Immunology and Molecular Biology
of Parasitic Infections

Third Edition

Immunology and Molecular Biology of Parasitic Infections

Edited by

KENNETH S. WARREN MD

Director for Science, Macmillan Inc.
New York

BOSTON

BLACKWELL SCIENTIFIC PUBLICATIONS

OXFORD LONDON EDINBURGH

MELBOURNE PARIS BERLIN VIENNA

© 1976, 1982, 1993 by
Blackwell Scientific Publications, Inc.
Editorial offices:
238 Main Street, Cambridge
    Massachusetts 02142, USA
Osney Mead, Oxford OX2 0EL, England
25 John Street, London WC1N 2BL
    England
23 Ainslie Place, Edinburgh EH3 6AJ
    Scotland
54 University Street, Carlton
    Victoria 3053, Australia

Other editorial offices:
Librairie Arnette SA
2, rue Casimir-Delavigne
75006 Paris
France

Blackwell Wissenschafts-Verlag
Meinekestrasse 4
D-1000 Berlin 15
Germany

Blackwell MZV
Feldgasse 13
A-1238 Wien
Austria

First published 1976
Second edition 1982
Third edition 1993

Set by Excel Typesetters Company,
Hong Kong
Printed and bound
in the United States of America
by BookCrafters, Chelsea, Michigan

93 94 95 96 5 4 3 2 1

DISTRIBUTORS

USA
    Blackwell Scientific Publications, Inc.
    238 Main Street
    Cambridge, Massachusetts 02142
    (*Orders*: Tel: 617 876-7000
                    800 759-6102)

Canada
    Times Mirror
    Professional Publishing, Ltd
    130 Flaska Drive
    Markham, Ontario L6G 1B8
    (*Orders*: Tel: 800 268-4178
                    416 470-6739)

Australia
    Blackwell Scientific Publications Pty Ltd
    54 University Street
    Carlton, Victoria 3053
    (*Orders*: Tel: 03 347-5552)

Outside North America and Australia
    Marston Book Services Ltd
    PO Box 87
    Oxford OX2 0DT
    (*Orders*: Tel: 0865 791155
                    Fax: 0865 791927
                    Telex: 837515)

Library of Congress
Cataloguing-in-Publication Data
Immunology and molecular biology of
    parasitic infections /
    edited by Kenneth S. Warren.—3rd ed.
        p.   cm.
    Rev. ed. of: Immunology of parasitic
        infections.
    2nd ed. 1982.
    Includes bibliographical references
        and index.
    ISBN 0-86542-095-5
    1. Parasitic diseases—Immunological
aspects.
    2. Parasitic diseases—Molecular
aspects.
    I. Warren, Kenneth S.
    II. Immunology of parasitic infections.
        [DNLM: 1. Parasitic Diseases—
immunology.
    2. Parasitic Diseases—metabolism.
    WC 695 I324]
    RC119.I49   1993
    616.9'6079—dc20

Contents

Contents

3

SECTION 3
HELMINTHS

4

SECTION 4
SYNOPSIS OF PARASITOLOGY

Contributors

Gordon L. Ada DSc
Division of Cell Biology, John Curtin School of Medical Research, The Australian National University, Canberra, Australia

Jenefer M. Blackwell BSc PhD
Department of Medicine, Addenbrooke's Hospital, Cambridge, UK

John C. Boothroyd PhD
Stanford University School of Medicine, Stanford, USA

James M. Burns Jr PhD
Seattle Biomedical Research Institute, Seattle, USA

André R. Capron MD
Centre d'Immunologie et de Biologie Parasitaire, Unité mixte INSERM U167-CNRS 624, Institut Pasteur, Lille Cedex, France

Stephen W. Chensue MD PhD
Department of Pathology, The University of Michigan Medical School, Ann Arbor, USA

Peter L. Chiodini BSc MBBS PhD FRCP
Hospital for Tropical Diseases, London, UK

Daniel G. Colley PhD
Vanderbilt Medical Center, Nashville, USA

Robert J. Dalgliesh BVSc MVSc DVSc
Animal Research Institute, 665 Fairfield Road, Yeerongpilly, Brisbane, Australia

Jean-Paul L. Dessaint MD
Centre d'Immunologie et de Biologie Parasitaire, Unité mixte INSERM U167-CNRS 624, Institut Pasteur, Lille Cedex, France

Peter Godfrey-Faussett BA MBBS MRCP DTM&H
Department of Clinical Sciences, London School of Hygiene and Tropical Medicine, University of London, London, UK

Bruce M. Greene MD (Deceased)
Division of Geographic Medicine, The University of Alabama at Birmingham, Birmingham, USA

William Harnett BSc PhD
Department of Immunology, University of Strathclyde, Glasgow, UK

Lloyd H. Kasper MD
Section of Neurology, Department of Medicine, Dartmouth Medical School, Hanover, USA

James W. Kazura MD
Division of Geographic Medicine, Department of Medicine, Case Western Reserve University School of Medicine, Cleveland, USA

Roberto Kretschmer MD
Biomedical Research Unit, Mexican Institute for Social Security, Facultad de Medicina, Edificio de Medicina Experimental, Ciudad Universitaria, Mexico City, Mexico

Steven L. Kunkel PhD
Department of Pathology, The University of Michigan Medical School, Ann Arbor, USA

Marshall W. Lightowlers BSc PhD
Molecular Parasitology Laboratory, University of Melbourne, Werribee, Australia

Jacques A. Louis PhD
WHO Immunology Research and Training Centre, Institute of Biochemistry, University of Lausanne, Lausanne, Switzerland

Keith P.W.J. McAdam MA MB BChiv FRCP
Department of Clinical Sciences, London School of Hygiene and Tropical Medicine, University of London, London, UK

Vincent McDonald BSc PhD
Department of Clinical Sciences, London School of Hygiene and Tropical Medicine, University of London, London, UK

Adel A.F. Mahmoud MD PhD
Department of Medicine, University Hospitals, Cleveland, USA

Adolfo Martínez-Palomo MD DSc
Center for Research and Advanced Studies of the National Polytechnic Institute, Mexico City, Mexico

Johanne Melancon-Kaplan PhD
Genzyme Corporation, Framingham, USA

Isaura Meza Gomez Palacio PhD
Center for Research and Advanced Studies of the National Polytechnic Institute, Mexico City, Mexico

Graham F. Mitchell BVSc RDA PhD FTS FAA
Royal Melbourne Zoological Gardens, Melbourne, Australia

Ralph Muller DSc PhD FIBiol
CAB International Institute of Parasitology, St Albans, UK

Peter J. Myler BSc PhD
Seattle Biomedical Research Institute, Seattle, USA

Theodore E. Nash MD
Laboratory of Parasitic Diseases, National Institute of Allergy and Infectious Diseases, National Institute of Health, Bethesda, USA

George R. Newport PhD
Intercampus Program in Molecular Parasitology, Department of Pharmaceutical Chemistry, University of California and San Francisco, Laurel Heights Campus, San Francisco, USA

Jaime Nina MD
Department of Clinical Sciences, London School of Hygiene and Tropical Medicine, University of London, London, UK

Thomas B. Nutman MD
Laboratory of Parasitic Diseases, National Institute of Allergy and Infectious Diseases, National Institute of Health, Bethesda, USA

R. Michael E. Parkhouse BSc PhD
Institute for Animal Health, Pirbright, UK

Richard D. Pearson MD
Department of Internal Medicine, University of Virginia Health Sciences Center, Charlottesville, USA

Wallace Peters MD DSc FRCP
CAB International Institute of Parasitology, St Albans, UK

S. Michael Phillips MD
Allergy and Immunology Division, School of Medicine, University of Pennsylvania, Philadelphia, USA

Michael D. Rickard BVSc PhD DVSc
Division of Animal Health, Commonwealth Scientific and Industrial Research Organization, Parkville, Australia

David L. Sacks PhD
Laboratory of Parasitic Diseases, National Institute of Allergy and Infectious Diseases, National Institute of Health, Bethesda, USA

Phillip A. Scott PhD
Department of Pathobiology, School of Veterinary Medicine, University of Pennsylvania, Philadelphia, USA

Alan Sher PhD
Immunology and Cell Biology Section, Laboratory of Parasitic Diseases, National Institute of Allergy and Infectious Diseases, National Institute of Health, Bethesda, USA

David Snary BSc PhD
Applied Development Laboratory, Imperial Cancer Research Fund, St Bartholomew's Hospital, London, UK

Kenneth D. Stuart BA MA PhD
Seattle Biomedical Research Institute, Seattle, USA

Garry B. Takle BSc PhD
Laboratory of Molecular Parastiology, The Rockefeller University, New York, USA

Rick L. Tarleton PhD
Department of Zoology, University of Georgia, Athens, USA

Akhil B. Vaidya PhD
Department of Microbiology and Immunology, School of Medicine, Hahnemann University, Philadelphia, USA

Contributors

Keith Vickerman DSc FRS PhD
Department of Zoology, University of Glasgow, Glasgow, UK

Derek Wakelin BSc DSc PhD
Department of Life Science, University of Nottingham, Nottingham, UK

H. Kyle Webster PhD
US Army Medical Component, Armed Forces Research Institute of Medical Sciences, Bangkok, Thailand

William P. Weidanz PhD
Department of Medical Microbiology and Immunology, School of Medicine, University of Wisconsin, Madison, USA

Dyann F. Wirth PhD
Department of Tropical Public Health, Harvard School of Public Health, Boston, USA

Stephen G. Wright MBBS MRCP MRCS DCMT
Department of Clinical Sciences, London School of Hygiene and Tropical Medicine, University of London, London, UK

Preface to the third edition

The opening sentence of the preface to the first edition (1974) of this book stated "In many tropical and subtropical countries parasitic infections exact a toll of human life and health grave enough to constitute a serious threat to economic and social development." The preface of the second edition (1982) noted that "during the intervening 8 years this serious situation has shown no significant improvement, the overall prevalence of parasitic infections increasing in parallel with the burgeoning growth of populations in the developing world." Sad to say, the 10-year hiatus between the second and third editions has seen little change in the socioeconomic status of most of the developing world, and a continuing stasis in the status of protozoan and helminth infections.

But the science of parasitology has taken two giant steps over the last 18 years, moving from a world dominated by the microscope and an obsession with immunodiagnosis, to a world concerned with the mechanisms of immunity and immunopathology, and now to a new world envisioned by Joshua Lederberg at a meeting on the future of parasitology in 1980. He prognosticated a "new wave" in the application of immunology, pharmacology, and molecular biology to parasitic infections, leading to an era of discovery of disease mechanisms, pharmaceutical agents, and vaccines that would "rival the great advances in microbiology of 40 years ago." This has been paralleled by new insights into the epidemiology and population biology of the host–parasite relationship, that should enhance the application of the products of laboratory research.

Over the last 15 years, research on the immunology and molecular biology of parasitic infections has been fostered by the development of several major new funding initiatives. These include the Tropical Diseases Research Progamme of the World Health Organization, the schistosomiasis and onchocerciasis strategic approaches of the Edna McConnell Clark Foundation, the Great Neglected Diseases network of the Rockefeller Foundation, the Biology of Parasitism Program of the MacArthur Foundation, and the Burroughs-Wellcome Fellowships in the Molecular Biology of Parasitic Infections. It is also heartening that the Wellcome Trust of the United Kingdom, which has for so long been a bulwark of parasitologic research, is now led by an eminent immunologist/parasitologist.

While the development of monoclonal antibodies and molecular probes have increased the sensitivity and specificity of immunodiagnosis, the methods have not as yet been refined enough for general use. Knowledge of the mechanisms of immunity and immunopathology continues its inexorable advance, greatly stimulated and accelerated by the new tools provided by genetic engineering. "The importance of immunoprophylactic measures in parasitic infections . . ." was stressed in the preface to the first edition, but those words were followed by, "Unfortunately this goal remains elusive." Let us hope that by the time the fourth edition of this seminal work is published the goal of parasite vaccines will not only be attained, but that they will be applied to the primary goal alluded to in the prefaces of both of the previous editions, the first reduction in the global prevalence of parasitic infections in recorded history.

Kenneth S. Warren
March 1992

Preface to the first edition

In many tropical and subtropical countries parasitic infections exact a toll of human life and health grave enough to constitute a serious threat to economic and social development. The magnitude of this problem can be illustrated by recent estimates that 300 million people are infected with amebiasis and 7 million with Chagas' disease; 400 million live in malarial regions with an annual morbidity of 50 million cases and mortality of about 1 million, while the current extension of irrigation projects and growing density of human populations have produced an explosive increase in the incidence of schistosomiasis. In these circumstances the importance of immunoprophylactic measures in parasitic infections need not be stressed. Unfortunately this goal remains elusive and practical success will demand a far deeper understanding of host–parasite interactions than exists at present.

All parasitic infections induce specific antibody synthesis and, as described in Section 2, immunodiagnostic tests are of great clinical and epidemiologic value. Only rarely, however, does the immune response lead to complete elimination of parasites. As illustrated in Section 3, acquired resistance is sometimes not manifest clinically and frequently is associated with persistent, low-grade infection. This state of affairs has been referred to by parasitologists as "premunition," and represents an equilibrium between host and parasite fundamental for evolutionary survival of parasitic species. Several mechanisms permitting parasite survival in the immunized host have been recognized including antigenic variation and disguise, production of soluble blocking antigens, intracellular location, and inhibition of various host defence mechanisms. It remains true, however, that the means whereby parasites evade acquired host immunity are not completely understood for any species. The similarities between continued parasite survival and progressive tumor growths in hosts manifesting potentially lethal immune responses are obvious. Parasitic infections can clearly provide models of great potential value to immunologists interested not only in mechanisms of acquired resistance, but also in the nature of immunopathologic complications of disease (Section 4).

These considerations encouraged Dr Elvio Sadun and myself to undertake the assembly of this book in the hope of promoting a greater interchange between parasitologists and immunologists. Towards this end the text includes some relevant basic immunology (Section 1) and summaries of the complex life-cycles of important human parasites (Appendix). The book was rendered viable by the generous cooperation of many distinguished investigators in the field of parasitic immunology. Dr Sadun himself wrote three invaluable chapters. His tragic death in April 1974 at the age of 55 was a profound shock which left an irreplaceable void for innumerable friends. Whatever merits this book may have are a reflection of Dr Sadun's deep knowledge of the subject, his kindly wisdom and the affectionate esteem which he engendered in colleagues throughout the world.

Sydney Cohen
June 1974

I Immune Responses to Parasitic Infections

1 Genetic variations in immunity to parasitic infections

Derek Wakelin & Jenefer M. Blackwell

INTRODUCTION

Immunogenetics has been one of the most rapidly developing fields of immunologic science in recent years. Growth in understanding the molecular basis of gene expression and gene regulation, and elucidation of the molecular structure of key cell-surface and secreted gene products and of their roles in cell–cell interactions, has provided a conceptual basis not only for the fundamental processes of the immune response itself but also for interpretation of the widespread polymorphism of the genes which regulate that response. A major challenge is to apply this knowledge to the problems of genetically determined variation in resistance and susceptibility to infectious disease, a challenge which has become more urgent as a result of the progress being made towards molecular vaccines and the possibility of using such vaccines to manipulate host immune responsiveness in a selective manner.

Individual, breed, or racial variation in response to infection with parasitic organisms has been recorded for many years, at least since the beginning of this century. Until recently much of this information was essentially observational and anecdotal rather than analytic, but the past 10–15 years have seen an explosive growth of interest in the experimental study of these phenomena. Three factors in particular have contributed to the development of this field of research:

1 the availability of genetically and immunologically well-defined inbred, congenic, recombinant, and mutant strains of mice;

2 the greater understanding of the mechanisms of innate and acquired immunity underlying resistance to infection;

3 the ease with which the techniques of modern molecular biology and molecular genetics may now be applied to immunoparasitologic problems.

In their recent review, Festing & Blackwell [1] identified six phases of study in the genetic analysis of host resistance. Much of the experimental literature currently available is concerned with Phases 1–3, i.e., definition of characters for analysis, search for a response pattern in inbred mouse strains, and analysis using classic crosses. In very few instances has progress to Phases 4–6 been attempted, i.e., mapping genes for resistance or susceptibility, producing congenic strains in which the activity of these genes can be analyzed, or cloning genes so that their gene products can be studied directly. The aim of this chapter is not to review the very extensive descriptive literature on variation in response to parasitic infection, for which a number of reviews are already available [2–8]. Our aim instead is to focus upon what we see as the most completely analyzed and most informative experimental systems, and to use these to point to the approaches which will be most productive in the future, especially with regard to the analysis of variation within human populations. Thus we will concentrate upon examples in which genetic variation in overall resistance or susceptibility to a particular parasite can be related to defined influences upon the initiation and expression of mechanisms mediating innate and acquired immunity. In a number of these examples it is possible to relate such variation to specific manifestations of host responsiveness, namely influences upon parasite development and survival, or to host pathology. Equal attention will be paid to those examples in which genetic variation has been associated with the expression of specific alleles, particularly where progress has allowed precise mapping and characterization. Finally, from this firm base of experimental data we shall discuss the application of proven concepts, approaches, and techniques to the more difficult, but ultimately more rewarding, problem of studying genetic variation in response to parasitic infections in humans.

VARIATION IN NATURAL RESISTANCE

Establishment of any parasitic infection in the vertebrate host is dependent upon a series of molecular–cellular interactions which occur prior to the development of specific antibody and/or T cell responses. The level of natural resistance in the host will reflect success or failure in preventing parasite invasion and/or early expansion of the parasite population. Conversely, the virulence or infectivity of the parasite will reflect its ability to gain entry into the host and evade early non-specific immune attack. In evolutionary terms, these early molecular–cellular interactions at the host–parasite interface provide the raw material on which natural selection can act, thus leading to the generation of transient or stable genetic polymorphism in both parasite and host populations. This review will examine only the evidence for genetic variation on the vertebrate host side of this interaction, but bearing in mind that selection for any molecular variant in the host which confers enhanced natural resistance may be matched in the parasite population by selection for genetic variation which promotes virulence.

The complement system

Parasites use a variety of ways to evade or withstand attack by the complement system of the vertebrate host. Rimoldi et al. [9] have shown, for example, that meta-cyclic trypomastigotes of *Trypanosoma cruzi* evade lysis by human alternative complement pathway because parasite molecules with decay-accelerating factor activity reduce the efficiency of binding of factor B to C3b on the parasite surface. Similarly, purified gp58/68, which acts as one of the fibronectin/collagen receptors of *T. cruzi*, has been shown to inhibit the formation of cell-bound and fluid-phase alternative pathway C3 convertase [10], although it is not able to enhance decay/dissociation of preformed alternative pathway C3 convertase sites. Other parasites, notably metacyclic promastigotes of *Leishmania major*, use complement activation to advantage in gaining entry via complement receptors (e.g., CR1 [11]) into their preferred host cell. To avoid lysis, the major C3 acceptor on the parasite surface, the lipophosphoglycan (LPG) molecule, undergoes developmentally regulated changes in glycosylation, allowing it to hold the membrane attack complex sufficiently far from the parasite surface to prevent insertion into the parasite membrane [12,13]. Another important molecule involved in binding leishmanial parasites to macrophages is the major surface glycoprotein gp63 [14,15]. Whereas it is known that there are differences in the relative expression and the structure of these two important molecular determinants of virulence between species of *Leishmania*, no one has looked to see how polymorphic they might be within species where marked differences in virulence have been observed [16,17] between different isolates. But what of the host side of this interaction?

The first reported [18] cloning of murine C3 in 1982 has been followed by rapid progress in molecular cloning and mapping of nearly all of the components of complement and the regulatory proteins and receptors associated with the system (reviewed in Reid [19]). Apart from the important contribution that these studies have made to our understanding of gene structure, organization, regulation, and function in the complement system, and the uncovering of important structural homologies between chains of the terminal attack complex and the perforin molecules found in the granules of cytotoxic T cells [20,21], the one very fascinating outcome of this work has been the demonstration that many complement proteins are highly polymorphic. In particular, the major histocompatibility complex (MHC) class III complement genes (C4, C2, and factor B), like the Class I and Class II genes, are highly polymorphic, although they show no structural similarity to the latter. Common electrophoretic variants of the non-MHC encoded complement component C3 (C3S and C3F) also occur, with different frequencies observed in all the major racial groups [19]. C3 is the most abundant of the complement proteins in plasma and plays a central role in activation of both classic and alternative pathways. Individuals with C3 and other complement component deficiencies nearly all show bacterial infections (e.g., Veitch et al. [22], Ross & Densen [23], Lachmann [24], and Goldstein & Marder [25]), as well as evidence of immune complex disease. Similarly, studies (e.g., Jarvinen & Dalmasso [26] and Ruppel et al. [27]) in C-deficient mice show quantitative differences in parasite loads following infection with a variety of parasitic organisms. However, no one to our knowledge has yet looked for associations between polymorphism at any of the loci-encoding proteins of the complement system and resistance or susceptibility to bacterial or parasitic infections.

Receptor interactions at the host–parasite interface

Molecules involved in receptor–ligand interactions between the parasite and cells of the host's immune system (e.g., polymorphonuclear leukocytes (PMN), eosinophils, macrophages) and/or target cells (e.g., erythrocytes) for parasite invasion are also likely to come under strong selective pressure. This is reflected in the complexity of the interactions, for example, between different species of *Plasmodium* and target erythrocytes (reviewed in Mitchell *et al.* [28]) and between *Leishmania* spp. and host macrophages (reviewed in Blackwell [29]). In both cases, no single species of parasite appears to have remained totally dependent on any single receptor–ligand interaction for entry into host cells. While rare genetic defects in expression of some of these receptor molecules (e.g., the LFA-1 family of integrin receptors [30] involved in leishmanial uptake by macrophages; the glycophorins [31] involved in entry of merozoites into erythrocytes) are known, no studies have yet reported on segregation of functionally relevant protein polymorphisms (i.e., sequence differences within the coding regions of these molecules) which might influence resistance/susceptibility to infection. Presence or absence of Duffy blood group antigens does, however, correlate with geographic distribution of malaria, with higher frequencies of Duffy-negative individuals occurring in areas of West Africa endemic for *Plasmodium vivax* infection (reviewed in Miller & Carter [32]). Laboratory studies by Miller *et al.* [33] have demonstrated that Duffy-negative erythrocytes are completely resistant to merozoite invasion.

Genetic regulation of intracellular survival

Once inside the host cell, survival of intracellular parasites will clearly be influenced by genetic factors which regulate host cell biochemistry/physiology. Malaria and leishmaniasis provide good examples of diseases where genetic polymorphism in cellular physiology leads to phenotypic variability in resistance and susceptibility to infection. It is now 40 years since Haldane [34] first proposed, for example, that the high carrier rates for lethal disorders such as sickle cell anemia and β-thalassemia in tropical and subtropical areas are maintained by heterozygous advantage in malaria-infected individuals. More recent molecular genetic analysis of the hemoglobinopathies, including α-thalassemia, provides compelling support for the Haldane hypothesis.

Genetics of hemoglobinopathies

The thalassemias and sickle cell anemia are caused by

genetic disorders in α- and β-globin genes (reviewed in Collins & Weissman [35] and Kazazian [36]). Since there is only one β-globin gene per haploid genome, severe mutations at this locus are readily identifiable in heterozygous carriers and can be studied at the population genetic level. The most common form of β$^+$-thalassemia in the Mediterranean, for example, is caused by a single nucleotide substitution (G to A) at position 110 of the first intervening sequence (IVS-1) of the β-globin gene which can be detected using oligonucleotide probes based on the normal and variant sequences [37]. Other mutations producing β-thalassemia in different populations are also well characterized [38,39], their geographic distributions suggesting that most have a single origin and have subsequently reached polymorphic frequencies by selection pressure [40]. Sickle cell anemia is caused by a single base-pair mutation in the sixth codon of the β-globin gene, which results in the loss of a DdeI (and MstII) restriction enzyme site [41]. Compound heterozygosity for sickle cell trait and β-thalassemias leads to sickling disorders of varying intensity, depending on the severity of the β-thalassemia defect [42].

α-Thalassemia (reviewed in Higgs *et al.* [43]) has posed more of a problem for population genetic analysis because normal individuals have two α-globin genes per haploid genome. The α-globin gene cluster resides on the short arm of human chromosome 16, the two α-globin genes coding identical globin chains and separated by less than 4 kb of DNA. Deletions of a single gene result in α$^+$-thalassemia. Homozygotes possess two α-genes (−a/−a) and have mild hypochromic anemia. Heterozygotes have three α-genes (−a/aa) and are often phenotypically indistinguishable from normal individuals. Deletion of the linked pair of α-globin genes results in α^o-thalassemia. Homozygotes (−−/−−) have no α-globin genes and are stillborn; heterozygotes are phenotypically indistinguishable from α$^+$-thalassemia, the phenotypic effect being the same whether the two α-globin genes are missing from the same (−−/aa) or opposite (−a/−a) pairs of chromosomes. In most populations, the α$^+$-haplotype (deletion of a single α-globin gene from one or both chromosomes) is the most common cause of α-thalassemia. α^o-Thalassemia (where both α-globin genes have been deleted from one chromosome) is rare. Seven varieties of nondeletion syndromes have also been described [44] but they appear to be responsible for a minority of α-thalassemias in all populations studied. This more complex genetic basis to the α-thalassemias has meant that, prior to the development of molecular tools (gene probes, molecular mapping, and restriction fragment length polymorphism (RFLP) analysis) for analysis at the DNA level, population genetic analysis of the α-thalassemias was extremely difficult. Detailed structural analysis of the α-thalassemia deletional types shows that they have involved a wide variety of DNA recombination events.

Mechanism of antimalarial effects of hemoglobinopathies

All of these variant globin genes influence the red cell environment and hence intracellular survival of the malarial parasite. For the hemoglobinopathies, the heterozygous advantage required to maintain variant globin genotypes means that the parasite must be at a disadvantage in this group compared to normal healthy individuals carrying wild-type globin genes. Studies *in vitro* show that the rate of merozoite invasion of red cells of both α- and β-thalassemia heterozygotes is the same as normal cells, and there are no differences in patterns of growth and development [45]. The red cell membrane in heterozygous thalassemia is, however, more sensitive to damage by oxidation [46], which could influence parasite survival *in vivo*. More recently [47], experimental studies to determine the protective effects of β-thalassemia against malaria have been carried out using transgenic mice infected with rodent malarial parasites, *Plasmodium chabaudi* and *Plasmodium berghei*, differing in their preference for mature erythrocytes vs. reticulocytes. In β-thalassemic C57BL/6 mice, *P. chabaudi* infection was inhibited and peak parasitemia was variably delayed. Mice corrected transgenically with the human β-A-globin gene showed the same pattern of infection as normal mice. For *P. berghei*, infection proceeded more rapidly in thalassemic mice but survival did not differ compared to control or transgenically corrected mice. These data provide direct evidence for the protective effect of the β-thalassemia gene for malarial parasites which invade mature erythrocytes but not for species which preferentially invade reticulocytes.

For sickle cell trait, Luzzatto *et al.* [48] found that *Plasmodium falciparum*-infected erythrocytes sickle more readily than nonparasitized cells under low oxygen tension. This could lead to more rapid clearance and destruction of parasitized erythrocytes. Other studies [49,50] show that invasion and growth of the parasite are inhibited under reduced oxygen conditions.

Studies in humans also provide evidence for associations between the hemoglobinopathies and other irregularities in erythrocyte physiology which might influence parasite survival. A recent study [51] in Saudi Arabia demonstrated a significant association between glucose-6-phosphate dehydrogenase (G6PD) deficiency and both the sickle cell gene and the β- and α-thalassemias. Similarly, in Papua New Guinea, the distribution of α+-thalassemia mimics that of G6PD deficiency and ovalocytosis [52]. An association between G6PD deficiency and resistance to malaria has been known for a long time [53], these cells also showing reduced parasite growth under reduced oxygen conditions [54]. Ovalocytic erythrocytes are resistant to invasion by merozoites [55]. In 18 β-thalassemia families from the Ferrara area [56] the incidence of an inherited low flavin mononucleotide (FMN)-dependent pyridoxine phosphate oxidase activity, a sensitive indicator of red-cell FMN deficiency, was found to be higher in related members of the families than in the unrelated spouses and control families. The authors suggest that the slower red-cell riboflavin metabolism in thalassemia families may be the result of coselection by malaria.

Population genetic analysis of the hemoglobinopathies and human malaria

At a gross level, the world distribution of sickle cell trait, the thalassemias, and G6PD deficiency mirrors that of past or present malaria infection [57], although different frequencies of each genetic variant occur in different geographic regions. Hence, in Africa, the sickle cell gene occurs at much higher frequency than β-thalassemia, probably owing to the severity of the clinical phenotype associated with compound heterozygosity for both β-gene disorders. The early epidemiologic studies of Allison [58], suggesting that hemoglobin (Hb) S heterozygotes are protected against *P. falciparum*, are supported by more recent studies [59] in Nigeria where population genetic analysis indicates that sickle cell trait confers a relative fitness of 0.2 compared to normal individuals in the same population. An interesting consequence of the protective effect of sickle cell in malaria is the possible run-on effect in protecting individuals against other diseases contracted through general ill-health associated with malarial infection. Colombo & Felicetti [60], for example, compared the frequency of Hb S heterozygotes in blood donors, outpatients, and inpatients of a general hospital in Maputo, Mozambique, where *P. falciparum* is endemic. The inpatient group showed a significantly lower percentage of Hb S heterozygotes suggesting that heterozygous individuals are not only protected against malarial infection but may be less prone to a wide spectrum of other diseases observed in the inpatient group. This situation should be borne in mind when carrying out any population survey for associations between disease phenotypes and particular genetic markers, since the causal association between genotype and phenotype may be indirect.

The early data of Siniscalco et al. [53], suggesting that high frequencies of β-thalassemia in Sardinia might have resulted from selective pressure due to malaria infection, have also received support from more recent studies [39,40,61,62]. In Papua New Guinea and Melanesia, β-thalassemia is rare but detailed analysis of α-thalassemia variants using RFLP analysis at the DNA level [63] has demonstrated a sharp cline in the frequency of α-thalassemia which matches the distribution of malaria as defined by parasite and spleen-rate data collected over many years.

Genetic (Lsh/Ity/Bcg) control of macrophage function

Human leishmanial and mycobacterial infections are characterized by a broad spectrum of disease pheno-types exhibited by individuals infected with the same species/strain of bacteria/parasite. Recent immuno-epidemiologic studies of leishmaniasis [64–66] highlight evidence for resistant individuals who become infected but develop no clinical symptoms of disease. Inter-estingly too, the human Bacille Calmette-Guerin (BCG) vaccination program has been plagued by marked variability in responsiveness both within and between populations of different ethnic origins (reviewed in Fine [67]). In studies designed to determine the underlying genetic basis to this variation, five major single gene (Lsh, Scl-1, Scl-2, H-2, H-11) responses controlling murine leishmaniasis have been identified (reviewed in Blackwell [68]), three of which (in addition to the MHC: H-2 in mice; HLA in humans—see section on MHC restriction, later) have been mapped to regions of known homology with human chromosomes. All of these genes operate directly or indirectly at the macrophage level, but one (Lsh) is of singular importance because of its parallel action in regulating the early phases of infection with Salmonella typhimurium (under the gene designation Ity [69]), Mycobacterium bovis (under the gene designation Bcg [70]), Mycobacterium lepraemurium [71,72], and Mycobacterium intracellulare [73], in addition to visceral infection with Leishmania donovani [74] and visceral-ization of Leishmania mexicana [75].

Mapping studies in mice and humans. The original mapping of Lsh/Ity/Bcg to a position between Idh-1 and ln on mouse chromosome 1 [70,76,77] has been followed by more detailed fine mapping of flanking genes [78–80], with estimates for the distance between Lsh and its nearest marker villin (Vil) ranging from zero [79] to 1.7 ± 0.8 cM [80], depending on the series of backcross mice employed. Current work in several laboratories is focusing on different ways of generating additional markers to saturate this region of mouse chromosome 1, thus facilitating a reverse genetic approach to cloning the gene [81]. One very positive observation which has been made [79,82,83] is that the region Idh-1 to Achrg of mouse chromosome 1 carrying the Lsh gene has been conserved onto the long arm (2q) of human chromosome 2. This not only provides one possible source for identifying additional marker loci in this region, but also forms the basis to family linkage studies currently in progress [82,84] which are designed to find a human homolog for Lsh.

Functional analysis of macrophage priming/activation. Work aimed at identifying a protein product for the Lsh gene has centered around in vivo and in vitro analysis of macrophage function in the context of L. donovani [84] or M. bovis [85] infections. Important in this has been the development [83,86,87] of congenic mouse strains (B10.L-Lsh; C.D2-Idh-1, Pep-3; C.D2-Pep-3) bearing the Lsh^r allele on susceptible B10 or BALB genetic back-grounds. Results obtained with these mice indicate that Lsh regulates receptor-mediated priming/activation of macrophages [84], the key factor in determining the ability of the three phylogenetically distinct groups of microorganisms to trigger the Lsh resistance mechanism, most likely relying on their ability to bind to one or more members of the activation/adhesion-promoting family of integrin receptors. The evidence available suggests that Lsh operates subsequent to ligand binding to these receptors, at the level of signal transduction and/or regulatory gene (DNA binding protein) control. Parameters/genes known to come under differential Lsh gene regulation include antimicrobial [88–91] and tumoricidal [85] activity, respiratory burst activity [83,92], lipopolysaccharide-elicited tumor necrosis factor pro-duction [84], phorbol myristate acetate plus ionophore-elicited c-fos expression [84], expression of the AcM.1 marker [85], and MHC Class II molecule expression [93–97]. In an attempt to identify functionally relevant protein differences associated with Lsh gene activity, protein phosphorylation during signal transduction has also been examined [84]: two-dimensional polyacrylamide gel electrophoresis (PAGE) analysis of proteins extracted 20 minutes after stimulation with leishmanial parasites, demonstrating phosphorylation of low-molecular-weight (10/12 kD and 14 kD; PI 5.9–6.1) proteins in susceptible B10 but not B10.L-*Lsh* macrophages. Whether these proteins represent putative *Lsh* gene products or are secondary to expression of the primary gene product awaits further investigation.

Influence of Lsh on accessory cell function. One finding pertinent to the quest for identification of an *Lsh* gene homolog in humans is the major influence that the gene has in regulating expression of MHC Class II molecules [93–97]. This, in turn, effects a dramatic difference in accessory cell function in macrophages from *Lsh*-resistant and congenic mouse strains [94], although it is not clear to what extent this might also be influenced by enhanced processing of antigen in resistant macrophages in addition to the higher levels of Class II expressed. The latter is a particularly prominent feature of both BCG [85,93] and *L. donovani* [94] infection models: *in vivo* studies [94] demonstrating massive differences in the interferon-γ (IFN-γ) generating capacity of splenic T cells isolated through 100 days of infection and stimulated with leishmanial antigen *in vitro*. The implications of these findings in the search for a human homolog for *Lsh* are twofold:

1 that resistant individuals in family linkage studies of mycobacterial or leishmanial infections might be identified through their enhanced T cell responses, despite no history of clinical disease (as in Ho *et al*. [64], Badaro *et al*. [65], and Sacks *et al*. [66]);

2 that *Lsh* gene-regulated differences in the macrophage response to BCG may have a profound impact on the ability to respond to BCG vaccination.

Since viable BCG [98] and *Salmonella* [99] are two of the most popular candidates for recombinant antigen vaccine vehicles, genetic regulation of responder/nonresponder phenotypes could clearly have a much more general impact on vaccine efficacy in humans.

VARIATION IN ACQUIRED IMMUNITY

Antigen recognition and T cell responses

Antigen recognition

The acquisition of immunity after infection depends initially upon the recognition of antigenic determinants (epitopes) by appropriate lymphocyte receptors. In the case of the eukaryotic parasites, it is almost always the case that recognition of those antigens that are involved in the generation of protective immunity is critically dependent upon presentation of epitopes to receptors on T lymphocytes (normally CD4$^+$ T helper cells), and this involves prior handling and processing of antigen by antigen-presenting cells (APC). A current view of this event is that the configuration of the epitope seen by the T cell receptor (TCR) is in fact determined by the way in which the processed antigen fragment binds to APC cell surface molecules that are coded by genes in the MHC. The entire sequence of events is therefore under tight genetic control, which is expressed through two major groups of cell surface molecules: those on the APC coded by MHC genes and those on the T cell coded by TCR genes. The polymorphism of the MHC genes and the rearrangements possible in the TCR gene repertoire allow the synthesis of a very large number of different molecules, capable of binding a very extensive range of antigens. There is, necessarily, a high degree of specificity in binding, which depends upon the molecular structure of each component and can be altered completely as a result of changes as small as single amino acid substitutions [100]. This specificity of binding, greater in the TCR than in the MHC molecules, has the consequence that some epitopes will not be recognized, either because the antigen fragment does not bind to the MHC molecule (or binds incorrectly) or because the epitope revealed by binding to the MHC molecule does not bind to the TCR. Examples of the former (so-called MHC restriction of recognition) are more common than those of the latter and several have been described in parasitic infections [8]. Relatively few of these examples are known to be related to differences in measurable resistance to infection; the majority relate only to differences in serologic or cellular immune responsiveness. The reason for this limited correlation with resistance arises from the antigenic complexity of parasites, which present their hosts with a wide array of antigenic molecules, each of which can carry a number of distinct epitopes. It is likely that several epitopes may be functionally involved in initiating protective responses, and therefore under conditions of natural infection it is unlikely that MHC restriction of the recognition of a given epitope will be reflected in a qualitative difference in resistance. Such restriction becomes significant, however, when attempts are made to use defined components of specific antigens as vaccines. Subunits of antigens, particularly low-molecular-weight peptides, are, by their nature, likely to carry only a limited number of epitopes. Under these circumstances restriction of T cell recognition is more likely to arise, because of the absence of the appropriate MHC molecule, and is more likely to be seen as a qualitative difference in ability to generate protective responses [101].

MHC restrictions of responses to malarial antigens

Vaccination strategies directed towards the immunologic control of malaria have three targets in the life-cycle: the sporozoite, merozoite, and gametocytes. Candidate vaccine antigens are now available for each of these targets and these have been studied extensively in rodent as well as primate hosts [102]. The substantial progress made with the first has been facilitated by the fact that the major surface antigen of the sporozoite is a single moiety, the circumsporozoite protein (CSP). The structure of this protein has been determined for a number of species, the coding genes sequenced, and both recombinant and synthetic peptides produced [103]. A characteristic of these molecules is the presence of extensive repetitive amino acid sequences. In *P. falciparum*, the central third of the molecule comprises a multiple repeat sequence of the tetrapeptide Asparagine–Alanine–Asparagine–Proline (NANP). The NANP repeat is a major target for antibody recognition, possessing an immunodominant B cell epitope. Antibodies directed against this epitope inhibit sporozoite invasion of hepatocytes and neutralize infectivity [104]. Protective activity of antibodies to the CSP and to epitopes located in repeat sequences has been demonstrated in other species of malaria as well [104].

Studies in mice have demonstrated a strict MHC restriction of the ability to make antibody responses to the CSP of *P. falciparum* or to the NANP repeat itself, which arises from an inability of some mouse strains to recognize the T cell epitopes necessary to allow stimulation of T helper cell populations [105,106]. Recognition of the T cell epitope within the NANP repeat occurs only in H-2^b haplotype mice and is restricted by the presence of the Class II gene product associated with the IA^b allele. Response to a second major T cell epitope (Th2R), which lies outside the NANP repeat, is similarly restricted by Class II gene products, occurring only in mice of the H-2^b and H-2^k haplotypes [107]. That the defect in nonresponder mice operates at the level of T cell epitope recognition rather

than at the level of recognition of the NANP-associated B cell epitope is amply demonstrated when NANP is linked to T cell epitopes that those strains can recognize, such as those in tetanus toxoid, keyhole limpet hemocyanin, or in a fusion peptide corresponding to part of the tetracycline resistance gene (tet$_{32}$). Under these conditions, anti-NANP antibody is freely produced [108].

Major histocompatibility complex restriction has also been shown to occur when mice are immunized with the dominant DPAPPNAN repeat of the CSP from *P. berghei* [109]. Recognition by T cells of the epitope concerned is IA^b restricted, but H-2D region Class I genes are also involved. Genetic restriction does not affect responses to irradiated sporozoites, however, implying the existence of other T cell epitopes on the intact molecule.

The results of a number of trials with peptides derived from CSP have produced strong evidence that MHC restriction of T cell epitopes occurs also in humans. This has been demonstrated indirectly from measurement of serologic recognition of the NANP repeat [110,111] and directly by proliferative responses to defined peptides [112,113]. For example, in a vaccine trial using a recombinant peptide containing the NANP repeat coupled to tet$_{32}$, only one out of 15 individuals developed significant anti-NANP titers [110]. Similarly, only limited serologic recognition of epitopes in a sequence of synthesized overlapping peptides could be demonstrated in individuals living in the Gambia, a *P. falciparum* holoendemic region [111]. Such restriction is to be predicted from current hypotheses of immune recognition, and is a difficulty likely to face all subunit vaccines. Nevertheless, some CSP-related T cell epitopes are apparently recognized by murine and human cells bearing widely different Class II MHC molecules [114]. A peptide of 20 amino acids from the CSP was recognized by human T cell clones in association with at least seven different DR antigens and by mice expressing seven different haplotypes. Evidence suggests that this may reflect separate recognition of closely overlapping epitopes within the sequence of 20 amino acids.

MHC restriction of in vivo responses to infection

Genetic control of responses to infection can be monitored by a variety of host- or parasite-related parameters. Some of these directly measure resistance *per se* (reductions in parasite growth, development, reproduction, survival) or the consequences of resistance (host growth and survival). Others reflect resistance only indirectly (serologic or cellular responses) or measure host susceptibility to induced immunopathologic change. MHC-linked effects have been recorded in many systems but it is clear in the majority that these relate to quantitative differences in host response rather than to the qualitative differences associated with MHC restriction of antigen presentation and recognition. Of the limited examples of qualitative differences in response, most concern MHC restriction of serologic recognition of parasite antigens, where the restriction of response is not clearly related to protection. For example, three studies in mice using *Ascaris suum*, *Schistosoma mansoni*, and *Trichuris muris* [115–117] have in each case shown that antibodies recognizing particular antigens appear only in mice of certain H-2 haplotypes, implying a restricted recognition of certain epitopes at the T or B cell level. This variation in response is apparent only when comparison is made between a number of different congenic strains and it is superimposed on a common recognition of other antigens in the same preparation.

In none of these examples is the significance of this selective recognition known and there is only limited evidence at present (in the case of *T. muris*) that it has any relevance to effective antiparasite resistance. The converse is true of those examples in which MHC-linked quantitative differences in resistance have been demonstrated. Here, the effect on the host–parasite relationship is known but, in the majority of cases, the underlying genetic or immunologic mechanisms are not. Some progress, however, has been made in examples where the MHC-linked influence involves selective antigen presentation, through particular Class II molecules, a phenomenon first described in a number of experimental systems using nonparasite antigens.

Leishmanial infections. Perhaps the most impressive evidence for the role of MHC in controlling parasitic infection has been found in murine models of visceral leishmaniasis caused by *L. donovani* (reviewed in Blackwell [68]). On a B10 genetic background, H-2^{r,s,b} mice show rapid resolution of liver and spleen parasite loads while H-2^{d,q,f} mice maintain high parasite loads over long periods (>130 days) of infection [118]. Considering the complex array of antigens that a whole parasite infection is likely to present to the immune system, one might hardly expect to observe such dramatic MHC "restriction" unless one particular immunodominant antigen is responsible for initiating protective or disease-promoting T cell responses. In fact, one-dimensional T cell blotting shows that T cells from curing strains proliferate and produce IFN-γ in response to a wide range of amastigote antigens of different molecular weights throughout infection [119]. Hence, if a particular antigen fails to bind to the MHC molecules of a given haplotype, it might be expected that many others could substitute in triggering protective T cell responses independently of MHC haplotype. How then does a change in H-2 haplotype exert such a dramatic all-or-none effect?

In attempting to answer this question one of the avenues investigated arose from the observation that the H-2^b and H-2^d pair of congenic strain mice differ not only in haplotype (i.e., in bearing different alleles at each of the loci encoded within the MHC) but also in expression of Class II molecules which present antigen to CD4$^+$ T cells. In mice, the heterodimeric Class II molecules are coded by genes at the IA and IE subregions, each subregion having α- and β-chains of the dimer. H-2^d mice express both IA and IE gene products, whereas H-2^b mice express only IA products. Treatment of noncuring H-2^d mice with monoclonal antibodies directed against these two molecules led to an interesting observation. Administration of anti-IE^d promoted resolution of liver and spleen parasite loads, whereas anti-IA^d resulted in exacerbation of disease [120]. One explanation for this might be that one (or a few) dominant antigen presented in the context of IE preferentially stimulates a disease-promoting CD4$^+$ T cell subset. This could, for example, result from preferential stimulation of CD4$^+$ T cells, which generate IL-4 rather than IFN-γ, this split in cytokine profiles having been associated with nonhealing vs. healing responses to cutaneous *L. major* infection in mice [121,122]. Again, T cell blotting in H-2^d noncure mice shows that T cells proliferate and produce IFN-γ in response to a broad array of antigens early (8 days) in infection, but become markedly restricted in their response (and eventually "shutdown") as *L. donovani* infection progresses, with no evidence *in vivo* or *in vitro* for the T helper 1 vs. T helper 2 split in cytokine profiles observed in the *L. major* infection model [123]. What causes this transition from early responder to late-phase nonresponder is not clear, but in attempting to follow up this story using a transgenic mouse model, other intriguing aspects of the interaction between MHC and TCR gene products have arisen.

In the NOD mouse model of insulin-dependent diabetes mellitus, onset of disease corresponds with generation of autoreactive T cells involved in β-cell destruction [124]. Like B10 H-2^b mice, NOD mice also fail to express IE. In this case, transgenic introduction of IE into NOD mice protects against onset of diabetes [125], the islet-specific T cells generated in normal NOD mice failing to develop. The traditional explanation for this was that IE preferentially stimulated suppressor T cells [126], a hypothesis which seemed consistent with the *L. donovani* infection model [120]. A more recent explanation is that the islet-specific T cells bear receptors encoded by a $V_{\beta}5$-gene segment [127] known to be deleted during development in IE-expressing mice [128]. This implies that it is not something detrimental which IE-restricted T cells do to "suppress" the immune response, but that there is a gap in the repertoire of T cells required to mount an anti-islet response. Again, this seems feasible in the context of an autoimmune response which might be directed against a defined autoantigen, but is difficult to integrate with an anti-genically complex infection model. Nevertheless, transgenic IE-expressing NOD mice infected with *L. donovani* do show enhanced parasite loads compared to control NOD mice at 30 and 50 days postinfection [129]. Both mouse strains show very good antigen-specific IFN-γ generating splenic T cell responses throughout infection, but a very clear difference is observed in the number of granulomas forming in the livers of NOD (high) vs. NOD-IE (low) mice as early as 15 days post-infection. $CD8^{+}$ MHC Class I-restricted T cells are known to play an important role in granuloma formation [130] and in resolution of *L. donovani* infection [131,132], thus offering the intriguing possibility that it is not the IE-mediated deletion of $CD4^{+}$ T cells bearing certain V_{β}-gene segments which is important, but the codeletion of $CD8^{+}$ T cells at the double positive phase in the thymus. The antigen specificity and V_{β}-gene expression in the TCR of $CD8^{+}$ T cells mediating granuloma formation early in *L. donovani* infection is currently under investigation. On examining HLA associations with disease phenotype in humans it should, however, be borne in mind that expression of certain MHC Class II molecules might have an indirect effect on Class I-regulated T cell responses because of their overall influence on thymic deletion of T cells bearing certain TCR gene rearrangements.

Nematode infections. The phenomenon of selective presentation involving sets of Class II molecules was first described in immunoparasitology in infections with *Trichinella spiralis* [133]. Data from experiments involving panels of congenic mice showed that relative resistance to infection, measured in terms of adult worm survival,

fecundity, and larval establishment, was correlated with IA-mediated presentation and relative susceptibility with IE-mediated events. In mice expressing both Class II molecules, IE-associated susceptibility dominated, suggesting preferential presentation of critical antigens by these molecules; this suggestion was supported by *in vitro* studies with T cell clones [134].

Similar associations between IA or IE presentation and resistance or susceptibility were also demonstrated in mice infected with *Heligmosomoides polygyrus* (*Nematospitoides dubius*), another intestinal nematode [133], suggesting that such associations may occur more widely in host–parasite systems, although it is clear that general statements cannot be made at the present time. For example, the presence of IE does not correlate with resistance to *T. muris* [135], and other workers using *H. polygyrus* have failed to find correlations similar to those originally reported [136]. As with *Leishmania* infections (see below), this variability in association between resistance and IE may indicate a complex interaction with background genetic influences and with parasite load. Indeed, in *T. spiralis*, Wassom *et al.* [133] has described modulation of IE influences by a gene located on chromosome 4 which, together with IE^{k} genes, controls IJ^{k} expression. When both genes are present, mice have a susceptible phenotype; when the chromosome 4 gene is absent, mice are more resistant. Similarly, expression of resistance alleles at IE can be modulated by genes present between the S and D loci.

Schistosome infections. Evidence for a direct association between the presence of particular HLA alleles and the immunopathologic consequences of schistosomiasis in humans has been accumulating steadily [137,138]. The fact that pathology is primarily the consequence of cell-mediated hypersensitivity reactions to antigens arising from the eggs of the parasite makes it logical to look for correlates between differential susceptibility and genes regulating T cell responsiveness. Sasasuki *et al.* in a series of papers (reviewed in Hirayama *et al.* [138]), have shown HLA linkage of *in vitro* lymphocyte proliferative responses to *Schistosoma japonicum* antigen, such that DQwl molecules (DQ being the HLA equivalent of IA) mediate antigen-specific suppressive responses, possibly through $CD8^{+}$ T cells, and DR2 molecules (HLA equivalent of IE) mediate proliferative responses. Patients with postschistosomal liver cirrhosis, a pathologic condition arising from antiworm and antiegg responses, showed a significant decrease in DQwl frequency, suggesting that their pathology reflected an unregulated high proliferative response. One interpretation of this is, again, that the Class II molecules involved in antigen presentation determine the CD4$^+$ T cell subset that responds, and this in turn determines collaborative interactions with other subsets.

The interplay of MHC-linked and background gene-associated effects upon *in vivo* responses

It is now well established that, in a majority of cases, the influences of MHC-linked genes upon resistance and susceptibility to infection will be expressed in the context of influences arising from genes located elsewhere in the genome. For this reason MHC influences are best identified by the use of MHC-congenic experimental animals, where only the alleles associated with the MHC are variable between strains, all other background genes remaining constant. Where the two groups of genes both influence a common endpoint of host resistance, e.g., a T cell-mediated inflammatory process, comparisons made between MHC-identical strains with different backgrounds may yield little useful data about the role of MHC-linked genes if it is the background genes that exert the major influence upon resistance. Such a situation occurs with *T. spiralis* infections in mice, where control of the intestinal phase of infection reflects the ability of mice to translate T cell responses into an intestinal inflammatory reaction [139]. The latter process reflects the influence of background genes upon myeloid cell responsiveness to cytokines, and this influence is dominant. Thus MHC-identical strains of mice may show opposite phenotypes because of background genetic effects. The converse situation can arise if background genes and MHC-linked genes influence quite different facets of resistance. Mouse strain variation in resistance to *L. donovani* is determined both by background gene control of intrinsic macrophage suitability for intracellular development and by MHC-linked control of T cell-mediated regulation of macrophage killing mechanisms [29]. Although the early response phenotype of mice is background-gene determined, the late-phase response phenotype is MHC-determined, and particular haplotypes (e.g., H-2^d, H-2^b) can generate a similar response phenotype on quite different (e.g., B10 vs. BALB) backgrounds.

The relationships between background- and MHC-linked genes in determining overall resistance or susceptibility are unpredictable. H-2^k mice on a B10 background (B10.BR) show a very variable response to *L. donovani*, some mice showing cure while others

maintaining high parasite loads for prolonged periods (130 days). Detailed studies with H-2 recombinant strains and with manipulation of IA/IE expression have led to the view that individual variations in parasite load, essentially determined by background genes, modulate the control that MHC-linked genes would be expected to exert, perhaps by tilting the balance between host-protective and parasite-protective T cell responses [68]. An analogous situation has been described recently in inbred and congenic mice infected with the nematode *T. muris*, where MHC-linked differences in rates of worm expulsion are seen within both BALB-background and B10-background congenic strains [135]. In the former, the BALB background determines a relatively rapid response and all congenics eliminate worms, although BALB/K (H-2^k) do so more slowly than BALB/c (H-2^d) or BALB/B (H-2^b). The slower response determined by the B10 background exaggerates the MHC-linked difference and results in survival of worms beyond 25 days. In some strains (B10.G–H-2^q; B10–H-2^b; B10.D2/n–H-2^d) this occurs in an increasing proportion of individuals, and in B10.BR–H-2^k mice it occurs in all individuals. All such individuals seem then to be immune-modulated by the surviving worms and fail completely to eliminate their parasite burden even after several weeks.

A cautionary note to add to such immunogenetic explanations of variable responsiveness is that variability *within* inbred strains might equally well arise from individual differences that have nothing to do with immune responsiveness to the parasite *per se*. This is thought to be the case with the intrastrain variations in response to *S. japonicium* seen in 129/J mice. Mitchell [140] has proposed that antigen crossreactivity, generated by cryptic viral infection perhaps at the level of T helper cell function, might influence the isotype-specific response to infection and thus contribute to individual variation in resistance. On the other hand, others have attributed individual differences in resistance to functional and morphologic variation in the portal supply to the liver [141,142]. In "resistant" mice, larval schistosomes are carried away from the liver and fail to survive. In mice with normal circulation, the worms locate in the liver and then mature.

GENETIC REGULATION OF EFFECTOR FUNCTION

Resistance to parasitic infection is, in many cases, mediated ultimately by cells of myeloid origin (monocytes, macrophages, eosinophils, neutrophils, basophils, and mast cells). Variation in the degree of resistance expressed may therefore originate in genetically determined variability in the development or activity of such cell populations. At the present time comparatively little is known about the control and expression of such variability, although there is sufficient experimental evidence to justify the conclusion that it comprises an important component of the mechanisms which determine resistance or susceptibility in individual hosts.

Macrophages (see also earlier section on genetic control of macrophage function)

Cells of the monocyte–macrophage series are involved as effectors of immunity against both protozoan and helminth infections. Their effector functions, including cytokine release, phagocytosis and intracellular killing, release of oxygen metabolites, and adherence to and destruction of extracellular targets, are mediated both by T cells and by antibodies. All of these activities are genetically regulated and inherently variable, as are the production and differentiation of the cells from their bone marrow-derived precursors. Although macrophage defects are known to occur in humans, these have not been formally related to parasitic infections; they have, however, been implicated in an increased susceptibility to other infectious organisms [143]. Detailed studies in experimental animals have analyzed a number of instances where macrophage defects profoundly influence levels of resistance. These can be illustrated primarily by work on P strain mice, which have a number of genetically determined abnormalities. Some of these abnormalities affect macrophage function [144], macrophage-mediated antibody-dependent cell-mediated cytotoxicity (ADCC) being an important component of resistance to infection [145].

Immunity to challenge infections with *S. mansoni* can be elicited in mice by vaccination with irradiated cerariae. Macrophage-mediated ADCC is thought to be an important effector in this vaccine-induced response. When panels of inbred mice are vaccinated, the level of protection achieved is strain-variable, but in P strain mice vaccination confers no immunity at all [146]. A major commponent of the defective response in this strain is related to cytokine (IFN-γ)-mediated activation of macrophages. The T lymphocytes of P mice release less cytokine in response to stimulation than do other strains, and P macrophages respond poorly to cytokines, being relatively inefficient at killing larval

worms *in vitro*. This defective response has been associated with a single gene, termed *Rsm-1* [147], but the expression of this gene has yet to be defined. In parallel studies it has been shown that P mice are also defective in their response to infection with *L. major*. Again, this defect is a consequence of deficient macrophage microbicidal activity and it is also under single gene control [148].

The genetic defect seen in macrophages of P strain mice is one of many known to influence resistance to infection. Indeed, the defective response of A strain mice to vaccination against *S. mansoni* is also macrophage-related, but under separate genetic control from *Rsm-1*. P × A hybrid mice show genetic complementation and exhibit a greater degree of resistance after vaccination than either parental strain [149]. Strain A mice are also known to exhibit impaired recruitment of macrophages into sites of infection, although this has been best studied to date in bacterial rather than parasitic infections. Other mouse strains are known to be poor at producing monocyte–macrophage cells from bone marrow precursors, despite adequate levels of the necessary factors, such as colony stimulating factor-1. Collectively, such deficiencies are likely to exert marked depressive influences on antiparasite responses where macrophages play important effector roles.

Eosinophils

Although eosinophils can act as microphages, *in vitro* data suggest that their antiparasite activities are expressed primarily through the release of mediators, enzymes, and a variety of cytotoxic factors [150]. Effective expression of resistance is therefore related to the ability of the host to generate eosinophils from bone marrow precursors, to mediate their localization around parasites, and to promote their parasiticidal activities. A number of experimental studies in mice have shown that the degree of eosinophilia after infection (or appropriate manipulation) is genetically determined, and varies markedly between different inbred strains [151,152]. More limited studies in other hosts (guinea pigs, sheep) have produced similar data, showing correlations between levels of eosinophilia and resistance to infection with intestinal nematodes [153]. *In vitro* studies of eosinophil differentiation, under culture conditions which provide optimal amounts of the necessary cytokine (IL-5), have shown that the variations in eosinophil response capacity seen between mouse strains reflects differences in precursor cell number and not differences in IL-5 production [154,155]. It remains to be seen whether the capacity of eosinophils to participate in cytotoxic events is also genetically variable and thus also likely to influence the expression of resistance.

Basophils and mast cells

Infiltration of amine-containing cells is a characteristic accompaniment of parasitic infections, particularly those involving blood-feeding arthropods, such as ticks, and those involving tissue-penetrating and intestinal-dwelling nematodes [156]. Correlations between expression of resistance and levels of both cell types in tissues have been firmly established in a number of experimental systems, although the functional bases of these correlations have yet to be fully understood. Much use has been made of mast cell-deficient mice (W/W^v strain) in analyzing the role of these cells in antiparasite responses. Clear-cut correlations with resistance were described in experiments using the tick *Haemaphysalis longicornis*, the genetically determined defect in response to infestation shown by W/W^v mice being corrected by grafting of skin from mast cell-sufficient litter mates [157]. In contrast, studies with intestinal nematodes in W/W^v mice have failed to produce a consistent picture regarding the role of mast cells in protective responses [158]. What has emerged from comparative studies of mast cell responses in rodents and larger animals is an overall positive correlation between the degree of mucosal mastocytosis and the ability to expel worms. Analysis of this correlation in mice has shown that mast cell-response phenotype in mast cell-sufficient strains is determined at the level of bone marrow precursor response to T cell-derived growth factors rather than at the T cell level itself [159]. This situation is clearly analogous to that seen with eosinophils, and similar *in vitro* studies using culture of bone marrow cells with optimum amounts of cytokine have confirmed that response phenotype is precursor-determined [160].

Variations in inflammatory responsiveness

The analyses described above lead to the conclusion that overall inflammatory responsiveness will be variable within genetically heterogeneous populations. This variation will be seen between individuals with similar levels of specific immune responsiveness as well as between individuals with different immune capacity because the origin of the variation will be different in each case. Many papers have discussed variations in inflammatory responsiveness in terms of resistance to infection but few have discussed its significance in terms of immunopathology. Certainly, it is well established from field studies that the pathologic sequelae of helminth infections, e.g., those involving schistosomes and filarial nematodes, vary considerably between individuals who live within endemic areas and who are presumably subjected to comparable levels of infection [161].

Variations in the immunopathologic consequences of schistosomiasis in humans have been linked with particular HLA haplotypes, and in the case of *S. japonicum* with altered T cell responsiveness [137,138]. However, the cellular composition of the schistosome granuloma, which involves macrophages, eosinophils, and mast cells, suggests that genetic influences directly affecting those cells may also contribute to the degree of variation. Similarly, the condition of tropical pulmonary eosinophilia, which appears in a minority of individuals exposed to infection with lymphatic filariases, has been associated with abnormal immune responses to microfilaria [162] but could equally well involve a component of abnormal eosinophil regulation. The complexity of the responses that lead to pathologic change is such that clear statements cannot always be made about the origins of differences between individuals, but the role of genetically determined variations in inflammatory competence *per se* should certainly be taken into account.

GENETIC VARIATION IN ANTIBODY-MEDIATED IMMUNITY

Antibodies contribute to antiparasite immunity through several different mechanisms, including direct interaction with functional parasite molecules, activation of complement, mediation of ADCC, and triggering of hypersensitivity reactions. Variations in the nature or level of antibody responses are likely, therefore, to influence resistance in many different ways. Because of the complex sequence of cellular interactions involved in the initiation and regulation of antibody production, genetic influences can act at a number of points but the effects may be detectable only by quantitative or qualitative differences in the final immunoglobulin response. For example, restriction at the T cell level of T epitope recognition can prevent antibody being made to a B cell epitope, even though B cells are perfectly capable of recognition and response (see earlier section on MHC restrictions of responses to malarial antigens). In this section, attention will be directed primarily to influences which affect B cells more directly, determining the amount or the isotype of antibody produced.

The most dramatic genetic influences on antibody response are those arising from stem cell deficiencies, which lead to agammaglobulinemia or hypogamma-globulinemia. The latter is associated with increased susceptibility to many infectious organisms, including protozoa [143]. Susceptibility to enteric protozoa is particularly marked in individuals with hypogamma-globulinemia or selective deficiency in IgA [163]. Under endemic conditions of multiple exposure to pathogens, individuals with these deficiencies may have impaired life expectancy. Less extreme variations in B cell response are more common but less easy to identify and analyze, other than through experimental models. One mutant mouse, the CBA/N mouse, has been particularly useful in such studies, having selective deficiencies in mounting IgM responses to certain antigens. Use of this strain has helped to clarify the role of antibody in resistance to organisms as diverse as malaria [164] and filarial nematodes [165].

Quantitative variation

The most comprehensive studies concerned with variation in amounts of antibody produced are those carried out with Biozzi high- and low-responder lines of mice [166]. These have shown a variety of responses to infection with protozoan and helminth infections (Table 1.1). In part, this variety reflects the contribution to resistance of macrophage activity, which differs between the lines of mice, but primarily it reflects the differences

Table 1.1 Differential resistance to parasitic infections in Biozzi high- and low-responder lines of mice. (From Biozzi et al. [166] and Wakelin [161])

High line, more resistant	Low line, more resistant
Trypanosoma cruzi*	Leishmania major
Plasmodium berghei*	Trichinella spiralis[†]
Plasmodium yoelii	Schistosoma mansoni
Plasmodium chabaudi	
Toxoplasma gondii	
Nematospiroides dubius[†]	
(Heligmosomoides polygyrus)	
Trichinella spiralis	
Mesocestoides corti	
Taenia taeniaeformis	

Resistance assessed after infection, vaccination,* or reinfection.[†]

in isotype, level, and rate of antibody production, known to be under polygenic control.

The consequences of differences in rate of antibody response have also been analyzed in detail in the mouse Taenia taeniaeformis model [167]. In susceptible strains, failure to make adequate amounts of antiparasite antibody (IgG) early enough in infection allows the larval cestodes to develop resistance to complement-mediated attack and thus to survive. After an initial infection, both susceptible and resistant strains of mice are able to resist a challenge infection because of enhanced secondary responsiveness, and resistance can be generated in susceptible mice by prior vaccination.

Qualitative variation

Variation in isotype-specific responses has been described in many infections, both in humans and in experimental animals. Work in the latter has shown that such variation can be genetically determined, but the degree to which such variation influences the level of resistance or susceptibility expressed is less clear cut, as can be illustrated by reference to some representative examples.

Marked differences in variant antigen-specific antibody responsiveness to infection with Trypanosoma brucei rhodesiense have been described in strains of mice with very different levels of resistance (time of survival) [168]. Resistant B10.BR produced pronounced IgM and IgG responses early in infection, intermediate CBA produced later, lower level responses, and susceptible C3H failed to respond. This apparent correlation between response and resistance was further investigated using radiation chimeras and by formal genetic analysis, from which it emerged that the genetic control of B cell responsiveness was unlinked with the control of survival time [169]. Indeed, response status was to a large extent parasite-determined. Although C3H mice were unresponsive to the variant antigen of the infecting clone during infection, after drug-cure IgM and IgG variant-specific antibodies were produced.

Genetic influences on antibody responses have also been studied during Plasmodium yoelii infections in mice [170]. In a panel of 13 different strains, three basic response patterns were identified based upon the time course of infection and the peak parasitemia reached after 18 days. In mice which cleared parasitemia most rapidly, total antibody response was quicker and higher than in the other strains. The most striking differences in isotype-specific response were seen in IgG_1, IgG_2, and

IgG$_3$, with production of the two latter isotypes correlating with ability to control infection. Variation in IgG isotypes has also been described in human infections with *P. falciparum* [171].

Strain-dependent differences in isotype responses have been reported in a number of experimental studies of helminth infections, including schistosomes, cestodes, and nematodes [172]. In few of these has there been clear-cut association with protection; indeed, in a number of cases an inverse correlation between antibody response and resistance has been demonstrated [173,174]. Such observations have led to the suggestion that production of "irrelevant," or of blocking, antibodies may down-regulate or prevent expression of protection, and there are now observations in human infections which point to the same phenomena. In schistosome infections, the isotype specificity of response is thought to determine the effectiveness of antilarval worm immunity, with IgG$_4$ antibodies having a blocking function [175]. One of the isotypes whose protective function may be blocked is IgE. IgM antibodies to carbohydrate components of schistosomulum surface antigens are also known to block killing mediated by IgG isotypes with specificity for other parts of the molecule [176]. The contribution of genetically determined variation in the production of blocking rather than protective isotypes to individual variation in resistance, and thus predisposition to heavy infection, is clearly of considerable theoretical and practical importance.

IgG blocking antibodies are known to modulate immediate hypersensitivity responses in patients infected with *Wuchereria bancrofti*, possibly by competition for the antigens that elicit this response [177]. Patients, especially those showing positive microfilaremia or tropical pulmonary eosinophilia, produce high levels of IgG$_4$ and this isotype may therefore be an important component of blocking activity [178].

Although the genetic basis and immunologic expression of these isotype variations has not been investigated in any detail in humans, data from experimental systems suggest that immunoregulatory control of B cell function may well be involved. Recent advances in knowledge of cytokines has clarified the ways in which isotype response can be controlled by the activity of soluble factors released by T cells. Genetic variation resulting in variation in isotype response may, therefore, be determined at several points in the overall response. In mice, where distinct subsets of T helper cells exist [179], there is a reciprocal relationship between levels of IgE and IgG$_1$, promoted by IL-4, and those of other IgG isotypes. IgE and IgG$_1$ hypergammaglobulinemias characterize a number of helminth infections and have been shown to be genetically variable in degree, e.g., in mice infected with *Nippostrongylus brasiliensis* or with *T. spiralis* [180,181]. High levels of IgE are also produced by susceptible BALB/c mice after infection with *L. major*. Manipulation of these mice to reduce IL-4 availability had the effect of reducing the IgE response and enhancing IFN-γ responses, i.e., of altering T helper subset activity [121]. Related phenomena have been described in mice infected with *N. brasiliensis* [182], where cells of IgE-nonresponder SJA/9 mice produced this isotype *in vitro* when supplied with IL-4. Collectively, these data supply strong evidence for direct genetic control of isotype specificity in response to parasitic infection, expressed through both T and B cells, and show that such control can markedly influence both resistance and pathology.

CONCLUSIONS

In this review we have attempted to examine the evidence for genetic variation in innate and acquired immune mechanisms and to relate these to observed variations in resistance and susceptibility to parasitic infections and/or in mounting specific immune responses to defined parasite antigens. In most cases, the hard evidence for genetic variation in these responses, and the identification of specific gene loci involved in their regulation, comes from the use of genetically defined murine models. The good news is that, whereas we might traditionally have been forced to "sell" our murine studies as merely "models" of infection *analogous* to the human disease, the recent increase in identification of synthenic relationships between murine and human genomes provides real evidence that *homologs* will be found in humans for resistance genes identified in mice. This is already true for most of the major genetic disorders (e.g., muscular dystrophy, retinoblastoma, chronic granulomatous disease, cystic fibrosis, Alzheimer's disease, etc.) of humans, although in this situation the order of events has usually been "from man to mouse" rather than the reverse. Since the molecular genetic tools are now available, the decade ahead should see much greater emphasis on determining the relative contribution of genetics as a risk factor for susceptibility to parasitic infection in humans, and in determining success or failure in response to vaccination. This in turn should provide a more rational basis to the development of new immunotherapeutic and immuno-prophylactic interventions for disease control.

ACKNOWLEDGMENTS

Work in the authors' laboratories on genetics of resistance to parasitic infections is supported by grants from the British Medical Research Council and the Wellcome Trust.

REFERENCES

1 Festing MFW, Blackwell JM. Determination of mode of inheritance of host response. In Wakelin D, Blackwell JM, eds. *Genetics of Resistance to Bacterial and Parasitic Infection.* London: Taylor & Francis, 1988:21–61.

2 Wakelin D. Genetic control of susceptibility and resistance to parasitic infection. Adv Parasitol 1978;16:219–308.

3 Skamene E, Kongshavn PAL, Landy M, eds. *Genetic Control of Natural Resistance to Infection and Malignancy.* New York: Academic Press, 1980.

4 Krco CJ, David CS. Genetics of immune response: a survey. Crit Rev Immunol 1981;1:211–257.

5 Mitchell GF, Anders RF, Brown GV. Analysis of infection characteristics and anti-parasite immune responses in resistant compared with susceptible hosts. Immunol Rev 1982;61:137–188.

6 Rosenstreich DL, Weinblatt AC, O'Brien AD. Genetic control of resistance to infection in mice. Crit Rev Immunol 1982;3:263–330.

7 Skamene E, ed. *Genetic Control of Host Resistance to Infection and Malignancy.* New York: Alan R Liss, 1985.

8 Wakelin D, Blackwell JM, eds. *Genetics of Resistance to Bacterial and Parasitic Infection.* London: Taylor & Francis, 1988.

9 Rimoldi MT, Sher A, Heiny S, Lituchy A, Hammer CH, Joiner K. Developmentally regulated expression by *Trypanosoma cruzi* of molecules that accelerate the decay of complement C3 convertases. Proc Natl Acad Sci USA 1988;85:193–197.

10 Fischer E, Ouaissi MA, Velge P, Cornette J, Kazatchkine MD. Gp58/68, a parasite component that contributes to the escape of the trypomastigote form of *T. cruzi* from damage by the human alternative complement pathway. Immunology 1988;65:299–303.

11 Da Silva RP, Hall BF, Joiner KA, Sacks DL. CR1, the C3b receptor, mediates binding of infective *Leishmania major* metacyclic promastigotes to human macrophages. J Immunol 1989;143:617–622.

12 Puentes SM, da Silva RP, Sacks DL, Hammer CH, Joiner KA. Serum resistance of metacyclic stage *Leishmania major* promastigotes is due to release of C5b-9. J Immunol 1989;143:3743–3749.

13 Pimenta PFP, da Silva RP, Sacks DL, da Silva PP. Cell surface nanoanatomy of *Leishmania major* as revealed by fracture-flip. A surface meshwork of 44 nm fusiform filaments identifies infective developmental stage promastigotes. Eur J Cell Biol 1989;48:180–190.

14 Russell DG, Wilhelm H. The involvement of the major surface glycoprotein (gp63) of *Leishmania* promastigotes in attachment to macrophages. J Immunol 1986;136:2613–2620.

15 Wilson ME, Hardin KK. The major concanavalin A-binding surface glycoprotein of *Leishmania donovani chagasi* promastigotes is involved in attachment to human macrophages. J Immunol 1988;144:265–272.

16 Kellina OI, Passova OM, Alekseev AN. Eksperimental'noe dokazatel'stvo geterogennosti sostava prirodnykh populiatsiii *Leishmania major* po priznaku virulentnosti. (English abstract.) Med Parazitol 1981;50:4–11.

17 Handman E, Hocking RE, Mitchell GF, Spithill TW. Isolation and characterization of infective and non-infective clones of *Leishmania tropica.* Mol Biochem Parasitol 1983;7:111–126.

18 Wiebauer K, Domday H, Diggelman H, Fey G. Isolation and analysis of genomic DNA clones encoding the third component of mouse complement. Proc Natl Acad Sci USA 1982;79:7077–7081.

19 Reid KBM. The complement system. In Hames BD, Glover DM, eds. *Molecular Immunology.* Oxford: IRL Press, 1988.

20 Tschopp J, Masson D, Stanley KK. Structural/functional similarity between protein involved in complement and cytotoxic T-lymphocyte-mediated cytolysis. Nature 1986;322:831–834.

21 Young JD, Cohn ZA, Podack ER. The ninth component of complement and pore-forming protein (perforin 1) from cytotoxic T cells: structural, immunological and functional similarities. Science 1986;233:184–190.

22 Veitch J, Love C, Chaudhuri AK, Whaley K. Complement deficiency syndromes and bacterial infections. Prog Brain Res 1983;59:69–80.

23 Ross SC, Densen P. Complement deficiency states and infection: epidemiology, pathogenesis and consequences of neisserial and other infections in an immune deficiency. Medicine (Baltimore) 1984;63:243–273.

24 Lachmann PJ. Inherited complement deficiencies. Philos Trans R Soc London Ser B 1984;306:419–430.

25 Goldstein IM, Marder SR, Infections and hypocomplementemia. Ann Rev Med 1983;34:47–53.

26 Jarvinen JA, Dalmasso AP. *Trypanosoma musculi* infections in normocomplementemic, C5-deficient, and C3-depleted mice. Infect Immun 1977;16:557–563.

27 Ruppel A, Rother U, Diesfeld HJ. *Schistosoma mansoni:* development of primary infection in mice genetically deficient or intact in the fifth component of complement. Parasitology 1982;85:315–323.

28 Mitchell G, Mitchell GH, Bannister LH. Malaria parasite invasion: interactions with the red cell membrane. CRC Crit Rev Oncol Hematol 1988;8:225–310.

29 Blackwell JM. Immunology of leishmaniasis. In Lachmann PJ, Peters DK, Rosen FS, Walport MJ, eds. *Clinical Aspects in Immunology,* 5th edn. Oxford: Blackwell Scientific Publications, 1993. (In press.)

30 Kishimoto TK, Hollander N, Roberts TM, Anderson DC, Springer TA. Heterogenous mutations in the beta subunit common to the LFA-1 Mac-1 and p150,95 glycoproteins cause leukocyte adhesion deficiency. Cell 1987;50:193–202.

31 Hadley TJ, Klotz FW, Pasvol G, *et al.* Falciparum malaria parasites invade erythrocytes that lack glycophorin A and B (MkMk). Strain differences indicate receptor heterogeneity and two pathways to invasion. J Clin Invest 1987;80:1190–1193.

32 Miller LH, Carter R. A review. Innate resistance in malaria. Exp Parasitol 1976;40:132–146.

33 Miller LH, Mason SJ, Dvorak JA, McGuiness MH, Rothman IK. Erythrocyte receptors for (*Plasmodium knowlesi*) malaria: Duffy blood group determinants. Science 1975;189:561–563.

34 Haldane JBS. The rate of mutation of human genes. Proc VIII Int Cong Gen Hered 1949;S35:367.

35 Collins FS, Weissman SM. The molecular genetics of human hemoglobins. Prog Nucleic Acid Res Mol Biol 1984;31:315–462.

36 Kazazian HH. Globin gene structure and the nature of mutation. Birth Defects 1987;23:77–92.

37 Kazazian HH, Orkin SH, Markham AF, Chapman CR, Youssoufian H, Waber PG. Quantification of the close association between DNA haplotypes and specific beta-thalassaemia mutations in Mediterraneans. Nature 1984; 310:152–154.

38 Kazazian HH, Boehm CD. Molecular basis and prenatal diagnosis of beta-thalassemia. Blood 1988;72:1107–1116.

39 Hill AV, Bowden DK, O'Shaughnessy DF, Weatherall DJ, Clegg JB. Beta thalassemia in Melanesia: association with malaria and characterization of a common variant (IVS-1 nt 5 G----C). Blood 1988;72:9–14.

40 Wainscott JS. The origin of mutant beta-globin genes in human populations. Acta Haematol Basel 1987;78:154–158.

41 Woodhead JL, Fallon R, Figueiredo H, Langdale J, Malcolm ADB. Alternative methods of gene diagnosis. In Davies KE, ed. *Human Genetic Diseases — a Practical Approach*. Oxford: IRL Press, 1986:51–64.

42 Atweh GF, Forget BG. Clinical and molecular correlations in the sickle/beta⁺-thalassemia syndrome. Am J Hematol 1987;24:31–36.

43 Higgs DR, Vickers MA, Wilkie AO, Pretorius IM, Jarman AP, Weatherall DJ. A review of the molecular genetics of the human alpha-globin gene cluster. Blood 1989;73:1081–1104.

44 Embury SH. The different types of alpha-thalassemia-2: genetic aspects. Hemoglobin 1988;12:445–453.

45 Pasvol G, Wilson RJM. The interaction of malaria parasites with red blood cells. Br Med Bull 1982;38:133–140.

46 Friedman MJ, Traeger W. The biochemistry of resistance to malaria. Sci Am 1981;244:154–155.

47 Roth EF, Shear HL, Costantini F, Tanowitz HB, Nagel RL. Malaria in beta-thalassemic mice and the effects of the transgenic human beta-globin gene and splenectomy. J Lab Clin Med 1988;111:35–41.

48 Luzzatto L, Nwachuku ES, Reddy S. Increased sickling of parasitized erythrocytes is the mechanism of resistance against malaria in the sickle trait. Lancet 1970;i:319–321.

49 Friedman MJ. Erythrocyte mechanism of sickle cell resistance to malaria. Proc Natl Acad Sci USA 1978;75:1994–1997.

50 Pasvol G, Weatherall DJ. A mechanism for the protective effect of haemoglobin S against *P. falciparum* malaria. Nature 1978;274:701–703.

51 Samuel AP, Saha N, Acquaye JK, Omer A, Ganeshaguru K, Hassounh E. Association of red cell glucose-6-phosphate dehydrogenase with haemoglobinopathies. Hum Hered 1986;36:107–112.

52 Yenchitsomanus P, Summers KM, Board PG, *et al.* Alpha-thalassemia in Papua New Guinea. Hum Genet 1986;74:432–437.

53 Siniscalco M, Bernini L, Filippi G, *et al.* Population genetics of haemoglobin variants, thalassaemia and glucose-6-phosphate dehydrogenase deficiency, with particular reference to malaria hypothesis. Bull WHO 1966;34:379–393.

54 Friedman MJ. Oxidant damage mediates variant red cell resistance to malaria. Nature 1979;280:245–249.

55 Kidson C, Lamont G, Saul A, Nurse GT. Ovalocytic erythrocytes from Melanesians are resistant to invasion by malarial parasites in culture. Proc Natl Acad Sci USA 1981;78:5829–5832.

56 Anderson BB, Perry GM, Clements JE, *et al.* Genetic and other influences on red-cell flavin enzymes, pyridoxine phosphate oxidase and glutathione reductase in families with beta-thalassaemia. Eur J Haematol 1989;42:354–360.

57 Weatherall DJ. Common genetic disorders of the red cell and the "malaria hypothesis". Ann Trop Med Parasitol 1987;81:539–548.

58 Allison AC. Protection afforded by sickle cell trait against subtertian malarial infection. Br Med J 1954;1:290–294.

59 Fleming AF, Storey J, Molineux L, Iroko EA, Attai EDE. Abnormal haemoglobins in the Sudan savanna of Nigeria. I. Prevalence of haemoglobins and relationships between sickle cell trait, malaria and survival. Ann Trop Med Parasitol 1979;73:161–172.

60 Colombo B, Felicetti L. Admission of Hb S heterozygotes to a general hospital is relatively reduced in malarial areas. J Med Genet 1985;22:291–292.

61 Willcox M, Bjorkman A, Brohult J. Falciparum malaria and beta-thalassaemia trait in northern Liberia. Ann Trop Med Parasitol 1983;77:335–347.

62 Livingstone FB. *Frequencies of Haemoglobin Variants*. New York: Oxford University Press, 1985.

63 Flint J, Hill AV, Bowden DK, *et al.* High frequencies of alpha-thalassaemia are the result of natural selection by malaria. Nature 1986;321:744–750.

64 Ho M, Siongok TK, Lyerly WH, Smith DH. Prevalence and disease spectrum in a new focus of visceral leishmaniasis in Kenya. Trans R Soc Trop Med Hyg 1982;76:741–746.

65 Badaro R, Jones TC, Lorenco BJC, *et al.* A prospective study of visceral leishmaniasis in an endemic area of Brazil. J Inf Dis 1986;154:639–649.

66 Sacks DL, Lal SL, Shrivastava SN, Blackwell J, Neva FA. An analysis of T cell responsiveness in Indian kala-azar. J Immunol 1987;138:908–913.

67 Fine PM. The BCG story: lessons from the past, implications for the future. Rev Infect Dis 1989;11:S353–359.

68 Blackwell JM. Protozoan infections. In Wakelin D, Blackwell JM, eds. *Genetics of Resistance to Bacterial and*

Parasitic Infection. London: Taylor & Francis, 1988:103–151.

69 Plant J, Glynn AA. Genetics of resistance to infection with *Salmonella typhimurium* in mice. J Infect Dis 1976;133:72–78.

70 Skamene E, Gros P, Forget A, Kongshavn PAL, St Charles C, Taylor BA. Genetic regulation of resistance to intracellular pathogens. Nature 1982;297:506–509.

71 Brown IN, Glynn AA, Plant J. Inbred mouse strain resistance to *Mycobacterium lepraemurium* follows the *Ity/Lsh* pattern. Immunology 1982;47:149–156.

72 Skamene E, Gros P, Forget A, Patel PJ, Nesbit MN. Regulation of resistance to leprosy by chromosome 1 locus in the mouse. Immunogenetics 1984;19:117–124.

73 Goto Y, Nakamura RM, Takahashi H, Tokunaga T. Genetic control of resistance to *Mycobacterium intracellulare* infection in mice. Infect Immun 1984;46:135–140.

74 Bradley DJ. Regulation of *Leishmania* populations within the host. II. Genetic control of acute susceptibility of mice to *Leishmania donovani* infection. Clin Exp Immunol 1977; 30:130–140.

75 Roberts M, Alexander J, Blackwell JM. Influence of *Lsh*, *H-2*, and an *H-11*-linked gene on visceralization and metastasis associated with *Leishmania mexicana* infection in mice. Infect Immun 1989;57:875–881.

76 Bradley DJ, Taylor BA, Blackwell JM, Evans EP, Freeman J. Regulation of *Leishmania* populations within the host. III. Mapping of the locus controlling susceptibility to visceral leishmaniasis in the mouse. Clin Exp Immunol 1979;37: 7–14.

77 Plant J, Glynn AA. Locating *Salmonella* resistance gene on mouse chromosome 1. Clin Exp Immunol 1979;37:1–6.

78 Mock B, Seldin M. A comparison of genetic linkage maps surrounding the *Lsh-Ity-Bcg* disease resistance locus. Res Immunol 1989;140:769–774.

79 Schurr E, Skamene E, Forget A, Gros P. Linkage analysis of the *Bcg* gene on mouse chromosome 1: identification of a tightly linked marker. J Immunol 1989;142:4507–4513.

80 Mock B, Krall M, Blackwell J, *et al*. A molecular map of mouse chromosome 1 near the *Lsh-Ity-Bcg* disease resistance locus. Genomics 1990;7:57–64.

81 Gros P, Malo D. A reverse genetic approach to *Bcg-Ity-Lsh* gene cloning. Res Immunol 1989;140:774–777.

82 Schurr E, Morgan K, Skamene E, Gros P. The search for a human homologue of the mouse *Bcg* host resistance locus. Res Immunol 1989;140:778–781.

83 Blackwell JM, Toole S, King M, Dawda P, Roach TIA, Cooper A. Analysis of *Lsh* gene expression in B10.L–*Lsh*[r] mice. Curr Top Microbiol Immunol 1988;137:301–309.

84 Blackwell JM, Roach TIA, Kiderlen A, Kaye PM. Role of *Lsh* in regulating macrophage priming/activation. Res Immunol 1989;140:798–805.

85 Buschman E, Taniyama T, Nakamura R, Skamene E. Functional expression of the *Bcg* gene in macrophages. Res Immunol 1989;140:793–797.

86 Potter M, O'Brien AD, Skamene E, *et al*. A BALB/c congenic strain of mice that carries a genetic locus (*Ity*[r]) controlling resistance to intracellular parasites. Infect Immun 1983;40: 1234–1235.

87 Mock B, Potter M. A molecular characterization of BALB/c congenic C.D1–*Idh-1*[b],*Lsh*[r],*Rep-1*[b],*Pep-3*[b] mice. Curr Top

Microbiol Immunol 1988;137:295–300.

88 Lissner CR, Swanson RN, O'Brien AD. Genetic control of the innate resistance of mice to *Salmonella typhimurium*: expression of the *Ity* gene in peritoneal and splenic macrophages isolated *in vitro*. J Immunol 1983;131:3006–3013.

89 Crocker PR, Blackwell JM, Bradley DJ. Expression of the natural resistance gene *Lsh* in resident liver macrophages. Infect Immun 1984;43:1033–1040.

90 Crocker PR, Davies EV, Blackwell JM. Variable expression of the natural resistance gene *Lsh* in different macrophage populations infected *in vitro* with *Leishmania donovani*. Parasite Immunol 1987;9:705–719.

91 Stach J-L, Gros P, Forget A, Skamene E. Phenotypic expression of genetically-controlled natural resistance to *Mycobacterium bovis* (BCG). J Immunol 1984;132:888–892.

92 Denis M, Forget A, Pelletier M, Skamene E. Pleiotropic effects of the *Bcg* gene. III. Respiratory burst in congenic *Bcg*[r] and *Bcg*[s] macrophages. Clin Exp Immunol 1988;73: 370–375.

93 Zwilling BS. MHC class II glycoprotein expression and resistance to mycobacterial growth. Res Immunol 1989; 140:806–809.

94 Kaye PM, Blackwell JM. *Lsh*, antigen presentation and the development of CMI. Res Immunol 1989;140:810–815.

95 Zwilling BS, Vespa L, Massie MS. Regulation of I-A expression by murine peritoneal macrophages: differences linked to the *Bcg* gene. J Immunol 1987;138:1372–1376.

96 Buschman E, Apt AS, Nickonenko BV, Moroz AM, Averbakh MH, Skamene E. Genetic aspects of innate resistance and acquired immunity to mycobacteria in inbred mice. Springer Semin Immunopathol 1988;10:319–336.

97 Kaye PM, Patel NK, Blackwell JM. Acquisition of cell-mediated immunity to *Leishmania*. II. *Lsh* gene regulation of accessory cell function. Immunology 1988;65:17–22.

98 Young RA. In Kohl H, LoVerde P, eds. *Vaccines: New Concepts and Developments*. London: Longman Group, 1988.

99 Brown A, Hormaeche CE, Demarco de Hormaeche R, *et al*. An attenuated aroA *Salmonella typhimurium* vaccine eliciting humoral and cellular immunity to cloned beta-galactosidase in mice. J Infect Dis 1987;155:86–92.

100 Schwartz RH. Immune response (Ir) genes of the murine major histocompatibility complex. Adv Immunol 1986;38: 31–201.

101 Good MF, Kumar S, Miller LH. The real difficulties for malaria sporozoite vaccine development: non responsiveness and antigenic variation. Immunol Today 1988;9:351–355.

102 Miller LH, Howard RH, Carter R, Good MF, Nussenzweig V, Nussenzweig RS. Research towards malaria vaccines. Science 1986;234:1349–1356.

103 Zavala F, Tam JP, Hollingdale MR, *et al*. Rationale for the development of a synthetic vaccine against *Plasmodium falciparum* sporozoites. Science 1985;228:1436–1440.

104 Hollingdale MR, Nardin EH, Tharavanji S, Schwartz AL, Nussenzweig RS. Inhibition of entry of *Plasmodium falciparum* and *P. vivax* sporozoites into cultured cells: an

in vitro assay of protective antibodies. J Immunol 1984;132: 909–913.

105 Del Guidice G, Cooper JA, Merino J, *et al*. The antibody response in mice to carrier-free synthetic polymers of *Plasmodium falciparum* circumsporozoite repetitive epitope is I-A[b] restricted: possible implications for malaria vaccines. J Immunol 1986;137:2952–2955.

106 Good MF, Berzofsky JA, Malay JA, *et al*. Genetic control of the immune response in mice to a *Plasmodium falciparum* sporozoite vaccine. J Exp Med 1986;164:655–660.

107 Good MF, Berzofsky JA, Miller LH. The T cell response to the malaria circumsporozoite protein: an immunological approach to vaccine development. Ann Rev Immunol 1988;6:633–688.

108 Wirtz RA, Ballou WR, Schneider I, *et al*. *Plasmodium falciparum*: immunogenicity of circumsporozoite protein constructs produced in *Escherichia coli*. Exp Parasitol 1987; 63:166–172.

109 Hoffman SL, Berzofsky JA, Isenbarger D, *et al*. Immune response gene regulation of immunity to *Plasmodium berghei* sporozoites and circumsporozoite protein vaccines. Overcoming genetic restriction with whole organism and subunit vaccines. J Immunol 1989;142:3581–3584.

110 Ballou WR, Sherwood JA, Neva FA, *et al*. Safety and efficacy of a recombinant DNA *Plasmodium falciparum* sporozoite vaccine. Lancet 1987;1:1277–1281.

111 Herrington DA, Clyde DF, Losonsky G, *et al*. Safety and immunogenicity in man of a synthetic peptide and malaria vaccine against *Plasmodium falciparum* sporozoites. Nature 1987;328:257–259.

112 Hoffman SL, Oster CN, Mason C, *et al*. Human lymphocyte proliferative response to a sporozoite T cell epitope correlates with resistance to *P. falciparum* malaria. J Immunol 1989;142:1299–1303.

113 de Groot AS, Johnson AH, Malay WL, *et al*. Human T cell recognition of polymorphic epitopes from malaria circumsporozoite protein. J Immunol 1989;142:4000–4005.

114 Sinigaglia F, Guttinger M, Kilgus J, *et al*. A malaria T cell epitope recognized in association with most mouse and human MHC class II molecules. Nature 1988;336:778–780.

115 Tomlinson LA, Kennedy MW, Christie JF, Fraser EM, McLaughlin D, McIntosh AE. MHC restriction of the antibody repertoire to secretory antigens, and a major allergen of the nematode parasite *Ascaris*. J Immunol 1989; 143:2349–2356.

116 Kee KC, Taylor DW, Cordingley JS, Butterworth AE, Munro AG. Genetic influence on the antibody response to antigens of *Schistosoma mansoni* in chronically infected mice. Parasite Immunol 1986;8:565–574.

117 Else KH, Wakelin D, Wassom DL, Hauda KM. MHC-restricted antibody responses to *Trichuris muris* excretory–secretory (E–S) antigens. Parasite Immunol 1990;12:509–527.

118 Blackwell J, Freeman J, Bradley D. Influence of H-2 complex on acquired resistance to *Leishmania donovani* infection in mice. Nature 1980;283:72–74.

119 Kaye PM, Blackwell JM, Roberts MB. Analysing the immune response to *L. donovani* infection. Ann Inst Pasteur/Immunol 1987;138:762–768.

120 Blackwell JM, Roberts MB. Immunomodulation of murine

visceral leishmaniasis by administration of monoclonal anti-I-A vs. anti-I-E antibodies. Eur J Immunol 1987;17: 1669–1672.

121 Heinzel FP, Sadick MD, Holaday BJ, Coffman RL, Locksley RM. Reciprocal expression of interferon or interleukin 4 during the resolution or progression of murine leishmaniasis. Evidence for expansion of distinct helper T cell subsets. J Exp Med 1989;169:59–72.

122 Scott P, Natovitz P, Coffman RL, Pearce E, Sher A. Immunoregulation of cutaneous leishmaniasis: T cell lines that transfer protective immunity or exacerbation belong to different T helper subsets and respond to distinct parasite antigens. J Exp Med 1988;168:1675–1684.

123 Kaye PM, Currey AJC, Blackwell JM. Differential production of TH1 and TH2 derived cytokines does not determine the genetically controlled or vaccine induced rate of cure in murine visceral leishmaniasis. J Immunol 1991;146:2763–2766.

124 Gepts W. Role of cellular immunity in the pathogenesis of type I diabetes. Curr Probl Clin Biochem 1983;12:86–107.

125 Nishimoto H, Kikutani H, Yamamura K-I, Kishimoto T. Prevention of autoimmune insulitis by expression of I-E molecules in NOD mice. Nature 1987;328:432–434.

126 Boitard C, Michie S, Serrurier P, Butcher GW, Larkins AP, McDevitt HO. *In vivo* prevention of thyroid and pancreatic autoimmunity in the BB rat by antibody to class II major histocompatibility complex gene products. Proc Natl Acad Sci USA 1985;82:6627–6631.

127 Reich E-P, Sherwin RS, Kanagawa O, Janeway CA. An explanation for the protective effect of the MHC class II I-E molecule in diabetes. Nature 1989;341:326–328.

128 Bill J, Appel VB, Palmer E. An analysis of T-cell receptor V-gene expression in MHC-disparate mice. Proc Natl Acad Sci USA 1988;85:9184–9188.

129 Kaye PM, Cooke A, Lund T, Kissousis D, Blackwell JM. Delayed cure from visceral leishmaniasis in mice expressing transgenic I-E molecules. Eur J Immunol 1992;22:357–364.

130 Stern JJ, Oca MJ, Rubin BY, Anderson SL, Murray HW. Role of L3T4[+] and Lyt-2[+] cells in experimental visceral leishmaniasis. J Immunol 1988;140:3971–3977.

131 McElrath JM, Murray HW, Cohn ZA. The dynamics of granuloma formation in experimental visceral leishmaniasis. J Exp Med 1988;167:1927–1937.

132 Roberts M, Kaye PM, Milon G, Blackwell JM. Studies of immune mechanisms in *H-11*-linked genetic susceptibility to murine visceral leishmaniasis. In Hart DT, ed. *Leishmaniasis: Current Status and New Strategies for Control*. New York: Plenum Publishers, 1989;259–266.

133 Wassom DL, Krco CJ, David CS. I-E expression and susceptibility to parasite infection. Immunol Today 1987;8: 39–43.

134 Krco CJ, Wassom DL, Abramson EJ, David CS. MHC-restricted T cell clones specific for a parasite antigen: *Trichinella spiralis*. Immunogenetics 1983;18:435–444.

135 Else KJ, Wakelin D, Wassom DL, Hauda KM. The influence of genes mapping within the major histocompatibility complex on resistance to *Trichuris muris* infections in mice. Parasitology 1990;101:61–67.

136 Behnke JM, Wahid F. Immunological relationships during

primary infection with *Heligmosomoides polygyrus* (*Nematospiroides dubius*). H-2 linked genes determine worm survival. Parasitology 1991;103:157–164.

137 Mahmoud AAF. Genetics of schistosomiasis. In Michal F, ed. *Modern Genetic Concepts and Techniques in the Study of Parasites*. Basel: Schwabe & Co., 1980.

138 Hirayama K, Matsushita S, Kikuchi I, Iuchi M, Ohta N, Sasazuki T. HLA-DQ is epistatic to HLA-DR in controlling the immune response to schistosomal antigen in humans. Nature 1987;327:426–430.

139 Wakelin D. Genetic control of immunity to helminth infection. Parasitol Today 1985;1:17–23.

140 Mitchell GF. Glutathione S-transferases—potential components of anti-schistosome vaccines? Parasitol Today 1989;5:34–37.

141 Coulson PS, Wilson RA. Portal shunting and resistance to *Schistosoma mansoni* in 129 strain mice. Parasitology 1989;99:383–389.

142 Elsaghier AAF, McLaren DJ. *Schistosoma mansoni*: evidence that vascular abnormalities correlate with the "nonpermissive" trait in 129/01a mice. Parasitology 1989;99:377–381.

143 Masur H. The compromised host: AIDS and other diseases. In Walzer PD, Genta RM, eds. *Parasitic Infections in the Compromised Host*. New York: Marcel Dekker, 1989:1–29.

144 Boraschi D, Meltzer MS. Defective tumoricidal capacity of macrophages from P/J mice: characterization of the macrophage cytotoxic defect after *in vivo* and *in vitro* activation stimuli. J Immunol 1980;125:771–776.

145 James SL, Skamene E, Meltzer MS. Macrophages as effectors of protective immunity in murine schistosomiasis. V. Variation in macrophage, schistosomulicidal and tumoricidal activities among mouse strains and correlation with resistance to reinfection. J Immunol 1983;131:948–533.

146 James SL, Correa-Oliveira R, Leonard EJ. Defective vaccine-induced immunity to *Schistosoma mansoni* in P strain mice. J Immunol 1984;133:1587–1593.

147 Correa-Oliveira R, James SL, McCall D, Sher A. Identification of a genetic locus, *Rsm-1* controlling protective immunity against *Schistosoma mansoni*. J Immunol 1986;137:2014–2019.

148 Fortier A, Meltzer MS, Nacy CA. Susceptibility of inbred mice to *Leishmania tropica* infection: genetic control of the development of cutaneous lesions in P/J mice. J Immunol 1984;133:454–459.

149 Correa-Oliveira R, James SL, Sher A. Genetic complementation of defects in vaccine-induced immunity against *Schistosoma mansoni* in P- and A- strain inbred mice. Infect Immun 1988;56:649–653.

150 Butterworth AE. Cell-mediated damage to helminths. Adv Parasitol 1984;23:144–235.

151 Vadas MA. Genetic control of eosinophilia in mice: gene(s) expressed in bone marrow-derived cells control high responsiveness. J Immunol 1982;128:691–695.

152 Wakelin D, Donachie AM. Genetic control of eosinophilia. Mouse strain variation in response to antigens of parasitic origin. Clin Exp Immunol 1983;51:230–246.

153 Handlinger JH, Rothwell TLW. Studies of the responses of basophil and eosinophil leukocytes and mast cells to the nematode *Trichostrongylus colubriformis*: comparison of cell populations in parasite resistant and susceptible guinea pigs. Int J Parasitol 1981;11:67–70.

154 Lammas DA, Mitchell LA, Wakelin D. Genetic influences upon eosinophilia and resistance in mice infected with *Mesocestoides corti*. Parasitology 1990;101:291–299.

155 Lammas DA, Mitchell LA, Wakelin D. Genetic control of eosinophilia. Analysis of production and response to eosinophil differentiating factor in strains of mice infected with *Trichinella spiralis*. Clin Exp Immunol 1988;77:137–143.

156 Askenase PW. Immunopathology of parasitic diseases: involvement of basophils and mast cells. Springer Semin Immunopathol 1979;4:1–59.

157 Matsuda H, Nakono T, Kiso Y, Kitmura Y. Normalization of anti-tick response to mast cell-deficient W/Wᵛ mice by intracutaneous injection of cultured mast cells. J Parasitol 1987;73:155–158.

158 Reed ND. Function and regulation of mast cells in parasite infections. In Galli SJ, Austen KF, eds. *Mast Cell and Basophil Differentiation and Function in Health and Disease*. New York: Raven Press, 1989:99–110.

159 Alizadeh H, Wakelin D. Genetic factors controlling the intestinal mast cell response in mice infected with *Trichinella spiralis*. Clin Exp Immunol 1982;49:331–337.

160 Reed ND, Wakelin D, Lammas DA, Grencis RK. Genetic control of mast cell development in bone marrow cultures. Strain-dependent variation in cultures from inbred mice. Clin Exp Immunol 1988;73:510–515.

161 Wakelin D. Helminth infections. In Wakelin D, Blackwell JM, eds. *Genetics of Resistance to Bacterial and Parasitic Infection*. London: Taylor & Francis, 1988:390–412.

162 Ottesen EA. Filariases and tropical eosinophilia. In Warren KS, Mahmoud AAF, eds. *Tropical and Geographical Medicine*, vol. 1. New York: McGraw-Hill Book Co., 1984.

163 Walzer PD, Genta RM, eds. *Parasitic Infections in the Compromised Host*. New York: Marcel Dekker, 1989.

164 Philipp M, Worms MJ, Maizel RM, Ogilvie B. Rodent models of filariasis. Contemp Top Immunobiol 1984;12:275–321.

165 Hunter KW, Finkelman FD, Strickland GT, Sayles PC, Scher I. Defective resistance to *Plasmodium yoelii* in CBA/N mice. J Immunol 1979;123:133–137.

166 Biozzi G, Mouton D, Siqueira M, Stiffel C. Effect of genetic modification of immune responsiveness on anti-infection and anti-tumor resistance. In Skamene E, ed. *Genetic Control of Host Resistance to Infection and Malignancy*. Prog Leukocyte Biol 3, New York: Alan R Liss, 1985:3–18.

167 Mitchell GF. Genetic variation in resistance of mice to *Taenia taeniaeformis*: analysis of host-protective immunity and immune evasion. In Flisser A, Willms A, Lachette JP, Ridaura C, Beltram F, Larralde C, eds. *Cysticercosis: Present State of Knowledge and Perspectives*. New York: Academic Press, 1982:575–589.

168 Levine RF, Mansfield JM. Genetics of resistance to the African trypanosomes: role of the H-2 locus in determining resistance to infection with *Trypanosoma rhodesiense*. Infect Immun 1981;34:573–578.

169 de Gee ALW, Levine RF, Mansfield JM. Genetics of resistance to the African trypanosomes. VI. Heredity of

resistance and variable surface glycoprotein-specific immune responses. J Immunol 1988;140:283–288.

170 Taylor DW, Pacheco E, Evans CB, Asofsky R. Inbred mice infected with *Plasmodium yoelii* differ in their antimalarial immunoglobulin isotype response. Parasite Immunol 1988;10:33–46.

171 Wahlgren M, Berzins K, Perlmann P, Persson M. Characterization of the humoral immune response in *Plasmodium falciparum* malaria: II: IgG subclass levels of anti-*P. falciparum* antibodies in different sera. Clin Exp Immunol 1983;54:135–142.

172 Parkhouse RME, Harrison LJS. Antigens of parasitic helminths in diagnosis, protection and pathology. Parasitology 1989;99:S5–S19.

173 Gibbens JC, Harrison LJS, Parkhouse RME. Immunoglobulin class responses to *Taenia taeniaeformis* in susceptible and resistant mice. Parasite Immunol 1986;8:491–502.

174 Else KJ, Wakelin D. Genetic variation in the humoral immune response of mice to the nematode *Trichuris muris*. Parasite Immunol 1989;11:77–90.

175 Butterworth AE, Bensted-Smith R, Capron A, *et al*. Immunity in human schistosomiasis mansoni: prevention by blocking antibodies of the expression of immunity in young children. Parasitology 1987;94:281–300.

176 Khalife J, Capron M, Capron A, *et al*. Immunity in human schistosomiasis mansoni. Regulation of protective immune mechanisms by IgM blocking antibodies. J Exp Med 1986;164:1626–1640.

177 Ottesen EA, Skvaril F, Tripathy SP, Poindexter RW, Hussain R. Prominence of IgG4 in the IgG antibody response to human filariasis. J Immunol 1981;127:2014–2020.

178 Ottesen EA, Skvaril F, Tripathy SP, Poindexter RW, Hussain R. Prominence of IgG4 in the IgG antibody response to human filariasis. J Immunol 1985;134:2707–2712.

179 Coffman RL, Seymour BWP, Lebman DA, *et al*. The role of helper T cell products in mouse B cell differentiation and isotype regulation. Immunol Rev 1988;102:5–28.

180 Pfeiffer P, Konig W, Bohn A. Genetic dependence of IgE antibody production in mice infected with the nematode *Nippostrongylus brasiliensis*. I. Modulation of the IgE antibody response *in vivo* by serum factors. Int Arch Allergy Appl Immunol 1983;72:347–355.

181 Rivera-Ortiz CI, Nussenzweig R. *Trichinella spiralis*: anaphylactic antibody formation and susceptibility in strains of inbred mice. Exp Parasitol 1976;39:7–17.

182 Azuma M, Hirano T, Miyajima H, *et al*. Regulation of murine IgE production by recombinant interleukin 4. J Immunol 1987;139:2583–2544.

2 Mononuclear phagocytes and resistance to parasitic infections

Adel A.F. Mahmoud

INTRODUCTION AND DEFINITIONS

The mononuclear phagocyte system is made up of a series of bone marrow, peripheral blood, and tissue cells [1]. In the bone marrow, the first recognizable cell of this series is a dividing pool of promonocytes. Mature monocytes are found in, as yet, a poorly defined storage compartment in the bone marrow but they quickly egress into peripheral blood. The production, maturation, and release of these cells is controlled by several factors, including IL-3, granulocyte macrophage colony stimulating factor (GM-CSF), and M-CSF [2–4]. Approximately 6 days are needed to produce a mature monocyte from its earliest bone marrow precursor. In peripheral blood, monocytes circulate for 1–3 days and then move into tissues to become macrophages where they remain for 4–12 weeks. Although kinetic information on different cell compartments of the mononuclear phagocyte system is incomplete, it is known that there are approximately 50 tissue macrophages for every mature monocyte circulating in peripheral blood. A final stage in cell development is the multinucleated giant cell, which may be found in some granulomatous inflammatory lesions, such as sarcoidosis and tuberculosis.

Migration of mature monocytes into tissues is a significant step in the differentiation of this cell system. There, the cells undergo a series of changes; some are tissue-specific. These include development of antigen-presenting capabilities and cell activation. Both changes are fundamental to the functional role of mononuclear phagocytes. In the context of this discussion, macrophage activation describes increased microbicidal activity in a general sense. This enhanced capability has been shown to play an important role in host defenses, not only against bacterial infections but in resistance to other infectious diseases, including protozoa, helminths, and fungi, and perhaps against tumor cells [5]. The concept of activated macrophages was originally introduced by Mackaness, based on changes in the effector role of these cells [6,7]. It is now appreciated that activated macrophages refer to and include morphologic, metabolic, and functional differentiation of these cells (Table 2.1) and that the process is dependent on several factors, among which interferon-γ (IFN-γ) is most prominent [1,5].

In parasitic infections, the mononuclear phagocytes are involved at multiple levels both in resistance against the invading organisms and in the pathogenesis of disease [8]. Central to this involvement is the immunologic role of mononuclear phagocytes in antigen processing and presentation and as part of the effector arm of immune responses. The focus of this chapter is on the role of mononuclear phagocytes in resistance to parasitic infections. For the purpose of clarifying some mechanistic concepts, the role of mononuclear phagocytes in each type of resistance mechanism to parasitic protozoa and helminths will be discussed separately. This approach is, by necessity, schematic but it is intended to relate phenomena to mechanisms in order to help in formulating a framework for understanding the central role of the cells. We will use examples

Table 2.1 Selected functional differentiation in activated macrophages

Enhanced function	Enhanced secretions
Microbicidal	Complement components
Tumoricidal	Acid phosphatase
Phagocytosis	Collagenase
Pinocytosis	Plasminogen activator
Killing of multicellular organisms	Arginase
	Interleukin-1
	Tumor necrosis factor
	Fibronectin
	Interferon

derived from only a few parasitic infections rather than exhaustively reviewing the details of many systems.

RESISTANCE TO PARASITIC INFECTIONS

We intentionally are using the term "resistance" rather than "immunity" to reflect the multiplicity of "types" of mechanisms contributing to the host's ability to protect itself against a specific microbial pathogen. A major differentiating factor between these "types" of resistance is our understanding of their components and mechanisms. In the following sections we will attempt to describe in detail three models of resistance to parasitic infections, where each is dissected and its effector mechanisms are described. The first type deals with species-related innate resistance which may not be related to the development of a specific immunologic response [9]. On the other hand, specific acquired resistance indicates the involvement of an immunologic mechanism, e.g., the development of specific protective antibody which adoptively transfers this protection to naïve recipients [10]. Similarly, cell-mediated immunity may contribute to protection through sensitization of a specific subset of T cells [11]. The term "cellular immunity" has unfortunately resulted in some confusion. The phenomenon originally was used to describe many initiating stimuli that lead to macrophage activation and acquisition of effector and protective capabilities against multiple organisms. The fact that macrophage activation also occurs during the course of a specific infection and may contribute to protection against the initiating organisms was also referred to as cellular immunity. To distinguish between these two types we will discuss the role of mononuclear phagocytes in specific acquired immunity separate from their role in nonspecific acquired resistance. This separation may be artificial in that both types of resistance mechanisms lead to a final common pathway, i.e., activation of macrophages to become more efficient effector cells. It is, however, an important mechanistic distinction as it relates to the specificity of the initiating stimulus. Finally, it has to be recognized that resistance against parasitic protozoa and helminths involves several other mediators and cells contributing to a complex set of mechanisms [12].

SPECIES-RELATED INNATE RESISTANCE

For decades, it has been known that susceptibility to *Schistosoma mansoni* infection is dependent on the

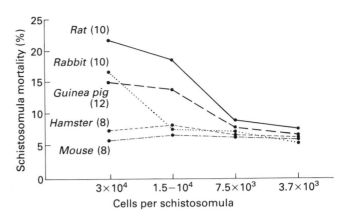

Fig. 2.1 Effect of various cell–parasite ratios on schistosomula killing by proteose peptone-induced peritoneal macrophages of five animal species. Adherent cell monolayers were cultured for 24 h with 200 schistosomula in supplemented RPMI 1640 medium with 10% fetal celf serum. Data are expressed as mean percentage killing; numbers of animals in each group are in parentheses.

species of the host animal. For example, while mice and hamsters are susceptible to cercariae of *S. mansoni*, rabbits are significantly more resistant and no patent infection can be established in fox or muskrat [9]. Conversely, humans are susceptible to five species of schistosomes and are resistant to many species of avian parasites. The mechanisms and mediators of this type of species-related resistance have been examined in our laboratories. Mononuclear phagocytes were selected for detailed examination because of previous studies showing their involvement in nonantibody-dependent *in vitro* killing of schistosomula of *S. mansoni* [13,14]. In our studies, peritoneal macrophages induced by proteose peptone were obtained from animals known to be either susceptible or resistant *in vivo*. Adherent peritoneal exudate cells from rats, guinea pigs, or rabbits, but not mice or hamsters, killed significant proportions of schistosomula following 24 h of incubation (Fig. 2.1). Killing was shown to be dependent on the species and on the cell–parasite ratio. Furthermore, peripheral blood mononuclear cells from a resistant animal species (rabbit) also showed significant ability to kill parasites *in vitro*. In Table 2.2 we summarize our data on the reproduction of parasite killing using culture supernatants when rat peritoneal exudates or rabbit peripheral blood mononuclear cells were used but not when mouse peritoneal exudates were incubated with the organisms. In these studies, as in previous investigations of the role of arginase (see below), the addition

Table 2.2 Role of mononuclear phagocytes from different animal species in innate resistance to *Schistosoma mansoni*: comparison of schistosomula killing by adherent cell monolayers or culture supernatants (200 schistosomula were incubated for 24 h with adherent monolayers; supernatants were collected, concentrated by PM-10 filter to one-fifth volume, and tested for cytotoxicity on fresh organisms)

Incubations	Killing by cell monolayers (%)	Killing by culture supernatants (mean ± SE)	Killing by supernatant effluent
Proteose peptone-induced rat peritoneal exudates + schistosomula	18.7 ± 2.8*	19.2 ± 0.8[†]	6.0 ± 1.15[‡]
Rabbit peripheral blood mononuclear cells + schistosomula	34.6 ± 5.8*	21.4 ± 4.2[†]	4.2 ± 1.2[‡]
Proteose peptone-induced mouse peritoneal exudates + schistosomula	6.1 ± 1.0	4.8 ± 0.7	5.2 ± 0.5
RPMI 1640 medium + schistosomula	4.23 ± 1.24*	5.74 ± 0.98[†]	ND

* Difference between parasite killing by rat peritoneal exudates or rabbit peripheral blood mononuclear cells and in controls incubated with medium alone is significant at the 0.1% level.
[†] Difference between killing by supernatants from rat peritoneal exudate culture or rabbit peripheral blood mononuclear cells and those from mouse peritoneal exudates or controls is significant at the 0.1% level.
[‡] Difference between killing by supernatant from rat peritoneal exudates or rabbit peripheral blood mononuclear cell cultures and their effluent from a PM-10 filter is significant at the 1% level.
ND, not determined.

of L-arginine in increasing concentration resulted in decreased killing (Fig. 2.2). These results show that the known variability of *in vivo* susceptibility or resistance to *S. mansoni* infection correlates with the *in vitro* degree of cytotoxicity exhibited by each species of mononuclear phagocyte. Furthermore, these observations confirm a role for arginase in parasite killing, as described below.

SPECIFIC ACQUIRED IMMUNITY

Acquired immunity to parasitic protozoa or helminths has been demonstrated in several animal species. In most models, however, the acquired immunity is partial and may be enhanced by adjuvants or other immuno-potentiators. For example, multiple mechanisms of resistance to schistosomiasis have been described involving antibodies, complement, and a host of effector cells [15]. We will briefly outline the role of mononuclear phagocytes in specific acquired immunity to schistosomiasis. James *et al.* [11,16] examined in detail the activity of mouse macrophages during the course of *S. mansoni* infection. They demonstrated that activation of these cells as a result of the helminth infection enhances their ability to kill the schistosomula stage *in vitro*. Their observations add to the multiple mechanisms previously described as potentially involved

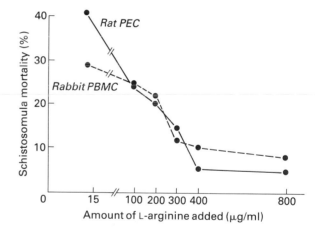

Fig. 2.2 Blocking of schistosomula killing by rat peritoneal exudate cells (PEC) or rabbit peripheral blood mononuclear cells (PBMC) (cell/target ratio $3 \times 10^4 : 1$) by various amounts of L-arginine. Adherent cell monolayers were cultured for 24 h with 200 schistosomula and 10% fetal calf serum; L-arginine was added at the initiation of incubations. Data are expressed as mean percentage killing of parasites based on three separate experiments.

in protection to infection and may dictate an approach to vaccine development that is based on using several antigens inducing multiple "types" of protective responses [17].

25

NONSPECIFIC ACQUIRED RESISTANCE

Protozoan infections

Nonspecific acquired resistance against several protozoan infections could be demonstrated in mice following injection of Bacille Calmette-Guerin (BCG), *Corynebacterium parvum*, or more recently by defined compounds such as muramyl dipeptide (MDP). Mice treated with these preparations have been shown to be significantly protected against a subsequent challenge with several protozoan species, such as *Plasmodium* or *Babesia*. Nonspecific resistance to protozoa can also be demonstrated *in vitro*: BCG increases the resistance of murine macrophages to *Trypanosoma cruzi*. Most of these observations can be reproduced using IFN-γ [5]. As can be seen, all studies on induction of nonspecific resistance to protozoan infections utilized intracellular parasites. The importance of localization of these parasites is unclear.

Induction of nonspecific resistance against murine infection with *Leishmania donovani* using BCG has been demonstrated both *in vivo* and *in vitro*. This enhanced activity of host macrophages has been studied extensively over the past decade. It is now recognized that "activation" of host macrophages may result not only from injection of BCG but following exposure of macrophages *in vitro* to soluble products of antigen or mitogenic stimulation of host lymphocytes. The requirements and characteristics of this protective process will be examined using *Leishmania* infection as a model. Davis *et al*. [18] recently reported a series of observations on the activation of mouse macrophages *in vitro* to kill increased numbers of *Leishmania major* (NIH strain 173). Enhanced resistance of macrophages was noted following exposure to an increased concentration of lymphokines obtained from culture supernatant of spleen cells of animals previously immunized with BCG (Table 2.3). In contrast, they observed that treatment of macrophages by increased doses of rIFN-γ showed no significant effect on the percentage of *L. major*-infected cells. This observation seems contradictory to what has been reported previously by Nacy *et al*. [19], demonstrating that IFN-γ is capable of activating resident macrophages to kill *Leishmania tropica*. In the recent study by Davis *et al*. [18] it appears that for macrophage activation to resist infection with *L. major*, more than one factor is required (Table 2.4). Although several earlier studies had demonstrated IFN-γ to be the main activating factor, the current hypothesis suggests the need for

Table 2.3 Treatment of macrophages with lymphokines and rIFN-γ for induction of resistance to *Leishmania* infection.* (From Davis *et al*. [18])

4-h Treatment of macrophages (dilution)	Percentage of infected cells	Percentage resistance to infection
Medium	35 ± 4	0
Lymphokines		
1/6	16 ± 3	55
1/18	17 ± 2	52
1/54	26 ± 3	27
rIFN-γ (U/ml)		
1000	34 ± 5	5
100	30 ± 3	16
10	36 ± 3	0
1	38 ± 6	0

* Macrophages were treated with lymphokines or rIFN-γ for 4 h, washed, and exposed to parasites at a multiplicity of infection of two amastigotes per macrophage for 2 h. Cells were examined microscopically and results were expressed as mean percentage of infected macrophages ± SEM and percentage resistance to infection, defined as the decrease in infected macrophages in treated cultures compared to cells in medium-treated controls. From 300 to 600 macrophages were observed in triplicate cultures for each data point. The 95% confidence limit for replicate control cultures were 21% and cultures with resistance to infection of >21% were significant at the 0.05 level.

IFN-γ and other lymphokines yet to be defined. This synergistic involvement of cytokines in the activation of other cells, such as T lymphocytes, has been demonstrated previously [20,21]. Furthermore, the interaction of mononuclear phagocytes and IFN-γ seems to depend on the order of cellular exposure to the cytokine or parasites, and probably on factors secreted by the activated cells themselves. Recently, Engelhorn *et al*. [22] have utilized human peripheral blood monocytes obtained from healthy volunteers to examine these phenomena. Prior exposure of the cells to rIFN-γ significantly decreased the number of promastigotes of *L. donovani* per monocyte. In contrast, the addition of INF-γ following monocyte infection with the protozoan failed to limit their multiplication in the cells (Table 2.5). This apparent contradiction was found to be due to a soluble factor which is released in supernatants of *Leishmania*-infected monocytes. The activity is dialyzable, and is destroyed by protease treatment. Production of this so-called activation suppressor factor is dependent on infection of human monocytes with viable *Leishmania* organisms and on the intact protein synthetic capabilities of the cells.

Table 2.4 Effect of sequential treatment of macrophages with lymphokines (LK) and rIFN-γ in the presence of protein synthesis inhibitors on resistance to *Leishmania* infection.* (From Davis *et al*. [18])

Treatment of macrophages

First signal (20 h)	Second signal (1 h)	Inhibitor (5 h)	Percentage of infected cells	Percentage resistance to infection
Medium	Medium	None	30 ± 3	0
LK	LK	None	13 ± 2	57
LK–IFN	Medium	None	28 ± 3	7
	Medium	Emetine	29 ± 2	3
	Medium	Cycloheximide	26 ± 3	13
	rIFN-γ	None	15 ± 2	50
	rIFN-γ	Emetine	25 ± 3	16
	rIFN-γ	Cycloheximide	23 ± 2	23
rIFN-γ	Medium	None	26 ± 3	13
	Medium	Emetine	30 ± 3	0
	Medium	Cycloheximide	30 ± 4	0
	LK–IFN	None	12 ± 4	60
	LK–IFN	Emetine	27 ± 2	10
	LK–IFN	Cycloheximide	25 ± 2	16

* Macrophages were treated with lymphokines and rINF-γ for 20 h, washed, pulsed with the opposite activation signal or medium for 1 h, washed, and incubated for 5 h with or without the protein synthesis inhibitors emetine (10 μg/ml) or cycloheximide (10 μg/ml). Cells were then exposed to parasites at a multiplicity of infection of two amastigotes per macrophage for 2 h. Cells were examined microscopically and results were expressed as mean percentage of infected macrophages ± SEM and percentage resistance to infection, defined as the decrease in infected macrophages in treated cultures compared to cells in medium-treated controls. From 300 to 600 macrophages were observed in triplicate cultures for each data point. The 95% confidence limit for replicate control cultures was 26%, and cultures with resistance to infection of >26% were significant at the 0.05 level. Results are from one of three similar experiments of similar design and indistinguishable results.

What has been learned from the detailed studies of one system indicates a complex series of mechanisms. Resistance of mononuclear phagocytes to infection with *Leishmania* is dependent on the cell source (mouse or human), on the parasite species, on the conditions prior to exposing the cells to parasites or activating molecules, such as INF-γ, on other lymphokines, and more recently on other regulatory factors secreted by parasitized cells.

Table 2.5 Effect of monocyte infection with *Leishmania donovani* before activation with IFN-γ on resistance to parasites. (From Engelhorn *et al*. [22])

	Total number of parasites/100 monocytes*			
	At 2 h		At 48 h	
Monocyte treatment	Exp. 1	Exp. 2	Exp. 1	Exp. 2
None†	73 ± 12	60 ± 13	281 ± 83	282 ± 23
IFN-γ prior to infection‡	96 ± 24	72 ± 18	87 ± 18‖	60 ± 6‖
Infection prior to IFN-γ§	73 ± 12	60 ± 13	251 ± 29	220 ± 35

* Values represent means of triplicates ± SE.
† Monocytes were infected with promastigotes of *L. donovani* for 2 h. In all experiments, coverslips were removed at 2 h and 48 h after infection.
‡ Monocytes were activated with 1000 U/ml IFN-γ for 72 h before infection.
§ Monocytes were infected with promastigotes of *L. donovani* for 2 h and then treated with 1000 U/ml IFN-γ for 48 h.
‖ $P \leqslant 0.005$.

Metazoan infections

Multicellular parasitic worms invade host tissues through skin or gastrointestinal mucosa; their adult stages vary in size from a few millimeters to several meters. Worms, therefore, represent a formidable challenge to host protective mechanisms [17]. Because worms usually do not multiply within their definitive host, they can serve as better models to define at least three types of resistance mechanisms. In the previous sections we introduced two "types": species-related innate resistance and specific acquired immunity. While studies on specific acquired immunity to worms have progressed considerably, attempts to define nonspecific resistance are just beginning. We learned of the feasibility of induction of nonspecific resistance to worm infections such as *S. mansoni*, *Trichinella spiralis* and others through several phenomena, as discussed below. Several new approaches had to be designed to allow better understanding of the interaction between the host and the multicellular invaders.

Induction of resistance to *S. mansoni* related to nonspecific mechanisms can be traced to early experiments employing a variety of substances, such as snail hemolymph, monkey erythrocytes, or killed *Escherichia coli* [23–25]. As the concept of nonspecific resistance was in its infancy, the significance of these observations was poorly understood. Over the past two decades, several attempts have been performed to examine systematically the role of mononuclear phagocytes in resistance against *S. mansoni* infection in mice and humans. These studies paved the way not only to the understanding of induced protection against this helminth but also developed our current appreciation of the phenomenon of macrophage activation, its mechanisms, and mediators. In earlier experiments, resistance was induced by *Toxoplasma gondii* [26]. Outbred Swiss albino mice were injected intraperitoneally with 10 cysts of the Gleadle strain of *T. gondii* (nonlethal in mice). At varying intervals thereafter, groups of animals were challenged with *S. mansoni* cercariae and adult worms were perfused from the portal circulation. Significant protection (35%) was demonstrated in *T. gondii*-infected mice. The protective effect was detected between 1 day and 4 weeks following the protozoan infection. In contrast, no resistance was noted in animals exposed to *T. gondii* following *S. mansoni* infection.

Bacille Calmette-Guerin was used in subsequent studies to allow comparison with its effect on protozoa and tumors and to define the characteristics of nonspecific resistance [27]. Following the intravenous injection of 2×10^7 colony-forming units (CFU) of BCG (Tice strain) into outbred mice, the animals were exposed to *S. mansoni*. Bacille Calmette-Guerin-treated mice were significantly more resistant to schistosomiasis, as evidenced by reduction of schistosomula recovery from the lungs or adult worms from the portal circulation (Table 2.6). Resistance was demonstrated in the treated

animals even if BCG administration was delayed for 3 days after cercarial challenge; mice were protected for at least 8 weeks following a single dose of BCG. The route of BCG injection proved just as important as in the case of protozoan infection; intravenous but not intraperitoneal or subcutaneous administration was protective. Another factor which proved equally relevant relates to the dose and species of *Mycobacterium*. In our studies, 2×10^7 CFU of BCG-treated mice obtained from the Chicago Research Foundation were used regularly to induce approximately 50% resistance in the recipient animals. Decreasing this dose to 10^6 CFU was associated with loss of protective activity. Similarly, other organisms obtained from commercial sources or live *Mycobacterium* showed marked variability in the degree of induced resistance.

The relationship of nonspecific resistance to the genetic background of hosts was subsequently examined [28]. Evidence to suggest differences in the degree of resistance observed upon vaccination of inbred strains of mice has been obtained from experiments using attenuated mycobacteria, *Salmonella* species, and *Listeria monocytogenes*. In our studies on BCG-induced protection against *S. mansoni* [28], it was found that resistance segregates in different inbred strains into high and low responders (Table 2.7). For example, significant protection was seen in the following strains: A/J(H-2ᵃ), C57BL/6 and C57BL/10(H-2ᵇ), DBA/2(H-2ᵈ), C57BR (H-2ᵏ), and SJL(H-2ˢ). In contrast, some strains with similar haplotypes were either not protected, such as BALB/c(H-2ᵈ), or acquired far less protection, such as C3H/He and CBA(H-2ᵏ). The ability to acquire nonspecific resistance following BCG vaccination was dominantly inherited and did not segregate with the H-2 complex as observed in several congenic strains. Bacille Calmette-Guerin induced resistance equally well in mice carrying the high-responder background with a substituted H-2 complex. On the other hand, mice carrying the non-responder background, such as BALB.B10, were not protected in spite of having the major histocompatibility complex (MHC) of responder animals. Protection induced by BCG against *S. mansoni* infection was, therefore, genetically controlled by genes not linked to the MHC.

In recent years, we have examined alternative methods for induction of resistance to *S. mansoni* infection in mice [29]. In these studies, freeze–thaw schistosomula or their extract resulted in significant resistance. The antibody response of immunized mice was limited; sera from vaccinated animals identified

Table 2.6 Protective effect of BCG against challenge with *Schistosoma mansoni**

Duration of *S. mansoni* infection	Schistosomula recovery from lungs		Adult worm recovery	
	Controls	BCG-treated	Controls	BCG-treated
3 days	18 ± 2	6 ± 1	–	–
5 days	166 ± 16	58 ± 6	–	–
7 days	177 ± 10	57 ± 6	–	–
8 weeks	–	–	40 ± 4	19 ± 1

*CG1 mice were each injected intravenously with 2×10^7 CFU of BCG-Tice. Two weeks later they were either exposed percutaneously to 500 cercariae for the lung recovery experiments or to 100 cercariae for adult worm recovery. Animals were individually exposed to *S. mansoni* cercariae.

Table 2.7 Role of genetic background of mice strains in nonspecific resistance to *Schistosoma mansoni**

Strain inbred	H-2 haplotype	Percentage protection	P
High responders			
A/J	a	39	<0.01
C57BL/6	b	50	<0.01
C57BL/10	b	49	<0.01
DBA/2	d	35	<0.01
C57BR	k	31	<0.01
SJL	s	65	
Low responders			
C3H/He	k	16	<0.05
CBA	k	19	<0.05
Nonresponders			
BALB/c	d	3	NS
Congenic			
High responders			
B10.A	a	53	<0.01
B10.D2	d	40	<0.01
Nonresponders			
BALB.B10	b	0.4	NS

* Mice were injected intravenously with 2×10^7 CFU of BCG-Tice. Two weeks later they were challenged percutaneously with 500 *S. mansoni* cercariae. Resistance was assessed by schistosomula recovery from the lungs on day 6. NS, not significant.

Table 2.8 Production of macrophage-activating lymphokine by splenocytes from intradermally immunized mice. (From James [16])

Spleen cell donors*	Antigen†	Dilution	Larvacidal activity‡ (%)
Control	None	1/10	13 ± 3
	SWAP	1/10	11 ± 4
	Freeze–thaw larvae	1/10	9 ± 3
BCG	None	1/10	11 ± 2
	SWAP	1/10	12 ± 5
	Freeze–thaw larvae	1/10	8 ± 1
	PPD	1/10	70 ± 11
		1/100	29 ± 6
		1/200	11 ± 2
SWAP + BCG	None	1/10	7 ± 1
	SWAP	1/10	100 ± 0
		1/100	34 ± 8
		1/200	15 ± 2
	Freeze–thaw larvae	1/10	100 ± 0
		1/100	32 ± 5
		1/200	28 ± 3
Freeze–thaw larvae + BCG	None	1/10	11 ± 6
	SWAP	1/10	100 ± 0
		1/100	43 ± 8
		1/200	36 ± 5
	Freeze–thaw larvae	1/10	100 ± 0
		1/100	46 ± 9
		1/200	32 ± 4

* Splenocytes were obtained from C57BL/6 mice maintained as untreated controls, or treated by intradermal injection of 5×10^6 CFU of BCG, 1 mg of soluble worm antigen preparation + BCG, or 10^4 freeze–thaw larvae + BCG several weeks previously.
† Splenocytes were stimulated *in vitro* with SWAP (500 µg/ml), freeze–thaw schistosomula (5000/ml), or PPD (50 µg/ml). Supernatant fluids were removed after 48 h and were used at the given dilutions to activate IC-21 cells to kill schistosomula in the larvacidal assay.
‡ Results represent the mean ± SD for two experiments.

three bands in schistosomula or adult worm extracts at 97, 68, and 43 kD [30]. One of these parasite antigens turned out to be schistosome paramyosin. Upon injection into mice, the purified paramyosin induced approximately 24% protection. When combined with BCG, a significant enhancement of resistance to 52% was achieved [30]. This set of observations as well as other reports indicate the existence of two additive mechanisms for resistance against schistosomiasis. One is induced specifically by several parasite antigens [12] and the other is nonspecifically induced by agents such as BCG. Simultaneously, James [16] examined the utilization of this model in exploring the mechanisms of resistance. In her studies, intradermal injection of freeze–thaw schistosomula and BCG was used. The acquired resistance in this model was associated with the development of delayed footpad swelling to parasite antigen. When soluble worm antigen was injected intraperitoneally, a population of activated macrophages was obtained which induced significant killing of schistosomula. Furthermore, control macrophages could be induced to kill schistosomula following incubation

with lymphokines obtained from antigen stimulation of spleen cells of immunized mice (Table 2.8).

Similar observation on the ability of human mononuclear phagocytes to kill schistosomula of *S. mansoni* have been made. Cells obtained from the peripheral blood of normal volunteers killed a significant percentage of organisms [13]. In addition, cells obtained from purified protein derivative (PPD)-positive individuals killed a higher proportion of targets [8,31]. Human mononuclear phagocytes could be induced to kill a significant proportion of incubated organisms following

Table 2.9 Killing of schistosomula of *Schistosoma mansoni* by human monocyte-derived macrophages cultured in presence or absence of recombinant cytokines. (From James *et al.* [32])

| Culture period* | Cytokine addition† | | Percentage of dead larvae‡ |
	Culture	Assay	
None		None	15 ± 7
		IFN-γ (200 U/ml)	10 ± 4
4 days	None	None	18 ± 2
	None	IFN-γ (60 U/ml)	20 ± 5
	None	IFN-γ (300 U/ml)	22 ± 0
	None	IFN-γ (1200 U/ml)	20 ± 15
1 week	None	None	23 ± 8
	None	IFN-γ (200 U/ml)	27 ± 17
	CSF-1 (1000 U/ml)	None	40 ± 5
	CSF-1 (1000 U/ml)	IFN-γ (200 U/ml)	58 ± 12
	IFN-γ (100 U/ml)	None	57 ± 5
	IFN-γ (100 U/ml)	IFN-γ (200 U/ml)	62 ± 11
2 weeks	None	None	33 ± 15
	None	IFN-γ (200 U/ml)	39 ± 11
	CSF-1 (1000 U/ml)	None	39 ± 14
	CSF-1 (1000 U/ml)	IFN-γ (200 U/ml)	68 ± 11
	IFN-γ (100 U/ml)	None	41 ± 1
	IFN-γ (100 U/ml)	IFN-γ (200 U/ml)	53 ± 4
3 weeks	None	None	36 ± 3
	None	IFN-γ (200 U/ml)	51 ± 13
	CSF-1 (1000 U/ml)	None	43 ± 9
	CSF-1 (1000 U/ml)	IFN-γ (200 U/ml)	71 ± 4

* Blood monocytes were obtained from normal donors and either used directly in the larvacidal assay or cultured *in vitro* for up to 3 weeks.
† Cells were cultured continuously in the presence or absence of recombinant human CSF-1 or IFN-γ at the concentrations shown. In some cases, fresh IFN-γ was added when the larvacidal assay was initiated.
‡ Larval viability was assessed after 48 h. Data represent the mean ± SD of two or three samples. Background larval mortality in these experiments was 12–18%.

exposure to IFN-γ and colony stimulating factor-1 [32]. These results are summarized in Table 2.9 and correlate with similar findings in mice.

In an attempt to define the active factors responsible for induction of nonspecific resistance, we examined the effects of several natural and synthetic mycobacterial extracts and products [33]. There is rapidly growing evidence that smaller, well-defined, synthetic molecules can reproduce most of the adjuvant and/or immunostimulant effects of BCG [34]. We evaluated the effects of these products on murine susceptibility to schistosomiasis. Natural cord factor was found to induce

significant protection in mice when given intravenously, intraperitoneally, or subcutaneously. This natural product of *Mycobacterium* can be solubilized in water and need not be suspended in oil or other vehicles. Examination of other mycobacterial products showed that some synthetic compounds, such as trehalose dipalmitate, induce a similar degree of resistance to that detected following BCG administration.

EFFECTOR MECHANISMS OF NONSPECIFIC RESISTANCE

The mononuclear phagocytes are armed with several effector mechanisms which can bring about target killing. Two systems have been described extensively: oxygen-dependent and oxygen-independent systems. The oxygen-dependent systems of the mononuclear phagocytes have been shown to play a crucial role in their antimicrobial effects. Peroxidase activity has been detected in some cells in this series; the mononuclear phagocyte peroxidase activity differs, however, from that of neutrophils by its greater inhibition by 3-amino-1,2,4-triazole and by its inability to catalyze peroxidative carboxylation of amino acids. Accumulation of peroxidase in monocytes may occur during their development in bone marrow, may be formed at extramedullary sites, or the cells may take the enzyme from the extracellular fluid by phagocytosis or pinocytosis [35]. The involvement of monocyte granule-associated peroxidase in their antimicrobial activity has been shown in studies measuring H_2O_2 production during phagocytosis and by the effect of enzyme inhibitors, as well as in cells from patients with chronic granulomatous disease and myeloperoxidase deficiency [1]. Oxygen-dependent but peroxidase-independent antimicrobial systems involving, for example, hydroxyl radicals have also been described in mononuclear phagocytes, but their relative importance is still unclear. On the other hand, several oxygen-independent systems have been described to explain the cytotoxic effect of mononuclear cells, particularly against tumor cells [36,37]. Particular models vary in regard to whether the oxygen-dependent or oxygen-independent mechanisms are of primary importance. Furthermore, dissociation of antimicrobial and cytotoxic effects of macrophage have been noted in several experimental models [38].

Killing of the intracellular parasites *T. cruzi* and *T. gondii* by activated macrophages has been studied by several investigators [5]. Injection of mice with BCG followed by intraperitoneal challenge with PPD-generated peritoneal macrophages capable of killing trypomastigotes [39]. In contrast, Murray & Cohn [40] found *T. gondii* to be more resistant to H_2O_2 than *T. cruzi*. These observations were limited to measuring one product of oxidative metabolism; the role of others and of the nonoxygen-dependent mechanisms in macrophage-mediated killing of protozoa remain to be explored.

The potential of activated macrophages to kill multicellular parasites *in vitro* was first evaluated using *S. mansoni* (Fig. 2.3) [41]. Peritoneal macrophages obtained from mice treated *in vivo* with BCG or *C. parvum* demonstrated increased *in vitro* killing of schistosomula. Resident peritoneal macrophages or those elicited with proteose peptone showed killing which was not different from background mortality of the parasites. The supernates of macrophages from *C. parvum*-treated mice cultured in the presence of schistosomula retained a significant degree of parasite cytotoxicity; upon incubation with fresh organisms, parasite mortality was demonstrable. Parasite killing in this system is nonspecific since the release of cytotoxic mediators in the culture supernates can be induced by several stimuli [41]. The cells involved in nonspecific resistance to schistosomiasis *in vitro* were evaluated using depletion and reconstitution studies. The cytotoxic effect was dependent on the presence of macrophages or their products in the final step. Activation of the macrophages, however, needed a T cell-dependent step which can be substituted for by supernatants of sensitized T lymphocytes cultured with a specific antigen [42] or with IFN-γ [5,32]. Furthermore, the ability of activated macrophages to kill *S. mansoni* schistosomula *in vitro* is a genetically restricted phenomenon similar to that previously demonstrated *in vivo* [31]. Macrophages from *C. parvum*-treated C57BL/6J mice showed increased killing of schistosomula, whereas those from BALB/CJ animals did not (Table 2.10).

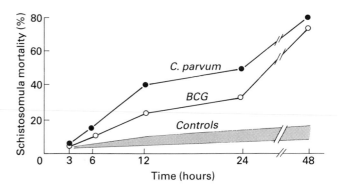

Fig. 2.3 Effect of peritoneal macrophage monolayers from *Corynebacterium parvum* on BCG-treated mice or normal animals on the viability of *Schistosoma mansoni* schistosomula.

Investigations of multicellular parasite killing by activated macrophages have addressed some of the questions posed by the nature and complexity of the target organisms. *In vitro*, it has been observed that killing of the multicellular targets occurs well before any gross evidence of parasite disintegration or phagocytosis by effector cells [13]. What mediates parasite cytotoxicity in these systems has been the subject of intensive investigations. The role of oxygen-dependent antimicrobial systems has been assumed to be similar to antibody-dependent eosinophil-mediated cytotoxicity [15]. It has to be realized, however, that no systematic examination of oxygen-dependent activated macrophage-mediated killing of multicellular targets has been performed. In contrast, several nonoxygen-dependent systems have been reported. In one set of studies, the genetic dependency on nonspecific resistance to schistosomiasis was used to explore the role of arginase. *Corynebacterium parvum* treatment of C57BL/6 mice resulted in activated macrophages which kill significant numbers of schistosomula *in vitro* and greater production of arginase by these cells (Table 2.10) [42]. The cytotoxic activity of

Table 2.10 Strain dependence of nonspecific resistance *in vitro*: effect of *Corynebacterium parvum*-treated macrophages on killing of *Schistosoma mansoni* schistosomula and on arginase and H_2O_2 production

Mouse strain	Treatment	Percentage schistosomula killing	Arginase (U/ml)	H_2O_2 (nmol/5 min/ 2×10^6 cells)
C57BL/6J	Controls	5 ± 4	0.9 ± 0.1	0.06 ± 0.01
	Corynebacterium parvum	30 ± 5	3.8 ± 0.8	0.23 ± 0.01
BALB/CJ	Controls	7 ± 1	1.1 ± 0.2	0.11 ± 0.01
	Corynebacterium parvum	6 ± 2	0.7 ± 0.2	0.39 ± 0.01

individual preparations of activated macrophages correlated significantly with the level of arginase measured in the culture supernatants (Fig. 2.4). In contrast, *C. parvum* treatment of BALB/CJ mice resulted in macrophages which were incapable of parasite cytotoxicity or of producing significant amounts of arginase in the culture medium. Confirmatory evidence for a central role for arginase was obtained from blocking experiments; cytotoxicity can be ablated by adding L-arginine and can be reproduced if exogenous arginase is used instead of activated macrophage. It is important to note that in these studies H_2O_2 production by *C. parvum*-treated macrophages of the two mouse strains tested was similar. Recent evidence obtained from mice immunized with schistosome paramyosin [30] supports a role for arginase in parasite killing.

Recently, another mediator system has been described in activated macrophages [43]. These cells are capable of forming reactive nitrogen intermediates from the terminal guanidionnitrogen atoms of L-arginine. An important role for these reactive nitrogen intermediates in the antitumor and antimicrobial effects of activated macrophage has been demonstrated [43,44]. Similar observations have now been reported using protozoa [45,46] or helminths [47] as targets.

ADJUVANTS AND IMMUNOPOTENTIATORS

Immunization against parasitic infections is an obvious goal to complement our armamentarium for controlling some of the most devastating diseases of humans. Experience over the past several decades indicates that resistance to parasites, whether induced or naturally occurring, is partial at best. Furthermore, the experience with some bacterial products, e.g., tetanus toxoid, indicates the necessity of using adjuvants such as aluminum hydroxide in order to achieve clinically meaningful degrees of protection. The recognition that, in parasitic infections, single antigens or subunit vaccines will not be immunogenic enough to lead to significant protection has resulted in examination of potentially useful adjuvants in humans [34]. These new substances and formulations are designed to enhance a specific subset of the host immune responses thought to be associated with protection. They also are structured in such a way to minimize the known side-effects of adjuvant therapy. Several preparations are currently being tested in humans, including MDP, pluronic polyals, liposomes,

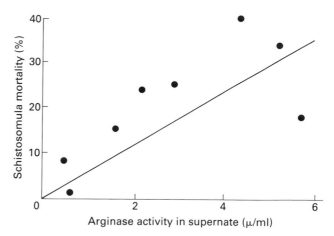

Fig. 2.4 Correlation of percentage dead *Schistasoma mansoni* schistosomula and arginase content of culture supernates of *Corynebacterium parvum*-treated peritoneal macrophages. Correlation coefficient = 0.72 ($P < 0.025$).

and immunostimulatory complexes (ISCOMS). Current vaccination efforts using malaria antigens or subunit vaccines or antigens from several other organisms in combination with these adjuvants show marked improvement in immunogenecity. Their role, however, in the formulation of vaccines for human use remains unclear.

REFERENCES

1 Johnston RB Jr. Monocytes and macrophages. N Engl J Med 1988;318:747.
2 Caracciolo D, Shirsat N, Wong GG, *et al.* Recombinant human macrophage colony stimulating factor (M-CSF) requires subliminal concentrations of granulocyte macrophage (GM) CSF for optimal stimulation of human macrophage colony formation *in vitro*. J Exp Med 1987;166: 1851.
3 Metcalf D. The granulocyte–macrophage colony stimulating factors. Science 1985;229:16.
4 Sieff CA. Hematopoietic growth factors. J Clin Invest 1987; 79:1549.
5 Murray HW. Interferon-gamma activated macrophages, and host defense against microbial challenge. Ann Intern Med 1988;108:595.
6 Mackaness GB. The monocyte in cellular immunity. Semin Hematol 1970;7:172.
7 Mackaness GB. Resistance to intracellular infection. J Infect Dis 1971;123:439.
8 Ellner JJ, Mahmoud AAF. Phagocytes & worms: David & Goliath revisited. Rev Infect Dis 1982;4:698.
9 Peck C, Carpenter M, Mahmoud AAF. Species-related innate resistance to schistosomiasis mansoni; role of mono-

nuclear phagocytes in schistosomula killing *in vitro*. J Clin Invest 1983;71:66.

10 Miller LH. Strategies for malaria control: realities, magic, and science. Ann NY Acad Sci 1989;569:118–126.

11 James SL, Scott P. Induction of cell-mediated immunity as a strategy for vaccination against parasites. In Englund PT, Sher A, eds. *The Biology of Parasitism*.

12 Mahmoud AAF. Strategies for vaccine development: schistosomiasis. Ann NY Acad Sci 1989;569:136.

13 Ellner JJ, Mahmoud AAF. Killing of schistosomula of *Schistosoma mansoni* by normal human monocytes. J Immunol 1979;123:949.

14 Ellner JJ, Mahmoud AAF. Cytotoxicity of activated macrophages for the multicellular parasite *Schistosoma mansoni*. In Pick E, ed. *Lymphokine Reports*. New York: Academic Press, 1981:231.

15 Capron A, Dessaint JP, Capron M, Ouma JH, Butterworth AE. Immunity to schistosomes: progress towards vaccine. Science 1987;238:1065.

16 James SL. Induction of protective immunity against *Schistosoma mansoni* by a nonliving vaccine III correlation of resistance with induction of activated larvicidal macrophages. J Immunol 1986;136:3872.

17 Mahmoud AAF. Parasitic protozoa and helminths: biological and immunological challenges. Science 1989;246:1015.

18 Davis CE, Belosevic M, Meltzer MS, *et al*. Regulation of activated macrophage antimicrobial activities: cooperation of lymphokines for induction of resistance to infection. J Immunol 1988;141:627.

19 Nacy CA, Fortier AH, Meltzer MS. Macrophage activation to kill *Leishmania tropica*: macrophages can be activated to kill amastigotes by both interferon and non-interferon lymphokines. J Immunol 1985;135:3505.

20 Karray S, Vazquez A, Merle-Beral H, *et al*. Synergistic effect of recombinant IL-2 and INF-γ on the proliferation of human monoclonal lymphocytes. J Immunol 1987;138:3824.

21 Aribia MHB, Leroy E, Lantz O, *et al*. rIL-2 induced proliferation of circulating NK cells and T lymphocytes: synergistic effects of IL-1 and IL-2. J Immunol 1987;139:443.

22 Engelhorn S, Buckner A, Remold HG. A soluble factor produced by inoculation of human monocytes with *Leishmania donovani* promastigotes suppresses IFN-γ-dependent monocyte activation. J Immunol 1990;145:2662.

23 Oliver-Gonzalez J. Our knowledge of immunity to schistosomiasis. Am J Trop Med Hyg 1967;16:565.

24 Oliver-Gonzalez J. Effect of hemolymph from *Biomphalaria glabrata* on *S. mansoni*. Proc Soc Exp Biol Med 1968;128:1029.

25 Smith MA, Clegg JA, Kusel JR, Webbe G. Lung inflammation in immunity to *Schistosoma mansoni*. Experientia 1975;31:595.

26 Mahmoud AAF, Warren KS, Strickland GT. Acquired resistance to infection with *Schistosoma mansoni* induced by *Toxoplasmosis gondii*. Nature 1976;263:56.

27 Civil RH, Warren KS, Mahmoud AAF. Conditions for Bacille Calmette-Guerin induced resistance to infection with *Schistosoma mansoni* in mice. J Infect Dis 1978;137:550.

28 Civil RH, Mahmoud AAF. Genetic differences in BCG-induced resistance to *Schistosoma mansoni* are not controlled by genes within the major histocompatibility complex of the mouse. J Immunol 1977;120:1070.

29 Lett RR. Induction of resistance against *Schistosoma mansoni*. MSc Thesis Edmonton, Alberta: University of Alberta, 1984:180.

30 Flanigan TP, King CH, Lett RR, Nanduri J, Mahmoud AAF. Induction of resistance to *Schistosoma mansoni* infection in mice by purified parasite paramyosin. J Clin Invest 1989;83:1010.

31 Olds GR, Ellner JJ, El Kholy A, Mahmoud AAF. Monocyte-mediated killing of schistosomula of *Schistosoma mansoni*: alterations in human schistosomiasis and tuberculosis. J Immunol 1981;127:1538.

32 James SL, Cook KW, Lazdins JK. Activation of human monocyte-derived macrophages to kill schistosomola of *Schistosoma mansoni in vitro*. J Immunol 1990;145:2686.

33 Olds GR, Chedid L, Lederer E, Mahmoud AAF. Induction of resistance to *Schistosoma mansoni* by natural cord factor and synthetic lower homologues. J Infect Dis 1980;141:473.

34 Bomford R. Adjuvants. In Liew FY, ed. *Vaccination Strategies for Tropical Diseases*. Boca Raton: CRC Press, 1989:93–103.

35 Klebanoff SJ. Oxygen intermediates and the microbicidal event. In van Furth R, ed. *Mononuclear Phagocytes, Functional Aspects*. The Hague: Martinus Nijhoff, 1980:1105.

36 Davies P, Allison AC. Secretion of macrophage enzymes in relation to the pathogenesis of chronic inflammation. In Nelson DS, ed. *Immunobiology of the Macrophage*. New York: Academic Press, 1976:427.

37 Otter WD. The effect of activated macrophages on tumor growth *in vitro* and *in vivo*. In Pick E, ed. *Lymphokine Reports*. New York: Academic Press, 1981:389.

38 Krahenbuhl JL, Remington JS, McLeod R. Cytotoxic and microbicidal properties of macrophages. In van Furth R, ed. *Mononuclear Phagocytes, Functional Aspects*. The Hague: Martinus Nijhoff, 1980:1631.

39 Nogueira N, Gordon S, Cohn ZA. *Trypanosma cruzi*: modification of macrophage function during infection. J Exp Med 1977;146:157.

40 Murray HW, Cohn ZA. Macrophage oxygen-dependent antimicrobial activity. I. Susceptibility to *Toxoplasma gondii* to oxygen intermediates. J Exp Med 1979;150:938.

41 Mahmoud AAF, Peters PAS, Civil RH, Remington JS. *In vitro* killing of schistosomula of *Schistosoma mansoni* by BCG and *C. parvum* activated macrophages. J Immunol 1979;122:1655.

42 Olds GR, Ellner JJ, Kearse LA, Kazura JW, Mahmoud AAF. Role of arginase in killing of schistosomula of *Schistosoma mansoni*. J Exp Med 1980;151:1557.

43 Hibbs JB Jr, Taintor RR, Vavrin Z. Macrophage cytotoxicity: role for L-arginine deaminase and imino nitrogen oxidation to nitrite. Science 1987;235:473.

44 Granger DL, Hibbs JB Jr, Perfect JR, Durack DT. Metabolic fate of L-arginine in relation to microbiostatic capacity of macrophages. J Clin Invest 1990;85:264.

45 Liew FY, Yun L, Millott S. Tumor necrosis factor-α synergizes with IFN-γ in mediating killing of *Leishmania major* through the induction of nitric oxide. J Immunol 1990;145:4306.

46 Corradin SB, Mauel J. Phagocytosis of Leishmania enhances macrophage activation by IFN-γ and lipopolysaccharide. J Immunol 1991;146:279.

47 James SL, Glaven J. Macrophage cytotoxicity against schistosomula of *Schistosoma mansoni* involves arginine-dependent production of reactive nitrogen intermediates. J Immunol 1989;143:4208.

3 Mechanisms of acquired immunity against parasites

Alan Sher & Phillip A. Scott

IMMUNITY TO PARASITES: SPECIAL PROBLEMS IMPOSED BY HOST–PARASITE ADAPTATION

The study of specific acquired immunity to parasites is of major importance both for understanding the epidemiology of parasitic disease and for designing vaccines to combat parasitic infections. Nevertheless, with certain exceptions, acquired resistance has been traditionally difficult to demonstrate in these diseases and the immune mechanisms underlying documented resistance phenomena have been even more elusive. The complexity of parasite life-cycles and transmission, as well as the often enormous intraspecies variations displayed by parasitic organisms, are important limitations encountered in such studies. However, the principal problem inherent in the demonstration and analysis of acquired immunity against parasitic disease is the remarkable evolutionary adaptation of parasites to the vertebrate immune system. These adaptations, which underlie and in part explain the chronicity of parasitic infections (see Chapter 5 for complete discussion), both severely limit and mask those immune responses otherwise capable of killing parasites. Thus, the identification of effector mechanisms of parasite immunity is very much a quest for responses which parasitic organisms cannot evade.

While acquired immunity is difficult to demonstrate in many host–parasite systems (e.g., human hookworm infection), it is prominent in others (e.g., infection with *Leishmania major*). In general, the more chronic the infection, the less evident the acquired immunity. Nevertheless, as discussed below, an important exception occurs in the case of the concomitant immunity which accompanies certain chronic parasitic infections (e.g., schistosomiasis). In addition, in situations where natural acquired resistance is clearly absent, it may be possible to stimulate protective immunity artificially with the appropriate vaccination strategy.

In this chapter, we will survey the different manifestations of acquired immunity against parasitic organisms and discuss the immunologic effector mechanisms currently implicated in both naturally acquired and vaccine-induced resistance. Since the last edition of this volume, enormous progress has been achieved in the molecular analysis of the immune response as well as in the identification of target antigens on parasites. While highlighting these important advances, we will also emphasize the continued importance of *in vivo* observation, from both epidemiologic studies and direct animal experimentation, in establishing the relevance of immunologic phenomena in parasitic infection.

DIFFERENT FORMS OF ACQUIRED IMMUNITY AGAINST PARASITES

Acquired immunity against parasites has been documented in a variety of different natural and experimental situations. These can be classified as follows.

Naturally acquired immunity

Immunity induced by previous infection

Poorly adapted parasites generally induce a strong immunity which not only clears most, if not all, of the initial infection but leaves the host solidly protected against subsequent challenge infection. Self-healing infections, such as those typically encountered in cutaneous leishmaniasis (Chapter 12), certain rodent malarias (Chapter 14), and intestinal nematode infestations (Chapter 19), are well-studied examples of this phenomenon. In general, the ability of these infections to resolve themselves is dependent on the particular host–parasite combination employed. For example, *Plasmodium berghei* produces a self-limited infection in rats but a lethal infection in mice. Similarly, whereas *L. major* produces a self-healing cutaneous lesion in most strains of mice, it causes a fatal visceralizing disease in BALB/c mice. Nevertheless, where resolution of the primary infection occurs, the host is usually left with a strong, long-lasting immunity which offers an excellent situation for the investigation of both effector mechanisms and target antigens potentially useful in vaccine design. Indeed, because of the absence of chronic pathogenicity, exposure to self-resolving parasitic infections can in itself be used as an effective form of live vaccination against subsequent disease. For example, in the Middle East, inoculation with live (unattenuated) promastigotes of *L. major* in inconspicuous anatomic sites has been used to prevent the cosmetic disfigurement (i.e., "Oriental Sore") associated with natural infections [1].

Acquired immunity can also result from drug-induced termination of chronic parasitic infections, such as malaria [2], African trypanosomiasis [3], visceral leishmaniasis [4], and amebiasis [5]. In these situations where live natural infection generally stimulates only a poor protective response, chemotherapy must result in improved antigen presentation or, alternatively, the abrogation of parasite-induced immunosuppression. In contrast, the resistance observed after chemotherapy of infections such as schistosomiasis [6] probably directly reflects the concomitant immunity (see below) already existing in the host at the time of treatment.

Premunition

Certain parasitic infections (e.g., human malarias, American trypanosomiasis, toxoplasmosis) induce a specific immune response, which results in clinical recovery and resistance to specific challenge, but they persist in the host in small numbers often as slow-replicating intracellular (tissue) forms. This situation, termed premunition by Sergent [7], promotes chronic survival of the parasite (and host) as a consequence of reduced pathology due to the lowered primary infection load and reinfection rate. As expected, premunition fails to develop or is reversed in immunocompromised hosts. An important clinical manifestation of this phenomenon is the activation of latent toxoplasmosis infection occurring in patients with acquired immunodeficiency syndrome (AIDS). Although the precise etiology of the human disease has not been elucidated (see Chapter 6 and 13), studies in murine animal models suggest that depletion of CD4$^+$ T lymphocytes [8] and interferon-γ (INF-γ) production [9] by these cells may be in part responsible for the burst of parasite growth encountered in human immunodeficiency virus (HIV)-infected immunodeficient patients. Curiously, there is as yet no evidence that AIDS activates latent clinical malaria infections, an observation which suggests that premunition in malaria may not depend on CD4$^+$ lymphocytes.

Concomitant immunity

In contrast to premunition in which immunity is maintained by the few parasites remaining after initial clearance, some parasitic infections induce resistance to subsequent challenge infection while apparently completely surviving the protective responses that they stimulate. Smithers & Terry [10] called this situation "concomitant immunity," a term which had previously been employed in cancer immunology to describe immunity to homologous tumor grafts induced by a concurrent malignancy. The phenomenon is most evident in schistosomiasis in permissive hosts where adult worms established from initial worms survive while inducing partial resistance to subsequent cercarial exposure. Concomitant immunity probably occurs in a number of other parasitic diseases (e.g., filariasis) but has not been formally documented because of difficulties in distinguishing primary infection from challenge infections.

The central paradox underlying concomitant immunity as well as premunition is how the initial infection escapes complete elimination by the protective immune responses that it triggers. In the case of schistosomiasis this is usually explained by stage specificity: the adult worms from the initial infection stimulate an effector mechanism which kills invading schistosomula but which fails to recognize the adult stages. Indeed, antibodies from infected mice react strongly with newly transformed schistosomula but bind poorly to adult parasites. The presence of a dense coat of host molecules (proteins and glycolipids) on the adult worm surface is the probable explanation of this phenomenon, although evidence for alternative mechanisms exist [11,12].

A major complication in the analysis of concomitant immunity is the presence of the initial infection and its accompanying pathology. Indeed, as discussed in Chapter 19, there is strong evidence in the murine *Schistosoma mansoni* model that circulatory changes due to egg pathology contribute significantly to the acquired resistance observed. Nevertheless, recent evidence indicates that concomitant immunity against schistosomes clearly exists in humans and does not appear to relate to the severity of disease induced by the primary infection. Instead, the current data argue that the balance of effector and blocking antibodies (see below) developed against the parasite determines the immune status of the infected individual [6].

Vaccine-induced immunity

Immunity stimulated by inoculation of attenuated parasites

As stated at the beginning of this chapter, the artificial induction of immunity against well-adapted parasites (those producing chronic infections) has been difficult to achieve. An important exception is the immunization against infective stages accomplished with living attenuated parasites. The induction of immunity by this procedure is generally highly effective and has served as an important model in many host–parasite systems for the analysis of effector mechanisms and antigens that could be utilized in a dead vaccine.

Working attenuated vaccines already exist for preventing a number of different parasitic infections of veterinary importance. These include canine hookworm (*Ancylostoma caninum*), bovine and ovine lungworm (*Dictyocaulus viviparus*, *Dictyocaulus filaria*) (see Chapter 19), and bovine *Babesia* (Chapter 15). The list of experimental attenuated vaccines is even larger, extending to plasmodia, *Leishmania*, *Trypanosoma cruzi*, filaria, and schistosomes. As mentioned above, many of these immunization protocols employ infective stages (e.g., sporozoites, cercariae, third-stage larval forms, promastigotes, metacyclic trypomastigotes) and utilize heavy doses of γ- or X-irradiation as the major attenuation procedures. The use of heavily irradiated infective stages probably stimulates effective immunity by creating a situation where the production of stage-specific antigen is prolonged in a fixed tissue site.

The effector mechanisms of attenuated vaccine-induced immunity have been extensively studied in rodent animal models. These will be discussed in detail below. However, certain generalizations have emerged from the existing data. Thus, in both the irradiated sporozoite (malaria) [13], cercariae (schistosomiasis) [14], and promastigote (*Leishmania*) [15] models, the attenuated vaccines appear to induce preferentially cell-mediated immunity and protection, with antibody-dependent immunity appearing only after hyper-immunization in the case of the malaria and schistosome

experiments. This capacity to elicit good cell-mediated responses may underly the unusual efficacy of this form of vaccination.

Immunization with native or recombinant antigens or peptide epitopes

The step from vaccination with attenuated parasites to immunization with defined parasite antigens has proved to be a difficult one in almost every system studied. A notable exception is the recent description of a recombinant vaccine against *Taenia ovis* in sheep [16]. Immunization with a fusion protein consisting of a recombinant oncosphere antigen linked to schistosome glutathione-S-transferase induced nearly complete protection against challenge infection. Nevertheless, the ease with which antibody-dependent immunity is induced against *Taenia* infections is highly unusual and the latter experiment merely recreates in recombinant form a vaccine which was already highly effective as a crude antigen mixture (Chapter 17).

The more typical result encountered in current vaccination experiments employing purified antigens is the induction of partial immunity. Whether the failure to achieve complete protection reflects a requirement for multiple antigens or the need for an antigen presentation method which more closely resembles that produced by attenuated living infection is not clear. In the case of antibody-dependent immunization, the induction of high titers or affinities of antibodies is usually essential for the protection observed. The requirements for the induction of optimal cell-mediated immunity are at present poorly understood, but hopefully will become more readily deciphered as our understanding of T cell subset regulation improves (see below).

Currently even more problematic is the use of subunit vaccines consisting of defined peptide epitopes. In addition to being poor immunogens, these constructs may be impractical for use on human or outbred animal populations because of major histocompatibility complex (MHC)-restricted host responsiveness [17]. Moreover, the recent demonstration of polymorphic T cell sites on the malaria circumsporozoite (CS) antigen further emphasizes the probable limitations associated with vaccines containing a low number of epitopes [17]. Thus, the use of vaccine constructs consisting of multiple full-length recombinant antigens would seem the most rational strategy for immunization with defined parasite antigens.

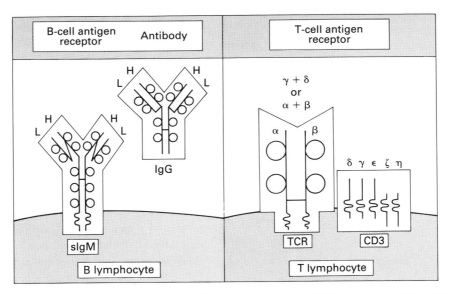

Fig. 3.1 Molecules responsible for immune recognition: B cell receptor/antibody and TCR proteins. Antigen recognition by B cells is mediated by cell surface antibody molecules usually belonging to the IgM (shown) or IgD isotypes. These molecules are anchored to the cell by transmembrane hydrophobic sequences. The secreted forms of antibody (IgG is shown) lack these regions. Antigen recognition by T cells is mediated by the TCR (α,β or γ,δ) and a cluster of closely associated proteins (CD3) thought to be involved in signal transduction. All of the proteins in the TCR complex are bound to the cell surface by transmembrane sequences. The antibody receptor and TCR molecules are encoded by genes belonging to the same supergene family yet recognize antigen in quite different contexts. The structures shown are schematic and do not conform to an accurate relative scale.

IMMUNOLOGIC EFFECTOR MECHANISMS

Molecular basis of immune recognition

Two different classes of molecules, both products of the immunoglobulin supergene family, are responsible for the recognition of foreign antigens by the immune system. These are the antibody proteins, which function as both antigen receptors on B lymphocytes as well as circulating effector molecules, and the T cell receptors (TCR) (Fig. 3.1). Both sets of molecules are constructed from two different sets of subunits, heavy and light chains in the case of antibodies, and α,β- or γ,δ-chains in the case of TCR. Each chain consists of multiple (2–5) disulfide-bridged domains which share a similar organization in the different molecules. Both the antibody (B cell) receptor and TCR are anchored to the cell by hydrophobic transmembrane regions. The TCR is closely associated with another molecular complex CD3, which is thought to be involved in signal transduction (Fig. 3.1).

The antigenic and idiotypic specificities of antibody and TCR are encoded by the variable regions of the different subunits, whereas functional specificity is determined by the constant regions. Among the antibody molecules these isotypic class and subclass differences associate with different biologic properties, such as fixation of complement, placental transfer, and binding to specific (e.g., mast cell, mononuclear cell) surface receptors. Of particular interest in parasite systems is IgE, which binds with high affinity to mast cells and basophils and with lower affinity to macrophage and eosinophil receptors. IgE antibodies are selectively induced by helminth infections and, as discussed below, are thought to play a role in acquired immunity to worms. While certain isotypes have been implicated in specific antiparasitic effector functions (e.g., IgA in control of intestinal infection), there are other situations where protective immunity can clearly be attributed to multiple antibody isotypes. In contrast, there is now convincing data that certain isotypes can block the function of others in promoting eosinophil-mediated damage to helminths (see below). Although TCR proteins also exist in two major isotypic forms (α,β; γ,δ), the function of the two types of receptors in unclear. Most T cells appear to express α,β-receptors. γ,δ-Receptors are not only infrequently expressed but may have less sequence diversity in their variable regions [18]. Recent evidence indicates that γ,δ-receptor-bearing T cells are elevated

Table 3.1 Differences in immune recognition by B and T lymphocytes

Property	B lymphocytes	T lymphocytes
Primary recognition molecule	B cell (IgH + IgL) receptor	T cell (α, β; γ, δ) receptor
Chemical structures recognized	Potentially all	Peptides only
Protein epitopes recognized	Linear and conformational	Linear only
Antigen processing	Not required	Required
Antigen presentation	Not required	Required
Receptor interaction	Directly with native antigen	With processed antigen in association with MHC molecule

in epithelial tissue (both epidermal and gut) and are further enriched in cutaneous lesions caused by *Mycobacterium leprae* and *Leishmania*. A portion of the γ,δ-T cells induced by mycobacterial infections appear to be directed against stress proteins and, in particular, against epitopes which crossreact with self-determinants [18,19]. The immunobiologic significance of this fascinating observation remains to be elucidated.

While antibodies and TCR are similar molecules, the manner in which they recognize antigens is quite different (Table 3.1). Antibody molecules are capable of binding specifically to a wide variety of different chemical structures and appear to distinguish readily conformational epitopes. In contrast, TCR bind linear peptide epitopes only, and thus require preprocessing (degradation) of most protein antigens. Moreover, TCR are able to recognize these peptide epitopes only when they are bound to a Class I or Class II MHC product (Fig. 3.2).

T cells can be subdivided on the basis of surface antigens into two major sets: CD4$^+$ and CD8$^+$ (Fig. 3.3). These populations have different biologic properties, CD4$^+$ cells functioning as "helper cells" for antibody production or cell-mediated immunity and CD8$^+$ cells functioning as cytotoxic killer cells, i.e., cytotoxic T lymphocytes (CTL), or possibly suppressor cells. This functional specificity may stem from the manner in which they recognize presented antigens. CD4$^+$ cells see peptides associated with Class II MHC molecules, whereas CD8$^+$ lymphocytes recognize antigen only in the context of Class I molecules. As discussed later in this chapter, the manner in which antigen is processed can influence its association with Class I or Class II molecules and, consequently, its ability to induce or be recognized by CD4$^+$ vs. CD8$^+$ cells.

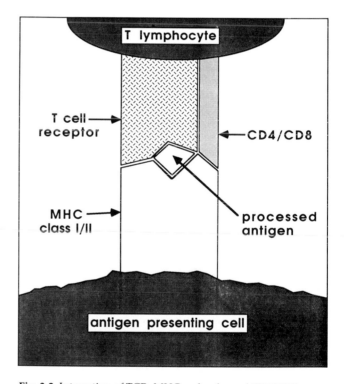

Fig. 3.2 Interaction of TCR, MHC molecule, and CD4/CD8 molecules in the recognition of processed antigen by T cells. Protein antigens are processed into smaller peptides which bind to specific regions on Class I or Class II molecules in intracellular compartments of APC (see Fig. 3.6). The MHC–peptide complex is transported to the cell surface where it is recognized by a specific TCR on a T lymphocyte. The CD4$^+$ and CD8$^+$ antigens which distinguish the two major classes of T cells also serve as recognition elements for MHC on APC, CD4$^+$ interacting specifically with Class II and CD8$^+$ with Class I MHC molecules. Thus, T cell recognition involves a minimum of three specific binding events: (1) peptide to MHC molecule; (2) MHC–peptide to TCR; and (3) CD4$^+$/CD8$^+$ to Class I and Class II molecules.

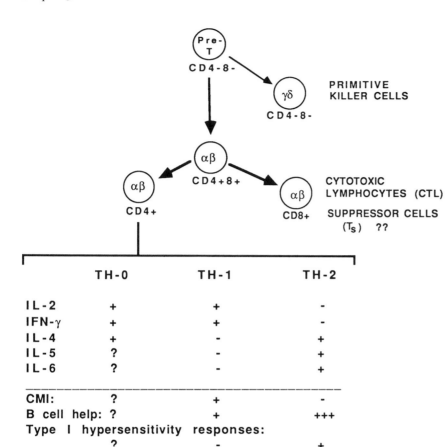

Fig. 3.3 T lymphocyte subsets. While CD4⁺ and CD8⁺ T cells are distinguished on the basis of cell surface markers, the TH-0, TH-1, and TH-2 CD4⁺ subsets (identified from studies on murine clones) are differentiated on the basis of the cytokine production patterns shown. T cells can also be distinguished on the basis of TCR isotypes. While most have TCR consisting of α,β-chains, a second subset bearing γ,δ-subunits has been identified which appears to display a different range of antigenic recognition. The pathway of differentiation shown is an approximation.

The specificity of the host response to parasites is governed by these recognition molecules. Nevertheless, while most parasitic infections induce a plethora of dfferent antibody and T cell responses, only a minority of the immune reactions generated have been shown to mediate killing of the organisms *in vitro*. Even fewer have been directly shown to be responsible for protective immunity *in vivo*. The major effector mechanisms which have been studied in host–parasite systems are summarized below.

Antibody-dependent effector mechanisms

Although most parasites stimulate significant humoral responses, there are relatively few instances in which antibody in the absence of other immune responses has been shown to be directly and solely responsible for protective immunity. Evidence for a role of antibody in protection has been obtained from passive transfer experiments in a variety of host–parasite models.

However, in most cases the protection transferred is partial and may involve cooperation with cell-mediated responses induced in the recipient animals by the challenge infection. On the other hand, the failure of immune serum to transfer high levels of protection may simply reflect a deficiency or active suppression of the appropriate antibody response in the donor host. In this regard, monoclonal antibodies have been reported in several models to be more effective than polyspecific immune sera in transferring protection against parasite challenge. For example, only low levels of resistance to malaria sporozoites are observed in recipients of sera from mice immunized with irradiated parasites, while solid sterilizing immunity is achieved by transfer of monoclonal antibodies directed against the major CS antigen [13,20,21]. Moreover, as already emphasized above, successful vaccination against sporozoite challenge with the major repeating epitope of the CS antigen was achieved only in mice producing extremely high titers of anti-CS antibodies [22]. Similar data in other parasite models (e.g., schistosomiasis) suggest that effective antibody-dependent immunity against surface antigens may require the induction of particularly potent humoral responses.

The major antibody-dependent effector mechanisms implicated in parasite immunity are summarized below.

Receptor blockade

Many intracellular parasites utilize distinct receptors on the surface of host cells for recognition and entry. By blocking binding to these receptors, antibodies may limit parasite invasion and infection (Fig. 3.4). In malaria, this is a major approach employed in vaccination against established blood stages. Malaria merozoites recognize a number of distinct receptors on the erythrocyte surface, including the Duffy blood group (*Plasmodium vivax*, *Plasmodium knowlesi*) glycophorin A, and a trypsin-sensitive sialic acid-independent molecule (*Plasmodium falciparum*). Antibodies against several different merozoite surface antigens or proteins present in rhoptries and micronemes (the organelles involved in erythrocyte penetration) block host cell infection. In the case of the 66/44/42 kD protein of *P. knowlesi*, the ability

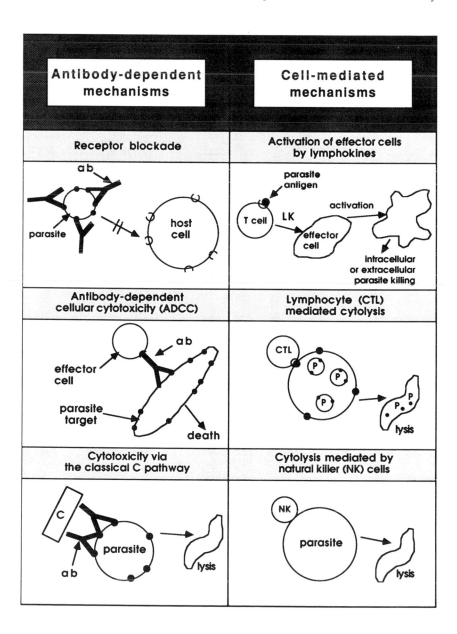

Fig. 3.4 Major immunologic effector mechanisms used against parasitic organisms. (See text for description.)

of Fab fragments to mediate the same inhibition of invasion supports the involvement of receptor blockade. These, as well as other putative receptor molecules (e.g., the 175 kD erythrocyte binding protein of *P. falciparum*), are being studied as important vaccine candidates (see Chapter 16) [21]. Inhibition of sporozoite receptor attachment to hepatocytes may also explain the protection induced by immunization with CS antigen [21].

Receptor blockade has also been described in *T. cruzi* infection. Antibodies to a major 85 kD surface antigen block invasion of fibroblasts, presumably by inhibiting binding to fibronectin receptors on the host cells [23,24]. Similarly, antibodies against either the 63 kD glycoprotein or excreted factor (EF) glycolipid of *Leishmania* block infection of macrophages [25]. In this case, the antibodies probably inhibit binding of the parasites to complement receptors which have been implicated in cell invasion [26,27].

In addition to blocking receptors involved in host cell invasion, antibodies may also limit parasite survival by blocking cytoadherence. Thus, as discussed in more detail in Chapter 14, erythrocytes infected with *P. falciparum* attach to endothelial cells as they mature from ring to trophozoite stage. This process of cytoadherence is thought to sequester the parasites, preventing their passage into and destruction by the spleen, and is apparently mediated by knob-like protrusions on the surface of the infected erythrocytes [21]. Antibodies directed against a 300 kD protein associated with knob expression block adherence to endothelial cells *in vitro* [21]. Thus, immunization against this or other knob-associated cytoadherence molecules could prevent sequestration, thereby inhibiting the development of malaria infections.

Antibody-dependent cell-mediated cytotoxicity (ADCC)

Probably the most extensively studied antibody-dependent protective responses against parasite, and, in particular, helminth targets, are those mediated by effector cells (Fig. 3.4). The major antibody isotypes and cells involved in antiparasitic ADCC reactions are summarized below.

Eosinophils. The prominence of eosinophils in both the peripheral blood of helminth-infected hosts and in the tissue responses to these parasites suggested that eosinophils might play a role in host defense against worm infections. This hypothesis was supported by the pioneer observation of Butterworth *et al.* [28], that schistosomula of *S. mansoni* can be killed *in vitro* by a combination of antibodies from infected patients and eosinophils. Subsequently, ADCC reactions involving antibodies and eosinophils were shown to be toxic to other helminths, including *Trichinella* and filariid larval stages [29].

The mechanism of antibody-dependent eosinophil killing has been studied in greatest detail in the *in vitro* schistosomulum model. Both IgG and IgE antibodies promote this reaction by binding to specific Fc receptors on the eosinophil surface [30]. Interaction with these receptors promotes degranulation, which in turn results in enhanced adherence of the effector cell to the target. Both the basic granular proteins and the peroxide–halide–peroxidase generating systems of the eosinophil have been implicated as the final mediators of parasite killing [29,30].

The antibody-dependent killing activity of eosinophils against schistosome larvae can be enhanced by a variety of mediators, including mast cell-produced chemotactic factors and cytokines, such as tumor necrosis factor (TNF), colony stimulating factors (CSF), and other undefined products of activated lymphocytes and monocytes [31]. The effects of these substances on eosinophils may explain the increased killing capacity of the cells when obtained from patients with eosinophilia [30,31].

While clearly a prominent cell in the immune inflammatory response to helminths and in *in vitro* killing reactions, the role of eosinophils in protective immunity *in vivo* is supported only by circumstantial evidence. Thus, while infected mice depleted of eosinophils by treatment with antisera produced against these cells show lowered resistance to both schistosome and *Trichinella* infections [29,30], it has not been established whether this loss of immunity results directly from the elimination of eosinophil effector cells or from some other consequence of the administration of the antibodies. Indeed, in recent experiments in which eosinophils were depleted from animals by treatment with monoclonal antibodies against IL-5, a cytokine which specifically controls the differentiation of eosinophils in the bone marrow, no reduction in vaccine-induced resistance against *S. mansoni* was observed [32].

Macrophages. Antibodies can promote the killing of parasites by macrophages via several distinct mechanisms. Firstly, antibody-opsonized protozoa are more readily phagocytosed. The latter phenomenon may play an important role in the clearance of parasites such as African trypanosomes or malaria blood stages from the circulation. Secondly, with extracellular targets such as helminths, antibody can facilitate cell contact with the parasite and thus augment killing by promoting Fc receptor attachment of activated macrophages (see below) [33]. Finally, IgE antibodies induce killing of schistosomula by rat macrophages both by activating the cells and by promoting their attachment to the target [30].

Neutrophils. While implicated as effector cells in ADCC reactions against helminths and certain protozoa, the neutrophil in general is thought to be less potent at killing parasites than the eosinophil. However, recent evidence in an *in vivo* schistosome model indicates that depletion of neutrophils by treatment with a monoclonal antibody can profoundly alter resistance to challenge infection [34]. Although similar caveats apply to this experiment, as do the original eosinophil depletion experiments employing polyvalent antisera discussed above, the results reopen the question of neutrophil involvement in parasite immunity.

Platelets. The most recently discovered and novel effector cells in ADCC reactions against parasites are the platelets. The killing function of these cells against schistosome larvae is dependent on IgE antibodies and ε-Fc receptors belonging to the same class (FcεRII) as those involved in IgE-dependent macrophage and eosinophil-mediated killing of schistosomula [30]. The *in vivo* relevance of this *in vitro* phenomenon is supported by experiments in which platelets from infected rat donors transferred protective immunity against schistosome infection to naïve recipients when injected locally [30].

Antibody-dependent killing mediated by the classic complement pathway

Perhaps the prototype of antibody-dependent effector mechanisms is the killing of targets opsonized with antibody by the classic complement pathway (Fig. 3.4). Indeed, a variety of different parasites can be killed by this reaction *in vitro* [35]. Nevertheless, in most cases the *in vivo* relevance of these reactions remains unconfirmed. For example, while African trypanosomes are readily lysed *in vitro* by antibodies against the variant-specific glycoprotein and complement, normal antibody-mediated clearance of the parasite occurs in

animals carefully depleted of complement by treatment with cobra venom factor [36].

Cell-mediated immunity (CMI)

Protection due to the recognition and killing of targets by cells in the absence of antibody is referred to as CMI. In recent years, CMI responses have been shown to play a major role in acquired immunity against both protozoa and helminths. Indeed, the induction of CMI has several attractive features as a strategy for vaccination against parasites [14]. Thus, while parasites are clearly highly adept at evading host antibody responses, evasion of CMI appears to be much less common (see Chapter 5). This is particularly true in the case of CMI involving activated macrophages where recognition of parasite surface antigens (the process interrupted by most evasion mechanisms) is not essential for effector cell function. In addition, most CMI responses involve T cell recognition of peptide as opposed to carbohydrate epitopes. This property permits potential production of the appropriate antigens using gene cloning or peptide synthesis techniques.

The major cell-mediated effector mechanisms documented to play a role in parasite immunity are outlined below.

Cytotoxic lymphocytes

While repeatedly invoked as a potential antiparasitic effector cell, the CTL (Fig. 3.4) has only recently been demonstrated to have defined activity. These cells, which are usually CD8[+], have been traditionally associated with the killing of Class I MHC bearing host cells infected with viruses or bacteria. The best-studied parasite targets are lymphoblastoid lines infected with *Theileiria parva*. This protozoan of cattle preferentially infects and transforms lymphocytes into stable cell lines. Class I-restricted CD8[+] T lymphocytes which lyse the infected lymphoblastoid lines *in vitro* develop in *Theileiria*-immunized animals [37]. The relevance of this effector mechanism (which often shows a high degree of strain specificity) to protective immunity is now being analyzed.

Since the activity of CTL depends on their recognition of Class I MHC products, schistosomes, which acquire MHC antigens from the host [38], are unique among the extracellular parasites in serving as possible targets for these effector cells. Indeed, alloreactive CTL generated against MHC antigens will adhere specifically to lung-stage schistosomula recovered from mice with the appropriate H-2 haplotype. Nevertheless, this interaction fails to damage the target [39], a finding which suggests that schistosomes are intrinsically resistant to killing by CTL.

The possibility that CTL play a role in immunity to malaria sporozoites is raised by the demonstration that the resistance to infection expressed in mice vaccinated with X-irradiated parasites is blocked in animals depleted of CD8+ cells by monoclonal antibody treatment [40,41]. This CD8+ cell requirement may be due to direct CTL lytic activity against sporozoite-infected hepatocytes or to IFN-γ production by the same effector cells (Fig. 3.4). Similarly, the effects of CD8+ lymphocytes on *Toxoplasma* infection [42] may also be due to their synthesis of IFN-γ. Indeed, the intracellular growth of both of the exoerythrocytic forms of malaria and *Toxoplasma* is inhibited by this cytokine, and depletion of IFN-γ *in vivo* reverses immunity against both parasites [40,43].

Cellular activation by lymphokines (LK)

The production of LK by Class II MHC-restricted CD4+ T helper cells is the form of CMI thought to play the most significant role in antiparasitic immunity (Fig. 3.4) [14]. These soluble mediators activate effector cells to kill parasite targets directly or, in the case of ADCC reactions (see above), in concert with antibody. In protozoan infections, the effects of LK activation are most dramatic with those parasites (e.g., *Leishmania*, *Trypanosoma cruzi*, *Toxoplasma*) which invade macrophages. Exposure of infected cells to LK-containing supernatants produced by mitogen or antigen-induced stimulation of T lymphocytes results in a marked inhibition of replication and/or killing of the intracellular parasite stages [44–46]. Furthermore, products of LK-activated macrophages are also able to inhibit the growth of malaria parasites in erythrocytes *in vitro* [21].

Effector cells can also be activated by LK to kill helminth targets. Lymphokine-activated mature tissue macrophages appear to be particularly potent at killing schistosome larvae *in vitro* [47,48]. Similarly, mouse eosinophils can be triggered to kill schistosome ova *in vitro* by LK (e.g., eosinophil stimulation promoters) which are distinct from those involved in macrophage activation [49]. Other cytokines, some of which are lymphocyte-derived (e.g., GM-CSF, TNF), can markedly enhance the antibody-dependent killing activity of eosinophils against schistosomula [31].

The principal LK responsible for cellular activation for both intracellular and extracellular parasite killing appears to be IFN-γ, although other non-IFN mediators yet to be identified have been noted in intracellular killing [50]. A variety of substances produced as a consequence of LK activation have been implicated as the final agents of parasite killing. These include oxidative metabolites, proteases, and TNF.

The role of LK-dependent cell-mediated responses in parasite immunity *in vivo* is supported by an ever-expanding body of evidence. Thus, inbred mice with defects in LK production and macrophage activation are more susceptible to infection with *L. major* [51] and are unable to develop vaccine-induced immunity against *S. mansoni* [52]. In other host–parasite models, such as murine malaria, a role for CMI is suggested by the failure of B lymphocyte depletion to ablate an immunity already known to be thymus-dependent [21] or, in other cases, by the loss of immunity observed after CD4+ T cell and/or IFN-γ depletion.

The role of CMI in parasite immunity is also strongly substantiated by adoptive transfer studies with helper T cells. Recently, T cell cloning technology has been employed to determine the phenotype of the protective lymphocytes. In both murine-acquired immunity to *Plasmodium chaubaudi* [53] and vaccine-induced immunity to *L. major* [54], the T cell clones or lines transferring protection belong to the CD4+ subset (i.e., TH-1), which uniquely produces IFN-γ (Fig. 3.3). The immunoregulatory significance of this finding will be discussed further below.

Natural killer cells (NK)

Natural killer cells are cytotoxic effectors believed to be of lymphoid origin which recognize a highly limited range of antigenic specificities and produce IFN-γ (Fig. 3.4). Enhanced NK cell activity against both tumor cell and homologous parasite targets has been noted in mice with *T. cruzi* infections [55,56]. Nevertheless, a formal link between NK cells and immunity has not yet been established in any host–parasite system.

REGULATION OF ACQUIRED IMMUNITY

Acquired immunity against parasitic infection is subject to strong immunoregulatory influences. An understanding of these mechanisms, which are often genetically controlled, is important both for the design of effective vaccines as well as for the interpretation of natural resistance, as it occurs in human and animal populations. The following is a brief summary of the immuno-regulatory elements which play a role in determining the expression of parasite immunity.

Regulation by antibodies

Antibodies have been postulated to exert a major immunoregulatory influence on acquired immunity in both animal models and man. While a number of different mechanisms have been proposed (e.g., anti-idiotypic regulation), the principal suppressive function of antibodies in protective immunity appears to be the binding of target antigens, thereby preventing their recognition by other effector antibodies or by cells. The major evidence is from studies of the role of antibody isotypes in immunity against helminths (see Chapter 17). Thus, whereas an IgG2a monoclonal antibody against a 38 kD surface glycoprotein of *S. mansoni* is protective in rats, an IgG2c antibody recognizing the same antigen fails to passively immunize and in fact blocks the protection conferred by the first antibody [57]. In humans, an inverse correlation has been described between levels of antischistosomulum antibodies of the IgM or IgG2 isotype and lowered reinfection rates after chemotherapy [6,30,58]. Since the antibodies with blocking isotypes bind only weakly to Fc receptors on eosinophils, they could act by preventing the interaction of these effector cells with schistosomula in the ADCC reaction described above [6,30,58]. Regardless of the precise mechanism, the data suggest that isotype switching during natural infection or vaccination could have a profound influence on the expression of immunity. Since antibody isotypes are known to be determined by CD4$^+$ helper cells, elucidation of the factors governing the induction of the different CD4$^+$ subsets (see below) may be the best approach to understanding how to manipulate this form of immunoregulation.

Antibodies (and/or immune complexes) have also been shown to block the induction of T cell responses to parasite antigens *in vitro* (e.g., Colley *et al.* [59]). Nevertheless, the *in vivo* relevance of this phenomenon, which can readily be explained by the removal by antibody of antigen before processing or presentation to T lymphocytes, has never been formally established.

Regulation by cells

Cells are the primary modulators of parasite immunity. Nevertheless, the immunoregulatory mechanisms involved are varied, complex, and, with several recent exceptions, poorly understood. The major cellular elements which have been implicated in the control of acquired immunity are discussed below.

Adherent cells

The removal of cells adherent to glass or plastic can cause a marked augmentation of *in vitro* responses to mitogens or specific antigens in lymphocytes recovered from different stages of parasitic infections (e.g., Todd *et al.* [60]). The adherent suppressors, which are likely to be monocytes or macrophages, may function by producing prostaglandins (e.g., Scott & Farrell [61]) or by acting as defective antigen-presenting cells. The *in vivo* role of these cells in the regulation of acquired immunity has never been formally established.

B lymphocytes

Studies in the murine cutaneous leishmaniasis model have revealed an interesting dual function for B lymphocytes in the regulation of infection (Chapter 12). Thus, *in vivo* depletion of B lymphocytes (by injection of anti-μ-chain antibodies) causes susceptible BALB/c mice to become resistant, whereas resistant C3H mice become susceptible to chronic infection after the same treatment [62,63]. Adoptive transfer experiments in this system suggest that B cells and/or their antibody products influence the development of the antileishmanial T cell repertoire. Given the increasing evidence for the role of B cells as antigen-presenting cells (APC), an alternative explanation of these results is that B cell depletion blocks the induction of the different CD4$^+$ cell populations (i.e., TH-2 and TH-1) which determine the expression of immunity in the two mouse strains (see below). Based on the above results, it is probable that variations in antigen-specific B cell responses could modulate the expression of T cell-dependent CMI in a number of different parasitic infections.

CD8$^+$ T lymphocytes

Until recently, the CD8$^+$ T lymphocyte, because of its initial description as a suppressor of immune responses, was thought to be the cell most likely to play a role in the regulation of parasitic infections. Indeed, in a number of different models, CD8$^+$ cells appear to modulate *in vitro* lymphocyte responses (e.g., Tarleton [64]). Nevertheless, in none of the systems has direct suppression of protective immunity by CD8$^+$ cells been formally established. Given current skepticism concerning the existence of T suppressor cells and suppressor factors, observations on CD8$^+$-mediated suppression phenomena must be viewed with caution. In this respect, it should be remembered that CD8$^+$ cells produce high levels of IFN-γ, a cytokine with potent immunoregulatory activity (see below).

CD4$^+$ T lymphocytes and their subsets

Recent data in murine model systems have indicated that a major form of immune regulation occurs within the CD4$^+$ helper population of T lymphocytes (Fig. 3.3). Studies on CD4$^+$ clones have revealed that the majority fall into either of two subsets based on the cytokines that they produce. T cells belonging to the TH-1 subset produce IFN-γ and IL-2 uniquely, whereas those belonging to the TH-2 subset produce IL-4, IL-5, and IL-6 but not IFN-γ or IL-2. Some clones, particularly those analyzed in early generations, produce both groups of cytokines and are referred to as TH-0 cells [65].

Of great interest is the observation that different immunobiologic activities can be ascribed to the different CD4$^+$ subsets. Thus, TH-1 cells, because of the cytokines that they produce, can transfer delayed-type hypersensitivity and induce CMI. In contrast, TH-2 cells, because of their IL-4, IL-5, and IL-6 activity, are the helper cells for the antibody isotypes (IgG1 and IgE) associated with immediate hypersensitivity and play a role in the induction of both mast cell and eosinophil differentiation [65]. These two groups of biologic activities often play opposing roles in parasitic infections. Thus, in infection with *L. major*, TH-1 cells mediate healing of cutaneous lesions while TH-2 cells are associated with disease exacerbation [66]. Similarly, helminths stimulate strong TH-2 responses during chronic infection while in certain cases (e.g., the *S. mansoni*-irradiated vaccine model) protective immunity is associated with TH-1 responses [66]. As explained below, the selective expression of either TH subset appears to be regulated by certain cytokines which suppress the proliferation or function of the opposing set.

Regulation by cytokines

The regulatory function of cells on protective immune responses is determined by the biologically active cytokine molecules that they produce. As summarized in Table 3.2 [67], the number of different cytokines which have been formally identified by gene cloning is now extensive. Since all of these mediators have effects on the differentiation or function of different cells in the

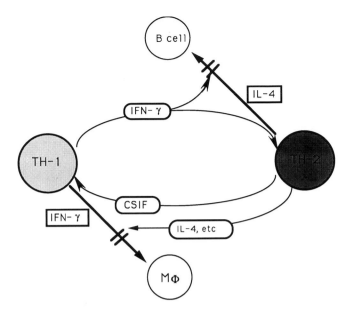

Fig. 3.5 Cross-regulation of TH-1 and TH-2 activities. The bold lines indicate activation pathways while the thin lines refer to inhibitory circuits. As shown, TH-1 cells produce IFN-γ which, in addition to activating macrophages and other cells, inhibits the proliferation of TH-2 lymphocytes as well the induction of B cell IgE responses by IL-4, a cytokine product of TH-2 cells. In turn, TH-2 cells produce IL-4 which, in addition to acting as a helper factor for antibody production, can (along with other cytokines such as GM-CSF) inhibit macrophage activation. Moreover, CSIF, a recently identified cytokine produced by TH-2 cells, inhibits the synthesis of IFN-γ and IL-2 by TH-1 lymphocytes. These regulatory circuits probably explain why, once triggered, the different TH responses tend to dominate each another.

immune system, each has potential immunoregulatory function. Nevertheless, certain cytokines appear to play a prominent function in governing the crucially important biologic activities of helper T lymphocytes. Thus, IFN-γ (produced by TH-1 or CD8$^+$ cells) can suppress the proliferation of TH-2 cells and at the same time blocks the effect of the TH-2 cytokine, IL-4, in stimulating IgE production by B cells (Fig. 3.5) [65]. In the opposite direction, IL-4 (along with several other cytokines produced by TH-2 cells) can inhibit activation of macrophages by IFN-γ [66,68]. Moreover, a recently identified mediator, cytokine synthesis inhibitory factor (CSIF), now designated IL-10, produced uniquely by TH-2 cells has been shown to block the synthesis of IFN-γ and IL-2 by TH-1 cells (Fig. 3.5) [69].

Evidence is rapidly accumulating for the role of cytokines in regulating parasite immunity and parasite-induced immune responses. Thus, *in vivo* depletion

Table 3.2 The major cytokines and their immunobiologic activities. (Adapted from Roitt *et al.* [67])*

Cytokine	Primary immune system source	Principal targets	Major effects
IL-1α IL-1β	Macrophages	T, B cells, macrophages, endothelium	Lymphocyte activation, macrophage stimulation, endothelial adhesion, pyrexia, acute-phase protein responses
IL-2	T cells	T cells	T cell growth factor
IL-3	T cells	Stem cells	Multilineage colony stimulating factor
IL-4	T cells	B, T cells	B cell growth and differentiation factor
IL-5	T cells	B cells, eosinophil precursors	B cell growth and differentiation, eosinophil differentiation and activation
IL-6	T, B cells	B cells, hepatocytes	B cell growth and differentiation, acute-phase response
IL-7	Stromal cells	B, T cell precursors	Early B, T cell differentiation, proliferation
IL-8	Monocytes	Neutrophil	Neutrophil chemotaxis, activation
TNF-α	Macrophages, lymphocytes	Macrophages, adipocytes, tumors	Activation of macrophages, granulocytes, cachexia, cytotoxicity
TNF-β (LT)	T cells	Tumor and allogeneic cells	Cytotoxicity
TNF-γ	T cells, NK cells	Lymphocytes, monocytic and tissue cells	Macrophage activation, MHC induction, T, B cell regulation, inhibition of intracellular microbial growth, adhesion of lymphocytes to endothelial cells
M-CSF	Monocytes	Stem cells, macrophages, granulocytes	Specific induction of stem cell division and differentiation to macrophages, cell activation
G-CSF	Macrophages	Stem cells, granulocytes, macrophages	Specific induction of stem cell division and differentiation to granulocytes, cell activation
GM-CSF	T cells, macrophages	Stem cells, granulocytes, macrophages	Specific induction of stem cell division and differentiation to granulocytes and macrophages, cell activation
MIF	T cells	Macrophages	Migration inhibition
TGF-β	Lymphocytes, monocytes, platelets	T cells, fibroblasts, endothelial	Immune suppression, fibrosis, angiogenesis, inflammatory cell recruitment
IL-10	TH-2 lymphocytes	TH-1 lymphocytes	Suppression of cytokine synthesis

* Since the preparation of this chapter four new interleukins (IL-9, IL-10, IL-11, IL-12) have been identified and functionally characterized.

of IL-4 results in the restoration of healing (CMI) to normally exacerbating *L. major* infections in BALB/c [70], while treatment with anti-IFN-γ antibodies blocks healing in resistant C3H mice [66,71]. Similarly, *in vivo* depletion of IL-4 and IL-5 ablates helminth-induced IgE and eosinophil responses [65,66], while IFN-γ, because of its effect on B cells [65], could be used to suppress IgE responses in worm infections (Fig. 3.5). These findings suggest new approaches for immunopotentiation of specific antiparasite immune responses, as well as for regulating parasite-induced immunopathology.

Genetic influences on protective immunity

The genetic status of the host can have a profound regulatory influence on the expression of protective immunity. As discussed extensively in Chapter 1, the most vivid examples of this phenomenon occur in parasitic infections in inbred mice, and in particular in protozoan models. It is interesting to note that many of the genes studied appear to exert their immuno-regulatory influence at the level of the macrophage by altering either antigen processing or presentation or by reducing the capacity of macrophages to be activated for intracellular or extracellular parasite killing.

THE ROLE OF ANTIGEN PRESENTATION IN THE INDUCTION OF PROTECTIVE IMMUNE RESPONSES AGAINST PARASITES

Antigen processing and presentation are fundamental steps in the generation of immune responses and, as expected, have a major influence on the induction of parasite immunity. Nevertheless, the mechanisms by which processing and presentation determine the outcome of antigenic stimulation are poorly understood and represent a major frontier for current research in immunology.

Antigen-presenting cells, which include macro-phages, B cells, Langerhans, and dendritic cells, fall into two major categories: those in which processed proteins are presented by surface Class I MHC mol-ecules, and those utilizing Class II molecules as present-ing elements. In each case, the processed peptide portion of the antigen binds to the MHC molecule, which in turn is presented to a T cell capable of interacting with the combined structure. Moreover, Class I-associated antigen is preferentially recognized by $CD8^+$ cells, whereas Class II-associated antigen is normally pre-sented to $CD4^+$ lymphocytes, presumably because of a direct interaction of the $CD8^+/CD4^+$ and Class I/Class II molecules (Fig. 3.2). Recent data suggest that in APC expressing both Class I and II MHC molecules, the manner in which antigen is processed determines its eventual MHC association and consequently the T cell subset induced. Thus, soluble antigens which are taken from the "outside" of the cell "in" by endocytosis tend to associate with Class II molecules and induce $CD4^+$ T cell responses, whereas cytoplasmically synthesized membrane antigens from most viruses are transported from "inside" the cell "outside," becoming associated in the endoplasmic reticulum of the infected APC with Class I molecules which induce $CD8^+$ responses preferentially (Fig. 3.6) [72].

Whereas soluble parasite antigens clearly can elicit Class II-restricted $CD4^+$ T cell responses, presumably because of their endocytosis by APC, the rules govern-

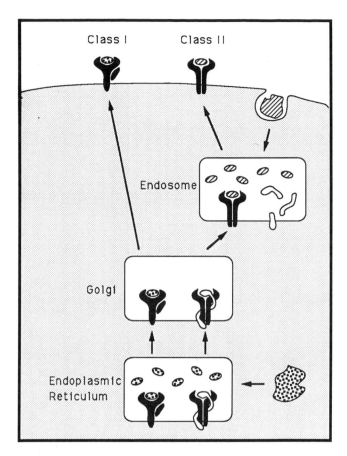

Fig. 3.6 Intracellular pathways of antigen processing in APC. As shown, cytoplasmically synthesized antigens associate with Class I molecules during their synthesis on endoplasmic reticulum and are transported via the Golgi to the cell surface. In contrast, exogenous soluble antigens are taken up by the APC and degraded. In the endosomes, the peptides bind to Class II molecules and are transported back to the external plasma membrane. An invariant chain (shown in white) associated with newly synthesized Class II molecules may be responsible for their selective transport to endosomes. (From Long [72].)

ing the induction of $CD8^+$ responses by parasites are poorly understood. Such responses are evident in malaria, *Toxoplasma*, and *Leishmania* infections (see Chapters 12, 13, and 14). Nevertheless, the manner in which antigens from these intracellular protozoa become associated with MHC molecules in the host cells they infect, in order to allow presentation to $CD8^+$ T lymphocytes, has yet to be elucidated.

Perhaps of greater importance for the induction of protective immunity against parasites is the role of antigen presentation in the selective elicitation of TH-1 or TH-2 CD4[+] T cell responses. While both TH subsets are Class II restricted and employ the same TCR genes, they are differentially induced under different conditions of antigenic stimulation (see above). Parasite models offer excellent examples of these presentation effects. For example, while *L. major* antigen inoculated intravenously or intraperitoneally into BALB/c mice induces a protective TH-1 response, the same material given subcutaneously exacerbates infection, presumably by the stimulation of a TH-2 response [65]. Such phenomena suggest that the different TH subsets may employ different APC (localized in different anatomic sites) or be triggered by different concentrations of processed antigen. Clearly the elucidation of the rules governing the selective induction of different T cell subsets, cytokines, and antibody isotypes will have a major impact on our understanding of how acquired immunity is stimulated naturally, as well as how to induce the most efficient protective responses by vaccination.

CONCLUSIONS

It should be evident from the information summarized above, as well as in the following chapters dealing with individual parasitic organisms, that major advances have been made in the definition of mechanisms of acquired immunity against parasites in recent years. Perhaps the most significant developments have occurred in the identification of *in vitro* effector mechanisms and target antigens of protective immunity, and in the testing of some of the first nonliving experimental vaccines against parasites. Nevertheless, it is clear from the imperfect immunity induced by most vaccine regimens employing dead antigens that there is still a lot more information needed about mechanisms of parasite killing operating *in vivo* and on the manipulation and optimization of antiparasite immune responses. Fortunately, many of the questions underlying these problems now form part of a current focus of basic immunologic research: the study of cytokines and their role in the induction and regulation of humoral and cellular responses. The exciting new information gained from this flourishing field should provide innovative new approaches for analyzing the basis of acquired immunity against parasites and for the rationale design of improved vaccination strategies.

REFERENCES

1 Greenblatt CL. The present and future of vaccination against cutaneous leishmaniasis. Prog Clin Biol Res 1980;47:259–285.

2 James SP, Nicol WD, Shute PG. A study of induced malignant Tertian malaria. Proc R Soc Med 1932;25:1153–1158.

3 Terry RJ. Immunity to African trypanosomiasis. In Cohen S, Sadeen EH, eds. *Immunology of Parasitic Infections*, 1st edn. Oxford: Blackwell Scientific Publications, 1976:203–226.

4 Manson PEC. Immunity in kala-azar. Trans R Soc Trop Med Hyg 1961;5:550–561.

5 Sepulveda B, Martinez-Palomo A. *Immunology of Parasitic Infection*, 2nd edn. Oxford: Blackwell Scientific Publications, 1982:120–191.

6 Butterworth AE. Control of schistosomiasis in man. In Englund PT, Sher A, eds. *The Biology of Parasitism*. New York: Alan R Liss, 1988:43–60.

7 Sergent E. Latent infection and premunition. Some definitions of microbiology and immunology. In Garnham PCC, Pierce AE, Roitt I, eds. *Immunity to Protozoa*. Oxford: Blackwell Scientific Publications, 1963:39–47.

8 Gazzinelli RT, Xu Y, Hieny S, Cheever AW, Sher A. Simultaneous depletion of CD4[+] and CD8[+] lymphocytes is required to re-activate chronic infection with *Toxoplasma gondii*. J Immunol 1992. (In press.)

9 Suzuki Y, Remington IS. Importance of endogenous IFN-γ for prevention of toxoplasmic encephalitis in mice. J Immunol 1989;143:2045–2050.

10 Smithers SR, Terry RJ. Immunity in schistosomiasis. Am NY Acad Sci 1969;160:826–839.

11 Smithers SR, Terry RJ, Hockley DH. Host antigens in schistosomiasis. Proc R Soc London Ser B 1969;171:483–490.

12 Pearce EJ, Sher A. Mechanisms of immune evasion in schistosomiasis. Contrib Microbiol Immunol 1987;8:219–232.

13 Nussenzweig V, Nussenzweig RS. Sporozoite malaria vaccines. In Englund PT, Sher A, eds. *The Biology of Parasitism: a Molecular and Immunologic Approach*. New York: Alan R Liss, 1988:183–199.

14 James SL, Scott PA. Induction of cell-mediated immunity as a strategy for vaccine production against parasites. In Englund PT, Sher A, eds. *The Biology of Parasitism: a Molecular and Immunologic Approach*. New York: Alan R Liss, 1988:249–264.

15 Howard JG. Immunological regulation and control of experimental leishmaniasis. Int Rev Exp Pathol 1986;28:79–116.

16 Johnson KS, Harrison GB, Lightowlers MW, *et al.* Vaccination against ovine cysticercosis using a defined recombinant antigen. Nature 1989;338:585–587.

17 Good MF, Kumar S, Miller LH. The real difficulties for malaria sporozoite vaccine development: non-responsiveness and antigenic variation. Immunol Today 1988;9:351–355.

18 Raulet DH. Antigens for γ,δ T cells. (News and Views.) Nature 1989;339:342–343.

19 Modlin RL, Pirmez C, Hoffman FM, *et al.* Lymphocytes bearing antigen specific γ,δ T cell receptors accumulate in human infectious disease lesions. Nature 1989;339:544–548.

20 Cochrane AH, Nussenzweig RS, Nardin EH. Immunization against sporozoites. In Kreir JP, ed. *Malaria*. New York: Academic Press, 1980:163–197.

21 Miller LH, Howard RJ, Carter R, Good MF, Nussenzweig V, Nussenzweig RS. Research toward malaria vaccines. Science 1986;234:1349–1356.

22 Zavala F, Tam JP, Barr PH, Romero PH, Nussenzweig RS, Nussenzweig V. Synthetic peptide vaccine confers protection against murine malaria. J Exp Med 1987;166:1591–1596.

23 Alves MJ, Abuin G, Kuwajima VY, Colli W. Partial inhibition of trypomastigote entry into cultured mammalian cells by monoclonal antibodies against a surface glycoprotein of *Trypanosoma cruzi*. Mol Biochem Parasitol 1986;21:75–82.

24 Ouaissi MA, Cornette J, Afchain D, Capron A, Grasmasse H. *Trypanosoma cruzi* infection inhibited by peptides modeled from a fibronectin cell attachment domain. Science 1986;234:603–607.

25 Louis J, Milon G. Immunobiology of experimental leishmaniasis. Ann Inst Pasteur/Immunol 1987;138:737–795.

26 Blackwell JM, Ezekowitz RAB, Roberts MB, Channon JY, Sim RB, Gordon S. Macrophage complement and lectin-like receptors bind *Leishmania* in the absence of serum. J Exp Med 1985;162:324–332.

27 Da Silva R, Hall BF, Joiner KA, Sacks DL. CR1, the C3b receptor mediates binding of infective *L. major* metacyclic promastigotes to human macrophages. J Immunol 1989;143:617–622.

28 Butterworth AE, Sturrock RF, Houba V, Mahmoud AAF, Sher A, Rees PH. Eosinophils as mediators of antibody-dependent damage to schistosomula. Nature 1975;256:727–729.

29 Butterworth AE. Cell-mediated damage to helminths. Adv Parasitol 1984;23:143–235.

30 Capron A, Desaint JP, Capron M, Duma JH, Butterworth A. Immunity to schistosomes: progress toward vaccine. Science 1987;238:1065–1072.

31 Silberstein DS, David JR. The regulation of human eosinophil function by cytokines. Immunol Today 1987;8:380–385.

32 Sher A, Coffman RL, Hieny S, Cheever AW. Ablation of eosnophils and IgE response with anti-IL-5 or anti-IL-4 antibodies fails to affect immunity against *Schistosoma mansoni* in the mouse. J Immunol 1990;145:3911–3916.

33 James SL, Natovitz PC, Farrar WL, Leonard EJ. Macrophages as effector cells of protective immunity in murine schistosomiasis: macrophage activation in mice vaccinated with radiation attenuated cercariae. Infect Immun 1984;44:569–578.

34 McLaren DJ, Strath M, Smithers SR. *Schistosoma mansoni*: evidence that immunity in vaccinated and chronically infected CBA/Ca mice is sensitive to treatment with monoclonal antibody that depletes cutaneous effector cells. Parasite Immunol 1987;9:667–682.

35 Joiner KA. Complement evasion by bacteria and parasites. Ann Rev Microbiol 1988;42:201–230.

36 Shirazi MF, Holman M, Hudson KM, Klaus GGB, Terry RJ. Complement (C3) levels and the effect of C3 depletion in infections of *Trypanosoma brucei* in mice. Parasite Immunol 1980;2:155–163.

37 Goddeeris BM, Morrison WI, Teale AJ, Bensaid A, Baldwin CL. Bovine cytotoxic T cell clones specific for cells infected with the protozoan parasite *Theileria parva*: parasite strain specificity and Class I major histocompatibility complex restriction. Proc Natl Acad Sci USA 1986;83:5238–5242.

38 Sher A, Hall BF, Vadas MA. Acquisition of murine major histocompatibility complex gene products by schistosomula of *S. mansoni*. J Exp Med 1978;148:46–52.

39 Butterworth AE, Vadas MA, Martz E, Sher A. Cytolytic T lymphocytes recognize alloantigens on schistosomula of *Schistosoma mansoni* but fail to induce damage. J Immunol 1979;122:1314–1321.

40 Schofield L, Villaquiran J, Ferreira A, Schellekens H. Gamma interferon, CD8+ T cells and antibodies required for immunity to malaria sporozoites. Nature 1987;330:664–666.

41 Weiss WR, Sedegah M, Beaudoin RL, Miller LH, Good MF. CD8+ T cells are required for protection in mice immunized with malaria sporozoites. Proc Natl Acad Sci USA 1988;85:573–576.

42 Suzuki Y, Remington JS. Dual regulation of resistance against *Toxoplasma gondii* infection by Lyt-2+ and Lyt-1+, L3T4+ T cells in mice. J Immunol 1988;140:3943–3946.

43 Suzuki Y, Oreiiana MA, Schreiber RD, Remington JS. Interferon-γ: the major mediator of resistance against *Toxoplasma gondii*. Science 1988;240:516–519.

44 Borges JS, Johnson WD. Inhibition of multiplication of *Toxoplasma gondii* by human monocytes exposed to T lymphocyte products. J Exp Med 1975;141:483–490.

45 Nacy CA, Meltzer MS, Leonard EJ, Wyler DJ. Intracellular replication and lymphokine induced destruction of *Leishmania tropica* in C3H/HEN mouse macrophages. J Immunol 1981;127:238–247.

46 Nathan CF, Murray HW, Wiebe ME, Rubin BY. Identification of interferon-γ as the lymphokine that activates human macrophage oxidative metabolism and anti-microbial activity. J Exp Med 1983;158:670–678.

47 Mahmoud AAF, Peters PA, Civil RH, Remington JS. *In vitro* killing of schistosomula of *Schistosoma mansoni* by BCG and *C. parvum* activated macrophages. J Immunol 1979;122:1655–1661.

48 Bout D, Joseph M, David JR, Capron A. *In vitro* killing of *S. mansoni* schistosomula by lymphokine activated mouse macrophages. J Immunol 1981;127:1–10.

49 Colley DG. Lymphokine-related eosinophil responses. Lymphokine Rep 1980;1:135–155.

50 Hoover DL, Finbloom DS, Crawford RM, Nacy CA. A lymphokine distance from interferon-γ that activates human monocytes to kill *Leishmania donovani in vitro*. J Immunol 1986;136:1329–1333.

51 Fortier AH, Meltzer MS, Nacy CA. Susceptibility of inbred mice to *Leishmania tropica* infection: genetic control of the development of cutaneous lesions in P/J mice. J Immunol 1984;133:454–462.

52 James SL, Correa-Oliveira R, Leonard EJ. Defective vaccine-induced immunity to *Schistosoma mansoni* in P strain mice II. Analysis of cellular responses. J Immunol 1984;133:1587–1593.

53 Brake D, Long CA, Weidanz WP. Adoptive protection against *Plasmodium chabaudi adami* malaria in athymic nude mice by a cloned T cell line. J Immunol 1988;140:1989–1993.

54 Scott PA, Natovitz P, Coffman R, Pearce EJ, Sher A. Vaccination against cutaneous leishmaniasis in a murine model. III. Protective and exacerbating T cell lines belong to different TH subsets and respond to distinct parasite antigens. J Exp Med 1988;168:1674–1684.

55 Hatcher FM, Kuhn RE, Cerrone MC, Burton RC. Increased natural killer cell activity in experimental American trypanosomiasis. J Immunol 1981;127:1126–1130.

56 Albright JW, Hatcher FM, Albright JF. Interaction between murine natural killer cells and trypanosomes of different species. Infect Immun 1984;44:315–319.

57 Grzych JM, Capron M, Dissous C, Capron A. Blocking activity of rat monoclonal antibodies in experimental schistosomiasis. J Immunol 1984;133:1988–2004.

58 Khalife J, Capron M, Capron A, et al. Immunity in human schistosomiasis mansoni. Regulation of protective immune mechanisms by IgM blocking antibodies. J Exp Med 1986; 164:1626–1640.

59 Colley DG, Hieny SE, Bartholomew RK, Cook JA. Immune responses during human schistosomiasis. III. Regulatory effect of patient sera on human lymphocyte blastogenic responses to schistosome antigen preparations. Am J Trop Hyg 1977;26:197–925.

60 Todd CW, Goodgame RW, Colley DG. Immune responses during human schistosomiasis. V. Suppression of schistosome antigen-specific lymphocyte blastogenesis by adherent/phagocytic cells. J Immunol 1979;122:1440.

61 Scott PA, Farrell JP. Experimental cutaneous leishmaniasis. I. Non-specific immunodepression in BALB/c mice infected with Leishmania tropica. J Immunol 1981;127:2395–2400.

62 Sacks DL, Scott PA, Asofsky R, Sher FA. Cutaneous leishmaniasis in anti-IgM treated mice: enhanced resistance due to functional depletion of a B cell-dependent T cell involved in the suppressor pathway. J Immunol 1984;132: 2072.

63 Scott PA, Natovitz P, Sher A. B lymphocytes are required for the generation of T cells that mediate healing of cutaneous leishmaniasis. J Immunol 1986;137:1017–1021.

64 Tarleton R. Trypanosoma cruzi induced suppression of IL-2 production. II. Evidence for a role of suppressor cells. J Immunol 1988;140:2769–2773.

65 Mosmann TR, Coffman RL. TH1 and TH2 cells: different patterns of lymphokine secretion lead to different functional properties. Ann Rev Immunol 1989;7:145–173.

66 Scott P, Pearce E, Cheever AW, Coffman RL, Sher A. Role of cytokines and CD4⁺ T cell subsets in the regulation of parasite immunity and disease. Immunol Rev 1989;112:161–182.

67 Roitt I, Brostoff J, Male D, eds. Immunology, 2nd edn. London: Gower Medical Publishing, 1989.

68 Lehn M, Weiser W, Engelhorn S, Gillis S, Remold HG. IL-4 inhibits H_2O_2 production and anti-leishmanial capacity of human cultured monocyte mediated by IFN-γ. J Immunol 1989;143:3020–3024.

69 Mosmann TR, Moore KW. The role of IL-10 in cross-regulation of TH1 and TH2. In Ash C, Gallagher RB, eds. Immunoparasitology Today. Cambridge: Elsevier Trade Journals, 1991:49–53.

70 Heinzel FP, Sadick MD, Holaday BJ, Coffman RL, Locksley RM. Reciprocal expression of interferon-γ or interleukin 4 during the resolution or progression of murine leishmaniasis. J Exp Med 1989;169:59–65.

71 Belosovic M, Finbloom DS, Van der Meide P, Slayter M, Nacy CA. Administration of monoclonal anti-IFN-γ antibodies in vivo abrogates natural resistance of C3H/HEN mice to infection with Leishmania major. J Immunol 1989;143: 266–271.

72 Long EO. Intracellular traffic and antigen processing. Immunol Today 1989;10:232–234.

4 Mechanisms of immunopathology in parasitic infections

S. Michael Phillips

INTRODUCTION

The pathology of parasitic disease represents an extremely complex spectral and unique phenomenon.

The clinical severity of disease is inversely related to contributions of both the parasite and host. Parasites are extremely variable in terms of metabolic demands, their production of inflammatory reactions, complexities of the intrahost life-cycles, and tissue tropisms. Examples of parasite-induced pathology include: iron deficiency anemia, secondary to blood loss caused by intestinal nematodes; hemolytic anemia, secondary to intracellular protozoans; and hepatobiliary fibrosis, secondary to products produced by intraductal trematodes. However, in parasitic conditions, there often is limited pathology directly attributable to the organism. Most morbidity is related to the immunoinflammatory response of the host to the parasite. This reaction is the principal subject of this chapter.

The host immune response to a parasite may be divided into two major categories (Fig. 4.1). The first is immunoprotective, which controls parasite growth and development and minimizes parasite-induced pathology. The second aspect of the immune response produces pathology. These two categories involve multiple immune mechanisms and need not be consonant, either kinetically or mechanistically. Clinical disease is most salient when immunopathologic mechanisms are dominant or effective control of the parasite is not obtained.

Teleologically the body seeks to maintain an optimal homeostatic balance of immune reactivity which will maximize resistance and minimize morbidity. This goal is met with varying success in each host–parasite interaction, depending upon the nature and intensity of the immune response. At one extreme, host pathology may occur from the relatively "hyperergic state," i.e., too strong an immune response to the parasite will result in immunopathology (A). Examples include fibrosis and granuloma formation in schistosomiasis or leishmaniasis; erythema nodosum, conversion reactions, and nerve destruction in leprosy; glomerulonephropathies in malaria; tropical eosinophilic syndromes and lymphatic obstruction in filariasis. Conversely, if the immune response is relatively "hypoergic" it is inadequate to control the consequences of parasite growth (C) and pathology is caused by the infectious agent. Clinical disease is optimally contained and pathology is minimized at some central point in the spectrum of immune reactivity (B), suggesting the importance of secondary immune regulation (Fig. 4.1).

Finally, the consequences of host–parasite interactions must be considered within the context of factors which are not solely related to the parasite or the immune response. These factors include host genetic influences, environmental consideration, nutrition, drugs, and interactions with coexistent pathologic conditions. Pathology may be related to direct effects of infection on the host or indirect effects upon food supply, domestic animals, and the host's ability to function in his/her environment.

To illustrate the spectrum of pathology resulting from parasitic disease, this chapter utilizes a "Mini-symposium" approach to examine three dissimilar diseases: schistosomiasis, trypanosomiasis, and leishmaniasis. It is the intent of the authors to describe the spectrum of pathology resulting from these parasites to illustrate the wide range of immunopathologic consequences which can be produced by parasites. Within the context of evolving immunologic and molecular biologic approaches, we will mechanistically explore illustrative elements of the host and parasite which induce and regulate immunopathology.

Pathology of schistosomiasis

Steven L. Kunkel, Stephen W. Chensue, & S. Michael Phillips

Disease spectrum

The study of schistosomiasis has clearly resulted in major advances in our understanding of immuno-pathology. The application of the tools of immunology and molecular biology have increasingly shown that schistosomiasis is an immunologic disease and the parasite *per se* contributes little direct pathology.

The pathologic manifestations of schistosomiasis are clearly related to the life-cycle of this parasite. Since schistosomiasis *per se* is the subject of another chapter and there are many reviews of the manifestation of this disease [1–6], extensive considerations of the life-cycle and clinical syndromes will not be duplicated. However, it is of interest that the stages of the disease are clearly related to the life-cycle of the parasite (Table 4.1).

Schistosomiasis begins with skin penetration by the cercariae, with the production of a local dermatitis as-cribed to multiple mechanisms, including anaphylactic IgE and IgG antibodies and cell-mediated immunity (CMI).

Seven to ten days after heavy re-exposure, when the parasites are actively migrating through the lung, transient pulmonary symptoms, eosinophilia, and chest X-ray infiltrates are observed. This pulmonary infiltra-tional eosinophilia syndrome has been ascribed to inter-actions of IgE, IgG, and T cells.

Subacutely a serum resistance-like syndrome (Katayama Fever) develops, characterized by fever, chills, headache, anorexia, diarrhea, hepatospleno-megaly, lymphadenopathy, urticaria, and diffuse vasculitic lesions. The syndrome is contingent upon the intravascular production of large quantities of parasite antigen at the onset of egg deposition, circulating immune complexes, and the complexes of deposition of antigen–antibody complexes in tissues.

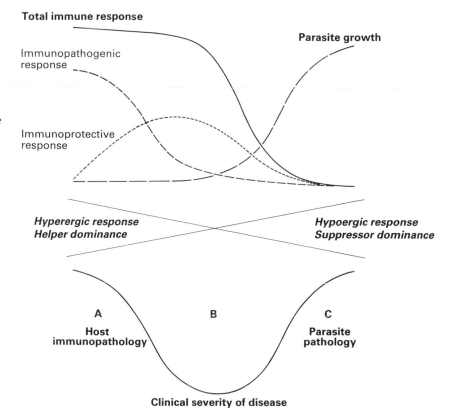

Fig. 4.1 Mechanistic model for the clinical spectrum of parasitic disease. The total immune response will be divided loosely into two categories: those which control disease and thus minimize parasite-associated pathology and those which produce pathology *per se*. The intensity of the immune response ranges from a relatively hyperergic state, characterized by strong immune responses, to a hypoergic state, characterized by relatively suppressed reactivity. The hyperergic state tends to be dominated by immunopathologic reactions secondary to host-dependent mechanisms, with relatively good containment of the parasite. The hypoergic state is often associated with relatively uncontrolled parasite growth. Clinical disease is most salient when the immunopathogenic responses or uncontrolled parasite growth are dominant.

Total immune response

Parasite growth

Immunopathogenic response

Immunoprotective response

Hyperergic response Helper dominance

Hypoergic response Suppressor dominance

A
Host immunopathology

B

C
Parasite pathology

Clinical severity of disease

Table 4.1 Spectrum of schistosomiasis

Stage of infection	Clinical manifestations	Pathologic manifestations	Immunologic mechanisms
Invasion			
Skin penetration	Swimmer's itch	Cutaneous inflammation	IgE, IgG, ADCC, CMI
Migration	PIE, cough, fever	Pulmonary/hepatic inflammation	IgE, IgG, ADCC
Maturation			
Initial oviposition	Katayama fever, serum sickness	Intense local and generalized vasculitic reaction	CIC, ADCC
Acute			
Intense oviposition	GI, Sx, hematuria	Large granulomata	CMI (S.m.)
Maximum egg production and excretion			B cell (S.j.)
Chronic			
Prolonged infection	Chronic disease, portal hypertension, nephropathy, CNS cor pulmonale	Modulated granulomata, Symmer's fibrosis, hepatosplenomegaly	CMI (S.m.)
Decreased egg production and excretion			B cell (S.j.)

ADCC, antibody-dependent cell-mediated cytotoxicity; CIC, circulating immune complexes; CMI, cell-mediated immunity; CNS, central nervous system; GI, gastrointestinal; PIE, pulmonary infiltrates and eosinophilia, S.j., *Schistosoma japonica*; S.m., *Schistosoma mansoni*; Sx, symptoms or signs.

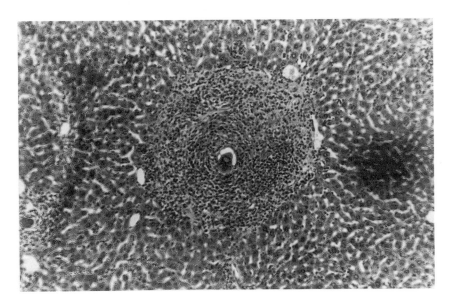

Fig. 4.2 The schistosome granuloma. The parasite ovum, containing germinating meracidial larva, can be seen at the nidus of a well-defined granulomatous reaction. Histologically the lesion is dominated centrally by mononuclear cells, lymphocytes, and macrophages. Cells toward the outer margins of the granuloma are mainly polymorphonuclear cells and predominantly eosinophils. At this stage the cytoarchitecture of the liver is relatively well maintained; however, in the chronic stages of the disease collagen and other matrix proteins are deposited, resulting in hepatic (Symmer's) fibrosis and ischemic hepatocellular damage.

Individuals with chronic schistosomiasis experience fatigue, a variety of bowel and bladder symptoms, and hepatic dysfunction. The local deposition of embolized eggs within the vessels, associated granulomatous reactions, and progressive fibrosis are observed histologically. Although some damage may result from egg products and worms *per se*, the major pathology appears to be contingent upon T cell-dependent immune responses.

The remainder of this section will discuss the generation and regulation of these lesions.

Pathology

There is compelling evidence to suggest that the granulomatous reaction and resulting fibrosis in chronic schistosomiasis mansoni is a result of T cell-mediated delayed-type hypersensitivity (DTH) immune responses

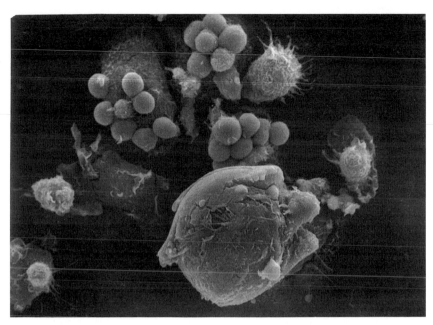

Fig. 4.3 *In vitro* granuloma formation: early events. Initial macrophage T cell lymphocyte interactions are evident. The first reactions involve the adherence of Ia⁺ macrophages to the surface of the egg or antigen-coated bead. Note particularly the three adjacent macrophage–lymphocyte clusters. These interactions are believed to be an essential component of the macrophage stimulation of lymphocytes and the development of an antigen-specific immune response. These cellular reactions occur before lymphocyte adherence to the bead. The initial lymphocytes adhering to these beads are CD4⁺, which rapidly express high-affinity IL-2 receptors.

to products of the egg (Fig. 4.2) [1–7]. Rectal biopsy granulomas are comprised of lymphocytes, macrophages, epithelioid cells, eosinophils, and fibroblasts [8]. These granulomatous reactions are decreased in immunosuppressed transplant recipients. The kinetics of cellular constituents, evolution, or regulation of the granulomatous lesions have not been studied in humans. Investigators utilizing *in vitro* granuloma formation by peripheral blood mononuclear cells, have demonstrated the progressive development of lymphocyte which can produce granulomas [9]. Experimentally, granulomas and clinical morbidity are reduced in congenitally athymic mice under pathogen-free conditions [10]. Immunosuppressed mice produce smaller granulomas, and granulomas do not occur in thymectomized chickens or T-depleted mice. Granulomatous hypersensitivity can be adoptively transferred through the use of selected subpopulations of T lymphocytes obtained from infected or soluble egg antigen-sensitized mice [1–6,11].

Granuloma formation can be initiated by cloned T cells, which are responsive to restricted antigenic epitopes presented by syngeneic Ia macrophages [12]. Monoclonal antibodies against Class II antigen CD4⁺ cells can block these Class II major histocompatibility complex (MHC)-restricted interactions of T cells and macrophages [13]. The CD4⁺ and CD8⁻ cells rapidly express

IL-2 receptors after antigenic excitation. Typical examples of granuloma formation *in vitro* are shown in Figs 4.3 and 4.4. These illustrations show the typical sequence of monocytic cell adherence to antigen-coated beads, subsequent interactive nests of macrophages and T cells, and the recruitment of various cells inducing macrophages, eosinophils, and fibroblasts; giant cell formation and matrix protein deposition is also observed [14].

Antigenic presentation is followed by the activation of T cells and macrophages by reciprocal stimulation with cytokines; IL-1 produced by macrophages, and interferon-γ (IFN-γ) produced by T cells. Further activation of T cells occurs by an autocrine pathway involving endogenous IL-2 generation and receptor expression. Once activated, the T cells and macrophages are the source of additional cytokines, such as MIF, granulocyte macrophage colony stimulating factor (GM-CSF), IFN-γ, IL-4, IL-5, IL-10, and transforming growth factor-β (TGF-β), tumor necrosis factor-α (TNF-α) and vasoactive intestinal polypeptide (VIP) [3,13,15–21]. These latter activities are apparently produced both locally and systemically, although comprehensive compartmental quantitations of production are only partially completed. These factors account for both granuloma-associated phenomena as well as indirect effects, such as eosinophil generation and mobilization.

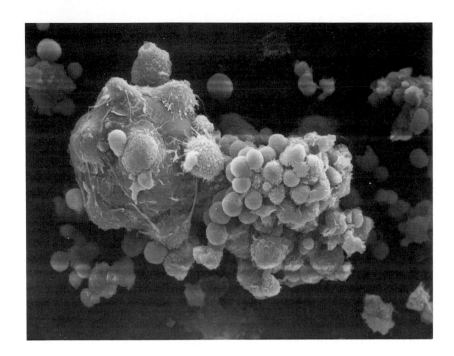

Fig. 4.4 *In vitro* granuloma: early and fully developed granulomas. Two soluble egg antigen-coated bead granulomas formed by *Schistosoma mansoni*-infected mouse splenocytes are shown. The granuloma on the left is at an early stage of development. Macrophages attached to the bead surface are the predominant cell type. The granuloma on the right demonstrates multiple additional layers of lymphocytes and other cells. Emigration of fibroblasts and fusion or palasading of macrophages to form giant cells is evident.

Regulation of granuloma formation: lymphocytic influences

Cellular constituents

The size of newly formed granulomas initially increases, reaches a maximum 8–10 weeks after infection, and then diminishes in size [22] owing to active suppression [23]. Experimental studies in *S. mansoni*-infected mice have defined participant helper T cells (L3T4$^+$), suppressor–inducer T cells, which also carry the helper determinants (L3T4$^+$, Lyt-1$^+$), and suppressor–effector T cells (Lyt-2$^+$), which are restricted by IJ/IC MHC molecules [3,7,11,13]. Analogous studies in humans utilizing *in vitro* granuloma formation have suggested that similar CD4 and CD8$^+$ helper and suppressor regulatory populations exist in the peripheral blood of infected humans [9,24]. *In vivo* depletion studies with anti-Lyt-1 and anti-L3T4 antisera diminish granulomatous hyper-sensitivity [3,7,11,13]. Conversely, depletion with anti-Lyt-2 can augment granulomatous hypersensitivity.

L3T4$^+$ or CD8$^+$ populations directly suppress lymphokine and granuloma formation by L3T4$^+$ and CD4$^+$ cells *in vitro* [3,7,11,13,24,25]. The *in vivo* co-adoptive transfer studies have confirmed the ability of the CD8$^+$ (Lyt-2$^+$) populations to suppress the ability of the CD4$^+$ (L3T4$^+$) cells to induce granulomatous hypersensitivity (Fig. 4.5).

The nature of the cellular infiltrates within the granulomas has also been examined [3,7,11,13]. In an elegant series of studies, subpopulations of L3T4$^+$, Lyt-2$^+$, and B lymphocytes were phenotypically isolated from the intact granulomas and studied functionally for lymphokine release and proliferation in response to egg antigen. Admixture experimental studies suggested active suppression of L3T4$^+$ cells by the Lyt-2$^+$ population and an inverse relationship to the intensity of the B cell response. Cells obtained from large granulomas produced approximately 10 times more IL-2 than those obtained from small lesions. The selective depletion of L3T4$^+$ cells led to suppression of the hepatic granulomas *in vivo*. This suppression could be reversed by recombinant IL-2 [7].

Cytokines and subsets

Additional studies have attempted to further subdivide the TH populations into two major subsets: TH-1 and TH-2. The eosinophilia and elevated IgE levels are at least in part dependent upon IL-4 and IL-5 [15–18]. Similarly, the immunoglobulin in schistosomiasis isotypes tends to be those associated with the TH-2 (IgG1, IgM, and IgA) but not with the TH-1 response (IgG2a) [26]. Depletion of IL-4 depresses IgE responses depletion with anti-IL-5 but results in granulomas nearly

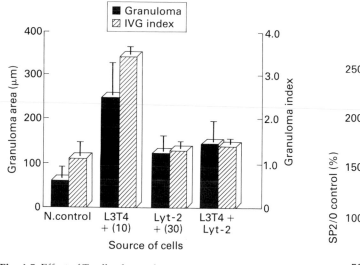

Fig. 4.5 Effect of T cell subpopulations upon granulomatous hypersensitivity. T lymphocytes are obtained from unexposed normal mice and mice exposed for 10 or 30 weeks to 25 *Schistosoma mansoni* cercariae. These cells were fractionated into L3T4$^+$ and Lyt-2$^+$ cells by panning techniques. The cells were injected into sublethally irradiated naïve recipients, which were challenged intravenously with freshly prepared *Schistosoma mansoni* eggs. Additional cells were maintained *in vitro* and assessed for cellular reactivity directed against antigen-coated latex beads. L3T4$^+$ cells, obtained from 10-week-infected animals, demonstrated the ability to transfer adoptively granulomas *in vivo* and produce large granulomas *in vitro*. Lyt-2$^+$ cells from chronically infected animals showed little reactivity. Admixture of the two populations resulted in suppression of the L3T4$^+$ cells' capabilities, both in terms of granuloma formation *in vivo* and *in vitro*.

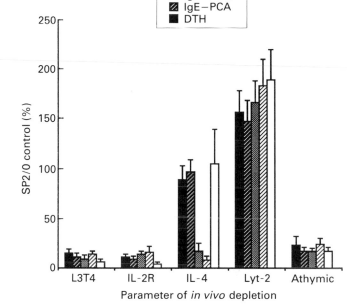

Fig. 4.6 Effect of *in vivo* depletion upon the development of immunity to *Schistosoma mansoni*. Animals were depleted of specific subpopulations of T lymphocytes and exposed to 500 irradiated *Schistosoma mansoni* cercariae. Parameters of the immune response included: IgM–Hem—total IgM-induced hemagglutination titers using soluble cercarial immunogen (SCI) antigens; IgG2a–E—IgG2a subclass antibody binding to SCI utilizing an enzyme-linked immunosorbent assay (ELISA); IgE–EoC—heat-labile antibody-dependent eosinophil-mediated cytotoxicity against schistosomule; IgE–PCA—heat-labile 24-h passive cutaneous anaphylaxis titer utilizing SCI; DTH—delayed-type hypersensitivity to SCI. The results indicated that there was a variable but significant suppression of all immune parameters in athymic and L3T4$^-$ or IL-2R$^+$-depleted animals. There was strong suppression of the IgE responses in the anti-IL-4-treated group. There was an augmentation of all immune responses in the Lyt-2$^+$-depleted group. To facilitate visual comparisons of the various parameters, the results are represented as a percentage change ±95% confidence interval vs. the value obtained for that parameter in SP2/0 injected animals.

devoid of eosinophils, whereas it causes only a marginal reduction of granuloma size and no alteration in hepatic fibrosis or egg deposition. Treatment with anti-IFN-γ has no effect on acute granuloma formation and IFN-γ responses are observed following antigen stimulation of acutely infected animal spleen cells. Interferon-γ will inhibit granuloma formation *in vitro*. Thus, exogenous IFN-γ may block the proliferation of TH-2 cells required for certain aspects of the response, and the granulomatous response involves both TH-2 populations. Assessment of IFN-γ and IL-2, IL-4, and IL-5 production which results from infection with normal or irradiated cercariae [18], suggests that cells of the TH-1 phenotype, secreting TH-1 cytokines, dominate prior to the onset of egg laying. After egg production, a strong TH-2 response is stimulated. A marked decrease of IFN-γ and IL-2 secretion coincides with elevated IL-4 and IL-5 production, suggesting a link between TH-2 responses and suppression of TH-1 function. The stimulation of TH-2 cells by egg antigen may also result in the production of IL-10 by these TH-2 cells [27], inhibiting the TH-1 population [28] and diminishing granulomatous hypersensitivity. The failure of exogenous IL-2 to promote IFN-γ production by antigen-stimulated spleen cell suspensions is also consistent with a role for IL-10, since this cytokine exerts an influence independent of the presence of IL-2 [28].

In vivo studies suggest that regulation may be multi-factorial. IL-2 reverses suppression by CD8+ cells or T suppressor–effector factor (TseF). In addition, IL-2R depletion leads to diminshed IgE and DTH reactivity. IL-4 depletion effects IgE production but not DTH, suggesting that IL-10 feedback may have limited effectiveness (Fig. 4.6).

Other regulatory substances may also play a role. For example, TGF-β, which may also interfere with IFN-γ production, is elevated in murine schistosomiasis [19] and may act independently of IL-10 [29]. In addition, TNF and IL-1β production are augmented in human disease, may be disease-stage specific, and contingent upon prior treatment [20,30]. Soluble IL-2 receptor also has been demonstrated in the serum of individuals infected with schistosomiasis [31]. The soluble IL-2 receptor may combine with circulating IL-2 and compete for the activation of TH-1- and IL-2-dependent cells. Since soluble CD8 antigen is also found in increased amounts in the serum of hepatosplenic schistosomiasis patients and is negatively correlated with CD3-induced mitogenesis, an imbalance of CD4 and CD8 ratios and increased IL-2 receptor/CD8 turnover also may reflect inhibitory circuits within the T cell compartment [32,33]. Many other humoral factors, such as circulating parasite antigen, immune complexes, idiotypic antibody, clono-typically derived T cells, and immunology products of the parasite, may play roles in both the specific and non-specific suppression observed in schistosomiasis [1–7]. Anti-idiotypic responses of T cells against idiotypes which recognize soluble egg antigen (SEA) have been extensively studied [34,35]. Similarly, the importance of autoanti-idiotypic regulation has been studied within the context of specific epitopes, paratopes, and anti-paratopic responses [36]. Specific anti-idiotypic cascades have been described and the importance of T cell receptor immunization upon anti-idiotypic responses has been studied at the global [37,38] and clonal [39] levels.

Receptor regulation

To investigate further the mechanisms whereby the subpopulations of T cells and their soluble cytokines might interact to induce granuloma formation and/or regulation, we have commenced study of the molecular basis of receptor-mediated regulation of granulomatous hypersensitivity [40–44]. A variety of functional activities of antigen-specific T suppressor cells has now been ascribed to the soluble products which they produce. Two or more distinct T cell populations interact sequen-

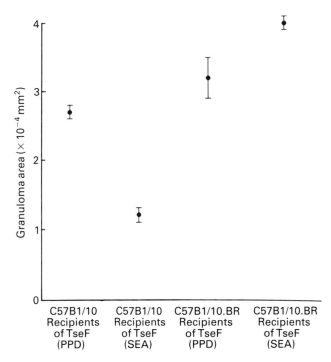

Fig. 4.7 Effective TseF administration on granuloma formation *in vivo*. The modulation of endogenous granuloma formation by TseF is genetically restricted. C57BL/10 (H-2b) and C57BL/10.BR (H-2k, congenic) mice, 6 weeks postexposure to 25 *Schistosoma mansoni* cercariae, were injected intraperitoneally with 0.5 ml of an IJ-B bearing TseF directed against either SEA or purified protein derivative (PPD); After 20 days the animals were sacrificed and the sizes of their liver granulomas were assessed. The isobars represent the 95% confidence interval. TseF suppressed granuloma formation only against the appropriate antigen and only in animals which were genetically compatible to the β-chain of the TseF.

tially through the production of regulatory factors; T suppressor–inducer cells do not suppress immune response directly; their product, T suppressor–inducer factor (TsiF), results in the generation of a second population of T suppressor–effector cells which produce a TseF capable of directly suppressing granuloma formation; the transfer of information may involve a population of suppressor transducers; the soluble mediators bear the same antigenic and genetic specificities as the cells which produce them, bear characteristic phenotypic markers, and have complex heterodimeric structures.

Factors are produced by incubating splenic lymphocytes from acutely infected animals with SEA resulting in the sequential production of TsiF and TseF and the suppression of granuloma formation both *in vivo* and *in vitro*. Antigenic and genetic restricted (Fig. 4.7) specificities or CD4$^+$ lymphocytes produce TsiF and are present in large numbers in early granulomas, decreasing dramatically during granulomatous regulation. Exogenous IL-2 will prevent the secretion of TsiF, suggesting regulation by a second population of IL-2 receptor-bearing cells. Once produced, the suppressor substances act to reduce granuloma formation even in the presence of exogenous IL-2. Interleukin-2 increases granuloma formation in chronically infected animals and reduces TseF activity *in vivo* and production. Evolving granulomas may be immunologic battlegrounds between conflicting helper and suppressor influences.

The structural analysis of TsiF and TseF suggests that they are heterodimeric molecules, which bear striking analogies to T cell receptors borne by the cells which produce them. The component chains are disulfide-linked, may be dissociated by reducing agents, and reassociated *in vitro*, thus restoring functional activity by complementation. The functional and phenotypic characterizations of the two chains has been published previously in detail [40–44]. The α-chain and β-chains are responsible for antigenic specificities and genetic restriction/function, respectively (Table 4.2).

T suppressor–effector factor suppresses granuloma formation through an interaction with the T cell receptor and the Ti complex. The target of TseF is an antigen-reactive CD4$^+$ T cell. Upon exposure to TseF these cells exhibit increased intracellular glutathione levels and ornithine decarboxylase production, abnormal signal transduction, and the failure to develop IL-high-affinity receptor following exposure to antigen.

Recently we have produced cloned T suppressor cells

which can suppress granuloma formation and DTH *in vivo* and produce TsiF *in vitro*. Hopefully, these advances will permit the molecular characterization of the suppressor substances and a more careful analysis of their modes of action.

Other soluble genetically restricted suppressor substances have also been described [45–47]. T suppressor factor (TsF) produced by T suppressor cells from chronically infected animals [48] is similar to the previously described TsiF and TseF. These factors, which function in an effector mode, are produced by cells obtained from both acutely and chronically infected mice and differ from the initial factors in that they are not genetically restricted, they lack IJ determinants, and they profoundly suppress antigen-mediated blast transformation. They are similar in that they are not cytotoxic and have a complex disulfide bond structure, they bear idiotypic determinants, and they apparently act upon a CD4$^+$ target population.

Regulation of granuloma formation: macrophage influences

Monocytes derived from granulomas have been shown to function in a highly activated state [49,50]. These cells are important in regulating the local milieu inside the granuloma.

Summary. In summary, the principal morbidity in schistosomiasis, granulomatous hypersensitivity, and fibrosis is clearly dependent upon multifactorial events. Although the generation and subsequent modulation of granulomatous hypersensitivity appears to be primarily under the aegis of T cell regulation, the importance of systemic and locally released cytokines suggests that the monocyte is also a strong contributor to pathology. Hopefully, newly developed techniques in educational analysis utilizing the tools of molecular biology and immunology will clarify many of the enigmas and provide constructive insights into the regulation of the prevention of disease through an attack upon pathogenic mechanisms.

The types and levels of inflammatory mediators, locally and/or systemically expressed, are likely to determine the course and outcome of disease. This spectrum of mediators, which include lipids, polypeptides, and metabolites of oxygen and arginine, are important in determining the pathophysiologic manifestation of the host's response to parasite antigens. Macrophages are critical effector cells in many parasitic diseases, serving

Table 4.2 Comparison of TsiF and TseF structural characteristics

TsiF	TseF
Disulfide-bonded heterodimer	Disulfide-bonded heterodimer
α-Chain has SEA receptor, 14–12 determinants	α-Chain has SEA receptor
β-Chain has IJ and TCR allotypic determinants	β-Chain has IJ, TCR allotypic, and 14–30 determinants
α-Chain determines antigenic specificity	α-Chain determines antigenic specificity
β-Chain imparts genetic restriction, suppressor–inducer mode	β-Chain imparts genetic restriction, suppressor–effector mode

SEA, soluble egg antigen; TCR, T cell receptor.

Table 4.3 Polypeptide mediators of inflammation include a heterogeneous group of cytokines synthesized from a variety of immune and nonimmune cells

Interleukins 1–8	Transforming growth factor
Tumor necrosis factor-α	Interferon-α, -β, -γ
Lymphotoxins	

both as phagocytic cells and as a source of immune mediators.

Synchronous granuloma formation

Macrophage mediators in granuloma formation

The most important cellular components of both foreign body and DTH granulomas are lymphocytes and macrophages. Lymphocyte-derived factors modulate the functional activity of macrophages, which in turn potentially direct the course of granuloma development via the production of inflammatory mediators. Macrophages secrete numerous biologically active agents, including polypeptides, proteases, arachidonate metabolites, and reactive oxygen intermediates [50–52]. The polypeptide cytokines can effectively modulate immune responses. Many of these polypeptide intercellular signals have been isolated, cloned, and expressed in recombinant form (Table 4.3).

IL-1 and TNF. There is increasing evidence that IL-1 and TNF may be important mediators in granulomatous diseases. Animal studies have shown that recombinant IL-1 bound to implanted ethyl vinyl disks [53] or to Sephadex beads [54] could induce a granulomatous response. In animal models of inflammation, IL-1 can be detected in aqueous extracts of murine pulmonary granulomas [55]. Peripheral blood monocytes of patients with either tuberculosis or leprosy spontaneously synthesize IL-1 [56,57]. Tumor necrosis factor-α has also been associated with granulomatous diseases, as neutralizing antibodies to TNF have been shown to reduce rat myobacterial granulomas [58]. However, as yet it is difficult to develop a clear causal relationship between these cytokines and the development of specific diseases.

The systematic analysis of the source and sequence of cytokine production by macrophages isolated from synchronously developing pulmonary granulomas has demonstrated an interesting kinetic pattern of IL-1 and TNF generation [52]. The pattern of IL-1 induction by

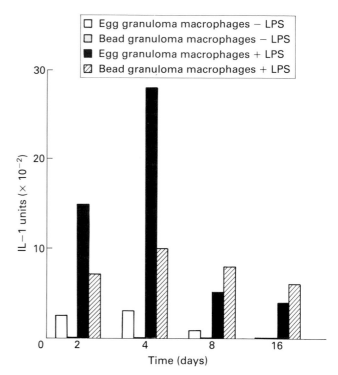

Fig. 4.8 Interleukin-1 production by macrophages isolated from synchronous foreign body (Sephadex bead) or DTH (schistosome egg) lesions. Analysis of the kinetics demonstrates an increase in IL-1 production by macrophages during early lesion development. LPS, lipopolysaccharide.

granuloma macrophages isolated at specific time points postembolization of Sephadex bead (foreign body-derived macrophages) or schistosome eggs (DTH-derived macrophages) is shown in Fig. 4.8. Interestingly, minimal IL-2 is spontaneously produced. Endotoxin- or zymosan-challenged granuloma macrophages isolated from early (day 2 and 4) *S. mansoni* egg-induced produced two- to three-fold more IL-1 than did macrophage pulmonary lesions isolated from early developing foreign body lesions. At days 16 and 32, macrophages from both types of lesions demonstrated similar capacities to generate IL-1. Macrophages from the foreign body lesions exhibited only minor changes in the induced synthesis of IL-1 over the entire 32-day study period. It is not known if the pattern of activity is related to decreased IL-1 synthesis or increased production of natural IL-1 antagonists. Teleologically, IL-1 may be needed for the initial recruitment and mobilization of inflammatory cells through effects on endothelial cell adhesion molecules, promotion of lymphocyte proliferation, etc. It should be noted that a membrane-bound or

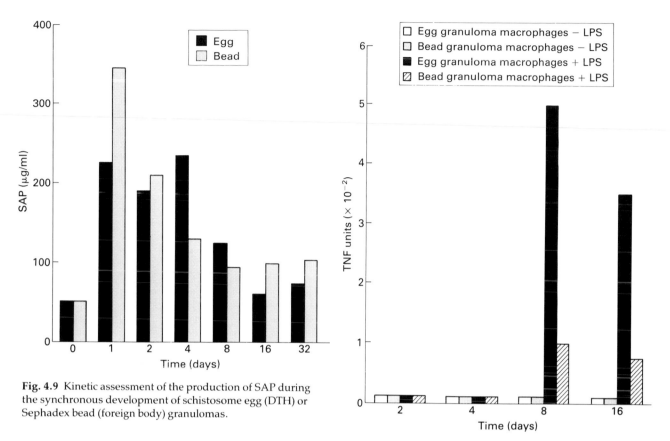

Fig. 4.9 Kinetic assessment of the production of SAP during the synchronous development of schistosome egg (DTH) or Sephadex bead (foreign body) granulomas.

Fig. 4.10 Tumor necrosis factor production by macrophages isolated from synchronous foreign body (Sephadex bead) or DTH lesions (schistosome egg). The kinetics in TNF production demonstrates an increased response of macrophages to an exogenous challenge.

cell-associated IL-1 may also be involved in granuloma formation.

IL-1 exists in two biologically active forms (IL-1α and IL-1β). Both are approximately 17 kD, represent cleavage products of an intercellular precursor, and share the same receptor. They only possess approximately 25% homology and IL-1α is a major cell-associated protein, while IL-1β is a secreted polypeptide [59]. Cell-free culture supernatants from stimulated macrophages isolated from 4-day egg granulomas demonstrated four-fold higher levels of IL-1β than IL-1α. This and other data suggest that IL-1β is the major form of IL-1 released by the granuloma macrophage.

In addition to contributing to local inflammatory events, IL-1 may influence systemic events. Increased IL-1 levels can lead to an increase in immature circulating neutrophils and a rise in acute-phase proteins. Murine serum amyloid P (SAP) can rise dramatically and is a true murine acute-phase protein. Hepatic acute-phase proteins are induced by many factors, including IL-1, IL-6, and TNF. The specific contribution of each cytokine to the acute-phase response during granuloma formation is not known. As depicted in Fig. 4.9, circulat-

ing levels of SAP were elevated during the early phases of the pulmonary granuloma development (days 2 and 4) and then slowly decreased over the next 12 days. A striking correlation exists between the elevation of SAP levels and the spontaneous production of granuloma macrophage IL-1. This observation suggests that IL-1 may contribute to the acute-phase response and have a positive impact on the subsequent evolution and growth of the T lymphocyte-mediated granuloma.

Another mononuclear cell-derived cytokine that may play a prominent role during the development of the schistosome egg granulomas is TNF. Although this cytokine possesses a number of overlapping functional activities with IL-1, it is not produced by stimulated granuloma macrophages until the growth and maintenance phase of the lesion, days 8–16 [52]. Little activity is detected during the initial 4 days of lesion

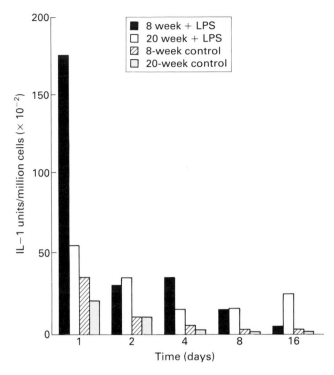

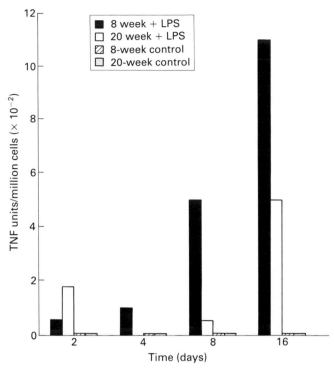

Fig. 4.11 Interleukin-1 production by pulmonary granuloma macrophages isolated 2, 4, 8, and 16 days postembolization from mice actively infected for 8 or 20 weeks.

Fig. 4.12 Kinetics of TNF production by pulmonary granuloma macrophages isolated 2, 4, 8, and 16 days postembolization in mice actively infected for 8 or 20 weeks.

development. Stimulated macrophages isolated from the *S. mansoni* granulomas produced fivefold more TNF than did cells isolated from synchronously developing foreign body granulomas (Fig. 4.10). Unstimulated macrophages did not spontaneously release TNF. Like IL-1, TNF promotes the recruitment and proliferation of inflammatory cells. The pattern of cytokine production indicates a shift from IL-1 to TNF synthesis and implies that TNF may be needed to sustain lesion development. The weaker capacity of the foreign body granulomas to produce TNF may explain the more ephemeral nature of these lesions.

The immunomodulated stage of schistosomiasis is an intriguing interaction of T lymphocyte subsets which ultimately leads to a diminished granulomatous response. Hyporesponsiveness may be attributed to direct cellular interactions, soluble mediators, and changes in lymphocyte-derived cytokines. To assess macrophage activation, IL-1 production by pulmonary granuloma macrophages induced in *S. mansoni*-infected mice at 8 and 20 weeks (vigorous and immunomodulated stages) was determined (Fig. 4.11). Maximum IL-1 produc-

tion was found early during lesion development and production was anamnestic. Maximum IL-2 production occurred earlier with macrophages derived from infected mice and correlated with the accelerated time course of lesion development. Granulomas induced during the modulated stage showed considerably reduced levels of IL-1, confirming that macrophage activity declined with modulated T cell function.

A similar pattern of reduced cytokine production was found with regard to TNF (Fig. 4.12). Macrophages isolated during the modulated stage (20 weeks) had a reduced capacity to generate TNF compared to the vigorous stage (8 weeks). Levels of macrophage-derived cytokines correlate with the kinetics and size of the *S. mansoni* egg-induced lesion. Large granuloma macrophages produced more cytokine than did macrophages from modulated lesions, suggesting a role for cytokines in granuloma formation.

Fibroblast growth factors. Since schistosomiasis terminates in hepatic fibrosis, the mediators of fibroblast growth may be important. Two heterogenous fibroblast

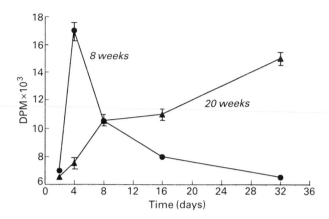

Fig. 4.13 Fibroblast growth factor production by granuloma macrophages during vigorous (8 weeks) and modulated (20 weeks) stages of active infection. Macrophages were isolated from synchronously induced pulmonary egg granulomas. Serum-free supernatants were collected and tested in a standard fibroblast proliferation assay.

growth/stimulation factors (FGF) of molecular weights 10–16 kD and 45–57 kD are produced by granulomas [60].

Figure 4.13 shows the spontaneous production of FGF by macrophages isolated from granulomas at intervals over a 32-day period. Macrophages from acute-stage lesions (8 weeks) produced maximal FGF activity at 4 days. Those from modulated lesions (20 weeks) did not show the peak of activity at 4 days, but rather showed a steady increase of FGF activity over the 32-day period. The FGF produced during the acute and modulated stages represent different polypeptides. Thus, regulation of FGF production may be related to the pathology observed at the different stages of disease and may provide insights into the interference of fibrosis.

Potential mechanisms of granuloma cytokine regulation. The immunomodulation associated with the regulation of macrophage cytokine expression is clearly multifactorial. T cells may indirectly regulate macrophage functional activity by alternating the activity of helper/effector T cells or directly through the production of IL-2, macrophage inhibition factor (MIF), and TsF.

In addition to antigen-specific suppressor factors, other molecules may regulate cytokine production or activity. For example, inhibitors of IL-1 activity have been demonstrated in plasma and serum [61,62] and in urine [63–65]. These glycoproteins have been shown to specifically inhibit IL-1 activity in the thymocyte assay to inhibit IL-2 activity. The role of IL-1 inhibitor proteins in the schistosome egg granuloma is still to be defined.

Corticosteroids and PGE_2 profoundly reduced schistosoma egg granuloma size, perhaps through inhibition of cytokine synthesis. Corticosteroids inhibit IL-1 production and expression of Ia antigen on murine macrophages [64]. Corticosteroids induce an inhibitor of phospholipase A and are elevated in chronic hepatic disease, potentiating their cytosuppressive role. The stimulation of macrophages with lipopolysaccharide (LPS) liberates IL-1 but also PGE_2. Prostaglandin E_2 inhibits the production of LPS-stimulated macrophage IL-1. Thus, products of arachidonic acid metabolism may endogenously regulate IL-1 production.

Tumor necrosis factor-α may play a central role in the activation of immune and nonimmune cells important for the successful growth and maintenance of the egg granuloma. The factors that regulate TNF production include glucocorticoids, which suppress the synthesis of TNF at both the transcriptional and post-transcriptional levels [65], and PGE_2, which modulates in an autocrine manner [66] via the intercellular signal cAMP [67]. The suppression of TNF by endogenous glucocorticoids and/or PGE_2 are potentially important *in vivo* processes that check the overproduction of a protein mediator and can have devastating physiologic consequences. Interferon-γ, in combination with an additional macrophage activator, such as LPS, has an enhancing effect on TNF expression. Since IFN-γ-induced augmentation of TNF mRNA is increased by cycloheximide, any augmentation of TNF may be regulated by short-lived repressor proteins.

Although IL-4 was already partially discussed within the context of lymphocytes, this T cell product is also a major endogenous modulator of several macrophage-derived mediators, including IL-1, TNF, IL-6, GM-CSF, and IL-8 [68,69]. Interleukin-8 is a potent chemoattractant for both neutrophils and lymphocytes, and is synthesized by macrophages, endothelial cells, epithelial cells, and fibroblasts. Since IL-4 is a late T cell product produced in an antigen-specific manner, its role during vigorous granuloma formation may be dominated by specific proinflammatory cytokines, such as IFN-γ, IL-1, and TNF. However, as the infection becomes chronic, IL-4 may contribute to the immunomodulation through effects on proinflammatory monocyte/macrophage-derived cytokines.

Chapter 4

Pathology of American trypanosomiasis

Rick L. Tarleton

Trypanosoma cruzi is an obligate intracellular parasite which infects 16–18 million people, with more than 45 million at risk of infection [70].

Disease spectrum

Three stages of infection have been described: acute, indeterminate, and chronic. The acute phase is characterized by an abundance of circulating parasites in the blood, often totally asymptomatic, and is terminated when the immune response of the infected host is sufficiently potent to control parasite growth. Hosts (5–10%) lacking an adequate immune response will fail to regulate parasite growth and succumb to an overwhelming parasitemia [71]. The indeterminate phase is an asymptomatic phase, characterized by positive serology for *T. cruzi* infection but low to undetectable parasitemia and the absence of clinical disease. An unknown percentage of infected individuals (estimated as high as 30%) eventually progress to the chronic phase, which is characterized by clinical disease usually focused in the heart and/or digestive tract. Chronic Chagas' disease is the leading cause of morbidity and mortality among middle-aged adults in many areas of Latin America.

Pathology

The pathology of *T. cruzi* infection and Chagas' disease have been well described [71–75]. The acute infection results in relatively high tissue parasite loads. The accompanying inflammatory response around infected cells at the site of entry of the parasite (chagoma or Romana's sign) or, more characteristically, in muscle, results in acute tissue damage. Chronic chagasic cardiac pathology includes enlargement of the chambers, muscle wall thinning with apical aneurysms, interstitial fibrosis, and inflammation and fibrosis of the conduction system. Digestive system involvement includes dilation and hypertrophy of the esophagus or colon (megaesophagus or megacolon), associated with inflammation, fibrosis, and denervation [72].

Mechanisms of pathogenesis

Chagas' disease is a spectral disease with the majority of infected individuals suffering no apparent clinical illness and the minority exhibiting a wide range and variety of disease symptoms. Unfortunately, ethical and technical reasons have prevented the systematic study of the progression of disease from the acute to chronic stage in humans. However, there are few data to argue that the chronic disease state is *not* the result of a slowly progressive pathologic process initiated by the lesions formed during the acute phase.

A number of mechanisms have been proposed to account for chronic chagasic pathology, including direct tissue destruction, loss of nervous tissue function, intravascular platelet aggregation, and generation of autoimmune reactivity [74,76–78].

The earliest proposed and still the simplest explanation for chronic-phase pathology is the cumulative damage due to parasite invasion of muscle cells, lysis of those cells, release of additional parasites, and a consequent inflammatory reaction which terminates in fibrosis. Sterilizing immunity and total clearance of *T. cruzi* is rarely reported, so it is presumed that most infected individuals harbor the parasite for life and may experience significant cumulative organ damage due to cycles of parasitosis.

The hypothesis that a parasite toxin is responsible for pathogenesis [79] has been challenged. Although, there is little doubt that the parasite load correlates to the degree of tissue damage during acute infection, the link between pathogenesis and parasite number during chronic *T. cruzi* infection is tenuous. Pathologic lesions and clinical disease do not increase in immunosuppressed patients experiencing reoccurrence of patent parasitemia [80]. Indeed, the relative absence of parasites during the chronic phase of infection, particularly in tissues exhibiting fibrotic and inflammatory lesions, has prompted other theories to explain chronic chagasic pathology.

The lack of evidence for a direct pathogenic effect of *T. cruzi* in chronic Chagas' disease does not imply that parasite levels are unrelated to pathogenesis. To the contrary, it is possible that acute-phase parasite levels determine both the severity of acute and of chronic disease. The initial parasitemia load might determine the severity of the initial lesion phase or the type of immune response elicited (see below). In addition, the methods for measuring chronic-phase parasitemia (xenodiagnosis and hemaculture) are inefficient and essentially non-quantitative, making evaluation of the connection between clinical disease and parasitemia in the chronic-phase nearly impossible to estimate. For example, the association between positive xenodiagnosis and abnormal electrocardiograms (ECG) is controversial [81,82]. Until a more sensitive and quantitative assay for *T. cruzi* is available, the question of parasite contribution to chronic pathology in Chagas' disease will remain open.

The reduction in the number of ganglion cells in the heart and gut of chronic chagasic patients prompted suggestions of a role for denervation in pathogenesis during chronic Chagas' disease [72,79,83]. Destruction of the myenteric plexus and cardiac vagal neurons may lead, respectively, to the gut megasyndrome and cardiac sympathetic abnormalities, autonomic imbalance, and compensatory ventricular dilatation [84]. A variety of direct and indirect (usually autoimmune) pathways have been proposed to explain neuronal destruction [72,85–91]. Data contrary to simple denervation hypothesis include the failure to observe denervation in chagasic hearts and the failure of vagectomy to produce similar pathology in rabbits or dogs [74]. The simultaneous measurement of cardiac autonomic innervation and ventricular function suggest that myocardial damage and mild ventricular dilatation *precedes* parasympathetic abnormalities [84].

The focal nature of chagasic heart pathology suggests that changes in cardiac microcirculation also may be a pathogenic mechanism in *T. cruzi* infection. Microcirculatory abnormalities have been found in both acute and chronic murine *T. cruzi* infection and etiologies have been suggested [92,93]. Conversely, autopsy data from a large number of human cases have demonstrated a low incidence of myocardial infarctions, arguing against either denervation or small coronary artery lesions as causative factors in chagasic heart pathology [94].

Autoimmune disease

The evidence for and against the extremely provocative and controversial autoimmune hypothesis for Chagas' disease has been extensively reviewed and debated [76,77,90,91]. Both autoantibodies [86,95–97] and autoreactive cellular responses [98–101] have been

documented. However, many of the original reports of autoreactivity used potentially alloreactive allogeneic cells to measure cellular autoreactivity or failed to demonstrate a cause-and-effect relationship between autoreactivity and pathogenesis (e.g., endocardium, vascular structures, and interstitia (EVI) autoantibodies [76,77]). While autoreactivity in Chagas' disease is "unequivocal" [78], its relationship to pathogenesis has not been firmly established. Similar to other infectious diseases with proported autoimmune etiologies [102], there is no "incontrovertible evidence that immunization with the crossreacting antigen(s), as well as passive transfer of either antibodies or lymphocytes specific for this antigen, result in the production of characteristic chagasic tissue lesions" [76].

Putative autoantigens and experiments examining the effects of passive transfers, particularly of lymphocytes, have been reported. These and previous studies suggest that the relationship between the immune system and *T. cruzi* is complex and that the immune response may contribute to pathology. Whether this contribution is "autoimmunity" in the classic sense, with crossreactive "autoimmune" epitopes on host and parasite initiating the pathology, or in a more subtle sense, with the immune system continuing a response which was initially parasite directed, is not clear. However, Chagas' disease does meet many of the criteria for classic autoimmune disease [102].

Susceptibility is polygenic and influenced by MHC Class I or Class II genes

Although the relationship between MHC genes and resistance to infection has been well studied in the mouse, few studies have examined the relationship between MHC genes and the severity of Chagas' disease in experimental models or in humans. Genetic differences among the human populations and/or a combination of other parasite, vector, or socioeconomic variables might explain variations in the severity of symptoms in different geographic regions of Latin America, as well as the variable relationship between seropositivity and clinical disease. An age-, sex-, and geographic area-matched study of the differences in pathology in southern Brazil showed a higher prevalence of ECG abnormalities in blacks than in whites. However, these findings were independent of seropositivity for *T. cruzi* infection and may be related to differences in socioeconomic conditions. Studies have shown no relationship between MHC genes and the presence of Chagas' disease [103] as well as a significant correlation between the presence of the combination of the B40 and Cw3 Class I MHC alleles and lack of cardiopathy in seropositive patients [104].

Susceptibility is multifactorial

Genetic predisposition alone does not assure the development of autoimmune disease, implying that environmental factors play major roles. Factors such as the strain, dosage, and route of infection influence disease severity in murine Chagas' disease [105]. Attempts to study the relationship of parasite genetics to human disease, by correlating differences in parasite zymodeme or schizodeme patterns with the presence or severity of pathology in human infections, have been largely unsuccessful. An association between humoral immunodepression in chronic *T. cruzi* infection and the zymodeme of *T. cruzi* isolated from particular patients is not supported by a correlation between the zymodeme and the pathologic state of the patients [106,107].

Onset of disease is age-related

The onset of classic autoimmune diseases often follows puberty or occurs around age 40–50 years [106]. The symptoms of chronic Chagas' disease generally occur in the third and fourth decades of life [71]. A recent longitudinal study of Chagas' disease [108] demonstrated a bimodal distribution of the development of ECG abnormalities in seropositive individuals. The greatest frequency of shifts from normal to abnormal ECG was in the 10–19-year age group (which would include the age of puberty) and the 30–39-year age group. Since most people living in endemic areas are infected in the first decade of life, these results might suggest at least two different pathways to disease development in human Chagas' disease. However, the factors which determine the timing and kinetics of clinical illness in Chagas' disease are not known.

Most autoimmune diseases exhibit "spontaneous exacerbations and remissions"

We are not aware of reports of exacerbation or remission of Chagas' disease symptoms; however, the relative insensitivity of methods for accessing disease onset and severity, as well as relative irreversibility of lesions, may account for this lack of information.

Most autoimmune diseases exhibit autoantibody

The production and patterns of autoantibodies are important criteria for autoimmune diseases. A large variety of autoantibodies are produced during Chagas' disease but none have been shown to directly induce pathology *in vivo*; thus, their role in pathogenesis is not clear. The autoantibodies may be epiphenomenological or involved in perpetuation rather than the initiation of tissue damage [102].

Thus, although meeting the broad criteria of an autoimmune disease, a cause-and-effect relationship between autoimmunity and pathogenesis in Chagas' disease has not been established. In addition, it is particularly difficult to separate the infectious agent-induced pathology from the immune-induced pathology.

If pathology in Chagas' disease is immune-mediated, what are the pathogenic immunologic components? Autoantibodies certainly have been the most studied and are the most controversial candidates. The proposed role for autoantibodies to heart tissue (EVI antibodies) in pathogenesis [95,109] have been challenged because of the heterophile nature of the antibodies, the lack of a correlation between the presence of these autoantibodies and disease status, and the presence of EVI antibodies in the circulation of patients with other parasitic diseases evidencing dissimilar pathology [76,77]. Antibodies to other common autoantigens (i.e., myosin, actin, tubulin, myoglobin, and DNA) have been reported in murine *T. cruzi* infections but the level of these antibodies correlates with the overall level of anti-*T. cruzi* antibodies but not with disease status [110].

Antimuscle sarcolemma antibodies induced during *T. cruzi* infection have been reported [96,111]. These antibodies may be related to anti-idiotypic antibodies specific for the idiotypes reactive with the membrane receptor on *T. cruzi* for muscle cells [112,113]. If the idiotypes on antibodies specific for *T. cruzi* attachment molecules express the internal image of the host cell receptor, anti-idiotypic antibodies could react with the *T. cruzi* receptor and induce myocarditis. A high correlation between the presence of these anti-idiotypic antibodies and severe cardiac pathology has been shown [114]. Both free and immune complex-bound anti-idiotypic antibodies were significantly higher in hospitalized chronic chagasic heart patients compared to asymptomatic seropositive and seronegative controls. The fact that the proposed pathogen-inducing antibody is measured as anti-idiotype (rather than as autoantibody to self-structure) minimizes the criticism that this autoimmune response is due to reaction to self-antigens which are normally sequestered and only released *following* host cell damage by *T. cruzi*. However, these anti-idiotypic antibodies have not been shown to mediate muscle cell destruction.

Antiacetylcholinesterase [115] and antimyocardial β-receptor antibodies [116,117] have also been described. There has been no correlation made between clinical status of patients or experimental animals and the presence of these antibodies; anti-idiotypic antibodies to anti-AChE are higher in symptomatic than in asymptomatic patients [115].

Studies of autoreactive cellular responses in Chagas' disease have also yielded mixed results. Original reports of cellular responses in human Chagas' patients to heart antigens used allogeneic or xenogenic antigens [76,77]. The proliferative responses of human patients to heart antigens has been correlated with prior exposure to streptococcal antigens and not to autoreactivity [118]. The response of cells from human chagasic patients to allogeneic tissue was found to be unrestricted to heart tissue (liver and kidney were also recognized) and not correlative with the clinical status of the patients [119]. Responses to rat heart extracts did not correlate with the clinical status in the patients [120]. Even in experimental animals where isogenicity can be attained, true auto-reactivity is somewhat elusive. Autoreactive cellular proliferative [101] and cytotoxic [100] responses have been reported in murine Chagas' disease. However, most of the "autoreactive" cellular activity of cells from *T. cruzi*-infected mice has been accounted for by allo-reactivity [121]. Without evidence of autoreactivity to syngeneic tissue, alloreactivity could be a byproduct of the well-documented polyclonal activation which occurs in *T. cruzi* infection.

Convincing evidence for cell-mediated immuno-pathology has come from murine *T. cruzi* infection. Immunoincompetent or immunosuppressed mice exhibit greatly reduced inflammatory response in the acute phase of infection and die of overwhelming parasitemias [122,123]. Effects of immunosuppression on chronic pathology has not been well studied. Re-activation of the infection (and thus an uncontrollable parasite load) makes these experiments difficult to perform.

The transfer of pathology by cells from infected mice would be a criteria to define Chagas' disease as a cellular autoimmune disorder [76]. CD8⁻ cells from chronically infected mice induce demyelination and inflammation in the sciatic nerves of normal mice [124]. The lesions induced by this transfer are histologically similar to those seen in chronically infected mice. CD8⁻ cells from chronically infected mice also transfer an inflammatory response to the liver (but not heart or skeletal muscle) when injected intravenously into normal recipients [125]. Unfortunately, the cells used in these latter studies were likely to be contaminated with parasites. However, long-term, parasite-free CD4$^+$ T cell lines cultivated with either *T. cruzi* antigens or peripheral nervous tissue antigen also transfer similar activity when injected locally instead of systemically. Although these studies are supportive of a role for autoreactive cells in pathogenesis during *T. cruzi* infection, they would be more convincing if they showed pathology in tissues other than the sciatic nerve and liver, which are not considered major foci of Chagas' disease.

The most convincing evidence for cellular immune pathogenesis in Chagas' disease comes from the murine syngeneic heart transplant model [126]. In this experimental system, neonatal hearts are implanted under the skin in the ear of adult mice. A syngeneic transplanted heart will vascularize and beat normally, whereas an allogeneic heart will be rejected within a few days. In chronic chagasic mice, however, syngeneic hearts are rejected as if they were allogeneic. This rejection phenomenon can be inhibited by depleting recipient mice of CD4$^+$ cells with anti-CD4 antibodies, and CD4$^+$ T cells from chronic mice can induce rejection in normal recipients of neonatal hearts [126]. This model should provide a powerful tool to examine the factors contributing to heart pathology and to test potential therapies for chagasic heart disease.

Other factors involved in immune-based pathology

The difficulty in demonstrating an autoimmune etiology for Chagas' disease by injection of putative autoantigens or serum/cell transfers is not necessarily surprising. The presence of autoantibodies does not imply autoimmune etiology of infectious disease, and autoantibodies or autoreactive cells are often insufficient to induce the disease which they mediate [127]. Other factors may be required to create the appropriate milieu for auto-reactivity to create immunopathology.

Cytokine production

Recognition of parasite antigens by the immune system of infected mice results in the release of lymphokines and monokines with a wide range of systemic and local activities. Besides cell recruitment, which results in an inflammatory focus, these cytokines also have effects on resident cells. For instance, IFN-γ and TNF induce MHC gene expression [128–130], which is critical for antigen presentation and T cell recognition. Chronic production of IFN-γ or TNF might lead to inappropriate expression of MHC gene products and consequent autoimmune reactivity. For example, ectopic expression of either Class II MHC or IFN-γ results in the spontaneous development of insulin-dependent autoimmune diabetes in transgenic mice [131]. Major histocompatibility complex antigens are poorly expressed on myocardium but can be induced by IFN-γ [130], are associated with autoimmune cardiomyopathology [132], and may contribute to immunopathology.

Trypanosoma cruzi infection induces heightened production of both IFN-γ and TNF [133] (G.S. Nabors & R.L. Tarleton, submitted). The production of both of these cytokines, although induced by the infection, is regulated *before* the peak parasitemia is reached. Interestingly, in the heart transplant model, rejection transferred to normal mice by CD4$^+$ T cell from chronically infected mice could be delayed by anti-IFN-γ treatment of the recipient mice [126]. Thus the level and nature of cytokine production may contribute to immunopathogenesis and regulate disease severity.

Polyclonal activation

Polyclonal activation of B and T cells has been well documented in murine Chagas' disease [134–136]. It has been proposed that the majority of responding clones are not directed against parasite antigens [137]. However, no parasite mitogen or superantigen has been identified and significant parasite-specific antibody is produced in mice during the burst of polyclonal activation. The multiclonal nature of the response to *T. cruzi* infection may represent an enhanced version of the polyclonal response that accompanies all antigen-specific responses [138]. The polyclonal nature of the response in *T. cruzi* is accentuated by the diversity of the antigen repertoire of the parasite (thereby stimulating multiple clones of lymphocytes). Heavy and persistent antigenic stimulation may not allow the selective maturation of high-affinity receptor-bearing cells. Chronic polyclonal activation initiated in response to antigen can lead to immunopathology through an autoimmune-type reaction [139]. Polyclonal activation in an inflammatory reaction, for example, might result

in the release of self-epitopes capable of driving the autoreactive response. In this way the specific response elicited by *T. cruzi* when accompanied by polyclonal activation would initiate self-perpetuating pathology. Subsequent immune regulation would determine the severity of resultant disease.

Ly-1-positive B cells and double-negative T cells

The preferential expansion of subpopulations of Ly-1$^+$ B cells and double-negative (CD4$^-$, CD8$^-$) T cells during the acute phase of murine *T. cruzi* infection has been reported [135]. Both cell types are relatively rare in adult animals. Although of unknown function, both Ly-1$^+$ B cells and double-negative T cells have been associated with autoimmune disorders [140,141].

Heat shock proteins (Hsp)

These highly conserved molecules carry out a variety of functions, from protection of the cell from stress to assisting in basic cell maintenance. Heat shock protein expression is induced by many stimuli [144], and Hsp are immune targets in a number of parasitic infections, including *T. cruzi* [143–147]. Because of their high immunogenicity, amino acid sequence conservation, and their differential expression under stress conditions, Hsp are prime candidates for targets of autoimmune responses [144]. Anti-*T. cruzi* Hsp antibodies are produced in both infected humans and experimental animals [146,147]. Despite the homology between *T. cruzi* and mammalian Hsp, the anti-*T. cruzi* Hsp antibodies produced in human patients do not react with host Hsp [147]. The cellular immune response to self- or *T. cruzi*-Hsp has not been studied.

Immunoregulation

If pathogenesis in Chagas' disease is largely immune-mediated, then immunoregulation would be expected to modify disease severity. Low-dose cyclophosphamide treatment to inhibit T suppressor cell function results in an increased inflammatory response in acutely infected mice [148] and an exacerbation of chronic myocarditis in infected dogs [149], suggesting that suppressor mechanisms normally regulate disease severity. These effects were seen without significant change in the parasite load of treated animals. Other attempts to modulate suppressor cell activity, however, have suffered from this latter problem [150]. Depletion of CD8$^+$ T cells (classic cytotoxic/suppressor) in acutely infected mice led to the surprising result of greatly enhanced parasite loads and decreased host longevity, suggesting that CD8$^+$ T cells also function to control *T. cruzi* infection [150]. CD8 depletion during the chronic stage of infection does not exacerbate the infection or affect infection-induced resistance to rechallenge. The effects of this treatment on disease severity have not been studied.

If immunoregulation limits immunopathology, then overcoming these regulatory effects by boosting the immune response would be expected to increase pathology. Chemical or biologic immunoenhancers failed to increase the severity of murine disease [151]. Enhancement of immunity by immunization with killed or attenuated parasites prior to infection did not increase pathology [152,153]. However, the potential multiple effects of immunoenhancing treatments also limit interpretation of experiments which employ them. Furthermore, these treatments decreased acute parasitemias which might influence *chronic* pathology.

Immunoregulatory mechanisms are activated as a result of *T. cruzi* infection. Immunosuppression in acute murine infection affects a wide range of immune responses to parasite and control antigens [154–157]. Immunosuppression also occurs in human *T. cruzi* infections [158–160]. However, the role of acute immunosuppression in limiting pathology in Chagas' disease is not clear and there are no reports of an association between the intensity of immunosuppression and the level of pathology among different mouse strains or in humans [106,119,161,162]. Proving a linkage between immunoregulation and pathology may require that patients be identified early in the acute-phase and that both immunologic and pathologic changes be evaluated over a period of many years.

The best example of correlations between the presence of immunoregulatory responses and pathology come from studies of idiotypic networks in chagasic patients. Both symptomatic and asymptomatic Chagas' patients produce parasite-specific antibodies and T cells which respond to the idiotype on these antibodies [163]. Antiepimastigote antibodies from cardiac patients are more highly stimulatory for T cells from either cardiac or indeterminate patients than are the antibodies from indeterminate patients [163]. The failure to express the T cell stimulatory antiepimastigote idiotype, perhaps due to active suppression, may correlate with the absence of disease. Interestingly, T cells in the cord blood of neonates born of mothers with Chagas' disease also respond to idiotypes on antiepimastigote antibodies. *In utero* sensitization has been proposed to account for lower pathology in endemic populations by priming neonates to develop protective immunity to *T. cruzi* or by inducing a regulatory suppressor cell population which prevents the elicitation of the "pathogenic" idiotype [164].

Both T cells and macrophages regulate immune responses in the acute phase of murine *T. cruzi* infection [155,159,165–172]. IL-2 production in infected mice correlates with the overall immunologic state of the mice, and some suppressed responses can be enhanced by the addition of exogenous IL-2 [172–174]. IL-2 is central to the generation of immune responses and influences the production and activity of other cytokines. In *T. cruzi* infection, the decreased IL-2 production correlates with the regulation of both TNF and IFN-γ production. The suppression of IL-2 production in *T. cruzi* infection is made more interesting in the context of immunopathology because suppressed IL-2 production has also been noted in a number of autoimmune disorders [175,176]. The similarities between the characteristics of IL-2 suppression in genetically autoimmune mice and murine *T. cruzi* infection is striking. In both cases the suppression is not due to a lack of T cells capable of making IL-2 or to the absence of cofactors such as IL-1. Suppressor cells have been identified in both cases and normal IL-2 production can be established by stimulating T cells with the combination of phorbal ester and ionophore or by "resting" the cells *in vitro* for 48 to 72 h.

The relationship between suppressed IL-2 production and autoimmunity is unclear. Suppressed IL-2 production may be the *result* of the autoimmune response and an attempt to regulate a response to a persistent antigenic stimulus (autoantigen or the infectious agent) [175]. Alternatively, suppressed IL-2 production may be an *in vitro* artefact of high *in vivo* production of IL-2 which is causing (or exacerbating) immune pathology [176]. Finally, low *in vivo* production of IL-2 may induce autoimmune syndromes by allowing selective expansion of autoreactive clones [177].

In summary, immunoregulatory circuits are activated in *T. cruzi* infection. Indeed, chronic infectious diseases may be the evolutionary force behind the development of immunoregulatory mechanisms [178]. If immunopathology drives disease development, then immunoregulation might ameliorate it.

General conclusions and conceptual model

1 The immune system has both a protective and pathogenic function in *T. cruzi* infection.

Without a fully competent immune system, infected hosts are unable to control parasite growth. Many components of the immune system play a role in this resistance, as shown in experiments using hosts depleted or deficient in B cells [179–181], CD4+ [123,181,182], CD8+ [82] or all T cells [180,184], natural killer cells [182], or macrophages [185]. Furthermore, some human patients and experimental animals who have controlled the acute infection and are thus chronically infected suffer reactivation of infection when immunosuppressed [123,186–188].

The antiparasite immune responses also determines the severity of the disease in humans. Investigations both support [119,161,162] and contradict [113,189] an association between proliferative responses to parasite antigens and clinical status of patients. A detailed longitudinal study must be completed before firm conclusions can be reached. Whether the parasite-specific immune response directly causes pathology is not clear; however, there is little question that the response set in motion by *T. cruzi* infection contributes.

2 Immunoregulatory mechanisms are activated during *T. cruzi* infection.

Immunoregulatory pathways are activated in *T. cruzi* infection. These result in the profound immunosuppression characteristic of the acute phase of experimental infections and the less dramatic, but perhaps equally important, idiotypic–anti-idiotypic regulation in chronic infection.

3 Immunoregulatory mechanisms control immunity and pathogenesis in *T. cruzi* infection.

Immunosuppression in the acute phase of infection affects the antiparasite response but does not totally handicap the host and prevent immune control of the infection. There is little evidence linking the intensity of immunosuppression and reduced ability to regulate parasite load. Since the immune response contributes to pathogenesis, the ability to regulate those responses is an advantage for the host.

4 Pathogenesis is multifactorial.

No single mechanism can account for the variability in the expression and course of Chagas' disease, suggesting that pathogenesis is both complex and multifactorial [71]. Factors which contribute to the course of the infection and disease include parasite genetics, host genetics, sex, age, and prior history of exposure (perhaps including *in utero* experiences). Pathogenesis may vary in etiology and timing [108] or tissue tropisms [188] in individual patients.

Trypanosoma cruzi may only serve as a catalyst for initiation of disease. The persistence of the parasites in the face of an apparently vigorous immune response sets in motion a series of events which can lead to pathogenesis in motion in genetically susceptible hosts. These events include polyclonal activation, production of immunomodulatory cytokines, and activation of immunoregulatory elements. Confounding factors may include autoreactive responses initiated by polyclonal activation or by ''shared'' antigens (including Hsp), enhanced MHC gene expression, and expansion of particular cell populations (Ly-1+ B cells and double-negative T cells) associated with autoimmunity. However, it is the antigenic complexity of *T. cruzi* and the parasite's ability to evade the immune response which ultimately lead to clinical disease.

This same scenario with similar constituent factors has been evoked as the etiology of diseases as varied as graft-vs.-host disease [175], systemic lupus erythmatosus, a variety of cardiac diseases [190,191], diabetes [192], and acquired immunodeficiency syndrome (AIDS) [193]. The variations in pathology in these disease states may be determined by multiple factors, including the site of tissue inflammation and shared antigens, existing between host and inciting agent. Thus, although Chagas' disease is certainly complicated, it is really not very different from any number of other chronic diseases. These diseases are characterized by the inability of the immune response to eradicate totally the infectious agent. The continued immune response leads to pathology unless the immune response is tightly regulated. Those individuals with the ability to regulate both the parasite load and the immune response, and in which the immune response to *T. cruzi* and the immunoregulatory machinery are appropriately balanced, survive infection with minimal long-term damage.

Unfortunately, our understanding of the basic cell biology, molecular biology, and immunology of *T. cruzi* infection has resolved virtually none of the major questions in the field, including the cause of chronic pathology. Perhaps we are cursed by the backward course charted by Carlos Chagas, first finding the vector, then the parasite, and finally the disease. Certainly we are limited by the complexity of the host–parasite interaction, the methodologies available to monitor parasite levels and the progression of disease, the model systems in which we work, and the ethical limitations of the study of the infection and disease in humans. Methods to assess parasite levels and quantify pathology are terribly insensitive. We are largely in the dark about which immune responses and target antigens are relevant to immune protection and immunopathogenesis. Although longitudinal studies of human patients have yielded new and important information, many questions remain concerning the relationship between *T. cruzi*, the immune system, and Chagas' disease.

ACKNOWLEDGMENTS

While not associating them with any of the ideas expressed in this book, RLT wishes to thank Drs Ray Kuhn and Ed Rowland for their critiques of parts of this chapter. The work in Dr Tarleton's laboratory was supported in part by NIH grant AI-22070, by the UNDP/World Bank/WHO Special Programme for Research and Training in Tropical Diseases and American Heart Association.

Pathology of leishmaniasis

Richard D. Pearson

Leishmaniasis has historically been divided into three major clinical syndromes: visceral, cutaneous, and mucosal (also known as mucocutaneous). *Leishmania donovani* and *Leishmania chagasi* are the species most commonly associated with visceral leishmaniasis. The other *Leishmania* species typically cause cutaneous diseases. *Leishmania braziliensis* is also responsible for mucosal leishmaniasis in Latin America [194,195].

The pathology, clinical manifestations, and course of leishmanial infections are dependent on a complex interaction between genetically determined virulence characteristics of the *Leishmania* species and the cell-mediated immune responses of the host. Interest in leishmaniasis has increased in recent years both because of the morbidity and mortality associated with leishmanial infections and the emergence of leishmaniasis as a model system for the study of CMI [196,197].

Leishmania

The *Leishmania* exist in two basic developmental forms: promastigotes and amastigotes [194]. Their detailed life-cycle is summarized in Chapter 20. Promastigotes multiply and differentiate in the gut of the sandfly and are inoculated into the skin of a susceptible mammal when the sandfly attempts to take a blood meal.

Promastigotes do not survive at 37°C [198]. They can bind directly to several receptors on human macrophages, including the Type 3 complement receptor (CR3) [199,200], mannose/fucose receptors [201–203], fibronectin receptors [204], and receptors for advanced glycosylation endproducts [205]. At least two types of parasite surface molecules, a surface acid protease (gp63) and lipophosphoglycans, are involved. After attachment, promastigotes are ingested by the phagocytic apparatus of the macrophage. Intracellular survival of the parasites depends in part on the oxidative potency of the ingesting mononuclear phagocyte and the oxidative response elicited by the parasite. Promastigotes can survive ingestion by mature macrophages which have not been "activated" by lymphokines, and they convert to the amastigote form within them [206].

Following parasite ingestion, the parasite-containing phagosome fuses with macrophage lysosomes. Amastigotes (Fig. 4.14) are ideally suited for intracellular survival at 37°C in the acid environment of the phagolysosome. Amastigotes multiply to high numbers within macrophages *in vitro* at 37°C without destroying them. *In vivo*, amastigotes multiply, are apparently released from infected macrophages, and gain access to mononuclear phagocytes that are recruited to the site of infection.

Macrophages activated by IFN-γ alone or in concert

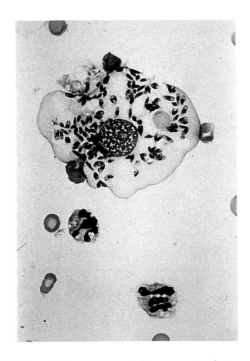

Fig. 4.14 Amastigotes are seen within a mononuclear phagocyte in this Wright–Giemsa-stained preparation of pleural fluid from a patient with visceral leishmaniasis and AIDS. (Photomicrograph kindly provided by Dr Robert Betts and Dr David M. Markowitz.)

immune T helper cells can also result in killing of intracellular amastigotes, but the mechanism [197] has not yet been determined [209].

Disease spectrum cutaneous leishmaniasis

The classic form of cutaneous leishmaniasis is the "oriental sore," which is caused by *Leishmania major* in rural areas of the Middle East, central Asia, and North Africa, and by *Leishmania tropica* in urban areas of the Middle East, the Indian subcontinent and the Mediterranean basin (Table 4.4) [194,195,210]. Cutaneous leishmaniasis in the Old World is usually sporadic, but epidemics have occurred when large groups of persons are exposed.

In the urban or dry form caused by *L. tropica*, lesions tend to be single, grow slowly, and last for a year or more. In the moist or wet form caused by *L. major*, lesions are often multiplied, expand rapidly, and heal after several months.

American cutaneous leishmaniasis is widespread in Central and South America. The manifestations of American cutaneous leishmaniasis range from single, small, dry, crusted lesions, which are more likely with *Leishmania mexicana* subspecies, to large, deep, mutilating ulcers, which are more common with *L. braziliensis* subspecies. Lesions may be single or multiple and are usually found on exposed areas of the skin (Fig. 4.15).

The lesions of cutaneous leishmaniasis develop at the site where promastigotes are inoculated by *Leishmania*-infected sandflies. Typically, lesions start as papules, enlarge, and then ulcerate. Ulcers heal after variable

with other cytokines, e.g., GM-CSF, can inhibit and/or kill *Leishmania* by the hydrogen peroxide–myeloperoxidase–halide or nitric oxide microbicidal mechanisms [207,208]. Direct contact of amastigote-infected macrophages with histocompatible, *Leishmania*-specific,

Table 4.4 Clinical manifestations and geographic distribution of leishmaniasis. (From Lainson & Shaw [212])

Clinical syndromes	Parasite	Geographic areas
New World cutaneous leishmaniasis		
Single or limited number of skin lesions	L. mexicana mexicana (chiclero ulcer)	Mexico, Central America (Texas?)
	L. m. amazonensis	Amazon basin and neighboring areas, Panama and Trinidad
	L. m. pifanoi	Venezuela
	L. m. garnhami	Venezuela
	L. m. venezuelensis	Venezuela
	L. mexicana sp.	Dominican Republic

continued

Table 4.4 *Continued*

Clinical syndromes	Parasite	Geographic areas
	L. braziliensis braziliensis	Brazil, Peru, Ecuador, Bolivia, Paraguay, Argentina
	L. b. guyanensis (pian bois, bush yaws)	Guyana, Surinam, northern Amazon basin
	L. b. peruviana (uta)	Peru, Western Andes, Argentinian highlands
	L. b. panamensis	Panama and adjacent areas
	L. donovani chagasi	Latin America
Diffuse cutaneous leishmaniasis	*L. m. amazonensis*	Amazon basin and neighboring areas
	L. m. pifanoi	Venezuela
	L. m. mexicana	Mexico, Central America
	L. mexicana sp.	Dominican Republic
Mucosal leishmaniasis	*L. b. braziliensis* (espundia)	Multiple areas in South America
	L. b. panamensis (rare)	Panama and adjacent areas
Old World cutaneous leishmaniasis		
Single or limited number of skin lesions	*L. major*	Middle East, central Asia, Africa, Indian subcontinent
	L. tropica	Mediterranean littoral, Middle East, west Asia, Indian subcontinent
	L. aethiopica	Ethiopian highlands, Kenya
Diffuse cutaneous leishmaniasis	*L. aethiopica*	Ethiopian highlands, Kenya
Visceral leishmaniasis (kala-azar)		
General involvement of the reticuloendothelial system (spleen, bone marrow, liver, etc.)	*L. donovani donovani*	Indian subcontinent, China
	L. d. infantum	Middle East, Mediterranean littoral, Balkans, western Asiatic area, north-western Iberia, China, subSaharan Africa
	L. d. chagasi	Latin America
Viscerotrophic leishmaniasis	*L. tropica*	Saudi Arabia
Post-kala-azar dermal leishmaniasis	*L. d. donovani*	Indian subcontinent
	L. donovani sp.	Kenya, possibly Ethiopia and Somalia

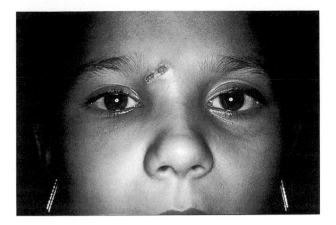

Fig. 4.15 A patient with uncomplicated cutaneous leishmaniasis due to *Leishmania b. braziliensis* infection in Fortaleza, Brazil. (Photograph kindly provided by Dr Anastacio de Q. Sousa.)

periods of time, leaving flat atrophic scars as evidence of disease.

Two relatively uncommon variants of cutaneous leishmaniasis demonstrate the extreme of this spectral disease. Diffuse cutaneous leishmaniasis starts as a papule that does not ulcerate. Chronic nonulcerative satellite lesions develop as amastigotes and metastasize to other areas of the skin. The parasite density is high, and those afflicted lack evidence of *Leishmania*-specific cell-mediated immune responses. Leishmaniasis recidivans is a relapsing, tuberculoid form of cutaneous leishmaniasis. The lesions slowly spread outward while healing in the center, persisting for 20 or more years. Affected patients have evidence of DTH to leishmanial antigens, and the parasite density in the lesions is low.

Following spontaneous resolution of cutaneous lesions, a small subset (2–3%) of people with *L. braziliensis braziliensis* infection develop chronic, mutilating mucosal lesions of the nose, face, or oral pharynx (Fig. 4.16) [211]. Amastigotes are scant in these mucosal lesions, and patients evidence vigorous DTH responses to leishmanial antigens.

Immunopathogenesis

Clinically, the spectrum of cutaneous leishmaniasis resembles that of leprosy. At one end lies diffuse cutaneous leishmaniasis, in which there is no evidence of cell-mediated immune responses and macrophages are heavily infected with amastigotes. This is similar to

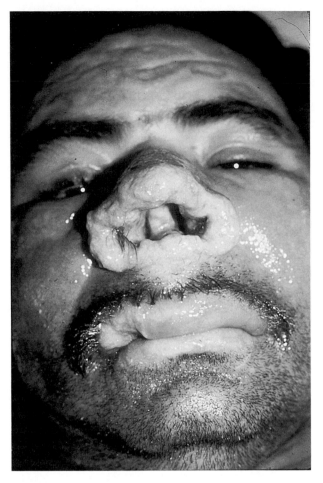

Fig. 4.16 A patient with mucosal leishmaniasis due to *Leishmania b. braziliensis* in Fortaleza, Brazil. (Photograph kindly provided by Dr Anastacio de Q. Sousa.)

lepromatous leprosy in which macrophages filled with *Mycobacterium leprae* predominate in the skin in the absence of cell-mediated immune responses. At the other end of the spectrum is mucosal leishmaniasis. Cell-mediated immune responses are vigorous and parasites are scant. Mucosal leishmaniasis is somewhat analogous to tuberculoid leprosy, in which there is an intense mononuclear cell infiltrate with few bacilli, and tissue damage comes as an apparent consequence of cell-mediated immune responses.

However, the analogy between cutaneous leishmaniasis and leprosy breaks down at the tissue level. Whereas the organization and cellular composition of the granulomatous response in leprosy are invariably characteristic of the position in the clinical spectrum, this is not true for human cutaneous leishmaniasis [212,213].

When amastigotes are numerous in human lesions the dominant cell type is nearly always the macrophage, but when amastigotes are scanty the characteristics of the granuloma, i.e., its composition and organization, are not predictable.

Although species-specific differences in virulence exist among the leishmanias, the genetically controlled cell-mediated immune responses of the host are also important determinants of the clinical manifestations of disease. In areas where *Leishmania aethiopica* and *L. mexicana* are endemic, the majority of persons develop cutaneous ulcers; but a subset goes on to have the anergic, nonulcerative variant, diffuse cutaneous leishmaniasis. In addition, racial factors appear to affect the severity of mucosal leishmaniasis. Extreme facial mutilation was observed almost exclusively among people of African ancestry in a study conducted in the eastern Andes of Bolivia, even though more cases of cutaneous leishmaniasis were observed in the region among indigenous Indians [214]. Infected people with African ancestry also had more vigorous DTH responses to leishmanial antigens than the infected Indians.

Murine cutaneous leishmaniasis has emerged as a model system for the study of the development and control of cell-mediated immune responses [196,197]. Susceptibility to *Leishmania* is genetically determined among inbred strains of mice. In resistant strains, amastigote multiplication is controlled at the macrophage level. Among susceptible animals, the eventual outcome of infection and the acquisition of resistance to reinfection are controlled by cell-mediated immune mechanisms and not humoral factors.

In vitro and *in vivo* data indicate that control of *Leishmania* by mononuclear phagocytes is dependent on activation of macrophage microbicidal mechanisms by IFN-γ [207,208] or by direct contact with *Leishmania*-specific T helper cells [209]. The outcome of murine infection depends on a complex interplay between protective *Leishmania*-specific T helper lymphocytes of the TH-1 type, which produce IFN-γ and IL-2, and nonprotective phenotypic T helper cells of the TH-2 type, which secrete IL-4 and other cytokines but not IFN-γ or IL-2 [215,216]. There is also evidence that *Leishmania*-infected macrophages may inhibit development of protective T cell-mediated immune responses by secreting suppressive prostaglandins [217,218]. The evolution of immune responses in murine leishmaniasis has been reviewed elsewhere [196,197]. The relevance of these studies to human disease is still conjectural.

Although circumstantial, the data suggest that cutaneous ulceration is a consequence of the host immune response. First, the sequence of events in cutaneous leishmaniasis has been studied following accidental inoculation of *L. tropica* into a laboratory worker. Cutaneous ulceration was coincidental with the development of blastogenic responses and the secretion of IL-2 by peripheral blood lymphocytes exposed to leishmanial antigen *in vitro* [219]. In other studies, people presenting with ulcerated cutaneous leishmanial lesions typically have peripheral blood mononuclear cells which proliferate and produce IL-2 and IFN-γ in response to leishmanial antigens.

Secondly, as noted earlier, tissue destruction is often severe in persons with mucosal leishmaniasis due to *L. b. braziliensis*. The parasite burden is low, and patients with mucosal leishmaniasis have vigorous T cell responses to leishmanial antigens, as measured by delayed-type hypersensitivity responses to intradermally administered leishmanial antigens (Montenegro test), and lymphocyte proliferation and secretion of IFN-γ and IL-2 in response to parasite antigens *in vitro*.

Conversely, cutaneous ulceration is not observed in diffuse cutaneous leishmaniasis, a syndrome in which the parasite density is high and in which there is no evidence of *Leishmania*-specific cellular immunity, as evidenced by the lack of DTH responses to leishmanial antigens *in vivo* and the lack of lymphocyte blastogenic responses or cytokine secretion to parasite antigens *in vitro*.

Finally, direct evidence that immune responses cause ulceration in cutaneous leishmaniasis comes from studies in which the skin of *Leishmania*-susceptible mice was exposed to ultraviolet B (UV-B) irradiation before inoculation of *L. major* [220]. In normal skin, parasites multiplied in macrophages and a granulomatous response developed, followed by cutaneous ulceration. In irradiated animals, parasites multiplied but there was no ulceration. Ultraviolet B irradiation of skin abrogated the induction of DTH to dinitrofluorobenzene or to leishmanial antigens. These data suggest that the cell-mediated immune responses of the host produce tissue destruction and ulceration, which are the hallmarks of cutaneous leishmaniasis.

Regulation of immunopathology

Early histologic studies of experimental cutaneous leishmaniasis in animals suggested that T lymphocytes play a central role in the healing of cutaneous lesions. After inoculation of *Leishmania enriettii* in guinea pigs, there was initially a heavy local infiltration of macrophages containing amastigotes [221]. A surrounding ring of lymphocytes and fibroblasts developed later, accompanied by development of DTH responses to leishmanial antigens. The number of infected macrophages subsequently decreased and healing ensued.

Immunocytochemical and electron microscopic studies using inbred strains of mice have also supported the role of T cells in the final resolution of cutaneous leishmaniasis [222]. Infected self-healing C57BL/6 mice initially displayed progressively enlarging cutaneous lesions with epidermal thickening, ulceration, and accumulation of eosinophils and Ia$^+$ infected macrophages. Subsequent healing was associated with a local influx of T helper (L3T4$^+$) cells and cytotoxic/suppressor (Lyt-2$^+$) cells into the dermis, and Ia antigen expression by epidermal keratinocytes. Intracellular parasites became scarce as T lymphocyte infiltration increased. In contrast, lesions in susceptible, nonhealing, BALB/c mice continued to enlarge and never healed. There was minimal T lymphocyte influx and no keratinocyte Ia expression. It is interesting to note that numbers of circulating helper and cytotoxic/suppressor lymphocytes were normal in the peripheral circulation of both mouse strains during infection.

Microscopically there is a mixed granulomatous inflammatory infiltrate with varying numbers of amastigote-containing macrophages, lymphocytes, and plasma cells in humans. Amastigotes are eliminated from human cutaneous lesions as an apparent result of destruction within infected macrophages. At times, infected macrophages appear to have been killed individually, in small clusters, or by focal necrosis of a mass of parasitized mononuclear cells [212,213]. The mechanism of necrosis of *Leishmania*-infected macrophages in humans remains uncertain. Cytotoxic/suppressor (Lyt-2$^+$) cells are involved in the protective immune responses that result in eradication of amastigotes in mice, but there is no direct evidence that they kill amastigote-infected macrophages *in vivo* and, furthermore, lymphocytes are not always seen in close proximity to necrotic-appearing macrophages. Tissue necrosis may be due to locally produced cytokines, such as possibly TNF/cachectin or lymphotoxin, to the effects of immune complexes [213], or to one or more populations of cytotoxic immune effector cells.

Despite their chronicity, cutaneous lesions eventually heal in immunocompetent humans. The major question is why the resolution of infection is not concomitant with the development of parasite-specific circulating T cell responses. The answer probably lies in complex interactions in the skin between potentially protective T helper cell subsets and nonprotective phenotypic T helper cells, other T cells, macrophages, and/or humoral factors such as immune complexes. Spontaneous resolution of cutaneous leishmaniasis is associated with lifelong immunity against the infecting *Leishmania* species.

For reasons that are not understood, the vigorous immune responses observed in people with mucosal leishmaniasis or leishmaniasis recidivans do not eradicate parasites from the lesions. The persistence of a small number of amastigotes seems to be the stimulus for the chronic, destructive granulomatous responses that characterize these syndromes.

Visceral leishmaniasis

Leishmania donovani in the Old World and *L. chagasi* in Latin America are the *Leishmania* species most commonly isolated from people with visceral leishmaniasis. They produce a spectrum of disease [223–226] ranging from asymptomatic self-resolving infection to progressive visceral leishmaniasis, which has been termed kala-azar, Dumdum fever, Assam fever, or infantile splenomegaly. The ratio of inapparent to apparent infections with *L. donovani* and *L. chagasi* depends on multiple factors, including the age and nutritional status of the host, but only a minority of infected persons develop classic kala-azar.

Amastigotes initially multiply locally in macrophages at the site of parasite inoculation and then spread through the lymphatics and bloodstream to infect the entire reticuloendothelial system.

Patients with kala-azar characteristically present with anorexia, malaise, fever, weight loss, and hepatosplenomegaly (Fig. 4.17) [227]. Anemia, neutropenia, thrombocytopenia, hypergammaglobulinemia, circulating immune complexes, and rheumatoid factors are prevalent among patients with progressive visceral leishmaniasis [228,229].

Immunopathogenesis

Patients with progressive visceral leishmaniasis, or kala-azar, fail to develop *Leishmania*-specific T cell responses. They lack cutaneous hypersensitivity to intradermally administered *Leishmania* antigens and their peripheral blood mononuclear cells neither proliferate nor secrete IFN-γ or IL-2 in response to leishmanial antigens [230].

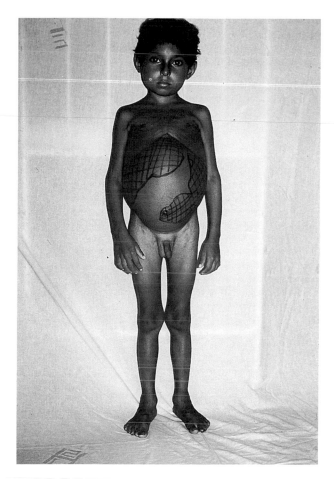

Fig. 4.17 A cachectic Brazilian child with visceral leishmaniasis due to *Leishmania chagasi*. The enlarged liver and spleen are outlined on the abdomen. (Photograph kindly provided by Dr Anastacio de Q. Sousa.)

Amastigote-containing macrophages crowd the bone marrow, liver, and spleen and are found in the gastro-intestinal tract, kidneys, and other organs. There is immune complex deposition, mesangial cell proliferation, and occasionally amyloid deposition in the kidney. Paradoxically, despite parasite-specific cell-mediated anergy, people with kala-azar produce high titers of antileishmanial antibodies [231].

Regulation of immunopathology

The resolution of *L. donovani* infection and protection against reinfection appear to be mediated by *Leishmania*-specific T cells. People with asymptomatic self-resolving infections most likely have early proliferation of pro-

tective T helper cells that are capable of activating macrophages to kill amastigotes. These patients evidence cutaneous DTH responses to leishmanial antigens and produce relatively low titers of antileishmanial antibodies.

The "oligosymptomatic" form of disease probably represents a tenuous balance between protective T helper cell populations and nonprotective T cell populations and/or the suppressive effects of infected macrophages or humoral factors [195,196]. If protective T helper cells come to dominate, then infection resolves. When nonprotective T cells or other suppressive factors dominate, the disease progresses.

Clinically apparent visceral leishmaniasis may first develop years after infection if people become immunosuppressed due to human immunodeficiency virus (HIV) infection, steroids, or neoplasm. Furthermore, visceral leishmaniasis has emerged as an opportunistic pathogen in people with AIDS [232,233].

Data from experimentally infected BALB/cJ mice indicate that the balance of *Leishmania*-specific TH-1 and TH-2 phenotypic T helper cell populations determine the course of murine *L. donovani* infection [234]. Cytotoxic/suppressor T cells (Lyt-2$^+$) also participate in the development of granulomas and eradication of amastigotes [235], but the precise interactions of the various mononuclear cell populations that result in protective immunity against *L. donovani* have not yet been defined.

Although the sequence of events following inoculation of promastigotes into humans who develop visceral leishmaniasis has not been characterized, data from Syrian hamsters, which are exquisitely sensitive to *L. donovani*, provide some insight into what may happen (Fig. 4.18) [236]. When hamsters were infected with 100 000 stationary-phase promastigotes, administered intradermally, a local mixed inflammatory response was observed within 1 h. Both polymorphonuclear and mononuclear phagocytes were present, and comparable numbers of parasites were observed within them. *Leishmania* appeared to be killed within polymorphonuclear leukocytes (Fig. 4.18a), but promastigotes converted to amastigotes (Fig. 4.18b) and subsequently multiplied within mononuclear phagocytes over the subsequent 2 weeks (Fig. 4.19).

Granulomas containing epithelioid cells, Langerhans' giant cells, and eosinophils formed between weeks 4 and 6, and amastigotes were progressively eliminated from the cutaneous lesions (Fig. 4.20). Although some amastigote-infected mononuclear phagocytes appeared

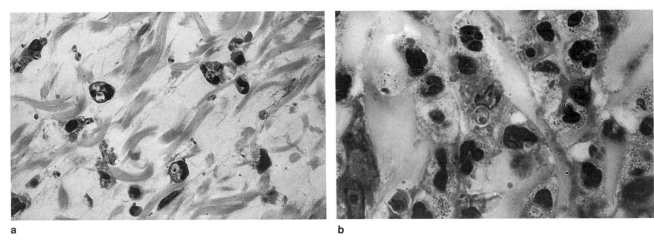

a   b

Fig. 4.18 Phagocytic responses following the intradermal inoculation of *L. donovani* promastigotes into the skin of hamsters. (a) Degenerating *L. donovani* within polymorphonuclear leukocytes 1 h after parasite inoculation. (b) *L. donovani* within a mononuclear phagocyte 6 h after inoculation. Notice the parasite's eccentrically located nucleus and dark staining kinetoplast. (Courtesy of Dr Mary E. Wilson.)

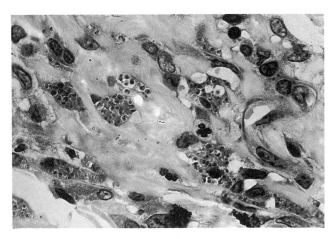

Fig. 4.19 Mononuclear phagocytes containing large numbers of intracellular amastigotes 1 week after inoculation of promastigotes. (Courtesy of Dr Mary E. Wilson.)

to have undergone necrosis in the granuloma, the overlaying skin did not ulcerate. Even though there was microscopic resolution of the local lesions, visceral dissemination of amastigotes occurred in some animals as evidenced by development of classic visceral leishmaniasis several months later.

The hamster model has also been used to study the mechanisms responsible for the profound wasting observed with experimental visceral leishmaniasis and with naturally acquired progressive human infections. Some people with visceral leishmaniasis are reminiscent of those with the "consumption" of miliary tuberculosis or the cachexia of neoplasms or AIDS. The presence of fever provides circumstantial evidence that IL-1 or TNF/cachectin, which is also a pyrogen and can trigger release of IL-1, or other yet-to-be identified endogenous pyrogens are produced by host cells during visceral leishmaniasis.

In *L. donovani*-infected hamsters, decreased food intake and weight loss correlated with the production of high levels of IL-1 and TNF/cachectin by splenic mononuclear phagocytes [237]. In contrast, inbred strains of mice infected with *L. donovani* do not lose weight or die early despite heavy parasite burdens, suggesting that the parasites themselves are not directly responsible for wasting. Studies in which the effects of IL-1 and TNF/cachectin are neutralized by antibodies or pharmacologic agents have not yet been performed, but TNF/cachectin has been shown to cause anorexia and weight loss.

In summary, the pathology and clinical manifestation of visceral leishmaniasis are a result of the interaction between the leishmanias and the complex CMI responses which they elicit. When protective T helper cells develop, infection is self-resolving. Evidence from murine models suggests that potentially protective T helper cell responses do not develop or are delayed in animals with progressive disease because they are blocked by *Leishmania*-specific, nonprotective T helper cell populations, infected mononuclear phagocytes,

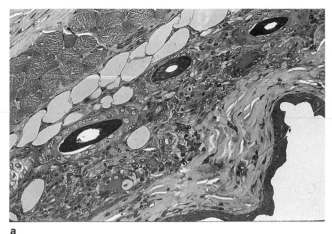
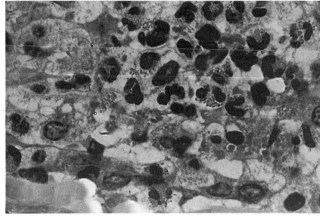

a   b

Fig. 4.20 Granulomata were observed 4 and 6 weeks after parasite inoculation. (a) Well-developed granuloma with Langerhans' giant cells seen at 4 weeks. No parasites are present. (b) A close up of a different granuloma revealed eosinophils (see left lower quadrant) and necrotic mononuclear cells, some apparently containing the remains of intracellular parasites (superior and right lower quadrants). (Courtesy of Dr Mary E. Wilson.)

and/or humoral factors. Finally, ulceration in cutaneous leishmaniasis and wasting in visceral leishmaniasis are probably due at least in part to the deleterious consequences of host immune responses.

SUMMARY

The work of many investigators has revealed that the pathology which is associated with parasitic diseases results from an intricate web of tangled immune responses, both cellular and humoral. In addition, the immune system is entangled with a variety of mechanisms by which the parasite evades the immune system and attacks the host. In order to unravel this Gordian knot of pathology, a variety of *in vivo* and *in vitro* "submodels" have been developed. The emerging powers of immunologic and molecular biologic dissection will help us slice through the enigmas of pathology.

ACKNOWLEDGMENT

The authors wish to thank Ms Anne Barr and Mrs Peggy Otto for their excellent secretarial support. This work was supported in part by NIH grants AI15193, HL31693, and HL35276.

REFERENCES

1 Phillips SM, Fox EG. The immunopathology of parasitic

disease. Clin Immunol Allergy 1982;2:667.

2 Phillips SM, Fox EG. The immunopathology of parasitic diseases: a conceptual approach. In Marchalonis JJ, ed. *Contemporary Topics in Immunobiology*, vol. 12. New York: Plenum Publishers, 1984:421.

3 Boros DL. Immunopathology of *Schistosoma mansoni* infection. Clin Microbiol Rev 1989;2:250.

4 Nash TE, Cheever AW, Ottesen EA, Cook JA. Schistosome infections in humans: perspectives and recent findings. Ann Intern Med 1982;97:740.

5 Colley DG. Immune responses and immunoregulation in experimental and clinical schistosomiasis. In Mansfield JM, ed. *The Immunology of Parasitic Diseases*. New York: Marcel Dekker, 1981:1–83.

6 Phillips SM, Colley DG. Immunologic aspects of host response to schistosomiasis: resistance, immunopathology and eosinophil involvement. Prog Allergy 1978;24: 49.

7 Boros DL. Immunoregulatory mechanisms active in the suppression of the schistosome egg granuloma. In Yoshida T, Torisu M, eds. *Basic Mechanisms of Granulomatous Inflammation*. Amsterdam: Elsevier, 1989:143.

8 Rocklin RE, Brown AP, Warren KS, *et al.* Factors that modify the cellular immune response in patients infected by *Schistosoma mansoni*. J Immunol 1980;125:1916.

9 Doughty BL, Ottesen EA, Nash TE, Phillips SM. Delayed hypersensitivity granuloma formation around *Schistosoma mansoni* eggs *in vitro*. III. Granuloma formation and modulation in human schistosomiasis mansoni. J Immunol 1984;133:993.

10 Phillips SM, DiConza JJ, Gold JA, Reid WA. Schistosomiasis in the congenitally athymic (nude) mouse. I. Thymic dependency of eosinophilia, granuloma formation, and host morbidity. J Immunol 1977;118:594.

11 Boros DL. Immunoregulation of granuloma formation

in murine schistosomiasis mansoni. Ann NY Acad Sci 1986;465:313.

12 Lammie PJ, Phillips SM, Linette GP, Michael AI, Bentley AG. In vitro granuloma formation using defined antigenic nidi. Tenth international conference on sarcoidosis and other granulomatous disorders. Ann NY Acad Sci 1986; 465:340.

13 Phillips SM, Lammie PJ. Immunopathology of granuloma formation and fibrosis in schistosomiasis. Parasitol Today 1986;2:296.

14 Bentley AG, Phillips SM, Kaner RJ, Theodorides VJ, Linette GP, Doughty BL. In vitro delayed hypersensitivity granuloma formation: development of an antigen-conjugated bead model. J Immunol 1985;134:4163.

15 Scott P, Pearce E, Cheever AW, Coffman RL, Sher A. Role of cytokines and CD4$^+$ T cell subsets in the regulation of parasite immunity and disease. Immunol Rev 1989;112:161.

16 Sher A, Coffman RL, Hieny S, Scott P, Cheever A. Interleukin-5 (IL-5) is required for the blood and tissue eosinophilia but not granuloma formation induced by infection with Schistosoma mansoni. Proc Natl Acad Sci USA 1990;87:61.

17 Secor WE, Stewart SJ, Colley DG. Eosinophils and immune mechanisms. VI. The synergistic combination of granulocyte–macrophage colony-stimulating factor and IL-5 accounts for eosinophil-stimulation promoter activity in Schistosoma mansoni-infected mice. J Immunol 1990;144:1484.

18 Pearce EJ, Caspar P, Grzych J, Lewis FA, Sher A. Down regulation of Th1 cytokine production accompanies induction of Th2 responses by a parasitic helminth, Schistosoma mansoni. J Exp Med 1991;173:159.

19 Czaja MJ, Weiner FR, Flanders KC, Giambrone M-A, Wind R, Biempica L, Zern MA. In vitro and in vivo association of transforming growth factor-B1 with hepatic fibrosis. J Cell Biol 1989;108:2477.

20 Zwingenberger K, Irschick E, Vergetti Siqueira JG, Correia Dacal AR, Feldmeier H. Tumour necrosis factor in hepatosplenic schistosomiasis. Scand J Immunol 1990;31:205.

21 Weinstock JV, Blum AM. Detection of vasoactive intestinal peptide and localization of its mRNA within granulomas of murine schistosomiasis. Cell Immunol 1990;125:291.

22 Domingo EO, Warren KS. Endogenous desensitization: changing host granulomatous response to schistosome eggs at different stages of infection with Schistosoma mansoni. Am J Pathol 1968;52:369.

23 Colley DG. Adoptive suppression of granuloma formation. J Exp Med 1976;143:696.

24 Doughty BL, Goes AM, Parra JC, et al. Granulomatous hypersensitivity to Schistosoma mansoni egg antigens in human schistosomiasis. I. Granuloma formation and modulation around polyacrylamide antigen-conjugated beads. Mem Inst Oswaldo Cruz Rio de Janeiro 1987;82(IV): 47.

25 Chensue SW, Wellhausen SR, Boros DL. Modulation of granulomatous hypersensitivity. II. Participation of Ly1$^+$ and Ly2$^+$ T lymphocytes in the suppression of granuloma formation and lymphokine production in Schistosoma mansoni infected mice. J Immunol 1981;127:363.

26 Sher A, McIntyre S, von Lichtenberg F. Kinetics and class

27 Sher A, Caspar P, Fiorentino DF, Mosmann TR. (Manuscript submitted.)

28 Fiorentino DF, Bond MA, Mosmann TR. Two types of T helper cells. IV. Th2 clones secrete a factor that inhibits cytokine production by Th1 clones. J Exp Med 1989;170:2081.

29 Espevik T, Figari I, Shalaby MR, Lackides G, Lewis G, Shepard H, Palladino MA. Inhibition of cytokine production by cyclosporin A and transforming growth factor B. J Exp Med 1987;166:571.

30 Chensue SW, Otterness IG, Higashi GI, Forsch CS, Kunkel SL. Monokine production by hypersensitivity (Schistosoma mansoni egg) and foreign body (sephadex bead)–granuloma macrophages. J Immunol 1988;142:281.

31 Josimovic-Alasevic O, Feldmeier H, Zsingenberger K, et al. Interleukin 2 receptor in patients with localized and systemic parasitic diseases. Clin Exp Immunol 1988;72:249.

32 Zwingenberger K, Harms G, Siqueira JGV, et al. T cell phenotype alterations in hepatosplenic schistosomiasis mansoni normalize after chemotherapy. Immunobiology 1989;179:342.

33 Colley DG, Katz N, Rocha RS, Abrantes W, Da Silva AL, Gazzinelli G. Immune responses during human schistosomiasis mansoni. 9. T-lymphocyte subset analysis by monoclonal antibodies in hepatosplenic disease. Scand J Immunol 1983;17:297.

34 Lima MS, Gazzinelli G, Nascimento E, Carvalho Parra J, Montesano MA, Colley DG. Immune responses during human schistosomiasis mansoni. Evidence for anti-idiotypic T lymphocyte responsiveness. J Clin Invest 1986; 78:983.

35 Colley DG, Parra JC, Montesano MA, et al. Immunoregulation in human schistosomiasis by idiotypic interactions and lymphokine-mediated mechanisms. Mem Inst Oswaldo Cruz Rio de Janeiro 1987;82(IV):105.

36 Phillips SM, Fox EG, Fathelbab NG, Walker D. Epitopic and paratopically directed anti-idiotypic factors in the regulation of resistance to murine Schistosomiasis mansoni. J Immunol 1986;137:2339.

37 Phillips SM, Perrin PJ, Walker DJ, Fathelbab NG, Linette GP, Idris MA. The regulation of resistance to Schistosoma mansoni by auto-anti-idiotypic immunity. J Immunol 1988;141:1728.

38 Phillips SM, Lin J, Gala N, Linette GP, Walker DJ, Perrin PJ. The regulation of resistance to Schistosoma mansoni by auto-anti-idiotypic immunity. II. Global qualitative and quantitative regulation. J Immunol 1990;144:4005.

39 Phillips SM, Lin J, Galal N, Linette GP, Walker DJ, Perrin PJ. The regulation of resistance to Schistosoma mansoni by auto-anti-idiotypic immunity. III. An analysis of effects on epitopic recognition, idiotypic expression and anti-idiotypic reactivity at the clonal level. J Immunol 1990;145:2272.

40 Perrin PJ, Phillips SM. The molecular basis of granuloma formation in schistosomiasis: I. T cell derived regulatory factors. J Immunol 1988;141:1714.

41 Perrin PJ, Prystowsky MB, Phillips SM. The molecular basis of granuloma formation in schistosomiasis. II.

Analogies of a T suppressor effector factor to the T cell receptor. J Immunol 1989;142:985.

42 Perrin PJ, Phillips SM. The molecular basis of granuloma formation in schistosomiasis. III. *In vivo* effects of a T cell derived suppressor effector factor and IL-2 on granuloma formation. J Immunol 1989;143:649.

43 Perrin PJ, Phillips RJ, Phillips SM. The molecular basis of granuloma formation in schistosomiasis. IV. T cell derived suppressor–inducer and suppressor–effector factor reactivities are regulated by a TCR β chain analog. Cell Immunol 1989;124:345.

44 Perrin PJ, Phillips SM. The molecular basis of receptor mediated regulation of granulomatous hypersensitivity. In Yoshida T, Torisu M, eds. *Basic Mechanisms of Granulomatous Inflammation*. Amsterdam: Elsevier, 1989:185.

45 Chensue SW, Boros DL, David CS. Regulation of granulomatous inflammation in murine schistosomiasis. II. T suppressor cell-derived, I-C subregionencoded soluble suppressor factor mediated regulation of lymphokine production. J Exp Med 1983;157:219.

46 Mathew RC, Boros DL. Regulation of granulomatous inflammation in murine schistosomiasis. III. Recruitment of antigen-specific I-J⁺ T suppressor cells of the granulomatous response by I-J⁺ soluble suppressor factor. J Immunol 1986;136:1093.

47 Abe T, Colley DG. Modulation of *Schistosoma mansoni* egg-induced granuloma formation. III. Evidence for an anti-idiotypic, I-J⁺, I-J restricted, soluble T suppressor factor. J Immunol 1984;132:2084.

48 Fidel PL, Jr, Boros DL. Antigen-induced suppressor T cells and soluble factors down regulate proliferation and IL-2 production in murine schistosomiasis mansoni. FASEB Abstr 1990:1905.

49 Wellhausen SR, Boros DL. Comparison of Fc, C3 receptors and Ia antigens on the inflammatory macrophage isolated from vigorous or immunomodulated liver granulomas of schistosome-infected mice. J Reticuloendothel Soc 1981; 30:191.

50 Chensue SW, Kunkel SL, Higashi GI, Ward PA, Boros DL. Production of superoxide anion, prostaglandins, and hydroxyeicosatetraenoic acids by macrophages from hypersensitivity-type (*Schistosoma mansoni* egg) and foreign body-type granulomas. Infect Immun 1983;42:1116.

51 Chensue SW, Ellul DA, Spengler M, Higashi GI, Kunkel SL. Dynamics or arachidonic acid metabolism in macrophages from delayed-type hypersensitivity (*Schistosoma mansoni* egg) and foreign-body type granulomas. J Leuk Biol 1985;38:671.

52 Chensue SW, Otterness IG, Higashi GI, Forsch CS, Kunkel SL. Monokine production by hypersensitivity (*Schistosoma mansoni* egg) and foreign body (Sephadex bead)-type granuloma macrophages: Evidence for sequential production of interleukin-1 and tumor necrosis factor. J Immunol 1989;142:1281.

53 Dunn CJ, Gibbons AJ. Human recombinant interleukin-1 induces chronic granulomatous inflammation. J Leuk Biol 1987;42:615.

54 Kasahara K, Kobayashi K, Shikama Y, *et al*. Direct evidence for granuloma-inducing activity of interleukin-1. Induction of experimental pulmonary granuloma formation in mice by interleukin-1 coupled beads. Am J Pathol 1988;130:629.

55 Kobayashi K, Allred C, Cohen S, Toshida T. Role of interleukin-1 in experimental pulmonary granuloma in mice. J Immunol 1985;135:358.

56 Ridel PR, Jamet P, Robin Y, Bach MA. Interleukin-1 released by blood-monocyte-derived macrophages from patients with leprosy. Infect Immun 1986;52:303.

57 Chensue SW, Davey MP, Remick DG, Kunkel SL. Release of interleukin-1 by peripheral blood mononuclear cells in patients with tuberculosis and active infection. Infect Immun 1986;52:314.

58 Kindler V, Sappino A, Grau G, Piguet P, Vassalli P. The inducing role of tumor necrosis factor in the development of bactericidal granulomas during BCG. Cell 1989;56:731.

59 Chensue SW, Shyr-Forsch C, Otterness IG, Kunkel SL. The beta form is the dominant interleukin-1 released by murine peritoneal macrophages. Biochem Biophys Res Commun 1989;150:404.

60 Wyler DJ, Stadecker MJ, Dinarello CA, O'Dea JF. Fibroblast stimulation in schistosomiasis. V. Egg granuloma macrophages spontaneously secrete a fibroblast-stimulating factor. J Immunol 1984;132:3142.

61 Cannon JC, Dinarello CA. Increased plasma interleukin-1 activity in women after ovulation. Science 1985;227:1247.

62 Dinarello CA, Rosenwasser LJ, Wolff SM. Demonstration of a circulating suppressor factor in thymocyte proliferation during endotoxin fever in humans. J Immunol 1981;127:2517.

63 Brown KM, Muchmore AV, Rosensteich DL. Uromodulin, an immunosuppressive protein derived from pregnancy urine, is an inhibitor of interleukin-1. Proc Natl Acad Sci USA 1986;83:9119.

64 Snyder DS, Unanue ER. Corticosteroids inhibit murine macrophage Ia expression and interleukin-1 production. J Immunol 1982;129:1803.

65 Beutler B, Krochin N, Milsark IW, Luedke C, Cerami A. Control of cachectin (tumor necrosis factor) synthesis: mechanisms of endotoxin resistance. Science 1986;232:977.

66 Kunkel SL, Spengler M, May MA, Spengler RM, Larrik J, Remick D. Prostaglandin E regulates macrophage-derived tumor necrosis factor gene expression. J Biol Chem 1988;263:5380.

67 Spengler RN, Spengler M, Lincoln P, Remick DG, Strieter RM, Kunkel SL. Dynamics of dibutyryl cyclic AMP and prostaglandin E mediated suppression of lipopolysaccharide-induced tumor necrosis factor alpha gene expression. Infect Immun 1989;57:2837.

68 Standiford TJ, Strieter RM, Chensue SW, Westwick J, Kasahara K, Kunkel SL. IL-4 inhibits the expression of IL-8 from stimulated human monocytes. J Immunol 1990;145:1435.

69 Hart PH, Vitti GF, Burgess DR, Whitty GA, Piccoli DS, Hamilton JH. Potential anti-inflammatory effects of IL-4: suppression of human monocyte tumor necrosis factor, IL-1, prostaglandin E. Proc Natl Acad Sci USA 1989;86:3803.

70 Tropical diseases in media spotlight. *TDR News* 1990;31:3.

71 Amorim D, de Souza JAA. Chagas' disease. In Yu PN,

Goodwin JF, eds. *Progress in Cardiology*. Philadelphia: Lea & Febiger, 1979:235.

72 Koberle F. Chagas' disease and Chagas' syndromes: the pathology of American trypanosomiasis. Adv Parasitol 1968;6:63.

73 Andrade ZA. Mechanism of myocardial damage in *Trypanosoma cruzi* infection. In *Cytopathology of Parasitic Disease* (Ciba Foundation symposium 99). London: Pitman Books, 1983:214.

74 Santos-Buch CA, Acosta AM. Pathology of Chagas' disease. In Tizard I, ed. *Immunology and Pathology of Trypanosomiasis*. Boca Raton: CRC Press, 1985:145.

75 de Rezende JM, Moreira H. Chagasic meaesophagus and megacolon. Historical review and present concepts. Arq Gastroenterol Sao Paulo 1988;25(special issue):32.

76 Kierszenbaum F, Hudson L. Autoimmunity in Chagas' disease: cause or symptom? Parasitol Today 1985;1:4.

77 Kierszenbaum F. Autoimmunity in Chagas' disease. J Parasitol 1986;72:201.

78 Takle GB, Hudson L. Autoimmunity and Chagas' disease. Curr Top Microbiol Immunol 1989;145:79.

79 Kobrele F. The causation and importance of nervous lesions in American Trypanosomiasis. Bull WHO 1970;42: 739.

80 Barousse AP, Costa JA, Esposto M, Laplume H, Segura EL. Enfermedad de Chagas e Inmunosupresion. Medicina (Buenos Aires) 1980;40(1):17.

81 Maekelt GA. Evaluacion estadistica de los resultados de encuestas epidemiologicas realizadas en Venezuela respecto a la etiologia Chagasica de las miocardiopatias cronicas rurales. Arch Venez Med Trop Parasitol Med 1973;5:107.

82 Maguire JH, Mott KE, Hoff R, *et al*. A three-year follow-up study of infection with *Trypanosoma cruzi* and electrocardiographic abnormalities in a rural community in northeast Brazil. Am J Trop Med Hyg 1982;31:42.

83 Koberle F. Pathogenesis of Chagas' disease. In *Trypanosomiasis and Leishmaniasis with Special Reference to Chagas' Disease* (Ciba Foundation symposium 20). London: Pitman Books, 1974:137.

84 Davila DF, Rossell RO, Donis JH. Cardiac parasympathetic abnormalities: cause or consequence of Chagas' heart disease. Parasitol Today 1989;10:327.

85 Oliveira JSM. A natural human model for intrinsic heart nervous system denervation: Chagas' cardiopathy. Am Heart J 1985;110:1092.

86 Khoury EL, Ritacco V, Cossio PM, *et al*. Circulating antibodies to peripheral nerve in American trypanosomiasis (Chagas' disease). Clin Exp Immunol 1979;36:8.

87 Ribeiro-dos-Santos R, Hudson L. *Trypanosoma cruzi*: immunological consequences of parasite modification of host cells. Clin Exp Immunol 1980;40:36.

88 Ribeiro-dos-Santos R, Hudson L. *Trypanosoma cruzi*: binding of parasite antigens to mammalian cell membranes. Parasite Immunol 1980;2:1.

89 Ribeiro-dos-Santos R, Hudson L. Denervation and the immune response in mice infected with *Trypanosoma cruzi*. Clin Exp Immunol 1981;44:349.

90 Hudson L. Immunopathogenesis of experimental Chagas' disease in mice: damage to the autonomic nervous system. In *Cytopathology of Parasitic Disease*. (Ciba Foundation symposium 99) London: Pitman Books, 1983:234.

91 Petry K, Eisen H. Chagas' disease: a model for the study of autoimmune diseases. Parasitol Today 1989;5:111.

92 Rossi MA, Goncalves S, Ribeiro-Dos-Santos R. Experimental *Trypanosoma cruzi* cardiomyopathy in BALB/c mice. The potential role of intravascular platelet aggregation in its genesis. Am J Pathol 1983;114:209.

93 Factor SM, Cho S, Wittner M, Tanowitz H. Abnormalities of the coronary microcirculation in acute murine Chagas' disease. Am J Trop Med Hyg 1985;34:246.

94 De Morais CF, Higughi HL. Chagas' heart disease and myocardial infarct. Incidence and report of four necropsy cases. Ann Trop Med Parasitol 1989;83:207.

95 Cossio PM, Laguens RP, Diez C, Szarfman A, Segal A, Arana RM. Chagasic cardiopathy: antibodies reacting with plasma membrane of striated muscle and endothelial cells. Circulation 1974;50:1252.

96 Acosta AM, Sadigursky M, Santos-Buch C. Antistriated muscle antibody activity produced by *Trypanosoma cruzi*. Proc Soc Exp Biol Med 1983;172:364.

97 Szarfman A, Terranova VP, Rennard SI, *et al*. Antibodies to laminin in Chagas' disease. J Exp Med 1982;155:1161.

98 Santos-Buch CA, Teixeira ARL. The immunology of experimental Chagas' disease. III. Rejection of allogeneic heart cells *in vitro*. J Exp Med 1974;140:38.

99 Teixeira ARL, Teixeira G, Macedo V, Prata A. *Trypanosoma cruzi*-sensitized T-lymphocyte mediated 51-Cr release from human heart cells in Chagas' disease. Am J Trop Med Hyg 1978;27:1097.

100 Acosta AM, Santos-Buch CA. Autoimmune myocarditis induced by *Trypanosoma cruzi*. Circulation 1985;71:1255.

101 Rizzo LV, Cunha-Neto E, Teixeira ARL. Autoimmunity in Chagas' disease: specific inhibition of reactivity of CD4[+]T cells against myosin in mice chronically infected with *Trypanosoma cruzi*. Infect Immun 1989;57:2640.

102 Sinha AA, Lopez MT, McDevitt HO. Autoimmune diseases: the failure of self tolerance. Science 1990;248:1380.

103 Ferreira F, Neva F, Gusmao R, *et al*. HLA and Chagas' disease. Abstr Congr Int Sobre Doenca de Chagas 1979:141.

104 Llop E, Rothhammer F, Acuna M, Apt W. HLA antigens in cardiomyopathic Chilean chagasics. Am J Hum Genet 1988;43:770.

105 Postan M, Bailey JJ, Dvorak JA, McDaniel JP, Pottala EW. Studies of *Trypanosoma cruzi* clones in inbred mice III. Histopathological and electrocardiographical responses to chronic infection. Am J Trop Med Hyg 1987;37:541.

106 Breniere SF, Carrasco R, Antezana G, Desjeux P, Tibayrenc M. Association between *Trypanosoma cruzi* zymodemes and specific humoral immunodepression in chronic chagasic patients. Trans R Soc Trop Med Hyg 1989;83:517.

107 Breniere SF, Carrasco R, Revollo S, Aparicio G, Desjeux P, Tibayrenc M. Chagas' disease in Bolivia: clinical and epidemiological features and zymodeme variability of *Trypanosoma cruzi* strains isolated from patients. Am J Trop Med Hyg 1989;41:521.

108 Mota EA, Guimaraes AC, Santana OO, Sherlock I, Hoff R, Weller TH. A nine year prospective study of Chagas' disease in a defined rural population in northeast Brazil. Am J Trop Med Hyg 1990;42:429.

109 Cossio PM, Diez C, Szarfman A, Kreutzer E, Candiolo B, Arana RM. Chagasic cardiopathy. Demonstration of a serum globulin factor which reacts with endocardium and vascular structures. Circulation 1974;49:13.

110 Ternynck T, Bleux C, Gregoire J, Avrameas S, Kanellopoulos-Langevin C. Comparison between auto-antibodies arising during *Trypanosoma cruzi* infection in mice and natural autoantibodies. J Immunol 1990;144:1504.

111 Sadigursky M, Acosta AM, Santos-Buch CA. Muscle sarcoplasmic reticulum antigen shared by *Trypanosoma cruzi* clone. Am J Trop Med Hyg 1982;31:934.

112 Von Kreuter BF, Sadigursky M, Santo-Buch CA. Complementary surface epitopes, myotropic adhesion and active grip in *Trypanosoma cruzi*-host cell recognition. Mol Biochem Parasitol 1988;30:197.

113 Von Kreuter BF, Santos-Buch CA. Modulation of *Trypanosoma cruzi* adhesion to host muscle cell membranes by ligands of muscarinic cholinergic and β adrenergic receptors. Mol Biochem Parasitol 1989;36:41.

114 Sadigursky M, von Kreuter BF, Ling P-Y, Santos-Buch CA. Association of elevated anti-sarcolemma anti-idiotype antibody levels with the clinical and pathologic expression of chronic Chagas' myocarditis. Circulation 1989;80:1269.

115 Ouaissi A, Cornette J, Velge P, Capron A. Identification of anti-acetylcholinesterase and anti-idiotype antibodies in human and experimental Chagas' disease: pathological implications. Eur J Immunol 1988;18:1889.

116 Sterin-Borda L, Leiros CP, Wald M, Cremaschi G, Borda E. Antibodies to β-1 and β-2 adrenoreceptors in Chagas' disease. Clin Exp Immunol 1988;74:349.

117 Sterin-Borda L, Gorelik G, Genaro A, Goin JC, Borda ES. Human chagasic IgG interacting with lymphocyte neurotransmitter receptors triggers intra-cellular signal transduction. FASEB J 1990;4:1661.

118 Todd CW, Todd NR, Guimaraes A. Do lymphocytes from chagasic patients respond to heart antigens? Infect Immun 1983;40:832.

119 De Titto EH, Braun M, Lazzari JO, Segura EL. Cell-mediated reactivity against human and *Trypanosoma cruzi* antigens according to clinical status in Chagas' disease patients. Immunol Lett 1985;9:249.

120 Mosca W, Plaja J, Hubsch R, Cedillos R. Longitudinal study of immune response in human Chagas' disease. J Clin Microbiol 1985;22:438.

121 Gattass CR, Lima MT, Nobrega AF, Barcinski MA, Dos Reis GA. Do self-heart-reactive T cells expand in *Trypanosoma cruzi*-immune hosts? Infect Immun 1988;56:1402.

122 Goncalves da Costa SC, Lagrange PH, Hurtrel B, Kerr I, Alencar A. Role of T lymphocytes in the resistance and immunopathology of experimental Chagas' disease. Ann Inst Pasteur/Immunol 1984;135C:317.

123 Russo M, Starobinas N, Minoprio P, Coutinho A, Hontebeyrie-Joskowicz M. Parasitic load increases and myocardial inflammation decreases in *Trypanosoma cruzi*-infected mice after inactivation of helper cells. Ann Inst Pasteur/Immunol 1988;139:225–236.

124 Said G, Joskowicz M, Barreira AA, Eisen H. Neuropathy associated with experimental Chagas' disease. Ann Neurol 1985;18:676.

125 Hontebeyrie-Joskowicz M, Said G, Millon G, Machal G, Eisen H. L3T4 cells from mice chronically infected by *Trypanosoma cruzi*. Eur J Immunol 1987;17:1027.

126 Ribeiro-dos-Santos R, Pirmez C, Savino W. T-cell function and autoimmunity in Chagas' disease. *Biology of Parasitism* (Abstract). AAAS Annual Meeting, New Orleans 1990.

127 Shoenfeld Y, Isenberg DA. Mycobacteria and auto-immunity. Immunol Today 1988;9:178.

128 Lipsky PE, Davis LS, Cush JJ, Oppenheimer-Marks N. The role of cytokines in the pathogenesis of rheumatoid arthritis. Springer Semin Immunopathol 1989;11:123.

129 Singer DS, Maguire JE. Regulation of the expression of class I MHC genes. Crit Rev Immunol 1990;10:235.

130 David-Watine B, Israel A, Kourilsky P. The regulation and expression of MHC class I genes. Immunol Today 1990;11:286.

131 Sarvetnick N, Liggitt D, Pitts SL, Hansen SE, Stewart TA. Insulin-dependent diabetes mellitus induced in transgenic mice by ectopic expression of class II MHC and interferon-γ. Cell 1988;52:773.

132 Caforio ALP, Stewart JT, Bonifacio E, *et al.* Inappropriate major histocompatibility complex expression on cardiac tissue in dilated cardiopathy. Relevance for autoimmunity? J Autoimmun 1990;3:187.

133 Tarleton RL. Tumor necrosis factor (cachectin) production during experimental Chagas' disease. Clin Exp Immunol 1988;73:186.

134 Minoprio P, Burlen O, Pereira P, *et al.* Most B cells in acute *Trypanosoma cruzi* infection lack parasite specificity. Scand J Immunol 1988;28:553.

135 Minoprio P, Bandeira A, Perreira P, Mota Santos T, Coutinho A. Preferential expansion of LY-1 and CD4⁻ CD8⁻ T cells in the polyclonal lymphocyte responses to murine *T. cruzi* infection. Int Immunol 1989;1:176.

136 Minoprio P, Andrade L, Lembezat M-P, Ozaki LS, Coutinho A. Indescriminate representation of Vh-gene families in the murine B lymphocyte response to *Trypanosoma cruzi*. J Immunol 1989;142:4017.

137 Minoprio PM, Eisen H, Forni L, D'Imperio Lima MR, Joskowicz M, Coutinho A. Polyclonal lymphocyte responses to murine *Trypanosoma cruzi* infection. Scand J Immunol 1986;24:661.

138 Rosenberg YL, Chiller JM. Ability of antigen-specific helper cells to effect a class-restricted increase in total Ig-secreting cells in spleens after immunization with the antigen. J Exp Med 1979;150:517.

139 Dziarski R. Autoimmunity: polyclonal activation or antigen induction? Immunol Today 1988;9:340.

140 Hayakawa K, Hardy RR. Normal, autoimmune, and malignant CD5⁺ B cells: the Ly-1 B linkage? Ann Rev Immunol 1988;6:197.

141 Holoshitz J, Koning F, Coligan JE, de Bruyn J, Strober S. Isolation of CD4⁻ CD8⁻ mycobacteria-reactive T lymphocyte clones from rheumatoid arthritis synovial fluid. Nature 1989;339:226.

142 Lindquist S, Craig EA. The heat-shock proteins. Annu Rev Genet 1988;22:631.

143 Young RA, Elliott TJ. Stress proteins, infection, and immune surveillance. Cell 1989;59:5.

144 Lamb JR, Bal V, Mendez-Samperio P, Mehlert A, *et al.* Stress proteins may provide a link between the immune

response to infection and autoimmunity. Int Immunol 1989;1:191.

145 Newport G, Culpepper J, Agabian N. Parasite heat shock proteins. Parasitol Today 1988;4:306.

146 Dragon EA, Sias SR, Kato EA, Gabe JD. The genome of *Trypanosoma cruzi* contains a constitutively expressed, tandemly arranged multicopy gene homologous to a major heat shock protein. Mol Cell Biol 1987;7:1271.

147 Engman DM, Dragon EA, Donelson JE. Human humoral response to hsp70 during *Trypanosoma cruzi* infection. J Immunol 1990;144:3987.

148 Silva JS, Rossi MA. Intensification of acute *Trypanosoma cruzi* myocarditis in BALB/c mice pretreated with low doses of cyclophosphamide or gamma irradiation. J Exp Pathol 1988;71:33.

149 Andrade ZA, Andrade SG, Sadigursky M. Enhancement of chronic *Trypanosoma cruzi* myocarditits in dogs treated with low doses of cyclophosphamide. Am J Pathol 1987;127:467.

150 Tarleton RL. Depletion of CD8⁺ T cells increases susceptibility and reverses vaccine-induced immunity in mice infected with *Trypanosoma cruzi*. J Immunol 1990;144:717.

151 Abath FGC, Coutinho EM, Montenegro SML, Gomes YM, Carvalho AB. The use of non-specific immunopotentiators in experimental *Trypanosoma cruzi* infection. Trans R Soc Trop Med Hyg 1988;82:73.

152 Rowland EC, Levy RL. Cardiac histology of mice with experimental Chagas' disease. J Parasitol 1987;73:240.

153 Cuneo CA, Molina de Raspi E, Basombrio MA. Prevention of electrocardiographic and histopathologic alterations in the murine model of Chagas' disease by preinoculation of an attenutated *Trypanosoma cruzi* strain. Rev Inst Med Trop Sao Paulo 1989;31:248.

154 Kuhn RE. Immunity against *Trypanosoma cruzi*. In Mansfield J, ed. *Parasitic Diseases: The Immunology*. New York: Marcel Dekker, 1983:137.

155 Kierszenbaum F. Immunological deficiency during experimental Chagas' disease (*Trypanosoma cruzi* infection): role of adherent, nonspecific esterase-positive splenic cells. J Immunol 1982;129:2202.

156 Cunningham DS, Kuhn RE, Rowland EC. Suppression of humoral responses during *Trypanosoma cruzi* infections in mice. Infect Immun 1978;22:155.

157 Tarleton RL, Scott DW. Initial induction of immunity, followed by suppression of responses to parasite antigens during *Trypanosoma cruzi* infection of mice. Parasite Immunol 1987;9:579.

158 Teixeira ARL, Teixiera G, Macedo VO, Prata A. Acquired cell-mediated immunodepression in acute Chagas' disease. J Clin Invest 1978;62:1132.

159 Cunningham DS, Grogl M, Kuhn RE. Suppression of antibody responses in humans infected with *Trypanosoma cruzi*. Infect Immun 1980;30:496.

160 Breniere SF, Poch O, Selaes H, *et al*. Specific humoral depression in chronic patients infected by *Trypanosoma cruzi*. Rev Inst Med Trop Sao Paulo 1984;26:254.

161 Mosca W, Plaja J. Delayed hypersensitivity to heart antigens in Chagas' disease as measured by *in vitro* lymphocyte stimulation. J Clin Microbiol 1981;14:1.

162 Gusmao R d'A, Rassi A, Rezende JM, Neva FA. Specific

and non-specific lymphocyte blastogenic responses in individuals infected with *Trypanosoma cruzi*. Am J Trop Med Hyg 1984;33:827.

163 Gazzinelli RT, Morato MJF, Nunes RMB, Cancado JR, Brener Z, Gazzinelli G. Idiotype stimulation of T lymphocytes from *Trypanosoma cruzi*-infected patients. J Immunol 1988;140:3167.

164 Eloi-Santos SM, Novato-Silva E, Maselli VM, Gazzinelli G, Colley DG, Correa-Oliveira R. Idiotypic sensitization in utero of children born to mothers with schistosomiasis or Chagas' disease. J Clin Invest 1989;84:1028.

165 Cunningham DS, Kuhn RE. *Trypanosoma cruzi*-induced suppression of the primary immune response in murine cell cultures to T cell-dependent and -independent antigens. J Parasitol 1980;66:16.

166 Tarleton RL. *Trypanosoma cruzi*-induced suppression of IL-2 production. II. Evidence for a role for suppressor cells. J Immunol 1988;140:2769.

167 Tarleton RL, Kuhn RE. Measurement of parasite-specific immune responses *in vitro*: evidence for suppression of the antibody response to *Trypanosoma cruzi*. Eur J Immunol 1985;15:845.

168 Reed SG, Roters SB, Goidl EA. Spleen cell-mediated suppression of IgG production to a non-parasite antigen during chronic *Trypanosoma cruzi* infection in mice. J Immunol 1983;131:1978.

169 Scott MT. Delayed hypersensitivity to *Trypanosoma cruzi* in mice: specific suppressor cells in chronic infection. Immunology 1981;44:409.

170 Ramos C, Schadtler-Siwon I, Ortiz-Ortiz L. Suppressor cells present in the spleens of *Trypanosoma cruzi*-infected mice. J Immunol 1979;122:1243.

171 Harel-Bellan A, Joskowicz M, Fradelizi D, Eisen H. T lymphocyte function during experimental Chagas' disease: production of and response to interleukin 2. Eur J Immunol 1985;15:438.

172 Tarleton RL, Kuhn RE. Restoration of *in vitro* immune responses of spleen cells from mice infected with *Trypanosoma cruzi* by supernatants containing interleukin 2. J Immunol 1984;133:1570.

173 Reed SG, Inverso JA, Roters S. Heterologous antibody responses in mice with chronic *T. cruzi* infection: depressed T helper function restored with supernatants containing interleukin 2. J Immunol 1984;133:1558.

174 Choromanski L, Kuhn RE. Interleukin 2 enhances specific and nonspecific immune responses in experimental Chagas' disease. Infect Immun 1985;50:354.

175 Via CS, Shearer GM. T-cell interactions in autoimmunity: insights from a murine model of graft-versus-host disease. Immunol Today 1988;9:207.

176 Kroemer G, Wick G. The role of interleukin 2 in autoimmunity. Immunol Today 1989;10:246.

177 Gutierrez-Ramos JC, Andreu JL, Revilla Y, Vinuela E, Martinez-A C. Recovery from autoimmunity of MRL/lpr mice after infection with an interleukin-2/vaccinia recombinant virus. Nature 1990;346:271.

178 Mitchison NA, Oliveira DBG. Chronic infection as a major force in the evolution of the suppressor T-cell system. Parasite Immunol 1986;2:312.

179 Rodriguez A-M, Santoro F, Afchain D, Bazin H, Capron A.

Trypanosoma cruzi infection in B-cell deficient rats. Parasite Immunol 1981;31:524.

180 Trischmann TM. Non-antibody mediated control of parasitemia in acute experimental Chagas' disease. J Immunol 1983;130:1953.

181 Trischmann TM. Role of cellular immunity in protection against *Trypanosoma cruzi* in mice. Parasite Immunol 1984;6:561.

182 Rottenberg M, Cardoni RL, Anderson R, Segura EL, Orn A. Role of T helper/inducer as well as natural killer cells in resistance to *Trypanosoma cruzi* infection. Scand J Immunol 1988;28:573.

183 Araujo F. Development of resistance to *Trypanosoma cruzi* in mice depends on a viable population of L3T4$^+$ (CD4$^+$) T lymphocytes. Infect Immun 1989;57:2246.

184 Burgess DE, Hanson WE. *Trypanosoma cruzi*: the T cell dependence of the primary immune response and the effects of depletion of T cells and Ig-bearing cells on immunological memory. Cell Immunol 1980;52:176.

185 Tanowitz HB, Rager-Zisman B, Wittner M. The effect of silica treatment on resistance to the "Brazil" strain of *Trypanosoma cruzi* in C57BL/10 (B10) mice. Trans R Soc Trop Med Hyg 1980;74:820.

186 Boullon F, Sinagra A, Riarte A, *et al.* Experimental cardiac transplantation and chronic Chagas' disease in dogs. Transplant Proc 1988;20:432.

187 Brener Z, Chiari E. The effects of some immunosuppressive agents in experimental chronic Chagas' disease. Trans R Soc Trop Med Hyg 1971;65:629.

188 Cabeza Meckert PM, Chambo JG, Laguens RP. Modification of the pattern of infection and evolution of cardiopathy in experimental Chagas' disease after treatment with immunosuppressive and trypanocidal drugs. Medicina (Buenos Aires) 1988;48:7.

189 Morato MJF, Brener Z, Cancado JR, Nunes RMB, Chiari E, Gazzinelli G. Cellular immune responses of chagasic patients to antigens derived from different *Trypanosoma cruzi* strains and clones. Am J Trop Med Hyg 1986;35:505.

190 Rose NR, Herskowitz A, Neumann DA, Nikolaus KD. Autoimmune myocarditis: a paradigm of post-infection autoimmune disease. Immunol Today 1988;9:117.

191 Maisch B. Retrospective and perspectives in the immunology of cardiac diseases. Springer Semin Immunopathol 1989;11:479.

192 Foulis AK. Does viral infection initiate autoimmunity in type I diabetes? J Autoimmun 1990;3:21.

193 Via CS, Morse III HC, Shearer GM. Altered immunoregulation and autoimmune aspects of HIV infection: relevant murine models. Immunol Today 1990;11:250.

194 Pearson RD, Sousa AQ. Leishmania species: visceral (kalaazar), cutaneous, and mucosal leishmaniasis. In Mandell GL, Douglas RG Jr, Bennett JE, eds. *Principles and Practice of Infectious Diseases*, 3rd edn. New York: Churchill Livingstone, 1990:2066–2077.

195 Jeronimo SMB, Pearson RD. Leishmaniasis: an update. Current Opinion Infect Dis 1989;2:631.

196 Pearson RD, Wilson ME. Host defenses against prototypical intracellular protozoans, the *Leishmania*. In Walzer PD, Genta RM, eds. *Parasite Infections in the Compromised Host. Immunological Mechanisms and Clinical Applications.*

New York: Marcel Dekker, 1988:31–81.

197 Wilson ME, Pearson RD. Immunology of leishmaniasis. In Wyler DJ, ed. *Modern Parasite Biology: Cellular, Immunological, and Molecular Aspects*. New York: WH Freeman, 1990:200–221.

198 Pearson RD, Navin TR, Sousa AQ, *et al.* Leishmaniasis. In Kass EH, Platt R, eds. *Current Therapy in Infectious Disease*, vol. 3. Toronto: BC Decker, 1990:384–389.

199 Wilson ME, Pearson RD. Roles of CR3 and mannose receptors in the attachment and ingestion of *Leishmania donovani* by human mononuclear phagocytes. Infect Immun 1988;56:363.

200 Wozencraft AO, Sayers G, Blackwell JM. Macrophage type 3 complement receptors mediate serum-independent binding of *Leishmania donovani*. J Exp Med 1986;164:1332.

201 Blackwell JM, Ezekowitz RAB, Roberts MB, *et al.* Macrophage complement and lectin-like receptors bind *Leishmania* in the absence of serum. J Exp Med 1985;162:324.

202 Blackwell JM. Receptors and recognition mechanisms of *Leishmania* species. Trans R Soc Trop Med Hyg 1985;79:606.

203 Wilson ME, Pearson RD. Evidence that *Leishmania donovani* utilizes a mannose receptor on human mononuclear phagocytes to establish intracellular parasitism. J Immunol 1986;136:4681.

204 Wyler DJ, Sypek JP, McDonald JA. *In vitro* parasite–monocyte interactions in human leishmaniasis: possible role of fibronectin in parasite attachment. Infect Immun 1985;49:305.

205 Mosser DM, Valassara H, Edelson PJ, *et al. Leishmania* promastigotes are recognized by the macrophage receptor for advanced glycosylation endproducts. J Exp Med 1987;165:140.

206 Pearson RD, Harcus JL, Roberts D, *et al.* Differential survival of *Leishmania donovani* amastigotes in human monocytes. J Immunol 1983;131:1994.

207 Murray HW, Rubin BY, Rothermel CD. Killing of intracellular *Leishmania donovani* by lymphokine-stimulated human mononuclear phagocytes. Evidence that interferon-gamma is the activating lymphokine. J Clin Invest 1983;72:1506.

208 Murray HW, Stern JJ, Welte K, *et al.* Experimental visceral leishmaniasis: production of interleukin-2 and interferon-gamma, tissue immune reaction, and response to treatment with interleukin 2 and interferon-gamma. J Immunol 1987;138:2290.

209 Sypek JP, Wyler DJ. Cell contact-mediated macrophage activation for antileishmanial defense; mapping of the genetic restriction to the I region of the MHC. Clin Exp Immunol 1985;62:449.

210 Lainson R, Shaw JJ. Evolution, classification and geographic distribution. In Peters W, Killick-Kendrick R, eds. *The Leishmaniases in Biology and Medicine*, vol. 1. London: Academic Press, 1987:1–120.

211 Jones TC, Johnson WD Jr, Barretto AC, *et al.* Epidemiology of American cutaneous leishmaniasis due to *Leishmania braziliensis braziliensis*. J Infect Dis 1987;156:73.

212 Ridley DS, Ridley MJ. The evolution of the lesion in cutaneous leishmaniasis. J Pathol 1983;141:83.

213 Ridley MJ, Ridley DS. Cutaneous leishmaniasis: immune complex formation and necrosis in the acute phase. Br J Exp Pathol 1984;65:327.

214 Walton BC, Valverde L. Racial differences in espundia. Ann Trop Med Parasitol 1979;73:23

215 Heinzel FP, Sadick MD, Holaday BJ, et al. Reciprocal expression of interferon-gamma or interleukin 4 during the resolution or progression of murine leishmaniasis. Evidence for expansion of distinct helper T cell subsets. J Exp Med 1989;169:59.

216 Scott P, Natovitz P, Coffman RL, et al. Immunoregulation of cutaneous leishmaniasis. T cell lines that transfer protective immunity or exacerbation belong to different T helper subsets and respond to distinct parasite antigens. J Exp Med 1988;168:1675.

217 Peterson EA, Neva FA, Barral A, et al. Monocyte suppression of antigen-specific lymphocyte responses in diffuse cutaneous leishmaniasis patients from the Dominican Republic. J Immunol 1984;132:2603.

218 Reiner NE, Melamud CJ. Arachodonic acid metabolism by murine peritoneal macrophages infected with Leishmania donovani: in vitro evidence for parasite-induced alterations in cyclooxygenase and lipoxygenase pathways. J Immunol 1985;134:556.

219 Sadick MD, Locksley RM, Raff HV. Development of cellular immunity in cutaneous leishmaniasis due to Leishmania tropica. J Infect Dis 1984;150:135.

220 Giannini MSH. Suppression of pathogenesis in cutaneous leishmaniasis by UV irradiation. Infect Immun 1986; 51:838.

221 Bryceson ADM, Bray RS, Wolstencroft RA, et al. Immunity in cutaneous leishmaniasis of the guinea pig. Clin Exp Immunol 1970;7:301.

222 McElrath MJ, Kaplan G, Nusrat A, et al. Cutaneous leishmaniasis. The defect in T cell influx in BALB/c mice. J Exp Med 1987;165:546.

223 Southgate BA, Manson-Bahr PEC. Studies in the epidemiology of East African leishmaniasis. 4. The significance of the positive leishmanin test. J Trop Med Hyg 1967; 70:29.

224 Pampiglione S, Manson-Bahr PEC, La Placa M, et al. Studies in Mediterranean leishmaniasis. 3. The leishmanin skin test in kala-azar. Trans R Soc Trop Med Hyg 1975;69:60.

225 Badaro R, Jones TC, Lorenco R, et al. A prospective study of visceral leishmaniasis in an endemic area of Brazil. J Infect Dis 1986;154:639.

226 Badaro R, Jones TC, Carvalho EM, et al. New perspectives on a subclinical form of visceral leishmaniasis. J Infect Dis 1986;154:1003.

227 Evans T, Reis FE, Alencar JE, et al. American visceral leishmaniasis (kala-azar). West J Med 1985;142:777.

228 Pearson RD, Alencar JE, Romito R, et al. Circulating immune complexes and rheumatoid factors in visceral leishmaniasis. J Infect Dis 1983;147:1102.

229 Carvalho EM, Andrews BS, Martinelli R, et al. Circulating immune complexes and rheumatoid factor in schistosomiasis and visceral leishmaniasis. Am J Trop Med Hyg 1983;32:61.

230 Carvalho EM, Badaro R, Reed SG, et al. Absence of gamma interferon and interleukin 2 production during active visceral leishmaniasis. J Clin Invest 1985;76:2066.

231 Pearson RD, Evans T, Naidu TG, et al. Humoral factors during South American visceral leishmaniasis. Ann Trop Med Parasitol 1986;80:465.

232 Rizzi M, Arici C, Bonaccorso C, et al. Visceral leishmaniasis in a patient with human immunodeficiency virus. Trans R Soc Trop Med Hyg 1988;82:565.

233 Fernandez-Guerrero ML, Aguado JM, Buzon L, et al. Visceral leishmaniasis in immunocompromised hosts. Am J Med 1987;83:1098.

234 Holaday B, Sadick MD, Pearson RD. Isolation of protective T cells from BALB/cJ mice chronically infected with Leishmania donovani. J Immunol 1988;141:2132.

235 Stern JJ, Oca MJ, Rubin BY, et al. Role of L3T4[+] and Lyt-2[+] cells in experimental visceral leishmaniasis. J Immunol 1988;140:3971.

236 Wilson ME, Innes DJ, Sousa AQ, et al. Early histopathology of experimental infection with Leishmania donovani in hamsters. J Parasitol 1987;73:55.

237 Pearson RD, Cox G, Evans T, et al. Wasting and macrophage production of tumor necrosis factor/cachectin and interleukin 1 in visceral leishmaniasis. Am J Trop Med Hyg 1990;43:640.

5 Survival strategies of parasites in their immunocompetent hosts

Jean-Paul L. Dessaint & André R. Capron

Survival and growth of infectious organisms in the face of the immune response mounted by their immuno-competent hosts is made possible through one of two strategies: one is to outrun rejection by fast replication and/or mutation, as observed in some bacterial or viral infections; the other is to impair development or expression of immunity so as to evade the deleterious consequences of the host's counter-attack. Indeed, it is now recognized that in most infections it is the infectious organism that determines whether or not to allow itself to be rejected by immune effectors. This is clearly the case of parasites, which in general induce chronic infections and are thus confronted for months and often years with the immune response that they have to maintain within limits compatible with both their own survival and that of their mammalian hosts. From an immunologic standpoint, the adaptation of parasites to vertebrates has been the result of selective pressure by the host's immune system on their evolution and biology. Detailed analyses of host–parasite relationships, which have been enriched in recent years by several molecular approaches, reveal in most systems that several aspects of the biology of parasites, as well as the remote and vague notions of infectivity and virulence, are indeed the result of immune restriction, and even that some parasites are able to utilize for their own benefit some components of the immune response. This leads in many instances to the refined elaboration in these organisms of a complex network of processes allowing them to escape the defense mechanisms of their host. Such a complexity gives the expression of immunity in parasitic diseases a dynamic aspect, reflecting the permanent balance between effector and regulatory mechanisms. A second important feature of parasitism concerns the biologic complexity of the parasite life-cycle, which involves in many instances a succession of developmental stages each characterized by differences in biologic niches, metabolic require-

ments, and expression of antigenicity.

It is not surprising, therefore, that infection by most parasites does not result in efficient and persistent sterilizing immunity, as in many other infectious diseases, but rather in a partial nonsterilizing immunity that reduces the parasite burden and pathologic manifestations without necessarily wiping out the invader. In helminthic infections, such as schistosomiasis, the basis for acquired resistance is the development of so-called concomitant immunity, a situation in which the adult worm provides the major antigenic stimulus without it being affected by the resulting effector mechanisms. The major target of immunity in this case is the invasive (larval) stage of the parasite, i.e., the schistosomula. A similar concept has been developed for malaria by the name of "immunité de prémunition" by Sergent.

Most of the host–parasite pairs demonstrate their own tactics, which in part reflect the striking diversity of parasites in their taxonomy, their localization, and their successive stages of development in their vertebrate hosts. Yet the strategy of parasite survival can be analyzed according to some general events and mechanisms that lead to impairment of immune function in parasitized hosts, be it generalized immunosuppression or specific unresponsiveness. Although a specific immune response can be detected against an array of parasite antigens, the important observation is that the response is made totally or partially inefficient by being deviated towards nonfunctionally important antigens. Thus, modification of the structure or topography of expression of parasite antigens, production by parasites of immunomodulatory molecules, isotypic selection, and restriction of the response are essential components of the immunodeviation that allows parasites to survive during chronic infections.

ESCAPE FROM IMMUNE EFFECTORS

Seclusion of parasites

Encystment of some parasites (e.g., ameba, *Echinococcus* spp.) might allow them to resist immune attack. Yet it is known that hydatid cysts, for instance, are penetrated by host macromolecules. Likewise, enteroparasites with strict intraluminal location are seemingly protected from attack by most immune effectors, but a so-called self-cure may occur and be facilitated by local anaphylaxis. A resistance mechanism common to several protozoan parasites is cell penetration and intracellular multiplication. Intracellular stages might thus be considered as protected, but it is known that infected cells bear parasite antigens on their surface and can be the target of antibodies or cellular responses, as demonstrated for instance in malaria [1].

In fact, we are far from a comprehensive knowledge of the various events that control cell penetration and intracellular resistance. If one accepts the schematic view that cell infection by a protozoan parasite proceeds in two major steps—recognition, attachment, and penetration on the one hand and intracellular survival and multiplication on the other hand—one is left in many cases with series of morphologic and biologic observations with little molecular support.

It is known that malaria parasites invade erythrocytes by interacting with specific ligands on the cell membrane, and that the erythrocyte structure involved both in the case of *Plasmodium knowlesi* and *Plasmodium vivax* is the Duffy blood group antigen. More recently, a 135 kD protein made synthetically by *P. knowlesi* and binding to Duffy-positive erythrocytes has been identified. A comparable receptor–ligand system for *Plasmodium falciparum* appears to involve a 170 kD parasite protein that binds erythrocyte sialic acid. Several other putative attachment sites, including glycophorin A and fibronectin receptors, have been suspected to provide alternative binding possibilities and it is likely, as emphasized by Miller *et al.* [2], that the malaria merozoites can bypass several steps or use several pathways for cell invasion.

Another example is provided by *Trypanosoma cruzi* infection. It was originally reported that *T. cruzi* trypomastigotes express on their membrane a high-affinity receptor for host fibronectin and that antibodies against either the parasite receptor (p85) or the attachment domain of fibronectin and derived synthetic peptides can block cell penetration and decrease infection both

in vitro and *in vivo* [3,4]. In this context, model peptides, including the Arg–Gly–Asp (RGD) sequence, once coupled to an appropriate carrier, have a protective effect in mice against experimental challenge. A similar observation can be made when synthetic peptides are constructed from the sequence of the surface protein p85, the putative receptor for fibronectin on *T. cruzi* trypomastigotes [5]. Fibronectin has also been reported to bind to *Leishmania mexicana* promastigotes and amastigotes as well as to *Leishmania tropica* amastigotes, and to enhance binding of the parasite to human monocytes. Moreover, major surface glycoproteins of *Leishmania* contain the RGD sequence as shown, for instance, for *Leishmania major* [6]. Parasite receptors for fibronectin may play a role directly in parasite attachment to host cells, or indirectly through the capacity of the protein to up-regulate complement receptors of mononuclear phagocytes [5].

Many protozoan parasites have been shown to activate complement on their surface, and C3 cleavage products have been claimed to be the principal ligand mediating uptake of intracellular parasites, including *Leishmania* promastigotes and *T. cruzi* trypomastigotes. C3-Binding molecules have been identified on the surface of the parasite: a 72 kD glycoprotein (gp72) on noninfective *T. cruzi* epimastigotes [7], a developmentally regulated lipophosphoglycan on *L. major* [8], and a 63 kD glycoprotein (gp63) on *L. mexicana* promastigotes [9]. Yet direct binding of gp63 through an RGD sequence to complement receptor Type 3 of host macrophages has recently been demonstrated [10]. Interaction of *Leishmania* with other receptors of host cells has also been suggested, including the mannosyl–fucosyl receptor of macrophages [11] and a receptor for nonenzymatically glucosylated moieties [12].

These examples show the diversity of interactions between protozoan parasites and extracellular and cell surface molecules, including adhesiotopes, complement components, and complement receptors. Since the complement cascade is known to be involved in some protective reactions, one has to consider how complement activation by parasites does not affect survival of the extracellular stage before it enters the host cell. Several escape mechanisms preventing complement-dependent attack have been described. In particular, infective metacyclic trypomastigotes of *T. cruzi* do not support efficient formation of the alternative C3 convertase, by failing to bind factor B. This appears to be due to a parasite molecule released by culture-derived metacyclic trypomastigotes, which functions in a similar way to a mammalian decay acceleration factor [13]. Infective-stage metacyclic promastigotes of all *Leishmania* species resist serum killing by another mechanism, interfering with the insertion of the complement membrane attack complex that is dependent on developmentally regulated synthesis of surface proteins and N-linked glycans [14]. Thus, resistance to complement-mediated killing may be mediated by parasite surface molecules at different stages in the complement cascade.

In fact, in some cases, controlled complement activation appears paradoxically as a survival mechanism for those parasites that infect cells endowed with phagocytic and microbicidal properties, such as macrophages. Indeed, in the case of *L. major*, infective metacyclic stages penetrate through the C3b receptor (CR1), which triggers no respiratory burst and thus allows them to evade killing by macrophages, whereas avirulent promastigotes enter macrophages by CR3-dependent endocytosis, which triggers cellular activation [15,16]. Likewise, *Toxoplasma gondii* elicits a little metabolic burst upon entering mononuclear phagocytes, which may be related to the nature of parasite surface ligands.

The prolonged survival of protozoan parasites in the hostile intracellular environment of phagocytic cells is often poorly understood. For instance, amastigotes of *Leishmania donovani* that reside in the phagolysosome of host macrophages can take up oxygen, incorporate nucleotides and aminoacids, and metabolize glucose optimally at pH 4.0–5.5 while their metabolic activity is sharply reduced at neutral pH, but the reverse is true for the extracellular form of the parasite [17]. However, the mechanism by which *T. gondii* blocks acidification of the phagosome is still unknown. In trypanosomes and in *Leishmania*, the levels of toxic oxidative metabolites such as H_2O_2 and OH^- are kept low by a complex of glutathione covalently linked to spermidine (trypanothione), acting as a cofactor for a parasite-specific trypanothione peroxidase [18].

Modulation of immune effectors

Another tactic has been developed by helminth parasites, which acts at the effector arm of the immune response. As illustrated in the case of schistosomes, this tactic is linked to the action of proteases secreted by transforming schistosomula.

Schistosome proteases have been shown to confer on schistosome larvae resistance to killing by complement [19]. Another cytotoxicity mechanism, involving cytotoxic macrophages activated by schistosome-specific IgE

antibodies [20], is also the target of modulation through parasite proteases. Convergent observations made on the existence of Fc receptors for IgG on the surface of schistosomula and the demonstration of the release of proteolytic enzymes by the parasite at different lifestages led to the observation that, shortly after binding of the Fc portion of IgG molecules to the membrane receptor on schistosomula, parasite proteinases, identified as a serine protease and an imidopeptidase, cleave the bound IgG molecules and lead to the release of peptides. These peptides significantly decrease many macrophage functions, including the release of lysosomal enzymes, glucosamine incorporation, superoxide anion generation, and phagocytosis. The production of IL-1 is also inhibited. Furthermore, IgE antibody-dependent macrophage cytotoxicity for schistosomula is dramatically decreased. The main inhibitory activity was attributed to a tripeptide (Thr–Lys–Pro), which has been made synthetically. The synthetic tripeptide, at concentrations down to 10^{-12} M, reproduces the biologic effects of the IgG cleavage peptides. Interestingly, macrophage activation by a stimulatory tetrapeptide from proteolysis of antibodies by host enzymes—tuftsin—differs from the schistosome-made inhibitory tripeptide only by one amino acid. This provides a remarkable example of how parasites utilize host molecules for their own survival by secreting an imidopeptidase rarely found in vertebrates [21].

There are also many examples showing that parasites can modulate the immune response of their host not only by acting at the effector level but also by causing immune suppression.

IMMUNE SUPPRESSION

Many parasites indeed suppress immune responses during, and sometimes after, their establishment in their mammalian host. Severe immunodeficiencies are associated mainly with infection by protozoan parasites but can also evolve in systemic helminthic infections [22]. It should be pointed out that in many instances susceptibility to a parasite may be increased by infection by another one. This has been shown, for instance, in *Plasmodium* infection, which augments trypanosome virulence in mice [23] and in schistosomiasis: rats are normally not susceptible to infection by the filarial worm *Dipetalonema viteae*. However, similar to athymic (nude) rats, normal rats previously infected by *Schistosoma mansoni* allow adult filarial worms to develop and to lay microfilariae which can be detected in the circulation after a challenge *D. viteae* infection [24]. Thus, multiple parasite infections may not only reinforce their respective immunosuppressive effect but also increase reciprocally host susceptibility.

Several mechanisms have been considered in the induction of immune suppression, although their contribution may vary from one parasite to another, and even at the different stages of the infection.

Antigenic competition and impaired presentation of antigens

Antigenic competition has been put forward in a variety of models. Indeed, parasites contain and liberate in the circulation an array of antigens that might compete with each other, as well as with nonparasite antigens, to decrease the response. Little supportive information has been provided; however, a decrease in the capacity of macrophages to present antigens has been observed in malaria, together with decreased uptake of particles and decreased microbicidal capacity of malarious macrophages [25]. Similar defects have also been observed in African trypanosomiasis.

Suppressor cells

Induction of suppressor cells is another mechanism common to many parasitic infections. The nature of the suppressor cell varies according to the model, including macrophages and T cells, these being either specific or nonspecific for parasite antigens.

Macrophages with suppressive activity have been evidenced in *Plasmodium* infection, in African trypanosomiasis, and in *T. cruzi*-induced immunosuppression. Generation of prostaglandin E2 (PGE$_2$) and reduction of IL-1 production could be responsible for this suppressive activity of macrophages in protozoan infections. Indeed, the production by T cells of T cell growth factors is markedly diminished during experimental infection with different trypanosomes. Recent experiments indicate that the addition of indomethacin, which blocks prostaglandin synthesis by macrophages, restores the potential of lymph node lymphocytes from *Trypanosoma brucei*-infected mice to secrete IL-2 [26]. Likewise, T cells from *T. cruzi*-infected mice fail to produce IL-2 [27], but the macrophage defect leading to impairment in helper T cell function can be restored by *in vivo* or *in vitro* administration of recombinant IL-1 [28]. However, the suppressive role of macrophages may be indirect in this model, by acting as inducers of sup-

pressor–effector CD8$^+$ cells [29]. In leishmaniasis, also, the suppression of IL-2 production appears to be mediated by macrophages, but only at the final stage of infection.

Suppressor T cells have also been incriminated in the antigen-specific or nonspecific inhibition of the immune response.

Suppressor T cell circuits have been investigated extensively in schistosomiasis in relation with the appearance of immune suppression, which coincides in mice with the onset of egg laying and with granulomatous modulation. The specificity of suppressor T cells in human schistosomiasis mansoni has been analyzed: a subset of such suppressors appears to be specific for soluble egg antigens (SEA), and their glycoprotein fraction (MSA$_1$, MSA$_2$) was found to be responsible for inducing nonspecific suppression when their protein (MSA$_3$) fraction was involved in antigen-specific suppression [30]. Another subset is anti-idiotype-specific T cells. These cells were detected by their proliferation in culture in the presence of antibodies to SEA collected from other patients, and the T cells recognize these idiotypes directly [31]. Interestingly, idiotypic differences in antibody preparations from individuals with different forms of schistosomiasis were evidenced: T cell stimulatory, immunoregulatory idiotypes on antiegg antibodies were found in sera from patients with asymptomatic or intestinal forms, but not in those with the severe acute or hepatosplenic forms, indicating that direct and indirect regulation of responsiveness to SEA plays a prominent role in determining the severity of the disease [32]. However, not all suppressor T cells are specific for egg antigens, and in the rat model where infection aborts before worms reach sexual maturity, suppressor T cells were also detected in relation to the appearance of unresponsiveness to adult worm antigens [33]. Nonantigen-specific suppressor cells were also characterized in the spleen of patients with hepatosplenic schistosomiasis [34].

Suppressor T cells have also been investigated in protozoan models. For instance, in the infection of susceptible BALB/c mice with *L. major*, the antigen-specific suppression of delayed-type hypersensitivity response to leishmanial antigen occurs before the appearance of suppressor macrophages responsible for nonantigen-specific suppression [35]. However, the suppressor T cells are remarkable since they have the uncommon CD4$^+$ phenotype. Accordingly, BALB/c mice subjected to adult thymectomy and repeated injection of low doses of anti-CD4 monoclonal antibody develop resistance and delayed footpad response to leishmanial antigen [36]. The preferential induction of the disease-promoting CD4[+] subset may be related to a defect in antigen presentation, since infected macrophages do not augment their expression of Class I and Class II major histocompatibility complex (MHC) gene products in response to interferon-γ (IFN-γ) [37].

Some of the above-mentioned mechanisms can also be found in microbial infections. It is indeed possible to attribute impairment of accessory cell function of macrophages to their infection by intracellular organisms, and induction of suppressor cells of the monocyte–macrophage or T cell lineages is common to many chronic infections. However, a remarkable feature of parasites is their direct interference in immunologic networks by releasing molecules with immunomodulatory function.

Immunosuppressive substances affecting lymphocytes

The mechanisms underlying the induction of suppression by parasites appear in some models to be related to intrinsic properties of parasite products. Many parasite-derived factors with suppressive activity have been described, but most of these are poorly characterized at the molecular level. A tryptophane metabolite, tryptophol (indole-3-ethanol), produced by trypanosomes inhibits antibody responses in mice immunized with red blood cells, but the cellular target, whether T or B cells or macrophages, was not characterized [38]. The ability to induce antigen-specific suppression that dominates the immune response in patients with diffuse cutaneous leishmaniasis is abrogated by drug treatment or removal of live parasites, contrary to lepromatous leprosy where the anergy is irreversible after successful treatment, which points to a direct effect of antigen-induced short-lived regulatory cells [39]. Coculture of the blood forms of T. cruzi and human peripheral lymphocytes induces a marked suppression of proliferative responses to different mitogenic lectins or anti-CD3. Contrary to the experimental model in mice, however, exogenous IL-2 does not restore responsiveness. The observation of a reduction in the proportion of human lymphocytes capable of expressing CD25 antigen has led to the demonstration that the parasite inhibits IL-2 receptor expression, thereby affecting an early event during lymphocyte activation. This inhibition of responsiveness to IL-2 was also observed when trypanosomes are separated from lymphoid cells by a filter, pointing to the release of a

trypanosome immunosuppressive factor, as yet not characterized at the molecular level, that blocks the expression of high-affinity IL-2 receptors [40]. Decreased responsiveness to IL-2 has also been reported in lymphocytes from T. cruzi-infected mice [26].

Part of the defect in interleukin-driven responses may also be attributed to the utilization by the parasite, for its own benefit, of so-called T cell growth factors. Indeed, in a series of experiments conducted with Leishmania mexicana americana and L. donovani we could provide evidence that lymphokines produced by activated T cells are directly involved in the stimulation of parasite growth. This effect could be reproduced by addition of recombinant IL-2. An anti-IL-2 antibody significantly reversed the stimulatory effect of IL-2. More strikingly, in situ treatment of infected mice with recombinant IL-2 led to an exacerbation of the lesions. The increased foodpad swelling after r-IL-2 treatment was correlated with a higher number of parasites per lesion [41]. While this effect of IL-2 could contribute to the immune pathogenesis of cutaneous lesions [35], IL-2 consumption by parasites also leads to a decrease in the level of paracrine messages delivered to host cells, which strengthens the concept that parasites not only are able to evade the host immune response but also have the ability to use it to their own advantage. It is of interest that cyclosporin A, an inhibitor of interleukin production by T cells, has been reported by several groups to exert a protective effect on the infection of mice by various parasites, including Leishmania, schistosome, etc., the exact mechanism underlying this protective effect being so far unknown. In the framework of our experiment, we could show that the protective effect of cyclosporin A against the development of Leishmania infection in mice could be almost totally reversed by r-IL-2 treatment. It is also known that the passive transfer of parasite-specific CD4-positive helper T cells leads to an exacerbation of Leishmania-induced lesions [42]. It is therefore tempting to assume, on the basis of our in vitro and in vivo observations, that among other possible mechanisms this paradoxical effect could be related to a direct role of IL-2 on parasite growth.

Sometimes, modulation by protozoan parasites can be induced directly by infection of lymphoid cells. Trypanosoma cruzi trypomastigotes can directly infect T cells of both the CD4 and CD8 subsets, as shown recently, but not B cells in similar experimental conditions. An interesting observation is a significantly increased susceptibility to infection of T cells collected from infected animals during the acute phase of the infection compared to normal T cells (17% vs. 2%). This observation has been confirmed by the finding that, *in vivo*, 3.5% of mouse T cells are seen infected. The study of the binding sites involved in penetration of human T cells shows that this mechanism is associated with CD3 and HLA-DR antigens, which might explain the increased susceptibility during the course of infection, since HLA-DR antigens are considered as activation markers of T cells. Besides having striking similarities with T cell retroviral infection, this so far unsuspected mechanism of T cell penetration and subsequent multiplication might provide a new basis for understanding the impairment of T cell response in Chagas' disease and more broadly the host–parasite relation in this model [43].

In helminthic infections as well, there are several examples indicating that parasite-derived molecules have immunosuppressive activities. Metabolic products of *S. mansoni* adult worms strongly inhibit lymphocyte proliferation *in vitro* and *in vivo*. This schistosome-derived inhibitory factor (SDIF) of low-molecular-weight selectively inhibits T cell proliferation [44], particularly the generation of cytotoxic T cells in mixed lymphocyte cultures, but without inhibiting IL-2 production. Purified SDIF preparations selectively inhibit the proliferation of IL-2-dependent as well as IL-2-independent T cell lines; in contrast, macrophage and B cell lines are not affected. Cytofluorographic studies have shown that SDIF has no effect on cells in the G_0 phase of the cell cycle; T cells progress through the G_1 phase, as shown by the normal production of IL-2 and the unaltered expression of IL-2 receptors [45]. By reducing or suppressing T cell responses, SDIF injected into rats transforms them into susceptible hosts in which *D. viteae* microfilariae are produced in similar numbers as in congenic nude rats, therefore giving a molecular basis to the reciprocal influences observed in mixed parasite infections [24]. The production of this suppressor of T cell function may also explain the inefficiency of cytotoxic T cells in damaging the helminth parasite. Similarly, exposure of mononuclear cells from uninfected humans to a soluble extract of the filarial worm *Onchocerca volvulus* results in the induction of suppressor T cells capable of inhibiting proliferative responses to unrelated antigens *in vitro* [46]. This was recently confirmed by showing that lymphoid cells from non-infected subjects in an *Onchocerca*-exposed population produce more IL-2 when pulsed with filarial antigen than did infected subjects [47,48]. Finally, a substance released by *Trichinella spiralis* and toxic for lymphoid

cells is thought to be responsible for lymphocyte depletion in the germinal centers observed in the infection [49].

Another series of regulatory molecules has been evidenced in the protozoan infections characterized by polyclonal hypergammaglobulinemia, contrasting with decreased specific antibody responses to parasite antigens. This has led to the investigation of parasite-derived polyclonal B cell activators, which have been found in supernatants from malarious red blood cells and in saline extracts from African trypanosomes. The release of polyclonal activators has generally been incriminated as a factor responsible for the occurrence of autoantibodies in many protozoan infections, whereas parasite-specific antibody unresponsiveness has been attributed to this mitogenic activity which, by driving polyclonal B cell expansion, might abrogate specific B cell responses [50]. In human *P. falciparum* malaria also, suggestive evidence has been brought to light showing the existence of a polyclonal T cell activator of 70 kD in schizonts, which stimulates the proliferation of CD4[+] and CD8[+] cells [51].

Parasites can evade the immune response by ever more subtle mechanisms, which allow them to reduce, modify, or disguise their antigenicity or to deviate the immune response towards nonaggressive and/or blocking pathways.

ALTERED PARASITE ANTIGENICITY OR IMMUNOGENICITY

Limited expression of surface antigens

In parallel with the successive developmental stages of parasites in their mammalian host, stage-specific antigens can be recognized. This suggests that the host is exposed to an array of antigenic entities. However, the antigenic complexity of parasites is reduced when we consider those functional antigens that are expressed at the parasite surface, and which represent the direct targets of the protective immune response. The circumsporozoite protein (CSP) of *Plasmodium*, the variant-specific glycoprotein (VSG) of African trypanosomes, and the surface glycoproteins of trematodes provide such examples of a limited repertoire of surface antigens.

Switch of antigen expression

Stage specificity appears linked to differences in the major surface antigens in the successive stages of the life-cycle of many protozoan parasites, but the structural and genetic bases of these developmentally regulated changes are poorly understood. In contrast, modifications in antigenicity of some metazoan parasites such as schistosomes are linked not to expression or absence of a given molecule but to the topography of its expression in or on the parasite. Indeed, crossreactivity can be observed between antigens that are expressed at different locations in different developmental stages. In the case of schistosomes, a shared carbohydrate moiety is found in a major surface glycoprotein of schistosomula (gp38), in a 115 kD antigen secreted by adult worms, and also in the eggs [52]. The switch from surface to internal —and secreted—expression of this polysaccharide may thus provide a molecular basis for the fundamental phenomenon of concomitant immunity. The 28 kD protein (p28) is also transiently expressed on the surface of schistosomula and is found in the metabolic products of adult worms [53].

Other parasites use an alternative escape strategy by permanently modifying their most expressed surface antigen. This antigenic variation is the hallmark of African trypanosomes, which use 10% (about 1000) of their genes to convert their surface antigenicity by duplicating a particular gene encoding the particular VSG and transposing the copy to a distant chromosomal region. This antigenic variation occurs spontaneously in culture, so that the immune response to the successive variants does not appear to trigger the induction of new variants but to serve a (negative) selective function and to reduce the probability of re-expressing early VSG genes as the infection progresses. In consequence, although a strong immune response to VSG is elicited, it is rendered totally ineffective by the development of variant parasite clones, since the surface coat appears to consist of closely packed VSG molecules. The potential for antigenic variation in an individual trypanosome appears to be only restricted by the time for which the infected host survives, and the genetic polymorphism may be increased by genetic exchange during sexual encounters in *T. brucei* [54,55]. Variation of parasite-derived antigens expressed on the surface of infected red blood cells has also been demonstrated in infection with some species of *Plasmodia*. Unlike trypanosome variant antigens, however, the schizont variant antigens constitute only a minor proportion of the total antigens produced by the parasite, and antigenic variation in malaria appears to be driven by the host antibody response [56,57]. Indeed, monkeys immunized against a conserved 140 kD merozoite surface antigen of *P. knowlesi*, upon challenge with a parasite clone expressing the same antigen, develop infections lacking the original 140 kD protein but displaying new antigens of different molecular weights [58].

Genetic polymorphism in T cell determinants from malaria CSP may also contribute to decreasing the efficiency of the immune response [59]. Again, polymorphism in the epitopes expressed by *Leishmania aethiopica* promastigotes may explain the development of diffuse vs. local cutaneous disease, since promastigotes derived from patients with diffuse cutaneous leishmaniasis were shown to stimulate suppressor cells and these determinants appear less (or not at all) expressed by promastigotes from self-healing individuals [39].

Reduced immunogenicity of major surface antigens

A remarkable finding has been the demonstration of repetitive epitopes in surface antigens of protozoan parasites. In malaria, for instance, several small oligopeptides are extensively repeated in tandem arrays in the unique surface antigen of *P. knowlesi* or *P. falciparum*. The CSP appears thus as a sheet with these immunodominant repetitive sequences exposed. The host immune system is therefore prevented from mounting a response against other (hidden) epitopes. Moreover, the repeat region alone does not elicit T cell responses in most humans from malaria endemic regions, and it is not immunogenic in mice [60]. Besides, there are extensive crossreactions amongst these repeats, and the interference they provide in the normal maturation of high-affinity antibody clones, by causing an abnormally high proportion of somatically-mutated B cells to be preserved, may contribute to the inefficiency of the antibody response in malaria [61] when, on the contrary, there is little crossreactivity between variant T cell epitopes from CSP [59]. Together with the turnover of the surface protein and genetic polymorphism of dominated epitopes, this might explain the inadequately unsuccessful attempts at vaccination with peptides from CS antigen of *Plasmodium* [62].

Mimicry

Another level of escape to strong immune response is antigenic mimicry [63,64]. Indeed, one of the simplest ideas when considering the host–parasite relation and its stability in terms of evolution, population dynamics, and individual infection is the likely existence of structures common to parasites and their hosts which would allow a precise adaptation of their respective metabolic requirements.

In the case of schistosomes, there are crossreactivities between the various developmental stages of the parasite and antigenic components of its invertebrate or vertebrate host. A parasite-derived surface molecule was shown to crossreact with host α_2-macroglobulin [65], and numerous parasite genes cloned so far show high levels of nucleotide sequence homology with mammalian genes, attributable to functional conservation since the corresponding proteins are potent immunogens [66,67]. Besides, antigenic disguise appears to be employed by schistosomes during their life in their vertebrate host. Indeed, within hours after their penetration through the skin they acquire a masking coat of host molecules—both glycolipids (e.g., A, B, and H blood group antigens) [68] and glycoproteins—and worm receptors for host molecules, such as the Fc portion of IgG, have been characterized [69]. An interesting observation is acquisition by schistosomes of Class I and Class II MHC products [69,70]. Although these acquired MHC products are recognizable on the surface of worms by alloantibodies or CD8$^+$ T cells [71,72], they do not appear to play a role in promoting parasite recognition by T cells or in immunoregulation. It is generally assumed that these host antigens limit access of immune effectors to target antigens, since their acquisition coincides with the capacity of transforming schistosomula to resist immune damage. Indeed, from this stage onward, disguised schistosomes are resistant to immune effectors, a phenomenon contributing to concomitant immunity. Sequence homology with human proteins was also demonstrated for the repeats of CS antigen of *Plasmodium* species [73], *T. cruzi* [74,75], and, very recently, *O. volvulus* [76].

In fact, the contribution of antigenic disguise or antigenic mimicry in the reduction of surface recognition of parasites is difficult to appreciate. For instance, in schistosomes, the acquisition of so-called host antigens proceeds concomitantly with intrinsic changes in membrane susceptibility to antibody-dependent killing and with the loss of expression in the worm membrane of the major target antigens that are expressed on the schistosomulum surface. Sharing of epitopes between host and parasite may accordingly appear to be related more to phyletic convergence and adaptation to common metabolic environments than to a strategy of evasion from immune clearance. However, foreign parasitic antigens may trigger an inappropriate immune response against host self-antigens through molecular mimicry, which can be involved in the autoimmunity often associated with parasitic diseases.

Deviation of immunity towards nonprotective responses

The circulating antigens excreted by parasites into the bloodstream of their mammalian host, by binding putative effector antibodies, may act as blocking factors. However, a remarkable feature of parasites is their direct interference in immunologic networks. Most parasite infections are characterized by some immunodeviation, with not only the absence of conventional cytotoxic T cell killing but also a striking selection of precise antibody classes or subclasses (isotypes) or T cell subsets.

Preferential elicitation of one subset of helper T cells

An illustration of such a selection is brought to light by the study of murine leishmaniasis, a model in which the induction of a particular CD4$^+$ T cell subset, i.e., Type 1 vs. Type 2 helper cells, governs resistance vs. susceptibility. Indeed, in this model, the protective CD4$^+$ subset mainly produced by resistant strains of mice can secrete IFN-γ upon leishmanial antigen stimulation *in vitro*, whereas the "suppressive" CD4$^+$ subset (mainly produced by susceptible BALB/c mice) produces IL-3 and IL-4 but not IFN-γ in these conditions [77]. However, BALB/c mice generate higher numbers of antigen-reactive T cells than do resistant strains of mice, with 80–90% of the specific T cells expressing the CD4$^+$ phenotype, whereas resistant mice produce both CD4$^+$ and CD8$^+$ T cells; CD8$^+$ T cells also appear to participate in the protection against *Leishmania* lesions [78]. This schematic view has, however, recently been challenged, since treatment of BALB/c mice by recombinant IL-4 not only reverses the usual course of *L. major* infection in this susceptible strain but also promotes the development of sterilizing immunity [79].

Blocking antibodies

In other parasitic diseases, a refined control of the production of precise antibody isotypes is demonstrated. In lymphatic filariasis, the main antibody complexed with circulating filarial antigen is IgM in asymptomatic patients, whereas it is mainly IgG in patients with swelling and lymphedema or elephantiasis [80]. However, levels of filarial-specific IgG4 are higher in asymptomatic individuals than in those with lymphatic obstructive pathology, and the spectrum of antigen profile is also broader in the former patients [81]. In schistosomiasis japonica, induction of specific IgG1 antibody to the eggs is responsible for suppression of granulomas [82].

Observations made in our laboratory have demonstrated that anaphylactic antibody isotypes play a prominent role in conjunction with nonlymphoid cells like macrophages, eosinophils, and platelets in antibody-dependent cell-mediated cytotoxicity (ADCC) against schistosomes [20]. In addition to this restricted range of protective antibody isotypes, other restricted antibody isotypes are involved in the escape strategy of schistosomes. Against the carbohydrate epitopes of the major surface glycoprotein of schistosomula (gp38), two antibody isotypes can indeed be produced during the experimental infection. One, IgG2a in the rat, is an anaphylactic antibody involved in protection [83]; the other, IgG2c in the rat, is a blocking antibody as shown both *in vitro* by inhibiting ADCC involving eosinophils and *in vivo*, since injection of a monoclonal IgG2c antibody inhibits the protection conferred by passive transfer of a protective IgG2a that is monoclonal-specific for the same gp38 antigen [84]. This observation led to the identification of such blocking antibodies in human mansoni schistosomiasis: the presence of antibodies to *S. mansoni* carbohydrate epitopes was shown to be highly correlated to a state of nonresistance to reinfection in human children [85], and blocking antibodies of IgM and IgG2 (sub)class against the gp38 antigen were characterized in the susceptible but not in the resistant population [85,86]. This indicates that in a chronic transmissible disease such as schistosomiasis, whereas the antibody profile of potentially protective (anaphylactic) antibodies does not differ significantly in immune and nonimmune populations, susceptibility is controlled by the development of a (nonanaphylactic) blocking antibody response, and thus it might be more important to characterize markers of susceptibility than putative indicators of protection [87]. Besides, the general concept that a given antigen might selectively be involved in the production of a defined effector isotype, and that the same molecule may elicit both effector and blocking isotypes, is obviously of considerable importance in the framework of vaccine strategy.

CONCLUSION

This survey of how parasites escape or resist the immune response they elicit in their immunocompetent hosts shows that the multiplicity of parasites is reflected by multiple adaptative mechanisms, ranging from absence of response to a critical target antigen to more subtle mechanisms of antigenic mimicry, immunodominance of nonessential target epitopes, variation in antigen structure or topographical expression, and immunodeviation towards nonprotective immune response at the expense of the putatively protective ones. This could be taken as an indication that the selective pressure of the host's immune system has prompted deviation in the biology of parasites, in particular by leading them to modify the expression of their surface antigens and the presentation of these to the immune system.

In the case of the carbohydrate of *S. mansoni* common to schistosomulum surface gp38, a putative target of immune attack, and to adult worm excretory–secretory antigen, but also found in molecules of miracidia and the intermediate host *Biomphalaria glabrata* [52,88], it is thought that it might represent an adaptive structure for osmotic regulation. Indeed, recent studies performed in plants, in a variety of microorganisms, and in marine invertebrates, have shown that membrane oligosaccharides play a major role in the mechanisms controlling osmotic regulation. It is tempting to speculate that this highly conserved glycannic structure has, during evolution, followed the adaptation of snails from deep marine to freshwater habitats. It is tempting to speculate further that schistosomes, which have at several critical stages of their life-cycle to overcome important changes in osmotic pressure, derive benefit from the expression of such glycannic structures. It is noticeable in this respect that the expression of this epitope is particularly abundant at the miracidium and cercarial stages, both of which confront the parasite with dramatic and rapid exposure to osmotic variations (110 mosmol in fresh water, 160 mosmol in snails, 340 mosmol in vertebrate hosts). Yet, from the standpoint of the parasite, the expression of such a conserved and highly functional structure is not a benefit with regard to the host immune response that it potently elicits.

In the face of this biologic constraint, a sophisticated escape strategy is mounted: rapid loss of surface expression by transforming schistosomula and induction of nonprotective blocking antibodies. Likewise, an immunodominant antigen of *S. mansoni* with 70–80% identity with heat shock protein (Hsp70) may be involved in adaptation of this parasite to temperature changes in transition between the snail and mammalian host [66]. The ingenuity of parasites can even reach so far as to utilize T cell growth factors for their own development or manipulate complement pathways to direct their penetration into host cells. It is therefore not surprizing that in contrast to a number of bacterial and viral infections, for which successful immunization of human populations has been achieved, no efficient vaccine has so far been developed against the major human parasitic diseases, such as malaria, schistosomiasis, trypanosomiasis, leishmaniasis, and filariasis. This is certainly due to the biologic complexity of the parasite life-cycle and the tight adaptation of parasites to their mammalian host, which points to the crucial importance of the selection of appropriate immunogens and thus to the analysis of the essential criteria leading to their definition.

Most attempts on the characterization of potential parasite immunogens have concerned surface protein or glycoproteins. However, the expression of essential structures at the parasite surface, which at this location would be prime targets of immune attack, would be a suicidal process which would have necessarily led, during evolution, to the disappearance of the parasite itself. The logical consequence is that either such surface compounds are not essential biologic structures or that the parasite has acquired the capacity of escaping, at their level, the deleterious effects of effector antibodies or cells. This hypothesis implies that in many cases the parasite components, which are the targets of protective immune responses, either represent minor constituents of the parasite surface or are not resident membrane components but are expressed transiently at the surface level. In fact, excretory–secretory antigens, which are commonly released by parasite in the circulation of their host, might represent more suitable candidates for immunization. Besides the protective epitope of the schistosomulum surface gp38, which is released in the host mainly by adult worms, major schistosome immunogens appear indeed to be nonsurface proteins or glycoproteins. This is the case of a schistosome glutathione *S*-transferase of *S. mansoni* (p28) [87,89] and *Schistosoma japonicum* [90], as well as of paramyosin (Sm97) [91], shown to induce protection in experimental models. There are also indications that excretory–secretory antigens might be involved in protection against protozoan parasites, such as *T. gondii* [92].

REFERENCES

1 Schofield L, Villaquiran J, Ferreira A, Schellekens H, Nussenzweig RS, Nussenzweig V. Gamma-interferon, CD8⁺ T cells and antibodies required for immunity to malaria sporozoites. Nature 1987;330:664–666.

2 Miller LH, Howard RJ, Carter R, Good MF, Nussenzweig V, Nussenzweig RS. Research toward malaria vaccines. Science 1986;234:1349–1356.

3 Alves MJ, Abuin G, Kuwajima VY, Colli W. Partial inhibition of trypomastigote entry into cultured mammalian cells by monoclonal antibodies against a surface glycoprotein of *Trypanosoma cruzi*. Mol Biochem Parasitol 1986;21:75–82.

4 Ouaissi MA, Cornette J, Afchain D, Capron A, Gras-Masse H, Tartar A. *Trypanosoma cruzi* infection inhibited by peptides modeled from a fibronectin cell attachment domain. Science 1986;234:603–607.

5 Ouaissi MA. Role of the RGD sequence in parasite adhesion to host cells. Parasitol Today 1988;4:169–173.

6 Rizvi FS, Ouaissi MA, Marty B, Santoro F, Capron A. The major surface protein of *Leishmania* promastigotes is a fibronectin-like molecule. Eur J Immunol 1988;18:473–476.

7 Joiner K, Hieny S, Kirchhoff LV, Sher A. gp72, the 72 kilodalton glycoprotein, is the membrane acceptor site for C3 on *Trypanosoma cruzi* epimastigotes. J Exp Med 1985;161:1196–1212.

8 Handman E, Goding JW. The *Leishmania* receptor for macrophages is a lipid containing glycoconjugate. EMBO J 1985;4:329–336.

9 Russell DG, Wilhelm H. The involvement of the major surface glycoprotein (gp63) of *Leishmania* promastigotes in attachment of macrophages. J Immunol 1986;136:2613–2620.

10 Russell DG, Wright SD. Complement receptor type 3 (CR3) binds to an Arg–Gly–Asp-containing region of the major surface glycoprotein, gp63, of *Leishmania* promastigotes. J Exp Med 1988;168:279–292.

11 Klempner MS, Cendron M, Wyler DJ. Attachment of plasma membrane vesicles of human macrophages to *Leishmania tropica* promastigotes. J Infect Dis 1983;148:377–384.

12 Mosser DM, Vlassara H, Edelson PJ, Cerami A. *Leishmania* promastigotes are recognized by the macrophage receptor for advanced glycosylation end products. J Exp Med 1987;165:140–145.

13 Rimoldi MT, Sher A, Hieny S, Lituchy A, Hammer CH, Joiner K. Developmentally regulated expression by *Trypanosoma cruzi* of molecules that accelerate the decay of complement C3 convertases. Proc Natl Acad Sci USA 1988;85:193–197.

14 Sher A, Hieny S, Joiner K. Evasion of the alternative complement pathway by metacyclic trypomastigotes of *Trypanosoma cruzi*: dependence on the developmentally regulated synthesis of surface protein and *N*-linked carbohydrate. J Immunol 1986;137:2961–2967.

15 Mosser DM, Edelson PJ. The third component of complement (C3) is responsible for the intracellular survival of *Leishmania major*. Nature 1987;327:329–331.

16 Da Silva R, Hall BF, Joiner KA, Sacks DL. CR1, the C3b

receptor mediates binding of infective *L. major* metacyclic promastigotes to human macrophages. J Immunol 1989;143:617–622.

17 Mukkada AJ, Meade JC, Glaser TA, Bonventre PF. Enhanced metabolism of *Leishmania donovani* amastigotes at acid pH: an adaptation for intracellular growth. Science 1985;229:1099–1101.

18 Tait A, Sacks DL. The cell biology of parasite invasion and survival. Parasitol Today 1988;4:228–234.

19 Marikovsky M, Arnon R, Fishelson Z. Proteases secreted by transforming schistosomula of *Schistosoma mansoni* promote resistance to killing by complement. J Immunol 1988;141:273–278.

20 Capron A, Dessaint JP. Effector and regulatory mechanisms in immunity to schistosomes: a heuristic view. Ann Rev Immunol 1985;3:455–476.

21 Auriault C, Joseph M, Tartar A, Capron A. Characterization and synthesis of a macrophage inhibitory peptide from the second constant domain of IgG. FEBS Letts 1983;153:11–15.

22 Dessaint JP, Capron A. Immunodeficiencies in parasitic diseases. Immunodeficiency Rev 1989;1:311–324.

23 Cox FEG. Enhanced *Trypanosoma musculi* infections in mice with concomitant malaria. Nature 1975;258:148–149.

24 Haque A, Camus D, Ogilvie BM, Capron M, Bazin H, Capron A. *Dipetalonema viteae* infective larvae reach reproductive maturity in rats immunodepressed by prior exposure to *Schistosoma mansoni* or its products and in congenitally athymic rats. Clin Exp Immunol 1981;43:1–9.

25 Murphy JR. Host defenses in murine malaria: analysis of plasmodial infection-caused defects in macrophage microbicidal capacities. Infect Immun 1981;31:396–407.

26 Sileghem M, Darji A, Remels L, Hamers R, de Baetselier P. Different mechanisms account for the suppression of interleukin-2 production and the suppression of interleukin 2 receptor expression in *Trypanosoma brucei*-infected mice. Eur J Immunol 1989;19:119–124.

27 Harel-Bellan A, Joskowicz M, Fradelizi D, Eisen H. T lymphocyte function during experimental Chagas' disease: production of and response to interleukin 2. Eur J Immunol 1985;15:438–442.

28 Reed SG, Pihl DL, Grabstein H. Immune deficiency in chronic *Trypanosoma cruzi* infection. Recombinant IL-1 restores Th function for antibody production. J Immunol 1989;142:2067–2071.

29 Tarleton RL. *Trypanosoma cruzi* induced suppression of IL-2 production. II. Evidence for a role for suppressor cells. J Immunol 1988;140:2769–2773.

30 Rocklin RE, Tracy JW, El Kholy A. Activation of antigen-specific suppressor cells in human Schistosomiasis *mansoni* by fractions of soluble egg antigen non-adherent to Con A-Sepharose. J Immunol 1981;127:2314–2318.

31 Lima MS, Gazzinelli G, Nascimento E, Carvalho-Parra J, Montesano MA, Colley DG. Immune responses during human Schistosomiasis *mansoni*. Evidence for antiidiotypic T lymphocyte unresponsiveness. J Clin Invest 1986;78:983–988.

32 Montesano MA, Lima MS, Correa-Oliveira R, Gazzinelli G, Colley DG. Immune responses during human schistosomiasis mansoni. XVI. Idiotypic differences in antibody preparations from patients with different clinical forms of infection. J Immunol 1989;142:2501–2506.

33 Camus D, Nosseir A, Mazingue C, Capron A. Immunoregulation by *Schistosoma mansoni*. Immunopharmacology 1981;3:193–204.

34 Ellner JJ, Olds GR, Kamel R, Osman GS, El Kholy A, Mahmoud AAF. Suppressor splenic T lymphocytes in human hepatosplenic schistosomiasis mansoni. J Immunol 1980;125:309–312.

35 Louis J, Milon G. Immunobiology of experimental leishmaniasis (20th Forum in Immunology), Ann Inst Pasteur/Immunol 1987;138:737–795.

36 Liew FY, Millott S, Lelchuck R, Cobbold S, Waldmann H. Effect of CD4 monoclonal antibody *in vivo* on lesion development, delayed-type hypersensitivity and interleukin-3 production in experimental murine cutaneous leishmaniasis. Clin Exp Immunol 1989;75:438–443.

37 Reiner NE, Ng W, Ma T, McMaster WR. Kinetics of gamma interferon binding and induction of major histocompatibility complex class II mRNA in *Leishmania*-infected macrophages. Proc Natl Acad Sci USA 1988;85:4330–4334.

38 Ackerman SB, Seed JR. The effects of tryptophol on immune responses and its implication toward trypanosome-induced immunosuppression. Experientia 1976;32:645–650.

39 Akuffo HO, Fehniger TE, Britton S. Differential recognition of *Leishmania aethiopica* antigens by lymphocytes from patients with local and diffuse cutaneous leishmaniasis. Evidence for antigen-induced immune suppression. J Immunol 1988;141:2461–2466.

40 Kierszenbaum F, Sztein MB, Beltz LA. Decreased human IL-2 receptor expression due to a protozoan pathogen. Immunol Today 1989;10:129–131.

41 Mazingue C, Cottrez-Detoeuf F, Louis J, Kweider M, Auriault C, Capron A. *In vitro* and *in vivo* effects of interleukin 2 on the protozoan parasite *Leishmania*. Eur J Immunol 1989;19:487–491.

42 Titus RG, Lima GC, Engers HD, Louis JA. Exacerbation of murine cutaneous leishmaniasis by adoptive transfer of parasite-specific helper T cell populations capable of mediating *Leishmania major*-specific delayed-type hypersensitivity. J Immunol 1984;133:1594–1600.

43 Velge P, Ouaissi MA, Kusnierz JP, Capron A. Infection des lymphocytes T par *Trypanosoma cruzi*. Rôle possible du CD3 et du HLA-DR. CR Acad Sci Paris 1989;309:93–99.

44 Dessaint JP, Camus D, Fischer E, Capron A. Inhibition of lymphocyte proliferation by factor(s) produced by *Schistosoma mansoni*. Eur J Immunol 1977;7:624–629.

45 Mazingue C, Walker C, Domzig W, Capron A, De Weck A, Stadler BM. Effect of schistosome-derived inhibitory factor on the cell cycle of T lymphocytes. Int Arch Allergy Appl Immunol 1987;83:12–18.

46 Ouaissi MA, Dessaint JP, Cornette J, Desmoutis I, Capron A. Induction of non-specific human suppressor cells *in vitro* by defined *Onchocerca volvulus* antigens. Clin Exp Immunol 1983;53:634–644.

47 Ward DJ, Nutman TB, Zea-Flores G, Portocarrero C, Lujans A, Ottesen EA. Onchocerciasis and immunity in humans: enhanced T cell responsiveness to parasite antigen in putatively immune individuals. J Infect Dis 1988;157:536–543.

48 Gallin M, Edmonds K, Ellner JJ, *et al*. Cell-mediated

immune responses in human infection with *Onchocerca volvulus*. J Immunol 1988;140:1999–2007.

49 Faubert GM, Tanner CE. The suppression of sheep rosette forming cells and the inability of mouse bone marrow cells to reconstitute competence after infection with the nematode *Trichinella spiralis*. Immunology 1974;27:501–505.

50 Greenwood BM. Possible role of a B cell mitogen in hypergammaglobulinaemia in malaria and trypanosomiasis. Lancet 1974;i:435.

51 Jaureguiberry G, Ogunkolade W, Bailly E, Rhodes-Feuillette A, Agrapart M, Ballet JJ. *Plasmodium falciparum* exoprotein stimulation of human T lymphocytes unsensitized to malaria. J Chromatogr 1988;440:385–396.

52 Dissous C, Capron A. *Schistosoma mansoni* and its intermediate host *Biomphalaria glabrata* express a common 39 kilodalton acidic protein. Mol Biochem Parasitol 1989;32:49–56.

53 Balloul JM, Grzych JM, Pierce RJ, Capron A. A purified 28 000 dalton protein from *Schistosoma mansoni* adult worms protects rats and mice against experimental schistosomiasis. J Immunol 1987;138:3448–3453.

54 Cross GAM. Cellular and genetic aspects of antigenic variation in trypanosomes. Ann Rev Immunol 1990;8:83–110.

55 Donelson JE, Rice-Ficht AC. Molecular biology of trypanosome antigenic variation. Microbiol Rev 1985;49:107–125.

56 Brown KN, Hills HI. Antigenic variation and immunity to *Plasmodium knowlesi*: antibodies which induce antigenic variation and antibodies which destroy parasites. Trans R Soc Trop Med Hyg 1974;68:139–148.

57 Howard RJ, Barnwell JW, Kao V. Antigenic variation in *Plasmodium knowlesi* malaria: identification of the variant antigen on infected erythrocytes. Proc Natl Acad Sci USA 1983;80:4129–4137.

58 Klotz FW, Hudson DE, Coon HG, Miller LH. Vaccination-induced variation in the 140 kd merozoite surface antigen of *Plasmodium knowlesi* malaria. J Exp Med 1987;165:359–367.

59 De la Cruz VF, Maloy WL, Miller LH, Lal AA, Good MF, McCutchan TF. Lack of cross-reactivity between variant T cell determinants from malaria circumsporozoite protein. J Immunol 1988;141:2456–2460.

60 Good MF, Berzofsky JA, Miller LH. The T cell response to the malaria circumsporozoite protein: an immunologic approach to vaccine development. Ann Rev Immunol 1988;6:663–688.

61 Anders RF. Multiple cross-reactivities amongst antigens of *Plasmodium falciparum* impair the development of protective immunity against malaria. Parasite Immunol 1986;8:529–539.

62 Egan JE, Weber JL, Ballou WR, *et al.* Efficacy of murine malaria sporozoite vaccines: implications for human vaccine development. Science 1987;236:453–456.

63 Damian RT. Molecular mimicry revisited. Parasitol Today 1987;3:263–266.

64 Capron A, Biguet J, Rose F, Vernes A. Les antigènes de *Schistosoma mansoni*. II. Etude immunoélectrophorétique comparée de divers stades larvaires et des adultes des deux sexes. Aspects immunologiques des relations hôte–parasite de la cercaire et de l'adulte de *S. mansoni*. Ann Inst Pasteur Paris 1965;109:798–810.

65 Damian RT, Greene ND, Hubbard WJ. Occurrence of mouse alpha2-macroglobulin antigenic determinants on *S. mansoni* adults with evidence of their nature. J Parasitol 1973;59:64–75.

66 Hedstrom R, Culpepper J, Harrison RA, Agabian H, Newport G. A major immunogen in *Schistosoma mansoni* infections is homologous to the heat-shock protein Hsp70. J Exp Med 1987;165:1430–1435.

67 Newport GR, Harrison RA, McKerrow J, Tarr P, Kallestad J, Agabian N. Molecular cloning of *Schistosoma mansoni* myosin. Mol Biochem Parasitol 1987;26:29–38.

68 Smithers SR, Terry RJ, Hockley DH. Host antigens in schistosomiasis. Proc R Soc London (Biol) 1969;171:483–490.

69 Torpier G, Capron A, Ouaissi MA. Receptor for IgG (Fc) and human β_2-microglobulin on *S. mansoni* schistosomula. Nature 1979;278:447–449.

70 Hedstrom R, Culpepper J, Harrison RA, Agabian H, Newport G. A major immunogen in *Schistosoma mansoni* infections is homologous to the heat-shock protein Hsp70. J Exp Med 1987;165:1430–1435.

71 Sher A, Hall BF, Vadas MA. Acquisition of murine major histocompatibility gene products by schistosomula of *Schistosoma mansoni*. J Exp Med 1978;148:46–52.

72 Butterworth AE, Vadas MA, Martz E, Sher A. Cytolytic T lymphocytes recognize alloantigens on schistosomula of *Schistosoma mansoni*, but fail to induce damage. J Immunol 1979;122:1314–1321.

73 McLaughlin GL, Benedik MJ, Campbell GH. Repeated immunogenic aminoacid sequences of *Plasmodium* species share sequence homologies with proteins from human viruses. Am J Trop Med Hyg 1987;37:258–262.

74 Szarfman A, Terranova VP, Rennard SI. Antibody to laminin in Chagas' disease. J Exp Med 1982;155:1161–1171.

75 Ouaissi MA, Cornette J, Velge P, Capron A. Identification of anti-acetyl-cholinesterase and anti-idiotype antibodies in human and experimental Chagas' disease: pathological implications. Eur J Immunol 1988;18:1889–1894.

76 McCauliffe DP, Lux FA, Lieu Tsu-San, *et al.* Molecular cloning, expression, and cChromosome 19 localization of a human Ro/SS-A autoantigen. J Clin Invest 1990;85:1379–1391.

77 Liew FY, Hodson K, Leichuk R. Prophylactic immunization against experimental leishmaniasis. VI. Comparison of protective and disease-promoting T cells. J Immunol 1987;139:3112–3117.

78 Farell JP, Muller I, Louis JA. A role for Lyt-2$^+$ T cells in resistance to cutaneous leishmaniasis in immunized mice. J Immunol 1989;142:2052–2056.

79 Carter KC, Gallagher G, Baillie AJ, Alexander J. The induction of protective immunity to *Leishmania major* in the Balb/c mouse by interleukin 4 treatment. Eur J Immunol 1989;19:779–782.

80 Lutsch C, Cesbron JY, Henry D, *et al.* Lymphatic filariasis: detection of circulating and urinary antigen and differences in antibody isotypes complexed with circulating antigen between symptomatic and asymptomatic subjects. Clin Exp Immunol 1988;71:253–260.

81 Hussain R, Grögl M, Ottesen EA. IgG antibody subclasses

in human filariasis: differential subclass recognition of parasite antigens correlates with different clinical manifestations of infection. J Immunol 1987;139:2794–2798.

82 Stavitsky AB. Immune regulation in schistosomiasis japonica. Immunol Today 1987;8:228–232.

83 Grzych JM, Capron M, Bazin H, Capron A. *In vitro* and *in vivo* effector function of rat IgG$_{2a}$ monoclonal anti-*S. mansoni* antibodies. J Immunol 1982;129:2739–2743.

84 Grzych JM, Capron M, Dissous C, Capron A. Blocking activity of rat monoclonal antibodies in experimental schistosomiasis. J Immunol 1984;133:998–1004.

85 Butterworth AE, Bensted-Smith R, Capron A, *et al.* Immunity in human schistosomiasis mansoni: prevention by blocking antibodies of the expression of immunity in young children. Parasitology 1987;94:281–300.

86 Khalife J, Capron M, Capron A, *et al.* Immunity in human Schistosomiasis *mansoni*. Regulation of protective immune mechanisms by IgM blocking antibodies. J Exp Med 1986;164:1626–1640.

87 Capron A, Dessaint JP, Capron M, Ouma JH, Butterworth AE. Immunity to schistosomes: progress toward vaccine. Science 1987;238:1065–1072.

88 Dissous C, Grzych JM, Capron A. *Schistosoma mansoni* shares a protective oligosaccharide epitope with fresh water and marine snails. Nature 1986;323:443–448.

89 Taylor JB, Vidal A, Torpier G, *et al.* The glutathione transferase activity and tissue distribution of a cloned Mr 28 K protective antigen of *Schistosoma mansoni*. EMBO J 1988;7:465–472.

90 Smith DB, Davern KM, Board PG, Tiu Wu, Garcia EG, Mitchell GF. Mr 26 000 antigen of *Schistosoma japonicum* recognized by resistant WEHI 129/J mice is a parasite glutathione S-transferase. Proc Natl Acad Sci USA 1986;83:8703–8707.

91 Pearce EJ, James SL, Hieny S, Lanar DE, Sher A. Induction of protective immunity against *Schistosoma mansoni* by vaccination with schistosome paramyosin (Sm97) a non-surface parasite antigen. Proc Natl Acad Sci USA 1988;85:5678–5682.

92 Darcy F, Deslee D, Santoro F, *et al.* Induction of a protective antibody-dependent response against toxoplasmosis by *in vitro* excreted/secreted antigens from tachyzoites of *Toxoplasma gondii*. Parasite Immunol 1988;10:553–567.

6 Parasites in the immunocompromised host

Peter Godfrey-Faussett, Stephen G. Wright, Vincent McDonald, Jaime Nina, Peter L. Chiodini, & Keith P.W.J. McAdam

INTRODUCTION

The mechanisms for the elimination of parasites from the normal host are a complex interaction that involve immunoglobulins of all classes and the full repertoire of the cellular immune system. Many parasites are able to complete their life-cycle despite the defences of the immune system and, in doing so, some cause disease while in others there is no significant pathology.

Following infection, "immunity" (in its most general usage) develops and in the absence of therapeutic intervention three possible outcomes may be observed. The parasite may be eliminated; it may be held at some equilibrium level, or it may overwhelm its host. The importance of the various constituents of the immune system and of factors intrinsic to the parasite in determining this outcome is the subject of much study. One approach is to study the effect on the parasite of suppressing the immune system of the host. In laboratory animals it is possible to create specific deficits, whereas in humans it is necessary to observe patients with immune deficits that are either congenital or acquired. In these immunocompromised hosts, differences are seen both in the relative frequency with which particular parasites are found and in the range of pathology they cause. With the advent of the acquired immunodeficiency syndrome (AIDS) epidemic, more attention has been focused on the opportunistic infections found in the syndrome. It has become clear that some of these diseases arise from reactivation of pathogens that had been inactive, presumably held in check by the immune system, while others arise from infection by organisms that are not generally pathogenic to the normal host. An understanding of why parasites that are closely related, such as *Cryptosporidia* and malaria, should be handled in such a different way will enhance our appreciation both of the normal immune response to these organisms and of the nature of the deficit observed in AIDS and other immunodeficient syndromes.

In this chapter we have focused on the protozoa, as these are the parasites in which the outcome of infection depends most critically on the state of the immune system of the host and whose presentation and management has changed as a result of the AIDS epidemic. Most helminths do not multiply in the definitive host so that although elimination of the parasite may be impaired in the immunocompromised, overwhelming infection is not usually seen. Indeed, in cases where pathology is mediated by the immune response, e.g., the delayed hypersensitivity granuloma seen in schistosomiasis, immunodeficiency might actually ameliorate the disease. The most important exception to this pattern is *Strongyloides stercoralis*, which is able to complete its life-cycle within one host and in which hyperinfection is well recognized in patients receiving organ transplants or undergoing immunosuppressive chemotherapy for hematologic malignances [1], although the diagnosis may be delayed if appropriate parasitologic tests are not performed [2].

Although specific cellular and humoral responses to filariform larvae develop in both patients and experimental animals, they appear to have little role in controlling autoinfection [3]. There have been several case reports of hyperinfection in patients with human immunodeficiency virus (HIV) infection [4–6] but, considering the prevalence of both infections in sub-Saharan Africa, the incidence is very low [7] and may represent coincidence—up to 15% of patients with severe strongyloidiasis have no demonstrable defect in cell-mediated immunity [1]. Patients with *S. stercoralis* infection are infected with HTLV-1 more commonly than local controls [8] and there is some evidence that patients with both infections have more severe strongyloidiasis, even in those patients without T cell leukemia/lymphoma [9]. The mechanisms for preventing hyperinfection, therefore, remain poorly understood but may be more related to local intestinal factors than to systemic immunity [10].

The infections seen in patients with HIV infection depend on the organisms to which the host has been exposed. The relative frequencies of AIDS-defining illnesses in patients reported to the Center for Disease Control (CDC) are shown by region of birth in Table 1 of Kreiss & Castro [11]. We will consider the most frequent protozoal pathogens (*Pneumocystis carinii*, despite the debate over its taxonomy as detailed below; *Toxoplasma gondii*; *Cryptosporidium* spp.; *Isospora* spp.) and then consider others that, whilst less common, are being increasingly recognized in immunodeficient hosts (*Microsporidia* spp.; *Leishmania* spp.). Finally, we will discuss organisms that seem not to cause problems in patients with HIV infection but are relevant to other immunodeficient states, such as pregnancy, corticosteroid therapy, and hypogammaglobulinemia (*Plasmodium falciparum*, *Entamoeba histolytica*, and *Giardia lamblia*).

PNEUMOCYSTIS CARINII

Introduction

The history of *P. carinii* infection is inextricably linked with immunodeficiency. Although it was described in the lungs of humans and guinea pigs in 1909, it was not recognized as a cause of disease for 40 years. Interstitial plasma cell pneumonitis occurred in epidemics among premature and malnourished infants in European orphanages in the 1930s but *P. carinii* was first implicated in 1951. Thereafter it was also seen as the cause of pneumonia in adults and children undergoing treatment with corticosteroids, which came into general usage in the 1950s, and with the new immunosuppressive anti-cancer drugs. Over the next 30 years, *P. carinii* pneumonia continued to be seen in patients whose immune system was deficient either congenitally or due to hematologic malignancies or drugs. So, in 1981, when clusters of cases of *P. carinii* pneumonia began to appear in the USA in fit young men, it was quickly realized that this was a new acquired immunodeficiency syndrome.

Despite its 80-year history and the renewed research interest engendered by the scale of the AIDS epidemic, remarkably little is known of the biology of the organism. The mode of acquisition and the infective stage of the organism remains uncertain. If there are indeed small numbers lurking in most people's lungs, being held in

check by the integrity of the immune system but never being eliminated, it is surprising that a technique as sensitive as the polymerase chain reaction (PCR) fails to detect DNA sequences from *P. carinii* in bronchoalveolar lavage from patients who do not have active disease [12]. Pneumonia and its recurrences may yet be shown to be due to infection and reinfection, which may make it necessary to isolate patients with cysts or trophozoites in their sputum. *Pneumocystis carinii* pneumonia is seen less frequently in patients with AIDS in African studies [13,14] but it is not known whether this is due to differences in host immunity or variation in the parasite. Does the patient with AIDS in Africa succumb to other more virulent infections before his or her defences are low enough to allow *P. carinii* to flourish or are there fewer infective stages of the parasite in the environment? Are there different strains of human *Pneumocystis* with differing virulence? If current dogma changes and reactivation is seen as less important, the role of different antigens in provoking useful humoral and cell-mediated defences will need to be studied further to design an effective vaccine for populations at risk. Until 1988 it was generally regarded as one of the protozoa but nucleotide sequences of RNA are phylogenetically closer to the fungi [15]. Advances in molecular biology have led to genes being sequenced [15] and chromosomes identified and mapped [16]. A greater understanding of the basic biology of the organism should result in more rational approaches to the treatment and prevention of disease.

Animal models

A major obstacle in the study of the organism has been the difficulty in its culture. A variety of cell-free media have been used unsuccessfully and in cell culture systems results are variable, with the best techniques producing a 10-fold growth over 1–3 weeks before the number of organisms reaches a plateau or declines [17].

Most studies of *P. carinii* have therefore relied on animal models. Since its original description, it has been recognized that the organism may be found sporadically in the lungs of a variety of mammals. In 1955, while trying to infect rats with a homogenate of lung from an infant who had died from *P. carinii* pneumonia, Weller discovered that both the control group and those inoculated developed *P. carinii* pneumonia if they had been immunosuppressed with corticosteroids. These observations were confirmed and form the basis of animal models in current use. Other animals have also been shown to develop *P. carinii* pneumonia when immunosuppressed, including rabbits, ferrets, and mice. The intensity of the infection and the duration of immunosuppression required vary from species to species and from strain to strain. Although it is not possible to distinguish *P. carinii* from different animal models by light or electron microscopy, they do not have identical antigenic determinants [18] so caution is required in extrapolating to human disease.

Although most rats and many other animals develop *P. carinii* when immunosuppressed, neither the infective stage nor the mode of acquisition of the organism is certain. It is not possible to infect germ-free rats by feeding them on homogenates of infected lung or by spraying finely minced lung through the environment [19].

In the rat model, infection is acquired via an airborne route either from the environment or from other infected rats. The organism then survives at a level that is not detectable by conventional techniques until the rat is immunosuppressed, when it flourishes and causes disease.

Whether *P. carinii* pneumonia is the result of acute infection or a resurgence of organisms that have survived for a prolonged latent period, it is likely that both cell-mediated and humoral defences are essential for the prevention of disease. Dietary protein deprivation leads to *P. carinii* pneumonia in rats and the course of the disease can be reversed by returning the animals to a normal diet [20]. Cutting the amount of protein in the feed from 23% to 8% resulted in 10% of rats developing pneumonia, while a protein-free diet led to an incidence of 88%.

Humoral and cell-mediated immunity

The importance of T lymphocytes in experimental models has been studied in athymic nude mice (the only model that does not require corticosteroids to produce a reliable persistent infection) [21]. Transfer of splenic T cells was effective in reducing the number of *P. carinii* in lung tissue and caused the disease to regress. In euthymic mice, intranasal inoculation results in a transient infection. *Pneumocystis carinii* can be isolated from the lungs for up to 4 weeks and histology shows an inflammatory response during the first 2 weeks, following which the infection is cleared and all changes resolve. The delayed-type hypersensitivity to a *P. carinii* antigen injected into a footpad also peaks at 2 weeks and this hypersensitivity can also be transferred with splenic T cells [22].

Humoral defences probably play a smaller part. Most animal studies have shown little or no protection from *P. carinii* pneumonia following transfer of immune serum. However, IgG and IgM antibodies are found in serum following infection in rats, and IgA rises in bronchial lavage [23]. Organisms retrieved at lavage are coated with antibodies of all three classes. Whether these immunoglobulins are useful in clearing the infection is not known. Previous immunization with *P. carinii* does not prevent disease when rats are treated with steroids. However, in one study, immunoprophylaxis with a monoclonal antibody raised against an epitope shared by *P. carinii* from various different species reduced the number of *P. carinii* cysts that could be detected in ferrets and rats following steroid treatment [24].

Human immunodeficiency states

The histopathologic features of the infection in humans are similar to those found in rats and the rat has become the principal animal model for human disease. It is generally accepted that humans also acquire the infection at a young age, a hypothesis supported by those serologic surveys in the USA and Europe that show the majority of the population to have antibodies against *P. carinii* by the age of 4 years [25,26]. The organism is then believed to survive in low numbers until the host becomes immunosuppressed, when it multiplies and causes pathology.

In contrast, postmortem series have only found *P. carinii* in a very small proportion of humans with no other evidence of immune suppression [27,28] and not all seroepidemiologic studies have found the same high prevalence in the general population [29,30]. There have also been clusters of cases in oncology units [31,32] and one report of a family in which person-to-person spread could have been implicated [33].

As in the rat model, nutrition also plays an important role in human disease. Most of the orphans in the first epidemics of *P. carinii* pneumonia were below the third centile for weight and height [34]. In South Africa, 8% of children dying with protein-calorie malnutrition were found to have *P. carinii* in their lungs, whereas none was found in a well-nourished control group. A comparison between leukemic patients with pneumonitis due either to *P. carinii* or to other causes showed that the *P. carinii* group weighed less and had lower serum albumin levels, although there was considerable overlap [20].

Humoral and cell-mediated immunity

Malnutrition probably has its major effect on cell-mediated immunity. The South African children had poor skin test reactivity to injected antigens and decreased lymphocyte transformation on stimulation with phytohemagglutinin. Other malnourished patients with *P. carinii* pneumonia have had decreased numbers of T lymphocytes and these have returned to normal with refeeding [35].

While it appears that cell-mediated immunity is the mainstay of small mammals' mechanism for preventing resurgence or reinfection with *P. carinii*, humoral defences also play a part. The situation in humans may well be analogous. In congenital immunodeficiency syndromes, *P. carinii* pneumonia is seen in both T cell alpasia and in agammaglobulinemia, although it occurs most commonly in severe combined immunodeficiency.

Clinical features

Diagnosis. Because many people have antibodies from a young age and antibody production in immuno-suppressed hosts may be ineffective, serology has little part to play in the diagnosis of *P. carinii* pneumonia. Current techniques depend upon the demonstration of organisms in the exudate from the alveolar spaces. This can be sampled most directly by bronchoscopy—either rigid or fiberoptic bronchoscopy—or by open lung biopsy. Fiberoptic bronchoscopy is the procedure most commonly performed with either transbronchial biopsy or bronchoalveolar lavage. Biopsy has the advantage of giving histologic information, particularly when the diagnosis is not *P. carinii* pneumonia, but the complication rate is appreciable and has been shown to be higher in patients with AIDS than other patients. Bronchoalveolar lavage has a higher yield for *P. carinii* but may cause hypoxia during the procedure [36]. Induction of sputum from the alveoli by inhaling a hypertonic solution of saline [37,38] has advantages in cost and safety (both for the patient and the physician), although attention to technique is important if an adequate specimen is to be achieved [39].

The alveolar exudate achieved by any of these procedures may be concentrated by a variety of methods that may increase the diagnostic yield [40], and the trophozoites and cysts are visualized using a variety of stains. More recently, monoclonal antibodies have been developed against some *P. carinii* antigens both from rats [41] and humans [42,43], and these may be used in immunofluorescent tests to improve the sensitivity, specificity, and speed of diagnosis.

Recombinant DNA clones carrying part of the genome of *P. carinii* have been produced and used to detect the organism by hybridization either with tissue sections or with alveolar secretions from patients with pneumonia [44,45]. The sensitivity of these probes can be greatly enhanced by amplifying specific sequences of DNA using the PCR [12].

Treatment and prophylaxis. If the diagnosis is made promptly, treatment with parenteral trimethoprim-sulfamethoxazole in high doses is usually effective [46]. Adverse reactions are more common in patients with AIDS and may necessitate a change of therapy, most commonly to pentamidine given as an intravenous infusion. The use of nebulized pentamidine is attractive since, by using a device that generates an aerosol of suitable size and density, it is possible to achieve good pulmonary deposition with minimal serum drug levels [47,48]. Larger randomized trials will establish whether such treatment can be considered as the first approach in patients with pneumonia.

In patients who deteriorate despite trimethoprim-sulfamethoxazole (Septrin or Bactrim) or pentamidine, salvage therapy with inhibitors of either dihydrofolate reductase, such as trimetrexate [49], or ornithine de-carboxylase, such as eflornithine [50], have been shown to be useful. In patients with severe pneumonitis, there is probably also a role for corticosteroids, although the correct dosage regimen has not been established [51,52].

In the absence of effective immunity, clearance of *P. carinii* is incomplete and relapses of pneumonia are common. Hughes [53] showed in 1974 that trimethoprim-sulfamethoxazole could be given as prophylaxis to prevent the development of pneumonitis in rats treated with steroids. In 1977 he showed that recurrence of pneumonitis could also be prevented in children with malignancies who had recovered from a first episode [54] and subsequently many centers confirmed that prophylaxis in high-risk patients is highly effective.

In patients with HIV infection, the risk of developing *P. carinii* pneumonia is related to the degree of immuno-suppression and correlates well with the number of CD4-positive lymphocytes present [55]. Recurrence of pneumonia is also common so that both primary (for patients with less than 200 CD4-positive lymphocytes per cubic milliliter) and secondary prophylaxis has been recommended [56]. A variety of drugs and regimens have been tried, including trimethoprim-sulfamethoxazole [57], dapsone [58], pyrimethamine–sulfadoxine [59], and pentamidine either parenterally [60] or by inhalation [61]. Most studies have been open and controls have been historical. Nebulized pentamidine is attractive because of its low incidence of serious toxicity, and a larger open trial of different dosage regimens is under way in San Francisco. The need for proper assessment and supervision of patients receiving prophylaxis in this way is illustrated by a recent outbreak of tuberculosis among patients and staff in a Florida clinic [62]. There have also been suggestions that the presentation of *P. carinii* pneumonia and the diagnostic yield of various tests may be altered, which might result in delays in starting treatment [63]. Studies are also under way comparing pentamidine with trimethoprim-sulfamethoxazole for prophylaxis.

TOXOPLASMA GONDII

Introduction

Despite the fact that defective immune responses have been recognized with increasing frequency in congential, acquired, and iatrogenic conditions over the past few decades, new infection or recrudescence of latent infection due to toxoplasmosis has been reported (reviewed by Luft & Remington [64]) but is a relatively uncommon event. The exception to this has been disseminated infection in the setting of cardiac transplantation [65,66], where the transplanted heart can be the source of infection. Four out of seven seronegative recipients of hearts from seropositive donors developed clinically apparent disease, whereas reactivation of latent infection due to iatrogenic immunosuppression in seropositive recipients was much less frequent (one out of 39 patients) [67] and appeared to be a consequence of increased doses of immunosuppressive drugs to counter rejection episodes.

In contrast, toxoplasmosis of the central nervous system (CNS) and, less often, disseminated toxoplasmosis are frequent opportunistic infections in AIDS due to HIV [68,69]. This is well illustrated by the experience of Remington, whose reference center received specimens and samples from six cases with CNS or disseminated toxoplasmosis over the previous 6-year period and material from nearly 100 cases from July 1981 through to December 1983 [68].

Toxoplasmosis ranks as the most common nonviral CNS infection in AIDS [64,69] and a very common cause of death [70,71]. Its frequency in AIDS patients is in

proportion to the frequency of *Toxoplasma* seropositivity in the general population in a given area. Most of the following discussion relates to toxoplasmosis in AIDS but reference is also made to other immune deficiency states.

Animal models

Animals that have undergone mutations causing severe defects in cell-mediated immune responses are most valuable in examining toxoplasmosis in the setting of defective immunity. The nude mouse and the mouse exhibiting severe combined immunodeficiency (SCID) are two well-recognized examples (reviewed by Shultz & Sidman [72]). The former has been used for studies of toxoplasmosis [73]. The latter is a relatively recent innovation and studies of this infection are awaited. Use of the SCID mouse presents an opportunity to study infections in the immunodeficient animal reconstituted with those components of the human immune system required by the investigator (SCID-hu) [74] and infected with HIV [75] if necessary.

In the past, surgical removal of the thymus or administration of drugs such as cytotoxics or steroids has been used to impair immune responses but it is now possible to deplete a specific population of lymphocytes using monoclonal antibodies directed to specific antigenic determinants of that population. Vollmer *et al.* [76] depleted T4+ lymphocytes in mice and found that toxoplasmic encephalitis occurred in chronically infected mice, while disseminated infection with relatively little cerebral involvement occurred in acute infections. The study of chronic infection showed that cerebral disease occurred despite high levels of circulating anti-*Toxoplasma* IgG. This provides murine infection akin to the two possible patterns of disease seen in humans with AIDS, representing recrudescence of latent infection on the one hand and newly acquired infection on the other. This experimental approach may prove to be useful for study of chemotherapy as well as pathogenesis.

Humoral and cell-mediated immunity

The role of the various arms of the immune response in toxoplasmosis has been examined in experimental studies with additional inferential studies in humans. In thinking about toxoplasmosis it must be remembered that tachyzoites actively invade cells of any tissue in the host and are not solely parasites of the phagocytic cells of the reticuloendothelial system, so that control of parasite populations must have its effects on either parasites within the intact cell [77] or parasites released from cells.

There are well-recognized differences in the outcome of this infection among different strains of mice. Immune responsiveness and immune reactivity of resistant (e.g., AJ) and susceptible (e.g., C57BL/6) strains of mice were examined in a detailed study [78]. Innate macrophage killing was the same, as was IgG antibody production. IgM antibody levels were higher in AJ mice. Depression of cellular reactivity was less in resistant mice and increments in unstimulated and stimulated production of interferon-γ (IFN-γ) were greater in susceptible mice, although serum levels were greater in AJ mice prior to infection. It therefore appears that a range of genetically determined manifestations of immune reactivity determine the outcome of this infection.

The presence of adequate levels of IFN-γ is thought to be a major factor in controlling intracellular proliferation. This was very clearly shown by Suzuki et al. [79], who injected a monoclonal antibody against IFN-γ into mice. The critical observations from this experiment were that mice receiving the monoclonal antibody developed toxoplasmic encephalitis, that many more brain cysts were found in the injected mice, and that these events occurred despite serum antibody levels and macrophage toxoplasmicidal activity, which were comparable with those of control animals. The effect of IFN-γ is mediated by induction of indoleamine, 2,3-dioxygenase, which enhances the degradation of tryptophan in host cells, so denying the amino acid to the parasite [80].

Antibody to the parasite plays a minor role in protection, as passive transfer of high-titer antibody gave no protection in nude mice [73]. The anti-*Toxoplasma* activity of natural killer cells activated by lymphokines in the presence of complement could be further enhanced by precoating the trophozoites with specific antibody [81].

Other regulatory cells, which include B cells [82], are concerned with modulating the expression of immunity through T cells. Overall susceptibility is determined by several genes [83]. The T cell subset which replicates in response to *Toxoplasma* antigen has the helper phenotype [82].

Most of the above discussion has concentrated on immunosuppression which allows unrestrained proliferation of *T. gondii*, but it must also be recognized that this infection is itself immunosuppressive [84]. The effects, if any, of this infection on the progression of HIV infection in AIDS and on the occurrence of other opportunistic infections have not been examined in great detail.

Human immunodeficiency states

Humoral and cell-mediated immunity

Experimental infections in animals suggest that organisms are widely disseminated in varying numbers before the host's immune response prevents further progression; the nervous system is infected during this process [64]. While innate, cellular, and humoral responses are elicited and each contributes to immunity, it seems likely that specific cellular responsiveness mediated by T cells exerts a major controlling influence [73,77]. Disease due to toxoplasmosis is not a problem in patients with impaired humoral immune responses but occurs in patients with defective cellular immunity seen in Hodgkin's disease, transplant recipients taking immunosuppressant drugs, and most commonly in AIDS [64].

Clinical features

The dominant clinical manifestations in immunodeficiency relate to neurologic involvement [64–66,68,69], although disease at one or more sites outside the CNS certainly does occur [66,70,71,85] but is less common or, perhaps, less commonly apparent. Diffuse encephalitis, space-occupying lesions—which can be single or multiple, affecting the cerebral or cerebellar hemispheres—and focal lesions occur. Spinal cord involvement is reported [86]. Meningeal involvement with appropriate signs is recognized [87] but appears to be uncommon. Choroidoretinitis, pulmonary disease, and cardiac involvement are sites of organ involvement outside the CNS [69,70,88] but virtually any tissue in the body can be affected.

An additional problem in AIDS is that in any patient there may be opportunistic infection at more than one site at the same time and there may be more than one pathogen infecting an organ such as the brain.

The main pathologic change seen in affected tissues is necrosis [89]. Quite why this occurs is not clear; it may be due to some effect of unrestrained parasite replication on the tissue affected, as parasites are often readily found in pathologic material from patients [89], although this is not invariable [90].

Diagnosis by standard techniques in neurologic investigation.
Computerized axial tomography (CT) scanning shows
abnormalities in toxoplasmic encephalitis. These comprise single or multiple ring-enhancing lesions, hypodense lesions, and nonenhancing lesions [91], but these
occur with other CNS infections known to occur in AIDS
and concurrently with toxoplasmosis of the CNS so
that these appearances are not diagnostic. Magnetic
resonance imaging (MRI) is becoming more widely
available and is particularly valuable in the CNS. Studies
of MRI have been published [92] and the technique
appears to be more sensitive in detecting focal abnormalities. Lesions are seen at the junction of cortex and
medulla and are multiple on MRI, although CT may
show only one so that when MRI scanning shows a
single lesion this is unlikely to be due to toxoplasmosis,
and biopsy is recommended. Both techniques can be
used to follow the response to treatment, in which there
may be complete resolution of small lesions and residual
changes at the sites of larger ones, where calcification
may be seen. The presence of calcification does not
indicate that parasitologic cure has occurred.

The cerebrospinal fluid (CSF) shows nonspecific
abnormalities, an increase in protein content, an
increase in cells, and usually a normal sugar content,
but these again are nonspecific and not diagnostic [87].

Diagnosis by serologic testing. Serologic testing, the usual
means of diagnosis in immunocompetent patients, does
not yield conclusive or consistent results when disease
is associated with immunodeficiency. Among a group
of seronegative recipients of hearts transplanted from
seropositive donors, seroconversion after surgery
occurred at 4–10 weeks (IgM) and 5–12 weeks (IgG)
[66]. Primary infection in most of a similar series of
patients reported from the UK occurred within 5 weeks
of transplantation [93]. Exceptions were one fatal case
who did not seroconvert, and disease and delayed seroconversion at 115 days after transplantation in another.

The Sabin–Feldman dye test has long been the main
serologic test employed in toxoplasmosis. Disease in
immunodeficient patients may be strongly suspected
with titers over 1:1024 or with rising titers in the
appropriate clinical situation [94], but titers at or above
1:1024 were found in eight out of 43 patients with
biopsy-proven toxoplasmic encephalitis [95]. Fourteen
of the forty-three had titers of 1:64 or less and one had a
negative test.

This lack of discrimination prompted studies of
alternative techniques. Detection of agglutinating antibodies to either acetone-fixed or formalin-fixed antigens
gave more useful results. Acetone-fixed antigen tests
had a sensitivity of 61% and a specificity of 97%, and,
when results of the two agglutination tests were combined, 70% and 93% [95] compared with 19% and 95%
for the dye test.

Detection of antibody in CSF is a further refinement
which might help in proving CNS disease. However,
finding IgG antibody in the CSF does not necessarily
mean that it was produced by antibody-secreting cells
within the neuraxis in response to infection by that
organism. It could have crossed inflamed meninges due
to some other infection and merely represent serum
antibody from past and inactive infection. Potasman
et al. [96], using a relationship which compares the ratio
of *Toxoplasma* antibody in CSF and serum to that of total
IgG in CSF and serum, found that a value over 1.0
indicated intrathecal production of antibody. Antibody
is only present in about 50% of cases, which limits the
application of this technique [97].

Diagnosis by antigen detection. Suzuki & Kobayashi [98]
found high levels of circulating antigen in nude mice
while others have found protein antigens in urine and
serum of mice [99]. Antigen was found in urine from five
out of 20 AIDS patients with toxoplasmic encephalitis
and in the serum of three out of 18 of these cases.
Antigen detection holds promise as a valuable diagnostic technique in this condition, although more
studies are needed. It should also be noted that
circulating immune complexes containing *Toxoplasma*
antigen have been found in healthy blood donors [100],
stressing the importance of interpreting laboratory data
in the context of the clinical situation for each patient.

Diagnosis by examination of biopsy material. Brain biopsy
is not undertaken lightly but has a definite place in the
management of CNS syndromes in immunosuppressed
patients because it may allow definitive and early recognition of one or more pathogens, with the possibility of
starting definitive treatment. Biopsies obtained using
stereotactic techniques may be less traumatic and
provide material from abnormal areas defined by CT or
MRI scanning.

Three patterns of pathologic changes were observed
in toxoplasmic encephalitis [89]: five cases showed
widespread ill-defined and often confluent areas of
necrosis associated with a weak inflammatory response
and large numbers of tachyzoites peripheral to the
necrotic areas; in 26 cases there were multiple

abscesses, which were well circumscribed and most often seen at the border between cortex and white matter and in the basal ganglia. The histologic features were central necrosis with a surrounding rim of lymphocytes, macrophages, and occasional polymorphs. Tachyzoites were present in variable numbers, both intra- and extracellularly, with some encysted bradyzoites. The third pattern was characterized by multiple microglial nodules with scattered encysted bradyzoites and a few tachyzoites within or around the nodules.

Parasites may be found by light microscopy examination of sections stained using routine techniques, but Wanke et al. [90] found parasites in three out of seven patients. Electron microscopy has also been used to identify the parasite in biopsy material [91]. Indirect immunofluorescence examination of sections stained with a monoclonal antibody against the P30 membrane antigen of the tachyzoite showed clear differentiation of tachyzoites from tissue debris in specimens from lungs [101]. The peroxidase–antiperoxidase, avidin–biotin and enzyme-linked immunosorbent assay (ELISA) methods can be applied to demonstrate *Toxoplasma* in tissues. Touch impression smears and cytocentrifuge specimens from biopsies and body fluids, e.g., CSF or bronchiolar lavage fluid, can be processed using these methods.

Diagnosis by culture of the organism. This can be done using uninfected mice or permissive cell cultures. Parasites can be isolated from infected immuno-competent humans [102]. In a recent study, trophozoites were readily isolated from bronchoalveolar lavage fluid in patients with AIDS [101] after culture for 48 h in fibroblast monolayers. The monolayers are examined for parasites using immunofluorscence. This technique has not been used extensively as yet in clinical laboratories but may be widely applicable using specimens from a range of sites.

Diagnosis by the PCR. This new technique is currently being applied in many infections and Burg et al. [103] applied the PCR to identify *T. gondii*. They used the B1 gene sequence as the sequence for amplification and positive results were obtained in all six strains tested in the initial studies and in strains isolated from patients with toxoplasmic encephalitis and three additional strains. It was possible to detect parasite DNA from one organism in a crude cell lysate which equates to 10 organisms in 100 000 human leukocytes. The authors showed that there was no crossreaction with a wide

range of other microbial pathogens which infect the CNS. Clinical applications of this technique have yet to be reported but the PCR has considerable potential for assisting diagnosis in this disease.

Treatment and prophylaxis

In most instances the immunocompetent host does not require chemotherapy, but in those who do, three weeks' treatment is usually enough. The same drugs, pyrimethamine with sulfadiazine, are the treatment of first choice and experience in AIDS patients shows it to be effective but, as with other infections in AIDS, treatment must be continued, perhaps indefinitely, as relapse occurs if therapy is discontinued [104].

Bearing in mind the high frequency of this infection in AIDS patients, a policy of empiric therapy for suspected toxoplasmic encephalitis after initial investigation [105] followed by brain biopsy in those failing to respond is reasonable. Unfortunately this drug combination is not free of toxicity. Daily pyrimethamine administration can lead to bone marrow toxicity [106], although this can be circumvented by folinic acid administration [107]. Sulfonamides are also known to evoke rashes, fever, and bone marrow suppression in AIDS patients [108]. Failure to respond to this drug combination in relapses raises the possibility that parasites may become drug-resistant, although it has been suggested that parasite populations are not high enough for spontaneous drug-resistant mutations to occur [109]. It is clear that new drugs are needed and efforts to find alternatives are continuing.

Chemoprophylaxis of toxoplasmosis is necessary in cardiac transplantation when the heart of a seropositive donor is given to a seronegative recipient in view of the high frequency of primary infections. Spiramycin has been tried but was not very effective [110], whereas pyrimethamine alone is effective [93]. Among AIDS patients who have had toxoplasmic encephalitis, maintenance chemotherapy with pyrimethamine and sulfonamide is essential after treatment to prevent relapse [106].

CRYPTOSPORIDIUM SPP. AND ISOSPORA BELLI

Introduction

Cryptosporidium spp. and *Isospora belli* are protozoan parasites belonging to the suborder Eimeriina. They develop intracellularly in epithelial cells of the intestine (although *Cryptosporidium* may be found in other organs) where they undergo asexual multiplication, gametogeny, and oocyst formation. Oocysts are released from the host in the feces; at this time, those of *Cryptosporidium* are already sporulated and infective—whereas those of *Isospora* sporulate after leaving the host. The disease patterns of both parasites are strongly dependent on the host's immune status: an acute self-limiting infection in an immunocompetent host contrasts with a severe, chronic, and life-threatening disease in the immuno-compromised host. In fact, this feature is so characteristic that both are included in the US Centers for Disease Control/World Health Organization AIDS case de-finition. In the absence of other causes of immuno-deficiency, to be classified as an AIDS case it is sufficient that there has been persistent diarrhea (>1 month) with a definitive diagnosis of isosporiasis (and an HIV positive test) or cryptosporidiosis (with an HIV positive test, in the absence of HIV testing, or even with an HIV negative test if with a T CD4$^+$ lymphocyte count <400/mm^3).

CRYPTOSPORIDIUM SP.

Cryptosporidium was recognized as an animal parasite in the early part of this century [111,112]. The taxonomy of the parasite has been poorly studied, but at present it is widely accepted that one species, *Cryptosporidium parvum*, is probably responsible for most reported disease in mammals. Its role as a human pathogen was not established until the late 1970s. It is interesting to note that the first two published reports showed a dichotomy in the clinical manifestations of the disease caused by this organism, which has become its hallmark. Nime *et al.* [113] described the case of an immuno-competent child with severe acute self-limiting, but life-threatening enterocolitis, whereas Meisel *et al.* [114] related the case of an adult who was therapeutically immunosuppressed (high doses of cyclophosphamide and prednisolone) with chronic severe dehydrating diarrhea which subsided 2 weeks after termination of the immunosuppressive therapy. The next few years saw a number of case reports, the majority involving immunocompromised patients. In 1980, diagnosis of cryptosporidiosis, until then requiring a biopsy, was simplified by the development of techniques for the detection of oocysts in fecal smears [115]. Subsequently a large number of published studies clearly demon-strated the high incidence of infection with this parasite:

between 1% and 4% of children's acute diarrhea in developed countries and up to 20% in developing countries [116,117].

Animal models

Cryptosporidium has been observed in a variety of vertebrates, including fish, reptiles, birds, and many mammalian species [116]. In mammals, two species, *Cryptosporidium muris* and *C. parvum*, have been described, which can be distinguished by the size of oocysts and the preferred site of development in the host [111,112]. Based on observations of oocyst morphology and/or endogenous development, it is at present generally believed that *C. parvum* is usually the cause of cryptosporidiosis in mammals.

In mammals other than humans, cryptosporidiosis is normally a disease of neonates [117], and older animals are resistant to infection. This age-related resistance has meant that a suitable adult small animal model has not been available for research.

Immunosuppression of adult rats with drugs has made them more susceptible to infection [118,119]. Neonatal mice are susceptible to infection, although little pathology is observed in these animals [120,121]. Mice infected as neonates and injected regularly with anti-CD4 or anti-CD4 plus anti-CD8 monoclonal anti-bodies remained infected as adults, but recovery occurred a number of weeks after the injections were stopped [122]. Neonatal nude mice had more protracted infec-tions than immunocompetent counterparts, and signs of pathology were observed with the immunocom-promised animals [121]. In one study adult nude mice were resistant to infection [121], but in another study in which larger numbers of oocysts were used when dosing inoculate mice, severe chronic and sometimes fatal cryptosporidiosis was observed [122]. Outbreaks of cyptosporidiosis have been observed occasionally in colonies of guinea pigs, and Vetterling *et al.* [123] proposed a separate species for a parasite which pro-duced enteritis in this host. Parasite isolates responsible for natural outbreaks of disease in colonies of guinea pigs also caused disease in experimental animals [124,125]. Few experimental studies have been made with *C. muris*, the first species of *Cryptosporidium* to be described [111]. Recently, a strain of this parasite was reported to produce heavy infections in weaned immunocompetent mice, and also infected guinea pigs, rabbits, cats, and dogs [126]. Infection in the chicken with the avian parasite *Cryptosporidium baileyi* may prove to be a useful model for studying immunity to cryptosporidiosis [127].

In the immunocompetent host the development of *C. parvum* occurs predominantly in the small intestine, although development may also be found in the colon and large intestine. In infected tissue there is cellular infiltration of the lamina propria by plasma cells, lymphocytes, neutrophils, and eosinophils [117], and crypt hyperplasia is a common feature. Partial villous atrophy has been observed and the lining epithelium may be flat or cuboidal. The pathologic findings in calves, lambs, deer calves, and goat kids were similar, and suggest that malabsorption due to villous atrophy and damage to epithelial cells may be responsible for the diarrhea associated with cryptosporidial infection. It has been proposed that a parasite-derived toxin may play a role in pathogenesis, but so far there is little evidence for this.

Whereas little pathology is observed in immunocompetent neonatal infected mice, neonatal [121] or adult nude mice [122] showed signs of pathology similar to those described above. Ungar *et al*. [122] also observed extraintestinal spread of infection in adult nude mice.

Humoral and cell-mediated immunity

Immunity to cryptosporidiosis may be acquired as a result of infection [125,127]. Infection of an immunocompetent host with *Cryptosporidium* produces a parasite-specific humoral response. Antibodies against *Cryptosporidium* have been found in all host species studied after parasitologically proven infection. The serologic tests mainly used for detecting cryptosporidial antibodies are immunofluorescence [128–132] and ELISA [133,134]. Each method shows a comparable sensitivity and also strong specificity when tested against the closely related parasites *Toxoplasma*, *Sarcocystis*, and *Isospora* [130]. Seroconversion appears usually in the second or third week [116,132,135–137].

Cryptosporidial antibodies in feces were detected in both experimentally infected lambs [132] and calves [134], appearing soon after the oocysts and increasing in time until the parasite was eliminated, usually in the third week postinfection. In calves the three classes of antibody IgG, IgM, and IgA were detected [134], but in lambs only IgA was found [132]. These limited studies may suggest a protective role for secreted intestinal immunoglobulin, particularly IgA.

Passive immunotherapy of calves or humans using colostrum from previously infected cows was unsuc-

cessful [138,139]. However, colostrum from cows hyperimmunized by injection with parasite antigen plus adjuvant protected calves from disease [140]. This hyperimmune colostrum could neutralize sporozoites *in vitro*, and treated mice were protected from infection [141]. The protective element in this colostrum was not identified.

Sera from animals immunized with cryptosporidial antigens have identified large numbers of oocyst polypeptide antigens in immunoblots following acrylamide gel electrophoresis of soluble oocyst material [142]. There is an important carbohydrate contribution to immune recognition [143]. Studies of the antigenic structure of the sporozoite also showed a large number of polypeptides, one of which had a molecular weight of 20 kD [144] or 23 kD [132,145]. This polypeptide was prominent and almost always recognized by convalescent sera of different animal species [144]. Antisera and a monoclonal antibody against this 20/23 kD antigen were found to react with sporozoites and merozoites [132,144], suggesting that it might be involved in the stimulation of protective immunity.

There is strong evidence from mouse studies that T cells are important in protective immunity. Heine *et al*. [121] found that while infections in normal neonatal mice were short-lived and caused little pathology, infections in nude counterparts were long-lasting, produced pathology, and were sometimes fatal. Infections introduced to neonates which were subsequently treated regularly with monoclonal antibodies recognizing CD4+ T cells were not eliminated until after the treatment was stopped in adulthood [122]. Observations made on cryptosporidial infections in animals with T cell function impaired by infection with other pathogens further suggest a role for this cell in immunity. Chronic diarrheal cryptosporidiosis similar to the disease in the immunocompromised human was reported in animals naturally infected with retrovirus, both in several rhesus monkeys with SIVmac (a virus closely related to HIV-2) [146] and in a cat infected with feline leukemia virus (a virus similar to HTLV-1 and -2) [147].

Human immunodeficiency states

Although epidemiologists generally agree that *Cryptosporidium* is a worldwide major cause of children's acute diarrhea, the driving force of research in human cryptosporidiosis comes from the parasite's prominent role in AIDS. Chronic cryptosporidial diarrhea in people with AIDS is very common in developed countries, with an incidence of 3.6% in American AIDS patients notified to the CDC up to April 1986 [148], 9% in New York City [135], and 11% in London [149]. In developing countries the incidence is even greater: in HIV-1, 31% and 38% of diarrheal AIDS patients from Zaire and Haiti, respectively [136,150], and in HIV-2, 18% of West African AIDS patients [137]. However, protracted infections have been observed in people with other immunodeficiency states (see below).

Development of *Cryptosporidium* in the immunocompetent individual usually takes place in the small intestine, particularly the ileum [117], and is also frequently found in the jejunum [151]. In rare cases, there is extraintestinal spread of infection. The intestinal architecture is normally only mildly or moderately affected. Histopathologic observations have included stunting and fusing of villi, with mild infiltration of the mucosa by lymphocytes, plasma cells, macrophages, and polymorphs, and a degree of crypt hyperplasia [117,151,152]. Similar observations have also been made in immunocompromised patients, including those with AIDS [153,154]. However, in sequential studies with two children, one with immunoglobulin deficiency [155] and the other with combined immunodeficiency [156], the mucosal damage as a result of cryptosporidial infection altered from mild to severe prior to death.

Infection of the colon and rectum has commonly been reported in immunocompromised patients [113,153,154]. In addition, extraintestinal spread of infection often occurs. Infection of the biliary tract has been a common finding since Pitlik *et al.* [157] first described it. *Cryptosporidium* has been reported as a cause of sinusitis [158], esophagitis [159], hepatitis [160], appendicitis [161], laryngotracheitis [162], and pancreatitis [163]. Infection of the respiratory tract in the immunocompromised person has also been documented [164–166], and this may be a common occurrence in AIDS patients [167,168].

Humoral and cell-mediated immunity

Seroconversion has been observed in both immunocompetent and AIDS patients but not in hypogammaglobulinemic patients [130]. The classic switch of immunoglobulin class was demonstrated in immunocompetent and, more variably, in AIDS patients, with an initially strong IgM response being replaced by IgG [133]. The study of Casemore [131] also showed a serum IgE response during the acute phase of infection.

However, differences in results of serologic tests may be obtained depending on whether oocysts or developmental stages of the parasite are used as antigens. In one study, when developmental stages were used the large majority of randomly screened human and animal populations were seropositive, with titers being stable or slowly descending long after infection [128]. This suggested that cryptosporidial infection is widespread, occurs early in life, and, following infection, anticryptosporidial antibodies are produced for a long time. In contrast, when oocysts were used as antigen, a shorter lived response was observed, with the antibody level becoming undetectable in humans about 1 year postinfection, and a low prevalence rate was found in a randomly screened population [130].

Indirect evidence for some protective role of humoral immunity in cryptosporidial infection comes from a number of clinical reports of chronic symptomatic infection in individuals whose sole immunologic defect has been in immunoglobulin production. In addition to the chronicity of diarrhea, extraintestinal infection was also documented in some of these patients, e.g., cholangitis [169] and sinusitis [158], in two hypogammaglobulinemic children. This extraintestinal spread has been rarely reported in immunocompetent patients. The passive transfer of serum from people with high *C. parvum* antibody titers to a patient with hypogammaglobulinemia and cryptosporidiosis did not affect the course of disease in this patient [138]. In the absence of reliable etiologic therapy, however, cryptosporidial diarrhea has been successfully treated with hyperimmune colostrum. Tzipori *et al.* [170] hyperimmunized cows with *Cryptosporidium* oocyst antigens in Freund's adjuvant and obtained colostrum with high titers of cryptosporidial antibodies. This colostrum appeared to induce clinical and, in some, but not all parasitologic improvement both in hypogammaglobulinemic and T cell immunodeficient patients [169,170]. Similar results have been reported by Ungar *et al.* [122].

T cells appear to be involved in the control of *Cryptosporidium* infection in humans. Chronic *Cryptosporidium* diarrhea has been reported in almost every clinical situation with depression of T cell function, including congenital immunodeficiencies [156], transitory postviral immunodepression [171], cancer chemotherapy-induced immunodepression [172–174], transplantation immunosuppression [166,175,176], AIDS, etc. The resolution of symptoms and the eradication of the parasite correlates well with the immune status of the host. In chemotherapeutically immunosuppressed patients with cryptosporidial chronic diarrhea, the suspension of therapy, when possible, was followed in 2 or 3 weeks by the clearance of the parasite [114,172,173]. Parasite clearance was also observed in bone marrow transplanted patients following the repopulation of bone marrow and peripheral blood by lymphocytes of donor origin [175]. In AIDS patients some correlation has been found between the severity of cryptosporidial symptoms and the absolute number of CD4$^+$ T cells, and a spontaneous long-lasting clinical remission has been observed in patients with only moderately reduced numbers of CD4$^+$ T lymphocytes [177,178]. Connolly *et al.* [149] were unable to eradicate *Cryptosporidium* with antimicrobial therapy in AIDS patients, but eradication was achieved in three patients treated with zidovudine. In an *in vitro* cell culture system zidovudine had no significant effect on *Cryptosporidium* growth [179], suggesting, as proposed by Connolly *et al.* [149], that in the clinical study efficacy was due to inhibition of HIV replication and HIV-induced CD4$^+$ T cell depletion.

Results of studies on the efficacy of transfer factor against cryptosporidiosis are contradictory. In one small clinical trial [180], AIDS patients were treated with transfer factor obtained from mesenteric lymph nodes of calves following *Cryptosporidium* infection or equivalent material from uninfected animals. The patients given transfer factor from infected animals experienced relief of symptoms and the parasite was eradicated in some. The same transfer factor material, however, failed to protect experimentally infected calves against acute disease [181]. Treatment of AIDS patients with IL-2 when suffering from cryptosporidiosis led to reduced symptoms and parasite numbers [149].

Clinical features

The clinical features of cryptosporidiosis may be completely different in the immunocompetent and immunocompromised patient. The most important symptom is diarrhea, which is often profuse, watery, and occasionally contains mucus but not blood or pus. The immunocompetent person may experience an acute diarrhea lasting between 3 and 10 days [116,151] before a complete recovery is made. There may be a strong correlation between the number of oocysts in the stools and severity of diarrhea [116]. In the immunocompromised patient (most observations have been made with AIDS patients) the diarrhea may be chronic and more severe, with up to 70 bowel movements and 20

liters of stool being produced daily [116]. In AIDS this chronic diarrhea is accompanied by wasting and a huge weight loss. *Cryptosporidium* is probably a contributory factor in "slim disease" of Central Africa as it has frequently been found to be the most common agent of African AIDS diarrhea [136,137,150].

Diagnosis. Diagnosis is usually performed by stool examination, and immunodeficiency is unlikely to affect this.

Treatment. Although many antimicrobial agents have been examined, an effective anticryptosporidial drug has yet to be found [116]. Immunotherapy with hyperimmune colostrum [169,182], IL-2 [149], or transfer factor [180] might be of value in cases involving immunocompromised individuals. Treatment with somatostatin has been of some value in a number of AIDS patients [183,184].

ISOSPORA BELLI

Introduction

Human isosporiasis is a poorly researched disease, probably because it is considered rare (or was, before the AIDS epidemic), is easily diagnosed from fecal smears, and is easily and efficiently treated with cheap, non-toxic, and widely available drugs [185].

Humoral and cell-mediated immunity

Little is known about the immune response to *I. belli* infection. The clinical data show a contrast between the typical (there are frequent exceptions) short, self-limiting diarrhea in the immunocompetent host and the chronic, wasting diarrhea in the immunocompromised patient, including those with AIDS or other immunodeficiencies, such as α-chain disease [186] or in HTLV-1-associated adult T cell leukemia [187].

Clinical features

In AIDS patients from some tropical countries *I. belli* is a very common parasitic cause of chronic diarrhea, usually second only to *Cryptosporidium*, e.g., in Zaire the parasite was found in 12% of AIDS patients with diarrhea [150], in Haiti in 13% [185], and in 53% of West African HIV-2 AIDS patients with diarrhea [145].

No studies have been published on the relationship between immunologic parameters and *I. belli* infection, and no serologic test has been reported.

Treatment. Both immunocompetent and immunocompromised patients can be easily treated for isosporiasis. But as AIDS patients have a tendency to have recurrence of *I. belli* diarrhea, it has been indicated that they should be maintained indefinitely on specific chemoprophylaxis [185].

MICROSPORIDIA

Introduction

Microsporidia are obligate intracellular parasitic protozoa which infect invertebrate and vertebrate hosts, and are the causative agents of microsporidiosis [188]. The genera of microsporidia which may infect humans are *Encephalitozoon*, *Enterocytozoon*, *Nosema*, and *Pleistophora* [189]. The parasites may have a preference for a particular type of host cell. Multiplication within the cell occurs by merogony, following which, spores (usually 1–2 μm in size) are formed which are able to infect other cells. Invasion of host cells may occur by phagocytosis or with the help of a specialized organelle of the spore, the polar filament. Under appropriate conditions, the polar filament extrudes from the spore, forming a tube-like structure through which the infective body of the spore is injected, and if the end of the filament has penetrated a host cell the body will emerge in the cytoplasm.

Diagnosis of microsporidiosis is usually obtained by identification of parasites in body fluids or tissue sections by light microscopy using appropriate staining [190], and unequivocal identification of the parasite is likely using electron microscopy [154]. Evidence of infection or previous infection may be found using serologic techniques (see below).

Animal models

Encephalitozoon cuniculi is a parasite commonly found in mammals which resides in a parasitophorous vacuole in a number of cell types, including macrophages and vascular endothelial cells. The disease produced in young infected canines or neonatal monkeys is often fatal [189,191,192], but infections in rabbits and mice are usually nonlethal and chronic [193,194]. Confirmation of *E. cuniculi* infection in humans may be obtained if mice injected with body fluid from a patient develop an infection [195].

Young infected canines may suffer severe and often fatal encephalitis or nephritis [191,192]. Infection appears to occur by transplacental transmission since disease has been found only in cubs with infected mothers. Damage to arterial walls may result in cerebral hemorrhage, a common cause of death in young animals. The signs of neurologic disease include ataxia, paralysis, convulsions, and blindness. Interstitial nephritis is commonly observed, and this may lead to renal disfunction.

Humoral and cell-mediated immunity

Infection in animals usually elicits a serologic response [196,197] and the presence of parasite-specific antibodies may correlate with pathologic findings, but in the study of Szabo & Shadduk [197] parasites were not always found in the tissues of experimentally infected dogs which had high antibody titers. Detection of microsporidial antibodies may be taken as evidence of infection or previous exposure to the parasites. Using spores as a source of antigen, antibodies are usually detected by any of a number of serologic tests, which include immunofluorescence [198], carbon immunoassay [199], or ELISA [200,201].

Antibodies may also be involved in immunity to *E. cuniculi*. Pretreatment of spores with serum from immune mice was found to reduce their viability [202]. In a canine study, immune serum increased the phagocytosis of spores *in vitro* by blood monocytes [197]. The titers of parasite-specific antibody in infected mice correlated with their immune status, but passively transferred serum from immune mice failed to protect nude mice from infection [203].

Susceptibility of mice to *E. cuniculi* varies with the strain; for example, BALB/c mice were shown to be more resistant than C57B1/6 mice [204]. The thymus is required for protective immunity, since euthymic BALB/c mice had chronic nonlethal infections while nude mice of this strain had acute and lethal infections [202]. T cells were shown to have a role in immunity in experiments in which T cell-enriched spleen cells from previously infected mice conferred protection against infection on nude mice, but T cell-depleted spleen cells had no protective effect [202,203]. There was no evidence of cytotoxic T cells being able to kill parasites, but macrophages, plus a few contaminating T cells (too small a number to be protective themselves), from immune mice were able to induce protection in nude mice. Supernatants from *E. cuniculi*-sensitized spleen cells were able to induce peritoneal exudate cells to kill *E. cuniculi* spores [203].

There was an increase in natural killer cell activity of the spleen in both BALB/c and C57B1/6 mice during infection, but increased activity was also obtained with BALB/c nude mice, indicating that this nonspecific mechanism may not play a prominent part in the elimination of infection [205].

Studies in the mouse model system, therefore, point to an important role for T cells in protective immunity, possibly acting as helper cells for antibody production and/or releasing cytokines which activate monocytes to kill parasites.

Human immunodeficiency states

Few cases of microsporidial infection in immuno-competent people have been reported, and most reports have concerned individuals who were immuno-suppressed. It is not clear whether the small number of reports in the literature reflects the actual incidence of infection or a failure to detect the presence of micro-sporidia. However, there has been an increase in the number of cases of microsporidiosis in relation to AIDS in recent years. *Enterocytozoon bienusi*, a parasite found in intestinal epithelium, has been described only in association with AIDS [153,154,190,206]. The absence or immaturity of spores in samples taken in some of these studies suggests that *E. bienusi* may not be a natural parasite of humans [206]. In an African study, two out of 22 autopsies of AIDS cases in Kampala revealed microsporidia in the intestine [207]. Two other micro-sporidians, *Pleistophora* sp. and *E. cuniculi*, have been observed in AIDS patients [208,209]. Infection with *E. cuniculi* was also reported in a child with an abnormally low T helper cell count [195]. Disseminated micro-sporidiosis, possibly caused by *Nosema connori*, was reported in a child with thymic dysplasia [210]. Microsporidian keratinoconjunctivitis in several AIDS patients has recently been documented [211,212].

Caution should be exercised when attributing pathologic observations in immunocompromised hosts to an infecting organism, as other pathogens may also be present.

In intestinal microsporidial infections of AIDS patients, mild increases in numbers of plasma cells and macro-phages in the local mucosa have been reported but there was little or no alteration to villous architecture [153,154].

Liver examination of an AIDS patient who developed fatal hepatitis due to *E. cuniculi* showed sinusoidal congestion, microgranuloma, and hepatocellular necrosis [208]. In a case of myositis in an AIDS patient [209], atrophic and degenerating deltoid muscle fibers were infiltrated with clusters of *Pleistophora* spores, and there was scarring and fibrosis of the muscle with an intense inflammatory reaction of lymphocytes, plasma cells, and histiocytes. In a case of disseminated micro-sporidiosis in a child with thymic dysplasia, parasites were found mainly in the smooth muscle of organs and arteries, and in nerve fibers [210]. Keratinoconjunctivitis caused by microsporidial infections was characterized by conjunctival inflammation and diffuse punctate keratopathy [211,212].

Humoral and cell-mediated immunity

No microsporidial antibodies were found in 116 blood donors in the UK [201], suggesting a low incidence of infection in the immunocompetent population. However, from a group of 28 European homosexuals considered to be at risk of AIDS, 32.1% had antibodies to *E. cuniculi* [213], and in another study 23.8% of 63 people at risk of AIDS were seropositive [189]. Homo-sexuals with or without HIV infection may, therefore, be predisposed to microsporidial infection. Interestingly, an unexpectedly high incidence of antibodies to *E. cuniculi* has been found in patients with certain tropical diseases which cause immunodepression [201,214]. Little is known about cell-mediated immune responses against microsporidia in humans.

Clinical features

The symptoms resulting from microsporidial infection may vary according to the site(s) of infection. Diarrhea in AIDS patients has been associated with gastrointestinal infection with microsporidia [153,154,190] and it has also been found in a case of disseminated microsporidiosis in a child with impaired thymic development [210]. An AIDS patient with infection of the muscles had progressive generalized muscle weakness with con-tractures, lymphadenopathy, sinusitis, and fever [209]. Microsporidial infection of the liver of an AIDS patient which proved to be fatal was accompanied with vomiting, nausea, abdominal pain, general fatigue, and weight loss [208]. AIDS patients with ocular micro-sporidiosis have commonly presented with either conjunctivitis or scleritis, foreign body sensation, blurred vision, and photophobia [211,212].

Treatment. There are some reports of apparently successful treatment of microsporidiosis with drugs in individual cases, but no antimicrobial agent has been proven to be effective in scientific trials.

LEISHMANIA SPP.

Introduction

The behavior of parasites of the *Leishmania donovani* complex as opportunistic infections suggests that the normal state of subclinical or asymptomatic *Leishmania* infections existing in endemic areas may be altered by HIV infection, with immunosuppression either facilitating acquisition of the disease or allowing reactivation of a latent infection [215,216]. However, immunosuppression may also be caused by visceral leishmaniasis (VL) patients who die usually succumbing to intercurrent infections, e.g., pneumococcal, tuberculous. Thus, advanced VL may itself contribute to and worsen the immunodeficiency and poor outlook for this condition in HIV-positive patients [217]. Since *Leishmania* acts as a stimulus to the activation of lymphocytes and macrophages, it has been suggested that stimulation of these cells, which may be latently infected with HIV, may increase viral replication and thus accelerate progression of the resulting immunodeficiency [218].

Leishmania is usually transmitted to humans by a sandfly vector. However, blood transfusion is a well-recognized means of infection in endemic areas. Montalban *et al.* [219], reporting 16 patients with VL in the presence of HIV infection, of whom 15 were intravenous drug abusers, suggested that this route of infection is more likely in AIDS. However, Berenguer *et al.* [216], in reporting nine HIV positive patients (eight of them intravenous drug abusers) with VL, felt that person-to-person transmission by hypodermic needles or blood was less convincing because no known contact among their patients has occurred and the transmission of VL among non-HIV-infected drug abusers had not been reported. It seems likely that different degrees of immunosuppression between HIV-positive and HIV-negative intravenous drug abusers provides the explanation.

Clinical features

Presenting clinical features may be atypical. Peters *et al.* [217] reviewed 47 reported cases of VL in patients who had AIDS or were HIV positive. In eight cases splenomegaly was absent. Since splenomegaly in VL is largely due to macrophage proliferation, the authors suggested that lack of splenic enlargement was due to failure of the macrophage response. Three patients had neither an enlarged spleen nor enlarged liver.

Other authors have recorded atypical presentations, for example prolonged fever and progressive severe pancytopenia in the absence of splenomegaly or hyperglobulinemia [220]. Infection of unusual sites, for example lung [217], cutaneous dissemination, infection of clinically unaffected skin [221], and gastroduodenal mucosa [222], is also recorded.

Diagnosis. Lack of clinical suspicion and atypical clinical presentation have led to late diagnosis of VL in HIV-positive patients [216,217].

Diagnosis by microscopy and culture. The definitive means of diagnosing VL is microscopic demonstration of amastigotes in tissue biopsies and culture of the same biopsies to the promastigote stage. Many of the cases of VL in HIV-positive patients have been diagnosed by bone marrow examination performed as a routine investigation of fever of unknown origin [217]. Amastigotes have been plentiful in the bone marrow samples obtained [223,224] (Chiodini, unpublished observations). Berenguer *et al.* [216], on the other hand, recorded only two microscopically positive bone marrow samples in his series of eight cases, although all eight marrows proved to be culture positive. Samples other than bone marrow found to be useful are splenic aspirates and liver or lymph node biopsies [221]. Amastigotes have also been found coincidentally in the skin of patients biopsied for Kaposi's sarcoma and in macrophages obtained by bronchoalveolar lavage [217]. It is important that biopsies are routinely cultured if leishmaniasis is suspected [219], as not only will sensitivity be increased but isoenzyme analysis of cultured material will identify the subspecies responsible [216].

Diagnosis by serology. Approximately 95% of immunocompetent patients with VL have positive serology [217,219]. In the case of AIDS or HIV-positive patients with AIDS, only 16 out of 45 (36%) patients from Peters' series had significant antibody titers to *Leishmania*. The authors suggested that patients without anti-*Leishmania* antibodies had contracted the infection while already immunosuppressed rather than reactivated a latent infection. Hyperglobulinemia, a useful indicator of VL in the immunologically intact, may be absent in HIV-positive patients with VL [220]. Thus atypical laboratory features have also contributed to late diagnosis.

Treatment

In immunocompetent individuals less than 2% show primary unresponsiveness to treatment [225]. In contrast, the prognosis of VL in HIV-positive patients has been poor so far. Eight of Berenguer's series in 1989 [216] received therapy with pentavalent antimony: four of them died, including two who had completed at least one 3-week course of N-methylglucamine antimonate; three patients improved and one showed no clinical change. Fernandez-Guerrero's 1987 series [224] of VL in the immunocompromised hosts included three with HIV infection or AIDS: one was cured and two relapsed. Overall figures from a review of 47 reported cases, including the two series mentioned above, showed that nine HIV-positive patients failed to exhibit a primary clinical response to antimonial therapy and eight of them had died within 5 months of starting treatment. Of those who did show a primary response, at least 13 relapsed. The mode of death was known for 13 of the 17 patients who died and was ascribed to respiratory complications in eight cases [217]. Montalban et al. [219] reported a 44% relapse rate in their series, in contrast to a rate of 10% for immunocompetent patients in the same country. On the basis of a single case report, Flegg et al. [226] have suggested that VL in HIV infection can conform to a more benign pattern and that only severe relapsing or disseminating disease should be classified as CDC group IVC-1 (AIDS).

MALARIA

Human immunodeficiency states

Early reports that antibodies to HIV were associated with acute malaria infections were later explained by a high antimalarial antibody titer causing nonspecific reactivity and false positive results in the HIV antibody assays then in use [227,228]. Later assays have shown no crossreaction between antibody to P. falciparum and anti-HIV antibody [228]. Simooya et al. [229] determined whether infection with HIV increased the risk or severity of infection with P. falciparum in 170 Zambian patients aged 12 years and above. Parasitemia was less common in those with HIV antibodies than those without (29% vs. 42%) but the difference was not significant. The mean parasite density in those with parasitemia was higher in those who were HIV seropositive, but again the difference was not significant. There was no significant difference in P. falciparum antibody titers between patients positive for HIV antibody and those negative, whether or not they had a patent parasitemia. The study included patients aged 12 years and above which, in a malaria endemic area, means that a very high proportion of them would have experienced previous malaria infections. The authors therefore suggested a study of the acquisition of malarial immunity in infants positive and negative for HIV antibody.

Nguyen-Dinh et al. [230] looked for an association between P. falciparum malaria and HIV seropositivity in children in Kinshasa, Zaire. In childen aged 9 months to 12 years, the HIV seropositivity rate (1.2% of 164) for those presenting with P. falciparum malaria was not significantly different from the rate (0.6% of 169) for healthy controls. The second part of the study, on 1046 children aged 1 month to 13 years, showed no association between P. falciparum slide positivity (51.6%) and HIV seropositivity (3.8%). Malarial antibodies were not measured.

Wabwire-Mangen et al. [231] studied a hospital-based population in Zambia and measured, by ELISA, antibody levels to the RESA-4, RESA-8, and RESA-11 synthetic ring-stage peptides of P. falciparum. HIV-1 positive patients with clinical features of AIDS had significantly lower antibody levels to RESA-8 than did a comparable group of seronegative patients. Antibody levels in the HIV-1-positive AIDS group were also lower for RESA-4 and RESA-11. In HIV-1-seropositive trauma (non-AIDS) patients antibody levels were higher than those found in HIV-1-seronegative trauma patients, although the differences were not statistically significant. The authors postulated that in HIV-1 infection, a polyclonal B cell stimulation may occur, for example in the HIV-1-seropositive trauma patients, but that in advanced AIDS B cell stimulation appears to be diminished as a result of increased immunosuppression, resulting in a decreased production of malarial antibody. Longitudinal studies of HIV-infected patients residing in malarious areas are required to test this hypothesis.

Leaver et al. [232] working in Lusaka, Zambia, studied 27 patients admitted to an intensive care unit with cerebral malaria; eight out of 27 (30%) were HIV seropositive. Two seropositive and two seronegative patients died. A control group of 396 inpatients and a group of 260 intensive care unit patients in the same hospital had

HIV seropositivity rates of 20.5% and 33%, respectively. The authors concluded that the outcome of cerebral malaria in the Lusaka area is not affected by HIV status.

So far there seems to be no difference in incidence or severity of malaria in HIV-seropositive versus HIV-seronegative patients. Studies in patients with advanced AIDS may partially alter this view [7].

ENTAMOEBA HISTOLYTICA

Human immunodeficiency states

While infection with *E. histolytica* is well recognized in homosexuals, symptomatic disease due to invasion by this parasite, causing colitis or amebic liver abscess, appears to be uncommon [233]. Studies in the tropics have not shown any increased incidence of invasive disease among those with HIV and AIDS [234]. Reports that indicate symptomatic improvement in AIDS patients after treatment with metronidazole for amebiasis do not contain clear evidence for a pathogenic role for this infection [235]. Three out of five improved, but in only one was *E. histolytica* the sole pathogen and two needed additional medication for other infections. One patient received metronidazole alone and improved. Diarrhea in AIDS and HIV infection is frequently caused by more than one pathogen at the same time [235,236].

There is anecdotal evidence to suggest that steroids have an adverse effect on the progression of hepatic amebiasis in humans [237] and amebic colitis does not improve when steroid therapy is given for a mistaken diagnosis of nonspecific inflammatory bowel disease [238]. There are instances in which intestinal amebiasis may progress much more rapidly with a fatal outcome in patients given steroids. There is also a widely perceived clinical impression that amebic liver abscess is more common among persons with a high alcohol intake and there is much evidence to suggest that alcohol has immunosuppressant actions [239]. The nutritional aspects of host immune responses are poorly understood. The Murrays [240] found that milk-drinking Somali nomads developed an increased incidence of amebiasis, i.e., cysts of the parasite, trophozoites, and positive amebic serology, among those given oral iron supplements. They ascribed this to a beneficial effect of iron on luminal proliferation of amebae. Pregnancy is another situation in which there are changes in immune responsiveness, and clinical experience suggests increased susceptibility to invasive amebic disease. Infants and malnourished persons show an increased incidence.

Clinical features

The methods used for diagnosis are the same in immunodeficient and normal hosts. The response to treatment is generally good, although those who have serious underlying disease, such as malnutrition, will need intensive supportive therapy in addition to anti-amebic therapy with the standard drugs. Patients who have received steroids and have amebic dysentery are frequently extremely sick and may develop complications, such as perforation, which can be fatal.

GIARDIA LAMBLIA

Introduction

Giardia lamblia is a common protozoan that infects the upper gastrointestinal tract of humans. This infection has a worldwide distribution and is not uncommon in the temperate parts of the world. Giardiasis is frequently asymptomatic but in symptomatic cases the severity can range from mild diarrhea to profuse diarrhea with significant weight loss. In immunocompetent patients the natural history of infection is for symptoms to improve with time and in asymptomatic persons for infection to be eradicated with time. Against this background, the effects of impaired immune responses on giardiasis can be considered in both patients and experimental animals.

Animal models

Giardia muris was found in the intestines of nude mice together with *Hexamita*; this observation raised the possibility that defective immune responses allowed these infections to persist and it was also suggested that these infections contributed to the increased mortality in these animals [241]. Subsequently nude mice with and without immune reconstitution [242], mice treated with steroids [243], and various strains of mice [244] have all been used for studies to examine the effects of different genetic constitutions on expression of the host response to *Giardia*.

Humoral and cell-mediated immunity

Trophozoites can be ingested and digested by macrophages [245] but the significance of this *in vivo* is not clear; it could be responsible for sensitizing the host to parasite antigens, although this would involve cellular migration into the gut lumen and then return to the mucosa, or mucosal invasion by the parasite, referred to above.

Antibody to *Giardia*, particularly IgA secreted from the mucosa and in the bile [246], would appear to be important and, in the rat experimentally infected with *G. lamblia*, appeared within 10 days of challenge. Trophozoite motility and morphology are adversely affected by anti-*Giardia* antibody [247]. Cellular reactivity appears to be important in the development of mucosal changes in enterocyte dynamics [248]. The nature of the antigens involved in the latter is unclear. The immunologic aspects of this infection are discussed in detail in Chapter 9.

Studies in murine giardiasis have indicated the importance of the host immune reactivity in both controlling trophozoite populations in the gut lumen and in contributing to the morphologic changes seen in the mucosa in infection. While the importance of T cell populations seems to have been demonstrated in the mouse, this does not seem to have a counterpart in the most severe and currently the most common immune deficiency state in humans, AIDS.

Human immunodeficiency states

The best-recognized association in humans occurs in patients with hypogammaglobulinemia [249]. These rather uncommon cases are often troubled by diarrhea and malabsorption, and giardiasis is a well-recognized cause. There have been suggestions that the more common selective IgA deficiency is associated with increased risk of giardiasis but large series of cases with symptomatic disease have not been reported. Impaired cellular immunity does not appear to predispose to giardiasis. Infection with HIV causing AIDS is currently one of the most common causes of defective cellular immune responses and diarrheal disease in a common clinical feature [250]. Among protozoal causes for this, cryptosporidiosis (see above) is the most common and *G. lamblia* seems to be relatively uncommon in both tropical and temperate environments (reviewed by Cook [251]), despite the observation that *Giardia* was a fairly common parasite among homosexuals [233]. Malnutrition is the most common cause of impaired immunity in the world today and there is some evidence to suggest that giardiasis is more common in malnourished children, although the extent to which this relates to impaired immunity rather than the contaminated environment in which they live [252] is not clear.

Clinical features

Diagnosis. The diagnosis is made by the same methods in immunocompromised and normal hosts. In humoral immunodeficiency, parasite populations may be persistently higher, allowing their ready detection. Antigen detection for diagnosis of giardiasis [253] may prove useful in some patients. Obviously serologic testing would be of little value in hypogammaglobulinemic patients.

Treatment. There is no evidence to suggest that treatment is likely to be any less effective than in immunocompetent hosts. However, 100% cure rates cannot be assured with any of the drugs presently used and so repeated dosing may be needed to eradicate infection where there is no immune response to augment the effects of therapy. There is now some *in vitro* evidence to suggest that *G. lamblia* can become resistant to drugs, which may add to the difficulty of managing some cases. In view of the enhanced susceptibility of patients with immunoglobulin deficiencies to giardiasis it is essential to ensure that family members and other close household contacts are free of infection, as reinfection from close contacts is a strong possibility.

CONCLUSIONS

This chapter has focused on some of the parasites on which the immunocompromised host can shed light. The past decade, since the description of AIDS, has seen a huge increase in the amount of work being done on all these organisms. Some, like *P. carinii*, have re-emerged into the limelight while others, like the microsporidia, have only been recognized as human pathogens since the AIDS epidemic. For many, the amount known of their basic biology and life-cycles has lagged behind clinical experience. However, with the application of new technologies in immunology, molecular biology and biochemistry, we are making steady progress and the time is fast approaching when results from the laboratories will be translated into novel therapies and a clearer understanding in the clinic.

REFERENCES

1 Grove DI. Clinical aspects. In Grove DI, ed. *Strongyloidiasis*. London: Taylor & Francis, 1989:155–174.
2 Cook GC. *Strongyloides stercoralis* hyperinfection syndrome: how often is it missed? Q J Med 1987;64:625–629.

3 Genta RM. Immunobiology of strongyloidiasis. Trop Geogr Med 1984;36:223–229.

4 Maayan S, Wormser GP, Widerhorn J, *et al. Strongyloides stercoralis* hyperinfection in a patient with the acquired immune deficiency syndrome. Am J Med 1987;83:945–948.

5 Vierya HG, Becerril CG, Padua GA, Jessurun J. *Strongyloides stercoralis* hyperinfection in a patient with the acquired immune deficiency syndrome. Acta Cytol 1988;32:277–278.

6 Schainberg L, Scheinberg MA. Recovery of *Strongyloides stercoralis* by bronchoalveolar lavage in a patient with acquired immunodeficiency syndrome. Am J Med 1989;87:486.

7 Lucas SB. Missing infections in AIDS. Trans R Soc Trop Med Hyg 1990;84 (Suppl. 1):34–38.

8 Nakada K, Kohakura M, Komoda H, Hinuma Y. High incidence of HTLV antibody in carriers of *Strongyloides stercoralis*. Lancet 1984;i:633.

9 Nakada K, Yamaguchi K, Furugen S, *et al*. Monoclonal integration of HTLV-1 proviral DNA in patients with strongyloidiasis. Int J Cancer 1987;40:145–148.

10 Genta RM. *Strongyloides stercoralis*: immunobiological considerations of an unusual worm. Parasitol Today 1986;2:241–246.

11 Kreiss JK, Castro KG. Special considerations for managing suspected human immunodeficiency virus infection and AIDS in patients from developing countries. J Infect Dis 1990;162:955–960.

12 Wakefield AE, Pixley FJ, Banerji S, *et al*. Detection of *Pneumocystis carinii* with DNA amplification. Lancet 1990;336:451–453.

13 McLeod DT, Neill P, Gwanzura L, *et al. Pneumocystis carinii* pneumonia in patients with AIDS in Central Africa. Resp Med 1990;84:225–228.

14 Lucas S, Sewankambo N, Nambuya A, *et al*. The morbid anatomy of African AIDS. In Giraldo G, Beth-Giraldo E, Clumeck N, *et al.*, eds. *AIDS and Associated Cancers in Africa*. Basel: Karger, 1988:124–133.

15 Edman U, Edman JC, Lundgren B, *et al*. Isolation and expression of the *Pneumocystis carinii* thymidilate synthetase gene. Proc Natl Acad Sci USA 1989;86:6503–6507.

16 Lundgren B, Cotton R, Lundgren J, *et al*. Identification of *Pneumocystis carinii* chromosomes and mapping of five genes. Infect Immun 1990;58:1705–1710.

17 Cushion MT, Ruffolo JJ, Linke MJ, *et al. Pneumocystis carinii* growth variables and estimates in the A549 and WI-38 VA13 human cell lines. Exp Parasitol 1985;60:43–54.

18 Walzer PD, Linke JM. A comparison of the antigenic characteristics of rat and human *Pneumocystis carinii* by immunoblotting. J Immunol 1987;138:2257–2265.

19 Hughes WT. Natural mode of acquisition for *de novo* infection with *Pneumocystis carinii*. J Infect Dis 1982;145:842.

20 Hughes WT, Price RA, Sisko F, *et al*. Protein-calorie malnutrition—a host determinant for *Pneumocystis carinii* infection. Am J Dis Child 1974;128:44.

21 Furuta T, Ueda K, Kyuwa S, Fujiwara K. Effect of T cell transfer on *Pneumocystis carinii* infection in nude mice. Jpn J Exp Med 1984;54:65–72.

22 Furuta T, Ueda K, Fujiwara K, *et al*. Cellular and humoral immune responses of mice subclinically infected with *Pneumocystis carinii*. Infect Immun 1985;47:544–548.

23 Walzer PD, Rutledge ME. Humoral immunity in experimental *Pneumocystis carinii* infection. J Lab Clin Med 1981;97:820–833.

24 Gigliotti F, Hughes WT. Passive immunoprophylaxis with specific monoclonal antibody confers partial protection against *Pneumocystis carinii* pneumonitis in animal models. J Clin Invest 1988;81:1666–1668.

25 Pfifer LL, Hughes WT, Stagno S, *et al. Pneumocystis carinii* infection: evidence for high prevalence in normal and immunosuppressed children. Pediatrics 1978;61:35.

26 Meuwissen JHE Th, Tauber I, Leeuwenberg ADEM, *et al*. Parasitologic and serologic observations of infection with *Pneumocystis carinii* in humans. J Infect Dis 1977;186:43.

27 Settnes OP, Genner J. *Pneumocystis carinii* in human lungs at autopsy. Scand J Infect Dis 1986;18:489–496.

28 Millard PR, Heryet AR. Observations favouring *Pneumocystis carinii* pneumonia as a primary infection: a monoclonal antibody study on paraffin sections. J Pathol 1988;154:365–370.

29 Shepherd V, Jameson B, Knowles GK. *Pneumocystis carinii* pneumonitis, a serological study. J Clin Pathol 1979;32:773.

30 Norman L, Kagan I. Some observations on the serology of *Pneumocystis carinii* infections in the United States. Infect Immun 1973;8:317–321.

31 Singer C, Armstrong D, Rosen PP, *et al. Pneumocystis carinii* pneumonia: a cluster of 11 cases. Ann Intern Med 1975;82:772.

32 Ruebush TK, Weinstein RA, Baehner RLS, *et al*. An outbreak of *Pneumocystis carinii* pneumonia in children with acute lymphocytic leukaemia. Am J Dis Child 1978;132:143.

33 Watanabe JM, Chinchinian H, Weitz C, *et al. Pneumocystis carinii* pneumonia in a family. J Am Med Assoc 1965;193:685.

34 Dutz W, Post C, Vessel K, *et al*. Endemic infantile *Pneumocystis carinii* infection: the Shiraz study. Symposium on *Pneumocystis carinii* infection. Natl Cancer Inst Monogr 1976;43:31.

35 Gleason WA, Roodman ST. Reversible T cell depression in malnourished infants with *Pneumocystis pneumonia*. J Pediatr 1977;90:1022.

36 Griffiths MH, Kocjan GC, Miller RF, Godfrey-Faussett P. The diagnosis of pulmonary disease in HIV infection: the role of transbronchial biopsy and broncho-alveolar lavage. Thorax 1989;44:554–558.

37 Pitchenik AE, Ganjei P, Torres A, *et al*. Sputum examination for the diagnosis of *Pneumocystis carinii* pneumonia in the acquired immunodeficiency syndrome. Am Rev Respir Dis 1986;133:226–229.

38 Bigby TD, Margolskee D, Curtis JL, *et al*. The usefulness of induced sputum in the diagnosis of *Pneumocystis carinii* pneumonia in patients with the acquired immunodeficiency syndrome. Am Rev Respir Dis 1986;133:515–518.

39 Leigh TR, Parsons B, Hume C, *et al*. Sputum induction for diagnosis of *Pneumocystis carinii* pneumonia. Lancet 1989;2:205–206.

40 Zaman MK, *et al*. Rapid noninvasive diagnosis of *Pneumocystis carinii* from induced liquefied sputum. Ann

Intern Med 1988;ii:7–10.

41 Matsumoto Y, Amagai T, Yamada M, *et al*. Production of a monoclonal antibody with specificity for the pellicle of *Pneumocystis carinii* by hybridoma. Parasitol Res 1987;73: 228–233.

42 Kovacs JA, Ng VL, Masur H, *et al*. Diagnosis of *Pneumocystis carinii* pneumonia: improved detection in sputum with use of monoclonal antibodies. N Engl J Med 1988;318:589–593.

43 Elvin KM, Bjorkman A, Linder E, *et al*. *Pneumocystis carinii* pneumonia: detection of parasites in sputum and bronchoalveolar lavage fluid by monoclonal antibodies. Br Med J 1988;297:381–384.

44 Wakefield AE, Hopkin JM, Burns J, *et al*. Cloning of DNA from *Pneumocystis carinii*. J Infect Dis 1988;158:859–862.

45 Tanabe K, Fuchimoto M, Egawa K, *et al*. Use of *Pneumocystis carinii* genomic DNA clones for DNA hybridization analysis of infected human lungs. J Infect Dis 1988;157:593–596.

46 Murray JF, Felton CP, Garay SM, *et al*. Pulmonary complications of the acquired immunodeficiency syndrome. N Engl J Med 1984;310:1682.

47 Montgomery AB, Debs RJ, Luce JM, *et al*. Aerosolized pentamidine as sole therapy for *Pneumocystis carinii* pneumonia in patients with the acquired immunodeficiency syndrome. Lancet 1987;1:480–483.

48 Miller RF, Godfrey-Faussett P, Semple SJG. Nebulised pentamidine as treatment for *Pneumocystis carinii* pneumonia in the acquired immunodeficiency syndrome. Thorax 1989;44:565–569.

49 Allegra CJ, Chabner BA, Tuazon CU, *et al*. Trimetrexate for the treatment of *Pneumocystis carinii* pneumonia in patients with the acquired immunodeficiency syndrome. N Engl J Med 1987;317:978–985.

50 Paulson TJ, Gilman TM, Helestine PNR, Sharma OP, Boylen CT. Eflornithine treatment of refractory *Pneumocystis carinii* pneumonia in patients with acquired immunodeficiency syndrome. Chest 1992;101:67–74.

51 MacFadden DK, Edelson JD, Hyland RH, *et al*. Corticosteroids as adjunctive therapy in treatment of *Pneumocystis carinii* pneumonia in patients with the acquired immunodeficiency syndrome. Lancet 1987;1: 1477–1479.

52 Miller RF, Mitchell DM. Management of respiratory failure in patients with the acquired immunodeficiency syndrome and *Pneumocystis carinii* pneumonia. Thorax 1990;45:140–146.

53 Hughes WT, McNabb PC, Makres TD, *et al*. Efficacy of trimethoprim and sulphamethoxazole in the prevention and treatment of *Pneumocystis carinii* pneumonitis. Antimicrob Agents Chemother 1974;5:289.

54 Hughes WT, Kuhn S, Chandhary S, *et al*. Successful chemoprophylaxis for *Pneumocystis carinii* pneumonitis. N Engl J Med 1977;297:1419.

55 Phair J, Munoz A, Detels R, *et al*. The risk of *Pneumocystis carinii* among men infected with human immunodeficiency virus type 1. N Engl J Med 1990;322:161–165.

56 CDC. Guidelines for prophylaxis against *Pneumocystis carinii* for persons infected with human immunodeficiency virus. Morbidity and Mortality Weekly Report 1989;38:S51.

57 Fischl MA, Dickinson GM, La Voie L. Safety and efficacy of sulfamethoxazole and trimethoprim chemoprophylaxis for *Pneumocystis carinii* pneumonia in AIDS. J Am Med Assoc 1988;259:1185–1189.

58 Martin MA, Cox PH, Beck K, Styer CM, Beall GN. A comparison in the prevention of *Pneumocystis carinii* pneumonia in human immunodeficiency virus-infected patients. Arch Intern Med 1992;152:523–528.

59 Gottlieb MS, Knught S, Mitsuyasu R, *et al*. Prophylaxis of *Pneumocystis carinii* infection in acquired immunodeficiency syndrome (AIDS) with pyrimethamine–sulfadoxine. Lancet 1984;ii:398–399.

60 Busch DF, Follansbee SE. Continuation therapy with pentamidine isethionate for prevention of relapse of *Pneumocystis carinii* pneumonia in AIDS. Proc II Int Conf AIDS Paris 1986;150:531–538.

61 Golden JA, Chernoff D, Hollander H, *et al*. Prevention of relapse of *Pneumocystis carinii* pneumonia by inhaled pentamidine. Lancet 1989;1:654–657.

62 CDC. *Mycobacterium tuberculosis* outbreak in a Florida health clinic. Morbidity and Mortality Weekly Report 1989; 38:256–264.

63 Jules-Elysee KM, Stover D, Zaman MB, *et al*. Aerosolized pentamidine: effect on diagnosis and presentation of *Pneumocystis carinii* pneumonia. Ann Intern Med 1990;112: 750–757.

64 Luft BJ, Remington JS. Toxoplasmosis of the central nervous system. In Remington JS, Swartz MN, eds. *Current Clinical Topics in Infectious Diseases*, vol. 6. New York: McGraw-Hill Book Co., 1985:315–358.

65 Britt RH, Enzmann DR, Remington JS. Intracranial infection in cardiac transplant recipients. Ann Neurol 1981;9:107–119.

66 Luft BJ, Naot Y, Araujo FG, Stinson EB, Remington JS. Primary and reactivated *Toxoplasma* infections in patients with cardiac transplants: clinical spectrum, and problems in diagnosis in a defined populating. Ann Intern Med 1983;99:27–31.

67 Hakim M, Esmore D, Wallwork J, English TAH, Wreghitt T. Toxoplasmosis in cardiac transplantation. Br Med J 1986;292:1108.

68 Luft BJ, Brooks RG, Conley FK, McCabe RE, Remington JS. Toxoplasmic encephalitis in patients with acquired immunodeficiency syndrome. J Am Med Assoc 1984;252: 913–917.

69 Levy RM, Bredesen DE, Rosenblum ML. Neurological manifestations of the acquired immunodeficiency syndrome (AIDS): experience at UCSF and review of the literature. J Neurosurg 1985;62:475–495.

70 Moskowitz L, Hensley GT, Chan JC, Adams K. Immediate cause of death in acquired immunodeficiency syndrome. Arch Pathol Lab Med 1985;109:735–738.

71 Tschirhart D, Klatt EC. Disseminated toxoplasmosis in acquired immunodeficiency syndrome. Arch Pathol Lab Med 1988;112:1237–1241.

72 Shultz LD, Sidman CL. Genetically determined murine models of immunodeficiency. Ann Rev Immunol 1987;5: 367–403.

73 Lindberg RE, Frenkel JK. Toxoplasmosis in nude mice. J Parasitol 1977;63:219–221.

74 McCune JM, Namikawa R, Kaneshimsa H, Shultz LD, Lieberman M, Weissman IL. The SCID-hu mouse: murine model for the analysis of human hematolymphoid differentiation and function. Science 1988;241:1632–1639.

75 McCune JM, Namikawa R, Chu-Chih Shi, Rabin L, Kaneshima H. Suppression of HIV in AZT-treated SCID-hu mice. Science 1990;247:564–566.

76 Vollmer TL, Waldor MK, Steinman L, Conley FK. Depletion of T-4$^+$ lymphocytes with monoclonal antibody reactivates toxoplasmosis in the central nervous system: a model of superinfection in AIDS. J Immunol 1989;138:3737–3741.

77 Chinchilla M, Frenkel JK. Specific mediation of cellular immunity to Toxoplasma gondii in somatic cells of mice. Infect Immun 1984;46:862–866.

78 McCleod R, Eisenhauer P, Mack D, Brown C, Filice G, Spitalny G. Immune responses associated with early survival after peroral infection with Toxoplasma gondii. J Immunol 1989;142:3247–3255.

79 Suzuki Y, Conley FK, Remington JS. Importance of endogenous interferon-gamma for prevention of toxoplasmic encephalitis in mice. J Immunol 1989;143:2045–2050.

80 Pfefferkorn ER, Eckel M, Guyre PM. Interferon-gamma suppresses the growth of Toxoplasma gondii in human fibroblasts through starvation for tryptophan. Mol Biochem Parasitol 1986;20:215–222.

81 Dannemann BR, Morris VA, Araujo EG, Remington JS. Assessment of human natural killer and lymphokine-activated killer cell cytotoxicity against Toxoplasma gondii trophozoites and brain cysts. J Immunol 1989;143:2684–2691.

82 Brinkmann V, Sharma SD, Remington JS. Different regulation of the L3T4-T cell subset by B cells in different mouse strains bearing the H-2^k haplotype. J Immunol 1986;137:2991–2997.

83 Williams DM, Grumet FC, Remington JS. Genetic control of murine resistance to Toxoplasma gondii. Infect Immun 1978;19:416–420.

84 Strickland GT, Pettit LE, Voller A. Immune depression in mice infected with Toxoplasma gondii. Am J Trop Med Hyg 1973;22:452–455.

85 Cook GC. Toxoplasma gondii infection: a potential danger to the unborn fetus and AIDS sufferers. Q J Med 1990;74:3–20.

86 Herskovitz S, Siegel SE, Schneider AT, Nelson SJ, Goodrich JT, Lantos G. Spinal cord toxoplasmosis in AIDS. Neurology 1989;39:1552–1553.

87 Koppel BS, Wormser GP, Tuchman AJ, Maayan S, Hewlett D Jr, Daras M. Central nervous system involvement in patients with acquired immune deficiency syndrome (AIDS). Acta Neurol Scand 1985;71:337–353.

88 Wong B, Gold JWM, Brown AE, et al. Central nervous system toxoplasmosis in homosexual men and parenteral drug abusers. Ann Intern Med 1984;100:36–42.

89 Lang W, Miklossy J, Deruaz JP, et al. Neuropathology of the acquired immune deficiency syndrome (AIDS): a report of 135 consecutive autopsy cases from Switzerland. Acta Neuropathol 1989;77:379–390.

90 Wanke C, Tuazon CU, Kovacs A, et al. Toxoplasma encephalitis in patients with acquired immune deficiency syndrome: diagnosis and response to therapy. Am J Trop Med Hyg 1987;36:509–516.

91 Tang TT, Harb JM, Dunne WM, et al. Cerebral toxoplasmosis in an immunocompromised host. Am J Clin Pathol 1986;85:104–110.

92 Paz R de la, Enzmann D. Neuroradiology of the acquired immune deficiency syndrome. In Rosenblum ML, Levy RM, eds. AIDS and the Nervous System. New York: Raven Press, 1988:121–154.

93 Wreghitt TG, Hakim M, Gray JJ, et al. Toxoplasmosis in heart and heart and lung transplant recipients. J Clin Pathol 1989;42:194–199.

94 Luft BJ, Remington JS. Toxoplasmic encephalitis. J Infect Dis 1988;157:1–6.

95 Suzuki Y, Israelski DM, Danneman BR, Stepick-Biek P, Thulliez P, Remington JS. Diagnosis of toxoplasmic encephalitis in patients with acquired immunodeficiency syndrome by using a new serologic method. J Clin Microbiol 1988;26:2541–2543.

96 Potasman I, Resnick L, Luft BJ, Remington JS. Intrathecal production of antibodies against Toxoplasma gondii in patients with toxoplasmic encephalitis and the acquired immunodeficiency syndrome. Ann Intern Med 1988;108:49–51.

97 Bishburg E, Eng RHK, Slim J, Perez G, Johnson E. Brain lesions in patients with acquired immunodeficiency syndrome Arch Intern Med 1989;149:941–943.

98 Suzuki Y, Kobayashi A. Presence of high concentrations of circulating Toxoplasma antigens during acute Toxoplasma infection in athymic nude mice. Infect Immun 1987;55:1017–1018.

99 Huskinson J, Stepick-Biek P, Remington JS. Detection of antigens in urine during acute toxoplasmosis. J Clin Microbiol 1989;27:1099–1101.

100 Van Trapen F, Panggbean SO, van Leusden J. Demonstration of Toxoplasma antigen containing complexes in active toxoplasmosis. J Clin Microbiol 1985;22:645–650.

101 Derouin F, Sarfati C, Beauvais B, Iliou M-C, Dehen L, Lariviere M. Laboratory diagnosis of pulmonary toxoplasmosis in patients with acquired immunodeficiency syndrome. J Clin Microbiol 1989;27:1661–1663.

102 Miller MJ, Aronson WJ, Remington JS. Late parasitemia in asymptomatic acquired toxoplasmosis. Ann Intern Med 1969;71:139–145.

103 Burg JL, Grover CM, Pouletty P, Boothroyd JC. Direct and sensitive detection of a pathogenic protozoan, Toxoplasma gondii, by polymerase chain reaction. J Clin Microbiol 1989;27:1787–1792.

104 Navia BA, Petito CK, Gold JWM, Cho ES, Jordon BD, Price RW. Cerebral toxoplasmosis complicating the acquired immune deficiency syndrome: clinical and neuropathological findings in 27 patients. Ann Neurol 1986;19:224–238.

105 Cohn JA, McMeeking A, Cohen W, Holzman RS. Evaluation of the policy of empiric treatment of suspected Toxoplasma encephalitis in patients with the acquired immune deficiency syndrome. Am J Med 1989;86:521–527.

106 Leport C, Raffi F, Matheron S, et al. Treatment of central nervous system toxoplasmosis with pyrimethamine/

sulphadiazine combination in 35 patients with the acquired immunodeficiency syndrome: efficacy of long-term continuous therapy. Am J Med 1988;84:94–100.

107 Webster LT Jr. Drugs used in the chemotherapy of protozoal infections. In Gilman AG, Goodman LS, Rall TW, Murad F, eds. *Goodman and Gilman's The Pharmacological Basis of Therapeutics*, 7th edn. New York: Macmillan Publishing, 1985:1037.

108 Gordon FM, Simon GL, Wofsby CB, Mills J. Adverse reactions to trimethoprim–sulfamethoxazole in patients with the acquired immunodeficiency syndrome. Ann Intern Med 1984;100:495–499.

109 Pfefferkorn ER. Anticoccidial agents in the treatment of toxoplasmic encephalitia. In McAdam KPWJ, ed. *New Strategies in Parasitology*. Proceedings of an international symposium. Edinburgh: Churchill Livingstone, 1989:237–254.

110 Sluiters JF, Balk AHMM, Essed CE, *et al*. Indirect enzyme-linked immunosorbent assay for immunoglobulin G and four immunoassays for immunoglobulin M to *Toxoplasma gondii* in a series of heart transplant recipients. J Clin Microbiol 1989;27:529–535.

111 Tyzzer EE. A sporozoan found in the peptic glands of the common mouse. Proc Soc Exp Biol Med 1907;5:12–13.

112 Tyzzer EE. *Cryptosporidium parvum* (sp. nov), a coccidium found in the small intestine of the common mouse. Arch Parasitol 1912;26:394–413.

113 Nime FA, Burek JD, Page DL, Holscher MA, Yardley JH. Acute enterocolitis in a human being infected with the protozoan *Cryptosporidium*. Gastroenterology 1976;70:592–598.

114 Meisel JL, Perera DR, Meligro C, Rubin CE. Overwhelming watery diarrhea associated with *Cryptosporidium* in an immunosuppressed patient. Gastroenterology 1976;70:1156–1160.

115 Tzipori S, Angus KW, Gray EW, Campbell I. Vomiting and diarrhea associated with cryptosporidial infection. N Engl J Med 1980;303:818.

116 Fayer R, Ungar BLP. *Cryptosporidium* spp. and cryptosporidiosis. Microbiol Rev 1986;50:458–483.

117 Tzipori S. Cryptosporidiosis in perspective. Adv Parasitol 1988;27:63–129.

118 Rehg JE, Hancock ML, Woodmansee DB. Characterization of a cyclophosphamide-rat model of cryptosporidiosis. Infect Immun 1987;55:2669–2674; Characterization of a dexamethasone-rat model of cryptosporidial infection. J Infect Dis 1988;158:1406–1407.

119 Brasseur P, Lemeteil D, Ballet JJ. Rat model for human cryptosporidiosis. J Clin Microbiol 1988;26:1037–1039.

120 Sherwood D, Angus KW, Snodgrass DR, Tzipori S. Experimental cryptosporidiosis in laboratory mice. Infect Immun 1982;38:471–475.

121 Heine J, Moon HW, Woodmansee DB. Persistent *Cryptosporidium* infection in congenitally athymic (nude) mice. Infect Immun 1984;43:856–859.

122 Ungar BLP, Burris JA, Quinn CA, Finkelman FD. New mouse models for chronic *Cryptosporidium* infection in immunodeficient hosts. Infect Immun 1990;58:961–969.

123 Vetterling JM, Jervis HR, Merrill TG, Sprinz H. *Cryptosporidium wrairi* (sp. n) from the guinea pig. J Protozool 1971;18:248–261.

124 Angus KW, Hutchinson G. Infectivity of a strain of *Cryptosporidium* found in the guinea pig (Cavia porcellus) for guinea pigs, mice and lambs. J Comp Pathol 1985;95:151–165.

125 Chrisp CE, Reid WC, Rush HG, Suckow MA, Bush A, Thomann MJ. Cryptosporidiosis in guinea pigs: an animal model. Infect Immun 1990;58:674–679.

126 Iseki M, Maekawa T, Moriya K, Uni S, Takada S. Infectivity of *Cryptosporidium muris* (strain RN66) in various laboratory animals. Parasitol Res 1989;75:218–222.

127 Current WL, Snyder DB. Development of and serologic evaluation of acquired immunity to *Cryptosporidium baileyi* by broiler chickens. Poult Sci 1988;67:720–729.

128 Tzipori S, Campbell I. Prevalence of *Cryptosporidium* antibodies in ten animal species. J Clin Microbiol 1981;14:455–456.

129 Van Opdenbosch E, Wellemans G. Detection of antibodies to *Cryptosporidium* by indirect immunofluorescence (drop method): prevalence of antibodies in different animal species. Vlaams Diergeneeskg Tijdschr 1985;54:49–54.

130 Campbell PN, Current WL. Demonstration of serum antibodies to *Cryptosporidium* sp. in normal and immunodeficient humans with confirmed infections. J Clin Microbiol 1983;18:165–169.

131 Casemore DP. The antibody response to *Cryptosporidium*: development of a serological test and its use in a study of immunological normal persons. J Infect 1987;14:125–134.

132 Hill BD. Immune responses in *Cryptosporidium* infections. In Angus KW, Blewett DA, eds. *Cryptosporidiosis: Proceedings of the First International Workshop*. Edinburgh: Animal Diseases Research Association, 1989:97–105.

133 Ungar BL, Soave R, Fayer R, Nash TE. Enzyme immunoassay detection of immunoglobulin M and G antibodies to *Cryptosporidium* in immunocompetent and immunocompromised persons. J Infect Dis 1986;153:570–578.

134 Williams RO, Burden DJ. Measurement of class specific antibody against *Cryptosporidium* in serum and faeces from experimentally infected calves. Res Vet Sci 1987;43:264–265.

135 Douglas RG, Roberts RB, Romano P, *et al*. Infectious complications in acquired immunodeficiency syndrome—experience at the New York Hospital Cornell Medical Center. In Gottlieb MS, Douglas RG, Roberts MS, Groopman JE, eds. *Acquired Immunodeficiency Syndrome*. Sympsosium on AIDS, Park City, Feb 5–10 1984. New York: Alan R Liss, 1984.

136 Malabranche R, Guerin JM, Laroche AC, *et al*. AIDS with severe gastrointestinal manifestations in Haiti. Lancet 1983;ii:873–877.

137 Clavel F, Mansinho K, Chamaret S, *et al*. Human immunodeficiency virus type 2 infection associated with AIDS in West Africa. N Engl J Med 1987;316:1180–1185.

138 Current WL. *Cryptosporidium* spp. In McAdam KPWJ, ed. *New Strategies in Parasitology*. Proceedings of an international symposium. Edinburgh: Churchill Livingstone, 1989:281–341.

139 Saxon A, Weinstein W. Oral administration of bovine colostrum anti-cryptosporidia antibody fails to alter the course of human cryptosporidiosis. J Parasitol 1987;73:

413–415.

140 Fayer R, Andrews C, Ungar BLP. Efficacy of hyper-immune bovine colostrum for prophylaxis of crypto-sporidiosis in neonatal calves. J Parasitol 1989;75:393.

141 Fayer R, Perryman LE, Riggs MW. Hyperimmune bovine colotrum neutralizes *Cryptosporidium* sporozoites and protects mice against oocyst challenge. J Pathol 1989;75:151–153.

142 Lazo A, Barriga OO, Redman DR, Nielsen SB. Identification by transfer blot of antigens reactive in ELISA in rabbits immunized and a calf infected with *Cryptosporidium* sp. Vet Parasitol 1986;21:151–163.

143 Luft BJ, Payne D, Woodmanse D, Kim CW. Characterization of the *Cryptosporidium* antigens from sporulated oocysts of *Cryptosporidium parvum*. Infect Immun 1987;55:2436–2441.

144 Mead JR, Arrowood MJ, Sterling CR. Antigens of *Cryptosporidium* sporozoites recognised by immune sera of infected animals and humans. J Parasitol 1988;74:135–143.

145 Ungar BLP, Nash TE. Quantification of specific antibody response to *Cryptosporidium* antigens by laser densitometry. Infect Immun 1986;53:124–128.

146 Blanchard JL, Baskin GB, Murphey-Corb M, Martin LN. Disseminated cryptosporidiosis in simian immuno-deficiency virus/delta-infected rhesus monkeys. Vet Pathol 1987;24:454–456.

147 Monticello TM, Levy MG, Bunch SE, Fairley RA. Crypto-sporidiosis in a feline leukemia virus-positive cat. J Am Vet Med Assoc 1987;191:705–706.

148 Navin TR, Hardy AM. Cryptosporidiosis in patients with AIDS. J Infect Dis 1987;155:150.

149 Connolly GM, Dryden MS, Shanson DC, Gazzard BG. Cryptosporidial diarrhoea in AIDS and its treatment. Gut 1988;29:593–597.

150 Colebunders R, Lusakumuni K, Nelson AM, *et al.* Persistent diarrhoea in Zairian AIDS patients: an endoscopic and histological study. Gut 1988;29:1687–1691.

151 Casemore DP. Human cryptosporidiosis. In Reeves DS, Geddes AM, eds. *Recent Advances in Infection*, vol. 3. Edinburgh: Churchill Livingstone, 1989:209–236.

152 Marcial MA, Madra JL. *Cryptosporidium*: cellular localization, structural analysis of absorptive cell–parasite membrane–membrane interactions in guinea pigs and suggestion of protozoan transport by M cells. Gastroenterology 1986;90:583–594.

153 Dobbins WO III, Weinstein WM. Electron microscopy of the intestine and rectum in acquired immunodeficiency syndrome. Gastroenterology 1985;88:738–749.

154 Modigliani R, Bories C, Le Charpentier Y, *et al.* Diarrhoea and malabsorption in acquired immune deficiency syndrome: a study of four cases with special emphasis on opportunistic protozoan infestations. Gut 1985;26:179–187.

155 Sloper US, Dourmashkin RR, Bird RB, Shaun G, Webster ADB. Chronic malabsorption due to cryptosporidiosis in a child with immunoglobulin deficiency. Gut 1982;23:80–82.

156 Kocoshis SA, Cibull ML, Davis TE, Hinton JT, Seip M, Banwell JG. Intestinal and pulmonary cryptosporidiosis in an infant with severe combined immunodeficiency. J Pediatr Gastroenterol Nutr 1984;3:149–157.

157 Pitlik SD, Fainstein V, Garza C, *et al.* Human crypto-sporidiosis, spectrum of disease. Arch Intern Med 1983;143:2269–2275.

158 Davis JJ, Heyman MB. Cryptosporidiosis and sinusitis in an immunodeficient adolescent. J Infect Dis 1988;158:649.

159 Kazlow PG, Shah K, Benkov KJ, Dische R, Leleiko NS. Oesophageal cryptosporidiosis in a child with acquired immune deficiency syndrome. Gastroenterology 1986;91:1301–1303.

160 Khan DG, Garfinkle JM, Klonoff DC, Pembrook LJ, Morrow DJ. Cryptosporidial and cytomegaloviral hepatitis and chole-cystitis. Arch Path Lab Med 1987;111:879–881.

161 Guarda LA, Stein SA, Cleary KA, Ordonez NG. Human crytosporidiosis in AIDS. Arch Pathol Med 1983;107:562–566.

162 Harari MD, West B, Dwyer B. *Cryptosporidium* as a cause of laryngotracheitis in an infant. Lancet 1986;1:1207.

163 Alonso JF, Andrade BE, Heras MMJ, Rodriquez JL, Rodriquez JMA, Leal JAL. Intestinal cryptosporidiosis with affection of biliopancreatic tract and bronchial tree in a child with AIDS. Med Clin 1987;89:335–338.

164 Forgacs P, Tarshis A, Ma P, *et al.* Intestinal and bronchial cryptosporidiosis in a immunodeficient homosexual man. Ann Interm Med 1983;99:793–794.

165 Brady EM, Margolis MC, Korzeniowski OM. Pulmonary cryptosporidiosis in acquired immune deficiency syndrome. J Am Med Assoc 1984;252:89–90.

166 Kibbler CC, Smith A, Hamilton-Dutoit SJ, Milburn H, Pattinson JK, Prentice HG. Pulmonary cryptosporidiosis occurring in a bone marrow transplant patient. Scand J Infect Dis 1987;19:581–584.

167 Hojlyng N, Jensen BN. Respiratory cryptosporidiosis in HIV positive patients. Lancet 1988;1:590.

168 Ma P, Villanueva TG, Kaufman D, Gillooley JF. Respiratory cryptosporidiosis in the acquired immune deficiency syndrome. J Am Med Assoc 1984;252:1298–1301.

169 Tzipori S, Robertson D, Cooper DA, White L. Chronic cryptosporidial diarrhoea and hyperimmune cow colostrum. Lancet 1987;2:344–345.

170 Tzipori S, Robertson D, Chapman C. Remission of diarrhoea due to *Cryptosporidium* in an immunodeficient child treated with hyperimmune bovine colostrum. Br Med J 1986;293:1276–1277.

171 De Mol P, Mukashema S, Bogaerts J, Hemelhof W, Butzler JP. *Cryptosporidium* related to measles diarrhoea in Rwanda. Lancet 1984;ii:42–43.

172 Miller RA, Holmberg RE Jr, Clausen CR. Life threatening diarrhea caused by *Cryptosporidium* in a child undergoing therapy for acute lymphocytic leukemia. J Pediatr 1983;103:256–259.

173 Lewis IJ, Hart CA, Bazby D. Diarrhoea due to *Cryptosporidium* in acute lymphoblastic leukaemia. Arch Dis Child 1985;60:60–62.

174 Mead GM, Sweetenham JW, Ewins DL, Furlong M, Lowes JA. Intestinal cryptosporidiosis: a complication of cancer treatment. Cancer Treat Rep 1986;70:769–770.

175 Martino P, Gentile G, Caprioli A, *et al.* Hospital-acquired cryptosporidiosis in a bone-marrow transplantation unit. J Infect Dis 1988;158:647–648.

176 Weisburger WR, Hutcheon DF, Yardley JH, Roche JC, Hillis WD, Charache P. Cryptosporidiosis in an immuno-

suppressed renal-transplant recipient with IgA deficiency. Am J Clin Pathol 1979;72:473–478.

177 Berkowitz CD, Seidel JS. Spontaneous resolution of cryptosporidiosis in a child with acquired immunodeficiency syndrome. Am J Dis Child 1985;139:967.

178 Just G, Nielsen F, Helm EB, Brodt HR, Stumer S, Stille W. Cryptosporidial infections in AIDS. Dtsch Med Wochensch 1987;112:378–381.

179 McDonald V, Stables R, Warhurst DC, et al. Cryptosporidium parvum: in vitro cultivation of and screening for anticryptosporidial drugs. Antimicrob Agents Chemother 1990;34:1498–1500.

180 McMeeking A, Borkowsky W, Klesius PH, et al. A controlled trial of bovine dialyzable leukocyte extract for cryptosporidiosis in patients with AIDS. J Infect Dis 1990;161:108–112.

181 Fayer R, Klesius PH, Andrews C. Efficacy of bovine transfer factor to protect neonatal calves against experimentally induced clinical cryptosporidiosis. J Parasitol 1987;73:1061–1062.

182 Ungar BLP, Ward DJ, Fayer R, Quinn CA. Cessation of Cryptosporidium-associated diarrhoea in an acquired immunodeficiency syndrome patient after treatment with hyperimmune bovine colostrum. Gastroenterology 1990;98:486–489.

183 Katz MD, Erstad BL, Rose C. Treatment of severe cryptosporidiosis related diarrhoea with octreotide in a patient with AIDS. Drug Intell Clin Pharm 1988;22:134–136.

184 Cook DJ, Kelton JG, Stanisz AM, Collins SM. Somatostation treatment for cryptosporidial diarrhoea in a patient with the acquired immune deficiency syndrome (AIDS). Ann Intern Med 1988;108:708–709.

185 Pape JW, Verdier R-I, Johnson WD. Treatment and prophylaxis of Isospora belli infection in patients with AIDS. N Engl J Med 1989;320:1044–1047.

186 Henry MC, de Clercq D, Lokombe B, et al. Parasitological observations of chronic diarrhoea in suspected AIDS adult patients in Kinshasa (Zaire). Trans R Soc Trop Med Hyg 1986;80:309–310.

187 Greenberg SJ, Davey MP, Zierdt WS, Waldman T. Isospora belli enteric infections in patients with HTLV-1-associated adult T cell leukemia. Am J Med 1988;85:435–438.

188 Canning EU, Lom J. The Microsporidia of Vertebrates. New York: Academic Press, 1986:201–207.

189 Canning EU, Hollister WS. Microsporidia of animals—widespread pathogens or opportunistic curiosities. Parasitol Today 1987;3:267–273.

190 Rijpstra AC, Canning EU, Van Kettel RJ, Eeftinck Schattenkerk JKM, Laarman JJ. Use of light microscopy to diagnose small-intestinal microsporidiosis in patients with AIDS. J Infect Dis 1988;157:827–831.

191 Mohn SF, Nordstoga K. Electrophoretic patterns of serum proteins in blue foxes with special reference to changes associated with nosematosis. Acta Vet Scand 1975;16:297–306.

192 Plowright W. An encephalitis – nephritis syndrome in the dog probably due to congenital Encephalitozoon infection. J Comp Pathol 1952;62:83–93.

193 Bywater JE, Kellett BS. Encephalitozoon cuniculi. Antibodies in a specific pathogen-free rabbit unit. Infect Immun 1978;21:360–364.

194 Shadduk JA, Watson WT, Pakes SP, Cali A. Animal infectivity of Encephalitozoon cuniculi. J Parasitol 1979;65:123–129.

195 Bergquist NR, Stintzing G, Smedman L, Waller T, Anderssen T. Diagnosis of encephalitozoonosis in man by serological tests. Br Med J 1984;288:902.

196 Wosu NJ, Shadduk JA, Pakes SP, Frenkel JK, Todd KS, Conroy JD. Diagonsis of encephalitozoonosis in experimentally infected rabbits by intradermal and immunofluorescence tests. Lab Anim Sci 1977;27:210–216.

197 Szabo JR, Shadduk JA. Immunologic and clinicopathologic evaluation of adult dogs inoculated with Encephalitozoon cuniculi. J Clin Microbiol 1988;26:557–563.

198 Cox JC, Gallichio HA. An evaluation of immunofluorescence in the serological diagnosis of Nosema cuniculi infection. Res Vet Sci 1977;22:50–52.

199 Waller T, Uggla A, Bergquist NR. Encephalitozoonosis and toxoplasmosis diagnosed simultaneously by a novel rapid test; the carbon immunoassay. Proc Third Int Symp of the World Association of Veterinary Laboratory Diagnosticians Ames, IA, June 13–15, 1983. 171–177.

200 Cox JC, Horsburgh R, Pye D. Simple diagnostic test for antibodies to Encephalitozoon based on enzyme immunoassay. Lab Anim 1981;15:41–43.

201 Hollister WS, Canning EU. An enzyme-linked immunosorbent assay (ELISA) for detection of antibodies to Encephalitozoon cuniculi and its use in determination of infections in man. Parasitology 1987;94:209–219.

202 Schmidt EC, Shadduk JA. Mechanisms of resistance to the intracellular protozoon Encephalitozoon in mice. J Immunol 1984;133:2712–2719.

203 Schmidt EC, Shadduk JA. Murine encephalitozoonosis model for studying the host–parasite relationship of a chronic infection. Infect Immun 1983;40:936–942.

204 Niederkorn JY, Shadduk JA, Schmidt EC. Susceptibility of selected inbred strains of mice to Encephalitozoon. J Infect Dis 1981;144:249–253.

205 Niederkorn JY, Brieland JK, Mayhew E. Enhanced natural killer cell activity in experimental murine encephalitozoonosis. Infect Immun 1983;41:302–307.

206 Desportes I, le Charpentier Y, Galian I, et al. Occurrence of a new microsporidian: Enterocytozoon bieneusi n. g., n. sp., in the enterocytes of a human patient with AIDS. J Protozool 1985;32:250–254.

207 Lucas SB, Wamukota W. HIV and the local African population. In Pounder RE, Chiodini PL, eds. Advances in Medicine, vol. 23. London: Baillière Tindell, 1987:102–111.

208 Terada S, Reddy KR, Jeffers LJ, Cali A, Schiff ER. Microsporidian hepatitis in the acquired immune deficiency syndrome. Ann Intern Med 1987;107:61–62.

209 Ledford DK, Overman MD, Gonzalvo D, Cali A, Mester SW, Lockey RF. Microsporidiosis myositis in a patient with acquired immunodeficiency syndrome. Ann Intern Med 1985;102:628–630.

210 Mergileth AM, Strano AJ, Chandra R, Neafie R, Blum M, McCully RM. Disseminated nosematosis in an immunologically compromised infant. Arch Pathol 1974;57:145–150.

211 Friedberg DN, Stenson SM, Orensten JM, Tierno PM, Charles NC. Microsporidial keratoconjunctivitis in acquired immune deficiency syndrome. Arch Opthalmol 1990;108:504–508.

212 Louder CY, Meisler DM, McMahon JT, Longworth DL, Rutherford T. Microsporidia infection of the cornea in a seropositive for human immunodeficiency virus. Am J Opthalmol 1990;109:242–244.

213 Bergquist NR, Morfeldt-Mansson L, Pehrson PO, Petrini B, Wasserman J. Antibody against *Encephalitozoon* in Swedish homosexual men. Scand J Infect Dis 1984;16:389–391.

214 Singh M, Kane GJ, Mackinlay L, *et al*. Detection of antibodies to *Nosema cuniculi* (protozoa: microsporidia) in human and animal sera by the indirect fluorescent antibody response. Southeast Asian J Trop Med Public Health 1982;13:110–113.

215 Alvar J, Blazquez J, Najera R. Association of visceral leishmaniasis and human immunodeficiency virus infections. J Infect Dis 1989;160:560–561.

216 Berenguer J, Moreno S, Cercenado E, Bernaldo de Quiros JCL, Garcia de la Fuente A, Bouza E. Visceral leishmaniasis in patients with human immuno-deficiency virus (HIV). Ann Intern Med 1989;111:129–132.

217 Peters BS, Fish D, Golden R, *et al*. Visceral leishmaniasis in AIDS: clinical features and response to therapy. Q J Med 1990;77:1101–1111.

218 Pinching AJ. Factors affecting the natural history of human immunodeficiency virus infection. Immunodef Rev 1988;1:23–38.

219 Montalban C, Martinez-Fernandez R, Calleja JL, *et al*. Visceral leishmaniasis (kala-azar) as an opportunistic infection with the human immunodeficiency virus in Spain. Rev Infect Dis 1989;11:655–660.

220 Grau JM, Bosch X, Salgado AC, Urbano-Marquez A. Human immunodeficiency virus (HIV) and aplastic anaemia. Ann Intern Med 1989;110:576–577.

221 Yebra M, Segovia J, Manzano L, Vargas JA, Bernaldo de Quiros L. Disseminated-to-skin kala-azar and the acquired immunodeficiency syndrome. Ann Intern Med 1988;108:490–491.

222 Datry A, Similowski T, Jais P, *et al*. AIDS-associated leishmaniasis—an unusual gastro-duodenal presentation. Trans R Soc Trop Med Hyg 1990;84:239–240.

223 Senaldi G, Cadeo G, Carnavale G, di Perri G, Carosi G. Visceral leishmaniasis as an opportunistic infection. Lancet 1986;1:1094.

224 Fernandez-Guerrero ML, Aguado JM, Buzon L, *et al*. Visceral leishmaniasis in immunocompromised hosts. Am J Med 1987;83:1098–1102.

225 Bryceson ADM, Chulay JD, Ho M, *et al*. Visceral leishmaniasis unresponsive to antimonial drugs. I. Clinical and immunological studies. Trans R Soc Trop Med Hyg 1985;79:700–704.

226 Flegg PJ, Brettle RP, Clarkson R, Whitelaw J, Bird AG. Visceral leishmaniasis in association with HIV infection; immunological changes demonstrating a lack of disease progression. Trans R Soc Trop Med Hyg 1990;84:168.

227 Biggar RJ, Gigase PL, Melbye M, *et al*. ELISA HTLV retrovirus antibody reactivity associated with malaria and immune complexes in healthy Africans. Lancet 1985;ii:520–523.

228 Biggar RJ. Possible non specific associations between malaria and HTLV III/LAV. N Engl J Med 1986;315:457.

229 Simooya OO, Mwendapole RM, Siziya S, Fleming AF. Relation between *falciparum* malaria and HIV seropositivity in Ndola, Zambia. Br Med J 1988;297:30–31.

230 Nguyen-Dinh P, Greenberg AE, Mann JM, *et al*. Absence of association between *Plasmodium falciparum* malaria and human immunodeficiency virus infection in children. Bull WHO, 1987;65:607–613.

231 Wabwire-Mangen F, Shiff CJ, Vlahov D, *et al*. Immunological effects of HIV-1 infection on the humoral response to malaria in an African population. Am J Trop Med Hyg 1989;41:504–511.

232 Leaver RJ, Haile Z, Watters DAK. HIV and cerebral malaria. Trans R Soc Trop Med Hyg 1990;84:201.

233 Quinn TC, Stamm WE, Goodell SE, *et al*. The polymicrobial origin of intestinal infections in homosexual men. N Engl J Med 1983;309:576–582.

234 Sewankambo N, Mugerwa RD, Goodgame R, *et al*. Enteropathic AIDS in Uganda—an endoscopic, histological and microbiological study. AIDS 1987;1:9–13.

235 Smith PD, Lane HC, Gill VJ, *et al*. Intestinal infections in patients with the acquired immunodeficiency syndrome. Ann Intern Med 1988;108:328–333.

236 Laughon BE, Druckman DA, Quinn TC, *et al*. Prevalence of enteric pathogens in homosexual men with and without acquired immunodeficiency syndrome. Gastroenterology 1988;94:984–993.

237 Stuiver PC, Gould TJLM. Corticosteroids and liver amoebiasis. Br Med J 1978;2:394–395.

238 Sanderson IR, Walker-Smith JA. Indigenous amoebiasis: an important differential diagnosis of chronic inflammatory bowel disease. Br Med J 1984;289:823.

239 Dunne FJ. Alcohol and the immune system Br Med J 1989;298:543–544.

240 Murray MJ, Murray A, Murray CJ. The salutary effect of milk on amoebiasis and its reversal by iron. Br Med J 1980;2:1351–1352.

241 Boorman GA, Lina PHC, Zurcher C, Nieuwerker HTM. *Hexamita* and *Giardia* as a cause of increased mortality in congenitally thymusless (nude) mice. Clin Exp Immunol 1973;15:623–627.

242 Stevens DP, Frank DM, Mahmoud AAF. Thymic dependency of host resistance to *Giardia muris* infection: studies in nude mice. J Immunol 1978;120:680–682.

243 Nair KV, Gillon J, Ferguson AM. Corticosteroid treatment increases parasite numbers in murine giardiasis. Gut 1987; 22:475–480.

244 Roberts-Thomson IC. Giardiasis: the role of immunological mechanisms to host–parasite relationships. In Marsh MN, ed. *Immunopathology of the Small Intestine*. Chichester: Wiley, 1987.

245 Owen RL, Nemanic PC, Stevens DP. Phagocytosis of *Giardia muris* by macrophages in Peyer's patch epithelium in mice. Infect Immun 1981;33:591–601.

246 Loftness TJ, Erlandsen SL, Wilson ID, Meyer EA. Occurrence of specific secretory immunoglobulin A in bile after inoculation of *Giardia lamblia* trophozoites into rat

duodenum. Gastroenterology 1984;87:1022–1029.

247 Farthing MJG. Host–parasite interactions in human giardiasis. Q J Med 1989;70:191–204.

248 MacDonald TT, Spencer J. Evidence that activated T-cells play a role in the pathogenesis of enteropathy in human intestine. J Exp Med 1988;167:1341–1349.

249 Ament ME, Rubin CE. Relations of giardiasis to abnormal intestinal structure and function in gastrointestinal immunodeficiency syndromes. Gastroenterology 1972;62: 216–226.

250 Quinn TC, Mann JM, Curran JW, Piot P. AIDS in Africa: an epidemiologic paradigm. Science 1986;234:955–963.

251 Cook GC. Opportunistic parasitic infections associated with the acquired immune deficiency syndrome (AIDS): parasitology, clinical presentation, diagnosis and management. Q J Med 1987;65:967–984.

252 Gilman RH, Marquis GS, Miranda E, Vestegui M, Martinez H. Rapid reinfection by *Giardia lamblia* after treatment in a hyperendemic Third World community. Lancet 1988;1:343–345.

253 Green EL, Miles MA, Warhurst DC. Immunodiagnostic detection of *Giardia* by a rapid visual enzyme-linked immunosorbent assay. Lancet 1985;1:691–692.

7 Vaccines and the challenge of parasitic infections

Gordon L. Ada

INTRODUCTION

The bicentenary of the first "modern" vaccine is due in 1996, when the crucial experiment of Edward Jenner, in inoculating James Phipps with material from cow pocks and then challenging him with smallpox material, will be celebrated. Jenner's procedure gained favor internationally and replaced the ancient practice of variolation quite quickly. The practice of vaccination, as Pasteur later termed it, became a generic term and Jenner's vaccine was the major weapon in the recent successful campaign to eradicate smallpox from the earth—a remarkable feat. The precise relationship between vaccinia virus, the basis of recent smallpox vaccines, and the cowpox virus of Jenner's day is not clear but the important principle was the use of an organism adapted to and pathogenic for one host to immunize another distantly related host.

The early success of the smallpox vaccine set the stage for the subsequent development of other vaccines: a vaccine aimed to protect the vaccinee from disease caused by the wild-type organism without itself causing a significant level of morbidity and mortality. Administration of the smallpox vaccine does cause appreciable levels of both morbidity and mortality but far less than the smallpox virus, variola [1]. The criteria of efficacy and safety became the crucial yardsticks for the acceptability of a vaccine. The smallpox vaccine possessed another great advantage—the vaccinee developed a scar which was a clear indication of a successful immunization and became a powerful epidemiologic tool. The net result was the development of many vaccines—there are about 50 in medical and veterinarian use in industrialized countries today [2]—in the absence of much information about the immune response of the host. A knowledge of the properties of the infectious agent, its pattern of replication, and the role of the different immune processes in the control and subsequent clearance of the infectious agent seemed to be

superfluous. In fact, some would still claim that much of this information is unnecessary for the development of many vaccines. One subsequent development was the measurement of seroconversion after vaccination; for some vaccines this proved to be a reasonable indication of the efficacy of the vaccine. Our knowledge of the different immune responses and their role in controlling an infection has been gained almost entirely by the study of model systems and by a relatively small group of investigators.

This pattern is changing. A glance at a list of vaccines for human use clearly shows that most, and certainly the more successful vaccines, are used to control acute infections, i.e., those in which the infection is controlled and cleared within 2–3 weeks in the very great majority of people. Most of the remaining unconquered infectious diseases are either caused by agents which show great antigenic variation, such as rhinovirus (a major etiological agent of the common cold) and potentially influenza virus, or those agents which cause a persistent and chronic infection. Among human infections in the latter category the great parasitic infections and, more recently, human immunodeficiency virus (HIV), the causative agent of the acquired immunodeficiency syndrome (AIDS) pandemic [3], are of major concern.

After an initial discussion of the main classes of vaccines currently available, a major thrust of this chapter will be the need to know in considerable detail the properties of the agent itself, its pattern of replication and tropism *in vivo*, and the different immune responses it generates in a relevant host if an effective vaccine is to be developed. This need has been recognized by many of those studying the parasitic infections and is now recognized by some for HIV [4].

THE TRADITIONAL VACCINES

The provision of proper sanitation facilities and safe drinking water were two of the improvements in personal and community health practices which led to a substantial reduction in the communicable disease burden in industrialized countries from the mid-1850s [5]. Vaccination against the common infectious diseases, especially the highly contagious agents, became increasingly important but extensive coverage of a population is a recent event. Thus, when the World Health Organization's expanded program of immunization began in the mid-1970s, only about 5% of the world's children in developing countries were being vaccinated against the six common childhood infections. Some 15 years later after a sustained effort involving, in addition, other agencies, including UNICEF, UNDP, and Save the Children Task Force, this figure approaches 70%. It is a sorry fact that the immunization programs in some of the developed countries still have poor records in this respect. The USA has set a proud example by requiring evidence of vaccination before children are admitted to school; exceptions are allowed for legitimate reasons. This led initially to a dramatic decrease in that country of the prevalence of measles and the unfortunate sequelae of infection by the wild-type virus. In recent years, however, the delivery of vaccines to the inner areas of some large cities has become inefficient.

Table 7.1 lists the major viral and bacterial vaccines in use. They are grouped under four headings—live agents, live attenuated agents, inactivated whole agents, and subunit preparations. The virus in the adenovirus vaccine is poorly attenuated but the vaccine is safe if it is administered orally in an enteric-coated capsule; this is an unusual route as this virus causes respiratory infections, but the success of the vaccine is due to the common pathway of the mucosal immune system [6]. The attenuated live viral vaccines have proved to be very effective, giving long-lasting (sometimes life-long) immunity after one or two administrations. Furthermore, most cause a very low incidence of side-effects [7]. It is for this reason that this approach to vaccine development to control viral infections is still popular (see below).

Several examples of inactivated whole organisms, both viral and bacterial, are quoted in Table 7.1. Being noninfectious, such vaccines are safer for immunodeficient or immunocompromised people than are infectious preparations. However, a much greater antigenic mass is required to achieve protection and, as has been recognized very recently, noninfectious vaccines poorly induce certain cell-mediated immune

Table 7.1 "Traditional" vaccines against viral and bacterial diseases of medical importance

Type	Viral	Bacterial
Live	Adeno	—
Live, attenuated	Vaccinia	BCG
	Polio (OPV)	
	Measles	
	Rubella	
	Yellow fever	
	Mumps	
Inactivated, whole	Polio (IPV)	*Vibrio cholerae*
	Influenza	*Salmonella typhi*
	Japanese encephalitis	*Bordetella pertussis*
	Rabies	*Streptococcus pneumoniae* (pneumococcus)
Subunit	Influenza	*Haemophilus influenzae*, type B (capsular polysaccharide)
	Hepatitis B	*Clostridium tetani*, tetanus toxoid
		Corynebacterium diphtheriae diphtheria toxoid

BCG, Bacille Calmette-Guerin; IPV, inactivated polio vaccine; OPV, oral polio vaccine.

(CMI) responses which are now known to be important in the recovery from many infections. In the absence of significant antigenic variation, inactivated preparations such as rabies and Japanese encephalitis viruses and *Bordetella pertussis* are generally effective, although these viral vaccines are more expensive. The pertussis vaccine, although effective [7], also causes side-effects.

To date, the only effective subunit viral vaccine is the hepatitis B surface antigen (HBsAg), initially prepared from infected plasma but now prepared from DNA-transfected yeast. A proportion of people respond poorly, even after several injections, and probably because of a paucity of T cell epitopes in the protein and the poorer response of elderly recipients. The influenza virus vaccine, consisting of the two surface glycoproteins, is less reactogenic than the whole virus but is even less effective than the latter in inducing a CMI response [8]. Apart from the two bacterial toxoid vaccines which are highly effective and safe but which require multiple injections to induce long-lasting immunity [9], the other bacterial subunit vaccines are the capsular polysaccharides of the encapsulated bacteria. These are of variable efficacy. Generally, children under 2 years of age respond poorly, probably because of an inability to present these antigens to T cells (discussed in detail elsewhere [10,11]). In addition, some polysaccharides are structurally so closely related to host carbohydrates that the host mounts a very poor immune response [10].

Table 7.2 lists some of the vaccines developed by traditional approaches and currently under trial. They illustrate the four approaches using live organisms as the basis of a vaccine. Examples of the "Jennerian" approach, i.e., using a virus pathogenic for another host, are the rota (monkey) and parainfluenza viral preparations. Another approach is to make reassortants, i.e., the combination of genes from two or more viral strains. This is only possible in the case of organisms with segmented genomes. The largest group is the attenuated preparations made by passage in other hosts or in cell culture. The derivation of temperature-sensitive strains requires eight mutations in order to achieve genetic stability. The last approach is to construct deletion mutants. This has not been successful with viral vaccines owing to stability, but recently derived bacterial preparations appear to be more stable.

Of the remaining preparations under trial, a subunit

Table 7.2 "Traditional-type" vaccines against viral and bacterial diseases of medical importance under trial

Type	Viral	Bacterial
Live, attenuated	Varicella zoster	*Vibrio cholerae*
	Cytomegalovirus	*Salmonella typhi* (ty21a) (Aro A mutant)
	Hepatitis A	
	Influenza (cold-adapted variant) (human–avian reassortant)	
	Dengue	
	Rotavirus (human)	
Live, pathogenic for another host	Rotavirus (Rhesus monkey) Parainfluenza (bovine), type 3	—
Inactivated, whole	—	*Mycobacterium leprae*
		V. cholerae (plus subunit)
Subunit	—	*S. typhi* (capsular polysaccharide)
		Haemophilus influenzae, type b (polysaccharide— diphtheria toxoid conjugate)

pertussis preparation has given moderate success in recent trials and conjugation of capsular polysaccharides with a carrier protein such as a bacterial toxoid appears promising. (For further details, see Bell & Torrigiani [11].)

Vaccines to protect against human parasitic infections are not yet available for a variety of reasons. Generally, the most successful approach used to combat viral infections, that of attenuation, is not feasible for parasites owing to the impracticability of growing sufficient organisms. A study of the successful vaccines can, however, tell us much about the general requirements if a parasite vaccine is to be effective. It is first necessary briefly to review the induction requirements of different immune responses and the role of each during an infection.

THE INDUCTION REQUIREMENTS OF DIFFERENT IMMUNE RESPONSES

The immune response has two main components—specific (adaptive) and nonspecific (nonadaptive). For brevity, this chapter will refer mainly to the former. Infectious agents generally present to the immune system either arrays of repeating antigenic molecules or repeating peptide (or carbohydrate) sequences. They are handled in three ways.

1 The agent or fractions thereof are nonspecifically endocytosed by phagocytic cells, such as a macrophage, and degraded to peptides; some of these may bind to Class II major histocompatibility complex (MHC) molecules and the complex is transported to the cell surface and recognized by specific receptors on an immunocompetent T cell (referred to as a Class II MHC-restricted cell). The antigen-presenting cell (APC) also produces soluble factors (interleukins) so that the recognizing T cell is activated. The T cell itself produces another interleukin (IL-2) and specific IL-2 receptors and proceeds to differentiate and replicate to form an effector cell, in this case a helper T cell. Current evidence suggests that another cell, called a dendritic cell, may act as an intermediate in antigen presentation between the phagocytic cell and the T cell.

2 The agent or a fraction thereof is also recognized by an immunocompetent B cell, this time by means of *specific* immunoglobulin (Ig) receptors. The antigen–Ig complex is endocytosed and degraded in the B cell. Peptide–MHC complexes are produced and expressed at the cell surface in much the same way as occurs in macrophages; if one or more of these resemble the complex recognized by the T cell, the latter, now activated, will recognize the peptide–MHC complex on the B cell and secrete other interleukins. These facilitate the replication and differentiation of the B cell to synthesize and secrete Igs of the same specificity as the Ig receptor originally expressed on the cell. This T cell thus "helps" the B cell to produce antibody; it may also have other activities, such as mediating delayed-type hypersensitivity (DTH) reactions.

3 The agent may infect the APC, sometimes via a receptor, and undergo either productive replication (i.e., produce infectious progeny) or an abortive infection (i.e., produce only viral proteins). In either case, some of the newly synthesized viral antigens are degraded to peptides (by a different route, now called the cytoplasmic pathway). Some of these peptides may bind to Class I MHC antigens and the complex, expressed at the cell surface, is recognized by receptors on a Class I-restricted immunocompetent T cell. This T cell is activated to become an effector cell, but its major functional activity is the ability to kill, by lysis, other cells expressing the peptide–MHC (Class I) complex on their surface. These cells are called cytotoxic T cells and the process is called cell-mediated lysis. Although cell-mediated lysis has been decisively shown *in vitro*, the evidence for this mechanism *in vivo*, although completely consistent with this activity, is circumstantial. Some evidence suggests that secretion of interferon-γ (IFN-γ), an antiviral agent, may be a major mechanism for the control of viral replication.

Although there is limited evidence that suppressor T cells may have a role in a few infections, there is little solid data on the pathway to their generation. For this reason they will not be discussed further.

There are two crucial general conclusions to be drawn from this outline of the immune response to infectious agents.

1 Generally, antibody to one or possibly two antigens of an agent is effective at preventing infection and, usually, only certain segments (epitopes) are important. Receptors on B cells preferentially see tertiary or quaternary structures (or conformations), i.e., structures formed by an "intact" protein or by aggregates of proteins. Antibodies can be made to simpler structures, such as peptides, but antibodies with important activities, such as neutralization of viral infectivity, are generally in the former category. The extent to which this generalization may apply to other agents, such as bacteria and parasites, remains to be seen.

2 The T cell receptor, which is closely related to the

B cell receptor, also recognizes tertiary structures but the contribution of the infectious agent to this structure is a short peptide, which itself has only secondary conformation. Cytotoxic T cell formation and activity was initially thought to be induced only during a viral infection [12] but it is now realized that *de novo* synthesis of a foreign protein (i.e., bacterial or parasitial, as well as viral) in a cell potentially may give rise to peptide–MHC complexes which initiate cytotoxic T cell formation [13]. Furthermore, in contrast to neutralization of infectivity by antibody, any protein of an agent, during synthesis, may in principle give rise to peptides which associate with Class I or II MHC antigens. The example of the influenza virus can be quoted [14]. Antibody to a single protein only (the hemagglutinin) neutralizes infectivity. Helper and cytotoxic T cells recognize epitopes in most, if not all, proteins of the virus. The importance of this strategy is discussed in the next section.

STAGES OF AN INFECTIOUS CYCLE AND THE ROLE OF DIFFERENT IMMUNE RESPONSES

During an acute infection, four stages of an immune response can clearly be discerned—infection (and prevention of infection), limitation of replication, recovery from infection (clearance of the agent), and the generation of immunologic memory.

Infection and its prevention

Infection of a cell is often mediated by the presence of a specific receptor for the agent on the cell and this determines the tropism of the agent. The only practical immunologic way to prevent infection is the presence of specific antibody, which is usually directed to a few epitopes on one or occasionally a few antigens. These epitopes are frequently discontinuous (i.e., nonlinear). Little is known about the method of neutralization. Antibody may block a receptor site, block a stage of replication after entry into the cell, or cause the agent to be diverted to a cell with Fc receptors, such as a macrophage: here, it may be destroyed, or if the agent replicates in that cell it may enhance the infection [15].

In the past it has been tacitly assumed that the major, if not the only, important function of a vaccine to an infectious agent was to produce specific antibody. Although unquestionably important, evidence that antibody completely prevents infection to any agent is generally not available. In fact, evidence has been accumulating for some time (and recently quite rapidly) that CMI responses may be very important.

Limitation of replication

Nonspecific factors either present initially, e.g., natural killer (NK) cells, complement, or rapidly induced, e.g., interferons, may limit the infection. These have been shown to be of some importance in a few cases but generally they have not been well studied in parasitic infections.

Recovery from infection

Antibody or the nonspecific responses mentioned above may contribute to the clearance of the infectious agent but evidence that these alone can achieve this goal has not yet been produced. Antibody of the appropriate isotype can mediate complemented dependent lysis of a cell expressing foreign antigens, or antibody can "arm" a killer cell and so mediate lysis of an infected cell. Effector T cell responses, either Class I or Class II restricted, are of crucial importance in many systems. In the case of many viral infections, there is overwhelming evidence that Class I cells which are cytotoxic for virus-infected target cells will clear the virus *in vivo*. In similar systems, Class II cells which express DTH reactions may cause enhanced pathology without viral clearance [16]. In some other systems, such as intracellular bacteria growing in macrophages, this class of response may be required to "activate" the cell and so liberate products which are harmful to and may destroy the agent [17]. In the case of at least some viral infections, there seems to be a relationship between the presence of infectious virus and Class I MHC-restricted T cell generation and persistence; cytotoxic T cell activity persists for only a few days after an infectious virus can no longer be detected [18,19].

Generation of memory cells

Current evidence from a model system [20] would suggest that long-lived immunity following infection by virus is dependent upon two factors:
1 the generation of a large pool of T and B memory cells within a few weeks of the infection;
2 the persistence of antigen in a form which can recruit the latter to form antibody-secreting cells, as such cells tend to have short half-lives.

It follows that a vaccine, to be effective, will need

to duplicate this effect. Unfortunately, although immunologic memory is the hallmark of the specific immune system, there is an almost total lack of information on the requirements for induction and in what way the pathway of activation differs from activation to produce effector cells. Furthermore, the process of affinity maturation of the antibody response also seems to occur mainly in the B memory cell pathway [21]. This area should therefore be of high priority for further research.

THE NEWER APPROACHES TO VACCINE DEVELOPMENT

There are no vaccines currently available to control the major human parasitic diseases. There are several reasons for this.

1 It is not feasible to grow the parasite in sufficient amount to produce a vaccine which could be used to immunize the great numbers of people at risk. It has been said that about one-quarter of the world's population suffers from infection by five parasites which cause the diseases malaria, schistosomiasis, filariasis, leishmaniasis, and trypanosomiasis. About one-third of the world's population is at risk from malaria alone.

2 These parasites cause a chronic and persistent infection and there are often several stages in the disease, because of morphologically distinct forms of the parasite. A vaccine, to be successful, may need to protect against more than one infectious stage.

There are four new approaches to vaccine development, some of which offer bright prospects for the development of vaccines to parasites. These are the synthesis of oligopeptides representing particular epitopes of antigens, the production of anti-idiotypes, and two methods based on recombinant DNA (rDNA) technology. The latter are transfection of cells with DNA coding for protective antigens, and the use of live agents, viruses, or bacteria as vectors for such DNA. They will be discussed briefly in turn.

Development of peptide-based vaccines

This approach has two origins.

1 The degradation of proteins and the demonstration that one or a few products would retain antigenic activity.

2 The synthesis of linear and branched amino acid polymers. The use of the latter in particular was of great importance since such preparations showed a more specific antibody response than native proteins, and their use led to the first clear-cut demonstration of immune response gene effects [22]. A rapid accumulation of protein sequence data occurred following the development of quick methods of DNA sequencing. Thus, the amino acid sequences of all the major proteins in HIV became available in a remarkably short time following isolation of the virus.

The success of the HBsAg vaccine in preventing infection by hepatitis B virus made this an obvious target for study and both T and B cell epitopes have been identified. The vaccine confers protection on about 90% of recipients. A pre-S region of the protein has been found to contain an immunodominant T cell epitope [23] and, recently, synthetic peptides corresponding to sequences in this region have protected chimpanzees from a hepatitis B virus challenge [24]. This is likely to develop into a "test" situation in which the feasibility of a peptide-based preparation as an alternate vaccine to an existing effective protein subunit vaccine will be tested. The prototype vaccine, prepared from the plasma of hepatitis B-infected people, was quite expensive (ca. US$100 for three doses). A similar product was subsequently prepared from DNA-transfected yeast and is cheaper. Recently, a plasma-derived product, produced in bulk in South Korea, has become available (for very large orders) at US$1 per dose; it may be difficult for a peptide-based preparation to compete with this figure.

Herpes simplex virus occurs as two serotypes, HSV-1 and HSV-2. After infection, the viral DNA may become integrated into the host cell genome, leading to latent infection. Both direct and indirect evidence shows that CMI responses are required to control infection. Immunization of rabbits with a peptide, 1–23 of the glycoprotein D (gD), coupled to keyhole limpet hemocyanin produced antibody which reacted with gD; the immunized rabbits were protected from infection following challenge with the virus. However, protection has not correlated with neutralizing antibody titers so there is a possibility that protection might correlate with effector T cell activity induced by the peptide [25].

These are only two of several examples of work in this area. There are two types of peptide-based vaccines at present under trial in humans. One is a fertility control vaccine based on the C-terminal 37 amino acid peptide from the B subunit of the human chorionic gonadotrophin hormone, coupled to diphtheria toxoid and administered with an adjuvant. This vaccine has undergone phase 1 trials successfully [26,27] and phase 2 trials are planned.

The second type of peptide-based vaccine is to control malaria. Two human vaccine trials have recently been carried out [28,29] in which the repetitive B cell epitope of the circumsporozoite protein (CSP) was fused to nonmalaria T cell epitopes. Both preparations failed to generate high antibody titers in humans and only a few vaccinated individuals were protected against challenge by the organism. It has since been found that the T cell epitopes within the CSP are not only severely restricted in their ability to interact with a variety of Class II MHC antigens but may also be variable in sequence [30]. It has also been reported that a cocktail of peptides from *Plasmodium falciparum* can protect humans [31], but this finding has not yet been repeated by other investigators.

Even under "optimum" conditions, i.e., infection by a natural pathogen such as influenza virus where there are many T cell epitopes on the different viral antigens, it seems that few, if any, people make antibody to all the important B cell epitopes on the protective antigen of the pathogen [32]. The question inevitably arises—Will it be possible to synthesize a "universal" T cell epitope, i.e., an amino acid sequence which binds to a wide range of Class II MHC haplotypes? If such sequences occur naturally or can be made, it would be a boost to this area.

Anti-idiotype antibodies

The antigen binding site of an antibody may be a unique structure (and hence is called an idiotype of the antibody), especially if it has arisen by somatic mutation from an original B cell receptor specificity. It would then be regarded by the body as foreign so that it may stimulate the formation of antibodies whose antigen combining site recognizes the idiotype. These are called anti-idiotypic antibodies and some may be the "mirror image" of the original antigenic epitope [33]. If so, immunization with this preparation may induce an immune response which protects the host against the pathogen itself. There are now many examples of model systems where this has been observed. Most of these are infections where vaccines can or have been made by standard procedures [34]. There are, however, two parasite examples. Anti-idiotype preparations have been made which have protective effects in model systems for trypanosomiasis [35] and for schistosomiasis [36]. There seems to have been little progress in developing this approach for parasite vaccines for human use. The anti-idiotype approach has been unsuccessful in some cases,

for example, where considerable antigenic variation occurs, such as with HIV in experimental systems [37]. Early results suggest that they may have a useful role in the control of certain autoimmune diseases in destroying effector T cells [38].

Approaches based on rDNA technology

The ability to manipulate DNA is perhaps the most important technical advance in biology this century. It has given rise to two techniques which have great potential for vaccine development.

The first is the ability to transfect both prokaryotic and eukaryotic cells with DNA coding for the protein of interest and obtain expression of the DNA in the cell. The protein may be expressed as such or as a polypeptide "fused" to a host cell protein. The latter approach has given rise to useful products from transfected prokaryotic cells, such as *Escherichia coli*.

There are, however, examples where this approach has yielded products without the appropriate biologic activity [39], owing to incorrect conformation of the expressed protein. In such cases, e.g., some viral glycoproteins, transfection of eukaryotic cells has been more successful. The first genetically engineered vaccine to become commercially available contains HBsAg, produced by transfected yeast cells [40]. These products may not have the "correct" glycosylation pattern, in which case transfection of mammalian eukaryotic cells may be a more appropriate approach. Three classes of cells are used:

1 Primary cells—uncultured cells from a tissue;
2 Cell strains—cells which have a finite capacity to replicate;
3 Continuous cell lines—cells that replicate indefinitely and are capable, under some circumstances, of producing tumors.

Vaccines for medical or veterinary use are made in all three cell types. There are possible causes for concern regarding vaccines for medical use and containing live agents, which are produced in continuous cell lines. These are the presence of heterogeneous contaminating DNA, of endogenous or latent viruses, or of transforming protein [41]. Their use, therefore, should be judged on a case-by-case basis.

Perhaps the greatest potential contribution of this technology to vaccine development is the development of recombinant live vectors. The basis of this approach is the incorporation of genes coding for foreign antigens into the genome of live agents (viruses or bacteria), particularly those which form the basis of existing medical vaccines: the bacteria *Salmonella* and Bacille Calmette-Guerin (BCG); the pox viruses, including vaccinia, orf, and avipox viruses; adenovirus; polio viruses. Each of these has particular advantages and disadvantages. For example, adenovirus, polio viruses, and *Salmonella*, given orally, would induce immunity at mucosal surfaces whereas BCG and vaccinia, administered parentally, would provide systemic immunity. The general requirements for the preparation of such recombinants are described elsewhere [42] but a brief description of the properties and use of one vector vaccinia, is given here as much more work has been done with this vector. Genes coding for foreign antigens have been inserted into vaccinia DNA, usually into the viral thymidine kinase gene.

The use of vaccinia or related viruses as vectors offers several advantages.

1 Both humoral and cell-mediated immunity to the foreign antigen is induced. In fact, the use of a live vector is probably the most effective means for generating a Class I MHC response to a soluble antigen [43].

2 DNA coding for several antigens can be incorporated into vaccinia virus and be expressed [44].

3 The DNA coding for the antigen can be modified so that the product is expressed in the cell cytoplasm, at the cell surface, or secreted into the medium [45].

4 The recombinant virus provides long-lasting immunity to the foreign antigen after a single injection.

5 The freeze-dried preparation is heat stable, which is an important attribute for Third World use as it bypasses the need for a cold-chain.

6 Successful vaccination results in a scar and this can be a very useful epidemiologic marker.

There are some potential disadvantages:

1 There is a significant level of side-effects; the mortality rate following vaccination during the eradication program was about one out of 200 000 vacinees [1]. Immunodeficient or immunocompromised individuals were particularly at risk.

2 The development of more attenuated strains to overcome this risk could result in poorly immunogenic preparations.

3 Because of the long-lasting immunity generated, the period which must elapse between vaccinations in order to generate a response to novel incorporated antigens might be several years.

Recent work suggests that some of these disadvantages might be ameliorated.

1 Cleavage of the thymidine kinase gene in the laboratory strain of vaccinia greatly decreased its neurovirulence for mice [46].

2 Administration of a recombinant vaccinia preparation which also contained the gene for IL-2 to athymic mice prevented death due to generalized vaccinia [47,48]. Perhaps a similar approach would afford protection to immunodeficient or immunocompromised individuals.

This latter finding has the great advantage that the interleukin is produced at the site of infection, i.e., where the factor is most required. In this sense, the technique can be called site-directed synthesis. It also raises the possibility that incorporation of genes coding for similar factors may boost the immune response and thus allow, if required, a further attenuation of the virus.

IMMUNOPOTENTIATION OF THE IMMUNE RESPONSE

The generation of an immune response requires the "activation" of different cell types and the production of a variety of interleukins. Antigenic preparations can achieve this to different extents and, in discussing the different requirements, it is convenient to classify them into four categories, as illustrated in Table 7.3. These are often considered together under the umbrella heading of adjuvant, but this may only involve a portion of the requirements.

Peptides generally need to be conjugated to a carrier protein, such as tetanus or diphtheria toxoid. The carrier is a source of T cell epitopes and the efficacy of proteins in this regard will vary, depending on several factors. An alternative approach is to couple the antigen to a T cell activator, such as purified protein derivative (PPD) [49]. In both cases, the carrier should stimulate T cells in hosts which had previously been immunized with the toxoid or with BCG, respectively, and hence would express T cell memory. Immunogens such as intact proteins (or anti-idiotypes) might provide the necessary T cell epitopes *per se*, but this could be quite a variable effect, depending on the size and amino acid sequence of the protein.

Adjuvants are now recognized as having broadly two effects. One of these is to "activate" the APC so that the appropriate factors (interleukins, such as IL-1) are synthesized and secreted. Table 7.4 lists some of the preparations with adjuvant activity and their possible mode of action. Clearly, a great variety of products have been or are used on an experimental basis. The only preparation licensed for general medical use is alum,

Table 7.3 Additional requirements for generating an immune response

Antigen	Adjuvant	T cell responses			Delivery system		
		Carrier		Activator	Depot	Controlled releases	Targeting
Peptide	+	+	or	+	+	or +	+
Protein	+	−		−	+	+	+
Anti-idiotype	+	−		−	+	+	+
Live vector	−	−		−	−	−	−

Table 7.4 Materials with adjuvant activity and possible sites of action. (From Ada [50])

Materials	Functions
Protein carriers; proteins containing appropriate T cell epitopes	Mobilization of T cell help
"Inert" carriers; alum, bentonite latex, acrylic particles	Aggregation of soluble antigens, induction of T cell help, focus for congregation of lymphoid cells in different areas of lymphoid organs
"Hydrophobic" antigen; addition of lipid tail to proteins, adding MDP to antigen in oil	Localization of antigen in T dependent areas, generation of effector T cells (DTH), formation of amphipathic structures?
"Depot" formers; water/oil and oil/water emulsions, some polysaccharides	Delayed release of antigen, leading to prolonged immune response, recruitment of memory cells?
Polyclonal activators of T cells; PPD, poly A : poly U	Provide additional T cell help, secondary activation of macrophage
B cell activators; antigen-polymerizing factors, B cell mitogena, e.g., LPS	Modulate Ig receptors on B cells
Surface-active materials; saponin, lysolecithin, retinal, Quil A, some liposomes, pluronic polymer formulations	Facilitate cell–cell interaction, aggregate antigen in specific way, e.g., ISCOMS
Macrophage (APC) stimulators; MDP and derivatives, LPS, various factors, see also below	Synthesis of factors—IL-1, IL-4, IFNs, CSFs, complement components
Alternative pathway; Complement activators: Inulin, zymosan, endotoxin, levamisole, *Corynebacterium parvum*, etc.	Focusing of antigen on and stimulation of leukocytes with C3 receptors, e.g., macrophages, B cells, follicular dendritic cells

APC, antigen-presenting cells; CSF, colony stimulating factor; DTH, delayed-type hypersensitivity; IFN, interferon; ISCOMS, immunostimulatory complexes; LPS, lipopolysaccharide; MDP, muramyl dipeptide; PPD, purified protein derivative.

which was first used over 50 years ago; this is an indication of the rate of progress in this field.

The second function of an adjuvant is to prevent the rapid disappearance of the antigen from the injection site, and the various preparations such as alum and Freund's adjuvants have this effect. It is only comparatively recently that an approach used successfully for many years in the therapeutic drug area for controlled release of the drug from the injection site is being investigated for antigen administration. These preparations are usually biodegradable microcapsules or spheres which allow for programmed release schedules, such as constant or pulsed patterns. The antigen and adjuvant are incorporated into the capsule and the mixture is injected, usually intramuscularly.

If successful experimentally, the application of this technology to vaccination programs may have a major impact. At present, the WHO/EPI schedule for immunization against diphtheria, tetanus, and pertussis is three injections of the diphtheria, pertussis, tetanus triple vaccine, involving three separate visits to the health center. The use of controlled-release formulations might reduce this to one or two injections.

The function of the adjuvant therefore is to provide a depot of antigen and to attract and activate appropriate cells of the immune system. The day may well be approaching when the use of many of the products listed in Table 7.4 will be replaced by an appropriate cocktail of growth factors to achieve a desired immune response pattern. These might either be incorporated into a controlled-release system or the DNA coding for them into a live vector.

A final aspect which indicates current trends is the ability to target the antigen to particular cell types. There are recent reports of targeting antigens to B cells [51] or to cells expressing Class II antigens [52]. This is a further step in the direction of mimicking a live agent, such as a virus, which infects those cells expressing a receptor specific for the agent.

ROUTE OF VACCINE ADMINISTRATION

Infectious agents enter the body mainly in one of two ways—via a mucosal surface or by penetrating the skin. Routes of administration of vaccines do not always mimic this pattern. Influenza virus and the cholera *Vibrio* infect via mucosal surfaces whereas the current vaccines for control of these infections are administered parenterally. Vaccines which aim to generate systemic immunity will continue to be administered parenterally but there is a growing interest in utilizing the mucosal route where this is most appropriate, and especially because of the interconnections of immune responses between such sites. The success of the adenovirus vaccine, which is given orally in an enteric-coated capsule and protects against a respiratory infection, not only illustrates the feasibility of this approach but also is greatly increasing interest in the possibility of exploiting this route for other vaccines. It has been established for example that oral administration of cholera toxin with an unrelated antigen greatly increases the immune response to the latter [53] so it is likely that this will

become a very active research area. It takes little imagination to assess the difference to worldwide immunization practices if some of the vaccines now administered parenterally could be given orally.

SELECTIVE INDUCTION OF PARTICULAR IMMUNE RESPONSES

As more is learnt about the immune response to different infectious agents and hence the way to proceed to vaccine development, the ability to manipulate the immune response to achieve particular responses would be a great achievement. Different adjuvant preparations are known to favor or not to facilitate certain responses; for example, alum is known to favor humoral rather than CMI responses.

There are now signs that this goal may be achieved by a more direct approach. It has recently been established that murine helper T cell clones fall into two subsets, as judged by the synthesis and secretion of different lymphokines. These in turn preferentially induce particular responses, e.g., different Ig isotypes (Table 7.5). The important aim now is to define the conditions of stimulation which lead to the preferential generation

Table 7.5 Two subsets of mouse T helper cell clones. By courtesy of Dr R.L. Coffman. (From Ada [50])

	TH-1	TH-2
Surface markers		
Ly1	+	+
L3T4	+	+
Lymphokines		
Interferon	+ +	−
Interleukin 2	+ +	−
Lymphotoxin	+ +	−
GM-CSF	+ +	+
Tumour necrosis factor	+ +	+
Ty5	+ +	+
Interleukin 3	+ +	+ +
Met-enkephalin	+	+ +
Interleukin 4	−	+ +
Interleukin 5	−	+ +
P600	−	+ +
B cell help		
IgG2a	+ +	+
IgE	−	+ +
Delayed type hypersensitivity	+ +	−

GM-CSF, granulocyte macrophage colony stimulating factor.

of either subset. Its achievement will signify that "the age of controlled immune regulation" will have been attained.

THE UNFOLDING DRAMA OF VACCINE DEVELOPMENT TO CONTROL PARASITIC DISEASES

Other chapters in this book will describe in detail the progress being made towards vaccine development for different parasites. This chapter will conclude with a brief description of recent events in two contrasting situations, malaria and schistosomiasis vaccine development.

Towards malaria vaccines

The life-cycle of the malaria parasite in humans begins when a person is bitten by a female *Anopheles* mosquito which is infected with one of the four *Plasmodium* species infectious for humans. The malarial sporozoites in the mosquito saliva enter liver cells within minutes of the bite and start to divide, forming schizonts. These contain a second form of parasite, the merozoites, which enter the bloodstream when the schizont ruptures. Within seconds, the merozoites infect red blood cells and replicate asexually. Each time an infected red blood cell bursts, the released progeny infect other red blood cells. A few merozoites differentiate into gametocytes, a sexual form which replicates in the mosquito to form sporozoites, and the stage is set for a fresh cycle of infection to begin.

Immunity to malaria occurs and, in 1961, immune γ-globulin antibody transfer was shown to decrease dramatically the parasite load in children [54]. The generation of antibody-mediated protection has remained the major research aim of many laboratories until quite recently, although evidence has been steadily accumulating, at least in experimental systems, that CMI responses might be equally, if not more, important (reviewed by Good *et al.* [30] and Good [55]). The evidence for nonantibody-mediated protective responses in animal models extends back for well over 10 years. The availability of particular specific reagents in the last few years has allowed crucial experiments to be performed. Three experimental findings which demonstrate this point will be reported briefly.

The first of these was the finding [56,57] that depletion of CD8+ but not CD4+ cells in mice previously immunized with X-irradiated sporozoites rendered them susceptible to challenge with sporozoties, thus implying a critical role for Class I MHC-restricted T cells (or products secreted by them). The second finding was the recent demonstration [58] that a single CD4+ T cell clone, on adoptive transfer, protected mice from *Plasmodium chabaudi*. A likely interpretation seems to be that the cloned T cells secrete IFN, which activates monocytes to destroy intraerythrocytic parasites. Indeed, supernatants from malaria antigen-activated human T cells were shown earlier [59] to be able to induce adherent mononuclear cells to kill parasites *in vitro*, in the absence of antibody. The parasites were in the intraerythrocytic crisis form.

A third example is the demonstration [60] that mice immunized with *Salmonella typhimurium* WR4017, an avirulent strain with impaired ability to multiply in macrophages, and transfected with DNA coding for the CSP of *Plasmodium berghei*, were protected when challenged with live sporozoites. The transfected bacteria induced specific CMI responses but no antibody was detected.

These three findings do not deny a role for antibody in protection from malaria. They illustrate "the importance of taking into account cellular mechanisms of immunity in vaccine developments [55]." It seems very likely that these findings will greatly stimulate similar approaches to determine the role of T cell responses in infections by other parasites.

Towards schistosomiasis vaccines

The life-cycle of this parasite begins when cercariae from infected snails penetrate the skin of people. The invading cercariae mature to become adult worms; male and female worms may reside in the blood for many years, are "permanently" coupled and produce prodigious numbers of eggs. It is the localization of the eggs at different sites which causes inflammation and granuloma formation with immunopathologic effects. Attempts to develop vaccines have two goals—to prevent infection and/or to limit egg production.

The first contrast between strategies to produce malaria and schistosomiasis vaccines is that in malaria infections, the parasite, once injected into humans, is extracellular for only very brief periods of time. Thus, much of the thrust of vaccine research is to prevent the parasite from infecting susceptible cells. In contrast, schistosomes may live in the bloodstream for many years, apparently impervious to the host's immune response.

In any infectious disease situation, if it can be demonstrated that immunity to reinfection occurs after a natural infection, this is a hopeful sign that a vaccine can be developed which will prevent reinfection by the cercariae [61]; a similar age-dependent pattern of resistance to reinfection was previously demonstrated with malaria. In addition, it was shown in the 1960s [54] that transfer of immune sera from malaria-resistant donors to young children gave partial protection. There is now epidemiologic evidence [62,63] that resistance in humans at least to *Schistosoma mansoni* correlates well with the level of IgE antibodies, less well but still significantly with the IgA response, and not with IgG levels, to unpurified adult worm antigens. In fact, blocking antibodies are of the IgG isotype. The value for vaccine development of such findings can now be tested directly by passive transfer studies into susceptible children of IgG-, IgE-, or IgA-enriched preparations from the serum of immune individuals, with subsequent assessment of the effect upon reinfection.

In malaria, it has been shown in model systems that Class I MHC-restricted T cells which secrete IFN-γ effectively prevent progress from the sporozoite to the merozoite stage of infection [64]. In contrast, there seems to be no reason to develop a schistosomiasis vaccine which would induce Class I MHC-restricted T cell responses, as the parasite does not infect cells. However, there is evidence from murine studies that IFN-γ-dependent CMI responses may be helpful [65].

Many schistosomal proteins have been isolated or synthesized and many are undergoing assessment as vaccine candidates. Three of these, the glutathione S-transferase enzyme family, Sm97 or paramyosin, and triose phosphate isomerase, have reached the stage at which trials in humans might be considered [62]. It is noteworthy that two of these candidates are enzymes which illustrate another current difference between the approaches to vaccine development to control malaria and schistosomiasis.

Finally, all attempts to immunize against malaria aim to induce systemic immunity, as this is a vector-borne infection and a bloodborne disease. If, however, the observed correlation between specific IgA levels and immunity to infection leads in due course to a demonstration that the transfusion of specific IgA-enriched preparations to susceptible praziquantel-treated children protects against reinfection, administration of a schistosomal vaccine via a mucosal rather than a parenteral route may increase efficacy.

These two examples indicate some differences and some similarities in the pathways to vaccine development to two major parasites. They illustrate how our better understanding of immune responses and our ability to produce candidate antigens at will are influencing in a positive way the prospects for successful vaccine development.

ACKNOWLEDGMENT

The author wishes to thank the Rockefeller Foundation for support while this manuscript was prepared.

REFERENCES

1 Fenner F, Henderson DA, Arita I, Jezek Z, Ladnyi LD. *Smallpox and its Eradication*. WHO, Geneva, 1988.
2 Willem JS, Sanders CR. Cost-effectiveness and cost-benefit analysis of vaccines. J Infect Dis 1981;114:486–493.
3 Fauci AS. Current issues in developing a strategy for dealing with the acquired immunodeficiency syndrome. Proc Natl Acad Sci USA 1986;83:9278–9283.
4 Ada GL. Prospects for HIV vaccines. J Autoimmun Def Syn 1988;1:295–303.
5 Fenner FJ. The effects of changing social organization on the infectious diseases of man. In Boyden SW, ed. *The Impact of Civilization on the Biology of Man*. Toronto: University of Toronto Press, 1970:48–68.
6 Bienenstock J, ed. *Immunology of the Lung and Upper Respiratory Tract*. New York: McGraw-Hill Book Co., 1984.
7 Galazka AM, Lauer BA, Henderson RH, Keja J. Indications and contraindications for vaccines used in the Expanded Programme of Immunization. Bull WHO 1984;62:357–366.
8 Ada GL, Jones PD. The immune response to influenza infection. Current Top Microbiol Immunol 1986;128:1–54.
9 World Health Organization, Expanded Programme of Immunization. Immunization policy. WHO/EPI/GEN/86/7 REV 1–23.
10 Kabat EA, Nickerson KG, Liao J, *et al.* A human monoclonal macroglobulin with specificity for (2–8) linked poly-N-acetyl neuraminic acid, the capsular polysaccharide of group B meningococci and *Escherichia coli* KI which cross-reacts with polynucleotides and with denatured DNA. J Exp Med 1986;164:642–654.
11 Bell R, Torrigiani G, eds. *Towards Better Carbohydrate Vaccines*. WHO. New York: Wiley, 1987.
12 Ada GL, Leung K-N, Ertl HCJ. An analysis of effector T cell generation and function in mice exposed to influenza A or Sendai viruses. Immunol Rev 1981;58:4–24.
13 Townsend ARM, Rothbard J, Gotch FM, Bahadur G, Wraith D, McMichael AJ. The epitopes of influenza nucleoprotein recognized by cytotoxic T cells can be defined with short synthetic peptides. Cell 1986;44:959–969.
14 Nestorowicz A, Laver G, Jackson DC. Antigenic determinants of influenza haemagglutinin. X. A comparison of the physical and antigenic properties of monomeric and trimeric forms. J Gen Virol 1985;65:1687–1695.
15 Peiris JSM, Porterfield JS. Antibody-mediated enhancement

of flavivirus replication in macrophagelike cell lines. Nature 1979;282:509–511.

16 Leung K-N, Ada GL. Different functions of subsets of effector T cells in murine influenza virus infection. Cell Immunol 1982;67:312–324.

17 Mackaness GB. Resistance to intracellular infection. J Infect Dis 1971;123:439–445.

18 Yap KL, Ada GL. Cytotoxic T cells in the lungs of mice infected with influenza A virus. Scand J Immunol 1978;7:73–80.

19 Blanden RV, Gardiner ID. The cell-mediated immune response to ectromelia virus infection. 1. Kinetics and characteristics of the primary effector T cell response *in vivo*. Cell Immunol 1976;22:271–282.

20 Jones PD, Ada GL. Persistence of influenza virus specific antibody-secreting cells and B memory after primary murine influenza virus infection. Cell Immunol 1986;109:53–64.

21 Allen D, Cumano A, Dildrop R, *et al*. Timing, genetic requirements and functional consequences of somatic hypermutation during B cell development. Immunol Rev 1987;96:3–22.

22 Benacerraf B, McDevitt HO. Histocompatibility-linked immune response genes. Science 1972;175:273–279.

23 Neurath ARS, Kent BH, Strick N. Location and chemical synthsis of a pre-S gene coded immunodominant epitope of hepatitis B virus. Science 1984;224:392–395.

24 Neurath ARS, Kent SBH, Strick N, *et al*. Antibodies to synthetic peptides from the pre-S1 and pre-S2 regions of one subtype of the hepatitis B virus (HBV) envelope protein recognizes all HBV subtypes. Mol Immunol 1987;24:975–980.

25 Eisenberg R, Cerini CP, Heilman CJ, *et al*. Synthetic protein D related peptides protect mice against herpes simplex challenge. J Virol 1985;56:1014–1027.

26 Stevens VC. Use of synthetic peptides as immunogens for developing a vaccine against human chorionic gonadotrophin. In Porter R, Whelan J, eds. *Synthetic Peptides as Antigens*. Ciba Foundation Symposium 119. Chichester: Wiley, 1986:200–215.

27 Jones, WR, Bradley J, Judd SJ, *et al*. Phase 1 clinical trial of a World Health Organization birth control vaccine. Lancet 1988;1:1295–1299.

28 Ballou WR, Hoffman SL, Sherwood JA, *et al*. Safety and efficacy of a recombinant DNA *Plasmodium falciparum* sporozoite vaccine. Lancet 1987;1:1277–1281.

29 Herrington DA, Clyde DF, Losonsky G, *et al*. Safety and immunogenicity in man of a synthetic peptide malaria vaccine against *Plasmodium falciparum* sporozoites. Nature 1987;328:257–259.

30 Good MF, Kumar S, Miller LH. The real difficulties for malaria sporozoite vaccine development: nonresponsiveness and antigenic variation. Immunol Today 1988;9:352–355.

31 Patarroyo ME, Amador R, Clarijo P, *et al*. A synthetic vaccine protects humans against challenge with asexual blood stages of *Plasmodium falciparum* malaria. Nature 1988;332:156–158.

32 Wang ML, Skehel JJ, Wiley DC. Comparative analysis of the specificities of anti-influenza hemagglutinin antibodies in human sera. J Virol 1986;57:124–128.

33 Jerne N. Towards a network theory of the immune system. Ann Inst Pasteur/Immunol 1974;125C:373–389.

34 Ada GL, Jones PD. Vaccines for the future—an update. Immunol Cell Biol 1987;65:11–24.

35 Sachs DL, Sher A. Evidence that anti-idiotype induced immunity of African trypanosomes is genetically restricted and requires recognition of combining site related idiotypes. J Immunol 1983;131:1511–1515.

36 Grzijch JM, Capron M, Lambert PH, Dissous C, Torres S, Capron A. An anti-idiotype vaccine against experimental schistosomiasis. Nature 1985;316:74–76.

37 Weiss RA, Clapham PR, McClure MD, *et al*. Human immunodeficiency viruses: neutralization and receptors. J Autoimmun Def Syn 1988;1:536–541.

38 Cohen IR. Regulation of autoimmune disease: physiologic and therapeutic. Immunol Rev 1989;94:5–13.

39 Davis AR, Nayak DP, Ueda M, *et al*. Expression of antigenic determinants of the hemagglutinin gene of a human influenza virus in *Escherichia coli*. Proc Natl Acad Sci USA 1981;78:5376–5380.

40 Jilg W, Schmidt M, Zouleck G, Larbeer B, Wilske B, Deinhardt F. Clinical evaluation of a recombinant hepatitis B vaccine. Lancet 1984;ii:1174–1175.

41 WHO Study Group. Acceptability of cell substrates for production of biologicals. WHO Tech Rep Ser 1987;747:3–29.

42 Smith GL, Moss B. Infectious poxvirus vectors have capacity for at least 25 000 base pairs of foreign DNA. Gene 1983;25:21–28.

43 Moss B, Flexner C. Vaccinia virus expression vectors. Ann Rev Immunol 1987;5:305–324.

44 Perkus ME, Piccini A, Lipinskas PR, Paoletti A. Recombinant vaccinia virus; immunization against multiple pathogens. Science 1985;229:981–984.

45 Langford CJ, Edwards SJ, Smith GL, *et al*. Anchoring a secreted *Plasmodium* antigen on the surface of recombinant vaccinia virus-infected cells increases its immunogenicity. Mol Cell Biol 1986;6:3191–3199.

46 Buller RML, Smith GL, Cremer K, Notkins AL, Moss B. Decreased virulence of recombinant vaccinia virus expression vectors is associated with a thymidine kinase-negative phenotype. Nature 1985;317:813–815.

47 Ramshaw IA, Andrew ME, Phillips SM, Boyle DB, Coupar BEH. Recovery of immunodeficient mice from a vaccinia/IL-2 recombinant infection. Nature 1987;329:545–546.

48 Flexner H, Hugin A, Moss B. Prevention of vaccinia virus infection in immunodeficient mice by vector-directed IL-2 expression. Nature 1987;330:259–262.

49 Lachmann PJ, Strangeways L, Vyakarnum A, Evans GI. Raising antibodies by coupling peptides to PPD and immunizing BCG-sensitized animals. In Porter R, Whelan J, eds. *Synthetic Peptides as Antigens*. Ciba Foundation Symposium 119. Chichester: Wiley, 1986:25–40.

50 Ada GL. Vaccines. In Paul WE, ed. *Fundamental Immunology*, 2nd edn. New York: Raven Press, 1989:985–1032.

51 Kawamura H, Berzofsky JA. Enhancement of antigenic potency *in vitro* and immunogenicity *in vivo* by coupling

the antigen to anti-immunoglobulin. J Immunol 1986; 136:58–65.

52 Carayamiotis G, Barber BH. Adjuvant-free IgG responses induced with antigen coupled to antibodies against class 2 MHC. Nature 1987;327:59–61.

53 Pierce NF, Cray WC, Sacci JB. Oral immunization of dogs with purified cholera toxin, crude toxin or subunit: evidence for synergistic protection by antitoxic and antibacterial mechanisms. Infect Immun 1982;37:687–695.

54 Cohen S, McGregor IA, Carrington S. Gamma globulin and acquired immunity to human malaria. Nature 1961;192:733–737.

55 Good MF. T cells, T sites and malaria immunity—further optimism for vaccine development. J Immunol 1988;140:1715–1716.

56 Schofield L, Villaquiran J, Ferreira A, Schellekens H, Nussenzweig RN, Nussenzweig V. Interferon, CD8$^+$ T cells and antibody are required for immunity to malaria sporozoites. Nature 1987;330:664–666.

57 Weiss WR, Sedegah M, Beaudoin RL, Miller LH, Good M. CD8$^+$ T cells (cytotoxic/suppressors) are required for protection in mice immunized with malaria sporozoites. Proc Natl Acad Sci USA 1988;85:573–577.

58 Brake DA, Long CA, Weidanz WP. Adoptive protection of nude mice by a cloned T cell line. J Immunol 1988;140:1989–1993.

59 Brown J, Greenwood BM, Terry RJ. Cellular mechanisms involved in recovery from acute malaria in Gambian children. Parasite Immunol 1986;8:551–558.

60 Sadoff JC, Ripley Ballou W, Bacon LS, et al. Oral *Salmonella typhimurium* vaccine expressing circumsporozoite protein protects against malaria. Science 1988;240:336–338.

61 Ada GL. What to expect of a good vaccine and how to achieve it. Vaccine 1988;6:77–79.

62 Berquist R. Prospects of vaccination against Schistosomiasis. Scand J Infect Dis Suppl 1991. (In press.)

63 World Health Organization. *Strategies for the Development of a Schistosomiasis Vaccine*. WHO, Geneva, 1991.

64 Romero PJL, Maryanski G, Conadin RS et al. Cloned cytotoxic T cells recognize an epitope on the CS protein and protect against malaria. Nature 1989;341:323–326

65 Sher A, Coffman RL, Hieny S, Cheever AW. Ablation of eosinophil and IgE responses with anti-IL-5 or anti-IL-4 antibodies fails to affect immunity against *Schistosomiasis mansoni* in the mouse. J Immunol 1990;145:3911–3916.

2 Protozoa

8 *Entamoeba histolytica* and amebiasis

Adolfo Martínez-Palomo, Roberto Kretschmer, & Isaura Meza Gomez Palacio

Amebiasis, i.e., the infection of humans with the protozoan parasite *Entamoeba histolytica*, has a worldwide distribution. The motile form of the parasite, the trophozoite, usually lives as a harmless commensal in the lumen of the large intestine, where it multiplies and differentiates into a cyst, the resistance form responsible for the transmission of the infection. The undisputed causal role of *E. histolytica* in symptomatic cases of amebiasis and its apparent innocuousness in many others has given rise to a still unsolved debate that has lasted for over a century. Even modern techniques of molecular biology have failed—so far—to solve the puzzling question of the pathogenesis of amebiasis. The challenge is to determine whether:

1 all strains of *E. histolytica* are potentially pathogenic and the main variable is the susceptibility of the host;
2 there are stable and genetically different pathogenic and nonpathogenic strains associated with invasive and asymptomatic forms of amebiasis, respectively;
3 local intestinal conditions produce interconversion between pathogenic and nonpathogenic forms of the parasite, allowing the differential expression of certain genes and their products. Clearly, knowledge of the natural and acquired immunity in amebiasis is also crucial in order to solve these questions.

As a commensal, *E. histolytica* produces no signs or symptoms in a condition known as luminal amebiasis. As a pathogen, it causes invasive amebiasis, which is prevalent in certain developing countries. Virulent amebae may invade the intestinal mucosa and produce mostly dysentery and occasionally ameboma, and through bloodborne spreading give rise to extraintestinal lesions, mainly liver abscesses. Most cases of intestinal invasive amebiasis manifest as diarrhea or dysentery and are usually self-limited. However, unless promptly diagnosed and properly treated, amebic liver abscess is a potentially lethal disease. Cerebral amebiasis, a rare but dreaded condition, is almost always fatal [1].

Invasive amebiasis is a major health problem in areas of Africa, Asia, and Latin America, where inadequate sanitary conditions and perhaps the presence of highly virulent strains of *E. histolytica* combine to produce a high incidence of symptomatic infections [2]. Estimates suggest that in 1984, 500 million people were infected with *E. histolytica*, 40 million of which developed colitis or liver abscesses. No less than 40 000 deaths that year may have been due to amebiasis, many as a consequence of liver abscesses. These figures do not include mainland China, where amebiasis is prevalent, yet information is scarce. Therefore, on a global scale, amebiasis comes second only to malaria among protozoan causes of death [3]. The number of severe cases of amebiasis in industrialized countries is much lower than in developing countries. Nevertheless, knowledge of the disease in those regions is important, since failure to identify an amebic infection may result in an unavoidable fatal outcome. Furthermore, high infection rates prevail among certain immigrant groups and a striking increase in luminal amebic infections—presumably with nonpathogenic strains—has been detected among male homosexuals in several large cities of North America, Japan, and Europe [4,5].

THE PARASITE

The trophozoite of *E. histolytica* is a highly dynamic, pleomorphic, essentially anaerobic protozoan, 10–60 µm in diameter. Surface and cytoplasmic movements are continuous in amebae, which often show pseudopodia, a tail or uroid, or even occasional slender filopodia. In addition, a prominent surface endocytic activity is reflected in the occurrence of a varying number of phagocytic openings and pinocytotic vesicles [6]. Trophozoites dwell in the colon, where they multiply and encyst, producing typically four-nucleated cysts, which are found in the formed stools of carriers as round or slightly oval hyaline bodies 8–20 µm in diameter, with a refractive wall. When ingested, the cyst wall is dissolved in the upper gastrointestinal tract and further down gives rise to eight potentially invasive uninucleated trophozoites. Cysts do not develop within tissues. The invasive form of the parasite, the trophozoite, can penetrate the intestinal mucosa and disseminate to other organs. Trophozoites are short-lived outside the body and do not survive passage through the upper gastrointestinal tract, hence they do not transmit the disease. In contrast, cysts may remain viable in a humid environment and thus stay infective for several days.

Strain differences

Nonpathogenic strains of amebae were first shown to differ from those obtained from invasive cases in certain surface properties, namely, in their reduced phagocytosis and low concanavalin A-induced agglu-tination [7]. Subsequently, the isoenzyme technique demonstrated that amebae cultured from samples of well-characterized cases of invasive amebiasis clustered around distinct isoenzyme patterns, or zymodemes, that are different from those of amebae isolated from asymptomatic carriers [8]. Although the technique is still too cumbersome and time-consuming to be used in clinical practice for the identification and classification of *E. histolytica*, it has been a useful tool in demonstrating that most amebae found in asymptomatic carriers, particularly in countries where invasive amebiasis is nonendemic, belong to "nonpathogenic" zymodemes. Modifications of the technique are currently being used [9]. Male homosexuals of these countries appear particularly prone to infection with nonpathogenic strains, although infections with pathogenic strains have been found in a recent screening of Japanese homosexuals [5,10]. Differentiation between pathogenic and nonpathogenic strains has also been reported with the use of monoclonal antibodies [11,12] and, more recently,

Table 8.1 Structural and molecular organization of *Entamoeba histolytica*

Structural components		Biochemical characterization molecules and activities [reference]	Characterized genes [reference]
Present	Absent		
Trilaminar plasma membrane	Golgi system	260 kD N-acetylgalactosamine-inhibitable adhesin [29]	Ribosomal sequences [17,53]
Vacuoles (phagocytic, endocytic)	Endoplasmic reticulum	220 kD N-acetyl-glucosamine-inhibitable lectin [31]	Actin [22,23]
Ribosomes organized as clusters or helices	Mitochondria	Amebapore (membrane pore-forming protein) [46,47]	Ferredoxin [58]
Condensations of actin-forming adhesion plates and endocytic and phagocytic invaginations	Actin stress fibers or other filamentous structures	Neutral protease [42]	Specific gene sequences for pathogenic and nonpathogenic strains [14]
	Primary and secondary lysosomes	Other thiol proteases [44,45]	125 kD glycoprotein [34]
Nucleus	Cytoplasmic microtubules	Collagenase [49]	29 kD surface molecule [35]
Nuclear lamina	Organized chromosomes with kinetochores	Glycosidases [50]	170 kD subunit of the 260 kD N-Acetyl-galactosamine-inhibitable adhesin [39]
Nuclear pores	Nucleolus	96 kD glycoprotein [32]	
Chromatin	Mitotic spindle	Phospholipase A [41]	Superoxide dismutase [92]
Nucleosomes		Cytotoxin [43]	Amebapore [48]
Microtubules in dividing nuclei		Fibronectin and laminin receptors [37]	Serine-rich E. histolytica protein [93]
		112 kD antigen [33]	
		Hemolysins [40]	
		Actin [21]	
		125 kD surface protein antigen [34]	

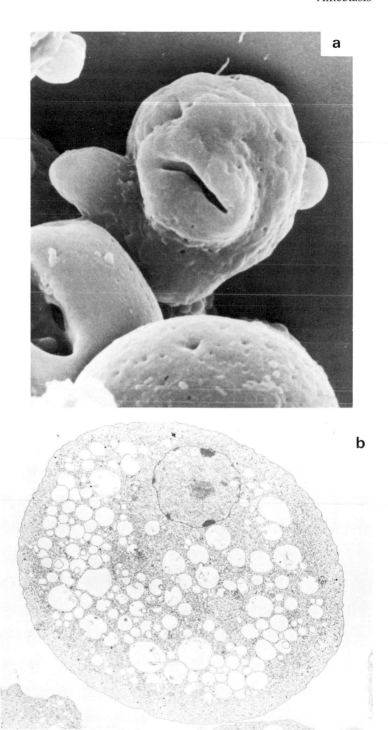

Fig. 8.1 Ultrastructure of *Entamoeba histolytica* trophozoites. (a) Scanning electron micrograph. Pseudopodia, a phagocytic stoma, and numerous pinocytotic vesicles are seen. Original magnification, ×2000. (b) Transmission electron micrograph of a thin section showing the cytoplasmic and nuclear components of a trophozoite. Original magnification, ×3000.

using DNA probes or rRNA sequences [13–18]. It has been found, however, that experimental changes in the bacterial flora associated with the parasite may result in the occasional conversion of a nonpathogenic strain into a pathogenic strain [19]. However, the possibility that these changes are due to the selection of a previously existing but undetectable subpopulation has not been ruled out yet. Thus, the question remains as to whether the zymodeme is a genotypically stable feature of *E. histolytica*.

Molecular biology techniques have provided what is probably the most important development in the differentiation between pathogenic and nonpathogenic strains. With the availability of genomic and expression libraries of *E. histolytica*, molecular probes have been developed that contain DNA sequences highly specific for pathogenic or nonpathogenic strains [14–16]. These studies have to be complemented with a better molecular characterization of the cellular components of both types of strains.

Cytoplasmic components

Entamoeba histolytica, being one of the most primitive eukaryotes, lacks the following cytoplasmic components: a structured cytoskeleton and cytoplasmic microtubules, a membranous system equivalent to the Golgi complex and endoplasmic reticulum of higher eukaryotic cells, mitochondria, and a system of primary and secondary lysosomes such as those regularly found in other eukaryotes (Table 8.1). Much of the cytoplasm consists of a vacuolar system mainly involved in endocytosis (Fig. 8.1). A lattice of tubules and vesicles only superficially resembling smooth endoplasmic reticulum may also be found by transmission electron microscopy. It remains to be shown if this membranous system is involved in the channeling of toxins and enzymes related to the destructive activity of the parasite. Trophozoites are facultative aerobes with peculiar glycolytic enzymes also found in certain bacteria. In many other aspects, amebae resemble more closely anaerobic and microaerophilic bacteria than typical eukaryotes, e.g., they lack glutathione metabolism [6,20].

The cytoskeletal components of the parasite are involved in motile processes (locomotion, adhesion, phagocytosis, and possibly the export of toxic substances through exocytosis) related to the cytopathic effect of pathogenic trophozoites. In spite of the presence of large amounts of cytoplasmic actin in trophozoites of *E. histolytica*, microfilaments are usually not visible under the electron microscope. With fluorescence microscopy, actin appears as dense patches in phagocytic and endocytic invaginations (Fig. 8.2a). The purified amebic actin polymerizes *in vitro* as typical 7-nm filaments and seems to be composed of a single isoform [21]. A single 1.5 kb RNA band has been identified using a cDNA actin clone [22]. Although at least four actin genes have been identified in the genome [22,23], it is possible that only one isoform is expressed in the parasite. Actin-regulating proteins remain to be described to provide a better understanding of the molecular bases of the ameboid movement and plasticity of the parasite. Microtubules have not been identified in

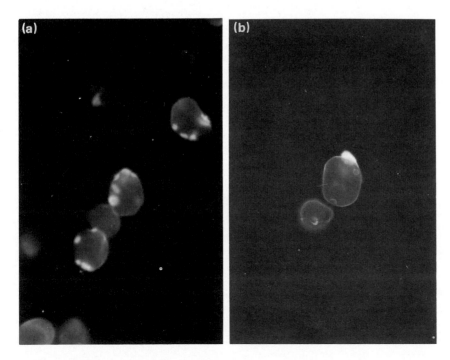

Fig. 8.2 Immunofluorescence micrographs of *Entamoeba histolytica* trophozoites. (a) Regions of endocytosis are strongly stained with rhodamine–phalloidin, a drug that specifically binds to structured filamentous actin. (b) Capping of antigen–antibody complexes induced with immune sera and revealed with fluorescent antibody against human immunoglobulins.

the cytoplasm and only dividing nuclei of *E. histolytica* show microtubular bundles [6].

Antigenic components

Early immunoelectrophoretic studies of aqueous extracts of axenically grown amebae showed an "antigenic skeleton" of *E. histolytica* made up of 20–32 antigenic components, the range and variations being explained as discrete periodic waves of emergence and hiding of antigens, a phenomenon apparently not correlated to virulence. Further analysis of soluble extracts of whole amebae revealed five fractions by gel sieve chromatography with molecular weights (MW) varying from 9 to 150 kD, 12 bands by autoradiography, seven polypeptides immunoadsorbing with human antiameba IgG, or eight bands by Western blotting using an antiamebic serum [24]. However, these studies did not take into consideration the extremely high proteoloytic activity of trophozoites manifested during the processing of protein samples. More recently it has been demonstrated that the processing of the sample will ultimately determine the electrophoretic profiles obtained, and hence the antigenic patterns observed. This explains, at least partially, the great variations in antigenic profiles reported by the different authors. It would be desirable, therefore, to standardize the conditions of sample treatment to unify criteria [25]. Recent experiments using sera from patients recovering from hepatic abscesses show that, for all its complexity, *E. histolytica* appears, however, antigenically quite uniform [26]; therefore, the characterization of surface antigens is of special interest.

Twelve glycoproteins ranging from 12 to 200 kD were identified by radiolabeling and autoradiography of external surface proteins [27], thus revealing a membrane composition more reminiscent of mammalian cells than of free-living amebae, such as *Acanthamoeba*. By using a variety of protease inhibitors and sodium dodecyl sulfate polyacrylamide gel electrophoresis (SDS-PAGE), surface proteins with MW ranging from 19 to 260 kD have been identified, several of the larger ones binding to concanavalin A and other lectins. Some of these antigens are more frequently recognized by sera from patients with amebic abscesses of the liver [26].

There is strong, passive, hydrophobic attachment of antigenic material, foremost of bovine serum albumin, from the axenic medium onto the amebae. Whether *E. histolytica* is capable of doing likewise with host antigens *in vivo*, as other parasites (i.e., *Schistosoma*) appear to do, and thus evade some mechanisms of

defense is not known. Interactions of antibodies with cell surface antigens elicit a rapid mobilization, capping (Fig. 8.2b), internalization, or shedding of surface antigen–antibody complexes [26]. This phenomenon together with the resistance to complement lysis [28] have been postulated as the main strategies employed by *E. histolytica* to evade the humoral response of the host.

Biochemically characterized surface molecules, probably involved in the adhesion to target cells or to intestinal mucus, include a 260 kD *N*-acetyl-D-galactosamine-inhibitable adhesin formed by two subunits of 170 and 35 kD [29,30] and an *N*-acetylglucosamine-inhibitable lectin of 220 kD [30]. Virtually all patients cured of invasive amebiasis have humoral antibodies against adherence lectins [26]. If such antibodies also occur in intestinal secretions, they may represent a resistance—probably short lived—factor to subsequent intestinal invasive disease. Antibodies against the 220 kD protein and the heavy subunit of the 260 kD lectin partially inhibit adhesion and phagocytosis of target cells *in vitro*, suggesting their participation in amebic adherence. Other monoclonal antibodies recently obtained against trophozoites have defined surface antigens of 96, 112, and 125 and 129 kD, corresponding to parasite glycoproteins [32–35]. Some of these antibodies inhibit adhesion and indirectly reduce the cytopathic effect *in vitro* [31,33].

One of these antigens, the 126 kD peptide, was first identified as one of the eight immunodominant antigens recognized by sera of patients with amebic liver abscesses [34]. From cDNA libraries constructed from the strain HM1-IMSS, two independent groups have now isolated positive clones containing the sequences for the 126 kD protein [14] (I. Meraz, personal communication). Both groups have identified homologous sequences in pathogenic and nonpathogenic strains of *E. histolytica*. Tannich *et al.* [14] found distinct Southern blot patterns, one characteristic for pathogenic amebae and one for nonpathogenic strains, suggesting, therefore, the existence of genetic differences between them. On the other hand, Edman·*et al.* [34] identified the 125 kD antigen as a highly glycosylated membrane protein containing a small sequence similar to the one present in β-integrins of typical eukaryotic cells. Monoclonal antibodies against the protein corroborated its presence in pathogenic and nonpathogenic strains isolated from cases of invasive and luminal amebiasis, respectively. Experiments with tissue culture cell mutants enriched for certain carbohydrates indicate a

Chapter 8

preferential adhesion of amebae to *N*-acetyllactosamine units on Asn-linked complex oligosaccharides [36].

Besides recognition of carbohydrates by amebic surface lectins, the initial contact and subsequent degradation of cells and extracellular matrix components possibly involve receptors to proteins, such as collagen, fibronectin, and laminin. A common receptor for the last two proteins has been identified in trophozoites of *E. histolytica* [37]. Binding of the parasite to these molecules, in soluble form or as solid substrate, is followed by their degradation. Binding to and degradation of collagen substrates follow a similar pattern, although a receptor for collagen has not been isolated. Moreover, binding to fibronectin or laminin substrates triggers a cytoplasmic response in trophozoites manifested as the formation of clearly defined actin adhesion plates. These surface amebic receptors for extracellular matrix components may thus be necessary for attachment and migration of the parasite into solid tissues.

Lytic activity

For descriptive purposes, the cytolytic action of *E. histolytica* on cultured cells can be conveniently divided into five stages: chemotaxis, adhesion, cytolysis following contact, phagocytosis, and intracellular degradation [38]. Even though most studies have focused on finding a single factor, such as aggression by a specific organelle, toxin, or enzyme action, the cytopathic effect of the amebae on mammalian tissues cannot be attributed to a single function of the parasite but is the result of combined chemical and mechanical actions.

Through chemotaxis, *E. histolytica*, even though devoid of cytoplasmic microtubules, is capable of efficient vectorial locomotion towards target cells [39]. After contact, the lysis of target cells requires mobilization and surface liberation of enzymes and toxins from the parasite. This translocation demands an intact cytoskeletal network, since disaggregation of actin caused by cytochalasin B blocks the cytopathic effect of the amebae. Lysis is also blocked by calcium-channel blockers, suggesting that calcium fluxes in the ameba and entry of calcium into the target cell participate in the cytopathic effect. Phospholipase A may be involved in the damage to the plasma membrane of host cells [40,41]. In addition, *E. histolytica* in culture contains and, under certain *in vitro* conditions, releases:

1 proteases that round up cultured cells [42–45];
2 pore-forming proteins that insert into natural or artificial membranes creating an ionic imbalance [46,48];
3 enzymes that degrade collagen and oligosaccharides in the extracellular matrix [49,50];
4 neurotransmitter-like compounds that induce water secretion in isolated intestinal loop models [51].

Despite these results and notwithstanding the apparent correlation between virulence and some of the above-mentioned factors, none of these alone have proven, so far, to be essential in the pathogenesis of amebiasis.

After contact-dependent cytolysis of target cells occurs, pathogenic amebae ingest the lysed cells, although they can also engulf living cells. In general, pathogenic strains show a high rate of erythrophagocytosis; in contrast, nonpathogenic amebae and those of attenuated virulence ingest few red blood cells. Further exploration of the relationship between high phagocytic rates and virulence has shown that trophozoites defective in phagocytosis simultaneously lose their virulence and erythrophagocytic activity, demonstrating that phagocytosis (Fig. 8.3) is one of the aggression mechanisms of the parasite [7].

Nuclear components

Little is known about nuclear organization and division in *E. histolytica*. The only reasonably settled issue is that nuclear division proceeds with the formation of microtubular bundles and without dissolution of the nuclear membrane [6]. Cytological evidence suggests that five to six nuclear condensations may represent chromatin organized as chromosomes, but pulse-field gel electrophoresis has so far failed to identify clearly the number of chromosomes in this protozoan. This technique has shown that the DNA of *E. histolytica* separates into six to nine bands [52]. DNA fibers contain nucleosomes; however, the structural unit is not formed by typical histones but by another type of DNA-binding protein. Recent estimates of the total DNA content of amebae is 0.5 pg per nucleus. All DNA is in the nucleus, as there are no mitochondria or other DNA-containing organelles. However, several copies of rDNA genes are present as extrachromosomal circular elements. The number of copies or the number of episomes in the different strains seems to be variable [53].

The first reports on DNA restriction patterns of several strains of *Entamoeba* revealed different repetitive fragments with prominent and specific bands of 0.6 and 0.7 kb in *E. histolytica* [54]. In lower eukaryotes, including the protozoa *Trypanosoma* and *Leishmania*, a large

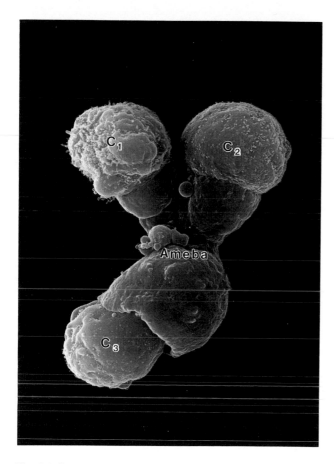

Fig. 8.3 Scanning electron micrograph of a single *Entamoeba histolytica* trophozoite (ameba) engulfing simultaneously three epithelial cells (C_1–C_3). Original magnification, ×2500.

number of highly repetitive sequences have also been detected. In *E. histolytica* the most conspicuous repetitive sequences correspond to ribosomal RNA genes. Recently, it has been established that the circular rDNA molecules contain two large repeat-regions, each at least 5.2 kb long. The inverted repeats are flanked by stretches of DNA which have tandemly reiterated sequences. These findings suggest an independent replication origin which allows the amplification of such molecules into a high copy number. The repetitive nature of these sequences and their polymorphism renders them excellent probes to identify *E. histolytica* in clinical samples and to differentiate between pathogenic and nonpathogenic strains [14,55]. *Entamoeba histolytica* mutants that are resistant to drugs such as emetine and colchicine [56] have allowed the identification of a family of mdr-like genes which may be involved in drug resistance, in which case a mechanism of multiple drug

resistance similar to that described for tumor cells could be envisaged for amebae.

The first gene described for *E. histolytica* was an actin gene [22,23]. Actin is the most abundant cytoskeletal protein in amebae and it participates in all the motility processes of the parasite. The inferred actin sequence shows 89% and 86% amino acid homology to human cytoplasmic and skeletal muscle actins, respectively. However, homology to the corresponding skeletal muscle gene is only 67%, indicating a peculiar codon usage in the making of actin. The codon usage for *E. histolytica* [57] actin shows great similarity to that present in other primitive eukaryotic actins, such as the *Saccharomyces* actin. The sequence analysis of a genomic actin clone revealed that the actin gene does not have intervening sequences. There are at least four actin genes [22].

A genomic ferredoxin clone has been isolated from *E. histolytica* [58]. Comparison of the genomic and cDNA sequences revealed that the ferredoxin gene is also unspliced. The deduced amino acid sequence of *E. histolytica* ferredoxin resembles a clostridial type of ferredoxin and reveals an arrangement of cysteines characteristic of the coordination of 2(4Fe–4S) centers. These genes are arranged in a family of at least two ferredoxin genes, one of which is marked by restriction-length polymorphism in different strains of *E. histolytica*.

Chemical mutagenesis of pathogenic amebae has already been used as a tool to explore the correlation of gene products and specific functions of the parasite. Several mutants show different degrees of defective adhesion and phagocytosis of target cells. The molecular changes leading to these properties have yet to be characterized [59,60]. Introduction of exogenous DNA into protozoan parasites either by electroporation or transformation and transfection using plasmids and viruses will certainly contribute in the near future to determine the role of specific molecules in the mechanisms of pathogenicity of amebiasis.

HOST–PARASITE RELATIONSHIP

Since the classic description of Councilman & Lafleur in 1891 [61], the lesions of invasive amebiasis were known for the paucity of its inflammatory reaction. However, early endoscopic biopsies and the evidence gathered from experimental models have established the existence of an intense, albeit ephemeral, acute inflammatory reaction in both intestinal and hepatic amebiasis. This reaction soon blends with the products of tissue necrosis, apparently caused by lysosomal enzymes released from host polymorphonuclear leukocytes (PMN) destroyed by the amebae [62–65]. Amebae may also produce anti-inflammatory factors capable of delaying the arrival of macrophages to the scene [66]. The paucity of late inflammatory elements could be related to the equally remarkable lack of scarring tissue and the complete regeneration of the affected organs (colon, liver, or skin) following recovery that is so striking of amebiasis [67].

Pathogenesis and pathology

Invasion of the colonic and cecal mucosa by *E. histolytica* begins in the interglandular epithelium (Fig. 8.4). Cell infiltration around invading amoebae leads to rapid lysis of inflammatory cells and tissue necrosis: thus, acute inflammatory cells are seldom found in biopsy samples or in scrapings of rectal mucosal lesions. Ulcerations may deepen and progress under the mucosa to form typical ''flask ulcers,'' which extend into the submucosa producing abundant microhemorrhages. This explains the finding of hematophagous amebae in stool specimens or in rectal scrapings, which are still the best indication of the amebic nature of a case of dysentery or bloody diarrhea. Macroscopically, the ulcers are superficial initially, with hyperemic borders, a necrotic base, and normal mucosa between the sites of invasion. Further progression of the lesions may produce loss of the mucosa and submucosa covering the muscle layers, and eventually lead to the rupture of the serosa.

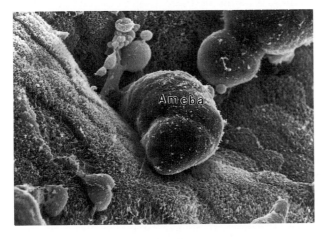

Fig. 8.4 Scanning electron micrograph of a trophozoite attached to epithelial cells of the large intestine near an interglandular region. Original magnification, ×2000.

Complications of intestinal amebiasis include perforation, direct extension to the skin, and dissemination, mainly to the liver. Amebae probably spread from the intestine to the liver through the portal circulation. The presence and extent of liver involvement bears no relationship to the degree of intestinal amebiasis, and these conditions do not necessarily coincide. The early stages of hepatic amebic invasion have not been studied in humans. In experimental animals, inoculation of *E. histolytica* trophozoites into the portal vein produces multiple foci of neutrophil accumulation around parasites, followed by focal necrosis and granulomatous infiltration. As the lesions extend in size, the granulomas are gradually substituted by necrosis until the lesions coalesce and necrotic tissue occupies progressively larger portions of the liver. Hepatocytes close to the early lesions show degenerative changes which lead to necrosis, but direct contact of liver cells with amebae is very rarely observed. The lesion can eventually develop large areas of liquefied necrotic material surrounded by a thin capsule of fibrous appearance [62–65]. These experimental results suggest that *E. histolytica* trophozoites do not produce amebic experimental liver abscesses through direct lysis of hepatocytes. Rather, tissue destruction is the result of the accumulation and subsequent lysis of neutrophils and macrophages surrounding the amebae. Human liver abscesses consist of areas in which the parenchyma has been completely substituted by semisolid or liquid material composed of necrotic matter and a few cells. Neutrophils are generally absent, and amebae tend to be located at the periphery of the abscess. Liver abscesses may heal, rupture, or disseminate [67].

Invasive amebic lesions in humans, whether localized in the large intestine, liver, or skin, almost invariably heal without the formation of scar tissue if properly treated. The absence of fibrotic tissue following necrosis is particularly striking in the liver. The complete anatomic and functional restitution of liver integrity after treatment of liver abscesses has been assessed by scintillography.

Interactions between amebae and inflammatory cells

Under *in vitro* conditions, when virulent amebae interact with human or mammalian neutrophils, the latter succumb after positive chemotaxis towards the parasite. Some of the lysed leukocytes are eventually phagocytized by the amebae which, in turn, emerge unscathed. Not even 3000 neutrophils per ameba or the presence of antibodies or complement succeed in killing virulent amebae. Only when less-virulent heat- or emetine-treated amebae are involved, and then only in PMN/ameba ratios in excess of 200:1, are leukocytes capable of destroying the parasite. It thus appears that the PMN is a remarkably incompetent effector cell against virulent *E. histolytica*. There are no differences in adherence between pathogenic and nonpathogenic amebae to human neutrophils [68]. It is hardly surprising, then, that humoral antibodies, the natural allies of PMN, have not established much of a protective reputation in clinical amebiasis. On the other hand, while non-stimulated lymphocytes also succumb to amebic cytolytic activity *in vitro*, immune T8 lymphocytes or lectin-stimulated nonimmune T lymphocytes kill virulent *E. histolytica* [69,70]. The supernatant fluid of stimulated lymphocytes inhibits amebic protein synthesis. Antibody-dependent cell-mediated cytotoxicity (ADCC), effective against other parasites, does not appear to operate against *E. histolytica*.

Interaction of virulent amebae with nonactivated normal macrophages culminates in contact-dependent, serum-independent lysis of macrophages without a decrease in amebic viability. Again, the presence of antiamebic antibody and complement fails to reverse this outcome. In contrast, lectin- or lymphokine-activated macrophages effectively kill virulent amebae through an extracellular, immunologically nonspecific, time- and contact-dependent, serum-independent process. Oxidative and nonoxidative mechanisms are at play in this process. This occurs at macrophage/amebae ratios as low as 10:1 and increases as the ratio approaches 100:1, but is usually followed by a decrease in macrophage viability due to toxic products released from the killed amebae [69]. The increased or decreased development of experimental amebiasis in laboratory animals in which macrophage function had been either depressed (using silica or antimacrophage serum) or enhanced (with Bacille Calmette-Guerin (BCG)), respectively, lends credence to this proposition [71,72].

The results suggest that the activated macrophage is a potentially competent effector cell against virulent *E. histolytica* and, therefore, possibly the only leukocyte that, following activation, possesses a phagocytic/cytolytic superiority over *E. histolytica* that may be critical in preventing invasive amebiasis [73]. The *in vivo* confrontation between macrophages and *E. histolytica* may, however, be a more precariously balanced duel, since in susceptible animal models, liver necrosis produced by virulent amebae occurs in spite of the recruitment of large numbers of macrophages around zones of trophozoite invasion [62]. Eosinophils have been studied recently in their *in vitro* interaction with virulent *E. histolytica*. Even aided by antiamebic antibodies and complement, the normal eosinophil undergoes the same fate as the neutrophil. However, when activated by lymphokines it behaves like an activated macrophage and is able to destroy the parasite.

HUMORAL IMMUNE RESPONSE

A prompt local secretory response followed by an equally rapid antibody response ensues upon intestinal invasion by *E. histolytica*. A mixture of IgA, IgG, and IgM coproantibodies has been found by indirect hemagglutination (IHA) in about 80% of cases of amebic dysentery, as opposed to 2% in healthy controls and 4% in nonamebic parasitic infections. Three weeks later this figure falls to 55%, just as serum antibodies make their appearance [74]. A comparable local antiamebic antibody production has been induced experimentally in the rat gut. Secretory IgA anti-*E. histolytica* antibodies have also been found in the bile of intracecally immunized rats [75] and in human milk, colostrum, and saliva. We do not know at the present time if IgA (and IgE) antiamebic coproantibodies play a defensive—if transient—role in amebiasis.

Circulating antibodies to *E. histolytica* can be demonstrated as early as 1 week after the onset of symptoms in humans and experimental animals. All immunoglobulin classes are involved, but there seems to be a predominance of IgG2 antibodies. Virtually all known serologic tests have been employed to this end, including immunofluorescent antibodies (IFA), indirect hemagglutination (IHA), radioimmunoassay (RIA), counterimmunoelectrophoresis (CIE), and enzyme-linked immunosorbent assay (ELISA), the latter being the most sensitive (no false negatives in cases of amebic liver abscess), specific (only 3.6% false positives in controls living in endemic areas), opportune (earliest detection < 1 week), and persistent (> 3 years) in measuring such antibodies (reviewed by Kretschmer [74]). Much of the earlier studies were done with IHA and CIE, which combined a reasonably high degree of specificity (only 6.6% and 5.8% positivity, respectively, in healthy controls living in endemic areas) and sensitivity (94.8% and 96.4% positivity, respectively, in proven cases of amebic liver abscess), coinciding in over 90% of both negative and positive cases. The IFA test also deserves to be mentioned because it is much simpler than all the other tests and when combined with IHA also reaches a 100% positivity in cases of liver abscesses produced by *E. histolytica*. The Center for Disease Control in Atlanta, Georgia, has chosen IHA as its standard serologic reference for amebiasis, with a 1:256 cutoff titer [76]. Because of its relative simplicity and efficiency, CIE, on the other hand, is particularly well suited for epidemiologic surveys: 19 442 nonselected individual serum samples revealed a 5.95% positivity in Mexico [77].

Antibody detection is an invaluable tool in the diagnosis of amebic abscess of the liver and of ameboma, where in short ascending order of sensitivity IHA, CIE, IFA, IHA + IFA, and ELISA give virtually no false negatives in very early sera. Serology is less useful in the diagnosis of invasive intestinal amebiasis, yielding only 60–90% positivity. Unfortunately, serologic tests cannot distinguish between present, recent, or past (> 3 years) amebic invasion and, furthermore, titers do not correlate with clinical severity in human amebiasis. The broad range of positivity that has been reported in *E. histolytica* cyst-passers (0–70%) and the background problem of subclinical amebic invasion casts further doubt on the usefulness of serology in intestinal amebiasis, specially in communities where amebiasis is endemic. Nevertheless, patients with suspected inflammatory bowel disease should be evaluated for amebiasis (i.e., stool examinations and serology), lest a potentially fatal steroid treatment be started [78].

Both elevated and decreased complement levels have been found in human and experimental invasive amebiasis. Such inconsistency contrasts with the observation that virulent and nonvirulent strains of *E. histolytica* are equally capable of activating both pathways of the complement system, the classic more vigorously than the alternative pathway, even in the absence of antibody. This activation is lethal for the nonvirulent strains, while virulent strains withstand lysis conditions [28]. *In vivo* experimental studies are perhaps more informative, since cobra venom factor-treated guinea pigs are significantly more susceptible to experimental amebic liver abscesses. Thus, it may well be that complement, notwithstanding some of the *in vitro* observations made so far, plays a role of defense against those intestinal amebae reaching the bloodstream and that many aborted amebic invasions remain silent. The significant increase in complotype SCO1 in Mexican mestizo patients with amebic abscess of the liver may be important in this respect, although a clear relationship between allelic forms of complement and

their function (i.e., complement activation) remains to be established.

High rates of intestinal amebic reinfection have been recorded in spite of elevated titers of antiamebic antibodies [73], and patients with agammaglobulinemia or B cell-immunosuppressed animals do not appear to be more susceptible to invasive amebiasis than the normal population [74]. This, and the apparent irrelevance of humoral antibodies and complement in *in vitro* lytic assays, has led to the consensus that circulating humoral antiamebic antibodies are not protective against intestinal and perhaps extraintestinal amebiasis. Even though no protective value is granted to circulating antiamebic antibodies, they do not appear to be harmful, as immune complex disease is not a feature in amebiasis, although immune complexes have been found [79] and in a small group of Indian patients amebiasis coexists with arthritis [80]. The protective role of secretory IgA and IgE antibodies in amebiasis remains to be clarified. Predictably, such secretory immunity will turn out to be ephemeral.

CELLULAR IMMUNE RESPONSE

Even though the basic ingredients for a local cellular immune response (mononuclear phagocytes, lymphocytes, etc.) are regularly present in early intestinal amebic lesions, their eventual role in the establishment or prevention of invasive infections is not well understood. It has been claimed that tissue invasion by *E. histolytica* must be preceded by, and associated with, some degree of T cell suppression, a condition that may be met by selection and/or induction [81]. The increased susceptibility of T cell-immunosuppressed patients or experimental animals to invasive amebiasis, the presence of malnutrition in over 90% of the autopsies of amebic liver abscess, the susceptibility of children to invasive amebiasis in spite of transplacental maternal antibodies, and the significant increase in HLA-DR3 [82] found in Mexican patients with amebic liver abscess support the selective proposition, while the inductive proposal is supported by the observation that cell-free extracts of the parasite can exhaust and thus suppress the cellular immune response of the host [83]. On the other hand, a state of acquired protective cell-mediated immunity may be responsible for the claimed rarity of recurrences of amebic liver abscesses in humans (i.e., 0.04% recurrences, vs. 0.2% first amebic liver abscess per year calculated in Mexico City) [73].

Patients with acquired immunodeficiency syndrome

(AIDS) surprisingly do not appear more susceptible to amebic disease than homosexual men without AIDS, and autopsies of AIDS patients with coincident invasive amebiasis carried out in Mexico have shown that the amebic lesions are not different from those seen in the pre-AIDS era.

Prospective selective immunization and immunosuppression of experimental animals, the few studies of passive transfer of immunity with cells, and the straightforward outcome of the *in vitro* interaction of virulent amebae with activated lymphocytes and macrophages as opposed to PMN (irrespective of the presence of antibodies or complement) favor the existence of an effective cellular, rather than humoral, immunity against extraintestinal amebiasis. While humoral antibodies have a more diagnostic than protective value, the reverse is true for cellular immune phenomena, since delayed hypersensitivity skin testing is of little diagnostic avail, except perhaps in epidemiologic surveys.

Important as cell mediated immunity may be in amebiasis, the actual defense strategy appears to gravitate around the macrophage. Depressing macrophage functions by using silica or antimacrophage serum, or enhancing them with BCG, increases or decreases the development of experimental amebiasis [71]. In fact, congenitally athymic nu/nu mice (devoid of T lymphocytes) and genetically susceptible mice (C57B1/6 and C3H/HeJ) only developed amebic liver abscesses and intestinal amebic disease, respectively, after macrophage blockade with silica [72].

IMMUNOLOGIC AND MOLECULAR DIAGNOSIS

A battery of polyclonal and monoclonal antibodies that can be used in ELISA assays has been gradually developed against *E. histolytica* antigens. Recent efforts have concentrated on the identification and characterization of highly specific amebic antigens. Yet, so far, only a few attempts to implement practical and specific diagnostic tools have appeared, such as the ELISA test for *E. histolytica* antigen detection in feces [84–86], specific IgA antibody detection in saliva [87,88], Western blots using a recombinant amebic protein [89], and the molecular DNA probe being developed by Samuelson *et al.* [55].

Human immune serum recognizes at least 15 different antigens, some of which are identified by most sera tested sera. In a recent survey of 108 patients' sera living in Mexico City [34], eight antigens were recognized by 90% of the sera. The antigenic peptides range in molecular weight from 46 to 220 kD and include glycoproteins of 37, 59, and 90 kD and the 220 kD lectin identified by other groups [26,31]. In addition, highly immunogenic 96 and 125 kD amebic surface antigens have been found with several monoclonal antibodies [32,34]. The first report on the selective identification of pathogenic and non-pathogenic amebae using monoclonal antibodies [11] has been mentioned earlier.

Furthermore, the characterization of several genes of *E. histolytica* has paved the way for the use of molecular probes to distinguish between pathogenic and non-pathogenic strains. Some of these probes are, in addition, complementary to repeated sequences of ribosomal DNA present in the amebae as extrachromosomal episomes, with internal polymorphism being characteristic of each strain [13]. The probes allow the detection of a small number of parasites in feces or in tissue samples [55,90] and some of them also distinguish between pathogenic and nonpathogenic strains of *E. histolytica* [14]. Further refinements of this methodology in the near future will certainly be of immense diagnostic and epidemiologic value. Characterization of *Entamoeba* genes will not only lead to the design of better probes but also to the production of recombinant DNA specific for *E. histolytica* antigens of potential use as immunogens.

PROSPECTS OF IMMUNOPROPHYLAXIS

The existence of acquired protective immunity in amebiasis rests essentially on two sets of basic observations. The first is the widely held belief that recurrent human amebic liver abscess is exceedingly rare, and the refractoriness of experimental animal models to amebic hepatic reinvasion with *E. histolytica* after spontaneous or therapeutic recovery. This argument calls for a better epidemiologic definition based on controlled prospective surveys.

The second line of arguments is based on successful immunization experiments with live trophozoites, crude antigens plus adjuvant, fractionated and chromatographed glycoproteins, and ribosomal or lysosomal amebic antigenic fractions [91], and more recently with purified amebic lectins (V. Tsutsumi, personal communication). The rise in morbidity and mortality due to *E. histolytica* with increasing age surely casts an uneasy doubt upon the concept of effective acquired protective immunity in amebiasis. Nevertheless, immunoprotection remains a reasonable proposition in human amebiasis, especially since public health control of the disease by other means would require radical socio-economic and political strategies that are easier phrased than implemented [2]. Several antigenic fractions of *E. histolytica* have been studied for this purpose, with different degrees of purity and a variety of adjuvants, animal models, and routes of administration.

The induction of protective secretory immunity to the adhesion lectins of *E. histolytica* appears to be a theoretically sound strategy, even though it would conceivably require a regular reactivation plan. Much remains to be learned before such a goal can be reached.

REFERENCES

1 Martínez-Palomo A, Ruíz-Palacios G. Amebiasis. In Warren KS, Mahmoud AAF, eds. *Tropical and Geographical Medicine*, 2nd edn. New York: McGraw-Hill Book Co., 1990:327–344.

2 Martínez-Palomo A, Martínez-Báez M. Selective primary health care: strategies for control of disease in the developing world. X. Amebiasis. Rev Infect Dis 1983;5:1093–1102.

3 Walsh J. Prevalence of *Entamoeba histolytica* infection. In Ravdin JI, ed. *Amebiasis. Human Infection by Entamoeba histolytica*. New York: Wiley, 1988:93–105.

4 Martínez-Palomo A. Amoebiasis. Clin Trop Med Commun Dis 1986;1:587–601.

5 Takeuchi T, Miyahira Y, Kobayashi S, Nozaki T, Motta SRN, Matsuda J. High seropositivity for *Entamoeba histolytica* infection in Japanese homosexual men—further evidence for the occurrence of pathogenic strains. Trans R Soc Trop Med Hyg 1990;84:250–251.

6 Martínez-Palomo A. *The Biology of Entamoeba histolytica*. Chichester: Wiley, 1982.

7 Martínez-Palomo A. Biology of ameiasis: progress and perspectives. In Englund PT, Sher A, eds. *The Biology of Parasitism*: a Molecular and Immunologic Approach. New York: Alan R Liss, 1988:61–76.

8 Sargeaunt PG, Jackson TFHG, Simjee AE. Biochemical homogeneity of *Entamoeba histolytica* isolates, especially those from liver abscess. Lancet 1982;i:1386–1388.

9 Meza I, De la Garza M, Meraz A, *et al*. Isoenzyme patterns of *Entamoeba histolytica* isolates from asymptomatic carriers: use of gradient acrylamide gels. Am J Trop Med Hyg 1986;35:1134–1139.

10 Nozaki T, Motta SR, Takeuchi T, Kobayashi S, Sargeaunt PG. Pathogenic zymodemes of *Entamoeba histolytica* in Japanese male homosexual population. Trans R Soc Trop Med Hyg 1989;83:525.

11 Stracham WD, Chiodini PL, Spice WM, Moody AH, Ackers JP. Immunological differentiation of pathogenic and non-pathogenic isolates of *Entamoeba histolytica*. Lancet 1988;1:561–562.

12 Petri WA, Jackson TFHG, Gathiram V, *et al*. Pathogenic and nonpathogenic strains of *Entamoeba histolytica* can be differentiated by monoclonal antibodies to the galactose-specific adherence lectin. Infect Immun 1990;58:1802–1806.

13 Garfinkel LI, Giladi M, Huber M, *et al*. DNA probes specific for *Entamoeba histolytica* possessing pathogenic and non-pathogenic zymodemes. Infect Immun 1989;57:926–931.

14 Tannich E, Horstmann RD, Knobloch J, Arnold HH. Genomic DNA differences between pathogenic and non-pathogenic *Entamoeba histolytica*. Proc Natl Acad Sci USA 1989;86:5118–5122.

15 Tachibana H, Ihara S, Kobayashi S, Kaneda Y, Takeuchi T, Watanabe Y. Differences in genomic DNA sequences between pathogenic and nonpathogenic isolates of *Entamoeba histolytica* identified by polymerase chain reaction. J Clin Microbiol 1991;29:2234–2239.

16 Tannich E, Burchard GD. Differentiation of pathogenic from nonpathogenic *Entamoeba histolytica* by restriction fragment analysis of a single gene amplified *in vitro*. J Clin Microbiol 1991;29:250–255.

17 Clark CG, Diamond LS. Ribosomal RNA genes of pathogenic and nonpathogenic *Entamoeba histolytica* are distinct. Mol Biochem Parasitol 1991;49:297–302.

18 Que X, Reed SL. Nucleotide sequence of a small subunit ribosomal RNA (16S-like rRNA) gene from *Entamoeba histolytica*—differentiation of pathogenic from nonpathogenic isolates. Nucleic Acids Res 1991;19:5438.

19 Mirelman D. Effect of culture conditions and bacterial associates on the zymodemes of *Entamoeba histolytica*. Parasitol Today 1987;3:37–40.

20 Fahey RC, Newton GL, Arrick B, Overdank-Bogart T, Aley SB. *Entamoeba histolytica*: a eukaryote without glutathione metabolism. Science 1984;224:70–72.

21 Meza I, Sabanero M, Cázares F, Bryan J. Isolation and characterization of actin from *Entamoeba histolytica*. J Biol Chem 1983;258:3936–3941.

22 Edman U, Meza I, Agabian N. Genomic and cDNA actin sequences from a virulent strain of *Entamoeba histolytica*. Proc Natl Acad Sci USA 1987;84:3024–3028.

23 Huber M, Garfinkel L, Gitler C, Mirelman D, Revel M, Rozenblatt S. *Entamoeba histolytica*: cloning and characterization of actin cDNA. Mol Biochem Parasitol 1987;24:227–235.

24 Aust-Kettis A, Thorstensson R, Utter G. Antigenicity of *Entamoeba histolytica* strain NIH-200: a survey of clinically relevant antigenic components. Am J Trop Med Hyg 1983;32:512–515.

25 Espinosa-Cantellano M, Martínez-Palomo A. The plasma membrane of *Entamoeba histolytica*—structure and dynamics. Biol Cell 1991;72:189–200.

26 Joyce MP, Ravdin JI. Antigens of *Entamoeba histolytica* recognized by immune sera from liver abscess patients. Am J Trop Med Hyg 1988;38:74–80.

27 Aley SB, Scott WA, Cohn ZA. Plasma membrane of *Entamoeba histolytica*. J Exp Med 1980;152:391–404.

28 Calderón J, Tovar R. Loss of susceptibility to complement lysis in *Entamoeba histolytica* HM1 by treatment with human sera. Immunology 1986;58:467–471.

29 Petri WA, Chapman MD, Snodgrass T, Mann BJ, Broman J, Ravdin JI. Subunit structure of the galactose and N-acetyl-D-galactosamine-inhibitable adherence lectin of *Entamoeba histolytica*. J Biol Chem 1989;246:3007–3011.

30 Tannich E, Ebert F, Horstmann RD. Primary structure of the 170-kDa surface lectin of pathogenic *Entamoeba histolytica*. Proc Natl Acad Sci USA 1991;88:1849–1853.

31 Rosales-Encina JL, Meza I, López de León A, Talamás-Rohana P, Rojkind M. Isolation of a 220 Kda protein with lectin properties from a virulent strain of *Entamoeba histolytica*. J Infect Dis 1987;156:790–794.

32 Torian BE, Lukehart SA, Stamm WE. Use of monoclonal antibodies to identify, characterize, and purify a 96 000-Dalton surface antigen of pathogenic *Entamoeba histolytica*. J Infect Dis 1987;156:334–343.

33 Arroyo R, Orozco E. Localization and identification of *Entamoeba histolytica* adhesin. Mol Biochem Parasitol 1987;23:151–158.

34 Edman U, Meraz MA, Rausser S, Agabian N, Meza I. Characterization of an immuno-dominant variable surface antigen from pathogenic and nonpathogenic *Entamoeba histolytica*. J Exp Med 1990;172:879–888.

35 Reed SL, Flores BM, Batzer MA, et al. Molecular and cellular characterization of the 29-kilodalton peripheral membrane protein of *Entamoeba histolytica*—differentiation between pathogenic and nonpathogenic isolates. Infect Immun 1992; 60:542–549.

36 Li E, Becker A, Stanley SL. Use of Chinese hamster ovary cells with altered glycosylation patterns to define the carbohydrate specificity of *Entamoeba histolytica* adhesion. J Exp Med 1988;167:1725–1730.

37 Talamás-Rohana P, Meza I. Interaction between pathogenic amebas and fibronectin: substrate degradation and changes in cytoskeleton organization. J Cell Biol 1988;106:1787–1794.

38 Martínez-Palomo A, González-Robles A, Chávez B, et al. Structural bases of the cytolytic mechanisms of *Entamoeba histolytica*. J Protozool 1985;32:166–175.

39 Bailey GB, Leitch GJ, Day DB. Chemotaxis by *Entamoeba histolytica*. J Protozool 1985;32:341–346.

40 Saíd-Fernandez S, López-Revilla R. Latency and heterogeneity of *Entamoeba histolytica* hemolysis. Z Parasitenkd 1983;69:435–438.

41 Long Krug SA, Fischer HJ, Hysmith RM, Ravdin JA. Phospholipase A enzymes of *Entamoeba histolytica*: description and subcellular localization. J Infect Dis 1985;152:536–541.

42 Keene WE, Petitt MG, Allen S, McKerrow JH. The major neutral proteinase of *Entamoeba histolytica*. J Exp Med 1986;163:536–549.

43 Lushbaugh WB, Hofbauer AF, Pittman FE. Proteinase activities of *Entamoeba histolytica* cytotoxin. Gastroenterology 1984;87:17–27.

44 Pérez-Montfort R, Ostoa-Saloma P, Velázquez-Medina L, Montfort I, Becker I. Catalytic classes of proteinases of *Entamoeba histolytica*. Mol Biochem Parasitol 1987;26:87–98.

45 Sholze H, Otte J, Werries E. Cysteine proteinase of *Entamoeba histolytica*. II. Identification of the major split position in bovine insulin β-chain. Mol Biochem Parasitol 1986;18:113–121.

46 Young JD-E, Young TM, Lu LP, Unkeless JC, Cohn ZA. Characterization of a membrane pore-forming protein from *Entamoeba histolytica*. J Exp Med 1982;156:1677–1690.

47 Rosenberg I, Bach D, Loew LM, Gitler C. Isolation, characterization and partial purification of a transferable membrane channel (amoebapore) produced by *Entamoeba histolytica*. Mol Biochem Parasitol 1989;33:237–248.

48 Leippe M, Ebel S, Schoenberger OL, Horstmann RD, Müller Eberhard HJ. Pore-forming peptide of pathogenic *Entamoeba histolytica*. Proc Natl Acad Sci USA 1991;88:7659–7663.

49 Muñoz ML, Calderón J, Rojkind M. The collagenase of *Entamoeba histolytica*. J Exp Med 1982;155:42–51.

50 Trissl D. Glycosidases of *Entamoeba histolytica*. Z Parasitenkd 1983;69:291–298.

51 McGowan K, Kane A, Asarkof N, et al. *Entamoeba histolytica* causes intestinal secretion: role of serotonin. Science 1983;221:762–764.

52 Valdés J, De la Cruz Hernández F, Ocádiz R, Orozco E. Molecular karyotype of *Entamoeba histolytica* and *Entamoeba invadens*. Trans R Soc Trop Med Hyg 1990;84:537–541.

53 Huber M, Koller B, Gitler C, et al. *Entamoeba histolytica* ribosomal genes are carried on palindromic circular DNA molecules. Mol Biochem Parasitol 1989;32:285–296.

54 Bhattacharya S, Battacharya A, Diamond LS. Comparison of repeated DNA from strains of *Entamoeba histolytica* and other *Entamoeba*. Mol Biochem Parasitol 1988;27:257–272.

55 Samuelson H, Acuña-Soto R, Reed S, Biagi F, Wirth D. DNA hybridization probe for clinical diagnosis of *Entamoeba histolytica*. J Clin Microbiol 1989;27:671–676.

56 Samuelson J, Ayala P, Orozco E, Wirth D. Emetine-resistant mutants of *Entamoeba histolytica* overexpress mRNA for multidrug resistance. Mol Biochem Parasitol 1990;38:281–290.

57 Tannich E, Horstmann RD. Codon usage in pathogenic *Entamoeba histolytica*. J Mol Evol 1992;34:272–273.

58 Huber M, Garfinkel L, Gitler C, Mirelman D, Revel M, Rozenblatt S. Nucleotide sequence analysis of an *Entamoeba histolytica* ferredoxin gene. Mol Biochem Parasitol 1988;31:27–34.

59 De la Garza M, Gallegos B, Meza I. Characterization of a cytochalasin D resistant mutant of *Entamoeba histolytica*. J Protozool 1989;36:556–560.

60 Rodríguez M, Orozco E. Isolation and characterization of phagocytosis and virulence deficient mutants of *Entamoeba histolytica*. J Infect Dis 1986;154:27–32.

61 Councilman WT, Lafleur HA. Amebic dysentery. Johns Hopkins Hosp Rep 1891;2:395–548.

62 Tsutsumi V, Mena R, Martínez-Palomo A. Cellular bases of experimental liver abscess formation. Am J Pathol 1984;117:81–91.

63 Tsutsumi V, Martínez-Palomo A. Inflammatory reaction in experimental invasive amebiasis. An ultrastructural study. Am J Pathol 1988;130:112–119.

64 Martínez-Palomo A. The pathogenesis of amoebiasis. Parasitol Today 1987;3:111–118.

65 Martínez-Palomo A, Tsutsumi V, Anaya-Velázquez F, González-Robles A. Ultrastructure of experimental intestinal amebiasis. Am J Trop Med Hyg 1989;41:273–279.

66 Krestchmer RR, Collado ML, Pacheco MG, et al. Inhibition of human monocyte locomotion by products of axenically grown E. histolytica. Parasite Immunol 1985;7:527–544.

67 Pérez-Tamayo R. Pathology of amebiasis. In: Martínez-Palomo A, ed. *Amebiasis*. Amsterdam: Elsevier Biomedical Publishers, 1986:45–94.

68 Burchard GD, Bilke R. Adherence of pathogenic and nonpathogenic *Entamoeba histolytica* strains to neutrophils. Parasitol Res 1992;78:146–153.

69 Salata RA, Pearson RD, Ravdin JI. Interaction of human leucocytes and *Entamoeba histolytica*. Killing of virulent amebae by the activated macrophage. J Clin Invest

1985;76:491–499.

70 Salata RA, Martínez-Palomo A, Murray HW, *et al*. Patients treated for amebic liver abscess develop cell-mediated immune responses effective *in vitro* against *Entamoeba histolytica*. J Immunol 1986;136:2633–2639.

71 Ghadirian E, Meerovitch E. Macrophage requirement for host defense against experimental hepatic amebiasis in the hamster. Parasite Immunol 1982;4:219–225.

72 Stern JJ, Graybill JR, Drutz DJ. Murine amebiasis: the role of the macrophage in host defense. Am J Trop Med Hyg 1984;33:372–380.

73 Kretschmer RR, López-Osuna M. Effector mechanisms and immunity to amebas. In Krestchmer RR, ed. *Amebiasis*. Boca Raton: CRC Press, 1990.

74 Kretschmer RR. Immunology of amebiasis. In Martínez-Palomo A, ed. *Amebiasis*. Amsterdam: Elsevier Biomedical Publishers, 1986:95–168.

75 Acosta G, Cote V, Isibasi A, Kumate J. Secretory IgA antibodies from bile of immunized rats reactive with trophozoites of *Entamoeba histolytica*. Ann NY Acad Sci 1983;409:760–765.

76 Jones JF. Serodiagnosis in parasitic infections. Clin Immunol Newslett 1984;5:103–105.

77 Gutiérrez G, Ludlow A, Espinosa G, *et al*. Encuesta serológica nacional. II. Investigación de anticuerpos contra *Entamoeba histolytica* en la República Mexicana. In Sepúlveda B, Diamond LS, eds. *Proceedings of the International Conference on Amebiasis*. Mexico: Instituto Mexicano del Seguro Social, 1976:599–608.

78 Krogstad DJ, Spencer HC, Healy GR. Current concepts in parasitology: Amebiasis. N Engl J Med 1978;298:262–265.

79 Pillai S, Mohimen A. A solid-phase sandwich radio-immunoassay for *Entamoeba histolytica* proteins and the detection of circulating antigens in amoebiasis. Gastroenterology 1982;83:1210–1216.

80 Jalan KN, Maitra TK. Amebiasis in the developing world. In Ravdin JI, ed. *Amebiasis*. New York: Wiley, 1988:535–555.

81 Harris WG, Bray RS. Cellular sensitivity in amoebiasis. Preliminary results of lymphocytic transformation in response to specific antigen and mitogen in carrier and disease states. Trans R Soc Trop Med Hyg 1976;70:340–343.

82 Arellano J, Granados J, Pérez E, Felix C, Kretschmer RR. Increased frequency of HLA-DR3 and complotype SCO1 in Mexican mestizo patients with amoebic abscess of the liver. Parasite Immunol 1991;13:23–29.

83 Diamanstein T, Klos M, Gold D, Hahn H. Interaction between *Entamoeba histolytica* and the immune system. I. Mitogenicity of *Entamoeba histolytica* extracts for human peripheral T lymphocytes. J Immunol 1981;126:2084–2086.

84 Grundy MS, Voller A, Warhurst D. An enzyme-linked immunosorbent assay for the detection of *Entamoeba histolytica* antigens in faecal material. Trans R Soc Trop Med Hyg 1987;81:627–632.

85 Del Muro R, Oliva A, Herion P, Capin R, Ortiz-Ortiz L. Diagnosis of *Entamoeba histolytica* in feces by ELISA. J Clin Lab Anal 1987;1:322–325.

86 Agarwal RK, Rawat R, Malaviya B, Das SR. An improved PVC strip ML-ELISA technique for diagnosis of recent cases of amoebiasis. Immunol Invest 1991;20:623–628.

87 Delmuro R, Acosta E, Merino E, Glender W, Ortiz L. Diagnosis of intestinal amebiasis using salivary IgA antibody detection. J Infect Dis 1990;162:1360–1364.

88 Aceti A, Pennica A, Celestino D, *et al*. Salivary IgA antibody detection in invasive amebiasis and in asymptomatic infection. J Infect Dis 1991;164:613–614.

89 Stanley SL, Jackson TFHG, Reed SL, *et al*. Serodiagnosis of invasive amebiasis using a recombinant *Entamoeba histolytica* protein. JAMA 1991;266:1984–1986.

90 Bracha R, Diamond LS, Ackers JP, Burchard GD, Mirelman D. Differentiation of clinical isolates of *Entamoeba histolytica* by using specific DNA probes. J Clin Microbiol 1990;28:680–684.

91 Sepúlveda B. Inducción de inmunidad protectora anti-amibiana con "nuevos" antígenos amibianos en el hamster lactante. Arch Invest Med (Mexico) 1978;9(Suppl. 1):309–310.

92 Tannich E, Bruchhaus I, Walter RD, Horstmann RD. Pathogenic and nonpathogenic *Entamoeba histolytica*—identification and molecular cloning of an iron-containing superoxide dismutase. Mol Biochem Parasitol 1991;49:61–72.

93 Stanley SL, Becker A, Kunz-Jenkins C, Foster L, Li E. Cloning and expression of a membrane antigen of *Entamoeba histolytica* possessing multiple TANDEM repeats. Proc Natl Acad Sci USA 1990;87:4976–4980.

9 *Giardia lamblia* and giardiasis

Theodore E. Nash

Giardia lamblia is a flagellated, binucleated protozoan which inhabits the small intestines of humans and other mammals [1]. Infection occurs following ingestion of a relatively resistant cyst stage which excysts forming motile trophozoites which multiply in the small intestine. Trophozoites adhere to the intestinal epithelium by means of a ventral adhesive disc [2], a process duplicated *in vitro* on almost any surface. Except in unusual instances, the parasite is limited to the lumen of the intestine [3]. Infections may be associated with varying degrees of villous atrophy. Acute and chronic inflammation occurs in the lamina propria and epithelium, which grossly correlates with the degree of symptoms [3,4]. Cysts form in the intestines and are excreted in the feces in concentrations approaching 10^7 per gram [5]. Because of the large numbers of cysts and the high infectivity (ingestion of as few as 10 resulted in infections in 30% of human volunteers [6–8]), infections are common wherever fecal contamination occurs.

In the USA, infections due to *Giardia* are particularly common among children attending daycare centers [9–11], travelers [12], homosexuals [13], and backpackers [14]. Large epidemics occur after the contamination of water supplies and, in fact, *Giardia* is the most common cause of defined waterborne diarrhea [15]. In developing countries, infections are usually more common, particularly in the young [16,17] where prospective studies have documented that almost everyone has been infected by the age of 2 years [18]. A number of studies associates giardiasis with malnutrition and diarrhea [19–21]. Reinfections are also common [22,23].

Most disease manifestations are related to the gastrointestinal tract [24,25], although a number of extraintestinal manifestations have been reported [25,26]. Symptoms range from asymptomatic carriage to fulminant diarrhea, nausea and vomiting, and malabsorption [24,25]. Infections may be transient or chronic, sometimes lasting for years [24,27]. Although the factors causing the variability of disease manifestations and duration of infections are incompletely understood, there is adequate evidence implicating both the immune responses of the host and the specific strain, type, or isolate of *Giardia*.

In model *Giardia* infections there is good evidence supporting the importance of immune responses in determining the nature of infection and the induction of immunity. In humans the evidence is less compelling, perhaps due to our poor understanding of the natural history of most infections: giardiasis is particularly common in patients with hypogammaglobulinemia [28,29] because of increased susceptibility and/or inability to mount an effective immune response. Life-threatening malabsorption is not unusual in these patients and standard, usually effective, courses of therapy many times fail to cure these individuals. In cross-sectional studies, *Giardia* is more common in children than in adults [22,30]. Although increased exposure and ingestion of cysts is likely in children and may explain the increased prevalence rates, infants may also be inherently more susceptible to infections. In model infections, infant mice [31] and rats [32,33] are clearly more susceptible to *G. lamblia* than adults. In addition, endemically exposed populations appear to have a decreased prevalence of infection compared to naïve newly exposed populations [34,35]. More direct evidence suggesting the development of immunity in humans was found in experimental human infections. Of the volunteers who became infected, 84% ceased shedding cysts a mean of 18.3 days following inoculation [6].

There is also evidence indicating that chronic and/or repeated infections are not uncommon. Apparently, immunologically normal individuals may be infected and have the disease for long periods of time [27]. This author documented *Giardia* infection in an asymptomatic carrier for 15 years. A number of studies in children, particularly in populations from developing countries, found that repeated and/or continued infections were common [22,23]. The experimentally infected population mentioned above, which showed self-cure in a majority of infections, also found chronic infections in the remaining 16% of volunteers [6].

Although current evidence suggests that *Giardia* infections produce both sterile immunity and chronic infections, little is known about the relative proportions of each and those factors which lead to one or the other.

As in most infections, there is an interaction between the parasite and host which determines the nature of the infection and disease. Previously, the study of *Giardia* infections in humans was hampered by the inability to culture *Giardia*. The availability of methods to analyze isolates and differentiate one isolate or strain from another allows an accurate determination of the host–parasite interaction. There is increasing evidence implicating the type of *Giardia* isolate or strain as an important determinant of the degree of infectivity.

THE PARASITE

Relatively little is known about the basic biology, biochemistry, and molecular biology of *Giardia*. It has two nuclei which are similar in size, but whether each nucleus is identical is unknown. Only recently have pulse-field gradient electrophoresis studies suggested that there are at least five chromosomes [36], but the ploidy of the parasite is still not definitely known. A number of studies indicate that *Giardia* are unusual parasites. For example, analysis of the small amount of RNA [37–39] indicated that *Giardia* are primitive eukaryotes and are not distantly related to some bacteria. Purine and pyrimidine metabolism is unusual [40,41] and *Giardia* require cysteine for survival [42,43].

There is no satisfactory classification for *Giardia* derived from different animals. Three morphologically distinct *Giardia* have been distinguished on the basis of the shape of the median body [1]. Only *Giardia* which possess a claw-hammer-shaped median body infect humans, but even this group is clearly heterogeneous [44]. *Giardia* that are morphologically identical to those in humans have been found in an increasing number of animals, including dogs [45], cats, guinea pigs, cows, horses, sheep, mice [46], and rats [46], and possibly beavers [47]. In addition, cysts derived from one species can infect some other species. Which *Giardia* carried by animals readily infect humans remains largely unstudied, but infected beavers have been found in

some contaminated water sources [15]. In one very brief description, cysts from both beaver and mule deer infected humans [48]. On the other hand, cysts from human sources are infectious for beavers [47], deer mice, gerbils [48,49], muskrats [47], and, in some parts of the world, mice [50] and rats [51]. There are conflicting reports concerning infectivity to dogs [33,52]. Cats appear to be resistant or largely resistant to human source cysts [33,53]. Gerbils [54] and deer mice [55] can be infected with cysts from a number of different animals. Beavers may also be infected with their own as well as human varieties of *Giardia* [47]; epidemics, if due to beavers, could be due to either type of *Giardia*.

Giardia isolated from humans have been compared using a variety of techniques, including endonuclease restriction banding patterns [56], pulse-field gradient electrophoresis [36], isoenzymes [57–60], ability to grow *in vitro*, sensitivity to chemotherapeutic agents [61], isoelectric focusing [62], and ability to become infected with an RNA virus found in some *Giardia* [63], and most studies indicate that *Giardia* are heterogeneous biochemically. In one series of studies three broadly classed groupings of organisms emerged [56,64]. One group had identical endonuclease banding patterns and a common surface epitope. The second group showed at least one band in common with the first group and there was antigenic similarity in surface antigens. The third group showed little similarity in surface antigens and no bands in common with the other group. More recent studies support these findings. A heat shock-like gene cloned from the first group hybridized to identical bands in Southern blots in the first and second groups of *Giardia* but failed to hybridize to any bands in the third group, indicating a lack of this gene [65]. This finding indicates substantial differences between Groups I and II and Group III. These studies show both major differences and similarities among isolates, but the exact relationship between various *Giardia* isolates has not been defined.

Biologic differences among isolates have also been noted in animals and humans. A number of investigators infected mice [48] and gerbils [66,67] with cysts from humans and suggested that the differences observed in cyst excretion were due to inherent differences among *Giardia*. When gerbils were infected with two well-defined biochemically divergent axenized isolates from Groups I and III defined earlier, two different patterns of infections and immune responses were noted [68]. In one infection with the Group I isolate, WB, the infection was self-limiting and no

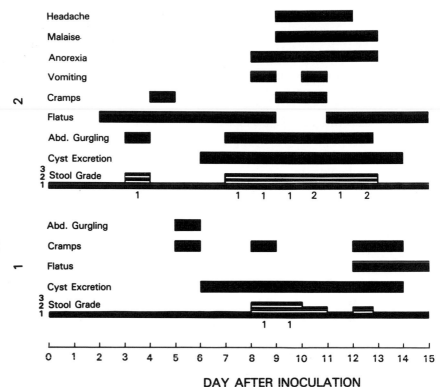

Fig. 9.1 Chart of symptoms and signs which developed following experimental infection of two normal volunteers with 50 000 trophozoites administered enterally. Grade 3 stools are defined as stools which take the shape of the container. The number under the horizontal line represents the number of grade 3 stools that occurred on a particular day. Typical signs and symptoms developed during infection [69].

trophozoites were detected in the intestines by day 28. These gerbils were refractory to homologous and heterologous challenge. In contrast, infections produced by a second Group III isolate (GS) were more chronic and only induced a partial protection after reinoculation with the homologous isolate and little resistance to heterologous challenge. Similar studies using two *G. lamblia*-like isolates derived from a rat and a mouse also produced different patterns of infection [46].

Differences in infectivity were noted in experimental human infections [69]. Humans were inoculated enterally with Group I and Group III isolates (Isr and GS, respectively). None of the five volunteers inoculated with Isr became infected compared to 12 out of 12 inoculated with GS (Fig. 9.1). In another series of experiments [70] volunteers were inoculated with two different *Giardia* clones (one from GS and the other most likely from a Group I isolate). One out of 13 became infected with one clone compared to four out of four inoculated with the GS-derived clone. Although the reasons for the differences in infectivity or course of infection in gerbils and humans remain speculative, they clearly indicate that the parasite itself is a major con-

tributor to the variability of infectivity, disease manifestations, chronicity, and immune response.

Giardia antigens have not been studied extensively and there is no general agreement about the antigenic makeup of *Giardia*. This may be due to differences in growth media, the parasite itself, or antigenic variation or instability [2]. In early studies most *Giardia* antigens appeared similar among *Giardia* isolates, although some were unique [71]. A few antigens have been described by more than one group of investigators and these include an 82 kD [72] and/or an 88 kD [73] surface antigen. A number of closely related somatic structural proteins, collectively termed "giardins," have recently been cloned and sequenced [74,75]. Some of these are dominant immunogens and readily induce humoral antibodies after immunization, but it is not clear how useful these are in detecting current *Giardia* infections. A trypsin-activated lectin that binds most avidly to mannose-6-phosphate has been described [76]. The authors believe that this may play a role in adherence of the parasite to intestinal villi, although *Giardia* adhere well to almost any surface. *Giardia* appear to contain limited kinds of complex carbohydrates. There are conflicting reports suggesting that the dominant sugar is either *N*-acetylglucosamine [77] or *N*-acetyl-galactosamine [78].

Although many *Giardia* proteins appear similar, recent studies show differences among isolates. Much of the variability of *Giardia* resides at the cell surface, as judged by a number of studies, including surface reactivities using isolate-specific polyclonal antibodies [79], antigen-specific monoclonal antibodies [80], surface radio-labeling [81], and polyacrylamide gel electrophoresis [64]. This is important because the surface is where biologically important host–parasite interactions are likely to occur. Many of the somatic antigens are similar [71,79].

The surface antigens of human isolates of *G. lamblia* are highly variable. Recent studies show that some of this variability is due to antigenic variation of the major surface antigens of *Giardia* [82]. Cytotoxic monoclonal antibodies (MAB) were produced to the major surface antigens of specific *Giardia* clones [80]. Exposure of the

clones to MAB *in vitro* caused immediate immobilization and death in all but a few *Giardia* [80]. The surviving *Giardia* became resistant to the effects of MAB and by a number of criteria, including surface radiolabeling, Western blots, and Northern blots [43], no longer expressed the major surface antigen [82]. A cytotoxic MAB was then produced to a newly appearing surface antigen on one of the resistant clones and the above procedure was repeated [83]. As before, the major surface antigen was lost and replaced by new surface antigens. Other important features are the rapid rate of change of the surface antigens, the cysteine-rich nature of the variant proteins [83,84], and the ability of a number of isolates to undergo similar changes [84].

Studies conducted on humans [70] and gerbils [85] show that antigenic variation occurs *in vivo* as well as *in vitro*, and that surface-specific humoral responses of the host are isolate dependent (Figs 9.2 and 9.3). Four volunteers [70] were enterally inoculated with a clone of GS that has a major 72 kD surface antigen specifically

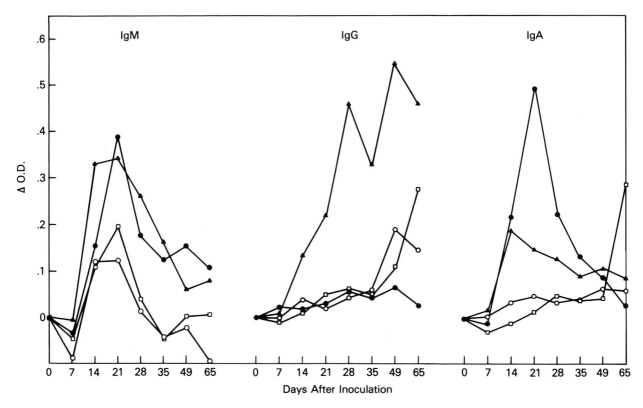

Fig. 9.2 Isotype-specific antibody titers as measured by enzyme-linked immunosorbent assay to homogenates (a measure of surface and somatic responses) of *Giardia* clone, GS/M-H7, in four individuals experimentally inoculated with the identical clone on day 0 and treated on day 22. All produced IgM antibodies but there were variable responses of IgG and IgA. Responses using a heterologous isolate were the same [70].

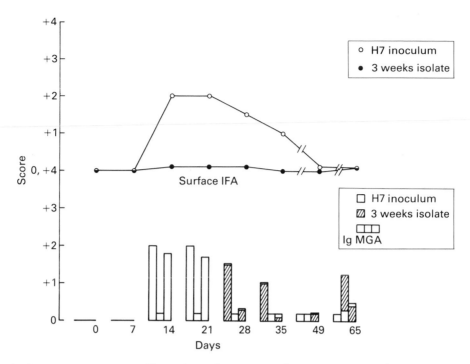

Fig. 9.3 Surface antibody responses as measured by surface indirect immunofluorescence (IFA) and serum cytotoxicity to the inoculating clone (see legend to Fig. 9.2) and the intestinal isolate recovered after 3 weeks of infection in one experimentally infected normal volunteer. Isolate-specific responses were seen using both techniques. Cytotoxicity responses developed to the inoculating clone but not the 3-week isolate. Surface antibody responses to the inoculating clone appeared by day 14. Immunofluorescent responses to the 3-week isolate were either delayed, weak, or absent in this volunteer as well as the others [70].

reactive with a MAB. The surface antigens of trophozoites reisolated from cysts and on trophozoites directly harvested from the jejunum on day 22 after inoculation were analyzed by surface labeling, indirect immunofluorescence, and Western blots. All four volunteers showed loss of the original major surface antigen and replacement by others. Analysis of isolates recovered serially from one volunteer revealed that surface antigen changes occurred after day 14 and was complete by day 22. Analysis of clones derived from the volunteer on day 17 showed a number of antigenic variants. One of the volunteers infected with another clone possessing another major surface antigen showed comparable findings.

Similar studies in gerbils [85] using a number of different clones derived from a single trophozoite also showed antigenic variation. However, the loss of the major antigen(s) occurred earlier and there was a strong tendency for a similar set of surface antigens to appear, even though the inoculating clones had different surface antigens. This suggests a nonrandom selection of

trophozoite surface antigens by the parasite, host, or both.

These studies clearly show that there are inherent differences among *Giardia* isolates and that individual organisms are able to vary their surface antigens both *in vitro* and *in vivo*. The biologic significance of antigenic variation is not known, but may play a role in the development of chronic or repeated infections.

An RNA virus has been found in some *Giardia* isolates [86,87] and could possibly alter important biologic characteristics [63,86,87]. As yet there is no clear evidence that the virus plays a role in altering infections in humans and is not required for antigenic variation to occur.

Bacteria have also been described in *Giardia muris* [88]. Their biologic function(s) are unknown.

INNATE RESISTANCE

In humans the occurrence of innate resistance to *Giardia* infection is difficult to assess because of the inability to determine the effect of potential prior exposure and infection on the host. On the other hand, when volunteers are given adequate numbers of cysts or trophozoites, infection is practically universal [6,69]. A number of nonimmunologic factors *in vitro* have been cytotoxic or protective to *Giardia*. For instance, free fatty acids formed in human milk kill *Giardia* [89–91] and intestinal mucus and bile protect *Giardia* from the lethal effects of human intestinal fluid [92]. Host genetic factors may contribute to prolongation of infection [93] or disease manifestations [94].

IMMUNE RESPONSES

There are a number of difficulties in interpreting the immune responses in *Giardia* infections. Although immune responses clearly occur in humans, our understanding of the relevant responses which may be responsible for self-cure or chronicity of infection, for example, are not known. In some aspects, the immune responses of mice to *G. muris* infection are better understood, but the use of different species of host and parasite limits the applicability of findings in human infections. In addition, *G. muris* cannot be cultured and the antigenic makeup, presence of surface antigen variability, and stability are not well studied. *Giardia lamblia* infections in gerbils have become a useful model of infection because of the ease of initiating infections with defined human isolates, but the immune system of the gerbil is not well studied and specific immunologic reagents are not generally available. Lastly, because there are no standard defined media employed to grow *Giardia* and no standard isolate or strain, results are not often comparable.

HUMORAL RESPONSES IN HUMANS

Systemic humoral responses have been detected by a variety of techniques [95]. These studies were of interest, in part, not only because of the relative ease of performing these assays but also in the hope that low or weak responses would indicate either susceptibility to infection or chronicity, and therefore might easily explain *Giardia* infections in patients with hypogammoglobulinemia. There is little evidence to support this idea. Early studies detected IgG or total immunoglobulin responses to fixed trophozoites [95–97] or cysts [98] using immunofluorescent antibodies (IFA), or to homogenates using enzyme-linked immunosorbent assay (ELISA) [99,100] and immunodiffusion [101]. Although

there was some variation among studies, about 80% of infected subjects showed elevated levels of anti-*Giardia* IgG compared to apparently uninfected controls [99,102]. Between 7% and 14% of normals from the USA had significant elevation of IgG antibodies [99,102]; however, antibodies were more common in persons from underdeveloped areas, presumably exposed to *Giardia* more frequently [103]. The duration of the antibody response varied, but persisted long enough to limit the usefulness of IgG-reactive antibodies to diagnose active infections [99,103,104]. Isotype-specific responses have only recently been studied and these responses may have greater clinical applicability (Fig. 9.1). Goka *et al.* [104] using an ELISA assay detected IgM-specific antibodies in 96% of infected patients. Nash *et al.* [69] experimentally infected 10 volunteers with isolate GS/M and measured the isotype-specific responses. The prepatent period ranged from 6 to 9 days and volunteers were treated on day 14 and followed up to 8 weeks. IgM, IgG, and IgA responses were detected in 100%, 70%, and 60% of volunteers, respectively. IgM and IgA responses appeared on day 14, peaked and then decreased over time. IgG responses were initially detected on day 14 but in some volunteers antibody levels continued to rise and peaked at 8 weeks, despite therapy at 2 weeks. Rises in IgM responses were not found in one out of two volunteers rechallenged 3 months later, suggesting that repeated infections may not be detected as readily. Additionally, both rechallenged volunteers became infected despite the presence of high levels of IgG antibodies in one of them. Specific IgM responses were not found in *Giardia*-infected acquired immunodeficiency syndrome (AIDS) patients [105].

Immune responses specific to the surface of the trophozoites may be important in the development of protective immune responses. Antibodies by themselves are capable of killing trophozoites *in vitro*. Monoclonal antibodies which recognize specific epitopes on the surface of *Giardia* are cytotoxic by complement-independent and complement-dependent mechanisms only to those *Giardia* possessing a particular epitope [80,106]. Rabbit antisera to different isolates showed varying degrees of complement-dependent cytotoxicity, which depended on the isolate and its surface antigens [84]. In addition, *in vitro* studies demonstrated antibodies (most likely IgG) which opsonize trophozoites, leading to increased phagocytosis by peripheral blood macrophages [96].

Cytotoxic antibodies are also found in humans

[70,108–110] and animals [68,106,107] following infection (Fig. 9.3). Complement-dependent cytotoxic antibodies were first detected by Hill *et al.* [108] in both infected and noninfected persons. Sera which showed increased degrees of cytotoxicity had the highest antibody titers to killed trophozoites, suggesting that it was antibody mediated. Subsequent studies in humans and gerbils [68] showed that cytotoxic antibodies developed in response to infection and cytotoxicity was isolate-specific (Fig. 9.3). In the human volunteer [70] studies described earlier, cytotoxic antibodies developed to the inoculating clone in four out of four infected volunteers but were not cytotoxic to any of the day 22 isolates (Fig. 9.3). These responses were most likely complement mediated.

The development of surface reactivity was further confirmed by estimating antibody responses to the surface of viable trophozoites [70]. Surface-specific IgM, IgG, and IgA responses measured by IFA were predominantly to the inoculum and were weak and or delayed to the day 22 isolates (Fig. 9.3). In contrast, antibodies as measured by ELISA to homogenates of either the inoculating or day 22 isolates were quantitatively similar (Fig. 9.2). These results show that immunologic responses to surface constituents are isolate dependent and highlight the potential importance of surface-reactive antibodies. In addition, they demonstrate how easily responses to surface antigens may be obscured by responses to somatic antigens. On a practical level, the use of heterologous trophozoites to measure immunologic responses may fail to detect important relevant responses in humans and most likely in other model systems.

Because *Giardia* is an intestinal parasite, attention has focused on the intestinal immune responses. In *G. muris* infections there is evidence suggesting that self-cure is IgA mediated (see below), but knowledge in human infections is primarily limited to measurement of intestinal IgA immune responses. Although early studies suggested depressed intestinal IgA concentrations in *Giardia*-infected patients [111], this finding has not been confirmed [112,113]. Increased numbers of IgA-containing plasma cells have been found in intestinal biopsies of infected [114] patients and intestinal trophozoites have IgA on their surface [115]. *Giardia*-specific intestinal IgA responses were detected in 50% of experimentally infected humans by day 14 [69]. Despite the presence of sometimes high levels of *Giardia*-specific IgA in the intestines, these patients remained infected and two out of two were reinfected after treatment.

Giardia-specific IgA antibodies are found in human milk [21,116] but it is not known if these antibodies play a role in preventing infection, although this may be the case in *Giardia* infections in the mouse.

CELLULAR IMMUNITY IN HUMANS

There is little information concerning the role of the cellular immune system in human *Giardia* infections. Clinically, except for the role of the cellular immune system in hypogammaglobulinemia, patients with profound cellular defects and AIDS are effectively treated with chemotherapeutic agents. Little evidence exists for suggesting that they are particularly susceptible to infection except that there is an increased prevalence of giardiasis in homosexuals, who are probably more frequently exposed to *Giardia*. The small intestines of infected individuals show a number of changes which may include varying degrees of inflammation and villous atrophy [3]. An increase in jejunal lymphocytic intraepithelial cells was found to correlate with the degree of malabsorption [4]. As mentioned earlier, a number of different types of cells are able to ingest alive [96,117] and/or degenerating *Giardia* [118–120] and this is enhanced by antibodies, but the role, if any, that these play in controlling infection is speculative.

IMMUNE RESPONSES IN *G. MURIS* INFECTIONS

Studies using *G. muris* infections in mice indicate both cellular and humoral mechanisms in the development of immunity [121–124]. The course of infection in *G. muris* infections of mice differs, depending on the strain of mice. Some infections result in a relatively limited infection or self-cure, while others develop chronic infections. Resistance to challenge also occurs to varying degrees [125–127]. Two general experimental approaches have been used. One approach correlates the loss or ablation of a particular response or type of cell with loss of immunity, while the other compares immunologic responses of strains of mice which develop chronic infections to those which develop self-limiting infections.

The strongest evidence implicating cellular immune mechanisms comes from experiments employing mice lacking T cells [125,128]. Nude mice are unable to self-cure *Giardia* infections, although they are able to limit infections after challenge. In contrast, their heterozygous littermates self-cure and fail to become infected after challenge. Nude mice given spleen cells from immune donors are partially able to resist infection [125]. The need for T cells, specifically L3T4+ cells, was further shown in mice depleted of these cells by administration of anti-L3T4 MAB [129]. These mice no longer had L3T4+ cells in their Peyer's patches and were no longer able to self-cure. There was no decrease in the numbers of *Giardia* in the intestines compared to phosphate buffered saline or anti-Ly-2 MAB-treated mice, which were able to self-cure. Both nude mice and anti-L3T4-treated mice lack L3T4+ T cells [129,130]. Nude mice fail to produce intestinal IgA [131,132] antibodies. In contrast, immunologically normal mice and untreated mice do produce intestinal IgA antibodies [133–136]. These results suggest that IgA and/or other antibodies may be needed to self-cure in addition to cellular mechanisms.

Although IgA antibodies are felt to be essential in the development of gastrointestinal tract immunity, the role of IgA or other antibodies in the development of immunity and self-cure in *Giardia* infections is unclear. *Giardia*-specific intestinal IgA [133–136] and possibly IgG [133] antibodies are produced during infection with *G. muris*, and IgA antibodies to surface antigens are detected in varying proportions on intestinal trophozoites [133,137]. As noted above, the lack of IgA intestinal antibodies in nude mice and anti-L3T4+-treated mice is associated with the inability to self-cure [131,132]. Neonatal mice treated with anti-μ antiserum are unable to mount systemic and intestinal IgG, IgM, or IgA responses and are also unable to self-cure [134]. Although this is suggestive of the importance of IgA or other antibodies, treatment with anti-μ not only ablates humoral responses but has effects on cellular immune responses as well. Experiments comparing the immune response of mouse strains which self-cure to those unable to self-cure have not shown biologically relevant differences in IgA responses [135,138] or cellular immune responses [139].

Factors in milk from immune mice, possibly IgA antibodies to *Giardia*, protect suckling pups from infection [136,140,141]. Pups from immune mothers suckled on normal mothers are not protected from infection whereas pups from normal mothers suckled on immune mothers are protected [140]. IgA antibodies to *Giardia* as well as other immunoglobulins are found in immune mouse milk and the precise factor or mechanism(s) involved in prevention of infection have not been defined. Although some IgG and IgM MAB to *Giardia* surface antigens are cytotoxic by complement- and noncomplement-mediated mechanisms, IgA-mediated cytotoxicity has not been described. In one *in vitro* study, milk containing *Giardia*-specific IgA antibodies enhanced phagocytosis by macrophages and neutrophils [142].

Other evidence suggesting a role of intestinal IgA antibodies in the development of immunity comes from experiments performed by Mayrhofer & Sharma [143] using rats infected with a *G. lamblia* type of parasite. Infusion of bile from immune rats into the intestines of infected rats led to a decrease in the number of parasites in the intestines.

A number of studies suggest that intestinal white cells play a role in controlling *Giardia* infections. Ultrastructural studies in mice showed lymphocytes and macrophages in contact with and macrophages sometimes ingesting trophozoites [144,145]. *Giardia in vitro* in the presence of macrophages, lymphocytes, or neutrophils adhered to, were phagocytosed by, or killed to various degrees, depending upon the system employed [139,142,146]. Most studies have shown enhanced cytotoxicity when *Giardia*-specific antibodies were included in these *in vitro* assays. Although antibody-mediated cellular cytotoxicity is an attractive effector mechanism, there is little direct evidence suggesting that this plays a direct role controlling *Giardia* infections *in vivo*.

Immune responses may also be responsible for disease. Nude mice reconstituted with spleen cells from immune mice developed greater loss of crypt height [125].

IMMUNODIAGNOSIS

Immunologically based methods have been developed to detect *Giardia* infections. Traditionally, the diagnosis of giardiasis depended on methods designed to visualize *Giardia* cysts or trophozoites in feces or duodenal fluid. Since the excretion of cysts is at times erratic and therefore unrewarding, as well as time consuming, and the sampling of duodenal fluid or small-bowel biopsies invasive, other methods were clearly needed. Craft & Nelson [147] were the first to detect *Giardia* antigens in stools using countercurrent immunoelectrophoresis. Ungar *et al.* [148] developed an ELISA test which was sensitive, specific, and easy to perform and a number of similar tests now exist. One assay based on detection of a 65 kD antigen found in trophozoites [149] is now commercially available.

Most of the important questions about *Giardia* and its relationship with its host(s) remain unanswered. The parasite itself is poorly understood, even though *Giardia* was most likely the first protozoan parasite described. Whether we should be considering one species or many is uncertain. Indeed, how *Giardia* cause diarrhea is not known. However, it is becoming increasingly clear that there are tricks that this parasite can play. The interplay between parasite and host is more intricate than previously believed.

REFERENCES

1 Feely DE, Erlandsen SL, Chase DG. Structure of the trophozoite and cyst. In Erlandsen SL, Meyer EA, eds. *Giardia and Giardiasis*. New York: Plenum Publishers, 1984:3–31.

2 Erlandsen SL, Freely DE. Trophozoite motility and the mechanism of attachment. In Erlandsen SL, Meyer EA, eds. *Giardia and Giardiasis*. New York: Plenum Publishers, 1984:33–63.

3 Gillon J, Ferguson A. Changes in the small intestinal mucosa in giardiasis. In Erlandsen SL, Meyer EA, eds. *Giardia and Giardiasis*. New York: Plenum Publishers, 1984: 163–183.

4 Wright SG, Tomkins AM. Quantification of the lymphocytic infiltrate in jejunal epithelium in giardiasis. Clin Exp Immunol 1977;29:408–412.

5 Danciger M, Lopez M. Numbers of *Giardia* in the feces of infected children. Am J Trop Med Hyg 1975;24:237–241.

6 Rendtorff RC. The experimental transmission of human intestinal protozoan parasites. II. *Giardia lamblia* cysts given in capsules. Am J Hyg 1954;59:209–220.

7 Rendtorff RC. The experimental transmission of human intestinal protozoan parasites. I. *Entamoeba coli* cysts given in capsules. Am J Hyg 1954;59:196–208.

8 Rendtorff RC. The experimental transmission of human intestinal protozoan parasites. IV. Attempts to transmit *Entamoeba coli* and *Giardia lamblia* by water. Am J Hyg 1954;60:327–338.

9 Keystone JS, Krajden S, Warren MR. Person-to-person transmission of *Giardia lamblia* in day-care nurseries. Can Med Assoc J 1978;119:241–248.

10 Pickering LK, Woodward WE, DuPont HL, Sullivan P. Occurrence of *Giardia lamblia* in children in day care centers. J Pediatr 1984;104:522–526.

11 Black RE, Dykes AC, Sinclair SP, Wells JG. Giardiasis in day-care centers: evidence of person-to-person transmission. Pediatrics 1977;60:486–491.

12 Brodsky RE, Spencer Jr HC, Schultz MG. Giardiasis in American travelers to the Soviet Union. J Infect Dis 1974; 130:319–323.

13 Schmerin MJ, Jones TC, Klein H. Giardiasis: association with homosexuality. Ann Intern Med 1978;88:801–803.

14 Barbour AG, Nichols CR, Fukushima T. An outbreak of giardiasis in a group of campers. Am J Trop Med Hyg 1976;25:384–389.

15 Juranek D. Waterborne giardiasis. In Jakubowski W, Hoff JC, eds. *Waterborne Transmission of Giardiasis*. Washington, DC: US Environmental Protection Agency, 1979:150–163.

16 Gilman RH, Brown KH, Visvesvara GS, *et al.* Epidemiology and serology of *Giardia lamblia* in a developing country: Bangladesh. Trans R Soc Trop Med Hyg 1985;79: 469–473.

17 Gupta MC, Urrutia JJ. Effect of periodic antiascaris and antigiardia treatment on nutritional status of preschool children. Am J Clin Nutr 1982;36:79–86.

18 Mata LJ. *The Children of Santa María Cauqué: a Prospective Field Study of Health and Growth*. Cambridge, MA: MIT Press, 1978.

19 Farthing MJG, Mata L, Urrutia JJ, Kronmal RA. Natural history of *Giardia* infection of infants and children in rural Guatemala and its impact on physical growth. Am J Clin Nutr 1986;43:395–405.

20 Pugh RNH, Burrows JW, Bradley AK. Malumfashi endemic disease research project, XVI. Ann Trop Med Parasitol 1981;75:281–292.

21 Islam A, Stoll BJ, Ljungstrom I, Biswas J, Nazrul H, Huldt G. *Giardia lamblia* infections in a cohort of Bangladeshi mothers and infants followed for one year. J Pediatr 1983; 103:996–1000.

22 Mason PR, Patterson BA. Epidemiology of *Giardia lamblia* infection in children: cross-sectional and longitudinal studies in urban and rural communities in Zimbabwe. Am J Trop Med Hyg 1987;37:227–282.

23 Gilman RH, Marquis GS, Miranda E, Vestegui M, Martinez H. Rapid reinfection by *Giardia lamblia* after treatment in a hyperendemic third world community. Lancet 1988;1:343–345.

24 Wolfe MA. Symptomatology, diagnosis, and treatment. In Erlandsen SL, Meyer EA, eds. *Giardia and Giardiasis*. New York: Plenum Publishers, 1984:147–161.

25 Webster BH. Human infection with *Giardia lamblia*. An analysis of 32 cases. Am J Dig Dis 1958;3:64–71.

26 Shaw RA, Stevens MB. The reactive arthritis of giardiasis. J Am Med Assoc 1987;258:2734–2735.

27 Chester AC, MacMurray FG, Restifo MD, Mann O. Giardiasis as a chronic disease. Dig Dis Sci 1985;30:215–218.

28 Hermans PE, Huizenga KA, Hoffman HN, Brown ALB Jr, Markowitz H. Dysgammaglobulinemia associated with nodular lymphoid hyperplasia of the small intestine. Am J Med 1966;40:78–89.

29 Ament ME, Ochs HD, Davis SD. Structure and function of the gastrointestinal tract in primary immunodeficiency syndromes a study of 39 patients. Medicine 1973;63:227–248.

30 Meleney HE, Bishop EL, Leathers WS. Investigations of *Entamoeba histolytica* and other intestinal protozoa in Tennessee III. State-wide survey of intestinal protozoa in man. Am J Hyg 1932;16:523–539.

31 Hill DR, Guerrant RL, Pearson RD, Hewlett EL. *Giardia lamblia* infection of suckling mice. J Infect Dis 1983;147(2): 217–221.

32 Craft JC. Experimental infection with *Giardia lamblia* in rats. J Infect Dis 1982;145(4):495–504.

33 Woo PTK, Paterson WB. *Giardia lamblia* in children in day-care centers in southern Ontario, Canada, and susceptibility of animals to *G. lamblia*. Trans R Soc Med Hyg 1986;80:56–59.

34 Moore GT, Cross WM, McGuire D, *et al.* Epidemic giardiasis at a ski resort. 1969;281:402–407.

35 Gleason NN, Horwitz MS, Newton LH, Moore GT. A stool survey for generic organisms in Aspen, Colorado. Am J Trop Med Hyg 1970;19:480–484.

36 Adam RA, Nash TE, Wellems TE. The *Giardia lamblia* trophozoite contains sets of closely related chromosomes. Nucleic Acids Res 1988;16:4555–4567.

37 Sogin ML, Gunderson JG, Elwood HJ, *et al.* Phylogenetic meaning of the kingdom concept: an unusual ribosomal RNA from *Giardia lamblia.* Science 1989;243:75–77.

38 Boothyrod JC, Wang A, Campbell DA, Wang CC. An unusually compact ribosomal DNA repeat in the protozoan *Giardia lamblia.* Nucleic Acids Res 1987;15:4065–4084.

39 Edlind TD, Chakraborty PR. Unusual ribosomal RNA of the intestinal parasite *Giardia lamblia.* Nucleic Acids Res 1987;15:7889–7901.

40 Wang CC, Aldritt S. Purine salvage networks in *Giardia lamblia.* J Exp Med 1983;158:1703–1712.

41 Aldritt SM, Tien P, Wang CC. Pyrimidine salvage in *Giardia lamblia.* J Exp Med 1985;161:437–445.

42 Gillin FD, Diamond LS. *Entamoeba histolytica* and *Giardia lamblia*: growth responses to reducing agents. Exp Parasitol 1981;51:382–391.

43 Gillin FD, Reiner DS. Attachment of the flagellate *Giardia lamblia*: role of reducing agents, serum, temperature, and ionic composition. Mol Cell Biol 1982;2:369–377.

44 Erlandsen SL, Meyer EA, Nash TE. Panel discussion on taxonomy of the genus *Giardia.* In Wallis PM, Hammond BR, eds. *Advances in Giardia Research.* Calgary: University of Calgary Press, 1988:287–289.

45 Levine ND. *Giardia lamblia*: classification, structure, identification. In Jakubowski W, Hoff JC, eds. *Waterborne Transmission of Giardiasis.* Washington, DC: US Environmental Protection Agency, 1979:2–8.

46 Sharma AW, Mayrhofer G. A comparative study of infections with rodent isolates of *Giardia duodenalis* in inbred strains of rats and mice and in hypothymic nude rats. Parasite Immunol 1988;10:169–179.

47 Erlandsen SL, Sherlock LA, Januschka M, *et al.* Cross-species transmission of *Giardia* spp.: inoculation of beavers and muskrats with cysts of human, beaver, mouse and muskrat origin. Appl Environ Microbiol 1988;54:2777–2785.

48 Davies RB, Hibler CP. Animal reservoirs and cross-species transmission of *Giardia.* In Jakubowski W, Hoff JC, eds. *Waterborne Transmission of Giardiasis.* Washington, DC: US Environmental Protection Agency, 1979:104–126.

49 Belosevic M, Faubert GM, MacLean JD, Law C, Croll NA. *Giardia lamblia* infections in Mongolian gerbils: an animal model. J Infect Dis 1983;147:222–226.

50 Aggarwal A, Bhatia A, Naik SR, Vinayak VI. Variable virulence of isolates of *Giardia lamblia* in mice. Ann Trop Med Parasitol 1963;77:163–167.

51 Anand BS, Kumar M, Chakravarti RN, Sehgal AK, Chhuttani PN. Pathogenesis of malabsorption in *Giardia* infection: an experimental study in rats. Trans R Soc Trop Med Hyg 1980;74:565–569.

52 Hewlett EL, Andres Jr JS, Ruffier J, Schaefer III FW. Experimental infection of mongrel dogs with *Giardia*

lamblia cysts and cultured trophozoites. J Infect Dis 1982; 145:89–93.

53 Kirkpatrick CE, Green IV GA. Susceptibility of domestic cats to infections with *Giardia lamblia* cysts and trophozoites from human sources. J Clin Microbiol 1985;21: 678–680.

54 Swabby KD, Hibler CP, Wegrzyn JG. Infection of Mongolian gerbils (*Meriones unquiculatus*) with *Giardia* from human and animal sources. In Wallis PM, Hammond BR, eds. *Advances in Giardia Research.* Calgary: University of Calgary Press, 1988:75–77.

55 Roach PD, Wallis PM. Transmission of *Giardia duodenalis* from human and animal sources in wild mice. In Wallis PM, Hammond BR, eds. *Advances in Giardia Research.* Calgary: University of Calgary Press, 1988:79–82.

56 Nash TE, McCutchan T, Keister D, Dame JB, Conrad JD, Gillin FD. Restriction-endonuclease analysis of DNA from 15 *Giardia* isolates obtained from humans and animals. J Infect Dis 1985;152:64–73.

57 Bertram MA, Meyer EA, Lile JD, Morse SA. A comparison of isozymes of five axenic *Giardia* isolates. J Parasitol 1983; 69:793–801.

58 Korman SEH, LeBlancq SM, Spira DT, El On J, Reifen RM, Deckelbaum RJ. *Giardia lamblia*: identification of different strains from man. Z Parasitenkd 1986;72:173–180.

59 Meloni BP, Lymbery AJ, Thompson RCA. Isoenzyme electrophoresis of 30 isolates of *Giardia* from humans and felines. Am J Trop Med Hyg 1988;38:65–73.

60 Baveja UK, Jyoti AS, Kaur M, Agarwal DS, Anand BS, Nanda R. Isoenzyme studies of *Giardia lamblia* isolated from symptomatic cases. Aust J Exp Biol Med Sci 1986;64: 119–126.

61 Boreham PFL, Smith NC, Shepherd RW. Drug resistance and the treatment of giardiasis. In Wallis PM, Hammond BR, eds. *Advances in Giardia Research.* Calgary: University of Calgary Press, 1988:3–7.

62 Isaac-Renton JL, Byrne SK, Prameya R. Isoelectric focusing of ten strains of *Giardia duodenalis.* J Parasitol 1988;74:1054.

63 Miller RL, Wang AL, Wang CC. Identification of *Giardia lamblia* isolates susceptible and resistant to infection by the double-stranded RNA virus. Exp Parasitol 1988;66:118–123.

64 Nash TE, Keister DB. Differences in excretory–secretory products and surface antigens among 19 isolates of *Giardia.* J Infect Dis 1985;152:1166–1171.

65 Aggarwal A, De la Cruz VF, Nash TE. A heat shock protein unrelated to Hsp70. Nucleic Acids Res 1990;18:3409.

66 Faubert GM, Belosevic M, Walker TS, MacLean JD, Meerovitch E. Comparative studies on the pattern of infection with *Giardia* spp. in Mongolian gerbils. J Parasitol 1983;69(5):802–805.

67 Visvesvara GS, Dickerson JW, Healy GR. Variable infectivity of human-derived *Giardia lamblia* cysts for Mongolian gerbils (*Meriones unguiculatus*). J Clin Microbiol 1988;26:837–841.

68 Aggarwal A, Nash TE. Comparison of two antigenically distinct *Giardia lamblia* isolates in gerbils. Am J Trop Med Hyg 1987;36:325–332.

69 Nash TE, Herrington DA, Losonsky GA, Levine MM. Experimental human infections with *Giardia lamblia.*

J Infect Dis 1987;156:974–984.

70 Nash TE, Herrington DA, Levine MM, Conrad JT, Merritt JW. Antigenic variation of *Giardia lamblia* in experimental human infections. J Immunol 1990;4362–4369.

71 Smith PD, Gillin FD, Kaushal NA, Nash TE. Antigenic analysis of *Giardia lamblia* from Afghanistan, Puerto Rico, Ecuador, and Oregon. Infect Immun 1982;36:714–719.

72 Einfeld DA, Stibbs HH. Identification and characterization of a major surface antigen of *Giardia lamblia*. Infect Immun 1984;46:377–383.

73 Edson CM, Farthing MJG, Thorley-Lawson DS, Keusch GT. An 88,000-M$_r$ *Giardia lamblia* surface protein which is immunogenic in humans. Infect Immun 1986;54:621–625.

74 Baker DA, Holberton DV, Marshall J. Sequence of a giardian subunit cDNA from *Giardia lamblia*. Nucleic Acids Res 1988;16:7177.

75 Aggarwal AA, Adam RD, Nash TE. Characterization of a 29.4 kilodalton structural protein of *Giardia lamblia* and localization to the ventral disk. Infect Immun 1989;57:1305–1310.

76 Lev B, Ward H, Keusch GT, Pereira MEA. Lectin activation in *Giardia lamblia* by host protease: a novel host–parasite interaction. Science 1986;232:71–73.

77 Ward HD, Alroy J, Lev BI, Keusch GT, Pereira MEA. Biology of *Giardia lamblia* detection of N-Acetyl-D-Glucosamine as the only surface saccharide moiety and identification of two distinct subsets of trophozoites by lectin binding. J Exp Med 1988;167:73–88.

78 Jarroll EL, Manning P, Lindmark DG, Coggins JR, Erlandsen SL. *Giardia* cyst wall-specific carbohydrate: evidence for the presence of galactosamine. Mol Biochem Parasitol 1989;32:121–132.

79 Ungar BLP, Nash TE. Cross-reactivity among different *Giardia lamblia* isolates using immunofluorescent antibody and enzyme immunoassay techniques. Am J Trop Med Hyg 1987;37:283–289.

80 Nash TE, Aggarwal A. Cytotoxicity of monoclonal antibodies to a subset of *Giardia* isolates. J Immunol 1986;136:2628–2632.

81 Nash TE, Gillin FD, Smith PD. Excretory–secretory products of *Giardia lamblia*. J Immunol 1983;131:2004–2010.

82 Nash TE, Aggarwal A, Adam RD, Conrad JT, Merritt JW Jr. Antigenic variation in *Giardia lamblia*. J Immunol 1988;141:636–641.

83 Adam RD, Aggarwal A, Lal AA, De la Cruz VF, McCutchan T, Nash TH. Antigenic variation of a cysteine-rich protein in *Giardia lamblia*. J Exp Med 1988;167:109–118.

84 Aggarwal A, Merritt JW Jr, Nash TE. Cysteine-rich variant surface proteins of *Giardia lamblia*. Mol Biochem Parasitol 1989;32:39–48.

85 Aggarwal A, Nash TE. Antigenic variation of *Giardia lamblia in vivo*. Infect Immun 1988;56:1420–1423.

86 Wang AL, Wang CC. Discovery of a specific double-stranded RNA virus in *Giardia lamblia*. Mol Biochem Parasitol 1986;21:269–276.

87 De Jonckheere JF, Gordts B. Occurrence and transfection of a *Giardia* virus. Mol Biochem Parasitol 1987;23:85–89.

88 Nemanic PC, Owen RL, Stevens DP, Mueller JC. Ultra-structural observations on giardiasis in a mouse model. II. Endosymbiosis and organelle distribution in *Giardia muris*

89 Reiner DS, Wang C-S, Gillin FD. Human milk kills *Giardia lamblia* by generating toxic lipolytic products. J Infect Dis 1986;154:825–832.

90 Hernell O, Ward H, Blackberg L, Pereira MEA. Killing of *Giardia lamblia* by human milk lipases: an effect mediated by lipolysis of milk lipids. J Infect Dis 1986;153:715–720.

91 Rohrer L, Winterhalter KH, Eckert J, Kohler P. Killing of *Giardia lamblia* by human milk is mediated by unsaturated fatty acids. Antimicrob Agents Chemother 1986;30:254–257.

92 Gillin FD. Killing of *Giardia lamblia* trophozoites by human intestinal fluid *in vitro*. J Infect Dis 1988;157:1257–1260.

93 Roberts-Thomson IC, Mitchell GF, Anders RF, *et al.* Genetic studies in human and murine giardiasis. Gut 1980;21:397–401.

94 Abaza H, Hilal G, Asser L, Abdo L, El-Sawy M. Histo-compatibility and immunological studies in giardiasis. Trans R Soc Trop Med Hyg 1988;82:437.

95 Visvesvara GS, Healy GR. Antigenicity of *Giardia lamblia* and the current status of serological diagnosis of giardiasis. In Erlandsen SL, Meyer EA, eds. *Giardia and Giardiasis.* New York: Plenum Publishers, 1984:219–232.

96 Radulescu S, Iancu L, Simionescu O, Meyer EA. Serum antibodies in giardiasis (Letter to the editor). J Clin Pathol 1976;29:863.

97 Visvesvara GS, Smith PD, Healy GR, Brown WR. An immunofluorescence test to detect serum antibodies to *Giardia lamblia*. Ann Intern Med 1980;93:802–805.

98 Ridley MJ, Ridley DS. Serum antibodies and jejunal histology in giardiasis associated with malabsorption. J Clin Pathol 1976;26:30–34.

99 Smith PD, Gillin FD, Brown WR, Nash TE. IgG antibody to *Giardia lamblia* detected by enzyme-linked immunosorbent assay. Gastroenterology 1981;80:1476–1480.

100 Wittner M, Maayan S, Farrer W, Tanowitz HB. Diagnosis of giardiasis by two methods. Immunofluorescence and enzyme-linked immunosorbent assay. Arch Pathol Lab Med 1983;107:524–527.

101 Vinayak VK, Jain P, Naik SR. Demonstration of antibodies in giardiasis using the immunodiffusion technique with *Giardia* cysts as antigen. Ann Trop Med Parasitol 1978;72(6):581–582.

102 Winiecka J, Kasprzak W, Kociecka W, Plotkowiak J, Myjak P. Serum antibodies to *Giardia intestinalis* detected by immunofluorescence using trophozoites as antigen. Trop Med Parasitol 1984;35:20–22.

103 Miotti PG, Gilman RH, Santosham M, Ryder RW, Yolken RH. Age-related rate of seropositivity of antibody to *Giardia lamblia* in four diverse populations. J Clin Microbiol 1986;24(6):972–975.

104 Goka AKJ, Rolston DDK, Mathan VI, Farthing MJG. Diagnosis of giardiasis by specific IgM antibody enzyme-linked immunosorbent assay. Lancet 1986;2:184–186.

105 Janoff EN, Smith PD, Blaser MJ. Acute antibody responses to *Giardia lamblia* are depressed in patients with AIDS. J Infect Dis 1988;157:798–804.

106 Belosevic M, Faubert GM. Lysis and immunobilization of *Giardia muris* trophozoites *in vitro* by immune serum from

susceptible and resistant mice. Parasite Immunol 1987;9: 11–19.

107 Butscher WG, Faubert GM. The therapeutic action of monoclonal antibodies against a surface glycoprotein of *Giardia muris*. Immunology 1988;64:175–180.

108 Hill DR, Burge JJ, Pearson RD. Susceptibility of *Giardia lamblia* trophozoites to the lethal effect of human serum. Immunology 1984;132:2046–2052.

109 Deguchi M, Gillin FD, Gigli I. Mechanism of killing of *Giardia lamblia* trophozoites by complement. J Clin Invest 1987;79:1296–1302.

110 Wallis PM, Lehmann DL. Differentiation of *Giardia duodenalis* from *Giardia muris* by immunobilization in various sera. In Wallis PM, Hammond BR, eds. *Advances in Giardia Research*. Calgary: University of Calgary Press, 1988:169–172.

111 Zinneman HH, Kaplan AP. The association of giardiasis with reduced intestinal secretory immunoglobulin A. Am J Dig Dis 1972;17:793–797.

112 Jones G, Brown WR. Serum and intestinal fluid immunoglobulins in patients with giardiasis. Dig Dis 1974;19:791–796.

113 Naik SR, Sehgal S, Krishnan S, Broor SL, Vinayak VK. Immunoglobulins in serum and duodenal juice and peripheral blood lymphocyte subpopulations in patients with giardiasis. Trop Geogr Med 1979;31:493–498.

114 Thompson A. Immunoglobulin-bearing cells in giardiasis (Letter to the editor). J Clin Pathol 1977;30:292–294.

115 Briaud M. Intestinal immune response in giardiasis (Letter to the editor). Lancet 1981;ii:358.

116 Miotti PG, Gilman RH, Pickering LK, Palacios-R G, Park HS, Yolken RH. Prevalence of serum and milk antibodies to *Giardia lamblia* in different populations of lactating women. J Infect Dis 1985;152:1025.

117 Hill DR, Pearson RD. Ingestion of *Giardia lamblia* trophozoites by human mononuclear phagocytes. Infect Immun 1987;55:3155–3161.

118 Smith PD, Elson CO, Keister DB, Nash TE. Human host response to *Giardia Lamblia*. I. Spontaneous killing by mononuclear leukocytes *in vitro*. J Immunol 1982;128:1372–1376.

119 Smith PD, Keister DB, Elson CO. Human host response to *Giardia lamblia*. II. Antibody-dependent killing *in vitro*. Cell Immunol 1983;82:308–315.

120 Aggarwal A, Nash TE. Lack of cellular cytotoxicity by human mononuclear cells to *Giardia*. J Immunol 1986;136:3486–3488.

121 Roberts-Thomson IC, Stevens DP, Mahmoud AAF, Warren KS. Giardiasis in the mouse: an animal model. Gastroenterology 1976;71:57–61.

122 Roberts-Thomson IC, Stevens DP, Mahmoud AAF, Warren KS. Acquired resistance to infection in an animal model of giardiasis. J Immunol 1976;117:2036–2037.

123 Belosevic M, Faubert GM. *Giardia muris*: correlation between oral dosage, course of infection, and trophozoite distribution in the mouse small intestine. Exp Parasitol 1983;56:93–100.

124 Duncombe VM, Bolin TD, Davis M, Fagan MR, Davis AE. The effect of iron deficiency, protein deficiency and dexamethasone on infection, re-infection and treatment of *Giardia muris* in the mouse. Aust J Exp Med Biol Sci 1980;58:19–26.

125 Roberts-Thomson IC, Mitchell GF. Giardiasis in mice. I. Prolonged infections in certain mouse strains and hypothymic (nude) mice. Gastroenterology 1978;75:42–46.

126 Underdown BJ, Roberts-Thomson IC, Anders RF, Mitchell GF. Giardiasis in mice: studies on the characteristics of chronic infection in C3H/He mice. Immunology 1981;126:669–672.

127 Belosevic M, Faubert GM. Temporal study of acquired resistance in infections of mice with *Giardia muris*. Parasitology 1983;87:517–524.

128 Stevens DP, Frank DM, Mahmoud AAF. Communications: thymus dependency of host resistance to *Giardia muris* infection: studies in nude mice. J Immunol 1978;120:680–683.

129 Heyworth MF, Carlson JR, Ermak TH. Clearance of *Giardia muris* infection requires helper/inducer T lymphocytes. J Exp Med 1987;165:1743–1748.

130 Carlson JR, Heyworth MF, Owen RL. T-lymphocyte subsets in nude mice with *Giardia muris* infection. Thymus 1987;9:189–196.

131 Heyworth MF, Kung JE, Caplin AB. Enzyme-linked immunosorbent assay for *Giardia*-specific IgA in mouse intestinal secretions. Parasite Immunol 1988;10:713–717.

132 Heyworth MF. Intestinal IgA responses to *Giardia muris* in mice depleted of helper T lymphocytes and in immunocompetent mice. J Parasitol 1989;75:246–251.

133 Heyworth MF. Antibody response to *Giardia muris* trophozoites in mouse intestine. Infect Immun 1986;52:568–571.

134 Snider SP, Gordon J, McDermontt MR, Underdown BJ. Chronic *Giardia muris* infection in anti-IgM-treated mice. Analysis of immunoglobulin and parasite-specific antibody in normal and immunoglobulin-deficient animals. J Immunol 1985;134:4153–4162.

135 Anders RF, Roberts-Thomson IC, Mitchell GF. Giardiasis in mice: analysis of humoral and cellular immune responses to *Giardia muris*. Parasite Immunol 1982;4:47–57.

136 Snider DP, Underdown BJ. Quantitative and temporal analyses of murine antibody response in serum and gut secretions to infection with *Giardia muris*. Infect Immun 1986;52:271–278.

137 Sharma AW, Mayrhofer G. Biliary antibody response in rats infected with rodent *Giardia duodenalis* isolates. Parasite Immunol 1988;10:181–191.

138 Snider DP, Skea D, Underdown BJ. Chronic giardiasis in B-cell-deficient mice expressing the *xid* gene. Infect Immun 1988;56:2838–2842.

139 Belosevic M, Faubert GM. Killing of *Giardia muris* trophozoites *in vitro* by spleen, mesenteric lymph node and peritoneal cells from susceptible and resistant mice. Immunology 1986;59:269–275.

140 Andrews JS, Hewlett EL. Protection against infection with *Giardia muris* by milk containing antibody to *Giardia*. J Infect Dis 1981;143:242–246.

141 Stevens DP, Frank DM. Local immunity in murine giardiasis is milk protective at the expense of maternal gut? Trans Assoc Am Physio 1978;91:268–272.

142 Kaplan BS, Uni S, Aikawa M, Mahmoud AAF. Effector mechanism of host resistance in murine giardiasis: specific IgG and IgA cell-mediated toxicity. J Immunol 1985;134: 1975–1981.

143 Mayrhofer G, Sharma AW. The secretory immune response in rats infected with rodent *Giardia duodenalis* isolate and evidence for passive protection with immune bile. In Wallis PM, Hammond BR, eds. *Advances in Giardia Research*. Calgary: University of Calgary Press, 1988:49–54.

144 Owen RL, Nemanic PC, Stevens DP. Ultrastructural observations on giardiasis in a murine model. I. Intestinal distribution, attachment, and relationship to the immune system of *Giardia muris*. Gastroenterology 1979;76:757–769.

145 Owen RL, Allen CL, Stevens DP. Phagocytosis of *Giardia muris* by macrophages in Peyer's patch epithelium in mice. Infect Immun 1981;33:591–601.

146 Radulescu S, Meyer EA. Opsonization *in vitro* of *Giardia lamblia* trophozoites. Infect Immun 1981;32:852–856.

147 Craft JC, Nelson JD. Diagnosis of giardiasis by counter-immunoelectrophoresis of feces. J Infect Dis 1982;145:499–504.

148 Ungar BLP, Yolken RH, Nash TE, Quinn TC. Enzyme-linked immunosorbent assay for the detection of *Giardia lamblia* in fecal specimens. J Infect Dis 1984;149:90–97.

149 Rosoff JD, Stibbs HH. Physical and chemical characterization of a *Giardia lamblia*-specific antigen useful in the coprodiagnosis of giardiasis. J Clin Microbiol 1986;24: 1079–1083.

10 African trypanosomiasis

Keith Vickerman, Peter J. Myler, & Kenneth D. Stuart

INTRODUCTION

No other disease has the stranglehold on development of a continent that trypanosomiasis has on Africa. The tsetse fly vector infests 11 million km^2 south of the Sahara, transmitting trypanosomes to both livestock and humans, causing nagana in the former and sleeping sickness in the latter. In cattle the result is slow growth, weight loss, poor milk yield, and impaired ability to work in transport animals. Infertility and abortion add to the toll. Some 50 million people are believed to be at risk from sleeping sickness, with 20 000 new cases reported annually. In the numerous active foci of this disease, communities are disrupted by depopulation and agriculture fails through depletion of human resources [1].

Trypanosomiasis, however, has another face. The antelopes and other indigenous mammals of Africa can harbor the same trypanosomes that cause disease but with no harmful effects; they have evolved resistance towards these parasites that enables them to survive in tsetse-ridden areas. By allowing survival of the native fauna and preventing overstocking of fragile land with cattle, and hence preventing widespread erosion, trypanosomiasis and the tsetse fly have kept Africa green.

While tsetse-borne trypanosomiasis is confined to Africa, the trypanosomes that cause the animal disease have extended their bounds further afield to southern Asia and South America. Transmission by tsetse involves lengthy development of the parasite in the fly, with the vector remaining infective for weeks or months. Outside the tsetse belt, such cyclic transmission is replaced by rapid syringe-like transfer of the parasite from the blood of one host to the blood of another by tabanid flies. Such noncyclic or mechanical transmission, though ephemeral, can cripple horses, camels, and other beasts of burden in regions where traction power is unmechanized.

Because these trypanosomes multiply in the blood and lymph of their mammalian hosts, they might be expected to provoke a strong immune response which would eliminate them from the body. Like most parasitic protozoa in immunocompetent hosts, however, African trypanosomes have evolved ways of dealing with such responses. What happens is that some parasites escape destruction and survive to continue the infection, so increasing the likelihood of their uptake by the blood-sucking insect vector for transmission to another mammal.

This ability to foil the mammalian host in its attempt to defend itself is imparted by a surface layer of glycoprotein which envelopes the entire plasma membrane of the parasite's body and flagellum. This surface coat serves as protective clothing, which not only enables the trypanosome to establish itself in the host in the face of nonspecific defense mechanisms but also provides the means whereby the parasite avoids the specific (lymphocyte-mediated) mechanisms which come into play later. This is because the trypanosome is able to change the antigenic nature of the single species of glycoprotein which composes the coat so that the host's acquired immune response is ineffectual, and response to the new antigen is called for. It is this antigenic variation—the tireless repetition by the trypanosome population of such changes of coat—which may lead the host literally to destroy itself in its vain attempt to overcome the infection. It is not so much the persistent presence of the parasite as the host's extravagent response to this presence which signals destruction.

And yet, within susceptible species, some hosts manage to live with their trypanosome infections without showing signs of suffering. Such trypanotolerance is reputedly a characteristic of the natural ungulate fauna of Africa, and is conspicuously absent from domestic stock, although certain trypanotolerant breeds of cattle provide an intriguing exception to this rule.

The urgent need to protect both animals and humans from trypanosomiasis has prompted a quest for some means of vaccinating susceptible hosts. This endeavor has become focused largely on understanding the process of trypanosome antigenic variation with a view to abrogating it: an alternative approach is to understand how the trypanotolerant host copes with its parasites.

The molecular basis of trypanosome antigen-switching has been a focus of avid attention over the last 20 years, with reviews currently appearing at the rate of two or more each year. As a result we now know more about the trypanosome as antigenic challenge than we do about the response of the natural host to infection with the parasite. This account attempts to put recent developments into the context of previous work. For historic development of the subject and further details the reader is referred to previous editions of this book [2–4], the reviews by Borst & Cross [5], Boothroyd [6], Borst [7], Clayton [8], Pays & Steinert [9], and Cross [10] for molecular aspects, and the symposium edited by Tizard [11] for immunology and immunopathology.

THE AFRICAN TRYPANOSOMES AND THEIR LIFE-CYCLES

Some knowledge of the intricacies of trypanosome life-cycles is essential for anyone trying to understand the development of immunity to these parasites. Detailed accounts are given by Hoare [12] and Vickerman [13].

African trypanosomiasis is a spectrum of diseases in humans and their domestic animals. It is caused by species of trypanosome—*Trypanosoma brucei*, *Trypanosoma congolense*, and *Trypanosoma vivax*—which are transmitted by bloodsucking tsetse flies (*Glossina* spp.) from one mammal to another. In the tsetse fly the parasites undergo an elaborate cycle of development with characteristic division phases and morphologic changes (Fig. 10.1). Similar diseases occur in ungulates outside the tsetse belt of Africa, but here the trypanosomes (*Trypanosoma evansi*, *T. vivax*) are transmitted mechanically, usually by tabanids; mechanical transmission may also occur when carnivores consume the flesh of trypanosome-infected prey. Somewhat exceptional life-cycles are found in vampire bat-transmitted *T. evansi* in South America (where trypanosomes pass from the blood of one host to that of another via the bat's mouth) and in *Trypanosoma equiperdum*, another *T. brucei*-like parasite, which is transmitted venereally between horses. Human sleeping sickness is caused by biologic races ("subspecies") of *T. brucei* designated *Trypanosoma brucei rhodesiense* and *Trypanosoma brucei gambiense*. The former causes the more acute East African illness and the latter the more chronic West African disease; *Trypanosoma brucei brucei* is morphologically identical to these two subspecies in all phases of its life-cycle but will not infect humans, being confined to wild and domestic ungulates and carnivores.

The African or salivarian trypanosomes, as their name implies, usually enter the mammal in the discharged saliva of their insect vector. In the cyclically-transmitted species, only the metacyclic stage (see Fig. 10.1) is known to be able to initiate infection. The fly's bite deposits metacyclic trypanosomes in dermal connective tissue and here a local inflammatory reaction—the "chancre"—develops. From the chancre the trypanosomes enter the draining lymphatics and then the bloodstream. *Trypanosoma congolense* multiplies in the tissue of the "chancre" as a morphologically distinct phase before it invades the bloodstream [14]. All three cyclically-transmitted species eventually undergo a change in form here to emerge with the characteristic morphology of the dividing bloodstream stages in the life-cycle.

The chancre phase completed, *T. congolense* and *T. vivax* remain largely intravascular parasites, the former localizing in small blood vessels where it attaches to the endothelium. Both species occur also in the lymphatics. *Trypanosoma brucei*, *T. evansi*, and *T. equiperdum* may secondarily escape from the bloodstream into the soft connective tissues and multiply in the tissue fluid; *T. equiperdum* is found principally in such tissues in its natural host. Some of these extravascular sites are "immunologically privileged," e.g., testis, cornea. Invasion of the brain and cerebrospinal fluid occurs in chronic *T. brucei* group infections and the brain has been implicated as a source of relapsing infections after chemotherapy [15,16]. Recently *T. vivax* has been found to behave similarly [17]. There are no intracellular stages in the life-cycles of the African trypanosomes as there are in the *T. cruzi* life-cycle (see Vickerman [13]).

All salivarian trypanosome infections in the natural host are characterized by an undulating parasitemia, each fall in trypanosome numbers (remission) corresponding to the destruction of a major antigenic type by the host's immune response and each ensuing increase in numbers (recrudescence) corresponding to the proliferation of trypanosomes of different antigenic type (Fig. 10.2).

Trypanosoma brucei is pleomorphic in the blood, multiplying by binary fission in the ascending parasitemia as a long slender flagellate and transforming to a nondivid-

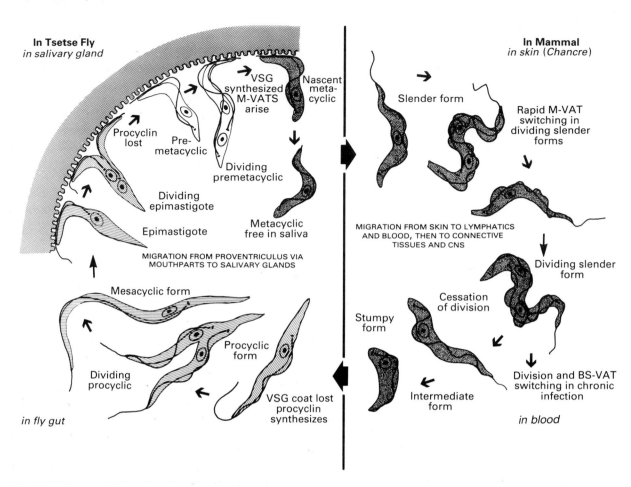

In Tsetse Fly
in salivary gland

VSG synthesized M-VATS arise

Nascent meta-cyclic

Procyclin lost

Pre-metacyclic

Dividing premetacyclic

Dividing epimastigote

Epimastigote

Metacyclic free in saliva

MIGRATION FROM PROVENTRICULUS VIA MOUTHPARTS TO SALIVARY GLANDS

Mesacyclic form

Procyclic form

Dividing procyclic

in fly gut

VSG coat lost procyclin synthesizes

In Mammal
in skin (Chancre)

Slender form

Rapid M-VAT switching in dividing slender forms

MIGRATION FROM SKIN TO LYMPHATICS AND BLOOD, THEN TO CONNECTIVE TISSUES AND CNS

Dividing slender form

Cessation of division

Stumpy form

Division and BS-VAT switching in chronic infection

Intermediate form

in blood

Fig. 10.1 Diagram of life-cycle of *Trypanosoma brucei* showing phases of multiplication and attachment and also changes in the surface immunodominant protein. The variant surface glycoprotein (VSG) coat (stippled) is present on all the trypomastigotes (kinetoplast behind nucleus) in the mammalian host. On entering the tsetse fly the trypanosomes replace the VSG largely with procyclin (cross-hatched), and subsequent stages in the vector up to the epimastigote (kinetoplast in front of nucleus) in the salivary gland bear this same surface glycoprotein. Premetacyclic trypomastigotes lack both procyclin and VSG, but VSG is synthesized after their division ceases. When fully coated with VSG the metacyclics detach and lie free in the fly's saliva. They are discharged into the skin of the mammalian host when the fly bites. (Courtesy of K. Vickerman.)

ing short stumpy trypanosome as the parasitemia passes through crisis into remission; the stumpy forms are more capable of surviving in the tsetse fly's blood meal and they initiate the cycle of development in the fly. The life-cycle of *T. brucei* is shown in Fig. 10.1. *Trypanosoma evansi* morphologically resembles *T. brucei* but rarely produces stumpy forms [12]. Lack of pleomorphism is also characteristic of *T. brucei* stocks which have been mechanically passaged through laboratory rodents; such monomorphic lines are not usually fly transmissible.

Although obvious gamete and zygote stages have not been demonstrated in trypanosomes, over the past 10 years evidence has accumulated that genetic exchange involving meiosis and syngamy occurs in *T. brucei*. Almost certainly, both events take place during development in the tsetse fly, most probably in the salivary gland stages, but unlike the sexual process in the life-cycle of malaria parasites, that of trypanosomes does not appear to be a necessary part of development in the vector (reviewed by Tait & Turner [19]).

Trypanosoma congolense and *T. vivax* are less markedly pleomorphic than *T. burcei* and show differences from that species in their developmental cycle in the vector. Thus, while both *T. brucei* and *T. congolense* undergo a multiplicative phase in the midgut of the fly, known as the so-called procyclic stage in the life-cycle, only

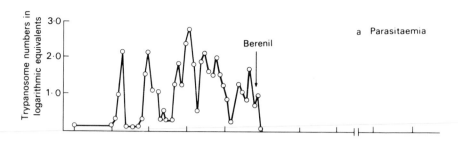

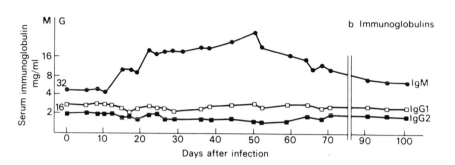

Fig. 10.2 Parasitemia and immunoglobulin levels in a zebu infected with *Trypanosoma congolense*. The parasitemia (a) shows characteristic peaks. (b) Serum IgM is elevated throughout the infection and declines slowly after treatment with Berenil (diminazene aceturate). IgG levels are scarcely affected by the infection. (From Luckins [18].)

T. brucei migrates to the fly's salivary glands for its final stages of development. The multiplicative epimastigote stage of *T. brucei* is attached by the flagellum to the salivary epithelium and differentiates into the non-multiplicative metacyclic trypomastigote which is liberated into the fly's saliva to be discharged during feeding. In *T. congolense*, epimastigote attachment and differentiation of the infective metacyclic form takes place on the chitinous wall of the proboscis, the metacyclic trypanosomes invading the hypopharynx for discharge. A procyclic phase in the vector's midgut does not occur in the life-cycle of *T. vivax*. Development of *T. vivax* is wholly in the fly's proboscis and corresponds to the latter part of the *T. congolense* cycle.

It is fortunate that the salivarian trypanosomes can, by syringe passage, be adapted to laboratory rodents for experimental work on the nature of antigenic variation and other aspects of host–parasite interaction. It is somewhat unfortunate, however, that at present our knowledge of the immunology of African trypanosomiasis is based largely on such mechanically transmitted infections of rodents rather than on cyclically transmitted infections of the natural host. Obvious changes take place in syringe-passaged trypanosome populations, e.g., in *T. brucei* loss of pleomorphism, increase in virulence, loss of ability to infect the vector, and stabilization of variable antigen type take place. As will become evident later, these changes are interrelated

and some are to the advantage of the experimentalist (see Turner [20]) in that they have assisted in the production of standardized materials which can be used to investigate the natural host's immune response. Increased virulence and stabilization of antigenic type, for example, have facilitated the harvesting of large numbers of antigenically characterized trypanosomes. The development of techniques for cryopreservation of trypanosome populations [21] as defined antigenic material has greatly helped these studies; sequential cryostabilates provide a means of monitoring changes in trypanosome antigenic character.

Within recent years, cultivation techniques for salivarian trypanosomes have advanced considerably and stages in the life-cycle previously obtainable with difficulty from either mammalian or insect host can now be produced *in vitro* (reviewed by Brun & Jenni [22], Gray *et al.* [23], and Gardiner [24]). Whereas the procyclic stage of *T. brucei* and *T. congolense* can be serially propogated with ease in a variety of cell-free media, the cultivation and maintenance of bloodstream stages *in vitro* is more difficult. Early success with feeder cells [25] has led to replacement of the feeder layer with mercaptoethanol [26] or cysteine [27], with use of bathocuproine sulfate to protect trypanosomes from cysteine toxicity [28]. Cloning of *in vitro*-produced bloodstream forms has also been achieved [29].

The unrewarding business of obtaining metacyclic

trypanosomes by infection of tsetse flies in the laboratory has deterred many from working on cyclically transmitted infections but the *in vitro* production of vector stages (including metacyclics) in quantity by Gray *et al.* [30] paved the way to eliminating such difficulties in the case of *T. congolense* and, eventually, *T. vivax* [31]. Whereas the metacyclics of these species differentiate from epimastigotes attached to inert (e.g., chitin, plastic) surfaces, *T. brucei* metacyclics arise from precursors attached to the tsetse salivary gland epithelium and a convenient and realistic *in vitro* system is still awaited. Kaminsky *et al.* [32] have reported some success in generating metacyclics by cultivating procyclic *T. brucei* together with *Anopheles gambiae* cells.

THE TRYPANOSOME AS ANTIGENIC STIMULUS

The antigens of African trypanosomes are classified as either stable (invariant) or variable antigens. The stable antigens (also called common antigens) are parasite constituents, such as structural proteins and enzymes, which can be isolated from the trypanosome population at any point in its infection in the mammal and may be continuously present through several stages in the lifecycle of the parasite. These antigens may occur in various stocks of the same species and be shared with other species [33]. Variable antigens, on the other hand, are developmentally regulated and confined to the metacyclic stage and trypanosome populations in the mammalian host. They are responsible for differences between serologic variants or variable antigen types (VATs) of the trypanosome.

Our knowledge of the structure, organization, and expression of trypanosome-variable antigens has expanded enormously in recent years, but the mechanisms that regulate antigenic variation on the part of the trypanosome population as a whole are still only imperfectly understood.

Variable antigens and the surface coat

The variable antigen makes up the surface coat of the trypanosome. Bloodstream forms of all the African trypanosomes have a 12–15 nm thick coat overlying the plasma membrane, as seen in electron micrographs of sections. While compact and dense in trypanosomes of the *T. brucei* group (*T. brucei* spp., *T. evansi*, *T. equiperdum*) and *T. congolense*, the coat is more diffuse in *T. vivax* (Fig. 10.3). The coat is lost along with the variable antigen

when the trypanosome embarks on cyclic development in the tsetse fly or undergoes transformation to the procyclic form *in vitro* [34]. The coat is reacquired during differentiation of the metacyclic form (reviewed by Vickerman [13], Gardiner *et al.* [35], and Tetley *et al.* [36]) so that trypanosomes already in possession of surface-variable antigen are inoculated into the mammalian host. Recent work (see Overath *et al.* [37] and Tetley *et al.* [38]) has confirmed the view that antigenic variation is associated with changes in the composition of the surface coat [39].

The coat of each trypanosome is made up of over 10^7 molecules of a single variant surface glycoprotein (VSG) [10,40] and the structure of this VSG determines the parasite's VAT. Antibodies produced against the VSG react only with parasites of the same VAT. Infection sera and also polyclonal and some monoclonal antibodies produced against purified VSG neutralize (i.e., prevent infection by) the homologous VATs. Similarly, immunization with purified VSG can protect only against infection with the homologous VAT.

Identification of VATs

Tests for indentification of VATs include *in vitro* tests, in which the results of the serologic reaction can be assessed directly, and *in vivo* tests, in which the results of the reactions are assessed indirectly by inoculation into animals. Of the *in vitro* tests, the agglutination reaction was much used in classic studies on antigenic variation (e.g., Gray [41,42] and Seed [43]) but it is now deemed too crude for the detailed quantitative comparisons of the individual VAT composition of trypanosome populations. For this, the tests of choice are the trypanolysis and immunofluorescence reactions (Plate 10.1 (facing p. 178)) [44,45]. Both polyclonal and monoclonal antibodies, some of which have been prepared against purified surface antigens, have been useful in these studies. Of the *in vivo* tests, protection against challenge and neutralization of infectivity have been most widely used.

Stable (invariant) antigens

Most of the trypanosome's stable antigens are not located at the parasite's surface and so are not exposed to the host's immune response against them. They may, however, be highly immunogenic and important in immunopathology of the infection (see section on immunopathology, pp. 194–198) or of potential use in

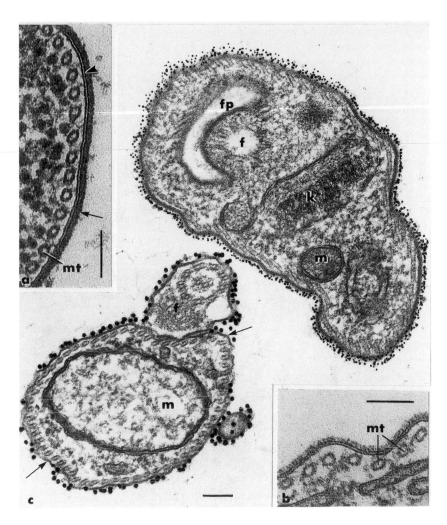

Fig. 10.3 Transmission electron micrographs of transverse sections of trypanosomes. (a) Detail of surface of bloodstream *Trypanosoma brucei* to show a dense 12 nm thick surface coat of VSG (arrow) on plasma membrane (arrowhead). (b) Similar section showing more diffuse coat on bloodstream *Trypanosoma vivax*. (c) Immunogold labeling of *T. brucei* bloodstream form (above) with 5 nm gold complexed with VSG-specific IgG antibody, and of procyclic (fly midgut) form (below) with 20 nm gold complexed with procyclin-specific monoclonal IgG antibody. Note lack of visible coat on procyclic trypanosome at arrow. Scale bars, 0.1 μm; f, flagellum; fp, flagellar pocket; k, kinetoplast; m, mitochondrion; mt, cortical microtubules. (a,b, Courtesy of K. Vickerman; c, Courtesy of Dr L. Tetley.)

serodiagnosis (see section on serodiagnosis, p. 194). Their high immunogenicity may be conferred by repetitive amino acid sequence motifs. Two such repetitive antigens have been recognized using early infection sera from cattle. These antigens have proved to be components of the trypanosome cytoskeleton (Fig. 10.4) [46].

In addition to the VSG, several other proteins are likely to be associated with the plasma membrane of African trypanosomes since they are required for functions such as nutrient import, response to external signals, and attachment to host surfaces. Receptors for low-density lipoproteins (LDL), ferritin, and glucose have been reported. The LDL and ferritin receptors appear to be localized in the flagellar pocket (Fig. 10.4), where endocytosis occurs [47]. The location of the glucose receptor [48] is unknown. The transferrin receptor

proteins belong to the VSG expression site-associated gene proteins (see section on VSG gene expression sites, pp. 181–182), which are probably all plasma membrane components [49–51]. The LDL receptor is produced in bloodstream forms but not in procyclic forms. The ferritin and glucose receptors are present in bloodstream forms but their presence in procyclic forms has not yet been demonstrated. Differential production of receptors for nutrient transport of proteins associated with surface composition is consistent with the environmental and metabolic changes that occur during the life-cycle of African trypanosomes (reviewed by Vickerman [13]). The organization of these membrane components relative to VSG is as yet unknown.

Most vector forms of *T. brucei* (Fig. 10.1) contain a dominant glycoprotein surface antigen called procyclin because it is present on procyclic but not bloodstream

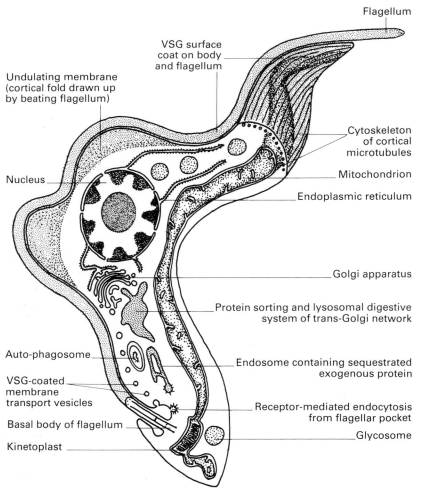

Flagellum

VSG surface
coat on body
and flagellum

Undulating membrane
(cortical fold drawn up
by beating flagellum)

Cytoskeleton
of cortical
microtubules

Mitochondrion

Nucleus

Endoplasmic reticulum

Golgi apparatus

Protein sorting and lysosomal digestive
system of trans-Golgi network

Auto-phagosome

Endosome containing sequestrated
exogenous protein

VSG-coated
membrane
transport vesicles

Receptor-mediated endocytosis
from flagellar pocket

Basal body of flagellum

Glycosome

Kinetoplast

Fig. 10.4 Diagram showing principal structures of *Trypanosoma brucei* (bloodstream intermediate form) in partial longitudinal section. Variant surface glycoprotein-coated membrane synthesized in the Golgi apparatus and trans-Golgi network is transported as vesicles to the flagellar pocket for insertion. The pocket is also the site of uptake of VSG-coated membrane in receptor-mediated endocytosis. After transport in clathrin-clothed vesicles to sequestering endosomes, endocytosed proteins and VSG are sorted in the trans-Golgi region; endocytosed proteins are digested in lysosomal compartments, from which VSG may be recirculated to the surface via the flagellar pocket. Autophagosomes and possibly other secondary lysosomes are also discharged into the pocket. (Courtesy of K. Vickerman.)

forms [52]. Procyclin is also called procyclic acidic-repetitive protein since it has glutamic acid–proline dipeptide repeats over half its length [53]. Procyclin appears to be synthesized and transported to the surface before the VSG is lost during transformation of bloodstream forms to procyclic forms [54]. Its function is a mystery because its exposed glycosylated N-terminal domain lays the trypanosome open to attack by the vector's lectin defenses during its development in the fly (see section on defense mechanisms of the vector, p. 187). Molecular modeling of procyclin and VSG molecules suggests that the procyclin repeat region interdigitates between the VSG molecules and extends beyond them during surface replacement of the latter by the former [54]. Procyclin does not form a visible coat, as seen by electron microscopy, but immunogold studies (Fig. 10.3)

have shown that it is present on all developmental stages in the tsetse fly up to the "naked" premetacyclic (trypomastigote intermediary between uncoated epimastigote and VSG-coated metacyclic stages, see Fig. 10.1) [55].

A variety of enzymes are also produced in a stage-specific fashion. While some enzymes of the glycolytic, Krebs cycle, and respiratory chain pathways are continuously produced, others are only produced in specific life-cycle stages [56]. Many glycolytic enzymes are localized in a specialized organelle, the glycosome (Fig. 10.4), and have cytosolic counterparts, while the Krebs cycle and respiratory chain enzymes are located in the mitochondrion (Fig. 10.1). The mitochrondion undergoes cyclic activation and repression during the course of the life-cycle—another chapter in trypanosome molecular biology which cannot be presented here (see Clayton [8]).

Variable antigen type repertoires: serodemes

Antigenic variation occurs in infections initiated by a single trypanosome. As will be discussed later, each VSG responsible for a VAT is the product of a single gene. The total number of VATs in the repertoire of a clone line is as yet undetermined for any trypanosome species, but has been shown to exceed a hundred in *T. equiperdum* [57]. Not all stocks of a given trypanosome species have the same repertoire of VATs. Different serodemes can be recognized, each with its own repertoire, though many serodemes have a few VATs in common [58,59]. These serologically crossreacting VATs are termed iso-VAT [60]. Estimates of VATs repertoire size based upon hybridization studies with probes for the conserved 3' region of VSG genes (see later section on VSG gene structure) give a figure of over 10^3 possible VATs for *T. brucei brucei* [61] but less than half this figure for *T. b. gambiense* [62].

Variant surface glycoprotein expression starts in the tsetse fly vector, at the final stage of development in the fly's salivary glands. During differentiation of the mammal-infective metacyclic stage, VSG synthesis commences just as parasite multiplication ceases (Fig. 10.1) [38], so that the trypanosomes reacquire the surface coat. The extruded metacyclic population is heterogeneous with respect to VAT (Fig. 10.1) but the range of VATs found is limited, forming a small subset of the parasite's complete repertoire; for *T. b. rhodesiense* this subset is no more than 27 VATs [63] but for *T. congolense* the number is as low as 12 [64]. Metacyclic VATs (M-VATs) appear with different frequencies, some M-VATs constituting over 30% of the population and others less than 1% [65]. The different M-VATs appear simultaneously during differentiation; as VAT switching appears to be a prerogative of dividing trypanosomes, it is unlikely that the nondividing metacyclic trypanosomes can undergo antigenic variation in the gland. Previous ideas that cyclically transmitted trypanosomes revert to a single "basic antigen" in the fly [41] have now been discarded, but finding reversion to a small subset characteristic of the serodeme raised hopes that vaccination of potential hosts against M-VATs might protect them from cyclically transmitted trypanosomiasis (see section on vaccines, pp. 193–194).

After entering the mammalian host, the first trypanosomes to appear in the blood continue to express M-VAT; rapid switching occurs from one M-VAT to another in the initial parasitemia in mice [66] and probably in the chancre in larger hosts. One of the first bloodstream VATs to appear in the initial parasitemia is that originally ingested by the transmitting fly (the I-VAT) [67]. Along with the I-VAT, a series of serodeme-characteristic "predominant VATs" appear in the blood as major VAT. In the ensuing relapsing parasitemia, a major VAT (or a few major VATs) tends to dominate each peak population of trypanosomes. For a given serodeme the major VAT appear in a characteristic order during the course of an infection. This order will be returned to later (see section on antigenic variation, p. 189).

At present it is difficult to estimate the number of serodemes which exist in the field for any trypanosome species or subspecies; it seems that for cattle trypanosome species there are several, even within one area [59,68]. Studies on sleeping sickness trypanosomes, however, suggest that within a given epidemiologic focus the number of serodemes is small. Thus, comparison of stocks of *T. b. gambiense* from Nigeria [42,69] or from several West African countries (D. Le Ray & E. Babiker, unpublished) indicate the presence of one major serodeme and very few minor ones. A similarly restricted range has been observed in the Lake Victoria region of East Africa, where a major serodeme of *T. b. rhodesiense* has been detected in recurring epidemics of acute sleeping sickness [66].

As might be expected for a host response–evasion mechanism which relies on the generation of antigenic diversity, VAT expression changes are evident in a serodeme with time. Constancy of M-VAT expression in the Lake Victoria *T. b. rhodesiense* serodeme does not appear to have been maintained over a period of 20 years [66], and whether this results from mutation or genetic recombination, or both, remains to be determined.

Preliminary studies on the inheritance of VAT repertoires following hybrid formation between *T. brucei* stocks in tsetse flies [70] have shown that each hybrid progeny clone contains some VATs from each parent, including (M-VATs), that most VSG genes are not allelic, and that progeny may inherit more than 50% of the parental repertoire. A variable antigen-type repertoire can therefore evolve through spontaneous alteration related to the VAT switching process (see below), through recombination within VSG genes leading to the production of chimeric genes [71], and through genetic exchange in hybrid formation. The last mechanism may result in the sudden appearance of new repertoires in the field.

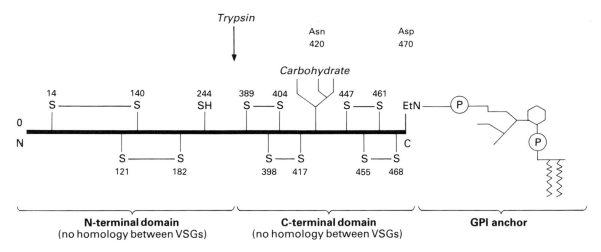

Fig. 10.5 Linear map of mature membrane form VSG (variant MITat 1.4) showing positions of the disulfide bonds, asparagine glycosylation site, glycolipid anchor, and the trypsin-sensitive hinge region between the N- and C-terminal domains. (From Ferguson & Homans [73].)

VSG structure

The structure of VSGs and their relationship with the plasma membrane have been studied in substantial detail [10,72,73]. Variant surface glycoproteins from different VATs have molecular weights ranging from 46 000 (*T. vivax* [74]) to 65 000 (*T. brucei* [75]). Although they are antigenically distinct within a species, VSG have numerous features in common (Fig. 10.5). The polypeptide chain of each VSG contains about 500 amino acids and is divisible into two domains defined by a trypsin cleavage site. The C-terminal domains of approximately 120 amino acids show some degree of homology, but the N-terminal domains are extraordinarily diverse in sequence. Substantial amino acid similarity occurs within the N-terminal 30 residues of some VSGs. The gene sequences can be more conserved than the amino acid sequence as a result of nucleotide additions and deletions that produce frameshift mutations. There are two general classes of VSG genes based largely on the features of the C-terminal sequence and the location of cysteine residues; in Class I, aspartic acid or asparagine is the C-terminal amino acid to which the glycolipid that anchors the VSG in the lipid bilayer is bound; in Class II the corresponding residue is serine.

Trypanosoma brucei VSGs contain between 7 and 17% carbohydrate [76], which plays no role in the antigenicity of the molecule. Several sites of glycosylation are conserved in the C-terminal region, while other sites are variable. Analysis of gene sequences has revealed that the primary polypeptide is synthesized with an N-terminal signal sequence that presumably functions in transport into the intracisternal space of the endoplasmic reticulum [77]. The C-terminal region contains a hydrophobic tail that is replaced in processing by glycolipid (Fig. 10.6). The glycolipid contains glycosyl-

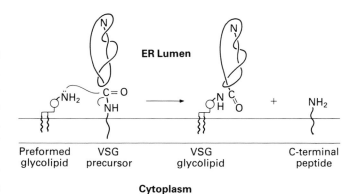

Fig. 10.6 Model of glycosylphosphatidylinositol (GPI; glycolipid anchor) addition to VSG in endoplasmic reticulum (ER). The VSG precursor is transported into the lumen of the ER by the hydrophobic N-terminal signal sequence, which is subsequently removed. The C-terminal GPI addition signal sequence is then recognized by an enzyme or enzyme complex which catalyzes the cleavage of the sequence and its replacement with the preformed GPI anchor. The processed VSG is then transported to the trypanosome surface along the route outlined in Fig. 10.4.

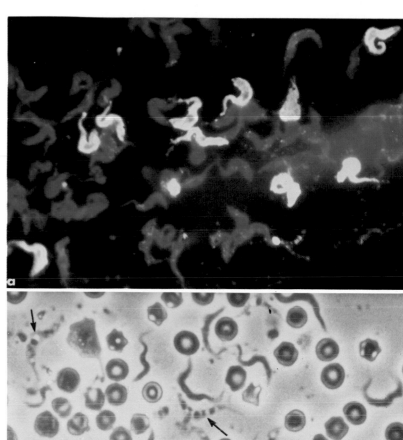

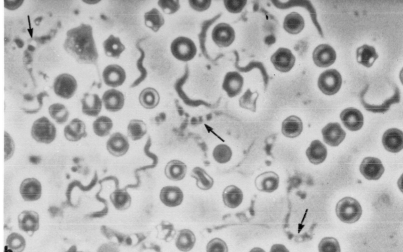

Plate 10.1 (a) Immunofluorescence reaction for identification of variable antigen types. Discharged metacyclics of *Trypanosoma brucei* in saliva of *Glossina morsitans*, stained with VAT AnTat 1.30-specific rabbit antiserum and FITC-conjugated antirabbit goat antiserum. AnTat 30 metacyclics fluoresce yellow; those belonging to other VATs are stained red by the Evans blue counterstain. Magnification, ×1700. (Courtesy of Dr S.L. Hajduk.) (b) Trypanolysis. Lysis of *T. brucei* as seen in withdrawn tail blood of mouse at onset of parasitemic remission. Lysing trypanosomes (arrowed) show vacuolation of both cytoplasm and nucleus. Those VATs to which the host has not mounted an immune response are unaffected. Phase contrast. Magnification, ×900. (Courtesy of Dr J.D. Barry.)

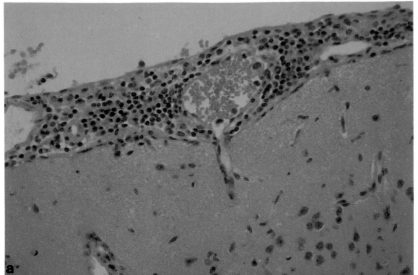

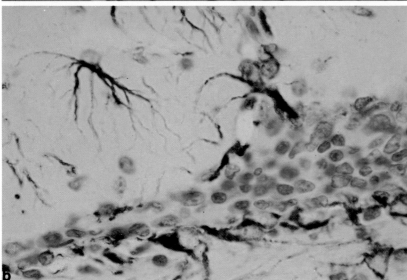

Plate 10.2 Pathology of *Trypanosoma brucei* infection in the brain of the experimental mouse. (a) Inflammatory reaction in the meninges. Trypanosomes can be seen in the blood vessel. Hematoxylin and eosin staining. Magnification, ×100. (b) Activation and proliferation of astrocytes in vicinity of blood vessels. Astrocytes are stained brown by a specific immunoperoxidase technique for glialfibrillary acid protein; inflammatory lymphocytes and macrophages, and gliotic scarring are evident below; hematoxylin counterstain. Magnification, ×300. (a, Courtesy of Dr C.J. Hunter; b, from Hunter *et al.* [310].)

phosphatidylinositol (GPI) and serves to anchor the VSG to the cell surface. The covalently linked glycolipid anchor of the African trypanosomes contains exclusively myristic acid whereas the GPI anchors of mammalian cells contain palmitic, stearic, and a variety of unsaturated fatty acids instead [78].

The three-dimensional structure of the VSG has been solved by X-ray crystallography [79,80]. The VSG exists as a dimer with a generally cylindrical structure that is oriented 90° to the cell surface. Interestingly, VSGs with very different amino acid sequences and no immunologic crossreactivity have very similar three-dimensional structures. Many antibodies, including the lytic antibodies, are directed at conformational epitopes. Most anti-VSG monoclonal antibodies react with acetone-fixed cells (i.e., unexposed epitopes); only a few react with live or formalin-fixed cells (i.e., with exposed epitopes) and generally these are the antibodies capable of neutralizing infectivity. Thus, the VSG appears to be densely packed on the cell surface with only terminal conformational epitopes exposed to the host's immune system (Fig. 10.7) [81,82].

Variant surface glycoprotein function appears to be to act as a barrier preventing host macromolecules from reaching the plasma membrane. Common portions of the VSG are also rendered inaccessible to antibodies. This includes the crossreacting determinant (CRD) [83, 84], a conserved immunologic feature of the glycolipid anchor. The N-glycans of VSG are not immunogenic, and they may protect the membrane from proteolytic

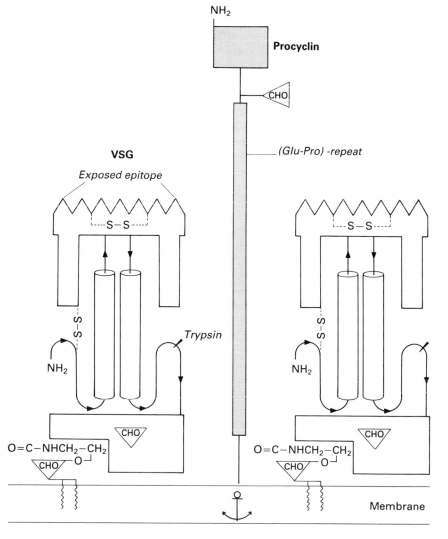

Fig. 10.7 Model for coexpression of variant surface glycoprotein and procyclin in the plasma membrane of *Trypanosoma brucei* differentiating from the bloodstream to the procyclic form. For simplicity the VSG is shown as a monomer (rather than a dimer). Its supporting core is formed by two 8 nm long α-helices in the NH$_2$-terminal domain (arrow denotes trypsin cleavage site). These uphold the globular exposed specific epitope-containing region made up of about 100 amino acid residues and rich in β-turn potential. CHO denotes carbohydrate chains, in VSG linked to the protein or as part of the COOH-terminal glycolipid which anchors the VSG in the surface membrane; they are not exposed on the trypanosome surface. In procyclin, the glutamic acid–proline repeat region forms an extended rod which can interdigitate with the VSG molecules. The 31 amino acid NH$_2$-terminal domain contains a glycosylation site so that carbohydrate groups are exposed on the living trypanosome for interaction with vector lectins. The nature of the procyclin anchor is not yet known. (From Roditi *et al.* [54]; VSG structure based on Jahnig *et al.* [81].)

attack or assist in dimerization of VSG [85]. Despite the dense packing of VSG molecules in the coat, they exhibit high lateral mobility within the plasma membrane [86].

The membrane-bound VSG molecule can be released as soluble VSG by the action of an endogenous GPI-specific phospholipase C [87]. Because this activity has been demonstrated only under nonphysiologic conditions, its biologic function is obscure. Such cleavage of the membrane form VSG may occur in the VAT switching process or in the recycling of VSG that takes place after its internalization by endocytosis.

The molecular basis of VSG switching

Genome organization

There are hundreds of VSG gene family members and they are distributed in the genome in a characteristic fashion. This distribution has been determined using pulse-field gradient electrophoresis (PFGE). African trypanosomes contain variable numbers of 50–150 kb mini chromosomes, depending on species. *Trypanosoma b. brucei* and *T. b. rhodesiense* contain about 100, *T. b. gambiense* contains about 10, *T. equiperdum* contains one and *T. vivax* none [88,89]. The genome also contains intermediate size chromosomes, which vary in number (1–8) and size (150–700 kb) among stocks and occasionally between variants (see below). In addition, trypanosome nuclei possess a set of larger chromosomes (0.7–6 Mb) that are difficult to resolve by PFGE. Bloodstream stages of syringe-passaged *T. brucei* spp. are basically diploid but the mini and intermediate size chromosomes probably do not occur as homologous pairs as in conventional diploids since hybrid trypanosomes initially contain both sets of parental mini and intermediate chromosomes and these diminish in number with population growth after cyclic transmission [90]. Recently, it has been shown in *T. brucei* stock 427 that the large chromosomes occur as 8–10 homologous pairs [91]. Homologous chromosomes contain different alleles of the diploid housekeeping genes, but often differ substantially in size. The variation in mini and intermediate chromosome number among stocks and species, especially the relative absence of these chromosomes in *T. equiperdum* and *T. gambiense*, as well as the primary location of housekeeping genes on the largest chromosomes, suggest that the largest chromosomes represent the core genome and the smaller chromosomes are associated with antigenic variation. Ploidy changes other than the chromosome diminution

following hybrid formation or meiosis have not been established but cannot be excluded at this point.

Chromosomal distribution of VSG genes

Variant surface glycoprotein genes are present in all classes of chromosomes and have two types of locations. They are present at many, if not all, telomeres (chromosome ends), as demonstrated by exonuclease sensitivity and restriction endonuclease mapping. Quantitative probe hybridization studies suggest that there are hundreds to a few thousand VSG genes in *T. brucei* [61]. About 200 of these genes are telomeric, assuming two telomeres on each of the approximately 100 chromosomes. The remaining VSG genes are intrachromosomal, some located on intermediate chromosomes but most on the largest chromosomes. Many VSG genes appear to occur in clusters of tandem repeats [61]. Even though VSG genes are found at the telomeres of all size classes of chromosomes, only those on the intermediate and large chromosomes are expressed *in situ* [92,93]. Mini chromosomes appear to serve largely as reservoirs for telomeric VSG genes as they appear to contain few sequences other than telomeric VSG genes or 76 bp and 177 bp repetitive elements (see below) [94].

VSG gene structure

All VSG genes have a similar nucleotide sequence organization although they have substantial sequence diversity in both the coding and flanking sequences. Telomeric VSG genes are associated with other characteristic genes and sequence elements (Fig. 10.8). The 5′ and 3′ termini of the coding sequence encode the signal peptide and hydrophobic tails, respectively, and a 13 nt sequence is also conserved at the 3′ end of the gene. No introns have been found in expressed VSG genes. Many VSG genes contain premature termination codons and consequently could not produce functional VSG if expressed *in situ*; they have thus been referred to as pseudogenes. However, segments of these pseudogenes may become incorporated into expressed VSG genes by recombination [95,96].

Sequences associated with VSG genes

Sequences are conserved in the 3′ flanks of both telomeric and nontelomeric VSG genes [97,98]. About 200 bp are conserved 3′ to telomeric VSG genes and

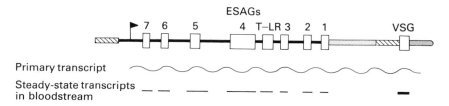

Fig. 10.8 Schematic representation of a VSG gene expression site. All *Trypanosoma brucei* VSG expression sites contain the following sequences moving upstream from the telomere (right): telomeric repeats; VSG gene (variable); cotransposed region; ''barren region'' containing 76 bp; ESAG 1–7 and ESAG 8 (T-LR); 50 bp repeat region. The putative promoter is indicated by the flag. The polycistronic primary transcript is processed in bloodstream forms to give the steady-state transcripts indicated.

beyond them are several subtelomeric segments with repeating motifs; then follow numerous repeats of the TTAGGG telomeric sequence. Variable amounts of the 200 bp sequence and some repeats are conserved among intrachromosomal VSG genes, although some have none of these conserved sequences. The immediate 5' flanks of VSG genes are diverse in size and sequence among both telomeric and intrachromosomal VSG genes. This region is also called the cotransposed region since copies of the sequences are transposed along with VSG genes. The sizes of the cotransposed regions vary between about 2 and 10 kb. Some cotransposed region sequences have homology to other VSG genes and may be pseudogenes. Transposon sequences called ''ingi'' or ''TRS'' occur in some cotransposed regions and multiple copies occur elsewhere in the genome, including some mini chromosomes [99–101]. These sequences are associated with VSG genes more frequently than would be expected by chance but the terminal segment of the ''ingi'' (called RIME) is associated with the ribosomal RNA genes [102]. Upstream of the cotransposed region of most VSG genes is a repeated sequence that has been called the 70–76 bp repeat or the V-sequence [103,104]. The repeat has characteristic sequence domains and occurs a variable number of times depending on the gene. Up to 11 tandem repeats of this sequence have been observed and some repeats contain an extended region of repeated (TAA)n sequence [105]. These repeats lack restriction sites and form the ''barren region'' upstream of all telomeric and some intrachromosomal VSG genes [106].

Thus, the VSG genes and their associated sequences as a class have an intrate pattern of homologous sequences scattered in and around the VSG genes. This is probably relevant to mechanisms of antigenic variation requiring homologous recombination (gene conversion).

VSG gene expression sites

While the majority of VSG genes occur at intrachromosomal locations, only those within telomeric ''expression sites'' located at the ends of intermediate and large chromosomes are expressed [92,93]. All *T. brucei* bloodstream stage expression sites examined to date contain eight genes, known collectively as expression site-associated genes (ESAGs), upstream of the barren region 5' to the expressed (telomeric) VSG gene (Fig. 10.8) [107–110]. *Trypanosoma equiperdum* expression sites appear to contain some of these ESAGs, in addition to a different ESAG [111]. *Trypanosoma brucei* metacyclic VSG expression sites studied to date lack most ESAGs and contain only short, if any, 76 bp repeats upstream of the VSG gene [112–114]. Expression site-associated-genes-related sequences also occur in genomic locations other than VSG gene expression sites. Most notably, an ESAG-2-related gene is found in the procyclin expression site [115], ESAG-4-related sequences are found elsewhere in the genome, and ESAG-1 sequences are found on mini chromosomes, which do not contain VSG expression sites. Possible functions for some of the ESAGs have been identified; ESAG-6 encodes a membrane-bound transferrin-binding protein [51], the ESAG-4 gene family has homology to mammalian adenylate/guanylate cyclase genes [50], and ESAG-1 encodes a membrane-associated glycoprotein [49]. Because of the similarity of ESAG sequences in different expression sites, it has been difficult to prove rigorously that ESAG and VSG gene transcription are coordinately regulated. However, nuclear run-on and transcription inactivation studies using ultraviolet light, as well as direct sequencing, suggest that the ESAGs and the VSG gene are members of the same transcriptional unit [116–118]. The promoter appears to be located upstream of the ESAG-7 gene,

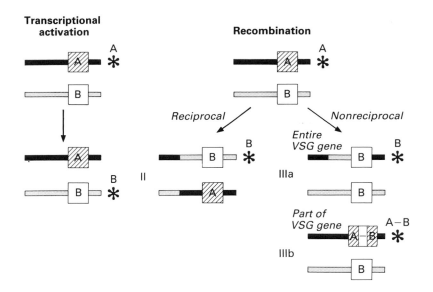

Fig. 10.9 Mechanisms for activation of VSG genes. Two telomeric expression sites contain VSG genes A and B. The expressed VSG gene is indicated by a star and letter. In the transcriptional activation mechanism (I), transcription is inactivated at site A and activated at site B by the antigenic switch. The recombination mechanism can entail either a reciprocal exchange of VSG genes A and B (and associated sequences) between expression sites (II), or gene conversion of all (IIIa) or part (IIIb) of VSG gene A by VSG gene B; transcription of the new VSG gene (B or A–B) continues in the same active expression site.

about 40–50 kb 5′ to the VSG gene [116,119], although these studies do not exclude the possibility of additional downstream promoters, and in some expression sites transcription also occurs upstream of this site [120,121].

Mechanisms of VSG gene expression

There are two general mechanisms of VSG gene expression (Fig. 10.9) and they operate independently of one another [122,123]. One entails switching transcription from one expression site to another [124], while the other entails genome rearrangement that places unexpressed VSG gene sequences in an expression site [125,126]. The only apparent genomic changes in the vicinity of either of the telomeric VSG genes associated with the transcriptional switch mechanism are the switch of transcription to a different VSG expression site, increased DNase I sensitivity of this gene [93], and partial modification of some nucleotides surrounding the inactivated VSG gene [127,128]. Aberrant changes in the length of the telomeres involved have also been observed [129,130]. Recently, genomic rearrangements upstream of the putative expression site promoter have been found during some transcriptional switches [120,121]. The role, if any, of these various changes in the molecular mechanism of transcriptional switching remains to be elucidated.

The genomic rearrangement mechanism probably entails homologous recombination, since the boundaries of the rearrangements occur at sites where sequences are conserved among VSG genes or where homology exists between the rearranged sequences. However, no studies have reported recombination intermediates such as heteroduplex molecules or recombinases, which might be specific for the VSG gene. There are two general types of genomic rearrangements; one resembles gene conversion, since a VSG gene in an active expression site is replaced (gene converted) by a copy of another VSG gene that occurs elsewhere in the genome; the second entails a reciprocal exchange between telomeric expression sites. The telomeric exchange process appears to be rare and in the two cases where it has been reported the exchange occurred within, or upstream of, the 5′ barren region [131,132].

The gene conversion can take several forms. Commonly, one entire VSG gene coding sequence is replaced by another. However, more extensive gene conversions may duplicate the VSG gene and upstream regions, including regions upstream of the barren region [133]. This has been most commonly observed for telomeric genes, which is to be expected since these regions are homologous for about 50 kb upstream of the VSG gene and these cases are referred to as telomeric gene conversions. Other cases of gene conversion only involve portions of VSG genes. This results in composites of two VSG genes or, more elaborately, mosaic VSG genes that are patchworks of segments from several VSG genes [95,134]. Two possible routes may create mosaic VSG genes: a series of sequential gene conversions as described above, or a more complex process involving the simultaneous interaction of several VSG genes or their transcripts. It has been found that "mosaic" genes are usually expressed late in an infection, suggesting that such a mechanism is important in increasing further the diversity of the VSG repertoire [96].

The gene conversions described above necessarily involve changes in expressed VSG genes since the antigenic switches require a change in VSG gene expression. However, the gene conversions may not be restricted to expressed VSG genes. Inactive telomeric VSG genes can be lost as a result of an antigenic switch [135,136]. It is likely that this is the result of a "silent" gene conversion, although gene deletion cannot be excluded. Detection of the donor for the gene conversion requires a probe, but since there are hundreds of VSG genes it is not surprising that the gene conversion has not been confirmed. As many as two gene conversions have been found to accompany a switch of transcription between different expression sites in single relapses [135,136]. It is possible that such an antigenic switch could occur by gene conversion of an inactive telomeric VSG gene followed by subsequent activation of the new VSG gene by switching of transcription from the active expression site. A second gene conversion replacing the VSG gene in the previously active expression site may have occurred during the same relapse. Gene conversions that alter nontelomeric genes are conceivable but have not been reported. The variety of gene conversions might serve continually to diversify the VAT repertoire by introducing new VSG genes into telomeric locations where they are more likely to be expressed. This diversification of repertoires may give selective advantage to the parasite.

Control of VSG expression site transcription

The processes that regulate transcription of the VSG gene expression site are unknown. Available evidence suggests that the expression site is a single transcription unit [116,117] and that the numerous steady-state transcripts from the ESAG and VSG genes result from processing of polycistronic transcripts [118]. In common with other trypanosome genes [137], VSG and ESAG transcripts contain no introns and are processed by mechanisms involving trans-splicing. This RNA processing entails cleavage of the transcripts and replacement of sequences 5' to the cleavage site by the 39 nt spliced leader sequence [138].

Although there are several expression sites within each trypanosome, only one is transcribed at a time. The simultaneous presence of two VSG has been observed in M-VATs [139,140] and in *T. equiperdum*

grown *in vitro* [141]. In the mammal, M-VAT are unstable and double expressers may be transcribing a single VSG gene but still contain mRNA from the previously expressed (metacyclic) VSG gene. Only one VAT resulted from infection of a mammalian host with the *T. equiperdum* double expressers [141]. Thus, only one expression site may have been transcribed, as in the M-VAT case, or two expression sites can be transcribed simultaneously, but such double expressers may not be viable *in vivo*. The latter possibility is strengthened by the observation that two expression sites are simultaneously transcribed in a VAT that contains a transposed sequence that terminates transcription before the VSG gene, resulting in production of a single VSG [142]. Thus, the system regulating transcription of the expression sites usually allows transcription of a single VSG gene but occasionally permits simultaneous expression of two sites.

The characteristics of the system that controls transcription of the VSG gene expression site not only ensures that only one expression site will (usually) be transcribed but it also enables transcription to be switched to a different expression site. Thus, transcription is inactivated at one expression site concomitant with activation of transcription at another expression site. Different VSG genes in their respective expression sites are transcribed in different VATs, as shown by a combination of PFGE and Northern blot experiments [135]. Sequence analysis of ESAGs expressed in different VATs indicate that they are transcribed only in those VATs where the downstream VSG gene is expressed [109,110]. This suggests that transcription of one entire expression site is inactivated when transcription of another entire expression site is activated in most antigenic switches, although further experimentation is required to substantiate this conclusion.

In procyclic forms, the VSG gene is no longer expressed from any expression site. Expression of ESAGs also ceases in procyclic forms, but some ESAG-related sequences not located in VSG expression sites are expressed at this stage [50,115]. However, recent results indicate that sequences close to the VSG expression site promoter are transcribed in procyclic forms [117], and this promoter is active in both life-cycle stages [119]. Thus, there appears to be post-transcriptional regulation of VSG gene expression in procyclic forms involving aberrant processing of VSG expression site transcripts [143]. Variant surface glycoprotein transcription is reinitiated in metacyclic forms but, interestingly, it appears to occur at expression sites that are M-VAT specific, lack barren regions, and are located on the largest chromosomes [112–114,144]. Intriguingly, the VSG gene that was inactivated upon conversion to the procyclic form is reactivated after the M-VAT switch antigenic type upon infection of the mammalian host [67,145], perhaps because the RNA transcribed from this expression site is no longer processed aberrantly (see p. 177).

These observations suggest the presence of an intricate regulatory system that controls the expression of VSG genes. It appears that separate processes are involved in the regulation of VSG expression during VAT switching, the transition to procyclic forms, and the expression of metacyclic VSG. Transcription of VSG genes is resistant to α-amanitin inhibition, unlike conventional RNA polymerase II-mediated transcription of mRNA [146]. This unusual feature suggests the existence of a special RNA polymerase that transcribes the VSG gene expression site and may also contribute to the complex regulatory properties of the system.

Antigenic variation at the population level

The molecular events surrounding the switch in expression of VSG genes at the individual trypanosome level represent one aspect of antigenic variation which has now been well explored. The other major aspect of this immune evasion mechanism is how these events are related to changes in VAT at the population level, and here we are on less certain ground (reviewed by Vickerman [147] and Barry & Turner [148]).

In order to obtain the maximum prolongation of infection, natural selection should adjust the rate of VSG gene switching so that one new VAT is generated in each of a series of regular parasitemic waves. Each peak would then represent an expanded clone of trypanosomes belonging to a particular VAT, the "homotype", against which the host would mount an immune response. When the homotype is sacrificed to the hosts's antibody response, the population would continue through the multiplication of VAT-switched individuals (heterotypes) present in the population. This simplistic view of antigenic variation is upheld by work on syringe-passaged stocks of monomorphic *T. brucei* in laboratory animals, but in cyclically transmitted infections of the natural host the situation is more complex, as can be seen from parasitemic profiles (Figs 10.2 and 10.11). These range from a few small sharp peaks, arising occasionally from a background of subpatency, to pronounced peaks which may run together over several days. Within the one infection peaks may vary in height,

width, frequency, and complexity, but the pattern is on the whole reproducible for a given trypanosome stock in a given host species.

Paced stimulation of the host's immune response and serial sacrifice of successive VATs are the essentials of antigenic variation at the population level. Screening of hosts for the appearance of antibodies against specific VAT suggests that particular major VATs stimulate the host's immune response (and therefore, presumably, dominate the bloodstream population) in a hierarchic order [41,57,149]. The hierarchy is not a linear sequence, however. Analysis of the mixtures of VATs arising in the second peak of cloned infections in mice [58,150] have shown that switching is divergent, i.e., trypanosomes of a particular VAT can give rise to several different VAT on switching. Divergent switching allows the infection to continue in the partially immune host, whereas a fixed linear order would not.

Variation in the switching rates between different combinations of VAT has been observed directly. The rate of switching is much higher in cyclically transmitted pleomorphic *T. brucei* infections (1×10^{-2} switches/cell/generation) [151] than in syringe-passaged monomorphic lines (1×10^{-5}–1×10^{-7} switches/cell/generation) [152,153]. This may explain why in monomorphic infections at peak the blood population may consist of over 99% of one VAT, whereas in pleomorphic infections a more pronounced mixture of VATs is usually present. The effective rate of VAT switching is progressively reduced from the basic rate as infection progresses and the host's immune responses weeds out an increasingly larger proportion of the VAT repertoire [151].

The various means by which the VSG genes are activated differentially may make a primary and stochastic contribution to VAT hierarchy. As discussed previously, telomeric genes tend to be expressed first, followed by chromosomal internal genes and then by genes activated through rare recombinational events. There is also a hierarchy among expression sites, with one site prevailing as the recipient for the duplicated internal genes [135,145,154]. Little is known about the activation hierarchy for chromosomal internal genes which make up most of the repertoire, but the extent of flanking homology between an internal gene and an expression site may be directly proportional to its activation frequency [155]. In two independent studies, however, trypanosomes belonging to the same VAT population were found to have activated the same internal VSG gene by independent routes [155,156].

Other factors which contribute to the VAT hierarchy

affect the ability of the newly switched trypanosome to multiply and dominate the bloodstream population. A seemingly obvious factor is the growth rate of VAT, once produced; those with the fastest growth should appear early in the infection; those with the longer doubling times should appear later [157]. This suggestion, however, has not received much support from experimental studies [158–160]. Factors other than VAT contribute to differences in growth rate, one being the tendency for trypanosomes to sidetrack into the G_0 phase of the cell cycle (e.g., in *T. brucei* by becoming stumpy forms; see section on control of parasitemia, pp. 200–201). There is also the interesting possibility that inhibitory interactions between trypanosomes of different VAT may influence VAT succession. The formation in spatially segregated sites of antigenically distinct subpopulations that can make sudden influxes into the bloodstream population, thus changing its VAT composition, has also been envisaged [161] but the high rate of interchange occurring between extravascular sites and the bloodstream [162] suggests that the VAT compositions of populations in different sites are not dissimilar. The possibility that the viability of certain inter-VAT switches may be limited by the incompatibility of VSG in forming a functional surface coat during the replacement of one VSG by another has also been raised [163].

Modeling approaches based on the logistic growth equation have been used to test hypotheses accounting for the hierarchic order of VAT. One approach incorporated VAT-specific growth rates into the model and concluded that this parameter could not explain hierarchic expression of VATs [152]. Another approach [163] also incorporated this parameter and the parameter associated with the simultaneous expression of two VSGs on the surface of individual trypanosomes; the conclusion was that the vulnerability of double expressors could be important in this respect.

At present, the most satisfactory explanation of VAT hierarchy is that based on differential switching rates [151]. Because actual rates of interswitching between VAT may vary by several orders of magnitude, the net rates of accumulation of different VATs to a point where they pass the threshold of immunostimulation may also vary, resulting in ordered VAT prevalence and paced stimulation of the antibody response. High switching rates in early tsetse-transmitted infections may delay the growth of certain variants to major VAT status until late in the infection.

The ability of the host to respond to the presence of trypanosomes, both in limiting the overall size of the

population in which switching is taking place and in responding to a particular VAT, is clearly important, yet is an area of ignorance (see section on trypano-tolerance, pp. 198–201). Given the high switching rates now demonstrated in cyclically transmitted infections, it is possible that the entire VAT repertoire is expressed in the first parasitemic wave in large mammals. Some explanation is needed as to why this plethora of VATs does not lead to early termination of the infection.

Antigenic disguise

From their work on the role of sheep serum supplement in adapting *T. vivax* to rodents, Desowitz & Watson [164] suggested that trypanosomes might adapt to new host species by binding specific plasma proteins to their surface and that these bound proteins might provide a shield against the host's defense mechanisms comparable to the "antigenic disguise" of schistosomes. Ketteridge [165] reported some evidence for adsorption of serum sialoglycoprotein on the surface of trypanosomes belonging to the same stock but subsequent workers failed to confirm the presence of host serum components on a ruminant-restricted stock of *T. vivax* naturally infective to mice [166]. The antigenic disguise hypothesis has, therefore, at present little to commend it in so far as salivarian trypanosomes are confirmed [24].

The adsorption of host serum components may occur for other reasons, however. Thus the detection of receptors for epidermal growth factor (EGF) on the surface of bloodstream and procyclic *T. brucei* [167] suggests that the trypanosome binds EGF and that binding may in turn control parasite proliferation. The relationship of the EGF receptor to the VSG coat has not yet been resolved. Major strides forward in our understanding of host–parasite relationships can be expected, shortly, from further studies on nonvariable membrane proteins, transmembrane signalling pathways, and control of trypanosome multiplication or differentiation (see section on trypanotolerance, pp. 198–201).

INNATE IMMUNITY IN MAMMALIAN HOST AND VECTOR

The glycoprotein coat of metacyclic and bloodstream forms of trypanosomes appears to render the parasites refractory to the usual mainstays of mammalian innate immunity—complement activation and opsonization for uptake by macrophages. Some potential hosts, however, have serum factors which can penetrate the barrier presented by the coat and lethally damage the trypanosomes.

Paradoxically, loss of the VSG coat in the midgut of the vector lays the parasite open to attack by the insect's natural defenses.

Complement activation

The VSG coat protects African trypanosomes from complement-mediated lysis in the nonimmune host. The uncoated stages in the life-cycle of T. brucei are readily lysed in fresh guinea pig serum but metacylic and bloodstream stages are not [168], unless deprived of the coat by trypsinization [169]. Such lysis probably results from activation of the alternative complement pathway [169]. Cultured bloodstream forms of T. congolense with a defect in coat synthesis were found by Ferrante & Allison [169] to be lysed by fresh human serum and by C2-deficient serum in the presence of EGTA but not by heat-inactivated serum or by fresh serum in the presence of ethylenediaminetetraacetic acid (EDTA), suggesting operation of the alternative pathway. The uncoated tsetse proboscis stages of T. congolense and T. vivax (epimastigote, premetacyclic), which have to withstand exposure to fresh plasma in the host's blood meal, are unaffected by fresh guinea pig serum and must have an additional mechanism for thwarting insertion of the membrane attack complex [36].

The coated stages of T. brucei are lysed by VAT-specific antibody with activation of complement by the classic pathway [171]. Bloodstream stages can also activate complement in the absence of a specific immune response; purified soluble-form VSG activates human complement via the classic pathway, depleting C1, C2, C4, and to some degree C3 [172]. Despite this activation the trypanosomes are not lysed. It is possible that the pathway, although initiated, is not completed [173,174]. Such activation may, however, trigger the initial inflammatory response in the chancre and contribute to the hypocomplementemia of T. congolense- and T. vivax-infected cattle [175]. The interaction of trypanosomes and complement has been reviewed by Nielsen [176].

In no case has it been shown for African trypanosomes that a mammal naturally resistant to infection destroys the parasite by complement activation. The complement pathways appear to be unimportant in nonspecific immunity to African trypanosomes.

Phagocytosis

The VSG coat appears to act as an efficient antiphagocytic capsule in the absence of VAT-specific immunity. Although the interaction of phagocytic cells (macrophages, polymorphs) with trypanosomes has not been investigated in detail, it is known that peritoneal macrophages from susceptible host species neither bind nor phagocytose bloodstream trypanosomes under such conditions [170,177].

The susceptibility of mice to African trypanosomes can be decreased nonspecifically following treatment with immunostimulants such as Bacille Calmette-Guerin, Propionibacterium acnes (= Corynebacterium parvum), and Bordetella pertussis [178], which are believed to act primarily by activating the mononuclear phagocyte system. In both tolerant and highly susceptible inbred strains of mice (see later section on serodiagnosis), the parasitemia takes longer to become established and is generally held at lower level. In view of the disinterest shown by macrophages in trypanosomes, this effect is hard to explain. More detailed investigation of the P. acnes effect on experimental T. brucei infections [179] suggests that the ability of treated mice to control the first parasitemic wave is not due to enhanced phagocytosis but to reduction in the number of dividing trypanosomes. Parasite growth-inhibiting molecules can be detected in the serum for 4 days after treatment. The cytokines produced by activated macrophages (IL-1, IL-2, IFN-α, IFN-β, IFN-γ, PGE$_1$, PGE$_2$, PGF$_2$) have no direct effect on trypanosome multiplication over a wide range of concentrations. The growth-inhibiting molecules may therefore be breakdown products of P. acnes itself. An additional possibility is that activated macrophages release killing agents (e.g., nitric oxide) [180] which effect local destruction of trypanosomes in spleen, liver, and bone marrow.

Trypanolytic serum factors and natural antibodies

There are several examples of a potential host's serum nonspecifically killing the trypanosome [181]; the best documented of these is the effect of normal human serum on T. b. brucei. The restriction of host range in this "subspecies" is explained by sensitivity of the parasite to lysis by toxic human high-density lipoprotein [182]. Lytic activity is associated with a minor subclass of very-high-density ($\rho_n = 1.24$ g/ml) lipoprotein of large size (15–21 nm diameter). It contains two unique apolipoproteins, LI (94 kD) and LII (49 kD), in addition to the typical HDL-associated lipoproteins A1 and A2. Monoclonal antibodies against the last two abolish trypanocidal activity [183]. The trypanosome lytic factor exerts its activity through a specific receptor located in the flagellar pocket: the bound factor is targetted to the lysosome at temperatures above 17°C. Lysis is prevented by agents (NH_4Cl, chloroquine) which block the acidification of lysosomes [184].

The blood incubation infectivity test (BIIT) [185], in which trypanosomes are incubated in fresh human blood for 5 h at 37°C and then injected into rats to test for retention of infectivity, has been much used in epidemiologic studies of trypanosomiasis, especially in identifying reservoir hosts [186]. It is possible that the capacity of some game animals to resist infection with human infective *T. b. rhodesiense* is due to innate serum factors. Thus, Mulla & Rickman [187] found trypanolytic activity for this parasite in waterbuck (*Kobus ellipsiprymni*) serum in which no antitrypanosome antibodies could be detected. The factor was active only against certain VATs, paralleling the variation in toxicity of normal human serum to different VATs of *T. b. rhodesiense* [188]. Whether steric hindrance of receptor function by particular VSGs occurs in such cases remains to be tested.

Natural antibodies are of uncertain significance. The reputed ability of cotton rat (*Sigmodon hispidus*) serum to agglutinate and lyse *T. vivax* (reviewed by Terry [189]) has been reinvestigated [190,191] and found to be restricted to one or two stocks only. The "natural antibody" appears to detect a surface-located invariant antigen in these stocks or, less likely, a stable VSG epitope [24].

Defense mechanisms of the vector

The tsetse fly vector has its own natural defense system against trypanosome infection in the form of lectins [192]. A midgut lectin with binding specificity for D(+)-glucosamine is reported to prevent colonization of the fly midgut by *T. brucei* or *T. congolense*. One might suppose that the lack of exposed carbohydrate groups on the surface of bloodstream *T. brucei* would favor retention of the VSG coat in the fly; by replacing the VSG with procyclin, the parasite makes itself more vulnerable. Perhaps VSG would sterically inhibit the novel transport mechanisms necessitated by transformation to the procyclic form. Possession of inherited rickettsia-like symbionts renders certain flies more susceptible to trypanosome midgut infection. The symbionts produce a chitinase which releases D(+)-glucosamine from the peritrophic membrane and the released sugar supposedly inhibits lectin action on the trypanosomes.

Curiously, the same lectin that prevents trypanosome development in the fly midgut is believed to enhance progress of the procyclic towards the metacyclic stage [193]. An agent which inhibits at one stage in the lifecycle and promotes at another in the same host would seem to pose problems for natural selection!

SPECIFIC IMMUNITY

Specific acquired immunity plays a vital role in determining the course and outcome of a trypanosome infection [193,194]. When a mammal is injected with trypanosomes belonging to a particular VAT, a strong and rapid humoral response develops leaving the host protected against that particular VAT but not against others. Antigenic variation enables the trypanosome to avoid this response and protective responses have to be mounted against all succeeding VATs. Infected cattle (and more rarely humans) occasionally undergo "self-cure": the immune system appears to eliminate the parasite completely and the host thereafter resists reinfection with the same serodome [195,196]. In overt clinical trypanosomiasis the outcome is more often a chronic infection leading to death of the host.

In the cyclically transmitted disease in ungulates, serodome-restricted protective immunity can be induced against metacyclic populations, especially if the parasite's ability to multiply is impaired. The evidence to date suggests that control of parasitemia and elimination of the infection is largely due to a series of VAT-specific T cell-independent IgM responses.

Induction of acquired immunity

Probably all VSG-coated stages can induce VAT-specific immunity in the immunocompetent host. Variant surface glycoprotein is not only the most abundant protein in the trypanosome body but it is also the most immunogenic. There is still uncertainty, however, over the manner of induction of acquired immunity.

Black *et al.* [197] postulated that trypanosome transformation from dividing to nondividing forms controls the kinetics of induction of host protective antibody responses, in addition to setting the level of parasitemic waves (see section on trypanotolerance, pp. 198–201). They reported that slender *T. brucei* parasites do not release VSG *in vitro*; only degenerating stumpy forms do this. The same group [198], however, found that antibody responses were induced in mice by exponentially growing *T. vivax*. Certainly, in *in vitro* culture, actively dividing bloodstream forms of *T. brucei*, surface-labeled with I^{125}, shed VSG into their surroundings (at a rate of $2.2 \pm 0.6\%/h$); it appears that the VSG coat is in a state of constant turnover, although much of it is recycled without degradation via the endocytotic pathway [199].

A variety of artificial immunization schedules, using formalin-fixed or living trypanosomes, soluble-released antigens, or purified VSG, have proved successful in protecting laboratory animals against African trypanosomes, but protection is specific only to the particular VAT used as immunogen (see Herbert & Lumsden [200]). African trypanosomiasis, however, is a cyclically transmitted disease of large animals and humans, rather than a syringe-transmitted disease of mice, and most current interest attaches to the induction of immunity against tsetse-transmitted infections of ruminants [193]. Such immunity is invariably serodome specific.

In cattle, establishment of a cyclically transmitted infection followed by drug treatment has been used to induce immunity, as also has inoculation with irradiated trypanosomes which have a limited lifespan in the host. Cattle primed with metacyclic *T. congolense* clones and treated with Berenil 3–4 weeks later are immune to homologous serodeme challenge 3–6 months afterwards [201,202]. Similar results have been obtained for this trypanosome species in mice [203] and rabbits [204], and for *T. brucei* in goats [205]. Culture-derived metacyclics are just as effective as tsetse-derived metacyclics in inducing immunity [202]. In all cases, challenge with a heterologous serodeme leads to reinfection. Protection against challenge coincides with the appearance of neutralizing M-VAT-specific antibodies in the blood of the host.

For *T. vivax* the picture is somewhat different. After priming with tsetse-transmitted clones and Berenil treatment, goats are still susceptible to homologous challenge [206]. It is possible that the small number of *T. vivax* metacyclics injected by the fly and their quick change act into dividing bloodstream trypomastigotes do not result in sensitization of the host to *T. vivax* M-VATs. *Trypanosoma vivax*-infected cattle which have undergone self-cure are refractory to reinfection with the same stock but may lack serum antibodies to M-VATs [196]. In syringe-induced *T. brucei* or *T. congolense* infections which have undergone self-cure, the cattle develop serum antibodies to M-VATs within 2 months of infection and are resistant to fly challenge with the same serodeme [195]. Protective immunity against metacyclic

populations induced by early drug treatment depends upon the trypanosome's inclination to loiter in the chancre (see below) before it invades the host's lymphatics [193]. *Trypanosoma vivax* has no such inclination.

Successful immunization of cattle against irradiated bloodstream trypanosomes (*T. b. rhodesiense*) was first achieved by Wellde *et al.* [207] but unfortunately the parasite stock utilized was uncharacterized with respect to VAT. Morrison *et al.* [208,209] found that a single intravenous inoculum of 10^7 irradiated *T. brucei* of defined clone (ILTat 1.3) protected cattle against challenge with 10^3 trypanosomes 14 days later; fewer trypanosomes in the primary inoculum resulted in delayed onset of parasitemia. Any protection achieved would appear to be VAT specific.

The use of irradiated trypanosomes to confer protection would seem to have an uncertain theoretic basis and an unpromising future. Both drug and ionizing radiation treatment impair trypanosome division and enhance both parasite mortality and the immunization process. By delaying drug treatment to allow for rapid switching from one M-VAT to another following introduction of metacyclic trypanosomes into the skin, the host becomes sensitized to the M-VAT repertoire and is then able to reject challenge infections with the same serodeme. Irradiation of metacyclics might be expected to inhibit M-VAT switching by preventing division.

Purified soluble VSG equivalent to 2.5×10^8 trypanosomes administered on its own does not induce VAT-specific protection in cattle but combined with adjuvants will confer immunity to challenge with 10^4 trypanosomes [208–210]. Invariant membrane antigens, on the other hand, do not usually induce immunity, even though high levels of antibody are induced [211]. But exceptions to this rule may be provided by flagellar pocket membrane antigens [212], possibly those acting as receptors for uptake of vital molecules. This particular approach to vaccination deserves follow-up.

The humoral response

The systemic immune response against Afran trypanosomes is mediated almost entirely by VAT-specific antibodies. The typal primary humoral response of mammals in which initial IgM synthesis is replaced by the synthesis of IgG antibodies is modified in trypanosome-infected hosts in that IgM production is greatly prolonged and enhanced.

Elevated serum IgM is found consistently in human sleeping sickness patients and IgM in spinal fluid is diagnostic of the disease [213]; 5–10% of this IgM is 7S, suggesting failure to assemble the 19S product. IgM levels are also raised in laboratory animals infected with *T. brucei* [214] and in experimental or natural infections of cattle or sheep with this trypanosome [18,215,216], where the increase over normal levels may be ninefold (Fig. 10.2). A similar situation occurs in cattle with *T. congolense* or *T. vivax* infections, with elevated serum IgM appearing about 10 days after infection, remaining high until animals are treated [18], and then falling. Temporary drops in IgM level are associated with remissions, possibly due to parasites mopping up specific IgM. Trypanosome-infected bushbuck (*Tragelaphus scriptus*) also show macroglobulinemia [18].

IgG (both IgG$_1$ and IgG$_2$) levels usually show little change in *T. congolense* or *T. vivax* experimentally infected cattle, although increases up to twofold have been recorded in natural infections of cattle and bushbuck [18,216]. In human sleeping sickness, IgG and IgA levels are in the normal range but IgE levels are raised [217].

IgM is more effective than early IgG at neutralizing, agglutinating, or lysing trypanosome *in vitro* and providing protection *in vivo* [18,218–220]. Variable antigen-type specific antibody is present in both IgM and IgG fractions. The specific humoral response is usually evident within a few days in rabbits and rats [218] and after about 10 days in cattle, where IgG$_1$ of low avidity usually arises simultaneously with IgM [220] or slightly later in *T. vivax* infections [45]. Humoral antibody against metacyclic VAT can be detected in the lymph glands of goats within 2–3 days of introduction of *T. brucei* by a single tsetse fly bite [221].

In cattle infected with *T. congolense* [222], *T. brucei* [223], and *T. vivax* [45], recurrent peaks of antibody activity against the infecting VAT have been found. Whether these responses were induced by recrudescence of the infecting VAT or by a VAT whose VSG bears crossreacting epitopes with the original is unclear. The production of VAT-specific IgG$_1$ of 25 times greater avidity several weeks after the first antibody peak has been interpreted as an anamnestic response to a recrudescence of the same VAT [220].

Owing to its lower molecular weight, IgG may be more effective at diffusing extravascularly. In *T. brucei* infections of rabbits, immunofluorescent antibody titers about one-fifth of those in the serum appear in tissue fluid 2 days after the initial serum rise, and IgG accounts for this activity; with increasing vascular permeability (a feature of trypanosome pathogenesis), IgM is also found [224].

It now seems likely that in humans and ungulates the sustained macroglobulinemia characteristic of African trypanosomiasis can be accounted for by antibodies produced against successive VAT. Although heterophile antibodies have been reported in *T. brucei*-infected monkeys and *T. vivax*-infected cattle [225], careful absorption with a wide range of trypanosome VATs has been found to remove most of the IgM from the serum of sleeping sickness patients or *T. brucei*-infected cattle [226,220], so the argument that much of the elevated IgM is nonspecific, resulting from polyclonal activation of B cells, must now be viewed with scepticism.

The efficacy of the humoral immune response, and of IgM in particular, in controlling trypanosome infections, and the fact that cell-mediated immunity against salivarian trypanosomes is either absent or unimportant have been shown using experimental infections in laboratory mice. Thus, specific protection against *T. brucei* spp. can be transferred with immune serum or with B lymphocytes but not with T lymphocytes. Suppression of the B cell response by treatment from birth with anti-µ serum results in much-reduced survival time. Homozygous athymic mice (nu/nu) which are devoid of mature T cells, and can mount an IgM but not an IgG response, control parasitemia better and survive longer than heterozygotes which mount both responses; moreover, the homozygotes can be immunized against a particular VAT [227,228].

Variable surface glycoprotein, then, may be regarded as a T-independent antigen (TIAg). It may be argued that the close-packing of the VSG molecules on the intact trypanosome surface results in the exposure of repeated antigenic determinants (the N-terminal epitopes) characteristic of TIAg. The relatively poor immune response to soluble VSG may be accounted for by lack of such packing. Membrane-form VSG behaves as a relatively atypical TIAg, however, in inducing a strong response; the primary response to TIAg is usually weak. Perhaps retention of the glycolipid anchor by the membrane-form VSG allows more effective line-up of the antigen on the presenting cell. T-independent antigens are usually presented to B cells by the macrophages of the marginal sinuses of spleen and lymph nodes. Details of VSG antigen presentation are still to be explored, however.

Antibodies of the IgG class are elicited by the invariant antigens of trypanosomes and can be demonstrated by immunofluorescence and complement fixation tests. A fall in complement-fixing antibody titer follows each parasitemic remission in sleeping sickness patients [229] and may be related to the release of invariant antigens at crisis. The temporary excess of such antigens stimulates IgG production. The circulating soluble immune complexes formed when antigen and antibody concentrations are equal may play a part in pathogenesis. As T helper cells are necessary for the IgG response it is interesting to note that mice devoid of T cells do not exhibit the pathologic changes of their immunocompetent counterparts (see section on immunopathology, pp. 194–198).

In cattle, goats, and mice, passive acquisition of maternal antibody via the colostrum or milk can afford temporary protection for young animals from challenge with trypanosomes of homologous serodeme [230–232]. Presumably the young are protected only against those VAT against which the mother has already mounted an immune response.

The cellular response

The chancre

When a tsetse fly injects metacyclic trypanosomes into the dermis of a susceptible host, an intense inflammatory reaction, the chancre, develops at the site of the bite within 7–11 days and then gradually subsides within 1 month of infection. Uninfected flies produce no such reaction. The chancre is characterized histologally by an initial infiltrate of neutrophils and small lymphocytes, and then at peak reaction of larger lymphocytes and

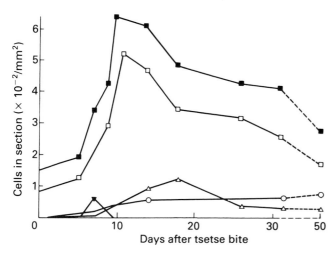

Fig. 10.10 Development of the chancre. Cellular response at the site of tsetse bite infecting a bovine with *Trypanosoma congolense*: (■) total cells; (□) lymphocytes; (▼) polymorphs; (○) macrophages; (△) plasma cells. The contribution of lymphocytes and plasma cells to the infiltrate is obvious. (From Emery *et al.* [205].)

plasma cells (Fig. 10.10) [205]. Numbers of the latter increase as the chancre subsides in the third week; macrophages also increase in number as the chancre regresses. Total cellularity increases in the chancre fourfold to tenfold. Dividing trypanosomes expressing M-VAT are found extensively throughout the chancre, among collagen fibers, and in dermal lymph spaces, from which they later spread via the dermal lymphatics to the circulation. Chancre development and accompanying enlargement of the draining lymph nodes in cattle precede the onset of parasitemia and fever [201].

Chancre size and severity depend upon the number of trypanosomes in residence and on the parasite species [201,233,234]. *Trypanosoma brucei* and *T. congolense* induce chancres up to 10 cm across in cattle, while *T. vivax* in the same hosts elicits smaller, quickly regressing (10–12 days) skin reactions, 2–3 cm across; some strains of this species may elicit no reaction at all [24]. The anomalous behavior of *T. vivax* may be due partly to the extremely small number of metacyclics injected and partly to the fact that metacyclics are less prone to linger in the skin, passing quickly to the lymph nodes draining the site of the bite and causing their enlargement [205].

Chancre formation has been reported in cattle, sheep, goats, game animals, and humans (reviewed by Morrison *et al.* [193]). Detailed accounts of the chancre in humans are sparse [16]. Experimentally, chancres have been studied in rabbits [235]. Mice, rats, and guinea pigs do not develop chancres because in these species the fly injects metacyclics into subcutaneous tissue. A local skin reaction is therefore not a prerequisite for metacyclic-induced infection, but the chancre appears to play an important part in the induction of serodeme-specific protective immunity.

The character of the cellular infiltrate in the chancre suggests that the reaction consists of two components: an initial inflammatory reaction followed by an immune response. Neutrophil infiltration is more prominent with *T. brucei* and *T. congolense* than with *T. vivax*. Morphometric analysis of changes in the mononuclear cell infiltrate in experimental sheep chancres [236] using monoclonal antibodies specific for ovine leukocyte subsets in immunoperoxidase staining shows that T lymphocyte subpopulations (CD5+, CD4+, CD8+) form a major component of the dense infiltrate for up to 15 days. Diffuse distribution of these cells in the dermis suggests passive recruitment but on a selective basis, as SBU-T19 T cells which form 15% peripheral blood leukocytes appear to be much rarer in the infiltrate. High numbers of CD45R+ major histocompatibility complex

(MHC) Class II$^+$ cells are abundant and represent the B cell component. These cells occur in aggregates, however, indicating local B cell proliferation to produce, eventually, antibody-secreting cells. Antibodies which neutralize M-VAT appear in the efferent lymph as early as 2 days after fly bite, and titers rise sharply within a few days [221].

As trypanosome numbers decline in the chancre, trypanolysis and phagocytosis of debris are evident [233], although phagocytosis of intact trypanosomes has not been observed [237]. In efferent lymph, large numbers of lysing trypanosomes can be observed.

In cyclically transmitted infections, protective immunity operates at the level of the skin. Animals that have been previously exposed to infection with metacyclic trypanosomes do not develop a chancre when challenged with metacyclics of the same serodeme [201, 205]. Challenge with metacyclics of a different serodeme can, however, result in a local skin reaction, although this may be delayed or even absent; in addition, the parasitemia may fail to rise and neutralizing antibody production to the challenge trypanosomes may be at a low level or undetectable [208,209,235,238]. These observations suggest operation of an interference effect which may even transcend parasite species boundaries, as observed in challenge of *T. congolense*-sensitized goats with *T. brucei* [239]. Interference does not appear to be dose dependent. Once again, *T. vivax* provides the exception: there is no delay in the establishment of cyclically transmitted superinfections with this species in goats already infected with either *T. congolense* or with a different serodeme of *T. vivax* [240].

The lack of development of the chancre after challenge with the homologous serodeme is presumably because the host has enough circulating antibody against M-VATs to neutralize the challenge dose. The interference phenomenon between serodemes or even species, however, has no obvious explanation. Interference does not occur if the two stocks are inoculated simultaneously and, because it is not dose dependent, antigenic competition seems an unlikely explanation. If immunosuppression was responsible for the reduced immune response, rapid exhaustion of the superinfecting parasite population might be expected. Differential growth rates are unlikely to play a part, as with *T. congolense* interference occurs if the order of administration of the stocks is reversed. Interference between trypanosome stocks can occur in syringe-induced bloodstream infections in mice [240] and this equally lacks a satisfactory explanation.

The systemic response

A gross proliferative response of B lymphocytes, first in the lymphoid system itself and later in the central nervous system, is characteristic of the systemic response to African trypanosomiasis.

Enlargement of lymph nodes and spleen are striking, with the spleen reaching 30 times its normal size in some infected animals. Histologic studies reveal a disproportionate increase in the size of the B cell-dependent follicular areas, and germinal centers are usually well developed. T cell-dependent areas (the paracortical zone of lymph nodes, periarteriolar regions of the spleen), moreover, become depleted of small lymphocytes and heavily infiltrated with plasma cells and macrophages, as do the thymus and red pulp of the spleen. Thymus size is reduced and its cortex and medulla become indistinct. A marked increase in the macrophage content of all lymphoid organs and in the number of blood monocytes and peritoneal macrophages is evident (for references, see Morrison *et al.* [193]).

The lymphoid proliferative phase starts with the first patent parasitemia and lasts for less than 2 weeks in experimental rodent infections but may last for 3–4 months in infected cattle. Changes in the lymph nodes occur later than in the spleen, their onset and severity depending on the species of both trypanosome and host. Other host species differences occur: for example, although germinal centers are prominent in hosts infected with *T. brucei* sspp. and in cattle infected with *T. congolense*, mice infected with the latter species lack germinal centers in spleen and lymph nodes [178].

After the proliferative phase in experimental rodent infections, a more protracted cellular depletion phase ensues, in which small lymphocytes disappear from the periarteriolar lymphatic sheaths (T dependent) in the spleen and then from the follicular areas. Although follicles contain predominantly large lymphocytes and lymphoblasts throughout the infection, they become progressively smaller and less cellular. Depletion of spleen white pulp is accompanied by expansion of the red; in lymph nodes, narrowing of the cortex and reduction in size of the follicles are not so marked as in the spleen. Similar depletion occurs in longstanding cattle infections [178].

Few studies have been conducted on the dynamics of characterized leukocyte populations in the systemic response to trypanosomes. A comparison of peripheral blood leukocyte changes in N'dama (trypanotolerant) and Boran (trypanosusceptible: see section on trypanotolerance, pp. 198–201) cattle following cyclical infection and homologous serodeme challenge with *T. congolense* has now been conducted using monoclonal antibodies to leukocyte subpopulation markers and flow cytometry [241,242]. In primary infections a transient leukopenia up to the first parasitemic peak (16 days postinfection) entailed a decrease in absolute numbers of T cells, B cells, and "null" cells, but not monocytes or granulocytes. In the ensuing lymphoproliferative response (after 22 days), the N'dama cattle showed significantly higher levels of B cells and null cells than Boran, but the CD4$^+$ (BoT4) and CD8$^+$ (BoT8) cell counts decreased in both breeds. On rechallenge, the numbers of circulating B cells increased in both groups after day 21; the CD4$^+$ population did not decline in the N'dama, although it did in the Boran; CD8$^+$ cell populations declined in both breeds. These results support the view that the trypanotolerance of the N'dama is correlated with a more efficient B cell response as compared with the Boran.

The decline of CD8$^+$ cell numbers in infected cattle is interesting in view of the extraordinary findings of Bakhiet *et al.* [243] that the selective depletion of CD8$^+$ cells in *T. brucei*-infected rats abrogated the rapid burst of interferon-α (IFN-α) production which normally follows infection, suppressed parasite growth and increased the survival time of the animals. The authors' conclusion that CD8$^+$ cell products may promote parasite growth is at variance with the potent macrophage-activating properties of IFN-α and the known role of CD8$^+$ cells in B cell differentiation and in antiparasitic activity.

Attempts have been made to relate changes in lymphocyte populations to the host's humoral response to the trypanosome, and to the pathology of trypanosomiasis, especially the profound macrogobulinemia and suppression of T cell-mediated responses (see section on autoimmunity, p. 197). The major question concerning the gross B cell response is whether it results from specific stimulation by a plethora of VSG or from non-specific polyclonal activation by VSG acting as a T-independent antigen. Other T-independent antigens (lipopolysaccharide, polymeric bacterial flagellin) are effective in polyclonal activation of B cells. The *in vitro* mitogenic effect of *T. brucei* and *T. congolense* on lymphocytes is well known [244], and many trypanosome immunobiologists now take it for granted that polyclonal activation occurs *in vivo* [245,246].

Effector mechanisms

The nature of the immune effector mechanisms operating in African trypanosomiasis is far from clear. The role of host antibody, especially IgM, in controlling trypanosome infections is well substantiated, and macrophages, especially those of the liver, are undoubtedly responsible for clearing trypanosomes from the bloodstream. Apart from the difficulties of extrapolating from *in vitro* observations to the *in vivo* situation, the investigator must beware of the temptation to extrapolate from the experimental rodent host to the infected ungulate.

The mechanisms involved in neutralization of trypanosome infectivity are perhaps least understood. Possibly disturbances of the surface coat induced by the binding of host antibody interferes with transport mechanisms (e.g., leucine) [247] or disorders metabolism, leading to the loss of parasite viability. Disruption of endocytosis, the lysosomal system, and the recycling of VSG seems a likely target (see section on trypanolytic serum factors, pp. 186–187).

One of the greatest mysteries surrounding clearance of trypanosomes from the blood is the role of complement in this process. Because an IgM response alone can control parasitemia, the most obvious mechanisms of parasite destruction would seem to be complement-mediated lysis and macrophage uptake of complement-opsonized trypanosomes. Immune lysis occurs readily *in vitro* in the presence of guinea pig serum as a source of complement and can occasionally be observed *in vivo* (Fig. 10.1). Long slender forms of *T. brucei* are lysed within 30 min whereas the short stumpy forms take up to 2 h [248]; the latter also appear to be more resistant to the host's immune response [249]. Similar differential lysis occurs in the BIIT, and as the first sign of trypanolysis is disruption of the endocytotic apparatus it is possible that the membrane attack complexes operate only after endocytosis. Immune lysis of *T. brucei* and *T. congolense* can be blocked by the presence of EDTA and EGTA, suggesting operation of the Ca^{2+}-dependent classic pathway of complement activation [172].

Complement levels are much reduced in the sera of *T. congolense*-infected cattle [176], *T. b. rhodesiense*-infected rhesus monkeys [250] and *T. b. gambiense* patients [251], although they are said to be unaffected in *T. b. brucei*-infected mice [251]. It is possible that antigen–antibody complexes are responsible for mopping up complement in chronically infected hosts. The essential role of complement in trypanosome clearance is disputed for the experimental mouse host. Cobra venom-treated (i.e., C3-reduced) mice, reputedly clear *T. brucei* infections as efficiently as untreated mice [252, 253] although reduced clearance of antibody-sensitized trypanosomes was noted by MacAskill *et al*. [254]. These last authors, however, point out that cobra venom does not eliminate C3 entirely but merely reduces complement levels to 10%, so the conclusion that complement is unnecessary for clearance may not be justified. They found that C5-deficient mice exhibited normal levels of clearance, and so concluded that in cobra venom-treated hosts it is phagocytosis, not immune lysis, that is depressed.

Murine macrophages avidly phagocytose trypanosomes *in vitro* in the presence of VAT-specific antibody. Attachment occurs within 5 min of the addition of antiserum, while ingestion requires 30–90 min. Because IgM is effective in trypanosome clearance, but reputedly does not bind to cells well, it might be expected that complement activated by antibody at the trypanosome surface would release C3b, opsonizing trypanosomes for macrophage uptake via the CR3 receptor. *In vitro*, however, heat decomplementation of serum does not inhibit trypanosome uptake in the presence of VAT-specific antibody [255], suggesting that an Fc receptor might be involved in uptake after all.

Support for participation of an IgM receptor in trypanosome phagocytosis also comes from work on bovine monocyte-derived macrophages [256]. Adherence of IgM-sensitized *T. brucei* occurs in the absence of bovine complement and is Ca^{2+}-dependent, suggesting that binding is not via the C3b receptor but via the Fc regions of the IgM molecule. Antibody (IgM or IgG)-sensitized trypanosomes will not bind to freshly isolated monocytes from noninfected animals, but such monocytes develop receptors for IgM- or IgG_1-sensitized trypanosomes on cultivation *in vitro* for 3 h and 7 days, respectively. Monocytes from trypanosome-infected blood develop the ability to bind IgM- or IgG_1-sensitized trypanosomes after 30 min and 24 h, respectively. Cattle neutrophils may also phagocytose sensitized trypanosomes. The number of neutrophils/mm^3 blood is higher in N'dama than in Boran cattle [257], and this difference may contribute to the trypanotolerance of the former breed (see section on trypanotolerance, pp. 198–201).

In wild bovids, the IgM receptor may be present on circulating peripheral blood leukocytes (PBL), as fresh PBL from wildebeest can bind *T. brucei* sensitized with cattle or wildebeest infection serum or IgM [258]. Circulating PBL (monocytes) from the trypanosome-infected buffalo are often engorged with ingested typanosomes

[259]. The faster and more efficient clearance of trypanosomes from the blood of native African bovids, as compared with domestic cattle, may be due to the former's high levels of neutralizing and phagocytosis-promoting antibody plus the available receptor for IgM on PBL [260].

Vaccines

The demonstration of M-VAT heterogeneity for a given serodeme and the limited number of VATs in the metacyclic repertoire [63,64], coupled with the demonstrable serodeme-specific protection induced in cattle by metacylic inoculation and drug treatment [201,202], have suggested the idea of a local vaccine based on a cocktail of metacyclic antigens. There is evidence, moreover, that livestock maintained under routine drug therapy in particular areas may acquire some resistance to trypanosomiasis [261]. This could be due to the limited number of serodemes circulating in such areas, the cattle becoming antigenically primed before the parasites are killed by the trypanocides [262].

Two major deterrents to the development of such vaccines are evident, however. First, the number of serodemes circulating in some areas may be high; for example, Frame *et al*. [68] found at least seven serodemes among 17 stocks of *T. congolense* isolated from an area in eastern Zambia; pastoral nomadism by shifting animals from one region to another may cause them to encounter new serodemes. Second, the M-VAT repertoire of a given serodeme changes with time as a consequence of the mechanism of VSG gene switching, as discussed above [66] and as a result of hybrid formation [70].

The generally bleak prospect for vaccine development based upon VSG, which are the protective antigens in trypanosome infections, has forced biologists to look to the surface invariant antigens as possible immunogens for immunoprophylaxis. Two categories deserve further attention [263]—the flagellar pocket receptors [212] and procyclic surface molecules [264]. Surface receptors for growth factors (located largely in the flagellar pocket), while representing a potentially attractive target, run the risk that host and parasite may use similar receptor structures and pathways of uptake, so that their use as immunogens could raise problems of tolerance and autoimmunity for the host. Procyclic surface immunogens (e.g., procyclin) would have to be used in a transmission blocking vaccine, i.e., to prevent infection of the vector. Such vaccines would seem to be more appropriate for cattle trypanosomiasis than sleeping sickness, although wild reservoir hosts would clearly pose a problem and use of the vaccine would have to be confined to restricted animal access areas, such as cattle ranches.

SERODIAGNOSIS

The true incidence and prevalence of African trypanosomiasis in both humans and livestock is unknown, because the available diagnostic tests are limited in their application. Traditional microscopic diagnosis from finding trypanosomes in blood or body fluids is unreliable on account of the fluctuating parasitemia and scarcity of parasites in some infections.

Antitrypanosome antibodies are often more readily detectable than the trypanosomes themselves, although such tests cannot distinguish between active and cured infections [265]. Tests for circulating antigens [266–269] certainly detect active infections, although at certain times in the infection antibody levels may result in rapid removal of all such antigen.

The indirect immunofluorescence antibody test (IFAT) has been most widely used in human sleeping sickness surveys. Blood collected on filter papers is eluted and used to detect invariant antigens in acetone-fixed trypanosome smears. A more readily performed test, widely used in surveys of Gambian sleeping sickness, is the card agglutination test for trypanosomiasis (CATT), in which a drop of fresh test blood is mixed on a card with a drop of fixed stained trypanosome suspension; trypanosomes expressing iso-VAT are commonly used. With positive sera, macroscopic agglutination occurs in 2 min [265]. Development of CATT to distinguish Gambian and Rhodesian sleeping sickness is awaited, as chemotherapy and pathogenesis of those forms of the disease differ. If *T. b. gambiense* and *T. b. rhodesiense* are restricted to only a few serodemes, as seems possible (see earlier section on serodemes) the use of M-VATs may permit early detection. Clinical diagnosis of early sleeping sickness is difficult owing to the lack of specificity of symptoms.

Enzyme-linked immunosorbent assay (ELISA) tests for circulating antigens in experimental infections of *T. evansi* and *T. congolense* have been described [270]. More recently monoclonal antibodies have been developed for use in sandwich ELISA tests for soluble antigens of cattle trypanosomes; specificity is sufficiently strict to allow distinction of *T. brucei* sspp. from *T. congolense* and *T. vivax* [269].

IMMUNOPATHOLOGY

There is little evidence that genesis of the diseased state associated with trypanosomiasis is due to toxin production by the parasites, some evidence that it is due to disturbance of the host's metabolism [271], and abundant support for the idea that the damage inflicted has its basis in hyperinflammatory reactions and in disorders of the host's immune response. Much of this support, however, comes from work on laboratory rodents (see Fig. 10.2) and caution must be exercised in extrapolating the conclusions to sleeping sickness patients or to nagana-infected domestic stock.

The outstanding pathologic effects of African trypanosome infections are lymphoid cell proliferation, anemia, circulatory disturbances (associated with increased vascular permeability and coagulopathy with increased fibrinolysis), tissue lesions, and immunosuppression. Anemia is of paramount importance in cattle trypanosomiasis [272]; it may be severe in Rhodesian sleeping sickness but is mild or absent in the Gambian disease [16]. Vascular disturbances [273] and disruption of tissue architecture (more importantly in heart and brain) are common in both human and animal trypanosomiasis [16,274]. Immunosuppression has been explored extensively in laboratory animals and demonstrated in both sleeping sickness patients and trypanosome-infected cattle; its relevance to the frequent incidence of intercurrent viral, bacterial, and parasitic infections in trypanosome-infected hosts and the possible failure of vaccination programs directed against these diseases is obvious [275]. The relation of the massive lymphoid cell proliferation of infected animals to the trypanosome on the one hand, and to the immunosuppression and serum protein changes on the other, has been the subject of extravagant speculation.

Our discussion here will be confined to those aspects of pathology which are relevant to the host's immune response to trypanosome infections. For further details and references to the classic literature the reader is referred to the symposia edited by Losos & Chouinard [276] and Tizard [11]. But immunopathology is currently undergoing a profound revolution as our understanding of cell communation in the immune system expands at an unprecedented rate. The polypeptide hormone-like mediators or cytokines responsible for much of this communation can act to protect the infected host, but if their production becomes imbalanced they can invoke damage or even death. At present our knowledge of cytokine production in African trypanosomiasis is rudimentary compared with our corresponding knowledge for other protozoan infections.

Cytokines and changes in the mononuclear phagocyte and lymphoid systems

The clinal features of African trypanosomiasis can largely be accounted for on the basis of a gross proliferative response of B lymphocytes, first in the lymphoid system itself and later in the central nervous system. The nature of this response has been discussed previously (see section on the cellular response, pp. 190–192). A notable accompaniment of B cell expansion is a decline in T cell numbers with depression of the host's T cell-dependent responses.

The mononuclear phagocyte system is also greatly expanded in trypanosomiasis, and in murine models it is the macrophages that play a pivotal suppressor role in the decline of T cell function [275]. Macrophages alone can offer the ability to inhibit T cell response to mitogen stimulation in vitro and in vivo; moreover, selective elimination of macrophages from spleen cell populations using L-leucine methyl ester, followed by reconstruction with naïve syngeneic macrophages, abrogates suppression to mitogen stimulation of spleen cells from infected animals in vitro [277].

What instigates suppressive activity on the part of macrophages is unknown. Specific antibody is not necessary [277], although it is for phagocytosis of coated trypanosomes (see section on phagocytosis, p. 186) and phagocytosis is reputedly necessary for the induction of suppressive activity [278]. Macrophages of T. brucei-infected mice appear to have reduced expression of mannose, Fc, and complement receptors but are in an activated state, i.e., have increased ability to secrete killing agents, such as superoxide anions and hydrogen peroxide [279]. It is possible that trypanosomes phagocytosed after destruction by complement-mediated lysis trigger macrophage suppressor activity, although the participation of soluble factors or another cell type cannot be ruled out.

Possible reasons for the failure of T cell stimulation include faulty processing of T-dependent antigens with inadequate presentation to T helper cells through MHC Class II molecules (Ia antigens) and changed secretion of macrophage-derived immunoregulatory cytokines. Class II molecule expression by macrophages has been reported to be both increased [280] and decreased [279] in T. brucei infections. The production of interleukin 1 (IL-1) by such macrophages is markedly increased, with

the increase due to release rather than synthesis [281]. In lymph node cells, however, interleukin 2 (IL-2) production is suppressed, as is the expression of IL-2 receptors [282]. Indomethacin treatment restores IL-2 synthesis, suggesting that macrophage-derived prostaglandins are responsible for inhibition, but does not restore IL-2 receptor synthesis in either CD4$^+$ or CD8$^+$ cells, suggesting that inhibition here entails a prostaglandin-independent pathway [283]. Inhibition of prostaglandin synthesis in macrophages also increases the output of IL-1 [284], so increased IL-1 output may inhibit IL-2 production and hence the proliferation of T helper cells.

Release of interferons is stimulated early in T. brucei infection [285], possibly from CD8$^+$ cells [243], and this may provoke activation of macrophages. Another macrophage-activating cytokine, but like IL-1 produced by macrophages themselves, is tumor necrosis factor (TNF). Overproduction of TNF has been held responsible for the wasting (cachexia) and the hypertrigliceridemia observed in trypanosomiasis [286]. Tumor necrosis factor causes endothelial cells to lose their anticoagulant properties, enhancing fibrin deposition. This may lead to the diffuse intravascular coagulation characteristic of trypanosomiasis [273]. By promoting adhesion and degranulation of polymorphs, TNF may be a major mediator in circulatory collapse in trypanosomiasis, along with the autocoids discussed below.

Immune complex formation and autocoid release

Immune complex formation and the consequent release of pharmacologly active substances (amines, peptides, lipids) called "autocoids" have been attributed a major role in the pathology of trypanosomiasis by Goodwin [224] and Boreham [287]. Such complexes could be formed during the repeated destruction of trypanosome populations by host antibody, and by the shedding of antigens from living trypanosomes into antibody-laden surroundings. The binding of invariant antigens to corresponding IgG antibodies has been regarded as particularly important; as T cell help is required for the production of such antibodies, it is interesting to note that T cell-deficient (nu/nu) mice can exhibit a relapsing parasitemia with a T. brucei infection and survive longer than their T-immunocompetent infected controls [228,288].

Circulating immune complexes have been detected in the blood and cerebrospinal fluid of sleeping sickness patients [289] and in the blood of various experimentally infected animals (see Boreham [287] for references).

Deposits have been reported in the tissues of monkeys [250], rabbits [290], and mice [291]. In *T. brucei*-infected rabbits these complexes absorb and activate Hageman factor (Factor VII), which in turn activates the kallikrein–kinin enzyme cascade, resulting in the formation of proteolytic enzymes and pharmacologically active peptides. Histamine and 5-hydroxytryptamine are released from platelets by immune complexes. A combination of blood vessel leakiness induced by kinins and the decrease in serum albumin level characteristic of trypanosomiasis produces a lower osmotic pressure within the circulation and results in edema. Other circulatory disturbances, such as stasis and microthrombus formation, may turn upon the increase in blood viscosity arising from elevated globulin and fibrinogen levels [292].

In addition to the release of autocoids, immune complexes probably lead to plasminogen activation associated with diffuse intravascular coagulation [273], and are undoubtedly implicated in the widespread complement activation leading to hypocomplementemia in trypanosomiasis. Activation of complement by immune complexes on the surface of various cell types is widely believed to result in lysis or opsonization of these cells, and red cell destruction by such means is believed to be important in the generation of anemia (see below).

Disruption of tissue architecture as a result of inflammatory reactions is particularly marked in *T. brucei* infections, especially in the myocardium, skeletal muscle, and brain [293]. In part the edema itself may be disruptive, for example in separating blood vessels of the choroid plexus from the overlying ependyma and so impairing the secretion of CSF, or Purkinje from myocardial fibers and interfering with the conducting system of the heart, but much tissue disruption takes place as a result of cellular infiltration [294].

Lymphocytes, plasma cells, macrophages, and, in some species, polymorphs invade the tissues attracted by chemotactic factors which result from complement activation by antigen–antibody complexes formed or deposited there. In experimentally infected mice, Ig and trypanosome antigen deposits can be detected by immunofluorescence reactions. Such lesions do not develop in sublethally irradiated or T cell-deficient mice unless syngeneic T cells are administered [287]; these observations again suggest that T cell-dependent antibodies or cytotoxic T cells are necessary for the infliction of damage and rule out direct damage by trypanosomes. The swarming of scavenging macrophages in pursuit of antigen–antibody complexes can be shown strikingly in regenerative ear chambers of infected rabbits [224]. Classic Arthus-type reactions are confined to *T. brucei*-infected animals where scavenging polymorphs releasing free-radicals during phagocytosis feature strongly (e.g., the dog). In cattle trypanosomiasis it is possible that the pronounced anemia is indirectly responsible for tissue lesions resulting from anoxia and, ultimately, degeneration.

Anemia and the transfer of VSG to other cells

In cattle infected with Afran trypanosomes, anemia as shown by packed cell volume is the overriding manifestation of damage to the host. Two phases in its development can be recognized [272]. In Phase I, increased red cell breakdown is associated with the development of parasitemia and is most rapid in the following 2–4 weeks. The rate of development and severity of anemia reflects the intensity and duration of parasitemia, and after drug treatment hemopoietic values return to normal. In Phase II, dyshemopoiesis sets in, the role of the trypanosome in the maintenance of anemia is less important, and drug treatment has little effect on the subsequent packed cell volume.

It is possible that Phase I has a multifactoral basis. Phagocytosis of red cells in this phase has been observed by several workers [272]. The expanded macrophage system, especially of spleen and liver, greatly facilitates red cell destruction. Evidence for intravascular hemolysis *in vivo* is lacking except in a few experimentally infected rodents. Although enzymes liberated by trypanosomes, and especially neuraminidase [295], have been shown to change the red cell surface, rendering it more prone to macrophage attack, immune reactions in which red cells become sensitized with host antibody and complement must take some of the blame. Immunoglobulin (both IgM and IgG) eluted from red cells of *T. vivax*-infected cattle reacts with soluble trypanosome antigen [296]. Two possible mechanisms of red cell sensitization have been postulated. First, erythrocytes could become coated with soluble trypanosome antigen and then opsonized by host antibody and complement; second, trypanosome antigen–antibody complexes could become passively attached to the erythrocyte membrane, inviting phagocytosis.

A more satisfactory version of the first explanation has emerged from the finding that the glycolipid membrane anchor of the trypanosome VSG antigen can migrate from the phospholipid bilayer of the parasite's surface to the membrane of the red blood cell [297]. Red blood cells carrying such artificially inserted VSG are susceptible to lysis by complement in the presence of VSG-specific antibody. The extent of erythrocyte lysis depends on the time of cell exposure to trypanosomes and upon trypanosomes concentration. No lysis is observed when trypanosomes are preincubated with VAT-specific antibody before adding red blood cells. Only purified membrane from VSG (which retains the glycolipid anchor) can sensitize cells to VAT-specific antibody and complement-mediated lysis; purified soluble VSG (which is deprived of the terminal diacyl glyceryl moiety) is ineffective [297].

These observations tie in well with the lack of anemia in hosts which control the first parasitemic wave; the sooner the host produces antibody, the more rapidly the rising parasitemia is quelled. Antibody and reduced parasite density prevent transfer of VSG from trypanosome to erythrocyte membrane. With each successive wave of antigenically distinct parasites, more and more red cells are destroyed, resulting in anemia. The degree of anemia in acute infections can be correlated with the number of circulating parasites. The rate, direction, and extent of intermembrane transfer of glycolipid anchors depend on the relative lipid composition and fluidity of donor and target cell membranes [298]. Not all host cells, then, would be expected to acquire trypanosome antigens, but the differential uptake by certain cells in the different sites invaded by the various salivarian trypanosome species may be sufficient to account for species differences in host pathology.

Autoimmunity

Although autoimmune responses are believed to play a major role in the pathology of chronic South American trypanosomiasis [299], there is no compelling evidence to support the suggestion of Seed & Gam [300] that autoimmunity underlies much of the damage to the host in the African disease. Antibodies against a wide variety of tissues and cell components (erythrocytes, liver, heart, brain, kidney, thymus cells intermediate filaments, DNA, and RNA) have been reported from both experimental and natural hosts of African trypanosomes (see Kobayakawa et al. [301], Mansfield & Kreier [302], and Anthoons et al. [303]). These antibodies have been popularly supposed to be the result of polyclonal B cell activation, rather than breakdown of tolerance of self. They appear before detectable tissue pathology, however, and as the host's health is rapidly restored after drug treatment of the infection it is unlikely that the antibodies lie at the root of tissue damage. The anti-intermediate filament antibodies produced by T. brucei-infected rats react with smooth muscle cells of blood vessel walls in indirect immunofluorescence tests, but as these antibodies would not normally penetrate the cell membrane, it seems improbable that they would elicit the vascular damage so characteristic of trypanosome infections [303]. The similar dynamics of the anti-trypanosome IgM antibody and the anti-intermediate filament autoantibody strongly suggests the presence of shared epitopes between trypanosome and host intermediate filaments, but not the production of autoantibodies as a reaction to tissue injury by the parasite. It is not inconceivable that the vast variety of VSG epitopes (including those not exposed on the living trypanosome) should include some that crossreact with host tissue components or heterologous host cells for that matter.

Immunosuppression

The immunosuppression described above begins to develop early in infection in laboratory rodents and the rapidity of its induction by virulent trypanosome stocks argues against host lymphocyte clonal exhaustion as a major mechanism in its operation. As indicated above, it appears that T cells become refractory to normal control signals and subject to defects in antigen presentation. IgM responses, including those to VSG, are not affected, so parasitemic control continues in chronic infections. A striking feature of immunosuppression is the rapid return to normal responsiveness of the host after chemotherapy of the infection; this observation suggests a direct role for the trypanosome in the maintenance of suppression. The suppressive activity of passively transferred macrophages lasts no more than 1 week [277]. Experimental supporting evidence for immunosuppression in laboratory animals is reviewed by Bancroft & Askonas [304].

An important question is whether such immunosuppression occurs in the natural host. In cattle the level of immunosuppression does not appear to be as extensive as in mice [193]. It seems probable that the relatively high parasite load borne by the infected mouse as compared with the cow is a significant factor. Cattle maintain scarcely affected levels of IgG in trypanosome infections (Fig. 10.2), although profound suppression of antibody response to Brucella abortus has been reported for cattle infected with T. congolense [175]. The increased catabolism of immunoglobulins observed in T. congolense-infected calves [174], suggests a possible explanation for reduced antibody levels in such cases. Nevertheless, even a slight reduction in the immune capability of cattle faced with poor feed and intercurrent infections may have serious consequences in the field [194]. The drop in levels of total hemolytic complement and C3 observed during spells of enhanced parasitemia with *T. congolense* and *T. vivax* infections in cattle may also encourage secondary infections.

Patients with Gambian sleeping sickness have "a modest general impairment of humoral and cellular immunity," as indicated by reduced response to typhoid vaccine and to skin test antigens [305]. These changes could be more marked in the more acute Rhodesian disease, where parasitemias are higher. It is possible that immunosuppression in cattle trypanosomiasis is more pronounced in late infections, as all experimental work has been conducted on infections of less than 6 weeks' duration. The underlying mechanisms of trypanosome-mediated immunosuppression in cattle and humans are not known to resemble those in mice.

Changes in the nervous system: advanced sleeping sickness

In the late stage of *T. brucei* infections the parasites penetrate into the brain via the choroid plexuses and other areas where the blood–brain barrier is incomplete. Infection of the brain and successful chemotherapy demand drugs that cross the blood–brain barrier; toxic organic arsenicals have been traditionally used for this purpose in human sleeping sickness [229].

Advanced sleeping sickness is characterized clinically by continuous daytime sleep, and pathologically by infiltration of the pia-arachnoid and perivascular Virchow–Robin spaces with lymphocytes (especially B cells), macrophages, plasma cells, and morula (Mott) cells [305,306]. B cells are present in the cerebrospinal fluid. Surprisingly there is general absence of nerve cell damage in the brain parenchyma itself (Plate 10.2 (facing p. 179)). Pentreath [307] has suggested that this is because astrocytes limit immune reactions to the perivascular "cuffs." Astrocytes are antigen-presenting immunoregulatory cells [308,309]. On activation (Plate 10.2) they synthesize MHC Class I molecules and release immune response modifiers [310], some of which are known to be somnogenic [311]. Thus Pentreath *et al.* [312] found prostaglandin D_2 (PGD_2) levels, as measured by radioimmune assay, to be strikingly elevated in late-stage (as compared with early-stage) Gambian sleeping sickness patients, although levels of IL-1 and PGE_2 (both believed to have a key role in the control of normal

sleep) [311] did not differ significantly between the two groups. PGD_2 receptors are concentrated in the preoptic area of the brain, which is the centre of sleep regulation [313]. The picture of advanced sleeping sickness which is emerging has thus more to do with wholesale imbalance in immune response modifiers throughout the neuraxis as a result of parasite–astrocyte–lymphocyte interactions than with specific or localized lesions in a particular part of the brain.

TRYPANOTOLERANCE

Trypanotolerance is the term used to describe animals which, despite infection with trypanosomes, show few clinical signs of disease. "Reduced susceptibility" is perhaps a more accurate description of this property, which is widely believed to be an attribute of the game animals that evolved in Africa but not of immigrant host species. Trypanotolerant hosts are of great importance epidemiologically. Trypanosomes are present in their blood in sufficient numbers to enable efficient transmission by the vector to susceptible hosts, and for prolonged periods, as the trypanotolerant hosts do not succumb. Certain breeds of cattle exhibit trypanotolerance as a genetic trait and so may be expected to figure increasingly in livestock development projects in tsetse-infested areas.

Game animals

Early experimental work on trypanotolerance was surveyed by Ashcroft *et al.* [314]. More recently, trypanotolerance has been demonstrated experimentally in wild Bovidae under experimental conditions [315] and the immune effector mechanisms of the wildebeest (*Connochaetes taurinus*), investigated in some detail [258] (see section on specific immunity, pp. 187–194). Some native African mammals (e.g., baboon), are totally refractory to infection (see section on innate immunity, pp. 185–187) while others are infectible and may maintain patent parasitemias for considerable periods (eland, duiker, bohor reedbuck, oribi, bushbuck, impala, and spotted hyena). Wild pigs support only scanty infections. The basis of trypanotolerance in game animals varies with the species and may involve either innate or specific immune mechanisms.

Domestic stock

Trypanotolerance has been investigated in more detail for domestic stock than for game animals. It is now

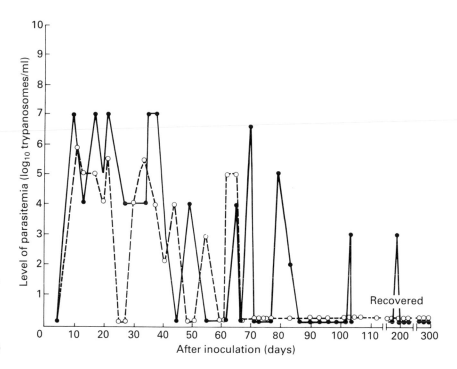

Fig. 10.11 Parasitemia profiles in an individual 4-year-old N'dama (○) and a 4-year-old zebu (●) inoculated with *Trypanosoma congolense*. The level of parasitemia is lower in the N'dama, as is the duration of the parasitemia. Both animals were aparasitemic for several months before the experiment was terminated, and both made a clinical recovery. (Data from Murray *et al.* [321].)

emerging as a genetic trait possessed by certain *Bos taurus* breeds of West and Central Africa, notably the N'dama longhorn and the West African shorthorn (muturu); *Bos indicus* (zebu, boran, brahmin) cattle do not exhibit this trait [1,316,317]. The latter have had a shorter period of residence in Africa (a millenium and a half as compared with 5–7 thousand years for the longhorn ancestors of the N'dama). Cattle of the trypanotolerant breeds are small in size (mean body weight 230 kg) but can be just as productive of meat and milk as the trypanosusceptible zebu. As yet, however, they have been little exploited. West African Guinea sheep and goats are also trypanotolerant [318]. Key factors in the trypanotolerant trait are the ability to control parasitemia and to resist anemia. In trypanotolerant breeds the intensity (peak height), prevalence, and duration of parasitemia are less pronounced than in the susceptible breeds (Fig. 10.11) [1,316]. Intensity of parasitemia has a pronounced effect on the subsequent anemia (see section on immune complex formation and autocoid release, pp. 195–196).

Until recently much of the experimental work on trypanotolerance has been conducted on cattle living in endemic areas and under suboptimally controlled conditions. Such cattle may have had previous experience of trypanosome infections or as calves received immune colostrum from their mothers. Paling *et al.*

[319], however, have shown that N'dama calves born of Boran surrogate mothers, although just as susceptible to cyclically transmitted *T. congolense* infections as Boran calves of identical upbringing, recover from the infection spontaneously, suffer less from anemia, and gain weight at the same rate as uninfected N'dama controls; moreover, infected females superovulate as normal. The ability to resist anemia, as shown by maintenance of packed red cell volume (PCV), has been found to be correlated with the capacity to be productive; indeed, PCV values represent the best indicator of trypanotolerance to date. The accelerated hemopoietic response of the N'dama in the face of trypanosome challenge complements the control of parasitemia in the trypanotolerant animal in resisting anemia [320]. The ability to control parasitemia and the ability to resist anemia may therefore be separate genetic traits.

Marker genes for trypanotolerance have not yet been identified but the MHC phenotype ECA121 and two common leukocyte antigens (IL-A37, IL-A39) appear to be associated with the maintenance of PCV values and animal performance under trypanosome challenge [321].

Trypanotolerant cattle resemble wild Bovidae in being more resistant to environmental constraints, including tickborne and helminthic diseases. They are better adapted to tropical conditions in terms of food utilization, heat tolerance, and water conservation than other

Bos taurus breeds [316]. Nevertheless, the stability of the trypanotolerant trait can be affected by overwork, intercurrent disease, repeated bleeding, pregnancy, and lactation. The single most important environmental constraint, however, is nutritional status. Groundnut meal food supplement can improve the PCV of infected N'dama cattle, and hence their productivity. After trypanocidal drug treatment N'dama recover their PCV in up to 1 month, whereas Boran cattle take much longer [1]. Considerable faith has been expressed by veterinarians in the future of trypanosomiasis control based on rational breeding programs exploiting the trypanotolerant trait.

Control of parasitemia

The precocity and amplitude of the protective antibody response play a crucial role in resistance to trypanosomiasis [318]. N'dama cattle have a greater ability to control the initial wave of parasitemia than zebu cattle (Fig. 10.11) [322] and exhibit a higher circulating B cell count during *T. congolense* infection [241]. Although it now seems certain that parasitemia control has a genetic basis, we are still uncertain as to whether this control is entirely the responsibility of the host's immune response. The expense and difficulty of conducting experimental studies on cattle have resulted once again in recourse to the laboratory mouse to shed light on this problem.

Different inbred strains of mice vary in their ability to handle homogeneous VAT infections with African trypanosomes (reviewed by Black *et al.* [197]). Susceptible mice, e.g., C3H/He, succumb to an initial parasitemia which does not go into remission. Resistant mouse strains, e.g., C57BL/6, control the first wave of parasites and develop a relapsing parasitemia as the trypanosomes undergo antigenic variation. Genetic analysis shows that the ability to cause remission is not linked to the MHC [323] and may be under the control of several genes [324]. Studies with chimeras constructed between H-2-compatible resistant and susceptible strains suggest that the remission-inducing ability is a property of B cells residing primarily in the spleen [325,326].

Susceptible mice produce little or no serologically detectable VAT-specific antibody during infection and show negligible change in overall serum Ig levels [327–329]. Yet both susceptible and resistant mouse strains mount similar antibody responses against lethally irradiated trypanosomes [328] and susceptible mice

show a rapid rise in VSG-specific antibody following trypanocidal drug treatment of a virulent trypanosome infection [330,331]. Plasma cells of susceptible mice do not suffer an impaired ability to secrete immunoglobulin; there are no structural differences between the plasma cells from infected mice belonging to the two strains [331]. The apparently negligible VSG-specific antibody production in susceptible mice may be due to parasites mopping up all that is produced so that the critical level for parasite clearance is not reached [331].

Black *et al.* [330,332] propose that the 2.5–10 times higher parasitemia level in C3H/He mice is due to failure of trypanosomes to differentiate to the nondividing stumpy form in this host strain. If the height of the first wave of parasitemia in C3H/He mice is controlled by the trypanostatic drug difluoromethyl ornithine (DFMO) (which supposedly promotes differentiation to stumpy forms), terminal development of antigen-induced plasma cell responses occurs [333]. The most significant difference between susceptible C3H/He and resistant C57BL/6 mice with respect to control of the first parasitemic wave according to Black's group lies in the nonimmunologic factors that regulate parasite differentiation [332].

The relationship between the trypanosome cell cycle, the host's immune response, and control of parasitemia is not entirely clear. In pleomorphic *T. brucei*, slender trypamastigotes divide and dominate the ascending parasitemia. They give rise to nondividing (G_0 phase) intermediate, and stumpy forms which are more abundant during parasitemic remission.

Neither VAT switching (which occurs only in dividing trypanosomes) nor differentiation to the stumpy form (which occurs only in nondividing trypanosomes) is induced by host antibody [249]. It is possible, then, that secular factors may regulate parasitemia. The change in relative abundance of slender and stumpy forms is due to reduction in the absolute abundance of slender forms [249,328]. Stumpy forms appear to be more resistant to destruction by host immune effector systems than slender forms (see section on effector mechanisms, pp. 192–193). Black *et al.* [197] ascribe the cessation of slender form division to depletion of a host-derived growth factor. They propose that exhaustion of this factor in the blood curbs parasite multiplication and induces transformation to the stumpy form.

The superimposition of a second infection of different VAT onto a primary infection at a time when stumpy forms predominate in the blood does not prevent multiplication of the challenge parasites as slender forms, however [334]. This finding suggests that transformation to the stumpy form is dependent upon neither depletion of a critical growth nutrient nor accumulation of an exogenous growth inhibitor, although Seed & Sechelski [335] found that plasma from infected rats inhibited thymidine incorporation into trypanosomes. Different VAT vary in the speed with which they transform to stumpy forms and certain agents, e.g., indomethacin [336] and the trypanostatic drug DFMO [337] can accelerate the change, suggesting that this part of the developmental cycle is not rigidly programmed. Curtailing the initial parasitemia seems to be vitally important in the establishment of a trypanotolerant relationship. Control of the cell cycle in bloodstream trypanosomes remains a problem urgently in need of further attention if we are to gain a better understanding of parasitemia control.

CONCLUSIONS

1 The immunology of the African trypanosomes is dominated by the presence on these parasites of a monomolecular glycoprotein surface coat which serves as a T-independent antigen in provoking the host's protective antibody (IgM) response. The trypanosomes can evade this response by replacing the single glycoprotein (VSG) with another of differing antigenic specificity, and so changing their VAT. At present we know more about the trypanosome as antigenic challenge than we do about the response of the natural host (human or ungulate) to infection. A great deal of our knowledge of immunity to trypanosomiasis is based upon experimental infections in laboratory animals.

2 The surface coat glycoprotein is attached to the plasma membrane by a glycolipid anchor. The exposed N-terminal portion of the glycoprotein contains the VSG-specific epitope(s). The parasite changes its VAT as a result of a spontaneous gene-switching process. Successive changes of VAT on the part of a subfraction of the trypanosome population before the host has produced antibody specific to the major VAT enables the trypanosome population to survive as a chronic infection which extends opportunities for transmission.

3 The VAT repertoire of a trypanosome clone depends upon 100–1000 nuclear VSG genes. Within each trypanosome species or subspecies several repertoires or serodemes occur; new repertoires may be formed by genetic recombination in the vector. All VSG genes may be expressed in the mammal. On entering the tsetse fly vector, the trypanosomes replace their various VSG coats with a common immunodominant protein, procyclin. Upon differentiation to the mammal-infective metacyclic stage in the fly, the VSG coat is reacquired and a limited number of M-VAT, characteristic of the serodeme, are injected into the skin of the new host.

4 Variant surface glycoprotein genes are either telomeric or intrachromosomal but can be transcribed only from certain (5–20) telomeric expression sites. One mechanism of novel VSG gene expression involves switching transcription from one telomeric expression site to another; the other mechanism entails genomic rearrangements that place unexpressed VSG gene sequences in an expression site. There are two types of genomic rearrangement: in the first (gene conversion), a VSG gene in an active expression site is replaced by a copy of another VSG gene (either telomeric or intrachromosomal); the second entails reciprocal exchange between telomeric expression sites and is rare.

5 Trypanosomes have large numbers of chromosomes (over 100 in T. brucei). Metacyclic variable antigen type VSG genes are located on approximately 10 (diploid) large (0.7–6 Mb) chromosome telomeres; expression sites are on large and intermediate (150–700 kb) size chromosomes, and the mini chromosomes (50–150 kb), whose numbers vary with the trypanosome species, contain only telomeric VSG genes. Intermediate and mini chromosomes are believed to be present as single copies only in the nucleus.

6 After the M-VAT, other VATs dominate the bloodstream population in a hierarchic order. This order is not linear, because VAT switching is divergent. The order may be dictated partly by the location of specific VSG genes in the genome and partly by other factors, especially differential switching rates between VATs. Because actual rates of interswitching between VATs may vary by several orders of magnitude (10^{-2}–10^{-7} switches/cell/generation), the net rates of accumulation of different VATs to a point where they pass the threshold of immunostimulation may also vary. Paced stimulation of the host's immune response and serial sacrifice of successive VAT subpopulations are important features of trypanosome antigenic variation.

7 Protective immunity to African trypanosomes is VAT specific. A satisfactory way of surmounting antigenic variation and vaccinating susceptible hosts has yet to be devised. Cocktail vaccines based on M-VAT of local serodemes represent one possibility, the number of M-VATs being low (12–30), although the number of serodemes may be high. Protective immunity has been shown to operate against M-VATs at the level of the

skin. Chemotherapy after multiplication of M-VAT in the chancre which forms at the site of fly bite has been used to induce immunity to challenge with the same serodeme. Change of the M-VAT repertoire with repeated cyclic transmission detracts from this idea. Other possibilities for vaccination include immune blockage of vital invariant trypanosome receptors which traverse the surface coat, and a vector transmission-blocking vaccine based on immunodominant surface proteins of the fly midgut (procyclic) stage of the trypanosome.

8 Immune effector mechanisms operating in African trypanosomiasis are still unclear, especially the role of complement in these processes. Phagocytosis and possibly lysis probably serve as clearance mechanisms. An IgM receptor on phagocytes (at least in ruminants) may account for the ability of an IgM response alone to control parasitemia. The VSG coat protects the parasite from complement-mediated lysis and phagocytosis in the nonimmune host. Serum factors may provide innate protection for some hosts against particular trypanosomes, e.g., a high-density lipoprotein protects humans from infection with *T. b. brucei*. The parasite's protein sorting and lysosomal system seems a likely target for disruption by such mechanisms.

9 Much of the pathology of African trypanosomiasis results from excessive inflammatory and immune responses to the parasite. Work with rodent models suggests polyclonal stimulation of B cells to account for the macroglobulinemia and production of heterophile and autoantibodies, but the IgM of human and cattle infections can be absorbed with a mixture of trypanosome VSG, suggesting that it is VAT-specific. Depression of T cell responses appears to be associated with macrophage defects; although IL-1 and TNF secretion are enhanced, production of IL-2 and IL-2 receptors by T cells is suppressed. Astrocytes may act as immunoregulatory cells in the brain and be stimulated to release somnogenic cytokines in human sleeping sickness. Immune complexes, especially those formed from invariant antigens, have been attributed a role in triggering production of pharmacologically active autocoids and, possibly through attachment to red cells, of promoting the hemolytic anemia characteristic of the cattle disease. Transfer of VSG along with its glycolipid anchor to the membrane of host cells may also signal their destruction.

10 The trypanotolerance of African game animals may be dependent upon either innate or acquired immune mechanisms. The trypanotolerance of certain cattle breeds is now emerging as a genetic trait depending on control of initial parasitemia and an accelerated hemopoietic response to compensate for induced destruction of red cells. Control of parasitemia is dependent primarily upon a superior acquired immune response, although other factors (e.g., induced transformation of trypanosomes from dividing to non-dividing forms) may play a part. The selection of trypanotolerant cattle is seen as a more realistic solution to the African trypanosomiasis problem than attempting to produce a vaccine for domestic stock.

REFERENCES

1 Murray M, Trail JCM, D'Ieteren GDM. Trypanotolerance in cattle and prospects for control of trypanosomiasis by selective breeding. Rev Sci Tech Off Int Epiz 1990;9:369–386.

2 Cohen S, Sadun EH, eds. *Immunology of Parasitic Infections*, 1st edn. Oxford: Blackwell Scientific Publications, 1976.

3 Cohen S, Warren KS, eds. *Immunology of Parasitic Infections*, 2nd edn. Oxford: Blackwell Scientific Publications, 1982.

4 Vickerman K, Barry JD. African trypanosomiasis. In Cohen S, Warren KS, eds. *Immunology of Parasitic Infections*, 2nd edn. Oxford: Blackwell Scientific Publications, 1982:204–260.

5 Borst P, Cross GMA. Molecular basis of antigenic variation. Cell 1982;28:291–303.

6 Boothroyd JC. Antigenic variation in African trypanosomes. Ann Rev Microbiol 1985;39:475–502.

7 Borst P. Discontinuous transcription and antigenic variation in trypanosomes. Ann Rev Biochem 1986;55:701–732.

8 Clayton CE. Molecular biology of the Kinetoplastidae. In Rigby PWJ, ed. *Genetic Engineering*, vol. 7. London: Academic Press, 1988:2–56.

9 Pays E, Steinert M. Control of antigen gene expression in African trypanosomes. Ann Rev Genet 1988;22:107–126.

10 Cross GAM. Cellular and genetic aspects of antigenic variation in trypanosomes. Ann Rev Immunol 1990;8:83–110.

11 Tizard I, ed. *Immunology and Pathogenesis of Trypanosomiasis*. Boca Raton: CRC Press, 1985.

12 Hoare CA. *The Trypanosomes of Mammals*. Oxford: Blackwell Scientific Publications, 1972.

13 Vickerman K. Developmental cycles and biology of pathogenic trypanosomes. Br Med Bull 1985;41:105–114.

14 Roberts CJ, Gray MA, Gray AR. Local skin reactions in cattle at the site of infection with *Trypanosoma congolense* by *Glossina morsitans* and *G. tachinoides*. Trans R Soc Trop Med Hyg 1969;63:620–624.

15 Jennings FW, Whitelaw DD, Holmes PH, Chizyuka HGB, Urquhart GM. The brain as a source of relapsing *Trypanosoma brucei* infection in mice after chemotherapy. Int J Parasitol 1979;9:381–384.

16 Poltera AA. Pathology of human African trypanosomiasis with reference to experimental African trypanosomiasis and infections of the central nervous system. Br Med Bull 1985;41:169–174.

17 Whitelaw DD, Gardiner PR, Murray M. Extravascular foci of *Trypanosoma vivax* in goats: the central nervous system and aqueous humour of the eye as potential sources of relapse infections. Parasitology 1988;97:51–61.

18 Luckins AG. The immune response of zebu cattle to infection with *Trypanosoma congolense* and *T. vivax*. Ann Trop Med Parasitol 1976;70:133–145.

19 Tait A, Turner CMR. Genetic exchange in *Trypanosoma brucei*. Parasitol Today 1990;6:70–75.

20 Turner CMR. The use of experimental artifacts in African trypanosome research. Parasitol Today 1990;6:14–17.

21 Lumsden WHR. Principles of viable preservation of parasitic protozoa. Int J Parasitol 1972;2:327–332.

22 Brun R, Jenni L. Salivarian trypanosomes: bloodstream forms (trypomastigotes). In Taylor AER, Baker JR, eds. *In vitro Methods for Parasite Cultivation*. London: Academic Press, 1987:94–117.

23 Gray MA, Hirumi H, Gardiner PR. Salivarian trypanosomes: insect forms. In Taylor AER, Baker JR, eds. *In vitro Methods for Parasite Cultivation*. London: Academic Press, 1987:118–152.

24 Gardiner PR. Recent studies of the biology of *Trypanosoma vivax*. Adv Parasitol 1989;28:229–317.

25 Hirumi H, Doyle JJ, Hirumi K. African trypanosomes: cultivation of animal infective *Trypanosoma brucei in vitro*. Science 1977;196:992–994.

26 Baltz T, Baltz D, Giroud C, Crockett J. Cultivation in a semi-defined medium of animal-infective forms of *Trypanosoma brucei*, *T. equiperdum*, *T. evansi*, *T. rhodesiense* and *T. gambiense*. EMBO J 1985;4:1273–1277.

27 Duszenko M, Feguson MAJ, Lamont GS, Rifkin MR, Cross GAM. Cysteine eliminates the feeder cell requirement for cultivation of *Trypanosoma brucei* bloodstream forms *in vitro*. J Exp Med 1985;162:1256–1263.

28 Yabu Y, Takayanagi T, Sato S. Long term culture and cloning system for *Trypanosoma brucei gambiense* bloodstream forms in semi-defined medium *in vitro*. Parasitol Res 1989;76:93–97.

29 Hirumi H, Hirumi K, Doyle JJ, Cross GAM. *In vitro* cloning of animal-infective bloodstream forms of *Trypanosoma brucei*. Parasitology 1980;80:371–382.

30 Gray MA, Cunningham I, Gardiner PR, Taylor AM, Luckins AG. Cultivation of infective forms of *Trypanosoma congolense* from trypanosomes in the proboscis of *Glossina morsitans*. Parasitology 1981;82:81–95.

31 Gumm ID. The axenic cultivation of insect forms of *Trypanosoma (Duttonella) vivax* and development to the infective metacyclic stage. J Protozool 1991;38:163–171.

32 Kaminsky R, Beaudoin E, Cunningham I. Studies on the development of metacyclic *Trypanosoma brucei* sspp. cultivated with insect cell lines. J Protozool 1987;34:372–377.

33 Le Ray D. Structures antigéniques de *Trypanosoma brucei* (Protozoa, Kinetoplastida). Analyse immunoélectrophorétique et étude comparative. Ann Soc Belg Med Trop 1975;55:129–311.

34 Vickerman K. On the surface coat and flagellar adhesion in trypanosomes. J Cell Sci 1969;5:163–193.

35 Gardiner PR, Webster P, Jenni L, Moloo SK. Metacyclic *Trypanosoma vivax* possess a surface coat. Parasitology 1986;92:75–82.

36 Tetley L, Vickerman K, Moloo SK. Absence of a surface coat from metacyclic *Trypanosoma vivax*: possible implications for vaccination against vivax trypanosomiasis. Trans R Soc Trop Med Hyg 1981;75:409–414.

37 Overath P, Czichos J, Stock V, Nonnongasser C. Repression of glycoprotein synthesis and release of surface coat during transformation of *Trypanosoma brucei*. EMBO J 1983;2:1721–1728.

38 Tetley L, Turner CMR, Barry JD, Crowe JS, Vickerman K. Onset of expression of the variant surface glycoproteins of *Trypanosoma brucei* in the tsetse fly studied using immunoelectron microscopy. J Cell Sci 1987;87:363–372.

39 Vickerman K, Luckins AG. Localization of variable antigens in the surface coat of *Trypanosoma brucei* using ferritin-conjugated antibody. Nature 1969;224:1125–1126.

40 Turner MJ, Cardosa de Almeida ML, Gurnett AM, Raper J, Ward J. Biosynthesis, attachment and release of variant surface glycoproteins of the African trypanosome. Curr Topics Microbiol Immunol 1985;117:23–55.

41 Gray AR. Antigenic variation in a strain of *Trypanosoma brucei* transmitted by *Glossina morsitans* and *G. palpalis*. J Gen Microbiol 1965;41:191–214.

42 Gray AR. A pattern in the development of agglutinogenic antigens in cyclically transmitted isolates of *Trypanosoma gambiense*. Trans R Soc Trop Med Hyg 1975;69:131–138.

43 Seed JR. Antigens and antigenic variability of the African trypanosomes. J Protozool 1974;21:639–646.

44 Van Meirvenne N, Janssens PG, Magnus E. Antigenic variation in syringe passaged populations of *Trypanosoma (Trypanozoon) brucei*. I. Rationalization of the experimental approach. Ann Soc Belg Med Trop 1975;55:1–23.

45 Vos GJ, Gardiner PR. Parasite-specific antibody responses of ruminants infected with *Trypanosoma vivax*. Parasitology 1990;100:93–100.

46 Muller N, Hemphill A, Imboden M, Duvallet G, Dwinger RH, Seebeck T. Identification and characterization of two repetitive non-variable antigens from African trypanosomes which are recognized early during infection. Parasitology. (In press.)

47 Coppens L, Opperdoes FR, Courtoy PJ, Baudhuin P. Receptor-mediated endocytosis in the bloodstream form of *Trypanosoma brucei*. J Protozool 1987;34:465–473.

48 Eisenthal R, Game S, Holman GD. Specificity and kinetics of hexose transport in *Trypanosoma brucei*. Biochim Biophys Acta 1989;985:81–89.

49 Cully DF, Gibbs CP, Cross GAM. Identification of proteins encoded by variant surface glycoprotein expression site-associated genes in *Trypanosoma brucei*. Mol Biochem Parasitol 1986;21:189–197.

50 Alexandre S, Paindavoine P, Tebabi P, et al. Differential expression of a family of putative adenylate/guanylate cyclase genes in *Trypanosoma brucei*. Mol Biochem Parasitol 1990;43:279–288.

51 Schell D, Evers R, Preis D, et al. A transferrin-binding protein of *Trypanosoma brucei* is encoded by one of the genes in the variant surface glycoprotein expression site. EMBO J 1991;10:1061–1066.

52 Richardson JP, Beecroft RP, Tolson DL, Liu MK, Pearson TW. Procyclin: an unusual immunodominant glycoprotein

surface antigen from the procyclic stage of African trypanosomes. J Immunol 1988;136:2259–2264.

53 Mowatt MR, Clayton CE. Polymorphism in the procyclic acid repetitive protein of *Trypanosoma brucei*. Mol Cell Biol 1988;8:4055–4062.

54 Roditi I, Schwartz H, Pearson TW, *et al*. Procyclin gene expression and loss of variant surface glycoprotein during differentiation of *Trypanosoma brucei*. J Cell Biol 1989;108: 737–746.

55 Vickerman K, Tetley L, Hendry KAK, Turner CMR. Biology of African trypanosomes in the tsetse fly. Biol Cell 1988;64:109–119.

56 Opperdoes FR. Biochemical peculiarities of trypanosomes, African and South American. Br Med Bull 1985;41:130–136.

57 Capbern A, Giroud C, Baltz T, Mattern P. *Trypanosoma equiperdum*: etude des variations antigeniques au cours de la trypanosomose experimentale du lapin. Exp Parasitol 1977;42:6–13.

58 Van Meirvenne N, Janssens PG, Magnus E, Lumsden WHR, Herbert WJ. Antigenic variation in syringe passaged populations of *Trypanosoma (Trypanozoon) brucei*. II. Comparative studies on two antigenic type collections. Ann Soc Belg Med Trop 1975;55:25–30.

59 Van Meirvenne N, Magnus E, Vervoort T. Comparison of variable antigenic types produced by trypanosome strains of the subgenus *Trypanozoon*. Ann Soc Belg Med Trop 1977;57:409–423.

60 World Health Organization. Proposals for the nomenclature of salivarian trypanosomes and for the maintenance of reference collections. Bull WHO 1978;56:467–480.

61 Van der Ploeg LHT, Valerio D, De Lange T, Bernards A, Borst P, Grosveld FG. An analysis of cosmid clones of nuclear DNA from *Trypanosoma brucei* shows that the genes for variant surface glycoproteins are clustered in the genome. Nucleic Acids Res 1982;10:5905–5923.

62 Dero B, Zampetti-Bosseler F, Pays E, Steinert M. The genome and the antigen gene repertoire of *Trypanosoma brucei gambiense* are smaller than those of *T. b. brucei*. Mol Biochem Parasitol 1987;26:247–256.

63 Turner CMR, Barry JD, Vickerman K. An estimate of the size of the metacyclic variable antigen repertoire of *Trypanosoma brucei rhodesiense*. Parasitology 1988;97:269–276.

64 Crowe JS, Barry JD, Luckins AG, Ross CA, Vickerman K. All metacyclic variable antigen types of *Trypanosoma congolense* identified using monoclonal antibodies. Nature 1983;306:389–391.

65 Hajduk SL, Cameron CR, Barry JD, Vickerman K. Antigenic variation in cyclically-transmitted *Trypanosoma brucei*. Variable antigen type composition of metacyclic trypanosome populations from the salivary glands of *Glossina morsitans*. Parasitology 1981;83:595–607.

66 Barry JD, Crowe JS, Vickerman K. Instability of the *Trypanosoma brucei rhodesiense* metacyclic variable antigen repertoire. Nature 1983;306:699–701.

67 Hajduk SL, Vickerman K. Antigenic variation in cyclically transmitted *Trypanosoma brucei*. Variable antigen composition of the first parasitaemia in mice bitten by try-

panosome-infected *Glossina morsitans*. Parasitology 1981;83:609–621.

68 Frame IA, Ross CAV, Luckins AG. Characterization of *Trypanosome congolense* serodemes in stocks isolated from Chipata District, Zambia. Parasitology 1990;101:235–242.

69 Jones TW, Cunningham I, Taylor AM, Gray AR. The use of culture-derived metacyclic trypanosomes in studies on the serological relationships of stocks of *Trypanosoma brucei gambiense*. Trans R Soc Trop Med Hyg 1981;75:560–565.

70 Turner CMR, Aslam N, Smith B, Buchanan N, Tait A. The effects of genetic exchange on variable antigen expression in *Trypanosoma brucei*. Parasitology 1991;103:379–386.

71 Thon G, Baltz T, Giroud C, Eisen H. Trypanosome variable surface glycoproteins: composite genes and order of expression. Genes Dev 1990;4:1374–1383.

72 Turner MJ. Trypanosome variant surface glycoprotein. In Englund PT, Sher A, eds. *The Biology of Parasitism: a Molecular and Immunologic Approach*. MBL Lectures in Biology, vol. 9. New York: Alan R Liss, 1988:349–369.

73 Ferguson MAJ, Homans SW. The membrane attachment of the variant surface glycoprotein coat of *Trypanosoma brucei*. In McAdam KPWJ, ed. *New strategies in Parasitology*. Edinburgh: Churchill Livingstone, 1989:121–140.

74 Gardiner PR, Pearson TW, Clarke MW, Mutharia LM. Identification and isolation of a variant surface glycoprotein from *Trypanosoma vivax*. Science 1987;235:774–777.

75 Cross GAM. Identification, purification and properties of clone-specific glycoprotein antigens constituting the surface coat of *Trypanosoma brucei*. Parasitology 1975;71:393–417.

76 Johnson JG, Cross GAM. Carbohydrate composition of variant-specific surface antigen glycoproteins from *Trypanosoma brucei*. J Protozool 1977;24:587–591.

77 McConnell J, Cordingley JS, Turner MJ. Biosynthesis of *Trypanosome brucei* variant surface glycoprotein 1. Synthesis, size and processing of an N-terminal signal peptide. Mol Biochem Parasitol 1981;4:226–242.

78 Ferguson MAJ, Williams AF. Cell surface anchoring of proteins via glycosyl-phosphatidylinositol structures. Ann Rev Biochem 1988;57:283–320.

79 Freymann D, Metcalf P, Turner MJ, Wiley D. 6 Å resolution X-ray structure of a variable surface glycoprotein from *Trypanosoma brucei*. Nature 1989;311:167–169.

80 Metcalfe P, Blum M, Freymann M, Turner M, Wyley DC. Two variant surface glycoproteins of *Trypanosoma brucei* of different sequence classes have similar 6 Å resolution X-ray structures. Nature 1987;325:84–86.

81 Jahnig F, Bulow R, Baltz T, Overath P. Secondary structure of the variant surface glycoproteins of trypanosomes. FEBS Lett 1987;221:37–42.

82 Masterson WJ, Taylor DW, Turner MJ. Topological analysis of variant surface glycoprotein of *Trypanosoma brucei*. J Immunol 1988;140:3194–3199.

83 Barbet AF, McGuire TC, Musoke AJ, Hirumi H. Cross-reacting determinants in trypanosome surface antigens. In Losos G, Chouinard A, eds. *Pathogenicity of Trypanosomes*. Ottawa: IDRC, 1979:38–43.

84 Zamze SE, Ferguson MAJ, Collins R, Dwek RA, Rademacher TW. Characterization of the cross-reacting

determinant (CRD) of the glycosyl-phosphatidylinositol membrane anchor of *Trypanosoma brucei* variant surface glycoprotein. Eur J Biochem 1988;176:527–534.

85 Reinwald E. Role of carbohydrate within variant surface glycoprotein of *Trypanosoma congolense*. Eur J Biochem 1985;151:385–391.

86 Bulow R, Overath P, Davoust J. Rapid lateral diffusion of the variant-specific glycoprotein in the coat of *Trypanosoma brucei*. Biochemistry 1988;27:2384–2388.

87 Cardosa de Almeida ML, Turner MJ. The membrane form of variant surface glycoproteins of *Trypanosoma brucei*. Nature 1983;302:349–352.

88 Gibson WC, Borst P. Size-fractionation of the small chromosomes of *Trypanozoon* and *Nannomonas* trypanosomes by pulsed field gradient gel electrophoresis. Mol Biochem Parasitol 1986;18:127–140.

89 Van der Ploeg LHT, Cornelissen AWCA, Barry JD, Borst P. Chromosomes of Kinetoplastida. EMBO J 1984;3:3109–3115.

90 Sternberg J, Tait A, Haley S, *et al*. Gene exchange in African trypanosomes: characterisation of a new hybrid genotype. Mol Biochem Parasitol 1988;27:191–200.

91 Gottesdiener K, Garcia-Anoveros J, Lee, MG-S, Van der Ploeg LHT. Chromosome organization of the protozoan *Trypanosoma brucei*. Mol Cell Biol 1990;10:6079–6083.

92 De Lange T, Borst P. Genomic environment of the expression-linked extra copies of genes for surface antigens of *Trypanosoma brucei* resembles the end of a chromosome. Nature 1982;299:451–453.

93 Myler PJ, Allison J, Agabian N, Stuart K. Antigenic variation in African trypanosomes by gene replacement or expression of alternate telomeres. Cell 1984;39:203–211.

94 Scholler JK, Myler PJ, Stuart KD. A novel telomeric gene conversion in *Trypanosoma brucei*. Mol Biochem Parasitol 1989;35:11–20.

95 Pays E. Pseudogenes, chimaeric genes and the timing of antigen variation in African trypanosomes. Trends Genet 1989;5:389–391.

96 Thon G, Baltz T, Eisen H. Antigenic diversity by the recombination of pseudogenes. Genes Dev 1989;3:1247–1254.

97 Majumder HK, Boothroyd JC, Weber H. Homologous 3′-terminal regions of mRNAs for surface antigens of different antigenic variants of *Trypanosoma brucei*. Nucleic Acids Res 1981;9:4745–4753.

98 Aline RF Jr, Stuart K. *Trypanosoma brucei*: conserved sequence organization 3′ to telomeric variant surface glycoprotein genes. Exp Parasitol 1989;68:57–66.

99 Kimmel BE, Ole-Moiyoi OK, Young JR. Ingi, a 5.2 kb dispersed sequence element from *Trypanosoma brucei* that carries half of a smaller mobile element at either end and has homology with mammalian LINEs. Mol Cell Biol 1987;7:1465–1475.

100 Smiley BL, Aline RF Jr, Myler PJ, Stuart K. A retroposon in the 5′ flank of a *Trypanosoma brucei* VSG gene lacks insertional terminal repeats. Mol Biochem Parasitol 1990;42:143–152.

101 Murphy NB, Pays A, Tebabi P, *et al*. *Trypanosoma brucei* repeated element with unusual structural and transcrip-tional properties. J Mol Biol 1987;195:855–871.

102 Hasan G, Turner MJ, Cordingley JS. Complete nucleotide sequence of an unusual mobile element from *Trypanosoma brucei*. Cell 1984;37:333–341.

103 Campbell DA, Van Bree MP, Boothroyd JC. The 5′-limit of transposition and upstream barren region of a trypanosome VSG gene: tandem 76 base-pair repeats flanking (TAA)90. Nucleic Acids Res 1984;12:2759–2774.

104 Aline RF Jr, Scholler JK, Nelson RG, Agabian N, Stuart K. Preferential activation of telomeric variant surface glycoprotein genes in *Trypanosoma brucei*. Mol Biochem Parasitol 1985;17:311–320.

105 Aline R Jr, MacDonald G, Brown E, *et al*. (TAA)*n* within sequences flanking several variant surface glycoprotein genes in *Trypanosoma brucei*. Nucleic Acids Res 1985;13: 3161–3177.

106 Florent I, Baltz T, Raibaud A, Eisen H. On the role of repeated sequences 5′ to variant surface glycoprotein genes in African trypanosomes. Gene 1987;53:55–62.

107 Crozatier M, Van der Ploeg LHT, Johnson PJ, Gommers-Ampt J, Borst P. Structure of a telomeric expression site for variant specific surface antigens in *Trypanosoma brucei*. Mol Biochem Parasitol 1990;42:1–12.

108 Alexandre S, Guyaux M, Murphy NB, *et al*. Putative genes of a variant-specific antigen gene transcription unit in *Trypanosoma brucei*. Mol Cell Biol 1988;8:2367–2378.

109 Pays E, Tebabi P, Pays A, *et al*. The genes and transcripts of an antigen gene expression site from *Trypanosoma brucei*. Cell 1989;57:835–845.

110 Stadnyk AW, Scholler JK, Myler PJ, Stuart KD. Ribonuclease protection determines sequences specific to a single variable surface glycoprotein gene expression site. In Agabian N, Cerami A, eds. *Parasites: Molecular Biology, Drug and Vaccine Design*. New York: Wiley–Liss, 1989:99–109.

111 Florent IC, Raibaud A, Eisen H. A family of genes related to a new expression site-associated gene in *Trypanosoma equiperdum*. Mol Cell Biol 1991;11:2180–2188.

112 Barnes DA, Mottram JC, Agabian N. Bloodstream and metacyclic variant surface glycoprotein gene expression sites of *Trypanosoma brucei gambiense*. Mol Biochem Parasitol 1990;41:104–114.

113 Jin Son H, Cook GA, Hall T, Donelson JE. Expression site associated genes of *Trypanosoma brucei rhodesiense*. Mol Biochem Parasitol 1989;33:59–66.

114 Matthews KR, Shiels PG, Graham SV, Cowan C, Barry JD. Duplicative activation mechanisms of two trypanosome telomeric VSG genes with structurally simple 5′ flanks. Nucleic Acids Res 1990;18:7219–7227.

115 Berberof M, Pays A, Pays E. A similar gene is shared by both the variant surface glycoprotein and procyclin gene transcription units of *Trypanosoma brucei*. Mol Cell Biol 1991;11:1473–1479.

116 Zomerdijk JCBM, Ouellette M, Ten Asbroek ALMA, *et al*. The promoter for a variant surface glycoprotein gene expression site in *Trypanosoma brucei*. EMBO J 1990;9:2791–2801.

117 Pays E, Coquelet H, Pays A, Tebabi P, Steinert M. *Trypanosoma brucei*: posttranscriptional control of the variable

surface glycoprotein gene expression site. Mol Cell Biol 1989;9:4018–4021.

118 Coquelet H, Tebabi P, Pays A, Steinert M, Pays E. *Trypanosoma brucei*: enrichment by UV of intergenic transcripts from the variable surface glycoprotein gene expression site. Mol Cell Biol 1989;9:4022–4025.

119 Pays E, Coquelet H, Tebabi P, *et al*. *Trypanosoma brucei*: constitutive activity of the VSG and procyclin gene promoters. EMBO J 1990;9:3145–3151.

120 Gottesdiener K, Chung H-M, Brown SD, Lee MG-S, Van der Ploeg LHT. Characterization of VSG gene expression site promoters and promoter-associated DNA rearrangement events. Mol Cell Biol 1991;11:2467–2480.

121 Zomerdijk JCBM, Kieft R, Duyndam M, Shiels PG, Borst P. Antigenic variation in *Trypanosoma brucei*: a telomeric expression site for variant-specific surface glycoprotein genes with novel features. Nucleic Acids Res 1991;19:1359–1367.

122 Myler P, Nelson R, Agabian N, Stuart K. Two mechanisms of expression of a predominant variant antigen gene of *Trypanosoma brucei*. Nature 1984;309:282–284.

123 Aline RF Jr, Stuart K. The two mechanisms for antigenic variation in *Trypanosoma brucei* are independent processes. Mol Biochem Parasitol 1985;16:11–20.

124 Williams RO, Young JR, Majiwa PA. Genomic environment of *Trypanosoma brucei* VSG genes: presence of a minichromosome. Nature 1982;299:417–421.

125 Hoeijmakers JH, Frasch AC, Bernards A, Borst P, Cross GAM. Novel expression-linked copies of the genes for variant surface antigens in trypanosomes. Nature 1980;284:78–80.

126 Pays E. Gene conversion in trypanosome antigenic variation. Prog Nucleic Acids Res Mol Biol 1985;32:1–26.

127 Pays E, Delauw MF, Laurent M, Steinert M. Possible DNA modification in GC dinucleotides of *Trypanosoma brucei* telomeric sequences; relationship with antigen gene transcription. Nucleic Acids Res 1984;12:5235–5247.

128 Bernards A, Van Harten-Loosbroek N, Borst P. Modification of telomeric DNA in *Trypanosoma brucei*; a role in antigenic variation. Nucleic Acids Res 1984;12:4153–4170.

129 Myler PJ, Aline RF Jr, Scholler JK, Stuart KD. Changes in telomere length associated with antigenic variation in *Trypanosoma brucei*. Mol Biochem Parasitol 1988;29:243–250.

130 Van Der Werf A, Van Assel S, Aerts D, Steinert M, Pays E. Telomere interactions may condition the programming of antigen expression in *Trypanosoma brucei*. EMBO J 1990;9:1035–1040.

131 Pays E, Guyaux M, Aerts D, Van Meirvenne N, Steinert M. Telomeric reciprocal recombination as a possible mechanism for antigenic variation in trypanosomes. Nature 1985;316:562–564.

132 Shea C, Glass DJ, Parangi S, Van der Ploeg LHT. Variant surface glycoprotein gene expression site switches in *Trypanosoma brucei*. J Biol Chem 1986;261:6056–6063.

133 Kooter JM, Winter AJ, De Oliveira C, Wagter R, Borst P. Boundaries of telomere conversion in *Trypanosoma brucei*. Gene 1988;69:1–11.

134 Roth CW, Longacre S, Raibaud A, Baltz T, Eisen H. The use of incomplete genes for the construction of a *Try-*

panosoma equiperdum variant surface glycoprotein gene. EMBO J 1986;5:1065–1070.

135 Myler PJ, Aline RF Jr, Scholler JK, Stuart KD. Multiple events associated with antigenic switching in *Trypanosoma brucei*. Mol Biochem Parasitol 1988;29:227–241.

136 Aline RF Jr, Myler PJ, Stuart KD. Frequent loss of a variant surface glycoprotein gene in *Trypanosoma brucei*. Exp Parasitol 1989;68:8–16.

137 Tschudi C, Young AS, Ruben L, Patton CL, Richards FF. Calmodulin genes in trypanosomes are tandemly repeated and produce multiple mRNAs with a common 5′ leader sequence. Proc Natl Acad Sci USA 1985;82:3998–4002.

138 Perry K, Agabian N. mRNA processing in the *Trypanosomatidae*. Experientia 1991;47:118–128.

139 Esser KM, Schoenbechler MJ. Expression of two variant surface glycoproteins on individual African trypanosomes during antigen switching. Science 1985;229:190–193.

140 Vickerman K, Tetley L, Turner CMR, Barry JD. Pattern of variant surface glycoprotein coating in nascent metacyclic *Trypanosoma brucei* in the salivary glands of *Glossina morsitans*. In Chang K-P, Snary D, eds. *Host–Parasite Cellular and Molecular Interactions in Protozoal Infections*. NATO ASI Series H, vol. 11. Berlin: Springer-Verlag, 1987:1–8.

141 Baltz T, Giroud C, Baltz D, Roth C, Raibaud A, Eisen H. Stable expression of two variable surface glycoproteins by cloned *Trypanosoma equiperdum*. Nature 1986;319:602–604.

142 Cornelissen AWCA, Johnson PJ, Kooter JM, Van der Ploeg LHT, Borst P. Two simultaneously active VSG gene transcription units in a single *Trypanosoma brucei* variant. Cell 1985;41:825–832.

143 Jefferies D, Tebabi P, Pays E. Transient activity assays of the *Trypanosoma brucei* variant surface glycoprotein gene promoter: control of gene expression at the posttranscriptional level. Mol Cell Biol 1991;11:338–343.

144 Lenardo MJ, Esser KM, Moon AM, Van der Ploeg LHT, Donelson JE. Metacyclic variant surface glycoprotein genes of *Trypanosoma brucei* subsp. *rhodesiense* are activated *in situ*, and their expression is transcriptionally regulated. Mol Cell Biol 1986;6:1991–1997.

145 Delauw MF, Pays E, Steinert M, Aerts D, Van Meirvenne N, Le Ray D. Inactivation and reactivation of a variant-specific antigen gene in cyclically transmitted *Trypanosoma brucei*. EMBO J 1985;4:989–993.

146 Kooter JM, Borst P. Alpha-amanitin-insensitive transcription of variant surface glycoprotein genes provides further evidence for discontinuous transcription in trypanosomes. Nucleic Acids Res 1984;12:9457–9472.

147 Vickerman K. Trypanosome sociology and antigenic variation. Parasitology 1989;99:S37–S47.

148 Barry JD, Turner CMR. The dynamics of antigenic variation and growth of African trypanosomes. Parasitol Today 1991;7:207–211.

149 Barry JD. Antigenic variation during *Trypanosoma vivax* infections of different host species. Parasitology 1986;92:51–65.

150 Miller EN, Turner MJ. Analysis of antigenic types appearing in first relapse populations of clones of *Trypanosoma brucei*. Parasitology 1981;82:63–80.

151 Turner CMR, Barry JD. High frequency of antigenic

variation in *Trypanosoma brucei rhodesiense* infections. Parasitology 1989;99:67–75.

152 Kosinski RJ. Antigenic variation in trypanosomes: a computer analysis of variant order. Parasitology 1980;80:343–357.

153 Lamont GS, Tucker RS, Cross GAM. Analysis of antigen switching rates in *Trypanosoma brucei*. Parasitology 1986;92:355–367.

154 Liu AY, Michels PA, Bernards A, Borst P. Trypanosome variant surface glycoprotein genes expressed early in infection. J Mol Biol 1985;182:383–396.

155 Timmers HTM, De Lange T, Kooter JM, Borst P. Coincident multiple activations of the same surface antigen gene in *Trypanosoma brucei*. J Mol Biol 1987;194:81–90.

156 Lee MGS, Van der Ploeg LHT. Frequent independent duplicative transpositions activate a single VSG gene. Mol Cell Biol 1987;7:357–364.

157 Seed JR. Competition among serologically different clones of *Trypanosoma brucei gambiense in vivo*. J Protozool 1978;25:526–529.

158 Barry JD, Le Ray D, Herbert WJ. Infectivity and virulence of *Trypanosoma (Trypanozoon) brucei* for mice. IV. Dissociation of virulence and variable antigen type in relation to pleomorphism. J Comp Pathol 1979;89:465–470.

159 Inverso JA, Mansfield JM. Genetics of resistance to African trypanosomes. II. Differences in virulence associated with VSSA expression among clones of *Trypanosoma rhodesiense*. J Immunol 1983;130:412–417.

160 Myler PJ, Allen AL, Agabian N, Stuart K. Antigenic variation in clones of *Trypanosoma brucei* grown in immune-deficient mice. Infect Immun 1985;47:684–690.

161 Seed JR, Effron HG. Simultaneous presence of different antigenic populations of *Trypanosoma brucei gambiense* in *Microtus montanus*. Parasitology 1973;66:269–278.

162 Turner CMR, Hunter CA, Barry JD, Vickerman K. Similarity in variable antigen type composition of *Trypanosoma brucei rhodesiense* populations in different sites within the mouse host. Trans R Soc Trop Med Hyg 1986;80:824–830.

163 Agur Z, Abiri D, Van der Ploeg LHT. Ordered appearance of antigenic variants of African trypanosomes explained in a mathematical model based on a stochastic switch process and immune selection against putative switch intermediates. Proc Natl Acad Sci USA 1989;86:9626–9630.

164 Desowitz RS, Watson HJC. Studies on *Trypanosoma vivax*. IV. The maintenance of a strain in white rats without sheep-serum supplement. Ann Trop Med Parasitol 1953;47:62–67.

165 Ketteridge D. *Trypanosoma vivax*: surface interrelationships between host and parasite. Trans R Soc Trop Med Hyg 1972;66:324.

166 De Gee ALW, Rovis L. *Trypanosoma vivax*: absence of host protein on the surface coat. Exp Parasitol 1981;51:124–132.

167 Hide G, Gray A, Harrison CM, Tait A. Identification of an epidermal growth factor receptor homologue in trypanosomes. Mol Biochem Parasitol 1989;36:51–60.

168 Ghiotto V, Brun R, Jenni L, Hecker H. *Trypanosoma brucei*: morphometric changes and loss of infectivity during transformation of bloodstream forms to procyclic forms *in vitro*. Exp Parasitol 1979;48:447–456.

169 Ferrante A, Allison AC. Alternate pathway activation of complement by African trypanosomes lacking a glycoprotein coat. Parasite Immunol 1983;5:491–498.

170 Mosser DM, Roberts JF. *Trypanosoma brucei*: recognition *in vitro* of two developmental forms by murine macrophages. Exp Parasitol 1982;54:310.

171 Balber AE, Bangs JD, Jones SM, Proia RL. Inactivation or elimination of potentially trypanolytic, complement-activating immune complexes by pathogenic trypanosomes. Infect Immun 1979;24:617–627.

172 Musoke AJ, Barbet AF. Activation of complement by variant-specific surface antigen of *Trypanosoma brucei*. Nature 1977;270:438–440.

173 Nielsen K, Sheppard J, Holmes W, Tizard I. Experimental bovine trypanosomiasis: changes in the catabolism of serum immunoglobulins and complement components in infected cattle. Immunology 1978;35:811–816.

174 Nielsen K, Sheppard J, Holmes W, Tizard I. Experimental bovine trypanosomiasis: changes in serum immunoglobulin, complement and complement component levels in infected cattle. Immunology 1978;35:817–826.

175 Rurangirwa FR, Musoke AJ, Nantulya VM, Tabel H. Immune depression in bovine trypanosomiasis: effects of acute and chronic *Trypanosoma congolense* and chronic *Trypanosoma vivax* infections on antibody response to *Brucella abortus* vaccine. Parasite Immunol 1983;5:207–276.

176 Nielsen KA. Complement in trypanosomiasis. In Tizard I, ed. *Immunology and Pathogenesis of Trypanosomiasis*. Boca Raton: CRC Press, 1985:134–144.

177 Lumsden WHR, Herbert WJ. Phagocytosis of trypanosomes by mouse peritoneal macrophages. Trans R Soc Trop Med Hyg 1967;61:142.

178 Morrison WI, Murray M. Lymphoid changes in African trypanosomiasis. In Losos G, Chouinard A, eds. *Pathogenicity of Trypanosomes*. Ottawa: IDRC, 1979:154–160.

179 Black SJ, Murray M, Shapiro SZ, et al. Analysis of *Propionibacterium acnes*-induced non-specific immunity to *Trypanosoma brucei* in mice. Parasite Immunol 1989;11:371–383.

180 Liew FY, Cox FEG. Non-specific defence mechanism: the role of nitric oxide. Parasitol Today 1991;7:A17–A21.

181 Rickman LR. The effects of some African game animal sera in the BIIT on *Trypanosoma (Trypanozoon) brucei* subspecies trypanosomes. Trans R Soc Trop Med Hyg 1981;75:122–123.

182 Rifkin MR. Identification of the trypanocidal factor in normal human serum: high density lipoprotein. Proc Natl Acad Sci USA 1978;75:3450–3454.

183 Hajduk SL, Moore DR, Vasudeva-Charya J, et al. Lysis of *Trypanosoma brucei* by a toxic subspecies of human high density lipoprotein. J Biol Chem 1989;264:5210–5217.

184 Hajduk SL, Hager K, Esko JD. High density lipoprotein-mediated lysis of trypanosomes. Parasitol Today 1992;8:95–98.

185 Rickman LR, Robson J. The blood incubation infectivity test: a simple test which may serve to distinguish *Trypanosoma brucei* from *T. rhodesiense*. Bull WHO 1970;42:650–651.

186 Hawking F. The action of human serum upon *Trypanosoma brucei*. Protozool Abstr 1979;3:199–206.

187 Mulla AF, Rickman R. Evidence for the presence of an innate trypanocidal factor in the serum of a non-immune African waterbuck (*Kobus ellipsiprymnus*). Trans R Soc Trop Med Hyg 1988;82:97–98.

188 Van Meirvenne N, Magnus B, Janssens PJ. The effect of normal human serum on Trypanosomes of distinct antigenic type (BTAT 1 to 12) isolated from a strain of *Trypanosoma brucei rhodesiense*. Ann Soc Belg Med Trop 1976;55: 55–63.

189 Terry RJ. Immunity to African trypanosomiasis. In Cohen S, Sadun EH, eds. *Immunology of Parasitic Infections*, 1st edn. Oxford: Blackwell Scientific Publications, 1976:203–221.

190 Cover B. *Trypanosoma vivax* and the action of cotton rat serum. Trans R Soc Trop Med Hyg 1984;78:140–141.

191 Gardiner PR, Thatthi R, King RC. Serum from cotton rat (*Sigmodon hispidus*) lacks lytic activity against *Trypanosoma vivax* stocks. Acta Trop 1988;45:187–188.

192 Maudlin, I, Welburn SC. Tsetse immunity and the transmission of trypanosomiasis. Parasitol Today 1988;4:109–111.

193 Morrison WI, Murray M, Akol GWO. Immune responses of cattle to African trypanosomes. In Tizard I, ed. *Immunology and Pathogenesis of Trypanosomiasis*. Boca Raton: CRC Press, 1985:103–131.

194 Urquhart GM, Holmes PH. African trypanosomiasis: In Soulsby EJL, ed. *Immune Responses in Parasitic Infections: Immunology, Immunopathology and Immunoprophylaxis*, vol. 3. Boca Raton: CRC Press, 1987:1–23.

195 Nantulya VM, Musoke AJ, Rurangirwa FR, Moloo SK. Resistance of cattle to tsetse transmitted challenge with *Trypanosoma brucei* or *Trypanosoma congolense* after spontaneous recovery from syringe-passaged infections. Infect Immun 1984;43:735–738.

196 Nantulya VM, Musoke AJ, Moloo SK. Apparent exhaustion of the variable antigen repertoires of *Trypanosoma vivax* in infected cattle. Infect Immun 1986;54:444–447.

197 Black SJ, Sendashonga CN, O'Brien C, *et al.* Regulation of parasitaemia in mice infected with *Trypanosoma brucei*. Curr Top Microbiol Immunol 1985;117:93–118.

198 Mahan SM, Black SJ. Differentiation, multiplication and control of bloodstream form *Trypanosoma (Duttonella) vivax* in mice. J Protozool 1989;36:424–428.

199 Seyfang A, Mecke D, Duszenko M. Degradation, recycling and shedding of *Trypanosoma brucei* variant surface glycoprotein. J Protozool 1990;37:546–552.

200 Herbert WJ, Lumsden WHR. Single dose vaccination of mice against experimental infection with *Trypanosoma (Trypanozoon) brucei*. J Med Microbiol 1968;1:23–32.

201 Akol GWO, Murray M. *Trypanosoma congolense*: susceptibility of cattle to cyclical challenge. Exp Parasitol 1983;55:386–393.

202 Akol GWO, Murray M. Induction of protective immunity in cattle by tsetse-transmitted cloned isolates of *Trypanosoma congolense*. Ann Trop Med Parasitol 1985;79:617–627.

203 Nantulya VM, Doyle JJ, Jenni L. Studies on *Trypanosoma (Nannomonas) congolense*. IV. Experimental immunization of mice against tsetse fly challenge. Parasitology 1980;80:133–137.

204 Luckins AG, Rae PF, Gray AR. Infection, immunity and the development of local skin reactions in rabbits infected with cyclically-transmitted stocks of *Trypanosoma congolense*. Ann Trop Med Parasitol 1983;77:569–582.

205 Emery DL, Akol GWO, Murray M, Morrison WI, Moloo SK. The chancre—early events in the pathogenesis of African trypanosomiasis in domestic livestock. In Van Den Bossche H, ed. *The Host–Invader Interplay*. Proceedings of the Third International Symposium on the Biochemistry of Parasites and Host–Parasite Relationships. Amsterdam: Elsevier/North Holland Biomedical Press, 1980:345–356.

206 De Gee ALW, Shah SD, Doyle JJ. An attempt to immunize against *Trypanosoma vivax* by cyclical infection followed by treatment. In De Gee ALW, ed. *Host-Parasite Relationships in Trypanosoma (Duttonella) vivax with Special Reference to the Influence of Antigenic Variation*. PhD Thesis. The Netherlands: Utrecht University 1980:113–136.

207 Wellde BT, Schoenbechler MJ, Diggs CL, Langbehn HR, Sadun EH. *Trypanosoma rhodesiense*: variant specificity of immunity induced by irradiated parasites. Exp Parasitol 1975;37:125–129.

208 Morrison WI, Black SJ, Paris J, Hinson CA, Wells PW. Protective immunity and specific antibody responses elicited in cattle by irradiated *Trypanosoma brucei*. Parasite Immunol 1982;4:395–401.

209 Morrison WI, Wells PW, Moloo SK, Paris J, Murray M. Interference in the establishment of superinfections with *Trypanosoma congolense* in cattle. J Parasitol 1982;68:755–764.

210 Wells PW, Emery DL, Hinson CA, Morrison WI, Murray M. Immunization of cattle with variant specific surface antigen of *Trypanosoma brucei*: the influence of different adjuvants. Infect Immun 1982;36:1–9.

211 Rovis LA, Musoke J, Moloo SK. Failure of trypanosome membrane antigens to induce protection against tsetse-transmitted *Trypanosoma vivax*. Acta Trop 1984;41:272–287.

212 Olecnick JG, Wolfe R, Nayman RK, McLaughlin J. A flagellar pocket membrane fraction from *Trypanosoma brucei rhodesiense*. Immunogold localization and non-variant immunoprotection. Infect Immun 1988;56:92–98.

213 Mattern P, Masseyeff R, Michel R, Peretti R. Étude immunochimique de la β₂-macroglobuline des sérums de malades atteints de trypanosomiase Africaine à *T. gambiense*. Ann Inst Pasteur Paris 1961;101:382–388.

214 Seed JR, Corville RL, Risby EL, Gam AA. The presence of agglutinating antibody in the IgM immunoglobulin fraction of rabbit anti-serum during experimental African trypanosomiasis. Parasitology 1969;59:283–292.

215 Luckins AG, Mehlitz D. Immunoglobulin levels in cattle trypanosomiasis. Ann Trop Med Parasitol 1976;70:479–480.

216 Bouteille B, Darde ML, Pestre-Alexandre M, *et al.* The sheep (*Ovis aries*) as an experimental model for African trypanosomiasis. II. A biological study. Ann Trop Med Parasitol 1988;82:149–158.

217 Masake RA, Morrison WI. Evaluation of the structural and functional changes in the lymphoid organs of Boran cattle infected with Trypanosoma vivax. Am J Vet Res 1981;42:1738–1746.

218 Seed JR. *Trypanosoma gambiense* and *T. equiperdum*: characterisation of variant-specific antigens. Exp Parasitol 1972;

31:98–108.

219 Takayanagi T, Enriquez GL. Effects of the IgG and IgM immunoglobulins in *Trypanosoma gambiense* infections in mice. J Parasitol 1973;59:644–647.

220 Musoke AJ, Nantulya VM, Barbet AF, Kironde F, McGuire TC. Bovine immune response to African trypanosomes: specific antibodies to variable surface glycoproteins of *Trypanosoma brucei*. Parasite Immunol 1981;3:97–106.

221 Barry JD, Emery DL. Parasite development and host responses during the establishment of *Trypanosoma brucei* infection transmitted by tsetse fly. Parasitology 1984;88:67–84.

222 Masake RA, Musoke AJ, Nantulya VM. Specific antibody responses to the variable surface glycoproteins of *Trypanosoma congolense* in infected cattle. Parasite Immunol 1983;5:345–355.

223 Nantulya VM, Musoke AJ, Barbet AF, Roelants GE. Evidence for re-appearance of *Trypanosoma brucei* variable antigen types in relapse populations. J Parasitol 1979;65:673–679.

224 Goodwin LG. The African scene: mechanisms of pathogenesis in trypanosomiasis. In *Trypanosomiasis and Leishmaniasis with Special Reference to Chagas' Disease*. Ciba Foundation Symposium No. 20 (new series). Amsterdam: Associated Scientific Publishers, 1974:107–119.

225 Parratt D, Herbert WJ. Heterophile antibodies in trypanosome infections. In Lumsden WHR, Evans DA, eds. *Biology of the Kinetoplastida*, vol. 2. New York: Academic Press, 1979:523–545.

226 Herbert WJ, Parratt D, Van Meirvenne N, Lennox B. An accidental laboratory infection with trypanosomes of a defined stock. II. Studies on the serological response of the patient and the identity of the infecting organism. J Infect 1980;2:113–124.

227 Campbell GH, Phillips SM. Adoptive transfer of variant-specific resistance to *Trypanosoma rhodesiense* with B lymphocytes and serum. Infect Immun 1976;14:1144–1150.

228 Campbell GH, Esser KM, Phillips SM. *Trypanosoma rhodesiense* infection in congenitally athymic (nude) mice. Infect Immun 1978;20:714–720.

229 De Raadt P. Immunity and antigenic variation: clinical observations suggestive of immune phenomena in African trypanosomiasis. In *Trypanosomiasis and Leishmaniasis*. Ciba Foundation Symposium 20 (new series). Amsterdam: Associated Scientific Publishers, 1974:199–216.

230 Mehlitz D, Heidrich-Joswig S, Fimmen H-O, Frehas EK, Karbe E. Observations on the colostral transfer of anti-trypanosome antibodies to N'dama calves and the immune response to infection with *Trypanosoma (Duttonella) vivax* and *T. (Nannomonas) congolense*. Ann Soc Belg Med Trop 1983;63:137–148.

231 Whitelaw DD, Jordt T. Colostral transfer of antibodies to *Trypanosoma brucei* in goats. Ann Soc Belg Med Trop 1985; 65:199–205.

232 Whitelaw DD, Urquhart GM. Maternally-derived immunity to *Trypanosoma brucei* in mice, and its potentiation by Berenil chemotherapy. Parasite Immunol 1985;7:289–300.

233 Gray AR, Luckins AG. The initial stage of infection with cyclically-transmitted *Trypanosoma congolense* in rabbits, calves and sheep. J Comp Pathol 1980;90:499–512.

234 Dwinger RH, Rudin W, Moloo SK, Murray M. Development of *Trypanosoma congolense*, *T. vivax* and *T. brucei* in the skin reaction induced in goats by infected *Glossina morsitans centralis*: a light and electron microscopical study. Res Vet Sci 1988;44:154–163.

235 Luckins AG, Gray AR. Interference with anti-trypanosome immune responses in rabbits infected with cyclically-transmitted *Trypanosoma congolense*. Parasite Immunol 1983;5:547–556.

236 Mwangi DM, Hopkins J, Luckins AG. Cellular phenotypes in *Trypanosoma congolense* infected sheep. The local skin reaction. Parasite Immunol 1990;12:647–658.

237 Emery DL, Moloo SK. The sequential cellular changes in the local skin reactions produced in goats by *Glossina morsitans morsitans* infected with *Trypanosoma (Trypanozoon) brucei*. Acta Trop 1980;37:137–149.

238 Dwinger RH, Luckins AG, Murray M, Rae P, Moloo SK. Interference between different serodemes of *Trypanosoma congolense* in the establishment of superinfection in goats following transmission by tsetse. Parasite Immunol 1986;8:293–305.

239 Dwinger RH, Murray M, Luckins AG, Rae P, Moloo SK. Interference in the establishment of tsetse-transmitted *Trypanosoma congolense*, *T. brucei* or *T. vivax* in goats already infected with *T. congolense* or *T. vivax*. Vet Parasitol 1989;30:177–189.

240 Herbert WJ. Interference between two strains of *Trypanosoma brucei*. Trans R Soc Trop Med Hyg 1975;69:272.

241 Ellis JA, Scott JR, Machugh NP, Gettinby G, Davis WC. Peripheral blood leucocyte subpopulation dynamics during *Trypanosoma congolense* infection in Boran and N'dama cattle: an analysis using monoclonal antibody and flow cytometry. Parasite Immunol 1987;9:363–378.

242 Williams DJL, Naessens J, Scott JR, Mgodimba FA. Analysis of peripheral leucocyte populations in N'dama and Boran cattle, following a rechallenge infection with *Trypanosoma congolense*. Parasite Immunol 1991;13:171–185.

243 Bakhiet M, Olsson T, Van Der Meide P, Kristensson K. Depletion of CD8$^+$ T cells suppresses growth of *Trypanosoma brucei brucei* and interferon gamma production in infected rats. Clin Exp Immunol 1990;81:195–199.

244 Mansfield JM, Craig SA, Steltzer GT. Lymphocyte function in African trypanosomiasis. II. Mitogenic effects of trypanosome extract *in vitro*. Infect Immun 1976;14:975–981.

245 Mansfield M. Immunology of African trypanosomiasis. In Wyler DG, ed. *Modern Parasite Biology*. New York: WH Freeman, 1990:222–246.

246 Pentreath VW. The search for primary events causing the pathology of African sleeping sickness. Trans R Soc Trop Med Hyg 1991;85:145–147.

247 Diggs C, Flemmings B, Dillon J, Snodgrass R, Campbell G, Esser K. Immune serum-mediated cytotoxicity against *Trypanosoma rhodesiense*. J Immunol 1976;116:1005–1009.

248 Barry JD, Vickerman K. Observations on short stumpy forms of *Trypanosoma brucei*. J Protozool 1977;24:42A.

249 Balber E. *Trypanosoma brucei*: fluxes of the morphological

variants in intact and X-irradiated mice. Exp Parasitol 1972;31:307–319.

250 Nagle RB, Ward PA, Lindsley HB, et al. Experimental infections with African trypanosomes. VI. Glomerulonephritis involving the alternate pathway of complement activation. Am J Trop Med Hyg 1974;23:15–26.

251 Greenwood B, Whittle HG. Complement activation in patients with Gambian sleeping sickness. Clin Exp Immunol 1976;24:133–138.

252 Shirazi MF, Holman M, Hudson KM, Klaus GGB, Terry RJ. Complement (C3) levels and the effect of C3 depletion in infections of Trypanosoma brucei in mice. Parasite Immunol 1980;2:155–161.

253 Dempsey WL, Mansfield JM. Lymphocyte function in experimental African trypanosomiasis. V. Role of antibody and the mononuclear phagocyte system in variant specific immunity. J Immunol 1983;130:405–411.

254 MacAskill JA, Holmes PH, Whitelaw DD, McConnell I, Jennings FW, Urquhart GM. Immunological clearance of ^{75}Se-labelled Trypanosoma brucei in mice. II. Mechanisms in immune animals. Immunology 1980;40:629–635.

255 Takayanagi T, Nakatake Y, Enriquez G. Trypanosoma gambiense: phagocytosis in vitro. Exp Parasitol 1974;36:106–113.

256 Ngaira JM, Nantulya VM, Musoke AJ, Hirumi H. Phagocytosis of antibody-sensitized Trypanosome brucei in vitro by bovine peripheral blood monocytes. Immunology 1983;49:393–400.

257 Kissling E, Karbe E, Freitas BK. In vitro phagocytic activity of neutrophils of various cattle breeds with and without Trypanosoma congolense infections. Tropenmed Parasit 1982;33:158–160.

258 Rurangirwa FR, Musoke AJ, Nantulya VM, et al. Immune effector mechanisms involved in the control of parasitaemia in Trypanosoma brucei-infected wild beast (Connochaetes taurinus). Immunology 1986;58:231–237.

259 Young AS, Kanhai GK, Stagg DA. Phagocytosis of Trypanosoma (Nannomonas) congolense by circulating macrophages in the African buffalo (Syncerus caffer). Res Vet Sci 1975;19:108–110.

260 Rurangirwa FR, Tabel H, Losos G, Tizard IR. Hemolytic complement and serum C_3 levels in Zebu cattle infected with Trypanosoma congolense and Trypanosoma vivax and the effect of trypanocidal treatment. Infect Immun 1980;27:832.

261 Wilson AJ, Paris J, Dar FK. Maintenance of a herd of breeding cattle in an area of high trypanosome challenge. Trop Anim Health Prod 1975;7:63–71.

262 Whitelaw DD, Bell IK, Holmes PH, et al. Isometamidium chloride (Samorin) prophylaxis against experimental Trypanosoma congolense challenge and its relationship to the development of immune responses in Boran cattle. Vet Rec 1986;118:722–726.

263 Barry JD. African trypanosomiasis. In Liew FY, ed. Vaccination Strategies for Tropical Diseases. Boca Raton: CRC Press, 1989:197:217.

264 Murray M, Hirumi H, Moloo SK. Suppression of Trypanosoma congolense, T. vivax and T. brucei infection rates in tsetse flies maintained on goats immunized with uncoated forms of trypanosomes grown in vitro. Parasitology 1985;91:53–66.

265 Van Meirvenne N, Le Ray P. Diagnosis of African and American trypanosomiases. Br Med Bull 1985;41:156–161.

266 Nantulya VM. Immunodiagnosis of rhodesian sleeping sickness: detection of circulating trypanosomal antigens in sera and cerebrospinal fluid by enzyme immunoassay using a monoclonal antibody. Bull Soc Pathol Exot Filial 1988;81:511–512.

267 Nantulya VM. An antigen detection enzyme immunoassay for the diagnosis of rhodesiense sleeping sickness. Parasite Immunol 1989;11:69–75.

268 Nantulya VM. Trypanosomiasis in domestic animals: the problems of diagnosis. Rev Sci Techn Off Int Epiz 1990;9:357–367.

269 Nantulya VM, Musoke AJ, Rurangirwa FR, Saigar N, Minja SH. Monoclonal antibodies that distinguish Trypanosoma congolense, T. vivax and T. brucei. Parasite Immunol 1987;9:421–431.

270 Rae PF, Luckins AG. Detection of circulating trypanosomal antigens by enzyme immunoassay. Ann Trop Med Parasitol 1984;78:587–596.

271 Seed JR, Hall JE. Pathophysiology of African trypanosomiasis. In Tizard I, ed. Immunology and Pathogenesis of Trypanosomiasis. Boca Raton: CRC Press, 1985:2–11.

272 Murray M, Dexter TM. Anaemia in bovine African trypanosomiasis. Acta Trop 1988;45:389–432.

273 Jenkins GC, Facer CA. Haematology of African trypanosomiasis. In Tizard I, ed. Immunology and Pathogenesis of Trypanosomiasis. Boca Raton: CRC Press, 1985:13–43.

274 Morrison WI, Murray M, Sayer PD, Preston JM. The pathogenesis of experimentally-induced Trypanosoma brucei infection in the dog. I. Tissue and organ damage. Am J Pathol 1981;102:168–181.

275 Askonas BA. Macrophages as mediators of immunosuppression in murine African trypanosomiasis. Curr Top Microbiol Immunol 1985;117:119–127.

276 Losos G, Chouinard A, eds. Pathogenicity of Trypanosomes. Ottawa: IDRC, 1979.

277 Borowy NK, Sternberg JM, Schreiber D, Nonnengasser C, Overath P. Suppressive macrophages occurring in murine Trypanosoma brucei infections inhibit T cell responses in vivo and in vitro. Parasite Immunol 1990;12:233–246.

278 Grosskinsky CM, Askonas BA. Macrophages as primary target cells and mediators of immune dysfunction in African trypanosomiasis. Infect Immun 1981;33:149–155.

279 Grosskinsky CM, Ezekowitz RAB, Berton G, Gordon S, Askonas BA. Macrophage activation in murine trypanosomiasis. Infect Immun 1983;39:1080–1086.

280 Bagasra O, Schell RF, Le Frock JL. Evidence for depletion of Ia$^+$ macrophages and associated immunosuppression in African trypanosomiasis. Infect Immun 1981;32:188–193.

281 Silighem M, Darji A, Remels L, Hamers R, De Baetselier P. Different mechanisms account for the suppression of interleukin 2 production and the suppression of interleukin 2 receptor expression in Trypanosoma brucei-infected mice. Eur J Immunol 1989;19:119–124.

282 Silighem M, Hamers R, De Baetselier P. Experimental Trypanosoma brucei infections selectively suppress both interleukin 2 production and interleukin 2 receptor expression. Eur J Immunol 1987;17:1417–1421.

283 Silighem M, Darji A, Hamers R, Van Der Winkel M, De Baetselier P. Dual role of macrophages in the suppression of interleukin 2 production and interleukin 2 receptor expression in trypanosome-infected mice. Eur J Immunol 1989;19:829–835.

284 Silighem M, Darji A, Hamers R, De Baetselier P. Modulation of IL-1 production and IL-1 release during experimental typanosome infections. Immunology 1989;68:137–139.

285 Bancroft GJ, Sutton CJ, Morris AG, Askonas BA. Production of interferons during experimental African trypanosomiasis. Clin Exp Immunol 1983;52:135–143.

286 Buetler B, Cerami A. Tumor necrosis, cachexia, shock and inflammation: a common mediator. Ann Rev Biochem 1988;57:505–518.

287 Boreham PFL. Autocoids: their release and possible role in the pathogenesis of African trypanosomiasis. In Tizard I, ed. *Immunology and Pathogenesis of Trypanosomiasis*. Boca Raton: CRC Press, 1985:45–66.

288 Galvao-Castro B, Hochmann A, Lambert PH. The role of the host immune response in the development of tissue lesions associated with African trypanosomiasis in mice. Clin Exp Immunol 1978;33:12–24.

289 Lambert PH, Berney M, Kazyumba G. Immune complexes in serum and in cerebrospinal fluid in African trypanosomiasis. J Clin Invest 1981;67:77–85.

290 Nagle RB, Dong S, Guillot JM, McDaniel KM, Lindsley HB. Pathology of experimental African trypanosomiasis in rabbits infected with *Trypanosome rhodesiense*. Am J Trop Med Hyg 1980;29:1187–1195.

291 Poltera AA, Hockmann A, Rudin W, Lambert PH. *Trypanosoma brucei brucei*: a model for cerebral trypanosomiasis in mice—an immunological, histological and electron-microscopical study. Clin Exp Immunol 1980;40:496–507.

292 Boreham PFL. The pathogenesis of African and American trypanosomiasis. In Levandowsky M, Hutner SH, eds. *Biochemistry and Physiology of Protozoa*, vol. 2. New York: Academic Press, 1979:429–457.

293 Morrison WI, Murray M, Whitelaw DD, Sayer PD. Pathology of infection with *Trypanosoma brucei*: disease syndromes in dogs and cattle resulting in severe tissue damage. Contrib Microbiol Immunol 1983;7:103–119.

294 Morrison WI, Murray M, Sayer PD. Pathogenesis of tissue lesions in *Trypanosoma brucei* infections. In Losos G, Chouinard A, eds. *Pathogenicity of Trypanosomes*. Ottawa: IDRC, 1979:171–177.

295 Esievo KAN, Saror DI, Kolo MN, Eduvie LO. Erythrocyte surface sialic acid in N'dama and zebu. J Comp Physiol 1986;96:95–99.

296 Facer CA, Crosskey JM, Clarkson MJ, Jenkins GC. Immune haemolytic anaemia in bovine trypanosomiasis. J Comp Pathol 1982;92:393–401.

297 Rifkin MR, Landsberger FR. Trypanosome variant surface glycoprotein transfer to target membranes: a model for pathogenesis of trypanosomiasis. Proc Natl Acad Sci USA 1990;87:801–805.

298 Cook SL, Bouma SR, Huestis WH. Cell to vesicle transfer of intrusive membrane proteins: effects of membrane fluidity. Biochemistry 1980;19:4601–4607.

299 Santos-Buch CA, Acosta AM. Pathology of Chagas' disease. In Tizard I, ed. *Immunology and Pathogenesis of Trypanosomiasis*. Boca Raton: CRC Press, 1985:145–183.

300 Seed JR, Gam AA. The presence of antibody to a normal liver antigen in rabbits infected with *Trypanosoma gambiense*. J Parasitol 1967;53:946–950.

301 Kobayakawa T, Louis J, Izui S, Lambert PH. Autoimmune response to DNA, red blood cells and thymocyte antigens in association with polyclonal antibody synthesis during experimental African trypanosomiasis. J Immunol 1979;122:296–301.

302 Mansfield JM, Kreier JP. Autoimmunity in experimental *Trypanosoma congolense* infections in rabbits. Infect Immun 1972;5:648–656.

303 Anthoons JAMS, Van Marck EAE, Gigase PLJ. Autoantibodies to intermediate filaments in experimental infections with *Trypanosoma brucei gambiense*. Z Parasitenkd 1986;72:443–452.

304 Bancroft GJ, Askonas BA. Immunobiology of African trypanosomiasis in laboratory rodents. In Tizard I, ed. *Immunology and Pathogenesis of Trypanosomes*. Boca Raton: CRC Press, 1985:75–101.

305 Greenwood BM, Whittle HC. The pathogenesis of sleeping sickness. Trans R Soc Trop Med Hyg 1980;74:716–725.

306 Adams JH, Haller L, Boa FY, Doua F, Dago A, Konian K. Human African trypanosomiasis (*Trypanosoma brucei gambiense*): a study of 16 fatal cases of sleeping sickness with some observations on acute reactive arsenical chemotherapy. Neuropathol Appl Neurol 1986;12:81–94.

307 Pentreath VW. Neurobiology of sleeping sickness. Parasitol Today 1989;5:215–218.

308 Fierez W, Fontana A. The role of astrocytes in the interaction between the immune and nervous system. In Federoff S, Vernadakis A, eds. *Astrocytes*, vol. 3. New York: Academic Press, 1986:203–230.

309 Hertting G, Seregi A. Formation and function of eicosanoids in the central nervous system. Ann NY Acad Sci 1989;559:84–99.

310 Hunter CA, Gow JW, Kennedy PGE, Jennings FW, Murray M. Immunopathology of experimental African sleeping sickness: detection of cytokine mRNA in the brains of *Trypanosoma brucei brucei*-infected mice. Infect Immun 1991;59:4636–4640.

311 Kreyger JM. Somnogenic activity of immune response modifiers. Trends Pharmacol Sci 1990;11:122–126.

312 Pentreath VW, Rees K, Owolabi OA, Philip KA, Doua F. The somnogenic T lymphocyte suppressor prostaglandin D_2 is selectively elevated in cerebrospinal fluid of advanced sleeping sickness patients. Trans R Soc Trop Med Hyg 1990;84:795–799.

313 Hayaishi O. Prostaglandin D_2 and sleep. Ann NY Acad Sci 1989;559:374–381.

314 Ashcroft MT, Burtt E, Fairbairn H. The experimental infection of some African wild animals with *Trypanosoma rhodesiense, T. brucei* and *T. congolense*. Ann Trop Med Parasitol 1959;53:147–160.

315 Grootenhuis JG, Varma Y, Black SJ, *et al*. Host response of some African wild bovidae to trypanosome infections. In Karbe E, Freitas EK, eds. *Trypanotolerance Research and Implementation*. Eschborn: 1982:37–42.

316 Murray M, Morrison WI, Whitelaw DD. Host suscepti-bility to African trypanosomiasis. Adv Parasitol 1982;21:1–68.

317 Roelants GE. Natural resistance to African trypano-somiasis. Parasite Immunol 1986;8:1–10.

318 Coulomb J, Gruvel J, Morel PC, Perreau P, Queval R, Tibayrence R, Provost A. *Trypanotolerance. Bibliographical Synthesis of Present Knowledge.* Maisons Alfort, France: Institut d'elevage et de Médecine veterinaire des pays tropicaux, 1978.

319 Paling RW, Moloo SK, Scott JR, *et al.* Susceptibility of N'dama and Boran cattle to tsetse-transmitted primary and rechallenge infections with a homologous serodeme of *Trypanosoma congolense.* Parasite Immunol 1991;13:413–425.

320 Paling RW, Moloo SK, Scott JR, Gettinby G, Mgodimba FA, Murray M. Susceptibility of N'dama and Boran cattle to sequential challenges with tsetse-transmitted clones of *Trypanosoma congolense.* Parasite Immunol 1991;13:427–445.

321 Murray M, Morrison WI, Murray PK, Clifford DJ, Trail JCM. Trypanotolerance—a review. World Anim Rev 1979;31:2–12.

322 Trail JCM, D'Ieteren GDH, Teale AJ. Trypanotolerance and the value of conserving lifestock genetic resources. Genome 1989;31:805–812.

323 Levine R, Mansfield JM. Genetics of resistance to the African trypanosomes. III. Variant specific antibody responses of H-2-compatible resistant and susceptible mice. J Immunol 1984;133:1564–1569.

324 De Gee AWL, Levine RF, Mansfield JM. Genetics of resistance to African trypanosomes. VI. Heredity of resistance and variable surface glycoprotein-specific immune responses. J Immunol 1988;140:283–288.

325 De Gee ALW, Mansfield J. Genetics of resistance to the African trypanosomes. IV. Resistance of radiation chimeras to *Trypanosoma rhodesiense* infection. Cell Immunol 1984;87:85–91.

326 Greenblatt HC, Diggs CL, Rosenstreich DL. *Trypanosoma rhodesiense*: analysis of the genetic control of resistance among mice. Infect Immun 1984;44:107–111.

327 Sendashonga CN, Black SJ. Analysis of B-cell and T-cell proliferation responses induced by monomorphic and pleomorphic *Trypanosoma brucei* parasites in mice. Parasite Immunol 1986;8:443–453.

328 Black SJ, Sendashonga CN, Lalor PA, *et al.* Regulation of growth and differentiation of *Trypanosoma (Trypanozoon) brucei brucei* in resistant (C57B1/6) and susceptible (C3H/He) mice. Parasite Immunol 1983;5:465–478.

329 Morrison WI, Murray M. The role of humoral immune responses in determining susceptibility of A/J and C57BL/b mice to infection with *Trypanosoma congolense.* Parasite Immunol 1985;7:63–79.

330 Black SJ, Sendashonga CN, Webster P, Koch GLE, Shapiro SZ. Regulation of parasite-specific antibody responses in resistant (C57BL/6) and susceptible (C3H/HE) mice infected with *Trypanosoma (Trypanozoon) brucei brucei.* Parasite Immunol 1986;8:425–442.

331 Mahan SM, Hendershot L, Black SJ. Control of trypano-destructive antibody responses and parasitaemia in mice infected with *Trypanosoma (Duttonella) vivax.* Infect Immun 1986;54:213–221.

332 Newson J, Mahan SM, Black SJ. Synthesis and secretion of immunoglobulin by spleen cells from resistant and sus-ceptible mice infected with *Trypanosoma brucei brucei* GUTat 3.1 Parasite Immunol 1990;12:125–139.

333 Seed JR, Sechelski JB. Immune response to minor variant antigen types (VATs) in a mixed VAT infection of the African trypanosomes. Parasite Immunol 1988;10:569–580.

334 McLintock L, Turner CMR, Vickerman K. A comparison of multiplication rates in primary and challenge infections of *Trypanosoma brucei* bloodstream forms. Parasitology 1990;101:49–55.

335 Seed JR, Sechelski JB. African trypanosomes: inheritance of factors involved in resistance. Exp Parasitol 1989;69:1–8.

336 Jack RM, Black SJ, Reed SL, Davis CE. Indomethacin promotes differentiation of *Trypanosoma brucei.* Infect Immun 1984;43:443–448.

337 Giffin BF, McCann PP, Bitonti AJ, Bacchi CJ. Polyamine depletion following exposure to DL-*a*-Difluoromethyl-ornithine both *in vivo* and *in vitro* initiates morphological alterations and mitochondrial activation in a monomorphic strain of *Trypanosoma brucei brucei.* J Protozool 1986;33:238–243.

11 South American trypanosomiasis (Chagas' disease)

Garry B. Takle & David Snary

THE PARASITE

The protozoan flagellate *Trypanosoma cruzi* causes Chagas' disease or American trypanosomiasis in humans. It is estimated that approximately 35 million people living in rural areas of Central and South America are at risk from infection, and that between 10 and 12 million have Chagas' disease [1].

Trypanosoma cruzi is a Stercorarian trypanosome, in contrast to the African trypanosomes which are Salivarian. It is transmitted as the infective metacyclic form to mammalian hosts via the insect vector's feces. The vectors of *T. cruzi* are the blood sucking reduviid bugs of the subfamily Triatominae. Following feeding by the vector, metacyclic trypanosomes enter the mammal either through the puncture wound or through nearby mucous membranes. Infection can occur congenitally and transmission during blood transfusion is still a major problem [2]. Once inside the body the metacyclics do not divide extracellularly but enter macrophages or nearby tissue cells in a polarized (Fig. 11.1) [3] and probably active manner, escape from the lysosomal compartment into the cytoplasm, and transform into rounded amastigotes. Escape from the phagolysosomes probably involves an acid-active molecule, or hemolysin, immunologically related to the C9 component of complement, that has pore-forming activity [4]. Intracellular division occurs several times by binary fission, the infected cells rupture, and the parasites are released as tissue-derived trypomastigotes which are capable of infecting surrounding cells or dissemination to other tissues via the bloodstream. Characteristically, muscle and neuronal tissues of all types are most heavily parasitized. The *T. cruzi* life-cycle is completed when trypomastigotes are taken up during feeding by the reduviid bugs and, in the insect midgut, these transform into the insect forms (or epimastigotes). The factors controlling differentiation into metacyclic forms in the insect midgut

are still unclear but the process is possibly mediated by an insect gut-wall lectin binding to parasite surface receptors [5]. Metacyclics finally migrate to the hindgut from where they can be voided in the feces.

The various *T. cruzi* strains that have been described are generally similar with regard to antigenicity [6] but show differences, particularly in susceptibility to host immune responses [7] and biochemical characteristics, such as isoenzyme profile (zymodeme) [8], kinetoplast DNA restriction maps (schizodeme) [9], and, to a lesser degree, morphology [10]. With the advent of recombinant DNA technology it is now possible to compare interstrain differences using cloned gene probes; this methodology has already revealed that there are restriction length polymorphisms between *T. cruzi* strains [11], and the breakthrough presented by polymerase chain reaction (PCR) technology has allowed the identification of specific probes for kinetoplast DNA, which can differentiate between major *T. cruzi* clonal populations [12]. However, a definitive description of possible subspeciation within *T. cruzi* has not been developed, although extensive population genetics studies using zymodemes and schizodemes have demonstrated a clonal structure to the natural population of *T. cruzi* [13,14]. The fine structure of *T. cruzi* has been reviewed in detail by Brener [10]; an important observation is that *T. cruzi* does not possess the clearly defined variant surface glycoprotein (VSG) coat characteristic of bloodstream forms of the African trypanosomes, and thus may not exhibit an identical form of semiprogrammed antigenic variation. Unlike African trypanosomes, *T. cruzi* can evade the host immune response by hiding within cells and dividing there unrestrained. However, an intriguing possible alternative is that by controlling the expression of multigene families, *T. cruzi* may be able to gain further advantage by modulating the expression of its surface antigens to present a constantly changing face to the host's immune system.

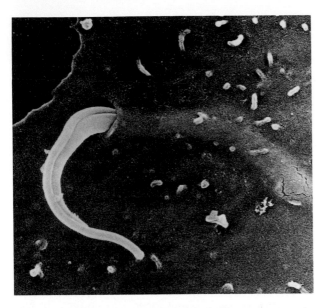

Fig. 11.1 Scanning electron micrograph of a *Trypanosoma cruzi* metacyclic trypomastigote invading a Mabin–Darby canine kidney cell. Courtesy of Dr S. Schenkman.

The entire *T. cruzi* life-cycle can be reproduced *in vitro* and this has obviously greatly facilitated the study of various aspects of the biology of the parasite. Epimastigotes grow rapidly in semidefined media containing hemin [15,16]. As the epimastigote cultures reach the stationary growth phase, metacyclic trypomastigotes start to appear and these can be separated from epimastigotes by centrifugation methods [17]. The use of an impoverished medium [18] increases the rate of metacyclogenesis. Bloodstream-form trypomastigotes are cultured by infecting monolayer cell cultures such as Vero cells [19] and allowing the intracellular amastigotes to multiply, finally releasing large numbers of trypomastigotes approximately 1 week after infection. During long-term laboratory maintenance it is routinely necessary to passage periodically the cultured bloodstream forms through mice to ensure that they retain infectivity.

CHAGAS' DISEASE

Following initial infection, the acute phase of Chagas' disease ensues; this is characterized by high blood parasitemia and extensive local tissue parasitism. Usually acute Chagas' disease is asymptomatic in adults but this is not always the case in children, where symptoms such as fever, vomiting, enlarged lymphoid tissue, and myocarditis are observable [20]. Swelling and skin lesions or "Chagoma" are frequently observed at the site of parasite entry. The high blood parasitemia subsides after 1–2 months, coincident with the development of a vigorous immune response, and patients enter a latent or indeterminate phase [21]. This phase is characterized by low-level parasitemia and is essentially asymptomatic. During the latent period, loss of immunity will lead to parasite proliferation [22]. Approximately 40% of *T. cruzi*-infected patients belong to the indeterminate or latent group. Some of these patients do show minor clinical symptoms of Chagas' disease, such as foci of inflammation and heart fibrosis [21].

The acute and latent phases of Chagas' disease thus appear to be relatively asymptomatic, and the more serious clinical manifestations of *T. cruzi* infection only occur much later in the progression of the disease. However, the acute and latent phases are not functionally benign, since muscle or neuronal damage that occurs during the period of high parasitemia is likely to be involved in the initiation of any autoimmune response [23]. The indeterminate phase of Chagas' disease lasts for an unspecified period. It may be a period of gradual enhancement of the symptoms characteristic of the chronic disease, possibly as a direct result of slowly developing autoimmunity.

The heart and the digestive tract are the organs specifically affected by chronic Chagas' disease. In the cardiac disease, most common in Brazil [24], enlargement of the heart, electrocardiogram (ECG) abnormalities, dilatation, and left ventricular apical aneurysms are found, and the whole syndrome is known as megacardia [25,26]. The digestive form of the disease involves aperistalsis and enlargement of regions of the alimentary canal, giving the characteristic megaesophagus and megacolon. Both cardiac and digestive forms of the disease appear to have a common cause associated with specific destruction of the autonomic nervous supply to the affected organs.

Frequently ganglion cells are destroyed, either by infection with the parasites or by the action of autoantibodies. Characteristic features of chronic disease, such as reduced control of heart beat and peristalsis, as a result of a reduction in muscle tone and muscular efficiency is caused by this loss of neurons. In addition to this, there is direct parasitization of cardiac and other muscle cells, and loss of musculature may contribute to loss of control and massive enlargement of these tissues. Sudden and unexpected death occurs in 60–70% of patients with Chagas' disease [26] and this is as a direct result of heart failure due to denervation and muscle block.

HOST–PARASITE RELATIONSHIP

The production of specific antibody and increased hypersensitivity reactions are evidence that immunity rapidly develops when animals are infected with *T. cruzi*. This immunity is required for continued resistance to parasite proliferation since animals immunocompromised by various methods, such as drug treatment [27], irradiation [28], and depletion of T cells [29], show markedly increased susceptibility to *T. cruzi* infection. Numerous studies have shown that *T. cruzi* infection stimulates nonspecific immunosuppression against unrelated antigens and in this way acute Chagas' disease is similar to many other parasite infections [30–32]. In Chagas' disease, immunosuppression may be a mechanism used by the parasites to establish an initial infection, and a variety of mechanisms have been proposed for the induction of immunosuppression by *T. cruzi*. This suppression may also be associated with the autoimmunity that characterizes the chronic phase of the disease, since suppressor T cells and T helper cells are likely to be involved in both processes [33,34]. The interplay between parasite-induced suppression and autoimmunity is complex and the possible roles of host responses in these phenomena are difficult to dissect. However, the possibility exists that autoimmunity could simply be the result of *T. cruzi* and selected host tissues having a range of similar and crossreactive surface antigens, and the appearance of parasite-induced immunosuppression, affecting control over autoimmune responses to these crossreactive antigens, could thus open the door for full-blown chronic chagasic syndromes.

The most intimate physical host–parasite relationship in Chagas' disease occurs by direct contact of the parasite with host cells (Fig. 11.1); this contact, leading to subsequent internalization, is essential to the survival of the parasite. Numerous parasite factors have been proposed to be involved in this contact and a number of surface antigens have been identified which may have a role in parasite attachment and penetration. Different groups have reported the identification of surface molecules, many with molecular weights of approximately 85 kD, antibodies to which prevent parasite internalization [35–39]. Many of these antigens can only be classified as "penetration–associated," while surface receptors which bind fibronectin [35] and collagen [40] may be involved more specifically in cell recognition and binding. It has been suggested that structures such as flagellar filopodia are used during attachment, and

although flagellum-specific surface proteins capable of acting as cell receptors have yet to be identified, cell penetration clearly takes place in a polarized manner [3]. Following receptor-mediated contact of the host cell by *T. cruzi*, it is possible that enzymes such as proteases [41] and glycosidases [42,43] may play a role in facilitating parasite entry. Many proteases, some life-cycle stage specific, with a variety of substrate specificities have been identified [41,44,45], and it is becoming evident that a major family of trypomastigote surface antigens contains regions of homology with bacterial sialidases [46]. There is, however, no evidence to suggest that host cell membrane proteins are digested during parasite penetration, and it may be that these surface enzymes use their active sites for specific binding to host cell surfaces. Once inside the host cell, *T. cruzi* parasites escape from membrane-bound structures into the cytoplasm, thus escaping the potentially destructive lysosomal enzymes. When inside the cytoplasm, *T. cruzi* rapidly transforms to the amastigote form and commences dividing; the mechanism by which *T. cruzi* achieves this intracytoplasmic habit is thought to involve a putative hemolysin, possibly related to the C9 component of complement [4].

Many tissues and cells have been reported to be successfully parasitized by *T. cruzi* [46]. For some strains there is a preference for muscle or neuronal cells, whereas others preferentially infect cells of the reticuloendothelial system. These differences are reproduced *in vitro*, but it is unclear whether these cell types possess greater numbers of parasite-specific receptors or provide higher concentrations of specific ligands for parasite binding. The promiscuity which *T. cruzi* demonstrates for a wide variety of host cell types *in vitro* could reflect the simultaneous expression of multiple members of a multigene family of closely related cell attachment proteins [46–48].

INNATE IMMUNITY

A wide variety of animals, including over a hundred different mammalian species, are known to be capable of being infected by *T. cruzi*. Amphibians and birds, however, are refractory to infection [10], and the resistance to infection of birds correlates directly with the lytic effects of normal serum on the parasite. Lysis of trypomastigotes in normal chicken serum is caused by the direct action of the C3 component of complement on the trypomastigote. The classic pathway (C142) is not involved and nor is antibody, since immunosuppression of chickens by various regimes does not influence their resistance to infection [49]. The epimastigote form of the parasite, which is the replicative form found in the gut of the insect vector and the form most easily grown in culture, is capable of activating normal mammalian sera to give lysis. The trypomastigote form is, however, not lysed in the absence of antibody. This differential sensitivity has been used to separate selectively *in vitro*-grown trypomastigotes and epimastigotes from each other [50]. Epimastigote sensitivity has been shown to be caused by activation of C3 after binding to a 72 kD cell surface glycoprotein [51]. Although this same glycoprotein is expressed by metacyclic trypomastigotes [52], lysis is not given; this resistance to lysis has been related to the presence of a 90 kD glycoprotein with a decay accelerating factor (DAF) activity on the surface of the parasite. This activity is shed into the medium on *in vitro* cultivation and inhibits the formation of C3 convertase activity formed by the interaction of factor B and C3b [53]. How these known parasite–complement interactions relate to events in the chicken is uncertain, but it may be that the parasite DAF activity does not interfere with the generation or accelerate the decay of C3 convertase in avian serum.

IMMUNITY ASSOCIATED WITH INFECTION

Infection with *T. cruzi* gives rise to an immune response in both humans and animals; this immune response is both humoral and cell mediated and is evident both as specific antibody and hypersensitivity reactions. The importance of immune mechanisms in establishing and maintaining the host–parasite relationship is clear from studies which have shown that immunosuppression by drugs [27], T cell depletion [29], and irradiation [28] have led to increased parasitemias and mortality. The presence of an immune response does not guarantee resistance to a further challenge, since infecting mice with one strain of parasite and rechallenging with a second strain, although giving apparent resistance to the challenge by the prevention of overt acute parasitemia, does not give an absolute block to infection, and both initial and rechallenge strains can be isolated from the blood of the mouse [7]. Furthermore, chronically infected mice which have been drug cured with Nifurtimox can still be reinfected with *T. cruzi* [54]. Therefore, although there is clear evidence for an immune response to the parasite, there is no evidence that this response can lead to a cure; rather, it maintains a host–parasite balance which lasts for the lifetime of the patient.

GENETIC BASIS OF IMMUNITY

All standard laboratory animals can be infected with *T. cruzi* and the course of infection is similar to that found in humans, provided a suitable infectious dose is found, with an acute phase when parasitemias are readily detected, followed by a chronic phase when parasitemias are subpatent. Their relative susceptibility of animals varies not only between but also within species, e.g., inbred strains of mice vary greatly in their susceptibility. The genetic basis for this susceptibility is unknown, but it is not linked to HLA in humans [55] or to H2 in mice [25,56]. Susceptibility appears to be complex and involves more than one genetic locus, although these loci have not been identified [57]. Studies have suggested that a slow or low initial antibody response during the early acute phase of infection is indicative of susceptibility [58]; however, these observations are not universally true, even for inbred mouse strains, and other additional factors also appear to be involved in the early control of parasite proliferation and related susceptibility [59,60].

HUMORAL IMMUNITY

Routine tests, e.g., immunofluorescence, hemagglutination, and complement fixation, readily detect antibody to *T. cruzi* in infected humans and animals. Specific antibody develops shortly after infection, initially during the acute phase as IgM and later as IgG and IgA [61].

Trypomastigotes recovered from the serum of infected animals are coated with immunoglobulin and many observations suggest that immunoglobulin could have a functional role in resistance to reinfection. These observations include: Biozzi mice selected for high or low antibody levels have differential sensitivity to infection [62], and the increased resistance found for the Biozzi "high" (BH) mouse has been ascribed to its ability to mount a more rapid, as well as a greater, antibody response to the parasite on infection [58]; repeated injection of anti-μ serum into rats to give B cell depletion also increases sensitivity to infection [63]; and mice recovered from infection with *T. cruzi* are resistant to reinfection and this resistance can be transferred to naïve mice by transfer of their serum [64]. Although the transferred protection is effective against homologous and heterologous challenge by different *T. cruzi* strains, it is possible that more than one mechanism of resistance can be effective, since passive transfer of serum is not effective against the CL strain of the parasite [7]. The importance of antibody has been confirmed by spleen cell transfer experiments; resistance can be transferred by spleen cells, but the level of resistance is reduced by depletion of B cells before transfer [65,66]. Not all classes of antibody are effective in transfer of immunity: IgM produced at an early stage of infection is not effective in transferring immunity; and later during infection the predominant isotypes produced, IgG2a and IgG2b, are effective whereas IgG1 is not [67–69]. Significantly, human chronic serum contains increased levels of IgG1 and IgG3, which are the equivalent of the mouse IgG2 antibody classes, i.e., they function in both antibody-dependent cell-mediated cytotoxicity (ADCC) and complement activation. This suggests that data derived from mouse studies are relevant to human infection [69].

Antibody will cause the parasite both to agglutinate and cap. Again the CL strain is the exception to the rule and CL blood-stage trypomastigotes are not agglutinated *in vitro* [70]. Immunofluorescence studies have shown that surface antigen caps into polar patches in the presence of antibody [71,72]. The ability to cap off surface antigens recognized by lytic antibody may have a role in resistance to immunity.

Complement will lyse blood trypomastigotes *in vitro* in the presence of sera from infected animals [73,74], but again the CL strain does not conform with these observations [75]. Lysis results from the activation of both alternative and classic pathways. That these *in vitro* observations reflect *in vivo* behavior was suggested by the finding that mice injected with cobra venom factor showed decreased resistance to infection [73]. However, contradictory observations were made with mice genetically deficient in C5 (unable to mount complement-mediated lysis) and in guinea pigs genetically deficient in C4 (lacking the classic C124 pathway of activation), both of which showed no increase in susceptibility to infection [76]. Further evidence for the absence of a role for complement was obtained when trypomastigotes were implanted into immune mice in Millipore chambers; the trypomastigotes were unaffected, even though the sera from the animals would protect on passive transfer [5,67]. The role of complement is further complicated by the observation that trypomastigotes express a molecule on their cell surfaces which has DAF activity [53]. The DAF-like activity, which is associated with a molecule of approximately 90 kD, inhibits the formation of C3 convertase and blocks complement-mediated lysis.

Significantly, lysis of trypomastigotes *in vitro* is only brought about by sera from infected animals and not by sera from animals immunized with lysed parasites or parasite antigen [77]. It has been postulated that infection sera could be blocking the DAF-like activity [78], and the sensitivity to lysis has been associated with another antigen of molecular weight 160 kD and not the 90 kD DAF molecule. Furthermore, a second anticomplement activity has been found associated with the fibronectin/collagen receptor of the parasite, which inhibits the action of C3 convertase in a manner distinctly different from that found for the DAF protein [79]. The role of complement is further complicated by observations that antibody fragments promote complement activation in the absence of an Fc portion of the immunoglobulin [80]. Therefore, the precise role for complement in *T. cruzi* infection is unclear since the parasite is capable of both activating the complement pathway in the absence of antibody Fc and of inhibiting complement activation [81,82].

Sera from individuals infected with *T. cruzi*, unlike sera from individuals with malaria and African trypanosomiasis, do not contain unusually high levels of immunoglobulin. However, polyclonal B cell activation has been reported in mice after *T. cruzi* infection. Increases in levels of plaque-forming cells to heterologous antigens and autoantigens have been observed, and the isotypes of the plaque-forming cells are predominately IgG2a, IgG2b, and IgG1. These effects appear to be related to both T helper-dependent regulation and a mitogenic effect of the parasite itself [83,84], and are persistent, still being present 6 months after infection when the isotype distribution of the plaque-forming cells has not changed [83]. Conversely, the changing population of spleen cells in mice on infection, and particularly the increase in B cells, has been shown to mirror the antibody response [85]. These increased cell levels were claimed to reflect the high and sustained antibody levels present in mice and not an exhaustive B cell polyclonal activation [86].

ANTIBODY-DEPENDENT CELL-MEDIATED IMMUNITY

In vitro studies have shown that antibody can be involved in cell-mediated immunity. Various cell types have been shown to be effective against antibody-coated blood trypomastigotes. When coated with mouse antibody, trypomastigotes are lysed or inactivated by lymph node cells, eosinophils, neutrophils, granulocytes, mononuclear cells, and platelets [87–90]. Human eosinophils and neutrophils are also effective in these *in vitro* tests and these cells can lyse the amastigote form of the parasite which replicates within mammalian cells [91,92]. The mechanism of killing of the parasite by these cells appears to be by both phagocytosis [93] and by the direct effect of hydrogen peroxide on the parasite [91], an effect that can be mimicked *in vitro* and inhibited by catalase. The binding of eosinophil basic protein to the parasite surface after phagocytosis may also be involved in parasite killing [92]. A further lytic mechanism involving antibody, complement, and platelets has been reported [90]; blood from C5 deficient mice will agglutinate trypomastigotes and platelets adhere to these clumps before parasite lysis. Recent studies have shown that whereas immune sera will agglutinate trypomastigotes *in vitro*, in the presence of platelets the parasites are killed; this phenomenum appears to be dependent upon the platelet C3b receptor since depletion of C3 abolished the lytic effects [90].

The relevance of ADCC to immunity *in vivo* remains to be determined but the resistance to lysis of trypomastigotes implanted in chambers into infected mice [5,67], and the apparent opsonization of radiolabeled parasites to the liver and spleen, accompanied by killing of these parasites, suggests that ADCC could be a significant factor in immunity [94]. This is a conclusion which is supported by the dominant role of antibody in protection and the uncompromised sensitivity of animals genetically deficient in complement components [76,90].

T CELL-MEDIATED IMMUNITY AND RESISTANCE TO INFECTION

The standard correlates of T cell-mediated immunity, delayed-type hypersensitivity (DTH) skin reactions, macrophage migration inhibition, and blastogenic responses are found in humans and animals infected with *T. cruzi*, and there is clear evidence for the existence of this type of immunity [95]. However, in contrast to the consistent appearance of antibody after infection, the expression of cell-mediated immunity is much more variable. In humans, skin reactions can be weak or absent [96,97] and this limits the usefulness of this type of test for diagnosis, especially since the presence of strong antibody-mediated immediate responses may require the use of a different antigen preparation to stimulate a delayed cell-mediated response and to discriminate this from the persistent immediate response [98,99].

Guinea pigs [100], monkeys [101], and rabbits [99] have all shown delayed skin reactions, which in the case of rabbits could be transferred to normal recipients by lymphoid cells. Mice undergoing an acute infection give a prolonged immediate skin reaction which can still be detected 24 h later and is undiminished from the 3-h reaction. However, no delayed skin reactions were detectable, and whereas the immediate reactions could be transferred by antibody and were probably of the Arthus type, neither the immediate nor any delayed reactions could be transferred by cells [66]. *Trypanosoma cruzi* induces suppression in antibody and cell-mediated responses [86,102], predominantly during the acute phase [102]. This suppression is evident in *in vitro* T cell responses, and an activity present in spleens from chronically infected mice suppresses delayed skin reactions [103] and mitogenic responses [26]. The suppression of delayed skin reactions has been associated with a glycoprotein of between 30 and 60 kD which is not IL-1, IL-2, or IL-3 and is not of parasite origin [104]. Although the suppressive activity for delayed skin reactions was found in chronically infected mouse spleens, mice still have functional T cells at this stage of infection since positive responses in macrophage migration inhibition assays have been observed [101, 105].

The weak or absent antiparasite cell-mediated immunity observed in humans and animals could have major significance to the development of Chagas' disease. Suppression of DTH responses by splenic factors present during the chronic phase of the disease could play a role in the suppression of the chronic pathology since lymphocytes of the appropriate phenotype for DTH positive T cells are found at the site of chronic lesions [106,107], as is persistent antigen [108]. Suppression of the activity of these cells could prevent or reduce the pathology associated with the chronic disease, since T cell clones stimulated by both *T. cruzi* and peripheral nerve tissue have been isolated from mice, and transfer of T cells depleted of Lyt-2 cells from infected mice transferred the DTH response and formed inflammatory lesions [106].

T cells are involved in resistance to infection, since mice congenitally [25,109] or experimentally [26] deficient in T cells are more susceptible to infection. These data do not discriminate between a direct T cell action or a helper role in antibody production, although it has been shown that neonatally thymectomyzed mice have a delayed antibody response [110]. Spleen cells from mice recovered from infection will transfer protection to normal individuals but depletion of T cells only marginally affects the ability to confer protection, and pure T cells do not transfer protection. These data suggest that immunity is predominately B cell dependent and antibody based [66]. A further study used mice immunized with epimastigotes and challenged with trypomastigotes, and although protection could be transferred with enriched T cells and cells depleted of B cells, on a per cell basis the depleted T cell population was more effective in providing protection [65]. The above data do not support a role for T cells in immunity; however, two studies—with spleen cells from infected mice [29] and from mice immunized with epimastigotes [111]—demonstrated that immunity can be transferred with T cells. Taken together, these data are difficult to reconcile but may relate to different parasite strains or experimental design; even when T cell transfers have reportedly given rise to immunity, the experimental design does not allow the discrimination between helper functions and direct cytotoxicity. However, treatment of mice with anti-L3T4 antibodies has shown that effector cell generation is predominantly T helper cell dependent [112]. Studies using cell surface markers have confirmed that T helper cells are found at the site of lesions in infected mice [107] and T helper cell clones have also been recovered from these mice [113]. These data, along with the increased sensitivity to infection and the reduced specific antibody production in mice specifically depleted of T helper/inducer cells [114], all support the concept of a major role for antibody in immunity and a primary role for T helper cells in the induction of this antibody.

Infected animals do not have cytotoxic T cells which act directly on the trypomastigote [61], although polyclonal cytotoxic T cell responses have been reported in mice [85]. Cytotoxic T cells against infected cells are present, but these cells, which probably recognize parasite antigens adhering to the surface of the infected cells, did not kill the parasite directly [115], although such a destruction of a host cell may render the parasite vulnerable to other forms of attack.

MACROPHAGES

The role of the macrophage in immunity to *T. cruzi* is complicated by the fact that this cell can both kill and support the intracellular growth of the parasite. Some strains of *T. cruzi* are reticulotrophic and will even preferentially invade spleen, liver, and bone marrow macrophages *in vivo* [46]. The reticulotrophic Y strain is also taken up *in vitro* by normal macrophages to a greater extent than the myotrophic CL strain [116,117].

Trypomastigotes can enter normal macrophages either by phagocytosis [118,119] or by active penetration [116]. Phagocytosis is mediated through pronase-sensitive macrophage surface proteins [117] and, whereas antibody to the parasite blocks invasion of fibroblasts, antibody potentiates infection of the macrophage presumably by interaction with Fc receptors [120,121]. The interaction is further complicated by the observation that, in order to be effective, polyspecific antiserum to *T. cruzi* needs to contain within it an antibody targeted to a specific glycoprotein found on the surface of living trypomastigotes [122], although it is possible to interpret this to mean that antibody on the surface of the target parasite is essential for Fc receptor interaction.

Although antibody enhances parasite uptake, it is not essential for phagocytosis. Trypomastigotes prepared from irradiated mice, with no evidence of surface immunoglobulin, are readily taken up by macrophages [117], and blockade of the Fc receptors by antimacrophage IgG does not impair ingestion [118]. The presence of antibody does not affect the ultimate fate of the parasite, since uptake is enhanced by antibody with normal macrophages *in vitro* and both uptake and killing are increased with activated macrophages [116,117]. After uptake, within 24 h, the parasite either escapes from the phagosome into the cytoplasm and commences its replicative cycle [118,119] or the parasite is killed within the phagolysosome, probably by mechanisms involving an oxidative burst [89,118,123,124]. Evidence to support the assumption that superoxides and hydrogen peroxide are involved in the killing process is provided by studies with eosinophils, where peroxidase from the eosinophil was shown to bind to the parasite surface and generate hydrogen peroxide, killing the trypomastigote in the process [124]. Epimastigotes are killed even by unactivated macrophages and do not survive to establish an infection [116,118], although this is not unexpected since this stage of the parasite is not adapted for mammalian intracellular development.

Available evidence suggests that activated macrophages play a role in immunity to *T. cruzi*. Selective destruction of macrophages in mice by silica treatment increased sensitivity to infection [28] whereas activation of macrophages by *Corynebacterium parvum* increased resistance [125,126] in the acute phase. Furthermore, histologic examination of inflammatory lesions in rats and mice have shown the presence during the course of infection of macrophages containing digested parasites [127,128]. Also, macrophages isolated from mice after Bacille Calmette-Guerin (BCG) treatment or *T. cruzi* infection were less supportive of *T. cruzi* intracellular growth than normal macrophages [129,130]. To kill the parasite the macrophage must become activated. Soluble factors elicited from T cells from mice infected with *T. cruzi* will activate macrophages to kill the parasite [113,131]. The T cells involved in this activation are helper cells [113] of the helper/inducer type and these could stimulate the production of numerous different factors which could be involved in macrophage activation. Interferons activate macrophages, and pretreatment of mice with the interferon inducer Tilerone enhanced resistance [132]. Interferon-α (IFN-α), IFN-β, and IFN-γ all potentiated *in vitro* killing of trypomastigotes by macrophages [118], although IFN-γ was the most effective at stimulating trypanocidal activity. Other cytokines, such as granulocyte macrophage colony stimulating factor (GM-CSF) and tumor necrosis factor (TNF), are also effective in inducing trypanocidal activity, although again none was as effective as IFN-γ [133]. However, IFN-γ alone is not as effective as the soluble supernatant from immune T cells [134], and these observations suggest that more than one factor is involved in the potentiation or activation of macrophages to kill *T. cruzi*. Although the factors involved could be enhancing the apparent killing activity by acting on other cell types, for instance natural killer cell activity is boosted by IFN-α and IFN-β, these cells will kill *T. cruzi* trypomastigotes [114,135].

The mechanisms of macrophage activation, although triggered by specific immune reactions, are in themselves nonspecific and capable of being stimulated by other agents, such as *C. parvum*, purified protein derivative (PPD), and BCG, all of which can induce trypanocidal activity [118,121,125]. However, whether this trypanocidal activity has a direct role in resistance to infection is unclear, since mouse strains sensitive to infection by the CL strain of parasite had greater levels of macrophage activation, evident as TNF and hydrogen peroxide release, than resistant mouse strains [136].

IMMUNOPATHOLOGY AND AUTOIMMUNITY

The discovery of autoantibodies to normal endocardium, vascular structures, and interstitia (EVI antibodies) [137,138] in chronic chagasic sera, thought to act as β-adrenergic agonists [139], led to the hypothesis that autoimmunity may be involved in the pathology of Chagas' disease. Further reports have described autoantibodies against Schwann cells [140], laminin

[141,142], myelin, striated muscle [138], and neurons [143] in chagasic sera. Similar antibodies can also be detected in acute Chagas' disease soon after initial infection [144,145]. In both chronic and acute sera, EVI activity can be removed by adsorption with *T. cruzi*, which implies that EVI serum contains antibodies which crossreact to both mammalian tissue and *T. cruzi* [137,138]; EVI antibodies can also be induced experimentally [146], but it is debatable whether the presence of EVI activity can be directly correlated with the presence or severity of chronic chagasic lesions [137,147,148]. Cell-mediated mechanisms of autoimmunity may also be involved in cardiac pathology in Chagas' disease since infected rabbits, or rabbits immunized with subcellular fractions of *T. cruzi*, that concurrently demonstrate cell-mediated immunity to *T. cruzi* and heart antigens develop chronic myocarditis [99]. Similar patterns are seen in human patients, where blood leukocytes are reactive to heart antigens and human cells *in vitro*, as well as to *T. cruzi* [149], and in mice, where T cells have been shown to bind *T. cruzi*-infected heart tissue *in vitro* [150]. Cytotoxic T cells have also been detected with dual reactivity to *T. cruzi* and heart tissue [149].

Autoimmune mechanisms are probably a major effector function in the generation of chronic Chagas' disease, and there are at least five possible routes by which these effects could be triggered:

1 crossreactive parasite surface antigens mimicking host antigens [151,152];

2 parasite antigens adsorbing to host cells [115];

3 host cell antigens adsorbing to parasites [153–155];

4 infected host cells expressing parasite antigen on their surface [156];

5 the appearance of crossreactive epitopes in host antigens by modification or alteration during the lysis of infected cells.

These routes are probably not mutually exclusive and more than one may be involved in the stimulation of autoimmunity.

It has been demonstrated that *T. cruzi* antigens are adsorbed to host tissues and cells [115,154] and that host material can adsorb to the parasites [155]; it is possible that antigen presented in this manner could give rise to an autoimmune response. Allied to these observations is the possibility that infected cells could actively express parasite antigens on their surface membrane, although there is controversy about whether this actually does occur since it is difficult to control for passively adsorbed antigens [7,67,157]. Available evidence, however, does suggest that infected host cells are able to incorporate parasite (glyco) protein(s) into their surface membrane [156,158]. Alternatively, the destruction of host neurons and other tissue cells, as well as being a direct cause of Chagas' disease syndromes, would release host antigen, some of which could be modified or altered and thus further stimulate or enhance autoimmunity. This hypothesis will be difficult to examine rigorously until a reliable *in vitro* or *in vivo* model system becomes available.

The most obvious way by which a parasite could induce autoimmunity is through shared or crossreactive surface antigens or epitopes common to both host and parasite. There is considerable evidence that this occurs in Chagas' disease and crossreactive surface molecules have been defined using chagasic sera, monoclonal antibodies, and recombinant DNA technology. Laminin is the major noncollagenous component of connective, muscle, and neuronal tissue, and it has been shown to be a target for chagasic EVI antibodies from both the acute and chronic phases of the disease. The activity contained within EVI serum directed towards basement membrane is specific for laminin and does not react with any other component of connective tissue [141]; a laminin-like molecule has been identified on the surface of *T. cruzi* trypomastigotes [159]. Crossreactive antibodies to this molecule specifically recognize terminal galactosyl-α-1,3-galactose groups on *N*-linked oligo-saccharides [160]. *Trypanosoma rangeli*, the closely related but nonpathogenic South American trypanosome, does not possess a similar surface laminin-like molecule [159].

The array of surface antigens on *T. cruzi* is becoming better characterized, and as information on possible crossreactive molecules or epitopes becomes available it is clear that *T. cruzi* trypomastigotes do express surface molecules which may be functionally or immuno-logically similar to their human counterparts. Examples are heat shock proteins [161] and the fibronectin and collagen receptors [35,40]. A 160 kD trypomastigote-specific recombinant antigen has also been identified which, by monoclonal antibody reactions, has been shown to contain epitopes that are crossreactive with a range of host tissues [162]. Since this antigen was pre-pared as an *Escherichia coli* recombinant protein, the crossreactive component must be encoded within the polypeptide sequence, the epitope probably being linear rather than conformational. Many other *T. cruzi* surface antigens contain crossreactive sulfated lipid or car-bohydrate structures [153,160] and these can be the

most immunogenic part of these molecules. The identity of the specific epitope recognized by another cross-reactive monoclonal antibody (CE5) that binds to mammalian neurones and to numerous *T. cruzi* antigens in Western blots [152] is not known, although it may be a common structure, such as a ubiquitous carbohydrate side-chain on glycoproteins, or a glycolipid structure, such as the glycosylphosphatidylinositol (GPI) tail found on several membrane proteins [163,164]. Numerous different *T. cruzi* surface molecules appear to be cross-reactive with host antigens and there does not seem to be a clearly identifiable unique structure common to these antigens which can be associated with crossre-activity. However, a complete examination of cross-reactive epitopes has only been carried out in the case of laminin, and further epitope mapping of other cross-reactive antigens is needed before the molecular basis of the reactions is fully understood.

Autoimmunity is thought to arise from the breakdown of self-tolerance [165]. Self-tolerance results from the interaction of a range of suppressor or contrasuppressor mechanisms, the regulated involvement of antiself T and B lymphocytes (clonal anergy) [166], clonal deletion of antiself lymphocytes, and complex idiotype net-works. Such a finely balanced multicomponent network is clearly susceptible to perturbation, and disturbance of any of these mechanisms may lead to autoimmune or antiself responses. Since infection by *T. cruzi* leads to immunosuppression, there is a clear possibility that this may have a causal effect in the autoimmunity associated with Chagas' disease. The initial acute phase of Chagas' disease is invariably characterized by a significant blastogenic response of both B and T lymphocytes and this is accompanied by severe immunodepression. These polyclonal responses may also be the cause of the various autoimmune phenomena, since polyclonal B cell activation could lead to autoantibody production, and autoreactive cytotoxic or T helper cell clones could be activated [149].

IMMUNOSUPPRESSION

Nonspecific immunosuppression towards unrelated antigens and tumors is a characteristic feature of many parasitic diseases, including Chagas' disease [167]. Numerous studies have been carried out to elucidate the causes of chagasic immunosuppression, and several factors have been implicated. Immunosuppression in Chagas' disease is evident by reduced antibody pro-duction [30,168], reduced IL-2 secretion by T lymphocytes [169–175], reduced T cell proliferation *in vitro*, [34,102,176,177], reduction of cytolytic T lymphocyte activity [31], reduced IL-2 receptor expression [178], and reduced DTH reactions to *T. cruzi* [179]. From these data it is not possible to attribute chagasic immunosuppression to a single cause or defect, although the use of different criteria for assessing immunosuppression has yielded conflicting results. Furthermore, there is evidence from *in vitro* studies that spleen cells from *T. cruzi*-infected immunosuppressed mice have greater T helper cell activity and produce higher levels of more interleukins than spleen cells from normal mice [180]. Additionally the presence of primed antigen-specific T helper cells can give rise to a sustained humoral immune response on further antigen challenge, even during *T. cruzi*-induced immunosuppression [181], although at the height of suppression during the peak of an acute-phase infection suppressor macrophages could even block this primed T helper response [182].

It appears that although macrophages have been shown to be involved as mediators of immunosuppression following initial *T. cruzi* infection [183], this is not due to defective antigen processing by the macrophages [32]. Reports have implicated L3T4+Ly-2⁻ T helper cells in immunosuppression and suggested their mode of action on effector cells to be by secretion of a suppressor substance [104,179]. Significantly, repeated antigenic stimulation can overcome the immunosuppressed antibody responses to unrelated antigens; this has been suggested to be by activation of small numbers of T helper cells, sufficient to induce specific B cell responses [184]. There is evidence for two possible routes to suppression, these routes involving the production of suppressor substances and a reduced synthesis and response to IL-2. These mechanisms are not mutually exclusive and elements of both are probably involved in the immunosuppression found after *T. cruzi* infection.

Host suppressor substances of molecular weights 14–15 kD [33], 30–60 kD [104,179], and 196–210 kD [185] have been identified, as has a parasite-derived suppressor factor found in epimastigote culture supernatants [58]. The 30–60 kD host activity, which is devoid of interleukin or IFN-γ activity, is a minor heterogeneous component of splenic T helper cell culture supernatants, and has not been purified. The 14–15 kD suppressor factor is found in splenocyte supernatants and infected mouse serum, and it is thought to exert its action through interaction with normal macrophages [33].

A common observation is that IL-2 production *in vitro*

is suppressed with cells from *T. cruzi*-infected animals [171,172], and restoration of lymphocyte functions can be achieved by using supernatants containing IL-2 activity [170,171]. It is possible that other, as yet unidentified, molecules in these crude culture supernatants are also responsible for this restoration of function, and purification of these is necessary to determine whether synergistic involvement of more than one factor is necessary to bring about immunosuppression. The reduced level of IL-2 production in mice appears to be brought about by a population of suppressor cells that are also capable of suppressing IL-2 production in cells from uninfected mice [174,175]. In humans, immunosuppression has also been related to an IL-1 independent reduction of IL-2 receptor expression on T cells, leading to deficient lymphocyte proliferation [178].

Interferon-γ has an enhancing effect on *T. cruzi* killing by mouse macrophages, and recombinant IFN-γ reduces immunosuppression in murine acute Chagas' disease [186]. This observation suggests that IFN-γ may be useful as part of a therapeutic regime for acute infections; however, since it has been suggested that immunosuppression may be a host defence mechanism to preventing autoimmunity in the chronic phase of the disease [179], its use in the latter stages of the disease may augment pathology.

ANTIGENS

Trypanosoma cruzi has a complex cell surface composed mainly of glycoprotein and glycolipid components, and this structural complexity reflects the diverse range of functions performed by the parasite membrane. As well as providing the normal functions associated with cell surface molecules, this array of antigens includes molecules involved in, for instance, attachment and penetration of host cells and resistance to the hostile environments experienced within the insect gut and mammalian host cell.

The various life-cycle stages of *T. cruzi* each express both common and unique surface molecules [187–190], and the range of surface molecules specific to a particular life-cycle stage must reflect an adaptation for existence in a particular host environment. Numerous methods have been employed to study the *T. cruzi* surface antigens, e.g., cell surface labeling, lectin binding, and the use of antibody probes. These methods, along with various methods used to purify membrane components, have contributed to our knowledge on several *T. cruzi* surface molecules. However, the range of different methodologies used has also led to some method-dependent variation in the identification of certain membrane components [191]. More recently the use of recombinant DNA methods has contributed to a greater understanding of the control of expression of some surface antigen genes and this will allow the preparation of realistic amounts of any recombinant polypeptides that may prove to be potential vaccine candidates.

All stages of *T. cruzi* are surrounded by the typical bilamellar membrane; in the epimastigote form this comprises 31% protein, 34% lipid, 16% carbohydrate, and 9% sterols, with a high level of sterol esters and low phosphonolipid content [192,193]. Most of the cell surface antigens characterized so far have been shown to be glycoconjugates and the carbohydrate components of some *T. cruzi* surface molecules can be used in their purification by lectin affinity chromatography or, because of their immunodominance, in antibody affinity chromatography [194]. Some *T. cruzi* glycoconjugates contain unusual sugars, for instance an insect stage-specific surface antigen with a molecular weight of 72 kD (gp72) contains not only xylose and rhamnose but also phosphorylated sugars [194,195] not normally found associated with eukaryotic glycoproteins. The possibility therefore exists that the glycosylation pathway in *T. cruzi* and, indeed, in other related pathogenic protozoa could contain some unique steps that may be potential targets for chemotherapy. For instance, in *T. cruzi*, unlike mammalian, avian and yeast systems, dolichol-bound oligosaccharides are not glucosylated and glucosylation takes place only after the oligosaccharides have been transferred to protein. The dolichol moiety itself is also unique, containing 13 isoprene units instead of the more usual 18 [196]. Also, *T. cruzi* is unable to incorporate *de novo* synthesized sialic acid, and is thought to utilize a trans-sialidase to incorporate sialic acid from exogeneous sources [197]. Although *T. cruzi* contains unusual surface glycosyl residues, autoimmune immunopathology of *T. cruzi* has been proposed to be intimately associated with parasite carbohydrate epitopes that mimic or are common with host molecules. Crossreactive autoantibodies from chagasic sera specifically recognize galactosyl-α-1,3-galactose epitopes present on host laminin and a *T. cruzi* surface antigen. Crossreactivity has also been shown to be directed towards sulfated lipid components of the *T. cruzi* membrane [198] and towards the polypeptide component of a 160 kD flagellar protein [161]. It is possible that further crossreactive parasite epitopes exist but have yet to be identified.

Six major glycoconjugates have been partially characterized and these have apparent molecular weights of 90, 85, 72, 25, 37–24, and 8–19 kD by sodium dodecyl-sulfate polyacrylamide gel electrophoresis (SDS-PAGE) [187–189,194,199–201]. When *T. cruzi* surface antigens are subjected to two-dimensional PAGE, these individual bands each separate into more than one spot and could represent either discrete glycoproteins or a single polypeptide species with variable degrees of post-translational modification, such as heterogeneous glycosylation or phosphorylation. In the case of the 85 kD band (a mammalian stage-specific band), this is composed of more than one discrete glycoprotein [36,202].

A major cell surface glycoprotein that has been identified by human chagasic sera is the 90 kD molecule present on all life-cycle stages of the parasite. This glycoprotein has a relatively simple structure, containing only 19% carbohydrate (high mannose) and no phosphate, and has been purified by lectin affinity chromatography [203]. Gp90 has no crossreactive determinants and is thus a possible candidate for vaccination [204]. Immunization with gp90 and adjuvant did confer some protection in mice, but protection was not absolute, and low but detectable parasitemias were observed [204]. Gp90 may have some degree of anti-phagocytic activity, since treatment of trypomastigotes with trypsin degrades gp90 and at the same time increases the rate of phagocytic uptake and complement-mediated lysis of the parasite [188]. A 90 kD metacyclic-specific antigen has been identified by a monoclonal antibody [189] but it is not clear whether this molecule is identical to, or related to, the original gp90, although premixing metacyclic trypomastigotes with monoclonal antibody to this metacyclic-specific glycoprotein also gives partial protection [190]. The gene for the polypeptide portion of a 90 kD glycoprotein has been cloned from an expression library using a monoclonal antibody probe [205]. The gene is part of a multicopy family organized in a tandem array of about 20 copies. The protein encoded by this gene appears to by trypomastigote specific and is not present on all strains of the parasite.

Several glycoproteins with apparent molecular weights of 85 kD by SDS-PAGE have been identified [35,55,160,188,202,206] and functions proposed for a number of them. Antibody interaction with molecules of this size reduced the extent of cell invasion *in vitro* and led to the hypothesis that some of these molecules could be involved in attachment to and/or penetration of host cells by the parasite [188,202]. One of the gp85 is known to be a fibronectin receptor [35] and incubation of parasites with the fibronectin receptor-binding peptide RGDS reduces parasite penetration of host cells *in vitro* [206]. The 85–90 kD molecule reported by Joiner *et al.* [53] is trypomastigote specific and is a C3-convertase DAF, and as such confers some protection to the parasite against complement mediated lysis. A gene for a further 85 kD surface antigen has been cloned and, from DNA sequence, shown to have homology with heat shock protein Hsp90 from *Saccharomyces* and chicken, and Hsp83 from *Drosophila* [161].

Genes encoding trypomastigote-specific surface antigens of 85 kD have been identified by three groups [46–48,207,208]. Each has reported that the 85 kD antigen genes are present as multiple, variable copies located throughout the genome, and it has recently been shown by DNA sequence comparison that these genes are from the same multigene family. The genomic arrangement of this multicopy variant family is similar to that of the African trypanosome VSG gene family, and by further analogy it has been shown that one preferentially transcribed member of the family is located at a telomere [209]. However, antigenic variation has not been described in *T. cruzi* [210,211], many of the gp85 gene copies are simultaneously transcribed [47,48], and some of these are not telomeric. Cell surface immunofluorescence studies using antisera to specific peptides designed from cDNA sequences have shown that at least three of the gp85 genes are expressed simultaneously [46], and PCR analysis of the 5' ends of the gp85 mRNA confirms that multicopies (at least 10) are simultaneously transcribed [47]. This would result in a heterogeneous cell surface composed of many different but related gp85. Many observations have shown that the 85 kD trypomastigote surface molecules are instrumental in parasite attachment; however, the specific mechanisms by which gp85 is involved in cell binding is not known.

Further analysis of the gp85 deduced amino acid sequence has revealed the presence of regions of homology with bacterial sialidases [212], and the presence in the gp85 of at least two separate eight amino acid "Asp boxes" (S–X–D–X–G–X–T–W) characteristic of the bacterial enzymes is striking. In the bacterial enzymes the Asp motifs are separated by amino acid stretches, the lengths of which are conserved between particular boxes and the size of these stretches is maintained in the gp85 sequences. This suggests that they may be involved in correct folding of the protein by allowing important conserved regions of amino acids to be orientated effectively. Although the structures of the influenza virus hemagglutinin and sialidase sialic acid binding sites are known [213,214], it is not possible to extrapolate to the bacterial enzyme by sequence comparison alone, and confirmation that Asp boxes are involved in sialic acid binding by both bacterial sialidases and gp85 awaits structural analyses. Asp boxes are not confined solely to sialidases; other bacterial glycosidases, T_{even} bacteriophage cell-attachment proteins, and laminin [215–217] also contain these sequences. It is therefore possible that Asp box-containing proteins are involved in binding to sugar residues, either prior to hydrolysis or during cell attachment processes.

The *T. cruzi* trypomastigote surface sialidase has been extensively characterized at the protein level [218–223]. Its role in the parasite penetration of host cells is still unclear, but desialylation of host cells does appear to enhance infection, although high- and low-density lipoproteins bind to *T. cruzi* and inhibit the sialidase activity and paradoxically enhance the infection of the host cells. Sialidase activity has been associated with proteins ranging in size from 70 to 200 kD, observations in keeping with the activity possibly being encoded by a heterogeneous gene family. Furthermore, monoclonal antibodies to *T. cruzi* neuraminidase identify subsets of trypomastigotes, and this may be due to the expression of different members of a multigene family by individual trypomastigotes. It is also possible that the variant nature of individuals of the gp85 family may lead to binding to a variety of different glycans and could explain how the parasite is able to infect such a diverse range of cells.

An epimastigote-specific surface antigen of 72 kD is the most extensively characterized at the biochemical level. It appears to play a role in the differentiation of the parasite in the insect midgut, possibly by interacting with insect lectin receptors [5]. However, analysis by binding of a monoclonal antibody specific for gp72 carbohydrate side-chains [224] appeared to demonstrate varying levels of gp72 on different *T. cruzi* isolates, and although the variation in level correlated with zymodeme classification [225] it was thought more likely to reflect heterogeneity in the structure of the carbohydrate side-chains than real differences in levels of gp72 expression [225]. Gp72 is an unusual glycoconjugate [194]; it contains 49% carbohydrate, comprising the pentoses xylose and rhamnose, in addition to fucose, mannose, galactose, and glucosamine. A high level of phosphate is also detectable. The carbohydrate side-chains are of two types; one type is mannose-rich and contains little phosphate, while the other contains all the pentoses, is rich in galactose, and is highly phosphorylated [194,195]. Immunization with gp72 partially protects mice from a lethal challenge by metacyclic trypomastigotes but not from blood trypomastigotes [52]. A candidate gene encoding gp72 has been identified by PCR amplification using oligonucleotides derived from amino acid sequences from cyanogen bromide-cleaved peptides isolated from the purified glycoprotein. The gene appears to be a single copy, in contrast to many surface antigens of *T. cruzi*, and the deduced amino acid sequence indicates a region that is rich in serine, threonine, and proline that is likely to be the site of *O*-glycosylation and linkage through serine and threonine to rhamnose and xylose in the highly phosphorylated side-chains [226]. The only other gene that has been identified which may have a function in metacyclogenesis was cloned using a subtractive hybridization strategy [227]. This gene is part of a multigene family and its transcription is markedly increased during metacyclogenesis, an increase that can be mimicked *in vitro* by incubation with cyclic AMP.

Another complex glycoconjugate group obtainable by phenol extraction is present on the surface of epimastigotes and comprises three glycoproteins and a glycolipid [228–230]. The three glycoproteins with molecular weights of 37, 31, and 25 kD have not been individually purified but the gross composition of the group is again unusual, with a high carbohydrate content and an atypical amino acid composition comprising high levels of aspartic acid, threonine, glutamic acid, and glycine. By contrast, the complex glycolipid has been isolated from the other glycoprotein components of the glycoconjugate by solvent extraction [199] and has undergone detailed analysis. It has been given the term lipopeptidophosphoglycan (LPPG) since it had a gross composition of lipid, carbohydrate, phosphate, and amino acids. Two groups have reported the composition of this molecule [230–232] and although both concur on the LPPG nature of molecule there are some differences between the two compositions, but this may relate to parasite strain differences or variation in growth conditions. Detailed structures for the LPPG have been reported, but these structures are those of a complex glycolipid [233,234] and the presence within the molecule of a peptide has not been confirmed. However, it has been suggested that the peptide could be attached through an ester link to the carbohydrate component of the LPPG [235], a possibility which is compatible with published data. The function of the LPPG is still unclear, but related molecules in *Acanthamoeba castellanii* [236] and *Leishmania* sp. [237,238] are proposed to protect these protozoa from harsh environments; in the case of *T. cruzi* epimastigotes, this would include the lytic enzymes of the insect midgut.

A glycoprotein of apparent molecular weight 25 kD expressed at all stages of the parasite's life-cycle was identified by surface radiolabeling, column chromatography, and immunoprecipitation using chronic chagasic serum [199]. Subsequent analyses using a monoclonal antibody revealed that gp25 is synthesized as a 57 kD molecule and that conversion to gp25 probably occurred as a result of the harsh conditions used during isolation [239]. This antigen has been proposed as a possible diagnostic reagent, since antibodies to it are detectable in the serum of 97% of chagasic patients [240].

Numerous other less prominent protein antigens of *T. cruzi* have been described. For instance, antigens of 67 kD [20] and 160 kD [161] that are serologically cross-reactive to host antigens may have a role in autoimmune disease pathogenesis. The 160 kD molecule has a flagellar location and may therefore be involved in cell motility. The gene for another flagellar protein [241], possibly also involving cell motility and possessing homology to a previously identified gene [242], has been cloned and the protein has a high affinity for calcium. A 58/68 kD collagen binding protein has been identified [40] that may have a role in binding to, and penetration of, host cells. Intriguingly this molecule may also have a role in preventing damage to the parasite by the alternative complement pathway [79]. Some antigens of *T. cruzi* show protease activity [41] and some, possibly all surface antigens, are GPI anchored [243].

Identification of surface antigens recognized by chagasic serum could be particularly important to our understanding of the immunopathology of the disease; the use of expression libraries and serum probes led, in one study, to the identification of 23 clones encoding antigens ranging from 85 kD to greater than 205 kD, some of which were stage specific [244,245]; in a further study another 59 clones were identified [246]. Many of these clones contain tandemly repeated sequence motifs and these repeats could confer high levels of immunogenicity to the antigens containing them. The repeat sequences identified within the genes cloned from *T. cruzi* are listed in Table 11.1. It is unclear if these short genomic sequences come from related genes that differ only in their repeats.

The discovery of systems for transfecting trypano-

Table 11.1 Repetitive sequences in *Trypanosoma cruzi* antigens

Protein size (kD)	Specificity*	Repeat†	Reference
>214	E, M, T	QKAAENRLADELE	[246]
>205	E, T	SMNARAQELAREKKLADRAFLDQRPECVPLRELPLDDDSDFVAMEQERRQQLE- KDPRRNAREIAALEE K	[245]
205/195/160	E	EKQKAAETKVAEA R M	[245]
205–165	T	DSSAHCTPSTPV T S A	[272]
150/140/125	E, T	NERLASVL P	[245]
105/75	E, M, T	PFGQAAAGDKPS	[246]
85	T	GDKPSPFGQAAA R PL GTV	[245]
85	T	KSAEP AG V E	[245]
85	E, T	ALPQEEQEDVGPRHVDPDHFRSTTQDAYRPVDPSAYKR H	[245]
85	T	DKKESGDSE	[48]
70	E, T	GNXG	[283]
–	?	SSASLSFSAAFCSS	[245]
–	E, M, T	LEELRVENEELRPEGEDKTRALQEVSEQAEDLEQ	[246]
–	E, M, T	QKAAEATKVAEAEK	[246]
–	E, M, T	AAPKKA	[246]
–	E, M, T	AAPAKA	[246]
–	E, M, T	HAHRAIHVLHCARHAFHLRTVLVVVLHIP	[246]

* E, epimastigote; M, metacyclic trypomastigote; T, trypomastigote.
† Single amino acid code, where X is any amino acid.

somatids with functional exogeneous genes [247,248] has led to a number of advances in the study of trypanosome molecular biology. Recently a system for transfecting *T. cruzi* epimastigotes has been reported [249], and variations on the constructs used in this study should significantly advance the study of *T. cruzi* cell surface antigens.

DIAGNOSIS

Diagnosis of *T. cruzi* infection relies either on the detection of the parasite or of the serologic responses to infection. Since infection of humans is not self-resolving in the absence of drug therapy, the presence of antibody to the parasite is evidence of an ongoing infection.

Direct detection of the parasite in blood is possible during the acute phase of the disease by examination of blood smears. The sensitivity of this method can be increased by concentration of the parasite along with the leukocytes after red cell lysis [250] or by enrichment using a density gradient centrifugation such as Ficoll–Triosil [251]. Lower levels of the parasite in the blood during the chronic phase of the disease make direct parasite detection more difficult and alternative procedures such as xenodiagnosis and hemoculture are needed. Xenodiagnosis is effected by feeding the natural

Triatomid vector on humans and examining the insect later for the presence of the parasites in its hindgut. Clearly this process has the disadvantage of time and takes 20–60 days; the likely success rate can also be low, and values as low as 40% [252] have been found, meaning that repeat diagnosis is often required even though allergic reactions at the site of the bite can occur [253]. Even allowing for the difficulties associated with this test, the definitive identification, by morphologic criteria, of *T. cruzi* can be an advantage. The sensitivity can also be increased by using larger insects, such as *Dipetalogaster maximus*, which can take up to 5 ml in a single blood meal [254].

An alternative to xenodiagnosis is hemoculture, where blood for testing is incubated in an appropriate medium and the growth of a potential parasite, in the epimastigote form, is monitored. The cultivation period again means that time is needed to achieve a result and contamination can be a problem [255]. Modern techniques using DNA probes should allow rapid and direct detection of parasites in the blood. Reports of parasite detection using whole DNA have been made [256] and the more promising approach of using cloned probes from the repetitive DNA sequences in the kinetoplast minicircles could enhance the detection limits and provide a greater degree of specificity and discrimination from other infectious agents [257].

Infection with *T. cruzi* gives rise to an antibody response, initially as IgM and then as IgG, 20–40 days after the onset of the acute phase of infection [258]. A number of different assays have been used to detect this antibody, including complement fixation [259,260], hemagglutination [261], latex agglutination [262], indirect immunofluorescence [263], and enzyme-linked immunosorbent assay (ELISA) [264]. These assays have various advantages and disadvantages based on, among other things, time taken, sensitivity, and suitability for field situations [265], but they all suffer from the same disadvantages of being unable to detect infection in the first few days when an antibody response has not yet been mounted, and the false positives caused by cross-reactivity with other parasites, particularly *Leishmania* spp. Monoclonal antibodies can be used as the basis for serologic tests to provide specificity and discrimination between *T. cruzi*, *Leishmania* spp., and *T. rangeli* [266–269]. Alternatively, purified proteins from *T. cruzi* can be used to provide more specific tests [244,270,271]. One such recombinant antigen (shed acute phase antigen), identified by screening an expression library with chagasic serum, is recognized by greater than 90%

of human acute sera. This 160–200 kD antigen or antigen family contains C-terminal repeats, a C-terminal hydrophobic region characteristic of a GPI anchor addition sequence, and is shed into the medium by trypomastigotes, possibly following cleavage of the GPI anchor. Its high reactivity with early acute sera makes it a candidate for diagnosis of newly infected cases [272].

Although there is no recorded case of an infection resolving itself, drug cure is possible. In many instances, on drug cure, serologic responses become negative [273]. In those instances when drug cure does not lead to a negative serology, it does lead to the loss of antibody reactivity to a 160 kD trypomastigote cell surface antigen. This antibody mediates complement-dependent killing, gives immunofluorescence with viable trypomastigotes, and its presence correlates with infection to such an extent that it has been proposed that it can be used as a marker to confirm the effectiveness of drug therapy [77,273].

VACCINATION

Vaccination studies have been undertaken in a number of different animal species by using a variety of different antigens, including lysed whole cells, subcellular fractions, and cell surface glycoprotein preparations [202,274–279]. The results in general show that these antigen preparations all reduce parasitemias during the acute phase of the disease, and in mice convert lethal to nonlethal infections. However, no vaccination study has given complete protection and the vaccinated animals all become infected. Included within these studies are experiments, analogous to those with the sporozoite stage of malaria when blocking immunity to the infecting sporozoite stage was induced [280], where immunization of mice with an antigen present on the surface of metacyclic trypomastigotes still allowed an infection to be established even though the reduced blood parasitemia did suggest that the immune response was partially effective [52]. The critical question to be answered is whether reducing the acute-phase parasitemia, and therefore presumably the total parasite load, reduces the incidence and severity of the chronic phase of the disease. The increased understanding of the molecular basis of the inflammatory responses and autoimmunity, together with the feasibility of measuring ECG [281] and histologic changes [282] on groups of rodents, should allow future studies to address the possible role for vaccination in chronic disease control in a more rational manner. Already studies with mice have shown that reduced blood parasite load is reflected in reduced tissue parasite burdens and reduced tissue damage [282].

NOTE

Sequence data from some of the *T. cruzi* surface antigens mentioned herein have now been reported. Comparison of the nucleotide sequences of these genes and their deduced amino acid sequences with the gp85 sequences described in the text, indicate that the genes all belong to the same mammalian stage-specific multigene family, and are homologous to bacterial sialidase genes, (SAPA) [284], *T. cruzi* sialidase [285], FL-160 [286], and SA-85 [287]. Recently, members of this gene family have also been shown to encode the *T. cruzi* trans-sialidase activity [288–290].

REFERENCES

1 World Heath Organization. *Chagas' Disease, Report of a Study Group.* WHO Tech. Rep. Ser. No. 202. Geneva: WHO, 1960.

2 Nickerson P, Orr P, Schroeder MI, Sekla L, Johnston JB. Transfusion-associated *Trypanosoma cruzi* infection in a non-endemic area. Ann Intern Med 1989;111:851.

3 Schenkman S, Andrews N, Nussensweig V, Robbins ES. *Trypanosoma cruzi* invade a mammalian epithelial cell in a polarised manner. Cell 1988;55:157.

4 Andrews NW, Abrams CK, Slatin SL, Griffiths G. A *T. cruzi* secreted protein immunologically related to the complement component C9: evidence for membrane pore forming activity at low pH. Cell 1990;61:1277.

5 Sher A, Snary D. Specific inhibition of the morphogenesis of *Trypanosoma cruzi* by a monoclonal antibody. Nature 1982;300:639.

6 Zingales B, Abuin G, Romanha AJ, Chiari E, Colli W. Surface antigens of stocks and clones of *Trypanosoma cruzi* isolated from humans. Acta Trop 1984;41:5.

7 Brener Z. Immunity to *Trypanosoma cruzi*. Adv Parasitol 1980;18:247.

8 Miles MA, Souza A, Provoa M, Shaw JJ, Lainson R, Toye PJ. Isozymic heterogeneity of *Trypanosoma cruzi* in the first autochthonous patients with Chagas' disease in Amazonia Brazil. Nature 1978;272:819.

9 Morel C, Chiari E, Camargo EP, Mattei DM, Romanha AJ, Simpson L. Strains and clones of *Trypanosoma cruzi* can be characterised by pattern of restriction endonuclease products of kinetoplast DNA minicircles. Proc Natl Acad Sci USA 1980;77:6810.

10 Brener Z. Biology of *Trypanosoma cruzi*. Ann Rev Microbiol 1973;27:347.

11 Engman DM, Reddy LV, Donelson JE, Kirchhoff LV. *Trypanosoma cruzi* exhibits inter and intra-strain heterogeneity in molecular karyotype and chromosomal gene location. Mol Biochem Parasitol 1987;22:115.

12 Veas F, Cuny G, Breniere SF, Tibayrenc M. Subspecific kDNA probes for major clones of *Trypanosoma cruzi*. Acta Trop 1991;48:79.

13 Tibayrenc M, Ward P, Moya A, Ayala FJ. Natural populations of *Trypanosoma cruzi*, the causative agent of Chagas' disease, have a complex multiclonal structure. Proc Natl Acad Sci USA 1986;83:115.

14 Tibayrenc M, Ayala FJ. Isoenzyme variability in *Trypanosoma cruzi*, the causative agent of Chagas' disease, genetical, taxonomical, and epidemiological significance. Evolution 1988;42:277.

15 Warren LG. Metabolism of *Schizotrypanum cruzi* Chagas I. Effects of culture age and substrate concentration on the respiratory rate. J Parasitol 1960;46:529.

16 Morel CM, ed. *Genes and Antigens of Parasites: A Laboratory Manual*, 2nd edn. Rio de Janeiro: Fundacao Oswaldo Cruz. WHO, 1984.

17 Deane MP, Moriearty PL, Thomaz N. Cell differentiation in trypanosomatids and other parasitic protozoa. In Morel CM, ed. *Genes and Antigens of Parasites: A Laboratory Manual*, 2nd edn. Rio de Janeiro: Fundacao Oswaldo Cruz. WHO, 1984:11–23.

18 Chiari E, Camargo EP. Culturing and cloning of *Trypanosoma cruzi*. In Morel CM, ed. *Genes and Antigens of Parasites: A Laboratory Manual*, 2nd edn. Rio de Janeiro: Fundacao Oswaldo Cruz. WHO, 1984:23–27.

19 Hudson L, Snary D, Morgan SJ. *Trypanosoma cruzi*: continuous culture with murine cell lines. Parasitology 1984; 88:283.

20 Santos-Buch CA. American trypanosomiasis: Chagas' disease. Int Rev Exp Pathol 1979;19:63.

21 Andrade ZA. *Mechanisms of Myocardial Damage in Trypanosoma cruzi Infection*. Ciba Foundation Symposium 99. London: Pitman Books, 1983:214.

22 Barousse AP, Costa JA, Epasto L, Plume H, Segura EL. Enfermedad de Chagas e immunosuppression. Medicina (Buenos Aires) 1980;40:17.

23 Takle GB, Hudson L. Autoimmunity and Chagas' disease. Curr Top Microbiol Immunol 1989;145:79.

24 Miles MA, Cedillos RA, Povoa MM, De Souza AA, Prata A, Macedo V. Do radically dissimilar *Trypanosoma cruzi* strains (zymodemes) cause Venezuelan and Brazilian forms of Chagas' disease? Lancet 1981;i:1338.

25 Anselmi A, Moliero F. Pathogenic mechanisms in Chagas' cardiomyopathy. Ciba Foundation Symposium 20. London: Pitman Books, 1974:125.

26 Koberle F. Pathogenesis of Chagas' disease. Ciba Foundation Symposium 20. London: Pitman Books, 1974: 136.

27 Kumar R, Kline IK, Abelmann WH. Immunosuppression in acute and subacute chagasic myocarditis. Am J Trop Med Hyg 1970;19:932.

28 Trischmann TM, Tanowitz H, Wittner M, Bloom BR. *Trypanosoma cruzi*: role of the immune response in natural resistance of inbred strains of mice. Exp Parasitol 1978;45: 160.

29 Burgess DE, Hanson WL. *Trypanosoma cruzi*: the T cell dependence of the primary immune response and the

effects of depletion of T cells and Ig bearing cells on immunological memory. Cell Immunol 1980;52:176.

30 Reed SG, Roters SB, Goidl EA. Spleen cell-mediated suppression of IgG production to a non-parasite antigen during chronic Trypanosoma cruzi infection in mice. J Immunol 1983;131:1978.

31 Plata F. Enhancement of tumour growth correlates with suppression of the tumour-specific cytolytic T-lymphocyte response in mice chronically infected by Trypanosoma cruzi. J Immunol 1985;134:1312.

32 Ramos C, Lamoyi E, Feoi M, Rodriquez M, Perez M, Ortiz-Ortiz L. Trypanosoma cruzi: immunosuppressed response to different antigens in the infected mouse. Exp Parasitol 1978;45:190.

33 Serrano LE, O'Daly JA. Protein fraction from Trypanosoma cruzi infected spleen cell supernatants with immuno-suppressive activity in vitro. Int J Parasitol 1987;17:851.

34 Kierszenbaum F, Budzko DB. Trypanosoma cruzi: deficient lymphocyte reactivity during experimental acute Chagas' disease in the absence of suppressor T cells. Parasite Immunol 1982;4:441.

35 Ouaissi MA, Cornette J, Capron A. Identification and isolation of Trypanosoma cruzi trypomastigote cell surface protein with properties expected of a fibronectin receptor. Mol Biochem Parasitol 1986;19:201.

36 Andrews NW, Katzin AM, Colli W. Mapping of surface glycoproteins of Trypanosoma cruzi by two dimensional electrophoresis. A correlation with cell invasion capacity. Eur J Biochem 1984;140:599.

37 Lima MF, Villalta F. Trypanosoma cruzi trypomastigote clones differentially express a parasite cell adhesion molecule. Mol Biochem Parasitol 1989;33:159.

38 Villalta F, Lima MF, Zhou L. Purification of Trypanosoma cruzi surface proteins involved in adhesion to host cells. Biochem Biophys Res Commun 1990;172:925.

39 Yoshida N, Blanco SA, Araguth MF, Russo M, Gonzalez J. The stage-specific 90-kilodalton surface antigen of metacyclic trypomastigotes of Trypanosoma cruzi. Mol Biochem Parasitol 1990;39:39.

40 Velge P, Ouaissi MA, Cornette J, Afchain D, Capron A. Identification and isolation of Trypanosoma cruzi trypomastigote collagen-binding proteins: possible role in cell–parasite interaction. Parasitology 1988;97:255.

41 Cazzulo JJ. Protein and amino acid catabolism in Trypanosoma cruzi. Comp Biochem Physiol 1984;79B:309.

42 Pereira MEA. A developmentally regulated neuraminidase activity in Trypanosoma cruzi. Science 1983;219:1444.

43 Harth G, Haidaris G, So M. Neuraminidase from Trypanosoma cruzi: analysis of enhanced expression of the enzyme in infectious forms. Proc Natl Acad Sci USA 1987; 84:8320.

44 Bontempi E, Franke de Cazzulo BM, Ruiz AM, Cazzulo JJ. Purification and properties of an acidic protease from epimastigotes of Trypanosoma cruzi. Comp Biochem Physiol 1984;77B:599.

45 Rangel HA, Araujo PMF, Camargo IJB, Bonfitto M, Repka D, Sakurada J, Atta AM. Detection of a protease common to epimastigotes, trypomastigotes and trypomastigotes of different strains of Trypanosoma cruzi. Tropenmed Parasitol 1987;32:87.

46 Takle GB, Cross GAM. An 85-kilodalton surface antigen gene family of Trypanosoma cruzi encodes polypeptides homologous to bacterial neuraminidases. Mol Biochem Parasitol 1991;48:185.

47 Kahn S, Van Voorhis WC, Eisen H. The major 85 kD surface antigen of the mammalian form of Trypanosoma cruzi is encoded by a large heterogeneous family of simultaneously expressed genes. J Exp Med 1990;172:589.

48 Peterson DS, Wrightsman RA, Manning JE. Cloning of a major surface antigen gene of Trypanosoma cruzi and identification of a nonapeptide repeat. Nature 1986;322: 566.

49 Kierszenbaum F, Ivanyi J, Budzko DB. Mechanism of natural resistance to Trypanosoma cruzi infection: role of complement in avian resistance to Trypanosoma cruzi infection. Immunology 1976;30:1.

50 Nogueira N, Bianco C, Cohn Z. Studies of the selective lysis and purification of Trypanosoma cruzi. J Exp Med 1975;142:224.

51 Joiner KA, Hieny S, Kirchhoff LV, Sher A. GP72, the 72 kilodalton glycoprotein is the membrane acceptor site for C3 on Trypanosoma cruzi. J Exp Med 1985;161:1196.

52 Snary D. Cell surface glycoproteins of Trypanosoma cruzi: protective immunity in mice and antibody levels in human chagasic sera. Trans R Soc Trop Med Hyg 1983;77:126.

53 Joiner K, Da Silva WD, Rimoldi MT, Hammer CH, Sher A, Kipnis TL. Biochemical characterisation of a factor produced by trypomastigotes of Trypanosoma cruzi that accelerates the decay of complement C3 convertases. J Biol Chem 1988;263:11327.

54 Meckert PC, Chambo JG, Laguens RP. Differences in resistance to reinfection with low and high inocula of Trypanosoma cruzi in chagasic mice treated with Nifurtimox and relation to immune response. Antimicrob Agents Chemother 1988;32:241.

55 Ferreira E, Neva F, Gusmoa R, et al. HLA and Chagas' disease. Abstracts of Congresso Internacional Sobre Doenca De Chagas, Rio de Janeiro, Brasil. Rio de Janeiro: Industria Grafica Cruzeiro Do Sul LTDA, 1979:181.

56 Wrightsman R, Krassner S, Watson J. Genetic control of responses to Trypanosoma cruzi in mice: multiple genes influencing parasitaemia and survival. Infect Immun 1982; 36:637.

57 Trischmann TM. Susceptibility of radiation chimeras to Trypanosoma cruzi. Infect Immun 1982;36:844.

58 Corsini AC, Costa MG, Oliveira OP, Camargo IJB, Rangel HA. A fraction (FAd) from Trypanosoma cruzi epimastigotes depresses the immune response in mice. Immunology 1980;40:505.

59 Andrade S, Andrade V, Brodskyn C, Magalhaes JB, Netto MB. Immunological response of Swiss mice to infection with three different strains of Trypanosoma cruzi. Ann Trop Med Parasitol 1985;79:397.

60 Trischmann TM. Trypanosoma cruzi: early parasite proliferation and host resistance in inbred strains of mice. Exp Parasitol 1986;62:194.

61 Texeira ARL, Santos-Buch CA. Immunology of experimental Chagas' disease: delayed hypersensitivity to Trypanosoma cruzi antigen. Immunology 1975;28:401.

62 Kierszenbaum F, Howard JG. Mechanism of resistance to

experimental *Trypanosoma cruzi* infections: the importance of antibodies and antibody-forming capacity in Biozzi high and low responder mice. J Immunol 1976;116:1208.

63 Rodriguez AM, Santoro F, Afchain D, Bazin H, Capron A. *Trypanosoma cruzi*: infection in B cell deficient rats. Infect Immun 1980;31:524.

64 McHardy N. Passive protection of mice against infection with *Trypanosoma cruzi* with plasma: the use of blood and bug-vector derived trypomastigote challenge. Parasitology 1980;80:471.

65 Trischmann TM, Bloom BR. *Trypanosoma cruzi*: ability to T cell enrich and deplete lymphocyte populations to passively protect mice. Exp Parasitol 1980;49:225.

66 Scott MT. The nature of immunity against *Trypanosoma cruzi* in mice recovered from acute infection. Parasite Immunol 1981;3:209.

67 Hanson WL. *Immune Response and Mechanism of Resistance to Trypanosoma cruzi*. PAHO Scientific Publication 347, Washington DC, 1977:22.

68 Takehara HA, Perini A, da Silva MH, Mota I. *Trypanosoma cruzi*: role of different antibody classes in protection against infection in the mouse. Exp Parasitol 1981;52:137.

69 Scott MT, Goss-Sampson M. Restricted isotype profiles in *T. cruzi* infected mice and Chagas' disease patients. Clin Exp Immunol 1984;58:372.

70 Krettli AU, Brener Z. Protective effects of specific antibodies in *Trypanosoma cruzi* infection. J Immunol 1976;116:755.

71 Schmunis GA, Szarfman A, Langenbach T, de Souza W. Induction of capping in the blood stage trypomastigote of *Trypanosoma cruzi* by human anti-*T. cruzi* antibody. Infect Immun 1978;20:576.

72 Leon W, Villalta F, Queiroz T, Szarfman A. Antibody induced capping of the intracellular stage of *Trypanosoma cruzi*. Infect Immun 1979;26:1218.

73 Budzko DB, Kierszenbaum F. Isolation of *Trypanosoma cruzi* from blood. J Parasitol 1974;60:1037.

74 Krettli AU, Nussenzweig RS. *Presence of Immunoglobulin on the Surface of Circulating Trypomastigotes of Trypanosoma cruzi Resulting in Activation of the Alternative Pathway of Complement and Lysis*. PAHO Scientific Publication 347, Washington DC, 1977:71.

75 Krettli AU, Weisz-Carpington P, Nussenzweig RS. Membrane bound antibodies to bloodstream *Trypanosoma cruzi* in mice. Strain differences in susceptibility to complement-mediated lysis. Clin Exp Immunol 1979;37:416.

76 Dalmasso AP, Jarvinen JA. Experimental Chagas' disease in complement-deficient mice and guinea pigs. Infect Immun 1980;28:434.

77 Krettli AU, Brener Z. Resistance to *Trypanosoma cruzi* associated to anti-living trypomastigote antibodies. J Immunol 1982;128:2009.

78 Schenkman S, Gunther MLS, Yoshida N. Mechanism of resistance to lysis by the alternate complement pathway in *Trypanosoma cruzi* trypomastigotes. Effect of specific monoclonal antibody. J Immunol 1986;137:1623.

79 Fischer E, Ouaissi MA, Velge P, Cornette J, Kazatchkine MD. Gp 58/68 a parasite component that contributes to the escape of the trypomastigote form of *T. cruzi* from damage by the human alternative complement pathway. Immunology 1988;65:299.

80 Kipnis TL, Krettli AU, Dias da Silva W. Transformation of trypomastigote forms of *Trypanosoma cruzi* into activators of the alternate complement pathway by immune Ig fragments. Scand J Immunol 1985;22:217.

81 Krettli AU, de Carvalho LCP. Binding of C3 fragments to the *Trypanosoma cruzi* surface in the absence of specific antibodies and without activation of the complement cascade. Clin Exp Immunol 1985;62:270.

82 Sher A, Hieny S, Joiner K. Evasion of the alternate complement pathway by metacyclic trypomastigotes of *Trypanosoma cruzi*: dependence on the developmentally regulated synthesis of surface protein and N-linked carbohydrate. J Immunol 1986;137:2961.

83 D'Imperio-Lima MR, Joskowicz M, Coutinho A, Kipnis T, Eisen H. Very large and atypical polyclonal plaque-forming cell responses in mice infected with *Trypanosoma cruzi*. Eur J Immunol 1985;15:201.

84 Minoprio PM, Coutinho A, Joskowicz M, d'Imperio Lima MR, Eisen H. Polyclonal lymphocyte responses to murine *Trypanosoma cruzi* infection: II cytotoxic T lymphocytes. Scand J Immunol 1986;24:669.

85 D'Imperio-Lima MR, Eisen H, Minoprio P, Joskowicz M, Coutinho A. Persistence of polyclonal B cell activation with undetectable parasitemia in late stages of experimental Chagas' disease. J Immunol 1986;137:353.

86 Tarleton RL, Kuhn RE. Changes in cell populations and immunoglobulin producing cells in the spleens of mice infected with *Trypanosoma cruzi*: correlation with parasite specific antibody response. Cell Immunol 1983;80:392.

87 Kierszenbaum F, Hayes MM. Mechanisms of resistance against experimental *Trypanosoma cruzi* infections: requirements for cellular destruction of circulating forms of *Trypanosoma cruzi* in human and murine *in vitro* systems. Immunology 1980;40:61.

88 Okabe K, Kipnis TL, Calich VLG, Dias da Silva W. Cell mediated cytotoxicity to *Trypanosoma cruzi*. I. Antibody dependent cell mediated cytotoxicity to trypomastigote bloodstream forms. Clin Immunol Immunopathol 1980;16:344.

89 Kipnis TL, James SL, Sher A, David JR. Cell mediated cytotoxicity of *Trypanosoma cruzi*. II. Antibody dependent killing of bloodstream forms by mouse eosinophils and neutrophils. Am J Trop Med Hyg 1980;30:47.

90 Umekita LF, Mota I. *In vitro* lysis of sensitized *Trypanosoma cruzi* by platelets: role of C3b receptors. Parasite Immunol 1989;11:561.

91 Villalta F, Kierszenbaum F. Role of polymorphonuclear cells in Chagas' disease. Uptake and mechanism of destruction of intracellular (amastigote) forms of *Trypanosoma cruzi* by human neutrophils. J Immunol 1983;131:1504.

92 Villalta F, Kierszenbaum F. Role of inflammatory cells in Chagas' disease. Uptake and mechanisms of destruction of intracellular (amastigote) forms of *Trypanosoma cruzi* by human eosinophils. J Immunol 1984;132:2053.

93 Rimoldi MT, Cardoni RL, Olabuenaga SE, de Bracco MM. *Trypanosoma cruzi*: sequence of phagocytosis and cytotoxicity by human polymorphonuclear leukocytes. Immunology 1981;42:521.

94 Scott MT, Moyes L. 75Se-methionine-labelled *Trypanosoma cruzi* blood trypomastigotes: opsonisation by chronic infection serum facilitates killing in spleen and liver. Clin Exp Immunol 1982;48:754.

95 Montufar OMB, Musatti CC, Mendes E, Mendes NF. Cellular immunity in chronic Chagas' disease. J Clin Microbiol 1977;5:401.

96 Barros MAMT, Neto VA, Mendes E, Mota I. *In vitro* cellular immunity in Chagas' disease. Clin Exp Immunol 1979;38:376.

97 Texeira ARL, Texeira G, Macedo V, Prata A. Acquired cell mediated immunodepression in acute Chagas' disease. J. Clin Invest 1978;62:1132.

98 Zeledon R, Ponce C. A skin test for the diagnosis of Chagas' disease. Trans R Soc Trop Med Hyg 1974;68:414.

99 Texeira ARL, Texeira ML, Santos-Buch CA. The immunology of experimental Chagas' disease: production of lesions in rabbits similar to those in chronic Chagas' disease in man. Am J. Pathol 1975;80:163.

100 Gonzalez-Cappa S, Schmunis GA, Traversa OC, Yanovski JF, Parodi AS. Complement fixation tests, skin tests and experimental immunisation with antigen from *Trypanosoma cruzi* prepared under pressure. Am J Trop Med Hyg 1968;17:709.

101 Seah SKK, Marsden PD, Voller A, Pettitt LE. Experimental *Trypanosoma cruzi* infection in Rhesus monkeys—the acute phase. Trans R Soc Trop Med Hyg 1974;68:63.

102 Maleckar JR, Kierszenbaum F. Inhibition of mitogen-induced proliferation of mouse T and B lymphocytes by bloodstream forms of *Trypanosoma cruzi*. J Immunol 1983;130:908.

103 Liew FY, Scott MT, Liu DS, Croft SL. Suppressive substance produced by T cells from mice chronically infected with *Trypanosoma cruzi*: preferential inhibition of the induction of delayed type hypersensitivity. J Immunol 1987;139:2452.

104 Gao XM, Schmidt JA, Liew FY. Suppressive substance produced by T cells from mice chronically infected with *Trypanosoma cruzi*. II. Genetic restriction and further characterisation. J Immunol 1988;141:989.

105 Ribeiro dos Santos R, Hudson L. Denervation and the immune response in mice infected with *Trypanosoma cruzi*. Clin Exp Immunol 1981;44:3490.

106 Hontebeyrie-Joskowicz M, Said G, Milon G, Marchal G, Eisen H. L3T4+T cells able to mediate parasite specific delayed type hypersensitivity play a role in the pathology of experimental Chagas' disease. Eur J Immunol 1987;17:1027.

107 Younes-Chennoufi AB, Said G, Eisen H, Durand A, Hontebeyrie-Joskowicz M. Cellular immunity to *Trypanosoma cruzi* is mediated by helper T cells. Trans R Soc Trop Med Hyg 1988;82:84.

108 Younes-Chennoufi AB, Hontebeyrie M, Tricottet V, Eisen H, Reynes M, Said G. Persistence of *Trypanosoma cruzi* antigens in the inflammatory lesions of chronically infected mice. Trans R Soc Trop Med Hyg 1988;82:77.

109 Kierszenbaum F, Pienkowski MM. Thymus dependent control of host immune defense mechanisms against *Trypanosoma cruzi* infection. Infect Immun 1979;24:117.

110 Schmunis GA, Gonzalez-Cappa SM, Traversa OC, Janovsky JF. The effect of immunodepression due to neonatal thymectomy on infection with *Trypanosoma cruzi*. Trans R Soc Trop Med Hyg 1971;65:89.

111 Reed SG. Adoptive transfer of resistance to acute *Trypanosoma cruzi* infection with T lymphocyte enriched spleen cells. Infect Immun 1980;28:404.

112 Minoprio PM, Eisen H, Joskowicz M, Pereira P, Coutinho A. Suppression of polyclonal antibody production in *Trypanosoma cruzi* infected mice by treatment with anti-L3T4 antibodies. J Immunol 1987;139:545.

113 Nickell SP, Gebremichael A, Hoff R, Boyer MH. Isolation and functional characterisation of murine T cell lines and clones specific for the protozoan parasite *Trypanosoma cruzi*. J Immunol 1987;138:914.

114 Rottenberg M, Cardoni RL, Andersson R, Segura EL, Orn A. Role of T helper/inducer cells as well as natural killer cells in resistance to *Trypanosoma cruzi* infection. Scand J Immunol 1988;28:573.

115 Ribeiro dos Santos R, Hudson L. *Trypanosoma cruzi*: immunological consequences of parasite modification of host cells. Clin Exp Immunol 1980;40:36.

116 Kipnis TL, Calich VLG, Dias da Silva W. Active entry of bloodstream forms of *Trypanosoma cruzi* into macrophages. Parasitology 1979;78:89.

117 Alcantara A, Brener Z. The *in vitro* interaction of *Trypanosoma cruzi* bloodstream forms and mouse peritoneal macrophages. Acta Trop 1978;35:209.

118 Nogueira N, Cohn Z. *Trypanosoma cruzi*: mechanism of entry and intracellular fate in mammalian cells. J Exp Med 1976;143:1402.

119 Milder R, Kloetzel J. The development of *Trypanosoma cruzi* in macrophages *in vitro*, interaction with lysosomes and host cell fate. Parasitology 1980;80:139.

120 Plata F, Wietzerbin J, Pons FG, Falcoff E, Eisen H. Synergistic protection by specific antibodies and interferon against *Trypanosoma cruzi in vitro*. Eur J Immunol 1984;14:930.

121 Nogueira N, Chaplan S, Cohn Z. *Trypanosoma cruzi*: factors modifying ingestion and fate of blood forms in macrophages. J Exp Med 1980;153:629.

122 Lages-Silva E, Ramirez LE, Krettli AU, Brener Z. Effect of protective and non-protective antibodies in the phagocytosis rate of *Trypanosoma cruzi* blood forms by mouse peritoneal macrophages. Parasite Immunol 1987;9:21.

123 Nathan C, Nogueira N, Juangbhanich C, Ellis J, Cohn Z. Activation of macrophages *in vivo* and *in vitro*: correlation between hydrogen peroxide release and killing of *Trypanosoma cruzi*. J Exp Med 1979;149:1056.

124 Nogueira N, Klebanoff S, Cohn Z. T. *cruzi*: sensitization to macrophage killing by eosinophil peroxidase. J Immunol 1982;128:1705.

125 Kierszenbaum F. Enhancement of resistance and suppression of immunisation against experimental *Trypanosoma cruzi* infection by *Corynebacterium parvum*. Infect Immun 1975;12:1227.

126 Abath GC, Coutinho A, Montenegro SML, Gomes YM, Carvalho AB. The use of non-specific immunopotentiators in experimental *Trypanosoma cruzi* infection. Trans R Soc Trop Med Hyg 1988;82:73.

127 Taliaferro WH, Pizzi T. Connective tissue reactions in normal and immunised mice to a reticulotrophic strain of *Trypanosoma cruzi*. J Infect Dis 1954;96:199.

128 Scorza C, Scorza JV. The role of the inflammatory macrophage in experimental acute chagasic myocarditis. J Reticuloendoth Soc 1972;11:604.

129 Nogueira N, Gordon S, Cohn Z. *Trypanosoma cruzi*: modification of macrophage function during infection. J Exp Med 1977;146:157.

130 Hoff R. Killing *in vitro* of *Trypanosoma cruzi* by macrophages from mice immunised with *Trypanosoma cruzi* or BCG and absence of cross immunity to challenge *in vivo*. J Exp Med 1975;142:299.

131 Nogueira N, Cohn Z. *Trypanosoma cruzi*: *in vitro* induction of macrophage microbial activity. J Exp Med 1978;148:288.

132 James SL, Kipnis TL, Sher A, Hoff R. Enhanced resistance to acute infection with *Trypanosoma cruzi* in mice treated with interferon inducer. Infect Immun 1982;35:588.

133 Reed SG, Nathan CF, Pihl DL, *et al*. Recombinant granulocyte macrophage colony stimulating factor activates macrophages to inhibit *Trypanosoma cruzi* and release hydrogen peroxide, comparison with interferon gamma. J Exp Med 1987;166:1734.

134 Alcina A, Fresno M. Activation by synergism between endotoxin and lymphokines of the mouse macrophage cell line J774 against infection by *Trypanosoma cruzi*. Parasite Immunol 1987;9:175.

135 Hatcher FM, Kuhn RE. Destruction of *Trypanosoma cruzi* by natural killer cells. Science 1982;218:295.

136 Russo M, Starobinas N, Ribeiro dos Santos R, Minopriori P, Eisen H, Hontebeyrie M. Susceptible mice present higher macrophage activation than resistant mice during infection with myotropic strains of *Trypanosoma cruzi*. Parasite Immunol 1989;11:385.

137 Cossio PM, Diez C, Szarfman A, Kreutzer E, Candiolo B, Arana RM. Chagasic cardiomyopathy: demonstration of a serum gamma globulin factor which reacts with endocardium and vascular structures. Circulation 1974;49:13.

138 Cossio PM, Laguens PR, Kreutzer E, Diez C, Segal A, Arana RM. Chagasic cardiomyopathy: antibodies reacting with plasma membrane of striated muscle and endothelial cells. Circulation 1974;50:1252.

139 Sterin-Borda L, Cossio PM, Gimeno MF, *et al*. Effect of chagasic sera on the rat isolated atrial preparation: immunological, morphological and functional aspects. Cardiovasc Res 1976;10:613.

140 Khoury EL, Ritacco V, Cossio PM, *et al*. Circulating antibodies to peripheral nerve in American trypanosomiasis (Chagas' disease). Clin Exp Immunol 1976;36:8.

141 Szarfman A, Terranova VP, Rennard SI, *et al*. Antibodies to laminin in Chagas' disease. J Exp Med 1982;155:1161.

142 Avila JL, Rojas M, Rieber M. Antibodies to laminin in American cutaneous leishmaniasis. Infect Immun 1987;43:402.

143 Ribeiro dos Santos R, Marquez JO, von Gal Furtado CC, Ramos de Oliviera JC, Martins AR, Koberle F. Antibodies against neurons in chronic Chagas' disease. Trop Med Parasitol 1979;30:19.

144 Szarfman A, Cossio PM, Schmunis GA, Arana RM. The EVI antibody in acute Chagas' disease. J Parasitol 1977;63:149.

145 Szarfman A, Cossio PM, Khoury EL, Ritaco V, Arana RM, Schmunis GA. Tissue reacting Ig in children parasitaemic with *Trypanosoma cruzi*. Trans R Soc Trop Med Hyg 1977;71:453.

146 Lenzi HL, Lenzi JG, Andrade ZA. Experimental production of EVI antibodies. Am J Trop Med Hyg 1982;31:48.

147 Laguens RP, Cossio PM, Diez C, *et al*. Immunopathologic and morphologic studies of skeletal muscle in Chagas' disease. Am J Pathol 1975;80:153.

148 Szarfman A, Luquetti A, Rassi A, Rezende JM, Schmunis GA. Tissue reacting immunoglobulins in patients with different clinical forms of Chagas' disease. Am J Trop Med Hyg 1981;30:43.

149 Laguens RP, Meckert PC, Chambo JG. Antiheart antibody dependent cytotoxicity in the sera of mice chronically infected with *Trypanosoma cruzi*. Infect Immun 1988;56:993.

150 Mortatti RC, Maia LCS, De Oliveira AV, Munk ME. Immunopathology of experimental Chagas' disease: binding of T-cells to *Trypanosoma cruzi* infected heart tissue. Infect Immun 1990;58:3588.

151 Cossio PM, Laguens RP, Kreutzer E, Diez C, Segal A, Arana RM. Chagasic cardiopathy: immunopathologic and morphologic studies in myocardial biopsies. Am J Pathol 1977;86:533.

152 Wood JN, Hudson L, Jessell TM, Yamamoto M. A monoclonal antibody defining determinants on subpopulations of mammalian neurones and *Trypanosoma cruzi* parasites. Nature 1982;296:34.

153 Petry K, Nudelman E, Eisen H, Hakomori S. Sulphated lipids represent common antigens on the surface of *Trypanosoma cruzi* and mammalian tissue. Mol Biochem Parasitol 1988;30:113.

154 Bretana A, O'Daly JA. Uptake of fetal proteins by *Trypanosoma cruzi* immunofluorescence and ultrastructural studies. Int J Parasitol 1976;6:379.

155 Williams GT, Fielder L, Smith H, Hudson L. Adsorption of *Trypanosoma cruzi* proteins to mammalian cells *in vitro*. Acta Trop 1985;42:33.

156 Peyrol S, Ouaissi MA, Capron A, Grimaud JA. *Trypanosoma cruzi*: ultrastructural visualisation of fibronectin bound to culture forms. Exp Parasitol 1987;63:112.

157 Araujo FG. *Trypanosoma cruzi*: expression of antigens on the membrane surface of parasitised cells. J Immunol 1988;135:4149.

158 Abrahamson IA, Kloetzel JK. Presence of *Trypanosoma cruzi* antigen on the surface of both infected and uninfected cells in tissue culture. Parasitology 1980;80:147.

159 Bretana A, Avila JH, Arias-Flores M, Contreras M, Tapia FJ. *Trypanosoma cruzi* and *Leishmania* sp: immunocytochemical localisation of a laminin like protein in the plasma membrane. Exp Parasitol 1986;61:168.

160 Towbin H, Rosenfelder G, Weislander J, *et al*. Circulating antibodies to mouse laminin in Chagas' disease, American cutaneous leishmaniasis and normal individuals recognise terminal galactosyl (alpha 1–3)-galactose epitopes. J Exp Med 1987;166:419.

161 Dragon EA, Roberts S, Kato EA, Gabe JD. The genome of *Trypanosoma cruzi* contains a constitutively expressed tandemly arranged multicopy gene homologous to a major heat shock protein. Mol Cell Biol 1987;7:1271.

162 Van Vorhis WC, Eisen H. FL–160: a surface antigen of

Trypanosoma cruzi that mimics mammalian nervous tissue. J Exp Med 1989;169:641.

163 Lowe MG. Biochemistry of glycophosphatidylinositol membrane protein anchors. Biochem J 1987;244:1.

164 Cross GAM. Glycolipid anchoring of plasma membrane proteins. Ann Rev Cell Biol 1990;6:1.

165 Shoenfeld Y, Isenburg DA. The mosaic of autoimmunity. Immunol Today 1989;10:123.

166 Rammensee HG, Kroschewski R, Frangoulis B. Clonal anergy induced in mature VB6117 T lymphocytes on immunising Mls-1b mice with Mls-1a expressing cells. Nature 1989;339:541.

167 Mitchell GF. Responses to infection with metazoan and protozoan parasites in mice. Adv Immunol 1979;28:451.

168 Cunningham DS, Kuhn RE, Rowland EC. Suppression of humoral responses during *Trypanosoma cruzi* infections in mice. Infect Immun 1978;22:155.

169 Harel-Bellan A, Joskowicz M, Fradelizi D, Eisen H. Modification of T cell proliferation and interleukin 2 production in mice infected with *Trypanosoma cruzi*. Proc Natl Acad Sci USA 1983;80:3466.

170 Tarleton RL, Kuhn RE. Restoration of *in vitro* human responses of spleen cells from mice infected with *Trypanosoma cruzi* by supernatants containing interleukin 2. J Immunol 1984;133:1570.

171 Reed SG, Inverso JA, Roters SB. Heterologous antibody responses in mice with chronic *T. cruzi* infection: depressed T helper function restored with supernatants containing interleukin 2. J Immunol 1984;133:1558.

172 Harel-Bellan A, Joskowicz M, Fradelizi D, Eisen H. T lymphocyte function during experimental Chagas' disease: production of and response to interleukin 2. Eur J Immunol 1985;15:438.

173 Choromanski L, Kuhn RE. Interleukin 2 enhances specific and non specific immune responses in experimental Chagas' disease. Infect Immun 1985;50:354.

174 Tarleton RL. *Trypanosoma cruzi*-induced suppression of IL-2 production. I. Evidence for the presence of IL-2 producing cells. J Immunol 1988;140:2763.

175 Tarleton RL. *Trypanosoma cruzi*-induced suppression of IL-2 production. II. Evidence for a role for suppressor cells. J Immunol 1988;140:2769.

176 Maleckar JR, Kierszenbaum F. Suppression of mouse lymphocyte responses to mitogens *in vitro* by *Trypanosoma cruzi*. Int J Parasitol 1984;14:45.

177 Rowland EC, Kuhn RE. Suppression of cellular responses in mice during *Trypanosoma cruzi* infections. Infect Immun 1978;20:393.

178 Beltz LA, Sztein MB, Kierszenbaum F. Novel mechanism for *Trypanosoma cruzi*-induced suppression of human lymphocytes. Inhibition of IL-2 receptor expression. J Immunol 1988;141:289.

179 Liew FY, Schmidt JA, Liu DS, *et al*. Suppressive substance produced by T cells from mice chronically infected with *Trypanosoma cruzi*. II. Partial biochemical characterisation. J Immunol 1988;140:969.

180 Cunningham DS, Benavides GR, Kuhn RE. Differences in the regulation of humoral responses between mice infected with *Trypanosoma cruzi* and mice administered *T. cruzi* suppressor substances. J Immunol 1980;125:2317.

181 Kierszenbaum F. Immunologic deficiency during experi-mental Chagas' disease (*Trypanosoma cruzi* infection): role of adherent, nonspecific esterase positive splenic cells. J Immunol 1982;129:2202.

182 Choromanski L, Kuhn RE. Augmentation of suppressed antibody response in mice during experimental Chagas' disease by helper T-cells activated in a time dependent mode of immunization. J Protozool 1990;37:388.

183 Ritter DM, Kuhn RE. Antigen specific T-helper cells abrogate suppression in *Trypanosoma cruzi* infected mice. Infect Immun 1990;58:3248.

184 Choromanski L, Kuhn RE. Repeated antigenic stimulation overcomes immunosuppression in experimental Chagas' disease. Immunology 1986;59:289.

185 Cunningham DS, Kuhn RE. *Trypanosoma cruzi*-induced suppressor substance. I. Cellular involvement and partial characterisation. J Immunol 1980;124:2122.

186 Reed SG. *In vivo* administration of recombinant IFN-gamma induces macrophage activation, and prevents acute disease, immune suppression, and death in experimental *Trypanosoma cruzi* infections. J Immunol 1988;140:4342.

187 Snary D, Hudson L. *Trypanosoma cruzi* cell surface proteins: identification of one major glycoprotein. FEBS Lett 1979;100:166.

188 Nogueira N, Chaplan S, Tydings JD, Unkless J, Cohn Z. *Trypanosoma cruzi*: surface antigens of blood and culture forms. J Exp Med 1981;153:629.

189 Mortara RA, Araguth MF, Yoshida N. Reactivity of stage specific monoclonal antibody 1G7 with metacyclic trypomastigotes of *Trypanosoma cruzi* strains: lytic properties and 90 000 mwt surface antigen polymorphism. Parasite Immunol 1988;10:369.

190 Araguth MF, Rodriguez MM, Yoshida N. *Trypanosoma cruzi* metacyclic trypomastigotes: neutralisation by the stage-specific monoclonal antibody 1G7 and immunogenicity of 90 kDa surface antigen. Parasite Immunol 1988;10:707.

191 Schechter M, Nogueira N. Variations induced by different methodologies in *Trypanosoma cruzi* surface antigen profiles. Mol Biochem Parasitol 1988;29:37.

192 Da Silveira JF, Colli W. Chemical composition of plasma membrane from epimastigote forms of *Trypanosoma cruzi*. Biochim Biophys Acta 1981;644:341.

193 Ferguson MAJ, Allen AK, Snary D. The detection of phosphonolipids in the protozoan *Trypanosoma cruzi*. Biochem J 1982;207:171.

194 Ferguson MAJ, Allen AK, Snary D. Studies on the struc-ture of a phosphoglycoprotein from a parasite protozoan *Trypanosoma cruzi*. Biochem J 1983;213:313.

195 Snary D. Cell surface glycoproteins of *Trypanosoma cruzi*. In Chang K-P, Snary D, eds. *Host–Parasite Cellular and Molecular Interactions in Protozoal Infections*. Berlin: Springer-Verlag, 1987;79.

196 Parodi AJ, de Lederkremer GZ, Mendelzon DH. Protein glycosylation in *Trypanosoma cruzi*, the mech-anism of glycosylation and structure of protein bound oligosaccharide. J Biol Chem 1983;258:5589.

197 Zingales B, Carniol C, de Lederkramer RM, Colli W. Direct sialic acid transfer from a protein donor to glycolipids of trypomastigote forms of *Trypanosoma cruzi*. Mol Biochem Parasitol 1987;26:135.

198 Petry K, Voisin P, Baltz T, Labouesse J. Epitopes common to *Trypanosoma cruzi, T. dionisii, T. vespertilionis (Schizotrypanum)*, astrocytes and neurones. J Neuroimmunol 1987;16:237.

199 Scharfstein J, Rodriguez MM, Alves CA, de Souza W, Previato JO, Mendonca Previato L. *Trypanosoma cruzi*: description of a highly purified surface antigen defined by human antibodies. J Immunol 1983;131:972.

200 De Lederkremer RM, Alves MJM, Fonseca GC, Colli W. A lipopeptidophosphoglycan from *Trypanosoma cruzi* (epìmastigote). Biochim Biophys Acta 1976;444:85.

201 Abuin G, Colli W, de Souza W, Alves MJM. A surface antigen of *Trypanosoma cruzi* involved in cell invasion (Tc 85) is heterogeneous in expression and molecular constitution. Mol Biochem Parasitol 1989;35:229.

202 Katzin AM, Colli W. Lectin receptors in *Trypanosoma cruzi*, an *N*-acetyl-D-glucosamine containing surface glycoprotein specific for the trypomastigote stage. Biochim Biophys Acta 1983;727:403.

203 Snary D. The cell surface of *Trypanosoma cruzi*. Curr Top Microbiol Immunol 1985;117:75.

204 Scott MT, Snary D. Protective immunisation of mice with cell surface glycoprotein from *Trypanosoma cruzi*. Nature 1979;282:73.

205 Beard CA, Wrightsman RA, Manning JE. Stage and strain specific expression of the tandemly repeated 90 kDa surface antigen gene family in *Trypanosoma cruzi*. Mol Biochem Parasitol 1988;28:227.

206 Ouaissi MA, Cornette J, Afchain D, Capron A, Gras-Masse H, Tartar A. *Trypanosoma cruzi* infection prevented by peptides modeled from a fibronectin cell attachment domain. Science 1986;234:603.

207 Manning JE, Peterson DS. Identification of the 85 kDa surface antigen gene of *Trypanosoma cruzi* as a member of a multigene family. In Turner MJ, Arnot D, eds. *Molecular Genetics of Parasitic Protozoa*. Cold Spring Harbor, NY: Cold Spring Harbor Laboratory, 1988;148–152.

208 Takle GB, Young A, Snary D, Hudson L, Nicholls SC. Cloning and expression of a trypomastigote-specific 85 kDa surface antigen gene from *Trypanosoma cruzi*. Mol Biochem Parasitol 1989;37:57.

209 Peterson DS, Fouts DL, Manning JE. The 85 kD surface antigen gene from *Trypanosoma cruzi* is telomeric and a member of a multigene family. EMBO J 1989;8:68.

210 Snary D. *Trypanosoma cruzi*: antigenic invariance of the cell surface glycoprotein. Exp Parasitol 1980;49:68.

211 Plata F, Pons FG, Eisen H. Antigenic polymorphism of *Trypanosoma cruzi*: clonal analysis of trypomastigote surface antigens. Eur J Immunol 1984;14:392.

212 Roggentin P, Rothe B, Kaper JB, Galen J, Lawrisuk L, Vimr ER, Schauer R. Conserved sequences in bacterial and viral sialidases. Glycoconj J 1989;6:349.

213 Weis W, Brown JH, Cusak S, Paulson JC, Skehel JJ, Wiley DC. Structure of the influenza virus haemagglutinin complexed with its receptor, sialic acid. Nature 1988;333:426.

214 Varghese JN, Laver WG, Colman PM. Structure of the influenza virus glycoprotein antigen neuraminidase at 2.9 A resolution. Nature 1983;303:35.

215 Martin I, Debarbouille M, Ferrari E, Klier A, Rapoport G. Characterisation of the levanase gene of *Bacillus subtilis*

216 Riede I, Drexler K, Eschbach ML, Henning U. DNA sequence of the tail fiber genes 37, encoding the receptor recognising part of the fiber of bacteriophage T2 and T3. J Mol Biol 1986;191:255.

217 Montell DJ, Goodman CS. *Drosophila* substrate adhesion molecule: sequence of laminin B1 chain reveals domains with homology with mouse. Cell 1988;53:463.

218 Prioli RP, Ordovas JM, Rosenberg I, Schaefer EJ, Periera MEA. Similarity of cruzin, an inhibitor of *Trypanosoma cruzi* neuraminidase, to high density lipoprotein. Science 1987;238:1417.

219 Prioli RP, Rosenberg I, Periera MEA. High and low density lipoproteins enhance infection of *Trypanosoma cruzi in vitro*. Mol Biochem Parasitol 1990;38:191.

220 Periera MEA, Hoff R. Heterogeneous distribution of neuraminidase activity in strains and clones of *Trypanosoma cruzi* and its possible association with parasite myotropism. Mol Biochem Parasitol 1986;20:183.

221 Harth G, Haidaris CG, So M. Purification and characterisation of stage-specific glycoproteins from *Trypanosoma cruzi*. Mol Biochem Parasitol 1989;33:143.

222 Cavallesco R, Periera MEA. Antibody to *Trypanosoma cruzi* neuraminidase enhances infection *in vitro* and identifies a subpopulation of trypomastigotes. J Immunol 1988;140:617.

223 Prioli RP, Santiago Mejia J, Periera MEA. Monoclonal antibodies against *Trypanosoma cruzi* neuraminidase reveal enzyme polymorphism, recognise a subset of trypomastigotes, and enhance infection *in vitro*. J Immunol 1990;144:4384.

224 Snary D, Ferguson MAJ, Scott MT, Allen AK. Cell surface antigens of *Trypanosoma cruzi*: use of monoclonal antibodies to identify and isolate an epimastigote specific glycoprotein. Mol Biochem Parasitol 1981;3:343.

225 Chapman MD, Snary D, Miles MA. Quantitative differences in the expression of a 72 000 molecular weight cell surface glycoprotein (gp72) in *Trypanosoma cruzi* zymodemes. J Immunol 1984;132:3149.

226 Cooper R, Inverso JA, Espinosa M, Nogueira N, Cross GAM. Characterisation of a candidate gene for gp72, an insect stage-specific antigen of *Trypanosoma cruzi*. Mol Biochem Parasitol 1991;49:45.

227 Heath S, Heiny S, Sher A. A cyclic AMP inducible gene is expressed during the development of infective stages of *Trypanosoma cruzi*. Mol Biochem Parasitol 1990;43:133.

228 Alves MJM, Colli W. Glycoproteins from *Trypanosoma cruzi*: partial purification by gel chromatography. FEBS Lett 1975;52:188.

229 Alves MJM, Da Silveira JF, De Paiva CHR, Tanaka CT, Colli W. Evidence for the plasma membrane localisation of carbohydrate containing macromolecules from epimastigote forms of *Trypanosoma cruzi*. FEBS Lett 1979;99:81.

230 Ferguson MAJ, Snary D, Allen AK. Comparative composition of cell surface glycoconjugates isolated from *Trypanosoma cruzi* epimastigotes. Biochim Biophys Acta 1985;842:39.

231 De Lederkremer RM, Tanaka CT, Alves MJM, Colli W.

Lipopeptidophosphoglycan from *Trypanosoma cruzi*, amide and ester-linked fatty acids. Eur J Biochem 1977;74:263.

232 De Lederkremer RM, Casal OL, Tanaka CT, Colli W. Ceramide and inositol content of the lipopeptido-phosphoglycan from *Trypanosoma cruzi*. Biochem Biophys Res Commun 1978;85:1268.

233 Previato JO, Gorin PAJ, Mazurek M, *et al*. Primary structure of the oligosaccharide chain of lipophosphoglycan of epimastigote forms of *Trypanosoma cruzi*. J Biol Chem 1990; 265:2518.

234 De Lederkremer RM, Lima C, Ramirez MI, Casal OL. Structural features of the lipopeptidophosphoglycan from *Trypanosoma cruzi* common with the glyco-phosphatidylinositol anchors. Eur J Biochem 1990;192:337.

235 De Lederkremer RM, Casal IL, Alves MJM, Colli W. *Trypanosoma cruzi*: amino acid and phosphorus linkages in the lipopeptidophosphoglycan. Biochem Int 1990;10:89.

236 Dearborn DG, Smith S, Korn ED. Lipophosphonoglycan of the plasma membrane of *Acanthamoeba castellanii*. Inositol and phytosphingosine content and general structural features. J Biol Chem 1976;251:2976.

237 El-On J, Bradley DJ, Freeman JC. *Leishmania donovani*: action of excreted factor on hydrolytic enzyme activity of macrophages from mice with genetically different resitance to infection. Exp Parasitol 1980;49:167.

238 McNeely TB, Turco SJ. Requirements of lipophosphoglycan for intracellular survival of *Leishmania donovani* within human monocytes. J Immunol 1990;144:3099.

239 Scharfstein J, Schechter M, Senna M, Peralta JM, Mendonca-Previato L, Miles MA. *Trypanosoma cruzi*: characterization and isolation of a 57/51 000 m.w. surface glycoprotein (gp57/51) expressed by epimastigotes and bloodstream trypomastigotes. J Immunol 1986;137:1336.

240 Scharfstein J, Luquetti A, Murta AC, *et al*. Chagas' disease: serodiagnosis with purified gp25 antigen. Am J Trop Med Hyg 1985;34:1153.

241 Engman DM, Krause KH, Blumin JH, *et al*. A novel flagellar Ca^{2+}-binding protein in trypanosomes. J Biol Chem 1989; 264:18627.

242 Gonzalez A, Lerner TJ, Huecas M, Sosa-Pineda B, Nogueira N, Lizardi PM. Apparent generation of segmented mRNA from two separate tandem gene families in *Trypanosoma cruzi*. Nucleic Acids Res 1985;13:5789.

243 Schenkman S, Yoshida N, Cardosa de Almeida ML. Glycophosphatydylinositol-anchored proteins in meta-cyclic trypomastigotes of *Trypanosoma cruzi*. Mol Biochem Parasitol 1988;29:141.

244 Ibanez CF, Affranchino JL, Frasch ACC. Antigenic determinants of *Trypanosoma cruzi* defined by cloning of parasite DNA. Mol Biochem Parasitol 1987;25:175.

245 Ibanez CF, Affranchino JL, Macina RA, *et al*. Multiple *Trypanosoma cruzi* antigens containing tandemly repeated amino acid sequence motifs. Mol Biochem Parasitol 1988; 30:27.

246 Hoft DF, Kim KS, Otso K, *et al*. *Trypanosoma cruzi* expresses diverse repetitive protein antigens. Infect Immun 1989;57: 1959.

247 Bellofatto V, Cross GAM. Expression of a bacterial gene in a trypanosomatid protozoan. Science 1989;244:1167.

248 Laban A, Wirth DF. Transfection of *Leishmania enrietta* and

expression of chloramphenicol acetyltransferase gene. Proc Natl Acad Sci USA 1989;86:9119.

249 Lu HY, Buck GA. Expression of an exogeneous gene in *Trypanosoma cruzi* epimastigotes. Mol Biochem Parasitol 1991;44:109.

250 Hoff R. A method for counting and concentrating living *Trypanosoma cruzi* in blood lysed with ammonium chloride. J Parasitol 1974;60:527.

251 Budzko DB, Pizzimenti MC, Kierszenbaum F. Effect of complement depletion in experimental Chagas' disease. Immune lysis of virulent blood forms of *Trypanosoma cruzi*. Infect Immun 1975;11:86.

252 Minter-Goedbloed E, Minter DM, Marshall TF de C. Quantitative comparison between xenodiagnosis and hemoculture in the detection of *Trypanosoma (Schizotrypanum) cruzi* in experimental and natural chronic infections. Trans R Soc Trop Med Hyg 1978;72:217.

253 Costa CHN, Costa MT, Weber JN, Gilks GF, Castro C, Marsden PD. Skin reactions to bug bites as a result of xenodiagnosis. Trans R Soc Trop Med Hyg 1981;75:405.

254 Cuba CC, Alvarenga NJ, Barreto AC, Marsden PD, Macedo V, Gama MP. *Dipetalogaster maximum* (Hemiptera, Triatominae) for xenodiagnosis of patients with serologically detectable *Trypanosoma cruzi* infection. Trans R Soc Trop Med Hyg 1979;73:524.

255 Minter-Goedbloed E. The primary isolation by haemo-culture of *Trypanosoma (Schizotrypanum) cruzi* from animals and man. Trans R Soc Trop Med Hyg 1978;72:22.

256 Ashall F, Yip-Chuck DA, Luquetti AA, Miles MA. Radiolabelled total parasite DNA probe specifically detects *Trypanosoma cruzi* in mammalian blood. J Clin Microbiol 1988;26:576.

257 Gonzalez A, Prediger E, Huecas ME, Nogueira N, Lizardi PM. Minichromosomal repetitive DNA in *Trypanosoma cruzi*: its use in a high sensitivity parasite detection assay. Proc Natl Acad Sci USA 1984;81:3356.

258 Schmunis GA, Szarfman A, Coarasa L, Guilleron C, Peralta JM. Anti-*Trypanosoma cruzi* agglutinins in acute human Chagas' disease. Am J Trop Med Hyg 1980;29:170.

259 Fife EH, Kent JF. Protein and carbohydrate complement fixing antigens of *Trypanosoma cruzi*. Am J Trop Med Hyg 1960;9:512.

260 Pereira CA, Longo IM, Ricci O, Silva AP. Automated complement fixation test for the detection of anti-*Trypanosoma cruzi* antibodies. Rev Inst Med Trop Sao Paulo 1980;22:180.

261 Camargo ME, Hoshino S, Correa NS, Peres BA. Hemagglutination test for Chagas' disease with chromium chloride, formalin treated erythrocytes, sensitised with *Trypanosoma cruzi* extract. Rev Inst Med Trop Sao Paulo 1971;13:45.

262 Minter-Goedbloed E, Franca S, Draper CC. The latex agglutination test for *Trypanosoma cruzi*: unsuitable for testing animals. J Trop Med Hyg 1980;83:157.

263 Cerisola JA, Alvarez M, de Rissio AM. Immunodiagnostico da doenca Chagas, evolucao serologica de pacientes com doenca de Chagas. Rev Inst Med Trop Sao Paulo 1970;12: 403.

264 Spencer HC, Allain DS, Sulzer AJ, Collins WE. Evaluation of one microenzyme linked immunosorbent assay for antibodies to *Trypanosoma cruzi*. Am J Trop Med Hyg

1980;29:179.

265 Lemesre JL, Afchain D, Orozco O, et al. Specific and sensitive immunological diagnosis of Chagas' disease by competitive antibody enzyme immunoassay using Trypanosoma cruzi specific monoclonal antibody. Am J Trop Med Hyg 1986;35:86.

266 Zicker F, Smith PG, Luquetti AO, Oliveira OS. Mass screening for Trypanosoma cruzi infections using the immunofluorescence, ELISA and haemagglutination tests on serum samples and on blood eluates from filter paper. Bull WHO 1990;68:465.

267 Dragon EA, Brothers VM, Wrightsman RA, Manning J. A Mr 90 000 surface polypeptide of Trypanosoma cruzi as a candidate for a Chagas' disease diagnostic antigen. Mol Biochem Parasitol 1985;16:213.

268 Constantine NT, Anthony RL. Antigenic differentiation of the kinetoplasts Leishmania brasiliensis from Trypanosoma cruzi by means of monoclonal antibodies. J Protozool 1983; 30:346.

269 Hudson L, Guhl F, Marinkelle CJ, Rodriguez J. Use of monoclonal antibodies for the differential detection of Trypanosoma cruzi and Trypanosoma rangeli in epidemiological studies and xenodiagnosis. Acta Trop 1987;44:387.

270 Schecter M, Flint JE, Voller A, Guhl F, Marinkelle CJ, Miles MA. Purified Trypanosoma cruzi specific glycoprotein for discriminative serological diagnosis of South American trypanosomiasis (Chagas' disease). Lancet 1983;ii:939.

271 Kirchhoff LV, Gam AA, Gusmao RA, Goldsmith RS, Rezende JM, Rassi A. Increased specificity of sero-diagnosis of Chagas' disease by detection of antibody to the 72- and 90-kilodalton glycoprotein of Trypanosoma cruzi. J Infect Dis 1987;155:561.

272 Affranchino JL, Ibanez CF, Luquetti AO, et al. Identification of a Trypanosoma cruzi antigen that is shed during the acute phase of Chagas' disease. Mol Biochem Parasitol 1989;34:221.

273 Martins MS, Hudson L, Krettli AU, Cancado JR, Brener Z. Human and mouse sera recognise the same polypeptide associated with immunological resistance to Trypanosoma cruzi infection. Clin Exp Immunol 1985;61:343.

274 McHardy N. Immunisation of mice against Trypanosoma cruzi: the effect of chemical treatment or immune serum on an epimastigote vaccine. Tropenmed Parasitol 1978;29:215.

275 Segura EL, Paulone I, Cerisola JA, Gonzalez-Cappa SM. Experimental Chagas' disease: protective activity in relation to subcellular fraction of the parasite. J Parasitol 1976; 62:131.

276 Scott MT, Neal RA, Woods NC. Immunisation of marmosets with Trypanosoma cruzi cell surface glycoprotein (gp90). Trans R Soc Trop Med Hyg 1985;79:451.

277 Texeira ARL. Immunoprophylaxis against Chagas disease. In Miller LH, Pino JA, McKelvey JJ, eds. Immunity to Blood Parasites of Animals and Man. New York: Plenum Publishers 1977:234.

278 Texeira ARL, Santos-Buch CA. The immunology of experimental Chagas' disease. I. Preparation of Trypanosoma cruzi and humoral antibody responses to these antigens. J Immunol 1974;113:859.

279 Gonzalez-Cappa SM, Bronzina A, Katzin AM, Golfera H, De Martini GW, Segura EL. Antigens of subcellular fractions of Trypanosoma cruzi. III. Humoral immune response and histology of immunised mice. J Protozool 1980;17:467.

280 Good MF, Berzofsky JA, Miller LH. The T cell response to the malaria circumsporozoite protein: an immunological approach to vaccine development. Ann Rev Immunol 1987;6:663.

281 De Oliveira JS, Bestetti RB, Soares EG, Marin-Neto JA. Ajmaline-induced electrocardiographic changes in chronic Trypanosoma cruzi-infected rats. Trans R Soc Trop Med Hyg 1986;80:415.

282 Postan M, Cheever AW, Dvorak JA, McDaniel JP. A histological analysis of the course of myocarditis in C3H/He mice infected with Trypanosoma cruzi clone Sylvio-X10/4. Trans R Soc Trop Med Hyg 1986;80:50.

283 Requena JM, Lopez MC, Jiminez Reis A, de la Torre JC, Alonso C. A head to tail tandem organisation of hsp 70 genes in Trypanosoma cruzi. Nucleic Acids Res 1988;16:1393.

284 Pollevick GD, Affranchino JL, Frasch ACC, Sanchez DO. The complete sequence of a shed acute phase antigen of Trypanosoma cruzi. Mol Biochem Parasitol 1991;47:247.

285 Pereira MEA, Mejia JS, Ortego-Barria E, Matzilevich D, Prioli RP. The Trypanosoma cruzi neuraminidase contains sequences similar to bacterial neuraminidases, YWTD repeats of the low density lipoprotein receptor, and Type III modules of fibronectin. J Exp Med 1991;174:179.

286 Van Voorhis WC, Schlekewy L, Le Trong H. Molecular mimicry by Trypanosoma cruzi: the FL-160 epitope that mimics mammalian nerve can be mapped to a 12 amino acid peptide. Proc Natl Acad Sci USA 1991;88:5993.

287 Kahn S, Colbert TG, Wallace JC, Hoagland NA, Eisen H. The major 85 kD surface antigen of the mammalian stage form of Trypanosoma cruzi is a family of sialidases. Proc Natl Acad Sci USA 1991;88:4481.

288 Schenkman S, Jiang MS, Hart GW, Nussenzweig VA. A novel cell surface trans-sialidase of Trypanosoma cruzi generates a stage-specific epitope required for invasion of mammalian cells. Cell 1991;65:1117.

289 Schenkman S, Pontes de Carvalho L, Nussenzweig V. Trypanosoma cruzi trans-sialidase and neuraminidase activities can be mediated by the same enzymes. J Exp Med 1991;175:567.

290 Parodi A, Pollevick GD, Mautner M, Buschiazzo A, Sanchez DO, Frasch ACC. Identification of the gene(s) coding for the trans-sialidase of Trypanosoma cruzi. EMBO J 1992;11:1705.

12 Leishmaniasis

David L. Sacks, Jacques A. Louis, & Dyann F. Wirth

INTRODUCTION

A cell biologist's and immunologist's approach to leishmaniasis is to understand the mechanisms whereby promastigotes are able to infect macrophages, transform into replicating amastigotes, and survive within an immunologically competent individual. A considerable degree of adaptation has taken place in the course of evolution to enable these protozoa to become established in the cell which constitutes the major antimicrobial effector arm of the cell-mediated immune response. The present review summarizes our current understanding of immune evasion by leishmanial parasites together with host immune mechanisms and strategies whereby these may be overcome.

DISEASE FORMS

The intimate association of the parasite with cellular components of the immune system requires that any immunologic discussion be set within the context of the histopathologic features of the principal forms of the disease. Depending mainly upon the species of leishmanial parasite, but also upon the immunologic status and response of the human host, expression of disease can be quite variable. Human infection may be entirely inapparent or subclinical or it may display a spectrum of manifestation from cutaneous involvement, through late destruction of mucous membranes, to generalized systemic disease with fatal outcome. The most important leishmanial species (and subspecies) known to produce these various forms of disease in humans are given in Table 12.1. While this classification is not universally accepted, the emphasis of this review will be in any case on the immunopathologic and immunologic aspects of the disease rather than the causative agent. The clinical features of the three main categories of human leishmaniasis are summarized below (reviewed by Neva & Sacks [1] and WHO [2]).

Cutaneous leishmaniasis (CL)

The basic clinical features of CL are very much the same wherever it occurs. Cutaneous lesions begin as small erythematous papules on exposed areas of the body where infected sandfly vectors have fed. The incubation period may be as short as 1–2 weeks up to as long as 1–2 months. The early lesion may be pruritic, but the ulcer, even the large ones, is not painful. The ulcer can remain relatively dry with a central crust (dry form) or it may exude seropurulent material (wet form). Multiple lesions may be present, depending upon the nature of exposure to infected sandflies. Cutaneous leishmanial lesions will eventually heal spontaneously, but they can persist for up to 1 year or more without treatment.

Diffuse cutaneous leishmaniasis (DCL) is a striking but uncommon complication of cutaneous disease associated with immunologic unresponsiveness to leishmanial antigens. It starts as a regular ulcerative cutaneous lesion but then progresses to involve multiple sites as nonulcerative nodules or plaques on cooler areas of the body, as in lepromatous leprosy. Another rare form of cutaneous disease associated most commonly with *Leishmania tropica* infections is the chronic relapsing or recidivans type. The lesions typically are papular and nonulcerating and they occur beyond the borders of the original scar as satellite lesions or in the center of a healed area.

Mucocutaneous leishmaniasis (MCL)

Most patients with MCL have a history and/or typical scar of previous cutaneous disease, and it is therefore thought to be due to metastasis of organisms to mucosal sites from a primary cutaneous lesion. The likelihood

Table 12.1 Geographic distribution and clinical disease caused by different species of *Leishmania*

Species	Geographic distribution	Clinical manifestations
Leishmania mexicana complex (*L. m. mexicana*, *L. m. amazonensis*, *L. m. venezuelensis*, ? others)	New World—from southern USA through Central America, northern and central South America, Dominican Republic	Cutaneous ulcers; small proportion of cases may develop DCL or MCL
Leishmania braziliensis complex (*L. b. braziliensis*, *L. b. guyanensis*, *L. b. panamensis*, *L. b. peruviana*)	New World—from Central America through various parts of South America, including Brazil, Venezuela, Bolivia, Peru to northern Argentina	Cutaneous ulcers; some cases may later develop MCL (probably more likely if cutaneous lesion not treated adequately)
Leishmania major	Northern Africa, Middle East, Central Africa, and south Asia	Cutaneous ulcers
Leishmania tropica	Middle East and southern Asia	Cutaneous ulcers and chronic relapsing cutaneous disease (recidivans form)
Leishmania aethiopica	Ethiopia	Cutaneous ulcers, rarely DCL
Leishmania donovani	Old World—East Africa and south of Sahara, southern Asia, including India and Iran	Visceral leishmaniasis; small proportion may develop post-kala-azar dermal leishmaniasis
Leishmania infantum (? separate species)	Old World—North Africa and southern Europe	Visceral leishmaniasis
Leishmania chagasi (? separate species)	New World—Foci in several areas of Brazil, Venezuela, and Colombia, and isolated cases in Central and South America	Visceral leishmaniasis

that MCL will later develop in a patient with regular CL cannot be predicted, except that it is encountered with greater frequency in certain geographic areas and it is more commonly associated with *Leishmania braziliensis braziliensis* than other species. Metastatic mucosal lesions run a slow but progessively necrotic course, resulting in irreparable damage to the soft and cartilaginous tissues of the nose, the palate, and the pharyngeal cavity.

Visceral leishmaniasis (VL)

In VL, or kala-azar, widespread parasitization of the mononuclear phagocyte system is especially prominent in the spleen and liver, with massive enlargement of these organs. The clinical aspects of VL caused by *Leishmania donovani*, *Leishmania donovani infantum*, and *Leishmania donovani chagasi* remain practically the same throughout the world, although they have quite different epidemiologic features. The incubation period of VL is usually long, 1–3 months. The onset of disease is generally insidious; fever, accompanied by sweating, weakness, and weight loss gradually become noticeable. In advanced cases edema and ascites can develop, and in untreated cases deaths are common due to secondary bacterial infections such as pneumonia, tuberculosis, or dysentery.

IMMUNODIAGNOSIS

The gold standard for the definitive diagnosis of leishmanial infection remains as the identification of parasites in biopsied material either by direct microscopic examination or by their growth in axenic culture. For the diagnosis of past or subclinical leishmanial infection or for those disease forms for which the parasite load is slight and/or parasite growth in culture is poor (e.g., mucosal disease) or for which biopsies are difficult to obtain in the field (e.g., visceral disease), immunodiagnostic techniques may need to be relied upon. Immunodiagnosis of leishmanial diseases has advanced relatively little in the last 10 years. The antigens used for skin testing or serologic assays are still crude heterogeneous preparations. They are therefore generally unable to distinguish between infections produced by different leishmanial species, particularly in the New World, and even more problematic is that they are often crossreactive in other parasitic infections, such as Chagas' disease and African trypanosomiasis. For the serodiagnosis of visceral leishmaniasis, however, crude antigen preparations have been used in a number of assays with extraordinary sensitivity and specificity. This is made possible by the fact that in VL the antileishmanial antibody titers are so extremely high that they permit the sera to be tested at high dilution, thus minimizing crossreactivity. A variety of enzyme-linked immunosorbent assays (ELISA) [3,4] and indirect fluorescent antibody tests (IFAT) [5,6] are in routine diagnostic and seroepidemiologic use for the detection of antileishmanial antibodies in VL. While the sensitivity of these tests approaches 100%, among the factors limiting the use of these methods in the field are the instability of reagents and the need for sophisticated equipment. A recently developed direct agglutination test (DAT) [7,8] has shown promise in fulfilling the requirements of an economic and simple field test. This test, which uses whole formalinized and trypsinized promastigotes, can be performed on whole blood at high dilution and has been applied successfully to the field diagnosis of VL in the canine reservoir as well as in humans [9]. Some recent progress has been made in the identification of antigens which are specifically diagnostic for VL. These include a cloned 60 kD membrane-associated antigen from *L. donovani* [10], an affinity-purified 70–72 kD glycoprotein from *L. donovani* [11], and a 62–63 kD protein from *L. chagasi* obtained by elution from sodium dodecyl sulfate polyacrylamide gels [12]. A highly specific, competitive, inhibition assay has also been developed using species-specific *L. donovani* monoclonal antibodies (MAB) [13]. Because assays which rely on antibody reactivity with specific antigens are better able to retain specificity even at low serum dilutions, it is possible that these assays might be able to detect the low serum antibody present in early or subclinical VL.

Because, generally, low serum antibody titers are also present in cutaneous infections, serodiagnosis of CL using IFAT, ELISA, or DAT has suffered from the inability to detect up to 30% of parasitologically proven cases [14,15]. A radioimmunassay displayed substantially better sensitivity when used to detect antibodies in Israeli CL patients, and its use with a carbohydrate–lipid fraction of the parasite improved the specificity problems associated with low serum dilutions [16]. Because of the strong cellular reactivity which accompanies CL and MCL, the delayed skin reaction or Montenegro test [17] remains an extremely useful tool for the immunodiagnosis of these diseases. The antigen commonly used is a suspension of promastigotes in phenolized saline (leishmanin) which is injected intradermally. A positive skin reaction does not necessarily indicate active infection, however, but may simply indicate some prior exposure to *Leishmania* species, especially in endemic areas. In addition, there is considerable crossreactivity at the skin test level with different leishmanial species, as well as with other chronic infections [18,19]. To date, molecularly defined skin test antigens which might elicit responses in a species- or even genus-specific manner have not been reported. Their use in *in vitro* assays for cellular reactivity will be discussed below.

In the event that clinical isolates can be obtained and successfully cultivated, a number of new techniques have been developed for their molecular characterization. The application of these techniques is useful not only for taxonomic and epidemiologic studies but the ability to distinguish between species likely to produce, for example, mucosal as opposed to simple cutaneous disease will influence the nature of treatment. The most widely used methods for identifying and classifying leishmanial field isolates have been enzyme electrophoresis (zymodeme analysis) [20,21] and serodeme analysis with monoclonal antibodies. The first MAB produced which could distinguish between species were specific for either *Leishmania mexicana* or *L. braziliensis* [22]. Since then, the considerable species or subspecies diversity of *Leishmania* in both the Old World and particularly the New World have been identified using monoclonal reagents [23–25]. Based on their reactivities with an extensive panel of MAB, over 1000 New World human isolates have been typed and assigned one of at least eight species designations, each with distinctive geographic and epidemiologic patterns [26]. Monoclonal antibodies have also been used for the *in situ* detection of amastigotes in cutaneous lesions [27,28], although this approach, which offers a potentially rapid and specific diagnosis of leishmanial infection at the species level, will undoubtedly be supplanted by far more sensitive DNA probe-based *in situ* detection systems, which will be discussed below.

DNA PROBE-BASED DIAGNOSIS OF LEISHMANIASIS

Leishmania species identification is based on a variety of ecologic, biologic, biochemical, and immunologic criteria [29]. In previous work, each cultured isolate of the parasite has been analyzed by one or more of these criteria and grouped into species and subspecies. There remain certain controversies as to whether organisms isolated in distant geographic locations but sharing certain common properties belong to the same or distinct subspecies of the genus *Leishmania*. Whether these organisms represent strains of the same subspecies or distinct subspecies cannot be resolved because there is no single generally accepted method for species identification in the genus *Leishmania*. There is no defined sexual stage so that traditional criteria for species identification cannot be applied. This complicates comparison of the disease epidemiology in distinct geographic locations and represents a potential limitation on the transfer of control measures from one geographic location to another. The WHO has addressed this problem by establishing a set of reference strains for the various species and subspecies which are to be used for comparison and classification of new isolates. In humans, the *Leishmania* parasite causes a range of diseases from simple cutaneous lesions to more serious forms of the disease, including MCL and DCL. The clinical manifestations of the disease are dependent at least in part on the infecting *Leishmania* species.

DNA probe-based diagnosis of leishmaniasis is an alternative to serodiagnosis and the Montenegro test. This method has the advantage of a direct assay of current infection. The initial DNA probes identified for leishmaniasis were based on the kinetoplast DNA of the parasite [30]. The kinetoplast DNA has two types of DNA molecules, the minicircle and the maxicircle. The maxicircle is equivalent to the mitochondrial genome in other eukaryotic organisms. It has several unique features, most notably the phenomenon of RNA editing in which transcripts from the maxicircle are post-transcriptionally modified by the addition or deletion of uridine residues within the transcript [31–35]. This process seems to require small "guide" RNA molecules which may act as templates or may participate in transesterification, resulting in the addition of uridines [36–38]. Until recently the function of minicircles was unknown but recent evidence has implicated the minicircles as templates for guide RNA necessary for the editing of some maxicircle transcripts [37,38].

The focus of the diagnostic work using DNA probes has been on the kinetoplast DNA minicircle. It has been studied extensively both in *Leishmania* spp. and *Trypanosome* spp. (both African and South American) [39–49]. It represents 10–25% of the total cell DNA and is repeated 10 000–100 000 times per kinetoplast.

The minicircles serve as ideal targets for DNA probe-based diagnosis in that they are highly repeated sequences which have a species-dependent divergence in DNA sequence. The recent suggestion that these minicircles serve as templates for guide RNA implies a certain constraint on the divergence of those sequences; however, the guide RNA identified thus far have been relatively short (30–50 nucleotides). DNA minicircles appear to undergo relatively rapid DNA sequence divergence, presumably because of the noncoding nature of the DNA. This observation has led to a useful application of molecular biology to the differentiation of *Leishmania* spp. and subspecies and subsequently to the direct diagnosis of leishmaniasis in patients [48]. In the New World *Leishmania* spp. the kDNA minicircle isolated from *L. mexicana* does not have significant sequence homology with those isolated from *L. braziliensis* [30]. This difference in DNA sequence has provided the basis for a DNA probe that can distinguish these two *Leishmania* spp. directly when material obtained from a lesion is applied to nitrocellulose [46,48]. This methodology has now been used to diagnose patients with CL, offering both a more rapid and specific diagnosis than is currently available with any other methodology. An alternative application of this methodology is that of *in situ* hybridization in which the lesion material is applied directly to a microscope slide followed by *in situ* hybridization with specific kDNA-based probes [50]. The advantage to this method is the ability to visualize directly the hybridization associated with the parasite.

Both work to develop subspecies-specific diagnostic probes and molecular analysis of kDNA had shown species-related restriction site heterogeneity within the kDNA and several groups have cloned species- or isolate-specific kDNA fragments from visceral and Old World cutaneous strains of *Leishmania* [39–48]. This work demonstrated that within the kDNA network, fragments of DNA existed with different taxonomic specificities. These cloned subspecies-specific DNA probes are important for the development of future diagnositic probes but are also very interesting with regard to the mechanism of kDNA sequence divergence. Within a single minicircle there are regions of conserved DNA sequence and regions of highly divergent

DNA sequence. The 10 000 minicircles within a single *Leishmania* organism are heterogeneous and fall into a number of different sequence classes. Within a given sequence class, the minicircle sequences are identical. The maintenance of homogeneous sequence classes in the face of rapid sequence variation presumably requires a mechanism whereby mutations in a single minicircle are transmitted to other minicircles. In addition, some mechanism must exist for the generation of rapid sequence divergence in one portion of the minicircle molecule with maintenance of relatively conserved regions within the same molecule.

The first phase of development of DNA probe-based diagnosis of leishmaniasis demonstrated the feasibility of the methodology and defined the limitations of the technology. Direct detection of parasites in human biopsies required an active infection with a minimum of a few hundred parasites. For certain infections, especially simple CL, the direct assay was adequate for routine diagnosis and epidemiologic surveys. However, application of the direct detection assay was of limited utility for detection of parasites in MCL or in VL because of the variable and often low number of parasites present in a clinical sample. Thus, methods to increase the sensitivity of the DNA probe-based assay were sought.

The polymerase chain reaction (PCR) is one such technique recently applied to the identification of infectious agents using DNA hybridization methods [51–54]. The PCR uses a thermostable DNA polymerase and a set of oligomer primers to direct the amplification of the hybridization target sequence through a series of priming–elongation–denaturation cycles [55]. Thirty such cycles is sufficient to amplify the target sequence 10^6-fold.

The PCR can be used to increase the sensitivity of detection of *Leishmania* parasites by DNA hybridization methods through the amplification of the minicircle target sequence [54]. The oligomer primers used are able to direct the amplification of all *Leishmania* strains tested. In addition, the PCR products from *L. mexicana* and *L. braziliensis* strains can be distinguished by hybridization with kDNA probes. The method is sensitive enough to detect the kDNA from a single organism and this sensitivity allows the use of nonradioactive hybridization methods. This method can be used to detect *Leishmania* from human biopsy material.

Methods are available to label hybridization probes with nonradioactive markers, such as biotin or digoxigenin. The detection of these probes is generally through an enzyme-linked system, such as alkaline phosphatase, which catalyzes a simple color development reaction. The practical introduction of these nonisotopic methods has long been anticipated by workers in developing countries and is greatly facilitated by the ability to increase the sensitivity of these probes using the PCR. The application of the PCR in conjunction with both ^{32}P-labeled and nonisotopic kDNA probes to detect and identify *Leishmania* from human biopsy material is now feasible.

The increase in the sensitivity of *Leishmania* kDNA detection will aid in solving several problems now faced in the diagnosis of leishmaniasis. Some lesions, such as those caused by *L. braziliensis* strains and those seen in leishmaniasis recidivans, have a characteristically low concentration of parasitized macrophages, which may be below the detection limit of direct hybridization methods [56]. In addition, parasites are frequently difficult to culture from these lesions, [57]. Given that *Leishmania* lesions frequently mimic other common bacterial and fungal skin infections, there is a real potential for misdiagnosis in such infections. In the case of MCL, an accurate identification of *L. braziliensis* parasites is critical, as a percentage of these infections will metastasize to the oronasopharyngeal mucosa months or even years after the initial inoculation, producing grossly disfiguring lesions. It has been reported that *L. braziliensis* infections which are aggressively treated were less likely to progress to the mucosal areas [58], making a prompt and accurate diagnosis of these infections essential.

Polymerase chain reaction analysis may also be useful in the treatment of those lesions where the taking of biopsy punches is to be avoided, such as facial lesions and those of small children. Scrapings and needle aspirates are the preferred samples from these types of lesions and the fact that they may yield fewer parasites for analysis is compensated for by the increased sensitivity provided by the PCR.

It is unclear whether the kDNA from killed parasites in biopsy samples is able to be amplified or how this amplification should affect the interpretation of a positive hybridization result. Experiments monitoring the parasitemia in a lesion over time, both by culturing or direct microscopic examination and PCR amplification of kDNA, are needed to determine the relationship of viable parasites to the presence of kDNA.

The increase in sensitivity produced by the amplification of the target sequence has allowed the use of nonradioactive DNA hybridization probes. In certain circumstances, hybridization may not be necessary, but instead, the PCR products can be visualized by staining with ethidium bromide after gel electrophoresis. While this method has certain limitations, it may provide a rapid means of identifying leishmaniasis, with subsequent species identification provided by hybridization. Leishmaniasis is primarily an endemic disease in developing countries, where radioisotopes are expensive and difficult to obtain and where the disposal of radioactive waste is a problem. Nonisotopic probes are easier to handle, have longer storage lives, and can yield quicker results compared to radioactive probes. These probes have, however, been associated with problems of low sensitivity and high background. Biotin-labeled probes frequently have a high background signal when hybridized to tissue samples. This is usually caused by the binding of the biotin detection molecule, avidin, to naturally occurring biotin present in various types of tissue [59]. Alternative nonradioactive labeling methods have been developed, including direct conjugation of enzymes to DNA oligonucleotide or the use of alternative haptens. Further, amplification of the target sequence reduces the amount of the tissue sample actually involved in the hybridization.

The method of preparing tissue biopsies for PCR amplification is quick and simple to perform. Only a small piece of tissue is required and provides material for several reactions. Recent work in *Trypanosoma cruzi* has demonstrated that increased sensitivity in detection of minicircle sequences by the PCR can be achieved by cleaving the minicircle DNA, thus releasing it from the network [60]. A similar application may be possible in the case of *Leishmania*.

The importance of methods aimed at detecting and identifying *Leishmania* species from natural samples cannot be overemphasized. Efforts to control leishmaniasis in endemic areas require a thorough knowledge of the ecology and epidemiology of *Leishmania*. The ability to amplify the hybridization target sequence will aid in studying the insect vectors and animal reservoirs of *Leishmania* by detecting parasites previously overlooked. Large-scale epidemiologic studies should also be enhanced by using this technique, given the potential to screen many samples at once. Further work is still needed concerning the development of more specific probes for all *Leishmania* species. The ability of the PCR to increase the sensitivity of these probes will make its application a necessity in the future.

The PCR methodology will also allow a new type of analysis in the future, namely the detection of specific genes directly from clinical or insect-derived samples. Until now, DNA probe-based diagnosis has been limited to highly repeated DNA [61] or RNA [62] sequences; however, the PCR has eliminated the need for such repeated sequences and should allow for the development of DNA probes specific for certain characteristics of the parasite, such as drug resistance or virulence.

ENCOUNTERS OF *LEISHMANIA* PARASITES WITH THE NONIMMUNE VERTEBRATE HOST

Complement

In order to initiate and sustain infection, sandfly-inoculated promastigotes and tissue amastigotes must survive microbicidal defense mechanisms to which they are exposed, even within nonimmune vertebrate hosts. These defenses include the lytic effects of fresh normal serum via alternative or classic complement pathway activation and the oxygen-dependent and -independent leishmanicidal activities of their host macrophages. The ability of fresh sera to cause the lysis of *Leishmania* was recognized as early as 1912 [63]. Later work with human and animal sera noted that heat-labile components in the serum were responsible for lysis, suggesting a role for complement [64]. The interaction of *Leishmania* with the complement system is now known to be influenced dramatically by the species and developmental stage of the parasite. Until recently it was believed that promastigotes of all species of *Leishmania* were susceptible to lysis by nonimmune serum because of their ability to activate complement, which for most species was shown to occur via the alternative pathway [65,66]. The exception is *L. donovani* promastigotes, which appear to bind natural or crossreacting IgM antibodies from normal human serum and activate complement via the classic pathway [67]. In contrast to promastigotes, *Leishmania* amastigotes of most species were found to be relatively serum resistant [68]. While their mechanism of complement resistance remains unexplained, it was shown, rather surprisingly, that even the most complement-resistant amastigotes (e.g., *L. donovani*) activate complement via the alternative pathway and efficiently fix C3 [68]. It was postulated that C3 fixation by tissue amastigotes may represent a mechanism seized upon by the parasite to facilitate re-entry into phagocytic cells.

The extreme susceptibility of promastigotes to complement-mediated lysis within nonimmune hosts raised the question as to how inoculated promastigotes survive serum exposure prior to their uptake by macrophages. The answer seems to be that the cultured promastigotes used in these complement studies were not representative of the forms which are actually inoculated by infected sandflies. What has emerged from a number of recent studies is that far from being uniform with respect to infectivity, *Leishmania* promastigotes undergo development within the sandfly midgut from a replicating serum-sensitive noninfective form to a nonreplicating serum-resistant infective or "metacyclic" form [69,70]. Similar developmental events have been shown to occur during the growth of most species of *Leishmania* promastigotes in axenic culture [71–73]. Study of culture-derived metacyclic promastigotes of *Leishmania major* and *L. donovani* has revealed that despite their serum resistance, these promastigotes also activate complement efficiently, resulting in extensive surface deposition of C3b and iC3b [74]. Their resistance to complement-mediated damage is thought to be due to acquisition of a surface coat which restricts the access of macromolecules (i.e., membrane attack complex) to the cell membrane [75,76]. The metacyclic surface thus provides a barrier against complement-mediated lysis but, as postulated for tissue amastigotes, because it activates complement efficiently it also promotes the attachment to and uptake by host macrophages.

Attachments to macrophages

Having survived their brief sojourn in the extracellular milieu, inoculated promastigotes or tissue amastigotes must then target themselves to macrophages and survive a barrage of intramacrophage microbicidal systems, which may include the lethal products of oxygen metabolism, lysosomal hydrolases, low pH, and cationic proteins. Studies involving *Leishmania*–macrophage interactions have focused most intensely in recent years on the identification of the parasite ligands and their complementary receptors on the macrophage surface as one possible means by which the parasite may influence the outcome of infection. Attachment to macrophages via receptors not normally associated with the triggering of hostile responses, such as the superoxide burst, may increase the chances for intracellular survival. The diversity of interactions that have been implicated to date is summarized below.

Early experiments from several laboratories indicated that the parasite–macrophage interaction was receptor mediated [77–79]. In the presence of an excess of parasites, promastigote ingestion by macrophages could be saturated and various exogenous saccharides could partially inhibit promastigote ingestion. In these and other studies, promastigote attachment was studied in the absence of fresh serum, thus the receptors binding directly to the parasite surface, known to be rich in carbohydrates, were revealed. The mannose/fucose receptor was implicated in the binding of *L. donovani* promastigotes to human and mouse macrophages [80–82], and the mannose-6-phosphate receptor was shown to be involved in the recognition of *L. mexicana* promastigotes [83]. A role for lectin-like receptors in promastigote recognition was further supported by the finding that the major surface glycoconjugate of promastigotes, the lipophosphoglycan (LPG), could bind directly to mouse macrophages and inhibit *L. major* promastigote attachment [84], and that mannan- and fucose-containing glycoconjugates from *L. donovani* promastigotes could inhibit internalization of these cells [85]. A role for the major surface glycoprotein of *Leishmania*, gp63, has also been established for the serum-independent binding of *L. mexicana* promastigotes to macrophages [86,87].

In the *in vivo* situation, promastigotes are likely to be exposed to components of the complement cascade present in extracellular fluid. Thus a number of investigators have examined *Leishmania*–macrophage interactions in the presence of fresh serum. As discussed above, both metacyclic promastigotes and tissue amastigotes are relatively serum resistant, yet it is known for *L. major* and *L. donovani* at least that they activate complement and fix C3 on their surface [72,74]. It is this interaction with complement that is postulated to be the key event in establishing *Leishmania* as intramacrophage parasites. In support of this view is the consistent observation that preincubation in fresh serum dramatically enhances the attachment and uptake of *L. major* and *L. donovani* promastigotes by C3 receptors expressed on the macrophage plasma membrane [88,89]. In addition to increasing parasite phagocytosis, C3 fixation enhances their subsequent intracellular survival [90], presumably because ligation of C3 receptors (CR3 and CR1) fails to trigger a respiratory burst [91].

The relative contribution of endogenous parasite ligands versus serum opsonins to the attachment and subsequent intracellular survival of *Leishmania* promastigotes may differ between species. Studies conducted on *L. mexicana* promastigotes indicated that the parasite's own ligands were sufficient to ensure successful infection of macrophages, and the effect of serum opsonization was minimal [92]. Interestingly, even the serum-independent macrophage binding of these promastigotes was found to be mediated by CR3, in this case binding not to C3 but to the major surface glycoprotein of *Leishmania* promastigotes, gp63 [93]. This observation probably explains the serum-independent CR3-mediated binding which has been reported for *L. donovani* and *L. major* [80,81,94], although opsonization by complement components produced locally by the target macrophages has not been ruled out [80]. Different leishmanial species undoubtedly differ in a number of endogenous ligands and in the manner and extent to which they interact with complement. It is therefore likely that differences, both qualitative and quantitative, exist between the manner in which these parasites bind to and enter macrophage subpopulations, which in turn determines the markedly different tissue tropism and disease syndromes which these species display. It is particularly important that in future studies these differences be considered in the context of amastigote–macrophage interactions, since it is this stage which is responsible for maintenance and amplification of infection.

Intramacrophage survival

Attachment of *Leishmania* amastigotes and metacyclic promastigotes to macrophages leads to extremely rapid phagocytosis, particularly when the organisms have been serum opsonized. Phagocytosis is generally accompanied by a respiratory burst and the consequent production of oxygen metabolites, such as superoxide and hydrogen peroxide, which have been shown to be toxic to all stages of the parasite [95]. Amastigotes and metacyclic promastigotes appear to evade oxygen-dependent destruction by triggering a minimal respiratory burst during attachment and ingestion [96,97]. This may be due, as already discussed, to the use of non-triggering receptors, such as CR1 and CR3, and/or to the presence of the LPG and a membrane-bound acid phosphatase, both of which have been shown to inhibit the oxygen burst [98,99]. After phagocytosis the parasite resides in a parasitophorous vacuole which then fuses with secondary lysosomes to form a phagolysosome where the organisms survive and replicate. Thus *Leishmania* challenge the very heart of the cell's defensive machinery, containing powerful hydrolytic and oxidative enzymes which kill and digest most living organisms. The phagosome is known to be functional because other material sharing the same vacuole can be seen undergoing digestion [100]. How the parasite sur-

vives within this environment has not been explained. Survival may be promoted by the resistance of the parasite's exposed surface to enzyme attack, perhaps because the presence of LPG makes it so highly negatively charged. Alternatively, the parasite may excrete inhibitors of lysosomal enzymes, such as the released form of LPG, or excretory factor (EF) as it was originally termed, which has been shown to display an anti-b-galactosidase activity *in vitro*, possibly linked to the presence of galactose in EF [101]. Recently the surface acid proteinase (gp63) of *L. mexicana* has been found to be a metalloenzyme capable of protecting liposome-encapsulated proteins from phagolysosomal degradation by macrophages [102]. The probability that the phagosomal membrane is modified in some way by the parasite is suggested by the different behavior of the membrane, depending upon the species. With *L. mexicana* the phagosome becomes greatly enlarged, and each parasite adheres to the membrane at one end. With *L. donovani* each parasite is usually surrounded by its own, closely opposed, individual membrane. Phagolysosomal swelling may be due to hyperosmolarity resulting from the failure of lysosomal enzymes to degrade high-molecular-weight solutes [103].

Since the parasite lies in a functioning phagosome, it must have available to it nutrients ingested by the host cell. The demonstration that extracellularly added electron-dense material is eventually detectable around the parasite indicates a continuous access to the extracellular environment via normal endocytic pathways [100]. Available nutrients are further digested by leishmanial enzymes, such as the membrane-bound and released acid phosphatases, which may play a nutritive role by hydrolyzing organic phosphates [104,105]. A similar function might be served by the unique nucleotidases present on the surface. *Leishmania*, as well as other trypanosomatids, cannot synthesize purines *de novo* and are therefore dependent upon an exogenous supply of preformed purines which can be utilized by leishmanial 3'- and 5'-nucleotidases [106]. Since normal phagolysosomal pH falls to 4.5–5.0, the parasite requires a way of maintaining its own intracellular pH. This is accomplished by the action of a membrane proton-translocating ATPase which is located on the cytoplasmic side of the parasite surface membrane and acts by coupling ATP hydrolysis to proton pumping activity [107]. This creates a proton electrochemical gradient across the membrane, which drives active transport of nutrients, such as glucose and proline, which are vital sources of energy.

USE OF MOLECULAR BIOLOGY TO UNDERSTAND PATHOGENESIS

Functional analysis of genes and gene products is necessary to fully understand their role in the infective process. Such analysis has been possible in the study of bacterial and viral pathogens through the use of molecular genetics. For example, in the case of listeria infection, the genes involved in the invasive process have been identified, cloned, and reintroduced into noninvasive organisms to demonstrate their function [108–113]. Multiple steps in the invasive process have been identified with this approach and because of the many similarities in the invasive process between listeria and leishmania, a similar multistep process is likely for *Leishmania* infection. Both in the identification of genes and in the final demonstration of their importance in the invasive process it is necessary to have functional assays. Functional analysis ideally involves the alteration of gene function by overexpression, mutation, or deletion. In the case of a loss of mutation function, the reintroduction of the gene to complement the defective gene is necessary to prove the essential role of any gene in the infective process.

The major development which was critical in continuing molecular genetic analysis of these parasites was the development of a transfection system which allowed the introduction and analysis of modified genes in the parasite. The first systems to be developed for the expression of foreign genes in *Leishmania* or related kinetoplastidae were transient expression systems [114,115]. Similar transient transfection systems are used frequently in the analysis of mammalian gene expression and offer the advantage of a rapid means to assay *cis*-acting expression signals. Successful transfection and transient expression of exogenously introduced DNA in *Leishmania enriettii* is the first step in this important technology [115]. A hybrid gene was constructed *in vitro* in which the intragenic region of the α-tubulin repeat was placed 5′ to the bacterial chloramphenicol acetyltransferase (CAT) gene in the pBluescript vector. This plasmid was then introduced into cells by electroporation and the CAT activity was assayed after 24–48 h. Expression of CAT required a portion of the α-tubulin intragenic region. This hybrid gene was active in *L. enriettii*, *L. tropica*, and *L. braziliensis*. Other plasmid vectors have been developed subsequently using the upstream regions of other transcriptional units, including the dihydrofolate–thymidylate synthase gene [116] and different reporter molecules. Again, in these vectors, there is a requirement for *Leishmania*-derived sequence. Similar types of experiments in trypanosomes have also resulted in the construction of transient transfection vectors for these parasites [117–119]; however, despite the relatedness of these organisms, vectors containing leishmania sequences are not active in trypanosomes [115]. A second important step in the development of functional assays for the parasite is to be able to introduce permanently modified genes into the parasite. This has now been accomplished both by the use of an extra-chromosomally replicating plasmid [120,121] and by the direct integration of selectable markers flanked by specific parasite sequences [122–125] and the subsequent "knock-out" of genes [123].

One of the first applications of these methods has been to identify functional sequences important in trans-splicing of mRNA. The *Leishmania* spp. have an unusual mechanism of mRNA processing. Each mRNA is a hybrid RNA molecule containing RNA sequences derived from two distinct parts of the genome, in fact from two different chromosomes. Each mRNA contains both a unique sequence encoding a structural protein and a common sequence which is an identical 39 nucleotide "cap" at its 5′ end. In *L. enriettii* and other trypanosomastids, many genes are arranged in tandem arrays and the intergenic sequences contain both the splice acceptor site for the addition of the Spliced Leader (SL) sequence and a putative polyadenylation site. There remains the question as to whether promoters or other transcriptional control elements are located in the intergenic regions or are upstream from the tandem arrays. Transient expression analysis (see above) demonstrated that the α-tubulin intergenic sequence was required for expression of CAT [115]. The expression of the CAT gene is dependent on the presence of a short sequence derived from the α-tubulin intergenic region. This sequence (less than 50 base pairs in length) contains both a trans-splice acceptor site and a stretch of polypyrimidines which have subsequently been demonstrated to be critical for the trans-splicing reaction (Lafaille, Laban, & Wirth, in press).

To answer the question as to whether the intergenic region provided transcriptional signals or only signals necessary for splicing, a stable transfection system was developed. The goal was to differentiate splicing signals and, from transcriptional signals, to define their relative roles in gene expression [120]. A selectable marker, *neo*[r], was cloned adjacent to the α-tubulin intergenic region and this plasmid was used to transfect *L. enriettii*. G418-

resistant parasites were isolated and analyzed. The neo[r] mRNA was processed (trans-spliced and poly-adenylated) using the signals of the α-tubulin gene present in the pALT-Neo plasmid. This work demonstrates the role of the α-tubulin intergenic region in trans-splicing and polyadenylation but not a role in transcription. In a parallel set of experiments in *L. major* [121], the upstream sequences from the dihydrofolate reductase–thymidylate synthetase (DHFR–TS) gene were also used in developing a stable transfection vector. Similar experiments using flanking regions from several *Leishmania* genes are now in progress. The next goal is to determine the specific signals necessary for transcription initiation. In trypanosomes, recent work has identified a potential "promoter" adjacent to the gene encoding procyclin, a major surface protein [117,118], but no analogous sequences have yet been identified in *Léishmania*. An additional question is also raised by these experiments; the plasmid sequences are found to be stable in the drug-resistant parasites but the mechanism of propagation of these sequences remains to be determined. The development of these methods allows functional analysis of sequences important for gene expression.

Another fascinating observation has arisen from the work to develop transfection systems, which has important practical applications. Homologous recombination of exogenously added DNA sequences occurs with a much higher frequency than nonhomologous recombination, thus facilitating site-directed mutagenesis of endogenous genes [122–126]. This is in contrast to mammalian systems in which exogenously added DNA is incorporated into nonhomologous DNA, thus hindering direct analysis of modified endogenous genes. In this regard, the trypanosomatids resemble yeast and other lower eukaryotes. Further experiments to define the mechanism of this homologous recombination gene replacement vs. reciprocal exchange are necessary. Application of this technology has just begun but it promises to provide the means to modify any gene by replacement or insertional mutagenesis. It also provides the potential for developing attenuated non-virulent parasites by deletion of genes essential for pathogenesis but not necessary for establishing infection [124]. This provides a novel approach to vaccine development in the parasite.

These developments have moved the field from descriptive experiments to functional analysis. There remain important and fundamental questions regarding the parasite and its pathogenicity, which now can be approached readily using transfection systems. These are listed below.

Genetic organization and mechanisms

Recent work has served to define the chromosomal organization of the parasite but there is no information on the exchange of genetic information in these organisms and the establishment of genetic systems for analysis. This area is just beginning to develop. No sexual stage has been defined for these parasites and there remains a controversy on even the most basic questions as to whether the organism is haploid or diploid and either promastigote or amastigote.

Developmental control of gene expression

One of the most interesting questions concerns the control of gene expression in the various stages of the parasite's life-cycle. The parasite exists in vastly different environments within the insect vector and the mammalian host. One major question which fascinates molecular biologists is what adaptive mechanisms has the parasite developed to maximally utilize the two environments. Thus a major focus of the current work and future work will be in understanding these adaptive mechanisms at the molecular level.

Novel biochemical and molecular processes

A third area of interest is that of novel biochemical/molecular pathways which are specific for the parasite. This is both of basic interest and has a direct practical application in the development of new points for chemotherapeutic attack. One very striking example is that of a novel mechanism of bimolecular splicing in the processing of RNA. A major area of interest both for genetic analysis and for the more practical consideration of drug therapy is the mechanisms of drug resistant in the *Leishmania* spp. Work in several laboratories has demonstrated that gene amplification plays a major role in the development of drug resistance [127–129]. At least one mechanism of resistance is the overproduction of the target gene product in the resistant parasite. This is similar to the mechanism found in many drug-resistant mammalian cells and may be an important *in vivo* step in the development of drug resistance. Gene amplification of enzyme structural genes appears to be at least one important mechanism in the development of drug resistance in *Leishmania* spp. One of the important questions which remains to be addressed is the molecular mechanism of this circular amplification and its role in the development of drug resistance in the field.

Mechanisms of pathogenesis

Finally, the detailed molecular mechanisms of pathogenesis remain a fundamental question. Several molecules have now been implicated in pathogenesis, including the gp63 glycoprotein. It should now be possible to assess directly its role by gene manipulation, although its multicopy nature increases the technical difficulty of the experiment.

IMMUNOLOGIC PARAMETERS AND IMMUNOPATHOLOGIC ASPECTS OF HUMAN LEISHMANIASIS

Simple cutaneous leishmaniasis

Evidence for naturally acquired immunity

The prototype of infections caused by *Leishmania* is simple CL (oriental sore), which is a self-healing disease. Immunologic studies of cutaneous infections in humans have focused primarily on the nature of the immune mechanisms responsible for control and resolution of infection and resistance to reinfection. Resistance to reinfection has been documented convincingly in studies of experimental leishmaniasis in humans in which individuals with naturally acquired infection have been challenged by experimental inoculation, individuals with experimental lesions at various stages of development have been experimentally challenged, and volunteers with experimentally induced lesions have been studied during subsequent exposure within an endemic area. In one of the earliest and largest studies of its kind involving almost 2000 subjects, Sokolova [130] found that experimental infection with *L. tropica* reduced the natural infection rate within an endemic area from 64% to 6%. In other studies, when individuals with healed natural or experimental *L. tropica* infection were challenged by inoculation or sandfly bite, a typical delayed-type hypersensitivity (DTH) reaction developed but not a lesion [131–133]. In the New World, past *L. mexicana* cutaneous infection was shown to confer protection against experimental challenge with both homologous and heterologous *L. mexicana* strains [134]. It is important to note that immunity to reinfection appears to be dependent on the stage of development of

the primary lesion. In another study by Sokolova [130], reinoculation of 61 subjects at various times after the appearance of their primary lesion resulted in superinfection when the primary lesion was active but not when it was healing or healed. Similar findings have been confirmed by others [133,135].

Correlates of cell-mediated immunity (CMI)

Self-cure and resistance to reinfection are due to the immune response which the infection evokes, and it was recognized as long as 60 years ago that CMI develops early during cutaneous infection and persists long after healing is complete [136]. Evidence for this cellular reactivity includes ulceration of the lesion, the histopathologic evidence, the positive DTH leishmanin skin test [136,137], and the ability of lymphocytes to respond *in vitro* to leishmanial antigens [138,139]. The onset of DTH positivity relative to the time of infection appears to vary from a few days to several months. The skin test is not necessarily an indication of immunity since in the early stages of the infection superinfection can occur [140–142]. Lymphoblast transformation (LT) *in vitro* is, like the skin test, a useful measure of immunologic stimulation at the T cell level but not necesssarily protection. The LT response has the advantage of being more quantitative than DTH, and in one series of studies the magnitude of the LT response was shown to correlate with the stage of the lesion, with the active phase of the nodule characterized by the highest mean stimulation index contrasted with the healed nodule group which had the lowest [143]. Lymphoblast transformation is especially quantitative when combined with limiting dilution to estimate the clonal frequency of *Leishmania*-specific T cells in a particular organ. These frequencies ranged from $1/10^5–1/10^3$ in the peripheral blood of 18 human cases of American cutaneous leishmaniasis (ACL) [144]. These relatively low frequencies might be explained by the light systemic involvement produced by localized cutaneous lesions.

In addition to providing a measure of *Leishmania*-specific T cell frequency, the activation of T cells *in vitro* can also be used to characterize the phenotypes of responding T cell subsets and to profile the cytokines which they produce. While there is only a minimal amount of clinical data which directly address these issues, it is none the less widely accepted that the *in vitro* response of peripheral blood cells from cutaneous patients can be characterized by $CD4^+$ T lymphocytes producing interferon-γ (IFN-γ), which has been shown to activate human macrophages to kill intracellular *Leishmania* parasites [145–148].

Another obvious advantage of LT as an *in vitro* correlate of CMI is that the antigens used to elicit T cell responses can be studied carefully. T lymphocytes from patients with active or cured CL are reactive to complex soluble antigens of both homologous and heterologous species, but specific T cell antigens have yet to be defined. In recent studies [148,149], the nature of the leishmanial antigens involved in eliciting T cell immunity in ACL was examined using a T cell immunoblotting method in which nitrocellulose-bound leishmanial antigens, resolved by one- or two-dimensional electrophoresis, were incorporated into lymphocyte cultures. The proliferative and IFN-γ responses were remarkably heterogeneous and occurred to as many as 50–70 distinct antigens. Control and resolution of cutaneous infection thus appears to be associated with a T cell response to a large and diverse pool of parasite antigens, and the possibility that many of these responses contribute to the immune status of the host has important implications in terms of vaccine development.

Immunohistologic findings

The early histologic findings in cutaneous disease indicate a macrophage granuloma in which the parasite multiplies and the antigenic load increases [150,151]. With an increase in the number of infected cells the inflammatory reaction in the dermis composed of lymphocytes, plasma cells, and macrophages becomes more prominent, and small foci of necrotic cells and scattered polymorphonuclear leukocytes may be seen. Cryostat sections of cutaneous lesions have been studied with immunofluorescence and immunoperoxidase staining methods for the phenotypic characterization of intralesional lymphocytes. In one study of 20 patients with ACL, the granulomas showed a mixture of T cell subpopulations with the CD4/CD8 ratio of phenotypes less than one [152]. The abundance of CD8 cells within these healing granulomas is particularly intriguing and has been confirmed in recent studies in which cell lines established from intralesional lymphocytes in ACL contained up to 30% CD8$^+$ cells (P. Melby, unpublished observations). The function of these CD8 cells is unknown, but they do not seem to down-regulate the CMI response since, in addition to these cells, IL-2-producing and IL-2 receptor-bearing cells (Tac$^+$) are present and the virtual elimination of the parasite can be seen [152,153]. The final histologic events are characterized by a postnecrotic granuloma of giant and epithelioid cells, marking the elimination of parasites and the onset of healing. The necrosis is most likely the consequence of the intense immunologic reaction occurring in the skin, but the exact mechanism of the process is not known.

Humoral immune response

The low levels of circulating specific antibodies and the lack of success of serum transfer experiments have led to the conclusion that humoral immunity plays a minimal role in protection against CL. Detection of antibodies in LCL has been reported by indirect immunofluorescence and complement fixation only when there is lymphatic involvement [154]. The antibody titers correlated with both lesion size and duration. A sensitive radioimmunoassay (RIA) has recently been developed employing both freeze–thawed promastigotes and a purified glycolipid from *L. major*, which despite their low titers was able to detect antibody in up to 90% of patients with parasitologically confirmed *L. major* infection [155]. In ACL, intralesional plasma cell numbers (primarily IgG) varied from less than 10% to more than 50% of cells in inflammatory infiltrates, in general with greater numbers in lesions of longer duration [156]. Interestingly, in this study there was a lack of correlation between the frequency of intralesional plasma cells and circulating titer, suggesting that serum antibody levels may not reflect the concentration of antibody produced locally, where they may play a role in protection or pathogenesis.

Diffuse cutaneous leishmaniasis

For many years it has been realized that CL presents a spectrum of disease similar to that seen in leprosy [157]. There is evidence that many of the clinical features of the disease in an individual patient are determined by the CMI response to the parasite. At one end of the spectrum lies lupoid or recidiva leishmaniasis in which the simple cutaneous lesion never quite heals and relapses. Histologically the lesion shows an intense cellular response in the form of a tuberculoid granuloma; parasites are scantly or absent and the condition is associated with marked DTH to leishmanin and sometimes an Arthus reaction. There have been no studies which have elucidated the immunologic events associated with persistent relapsing infection in this rare form of CL.

At the other end of the spectrum lies DCL which has been described from South America, notably by Convit *et al.* [158] in Venezeula, and from Ethiopia by Bryceson [159]. In DCL, the primary lesion does not usually ulcerate or heal but spreads locally after a period of weeks or months and also metastasizes to other parts of the skin. Patients do not react with DTH to leishmanin, and this unresponsiveness is usually specific. On histologic study their lesions show a mass of heavily parasitized macrophages and a few small lymphocytes and epithelioid cells, although plasma cells are usually present. In more recent immunohistologic studies of intralesional cells [152], DCL granulomas from patients in Venezuela showed a mixture of T lymphocyte sub-populations with a CD4/CD8 ratio of less than one, similar to what was found in ACL granulomas. In contrast, IL-2-containing cells were an order of magnitude less in DCL compared to ACL lesions, while cells expressing Tac were present in approximately equal numbers. The authors concluded from these studies that the immunologic effectiveness of the granulomas appeared to be related less to the numbers and location of T cell phenotypes than to the functional aspects of these cells, particularly the ability to generate lymphokines. In similar studies conducted on DCL patients in Ethiopia [153], IL-2-containing cells were also extremely few or absent, and in this case there were also lower numbers of Tac+ cells than in ACL lesions. Furthermore, epidermal keratinocytes above the ACL but not the DCL lesions expressed HLA-DR antigen, suggesting a lower IFN-γ production in DCL granulomas.

There have been a number of studies which have attempted to explain the evolution of antigen-specific unresponsiveness in DCL. Based on their study of four DCL patients in the Dominican Republic, Petersen *et al.* [160,161] implicated a population of adherent suppressor cells, since their removal on nylon wool or treatment of peripheral blood leukocytes (PBL) with indomethacin partially restored the lymphoproliferative response. It is, however, difficult to reconcile the antigen-specific nature of the immunosuppression with the activities of an adherent cell. Suppressor cell activity was also apparent in the studies of Castes *et al.* [162], in which costimulation of PBL from DCL patients with concanavalin-A (Con-A) plus leishmanial antigen suppressed the proliferative response of these cells but not cells from ACL or MCL patients. The nature of these suppressor cells was not explored. The clearest indication that suppressor cells might be involved in

the pathogenesis of DCL comes from recent work in Ethiopia in which it was found that lymphocytes from DCL patients could respond to *Leishmania aethiopica* promastigotes isolated from ACL patients but not to those from DCL patients [163]. The evidence suggests that promastigotes derived from DCL patients express epitopes which preferentially stimulate suppressor activities in DCL patients and that these suppressive cells disappear after treatment. These results are important because they imply that the development of DCL is not solely determined by host-related factors, such as genetic predisposition or previously existing defects in immune function, and that some differences in the parasite may contribute to the outcome of clinical infection with *L. aethiopica*.

Mucocutaneous leishmaniasis

Virtually all patients with MCL have a history and/or a typical scar of previous cutaneous disease. Two characteristic features of MCL are the presence of very few parasites in the mucosal lesions and a strong CMI response to leishmanial antigens. In fact, the CMI response in mucosal patients, as assessed by lymphocyte transformation, DTH, and even IFN-γ production, has often been found to be augmented compared to the response in patients with ACL [162,164,165]. The critical factors in the pathogenesis of mucosal disease, in which a low level of infection persists in the face of very strong specific cellular reactivity, remain poorly defined. In a recent quantitative study of T cell frequency in ACL and MCL, the correlation between high antigen-specific T cell frequencies and MCL was seen not in the peripheral blood but within intralesional sites containing fibrinoid necrosis, arguing that these cells might have a deleterious effect [166]. The enhanced lymphoproliferative response and increased T cell frequency in mucosal patients might also be a function of the long duration of active disease in this population and unrelated to the pathogenesis of their mucosal lesions. The possibility, however, that the hyperreactivity contributes to the severity of the lesions is supported by the findings of a large cross-sectional study in Columbia that mucosal disease and *L. b. braziliensis* infection are accompanied by significantly greater antibody and DTH responses and that these effects are independent of the time of lesion evolution [167]. According to this view, the inherent "reactogenicity" or sensitizing capacity of mucosal strains contribute to their pathogenicity. It has also been argued that the augmented CMI and resulting tissue destruction are due to a suppressor cell defect in MCL patients [164], i.e., an absence of cells which normally serve to down-regulate immune responses. It is extremely interesting that despite their apparent hyper-reactivity, 11 Venezuelan MCL patients who were treated with Bacille Calmette-Guerin (BCG) and killed leishmanial antigen in order to yet further immuno-potentiate their response all demonstrated cure or clinical improvement [168].

Unfortunately, because of the inaccessibility of tissue specimens in mucocutaneous disease, immunohistologic studies have not been undertaken, and the cellular events that might help to explain the noncuring nature of mucosal infections or the events promoted by immunotherapy have not been elucidated. Additional mechanisms that have been proposed to explain noncure include the possibility that macrophages within mucosal lesions are unable to be activated to kill *Leishmania* amastigotes, although they are clearly able to control parasite growth, as evidenced by the low numbers of parasites within MCL lesions. Several factors initially described in murine models might contribute to such a defect. These include the fact that some macrophage subpopulations are unresponsive to activation signals, that macrophages exhibit a decreased capacity to kill amastigotes at cutaneous temperatures [169], and that some leishmanial strains are resistant to activated macrophage killing [170].

Visceral leishmaniasis

Immunodepression in visceral disease

The critical immunologic feature of VL is the complete absence of CMI to leishmanial antigens. This results in uncontrolled parasitization of the mononuclear phagocyte system, which is especially prominent in the spleen. Patients have been shown to have negative intradermal skin tests to leishmanin [137], absent lymphocyte blastogenesis [171], and decreased IL-2 and IFN-γ production in response to parasite antigens [172]. The profound impairment of CMI is usually specific, but may extend to other antigens, although this remains a controversial point. In two studies of Indian kala-azar [173,174], the *in vitro* T cell unresponsiveness was found to be antigen specific. Studying 14 Brazilian patients *in vitro*, Carvalho *et al.* [171] characterized the immuno-suppression in American visceral leishmaniasis (AVL) as also being specific as well as short lived. In contract, Levy & Mendes [175] demonstrated impaired DTH to

leishmanin and other unrelated antigens in 10 Brazilian patients. Generalized immunodepression was also reported for 15 VL patients in Kenya [176]. In all of these studies, responses to mitogens appeared to remain generally intact.

Mechanisms of unresponsiveness

Visceral leishmaniasis is a complex disease associated with many features that could cause immunodepression. For example, immune complexes are present [177] and these have been associated with immune dysfunction in other diseases. In addition, sera from patients with active VL have been found to suppress mitogen-induced lymphocyte proliferation of normal cells [178]. This would not, however, explain the specific unresponsive-ness seen in most VL patients. It may be due simply to a reduction in the number of circulating T cells, which has been reported in Iranian and Mediterranean kala-azar patients [179], and/or a depletion of lymphocytes in the T-dependent areas of lymph nodes and the spleen, which has also been reported [180]. The significance of these observations is not clear, however, since in another study those AVL patients with normal total lymphocyte counts were equally as unresponsive to *L. d. chagasi* antigens as those patients who had reduced numbers of peripheral T cells [171]. It seems more likely that if there is a depletion of T cells during VL, then this will be restricted to antigen-specific clones. In an extensive study of T cell unresponsiveness in Indian kala-azar, the failure to reverse unresponsiveness by either CD8 depletion or the addition of human IL-2 to the cultures was interpreted as an absence of *Leishmania*-reactive T cells in the periphery [174]. The possibility that reactive cells might be found *in situ* could not be excluded by these or, for that matter, any studies to date. In the only other study to explore the cellular basis of antigen unresponsiveness in VL, Carvalho *et al.* [181] were also unable to restore responses by depletion of macrophages, B cells, Fc receptor high-avidity cells, or CD8 and CD4 cells. They did determine, however, that frozen cells obtained from unresponsive patients before therapy could abolish the ability of those same patients' cells to respond to antigen after therapy. To date, these are the only studies to implicate cell-mediated suppression in the evolution of antigen-specific un-responsiveness in VL. It needs to be emphasized that the immunologic lesion in VL does not appear to extend to a subset of T cells, namely helper cells, since extremely high titers of antileishmanial antibodies, including IgG antibodies, are always present in VL. It is clear that the analyses of T cell responsiveness in VL must somehow be modified so as to be able to detect the potential activation of these cells.

Evidence for naturally acquired immunity

There is general agreement that after successful treatment there is a gradual recovery of antigen-specific responses between 3 and 12 months [137,176,181,182]. This has prompted the belief that successfully treated VL patients will be naturally immune to reinfection. In the only study to address directly this question [137], three volunteers who had been cured of naturally acquired VL by chemotherapy were intradermally inoculated with *L. donovani* promastigotes. A transient nodule developed at the inoculation site, but no visceral disease, was noted. In these experiments there were no positive controls and the length of observation was not recorded. In another study [183], volunteers were inoculated with a rodent strain of *L. donovani*, developing a leishmanioma but no visceral disease over an 18–24-month followup. When six of these individuals were challenged with a human *L. donovani* isolate, none developed evidence of disease over an 18-month followup. The correlation of T cell responses with recovery from, and resistance to, VL provides a strong motivation to define the antigens recognized by T cells from recovered patients. In preliminary studies of former visceral patients in Brazil, *L. chagasi* antigens selected by reactivity with patient antibodies were used to induce proliferative responses in peripheral blood lymphocytes. Two purified glycoproteins, 30 and 42 kD, were consistently among the most effective in eliciting high proliferative and IFN-γ responses, and T cell clones specific for these antigens were successfully generated from cured visceral patients [184].

The development of naturally acquired immunity to visceral infection is consistent with a number of reports, suggesting that the usual outcome of *L. donovani* exposure is not visceral disease but infections which are subclinical or asymptomatic, and in either case self-healing. Ho *et al.* [176] estimated that the ratio of disease to asymptomatic cases in Kenya was 1:5. In an extensive prospective survey in Brazil which used both serology and skin testing, Badaro *et al.* [185] reported that the ratio of clinical to subclinical cases in a highly endemic area was 1:6.5. Of 25 family members of active cases of Indian kala-azar who themselves had no history of disease, almost half responded well by *in vitro* LT

and leishmanin skin testing, again suggesting that the number of *L. donovani*-exposed but -resistant individuals within an endemic area is quite large [174].

Visceral leishmaniasis as an opportunistic infection

There is evidence that even in healthy people the organism can persist and remain viable for long periods of time. For one thing, the DTH response generally remains positive many years after exposure [186] and for another, there are a few cases of disease that have developed in individuals long after leaving endemic areas [187]. There is now solid evidence that VL can be added to the growing list of opportunistic microbial infections since progressive disease has been documented in immunocompromised hosts, including human immunodeficiency virus (HIV)-infected patients as well as patients receiving corticosteroids or prednisone [188,189]. It can of course be argued that the manifestation of VL in primarily exposed individuals is also opportunistic in nature in that those risk factors which have been identified, such as young age, malnutrition, and concurrent infection, are all associated with suboptimal or compromised immune responses [185].

Human vaccination

The only immunization strategy against leishmaniasis used with any success so far in humans has been restricted to cutaneous disease (reviewed by Greenblatt [190]). Vaccines against cutaneous disease have historically progressed from the use of living virulent organisms to attenuated and then to killed promastigotes. The use of living vaccines, or leishmanization, is based on the convalescent immunity which is acquired following induction of a lesion at a selected site with small doses of a cutaneous strain. Leishmanization has been practised for hundreds of years, especially in the Middle East and parts of the Soviet Union. It is a currently used practise in Iran, where a massive program was begun in 1982 after outbreaks of CL among military personnel. As shown in the Soviet Union and Iran, leishmanization can protect 90% or more of the population from multiple cutaneous lesions or lesions on the face [130,191]. Unfortunately, leishmanization using virulent strains can be associated with several problems, including the development of large slowly healing lesions, lesion contamination, and exacerbation of other dermatologic conditions [192,193]. For these reasons the practice has been restricted in the Soviet Union and discontinued in Israel.

Several strains of attenuated promastigote vaccines have been tried. Salazar [194] injected 204 individuals with a laboratory-attenuated strain of *L. braziliensis*, and while no subsequent protection was studied, 58% of the recipients converted to leishmanin positivity. Manson-Bahr [195] used a naturally attenuated *L. donovani* strain for immunization against VL. A large study of 3000 individuals, divided between controls and test subjects, revealed no difference in the natural incidence of disease [195]. These disappointing results are consistent with the notion, discussed above, that no immunity to CL develops unless the infection progresses to the appearance and ulceration of a cutaneous lesion.

The prospects for a killed and even a subunit vaccine against cutaneous disease was revitalized by animal studies, which conclusively demonstrated that dead parasites or soluble fractions derived from them could induce protection if administered with certain adjuvants. These studies will be discussed in detail below. Almost coinciding with the stimulus of animal studies, human trials of a killed vaccine which had been used in Brazil in the 1930s were renewed in Brazil, beginning in 1979. The polyvalent vaccine, containing sonicated and merthiolate-treated promastigotes of up to five strains, yielded a large proportion of positive skin tests in the initial trial but the disease disappeared from the area, making an evaluation of protection impossible [196]. In subsequent randomized trials amongst vaccinees whose skin test converted, the protection rate was 67% and 86%, although because of low incidence rates these results could not be considered significant [197,198]. At least it appeared that immunization with whole promastigotes did not increase susceptibility to the disease and no severe side-effects were noted. Further Brazilian trials are currently in progress, as are trials in Venezuela and Iran using killed parasites and BCG. These immunoprophylactic trials are based partly on the success of immunotherapy of patients with CL using killed *L. mexicana* and BCG, in which three vaccinations over a course of 32 weeks gave a similar rate of cure as was achieved with chemotherapy but with considerably fewer severe side-effects [168]. This work has been confirmed by other Latin American groups and has given further impetus to killed vaccine studies.

ANIMAL MODELS OF INFECTION WITH *LEISHMANIA*

The disease pattern produced in animals infected experimentally with *Leishmania* differs widely according to both the animal and the *Leishmania* species involved. The guinea pig model of infection with *L. enriettii* has been studied for some time and the features of the different types of lesions that can be produced as well as the characteristics of the corresponding immune responses have been reviewed recently [199]. Experimental infections of hamsters and mice with *L. donovani* have been used as experimental models for human VL. In particular, studies of the murine model of infection with *L. donovani* has provided important information regarding the genetic aspects of resistance and susceptibility to infection with this parasite (see Chapter 1). Upon infection of mice with *L. major*, almost the entire spectrum of disease patterns seen in human leishmaniasis can be observed [200,201]. The severity of disease resulting from experimental infection of mice with *L. major* has been shown to be determined genetically [202]. For example, BALB/c mice exhibit severe lesions at the site of inoculation and fatal visceralization of infection, leading to death, does occur. Mice from other strains (e.g., CBA, C57BL) are resistant to infection, developing only small lesions at the site of inoculation which heal after a few weeks. These mice are immune to further challenge with virulent parasites. Similar patterns of disease have been observed in mice infected with *L. mexicana*. These animal models have been used recently by several research groups for the dissection of the mechanisms implicated in resistance (protection) and susceptibility to infection. In this chapter, mainly the recent results obtained by several research groups using the murine model of infection with *L. major*, pertaining to the understanding of the role of the various parameters of the specific immune responses triggered during infection on the pattern of disease, will be summarized. Some of the cellular parameters accounting for the anergy state which characterizes visceral infections with *L. donovani* in hamsters will also be discussed.

Role of humoral response

It is now believed that healing of an established infection in mice is independent of the specific antileishmanial antibody response elicited during infection (reviewed by Liew [203]). This contention does not preclude the possibility that, if present at the start of infection, specific antibodies could hamper the initiation of infection by preventing the attachment of *Leishmania* promastigotes to their host cells, i.e., the macrophages. Indeed, it has been demonstrated that both passive administration of antibodies directed against a glyco-protein of the membrane of *Leishmania amazonensis* and immunization of mice with this molecule prevented successful infection [204,205]. Furthermore, MAB specific for lipid-containing glycoconjugates of *L. major* have been shown to prevent parasitization of macrophages *in vitro* by inhibiting their binding to the macrophage membrane [206]. Since parasites not rapidly internalized by macrophages are destroyed within a few minutes (cited in Handman & Mitchell [207]), antibodies interfering with the access of parasites to their host cells could represent an important means by which to prevent the establishment of an infection.

Role of macrophages in resolution of lesions

Macrophages are the cells in which *Leishmania* grow in their mammalian host, and the basis for the capacity of these parasites to survive and multiply in an environment normally hostile for microorganisms is not yet known. Since activation of macrophages by T cell-derived lymphokines appears to be the main effector mechanism for the elimination of these parasites, these cells have a central role in the outcome of infection. In this context, several observations indicate that some impairments in macrophage function could be related to susceptibility to infection. For example, Class II and Class I major histocompatibility complex (MHC) molecules are reduced at the surface of macrophages from the genetically susceptible mice harboring intracellular *Leishmania*, and these cells are unresponsive to IFN-γ for the expression of these molecules [208]. Furthermore, compared to resistant mice, macrophages from susceptible mice infected *in vitro* with *L. major* are relatively resistant to activation by T cell-derived lymphokines [209]; compared to macrophages obtained from noninfected mice, macrophages isolated from susceptible mice 2–3 weeks after infection are impaired in their response to activation signals [210].

Role of T cells in determining the outcome of infection with *L. major*

Clear evidence exists that T cell immunity, in contrast to humoral responses, determines the pattern of disease resulting from infection of mice with *L. major*. There is a consensus of opinion that acquired T cell immunity is not only extremely important for resistance but can also contribute to susceptibility of the host to infection with *Leishmania*.

CD4+ T cells and resistance to infection with L. major

Numerous observations have demonstrated the importance of CD4+ T cells for the resolution of lesions induced by *L. major* in mice. The now classic observations by Mitchell *et al.* [211], showing that nu/nu (athymic) mice were extremely susceptible to infection and that transfer of small numbers of CD4+ T cells rendered them relatively resistant, have greatly contributed to the recognition of the importance of CD4+ T cells in resistance to *L. major* infection [211]. Observations which have shown that adoptive transfer of CD4+ spleen cells obtained from resistant mice which had recovered from a primary infection could protect normal mice against infection [212], and that elimination of CD4+ T cells by administration of anti-CD4 MAB *in vivo* led to the development of severe lesions in normally resistant mice, have further demonstrated the essential role of CD4+ T cells in resistance to infection with *L. major* [213].

This effect of CD4+ cells on the resolution of CL in mice has been shown to be the consequence of the macrophage-activating lymphokines that they release upon specific activation. Inasmuch as IFN-γ represents the main macrophage-activating lymphokine resulting in the destruction of intracellular *Leishmania in vitro* [214,215], the importance of IFN-γ in the resolution of lesions has been investigated by several research groups. Results so far accumulated strongly indicate that IFN-γ is important for the resolution of CL *in vivo*. First, the release of IFN-γ by lymphocytes from resistant mice was clearly related to the capacity of these mice to eliminate the parasites [216]. Second, lymphoid tissues of resistant mice contained, after infection, significantly higher amounts of IFN-γ mRNA than susceptible mice [217,218]. Third, administration of anti-IFN-γ MAB to resistant mice during the course of infection prevented the healing of lesions [219,220]. These results clearly indicate that the effect of CD4+ T cells on resolution of lesions is performed by the IFN-γ that they release. Inasmuch as, at least in the mouse, the CD4+ T cell subpopulation is composed of two distinct subsets releasing different lymphokines upon specific activation (e.g., TH-1 cells releasing IFN-γ and TH-2 cells producing IL-4 and IL-5 [221]), it appears that the anti-*Leishmania* effector function of CD4+ T cells is performed by TH-1 cells. The importance of TH-1 cells for the development of protective immunity against infection with *L. major* has been confirmed by studies using clonally derived parasite-specific T cells. CD4+ T cell lines and clones derived from lymphoid tissues of susceptible mice vaccinated with an antigenic fraction separated from a soluble extract of *L. major* promastigotes were shown to protect, after adoptive transfer, sublethally irradiated (200 rads) syngeneic BALB/c mice. Analysis of the pattern of lymphokines produced by these T cells, upon specific and mitogenic stimulation, showed that they belong to the TH-1 subset [210,222]. Results from our laboratory also indicate that normal BALB/c susceptible mice could be protected against infection with virulent *L. major* by adoptive transfer of parasite-specific CD4$^+$ T cells [223]. Data showing that the administration of anti-IFN-γ MAB to mice adoptively transferred with these T cell clones abrogated their protective capacity further demonstrate the importance of IFN-γ in protection against *L. major* infection. Furthermore, after stimulation *in vitro*, these clones also release IFN-γ and IL-2, indicating that they belong to the TH-1 subset. It is important to emphasize that the protective T cells derived in our laboratory recognize *in vitro* antigen(s) associated only with live parasites [224].

Tumor necrosis factor (TNF) has also been suggested to play an important role in mediating host protection against experimentally induced CL [225,226]. This contention derives from observations showing that:
1 lymph node cells obtained from resistant mice during the course of infection produced more TNF *in vitro* than susceptible mice;
2 repeated injections of TNF to infected mice had some therapeutic effect;
3 administration of anti-TNF antibodies enhanced the course of infection.

It should be mentioned, however, that although we have observed that, compared to resistant mice, the production of TNF in supernatant of specifically activated lymph node cells draining the lesions of BALB/c mice was delayed by 1–2 weeks, cells from infected mice of both susceptible and resistant strains produce similar amounts of TNF around 21 days after infection. Furthermore, these results were corroborated by observations showing similar amounts of TNF mRNA in lymph node cells from resistant and susceptible mice (Cossando, Grau, Louis, & Miller, in preparation). It is of interest that TNF was recently reported to be unable to activate parasitized macrophages for the killing of intracellular *L. major* unless IFN-γ was provided as a costimulatory signal [227].

Although, as described above, the body of evidence accumulated by several laboratories indicate that the anti-*Leishmania* effector function of T cells is performed by the IFN-γ that they release, it is important to mention that, under some circumstances, the ability of mice to produce IFN-γ in response to infection with *L. major* does not correlate with resistance. For example, using a sensitive limiting dilution assay we were able to show that residual viable parasites were present in resistant mice after complete resolution of cutaneous lesions, i.e., at a time when these mice are immune to rechallenge. Upon depletion of CD4$^+$ T cells by repeated injections of anti-CD4 MAB to these mice, cutaneous lesions reappeared at the site of primary infection, progressed inexorably, and dissemination of parasites was observed [228]. An analogous situation exists in humans where VL has been reported to develop as opportunistic infections in immunocompromised patients, including acquired immunodeficiency syndrome (AIDS), several years after sojourns in an area where transmission of disease does occur [188,189]. These observations strongly suggest that CD4$^+$ T cells precluded the unrestricted growth of residual parasites both in resistant mice and in humans who had recovered from infection. Interestingly, a significant amount of mRNA specific for IFN-γ could be detected using a sensitive RNase protection assay in mRNA isolated from lymphoid tissues and even lesions of mice developing uncontrolled lesions as a result of treatment with anti-CD4$^+$ MAB after the resolution of a primary infection [228].

Furthermore, although the administration of neutralizing anti-IFN-γ MAB to resistant mice during the course of infection prevented resolution of lesions [219,220], our results show that the size of lesions observed in resistant mice treated with anti-IFN-γ MAB never reached the magnitude of those seen in similarly infected susceptible mice [219]. Finally, using a sandwich ELISA method, titrable levels of IFN-γ were present in supernatant of specifically activated spleen cells from susceptible mice as soon as 3 weeks after initiation of infection (Miller & Louis, unpublished). The production of IFN-γ in susceptible mice during infection was confirmed by directly enumerating T cells secreting IFN-γ in lymphoid tissues using an ELISA spot assay [229] (Hug & Louis, unpublished). Compared to resistant mice, however, the production of IFN-γ in infected susceptible mice was delayed. Since, in the experiments which have demonstrated a greatly impaired IFN-γ production by lymphoid cells during the course of infection of BALB/c mice, IFN-γ was measured using a biologic assay, the possibility exists that the IFN-γ produced by BALB/c lymphocytes detected in an ELISA assay is biologically inactive. Taken together, these observations would also suggest that IFN-γ is not the only effector mechanism involved in the elimination of parasites from infected hosts and/or that the inactivation of its biologic activity accounts for disease progression.

CD8+ T cells and resistance to infection with L. major

Results accumulated over the last 3 years indicate that CD8+ T cells are also involved in the resolution of lesions induced in mice by *L. major*, even though it appears that their participation in the elimination of parasites is smaller than that of CD4+ T cells. The first indirect evidence for a role of CD8+ T cells in the healing of cutaneous lesions derives from observations that, in comparison to susceptible mice, resistant mice, during infection produced in the lymph nodes draining the lesions, had three times more parasite-specific CD8+ T cells which were able to transfer specific DTH reactions to naïve recipients [213]. The higher frequencies of specific CD8+ cells were observed at the time when lesions started to heal. Furthermore, depletion of CD8+ cells by treatment of infected resistant mice with anti-CD8 MAB led to the development of more severe lesions which eventually resolved [213].

Recent evidence indicates that CD8+ T cells could play a more important role in controlling the lesions that develop upon infectious challenge of immune mice. Indeed, depletion *in vivo* of CD8+ T cells abrogated the induction of resistance which is normally seen following immunization with killed promastigote, indicating that CD8+ T cells may represent an important component of this immune response [230]. The small lesions which normally develop upon infectious challenge of resistant mice which have completely resolved a primary infection were significantly enhanced as a result of repeated administration of anti-CD4 or anti-CD8 MAB starting at the time of secondary infectious challenge [228]. Although the secondary lesion of mice receiving either anti-CD4 or anti-CD8 MAB eventually healed, those of mice receiving both anti-CD4 and anti-CD8 MAB progressed without tendency to resolve. These results provide indirect evidence that, in immune mice, the course of secondary infection is at least partly controlled by CD8+ T cells. In addition, the DTH response which can be elicited upon challenge with viable *L. major* of resistant mice having recovered from a primary infection requires the participation of CD8+ T cells [228]. It appears that BALB/c-susceptible mice have the cellular components normally involved in the resolution of

L. major-induced lesions but that these cells are either not appropriately activated or their function is inhibited during infection. This contention derives from observations which have shown that several immune interventions are able to render normally susceptible mice resistant to infection with *L. major*. For example, normal BALB/c mice treated at the beginning of infection with a single dose of anti-CD4 MAB are able to resolve completely the cutaneous lesions at around 60 days after inoculation with *L. major*. These mice become resistant to further challenge with virulent parasites [219,231,232]. This therapeutic effect of administration of anti-CD4 MAB to BALB/c mice has been related to the appearance of IFN-γ mRNA in lymphoid tissues which is lacking in infected unmanipulated BALB/c mice [218]. Recent studies of the cellular parameters of the specific anti-*Leishmania* cellular response generated during infection in these BALB/c mice rendered resistant to infection following treatment with anti-CD4 MAB have also provided evidence for a role of CD8+ cells in the immunologic control of infection with *Leishmania*. Administration of anti-CD4 MAB to BALB/c mice at the beginning of infection resulted in a transient depletion of CD4+ T cells in lymphoid tissues, which subsequently were gradually repopulated with CD4+ T cells. Fifty to sixty days after infection these mice contained numbers of CD4+ T cells equivalent to those seen in normal mice. At that time, the DTH response that could be induced in these mice following challenge with viable promastigotes was partly dependent upon CD8+ T cells, since administration of anti-CD8 MAB 24 h prior to antigenic challenge led to an important inhibition of this response. Furthermore, compared to control untreated infected mice, the frequency of parasite-specific CD4+ T cells in lymph node draining of the lesions of BALB/c mice cured as a result of treatment with anti-CD4 MAB at the initiation of infection was drastically reduced whereas the frequencies of parasite-specific CD8+ T cells were increased. It also appears that CD8+ T cells contribute significantly to the important IFN-γ production, which is seen after specific stimulation *in vitro* with viable parasites of lymphoid cells from BALB/c mice cured as a result of administration of anti-CD4 MAB at the beginning of infection. The resistance to reinfection which is normally seen in BALB/c mice treated with anti-CD4 MAB at the time of a primary infection was greatly inhibited by treatment of mice with anti-CD8 MAB 24 h prior to reinfection (Milon, Miller, & Louis, in preparation). Finally, elimination of CD4+ T cells in thymectomized susceptible mice has also been shown to result in a reduction in the size of lesions developing at the site of parasite inoculation, and this protective effect has been attributed to the activity of CD8$^+$ T cells released in these mice free of CD4$^+$ T cells [233].

The induction of specific CD8$^+$ T cells during *Leishmania* infection and their possible anti-*Leishmania* effector function would suggest that some antigen(s) of this intracellular parasite, normally residing in the phagosomes of its host macrophages, follow the MHC class I pathway of antigen processing and presentation.

CD4$^+$ T cells and susceptibility to infection with L. major

Evidence for a contribution of CD4$^+$ T cells to the susceptibility of mice to infection with *L. major* was obtained several years ago. For example:

1 Adoptive transfer of high numbers of T cells from normal susceptible BALB/c mice to syngeneic nu/nu mice did not confer resistance and, moreover, injection of T cells from infected BALB/c mice rendered nu/nu mice more susceptible to infection with *L. major* [234];

2 Subcutaneous immunization of mice with crude extracts of *L. major* enhanced the development of lesions seen after infectious challenge, an effect which was related to the triggering of CD4$^+$ T cells [235,236]. Observations which have documented the expansion of excessive numbers of specific CD4$^+$ T cells in lymphoid tissues of genetically susceptible mice following experimental infection strongly supported the notion that CD4$^+$ T cells were somehow involved in susceptibility to infection with *L. major* [237]. Several experimental results indirectly indicated that the susceptibility of BALB/c mice was, at least partly, the consequence of the rapid expansion of specific CD4$^+$ T cells. Administration of anti-CD4 MAB at the beginning of infection rendered these mice resistant to infection and to further infectious challenge [219,231,238]. Furthermore, BALB/c mice treated with cyclosporin A also become resistant to infection and this effect was shown to be at least partly the consequence of the modulation of the T cell responses by cyclosporin A [239].

The elegant work from Locksley *et al.* [217] has recently provided a rational basis for understanding the cellular parameters of the T cell responses involved in the susceptibility of BALB/c mice to infection with *L. major*. It was found that, compared to resistant mice, lymph node cells from infected susceptible mice contained 50-fold more IL-4 mRNA and 50-fold less IFN-γ mRNA [218]. Furthermore, compared to infected mice, lymphoid cells from BALB/c mice rendered resistant to

infection as a result of treatment with anti-CD4 MAB at the time of parasite inoculation contained several-fold more IFN-γ mRNA and several-fold less IL-4 mRNA. Inasmuch as IL-4 and IFN-γ have been shown to be produced by distinct CD4$^+$ T cells [221], these results indicate not only the triggering of two functionally distinct parasite-specific CD4$^+$ T cells during infection but strongly suggest that the outcome of infection is dependent upon the proportions of these functionally different T cells expanding during infection. Thus, TH-1 IFN-γ-producing and TH-2 IL-4-producing CD4$^+$ T cells are preferentially induced during infection in resistant and susceptible mice, respectively.

The disease-promoting activity of TH-2 CD4$^+$ T cells was confirmed in studies using cloned T cell lines, since from a panel of T cell clones specific for antigen(s) in a fraction isolated from a soluble extract of *L. major* only T cells with functional properties of TH-2 T cells were found to enhance susceptibility [210,222].

It appears that the IL-4, produced by TH-2 cells triggered during infection of BALB/c mice, plays a causal role in disease progression since the administration of anti-IL-4 MAB *in vivo* to BALB/c mice at the time of infection not only significantly inhibited the development of lesions but also led to the establishment of protective immunity to reinfection [240]. Interestingly, the regression of lesions observed after neutralization of IL-4 appeared independent of IFN-γ since administration of anti-IFN-γ MAB to these mice did not abrogate the therapeutic effect of anti-IL-4 MAB. Furthermore, evidence was obtained that treatment with anti-IL4 MAB inhibited the development of TH-2 cells during infection and led to the development of protective immunity residing in both CD4$^+$ and CD8$^+$ T cell subpopulations [240]. Other studies have suggested that IL-4 was able to interfere with the ability of IFN-γ to activate macrophages (which normally leads to the destruction of intracellular *Leishmania*) [241]. However, it should be emphasized that this regulation of IFN-γ-mediated macrophage activation by IL-4 is strictly dependent upon the timing of lymphokine addition *in vitro* [210] and that other reports have shown that resistance of macrophages to infection with *L. major* amastigotes required IFN-γ in conjunction with other lymphokines, including IL-4 [242]. In addition, the killing of another protozoan parasite, *T. cruzi*, by macrophages was enhanced by IL-4 [243]. Given these apparently contradictory results, it is clear that the understanding of the mechanisms by which IL-4 enhance *L. major* induced lesions *in vivo* requires further investigation. Finally, it should be mentioned that the administration of rIL-4 subcutaneously in the site of parasite inoculation has been reported recently to interfere with the development of lesions and to lead to a state of protective immunity to reinfection [244]. Thus, it is possible that IL-4 may have different effects on the disease process, depending upon its concentration at the lesion site.

Other lymphokines have also been implicated in disease exacerbation. Direct evidence that IL-3 can favor the development of lesions induced by *L. major* has been obtained [245]. Granulocyte macrophage colony stimulating factor (GM-CSF) was found to have a similar effect [246]. Since both of these lymphokines promote the multiplication and differentiation of cells from various lineages, including macrophages, their effect on experimentally induced leishmaniasis has been explained by their capacity to increase the pool of host cells normally permissive for parasite growth, at least in the absence of macrophage-activating lymphokines. Other results have shown that IL-3 could inhibit the IFN-γ-mediated activation of macrophages, thus providing another mechanism by which IL-3 could favor the growth of *L. major* [210,247]. It should be mentioned that IL-3 has been shown to be released by both TH-1 and TH-2 T cells.

IL-2 has also been shown to enhance the cutaneous lesions produced in mice experimentally infected with *L. mexicana* upon administration in the vicinity of the developing lesions and also the multiplication of *L. mexicana* cultivated *in vitro* under suboptimal conditions. Furthermore, since IL-2 was also found to reverse the resistance to infection seen in mice administered cyclosporin A at the time of parasite inoculation, it is likely that the effect of IL-2 *in vivo* on the course of disease was a consequence of the activity of host lymphocytes stimulated by this lymphokine [248].

Early studies have shown that, compared to resistant mice, the frequency of CD4+ T cells triggered in lymphoid tissues was several-fold higher in susceptible mice during infection [237]. Inasmuch as these CD4+ T cells were functionally identified by their ability to transfer DTH to naïve syngeneic recipients, these results would indicate that the triggering of some TH-1 cells does not correlate with protection. A causal relationship between the expansion of these DTH-mediating CD4+ T cells and susceptibility to disease is indirectly supported by recent observations showing that the numbers of CD4+ T cells, found in lymphoid tissues of susceptible mice cured as a result of anti-CD4 MAB administered at the time of parasite inoculation, that are capable of

mediating *Leishmania*-specific DTH is drastically reduced compared to similarly infected, not treated, BALB/c mice (Milon, Miller, & Louis, in preparation).

Results obtained using *L. major*-specific T cell lines and clones also support the notion that some parasite-specific TH-1 T cells are not protective and even exacerbate the disease process. *Leishmania major*-specific CD4+ T cell lines were derived, in our laboratory, from lymph node cells of infected or immunized mice upon *in vitro* selection with a crude extract of *L. major* promastigotes as antigen. These specific T cells could transfer DTH reactivity to naïve syngeneic recipients, and released, upon stimulation *in vitro*, substantial amounts of IFN-γ, IL-3, and GM-CSF. They also significantly exacerbated the course of disease after adoptive transfer to syngeneic recipients [228,235]. Recent data show that the exacerbative effect of these T cell lines was a titrable phenomenon, i.e., was always observed but was proportional to the number of T cells transferred. This effect required the localization of the T cells transferred in the site of lesions, and their *in situ* activation was independent of any host-specific T and B cell responses. It is likely that this exacerbation of experimentally induced CL is the consequence of interactions between host-derived macrophages and the adoptively transferred parasite-specific CD4+ T cells [249] (Louis *et al.*, submitted). This hypothesis is strongly supported by recent observations showing that an important increase of parasite growth in lesions of lethally irradiated mice is observed 4 days after adoptive transfer of these CD4+ parasite-specific T cell lines, provided that these mice are given a source of circulating monocytes/macrophages (Mendonia, Titus, & Louis, submitted). Several T cell clones derived from these homogeneous exacerbative CD4+ T cell populations were also found to promote significantly the development of lesions. All these parasite-specific T cell clones expressed the CD4+ cell surface phenotype, mediated parasite-specific DTH reactivity, and produced IL-2 and IFN-γ but no IL-4 upon specific activation *in vitro* [228]. The reasons for the inability of these CD4+ T cell clones to eliminate parasites *in vivo* in spite of their production of IFN-γ is currently being investigated. Our current hypothesis is that the inability of these cells to eliminate the parasites might be related to their fine specificity.

It is likely that the ability of T cells to eliminate *L. major* from infected host will depend upon their recognition of parasite-derived antigens on the surface of macrophages containing living microorganisms, i.e., the target cells of protective immune mechanisms. Since these exacerbative T cell clones were selected *in vitro* using a crude extract of *L. major* promastigotes and the protective CD4+ TH-1 T cell clones (obtained above) did not respond to this antigen preparation but rather recognized antigen(s) only associated with living parasites, it is possible that macrophages containing multiplying parasites do not present their specific epitopes to the nonprotective T cells [224]. Thus, even though they secrete IFN-γ, these T cells would be unable to focus this macrophage-activating lymphokine on the parasitized macrophages.

CELLULAR PARAMETERS ASSOCIATED WITH IMMUNOLOGIC UNRESPONSIVENESS IN VISCERAL *L. DONOVANI* INFECTION

The hamster model of infection with *L. donovani* has been utilized in an attempt to delineate the cellular parameters responsible for the apparent anergic state which accompanies human kala-azar. Indeed, similarly to the human disease, hamsters inoculated intracardially with *L. donovani* develop uncontrolled growth of parasites in various visceral organs, which is accompanied by hepatosplenomegaly, anemia, and cachetia [250]. In contrast, after intradermal inoculation, some animals develop only self-limiting infections [251]. Hamsters with visceralized infections exhibit a state of anergy to *L. donovani* antigens, as evidenced by their failure to exhibit DTH reactions after specific antigenic challenge and the inability of their spleenic, blood, and lymph node lymphocytes to proliferate *in vitro* in the presence of parasite antigens [252]. In addition, spleen cells from these animals suppressed the specific proliferative response which is normally seen after antigenic challenge *in vitro* of lymphocytes obtained from animals with self-limiting infections. Interestingly, removal of adherent cells from unresponsive spleen cell populations restored their capacity to respond *in vitro* to *L. donovani* antigens, as well as their capacity to transfer specific DTH responses to normal recipients [252]. In this vein, adherent cells capable of suppressing lymphocyte proliferation have been documented also in mice infected with *L. major* and *L. donovani*, as well as in human DCL [160,253,254]. Since these adherent cells with suppressive activity have not been observed in blood of hamsters with VL, it has been hypothesized that the unresponsiveness of blood lymphocytes to leishmanial antigens could be due to a decrease in

specific lymphocytes as a result of their recruitment to visceral organs, i.e., spleens with high parasite load or antigens [252]. In addition to adherent cells, the immunosuppression associated with VL of hamsters has also been related to the activity of T cells able to suppress antigen-specific proliferative responses [255].

Vaccination against experimentally induced infection with *L. major* in mice

In recent years, numerous investigators have used the murine model of infection with *L. major* in an attempt to devise efficient immunoprophylactic measures. Observations which have shown that even genetically susceptible BALB/c mice are able to develop resistance to *L. major* infection following some vaccination regimens are particularly important. For example, immunization of BALB/c mice with repeated parenteral injection of killed promastigotes induced substantial resistance to an infectious challenge which was mediated by CD4+ T cells producing IFN-γ but not IL-3 and IL-4, and also not capable of transferring a DTH reaction (reviewed by Liew [241]). Similarly, BALB/c mice were completely protected against an otherwise fatal challenge by intraperitoneal immunization with a soluble leishmanial antigen (SLA) extract of promastigotes [256]. A fraction (fraction 9) obtained upon an exchange chromatography of SLA was observed to induce equivalent protection to that obtained with the whole extract, whereas another fraction (fraction 1) was not able to induce protection [256]. All T cell clones derived from mice immunized with fraction 9 expressed the TH-1 functional phenotype and could protect sublethally irradiated BALB/c mice against virulent challenge, whereas T cell clones derived from mice imunized with fraction 1 belong to the TH-2 subset and had no protective potential [210]. These important data suggest that different antigenic fractions have a differential ability to stimulate T cells of the two functionally different CD4+ T cell subsets. Scott *et al.* [257], using a protective TH-1 T cell clone responding to fraction 9 in an attempt to identify a protective immunogen, have identified a 10 kD protein which, because of its size, could represent an ideal immunogen for the construction of a molecularly defined vaccine.

Avirulent *L. major* has also been used as an immunogen in an attempt to vaccinate mice against virulent *L. major*. Using noninfective clones of *L. major*, Mitchell & Handman [258] could induce protective immunity in mice. Other studies have shown that avirulent parasites derived by chemical mutagenesis were able to immunize either susceptible or resistant mice against challenge with virulent *L. major* [259]. Immunization with these mutants triggered CD4$^+$ T cells that could transfer protective immunity in syngeneic normal recipient but could not transfer DTH reactions [260]. Interestingly, these cells appear to respond only to live promastigotes *in vitro*, a situation analogous to that described above with protective CD4$^+$ T cells isolated from infected susceptible mice.

Several studies have shown that immunization with purified material from *L. major* protect mice against infectious challenge. Immunization with a lipid-containing glycoconjugate present on the promastigote membrane, termed LPG, has been shown to protect mice against challenge with virulent parasites [261]. The cellular parameters of the protective immune response which is induced following immunization with this LPG have not been delineated. It should be stressed that molecules other than LPG are able to trigger protective responses, since immunization with avirulent parasites devoid of LPG protects mice against infectious challenge [258]. Specifically, some integral membrane proteins of *L. major* have been demonstrated to be good immunogens for immunoprophylaxis of experimentally induced leishmaniasis in mice [262]. A 63 kD glycoprotein (gp63) on the surface of *Leishmania*, known to be a metallo-protease, has also been shown to protect mice against virulent parasites using *L. mexicana* as a model system. In these successful vaccination experiments, native gp63 incorporated into liposomes were used [257]. Even though some studies have failed to demonstrate a successful outcome with gp63 immunization [263,264], other more recent and exciting results have revealed that immunization with synthetic peptides representing gp63 residue 154–168 could protect susceptible mice against *L. major* [265]. Since in these experiments protection was only seen when the peptides were administered together with a novel temperature-dependent sol–gel transition adjuvant, it is possible that this adjuvant was necessary for the triggering of a protective TH-1 T cell response [265].

Finally, immunization of mice with a protein (gp72) purified from *L. donovani* has been shown recently to protect significantly mice against a challenge with virulent *L. donovani* amastigotes [266]. As in most experimental vaccination studies so far, the precise effector mechanisms implicated in the establishment of this protective state have not been characterized in detail. Indeed, it is not known whether or not immunization with purified antigens generates antibodies that will prevent invasion of host cells by the parasites, thus reducing the size of the initial parasite inoculum, or specific T cells that will activate the parasitized macrophages, resulting in the destruction of intracellular microorganisms.

CONCLUSIONS

From the results summarized above, it appears that resistance and susceptibility to experimental infection with *L. major* are influenced by distinct subsets of parasite-specific CD4$^+$ T cells which exert their activity through the lymphokines that they release. The weight of the evidence to date also supports a role for CD4$^+$ IFN-γ-producing T cells in the control of human leishmanial infections. Whether or not distinct CD4$^+$ T cells also control susceptibility leading to chronic disease in humans has not been determined.

Studies on the mouse model of infection with *L. major*, which have demonstrated that susceptibility to disease can be overcome by immunologic interventions, clearly show that even genetically susceptible mice have the functional T cell repertoire necessary to resolve infection [267]. This situation is analogous to humans with generalized infections, where the capacity to generate immune responses earmarking resistance to infection is restored after successful chemotherapy. Thus, although major advances have been made in deciphering the role of T cell responses in resistance and susceptibility, much work remains to be done on delineating the precise mechanisms whereby certain parasites in certain hosts lead to the expansion of functionally distinct T cell phenotypes.

Encouraging results are also emerging which give good hope for possible control of leishmaniasis through vaccination using killed and molecularly defined immunogens. The findings that the fine specificity of T cells, even those belonging to the subset normally mediating resistance (i.e., TH-1), influences their effect on the outcome of infection has important implications in vaccine development.

ACKNOWLEDGMENTS

Work cited from the author's laboratory has been supported by the Swiss National Science Foundation and the UNDP/World Bank/WHO Special Program for research on tropical diseases.

REFERENCES

1 Neva FA, Sacks DL. Leishmaniasis. In Warren KS, Mahmoud AAF, eds. *Tropical and Geographical Medicine*, 2nd edn. New York: McGraw-Hill Book Co., 1990.

2 Report of World Health Organization Expert Committee. *The Leishmaniasis Technical Report Series 701*. WHO, Geneva, 1984.

3 Hommel M, Peters W, Ranque J, Quilici M, Lannotte G. The micro-ELISA technique in the serodiagnosis of visceral leishmaniasis. Ann Trop Med Parasitol 1978;29:213–218.

4 Anthony RL, Christensen HA, Johnson CM. Micro-ELISA for the serodiagnosis of New World leishmaniasis. Am J Trop Med Hyg 1980;29:190–194.

5 Duxbury RE, Sadun EH. Fluorescent antibody test for the serodiagnosis of visceral leishmaniasis. Am J Trop Med Hyg 1964;13:525–529.

6 Badaro R, Reed SG, Carvalho EM. Immunofluorescent antibody test in American visceral leishmaniasis: sensitivity and specificity of different morphological forms of two *Leishmania* species. Am J Trop Med Hyg 1983;32:480–484.

7 Allain DS, Kagan IG. A direct agglutination test for leishmaniasis. Am J Trop Med Hyg 1975;24:232–236.

8 Harith AE, Kolk AHJ, Leeuwenberg R, Muigai R, Kiugu S, Laarman JJ. A simple and economical field test for serodiagnosis and seroepidemiological studies of visceral leishmaniasis. Trans R Soc Trop Med Hyg 1986;80:583–587.

9 Harith AE, Slappendel RJ, Reiter I, *et al.* Application of a direct agglutination test for detection of specific anti-*Leishmania* antibody in the canine reservoir. J Clin Microbiol 1989;27:2252–2257.

10 Blaxter ML, Miles MA, Kelly JM. Specific diagnosis of visceral leishmaniasis using a *Leishmania donovani* antigen identified by expression cloning. Mol Biochem Parasitol 1988;30:259–270.

11 Jaffe CL, Zalis M. Use of purified proteins from *Leishmania donovani* for the rapid serodiagnosis of visceral leishmaniasis. J Infect Dis 1988;157:1212–1220.

12 Reed SG, Badaro R, Lloyd RM. Identification of species and cross-reactive antigens of *Leishmania donovani chagasi* by human infection sera. J Immunol 1987;138:1596–1601.

13 Jaffe CL, McMahon-Pratt D. Serodiagnostic assay for visceral leishmaniasis employing monoclonal antibodies. Trans R Soc Trop Med Hyg 1987;81:587–594.

14 Evans DA. A comparison of the direct agglutination test and ELISA in the sero-diagnosis of leishmaniasis in Sudan. Trans R Soc Trop Med Hyg 1989;83:334–337.

15 Edrissian GH, Darabian P. A comparison of ELISA and indirect immunofluorescent antibody test in the seridiagnosis of cutaneous and visceral leishmaniasis in Iran. Trans R Soc Trop Med Hyg 1979;73:289–292.

16 Rosen G, Londner MV, Greenblatt CL, Morsy TA, El-On J. *Leishmania major*: solid phase radioimmunoassay for anti-body detection in human cutaneous leishmaniasis. Exp Parasitol 1986;62:79–84.

17 Montenegro J. Cutaneous reaction in leishmaniasis. Arch Dermatol Syphilol 1926;13:187.

18 Dostrovsky A, Sagher F. The intra-cutaneous test in cutaneous leishmaniasis. Ann Trop Med Parasitol 1946;40:265.

19 Koufman Z, Egoz N, Greenblatt CL, Handman E, Montillo B, Even-Paz Z. Observations on immunization against cutaneous leishmaniasis in Israel. Israel J Med Sci 1978;14:218–222.

20 Kilgour V, Gardener PJ, Godfrey DG, Peters W. Demonstration of electrophoretic variation of two aminotransferases in *Leishmania*. Ann Trop Med Parasitol 1974;68:245–246.

21 Kreutzer RD, Christensen HA. Characterization of *Leishmania* spp. by isozyme electrophoresis. Am J Trop Med Hyg 1980;29:199–208.

22 Pratt DM, David JR. Monoclonal antibodies that distinguish between New World species of *Leishmania*. Nature 1981;291:581–583.

23 Pratt DM, Bennett E, David JR. Monoclonal antibodies that distinguish subspecies of *Leishmania braziliensis*. J Immunol 1982;129:926–927.

24 Jaffe CL, Pratt DM. Monoclonal antibodies specific for *L. tropica*. I. Characterization of antigens associated with stage- and species-specific determinants. J Immunol 1983;131:1987–1993.

25 Jaffe CL, Bennett E, Grimaldi G, Pratt DM. Production and characterization of species-specific monoclonal antibodies against *Leishmania donovani* for immunodiagnosis. J Immunol 1984;133:440–447.

26 Grimaldi G, Tesh RB, Pratt DM. A review of the geographic distribution and epidemiology of leishmaniasis in the New World. Am J Trop Med Hyg 1989;41:687–725.

27 Lynch NR, Malave C, Ifante RB, Modlin RL, Convit J. *In situ* detection of amastigotes in American cutaneous leishmaniasis using monoclonal antibodies. Trans R Soc Trop Med Hyg 1986;80:6–9.

28 Handman HE, Mitchell GF, Goding JW. Leishmania major: a very sensitive dot-blot ELISA for detection of parasites in cutaneous lesions. Mol Biol Med 1987;4:377–383.

29 Chance ML, Walton BC, eds. *Biochemical Characterization of Leishmania*. UNDP/World Bank/WHO, Geneva, 1982.

30 Wirth DF, McMahon-Pratt D. Rapid identification of *Leishmania* species by specific hybridization of kinetoplast DNA in cutaneous lesions. Proc Natl Acad Sci USA 1982;79:6999–7003.

31 Benne R, Van DBJ, Brakenhoff JP, Sloof P, Van BJH, Tromp MC. Major transcript of the frameshifted coxII gene from trypanosome mitochondria contains four nucleotides that are not encoded in the DNA. Cell 1986;46:819–826.

32 Decker CJ, Sollner WB. RNA editing involves indiscriminate U changes throughout precisely defined editing domains. Cell 1990;61:1001–1011.

33 Feagin JE, Abraham JM, Stuart K. Extensive editing of the cytochrome c oxidase III transcript in *Trypanosoma brucei*. Cell 1988;53:413–422.

34 Shaw JM, Feagin JE, Stuart K, Simpson L. Editing of kinetoplastid mitochondrial mRNAs by uridine addition and deletion generates conserved amino acid sequences and AUG initiation codons. Cell 1988;53:401–411.

35 Simpson L, Shaw J. RNA editing and the mitochondrial cryptogenes of kinetoplastid protozoa. Cell 1989;57:355–366.

36 Blum B, Bakalara N, Simpson L. A model for RNA editing in kinetoplastid mitochondria: "guide" RNA molecules transcribed from maxicircle DNA provide the edited information. Cell 1990;60:189–198.

37 Blum B, Sturm NR, Simpson AM, Simpson L. Chimeric gRNA–mRNA molecules with oligo(U) tails covalently linked at sites of RNA editing suggest that U addition occurs by transesterification. Cell 1991;65:543–550.

38 Pollard VW, Hajduk SL. Trypanosoma equiperdum minicircles encode three distinct primary transcripts which exhibit guide RNA characteristics. Mol Cell Biol 1991;11:1668–1675.

39 Barker DC. Molecular approaches to DNA diagnosis. Parasitology 1989;1:46–54.

40 Dasgupta S, Adhya S, Majumder HK. A simple procedure for the preparation of pure kinetoplast DNA network free of nuclear DNA from the kientoplast hemoflagellate Leishmania donovani. Anal Biochem 1986;158:189–194.

41 Greig SR, Akinsehinwa FA, Ashall F, et al. The feasibility of discrimination between Leishmania and Endotrypanum using total parasite DNA probes. Trans R Soc Trop Med Hyg 1989;83:196.

42 Jackson PR, Lawrie JM, Stiteler JM, Hawkins DW, Wohlhieter JA, Rowton ED. Detection and characterization of Leishmania species and strains from mammals and vectors by hybridization and restriction endonuclease digestion of kinetoplast DNA. Vet Parasitol 1986;20:195–215.

43 Lopes UG, Wirth DF. Identification of visceral Leishmania species with cloned sequences of kinetoplast DNA. Mol Biochem Parasitol 1986;20:77–84.

44 Lopez M, Montoya Y, Arana M, et al. The use of nonradioactive DNA probes for the characterization of Leishmania isolates from Peru. Am J Trop Med Hyg 1988;38:308–314.

45 Rogers WO, Wirth DF. Kinetoplast DNA minicircles: regions of extensive sequence divergence. Proc Natl Acad Sci USA 1987;84:565–569.

46 Rogers WO, Burnheim PF, Wirth DF. Detection of Leishmania within sand flies by kinetoplast DNA hybridization. Am J Trop Med Hyg 1988;39:434–439.

47 Smith DF, Searle S, Ready PD, Gramiccia M, Ben IR. A kinetoplast DNA probe diagnostic for Leishmania major: sequence homologies between regions of Leishmania minicircles. Mol Biochem Parasitol 1989;37:213–223.

48 Wirth DF, Rogers WO, Barker RJ, Dourado H, Suesebang L, Albuquerque B. Leishmaniasis and malaria: new tools for epidemiologic analysis. Science 1986;234:975–979.

49 Kidane GZ, Hughes D, Simpson L. Sequence heterogeneity and anomalous electrophoretic mobility of kinetoplast minicircle DNA from L. tarentolae. Gene 1984;27:265–277.

50 Van EGJ, Schoone GJ, Ligthart GS, Laarman JJ, Terpstra WJ. Detection of Leishmania parasites by DNA in situ hybridization with non-radioactive probes. Parasitol Res 1987;73:199–202.

51 Sturm NR, Degrave W, Morel C, Simpson L. Sensitive detection and schizodeme classification of Trypanosoma cruzi cells by amplification of kinetoplast minicircle DNA sequences; use in diagnosis of Chagas' disease. Mol Biol Parasitol 1989;33:205–214.

52 Moser DR, Cook GA, Ochs DE, Bailey CP, McKane MR, Donelson JE. Detection of Trypanosoma congolense and Trypanosoma brucei subspecies by DNA amplification using the polymerase chain reaction. Parasitology 1989;1:57–66.

53 Moser DR, Kirchhoff LV, Donelson JE. Detection of Trypanosoma cruzi by DNA amplification using the polymerase chain reaction. J Clin Microbiol 1989;27:1477–1482.

54 Rodgers MR, Popper SJ, Wirth DF. Amplification of kinetoplast DNA as a tool in the detection and diagnosis of Leishmania. Exp Parasitol 1990;71:267–275.

55 Saiki R, Gelfand D, Stoffel S, et al. Primer directed enzymatic amplification of DNA with a thermostable DNA polymerase. Science 1988;239:487–491.

56 Wyler DJ, Marsden PD. Leismaniasis. In Warren KS, Mahmoud AAF, eds. Tropical and Geographical Medicine, vol. 1. New York: McGraw-Hill Book Co., 1984:270–280.

57 Cuba Cuba CA, Netto EM, Marsden PD, de C Rosa A, Llanos Cuentas EA, Costa JLM. Cultivation of Leishmania braziliensis braziliensis from skin ulcers in man under field conditions. Trans R Soc Trop Med Hyg 1986;80:456–457.

58 Llanos Cuentas EA, Cuba CC, Barreto AC, Marsden PD. Clinical characterization of human L.b.b. infections. Trans R Soc Trop Med Hyg 1984;78:845–846.

59 Wood GS, Warnke R. Suppression of endogenous avidin binding activity in tissues and its relevance to biotin–avidin detection systems. J Histochem Cytochem 1981;29:1196–1204.

60 Avila HA, Sigman DS, Cohen LM, Millikan RC, Simpson L. Polymerase chain reaction amplification of Trypanosoma cruzi kinetoplast minicircle DNA isolated from whole blood lysates: diagnosis of chronic Chagas' disease. Mol Biomed Parasitol 1991;48:211–222.

61 Ashall F, Yip CDA, Luquetti AA, Miles MA. Radiolabeled total parasite DNA probe specifically detects Trypanosoma cruzi in mammalian blood. J Clin Microbiol 1988;26:576–578.

62 Ramirez JL, Guevara P. The ribosomal gene spacer as a tool for the taxonomy of Leishmania. Mol Biochem Parasitol 1987;22:177–183.

63 Patton WS. The development of the parasite of Indian kala-azar. Scientific Memoirs, Government of India 1912;27,31,53. In Hindle EP, Hou PC, Patton WS, eds. Serological Studies in Chinese kala azar. Proc R Soc London 1926;100:368.

64 Ulrich M, Ortiz D, Conuit, J. The effect of fresh serum on leptomonads of Leishmania. Trans R Soc Trop Med Hyg 1968;62:825.

65 Mosser DM, Edelson PJ. Activation of the alternative complement pathway by Leishmania promastigotes: parasite lysis and attachment to macrophages. J Immunol 1984;132:1501–1505.

66 Mosser DM, Burke SK, Coutavas EE, Wedgewood JF, Edelson PJ. Leishmania species: mechanisms of comple-

ment activation by five strains of promastigotes. Exp Parasitol 62:394–404.

67 Pearson RD, Steigbigel RT. Mechanism of lethal effect of human serum upon *Leishmania donovani*. J Immunol 1980; 125:2195–2201.

68 Mosser DM, Wedgewood JF, Edelson PJ. *Leishmania* amastigotes: resistance to complement-mediated lysis is not due to a failure to fix C3. J Immunol 1985;134:4128–4131.

69 Sacks DL, Perkins PV. Identification of an infective stage of *Leishmania* promastigotes. Science 1984;223:1417.

70 Sacks DL, Perkins PV. Development of infective stage *Leishmania* promastigotes within phlebotomine sandflies. Am J Trop Med Hyg 1985;34:456.

71 Giannini MS. Effects of promastigote growth phase, frequency of subculture, and host age on promastigote-initiated infections in *Leishmania donovani* in the golden hamster. J Protozool 1974;21:521–527.

72 Franke ED, McGreevy PB, Katz SP, Sacks DL. Growth cycle dependent generation of complement resistant *Leishmania* promastigotes. J Immunol 1985;134:2713–2718.

73 Howard MK, Sayers G, Miles MA. *Leishmania donovani* metacyclic promastigotes: transformation *in vitro*, lectin agglutination, complement resistance and infectivity. Exp Parasitol 1987;64:147–156.

74 Puentes SM, Sacks DL, da Silva RP, Joiner KA. Complement binding by two developmental stages of *Leishmania major* promastigotes varying in expression of a surface lipophosphoglycan. J Exp Med 1988;167:887–902.

75 Sacks DL, da Silva R. The generation of infective stage *Leishmania major* promastigotes is associated with the cell-surface expression and release of a developmentally regulated glycolipid. J Immunol 1987;139:3099.

76 Pimenta PF, da Silva R, Sacks DL, da Silva P. Cell surface nanonanatomy of *Leishmania major* as revealed by fracture-flip. A surface meshwork of 44 nm fusiform filaments identifies infective developmental stage promastigotes. Eur J Cell Biol 1989;48:180–190.

77 Chang KP. *Leishmania donovani* macrophage binding mediated by surface glycoproteins/antigens. Mol Biochem Parasitol 1981;4:67–76.

78 Klempner MS, Cendron M, Wyler DJ. Attachment of plasma membrane vesicles of human macrophages to *Leishmania tropica* promastigotes. J Infect Dis 1983;148:377–384.

79 Channon JY, Roberts MB, Blackwell JM. A study of the differential respiratory burst activity elicited by promastigotes and amastigotes of *Leishmania donovani* in murine resident peritoneal macrophages. Immunology 1984;53:345–355.

80 Blackwell JM, Ezekowitz RAB, Roberts MB, Channon JY, Sim RB, Gordon S. Macrophage complement and lectin-like receptors bind *Leishmania* in the absence of serum. J Exp Med 1985;162:324–331.

81 Wilson ME, Pearson RD. Evidence that *Leishmania donovani* utilizes a mannose receptor on human mononuclear phagocytes to establish intracellular parasitism. J Immunol 1986;136:4681–4688.

82 Wilson ME, Pearson RD. Roles of CR3 and mannose receptors in the attachment and ingestion of *Leishmania donovani* by human mononuclear phagocytes. Infect Immun 1988;56:363–369.

83 Saraiva EMB, Andrade AFB, de Souza W. Involvement of the macrophage mannose-6-phosphate receptor in the recognition of *Leishmania mexicana amazonensis*. Parasitol Res 1987;73:411–416.

84 Handman E, Goding JW. The *Leishmania* receptor for macrophages is a lipid containing glycoconjugate. EMBO J 1985;4:329–336.

85 Palatnik CB, Borojevic R, Previato JO, Mendonca-Previato L. Inhibition of *Leishmania donovani* promastigote internalization into murine macrophages by chemically defined parasite glycoconjugate ligands. Infect Immun 1989;57:754–763.

86 Russell DG, Wilhelm H. The involvement of the major surface glycoprotein (gp63) of *Leishmania* promastigotes in attachment to macrophages. J Immunol 1986;136:2613–2620.

87 Chang CS, Chang KP. Monoclonal antibody affinity purification of a *Leishmania* membrane glycoprotein and its inhibition of *Leishmania*-macrophage binding. Proc Natl Acad Sci USA 1986;83:100–104.

88 Mosser DM, Edelson PJ. The mouse macrophage receptor for C3bi (CR3) is a major mechanism in the phagocytosis of *Leishmania* promastigotes. J Immunol 1985;135:2786–2789.

89 da Silva RP, Hall BF, Joiner KA, Sacks DL. CR1, the C3b receptor, mediates binding of infective *Leishmania major* metacyclic promastigotes to human macrophages. J Immunol 1989;43:617–622.

90 Mosser DM, Edelson PJ. The third component of complement (C3) is responsible for the intracellular survival of *Leishmania major*. Nature 1987;327:329–331.

91 Wright SD, Silverstein SC. Receptors for C3b and C3bi promote phagocytosis but not release of toxic oxygen from human phagocytes. J Exp Med 1983;159:2016–2022.

92 Russell DG, Talamas-Rohana P. *Leishmania* and the macrophage: a marriage of inconvenience. Immunol Today 1989;10:328–334.

93 Russell DG, Wright SD. Complement receptor type 3 (CR3) binds to an Arg-Gly-Asp-containing region of the major surface glycoprotein, gp63, of *Leishmania* promastigotes. J Exp Med 1988;168:279–292.

94 Cooper AM, Rosen H, Blackwell JM. Monoclonal antibodies which recognize distinct epitopes of macrophage type three complement receptor differ in their ability to inhibit binding of *Leishmania* promastigotes harvested at different phases of their growth cycle. Immunology 1988; 65:511–514.

95 Murray HW. Susceptibility of *Leishmania* to oxygen intermediates and killing by normal macrophages. J Exp Med 1981;153:1302–1315.

96 Murray HW. Cell-mediated immune response in experimental visceral leishmaniasis. II. Oxygen-dependent killing of intracellular *Leishmania donovani* amastigotes. J Immunol 1982;129:351–357.

97 Pearson RD, Harcus JL, Sumes PH, Romito R, Donowitz GR. Failure of the phagocytic oxidative response to protect human monocyte-derived macrophages from infection by *Leishmania donovani*. J Immunol 1982;129:1282–1287.

98 Truco SJ. The lipophosphoglycan of *Leishmania*. Parasitol Today 1988;4:255–257.

99 Remaley AT, Kuhns DB, Basford RE, Glew RH, Kaplan SS. Leishmanial phosphatase blocks neutrophil O_2 production. J Biol Chem 1984;259:11173–11175.

100 Chang KP, Dwyer DM. *Leishmania donovani* hamster macrophage interaction *in vitro*. Cell entry, intracellular survival, and multiplication of amastigotes. J Exp Med 1978;147:515.

101 El-On J, Bradley DJ, Freeman JC. *Leishmania donovani*: action of excreted factor on hydrocytic enzyme activity of macrophages from mice with genetically different resistance to infection. Exp Parasitol 1980;49:167.

102 Chawdhuri G, Chaudhuri M, Pan A, Chang KP. Surface acid proteinase (gp63) of *Leishmania mexicana*. J Biol Chem 1989;264:7483–7489.

103 Steinman RM, Brodie SE, Cohn ZA. Membrane flow during pinocytosis, a stereologic analysis. J Cell Biol 1976; 68:665.

104 Gottlieb M, Dwyer DM. Identification and partial characterization of an extracellular acid phosphatase activity of *L. donovani* promastigote. Mol Cell Biol 1982;2:76–81.

105 Gottlieb M, Dwyer DM. *Leishmania donovani*: surface membrane acid phosphatase activity of promastigotes. Exp Parasitol 1981;52:117–128.

106 Dwyer DM, Gottlieb M. Surface membrane localization of 3′ and 5′-nucleotidase activities in *Leishmania donovani* promastigotes. Mol Biochem Parasitol 1984;10:139–150.

107 Zilberstein D, Dwyer DM. Proton motive force driven active transport of d-glucose and l-proline in the protozoan parasite *Leishmania donovani*. Proc Natl Acad Sci USA 1985; 82:1716–1720.

108 Betley MJ, Miller VL, Mekalanos JJ. Genetics of bacterial enterotoxins. Annu Rev Microbiol 1986;40:577–605.

109 Cianciotto N, Eisenstein BI, Engleberg NC, Shuman H. Genetics and molecular pathogenesis of *Legionella pneumophila*, an intracellular parasite of macrophages. Mol Biol Med 1989;6:490–524.

110 Conway B, Ronald A. An overview of some mechanisms of bacterial pathogenesis. Can J Microbiol 1988;34:281–286.

111 Cossart P, Mengaud J. *Listeria monocytogenes*. A model system for the molecular study of intracellular parasitism. Mol Biol Med 1989;6:463–474.

112 Dallmier AW, Martin SE. Catalase, superoxide dismutase, and hemolysin activities and heat susceptibility of *Listeria monocytogenes* after growth in media containing sodium chloride. Appl Environ Microbiol 1990;56:2807–2810.

113 Sun AN, Camilli A, Portnoy DA. Isolation of *Listeria monocytogenes* small-plaque mutants defective for intracellular growth and cell-to-cell spread. Infect Immun 1990; 58:3770–3778.

114 Bellofatto V, Cross GA. Expression of a bacterial gene in a trypanosomatid protozoan. Science 1989;244:1167–1169.

115 Laban A, Wirth DF. Transfection of *Leishmania enriettii* and expression of chloramphenicol acetyltransferase gene. Proc Natl Acad Sci USA 1989;86:9119–9123.

116 LeBowitz JH, Coburn CM, McMahon PD, Beverley SM. Development of a stable *Leishmania* expression vector and application to the study of parasite surface antigen genes. Proc Natl Acad Sci USA 1990;87:9736–9740.

117 Clayton CE, Fueri JP, Itzhaki JE, *et al*. Transcription of the procyclic acidic repetitive protein genes of *Trypanosoma brucei*. Mol Cell Biol 1990;10:3036–3047.

118 Rudenko G, Le BS, Smith J, Lee MG, Rattray A, Van DPLH. Procyclic acidic repetitive protein (PARP) genes located in an unusually small alpha-amanitin-resistant transcription unit: PARP promoter activity assayed by transient DNA transfection of *Trypanosoma brucei*. Mol Cell Biol 1990;10:3492–3504.

119 Lu HY, Buck GA. Expression of an exogenous gene in *Trypanosoma cruzi* epimastigotes. Mol Biochem Parasitol 1991;44:109–114.

120 Laban A, Tobin JF, Curotto DLMA, Wirth DF. Stable expression of the bacterial neor gene in *Leishmania enriettii*. Nature 1990;343:572–574.

121 Kapler GM, Coburn CM, Beverley SM. Stable transfection of the human parasite *Leishmania major* delineates a 30-kilobase region sufficient for extrachromosomal replication and expression. Mol Cell Biol 1990;10:1084–1094.

122 Ten AAL, Ouellette M, Borst P. Targeted insertion of the neomycin phosphotransferase gene into the tubulin gene cluster of *Trypanosoma brucei* [see comments]. Nature 1990; 348:174–175.

123 Cruz A, Beverley SM. Gene replacement in parasitic protozoa [see comments]. Nature 1990;348:171–173.

124 Lee MG, Van DPLH. Homologous recombination and stable transfection in the parasitic protozoan *Trypanosoma brucei*. Science 1990;250:1583–1587.

125 Eid J, Sollner WB. Stable integrative transformation of *Trypanosoma brucei* that occurs exclusively by homologous recombination. Proc Natl Acad Sci USA 1991;88:2118–2121.

126 Tobin JF, Laban A, Wirth DF. Homologous recombination in *Leishmania enriettii*. Proc Natl Acad Sci USA 1991;88:864–868.

127 Kapler GM, Beverley SM. Transcriptional mapping of the amplified region encoding the dihydrofolate reductase–thymidylate synthase of *Leishmania major* reveals a high density of transcripts, including overlapping and anti-sense RNAs. Mol Cell Biol 1989;9:3959–3972.

128 Kink JA, Chang KP. Tunicamycin-resistant *Leishmania mexicana amazonensis*: expression of virulence associated with an increased activity of N-acetylglucosaminyl-transferase and amplification of its presumptive gene. Proc Natl Acad Sci USA 1987;84:1253–1257.

129 Katakura K, Chang KP. H DNA amplification in *Leishmania* resistant to both arsenite and methotrexate. Mol Biochem Parasitol 1989;34:189–191.

130 Sokolova AN. Preventive vaccination with the living parasites of cutaneous leishmaniasis. Trans Turkmen Cutan-Venereol Inst Ashkhabad 1940;11–44.

131 Adler S, Theodor O. The transmission of *L. tropica* from artificially infected sandflies to man. Ann Trop Med Parasitol 1927;21:89–111.

132 Berberian DA. Vaccination and immunity against oriental sore. Trans R Soc Trop Med Hyg 1939;33:87–94.

133 Senekji HA, Beattie CP. Artificial infection and immunization of man with cultures of *L. tropica*. Trans R Soc Trop Med Hyg 1941;34:415–419.

134 Lainson R, Shaw JJ. Studies on the immunology and serology of leishmaniasis. III. On the cross immunity

between Panamanian cutaneous leishmaniasis and *L. mexicana* infection in humans. Trans R Soc Trop Med Hyg 1966;60:533–535.

135 Berberian DA. Cutaneous leishmaniasis (oriental sore). I. Time required for development of immunity after vaccination. Arch Dermatol Syphilol 1944;49:433–435.

136 Montenegro J. Cutaneous reactions in leishmaniasis. Arch Dermatol Syphilol 1986;13:187.

137 Manson-Bahr PEL. Immunity in kala-azar. Trans R Soc Trop Med Hyg 1961;55:550–555.

138 Witztum E, Spira DT, Zuckerman A. Blast transformation in different stages of cutaneous leishmaniasis. Israel J Med 1978;14:244–247.

139 Wyler DJ, Weinbaum FI, Herrod HR. Characterization of *in vitro* proliferative responses of human lymphocytes to leishmanial antigens. J Infect Dis 1979;140:215–221.

140 Koufman Z, Egoz N, Greenblott C, *et al.* Observations on immunization against cutaneous leishmaniasis in Israel. Israel J Med Sci 1978;4:218–222.

141 Guirges SY. Natural and experimental reinfection of man with oriental sore. Ann Trop Med Parasitol 1971;65:197–205.

142 Dostrovsky A. The incubation in experimental cutaneous leishmaniasis. Acta Med Orient 1945;4:303–305.

143 Green MS, Kark JD, Greenblatt CI, *et al.* The cellular and humoral immune response in subjects vaccinated against cutaneous leishmaniasis using *Leishmania major* promastigotes. Parasite Immunol 1983;5:337–344.

144 Dorea RCC, Coutinho SG, Sabroza PC, *et al.* Quantification of *Leishmania*-specific T cells in human American cutaneous leishmaniasis by limiting dilution analysis. Clin Exp Immunol 1988;71:26–31.

145 Muray HW, Rubin BY, Rothermel CD. Killing of intracellular *Leishmania donovani* by lymphokine-stimulated human mononuclear phagocytes. Evidence that interferon-gamma is the activating lymphokine. J Clin Invest 1983;72:1506–1510.

146 Rada E, Trujillo D, Castellanos PL, Convit J. Gamma interferon production induced by antigens in patients with leprosy and American cutaneous leishmaniasis. Am J Trop Med Hyg 1987;37:520–524.

147 Castes M, Cabrera M, Trujillo D, Convit J. T cell subpopulations, expression of interleukin-2 receptor, and production of interleukin-2 and gamma interferon in human American cutaneous leishmaniasis. J Clin Microbiol 1988;26:1207–1213.

148 Melby PC, Neva FA, Sacks DL. Profile of human T cell response to leishmanial antigens: analysis by immunoblotting. J Clin Invest 1989;83:1868.

149 Melby PC, Sacks DL. Identification of antigens recognized by T cells in human leishmaniasis: analysis of T cell clones by immunoblotting. Infect Immun 1989;57:2971–2976.

150 Ridley MJ, Wells CW. Macrophage–parasite interactions in the lesions of cutaneous leishmaniasis. Am J Pathol 1986;123:79–89.

151 Ridley DS, Ridley MJ. The evolution of the lesion in cutaneous leishmaniasis. J Pathol 1983;141:83–96.

152 Madlen RL, Tapia FJ, Bloom BR, *et al. In situ* characterization of the cellular immune response in American cutaneous leishmaniasis. Clin Exp Immunol 1985;60:241–248.

153 Nelson R, Mshana RN. *In situ* characterization of the cutaneous immune response in Ethiopian cutaneous leishmaniasis. Scand J Immunol 1987;26:503–512.

154 Menzel S, Bienzle V. Antibody response in patients with cutaneous leishmaniasis of the Old World. Tropenmed Parasitol 1978;29:194.

155 Rosen G, Londner M, Greenblatt C, *et al. Leishmania major*: solid phase radio immunoassay for antibody detection in human cutaneous leishmaniasis. Exp Parasitol 1986;62:79–84.

156 Moriearty PL, Gumaldi G, Galvao-Castro B, *et al.* Intralesional plasma cells and serological responses in human cutaneous leishmaniasis. Clin Exp Immunol 1982;47:59–64.

157 Bryceson ADM. Immunological aspects of clinical leishmaniasis. Proc R Soc Med 1970;63:40–44.

158 Convit J, Pinardi ME, Rondon A. Diffuse cutaneous leishmaniasis. A disease due to an immunological defect. Trans R Soc Trop Med Hyg 1972;66:603–610.

159 Bryceson ADM. Diffuse cutaneous leishmaniasis in Ethiopia. III. Immunological studies. Trans R Soc Trop Med Hyg 1970;64:380–387.

160 Petersen EA, Neva FA, Oster CN, Bogaert-Diaz H. Specific inhibition of lymphocyte-proliferation responses by adherent suppressor cells in diffuse cutaneous leishmaniasis. N Engl J Med 1982;306:387–392.

161 Petersen EA, Neva FA, Barral A, *et al.* Monocyte suppression of antigen-specific lymphocyte responses in diffuse cutaneous leishmaniasis patients from the Dominican Republic. J Immunol 1984;132:2603–2606.

162 Castes M, Agnelli A, Rondon AJ. Mechanisms associated with immunoregulation in human American cutaneous leishmaniasis. Clin Exp Immunol 1984;57:279–286.

163 Akuffo HO, Fehniger TE, Britton S. Differential recognition of *Leishmania aethiopica* antigens by lymphocytes from patients with local and diffuse cutaneous leishmaniasis. Evidence for antigen-induced immune suppression. J Immunol 1988;7:2461–2466.

164 Castes M, Agnelli A, Verde O, Rondon AJ. Characterization of the cellular immune response in American cutaneous leishmaniasis. Clin Immunol Immunopathol 1983;27:176–186.

165 Carvalho EM, Johnson WD, Barreto E, *et al.* Cell mediated immunity in American cutaneous and mucosal leishmaniasis. J Immunol 1985;135:4144–4148.

166 Conceicao-Silva F, Dorea RCC, Pirmez C, Schubach A, Coutinho SG. Quantitative study of *Leishmania braziliensis braziliensis* reactive T cells in peripheral blood and in the lesions of patients with American mucocutaneous leishmaniasis. Clin Exp Immunol 1990;79:221–226.

167 Saravia NG, Valderama L, Labrada M, *et al.* The relationship of *Leishmania braziliensis* subspecies and immune response to disease expression in New World leishmaniasis. J Infect Dis 1989;159:725–735.

168 Convit J, Castellanos PL, Ulrich M, *et al.* Immunotherapy of localized, intermediate, and diffuse forms of American cutaneous leishmaniasis. J Infect Dis 1989;160:104–115.

169 Scott P. Impaired macrophage leishmanicidal activity at

cutaneous temperature. Parasite Immunol 1989;7:277–288.

170 Scott P, Sacks D, Sher A. Resistance to macrophage-mediated killing as a factor influencing the pathogenesis of chronic cutaneous leishmaniasis. J Immunol 1983;131:966–971.

171 Carvalho EM, Teixeira RS, Johnson WD. Cell-mediated immunity in American visceral leishmaniasis: reversible immunosuppression during acute infection. Infect Immun 1981;33:498–502.

172 Carvalho EM, Badaro R, Reed SG, Jones TC, Johnson WD. Absence of gamma interferon and interleukin-2 production during active visceral leishmaniasis. J Clin Invest 1985;76:2066–2069.

173 Haldar JP, Ghose S, Saha KC, Ghose AC. Cell-mediated immune response in Indian kala-azar and post-kala-azar dermal leishmaniasis. Infect Immun 1983;42:702–707.

174 Sacks DL, Lal DL, Shrivastava SN, Blackwell J, Neva FA. An analysis of T cell responsiveness in Indian kala-azar. J Immunol 1987;138:908–913.

175 Levy LH, Mendes E. Impaired cell-mediated immunity in patients with kala-azar. Allergol Immunopathol 1981;9:109.

176 Ho M, Koech DK, Iha DW, Bryceson ADM. Immunosuppression in Kenyan visceral leishmaniasis. Clin Exp Immunol 1983;51:207–214.

177 Galvao-Castro B, Ferreira JA, Marzochi MC, et al. Polyclonal B cell activation, circulating immune complexes and autoimmunity in human visceral leishmaniasis. Clin Exp Immunol 1984;56:58–66.

178 Barral A, Carvalho EM, Badaro R, Barral-Netto M. Suppression of lymphocyte proliferative responses by sera from patients with American Visceral leishmaniasis. Am J Trop Med Hyg 1986;35:735–742.

179 Rezai HR, Ardehali SM, Amirhakimi G, Kharazmi A. Immunological features of kala-azar. Am J Trop Med Hyg 1987;27:1079–1083.

180 Veress B, Omer A, Sater A, El Hassam AM. Morphology of the spleen and lymph nodes in fatal visceral leishmaniasis. Immunology 1977;33:605–610.

181 Carvalho EM, Vacellar O, Barral A, Badero R, Johnson W. Antigen-specific immuno suppression in visceral leishmaniasis is cell-mediated. J Clin Invest 1989;83:860–864.

182 Neogy AB, Nandy A, Dastidar BG, Chowdhury AB. Leishmanin test in Indian kala-azar. Trans R Soc Trop Med Hyg 1986;80:454–455.

183 Manson-Bahr PEC, Southgate BA, Harvey AEC. Development of kala-azar in man after inoculation with Leishmania from a Kenya sandfly. Br Med J 1963;1208–1210.

184 Reed SG, Carvalho EM, Sherbert CH, et al. In vitro responses to Leishmania antigens by lymphocytes from patients with leishmaniasis or Chagas' disease. J Clin Invest 1990;85:690–696.

185 Badaro R, Jones TC, Lorenco FJC, et al. A prospective study of visceral leishmaniasis in an endemic area of Brazil. J Infect Dis 1986;154:639–649.

186 Reed SG, Badaro R, Masur H, et al. Selection of a specific skin test antigen for American visceral leishmaniasis. Am J Trop Med Hyg 1986;35:74–89.

187 Ma DDF, Concannon AJ, Hayes J. Fatal leishmaniasis in renal transplant patients. Lancet 1979;ii:311–312.

188 Badaro R, Carvalho EM, Rocha H, Wueiroz AC, Jones TC. Leishmania donovani: an opportunistic microbe associated with progressive disease in three immunocompromised patients. Lancet 1986;1:647–650.

189 Senalid G, Cadeo G, Carnevale G, et al. Visceral leishmaniasis as an opportunistic infection (Letter). Lancet 1986;1:2094.

190 Greenblatt CL. Cutaneous leishmaniasis: the prospects for a killed vaccine. Parasitol Today 1988;4:53–54.

191 Hazrati SM. Evaluation of leishmanization in the south of Iran. In Kager PA, Polderman AM, Goudsmit J, Laarman JJ, Vogel LC, eds. XIIth Int Congress Tropical Medicine and Malaria, Amsterdam, The Netherlands, Sept. 18–23, 1988, Vol. MoS-3-6.

192 Katzenellenbogen I. Vaccination against oriental sore: report of results of 555 inoculations. Arch Dermatol Syphild 1944;50:239–242.

193 Iarmukhamedov MA. On the question of counter-indications to vaccination to cutaneous leishmaniasis. Med Parasitol (Moskow) 1971;50:549–552 (in Russian).

194 Salazar J. Aspectos immunologicos de la leishmaniasis tegumentaria Americana. Ensayos profilacticos mediante vacunaciones con formas leptomonas vivas de cultivas de Leishmania braziliensis Vianna 1911. Arch Venez Med Trop Parasitol Med 1965;5:365–384.

195 Manson-Bahr PEC. Active immunization in leishmaniasis. In Garnham PCC, Pierce AE, Roitt I, eds. Immunity to Protozoa. Oxford: Blackwell Scientific Publications, 1963:246–252.

196 Mayrink W, Da Costa CA, Magalhaes PA, et al. A field trial of a vaccine against American dermal leishmaniasis. Trans R Soc Trop Med Hyg 1979;73:385–387.

197 Antunes CMF, Mayrink W, Magalhaes PA, et al. Controlled field trials of a vaccine against New World cutaneous leishmaniasis. Int J Epidemiol 1986;15:572–580.

198 Mayrink W, Antunes CMF, Da Costa CA, et al. Further trials of a vaccine against American cutaneous leishmaniasis. Trans R Soc Trop Med Hyg 1986;80:1001.

199 Mauel J, Behin R. In Peters W, Killick-Kendrick R, eds. The Leishmaniases in Biology and Medicine. New York: Academic Press, 1987;731:791.

200 Behin R, Mauel J, Sordat B. Leishmania tropica: pathogenicity and in vitro macrophage function in strains of inbred mice. Exp Parasitol 1979;48:81–91.

201 Handman E, Ceredig R, Mitchell GF. Murine cutaneous leishmaniasis: disease patterns in intact and nude mice of various genotypes and examination of some differences between normal and infected macrophages. Aust J Exp Biol Med Sci 1979;57:9–30.

202 Howard JG, Hale C, Chan-Liew WL. Immunological regulation of experimental cutaneous leishmaniasis. 1. Immunogenetic aspects of susceptibility to Leishmania tropica in mice. Parasite Immunol 1980;2:303–314.

203 Liew FY. Cell-mediated immunity in experimental cutaneous leishmaniasis. Parasitol Today 1986;2:264–270.

204 Anderson S, David JR, McMahon-Pratt D. In vivo protection against Leishmania mexicana-mediated by monoclonal antibodies. J Immunol 1983;131:1615–1618.

205 Champsi J, McMahon-Pratt D. Membrane glycoprotein M2

protects against *Leishmania amazonensis* infection. Infect Immun 1988;56:3272–3279.

206 Handman E, Goding JW. The *Leishmania* receptor for macrophages is a lipid-containing glycoconjugate. EMBO J 1985;4:329–336.

207 Handman E, Mitchell GF. Immunization with *Leishmania* receptor for macrophages protects mice against cutaneous leishmaniasis. Proc Natl Acad Sci USA 1985;82:5910–5914.

208 Reiner NE, Ng W, McMaster WR. Parasite–accessory cell interactions in murine leishmaniasis. II. *Leishmania donovani* suppresses macrophages expression of class I and class II major histocompatibility complex gene products. J Immunol 1987;138:1926–1932.

209 Nacy CA, Fortier AH, Pappas MG, Henry RR. Susceptibility of inbred mice to *Leishmania tropica* infection: correlation of susceptibility with *in vitro* defective macrophages microbicidal activities. Cell Immunol 1983;77:298–307.

210 Scott PH, Pearce E, Cheever AW, Coffman RL, Sher A. Role of cytokines and CD4+ T-cell subsets in the regulation of parasite immunity and disease. Immunol Rev 1989;112:161–182.

211 Mitchell GF, Curtis JM, Handman E, McKenzie IFC. Cutaneous leishmaniasis in mice: disease pattern in reconstituted nude mice of several genotypes infected with *Leishmania tropica*. Aust J Exp Biol Med Sci 1980;58:521–532.

212 Liew FY, Hale C, Howard JC. Immunologic regulation of experimental cutaneous leishmaniasis. V. Characterisation of effector and specific suppressor T cells. J Immunol 1982;128:1917–1922.

213 Titus RG, Milon G, Marchal G, Vassalli P, Cerottini JC, Louis JA. Involvement of specific Lyt-2+ T cells in the immunological control of experimentally induced murine cutaneous leishmaniasis. Eur J Immunol 1987;17:1429–1433.

214 Titus RG, Kelso A, Louis JA. Intracellular destruction of *Leishmania tropica* by macrophages activated with macrophage activating factor interferon. Clin Exp Immunol 1984;55:157–165.

215 Ralph P, Nacy CA, Meltzer MS, Williams N, Nakoinz I, Leonard EJ. Colony-stimulating factors and regulation of macrophage tumoricidal and microbicidal activities. Cell Immunol 1983;76:10–21.

216 Sadick MD, Locksley RM, Tubbs C, Raff HV. Murine cutaneous leishmaniasis. Resistance correlates with the capacity to generate IFN-γ in response to Leishmania antigens *in vitro*. J Immunol 1986;136:655–661.

217 Locksley RM, Heinzel FP, Sadick MD, Holaday BJ, Gardner KD. Murine cutaneous leishmaniasis: susceptibility correlates with differential expansion of helper T cell subsets. Ann Inst Pasteur/Immunol 1987;138:744–749.

218 Heinzel FP, Sadick MD, Holaday BJ, Coffman RL, Locksley RM. Reciprocal expression of Interferon-γ or Interleukin-4 during the resolution or progression of murine leishmaniasis. J Exp Med 1989;169:59–72.

219 Miller I, Pedrazzini TH, Farrell JP, Louis JA. T cell responses and immunity to experimental infection with *Leishmania major*. Ann Rev Immunol 1989;7:561–578.

220 Belesovic M, Finbloom DS, Van der Meide P, Slayter MZ, Nacy CA. Administration of monoclonal anti-IFN-γ antibodies *in vivo* abrogates natural resistance of C3H/HeN mice to infection with *Leishmania major*. J Immunol 1989;143:266–274.

221 Mosmann TR, Cherwinski H, Bond MW, Giedlin MA, Coffman RL. Two types of murine helper T cell clone. I. Definition according to profiles of lymphokine activities and secreted proteins. J Immunol 1986;136:2348–2357.

222 Scott P, Natovitz P, Coffman RL, Pearce E, Sher A. Immunoregulation of cutaneous leishmaniasis. T cell lines that transfer protective immunity or exacerbation belong to different T helper subsets and respond to distinct parasite antigens. J Exp Med 1988;168:1675–1684.

223 Miller I, Pedrazzini Th, Louis JA. Experimentally induced cutaneous leishmaniasis: are L3T4+ T cells that promote parasite growth distinct from those mediating resistance? Immunol Lett 1988;19:251–260.

224 Miller I, Louis JA. Immunity to experimental infection with *Leishmania major*: generation of protective L3T4+ T cell clones recognizing antigen(s) associated with live parasites. Eur J Immunol 1989;19:865–871.

225 Titus RG, Sherry B, Cerami A. Tumor necrosis factor plays a protective role in experimental murine cutaneous leishmaniasis. J Exp Med 1989;170:2097–2104.

226 Liew FY, Parkinson C, Millott S, Severn A, Carrier M. Tumor necrosis factor (TNFa) in leishmaniasis. I. TNFa mediates host protection against cutaneous leishmaniasis. Immunology 1990;69:570–573.

227 Bogdan C, Moll H, Solbach W, Rollinghoff M. Tumor necrosis factor and in combination with interferon γ, but not with Interleukin 4 activates murine macrophages for elimination of *Leishmania major* amastigotes. Eur J Immunol 1990;20:1131–1135.

228 Miller I, Garcia-Sanz JA, Titus R, Behin R, Louis JA. Analysis of the cellular parameters of the immune responses contributing to resistance and susceptibility of mice to infection with the intracellular parasite, *Leishmania major*. Immunol Rev 1989;112:95–113.

229 Skidmore BJ, Stamnes SA, Towsend K, *et al*. Enumeration of cytokines secreting cells at the single-cell level. Eur J Immunol 1989;19:1591–1597.

230 Farrell J, Miller I, Louis JA. A role for Lyt-2+ T cells in resistance to cutaneous leishmaniasis in immunized mice. J Immunol 1989;142:2052–2056.

231 Titus RG, Ceredig R, Cerottini JC, Louis JA. Therapeutic effect of anti-L3T4 monoclonal antibody GK1.5 on cutaneous leishmaniasis in genetically susceptible BALB/c mice. J Immunol 1985;135:2108–2114.

232 Sadick MD, Heinzel FP, Shigekane M, Fisher WL, Locksley RH. Cellular and humoral immunity to *Leishmania major* in genetically susceptible mice after *in vivo* depletion of L3T4+ T cells. J Immunol 1987;139:1303–1309.

233 Hill JO, Auwad M, North RJ. Elimination of CD4+ suppressor T cells from susceptible BALB/c mice releases CD8+ T lymphocytes to mediate protective immunity against *Leishmania*. J Exp Med 1989;169:1819–1827.

234 Mitchell GF. Host-protective immunity and its suppression in a parasitic disease: murine cutaneous leishmaniasis. Immunol Today 1984;5:224–226.

235 Titus RG, Lima GC, Engers HD, Louis JA. Exacerbation

of murine cutaneous leishmaniasis by adoptive transfer of parasite-specific helper T cell populations capable of mediating *Leishmania major*-specific delayed type hypersensitivity. J Immunol 1984;133:1594–1600.

236 Liew FY, Singleton A, Cilari E, Howard JG. Prophylactic immunization against experimental leishmaniasis. V. Mechanism of the anti-protective blocking effect induced by subcutaneous immunization against *Leishmania major*. J Immunol 1985;135:2101–2107.

237 Milon G, Titus RG, Cerottini J, Marchal G, Louis JA. Higher frequency of *Leishmania major*-specific L3T4⁺T cells in susceptible BALB/c mice than in resistant CBA-mice. J Immunol 1986;136:1467–1471.

238 Sadick MD, Heinzel FPO, Shigekane VN, Fisher WL, Locksley RM. Cellular and humoral immunity to *Leishmania major* in genetically susceptible mice following *in vivo* depletion of T3T4⁺ T cells. J Immunol 1987;139:1303–1309.

239 Solbach W, Forberg K, Kammerer E, Bogdan C, Rollinghoff M. Suppressive effect of cyclosporin A on the development of *Leishmania tropica* induced lesions in genetically susceptible BALB/c mice. J Immunol 1986;137:702–707.

240 Sadick MD, Heinzel FP, Holaday BJ, Pu RT, Dawkins RS, Locksley RM. Cure of murine leishmaniasis with anti-Interleukin 4 monoclonal antibody. Evidence for a T cell-dependent, Interferon-γ independent mechanism. J Exp Med 1990;171:115–127.

241 Liew FW. Functional heterogeneity of CD4⁺ T cells in leishmaniasis. Immunol Today 1989;10:40–45.

242 Belesovic M, Davis CE, Meltzer M, Nacy CA. Regulation of activated macrophage antimicrobial activities. Identification of lymphokines that cooperate with IFN-γ for induction of resistance to reinfection. J Immunol 1988;141:890–896.

243 Wirth JJ, Kierszenbaum F, Zlotnik A. Effects of IL-4 on macrophage functions increased uptake and killing of a protozoan parasite (*Trypanosoma cruzi*). Immunology 1989;66:296–301.

244 Carter KG, Gallagher G, Baillie AJ, Alexander J. The induction of protective immunity to *Leishmania major* in the BALB/c mouse by Interleukin 4 treatment. Eur J Immunol 1989;19:779–782.

245 Feng ZG, Louis JA, Kindler V, *et al.* Aggravation of experimental cutaneous leishmaniasis in mice by administration of interleukin 3. Eur J Immunol 1988;18:1245–1251.

246 Solbach W, Greil J, Rollinghoff M. Anti-infectious responses in *Leishmania major*-infected BALB/c mice injected with recombinant granulocyte-macrophage colony-stimulating factor. Ann Inst Pasteur/Immunol 1987;138:759–762.

247 Liew FY, Millott S, Li Y, Lechluk Chan WL, Ziltener H. Macrophage activation by interferon-γ from host-protective T cells is inhibited by Interleukin (IL)3 and IL-4 produced by disease-promoting T cells in leishmaniasis. Eur J Immunol 1989;19:1227–1232.

248 Mazingue C, Cottrez-Detoeuf F, Louis JA, Kweider M, Auriault C, Capron A. *In vitro* and *in vivo* effects of Interleukin 2 on the protozoan parasite *Leishmania*. Eur J Immunol 1989;19:487–491.

249 Louis JA, Mendonca S, Titus RG, *et al.* The role of specific T cell subpopulations in murine cutaneous leishmaniasis. In Cinader B, Miller RG, eds. *Progress in Immunology*, vol.VI.

New York: Academic Press,1990:762–769.

250 Campos-Neto A, Bunn-Moreno M. Polyclonal activation in hamsters infected with parasites of the genus *Leishmania*. Infect Immun 1982;43:1033–1040.

251 Farrell JP. *Leishmania donovani*: acquired resistance against visceral leishmaniasis in the golden hamster. Exp Parasitol 1976;40:89–94.

252 Gifawesen C, Farrell JP. Comparison of T cell responses in self-limiting versus progressive visceral *Leishmania donovani* infections in golden hamsters. Infect Immun 1989;57:3091–3096.

253 Scott P, Farrell JP. Experimental cutaneous leishmaniasis. I. Non-specific immunodepression in BALB/c mice infected with *Leishmania tropica*. J Immunol 1981;127:2395–2400.

254 Murray HW, Mosur H, Keithly JS. Cell-mediated immune response in experimental visceral leishmaniasis. I. Correlation between resistance to *L. donovani* and lymphokine generating capacity. J Immunol 1982;129:344–350.

255 Nickol AD, Bonventre PF. Immunosuppression associated with visceral leishmaniasis of hamsters. Parasite Immunol 1985;7:439–449.

256 Scott P, Pearce E, Natovitz P, Sher A. Vaccination against cutaneous leishmaniasis in a murine model. II. Immunologic properties of protective and non-protective subfractions of a soluble promastigote extract. J Immunol 1987;139:3118–3125.

257 Scott P, Casfar P, Sher A. Protection against *Leishmania major* in BALB/c mice by adoptive transfer of a T cell clone recognizing a low molecular weight antigen released by promastigotes. J Immnol 1989;144:1075–1079.

258 Mitchell GF, Handman E. Heterologous protection in murine cutaneous leishmaniasis. Immunol Cell Biol 1987;65:387–392.

259 Marchand M, Daoud S, Titus RG, Louis JA, Boon Th. Variants with reduced virulence derived from *Leishmania major* after mutagen treatment. Parasite Immunol 1987;9:81–92.

260 McGurn M, Boon Th, Louis JA, Titus RG. *Leishmania major*: nature of immunity induced by immunization with a mutagenized avirulent clone of the parasite in mice. Exp Parasitol 1990;71:81–89.

261 Handman E, Mitchell GF. Immunization with *Leishmania* receptor for macrophages protects mice against cutaneous leishmaniasis. Proc Natl Acad Sci USA 1985;82:5910–5914.

262 Murray PJ, Spithill TW, Handman E. Characterization of integral membrane proteins of *Leishmania major* by Triton X-114 fractionation and analysis of vaccination effects in mice. Infect Immun 1989;57:2203–2209.

263 Russell DG, Alexander J. Effective immunization against cutaneous leishmaniasis with defined membrane antigens reconstitutes into liposomes. J Immunol 1988;140:1274–1279.

264 Handman E, Button LL, McMaster RW. *Leishmania major*: production of recombinant gp63, its antigenicity and immunogenicity in mice. Exp Parasitol 1990;70:427–435.

265 Jardim A, Alexander J, Siafeh H, On D, Olafson W. Immunoprotective *Leishmania major* synthetic T cell epitopes. J Exp Med 1990;172:645–648.

266 Jaffe Ch, Rachamin N, Sarfstein R. Characterization of

two proteins from *Leishmania donovani* and their use for vaccination against visceral leishmaniasis. J Immunol 1990; 144:699–706.

267 Locksley RM, Sadick MD, Holaday BJ, *et al*. CD4$^+$ T cell subsets in murine leishmaniasis. In Keith PWJ, McAdam KPWJ, eds. *New Strategies in Parasitology*. Edinburgh: Churchill Livingstone, 1989:147–157.

13 *Toxoplasma gondii* and toxoplasmosis

Lloyd H. Kasper & John C. Boothroyd

In recent years, the obligate intracellular parasite, *Toxoplasma gondii*, has emerged as an important opportunistic pathogen in humans and animals. In humans, this ubiquitous parasite is the etiologic agent for toxoplasmosis, an infection that is frequently encountered among two groups of high-risk individuals. In human fetuses, *Toxoplasma* infection is associated with severe congenital defects when the primary infection is acquired during the first trimester of pregnancy. In the USA alone there are several thousand infants born each year with congenital toxoplasmosis: the yearly health costs for these individuals with neurologic sequelae of congenital infection has been estimated to be several hundred millions of dollars [1]. The other group of high-risk individuals are those immunosuppressed persons that develop fatal *Toxoplasma* meningoencephalitis. By far the most frequent group among these individuals are those with acquired immunodeficiency syndrome (AIDS). The range of incidence for infection in those with AIDS is between 5 and 33% [2,3]. Thus, with the ever-increasing number of AIDS patients in our society, it is important to develop an understanding of the immunology and molecular biology of this organism.

THE PARASITE

In order to understand better the immune aspects of this parasite, a general review of the parasite life-cycle is necessary. *Toxoplasma gondii*, an intracellular coccidian, infects both birds and mammals [4,5]. There are two distinct aspects of the life-cycle of *T. gondii*: a nonfeline and a feline cycle. In 1908, Nicole & Manceau first described the asexual tachyzoite form of this parasite. In this extraintestinal cycle within intermediate hosts, including humans, mice, sheep, and cattle, *T. gondii* enters the host cell and begins to divide by endodyogeny, a process akin to binary fision within a

parasitophorous vacuole. Division time is usually 6–8 h for the virulent RH strain tachyzoite. When the number of parasites within the cell approaches 64–128 (in tissue culture), the cell ruptures, releasing tachyzoites that infect adjoining cells. In this manner, an infected culture or organ soon shows evidence of a cytopathic process. Most of the tachyzoites are eliminated by humoral and cell-mediated immune responses. Seven to ten days after the systemic tachyzoite infection, tissue cysts that contain bradyzoites develop. These tissue cysts occur in a variety of host organs but principally within the central nervous system and muscle, where they may exist for the lifetime of the host. When cysts are ingested (i.e., humans eat undercooked meat products), the cyst membrane undergoes rapid digestion in the presence of acidic pH gastric secretions.

In nonfeline hosts, the ingested bradyzoites enter the small intestine epithelium and transform into rapidly dividing tachyzoites. This acute systemic tachyzoite infection is followed by the formation of chronic bradyzoite cysts, usually detectable 10–14 h following acute infection. Once cysts containing bradyzoites are formed, the nonfeline cycle is completed.

The sexual life-cycle of the parasite is defined by the formation of oocyst/sporozoite within the feline host. In 1970, four independent groups of investigators described the enteroepithelial cycle in the cat [6–10]. This life-cycle begins with the ingestion of the bradyzoite tissue cysts and culminates after several intermediate stages in the production of micro- and macrogametes. The microgamete is flagellated, allowing the parasite to seek a macrogamete. Gamete fusion produces a zygote which envelops itself in a rigid cyst membrane and is secreted as an unsporulated oocyst. During acute infection, a cat may excrete as many as 100 million parasites per day. After 2–3 days of exposure to air at ambient temperature, the noninfectious oocyst sporulates to produce eight sporozoite progeny. These very stable oocysts are now highly infectious and may remain viable for many years in the soil. They can be ingested by an intermediate host, such as a pregnant woman emptying a litter box or a pig rummaging in the barn yard, or perhaps a mouse. Once digested, the released sporozoites infect the intestinal epithelium, producing rapidly growing asexual tachyzoites and ultimately bradyzoites. This completes the feline entero-epithelial cycle of *T. gondii*.

Tachyzoite

The asexual tachyzoite is the most frequently studied stage of this parasite, primarily because of the ease with which it can be grown in the laboratory. The tachyzoite has no strong predilection for specific cell type or organ. These parasites are capable of infecting a wide range of phagocytic and nonphagocytic cells by an energy-requiring process [11]. Even erythrocytes can be targets for invasion, although the invasion process is rarely completed and no parasite growth occurs. Macrophages and other monocyte lines, however, are fully susceptible to productive infection by the parasite.

An unusual spiraling locomotion, the probing movement of the anterior pole (conoid), and the secretion of factors from the rhoptries allow the parasite to penetrate the host cell. The conoid is composed of microtubules that extend in a counter-clockwise spiral [12,13]. It is postulated that as the conoid moves, the polar ring moves along the spiral pathway of the subunit. When the conoid is retracted the polar ring rotates, giving rise to the characteristic torsion movement of the parasite. Motility is dependent on the pH gradient determined by extracellular ions. Internal pH greater than external pH induces motility and involves myosin [14], actin [15], and microtubules [16,17].

A factor which can be extracted from either intact parasite or the culture medium in which the parasite grows enhances the entry of tachyzoites into mammalian host cells [18,19]. This factor has been termed penetration enhancing factor (PEF). Although other enzymes enhance penetration, PEF is more active and has been considered to be a specific protein, probably secreted from the anterior organelles during the process of cell invasion. Although, in high concentrations, PEF can damage host cell membranes, its activity can be observed at low concentrations that fail to induce morphologic changes. Recently Schwartzman [19] has produced a library of monoclonal antibodies (MAB) that inhibit the activity of PEF as determined by reduction

in parasite plaque formation using human fibroblasts. Penetration enhancing factor can give a 10-fold increase in the efficiency of invasion in culture. Monoclonal antibody reactive to PEF can substantially block this enhancing activity: a sixfold enhancement of invasion with PEF extract can be specifically reduced to about twofold when the PEF extract is preincubated with MAB TG49, which apparently stains the rhoptries of fixed cells and recognizes a protein designated rhoptery-1 (ROP-1). Schwartzman has determined that TG49 reacts with the 60 kD protein in the presence of protease inhibitors. Trypsin digestion destroyed the antigenic reactivity of this molecule, whereas treatment with periodic acid and failure to react with Schiff's reagent supported the absence of significant carbohydrate.

The gene encoding the protein described by Schwartzman has recently been cloned and sequenced. The product shows no significant homology to any known protein except in being similar in motif to the proline-rich repeated proteins described by Bennick [20]. It has two regions of extreme charge, one being highly acidic (25 acidic and two basic residues in a region of about 100 residues), the other highly basic (25 basic and two acidic residues in a region of 100 residues). The role of these two regions in PEF function is not known. This and further studies await analyses of the recombinant protein in PEF assays.

Kimata & Tanabe [21] have also identified a secreted protein of molecular weight 66 kD which localizes to the anterior portion of the parasite and thus is possibly located in the rhoptries. Following invasion, this protein is apparently associated with the parasitophorous vacuole membrane. It is not yet known whether this protein is related to the other proteins of similar size and localization described above.

Similar observations were made by Sadak *et al*. [22]. They produced a MAB that reacted by immuno-fluorescence with rod-like organelles located in the anterior aspect of the parasite. By immunoblot, two proteins of 55 and 60 kD were identified. This protein was recognized on the three major life-cycle stages (bradyzoite, tachyzoite, and sporozoite) of the parasite. For tachyzoites, it appeared that the protein was derived from a 66–68 kD doublet which was processed 30 min after biosynthesis. This protein is apparently distinct from the ROP-1 of Schwartzman (J.F. Dubremetz & J.D. Schwartzman, personal communication).

Invasion begins with the attachment of the parasite to the host cell through unknown receptors and ligands. Recently, Joiner *et al*. [23] have implicated laminin as a possible host cell receptor for the parasite.

Recent work by Grimwood & Smith [24] as well as in one of our laboratories (LHK) suggests that P30 (a major surface protein of the parasite described below) may be an important attachment ligand for the parasite. Monoclonal and polyclonal, monospecific antibodies to this protein inhibit infection of human fibroblasts and murine enterocytes. Fabs prepared from polyclonal, monospecific antibody to P30 also have this inhibitory effect on invasion. Heat inactivated antisera derived from mice infected with either the RH or PTg strain (P30$^+$) of *T. gondii* significantly inhibit infection of fibroblasts by autologous wild type parasites whereas these antisera have little inhibitory effect against a P30-deficient mutant (PTgB). Antisera raised to the P30$^-$ mutant had no significant effect on infection by the wild strains (J.R. Mineo, D. Mack, R. McLeod, and L.H. Kasper, unpublished results). Further evaluation of the parasite attachment ligand indicates that it can be blocked by neoglycoproteins [25]. In our studies, neoglycoproteins (BSA-glucosamide) blocked P30$^+$ but not P30$^-$ tachyzoite infection of human fibroblasts. These data indicate that P30 has an important functional role in *T. gondii* infection and is perhaps a parasite ligand responsible for attachment to host cells.

Orientation of the parasite is such that the contents of the anterior organelles are displaced into the space between the parasite and host plasma membrane [13,26]. A tight junction then forms at the ring-like apposition of parasite and host plasma membrane. Exogenously added materials (e.g., antibodies bound to the surface) are thus excluded from the evolving vacuole. The parasite enters the host cell with the generation of a membrane-limited parasitophorous vacuole. This process appears to be active in that killed organisms are not taken up efficiently.

Upon entry into the host cell environment, *T. gondii* blocks acidification of the phagosome. Using cultured macrophages, Sibley *et al.* [27] showed that vacuoles containing *T. gondii* fail to acidify in normal (non-activated) macrophages. In contrast, antibody-opsonized parasites induced rapid phagosome acidification when entering normal macrophages, culminating in lysosome fusion. Since extracellular parasites are highly sensitive to pH, the ability to block acidification of the phagosome may be important for survival of this obligate intracellular pathogen. This block in phagosome acidification may be independent of other defensive maneuvers by the parasite, such as the lack of fusion with lysosomes.

Modification of the phagosome begins shortly after formation, with secretion of membranous vesicles that form a reticular network. The modified compartment is able to resist normal endocytic processing and digestion. There are at least six parasite-derived proteins within this network, including the surface antigens P43, P35, P30, and P22 and an internal 28 kD protein that is excreted from the posterior pole of the parasite shortly after invasion [28]. This work indicates that the membranous network is a unique structural modification of the host cell phagosome and includes most of the major *Toxoplasma* surface proteins.

One of the proteins formed within the parasitophorous vacuole is a 23 kD calcium-binding protein [27,29]. This is one of several excreted–secreted antigens (ESA) [30] that may have potential for protection in immunization. The gene for this 23 kD protein has been cloned and sequenced and has all of the properties predicted for a calcium-binding protein [29].

Joiner *et al.* [31] observed phagocytosis of tachyzoites by Chinese hamster ovary cells possessing the Fc receptor. In their studies, live, but not dead, tachyzoites manifest a generalized phagosome–lysosome fusion block. When tachyzoites are presensitized with antibodies directed against major surface proteins P30 or P22 prior to infection, attachment and entry of the parasite was increased by 2.5-fold. In a 24-h period, the percentage of infected cells and number of tachyzoites per 100 cells decreased by 68% when compared to tachyzoites that had not been antibody sensitized. These experiments suggested that tachyzoites normally form a fusion-incompetent parasitophorous vacuole. When a signal is provided for phagocytosis, the phagosome–lysosome fusion block is overcome and parasite destruction occurs.

Bradyzoite

The tissue cysts containing bradyzoites are formed within the host cell cytoplasm and may contain several thousand organisms. They may be present as early as 6–8 days following infection in mice and persist as viable parasites throughout the life of the host.

Much less is known about the bradyzoite stage of this parasite. During the feline enteroepithelial cycle, it is the bradyzoite that undergoes sexual differentiation and transforms into infectious sporozoites [32,33]. The parasites in this stage contain carbohydrate intracytoplasmic vacuoles that may be a source of energy during the latent phase of infection [34]. Bradyzoites are more resistant to pepsin, allowing them to escape the normal host digestive process. The contribution of the neuronal cytoskeletal to the structural wall of the *in vivo* cyst has been described recently [35]. Immunologically, parasite antigen can be detected in the cyst wall [36]. Although tachyzoites do share a number of common antigens with bradyzoites, each stage expresses unique antigens as well [37]. The bradyzoite is the presumed source of recrudescent infection in the immunosuppressed. Beyond these observations, our knowledge of the bradyzoite is limited. Little is known about the biochemical composition and metabolism of this stage. Transformation from the tachyzoite to the bradyzoite stage is not understood, although interferon-γ (IFN-γ) [38] as well as anti-*Toxoplasma* sera [39] have been reported to induce cyst-like structures *in vitro*. Gene products that may regulate transformation or selection of the bradyzoite have not yet been identified. Cyst rupture is believed to be the major causative factor for recrudesence of toxoplasmosis in patients with AIDS. In recent years, increasing attention has been directed towards the bradyzoite because of its infectious potential. The development of pseudocysts in culture by avirulent strains of parasites is well recognized. Jones *et al.* [38,40] have shown that long-term cultivation of *Toxoplasma* cysts can be accomplished *in vitro* using murine astrocytes and IFN-γ in the medium. In his studies, the "pseudocysts" appeared to be composed of a single trilaminar membrane containing amorphous electron-dense material. The observed bradyzoites were smaller and contained numerous electron-dense vacuoles. Although IFN-γ was not necessary for cyst formation, its presence appeared to control the division of tachyzoites and allowed for the development of cysts over a prolonged period without rupturing. These findings were similar to those of Shimada *et al.* [41], who demonstrated the development of *in vitro* cysts following treatment of infected cultures with anti-*Toxoplasma* sera and complement. They too observed the development of pseudocysts during a prolonged *in vitro* cultivation of infected host cells. The type of *in vivo* host cells responsible for supporting cyst development has never been identified. The wall of intact tissue cysts in the brains of mice with congenital toxoplasmosis was investigated by Sims *et al.* [35]. Their studies indicate that this wall may be composed, at least in part, by components derived from the neuronal cytoskeleton. Their suggestion is that this neuronal-derived membrane may allow protection of the *Toxoplasma* cyst from the host immune response. Frenkel has shown that intact cysts in sequential stages of disintegration can

be identified immunohistologically within microglial nodules. This supported the hypothesis that glial nodules may represent the original specific site of *Toxoplasma* cysts [42].

HOST–PARASITE RELATIONSHIP

Because of its obligate intracellular dependence, the interaction between the parasite and host cell is paramount for survival. *Toxoplasma* multiplies intracellularly in host cells, following which it destroys that cell and invades adjacent cells spreading throughout the body via lymphatics and the blood stream. With the induction of immunity the parasites are reduced in number, making histologic identification of the tachyzoite unusual. At this stage, only the bradyzoite-containing tissue cyst can be observed. The predilection of the bradyzoite for the central nervous system (CNS) has been attributed to a barrier of passive diffusion of antibody. Thus, the infection persists within the CNS whereas it is cleared from extraneural tissues. This explanation, however, is not complete in that the parasites persist in a number of organs, including skeletal muscle and heart, where antibody is not blocked. Furthermore, elevated antibody titers of the IgG and IgA isotype can be detected readily in the cerebrospinal fluid (CSF) of both humans and animals. Other factors, including a predilection of the parasite for specific host cell receptors within the CNS, should also be considered.

A mouse model for toxoplasmic encephalitis has demonstrated that normal mice are able to survive intracerebral infusion with *T. gondii* tachyzoites whereas immunosuppressed mice died from progressive disease. Cortisone treatment resulted in reduced inflammation but the presence of many tachyzoites. Mice treated with cyclophosphamide showed attenuated inflammation and the presence of cysts [43]. Persistence of low-grade infection in laboratory animals and humans has been demonstrated during chronic infection, even with high levels of neutralizing antibody present.

Cyst rupture has been attributed as a possible mechanism of toxoplasmic encephalitis [42]. The presence of small satellite cysts within the area of large cysts of chronically infected mice also supports this observation. By indirect fluorescent antibody analysis, the presence of antigen leakage from cysts has been observed. Antiperixodase stains have also been used to demonstrate the presence of toxoplasma organisms in brain biopsies of human toxoplasmosis. Further, *Toxoplasma* antigen can be detected in the form of tachyzoites and tissue cysts as well as within the particulate extracellular and amorphosis intracellular material of brain tissue sections from acutely infected mice.

Parasites also can be detected occasionally in the CSF from AIDS patients with toxoplasmic encephalitis (J.S. Remington, personal communication). The generality of this phenomenon and its potential exploitation for diagnosis are currently being actively explored [44–46].

HUMORAL RESISTANCE

Antigens and antibodies

Over the past 5–10 years, there have been a number of reports using a variety of immunologic and molecular biologic tools to separate and identify the variety of *Toxoplasma* antigens that stimulate the host immune response. Most work to date has concentrated on the tachyzoite stage of *T. gondii*. This is due primarily to the ease with which the asexual stage can be grown both in tissue culture and the ascites of infected mice. For the most part, these antigen preparations have been crude extracts, containing both parasites and host cell debris.

More recently surface radioiodination has been utilized to define the antigenic character of the parasite. At least five major tachyzoite membrane antigens can be identified by this technique, ranging in molecular weight from 6 to 95 kD [47–54]. The predicted molecular weights of these antigens shows variability among the different laboratories. This may be due in part to different methods of iodination, parasite strain variability, or perhaps technical differences in sodium dodecyl sulfate polyacrylamide gel electrophoresis (SDS-PAGE) procedures. By immunoprecipitation, a number of radioiodinated proteins were detected by infected mouse sera and the sera from acutely infected humans. Although the predicted sizes of these antigens as reported by different groups vary, there is no reason to believe that those differences are real. For simplicity we will use the nomenclature of Kasper *et al.* [53,55] based on predicted molecular weights of 45, 30, 22, and 14 kD, designated P45 through P14.

The advent of MAB has been of paramount importance in defining the antigenic composition of *T. gondii*. Because of the parasite's obligate nature, host cell contamination from tissue culture or the peritoneal cavity is potentially problematic. Monoclonal antibodies have circumvented this problem. A number of laboratories have now produced panels of MAB that are reactive toward the surface of the tachyzoite stage. In general, immunoprecipitation with MAB followed by SDS-PAGE analysis has provided a limited but important assessment of the critical antigens of *T. gondii* that elicit humoral immunity. Most studies to date confirm the presence of MAB-reactive antigens that were recognized by immune sera.

A more thorough analysis of reactive antigens has been achieved over the past several years using Western blot. Sera from humans that have been naturally infected with *Toxoplasma* identify a broad range of antigens by Western blot. These antigens range in molecular weight from 22 to 67 kD and were recognized by the IgG fraction from acutely or chronically infected patients and may differ somewhat from antigens recognized by the IgM fraction. These differences between the IgG and IgM fractions in acutely and chronically infected patients have been noted by others [56]. It was found that the IgM was directed primarily against a 6 kD antigen, whereas the IgG response identified this antigen as well as a variety of others of higher molecular weight. Interestingly, neonates in this group did not contain an IgM antibody directed against this 6 kD antigen.

A MAB raised against this 6 kD antigen reacted with a water-extracted parasite cytoplasmic component. Sharma's studies found that the sera of three individuals with *Toxoplasma*-specific IgM recognized this 6 kD antigen as well as antigens of approximate molecular weight of 22 and 32 kD. One of the three patient's sera also recognized a 45 kD antigen. Reactivity of the IgM fraction to the 45 and 22 kD antigens was completely abolished by periodate treatment, suggesting that the major epitopes may contain immunoreactive carbohydrates. Treatment of the preparations with trypsin and pronase eliminated reactivity to the 32 and 22 kD antigens. Lipase treatment did not have obvious effects on any of the antigens recognized by the *Toxoplasma* IgM-specific antibody. Ogata *et al.* [57] developed a panel of three MAB against the 43 kD antigen. These antibodies were found to react with different patterns on the tachyzoite, suggesting that this antigen was dissimilar to that reported by others.

In the past several years there have been an increasing number of reports on the role IgA antibody response in toxoplasmosis. McLeod & Mack [58] demonstrated the presence of a secretory IgA specific for *T. gondii* in the intestinal secretions of mice infected perorally with bradyzoites. Crossreactive immunoabsorption with tachyzoites reduced the bradyzoite activity. Specificity of this reaction was shown using cholera toxin, which produced high levels of intestinal IgA that failed to react with *T. gondii*. Other investigators have concentrated on determing the serum IgA concentrations by various immunodiagnostic assays. Most recent reports demonstrate that an IgA antibody response can be seen in individuals positive for *Toxoplasma* antibody. This response appears during the acute phase of infection and persists after the IgM titer has fallen. The importance of persistently elevated IgA titers against P30 was demonstrated by Decoster *et al.* [59], who found that IgA antibody was a sensitive marker of congenital and acute toxoplasmosis in humans. Ridel *et al.* [60] have suggested that there is a protective role of IgA in the immunocompromised rat infected with toxoplasmosis. The profile of those parasite antigens recognized by acute and chronic immune human sera suggests that most of these antigens are of similar molecular weight to those recognized by the IgG fraction (J.S. Remington, unpublished observation).

The IgA antibody response in an experimental model was further studied by Chardes *et al.* [61]. Following single oral infection, IgA production began during week 2 in serum and milk. IgG antibody was detected in the intestine after the rise in IgA titer was noted. Serum and mild IgG and IgM production began at the same time as IgA. By Western blot, the intestinal IgA response was directed at a number of different antigens, whereas the milk IgA reacted with P30 and P43. Most antigens recognized by IgA were also recognized by IgG.

Most work to date on the various *Toxoplasma* antigens recognized by Western blot has concentrated on either P30 or P22. These antigens are the principal radio-iodinated antigens recognized by human antisera as well as sera from immunized mice and rabbits. We utilized MAB to isolate and characterize these two principal iodinatable surface proteins [53,62]. Of the two, P30 is the major iodinated protein on a number of geographically diverse human and animal isolates of *Toxoplasma* [53,63]. It comprises 3–5% of the total parasite protein and is the major component of the vesicular network within the parasitophorous vacuole [27]. Monoclonal antibody raised against P30 is parasiticidal to extracellular *T. gondii* in the presence of serum complement. Similarly, antisera raised against purified antigen, P30, induced parasite lysis *in vitro* in the presence of complement.

Dubremetz *et al.* [26] has studied the distribution of MAB directed at P30 that occurs during host cell invasion. Monoclonal antibody against this antigen will bind the surface of isolated tachyzoites at all stages of their development. Living tachyzoites showed an equal distribution of anti-P30 on the surface membrane. Upon invasion of a host cell, however, most of the coat of antibody against P30 was shed from the tachyzoite [26]. Rodriguez *et al.* [64] raised a panel of four MAB that were directed against P30. By competition binding assay, it was shown that a single region of P30 was recognized by all four of the MAB. A single MAB directed against P30 inhibited 25–50% of the specific binding of antibodies of patients with toxoplasmosis to the antigenic extract of tachyzoites. Their conclusion was that P30 was the most immunogenic constituent of the tachyzoite and that a single region of this molecule contained most of the immunogenic activity.

The gene encoding P30 has been cloned and sequenced [65]. The predicted amino acid sequence reveals the expected hydrophobic N-terminal signal peptide, one potential N-linked glycosylation site (not yet known to be functional *in vivo*), no apparent internal repeats (contrary to the report of Santoro *et al.* [66], which may have been due to the ready dimerization of P30 in solution), and a C-terminal hydrophobic tail peptide.

The P30 gene is highly conserved between strains of *T. gondii*. Only 10 nucleotide differences are found in the coding region (just over 900 bp) when comparing RH and C strains, of which eight lead to amino acid substitutions [67].

The P30 sequence data suggest a glycolipid anchoring mechanism similar to that seen in many protozoa and higher enkaryotes. This has been confirmed by Nagel & Boothroyd [68], who have shown P30 to be anchored by a phosphatidylinositol-linked glycan which shows antigenic similarity to the crossreacting determinant of trypanosome variant surface glycoprotein (VSG) [68]. Tomavo *et al.* [69] have recently shown that this anchoring is apparently functioning for the other major surface antigens, in addition to P30.

There have been a variety of studies using lectins that indicate the absence of carbohydrates on the surface of tachyzoite membranes [50]. Most of these studies, however, have used the RH highly virulent strain of *Toxoplasma* which has been grown either in mouse peritoneum or tissue culture for many decades. In contrast to tachyzoites, cysts appear to bind to concanavalin A, wheatgerm, and soybean agglutinin [70]. This binding, however, may be related to the cyst wall rather than the parasite membrane. It has been shown that the cyst wall is *p*-aminosalicylic acid (PAS) positive, suggesting the presence of carbohydrates, and may be of host cell origin.

Circulating Antigens

A number of laboratories have reported the presence of circulating antigens during acute *Toxoplasma* infection. By enzyme-linked immunosorbent assay (ELISA) testing it was found that over 50% of the sera from patients with recently acquired toxoplasmosis contained circulating antigen. This antigenemia was not detected in the sera from seronegative individuals or from individuals with chronic infection. By using the Fab-2 fragment, the incidence of false-positives secondary to rheumatoid factor could be avoided. Monoclonal antibodies have also been used to detect antigenemia [56]. In these studies it was found that the IgG Fab-2 fragment of polyvalent sera is more sensitive than MAB in detecting antigenemia. Suzuki & Kobayashi [71] found that circulating *Toxoplasma* antigens were present in the sera 1–3 weeks postinfection. This suggested to these workers that the detection of circulating *Toxoplasma* antigens may be a reasonable approach to the diagnosis of acute toxoplasmosis. Similar findings in rabbits was reported by Ise *et al.* [41]. In their experiments they found that an ELISA that utilized a biotin–avidin assay was very beneficial as a diagnostic tool. Using this assay, antigen was detected as low as 4 ng/ml of sera. Within 3 days following infection, rabbits inoculated subcutaneously with tachzyoites exhibited antigenemia. When evaluated by liquid gel chromotography, these antigens appear to have molecular weights of >400, 220, 130, and 45 kD.

Another approach to detect circulating antigen was developed by Asai *et al.* [72,73]. These investigators observed that nucleoside triphosphate hydrolase was detected as a circulating antigen in the sera of mice infected with both the RH and the avirulent Beverly strain. This enzyme is one of the primary proteins that constitute the tachyzoite. It is their estimate that almost 8% of the total parasite protein in the tachyzoite was found to be nucleoside triphosphate hydrolase. Using a MAB they found that their assay for detection of this enzyme was sensitive to 0.3 ng/ml. Mice inoculated with RH-strain tachyzoites demonstrated circulating antigen within 24 h postinfection. The peak values were reached within several days in mice inoculated with either tachyzoites or cysts containing bradyzoites. Circulating antigen disappeared by the end of 2 months.

Complement

The role of complement activation in *T. gondii* was first described in the Sabin–Feldman dye test. These authors focused on accessory factors mediating the lysis of *T. gondii* tachyzoites in the presence of antibody. This factor has been identified as complement component C2, implicating that activation occurred by the classic pathway.

The mechanisms of resistance to complement-mediated killing have recently been addressed by Fuhrman & Joiner [74]. These authors examined serum resistance in the RH- and P-strain tachyzoites, which differ in virulence but are equally resistant to serum killing. In nonimmune human serum, activation of the alternative complement pathway occurred with deposition of C3. This complement component bound covalently to parasite receptor molecules by an ester linkage; iC3b was the primary component of C3. iC3b is unable to form a lytic C5B through C9 complex. When tachyzoites were presensitized with anti-P30 MAB, a new amide-linked C3 receptor complex formed. Threefold enhancement of radiolabeled C9 binding occurred when anti-P30-presensitized tachyzoites were compared to unsensitized organisms. Their data suggested that tachyzoites of *T. gondii* are serum resistant because of the inability to activate complement efficiently. Antibody presensitization altered the site of complement deposition and augmented the C5B-9 formation.

Strain variation

Toxoplasma strains isolated in nature exhibit different virulence when compared to laboratory strains. Attempts to detect distinctive antigens in relatively avirulent and highly virulent strains of parasite have, for the most part, been unsuccessful. It is generally accepted that virulence increases after repeated passage during routine laboratory maintenance. This change in virulence patterns is host dependent in that avirulent strains passed in mice show progressively increased pathogenicity whereas the same strain maintained in chick embryos fails to exhibit this change. By far the most commonly used virulent strain of *T. gondii* is RH. This strain, originally obtained from infected human tissue, has been maintained in the laboratory since 1941 by continuous passage in either mouse peritoneal cavities or cultured human fibroblasts. Over the years many investigators found the strain to be highly virulent in susceptible mice, frequently with a 90% lethal dose of less than 100 viable organisms. Because of its aggressive nature and ease of growth both *in vivo* and *in vitro*, it has been the predominantly studied strain of *T. gondii*. *In vivo*, the RH strain is unable to produce tissue cysts containing bradyzoites or oocysts/sporozoites in the cat, suggesting that it may not be able to undergo sexual reproduction in the cat intestine. Treatment with sulfa drugs can induce cyst formation with the RH strain.

Some reports have tried to define antigenic strain variation by immunologic means. Suggs *et al.* [75] compared four strains of *T. gondii* with two *Besnoitia* strains by using immunologic tests such as immunofluorescence and complement fixation. They concluded that there are strain variations possibly brought on by prolonged passage in mice. In an attempt to correlate virulence with strain variation, Handman & Remington [50] infected mice with either the virulent RH strain or the less virulent C37 and C56 strains, and characterized the antibodies that developed. They concluded that antisera to these strains recognize similar membrane antigens. Ware & Kasper [63] performed a detailed immunologic assessment of strain-specific antigens of *T. gondii*. Rabbit antisera against three strains of *T. gondii* obtained from divergent sources were generated. These strains included the RH strain, strain C obtained from a naturally infected kitten, and strain P which is maintained by passage in mice. Rabbit antisera were raised against these three strains and used to identify unique strain-specific and commonly shared tachyzoite antigens by radioiodination followed by immunoprecipitation. Both qualitative and quantitative differences of a number of the major tachyzoite antigens were found in these assays. These antigenic differences were confirmed by Western blot analysis. A parasite plaque reduction assay using parasiticidal MAB showed marked variation in the ability to kill these three tachyzoite subtypes, further supporting antigenic distinction among *T. gondii* strains.

Darde *et al.* [76] used isoelectric focusing to study strain variation in enzyme content. Fourteen enzymes had similar patterns in all strains tested. Four enzymes had patterns which had an approximate relationship between pathogenicity and oocyst production. These enzymes could be useful as markers in future genetic linkage studies.

Restriction fragment length polymorphisms (RFLP) have been used to characterize strains and particularly to determine if there is any correlation between genotype and virulence (usually as measured in mice), host or

geographic origin. In one study, Cristina *et al.* [77] used a DNA probe that detects about 15 fragments (i.e., the probe represents a middle-repetitive element) and found no obvious correlation between virulence and RFLP pattern in the six strains tested. Insufficient numbers of strains were examined to reach any conclusion on similarity based on host or geographic origin.

In a much larger study, L.D. Sibley & J.C. Boothroyd (submitted) have examined 26 strains using nine RFLP probes, most of which are single copy genes and about half of which encode known proteins. Their results can be summarized as follows.
1 There is relatively little polymorphism in the total gene pool of *T. gondii*, worldwide.
2 There is no apparent relatedness between strains from a given region or host.
3 There is a remarkable relatedness between strains of a similar virulence type such that all 10 virulent strains (originating from three continents) appear to be derived from a single clonal lineage. That is, they share the same genetic make-up except when a hyper-polymorphic locus is examined. This latter probe serves to distinguish or "fingerprint" the strains thus excluding lab contamination as an explanation for their near identity.

One example of the polymorphism seen is in the P30 gene where three polymorphic restriction sites are known. Interestingly, in the 26 strains, there are only two alleles identifiable by RFLP analysis (compared with the eight theoretically possible). Most importantly, all 10 virulent strains, regardless of their origin, apparently possess the same P30 allele which is distinct from that present in all 16 nonvirulent strains. It is not yet known if virulence as defined in mice correlates with any particular disease manifestation in humans but clearly the tools are now available to examine this.

Another approach for the isolation of variants of *T. gondii* has also been employed. Some parasitic protozoa, notably the African trypanosomes, exhibit marked antigenic variation. The mammalian immune response is probably effective in selecting antigenic variants of these trypanosomes because the parasite surface consists largely of a single glycoprotein. Although antigenic differences between *T. gondii* strains were demonstrated as previously described, there is no evidence for emergence of antigenic variance during the course of an infection. Such variance would probably have only a limited selective advantage because the parasite has multiple surface antigens as previously noted, and the polyclonal immune response would react nearly as well with the variant and the wild type. Monoclonal antibodies, however, are specific for a single antigenic determinant and are able to exert suitable selective pressure for the emergence of antigenic mutants of *T. gondii*. We used such parasiticidal MAB to isolate and characterize antibody-resistant mutants of *Toxoplasma* [62,78]. Two MAB were selected for these experiments. One, 7G1, identifies the surface protein P22 with an approximate molecular weight of 22 kD. The other, 2F2, is directed against P30. Both of these MAB exhibit parasiticidal activity in the presence of serum complement. For mutant selection, parasites were first treated with the mutagen ethylnitrosourea. Mutagenized parasites were then exposed to lytic MAB plus complement for a short incubation period. This treatment was repeated five to seven times until a resistant mutant that no longer was susceptible to the antibody plus complement emerged. These resistant mutants were then characterized by PAGE and Western blot. The P22-resistant mutant showed almost a complete absence of this band by SDS-PAGE. By immunoprecipitation of wild-type P-strain tachyzoites, MAB 7G1 clearly immunoprecipitated a band of 22 kD whereas the resistant mutant showed no precipitable band of this molecular weight. Further analysis by two-dimensional electrophoresis confirmed the absence of P22 on the resistant mutant.

Analysis of the P30-resistant mutant showed similar findings, in that there is no observable or reactive epitope by Western blot. Further analysis using rabbit antisera raised against P30 from the wild-type P strain showed no detectable P30 in these mutants. Sequence analysis of the P30 from one of these mutants has been done and it shows a nonsense mutation (giving a stop codon) about 600 bp from the start-codon, predicting a truncated protein of about 17 kD. The absence of a detectable protein by Western blot analysis implies that this truncated product is highly unstable (it lacks the C-terminal tail necessary for glycolipid anchoring) and/or it is missing most or all of the reactive epitopes normally found on P30. The former explanation seems the more likely. Both of these studies indicate that with sufficient environmental pressure, resistant mutants occur and can be identified. Such mutants should prove very useful in dissecting the function of these proteins.

Another approach has been adopted by Pfefferkorn & Pfefferkorn [79], who isolated a temperature-sensitive mutant of *T. gondii*. Some of these mutants, isolated by chemical mutagenesis of the parasite, have been shown to be avirulent in mice. This is remarkable since they were isolated from the highly virulent RH strain, which has an LD_{100} of less than 10 parasites in a highly sus-

ceptible strain. One such temperature-sensitive mutant has been used to design a veterinary vaccine (discussed below). In summary, antigenic mutants may be important in further understanding the pathogenesis of *T. gondii*.

Stage-specific antigens

Although the asexual rapidly multiplying tachyzoite has been most investigated, this stage is only one of several in the life-cycle of *T. gondii*. Stage-specific antigens have been described in other intracellular parasites, most notably *Plasmodium*. Antigens specific to the *Toxoplasma* bradyzoite, tachyzoite, and oocyst/sporozoite stage do occur.

Oocyst/sporozoite

The oocyst/sporozoite has been the least investigated of all three major life-cycle stages of *T. gondii*. As discussed earlier, this stage represents the highly infectious parasite that is excreted by the cat following sexual reproduction in the feline intestine. Identification of unique antigens to this stage would be important for improved serodiagnosis and potential vaccine development against *T. gondii* as well as epidemiologic studies aimed at determining whether oocysts or bradyzoites are the major sources of human infection.

Kasper *et al.* [80] developed a panel of MAB that identified unique surface membrane antigens on the sporozoite and oocyst wall of *T. gondii*. These MAB failed to react with *T. gondii* tachyzoite surface membrane antigens in several immunologic assays. A comparison of surface membrane radioiodinated tachyzoites and sporozoites by one- and two-dimensional electrophoresis showed that *T. gondii* sporozoites have two major membrane proteins of approximate molecular weight 67 and 25 kD that are not present on the tachyzoite stage. Two of the stage-specific MAB described were directed against the sporozoite 67 kD protein. Additionally, *T. gondii* sporozoites appeared to be deficient in the major radioiodinated tachyzoite protein P30.

Epidemiologic importance for the role of stage-specific oocyst/sporozoite antigens was further provided by Kasper & Ware [81]. These authors showed that human antiserum was able to identify stage-specific oocyst/sporozoite antigens. Acute and convalescent *Toxoplasma* sera were obtained from patients in an epidemiologically well-documented outbreak of oocyst-transmitted infection [82]. An ELISA comparing equal numbers of tachyzoites and oocyst/sporozoites indicated that these antisera recognized antigens from both stages. Adsorption of pooled antisera with purified oocysts reduced both the antioocyst IgG and IgM titers but only had minimal effect on the antitachyzoite titer. Conversely, adsorption of the antisera with tachyzoites reduced both antioocyst and antitachyzoite IgG and IgM. Sodium dodecyl sulfate polyacrylamide gel electrophoresis of radioiodinated oocysts revealed that the principle stage-specific surface proteins of the oocyst/sporozoite have approximate molecular weights of 67 and 25 kD. Periodic acid and silver stain of purified oocyst/sporozoite identified bands of similar molecular weight not present in the tachyzoite preparation. Western blot analysis of purified parasites assayed with human antioocyst antisera identified specific oocyst/sporozoite antigens not present on the tachyzoites. Reaction to these oocyst/sporozoite antigens was seen primarily in the IgM fraction of the acute phase and the IgG fraction of convalescent phase antisera. Neither adsorption of the antisera with tachyzoites nor periodate treatment of the oocyst/sporozoites reduced the antibody recognition of these stage-specific antigens. This report suggested that some individuals believed to be infected by oocysts develop antibodies against unique stage-specific oocyst/sporozoite antigens.

Bradyzoites

Antigenic specificity of bradyzoites has also been described. Lunde & Jacobs [83] showed that bradyzoites (cystozoites) were found to exhibit stage specificity by immunofluorescence. Similarly, Kasper et al. [81,84] have shown that some MAB directed at tachyzoite antigens failed to react toward bradyzoites. Recently, Kasper & Ware [35] performed an immunologic evaluation of the surface antigens of the three major life-cycle stages of T. gondii. Mouse antisera were raised against these stages, which included the oocyst/sporozoite, bradyzoite, and tachyzoite. The antisera were used in an ELISA test and Western blot to demonstrate the presence of stage specificity. These antibodies were able to distinguish their corresponding stages far better than they could distinguish noncorresponding stages. For example, MAB raised against two major tachyzoite antigens, P30 and P22, were 100-fold more reactive with the tachyzoite stage than with the oocyst/sporozoite and bradyzoite stages. Similarly, high titer (greater than 1:50 000), antioocyst/sporozoite MAB reacted poorly with bradyzoites and tachyzoites. The development of

an antibradyzoite MAB library [85] supports the previous observations that antigenic specificity of this stage also exists. Another recent study by Tomavo et al. [85] further supports the presence of bradyzoite stage specific antigens. This group developed a panel of MAB against the bradyzoite stage and were able to identify four pellicular antigens (36, 34, 21, and 18 kD). Three of these antigens were exposed on the surface of the organism and accessible to either antibody or trypsin cleavage. Moreover, these antigens were identified on human isolates of T. gondii. Since current epidemiologic evidence suggests that acute toxoplasmosis in patients with AIDS may represent recrudescence of previous infection, the identification, characterization, and purification of bradyzoite antigens in the future may be of clinical importance to the serodiagnosis of toxoplasmosis.

Tachyzoites

There also appears to be stage specificity of the tachyzoite. Neither bradyzoites nor oocyst/sporozoites appear to express the major tachyzoite antigens P30 and P22 as determined by MAB analysis. Interestingly, all three stages observed do exhibit crossimmunoreactivity by Western blot for a group of proteins with an approximate molecular weight of 24–32 kD. By this test, all three stages express antigens within this range. These antigens may represent a group of proteins that may perhaps be of importance in the development of a protective vaccine against T. gondii. Further discussion regarding tachyzoite antigens and their corresponding antibody response has been described previously in this chapter. Epitopes corresponding to tachyzoites and bradyzoites may be useful for vaccine development.

Cellular immunity

The cell-mediated response to T. gondii represents the major component of host immunity to this organism. Delayed-type hypersensitivity (DTH) has been shown in chronically infected individuals. The DTH response can be detected as early as 1 week postinfection in guinea pigs and 3 weeks in mice [50]. A number of reports indicate that a direct correlation exists between the development of DTH and the antibody response. In humans, DTH occurs following the initial rise in antibody titer. In spite of this, skin testing for DTH has never been a useful diagnostic technique for establishing chronic Toxoplasma infection in humans. Pavia [86,87] demonstrated that serum or spleen and lymph node cells from guinea pigs immune to infection with RH strain were able to produce partial protection against symptomatic disease in recipient guinea pigs. This cutaneous infection model was consistent with the earlier findings demonstrating that hamsters receiving unfractionated spleen and lymph node cells from chronically infected animals were protected against lethal challenge.

The induction of a lymphocyte proliferative response by *Toxoplasma* antigens is now well established. Presensitized lymphocytes from *Toxoplasma*-infected rabbits undergo blast transformation in the presence of parasite antigen. In humans, a lymphoproliferative response from sero-positive individuals has been observed routinely in many laboratories. The time between the induction of antigen-specific lymphocyte transformation and clinical illness is variable. Lymphocytes from almost one-third of those individuals with less than a 12-month history of chronic infection failed to undergo proliferation in response to *Toxoplasma* antigen. Eventually, antigen-specific lymphoproliferation could be observed in the majority of individuals who manifested clinical disease.

Wilson & Remington [1] reported that *Toxoplasma* antigen-specific proliferation may in fact be a reasonable indicator of congenital infection in both symptomatic and asymptomatic individuals. Lymphocyte blastogenic response to *T. gondii* antigens is impaired in a certain number of patients with chronic latent *Toxoplasma* infection. However, the degree of impairment was insufficient to cause reactivation of chronic latent *T. gondii* infection. A similar attenuated response to parasite antigen was observed in patients with Hodgkin's disease [88]. In these people, therapy of the underlying Hodgkin's disease failed to improve the depressed lymphocyte transformation in response to parasite antigens. Thus, lymphocyte blastogenesis in response to parasite antigen may not be crucial to protect against reactivation of chronic latent infection in humans.

Cellular response

T cell subsets

Over the past 5 years, there has been increasing interest in the identification of those T cell subsets associated with the cellular immune response to *T. gondii*. Identification of various T cell receptors and their role in host immunity has been the subject of much speculation in the field of immunoparasitology. Some of the earlier studies attempting to subtype phenotypically the T cell response to *Toxoplasma* are those of Macario *et al.* [89]. In those studies, the lymphocyte subpopulation of mice during chronic *Toxoplasma* infection was assessed. Infected mice exhibit involution of the thymus and other lymphoid organs in association with a generalized depletion of the Thy-1$^+$ lymphocyte subpopulation. This immune suppression during earlier stages of infection was attributed either to the induction of suppressor cells or perhaps the selective imbalance of lymphocyte subsets (see next section on suppression of immunity).

Natural killer (NK) cells have been postulated as a possible mechanism for cytotoxic activity [90,91]. Natural killer cells obtained from the spleens of normal and *Toxoplasma*-infected mice were assayed for reactivity against RH-strain tachyzoites *in vitro*. Presensitized effector NK cells from 3-day *Toxoplasma*-infected mice were incubated with target tachyzoites. In this experiment, these effector cells had significantly greater parasite cytotoxicity then the control cells from uninfected mice. Toxoplasmicidal activity was significant at all effector/target ratios and was mediated by direct contact between the host cell and the parasite.

These authors and others have demonstrated that acute *Toxoplasma* infection of mice results in an increased enhancement of NK cell cytotoxicity for a variety of tumors. This activity was observed in the first 24 h, peaking at 72 h postinfection. The NK function is independent of host mouse strain, age of mice, parasite strain, and route of infection. The NK activity was associated with a concomitant rise in serum IFN titers. Interestingly, Hauser & Tsai [90] found that NK cells obtained from normal mice were not as cytotoxic for extracellular RH tachyzoites as presensitized NK cells. This would perhaps contradict them being NK in phenotype, since NK cells need not be presensitized in order to exert their immune function.

Contrasting the reports of Hauser & Tsai [90] and Kamiyama [91] is that of Hughes *et al.* [92]. In this study, the absence of a role for NK cells in control of acute infection by *T. gondii* was assessed. Primary oral infection of inbred mice with P-strain oocysts resulted in the rapid increase of serum IFN titers followed by augmented NK activity against YAC-1 target cells. In subsequent challenge infections, NK was not augmented and IFN titers rose only if a high dose of oocysts was given. Inhibition studies indicated that *T. gondii* did not bind to NK cells and nor was there evidence for parasiticidal activity *in vitro* by cells exhibiting NK function. Genetic studies using inbred mice deficient in NK activity showed that there was no difference in time to death after administration of an oral lethal dose of parasites. A recent report by Dannemann *et al.* [93] indicates that human NK cells have no cytotoxic activity against radiolabeled tachyzoites. From these and the previous studies it appears that *T. gondii* increase NK cell production but the role of these cells in host immunity remains unclear.

The role of both helper and cytotoxic T cells has been investigated in toxoplasmosis. The cell interaction between tachyzoites and the human immune system suggests that immune seropositive cells proliferate in response to cocultivation with live attenuated tachyzoites. These stimulated cells are able to reduce *in vitro* growth of tachyzoites. In contrast, cocultivation of tachyzoites with peripheral blood mononuclear cells (PBMC) from seronegative donors fails to induce a proliferative response. All isolated clones produced IL-2 and IFN-γ consistent with their CD4 surface phenotype.

The importance of the CD4$^+$ T helper cell subset in experimental toxoplasmosis was also described by Vollmer *et al.* [94]. By using a mouse model of toxoplasmosis, his group found that depletion of the CD4$^+$ lymphocytes produced an overwhelming infection and death in both acute and chronic toxoplasmosis. In chronically infected mice, depletion of CD4 cells resulted in death associated with severe *Toxoplasma*-related central nervous system damage and minor systemic involvement. Reactivation of the *Toxoplasma* infection in the CD4-depleted chronically infected mice, occurs despite high titers of circulating anti-*Toxoplasma* antibody.

Others have suggested that *Toxoplasma* induces the propogation of CD8$^+$ cytotoxic T cells. Two reports demonstrate an increase in the number of CD8$^+$ cells in peripheral blood during *Toxoplasma* infection. A more recent study by Sklenar *et al.* [95] further supports these findings. Cytofluorometric analysis during recent toxoplasmosis revealed an increased absolute number of CD8$^+$ cytotoxic cells, NK cells and monocytes. The absolute number of CD4 helper cells was not significantly changed. CD4/CD8 ratios during symptomatic toxoplasmosis showed an increase in relative numbers of CD8$^+$ cells. Stimulation with *Toxoplasma* antigen induced higher numbers of CD8$^+$ from people with symptomatic infection. Similar results were reported in a study of 17 adults with acute toxoplasmosis. In these studies, the absolute numbers of CD3$^+$, CD8$^+$ circulating lymphocytes were increased. These changes were most frequently found in people with lymphadenopathy, and less frequently found in those without clinical

symptoms. Even though antigenemia resolved early following the onset of symptoms, the reversed CD4/CD8 persisted for 2–4 months.

McLeod *et al.* [96] has demonstrated that infection reduced percentages of spleen cells with Lyt-2$^+$ phenotype in susceptible but not resistant mice. Infection decreased percentages of spleen cells with L3T4$^+$ similarly in both strains of mice tested (C57 and A/J) [96]. The induction of both CD4$^+$ and CD8$^+$ subsets during *Toxoplasma* infection can perhaps be explained by the use of whole parasite extract as the immunogen or stimulant *in vitro*. Observations that a variety of lymphocyte subsets can be stimulated by parasite extract would indicate that antigens responsible for the induction of cellular immunity in this disease are complex. That is, different antigens perhaps stimulate different T cell subsets. In view of the complex biochemical nature of this eukaryotic cell, it has been impossible to determine which antigens are most important for stimulation of protective immunity. The existence of some of the major *Toxoplasma* antigens in recombinant form should help determine the relative role of each in stimulating a cellular immune response.

To approach the question of the relevant immunogenicity of specific *Toxoplasma* antigens, we chose to investigate P30 [97,98]. In these experiments, we found that P30 immune mouse splenocytes obtained from inbred BALB/c mice reduced extracellular *T. gondii* plaque forming units by more than 50% when incubated in an effector/target ratio of 10:1 or greater. By using a ^{3}H-uracil radioisotope release assay, the effect of the immune splenocytes was determined to be directly parasiticidal. The cytotoxic immune splenocytes were P30 antigen specific and of the Thy-1.2 Lyt-2,3$^+$ (CD4$^-$, CD8$^+$) phenotype specific for mouse cytotoxic T cells. Opsonization of the parasites with monoclonal anti-P30 reactive antibody did not enhance parasiticidal activity. Culture supernatants obtained during the 2 h cytotoxic assay were not parasiticidal and antiasialo GM1 antibody (antimouse NK cell) plus complement had no effect on the parasiticidal activity of the P30 responder cells. Thus, we identified an antigen-specific T cell subset that are directly parasiticidal to extracellular *T. gondii* and exhibit cytotoxicity independent of antibody opsonization, cytokine secretion, and NK activity.

In order to better understand the cellular immune response in humans to individual parasite antigens we investigated the induction of antigen-specific human T cells by *T. gondii* [99]. Antigen-specific T cell clones were generated using PBMC from seropositive individuals. Whole parasite extract was used to stimulate a proliferative expansion of antigen-reactive cells followed by limiting dilution cloning in the presence of irradiated autologous PBMC and recombinant IL-2. Parasite antigen-specific T cell clones expressing the CD3$^+$ phenotype were selected. The clones representing the highest stimulation index in response to parasite antigen were further analyzed with MAB reactive to CD8 or CD4 phenotype. When tested in a proliferation assay using Western blot T cell analyses of different *T. gondii* proteins, one clone reacted with a single large protein of greater than 180 kD as well as a smaller component of approximate molecular weight of 12–14 kD. Clone 2 reacted with a protein of 28 kD and clone 3 reacted with proteins of 116 and 12 kD. Only the 28 kD protein recognized by clone RTg2 was reactive with the donor antiserum. All three clones produced high titers of IFN-γ and IL-2. Interestingly, clone RTg1 exhibited direct parasite cytotoxicity. This clone when incubated at an effector/target ratio of 40:1, inhibited extracellular *T. gondii* growth by over 75%. The RTg1 cloned cytotoxic T cells were αβ dimer-TCR positive and exhibited cytotoxicity independent of MHC restriction.

In summary, all of these reports suggest that *T. gondii* is able to induce a variety of T cell subsets that exhibit both helper and suppressor cytotoxic function. Although some of these parasite antigens may be unable to induce an observable humoral response they are of potential interest as candidate T cell vaccines.

Cellular response/cytokine response

As discussed previously, cell-mediated immunity has been shown to be an important mechanism in recovery from *Toxoplasma* infection in both mice and humans. The role of secreted cytokines by these immune cells has been investigated over the past several years.

An important T cell cytokine is IFN-γ which activates macrophages to kill *T. gondii* [38,40,95,100,101]. Much of the work regarding the function of IFN-γ deals with its role in macrophage activation. *In vivo*, IFN-γ is an important factor responsible for activating macrophage oxidative metabolism and anti-*Toxoplasma* activity [102–106]. Following parenteral injection of recombinant IFN-γ, peritoneal macrophages exhibited increased H_2O_2 production and marked anti-*Toxoplasma* activity. In a similar study using humans [107,108] the immunotherapeutic role of IFN-γ as a macrophage phagocyto-activating agent was determined. In these

studies, human monocytes could be induced by recombinant IFN-γ to express immediate and persistent activation implying that the cytokine may be an important activator of mononuclear phagocytosis. When assayed following a 24 hour *in vitro* pulse with IFN-γ both normal and AIDS monocytes behaved similarly exhibiting little enhancement of their intrinsically high levels of H_2O_2 release and anti-*Toxoplasma* activity. Enhancement of this effect by IFN-γ was reported by Mellors *et al.* [109]. In this study, the ability of free and liposome-incorporated recombinant IFN-γ was compared. They found that the concentration of liposomal recombinant IFN-γ required to enhance macrophage H_2O_2 release and maximally inhibit *T. gondii* growth was 1/10th the concentration required for free IFN-γ.

The ability of IFN-γ to improve host immunity to *Toxoplasma* is not limited to macrophages. In 1984, Pfefferkorn clearly demonstrated that the growth of *T. gondii* in cultured human fibroblasts could be inhibited by human recombinant IFN-γ. Using a ^{3}H-uracil incorporation assay, he was able to demonstrate that low concentrations of IFN-γ effectively blocked *in vitro* growth of the parasite. Further studies by Pfefferkorn & Guyre [110], revealed that IFN-γ blocks the growth of *T. gondii* by inducing the host cells to degrade tryptophan. In 1986 Pfefferkorn *et al.* [111–113] was able to characterize the enzyme indolamine 2–3 dioxygenase as the mediator of this host-cell resistance. In the presence of IFN-γ host-cell fibroblasts utilize indolamine-2,3-dioxygenase to degrade intracellular tryptophan to *N*-formyl kynurenin. A constitutive host-cell formamidase then further degrades this product to kynurenine, which leaks from the host cell back into the mediun. By studying the fate of 14C tryptophan, the effect of this enzyme was determined. Confirmation of this process was performed using neutralization of antiviral activity by antibodies specific for IFN-γ.

Because of the complicated nature of the cellular response as described previously, we felt that analysis of the IFN-γ response to a purified parasite membrane protein would be important. Accordingly, we used P30 to induce proliferation of PBMC from seropositive individuals. The culture supernatants from stimulated cells blocked the growth of *T. gondii* in human fibroblasts, whereas those from antibody negative individuals failed to do so. The antitoxoplasmic effect of culture supernatants correlated with the induction of indolamine-2,3-dioxygenase and the destruction of tryptophan as previously described. Further, the antitoxoplasmic effect was blocked by MAB to IFN-γ. Thus, the level of IFN-γ appears to be an important immune factor in protection against toxoplasmosis in humans [97,98].

Further confirmation for the role of IFN-γ in experimental toxoplasmosis was reported by Suzuki *et al*. [114]. In their study, MAB directed against mouse IFN-γ dramatically reduced the ability of mice to resist infection with *Toxoplasma*. Suppression of chronic immunity to *Toxoplasma* by this antibody was not reported but would certainly be of interest in that Jones *et al*. [38] found that IFN-γ assists in the production of tissue cysts *in vitro*. Endogenous IFN appears to play a role in preventing encephalitis in chronically infected mice [115].

Aside from IFN-γ there are other cellular cytokines which appear to be important in host immunity to *Toxoplasma* infection. IL-2 has been demonstrated to induce a number of cellular immune functions. Sharma *et al*. [116] showed that administration of recombinant IL-2 resulted in a significant decrease in mortality in lethally infected mice. Additionally, these mice had significantly fewer cysts in their brains. When assayed for increased antibody synthesis or macrophage killing, it appeared that the effects of recombinant IL-2 was independent of either of these immune factors. He did find that mice treated with recombinant IL-2 showed increased NK cell activity. The recombinant IL-2 did not reverse the suppressed lymphoproliferative response in acutely infected mice when stimulated with either concanavalin A or lipopolysaccharide (LPS). In our own investigations (unpublished observations) we have found that IL-2 also enhances the ability of various purified parasite antigens, in particular P30, to induce protective immunity in mice. These and other studies related to IL-2 should be of long-term interest, especially in regard to potential therapy for those with AIDS.

The role of other cytokines has also been investigated in experimental toxoplasmosis. Activated macrophages are known to produce tumor necrosis factor-α (TNF-α) (or cachectin), a cytokine that demonstrates both anti-tumor and antimalarial activities. DeTitto *et al*. [117] investigated the role of recombinant TNF in the intra-cellular multiplication of *T. gondii*. Through their studies they found that this cytokine had no effect on either the intracellular multiplication of tachyzoites or bradyzoites in normal mouse peritoneal macrophages. There was no effect on parasite growth in cultured human fibroblasts. The killing of *Toxoplasma* tachyzoites by activated macrophages was not enhanced by recombinant TNF.

Recently, Remington *et al*. [118] have been able to identify polymorphisms in the TNF-α gene that correlate with murine resistance to the development of toxo-plasmic encephalitis in the brains of infected mice. These authors were able to correlate resistance to toxoplasmic encephalitis with RFLP and microsatellite variants in the TNF-gene. Susceptible mice expressed elevated levels of TNF mRNA in brain tissue 6 weeks after infection. Resistant and uninfected mice showed no detectable mRNA in brain tissue. These differences were localized to the first intron, promoter, and 3' region of the TNF gene. Their finding suggested that TNF regulation in brain tissue may be an important factor in determing host resistance to chronic infection.

The role of TNF-α in *T. gondii* infection has also been investigated by Sibley *et al*. [119]. Their studies indicate that TNF-α is able to regulate enhanced antimicrobial activity by triggering IFN-γ primed macrophages to kill or inhibit intracellular parasites. Mouse macrophages failed to display antimicrobial activity with rIFN-γ. However, in the presence of low concentrations of LPS (0.1 ng/ml) these cells became activated. rTNF-α alone was unable to trigger macrophage antitoxoplasmacidal activity. Significant antitoxoplasmacidal activity was observed when rTNF-α was used to stimulate IFN-γ treated macrophages cultured under low endotoxin conditions. The antitoxoplasmacidal activity was blocked by anti-TNF-α antisera. These findings further support the importance of TNF in the antimicrobial effect of macro-phages *in vitro*.

The *in vivo* role of TNF has also been evaluated in mice [120]. Mice resistant to ip infection (CB6F) were treated with anti-TNF antibody which caused a sig-nificant increase in the numbers of intraperitoneal tachyzoites. Peroral infected mice treated with anti-TNF resulted in increased mortality of the parent C57, but not the BALB/c F1 cross suggesting that TNF is perhaps an important modulator of protection in mice.

We have investigated the role of a cytotoxic lym-phokine present in the supernatant from cloned mouse and human immune T cells [37]. Supernatants from these antigen-specific T cell cultures inhibited the growth of L929 mouse fibroblasts *in vitro*. The immune supernatant had no effect on the viability of extracellular parasites, suggesting that other cell-mediated lym-phokines may potentially modulate cellular immunity against *T. gondii* infection.

Reyes & Frenkel [121] studied the specific and non-specific mediation of protective immunity by *Toxoplasma* immune lymphocytes and their supernatants. The specific protective mediator in the immune supernatants was characterized as having a molecular weight between 3 and 12 kD by both dialysis and molecular chromotography. This activity was destroyed by exposure to pH 2 treatment. *In vitro* production of cytokine from the immune lymphocytes was first noted between 7 and 10 days following vaccination of hamsters.

Another potentially important cytokine involved in the immune response to *T. gondii* is IL-10. Recently, Gazzinelli *et al.* [122] have demonstrated the potential importance of this molecule in macrophage activation against the parasite. Interleukin-10 is produced by CD4$^+$ T cells usually belonging to the TH-2 subset and has been shown to inhibit the synthesis of IFN-γ by both T cells and NK cells. In their study, these authors demonstrate that IL-10 can down regulate immunity by blocking the ability of the IFN-γ to activate macrophages. This activity is dose dependent and correlates with the inhibition of IFN induced toxic nitrogen oxide metabolites which is an important macrophage effector function in toxoplasmacidal activity. IL-10 requires at least 12 h exposure for maximal effect and must be present before or together with the IFN-γ stimulus. These observations suggest that IL-10 may be an important mechanism by which *Toxoplasma* evade IFN-γ dependent cell-mediated immunity.

Another group of immune modulators are the leukotrienes. These mediators have been identified as important factors involved in host defense. These immune modulators enhance permeability across capillary venuoles, promote vasoconstriction, and are a chemoattractant and stimulus for neutrophils. Locksley *et al.* [123] examined the ability of macrophages to generate leukotrienes following infection with *T. gondii*. In their study, *Toxoplasma*-infected peritoneal macrophages lead to the formation of 11-, 12-, and 15-hydroxyicosatetraenoic acids. Together with an unidentified compound, designated X, each of the compounds incorporated tritiated arachidonic acid during macrophage phagocytosis of *Toxoplasma*. If the parasites were previously opsonized with antibody or killed, they induced the production of LTD-4 (a leukotriene released in the early stages of inflammation) but not compound X. This absence of leukotriene production by macrophages may account in part for the lack of immune inflammatory response that has been well described during chronic *Toxoplasma* infection.

In summary, *T. gondii* induces a variety of immune cytokine responses in the infected host. These cytokines appear to play a critical role in host protection against *Toxoplasma* infection. Alteration of this cytokine response, such as that seen in AIDS patients, may result in increased susceptibility to infection.

Suppression of immunity

Chronic *Toxoplasma* infection in mice appears to alter host immunity, as determined by a variety of immunologic assays. Strickland *et al.* [124] first reported immunodepression in mice infected with *T. gondii*. Huldt [125] described immunosuppression of the primary antibody response to sheep red blood cells (SRBC) as well as polio virus in chronically infected mice. These findings were further confirmed and extended by Hibbs *et al.* [126], who observed that chronic *Toxoplasma* infection of mice suppresses specific immunologic reactions. In their studies, chronically infected mice were unable to mount an effective primary antibody response to tetanus toxoid, although the secondary antibody response to SRBC appeared to be normal. These workers also observed impaired rejection of primary skin grafts when normal mouse donors and *Toxoplasma*-infected recipients differed at the H-2 locus. Strickland *et al.* [127] assessed the blastogenic response of spleen cells to various T and B cell mitogens. Reduced proliferation to all mitogens occurred during the acute stages of infection. During chronic infection, persistent depression of the blastogenic response to the T cell mitogens was also noted. Suzuki & Kobayashi [128] examined the effect of infection with *T. gondii* on antibody responses to unrelated antigens. Their studies indicated that infection with low doses of organisms had no effect on antibody response to either SRBC or dinitrophenyl (DNP). By increasing the infectious dose, however, suppression of the secondary anti-SRBC antibody response was observed. By contrast, in primary infection anti-SRBC and anti-DNP responses were suppressed by injection with low-dose *T. gondii*. These results demonstrated that a nonspecific suppression of antibody response is provoked by *T. gondii* infection when mice receive a large number of organisms. The suppressed antibody response appeared to be caused by activation of suppressor cells.

Suppressed T cell lymphocyte proliferation during chronic infections has been reported [129]. Chan *et al.* [130] investigated T cell dysfunction during acute *Toxoplasma* infection. In their studies they found that the proliferative response of splenocytes and enriched T cells from infected mice was depressed in response to concanavalin A stimulation. Removal of macrophages had no effect on the depressed proliferative response.

Interleukin-2 production was decreased in all infected mice. Reconstitution of the depressed proliferative response was done by the addition of exogenous IL-2. Their conclusions were that an active suppressor factor could be identified and perhaps may be related to strain specificity. Further studies of the mononuclear response from patients during chronic infection has been reported. Adherent mononuclear cells can be induced by preincubation with antigen. These cells suppress the proliferative response of autologous mononuclear cells to *Toxoplasma* antigen. Induction of the adherent suppressor cell was T cell-dependent (CD4$^+$), suggesting that the macrophage and CD4$^+$ T cells may exert suppressor activity on the immune system. In further studies, antigen-specific cells were found to be able to suppress lymphocyte proliferation. The induction of suppressor cell activity and the persistent increase in Leu-2$^+$ suppressor T cells correlated with more severe and persistent clinical symptoms. In subjects who were acutely or chronically infected but asymptomatic, the number of Leu-2$^+$ suppressor T cells was normal and the induction of suppressor cell activity was minimal.

Yano *et al.* [131] reported that *Toxoplasma*-specific T cells failed to proliferate in response to antigen in a patient with symptomatic acute toxoplasmosis. The immunosuppression was mediated by suppressor T cells. These cells suppressed both *Toxoplasma*-specific and purified protein derivative (PPD)-specific T cell responses from a patient with chronic disease. Antibody to HLA-DQ altered the suppressive activity of *Toxoplasma*-specific inhibition. By contrast, *Toxoplasma*-specific T cell responses were activated by antibody to HLA-DR.

IMMUNOPATHOLOGY

As mentioned previously, *Toxoplasma* infection in both the experimental host and in humans has been associated with a transient reduction in delayed hypersensitivity. These investigations have demonstrated varying degrees of proliferation in response to *Toxoplasma* antigens during the first month of illness. Jones *et al.* [132] showed that cyclophosphamide in a single dose of 100 mg/kg given prior to infection resulted in an augmented response and decrease in the number of *Toxoplasma* brain cysts. During the first few weeks of *Toxoplasma* infection there was a decrease in CD4 T lymphocyte response in spleens and lymph nodes. Further, a reduced lymphoproliferative response of cultured spleen cells to *Toxoplasma* antigen and decreased

levels of IFN-γ production occurred. In other studies evaluating the importance of T cell subsets in the inflammatory response during toxoplasmic encephalitis, Israelski *et al.* [133] found that mice chronically infected with *T. gondii* and treated with anti-CD4$^+$ MAB exhibited significantly less inflammation in their brains when compared to controls. Histopathologic sections of the brain were examined at regular intervals and also revealed markedly decreased inflammatory response. Recrudescence of the inflammatory process occurred after discontinuation of treatment with MAB. These experiments were observed using different strains of mice as well as different lines of *Toxoplasma* and imply that CD4$^+$ T cells may play a significant role in the inflammatory response in the brains of animals with toxoplasmic encephalitis.

EFFECTS OF SPECIFIC THERAPY AND IMMUNOSUPPRESSION ON PATHOGENESIS OF INFECTION

Toxoplasma infection in the normal host can be modified by treatment with agents such as sulfadiazine and pyrimethamine. Conversely, immunosuppressive therapy can enhance *Toxoplasma* infection. *Toxoplasma gondii* is an important opportunistic infection in patients with compromised immunity, especially those with AIDS. The addition of immunosuppressive agents, such as corticosteroids or other cytotoxic agents, can result in an increased frequency of infection in the compromised host. Presumably most infections of this nature are due to recrudescence of latent infection.

Toxoplasmosis complicating malignancy and organ transplantation has been reviewed [134]. A variety of reports concerning life-threatening infections with *Toxoplasma* in cancer patients have been reported. Frenkel [135] studied the pathogenesis of recrudescent toxoplasmic encephalitis in hamsters. His studies suggest that relapse is due to the rupture of a single cyst in the center of each lesion. Treating chronically infected hamsters with either cortisone or total body irradiation resulted in a marked exacerbation of infection with fatal encephalitis. Later, Frenkel & Piekarski. [136] demonstrated that cortisone or cyclophosphamide treatment was associated with reactivation of chronic *Toxoplasma* infection in hamsters. Recently, Suzuki *et al.* [114] have demonstrated recrudescence in a mouse model using antibody that inhibits IFN-γ. Similarly, as previously discussed, treatment with antibody to the CD4$^+$ receptor results in reactivation of toxoplasmosis. In general, these and earlier studies suggest that *T. gondii* infection can be reactivated by a variety of immunosuppressive agents.

RESISTANCE TO *TOXOPLASMA*

Natural resistance

There appear to be several critical variables that govern the pathogenicity of *Toxoplasma* infection in the normal host. Different host species as well as host strains vary in their susceptibility to *T. gondii*. Further, different strains of *Toxoplasma* are markedly different in their virulence for given species of animal. The most commonly used laboratory animals, including hamster, mice, and rabbits, are susceptible to infection with *Toxoplasma*. Rats, guinea pigs, and nonhuman primates do appear to possess some degree of innate resistance. Rats have been found to exhibit little clinical effects after infection with large numbers of highly virulent parasite strains. Differences in susceptibility among different inbred and outbred strains of mice is well recognized. In general, most inbred strains of mice are more susceptible to infection than outbred strains, i.e., BALB/c and DBA-2 were the least resistant strains of mouse, whereas C57BL and C₃H mice, although more resistant than the BALB/c mice, did not exhibit the same degree of resistance as that of the SW outbred. Recently, McLeod *et al.* [96] have shown that there exists a marked difference in survival of inbred mouse strains following peroral infection with toxoplasma. In their studies, they found that A/J mice were resistant to infection whereas C57BL mice were highly susceptible to oral infection. Other mouse strains assayed in their studies exhibited different cumulative mortality following oral infection. Among those resistant strains tested were C3H/HEJ and DBA2/J, those showing moderate resistance were DBA1/J, BALB/c, and BYJ, and those exhibiting the least resistance were C57LJ, SJLJ, SBLE, and C57BL/6J.

Other studies employing congenic mice showed that at least one of the murine genes that determine resistance to the parasite is linked to the H-2 locus [137]. Brinkmann *et al.* [138] have assessed the production of antigen-specific IgM and IgG in immunocompetent CBA/J (H2K), C3H/HE, and B cell-deficient CBA/N mice. Most of the C3H/HE mice survived infection with 20 cysts of ME49 strain. This was compared to the CVA/J mice (71% survival) and CVA/N (53% survival) following infection with a similar number of cysts. After sulfadiazine treatment they observed decreased mortality in the CVA/J and C3H/HE mice, which was associated with reduced formation of brain cysts. *Toxoplasma*-specific IgM and IgG levels were lower in sulfadiazine-treated CVA/J and CVA/N mice. Although the CVA/N mice developed practically no humoral response, they produced fewer brain cysts than normal CVA/J mice.

This same group [138] has also reported on the regulation of the L3T4 T cell subset by B cells in different mice bearing the H2K haplotype. In this study, splenic L3T4 cells from *Toxoplasma*-infected CVA/J, but not C3H, mice proliferated in response to parasite antigen. The proliferative response of the C3H/HE spleen cells could be restored in part by adding recombinant IL-2. It was felt that the unresponsiveness of the C3H/HE spleen cells was due to IL-2 inhibiting factors present in the culture supernatant of the antigen-stimulated splenocytes. Splenocyte culture supernatant blocked the growth of the IL-2-dependent T cell line in the presence of optimal concentrations of IL-2. These results indicated that wide variations in T cell proliferation and lymphokine production in mice bearing the same H2K haplotype may exist. Studies by Suzuki *et al.* [139] further support genetic restriction in the development of toxoplasmic encephalitis. Using congenic mice, these investigators were able to conclude that genes within the H-2D region are involved in the development of toxoplasmic encephalitis.

Acquired resistance

Acquired resistance to *T. gondii* can be obtained in a variety of ways, including infecting with an avirulent strain of parasite, using a host that is naturally more resistant, or by altering the course of infection with other anti-*Toxoplasma* agents. Once immunity has been acquired it persists for prolonged periods, usually the life of the host. Acquired resistance begins shortly after infection in both guinea pigs and mice and may occur as early as 1–2 weeks following primary infection.

Although survival from acute *Toxoplasma* infection leads to resistance against reinfection, the immunity associated with chronic infection does not result in the absence of parasites from the infected host. As early as 1943, Fineman observed the persistence of organisms in tissues for the life of the host. Since that time, a variety of reports have shown that immunity to *Toxoplasma* in mice protects against disease but does not necessarily prevent reinfection. Although the rapid disappearance of an infecting dose can be observed in the peritoneal fluids of mice, prolonged persistence of the parasite in their brains is apparent. Even when animals are successfully immunized with either killed *Toxoplasma* vaccines or purified antigen, viable parasites can be obtained from the tissues of these immunized animals after challenge. This incomplete immunity can be observed in the offspring of these infected animals. Second challenge of mice chronically infected with an avirulent strain bore infected offspring that had properties attributable to the primary innoculum. Thus, immunologically intact but chronically infected mice give birth to congenitally infected offspring that carry the original infecting strain of the mother.

Interestingly, the immunity to oocyst excretion in the infected cat is also incomplete. Dubey *et al.* [6,7] showed that immunity to intestinal infection resulted in the failure to produce oocysts several months later. Others, however, have observed a relationship between the titers of infected cats and their renewed oocyst secretion following reinfection [139]. This was first shown by Kuhn & Weiland [140], who demonstrated that a second infection in cats results in renewed oocyst shedding.

Humoral antibody in acquired resistance

Various attempts have been made to induce resistance to infection by passive transfer of immune serum. The earliest studies performed by Eichenwald [141] show that immune serum passively transferred to chronically infected mice temporarily reduced parasitemia. The passive transfer of immune rabbit serum failed to protect against challenge with RH strain [142]. Johnson *et al.* [47] have shown that MAB can induce partial protection against virulent parasite challenge. Passive transfer of serum containing IgE from Fischer rats vaccinated with excreted/secreted antigen (ESA) of *T. gondii* conferred protection to the highly susceptible nu/nu strain [60]. Survival times following challenge were twice those of their controls.

Recently, Eisenhauer *et al.* [143] showed that the intramuscular administration of *T. gondii* lysate antigens produced high-titer antigen-specific sera. Mice receiving intravenous heat-killed *Propionibacterium acnes* produced peritoneal macrophages with enhanced toxoplasmacidal activity. These mice had reduced numbers of brain cysts 30 days after peroral challenge. In their study, both antibody and activated macrophages were necessary, since high antibody titer alone did not protect against congenital infection. Treatment that produced both high-titer antibody and activated macrophages provided protection against congenital infection. We have demonstrated that the passive transfer of MAB specific

for P30 in BALB/c mice had an antagonistic effect [81,84]. Normal recipients of MAB exhibited higher mortality and increased morbidity, as determined by the number of intracerebral brain cysts.

Antibody capping of the tachyzoite has been demonstrated. Antibody first moves to the anterior pole of the tachyzoite where a cap is formed. This cap is subsequently discarded upon infection of the host cell. Antibody opsonization may influence intracellular killing by normal mouse peritoneal macrophages. Jones *et al.* [144] has shown that antibody stimulates the fusion of lysosomes with phagosomes containing *Toxoplasma*. These studies suggest that antibody opsonization alters normal macrophage function. Recent work by Joiner *et al.* [31], as previously described, shows that the route of entry of *T. gondii* determines whether the parasitophorous vacuole containing the parasite fuses with other intracellular compartments. These results suggest that the native parasite actively forms a vacuole, devoid of signals for fusion, that is altered upon antibody opsonization.

Role of macrophage

Cell-mediated immunity is critical for specific resistance to *Toxoplasma*. His studies demonstrated that the adoptive transfer of spleen and lymph node cells from chronically infected hamsters protected normal recipients from lethal challenge with the highly virulent RH strain. Previous work had failed to transfer adoptive immunity to this strain in recipient mice using donor cells from mice infected with the avirulent Beverly strain.

Nonspecific enhancement as opposed to previously described immunosuppression of the cellular immune response has been associated with *Toxoplasma* infection. *Toxoplasma*-infected animals show increased resistance to a wide variety of phylogenetically unrelated and related organisms, such as *Listeria monocytogenes*, *Brucella*, *Salmonella*, *Cryptococcus*, *Mycobacterium leprae*, *Besnoitia*, and trypanosomes. Following infection, nonspecific enhanced resistance may be due to the activation of macrophages. However, as previously discussed, the role of NK cells, or perhaps the induction of cytotoxic T cells, may also play a role in this nonspecific enhanced resistance.

Activated macrophages probably do in fact play an important effector mechanism in the acquired resistance to *Toxoplasma*. The ability of *Toxoplasma* to undergo multiplication in normal mouse macrophages was first reported as early as 1954. These findings were confirmed by Nakayama, who was able to infect peritoneal macrophages in tissue culture. The ability of *Toxoplasma* to replicate within macrophages *in vitro* is well recognized [145]. The duration of chronic infection does not influence the capability of macrophages to kill intracellular *Toxoplasma*. Contrasting this view, Jones *et al.* [144] showed that macrophage stimulation appeared to vary as a function of time. In their studies, they showed that macrophages from acutely infected mice were able to inhibit parasite replication, whereas mice infected for more than 3 months had little *in vitro* inhibitory effect. Macrophages can be activated *in vitro* in the presence of lymphocytes and parasite antigen. In the presence of parasite antigen and sensitized lymphocytes, normal guinea pig peritoneal macrophages can be stimulated to resist challenge with lysteria [146]. In the absence of either parasite antigen or sensitized spleen cells, normal spleen cells are unable to show resistance to lysteria. Those cells responsible for activating macrophages have been shown to be sensitized T lymphocytes [147].

Alveolar macrophages play a key role in the defense of the host against pulmonary infection with *T. gondii*. *Toxoplasma* causes pneumonia in immunosuppressed patients. Rying *et al.* [148] followed chronic infection in mice and noted a transient period in which there was a large increase in the number of alveolar macrophages. These studies support activated alveolar macrophages as an effector in the resistance of lungs to infection with *Toxoplasma*.

Macrophages can also be activated *in vitro* when stimulated with lymphocyte mitogens, such as SKSD, conconavalin A or *Toxoplasma* antigen. It has been demonstrated that over 90% of human peripheral blood monocytes phagocytized *Toxoplasma*, degraded the organisms, and retarded the multiplication of those organisms that were not destroyed [149]. Freshly isolated human monocytes from sero-positive individuals were compared for toxoplasmacidal capacity to spleen mononuclear phagocytes by McLeod *et al.* [150]. Their study indicated that almost all the mononuclear phagocytes in peripheral blood have the capacity to eliminate *Toxoplasma*.

Mechanisms of monocyte activity

Studies of monocyte activation by *T. gondii* suggest that both oxygen-dependent and oxygen-independent mechanisms are employed. There are sufficient data supporting a role for the induction of both of these processes during infection with *T. gondii*. McLeod &

Remington [151] described activated macrophages that inhibit or kill *Toxoplasma* by an active process rather than by deprivation of substances necessary for replication. Their studies showed that cyclic AMP is important in the toxoplasmacidal activity of activated macrophages. The ability of TLCK to inhibit this killing suggested that serine esterases may be an important effector mechanism. Murray *et al.* [102–106,152] have done a number of investigations to explore the oxygen-dependent toxoplasmacidal activity of macrophages. In earlier studies, these investigators found that toxic oxygen intermediates mediate macrophage resistance to *Toxoplasma*. It appears that *Toxoplasma* contain high concentrations of catalase and are resistant to peroxide [25]. Inhibition of microbicidal activity by superoxide dismutase and hydrogen peroxide catalase indicated that neither superoxide nor peroxide alone were toxoplasmacidal and that the interaction of the two is required for parasite killing. These investigators went on to determine the intracellular fate of *T. gondii* in normal mouse peritoneal macrophages. In this situation normal macrophages released small quantities of oxide and allowed unrestricted multiplication of intracellular parasites. Chronically infected mouse macrophages released several-fold greater peroxide and displayed toxoplasmacidal activity against the parasite. Macrophages from immunized mice released 25 times more peroxide than normal macrophages and killed ingested *T. gondii* in less than 60 min. The *in vitro* studies suggested that OH and O^{2-} were the major toxic agents.

Wilson & Remington [1] studied the survival of tachyzoites in human monocyte-derived macrophages and normal mouse peritoneal macrophages. Failure of the parasite to stimulate an oxidative burst that normally occurs upon phagocytosis was perhaps responsible for persistence within these cells. In contrast, parasites were rapidly destroyed upon phagocytosis by monocytes. These monocytes exhibited an oxidative burst, as determined by nitro blue tetrazonime reduction, chemiluminescence, and carbon dioxide production. Monocytes from a child and its mother with X-linked chronic granulomatous disease, which is associated with an oxidative burst deficit in monocytes, were both impaired in their ability to kill *Toxoplasma*.

Murray *et al.* [105] determined whether oxygen-dependent or oxygen-independent mechanisms operate against *Toxoplasma*. In this study, fresh monocytes and lymphokine-activated macrophages from normal individuals killed 35–50% of *T. gondii* within 6 h. This activity was associated with release of large amounts of peroxide and respiratory burst activity. Inhibiting the ability to generate oxygen intermediates by glucose deprivation, treatment with superoxide dismutase, catalase, or mannitol inhibited toxoplasmacidal activity by over 80%, allowing a several-fold increase in the number of intracellular parasites. This was compared with fresh monocytes from individuals with chronic granulomotous disease which killed less than 8% of the parasites and exerted 50% less toxoplasmastatic activity. Stimulation with lymphokines (including IFN-γ) induced these granulomatous monocytes and macrophages to display normal levels of toxoplasmastatic activity. This enhancement of activity could be blocked by antibody to IFN-γ and reconstituted with the introduction of recombinant IFN-γ alone. Granulomatous monocytes lost all anti-*Toxoplasma* activity after 2 days in culture, whereas normal monocytes continued to exhibit effective inhibition of *T. gondii* growth. These studies suggest that human mononuclear phagocytes possess an oxygen-independent antiprotozoal mechanism that can be stimulated by IFN-γ. Similarly, Catterall *et al.* [153] explored human alveolar and peritoneal macrophages and their ability to kill *T. gondii*. Unstimulated alveolar macrophages were able to inhibit markedly *T. gondii* replication *in vitro*. Assessment of superoxide released by these alveolar macrophages showed markedly increased amounts when the cells were exposed to phorbal myristate acetate (PMA) or *Candida*. In contrast, these macrophages when incubated when *T. gondii* released no more superoxide than in medium alone. Normal human peritoneal macrophages were able to kill *T. gondii* in the absence of an oxidative burst. Thus, these data in humans imply that normal alveolar and peritoneal macrophages are able to kill intracellular parasites by a nonoxidative mechanism.

Macrophage peroxide released by resident thymocytes was studied by Mellors *et al.* [154]. They found that the peroxide-stimulating factor in thymocyte supernatant was distinct from IFN-γ. Macrophages cultured for several days with cell prethymocyte supernatant released several times more peroxide in response to PMA or opsonized thymocyte than did control macrophages. The factor within the thymocyte supernatant was of approximate molecular weight 2.4 kD and exhibited no IFN activity in either antiviral or enzyme-linked assays. Thymocyte supernatant-treated macrophages failed to exhibit killing of intracellular *T. gondii*, suggesting that the thymocyte factor did not fully activate macrophage microbicidal mechanisms. In this experimental model,

thymocytes can increase the respiratory burst capacity of macrophages in the absence of an antigen-specific immune response.

McCabe *et al.* [155] studied rats which are naturally resistant to *T. gondii* infection. This mode of resistance is believed to be mediated by macrophages. In their studies, peritoneal macrophages were able to kill greater than 90% of ingested *Toxoplasma in vitro*. Incubation with PMA, which exhausts the respiratory burst of the macrophages, impaired the ability of the parasite to replicate.

The role of oxidative metabolism on killing of *T. gondii* was evaluated by Sibley *et al.* [156]. In their studies, they found that the toxoplasmacidal activity of J774G8 cells and peritoneal macrophages was blocked by adding oxygen-intermediate scavengers (catalase or superoxide dismutase) during culture. Activated J774G8 cells produced low levels of oxygen intermediates but, in the presence of the appropriate lymphokine, elevated respiratory burst activity and increased cell killing was noted. These activated cells inhibited the replication of *Toxoplasma*, although *Toxoplasma* replication was not affected by the addition of exogenous catalase or superoxide dismutase. Peritoneal macrophages from immune mice showed that activation leads to destruction of intracellular parasites by perhaps two different mechanisms. Their conclusion was that the capability of *in vitro*-activated J774G8 cells and mouse peritoneal macrophages to inhibit *T. gondii* replication was independent of oxygen intermediates and not affected by exogenous oxygen-intermediate scavengers. When *T. gondii* organisms are protected by exogenous scavengers, they survive the initial microbicidal effects of activated macrophages and reside in vacuoles which resist lysosomal fusion.

DIAGNOSIS OF *TOXOPLASMA* INFECTION

Diagnosis of *Toxoplasma* infection can be determined by a number of methods. Infection may be established by identification of parasites from blood, body fluids, or other tissue. The presence of cysts in the placenta as well as identification of organisms may assist in this diagnosis. The most common method for diagnosis of *T. gondii* infection is serologic tests for antibody against *T. gondii* or the demonstration of either circulating antigen in serum. Identification of antigen or parasites in histologic specimens is the absolute test for parasite infection. The methodology for indirect or direct fluorescent microscopy to determine the presence of organisms in body fluids are well established.

Demonstration of antibodies in serum and body fluids

Detection of circulating antibody in the serum and body fluids of individuals infected with *T. gondii* is by far the most standard approach to serologic evaluation. There have been many articles written comparing the various antibody tests for this organism. In general, detection of rising antibody titer is the most frequent guideline for active infection.

Sabin–Feldman dye test

This test is the classic and gold standard for *Toxoplasma* infection. Sabin & Feldman in 1948 [157] first described the observation that living organisms when incubated with normal serum for 1 h at 37°C became swollen and stained blue. Parasites exposed to antibody-containing serum under the same conditions appear thin and distorted, are not stained when the dye is added, and die as a result of antibody complement lysis. This test is primarily a neutralization assay for IgG antibody and has been used over the years for testing antibody in serum as well as CSF. The reaction is dependent on the complement system [158]. The titer is reported as the dilution of serum at which half of the organisms are not killed (stained) and the other half are killed (unstained). These titers can be expressed in IU/ml of serum as compared to an international standard reference serum (i.e., IU/ml in the dye test corresponds to the following approximate titers expressed in dilution, 2 IU equals 1:8; 8 IU equals 1:32; 15 IU equals 1:64; 60 IU equals 1:256; 300 IU equals 1:1000; 3000 IU equals 1:10 000). Titers of less than or equal to 1:4 are not considered positive except in cases of ocular involvement. Generally, dye test antibodies appear several weeks after infection and rise rapidly to the highest titers over the next few weeks. There is no correlation between antibody titer and the severity of illness. Asymptomatic individuals may have titers of greater than 1:32 000. Persistence of titers in the range of 1:1000 or 1:4000 may continue for months to years, with titers as high as 1:64 remaining throughout the life of the host.

Immunofluorescence test

This test reportedly has about the same sensitivity as the Sabin–Feldman dye test. The advantage to this test is that it does not require living organisms or accessory factor. To perform this test, killed tachyzoites are incubated with serial dilutions of serum and antibody followed by fluorescein-labeled antiserum. Immuno-fluorescent antibody (IFA) titers greater than 1:8 are considered to be consistent with infection and titers greater than 1:1000 suggest acute infection.

The IFA test for IgM antibodies is helpful in the diagnosis of congenital infection. Earlier work by Eichenwald & Shinefield [159] determined that the fetus is able to produce IgM-specific antibody. By this assay, IgM antibodies appear earlier and disappear sooner than IgG antibody. The presence of IgM antibody in the neonate represents antibodies produced by the fetus *in utero* since maternal IgM does not normally pass through the placental barrier. This test attempts to make an early diagnosis of congenital infection by distinguishing passively transferred maternal antibody and the neonatal response to infection. Generally, IgM titers of greater than 1:10 in an adult and 1:2 in a neonate is considered to be positive. A low titer does not eliminate the possibility of active or recent infection. The titer usually declines to 1:10 or 1:40 within several months. However, in some patients the IgM IFA test may remain positive at low titer for years. Some individuals who are immunodeficient may not be able to mount a demonstrable IgM antibody response. Further, a false-positive IgM IFA test may occur in individuals that are rheumatoid factor positive or antinuclear antibody (ANA) positive [160,161]. This false positive is not limited to adults and may be found in newborns as well, perhaps as a result of an IgM immune response *in utero* to passively transferred maternal IgG. Hyde reports that treatment of heat-aggregated IgG can be used to differentiate false-positive IgM IFA tests due to rheumatoid factor from those due to specific IgM antibody to *Toxoplasma*. Failure to demonstrate IgM antibody in this assay may result from high titers of IgG antibody to *Toxoplasma* in the sera [162].

The indirect hemagglutination test and complement fixation test have been reported. In general these tests are not satisfactory as screening methods for pregnant women and AIDS patients, and are not available to most physicians.

ELISA test

The ELISA test has become the mainstay for clinical serologic evaluation of toxoplasmosis. This test is effective in measuring IgG antibody response to *Toxoplasma* infection in humans and other animals. There are a variety of preparative methods for the IgG ELISA. Most tests require the use of a solid substrate to which parasite antigen is bound. Test sera are then incubated with the parasite extract followed by a second antibody directed at the first that is linked to a marker, such as alkaline phosphatase or peroxidase. In general, the sensitivity of this assay equals that of Sabin–Feldman. Practically, the ELISA is much easier to perform and does not require the maintenance of viable parasites.

The use of a double-sandwich IgM for detection of antibodies has been beneficial [163]. This double sandwich avoids false-positive results due to rheumatoid factor and false-negative results due to competition from high levels of maternal IgG antibody. Increased sensitivity of this IgM ELISA was recently reported by Joss *et al.* [164]. They used biotin-labeled *Toxoplasma* antigen with avidin peroxidase, resulting in 100% sensitivity for antibody detection in their sera sampling. Their assay avoided false-positive results from rheumatoid factor or ANA. Balfour *et al.* [165] developed a MAB IgM ELISA. In this assay, MAB directed at human IgM was used on the solid phase followed by *Toxoplasma* antigen that was detected using another MAB coupled to peroxidase.

All of these assays have been hindered by the use of either water lysates or a crude parasite preparation as the specific antigen. Recently a number of investigators have attempted to use specific parasite antigens as a serodiagnostic source. Santoro *et al.* [64,66,166] used the major surface protein P30 of *T. gondii*. In their study, 37 patients with acute toxoplasmosis were evaluated for antibody against P30. Those tested had significantly elevated levels of IgM anti-P30 antibody. Additionally, 40 individuals with chronic *Toxoplasma* infection showed elevated IgG anti-P30 titers. Potasman *et al.* [167,168] examined antibody response to *T. gondii* in 12 congenitally infected infants and seven mothers. They report the presence of antibody to a 60 kD antigen in five out of six cases of neonatal toxoplasmosis. They also indicate that there is a 35 kD antigen that "provokes the strongest and most consistent antibody response following infection with *T. gondii*." In their studies, the 35 kD band differs from P30, although both are frequently noted on the same blot. Identification of IgA antibody directed at P30 may also be an important factor for serodiagnosis of congenital toxoplasmosis [59]. The genes for many antigens with a potential use in diagnosis have been cloned recently [65,169,170]. Recombinant protein is likely to be the basis of future diagnostic kits.

Circulating antigen tests

The presence of circulating *Toxoplasma* antigen in infected subjects has been investigated. Demonstration of *Toxoplasma* antigen in the circulation would be of paramount importance in connection with those individuals who are immunodeficient and are unable to mount a protective immune response. An assay for the direct identification of antigens of *T. gondii* would be valuable to aid in specific diagnosis. Many laboratories have pursued the identification of circulating toxoplasma antigens by various immunologic techniques, including counter-current electrophoresis, dot immunoblotting assay [171], radioimmunoassay, and latex agglutination [172], demonstrating that the lower limit of the ELISA test was sensitive to 4 ng/ml antigen. *Toxoplasma* antigen became demonstrable in the circulation 3 days after infection, prior to the emergence of an antibody response and the development of systemic parasitemia.

In addition to these circulating antigens, there have been several studies that have focused on the secreted antigens of *T. gondii* [173,174]. These exoantigens may constitute as much as 90% of the *Toxoplasma* circulating antigens [174]. Most recently, Decoster *et al.* [175] studied and characterized the various ESA antigens of tachyzoites in cell-free incubation medium. This study focused primarily on the immune response towards antigens which were released by the parasites in the cell-free medium. Sera from acute or chronic infection were used to perform immunoprecipitation studies on these ESA proteins. A variety of ESA antigens were recognized by the IgG fraction from individuals with chronic toxoplasmosis. These antigens ranged in molecular weight from 108 to 28.5 kD. In particular, the 28.5 kD and 108 kD doublets were most characteristic of chronic infection. IgM antibodies in acute infection recognized a 97 kD antigen first. During chronic infection, human sera recognized several antigens. With acute toxoplasmosis the principal parasite ESA antigens recognized by immune sera were of 43 and 30 kD molecular weight. They concluded that in human toxoplasmosis a characteristic antibody pattern could be identified that recognizes specific ESA proteins which may add to the improved serodiagnosis of this condition.

Diagnosis in the immunocompetent host

In immunocompetent adults, testing by ELISA or immunofluorescence virtually concludes the diagnosis. Confirmation of the diagnosis is made if there is a seroconversion from negative to positive titer or a serial twofold rise in antibody drawn at 3-week intervals. A single high titer in any test is not diagnostic although may be suggestive of active infection. In general, an IFA or dye test titer of greater than 1:1000 in the presence of an elevated IgM test (IFA greater than 1:80, IgA ELISA greater than 1:6) is diagnostic of acute infection in the presence or absence of symptoms in individuals with positive titers in the dye test or IFA test. The absence of IgM antibody by IgM ELISA excludes, for practical purposes, the diagnosis of acute infection. However, as mentioned earlier, a markedly elevated IgG titer may in fact cause a false-negative IgM assay. This can be avoided using IgM ELISA techniques as described in the previous section.

Diagnosis of ocular toxoplasmosis in children and adults is more difficult in that serum titer may not correlate with the presence of active lesions. Active toxoplasmic choreoretinitis may occur in the presence of low dye test or conventional immunofluorescent test titers of 1:4 or 1:64. Additionally, IgM antibodies may not be demonstrable in this setting. If retinal lesions are characteristic and serologic tests are confirming, diagnosis may be made with high confidence. If the retinal lesion is atypical and the serologic test is positive, or the retinal lesion is typical and the serologic test negative, the diagnosis is only presumptive.

Diagnosis in immunocompromised patients

Patients receiving high-dose immunosuppressive therapy will usually exhibit a detectable IgM and IgG antibody response to acute infection. It is recommended that *Toxoplasma* serology be performed to screen immunocompromised individuals to identify those who are at risk of primary infection or reactivation of latent disease.

Unique problems in serodiagnosis of toxoplasmosis are those individuals with AIDS. In a recent review, Luft *et al.* [2,129,176] analyzed the data for the diagnosis and treatment of toxoplasmic encephalitis in patients with AIDS in the USA. It is their feeling that those data support the reactivation of latent infection rather than the development of acute acquired infection in these individuals. IgM antibody is rarely present in the sera from these individuals, who almost all have IgG antibody before the onset of their neurologic symptoms. In France, which has a much higher incidence of infection, approximately 80% of the general population and 25% of all patients with AIDS have antibody to *Toxoplasma*. IgM antibody to *T. gondii* can be observed in as many as 20% of patients with acute toxoplasmic encephalitis and

AIDS. Fewer than one-third of all patients with AIDS in the USA have an IgG antibody titer of greater than 1:1000, as determined by either the dye test or conventional IFA. This is similar to the prevalence of toxoplasmosis in the USA.

Intrathecal production of antibody against *Toxoplasma* in AIDS patients has been reported [176]. In one study a CSF rise in antibody occurred, independent of serum antibody titer. Potasman [176] studied 37 patients with AIDS and found 23 had detectable antibody to the parasite in their CSF. Eleven out of 16 patients with AIDS and toxoplasmic encephalitis had production of antibody in the CNS. Thus, although a majority of patients with toxoplasmic encephalitis do not develop serum titers, many of these individuals can produce antibody at the site of infection. Although this may be a specific marker for infection, not all patients have detectable antibody to *Toxoplasma* in their CSF or evidence of local production. Although as little as 3 ng of *Toxoplasma* antigen can be detected in body fluids, current methods are not yet useful for predicting the response to therapy or development of disease.

The recent development of methods for *in vitro* DNA amplification (the "polymerase chain reaction" (PCR)) may be especially useful for the diagnosis of toxoplasmosis. The PCR requires knowledge of a sequence specific to the relevant organism. This then becomes the target for replication or amplification of a short stretch of the gene by repeated cycles of primed extension. The final product can be a several million-fold (or more) amplification of a specific sequence, which can then be detected by standard gene-probing methodologies which can otherwise lack the sensitivity to detect unamplified material.

Several target genes have recently been identified in *Toxoplasma* [177]. One such, arbitrarily referred to as "B1" (coding function unknown), is present in 35 copies in the *Toxoplasma* genome, giving it added advantage in amplification strategies. It is specific to *Toxoplasma* spp. (it is not found in even the closely related *Neospora*) and is conserved in all of over 25 isolates examined from around the world. Using this gene as a target, a single parasite can be detected in idealized laboratory conditions. These probes have been tested in the diagnosis of *Toxoplasma* infection in humans. In acutely infected women, PCR on amniotic fluid has shown unmatched sensitivity and specificity in diagnosing fetal infection [177]. In AIDS patients, PCR has successfully detected the parasite in many tissues [44–46], including cerebrospinal fluid but it is not yet clear how useful the technique will prove in routine management of the infected individual: it is often difficult to obtain a biopsy specimen containing the parasite and the sensitivity and specificity are not yet known because of the difficulty of doing a large-scale prospective study in this latter patient group.

TOXOPLASMA INFECTION IN THE PREGNANT WOMAN

If during routine serologic testing a pregnant woman is found to be positive, an IgM antibody test should be performed. If the IgM assay is unavailable, serology should be repeated in 2–3 weeks. This should be done in order to determine if the titer is changing. If the IgM antibody titer is negative, and/or their is no rise in the IgG titer, then no further testing is necessary. Titers stabilize at 1:1000 or greater within 2 months of infection, regardless of the IgM response. This would indicate that infection was acquired at least 1–2 months prior to the serum sample. Thus, if the IgG titer is greater than or equal to 1:1000 and stable when measured in the first 2 months of pregnancy, the fetus is apparently not at risk.

Congenital infection

Diagnosis of acute *Toxoplasma* infection in the newborn will hopefully be improved using more defined antigen-specific assays and the PCR, as described above. However, diagnosis currently is based on finding either persistent or rising titers in a conventional assay of neonatal serum or a positive IgM antibody in the absence of a placental leak. With only two qualifications, that being the presence of rheumatoid factor or a placental leak, demonstration of IgM antibody in the serum of the newborn is diagnostic of congenital toxoplasmosis. If the mother's serum is negative and the infant's serum is positive, the infant has congenital toxoplasmosis. If both mother and infant are positive, the infant should be tested again. A marked falloff in the IgM titer of the infant after several weeks will occur if the IgM was maternally acquired, as the half-life of IgM is only several days. If the IgM of the infant remains high or continues to rise, this is diagnostic of infection. Synthesis of *Toxoplasma* antibody by the infected infant is usually demonstrable by the third month in an un-treated situation, but may be delayed for 6–9 months if the newborn is treated.

The ability to discriminate between active and chronic infection is especially important in diseases like toxoplasmosis where the infection is common but the disease state restricted to only certain situations. For example, a positive serology in a woman can indicate essentially no risk to the infant (if due to a longstanding chronic infection) or extreme risk if indicative of an acute infection. Currently two serologic analyses separated by often precious weeks are needed to make the diagnosis. The PCR could offer a method for immediate (1–2 days) definitive diagnosis of congenital infection, saving not only time but establishing the diagnosis for the developing fetus directly rather than indirectly assessing risk through diagnosis of the mother.

GENOME AND ITS ORGANIZATION

Composition

Throughout the vast part of its life-cycle, including all of its asexual division in the intermediate hosts, the *T. gondii* nucleus is haploid [178]. The complexity of the genome is estimated at about 8×10^7 bp, which is about 20 times the size of the *Escherichia coli* genome and twice that of *Trypanosoma brucei* [179].

The number and sizes of the chromosome complement in *T. gondii* have been estimated through physical [180] and genetic (L.D. Sibley, E.R. Pfefferkorn, and J.C. Boothroyd, submitted) means. The results indicate that there are probably 11 chromosomes. The 10 smallest span from about 2 through 8–10 Mbp. The largest is of unknown size but by subtracting the combined sizes of the smaller chromosomes from the total genetic complement, it could be as big as 40 Mbp. Using chromosome-specific markers (see below) in comparisons of three strains of *T. gondii*, the chromosomes have been shown to vary in size (between strains) by no more than about 20%. Comparisons of a cloned population of the RH strain sampled (and frozen) in 1977 and in 1988 (following 11 years of continuous passage in culture) showed no detectable alteration in the molecular karyotype over the 11-year period.

An extrachromosomal DNA of *T. gondii* has been partially characterized [181] and shown to comprise apparent homogeneous circular molecules of 36 kb in size. These molecules have a substantial inverted repeat (10 kb). Details of the gene organization and origin of this extrachromosomal element are not known.

Others [182] have shown the presence of an extra-chromosomal element in *T. gondii* with homology to the 6 kb linear element in *Plasmodium* where it has been shown to possess many of the hallmarks expected of a mitochondrial genome although it is exceptionally small.

The form of the homologous element in *Toxoplasma* is not known.

GENE EXPRESSION

Transcription

Nothing is currently known of the RNA polymerases in *T. gondii*. However, several protein-coding genes have been analyzed, as have the ribosomal RNA genes.

The transcription start site for two protein-coding genes, P30 [59] and α-tubulin [183], have been identified unambiguously. The sequences in the immediate vicinity of this region show no striking similarities to each other or to consensus sequences from higher eukaryotes (e.g., TATA and CCAAT boxes). However, such sequences are unlikely to be readily apparent until more genes are analyzed. Interestingly, in the case of P30, about 60 bp upstream of the transcriptional start site, a series of at least five repeats of 27 bp each are found. The role of these repeats, if any, in expression of the P30 gene is unknown.

The start site for the rRNA genes has not been determined. However, there is no reason to suppose that *T. gondii* is unconventional and so the working hypothesis must be that it too will have three RNA polymerases responsible for transcription of the rRNA (polymerase I), protein-coding (polymerase II), and tRNA and 5SRNA genes (polymerase III).

Splicing

Of the six genes analyzed so far in sufficient detail, four have been found to contain one or more introns. In the case of the α- and β-tubulin genes, there are two and three introns, respectively [183]. This situation contrasts with the Kinetoplastida, where no conventional introns have been found in the many genes so far examined [184].

Recombinants containing portions of two exons flanking an intron have been constructed and used to generate RNA molecules *in vitro*. These have been added to a HeLa cell-splicing extract and shown to be spliced accurately and efficiently, indicating that the splicing signals have been conserved between *T. gondii* and mammals [183].

In the Kinetoplastida there exists a phenomenon known as trans-splicing, which describes the splicing together of two discrete transcripts to yield a single chimeric transcript. In all such cases, one of the transcripts encodes a short "mini-exon" or "spliced-leader,"

representing about 39 nucleotides of 5′ untranslated sequence [164]. Definitive analysis of the mRNA-encoding α-tubulin [183] and P30 [65] shows that they are not trans-spliced, i.e., the 5′ end of the mRNA is contiguously encoded with the protein-coding segment.

Also in contrast to the Kinetoplastida, the protein-coding genes of *T. gondii* are, with one exception, not tandemly reiterated, each gene being found as a single copy within the haploid genome. The coding function of the one exception, the so-called B1 gene, is not known [177].

Translation

The nucleotide sequences which have been determined strongly suggest that the universal genetic code is used in *T. gondii*. In the absence of direct amino acid sequence data and corresponding nucleotide sequences, however, an absolute conclusion on this point cannot be made.

The one case where N-terminal sequence data have been determined empirically (P30) is for a protein whose primary translation product is almost certainly processed by the removal of a signal peptide. Hence, the amino acid sequence data are for a processed product and the definitive start site for translation cannot be assigned unambiguously. However, given precedent for the size of signal peptides in other eukaryotes [185] and the presence of an in-frame Met (ATG) codon 30 codons upstream of the processing site, a probable guess can be made for the likely start codon. Based on comparisons with highly conserved sequences in other organisms, the start codon for the tubulin genes can likewise be predicted with a high degree of accuracy. These three sites together show that *T. gondii* appears to conform to "Kozak's rule" on start codon selection [186], i.e., they have a purine at position −3 and a G at position +4 relative to the ATG codon (defined as positions +1, +2, and +3, respectively).

Analysis of several abundantly expressed genes reveals a markedly skewed codon usage. In general, codons ending in C or G are much preferred over those ending in T or A, with some codons used rarely, if ever. (A similar conclusion has recently been reached by Johnson [4,5].) Analysis of less abundantly expressed genes, such as that encoding a putative component of the PEF, reveals little, if any, bias.

Biases in codon usage in abundantly expressed genes have been commonly observed in a wide variety of organisms. The use for this information comes in allowing predictions to be made of a likely gene sequence based on a known amino acid sequence. It can also become important when attempting to express such genes in heterologous systems where final yields are important.

VACCINATION AGAINST TOXOPLASMOSIS

An obvious long-term goal of many laboratories investigating *T. gondii* would be the development of an effective vaccine to protect against both acute and chronic infection. In humans, a vaccine of this nature could provide effective protection against congenital infection as well as infection in those who are immunosuppressed. In domestic animals (i.e., sheep and pigs) this vaccine could prevent spontaneous abortion as well as reduce a major epidemiologic vector for human infection, cysts containing bradyzoites in undercooked meat. Vaccine development could be directed at immunization of intermediate hosts, such as humans or domestic animals, or immunization of domestic cats to prevent sporogeny of the parasite may be possible. In general, most vaccines that utilize killed organisms (heat, formalin, etc.) have been unsuccessful in producing effective immunity. These earlier studies have shown that partial protection was obtainable, but complete protection against challenge with even an avirulent strain of parasite was not obtainable. Much of the earlier and current work has been reviewed in the excellent article by Johnson [187].

Investigation of this problem in our laboratory, as well as by others in recent years, has provided a new framework for the development of an effective vaccine against *Toxoplasma*. Capron *et al.* have shown that rats immunized with ESA are able to confer passively an immunity to the highly susceptible nu/nu strain [30,60]. Upon challenge, the passively transferred recipients were able to survive 30 days, compared to control rats which survived less than half that time. Depletion of the IgE in the sera greatly reduced its ability to induce passive protection in the recipient rats. Interestingly, both platelets bearing surface IgE as well as eosinophils from the donor Fischer +/+ rats exhibited cytotoxicity for the parasite *in vitro*. These platelets when adoptively transferred into recipient nu/nu rats confered significant protection against a lethal challenge of RH-strain parasite. The importance of humoral immunity in protection was also shown by the studies of Eisenhauer *et al.* [143] In their investigations, high titers of anti-*Toxoplasma* antibody acted in a synergistic fashion with *P. acnes*-treated macrophages. Under these conditions, significant protection against congenital transmission was observed [126].

Another approach to the development of a *Toxoplasma* vaccine has been the use of attenuated parasites. Use of irradiation to attenuate both virulent and less virulent parasite strains has resulted in partial protection against acute challenge. Another approach has been the use of ts-4, a temperature-sensitive mutant first isolated by Pfefferkorn *et al.* [79] This attenuated parasite, described earlier in this chapter, is unable to complete the life-cycle in infected cats, which has allowed for further evaluation of vaccine potential. Mice vaccinated with the ts-4 temperature-sensitive mutant have been shown to be highly resistant to challenge with virulent parasite strains. This immunity is dependent on the synthesis of IFN-γ. A recent analysis by Gazzinelli *et al.* [188] has shown that CD4$^+$ splenocytes from BALB/c mice vaccinated with this parasite produce high levels of TH-1 cytokines (IL-2 and IFN-γ) but not TH-2 cytokines (IL-4 and IL-5) when stimulated *in vitro*. CD8$^+$ cells produced lesser quantities of IFN-γ and no IL-2. *In vivo* treatment with MAB directed at CD4$^+$, CD8$^+$, and IFN-γ decreased resistance to infection. Treatment with anti-CD4$^+$ had no effect whereas treatment with anti-CD8$^+$ decreased vaccine-induced resistance. These results suggested that IFN-γ-producing CD8$^+$ cells are the major effectors of immunity *in vitro* in this experimental model.

Recent studies indicate that the cytolytic activity of these CD8$^+$ T cells is genetically restricted [188,189]. Sher *et al.* [188] using the ts-4 mutant have shown that when the target cell (bone marrow macrophages) is exposed to immune T cell from mice immunized with the ts-4 mutant, efficient killing is observed but is MHC Class I restricted. Similar observations have been made by Remington and his group. In their studies, cytotoxic T lymphocyte activity by CD8$^+$ mouse immune cells against target was genetically restricted [189]. In our laboratory, we have observed similar genetic restriction of CD8$^+$ immune cells against peritoneal macrophage infected with *T. gondii*. This cytolytic activity was P30 antigen specific as determined by ineffectual killing of macrophage infected with a P30 deficient mutant, PTgB [85].

The use of individual parasite antigens as effective immunogens has also been reported. Capron *et al.* have expressed a tachyzoite antigen *in vitro* that has an approximate molecular weight of 24 kD [29]. This protein, considered by these workers to be an important component of ESA, was found to be crossreactive with bradyzoites. Antibody directed at this protein was felt to correlate with chronic human infection. The gene expressing this protein was expressed in *E. coli* and eukaryotic cells. Immunocytochemical analysis of this protein located it within the dense granules of both tachyzoite and bradyzoites. In addition, this protein bound calcium, indicates a physiologic function. When used to immunize susceptible rats, this protein was able to induce partial protection by increasing the survival time for several days, although all of the rats eventually succumbed to infection.

In our laboratories we have used P30 to investigate its potential as an effective modulator of protective immunity against acute and chronic toxoplasmosis. When this protein is administered in the presence of either the saponin adjuvant, Quil A, or liposomes we have found that nearly 100% protection can be achieved in mice against acute infection [190,191,192]. Six weeks following challenge, none of the surviving mice were found to have brain cysts containing bradyzoites, suggesting similar protection against chronic infection. Similar results were found in both inbred A/J and C57BL mice. Analysis of the phenotype induced by immunization found an increase in the absolute $CD8^+$ T cell numbers as well as the $CD4^+/CD8^+$ ratio. Adoptive transfer of $CD8^+$ T cells conferred protection in naïve hosts. Moreover, these $CD8^+$ effector cells were found to be parasiticidal to extracellular parasites *in vitro*. Thus, our data suggest that P30 is a potentially effective antigen for immunization against acute and chronic toxoplasmosis.

CONCLUSION

Much has been learned over the past 5–10 years regarding the immunology and molecular biology of *T. gondii*. The modern tools of immunology—MAB, T cell phenotyping and cloning, assessment of cytokine production—have allowed us to identify and characterize those parasite antigens that are important in the host immune response. The reagents produced against these products have further allowed us to initiate studies on the molecular biology of the parasite. These approaches should enhance our ability to diagnose and treat this opportunistic pathogen. Obviously, the overall direction of all these approaches would be to produce an effective vaccine for both animals and humans. Knowledge of the parasite biochemistry, genetics, and physiology will undoubtedly play an important role in achieving this goal. With the advent of AIDS, toxoplasmosis has become an important pathogen and will receive increasing scientific attention in the future.

REFERENCES

1 Wilson CB, Remington JS. What can be done to prevent congenital toxoplasmosis? Am J Obstet Gynecol 1980; 138:357.

2 Luft BJ, Remington JS. AIDS commentary. Toxoplasmic encephalitis. J Infect Dis 1988;157:1–6 (a review).

3 Levy RM, Bredesen DE, Rosenblum ML. Neurological manifestations of the acquired immunodeficiency syndrome (AIDS): experience at UCSF and review of the literature. J Neurosurg 1985;62:475–495 (a review).

4 Johnson AM. Comparison of dinucleotide frequency and codon usage in *Toxoplasma* and *Plasmodium*: evolutionary implications. J Mol Evol 1990;30:383–387.

5 Johnson AM. Toxoplasma: biology, pathology, immunology and treatment. In Long PL, ed. *Coccidiosis of Man and Domestic Animals*. Boca Raton: CRC Press, 1989.

6 Frenkel JK, Dubey JP, Miller NL. *Toxoplasma gondii* in cats: fecal stages identified as coccidian oocysts. Science 1970;167:893–896.

7 Dubey JP, Miller NL, Frenkel JK. The *Toxoplasma gondii* oocyst from cat feces. J Exp Med 1970;132:636–662.

8 Hutchinson WM, Dunachie JF, Siim JC, Work K. Coccidian-like nature of *Toxoplasma gondii*. Br Med J 1970;1:142–144.

9 Overdulve JP. The identity of *Toxoplasma gondii*. Proc K Ned Akad Wet Ser C Biol Med 1970;73:129–141.

10 Sheffield HG, Melton ML. *Toxoplasma gondii*: the oocyst, sporozoite, and infection of cultured cells. Science 1970;167:892–893.

11 Werk R, Dunker R, Fischer S. Polycationic polypeptides: a possible model for the penetration-enhancing factor in the invasion of host cells by *Toxoplasma gondii*. J Gen Microbiol 1984;130:927–933.

12 Nichols BA, Chiappino ML. Cytoskeleton of *Toxoplasma gondii*. J Protozool 1987;34:217–226.

13 Nichols BA, Chiappino ML, O'Connor GR. Secretion from the rhoptires of *Toxoplasma gondii* during host-cell invasion. J Ultrastruct Res 1983;83:85–98.

14 Schwartzman JD, Pfefferkorn ER. Immunofluorescent localization of myosin at the anterior pole of the coccidian *Toxoplasma gondii*. J Protozool 1983;30:657–661.

15 Endo T, Yagita K, Yasuda T, Nakamura T. Detection and localization of actin in *Toxoplasma gondii*. Parasitol Res 1988;75:102–106.

16 Schwartzman JD, Krug EC, Binder LJ, Payne MR. Detection of the microtubule cytoskeleton of the coccidian *Toxoplasma gondii* and the hemoflagellate *Leishmania donovani* by monoclonal antibodies specific for Beta tubulin. J Protozool 1985;32:747.

17 Werk R. How does *Toxoplasma gondii* enter host cells? Rev Infect Dis 1985;7:449–457. (A review.)

18 Lycke E, Norrby R. Demonstration of a factor of *Toxoplasma gondii* enhancig the penetration of toxoplasma parasites into cultured host cells. Br J Exp Pathol 1966;47:248.

19 Schwartzman JD. Inhibition of a penetration-enhancing factor of *Toxoplasma gondii* by monoclonal antibodies specific for rhoptries. Infect Immun 1986;51:760–764.

20 Ossorio PN, Schwartzman JD, Boothroyd JC. A *toxoplasma gondii* rhoptry protein associated with host cell penetration

has unusual charge symmetry. Mol Biochem Parasitol 1992;50:1–16.

21 Kimata I, Tanabe K. Secretion by *Toxoplasma gondii* of an antigen that appears to become associated with the parasitophorous vacuole membrane upon invasion of the host cell. J Cell Sci 1987;88:231–259.

22 Sadak A, Taghy Z, Fortier B, Dubremetz JF. Characterization of a family of rhoptry proteins of *Toxoplasma gondii*. Mol Biochem Parasitol 1988;29:203–211.

23 Joiner KA, Furtado G, Mellman I, *et al*. Cell attachment and invasion by tachyzoites of *Toxoplasma gondii*. J Cell Biochem 1989;13E:64.

24 Grimwood J, Smith JE. *Toxoplasma gondii*: The role of a 30-kDa surface protein in host cell invasion. Exp Parasitol 1992;74:106–111.

25 Robert R, de la Jarrige PL, Mahaza C, Cottin J, Marot-Leblond A, Senet JM. Specific binding of neoglycoproteins to *Toxoplasma gondii* tachyzoites. Inf Immunity 1991;59: 4670–4673.

26 Dubremetz JF, Rodriguez C, Ferreira E. *Toxoplasma gondii*: redistribution of monoclonal antibodies on tachyzoites during host cell invasion. Exp Parasitol 1985;59:24–32.

27 Sibley LD, Weidner E, Krahenbuhl JL. Phagosome acidification blocked by intracellular *Toxoplasma gondii*. Nature 1985;315:416–419.

28 Sibley LD, Krahenbuhl JL. Modification of host cell phagosomes by *Toxoplasma gondii* involves redistribution of surface proteins and secretion of a 32 kda protein. Eur J Cell Biol 1988;47:81–87.

29 Cesbron-Delauw MF, Guy B, Torpier G, *et al*. Molecular characterization of a 23-kilodalton major antigen secreted by *Toxoplasma gondii*. Proc Natl Acad Sci USA 1989;86: 7537–7541.

30 Darcy F, Deslee D, Santoro F, *et al*. Induction of a protective antibody-dependent response against toxoplasmosis by *in vitro* excreted/secreted antigens from tachyzoites of *Toxoplasma gondii*. Parasite Immunol 1988;10:553–567.

31 Joiner KA, Fuhrman SA, Miettinen HM, Kasper LH, Mellman I. *Toxoplasma gondii*: fusion competence of parasitophorous vacuoles in Fc receptor-transfected fibroblasts. Science 1990;249:641–646.

32 Freyre A, Dubey JP, Smith DD, Frenkel JK. Oocyst-induced *Toxoplasma gondii* infection in cats. J Parasitol 1989;75:750–755.

33 Hoff R, Dubey JP, Behbehain AM, Frenkel JK. *Toxoplasma gondii* cysts in cell culture: new biologic evidence. J Parasitol 1977;63:1121–1124.

34 Beyer T, Sim V, Hutchinson W. Cytochemistry of *Toxoplasma gondii*. VI. Polysaccharides, lipids and phosphatase on cyst forms. Tsitogia 1987;19:979.

35 Sims TA, Hay J, Talbot IC. Host–parasite relationship in the brains of mice with congenital toxoplasmosis. J Pathol 1988;156:255–261.

36 Carver BK, Goldman M. Staining *Toxoplasma gondii* with fluoroscein labeled antibody. Am J Clin Pathol 1959;32: 159–164.

37 Kasper L, Ware PL. Identification of stage-specific antigens of *Toxoplasma gondii*. Infect Immun 1989;57:668–672.

38 Jones TC, Bienz KA, Erb P. *In vitro* cultivation of *Toxoplasma gondii* cysts in astrocytes in the presence of gamma

interferon. Infect Immun 1986;51:147–156.

39 Shimata K, O'Connor R, Yoneda C. Cyst formation by *Toxoplasma gondii* (RH strain) *in vitro*. Arch Ophthalmol 1974;92:496–500.

40 Jones TC, Alkan S, Erb P. Spleen and lymph node cell populations, *in vitro* cell proliferation and interferon-gamma production during the primary immune response to *Toxoplasma gondii*. Parasite Immunol 1986;8:619–629.

41 Ise Y, Iida T, Sato K, Suzuki T, Shimada K, Nishioka K. Detection of circulating antigens in sera of rabbits infected with *Toxoplasma gondii*. Infect Immun 1985;48:269–272.

42 Frenkel JK, Escajadillo A. Cyst rupture as a pathogenic mechanism of toxoplasmic encephalitis. Am J Trop Med Hyg 1987;36:517–522.

43 Hofflin JM, Conley FK, Remington JS. Murine model of intracerebral toxoplasmosis. J Infect Dis 1987;155:550–557.

44 Holliman RE, Johnson JD, Gillespie SH, Johnson MA, Squire SB, Savva D. New methods in the diagnosis and management of cerebral toxoplasmosis associated with the acquired immune deficiency syndrome. J Infect 1991;22: 281–285.

45 Holliman RE. Clinical and diagnostic findings in 20 patients with toxoplasmosis and the acquired immune deficiency syndrome. J Med Microbiol 1991;35:1–4.

46 Cristina N, Derouin F, Pelloux H, Pierce R, Cesbron-Delauwn MF, Ambroise-Thomas P. Detection of *Toxoplasma gondii* by "polymerase chain reaction" (PCR) technique in AIDS infected patients using the repetitive sequence TGRIE. Pathol Biol 1992;40:52–55.

47 Johnson AM, McDonald PJ, Neoh SH. Monoclonal antibodies to Toxoplasma cell membrane surface antigens protect mice from Toxoplasmosis. J Protoxool 1983;30:351–356.

48 Johnson AM, McDonald PJ, Neoh SH. Molecular weight analysis of soluble antigens from *Toxoplasma gondii*. J Parasitol 1983;69:459–464.

49 Handman E, Goding JW, Remington JS. Detection and characterization of membrane antigens of *Toxoplasma gondii*. J Immunol 1980;124:2578–2583.

50 Handman E, Remington JS. Serological and immunochemical characterization of monoclonal antibodies to *Toxoplasma gondii*. Immunology 1980;40:579.

51 Johnson AM, McNamara PJ, Neoh SH, McDonald PJ, Zola H. Hybridomas secreting monoclonal antibody to *Toxoplasma gondii*. Aust J Exp Biol Med Sci 1981;59:303.

52 Johnson AM, McDonald PJ, Neoh SH. Molecular weight analysis of the major polypeptides and glycopeptides of *Toxoplasma gondii*. Biochem Biophys Res Commun 1981; 100:934–943.

53 Kasper LH, Crabb JH, Pfefferkorn ER. Purification of a major membrane protein of *Toxoplasma gondii* by immunoabsorption with a monoclonal antibody. J Immunol 1983; 130:2407–2412.

54 Couvreur G, Sadak A, Fortier B, Dubremetz JF. Surface antigens of *Toxoplasma gondii*. Parasitology 1988;97:1–10.

55 Pfefferkorn ER, Kasper LH. *Toxoplasma gondii* genetic crosses reveal phenotypic suppression of hydroxyurea resistance by fluorodeoxyuridine resistance. Exp Parasitol 1983;55:207–218.

56 Sharma SJ, Mullenax J, Araujo AG, Erlich HA, Remington

JS. Western blot analysis of the antigens of *Toxoplasma gondii* recognized by human IgM and IgG antibodies. J Immunol 1983;131:977–983.

57 Ogata K, Kasahara T, Shioiri-Nakano K, Igarashi I, Suzuki M. Immunoenzymatic detection of three kinds of 43 000 molecular weight antigens by monoclonal antibodies in the insoluble fraction of *Toxoplasma gondii*. Infect Immun 1984;43:1047–1053.

58 McLeod R, Mack DG. Secretory IgA specific for *Toxoplasma gondii*. J Immunol 1986;136:2640–2643.

59 Decoster A, Darcy F, Caron A, Capron A. IgA antibodies against P30 as markers of congenital and acute toxoplasmosis. Lancet 1988;2:1104–1107.

60 Ridel PR, Auriault C, Darcy F, *et al*. Protective role of IgE in immunocompromised rat toxoplasmosis. J Immunol 1988; 141:978–983.

61 Chardes T, Bourguin I, Mevelec MN, Dubremetz JF, Bout D. Antibody responses to *Toxoplasma gondii* in sera, intestinal secretions, and milk from orally infected mice and characterization of target antigens. Infect Immun 1990; 58:1240–1246.

62 Kasper LH, Crabb JH, Pfefferkorn ER. Isolation and characterization of a monoclonal antibody resistant antigenic mutant of *Toxoplasma gondii*. J Immunol 1982;129: 1694–1699.

63 Ware PL, Kasper LH. Strain-specific antigens of *Toxoplasma gondii*. Infect Immun 1987;55:778–783.

64 Santoro F, Rodriguez C, Afchain D, Capron A, Dissous C. Major surface protein of *Toxoplasma gondii* (P30) contains an immunodominant region with repetitive epitopes. Eur J Immunol 1985;15:747–749.

65 Burg JL, Perelman D, Kasper LH, Ware PL, Boothroyd JC. Molecular analysis of the gene encoding the major surface antigen of *Toxoplasma gondii*. J Immunol 1988;141:3584–3591.

66 Santoro F, Afchain D, Pierce R, Cesbron JY, Ovlaque G, Capron A. Serodiagnosis of *Toxoplasma* infection using a purified parasite protein (P30). Clin Exp Immunol 1985;62: 262–269.

67 Buelow R, Boothroyd JC. Protection of mice from *Toxoplasma* infection with P30 antigen in liposomes. J Immunol 1991;147:3496–3500.

68 Nagel SD, Boothroyd JC. The major surface antigen, P30, of *Toxoplasma gondii* is anchored by a glycolipid. J Biol Chem 1989;264:5569–5574.

69 Tomavo S, Schwarz RT, Dubremetz JR. Evidence for glycosylphosphatidylinositol anchor of *Toxoplasma gondii* major surface antigens. Mol Cell Biol 1989;9:4576–4578.

70 Sethi KK, Rahman A, Pelster B, Brandis H. Search for the presence of lectin-binding sites on *Toxoplasma gondii*. J Parasitol 1977;63:1076–1080.

71 Suzuki Y, Kobayashi A. Presence of high concentrations of circulating toxoplasma antigens during acute *Toxoplasma* infection in athymic nude mice. Infect Immun 1987;55: 1017–1018.

72 Asai T, Kim TJ, Kobayashi M, Kojima S. Detection of nucleoside triphosphate hydrolase as a circulating antigen in sera of mice infected with *Toxoplasma gondii*. Infect Immun 1987;55:1332–1335.

73 Asai T, Kim T. Possible regulation mechanism of potent

74 Fuhrman SA, Joiner KA. *Toxoplasma gondii*: mechanism of resistance to complement-mediated killing. J Immunol 1989;142:940–947.

75 Suggs M, Walls KW, Kagan IG. Comparative antigenic study of *Besnoitia jellisoni*, *B. panamensis*, and five *Toxoplasma gondii* isolates. J Immunol 1968;101:166–175.

76 Darde ML, Bouteille B, Pestre-Alexander M. Isoenzymic characterization of seven strains of *Toxoplasma gondii* by isoelectrofocusing in polyacrylamide gels. Am J Trop Med Hyg 1988;39:551.

77 Cristina N, Oury B, Ambroise-Thomas P, Santoro F. Restriction-fragment-length polmorphisms among *Toxoplasma gondii* strains. Parasit Res 1991;77:266–268.

78 Kasper LH. Isolation and characterization of a monoclonal anti-P30 antibody resistant mutant of *Toxoplasma gondii*. Parasite Immunol 1987;9:433–445.

79 Pfefferkorn ER, Pfefferkorn LC. *Toxoplasma gondii*: isolation and preliminary characterization of temperature-sensitive mutants. Exp Parasitol 1976;39:365–373.

80 Kasper LH, Bradley MS, Pfefferkorn ER. Identification of stage-specific sporozoite antigens of *Toxoplasma gondii* by monoclonal antibodies. J Immunol 1984;132:443–449.

81 Kasper LH, Ware PL. Recognition and characterization of stage-specific oocyst/sporozoite antigens of *Toxoplasma gondii* by human antisera. J Clin Invest 1985;75:1570–1577.

82 Beneson MW, Takafuji ET, Lemon SM, Greenup RL, Sulzer AJ. Oocyst transmitted toxoplasmosis associated with the ingestion of contaminated water. N Engl J Med 1982;307:666–669.

83 Lunde MN, Jacobs L. Antigenic differences between endozoites and cystozoites of *Toxoplasma gondii*. J Parasitol 1983;69:806–811.

84 Kasper LH, Currie KM, Bradley MS. An unexpected response to vaccination with a purified major membrane tachyzoite antigen (P30) of *Toxoplasma gondii*. J Immunol 1985;134:3426–3431.

85 Tomavo S, Fortier B, Soete M, Ansel C, Camus D, Dubremetz JF. Characterization of bradyzoite-specific antigens of *Toxoplasma gondii*. Infect Immun 1991;59:3750–3753.

86 Pavia CS. Enhanced primary resistance to *Treponema pallidum* infection and increased susceptibility to toxoplasmosis in T-cell-depleted guinea pigs. Infect Immun 1986;53:305–311.

87 Pavia CS. Protection against experimental toxoplasmosis by adoptive immunotherapy. J Immunol 1986;137:2985–2990.

88 Krahenbuhl JL, Gaines JD, Remington JS. Lymphocyte transformation in human toxoplasmosis. J Infect Dis 1972; 125:283–288.

89 Macario A, Stahl WRM. Lymphocyte subpopulations and function in chronic murine toxoplasmosis. Clin Exp Immunol 1980;41:425.

90 Hauser WEJ, Tsai V. Acute toxoplasma infection of mice induces spleen NK cells that are cytotoxic for *T. gondii in vitro*. J Immun 1986;136:313–319.

91 Kamiyama TTH. Augmented followed by suppressed

levels of natural cell mediated cytotoxicity in mice infected with *Toxoplasma gondii*. Infect Immun 1982;36:628.

92 Hughes HP, Kasper LH, Little J, Dubey JP. Absence of a role for natural killer cells in the control of acute infection by *Toxoplasma gondii* oocysts. Clin Exp Immunol 1988;72: 394–399.

93 Dannemann BR, Morris VA, Araujo FG, Remington JS. Assessment of human natural killer and lymphokine-activated killer cell cytotoxicity against *Toxoplasma gondii* trophozoites and brain cysts. J Immunol 1989;143:2684–2691.

94 Vollmer TL, Waldor MK, Steinman L, Conley FK. Depletion of T-4+ lymphocytes with monoclonal antibody reactivates toxoplasmosis in the central nervous system: a model of superinfection in AIDS. J Immunol 1987;138: 3737–3741.

95 Sklenar I, Jones TC, Alkan S, Erb P. Association of symptomatic human infection with *Toxoplasma gondii* with imbalance of monocytes and antigen-specific T cell subsets. J Infect Dis 1986;153:315–324.

96 McLeod R, Eisenhauer P, Mack D, Brown C, Filice G, Spitalny G. Immune responses associated with early survival after peroral infection with *Toxoplasma gondii*. J Immunol 1989;142:3247–3255.

97 Khan IA, Smith KA, Kasper LH. Induction of antigen-specific parasiticidal cytotoxic T cell splenocytes by a major membrane protein (P30) of *Toxoplasma gondii*. J Immunol 1988;141:3600–3605.

98 Khan IA, Eckel ME, Pfefferkorn ER, Kasper LH. Production of gamma interferon by cultured human lymphocytes stimulated with a purified membrane protein (P30) from *Toxoplasma gondii*. J Infect Dis 1988;157:979–984.

99 Khan IA, Smith KA, Kasper LH. Induction of antigen-specific human cytotoxic T cells by *Toxoplasma gondii*. J Clin Invest 1990;85:1879–1886.

100 Shirahata T, Shimizu K. Production and properties of immune interferon from spleen cell cultures of *Toxoplasma*-infected mice. Microbiol Immunol 1980;24:1109–1120.

101 Sun T, Greenspan J, Tenenbaum M, *et al*. Diagnosis of cerebral toxoplasmosis using fluorescein-labeled anti-*Toxoplasma* monoclonal antibodies. Am J Surg Pathol 1986; 10:312–316.

102 Lepay DA, Nathan CF, Steinman RM, Murray HW, Cohn ZA. Murine kupffer cells. Mononuclear phagocytes deficient in the generation of reactive oxygen intermediates. J Exp Med 1985;161:1079–1096.

103 Lepay DA, Steinman RM, Nathan CF, Murray HW, Cohn ZA. Liver macrophages in murine listeriosis. Cell-mediated immunity is correlated with an influx of macrophages capable of generating reactive oxygen intermediates. J Exp Med 1985;161:1503–1512.

104 Murray HW, Gellene RA, Libby DM, Rothermel CD, Rubin BY. Activation of tissue macrophages from AIDS patients: *in vitro* response of aids alveolar macrophages to lymphokines and interferon-gamma. J Immunol 1985;135: 2374–2377.

105 Murray HW, Rubin BY, Carriero SM, Harris AM, Jaffee EA. Human mononuclear phagocyte antiprotozoal mechanisms: oxygen-dependent vs. oxygen-independent

activity against intracellular *Toxoplasma gondii*. J Immunol 1985;134:1982–1988.

106 Murray HW, Spitalny GL, Nathan CF. Activation of mouse peritoneal macrophages *in vitro* and *in vivo* by interferon-gamma. J Immunol 1985;134:1619–1622.

107 Kelly CD, Welte K, Murray HW. Antigen-induced human interferon-gamma production. Differential dependence on interleukin 2 and its receptor. J Immunol 1987;139:2325–2328.

108 Murray HW, Scavuzzo D, Jacobs JL, *et al*. *In vitro* and *in vivo* activation of human mononuclear phagocytes by interferon-gamma. Studies with normal and AIDS monocytes. J Immunol 1987;138:2457–2462.

109 Mellors JW, Debs RJ, Ryan JL. Incorporation of recombinant gamma interferon into liposomes enhances its ability to induce peritoneal macrophage anti-*Toxoplasma* activity. Infect Immun 1989;57:132–137.

110 Pfefferkorn ER, Guyre PM. Inhibition of growth of *Toxoplasma gondii* in cultured fibroblasts by human recombinant gamma interferon. Infect Immun 1984;44:211–216.

111 Pfefferkorn ER. Interferon gamma and the growth of *Toxoplasma gondii* in fibroblasts. Ann Inst Pasteur Microbiol 1986;137a:348–352.

112 Pfefferkorn ER, Eckel M, Rebhun S. Interferon-gamma suppresses the growth of *Toxoplasma gondii* in human fibroblasts through starvation for tryptophan. Mol Biochem Parasitol 1986;20:215–224.

113 Pfefferkorn ER, Rebhun S, Eckel M. Characterization of an indoleamine 2,3-dioxygenase induced by gamma-interferon in cultured human fibroblasts. J Interferon Res 1986;6:267–279.

114 Suzuki Y, Orellana MA, Schreiber RD, Remington JS. Interferon-gamma: the major mediator of resistance against *Toxoplasma gondii*. Science 1988;240:516–518.

115 Suzuki Y, Conley FK, Remington JS. Importance of endogenous IFN-γ for prevention of toxoplasmic encephalitis in mice. J Immunol 1989;143:2045–2050.

116 Sharma SD, Hofflin JM, Remington JS. *In vivo* recombinant interleukin 2 administration enhances survival against a lethal challenge with *Toxoplasma gondii*. J Immunol 1985; 135:4160–4163.

117 DeTitto E, Catterall JR, Remington JS. Activity of recombinant tumour necrosis factor on *Toxoplasma gondii* and *Trypanasoma cruzi*. J Immunol 1986;137:1342.

118 Freund YR, Sgarlato G, Jacob G, Suzuki Y, Remington JS. Polymorphisms in the tumor necrosis factor gene correlate with murine resistance to development of toxoplasmic encephalitis and with levels of TNF mRNA in infected brain tissue. JE p Med 1992;175:683–688.

119 Sibley LD, Adams LA, Fukutome Y, Krahenbuhl JL. Tumor necrosis factor-alpha triggers antitoxoplasmal activity of IFN-γ primed macrophages. J Immunol 1991;147:2340–2345.

120 Johnson LL. A protective role for endogenous tumor necrosis factor in *Toxoplasma gondii* infection. Inf Immun 1992;60:1979–1983.

121 Reyes L, Frenkel JK. Specific and nonspecific mediation of protective immunity to *Toxoplasma gondii*. Infect Immun 1987;55:856–863.

122 Gazzinelli RT, Oswald IP, James SL, Sher A. IL-10 inhibits

parasite killing and nitrogen oxide production by IFN γ activated macrophages. J Immunol 1992;148:1792–1796.

123 Locksley RM, Fankhauser J, Henderson WR. Alteration of leukotriene release by macrophages ingesting *Toxoplasma gondii*. Proc Natl Acad Sci USA 1985;82:6922–6926.

124 Strickland GT, Pettitt LE, Voller A. Immunodepression in mice infected with *Toxoplasma gondii*. Am J Trop Med Hyg 1973;22:452.

125 Huldt G. Effect of *Toxoplasma gondii* on the thymus. Nature 1973;244:301.

126 Hibbs J Jr, Remington J, Stewart C. Modulation of immunity and host resistance by micro-organisms. Pharmacol Ther 1980;8:37.

127 Strickland G, Ahmed A, Sell K. Blastogenic response of *Toxoplasma*-infected mouse spleen cells to T- and B-cell mitogens. Clin Exp Immunol 1975;22:167.

128 Suzuki Y, Kobayashi A. Suppressive effect of secondary *Toxoplasma gondii* infection on antibody responses in mice. Infect Immun 1985;48:686–689.

129 Luft BJ, Pedrotti PW, Remington JS. *In vitro* generation of adherent mononuclear suppressor cells to toxoplasma antigen. Immunology 1988;63:643–648.

130 Chan J, Siegel JP, Luft BJ. Demonstration of T-cell dysfunction during acute toxoplasma infection. Cell Immunol 1986;98:422–433.

131 Yano A, Norose K, Yamashita K, *et al*. Immune response to *Toxoplasma gondii*—analysis of suppressor T cells in a patient with symptomatic acute toxoplasmosis. J Parasitol 1987;73:954–961.

132 Jones TC, Alkan S, Erb P. Murine spleen and lymph node cellular composition and function during cyclophosphamide and splenectomy induced resistance to *Toxoplasma gondii*. Parasite Immunol 1987;9:117–131.

133 Israelski DM, Araujo FG, Conley FK, Suzuki Y, Sharma S, Remington JS. Treatment with anti-L3T4 (CD4) monoclonal antibody reduces the inflammatory response in toxoplasmic encephalitis. J Immunol 1989;142:954–958.

134 Ruskin J, Remington J. Toxoplasmosis in the compromised host. Ann Intern Med 1976;84:193.

135 Frenkel JK, Nelson BM, Arias-Stella J. Immunosuppression and toxoplasmic encephalitis: clinical and experimental aspects. Hum Pathol 1975;6:97.

136 Frenkel JK, Piekarski G. The demonstration of *Toxoplasma* and other organisms by immunofluorescence: a pitfall. J Infect Dis 1978;138:265.

137 Suzuki Y, Joh K, Oreilana MA, Conles Fk, Remington JS. A gene(s) within the H-2D region determines the development of toxoplasmic encephalitis in mice. Immunology 1991;74:732–739.

138 Brinkmann V, Sharma SD, Remington JS. Different regulation of the 13T4-T cell subset by B cells in different mouse strains bearing the H-2k haplotype. J Immunol 1986;137:2991–2997.

139 Piekarski G, Witte HM. Experimentelle und histologische studien zur *Toxoplasma*-Infection der Hauskatze. Z Parasitenkd 1971;36:95.

140 Kuhn D, Weiland G. Experimentelle Toxoplasma-Infektionen bei der Katze. I. Wiederholte ubertragung von *Toxoplasma gondii* durch kot von mit Nematoden

infizierten Katzen. Berl Muench Tieraerztl Wochenschr 1969;82:401.

141 Eichenwald HF. Experimental toxoplasmosis. II. Effect of sulfadiazine and antiserum on congenital toxoplasmosis in mice. Proc Soc Exp Biol Med 1949;71:45.

142 Gill HS, Prakash O. Chemotherapy of experimental *Toxoplasma gondii* (RH strain) infection. Indian J Med Res 1970; 58:1197.

143 Eisenhauer P, Mack DG, McLeod R. Prevention of peroral and congenital acquisition of *Toxoplasma gondii* by antibody and activated macrophages. Infect Immun 1988;56:83–87.

144 Jones TC, Len L, Hirsch JG. Assessment *in vitro* of immunity against *Toxoplasma gondii*. J Exp Med 1975;141: 466.

145 Nicoble C, Manceaux L. Sur une infection á corp de Leishman (on organism voisins) du gondi. CR Acad Sci 1908;147:763–766.

146 Krahenbuhl JL, Remington JS. *In vitro* induction of nonspecific resistance in macrophages by specifically sensitized lymphocytes. Infect Immun 1971;4:337.

147 Krahenbuhl JL, Rosenberg L, Remington JS. The role of thymus-derived lymphocytes in the *in vitro* activation of macrophages to kill *Listeria monocytogenes*. J Immunol 1973; 111:992.

148 Rying FW, Remington JS. Effect of alveolar macrophages on *Toxoplasma gondii*. Infect Immunol 1977;18:746.

149 Anderson SE Jr, Bautista S, Remington JS. Induction of resistance to *Toxoplasma gondii* in human macrophages by soluble lymphocyte products. J Immunol 1976;117:381.

150 McLeod R, Bensch KG, Smith SM, Remington JS. Effects of human peripheral blood monocytes, monocyte-derived macrophages, and spleen mononuclear phagocytes on *Toxoplasma gondii*. Cell Immunol 1980;54:330.

151 McLeod R, Remington JS. Inhibition or killing of an intracellular pathogen by activated macrophages is abrogated by TLCK or aminophylline. Immunology 1980;39:599.

152 Murray HW. Cellular resistance to protozoal infection. Annu Rev Med 1986;37:61–69.

153 Catterall JR, Black CM, Leventhal JP, Rizk NW, Wachtel JS, Remington JS. Nonoxidative microbicidal activity in normal human alveolar and peritoneal macrophages. Infect Immun 1987;55:1635–1640.

154 Mellors JW, Bartiss AH, Coleman DL. Stimulation of macrophage H_2O_2 release by resident thymocytes: effect of a soluble factor distinct from interferon-gamma. Cell Immunol 1987;110:391–399.

155 McCabe RE, Catterall JR, Remington JS. Unique differences in infectivity and seroreactivity of *Toxoplasma* harvested from mice infected for different lengths of time. J Parasitol 1987;73:1152–1157.

156 Sibley LD, Krahenbuhl JL, Weidner E. Lymphokine activation of J774G8 cells and mouse peritoneal macrophages challenged with *Toxoplasma gondii*. Infect Immun 1985;49: 760–764.

157 Sabin AB, Feldman HA. Dyes as microchemical indicators of a new immunity phenomenon affecting a protozoan parasite (*Toxoplasma*). Science 1948;108:660.

158 Schreiber RD, Feldman HA. Identification of the activator system for antibody to *Toxoplasma* as the classical complement pathway. J Infect Dis 1980;141:366.

159 Eichenwald HF, Shinefield HR. Antibody production by the human fetus. J Pediatr 1963;63:870.

160 Hyde B, Barnett EV, Remington JS. Method for differentiation of nonspecific from specific *Toxoplasma* IgM fluorescent antibodies in patients with rheumatoid factor. Proc Soc Exp Biol Med 1974;148:1184.

161 Araujo FG, Barnett EV, Gentry LO, Remington JS. False-positive anti-*Toxoplasma* fluorescent-antibody tests in patients with antinuclear antibodies. Appl Microbiol 1971; 22:270.

162 Filice GA, Yeager AS, Remington JS. Diagnostic significance of immunoglobulin M antibodies to *Toxoplasma gondii* detected after separation of immunoglobulin M from immunoglobulin G antibodies. J Clin Microbiol 1980;12: 336.

163 Naot Y, Remington JS. An enzyme-linked immunosorbent assay for detection of IgM antibodies to *Toxoplasma gondii*: use for diagnosis of acute acquired toxoplasmosis. J Infect Dis 1980;142;757.

164 Joss AW, Skinner LJ, Moir IL, Chatterton JM, Williams H, Ho YDO. Biotin-labelled antigen screening test for toxoplasma IgM antibody. J Clin Pathol 1989;42:206–209.

165 Balfour AH, Harford JP, Goodall M. Use of monoclonal antibodies in an ELISA to detect IgM class antibodies specific for *Toxoplasma gondii*. J Clin Pathol 1987;40:853–857.

166 Santoro F, Cesbron JY, Capron A, Ovlaque G. Use of a monoclonal antibody in a double-sandwich ELISA for detection of IgM antibodies to *Toxoplasma gondii* major surface protein (P30). J Immunol Methods 1985;83:151–158.

167 Hofflin JM, Potasman I, Baldwin JC, Oyer PE, Stinson EB, Remington JS. Infectious complications in heart transplant recipients receiving cyclosporine and corticosteroids. Ann Intern Med 1987;106:209–216.

168 Potasman I, Araujo FG, Thulliez P, Desmonts G, Remington JS. *Toxoplasma gondii* antigens recognized by sequential samples of serum obtained from congenitally infected infants. J Clin Microbiol 1987;25:1926–1931.

169 Tenter AM, Johnson AM. Recognition of recombinant *Toxoplasma gondii* antigens by human sera in an ELISA. Parasit Res 1991;77:197–203.

170 Tenter AM, Johnson AM. Human antibody response to the nucleoside triphosphate hydrolase of *Toxoplasma gondii*. J Immunoassay 1990;11:579–590.

171 Brooks RG, Sharma SD, Remington JS. Detection of *Toxoplasma gondii* antigens by a dot-immunobinding technique. J Clin Microbiol 1985;21:113–116.

172 Pallangyo KJ, Suzuki H, Fukumoto Y, Matsumoto K. One point dilution enzyme-linked immunosorbent assay (ELISA) for *Toxoplasma gondii* seroepidemiological surveys. Tohoku J Exp Med 1985;147:349–356.

173 Chumpitazi B, Ambroise TP, Cagnard M, Autheman JM. Isolation and characterization of *Toxoplasma* exo-antigens from *in vitro* culture in MRC5 and VERO cells. Int J Parasitol 1987;17:829–834.

174 Hughes HPA, VanKnapen F, Atkinson HJ, Balfour AH, Lee DH. A new soluble antigen preparation of *Toxoplasma gondii* and its use in serological diagnosis. Clin Exp Immunol 1982;49:239–246.

175 Decoster A, Darcy F, Capron A. Recognition of *Toxoplasma gondii* excreted and secreted antigens by human sera from acquired and congenital toxoplasmosis: identification of markers of acute and chronic infection. Clin Exp Immunol 1988;73:376–382.

176 Potasman I, Resnick L, Luft BJ, Remington JS. Intrathecal production of antibodies against *Toxoplasma gondii* in patients with toxoplasmic encephalitis and the acquired immunodeficiency syndrome (AIDS). Ann Intern Med 1988;108:49–51.

177 Grover CM, Thylliez P, Remington JS, Boothroyd JC. Rapid prenatal diagnosis of congenital *Toxoplasma* infection from amniotic fluid by polymerase chain reaction. J Clin Micro 1990;28:2297–2301.

178 Pfefferkorn EP, Pfefferkorn LC, Colby ED. Development of gametes and oocysts in cats fed cysts derived from cloned trophozoites of *Toxoplasma gondii*. J Parasitol 1977; 63:158–159.

179 Cornelissen AWCA, Overdulve JP, Van der Pleog M. Determination of nuclear DNA of five Eucoccidian parasites, *Isospora* (*Toxoplasma*) *gondii, Sarcocystis cruzi, Eimeria tenella, E. acervulina* and *Plasmodium berghei*, with special reference to gamontogenesis and meiosis in *I.* (*T.*) *gondii*. Parasitology 1984;88:531–553.

180 Sibley LD, Boothroyd JC. Construction of a molecular karyotype for *Toxoplasma gondii*. Mol Biochem Parasitol (in press).

181 Borst P, Overdulve JP, Weijers PJ, Fase-Fowler F, Van den Berg M. DAN circles with cruciforms from *Isospora* (*Toxoplasma*) *gondii*. Biochem Biophys Acta 1984;781:100–111.

182 Joseph JT, Aldritt SM, Unnasch T, Puijalon O, Wirth D. Characterization of a conserved extrachromosomal element isolated from the avian malarial parasite *Plasmodium gallinaceum*. Mol Cell Biol 1989;9:3621–3629.

183 Nagel SD, Boothroyd JC. The alpha- and beta-tubulins of *Toxoplasma gondii* are encoded by single copy genes containing multiple introns. Mol Biochem Parasitol 1988;29: 261–273.

184 Boothroyd JC. Trans-splicing of RNA. In Eckstein F, Lilley DMJ, eds. *Nucleic Acids and Molecular Biology*, vol. 3. Berlin: Springer-Verlag, 1989;216–230.

185 Walter P, Lingappa VR. Mechanism of protein translocation across the endoplasmic reticulum membrane. Annu Rev Cell Biol 1986;2:499–516.

186 Kozak M. The scanning model for translation: an update. J Cell Biol 1989;108:229–241.

187 Johnson AM. *Toxoplasma* vaccines. In Wright IG, ed. *Veterinary Protozoan and Hemoparasite Vaccines*, Chapter 9. Boca Raton: CRC Press, 1989.

188 Gazzinelli RT, Hakim FT, Hieny S, Shearer GM, Sher A. Synergistic role of CD4$^+$ and CD8$^+$ T lymphocytes in IFN-gamma production and protective immunity induced by an attenuated *Toxoplasma gondii* vaccine. J Immunol 1991; 146;286–292.

189 Subauste CS, Koniaris AH, Remington JS. Murine CD8$^+$ cytotoxic T lymphocytes lyse *Toxoplasma gondii*-infected cells. J Immunol 1991;147:3955–3959.

190 Khan IA, Ely K, Kasper LH. A purified parasite antigen (P30) mediates CD8$^+$ T cell immunity against fatal *Toxo-*

plasma infection in mice. J Immunol 1991;147:3501–3506.

191 Buelow E, Boothroyd JC. Protection of mice from fatal *Toxoplasma* infection by immunization with P30 antigen in liposomes. J Immunol 1991;147:3496–3502.

192 Kasper LH, Boothroyd JC, Buelow E, Ely KH, Khan IA. CD8⁺ T cells from mice immunized with a purified parasite antigen (P30) against fatal *Toxoplasma* infection exhibit MHC restricted cytolysis. J Immunol 1992;148:1493–1498.

14 Malaria

Johanne Melancon-Kaplan, James M. Burns Jr, Akhil B. Vaidya, H. Kyle Webster,
& William P. Weidanz

INTRODUCTION

Malaria remains one of the world's major killing diseases, with an estimated annual mortality of 2.5 million victims [1]. Recent World Health Organization statistics [2] suggest that more than 250 million people are infected with plasmodia and that some 2.1 billion people—half the world's population—live in areas where malaria is common. While the accuracy of these figures can be challenged [3], it is possible that morbidity and mortality due to malaria are greater today than ever before. In addition to the enormous health problems caused by malaria, the social and economic consequences of this disease continue to have a tremendous impact on development in Third World nations.

Following World War II it was thought that malaria could be eradicated through the use of antiparasite drugs and insecticides (see Bruce-Chwatt [4]). At first, these attempts appeared to be successful, with malaria being eliminated from many parts of Europe, Asia, and North America, but then resistance to both antimalarials and insecticides developed, and many of the poorer countries of the world where malaria is prevalent were faced with increasing hardships brought on by economic and political instability. Left without adequate resources, many of these countries have been unable to develop, implement, and maintain appropriate programs to eradicate or even control malaria [5].

PARASITE

The genus *Plasmodium* is a member of the protozoan phylum Apicomplexa. Its line of classification is: class: Sporozoa; subclass: Coccidia; order: Eucoccidiida; suborder: Haemosporina. All members of Apicomplexa are parasitic, and the invasion of host cells is believed to be assisted by the contents of a complex set of organelles present at the apical end of invading stages of parasites;

hence the name Apicomplexa.

More than 100 species of the genus *Plasmodium* are found in the blood of reptiles, birds, and mammals (reviewed by Garnham [6]). Of these, four species—*Plasmodium falciparum, Plasmodium vivax, Plasmodium ovale,* and *Plasmodium malariae*—cause malaria in humans. *Plasmodium falciparum* produces the most severe disease with the greatest mortality. Rarely, malaria results from infection with simian species of *Plasmodium* introduced accidentally or by natural means. In most cases, malaria is transmitted to humans via the bite of an infected female anopholine mosquito; however, the parasites may also be transferred congenitally, through blood transfusions, and by means of contaminated syringes. The infected mosquito injects <1000 sporozoites which pass via blood to the liver where, within minutes, they invade hepatocytes through some mechanism that is unknown at present but thought to involve a receptor–ligand interaction. Exoerythrocytic schizogony then follows wherein the sporozoite changes its morphology and proceeds to differentiate and divide, producing thousands of exoerythrocytic merozoites which, upon discharge from rupturing hepatocytes, invade erythrocytes. In erythrocytes, the merozoite proceeds through erythrocytic schizogony, a maturational process yielding morphologically distinct parasite stages and division which culminates in the destruction of the host erythrocyte and the release of merozoites which invade new erythrocytes. This erythrocytic cycle gives rise to the signs and symptoms of malaria and continues unabated until the host either dies or, more often, activates an immune response(s) capable of killing the parasites or suppressing their growth. Unfortunately, we do not know how this is accomplished in human or experimental malaria. Instead of proceeding through erythrocytic schizogony, certain merozoites differentiate into sex cells or gametes which the female mosquito ingests during a blood meal. Subsequently, the gametes

Table 14.1 Characteristics of infection with human plasmodia. (From Cohen & Lambert [20])

	Plasmodium vivax	Plasmodium falciparum	Plasmodium malariae	Plasmodium ovale
Incubation period (days)	12–17 (up to 12 months)	9–14	18–40	16–18
Approximate no. of merozoites per hepatic schizont	10 000	40 000	15 000	15 000
Duration erythrocytic cycle (h)	48	48	72	50
Parasitemia per µl (average)	20 000	20 000–500 000	6 000	9 000
Clinical severity	Mild	Severe	Mild	Mild
Duration of untreated infection (years)	1–3	1–2	3–50	1–2

are released in the gut of the mosquito and sexual reproduction follows, with infectious sporozoites eventually reaching the salivary glands of the mosquito which can now transmit malaria to a new host. The nature of malaria caused by human plasmodia varies according to the infecting species. Selected characteristics of infection are presented in Table 14.1.

During most of their life-cycle, malarial parasites exist with a haploid genome. The diploid state occurs fleetingly after fertilization of the macrogamete in the mosquito midgut. This is quickly followed by meiosis, so that the ookinette, the invading stage that traverses through the insect intestinal epithelium, is again a haploid. Since genetic variations among *Plasmodium* species are of critical importance in understanding the immune response against the parasite, as well as in designing strategies to control malaria, it may be important to bear in mind some of the biologic features of the parasite that drive such diversity.

Diversity and variations among populations of eukaryotes are driven mainly by genetic drift due to various mutations, reassortment of chromosomes at meiosis, and genetic recombination during meiosis. Whereas genetic drift is often a function of the rate of replication of the genome, such changes can be amplified and spread through the population by genetic reassortment and recombination during meiosis. Hence, variations in the malarial parasite population during erythrocytic stages are mostly due to genetic drift occurring in rapidly dividing cells, whereas shifts in genetic composition of the parasite population will result from chromosomal reassortment and recombination during meiosis in the mosquito midgut. Growth during asexual stages of malarial parasites is substantial: each successful sporozoite developing in a hepatocyte

may give rise to 10 000–50 000 merozoites, and each successful merozoite can yield 10–20 progeny at each subsequent erythrocytic infection. Because of the inherent propensity of the cell replication machinery to generate various forms of mutations, this rapid expansion of the malarial parasite population is the likely reason for variations that occur during asexual stages. These variations, of course, will have to pass through the test of selective pressures: silent mutations will go unnoticed, mutations deleterious to the parasite will be selected against, and those mutations helpful to the parasite will be selected for. The main selective pressure in untreated malaria is the immune response of the host, and variations that result in evasion of such responses will obviously be of selective advantage to the parasite. A case in point is the report by Hudson et al. [7] in which a rhesus monkey vaccinated with a 143/140 kD merozoite surface protein of *Plasmoduim knowlesi* was found, upon challenge, to be infected with *P. knowlesi* that failed to express the 143/140 kD protein. Apparently, the immune response against this polypeptide resulted in selection of *P. knowlesi* clones that were defective, through frame-shift and deletion mutations, in expressing the 143/140 kD protein. The possibility of more such occurrences should be kept in mind when proteins are evaluated as potential immunogens in vaccines.

A feature of malarial parasites that has considerable impact on their biology is the substantial size variations seen for their chromosomes [8–12]. These chromosomes can be resolved by means of pulse-field gradient gel electrophoresis, and, through Southern blot hybridizations, linkage maps assigning various cloned genes to specific chromosomes have been developed. Homologous chromosomes of different isolates of *P. falciparum* have been shown to have great variations in their sizes.

In some instances, chromsomal size variation has also been observed during clonal expansion of a single isolate [12,13]. Hence, it appears likely that chromosomal variations can occur during both mitotic and meiotic divisions. Two mechanisms have been proposed for this phenomenon. In one, telomeric and subtelomeric repeated sequences are proposed to be involved in recombination and unequal crossing-over to generate variations in chromosomal sizes [14]. The second proposal suggests breakage of chromosomes with subsequent telomeric healing [15]. In the second scenario, distal portions of the chromosomes that are now separated from their centromeres will be lost due to segregation. In fact, such losses have been observed in clones of *P. falciparum* that do not express either the knob-associated histidine-rich protein (KAHRP) or the ring-infected erythrocyte surface antigen (RESA) [15,16]. In *P. falciparum*, the knob structures are seen on the surface of parasite-infected erythrocytes and are believed to assist in cytoadherence and sequestration of mature parasites. A knob-minus parasite has a disadvantage during *in vivo* growth since maturing parasites will not be sequestered on capillary endothelial cells and will be forced to pass through the spleen with its parasitocidal consequences. The KAHRP is a component of the knob structure and its loss results in a knob-minus phenotype. It is interesting to note that during *in vitro* cultivation knob-minus KAHRP-minus *P. falciparum* arise quite readily, suggesting that molecular events underlying such mutations may be relatively common during mitotic growth of parasites, but only those events with neutral or positive consequences will survive *in vivo*.

Many antigens of malarial parasites contain regions composed of tandem repeats of amino acid sequences [17]. These regions of the antigens often provide immunodominant epitopes. Antigens bearing repeated sequences, in general, seem also to have regions that are highly conserved, and it is believed that domains important for physiologic functioning of the proteins are located in these regions. Functions of repeated sequences are not entirely clear, but it has been proposed that they serve as "decoys" or "smoke-screens" for the immune response: while the host is actively responding to the repeated immunodominant epitopes, the important portions of the protein are somehow not recognized as effectively [17]. It is also not clear how the repeated epitopes are generated and maintained by the parasites. There appears to be great diversity in the sequence, length, and organization of repeated epitopes of various antigens. While certain proteins, such as the circumsporozoite antigen of *P. falciparum*, show very limited allelic diversity, other proteins, such as the S-antigen, have numerous allelic forms [18]. Recombination between different alleles can further increase the number of variants observed for a given antigen.

Malarial parasites follow Mendelian patterns of inheritance, and reassortment and recombination occur during meiotic division. For a species of parasite with, say, 14 chromosomes, mere reassortment has the potential to generate $2^{14} = 16\,384$ different combinations at meiosis. The actual number can, of course, be much higher because of recombination and chromosomal crossing-over. Hence, allelic variations will be greatly amplified as the parasite population expands. The passage of malarial parasite through a vertebrate host can also be viewed as testing grounds for various combinations of alleles, as well as the means for generating diversity. It is through such dynamic and complex interactions that malarial parasites are able to survive and evolve.

HOST–PARASITE RELATIONSHIPS

While capable of producing severe disease and death, particularly in the young, plasmodia characteristically develop stable long-lasting relationships with their vertebrate hosts, often in the absence of clinical disease [19]. As will be discussed in subsequent sections, these host–parasite associations have a tremendous impact on the immune apparatus of the host, leading to the production of antibodies and cell-mediated immune reactivity. Such responses have been studied intensively to determine the nature of protective immunity in malaria as well as to identify immune responses which may have diagnostic or epidemiologic value. Why plasmodia survive in the immunologically active hosts remains a mystery, but it appears that much of the immunologic activity stimulated by plasmodia has little effect on the survival of the parasite. While its predominantly intracellular existence in the vertebrate host was thought to protect the parasite from host immune responses [20], it is well known that intraerythrocytic parasites die during crisis [21]. Also, CD8[+] T cells have been shown to congregate in the vicinity of infected hepatocytes *in situ* and have been shown to be capable of killing intrahepatocytic stages *in vitro* [22]. Other explanations which might account for "immune evasion" include antigenic variation and immunosuppression [23]. While both phenomena have been documented in human as well as experimental malaria, there is no agreement that either fully explains the ability of the parasite to survive in the immunocompetent host. It is also possible that parasite survival is achieved through a very different mechanism, i.e., instead of activating mechanisms to evade protective immune responses (see Weidanz et al. [24]), it is possible that through the expression of certain immunogenic epitopes, plasmodia are able to regulate the magnitude of protective immune responses. This would allow the parasite to control its numbers in blood so that pathology would be minimized in most hosts while allowing for the production of adequate numbers of gametocytes to ensure transmission to new hosts.

Aspects of the host–parasite relationship where our understanding of events appears to be further developed and where immunologic intervention may be expected to prevent disease include erythrocyte invasion by merozoites and the sequestration of certain species of plasmodia from the peripheral circulation. Much effort has centered on the identification of merozoite surface receptors and erythrocyte surface ligands, whose interaction initiates the red cell cycle of malarial infection. While polymorphism has been noted in many malarial proteins, it is expected to be more limited where molecules important in receptor–ligand interactions are involved. Early in vitro studies showed that antibodies inhibited the invasion of red cells by P. knowlesi [25,26] and P. falciparum [27,28]. Although this inhibition was most likely due to merozoite agglutination and not specifically receptor blockade [26,29], these studies demonstrated that appropriate immune responses to merozoite surface antigens could interrupt the erythrocytic cycle of infection and that vaccine intervention might be feasible.

The invasion of erythrocytes by malarial parasites is a specific, complex, and multistep process [30–34]. The initial attachment of the merozoite to the erythrocyte appears to be a random event, involving any two points on the surface of the two cells. Following contact, a reorientation of the apical protuberance of the merozoite toward the erythrocyte occurs, while the red cell undergoes a wave of deformation. These apical protrusions contain a pair of membrane-bound organelles called rhoptries, with associated micronemes, the contents of which are released into or onto the erythrocyte membrane. Merozoite reorientation results in the formation of a junction between the two cells and an invagination of the erythrocyte membrane. As the merozoite is internalized, its surface coat is lost, being shed and/or proteolytically degraded. The parasitophorous vacuole forms during merozoite entry and at completion, and

red cell membrane reseals and undergoes a second wave of deformation.

In the case of P. falciparum, glycophorins represent the major erythrocyte receptors, with terminal sialic acid residues playing a critical role in the interaction between the parasite and the host cell [35–37]. Although poorly understood, the actual invasion process appears to involve ligand–receptor interactions in addition to those involving sialic acid [38]. Other species of plasmodia use different erythrocyte receptors, with the Duffy blood group glycoproteins being the major receptor for P. vivax and P. knowlesi [39,40]. The significance of these receptors in innate immunity will be discussed in the section on innate resistance. Studies employing erythrocyte variants deficient in sialic acid or glycophorins have demonstrated that certain isolates of P. falciparum utilize alternative receptors when invading host cells [37,41]. Similarly, other species of plasmodia have been found capable of entering host cells via secondary receptors (see Perkins [42]).

The ability of merozoites to recognize erythrocyte receptors is dependent upon parasite proteins which have been termed receptor-binding proteins (RBP), which appear to be distinct for the various species of plasmodia that have been studied thus far [42]. For example, P. falciparum can bind to the major erythrocyte receptor via Pf200 (gp195), the precursor of the major merozoite surface antigen [36]. They may also recognize the major erythrocyte receptor by a 175 kD protein that binds to sialic acid [43]. Additional RBP must be utilized by P. falciparum strains binding to erythrocytes in a sialic acid-independent manner. Recent findings summarized by Perkins [42] suggest that sialic acid-independent binding is achieved through complexes of rhoptry proteins which range in size from 110 kD to 140 kD.

The Duffy receptor-binding proteins of P. vivax and P. knowlesi have also been identified [44]. Both parasites recognize the Duffy receptor by 135 kD RBP which, while related functionally and immunologically, differ in their ability to bind to rhesus erythrocytes. Plasmodium knowlesi also recognizes the Duffy receptor via a 155 kD protein [45]. Since polyclonal and monoclonal antibodies reactive with RBP have been shown to inhibit attachment and/or invasion of erythrocytes in vitro [42], these molecules are being considered as future immunizing agents. Also, RBP and the receptor molecules they combine with help to restrict the specificity of the host–parasite relationships commonly associated with malaria in different species.

Sequestration of erythrocytes in patients infected

with *P. falciparum* was first described prior to the turn of the century by Bignami & Bastiamelli [46]. Subsequently, it was realized that erythrocytes parasitized by certain plasmodial species acquired adhesion properties which allowed the parasitized erythrocytes to adhere to the endothelia of blood vessels in various tissues of the infected host [47]. *Plasmodium falciparum* is the only human malarial parasite to sequester and infections with this parasite are typified by the lack of trophozoites and schizonts in peripheral blood films of malarious patients. Sequestration appears to have several functions. It provides an environment with reduced oxygen tension which appears to favor the development and replication of *P. falciparum* and it appears to be a protective mechanism whereby plasmodia avoid passage through the spleen at a stage in their life-cycle when they would be most susceptible to parasite-destructive immune mechanisms activated in the spleen. In splenectomized patients and monkeys, all stages of erythrocytic *P. falciparum* are detected whereas only ring stages and occasional immature trophozoites are present in the peripheral blood of splenic-intact hosts [48] (M. Ho, personal communication). Additional evidence for sequestration having a parasite-protective role have come from studies utilizing laboratory variants of *P. falciparum* lacking the ability to sequester (see Howard & Gilladoga [49]). These parasites produced avirulent infections that healed spontaneously. The fact that different isolates of a single species of plasmodia differ in their ability to sequester and that many species of malaria parasites fail to sequester at all suggest that during the course of evolution many malarial parasites have learned to pass safely through the spleen.

More recently the idea that sequestration of erythrocytes parasitized with *P. falciparum* plays an essential role in the pathogenesis of malaria has provided a major impetus for understanding the molecular basis of sequestration, with the aim that the identification of both host and parasite molecules involved in the event could provide targets for therapeutic intervention utilizing blocking antibodies or soluble receptor molecules themselves [50]. Initial studies on sequestration were conducted in primate models using blood-stage infections initiated with *P. falciparum* or monkey malarias (reviewed by Howard *et al.* [51]). These studies demonstrated that sequestration involved the specific attachment of parasitized erythrocytes to endothelium associated with knob-like protrusions on the surface of the infected erythrocyte. In addition, they showed that passively transferred monkey antibody reversed *in vivo*

sequestration by *P. falciparum* in Saimiri monkeys, indicating the clinical potential of antibodies that could make the parasite more susceptible to host defense mechanisms.

In addition, *ex vivo* and *in vitro* models have been used to investigate various aspects of the sequestration phenomenon, including the requirement for knobs on the parasitized erythrocyte surface and the identification of molecules involved in the attachment phenomenon [51]. The relationship of findings obtained using these *in vitro* models to sequestration *in vivo*, for the most part, remains to be determined [52]. None the less, they provide a starting point and may yield findings of considerable significance. To date, three protein molecules have been identified as receptors to which erythrocytes bind [51, 53]. CD36, also known as glycoprotein IV is an 88 kD integral membrane protein molecule found on endothelial cells, platelets, and monocytes. Thrombospondin, a high-molecular-weight protein consisting of three peptide chains of 180 kD each connected by disulfide bonds, is a "bridge protein" secreted by many cells and thought to function by linking membrane-associated proteins. Intercellular adhesion molecule-1 (ICAM-1), CD54, is an 84–110 kD single-chain glycoprotein found on multiple cell types where it functions as a ligand for the integrin LFA-1. The expression of ICAM-1 is regulated by various cytokines [54], including tumor necrosis factor-α (TNF-α), IL-1, and interferon-γ (IFN-γ), which are produced during malarial infection (see section on cell-mediated responses). All three receptors expressed on transfected CV1 origen-defective SV40 cells or immobilized on plastic bind *P. falciparum*-parasitized erythrocytes (reviewed by Howard & Gilladoga [49] and Chulay & Ockenhouse [53]). Binding can be prevented by selected antibodies as well as soluble receptors. Whether these receptor molecules function as three separate entities or as a single cell surface molecule bearing three distinct receptor sites remains to be determined (see postulated mechanisms in Fig. 14.1). Also, the binding sites in the receptor molecules themselves have not been identified. The possible modulation of these receptor molecules by cytokines resulting from the activation of immune cells by malarial antigens might be expected to influence the pathogenesis of infections where sequestration appears to play an important role and might represent a mechanism whereby the parasite could regulate its own numbers in blood by activation of immune mechanisms altering the number of available receptors on endothelial cells.

As indicated above, erythrocytes infected with ring or

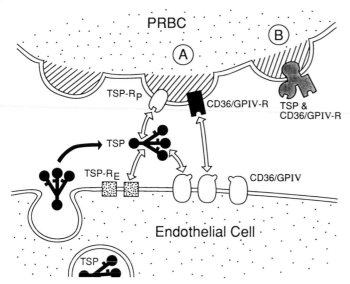

Fig. 14.1 Schematic of the receptor interactions involved in adherence of *Plasmodium falciparum*-infected red blood cells to endothelial cells, based on *in vitro* experiments with CD36, thrombospondin (TSP), or ICAM-1. Knob protrusions on the mature parasitized red blood cell (PRBC) are shown bearing a specific PRBC receptor for TSP (TSP-R$_P$), a specific receptor for CD36 (CD36-R), and a specific receptor for ICAM-1 (ICAM-1-R). These three receptors are shown alternatively as three different molecular entities or as one cell surface molecule bearing three distinct receptor sites (TSP, CD36, and ICAM-1-R). Secretion of TSP by the endothelial cell (EC) is shown, with subsequent binding of TSP to EC membrane-bound components (TSP-R$_E$). TSP on the EC surface may also derive from circulating platelets or other cells that synthesize and release this protein into plasma. TSP could therefore bridge the TSP-R$_P$ on the PRBC with CD36, which binds TSP *in vitro*, or other TSP receptors on the EC surface. Adherence of PRBC to EC could also be mediated independently of TSP by direct interaction of CD36 with the CD36-R on PRBC or by direct interaction of ICAM-1 on PRBC. (From Howard & Gilladoga [49].)

early trophozoite stages of *P. falciparum* fail to sequester and are found in peripheral blood. Why then do erythrocytes parasitized with more mature forms sequester? Early studies [51] showed a close relationship between the expression of knobs on the surface of parasitized erythrocytes and adherence to host cells, i.e., knob-bearing K$^+$ strains of *P. falciparum* sequestered well but K$^-$ strains did not. Similar results were obtained with K$^+$-parasitized erythrocytes infected with either *Plasmodium fragile* or *Plasmodium coatneyi*. Also, all wild isolates of *P. falciparum* have been K$^+$ and cytoadhered. On the other hand, the K$^+$-parasitized erythrocytes of *P. malariae* and *Plasmodium brasilianum* did not adhere to

host cells and K$^-$ parasites from other species of plasmodia did sequester *in vivo*. The dissociation between the K$^+$ phenotype and cytoadherence was further supported by the observation that certain K$^+$ *P. falciparum* failed to adhere to host cells [55,56] and that culture-adapted *P. falciparum* parasites which were K$^-$ adhered to C32 melanoma cells [57,58].

A large number of K$^+$ and K$^-$ *P. falciparum* clones have been generated from experimentally infected monkeys and *in vitro* cultures. These parasites have been examined for their ability to adhere to a variety of host cells and to induce on the erythrocyte surface the expression of parasite molecules responsible for adherence to host cells. Thus far, two large malarial proteins PfEMP-1 (*P. falciparum* erythrocyte membrane protein 1) and PfHRP-2 (*P. falciparum* histidine-rich protein 2), which are expressed on the surface of parasitized erythrocytes, have been associated with the cytoadherence phenotype (C$^+$) (see Howard & Gilladoga [49] and Howard *et al.* [51]). In addition, two altered host proteins of >240 kD and 85 kD which appear to be closely related to the erythrocyte anion transport protein band 3 may be involved in cytoadherence phenomena [59,60]. How these and other ligands on the surface of infected erythrocytes mediate or contribute to cytoadherence remains to be determined. Different isolates of *P. falciparum* can use the same or different molecules on homogeneous target cells. Similarly, a single clone can bind to the same or different receptors on different target cells, suggesting that a multiplicity of surface molecules on the parasitized erythrocyte mediate adherence phenomena [51]. Despite the complexity of studying multiple ligand–receptor interactions, the possibility that certain of these interactions leading to cytoadherence may be associated with the virulence of *P. falciparum* isolates or the pathology they induce makes this area of research particularly exciting.

HUMORAL RESPONSE

Nonspecific Ig

Antibody production is an important component of the immune response to plasmodia. The nature and function of the humoral response in malaria has been reviewed by Deans & Cohen [61] and more recently by Taylor [62]. Malarial infection induces a dramatic increase in the synthesis of antibodies by the host. The production of IgM and IgG in particular is markedly elevated, with IgA being increased to a smaller extent

[63]. Interestingly, only a small fraction of the total Ig produced (approximately 6–11%) is actually specific for malarial antigens [64]. The vast majority of antibodies generated are not reactive with the parasite and are thought to result from the induction of polyclonal activation by the parasite. Greenwood [65] originally proposed that a parasite-derived B cell mitogen was responsible for the hypergammaglobulinemia associated with human malaria. It has since been shown that *P. falciparum*-derived antigen(s) do not activate B lymphocytes directly but rather appear to stimulate T lymphocytes which are then induced to produce factors promoting the differentiation of B cells into Ig-secreting plasma cells [66,67].

Polyclonal Ig synthesis is believed to be at least partially responsible for the presence of autoantibodies in humans and animals infected with plasmodia [68]. Alternative possibilities that have been suggested to explain the generation of self-reactive antibodies include:

1 modification of host antigens by the parasite, possibly through the release of enzymes or through the binding of soluble parasite antigens to host cells;

2 crossreactivity between plasmodial antigens and host antigens;

3 malaria-induced inhibition of T suppressor cell activity.

A variety of autoantibodies have been detected in malaria-infected hosts, including antibodies directed against normal erythrocytes [69,70], lymphocytes [71], macrophages [71], serum globulins [72], heart, thyroid [73], nuclear factors (antinuclear antibodies, ANA [74]), and a variety of intracellular proteins [75]. The role played by such autoantibodies remains unclear. It has been hypothesized that autoantibodies may be involved in immunopathology. For example, immune complexes consisting of DNA–anti-DNA antibodies have been implicated in the development of glomerulonephritis by patients with nephrotic syndrome [74]. It was also suggested that antierythrocyte antibodies may be involved in the anemia associated with malaria since the degree of anemia observed is greater than can be accounted for by the loss of erythrocytes due to rupture by maturing parasites [70]. However, some investigators have argued that antierythrocyte activity directed at altered RBC determinants may also be important in protection against malaria [76]. Autoantibodies directed against lymphocytes and macrophages could conceivably be involved in immunoregulation, an area that remains to be explored.

Specific Ig

During the course of malaria, parasite-specific antibodies develop with the same kinetics as the total Ig and nonspecific Ig response. In both human and mice infected with plasmodia, specific antimalarial antibodies can generally be detected by indirect immunofluorescence a few days after the appearance of parasites in the blood. Thereafter, antibody titers increase rapidly. The increase in titer is due to the synthesis of specific antibodies of the IgG, IgM, and IgA isotypes while IgD and IgE levels remain essentially the same [63,77–79]. In humans, antibody titers rapidly decline following recovery from infection [79]. A drop in antibody titer is also observed in immune individuals moving to nonmalarious areas.

While immunity to plasmodia appears to be species- and stage-specific, a significant proportion of antigen-reactive antibodies produced by the host recognize antigens shared by different species and stages of the parasite [61]. This observation, taken together with the reported lack of correlation between antigen-specific antibody titers and immune status [79], suggests that a fraction of the antimalarial antibodies produced during infection has little or no protective value. Such nonprotective antibodies could, for example, be directed against metabolites, degradation products, or constituents of the parasite which do not represent effective targets for humoral immunity [61]. Nevertheless, the participation of parasite-specific antibodies in protective immunity has been clearly established.

Evidence for the protective activity of parasite-specific antibodies

The role of antimalarial antibodies in protection against malaria was demonstrated by various experimental approaches, including: passive immunization of naïve recipients with hyperimmune serum or monoclonal antibodies, adoptive transfer of immunity with B-cells, and increased susceptibility of B cell-deficient hosts.

Passive immunization with polyclonal and monoclonal antibodies

Passive immunization in humans was accomplished by Cohen *et al.* [80], who reported that administration of purified Ig from immune adults to 12 Gambian children with acute malaria reduced parasitemia to low levels and cured their disease. Passively transferred immune serum was also shown to confer at least some degree of protection in other host–parasite models, including rhesus monkeys infected with *P. knowlesi* [81], as well as rats and mice infected with *Plasmodium berghei* [82,83], *Plasmodium yoelii* [84,85], or *Plasmodium chabaudi* [86]. Erythrocytic rodent malaria has been the most extensively studied animal model and results from numerous studies have shown that, depending on the amount of hyperimmune serum transferred, the dose of challenging parasites, and the timing of serum administration relative to parasitic challenge, antimalarial antibodies are able to delay or prevent the onset of parasitemia [82,84, 86], reduce peak parasitemia [82,85,86], and enhance survival [81,82] of naïve recipients. The highest levels of passively transferred immunity were achieved with large doses of hyperimmune serum prior to challenge with small numbers of parasites. Antibodies of the IgG1 and IgG2 isotypes were found, for the most part, to be responsible for passive immunization [82].

Antibody-mediated immunity to the sexual stages of malaria parasites was demonstrated by experiments in which sera from humans naturally infected with *P. vivax* [87] or from animals immunized with gametocytes [88] or gametes [89] were shown to suppress the infectivity of gametocytes of the same species when administered to mosquitoes in a membrane-feeding apparatus. A protective role for antibodies against the sporozoite stage was shown by experiments in which passively transferred immune sera increased the rate of sporozoite clearance in naïve recipient mice [90]. However, complete protection against sporozoites was rarely obtained with the transfer of antibodies [91]. This may in part have been due to insufficient amounts of antibodies with the appropriate specificity and affinity in the sera used since small amounts of a monoclonal antibody against the circumsporozoite (CS) protein [92] or Fab fragments of this antibody [93] passively protected mice against sporozoite challenge.

Monoclonal antibodies directed against the erythrocytic stage of *P. yoelii* [94,95] and *P. chabaudi adami* [96] have also been produced that were shown to transfer passively immunity *in vivo*. Such antibodies displayed T cell-dependent isotypes. Monoclonal antibodies against certain surface antigens of gametes and zygotes of *Plasmodium gallinaceum* [97], *P. falciparum* [98,99], *P. yoelii nigeriensis* [100], and *P. vivax* [101] have also been shown to suppress infectivity of the parasites to mosquitoes, resulting in transmission-blocking immunity. The target antigens of monoclonal antibodies offering passive protection against the various life-stages of the parasite are discussed in further detail in the section on vaccines.

Adoptive transfer of immunity with B lymphocytes

In order to study the contribution of B and T lymphocytes to the immune response against malarial parasites, various investigators have performed transfer experiments in which immune spleen cells or selected spleen cell populations were administered to nonimmune recipient animals. Most studies examined blood-stage *P. yoelii* or *P. chabaudi* infections in mice and *P. berghei* in rats. In every instance, unfractionated immune spleen cells were able to confer resistance when transferred into naïve animals [102–106]. Cell fractionation experiments undertaken to determine the identity of the protective spleen cell population(s) revealed that, in these models, both immune B and T lymphocytes possessed the ability to transfer some level of protection to the recipient [86–91] and some investigators reported that, when transferred together, the two cell populations synergized to generate significantly enhanced levels of protective activity [103,104,107]. Such results suggest that antibody-producing immune B lymphocytes significantly contribute to malarial immunity and that their protective activity can be enhanced by the presence of immune T helper cells.

Immunity to malaria in B cell-deficient hosts

B cell-deficient animals, which are unable to produce a normal antibody response, have been shown to be more susceptible to malarial infection, thereby underlining the importance of antibodies in immunity to malaria. Chickens rendered B cell-deficient through surgical or hormonal bursectomy developed more severe infections than intact animals when infected with *Plasmodium lophurae* or *P. gallinaceum*, the latter being fatal for B cell-deficient chickens [108,109]. Similarly, mice rendered B cell-deficient by administration of antimouse μ-chain sera died when infected with blood-stage parasites of the normally avirulent 17X stain of *P. yoelii* [110,111]. CBA/N mice, which lack the Lyb-5[+] subset of B lymphocytes, have also been reported to suffer more severe *P. yoelii* malaria of prolonged duration when compared to normal mice [112].

Even though these findings establish that antibody-producing B cells play an important role in resistance to malaria, a role for antibody-independent T cell-mediated

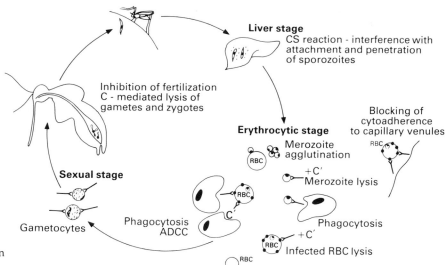

Fig. 14.2 Possible mechanisms of action of "protective antibodies."

immunity was also uncovered through the use of B cell-deficient animals, as will be discussed in the next section.

Possible modes of action of "protective" antibodies

While several possible mechanisms by which antibodies may help control malarial infection can be envisaged (Fig. 14.2), their actual occurrence and relative importance *in vivo* remains largely to be determined. In the case of sporozoites, it has been shown *in vitro* that serum antibodies [113] or monoclonal antibodies [77] directed against the CS protein can bind to the surface of the parasite and produce the formation of a thread-like precipitate which extends from the parasite. This phenomenon is called the CS reaction. Protective antibodies which induce the CS reaction have been shown to interfere with the attachment and penetration of *P. berghei* sporozoites into W138 human lung cells [114] and of *P. falciparum* and *P. vivax* sporozoites into human hepatoma cells *in vitro* [115]. Once inside, the development of sporozoites did not appear to be affected by the presence of anti-CS protein antibodies, leading to the conclusion that protective antibodies probably function primarily to prevent the attachment and entry of sporozoites into host liver cells.

Antibodies are also thought to prevent invasion of host erythrocytes by the merozoite stage of the parasite by binding to surface antigens and causing agglutination of merozoites or blocking of parasite surface molecules involved in the penetration of erythrocytes [25,28,29]. The identification of plasmodial proteins involved in the invasion process represents a major focus of research aimed at the development of a blood-stage malaria vaccine and is discussed further in the section on vaccines. As observed with sporozoites, once inside the host cell the merozoites appear to develop normally in the presence of antibodies [25]. Antibodies directed against mature schizonts, however, have been reported to inhibit the release of free merozoites [29].

Conceivably, complement-mediated lysis could take place following the binding of antibodies to parasite antigens expressed on the surface of infected erythrocytes or free merozoites. However, there is no evidence that such a phenomenon contributes to malarial immunity. In mouse models, the *in vivo* growth of malaria parasites did not appear to be significantly affected by depletion of C3 with cobra venom factor [116] or genetic defects in complement components [117]. Antibodies may also exert protective activity through the opsonization of free merozoites and parasitized erythrocytes, resulting in the enhancement of phagocytosis by macrophages [118–120] and polymorphonuclear leukocytes [119,121,122]. However, it is unclear to what extent phagocytosis contributes to resistance. For example, recovery from infection with blood-stage *P. berghei* in rats and *P. chabaudi* in mice does not appear to be accompanied by enhanced phagocytosis [123].

It has been suggested that antibody-dependent cell-mediated cytotoxicity (ADCC) mediated by K cells, mononuclear cells, and granulocytes binding to antibody-coated infected erythrocytes through their Fc receptors may also contribute to malarial immunity. In *in vitro* experiments, human peripheral blood lymphocytes were able to lyse *P. falciparum*-infected erythrocytes in the presence of immune serum [124] and spleen cells from *P. berghei*-infected mice exhibited ADCC activity against infected erythrocytes sensitized with antibodies against mouse erythrocytes [125]. Nevertheless, the actual occurrence and contribution of ADCC to malarial immunity *in vivo* remains to be determined.

Antibodies directed against the sexual stage of the malaria parasite are believed to be involved in transmission-blocking immunity. When taken up by the mosquito vector during a blood meal, host antibodies to gamete surface antigens appear to block parasite fertilization and promote complement-mediated lysis of gametes and newly fertilized zygotes in the mosquito midgut [126]. Finally, antibodies may act to reduce pathology by preventing the cytoadherence of infected erythrocytes to the endothelium of capillary venules, a phenomenon which is believed to contribute to cerebral malaria. Antibodies in immune sera have been shown to prevent specifically or reverse the binding of *P. falciparum*-infected erythrocytes to melanoma cells *in vitro* [127] and cause the release of *P. falciparum*-infected erythrocytes sequestered in the deep vasculature of squirrel monkeys *in vivo* [127]. Identification of the parasite antigens involved in cytoadherence, as well as the host molecules that they bind, is being pursued for the purpose of vaccination against cerebral malaria (see section on vaccines pp. 326–333).

CELL-MEDIATED RESPONSES

T cell activation

Malarial infection in humans and experimental animals activates T cells, as assessed by cellular proliferation, lymphokine secretion, the appearance of T cell-dependent isotypes of antibody, and the release of membrane surface molecules such as IL-2 receptor and CD8 [20,128–130]. Similarly, T cells obtained from hosts infected previously with malaria become activated when exposed to malarial antigens *in vitro*, and the injection of malarial antigens into hosts which have recovered from malaria or which have been immunized with malarial antigens produces classic delayed-type hypersensitivity reactions known to be mediated by T cells [131, 132].

T cell proliferation

T cell proliferation studies have been utilized by many investigators to measure the cellular immune status of hosts exposed to malarial infection (reviewed by Melancon-Kaplan & Weidanz [130]). Early studies involved the use of crude antigenic extracts which in high concentration were often mitogenic [133]. More recently, well-characterized antigens produced by recombinant DNA techniques or synthesized biochemically have been utilized [134,135]. While these well-defined antigens have been ideal for the mapping of T cell epitopes in malaria vaccines, as well as the individual host responses to particular T cell epitopes, they may be less useful in assessing the overall pattern of T cell responsiveness of individuals living in areas of malaria transmission because of host genetics and because of antigenic polymorphism restricting host recognition of particular T cell epitopes. Riley & Greenwood [136] have recently summarized various external factors which influence proliferative responsiveness to malarial antigens. They include the physiologic status of the host, the quality and quantity of exposure to malarial antigens, and infection-related immunosuppression. As will be discussed later, *in vitro* proliferative responses of peripheral blood T cells to malarial antigens are markedly suppressed during acute infection, despite the fact that activation molecules such as soluble IL-2r, CD8 antigen, and IFN-γ are increased over normal levels in these patients [137,138]. Also, increased numbers of proliferating T cells have been demonstrated in the spleens of mice with murine malaria at a time when the mice displayed generalized immunosuppression (reviewed by Weidanz [139]). While antigenic preparations from all life stages of plasmodia have been shown to cause T cell proliferation, recent studies with synthetic peptides of the *P. falciparum* blood-stage antigen Pf155/RESA demonstrated that the *in vitro* induction of IL-4 mRNA, IFN-γ secretion, proliferation, and B cell help could not be correlated to individual immune donors [140]. This finding suggests that the T cell proliferation should not be used alone as an indication of cellular immune responsiveness to a selected malarial antigen but rather T cell responsiveness should be established by examining multiple parameters of T cell activation.

Lymphokine production

The activation of T cells during malaria is accompanied by increases in the serum levels of IFN-γ and IL-6

[141,142]. Similarly, when T cells obtained from humans living in malaria endemic areas or from the spleens of "immune" mice are stimulated *in vitro* with malarial antigens, they secrete a broad spectrum of lymphokines, including IFN-γ, TNF-β, IL-2, IL-4, and IL-6 [143]. Moreover, studies in mice in which antilymphokines have been used to alter the functional capacity of T cells indicate that lymphokines in addition to the ones listed above are produced in response to stimulation with parasite antigens [143–145]. What remains to be determined is the *in situ* localization of these lymphokine-secreting cells as well as their physical association with malarial parasites in hepatocytes or erythrocytes, the concentration of lymphokines that they can produce locally in host tissues, and the biologic significance of their presence. Lymphokines are potent molecules which activate other cell types, including lymphocytes, monocytes, macrophages, and neutrophils. Activated macrophages produce IL-1 and TNF-α, which are increased in the sera of infected humans and experimental animals [146–149]. It seems quite likely that these molecules produced by activated T cells and macrophages play an important but poorly understood role in resistance to malaria as well as in the development of immunopathology associated with this disease, as will be discussed in subsequent sections.

Populations of T cells activated by malaria infections

During experimental malaria, T cells increase in number in the spleens of infected hosts. Both CD4+8− and CD4−8+ T cells show marked increases, with a disproportionate increase in CD4−8+ cells [24]. In human malaria, an inversion of the CD4+/CD8+ ratio of peripheral blood T cells has been recorded in patients who have recently recovered from an attack of malaria [150, 151]. While an immunoregulatory role has been sought for CD8+ cells, it should be realized that activated CD8+ cells also secrete an array of lymphokines similar to those secreted by CD4+ cells.

Recently, several laboratories have shown that γδ T cells increase in number during acute malaria. Increased numbers of CD4−8+ γδ T cells were enumerated in the spleens of mice infected with *P. chabaudi* [152]. Also, CD4−8− γδ T cells were elevated in the blood of patients infected with *P. falciparum* and remained so during the convalescent period [153,154]. The significance of these observations remains to be determined but it is known that γδ T cells participate in major histocompatibility complex (MHC)-nonrestricted cytotoxicity [155] and are

activated by heat shock proteins obtained from a variety of microbial agents [156]. Malarial parasites, including *P. falciparum*, are known to express heat shock proteins [157–160] and the ability of a cytotoxic T cell to lyse targets in an MHC nonrestricted manner might permit the destruction of parasitized erythrocytes which do not express MHC molecules on their membrane surfaces.

T cell lines and clones

A number of human and rodent continuous T cell lines and clones reactive with malarial antigens of different stages from plasmodia of different species have been generated during the past 6 years (reviewed by Weidanz & Long [128] and Melancon-Kaplan & Weidanz [130]). These cells were used in proliferation studies to determine antigenic specificity and in functional studies to measure their ability to provide help for antibody formation, mediate cytotoxicity, and secrete lymphokines. Some T cell clones were derived from the peripheral blood of individuals who had never experienced malaria and approximately half of all the clones generated reacted with crossreacting antigens found in diverse species of plasmodia [161–164]. Most antigen-reactive T cell lines and clones isolated were of the CD4+ phenotype, but CD8+ clones were also isolated by modifying the cloning procedures [165].

As will be discussed subsequently, continuous T cell lines and clones have proven most useful in adoptive transfer studies using rodent models of malaria to establish and characterize a protective role for T cells (reviewed by Weidanz & Long [128] and Melancon-Kaplan & Weidanz [130]). They have also been used successfully to map T cell epitopes in recombinant and synthetic peptides. For example, Nardin *et al.* [166] have recently isolated T cell lines and clones from a sporozoite-immunized human volunteer. These CD4+ T cells recognized both natural and recombinant CS antigens of *P. falciparum* via a T cell epitope contained in a 12-mer sequence, NANPNVDPNANP. In similar studies, CS-specific T cells isolated from a sporozoite-immunized chimpanzee were used to identify a T cell epitope within a repeat region of the *P. vivax* CS protein [167].

Major problems confronting investigators who are attempting to construct subunit vaccines to protect against malaria include the choice of antigens, antigenic polymorphism, and MHC restriction. Sinigaglia *et al.* [168] determined that the 378–398 sequence of the *P. falciparum* CS protein is conserved in different parasite isolates. They subsequently used a synthetic peptide,

CS.T3, corresponding to this sequence but with two cysteines replaced by alanines to generate T cell clones from several donors. CS.T3-reactive clones proliferated in response to native CS protein and provided T cell help which was boosted upon immunization with whole sporozoites. In addition, the CS.T3 peptide was shown to be capable of binding to and being recognized by most human Class II molecules. Hopefully, as other candidate molecules are examined, this type of analysis will prove helpful in defining which epitopes should be included in subunit vaccines.

Protective role of T cells in immunity to malaria

Malaria in immunodeficient hosts

There is no doubt that T cells play an essential role in the development of immunity to malaria. Evidence obtained from studies using animal models with thymic deficiencies induced experimentally or congenital in origin showed clearly that the presence of an intact functional thymus was essential for immunity to develop against erythrocytic infections (reviewed by Russo & Weidanz [132]). Similarly, T cell-deficient mice could not be immunized with irradiated sporozoites whereas immunologically intact mice were protected against challenge infection by the same procedure. The question arose as to how T cells functioned in the development of protective immunity. Was immunity achieved through the activation of T helper cells which were essential for the production of protective antibodies or did T cells participate in the development of cell-mediated immunity (CMI)?

The use of B cell-deficient animal models helped to clarify this issue. When B cell-deficient mice and chickens were infected with relatively avirulent plasmodia, they usually developed unremitting parasitemia and died, whereas immunologically intact controls became infected but then cleared parasites from their blood (reviewed by Weidanz & Long [128]). When acute malaria in B cell-deficient hosts was controlled with chemotherapy, the animals resisted exogenous challenge infection with homologous parasites despite the absence of detectable antibodies in their blood. In contrast, athymic hosts treated identically developed malaria upon challenge infection. Together, these results suggested that protection against reinfection with malaria could be achieved by T cell-dependent mechanisms of immunity in the absence of protective antibodies but that B cells and presumably antibodies were necessary for resis-

tance to acute disease. Additional studies utilizing other host–parasite combinations revealed that B cell-deficient chickens infected with *P. lophurae* and B cell-deficient mice infected with *P. chabaudi adami* cleared acute blood-stage infections spontaneously in the absence of chemotherapy [108,169]. These findings indicate that certain parasites may be more susceptible to T cell-dependent CMI than others or that these parasites, which produce self-curing infections in immune B cell-deficient mice, provoke quantitatively stronger responses than others. Since B cell-deficient mice immune to *P. yoelii* were also resistant to *P. chabaudi* but not vice versa [170], the former explanation probably accounts for the reported findings. Acute infections caused by *Plasmodium vinckei petteri*, *P. chabaudi chabaudi*, and *Babesia microti* also cured spontaneously in B cell-deficient mice whereas *P. yoelii* and *P. berghei* infections did not [171]. The fact that *P. chabaudi adami* infections were suppressed by a monoclonal antibody reactive with a 250 kD merozoite antigen indicates that multiple mechanisms of immunity can be utilized by a host to destroy parasites [96]. Immunization of B cell-deficient or athymic mice with irradiated sporozoites or CS antigen expressed by salmonella suggested that T cell-dependent CMI gave protection against sporozoites as well [172,173]. As indicated above, malaria has profound effects on human T cells but whether T cell-dependent CMI contributes to protection against human malaria remains to be determined.

Adoptive transfer of immunity with T cells

Numerous adoptive transfer studies have been undertaken for the purpose of establishing the identity of the cell types responsible for protective immunity against malaria (reviewed by Weidanz & Long [128] and Melancon-Kaplan & Weidanz [130]). Initially, immunity against experimental blood-stage malaria was transferred with immune spleen cells, but in subsequent studies utilizing fractionated spleen cells high levels of protection were achieved in most instances only when T and B cells were transferred together into recipient hosts. This suggested that the function of the transferred T cells was primarily to provide help to B cells making protective antibodies. The inability of these studies to indicate a role for CMI in resistance to malaria may have several explanations. Many of the adoptive transfer experiments utilized *P. yoelii* or *P. berghei* infection models, which appear to be controlled primarily by antibody-mediated mechanisms. In addition, the adoptive transfer of selected populations of lymphocytes was often complicated by the inability of the fractionation techniques employed to yield pure populations of desired cells. Also, the adoptive transfers utilized immunosuppressed recipients which, during the recovery period, may have contributed their own cells or factors to the developing immune response.

In part, some of these problems were overcome by utilizing the *P. c. adami* model where resistance to acute blood-stage infection was known to be mediated by T cell-dependent immune mechanisms and by using a well-characterized recipient, the nude mouse. Cavacini *et al.* [105] employed this system to show that immune T cells, but not B cells, transferred immunity to acute infection with this parasite. Moreover, immunity was transferred best with CD4$^+$ T cells. In other studies, antigen-reactive T cell lines and clones derived from mice immune to *P. c. adami* were shown to transfer protection against homologous but not heterologous plasmodia [174,175]. These T cells displayed the CD4$^+$8$^-$ phenotype, proliferated in response to homologous antigen *in vitro*, and secreted IL-2 and IFN-γ, suggesting that they belonged to the TH-1 subset of T cells [176]. The parasite antigens bearing the T cells epitopes recognized by these protective T cells have not been identified, nor is it known how T cells function in the development of protective immunity to blood-stage infections.

The adoptive transfer of immunity to sporozoite and sexual stages of plasmodia has also been accomplished using enriched T cell fractions from the spleens of mice immunized with either CS antigens of *P. berghei* or microgametes of *P. yoelii nigeriensis*, respectively (reviewed by Melancon-Kaplan & Weidanz [130]). More recently, the adoptive transfer of species-specific immunity to sporozoites has been achieved with cytolytic cloned T cells isolated from mice immunized with irradiated sporozoites of either *P. berghei* or *P. yoelii* [177,178]. The cloned T cells displayed the CD8$^+$ phenotype, recognized synthetic peptides corresponding to a homologous region in the CS proteins of both malaria species in the context of the same MHC Class I molecule, and in several instances adoptively conveyed a high degree of protection against challenge with homologous sporozoites. The epitopes recognized by cytolytic T cell lines were defined by a synthetic peptide comprising residues 249–260 of the CS protein of *P. berghei* and residues 277–288 of the CS protein of *P. yoelii*. These regions show a high degree of homology with only three amino acid differences. None the less, protective T cell clones were capable of transferring species-specific protection. Other cytolytic T cell clones recognizing the same epitopes were not protective in adoptive transfer studies. Whether this was due to differences in *in vivo* homing properties or survival upon transfer is unknown. Additional explanations given by the authors to account for differences in the protective activity of T cell clones recognizing the same epitopes included the fine specificity of the epitopes being recognized as well as possible differences in lymphokine secretion following activation by malarial antigens.

CD4$^+$ T cell clones have been generated from mice immunized with a 21-mer peptide corresponding to the amino acid positions 59–79 of the CS protein of *P. yoelii* and referred to as PyI [179]. These clones proliferated in an MHC-restricted manner in response to stimulation with the same peptide and produced IFN-γ and IL-2 or IL-4, IL-5, and IL-6. They recognized PyI via T cell receptors (TCR)-αβ, and in preliminary studies some clones, either TH-1 or TH-2, provided protection when adoptively transferred into mice which were then challenged with viable *P. yoelii* sporozoites. It is possible that these cells were cytotoxic for pre-erythrocyte parasites in hepatocytes or secreted lymphokines capable of activating killing mechanisms.

Deletion of selected T cell populations

Another approach to identify T cells participating in protective immune responses against malaria has been the *in vivo* depletion of subsets of T lymphocytes utilizing specific antibodies. This procedure was successfully employed to demonstrate that antisporozoite immunity induced by immunization with irradiated sporozoites was mediated by CD8$^+$ T cells [180,181]. More recently, it has been shown that the role of CD8$^+$ T cells in sporozoite immunity is variable, depending on host genetics as well as the strain of parasites being utilized [182]. While the reasons for this observation remain to be determined, it may be worth considering that γδ T cells may or may not express the CD8 molecule on their surface. Depletion of CD8$^+$ cells in the latter instance would have little effect on functional (CD8$^-$ γδ) T cells.

As might be expected from the previous discussion, depletion of CD4$^+$ cells by anti-CD4 treatment *in vivo* suppressed immunity against blood-stage infection, supporting the pivotal role played by this subset of lymphocytes [22,183,184]. Interestingly, the time of CD4$^+$ T cell depletion appeared critical. When delayed, CD4$^+$ T cell depletion had little or no effect, indicating that the essential role of these cells in developing immunity had already been accomplished. Treatment with anti-CD8⁺ antibody during blood-stage infection has produced conflicting results. But in one case, late treatment with anti-CD8⁺ antibody prevented a significant number of infected mice from clearing parasites from their blood [22]. This suggests that CD8⁺ cells activated by CD4⁺ helper cells may be involved in a cascade of effector events leading to erythrocytic parasite destruction.

Mechanisms of T cell immunity

Having established that T cells activated specifically by malarial antigens play an important role in CMI to plasmodia, we now turn to the question of how such responses lead to parasite death or growth inhibition. Sporozoites are in all likelihood poor targets for CMI, but from the previous discussion we can submit that exoerythrocytic stages in hepatocytes can be killed by CD8⁺ T cells via direct cytotoxicity or by CD8⁺ and/or CD4⁺ T cells secreting lymphokines. CD8⁺ T cells along with other inflammatory cells have been shown to accumulate adjacent to infected hepatocytes *in situ*, and CD8⁺ cells capable of recognizing a 16-mer peptide, PyCTL-1, from the *P. yoelii* CS protein were directly cytotoxic for exoerythrocytic schizonts *in vitro* [22]. These observations and the fact that the adoptive transfer of cloned cytotoxic T lymphocytes (CTL) capable of recognizing a homologous epitope on the CS protein of *P. berghei* protected mice against sporozoite challenge lend support to the hypothesis that CTL participate in the destruction of exoerythrocytic parasites [177]. However, the recent finding that the immunity to *P. yoelii* sporozoites in certain strains of mice was not dependent upon the presence of CD8⁺ CTL indicates that other mechanisms of immunity are also involved.

IFN-γ, IL-1 and TNF-α have been shown to prevent the development of exoerythrocytic stages and it now appears that they are part of a cytokine (lymphokine) cascade, possibly involving multiple cell types, leading to the destruction of exoerythrocytic schizonts in hepatocytes (reviewed by Mazier *et al.* [185]). Components of this cascade include the recruitment of macrophages by IL-4 and granulocyte macrophage colony stimulating factor (GM-CSF), the activation of macrophages and possibly natural killer (NK) cells by IFN-γ, the production of TNF-α by macrophages, and the induction of IL-6 synthesis followed by the activation of a killing mechanism by IL-6. The recent observation that the inhibition of nitric oxide synthesis from arginine by

N-monomethyl arginine in macrophages blocked the killing of *P. berghei* schizonts in cultures of hepatocytes indicates that the killing mechanism may be dependent upon the well-established toxicity of nitrogen oxides [185,186]. The significance of these findings in the development of future chemotherapeutic agents and vaccines remains to be determined, but they do provide potential targets for both types of agents.

The situation relative to the destruction of erythrocytic-stage parasites by T cell-dependent CMI is less certain. While phagocytosis of infected erythrocytes by splenic and liver macrophages as well as blood monocytes occurs during malaria, phagocytosis does not explain the presence of "crisis forms" circulating in the blood of infected hosts. Similarly, phagocytosis does not account for parasite destruction following treatment with various immunomodulating agents (reviewed by Weidanz & Long [128]). Is parasite destruction mediated by cytotoxic T cells and/or other cells of mesenchymal origin? Despite sporadic reports suggesting that T cells and/or other mononuclear cells are capable of killing intraerythrocytic parasites (see Weidanz & Long [128]), it is generally accepted that T cells cannot be directly cytotoxic for erythrocytic-stage parasites because the host erythrocytes usually do not express class I MHC molecules, which CTL require for antigenic recognition [187]; however, this assumption may be premature in view of recent findings that, during malaria, γδ T cells increase in the peripheral blood and spleens of humans and mice, respectively, as indicated above. These cells do not require the recognition of restriction elements in order to be cytotoxic and they are known to accumulate in the red pulp of the spleen [188,189] where they would be in intimate contact with parasitized erythrocytes. Whether γδ T cells are capable of killing erythrocytic-stage parasites directly or via the secretion of lymphokines remains to be determined.

T cell receptor 22 (αβ) cells produce an array of lymphokines when activated specifically by malarial antigens. However, none of the lymphokines secreted by these cells show direct cytotoxicity for erythrocytic-stage parasites [190]. Treatment with IFN-γ has been shown to lessen the severity of rodent malaria but appears to do so indirectly, possibly through the activation of other cell types, including macrophages [191, 192]. Similarly, while TNF-α has been shown to suppress parasitemia in experimentally infected mice, it did not kill cultured parasites [191,193,194].

However, there is a rather long list of agents derived from various cell types which have the capacity to kill blood-stage plasmodia *in vitro* and in some instances *in vivo* (see Playfair *et al.* [190]). Convincing arguments have been made for the role of reactive oxygen species in resistance to blood-stage malaria. Products of the respiratory burst, including hydrogen peroxide and singlet oxygen, have been reported to kill plasmodia both *in vitro* and *in vivo*. Moreover, the killing of plasmodia by tumor necrosis serum appears to be mediated by lipid peroxides. The fact that macrophages and granulocytes from patients with chronic granulomatous disease were able to kill intraerythrocytic parasites [195, 196] and that mice deficient in their ability to produce an oxidative burst were capable of clearing parasites from their blood [197] suggests that factors other than reactive oxygen species may be involved in the destruction of blood-stage parasites. In fact, Playfair has suggested that oxides of nitrogen are capable of killing cultured *P. falciparum* [198], and oligomers of prostaglandins have been shown to inhibit the growth (kill) of plasmodia both *in vivo* and *in vitro* [199,200].

In summary, aside from a possible cytotoxic role being played by $\gamma\delta$ T cells, one might visualize that the destruction of erythrocytic-stage parasites involves the production of lymphokines by selected subsets of T lymphocytes. Certain of these lymphokines may be involved in the recruitment and activation of other cells, which in turn produce various cytokines, including IL-1 and TNF-α, that influence other cells and activate mechanisms leading to parasite destruction.

Role of the spleen in immunity to malaria

For many years hepatosplenomegaly has been considered the hallmark of malaria and the results of numerous splenectomy studies have shown that usually malaria is most severe in hosts lacking a spleen [201]. Why this is so remains to be determined. The spleen is known to increase in cellularity during malaria and it appears that the accumulation of activated T cells, B cells, and macrophages within the spleen is due to the secretion of lymphokines by T cells activated by malarial antigens [202,203]. The significance of this cellular response is evidenced by the effects of splenectomy on B cell-deficient mice immune to *P. c. adami*. Once immunized, these mice resist reinfection with fresh parasites but frequently have low levels of endogenous blood parasites which they control by T cell-mediated immunity. Splenectomy activates these infections, resulting in unremitting high-grade parasitemias which eventually are lethal, indicating that the spleen is essen-

tial for the expression of CMI [204]. Whether malaria parasites are actually killed in the spleen or elsewhere, for example in the liver as suggested by Playfair *et al.* [205] in the case of *P. yoelii*, the spleen is thought to be essential for the development and maintenance of immunity.

Aside from the accumulation of immune cells and the cytokines that they secrete within the spleen, an intact splenic architecture also appears to be essential for immunity since a functioning spleen cannot be replaced by an equivalent number of dispersed spleen cells [204]. The spleen is a filtering organ and changes in the microcirculation of the spleen during infection have been reported. In fact, Weiss *et al.* [206] have shown that the reticular cells become activated during infection and, in doing so, function as barrier cells to restrict the circulation of blood through the spleen. It is possible that circulating parasitized erythrocytes are brought into close proximity of cells producing toxic molecules or perhaps are held within the spleen by activated cells expressing adhesion molecules which facilitate the adherence of parasites and/or parasitized erythrocytes to cell membranes, where they can then be phagocytized or destroyed by toxic molecules. Antibodies reacting with parasite antigens on the red cell membrane might enhance these interactions between effector cells and their targets. There is no doubt that the spleen has a vital function in immunity to malaria and a clear understanding of how this is accomplished could prove useful in the design of future immunizing agents.

IMMUNOPATHOLOGY

Infection with malarial parasites induces protective humoral and cell-mediated immune responses. However, the immunologic mechanisms triggered can sometimes be harmful to the host and actually contribute to the pathologic manifestations of the disease. Some of the complications of malaria, such as anemia, nephropathy, and splenomegaly, are believed to have an immunologic basis and to be a direct result of the immune response to the parasite. In addition, it has recently been suggested that TNF, a nonspecific mediator produced by activated macrophages, may be responsible for many of the clinical symptoms of malaria. This section reviews the evidence for the involvement of immune reactivity in the pathology of human and experimental malaria.

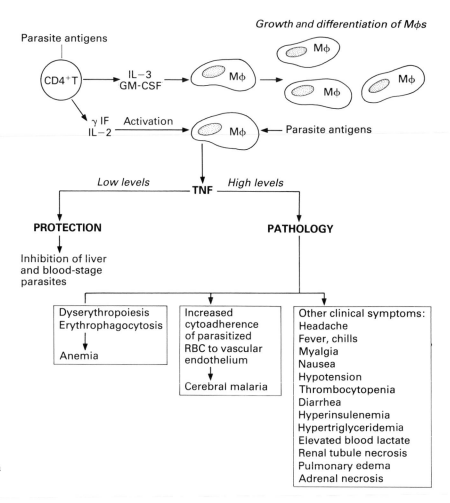

Growth and differentiation of Mϕs

Fig. 14.3 Postulated mechanism of action of TNF in the pathology of malaria.

Tumor necrosis factor

It has been proposed by Clark [207] and Clark & Chaudhri [208] that TNF may play a pivotal role in malarial pathophysiology and be responsible for much of the pathology which accompanies infection. Indeed, recombinant TNF administered to humans and animals reproduces many of the symptoms of clinical malaria (Fig. 14.3) (reviewed by Clark [207]). Tumor necrosis factor is produced by macrophages during the course of malaria, as evidenced by reports of elevated levels of TNF in the serum of patients suffering from infection [149,209]. Moreover, Grau *et al.* [209] reported that, in Malawian children with falciparum malaria, the mortality rate increased with the serum concentration of TNF. In mice infected with *P. vinckei*, plasma levels of TNF were found to increase exponentially at the time of the infection when pathologic changes started developing [208].

Parasite antigens presumably act as a trigger for TNF release by macrophages as it has recently been shown that *Plasmodium*-infected erythrocytes, as well as soluble heat-stable exoantigens from culture supernatants of parasitized erythrocytes, are able to stimulate directly TNF production by human [210] and mouse [211] macrophages. In addition, products of activated T cells, such as IFN-γ, which is present in the serum of patients with falciparum malaria, as well as IL-2 have been shown to enhance the production of TNF [208,212,213] and may also contribute to malarial pathology. This is supported by reports that splenomegaly and anemia in *P. yoelii*-infected mice [214] as well as cerebral malaria and thrombocytopenia in *P. berghei* infections [215] are T cell-dependent manifestations of malaria.

Playfair *et al.* [216,217] have suggested that, early in infection, induction of low levels of TNF by small numbers of infecting parasites may be beneficial since it has been shown that the *in vivo* administration of recombinant TNF inhibits the growth of both liver and blood-stage parasites in mice [180,191,193,218]. However, as the infection progresses and the antigenic load increases, the consequent increase in TNF production may result in pathology. In accordance with this, Playfair *et al.* [216] recently proposed that an "anti-disease" vaccine directed against TNF-inducing antigens rather than an "antiparasite" vaccine aimed at parasite elimination might be more effective in preventing the pathologic complications of malaria. The feasibility of such an approach is supported by a series of experiments conducted in mice. It was shown that the injection of heatstable soluble exoantigens of *P. yoelii* or *P. berghei* into mice previously primed with the macrophage-activating agent *Propionibacterium* induced the release of TNF in the serum [217]. Prior vaccination of the mice with boiled soluble parasite antigen prevented the release of TNF. Furthermore, soluble exoantigens were shown to cause the dose-dependent death of mice rendered hypersensitive to TNF by pretreatment with D-galactosamine, an effect which again could be prevented by prior vaccination with the antigens [217]. Characterization of the TNF-inducing exoantigens indicated that they are probably glycolipid in nature and behave as T-independent antigens [219]. Consequently, to be effective as vaccinating agents these molecules would have to be modified so as to render them T cell-dependent and elicit immunologic memory.

As suggested by Clark & Chaudhri [208], the involvement of TNF as a primary mediator in the pathophysiology of malaria would also explain the phenomenon of acquired malarial tolerance, i.e., the absence of clinical symptoms despite the presence of a high parasitemia in children from malaria endemic areas. Repeated exposure to TNF has been shown to render humans [220] and animals [221] refractory to its effects. By extension, repeated exposure to malaria parasites, which act as a trigger for TNF production, would render them tolerant to TNF and its pathophysiologic manifestations. This concept is supported by reports that subjects with acquired malarial tolerance are also resistant to the effects of endotoxin, a powerful inducer of TNF.

Cerebral malaria

Cerebral malaria, which often has a fatal outcome, is the most severe complication of acute falciparum malaria in young children and nonimmune adults. Autopsy findings show that, in individuals with this condition, blood vessels of the cerebral vascular bed are characteristically plugged with heavily parasitized erythrocytes. Petechial hemorrhages and various degrees of edema are also observed. The events involved in the pathogenesis of cerebral malaria remain unclear. Various groups have suggested that neurologic symptoms may be caused by:
1 the restriction of blood flow due to the adhesion of parasitized erythrocytes to capillary endothelium [222, 223];
2 a hyperergic reaction of the nervous system to the parasite [224];
3 circulating immune complexes and activation of complement [225];
4 T cell-mediated immune reactions [226].

The evidence for the involvement of T lymphocytes in the pathology of cerebral malaria is especially compelling. It was observed that, in humans, cerebral symptoms rarely developed in malnourished children who typically suffer from thymic atrophy [227]. Similarly, mice infected with *P. berghei*, athymic animals [226] or animals treated with anti-T cell serum [228] showed decreased cerebral pathology when compared to immunologically intact controls.

Recently, Grau *et al.* [215] developed a mouse model to investigate the pathology of cerebral malaria. Their model utilized genetically susceptible CBA mice which, when infected with the ANKA strain of *P. berghei*, develop neurologic lesions and succumb 7–15 days after infection. Human disease is not exactly reproduced by this model since adherence of parasitized red blood cells to cerebral vessels does not occur but rather capillary vessels become plugged with macrophages. This murine model has nevertheless yielded important information: the L3T4$^+$ subset of T lymphocytes was shown to play a significant role in the pathogenesis of murine cerebral malaria since:
1 The development of cerebral malaria in *P. berghei*-infected mice was prevented by *in vivo* depletion of L3T4$^+$ but not Lyt-2$^+$ T cells;
2 Adult CBA mice that were thymectomized, lethally irradiated, and bone marrow-reconstituted failed to develop neurologic lesions when infected with *P. berghei*. Reconstitution with CD4$^+$ but not CD8$^+$ T cells from normal mice restored their susceptibility to the development of cerebral malaria;
3 Adoptive transfer of L3T4$^+$ T cells from mice with cerebral malaria into infected euthymic mice resulted in the worsening of neurologic symptoms and earlier mortality.

The mechanisms by which L3T4$^+$ T lymphocytes influence the development of cerebral malaria remain to be defined. However, recent studies by Grau *et al.* [229] indicate that T cells may act by producing lymphokines (IFN-γ, IL-3, GM-CSF) which activate macrophages to release toxic factors such as oxygen radicals and, more importantly, TNF. Tumor necrosis factor levels were found to be significantly higher in susceptible CBA mice with cerebral malaria compared to infected CBA mice which did not develop neurologic symptoms or infected mice from genetically resistant strains. Depletion of L3T4+ T cells, which prevents cerebral malaria, was also found to prevent the increase in serum TNF. More direct evidence for the involvement of TNF in cerebral malaria was obtained in the following series of experiments:

1 92% of infected CBA mice injected with neutralizing anti-TNF antibody were protected against cerebral malaria;

2 infected mice genetically resistant to cerebral malaria developed neurologic symptoms characteristic of cerebral malaria when injected with recombinant TNF;

3 susceptible CBA mice developed neurologic lesions in the absence of malarial infection when injected with large doses of TNF [229,230].

Experiments involving the injection of neutralizing antibodies specific for various T cell lymphokines showed that a "cytokine cascade" involving IL-3, GM-CSF, and IFN-γ is involved in stimulating the overproduction of TNF by macrophages and the consequent development of cerebral malaria [143,197]. Treatment of CBA mice with a combination of anti-IL-3 and anti-GM-CSF (but not either one alone) or with anti-IFN-γ alone resulted in the inhibition of TNF production and prevention of cerebral malaria. Similar studies indicated that IL-1, platelet-activating factor, and IL-6 are not likely to play a major role in the pathogenesis of cerebral malaria [145,148]. The authors speculated that IL-3 and GM-CSF, which stimulate the growth and differentiation of hemopoietic cells, act by increasing the macrophage pool while IFN-γ is responsible for activating them to produce TNF. Tumor necrosis factor may then cause pathology through direct cytotoxic effects on endothelial cells (EC) and by increasing the EC adhesiveness for leukocytes and parasitized erythrocytes leading to local cell damage with consequent perivascular hemorrhages [229,230].

As already mentioned, three molecules on the surface of EC have been reported to act as receptors for *P. falciparum*-parasitized red blood cells: CD36, which is also found on the surface of platelets (glycoprotein IV), ICAM-1, and thrombospondin (reviewed by Howard

& Gilladoga [49]). Interestingly, ICAM-1 expression by human umbilical EC has been reported to be upregulated by IL-1, IFN-γ, and TNF, resulting in enhanced adherence of parasitized erythrocytes to the EC [54]. ICAM-1 is also a ligand for the leukocyte LFA-1 molecule and its upregulation by TNF (or IL-1, IFN-γ) may promote the adherence of monocytes to the EC of brain vessels such as observed in murine cerebral malaria. In accordance with this, Grau *et al.* [148] reported that administration of an anti-LFA-1 monoclonal antibody to *P. berghei*-infected CBA mice prevented the development of cerebral malaria, presumably by preventing adherence of leukocytes to the endothelium of blood vessels. The effect of cytokines on the expression of the CD36 and thrombospondin receptors remains to be determined, but it is possible that, as appears to be the case with ICAM-1, upregulation of these molecules by high levels of TNF may result in increased adherence of leukocytes (as in murine cerebral malaria) or of parasitized erythrocytes (as in human cerebral malaria) to EC, leading to brain capillary occlusion, local release of toxic factors, and consequent pathology. The fact that the immunopathologic process mediating cerebral malaria selectively occurs in blood vessels of the brain and not in the periphery may be explained by the phenotypic diversity of EC in different organs and tissues and the variability in their response to immunomodulators [49]. Finally, it should be mentioned that *P. falciparum*-parasitized red blood cells have also been shown to form rosettes with uninfected erythrocytes, a process which may further contribute to the occlusion of brain capillaries. However, the nature of the molecules involved in the binding of normal and infected red blood cells, as well as the effect of cytokines on the expression of such molecules, remain to be determined [49].

Anemia

Anemia is a common complication of malaria and can be especially severe in *P. falciparum* infections. Destruction of parasitized erythrocytes cannot entirely account for the degree of anemia generally observed. In fact, severe anemia can be found in the presence of low levels of parasitemia and can persist after complete eradication of the parasite [231,232]. It is believed that both an increased erythrocyte destruction and decreased production contribute to the anemia.

Parasitized erythrocytes are destroyed by the parasite itself at the time of merozoite release but can also be targets of the immune system. The binding of antibodies against plasmodial antigens expressed on the surface of infected red blood cells can lead to complement-mediated lysis, ADCC, or enhanced clearance by the reticuloendothelial system (see section on humoral response pp. 306–309). An increased destruction of normal red blood cells is also observed during the course of malaria, as evidenced by the decreased survival time of ^{51}Cr-labeled erythrocytes in both humans [233] and rodents [234]. The lysis of normal red blood cells may in part be due to sensitization with Ig and complement fragments, as a majority of individuals suffering from malaria were found to have a positive direct Coombs test [70,235]. Some of the IgG eluted from the surface of erythrocytes was found to be specific for *P. falciparum* schizont antigen and it was suggested that erythrocyte sensitization may result from the adsorption of circulating malaria antigen–antibody immune complexes [236]. Autoantibodies directed against red blood cells, which are found in malaria-infected hosts [69,70], may also contribute to normal red blood cell destruction.

Defective erythropoiesis is also believed to be a contributing factor in the anemia of malaria, as a decrease in the incorporation of ^{59}Fe by red blood cells as well as a reduction in the number of erythroblasts and reticulocytes has been reported in humans with malaria [231, 237,238]. Dyserythropoiesis, erythrophagocytosis, and decreased numbers of pluripotent stem cells and erythroid precursors have been observed in malarious hosts [239,240]. As in the case of cerebral malaria, recent evidence suggests that these pathologic changes may in part be due to the production of TNF by activated macrophages [239,241,242]. Clark & Chaudhri [239] showed that the injection of recombinant human TNF-α into mice with a low-grade *P. vinckei* infection reproduced the erythrophagocytosis and dyserythropoiesis observed in the bone marrow of mice with severe *P. vinckei* infections. Miller *et al.* [242] subsequently showed that both mice infected with *P. berghei* and mice given recombinant TNF-α via implanted osmotic pumps showed the same defects in erythropoiesis, i.e., reduction in bone marrow pluripotent stem cells, a decrease in erythroid progenitor cells, and a reduced incorporation of ^{59}Fe into erythrocytes. The observed changes were partially reversed in *P. berghei*-infected mice treated with antiserum to recombinant TNF-α [242]. The mechanism by which TNF-α is able to inhibit erythropoiesis remains to be determined but contributing factors may involve the demonstrated ability of TNF to:
1 inhibit the growth of erythroid precursors [243];

2 enhance phagocytic activity and hence erythrophagocytosis [244];
3 stimulate the release of toxic oxygen species by activated phagocytes, possibly resulting in damage to erythroblasts in bone marrow sinusoids [245].

In addition, as suggested by Miller *et al.* [242], TNF may promote monocyte differentiation at the expense of erythrocytes since it has been shown to induce differentiation of human myeloid cell lines [246].

Thrombocytopenia

Thrombocytopenia is a common complication of malarial infection and is believed to be due to increased peripheral platelet destruction rather than defective platelet production [247,248]. Malarial parasites have been observed inside blood platelets during malaria infection [249] but direct destruction of platelets by invading parasites is unlikely to be a major contributing factor in thrombocytopenia. Rather, it has been proposed that immune mechanisms may be involved in platelet destruction. This is supported mostly by the observation that levels of platelet-associated IgG are considerably increased during the course of human malaria [250]. Platelet-associated IgG levels, as well as platelet counts, return to normal following clearance of the parasites. Platelet-associated IgG may originate from circulating immune complexes, which are present in the circulation of malaria patients [251], or may represent autoantibodies directed against platelet surface antigens. In either case, platelet-associated IgG is likely to promote the rapid clearance of circulating platelets by the reticuloendothelial system, hence leading to thrombocytopenia [252]. Such a scenario is supported by the observation that CBA mice acutely infected with *P. berghei* developed an early thrombocytopenia accompanied by an increase in bone marrow megakaryocytes and a decreased survival of ^{111}In-labeled platelets [248]. Therefore, as in the human system, the *P. berghei*-induced thrombocytopenia was attributed to increased peripheral platelet destruction. Infected animals also showed increased levels of platelet-associated IgG which were inversely correlated with platelet counts. Moreover, passive transfer of serum from thrombocytopenic mice resulted in an immediate decrease of platelet counts in normal recipients, thereby supporting a role for antibody-mediated platelet destruction [248]. Interestingly, the occurrence of thrombocytopenia in the infected mice appeared to be dependent on the presence of CD4$^+$ T cells, since *P. berghei*-infected mice depleted of CD4$^+$ T cells by treatment with an anti-CD4 monoclonal antibody (GK1.5) failed to develop thrombocytopenia. Similarly, adult-thymectomized, irradiated mice reconstituted with bone marrow were also protected against thrombocytopenia but regained susceptibility upon receiving $CD4^+$ T cells. The mechanism by which $CD4^+$ T cells affect platelet levels is unclear but the authors speculate that it may involve helper activity for the production of anti-platelet antibodies and/or circulating immune complexes. T cell release of lymphokines having an effect on hemato-poiesis and/or platelet function may also play a role in the induction of thrombocytopenia.

Nephropathy

Both acute and chronic forms of kidney lesions associated with human malaria have been attributed to immunopathologic events involving the deposition of immune complexes in the glomeruli. Acute glomerulone-phritis observed in human and experimental malaria is characterized by the deposition of IgG, IgM, and complement in the mesangium and capillary loops of glomeruli. Malarial antigens have also been detected within the deposits in some cases. The glomerular lesions incurred are reversible and regress following antimalarial therapy [253–255]. In contrast, chronic glomerular lesions, which are typically associated with *P. malariae* (quartan) malaria, fail to respond to antima-larial chemotherapy and lead to the progressive development of a nephrotic syndrome. Serial biopsies showed that, as in acute glomerulonephritis, the majority of individuals with chronic nephropathy presented with granular deposits of IgG, IgM, and C3 in the glomeruli. *Plasmodium malariae* antigens were also detected in 25% of cases [256]. The mechanism behind the perpetuation of renal damage after treatment and elimination of *P. malariae* parasites remains unknown.

Immune complexes are generally recognized to play a causal role in the pathogenesis of both acute and chronic renal lesions. Soluble serum immune complexes have been detected in human malaria [251] as well as in mice infected with *P. berghei* [257], supporting the hypothesis that preformed circulating immune complexes become trapped in glomerular capillaries, leading to inflammation and tissue injury. Alternatively, it has been suggested that immune complexes may be formed locally following the deposition of free malarial antigen in kidney glomeruli. This is supported by the observation that, in the *P. berghei* model, antigens released from parasitized erythrocytes during their passage through glomerular capillaries appear to accumulate along the vessel walls and later become associated with Ig and complement [258,259]. A similar mechanism could conceivably operate in humans.

Houba [253,255] has suggested that, in quartan malaria, the progression of chronic renal lesions may be triggered initially by the deposition of malarial antigen–Ig immune complexes but that, after disappearance of malarial antigens from the body, other pathogenic events become involved. One possible mechanism involves the formation of autoantibodies. For example, sera from patients with malarial nephropathies have been reported to contain autoantibodies directed against host DNA and Ig [256,260,261]. Such antibodies could conceivably lead to the chronic formation and deposition of immune complexes in glomeruli.

Hyperreactive malaria splenomegaly

Splenomegaly is a consistent clinical feature of acute malaria infection and is sufficiently common in children so as to provide an index of malaria endemicity [262]. However, persistent splenomegaly in adults with acquired immunity is abnormal [263]. The prevalence of tropical splenomegaly closely parallels the geographic distribution of malaria, although idiopathic causes may be as frequent as malaria infection itself [264]. The term "tropical splenomegaly syndrome" has most frequently been used to describe the clinical condition associated with endemic malaria. More recently, the term "hyper-reactive malarial splenomegaly" (HMS) was proposed to distinguish the clinical, histopathologic, and immunologic condition due to malaria based on positive diagnostic criteria [265]. A nontransient reduction in spleen size upon antimalarial treatment is the essential clinical feature of HMS [266].

Hyperreactive malarial splenomegaly is characterized by massive and persistent splenomegaly, lymphocytic infiltration of hepatic sinuses, anemia, serum polyclonal IgM, and increased levels of antimalarial antibodies, serum antiglobulins, cryoglobulins, and immune complexes. The increase in IgM is due to increased synthesis rather than reduced catabolism [267]. Splenectomy does not correct the elevation in IgM [268]. It is believed that the immunopathology of splenomegaly may result from immune complex formation following the overproduction of serum IgM. During persistent malarial infection, the massive uptake of macromolecular aggregates of immune complexes by splenic mononuclear phagocytes, leading to hypertrophy of the reticuloendothelial system, is likely to contribute to the development of splenomegaly [269,270]. Genetic factors may be important in determining predisposition to HMS and may underlie familial and population patterns that have been observed [271]. For example, in areas of Papua New Guinea, HMS occurs in over half the adult population [272] whereas HMS in West Africa appears infrequent [263].

The precise immunologic mechanisms underlying the B cell overproduction of IgM remain unknown. However, it has been suggested that functional defects in CD8$^+$ T suppressor cells may lead to a loss of control over B cell activation and the consequent overproduction of IgM antibodies [273,274].

Immunosuppression

Suppression of immune responses to various antigens and mitogens has been observed in humans and experimental animals infected with plasmodia. McGregor & Barr [275] first reported that the antibody response to immunization with tetanus toxoid was greater in children receiving malaria prophylaxis than in untreated children. Later studies confirmed these findings and showed that children with acute falciparum malaria have depressed responses to a variety of vaccine antigens (e.g., meningococcal C polysaccharide, *Salmonella typhi* O antigen, and typhoid and poliomyelitis vaccines) while responses to other antigens remain unaffected (e.g., *S. typhi* H antigen, measles, and diphtheria, pertussis, tetanus, and Bacille Calmette-Guerin vaccines) [276–278]. Immune suppression may extend to specific responses to malarial antigens since decreased lymphocyte responsiveness to plasmodial antigens *in vitro* has been observed by several groups in subjects undergoing acute infection with *P. falciparum* [279–281]. In addition, the level of antiplasmodial antibodies was found to be significantly lower in patients who died from severe malarial infection compared to those who recovered [282]. The cellular basis for immunosuppression in acute falciparum malaria remains to be established. An observed defect in the production of IL-2 and expression of IL-2 receptors may be involved in the depression of immune responses [138]. High levels of circulating soluble IL-2 receptors were also reported to be present during acute infection and may also act to modulate immune reactivity [138].

Experimental animals have provided the opportunity for more detailed studies. Several workers have reported on the depression of humoral responses to a variety of

antigens administered to plasmodia-infected mice [23]. The inability of macrophages from infected mice to function as accessory cells appeared to be a contributing factor [283–285]. Macrophage dysfunction in malarious mice was attributed to:

1 a "dilution" of functional macrophages due to recruitment of large numbers of immature macrophages during infection [285];

2 the suppressive activity of macrophages from infected mice [284];

3 a defect in the handling and processing of antigen by macrophages [286,287].

Regarding the latter hypothesis, a recent study showed that overloading of macrophages with hemozoin may impair their function since hemozoin-laden macrophages from the liver and spleen of *P. berghei*-infected mice were defective in their ability to serve as accessory cells in the generation of an antisheep red blood cell response [287]. Moreover, the addition of purified hemozoin to cultures of normal spleen cells inhibited their response to sheep red blood cells. In addition to macrophage dysfunction, a defect in IL-2 synthesis in malaria-infected mice has been reported and may also contribute to immunosuppression [288].

Malaria-induced immunosuppression could conceivably render the host more susceptible to secondary infections. There are, however, few reports of this occurrence in the literature. Another cause for concern is the effect of malaria on the effectiveness of vaccines administered to infected individuals. The identification of the plasmodial antigens responsible for inducing immunosuppression may lead to a better understanding of the cellular events leading to immunosuppression and may also help prevent potentially deleterious effects of malaria vaccines.

INNATE RESISTANCE

Resistance to malarial infection is dependent on the development of an immune response by the host and, to a varying extent, on certain innate characteristics possessing protective value against infection. For example, individuals whose red blood cells lack Duffy blood group determinants appear to be completely resistant to infection with *P. vivax* [289,290]. Innate resistance, however, is not always absolute, and genetically determined factors can instead act to reduce the severity of infection and improve survival. It was first proposed by Haldane in 1949 [291] that the high fre-

Table 14.2 Erythrocyte variants linked to innate resistance to malaria

Variant	Resistance to malaria	Degree of protection	Postulated mechanism(s)
1 RBC membrane Lack of Duffy Ag (Fyb determinant)	*Plasmodium vivax*	+++	Lack of surface receptor for merozoites
Lack of glycophorin A	*Plasmodium falciparum*	+/−	Lack of surface receptor for merozoites
Ovalocytosis/elliptocytosis	*Plasmodium falciparum* *Plasmodium vivax* *Plasmodium malariae*	+	Invasion of RBC inhibited by: – reduced membrane deformability – restricted lateral mobility of membrane proteins
2 RBC enzyme deficiencies G6PD deficiency	*Plasmodium falciparum*	− (Homozygotes) − (Hemizygotes) + (Heterozygotes)	Decreased levels of GSH causing: – sensitivity to oxidant damage – impaired ribose metabolism of parasite
Pyridoxal kinase deficiency	*Plasmodium falciparum*	?	Decreased synthesis of the cofactor pyridoxal-5'-phosphate
3 Hemoglobinopathies Hemoglobin S (sickle cell trait)	*Plasmodium falciparum*	++	– Loss of intraerythrocytic K^+ under low O_2 tension (e.g., deep tissues) – Increased phagocytosis of parasitized erythrocytes
Thalassemia (α and β)	*Plasmodium falciparum*	+	– Sensitivity of RBC to oxidant damage – Decrease in number of Hb chains that can be used as a source of aa by parasite – Delay in replacement of fetal Hb in infants with β-thalassemia
Fetal hemoglobin	*Plasmodium falciparum*	?	?
Hemoglobin C, D, E (β-chain structural variants)	*Plasmodium falciparum*	?	– Delay in replacement of fetal Hb in infants

Hb, hemoglobin; G6PD, glucose-6-phosphate dehydrogenase; aa, amino acid; RBC, red blood cell.

quencies of thalassemia and other hemoglobinopathies observed in malaria endemic areas resulted from the fact that these genetic disorders offered some degree of protection against the parasite. Convincing evidence that malaria could indeed select for certain phenotypes and determine gene frequencies in human populations was subsequently presented by Allison with regard to the sickle cell trait [292,293]. It was demonstrated that the sickle cell trait existed in a state of "balanced polymorphism" in malaria endemic areas where the disadvantage of a defective hemoglobin was balanced by the improved survival that it conferred to the carrier [294]. The concept that genetic variation in human red cells can affect susceptibility to malaria has since been referred to as the "malaria hypothesis."

Even though most studies on innate resistance have been concerned with the ability of host erythrocyte variants to inhibit parasite growth, innate resistance can potentially affect the growth of the parasite at any of the stages in its life-cycle (Table 14.2).

Interference at the erythrocyte membrane level

The Duffy blood group system

It was shown in 1975 by Miller *et al.* [295] that human erythrocytes which do not express the Duffy blood group antigen on their surface (Fya⁻b⁻) are resistant to *in vitro* infection with *P. knowlesi*, the "monkey equivalent" of human *P. vivax*. In the absence of the Duffy antigen, *P. knowlesi* merozoites were able to attach to erythrocytes but failed to establish a junction with the membrane and became detached [34,295]. At the same time, anti-Duffy antibodies were shown to inhibit partially the invasion of Duffy positive erythrocytes [295], leading to the notion that the Duffy blood group antigen was the "receptor" for simian *P. knowlesi* and presumably *P. vivax* in humans. Direct support for this hypothesis was provided by experiments in which human volunteers were exposed to the bite of *P. vivax*-infected mosquitoes; Duffy-positive individuals developed malaria while Duffy-negative individuals invariably remained unaffected [289,290]. Nevertheless, direct ligand studies could not be performed with *P. vivax* since this parasite was not easily amenable to culture *in vitro*. Recently, Barnwell *et al.* [296] developed a short-term *in vitro* assay for the invasion of human erythrocytes by *P. vivax* parasites obtained from squirrel monkeys (*Saimiri sciureus*). Their results suggest that *P. vivax* merozoites do indeed use the Duffy glycoprotein, more specifically the newly characterized Fy-6 determinant [297], as a ligand for the invasion of red blood cells. However, the authors caution that the Duffy antigen is probably not the only ligand involved in the process of invasion, since *P. vivax* merozoites preferentially infect reticulocytes [298] even though the Duffy antigen is present on both reticulocytes and mature red blood cells. The authors suggest that a separate determinant peculiar to reticulocytes may act to facilitate the initial attachment of *P. vivax* merozoites while the Fy-6 epitope is involved in junction formation [296]. A low-affinity receptor on the surface of reticulocytes could conceivably be used for initial attachment by *P. falciparum* and *P. ovale* as well, since these parasites also exhibit a predilection for reticulocytes over mature red blood cells [299].

Glycophorin

Duffy-negative erythrocytes are resistant to infection with *P. vivax* but remain fully susceptible to invasion by *P. falciparum*, indicating that this parasite utilizes a different receptor on the surface of the red blood cell. Miller *et al.* [300] first noted that Ena⁻ human erythrocytes which lack glycophorin A, the major erythrocyte sialoglycoprotein, were less susceptible to invasion by *P. falciparum*. Several lines of evidence suggest that glycophorin A acts as a receptor in the invasion of

erythocytes by *P. falciparum*. It has been reported that anti-Ena antisera [301], Fab' fragments from antiglycophorin A antibodies [302], and more recently monoclonal antibodies to glycophorin A [296] all acted to inhibit invasion of human red blood cells by *P. falciparum* merozoites. Invasion of erythrocytes also appears to be inhibited by soluble glycophorin A [302] or treatment with trypsin, which cleaves glycophorin A from the erythrocyte surface [300]. It is unlikely that glycophorin A is the sole receptor for *P. falciparum* invasion, since Ena⁻ erythrocytes can still be invaded, although at a reduced rate. Other potential ligands for merozoites remain to be identified.

Ovalocytosis/elliptocytosis

Hereditary ovalocytosis, sometimes also referred to as hereditary elliptocytosis, occurs at a high frequency (up to 30%) in aboriginal populations throughout Southeast Asia, especially among Melanesians living in Papua New Guinea. Epidemiologic data suggest that, in these regions, individuals with ovalocytosis are less susceptible to infection with *P. falciparum*, *P. vivax*, and *P. malariae*, as they exhibit both a reduced parasitemia and frequency of infection with these parasites [303,304]. *In vitro* experiments have shown that ovalocytic erythrocytes from Melanesians are resistant to invasion by merozoites of both *P. falciparum* and *P. knowlesi* [305,306]. Therefore, it was suggested that ovalocytes offer resistance to malarial infection by interfering with a step(s) of the invasion process common to all malarial species. Ovalocytes display a marked reduction in membrane deformability and it has been suggested that this property is responsible for their resistance to invasion by malaria parasites [305–307]. In accordance with this hypothesis, Mohandas *et al.* [307], using an ektacytometer, showed that the ability of *P. falciparum* merozoites to infect ovalocytes from a given individual was inversely related to the degree of membrane rigidity. Moreover, when normal erythrocytes were treated with graded concentrations of glutaraldehyde to crosslink membrane proteins and increase membrane rigidity, susceptibility of *P. falciparum* infection decreased as the deformability of normal cells progressively decreased. It has also been proposed by other workers that restricted lateral mobility of certain membrane proteins, rather than physical indeformability of ovalocytes, may represent an important factor in the inhibition of the binding and penetration of merozoites [308].

Interference at the intracellular level

Enzyme deficiencies

Glucose-6-phosphate dehydrogenase deficiency. Glucose-6-phosphate dehydrogenase (G6PD) deficiency is an X-linked condition occurring in up to 10–20% of the population in malarious regions, such as Sardinia or tropical Africa [309]. Inhibition of *P. falciparum* growth in G6PD-deficient erythrocytes has been observed both *in vivo* [310] and *in vitro* [311]. Interestingly, heterozygote individuals whose blood contains both normal and G6PD-deficient erythrocytes are relatively protected against malaria, while female homozygotes or male hemizygotes whose blood contains only G6PD-deficient cells appear to be as susceptible to malaria as normal individuals [312]. It has been shown *in vitro* that, when faced with an erythrocyte population in which every cell is G6PD-deficient, the parasite is induced to produce its own enzyme and, after a few cycles of multiplication, gradually "adapts" to growth in the enzyme-deficient cells [313,314]. It is speculated that the same phenomenon may also take place *in vivo*. In homozygous females and hemizygous males, the parasite would presumably adapt to growth in the G6PD-deficient red blood cells of the host and ultimately induce the same levels of parasitemia as in normal individuals. In contrast, in heterozygotes, the merozoites would be cycling between both normal and deficient erythrocytes; merozoites developing in normal red blood cells would not be induced to produce their own enzyme and their growth would be inhibited upon subsequent invasion of G6PD-deficient red blood cells, thereby resulting in lower parasitemias.

Glucose-6-phosphate dehydrogenase-deficient erythrocytes have decreased levels of reduced glutathione (GSH) and it has been suggested that their ensuing sensitivity to oxidant damage is responsible for the inhibition of parasite growth. Indeed, Friedman [315] showed that the *in vitro* growth of *P. falciparum* in G6PD-deficient erythrocytes was most severely affected under oxidant stress (30% O_2). Under *in vivo* conditions, oxidant damage to the host erythrocyte or the parasite within could result from the production of H_2O_2 by the parasite itself, from the release of reactive oxygen species by polymorphonuclear leukocytes and macrophages undergoing a respiratory burst [316], or even from the dietary habits of the host, such as the consumption of fava beans which are known to generate strong oxidants causing hemolysis of G6PD-deficient red blood cells [317]. It has been argued that the low level of GSH in G6PD-deficient erythrocytes is deleterious to the parasite not only because of increased susceptibility to oxidant damage but also because of its effect on ribose metabolism and nucleic acid synthesis. Roth *et al.* [318] showed that, in parasitized G6PD-deficient red blood cells, 5'-phosphoribosyl-1-pyrophosphate (PRPP) synthetase, an enzyme which requires GSH for full activity, showed a dramatic decrease in function compared to normal infected red blood cells. The resulting decrease in the production of ribose compounds could undoubtedly affect the ability of merozoites to synthesize RNA and DNA and multiply within the host erythrocyte.

Pyridoxal kinase. Low erythrocyte pyridoxal kinase activity has been reported to occur at a relatively high frequency in Black Americans [319] and Africans [320], suggesting that reduced enzyme activity may present a selective advantage against malaria. In accordance with this hypothesis, Martin *et al.* [321] showed that the pyridoxal kinase activity of malaria patients with high *P. falciparum* parasitemia was significantly higher than that of a control group. Pyridoxal phosphate catalyzes the major step in the synthesis of the cofactor pyridoxal-5'-phosphate. At present, it is unclear what role this red blood cell enzyme might play in parasite development and to what extent its level contributes to resistance against malaria.

Hemoglobinopathies

Sickle cell trait (hemoglobin S). The sickle cell gene (HbS) occurs at a high frequency in areas of the world where *P. falciparum* malaria is prevalent [293], with 20–50% of West Africans being heterozygotes (HbS/HbA). The gene is believed to exist in a state of "balanced polymorphism;" homozygotes (SS) develop sickle cell anemia, an often fatal condition, but this disadvantage is balanced by the markedly improved survival of heterozygotes (AS; sickle cell trait) over normal individuals when infected with *P. falciparum* [294]. *In vitro* studies have shown that, under aerobic conditions (18% O_2), the growth of *P. falciparum* in AS and SS cells is essentially the same as in normal AA erythrocytes [322,323]. In contrast, at low oxygen tensions (1–5% O_2), a dramatic inhibition of parasite growth was observed [304,305]. In SS cells, extensive sickling of the host erythrocytes, accompanied by the formation of paracrystalline needles by the aggregated HbS, was considered responsible for physical destruction of the parasite. In AS cells, the parasites were not lysed but rather showed an arrested development inside both sickled and unsickled cells [322,323]. A possible explanation for this phenomenon was provided by Friedman who showed that, under low O_2 tension, intracellular K^+ decreased in AS erythrocytes, thereby inhibiting the development of erythrocytic parasites which require high levels of K^+ for growth [324]. *In vivo*, the growth inhibitory effect of low O_2 tension on parasites within AS erythrocytes is presumably reproduced in the later stages of the *P. falciparum* life-cycle when parasitized erythrocytes leave the circulation and become sequestered in the capillaries of deep tissues [322,323]. Phagocytosis may also play a role in the resistance of AS heterozygotes to *P. falciparum*, as Luzatto *et al.* [325] reported that parasitized AS erythrocytes have a greater tendency to sickle, thereby making them more susceptible to removal by the phagocytic elements of the RES.

Thalassemia, fetal hemoglobin (HbF), and β-chain variants (HbC, -D, -E). Thalassemia is characterized by the absence or deficiency in the synthesis of one or more of the (α- or β-thalassemia) chains of hemoglobin. The relatively high frequency of thalassemia in malaria endemic areas suggests that it may offer a selective advantage against malaria [326]. In the mouse, β-thalassemia was recently shown to be directly responsible for protection against *P. chabaudi adami*; thalassemic mice showed less severe infection with delayed peak parasitemia, while mice in which the thalassemia had been transgenically corrected with the human β-globin gene showed the same pattern of infection as normal mice [327]. In humans, erythrocytes from thalassemic patients appear to support the *in vitro* growth of *P. falciparum* normally unless an oxidant stress is applied [315] or an amino acid-poor medium is used to culture the parasitized erythrocytes [328]. The mechanism behind the observed susceptibility of parasitized thalassemic erythrocytes to oxidant damage is unclear but, as in the case of G6PD-deficient erythrocytes, it would render the host erythrocytes susceptible to damage by products of the respiratory burst released by phagocytes or H_2O_2 production by the parasite itself.

The inability of thalassemic erythrocytes to support growth of the parasite in amino acid-poor culture medium was attributed to the fact that *P. falciparum* is believed to depend upon normal host hemoglobin (Hb) as a source of amino acids, thereby making Hb-deficient thalassemic erythrocytes a poor source of nutrients.

Culture of parasitized thalassemic erythrocytes in one of the commonly used tissue culture media with very high amino acid contents would mask any detrimental effect of thalassemia since the parasite could obtain sufficient amino acids from the medium. In contrast, culture in an amino acid-poor medium allowed the detection of growth inhibition in Hb-deficient thalassemic red blood cells [328].

In the case of β-thalassemia, an additional factor may be involved in protection against *P. falciparum*. Infants heterozygous for β-thalassemia show a delay in the replacement of fetal hemoglobin (HbF) by adult Hb (HbA). Pasvol *et al.* [329] showed that, *in vitro*, erythrocytes containing HbF inhibited the development of *P. falciparum* even though there was no effect on the rate of invasion by the parasite. The mechanism responsible for the inhibitory effect of HbF remains to be determined. Nevertheless, retardation in the rate of decline of HbF production in infants heterozygote for β-thalassemia may confer some degree of protection against *P. falciparum* until active immunity is established [44]. The replacement of HbF with adult HbA is also delayed in infants with β-chain structural variants, such as HbC, -D, -E, and may also offer some degree of protection to these individuals.

IMMUNO- AND MOLECULAR DIAGNOSIS OF MALARIA

For the past 100 years, the most reliable means for the accurate diagnosis of malaria has been the detection of plasmodial parasites by light microscopy in the blood of infected individuals. This technique allows for the quantitation of blood parasitemia levels and species identification of the infecting malarial parasite with a sensitivity (20 parasites/μl or 0.0004%) which remains unsurpassed by alternative methods [330]. The technology is relatively simple and, with trained personnel, applicable under field conditions. Although time consuming when low parasitemias are encountered, light microscopy will most likely not be replaced as the method of choice for the individual diagnosis of malaria. However, malaria diagnosis in clinics and fieldstations is only one level of application for a diagnostic test. More efficient ways of detecting parasites are necessary in epidemiologic studies that attempt to evaluate the prevalence of malarial infection and the effectiveness of control programs, as well as in the screening of blood banks in endemic areas. To meet the demand for diagnostic tests at this level, the tools of modern immunology and molecular biology are being applied towards the development of alternative assays. Considering the current vaccine efforts, it seems likely that these new tests will find broad application in the near future.

Immunologic diagnostic assays

The detection of *P. falciparum*-specific antibodies in the sera of infected individuals can be readily accomplished by several methodologies. Although effective, their use as diagnostic tools is limited, as a distinction between past and present infections is not always clear. Alternative approaches have been tested that involve the diagnosis of acute *P. falciparum* infection by demonstrating the presence of parasite antigens in infected peripheral blood. Both solid-phase radioimmunoassays [331,332] and enzyme-linked immunosorbent assay (ELISA) [333] have been used to detect *P. falciparum* antigens of lysed erythrocytes, by their ability to block the binding of specific antibodies to wells coated with crude parasite antigen preparations. Although these assays have been adapted to avoid false-positive results due to antimalarial antibodies present in the test sample, the standardization of the crude antigen and antibody preparations may be problematic. To detect defined *P. falciparum* antigens in patient blood during acute infection, alternative antigen capture assays have employed monoclonal antibodies directed against a heat-stable antigen Pf93 [5], a histidine-rich protein PfHRP-2 [334,335], or a 50 kD exoantigen [336]. Although standardization of reagents appears to be less of a problem, the sensitivity of these assays needs to be enhanced. Due to the antigenic diversity between different plasmodial isolates, difficulties may also arise with tests based on the binding of a single antibody to a single antigenic epitope.

The most promising use of immunologic tests as malaria diagnostic tools involves the identification of sporozoites in infected mosquitoes. Traditionally, the determination of sporozoite rates in a given vector population has relied on the tedious dissection of mosquito salivary glands and the microscopic identification of sporozoites. With the availability of monoclonal antibodies specific for the CS protein of several species of *Plasmodium*, immunoradiometric assays have been developed to detect malaria parasites in mosquito vectors [337]. These assays have been adapted to take advantage of nonradioactive ELISA technologies [338–340], with sensitivities comparable to that achieved by mosquito dissection. Moreover, exquisite specificities

of the antibodies involved has allowed for the identification of the species of plasmodia present in a given vector population. This provides a considerable advantage over mosquito dissection methods. Both the immunoradiometric assays and ELISA have performed well in field evaluations in the identification of *P. falciparum*- and *P. vivax*-infected mosquitoes [341–343] and should continue to be of value in epidemiologic studies.

Molecular diagnostic assays

In 1984, Franzen *et al.* [344] reported the isolation and characterization of a repetitive DNA probe to identify *P. falciparum*-infected blood in a spot-hybridization assay. The identified recombinant clone contained a 21 bp tandemly repeated sequence that was specific for *P. falciparum*. The genomic organization of this noncoding highly reiterated element has since been characterized in detail and appears to comprise approximately 1% of the *P. falciparum* genome [345,346]. Other laboratories using similar experimental strategies have likewise identified diagnostic repetitive DNA probes for *P. falciparum*, most of which are based on the same 21 b p repeat [347–350]. The usefulness of these probes in the diagnosis of *P. falciparum* infections in geographically diverse locations has been evaluated using radiolabeled plasmid DNA [344,351,352] and synthetic oligonucleotides that are isotopically [352–355] or enzymatically [356–358] labeled. The findings in these studies indicate that hybridization assays based on this repetitive DNA probe are highly specific for *P. falciparum* but sensitivities parallel that of direct microscopic examination only in the best of circumstances. Difficulties in the detection of low parasitemias in partially immune individuals from malaria endemic areas were particularly noted [359]. No correlation between the intensity of the hybridization signal and the level of parasitemia could be established consistently. Nevertheless, the results from these initial trials were encouraging. In future studies it may be possible to develop additional probes to identify the other human malarial parasites and to increase the sensitivity of these assays by signal amplification using polymerase chain reaction technologies.

A new and more recent effort in the area of molecular diagnosis of malaria infection has focused on the detection of plasmodial ribosomal RNA [360,361]. This test employs radiolabeled oligonucleotides as probes which are complementary to species-specific regions of plasmodial RNA of the small ribosomal subunit. The sensitivity of the assay appears to be significantly better than that of repeat-based DNA probes, most likely because of the high abundance of target RNA. Like detection by light microscopy, this test permits the speciation of the plasmodial parasite in question and can possibly be used to quantitate parasitemia. It is likely that this new approach will yield useful information in field evaluations, as it is relatively quick and inexpensive and will use patient blood obtained from a finger stick directly.

VACCINES

The emergence and spread of insecticide-resistant mosquito vectors and drug-resistant plasmodia, despite major control efforts, have resulted in the continued prevalence of malaria throughout the world. Owing to the seriousness of this health problem, much effort has focused on the development of an antimalarial vaccine. The limited ability to produce native malarial antigen in sufficient quantities for vaccine use necessitates the construction of a subunit vaccine which will rely on recombinant DNA and/or peptide synthesis technologies.

The successful development of a defined malaria vaccine rests on the attainment of a clear understanding of both the cellular and humoral components of the immune responses elicited during malarial infection and the identification of the plasmodial antigens which trigger protective responses. Protective T cell and B cell epitopes of these antigens, once characterized, might then be incorporated into recombinant or synthetic subunit vaccines. Many factors, such as the polymorphisms inherent to the protective determinants, the most effective means of immunization, and the potential for natural boosting of protective responses upon subsequent exposure, must also be considered. Extensive efforts have been made towards the realization of this goal in examining the sporozoite, erythrocytic, and gametocyte stages of malarial parasites. Since the expression of plasmodial antigens is stage-specific, the use of a multicomponent vaccine containing protective antigens of each of these developmental stages is most desirable.

Sporozoites

Considering the course of plasmodial growth and development, a vaccine directed against the infective form of the parasite, the sporozoite, is the first possible target for vaccine intervention. An antisporozoite vaccine would be designed to prevent the initial parasite infection of, and development within, liver hepatocytes. Since infection with a single sporozoite can lead to the development of a severe blood-stage infection, an antisporozoite vaccine must be 100% effective, exerting its protective effect over a relatively short period of time.

Attenuated sporozoite vaccines

Early studies showed that it was possible to protect animals [362,363] and humans [364,365] from malarial infection by immunization with attenuated X-irradiated sporozoites. The protection appeared to involve a humoral component in that immunization induced antibodies which mediated the shedding of the sporozoite surface coat (CS protein reaction), rendering the parasites noninfectious both in vivo [113] and in vitro [114,366]. Likewise, the passive transfer to Fab fragments of a monoclonal antibody directed against the surface coat of P. berghei sporozoites protected mice from sporozoite challenge (93). However, the protection observed upon immunization with irradiated sporozoites clearly involved antibody-independent cell-mediated immune responses as well. As described earlier, the protection of B cell-deficient mice immunized with irradiated sporozoites was comparable to that observed with immunologically intact animals, while T cell-deficient animals similarly immunized were fully susceptible to sporozoite infection [172]. In later studies, the adoptive transfer of T cells but not B cells from sporozoite-immunized mice to naïve recipients conferred protection against infection [367].

Circumsporozoite proteins

Attempts to identify sporozoite antigens which elicit protective immune responses led to the characterization of the major surface antigen of sporozoites, the CS protein. The CS proteins are one of the major biosynthetic products of sporozoites [368]. Ranging in size from 40 to 60 kD, depending on the species being studied, these proteins form a coat covering the surface of mature sporozoites [369–372]. Immunologic characterization of the CS revealed the presence of a strikingly dominant B cell epitope, present multiple times within a single molecule [373,374]. In competitive binding studies, this determinant was recognized by all monoclonal antibodies raised against sporozoites, as well as most antisporozoite antibodies present within polyclonal antisera.

Through gene cloning and nucleic acid sequence analysis, the primary amino acid sequences of the CS proteins of *P. knowlesi* [375], *P. falciparum* [376], *P. vivax* [377], *P. berghei* [378], *P. cynomolgi* [379], *P. yoelii* [380], *P. brazilianum* [381], and *P. malariae* [382] have been deduced. The overall structure of the CS proteins appears similar among these parasites, and includes an N-terminal signal sequence and a C-terminal hydrophobic membrane anchor. The most striking feature of the CS proteins is the presence of a repeat domain, comprising the central third of the molecule. This region contains a series of tandemly repeated amino acid residues differing substantially among species in composition and number. To varying degrees, polymorphism is also apparent within the repeat domains of different strains and isolates of the same species, most notably with *P. cynomolgi* [383], *P. knowlesi* [384], and *P. vivax* [385]. In *P. falciparum*, however, DNA hybridization studies indicate that the variability in this domain of the CS protein is limited, and the characteristic NANP tetrapeptide repeat sequence is present in 18 isolates of *P. falciparum* examined thus far [386]. Nevertheless, the function of this extensive repeat domain in parasite infectivity is as yet unclear.

By sequence comparison, similarity in two additional CS protein domains was noted, with both containing a large number of charged amino acid residues. The first immediately flanks the repeat domain on the N-terminal side while the second is found among the cysteine residues at the C-terminus [375–382]. Within the N-terminal charged sequences, nine out of 15 amino acids (region I) are conserved between the CS proteins of *P. falciparum* and *P. knowlesi*. Studies employing synthetic peptides indicate that this region of the molecule may be involved in the interaction between sporozoites and hepatocytes [387–392]. A second region of similar sequence (region II), displaying identity at 12 out of 13 residues in most CS molecules, is present preceding the C-terminal charged area. Although direct evidence for the role of this region in sporozoite infectivity has not been clearly established, its similarity to sequences of thrombospondin and properdin suggests a possible role in cell adhesion or receptor–ligand interactions [388,389].

Subunit sporozoite vaccines

In the development of a subunit vaccine against *P. falciparum*, attention has focused on the repeat domain of the CS antigen. This region contains the immunodominant B cell determinant of the CS protein recognized by monoclonal and polyclonal antibodies which neutralize parasite infectivity. Although a large domain, the dominant epitope is present within as few as three repeats of the NANP tetrapeptide [390]. Unlike antibodies to synthetic peptides derived from regions I and II, antibodies raised against synthetic peptides containing the *P. falciparum* NANP repeat unit react with the native CS molecule and inhibit the sporozoite invasion of hepatoma cells [391]. To test the vaccine potential of this repeat domain, a synthetic antimalarial vaccine containing (NANP)3 conjugated to tetanus toxoid was constructed [392]. A second vaccine was produced concurrently, through the recombinant expression of the *P. falciparum* CS protein in *Escherichia coli* [393]. The recombinant peptide, designated R32tet32, contained two copies of the repeat unit [(NANP)16(NVDP)1] fused to 32 residues of the tetracycline resistance gene read out of frame.

In preclinical studies in animals, both antisporozoite vaccines were immunogenic and induced the production of high titers of anti-NANP antibodies. In safety and efficacy trials in humans, the candidate vaccines when administered with aluminum hydroxide as an adjuvant were safe, well tolerated, and elicited IgM and IgG antisporozoite antibodies [392,394]. This rise in antibody titer, however, did not correlate well with the observed T cell proliferative responses. Upon challenge with *P. falciparum* sporozoites, only two of nine immunized individuals were protected and failed to develop a blood-stage infection. The remaining immunized and control subjects in both vaccine trials developed clinical malaria within 2 weeks postchallenge. Vaccinated volunteers displayed a slight delay in the onset of parasitemia relative to controls, but this did not alter the severity of the ensuing blood-stage infection. Although some correlation between vaccine dose and antibody levels was apparent, no correlation existed between T cell responses and antibody levels or protection. This, combined with the inability to significantly boost either B or T lymphocyte responses revealed the poor immunogenicity in humans of both candidate vaccines and limited their further use in expanded field trials.

Several studies attempted to enhance the immunogenicity of these CS repeat-based synthetic and recombinant antisporozoite subunit vaccines, with the hope of significantly increasing the levels of anti-NANP antibodies. Successful strategies tested in experimental model systems included alternative routes of immunization [173], the use of novel adjuvants such as cytokines [395], the incorporation of vaccine peptides into proteosomes [396], and the selection of additional carrier proteins [397]. It is questionable, however, whether the induction of higher antibody levels to the NANP repeat will be protective. The level of naturally acquired antisporozoite antibodies in patient sera from highly endemic malaria areas does not correlate with protection [398,399]. Patients with serum anti-NANP antibody levels in excess of those elicited in either vaccine trial are fully susceptible to *P. falciparum* infection. Furthermore, *in vivo* studies involving the *P. berghei* [22,180,400,401] and *P. yoelii* [181,402] murine models stress the importance of cell-mediated components of the immune response in the development of protective immunity against malaria sporozoites. It is clear that the stimulation of effector T cells which mediate cytotoxicity or secrete parasiticidal lymphokines, such as IFN-γ, must not be overlooked.

T cell responses to CS proteins

Efforts characterizing T cell responses to the CS protein and specifically to the candidate vaccines have been extensive. Mapping of T cell epitopes of the *P. falciparum* CS protein highlighted the difficulties in developing a subunit malarial vaccine. Studies utilizing H-2 congenic mice revealed the limited recognition of T cell epitopes lying within the NANP repeat domain as well as those in the flanking nonrepeat regions [403–405]. Human T cell recognition of CS protein determinants also appears limited. Peripheral blood lymphocytes from 14 out of 35 adults living in a malaria endemic region of the Gambia did not proliferate in response to any of a set of overlapping synthetic peptides covering the entire *P. falciparum* CS molecule [406]. In addition, no relationship was observed between T cell proliferative responses to any peptide and the level of anti-NANP antibodies. Of those patients responding, three immunodominant T cell domains outside of the repeat region were mapped. However, comparison of CS protein sequences from three *P. falciparum* strains indicated that these T cell sites were located within polymorphic segments of the antigen [407,408] which lacked immunologic cross-reactivity [409].

The restricted recognition and polymorphic nature of the CS protein T cell sites present a considerable obstacle to the development of a subunit sporozoite vaccine. Appropriate sporozoite-derived T cell epitopes must be incorporated into such a vaccine if boosting of B and T lymphocyte responses upon natural exposure is to be expected. It is encouraging that additional studies have identified a C-terminal *P. falciparum* CS protein T cell epitope recognized in association with most mouse and human MHC class II molecules [410]. The recognition of this epitope may correlate with resistance to malarial infection [411]. A greater understanding of the protective immune responses leading to the establishment of antisporozoite immunity may be necessary before additional vaccines can be developed and tested.

Erythrocytic stages of malaria

The second potential target for malaria vaccine intervention deals with the asexual blood stages of plasmodial development. It is during the erythrocytic cycle of malarial infection that clinical disease is apparent, with the severity of the disease dependent upon the level of blood parasitemia. An effective blood-stage vaccine leading to the suppression of parasite growth and development in erythrocytes could alleviate the clinical manifestations of malaria infection. Unlike an antisporozoite vaccine, a blood-stage malarial vaccine would be beneficial if only a partial reduction in the parasite burden is achieved. It is also possible that antibodies reactive with certain parasite antigens may prevent clinical malaria without diminishing parasitemia. The development of a system for the *in vitro* culture of *P. falciparum* blood-stage parasites [412] has led to the identification and characterization of many proteins of malarial parasites as well as the genes encoding them. Several of these antigens, some secreted, some expressed on the parasite surface, and others associated with the surface of the infected erythrocyte, have potential use as vaccine components.

Blood-stage vaccines

Studies involving the merozoite invasion of erythrocytes have provided the basis for one major area of research directed at the development of a blood-stage malaria vaccine. The search for plasmodial proteins directly involved in the invasion process has led to the characterization of several proteins of the asexual stages of plasmodial parasites. A second major malaria vaccine development of effort has focused on the immunologic properties of malarial antigens evaluated, based on their reactivities with immune sera or immunizing potential in various animal model systems. These antigens can be classified into four major groups, namely, secreted antigens, merozoite surface antigens, rhoptry antigens, and erythrocyte surface antigens. Many of these molecules have been identified only in *P. falciparum*, while others have been studied in additional species of plasmodia as well. In general, these proteins are believed to play important biologic roles during the erythrocytic stages of plasmodial development and may serve as targets for vaccine intervention.

Secreted antigens. The secreted antigens of plasmodial parasites include the S-antigens, the glycophorin-binding proteins (GPB), and the erythrocyte-binding antigens (EBA). The S-antigens are a group of soluble, heat-stable, antigenically diverse molecules found in the sera of infected individuals [413–415]. Sequence analysis of the genes encoding four antigenically distinct S-antigens of *P. falciparum* revealed the presence of extensive, but unrelated, blocks of tandemly repeated amino acids [416–418]. At present, the function of these S-antigens is unknown. Although a monoclonal antibody directed against a *P. falciparum* S-antigen inhibited the merozoite invasion of red blood cells *in vitro* [419], little additional evidence exists to support their use as vaccine components.

In light of the biochemical and biologic evidence for the role of the erythrocyte glycophorins in malarial invasion of red cells, attempts were made to identify the plasmodial GPB. Using immobilized glycophorins as an affinity column, *P. falciparum* GBP of 140, 70, and 35 kD in one study [420] and 155 and 130 kD in another study [421] were identified. The 130 kD GBP has been cloned and sequenced and shown to contain 11 repeats of a 50 amino acid sequence [422] recognized by antibodies which inhibit merozoite invasion [423]. It is present in a number of geographically diverse isolates of *P. falciparum*, and appears to be released at the time of schizont rupture [424,425]. Recent evidence, however, suggests that the binding of this 130 kD GBP to immobilized glycophorin may be nonspecific and that binding to intact erythrocytes does not occur [425,426]. Consequently, its role as a parasite receptor for an erythrocyte surface ligand remains questionable and its use as a vaccine component uncertain.

For vaccine purposes, the most promising of the secreted antigen group are the EBA. By assaying for the binding of soluble plasmodial antigens to intact erythrocytes, *P. falciparum* EBA of 175, 120, 65, and 46 kD have been identified [43]. Using erythrocytes treated with various enzymes, as well as those naturally re-sistant to invasion, the 175 kD EBA was shown to bind to red blood cells with receptor-like specificity in a sialic acid-dependent manner. Interestingly, this *P. falciparum* EBA also binds to merozoites in a strain-specific manner. It has been suggested that, once secreted, the dual binding of this EBA may facilitate the invasion of red blood cells by forming a bridge between the merozoite and erythrocyte surfaces. Using similar erythrocyte binding assays, EBA of 135 kD [427] and 155 kD [45] have likewise been identified in the culture supernatant of *P. knowlesi*-infected cells. The 135 kD putative parasite receptor was shown to specifically interact with the 35–45 kD erythrocyte glycoprotein carrying the Duffy determinant previously shown to be involved in *P. knowlesi* invasion of red blood cells. The second 155 kD erythrocyte-specific binding protein of *P. knowlesi* does not appear to interact with this same Duffy determinant, but may mediate the initial merozoite attachment to red blood cells. Further characterization at the molecular level of these EBA, as well as those of other malarial parasites, will allow their evaluation as potential vaccine components.

Merozoite surface antigens. A limited number of antigens have been shown to be expressed on the merozoite surface [428,429], some of which have drawn considerable interest as targets for vaccine intervention. A number of antibodies exist, both monoclonal and polyclonal, which recognize various merozoite surface components. Some of these can inhibit, *in vitro*, the merozoite invasion of erythrocytes. Despite these observations, however, a membrane-bound plasmodial receptor for erythrocytes has not been clearly defined. As mentioned above, this has mainly been due to the difficulties in distinguishing antireceptor antibodies from those which agglutinate merozoites. Only one monoclonal antibody possessing antireceptor activity was effective as a Fab fragment in inhibiting the merozoite invasion of erythrocytes [44,430]. This antibody recognizes a 66 kD antigen of *P. knowlesi*, which appears on the merozoite surface as well as in supernatants of parasite cultures [431].

A class of high-molecular-weight merozoite surface antigens has been identified in rodent, simian, and human plasmodial parasites (reviewed by Holder [432]). These antigens, referred to as the precursors to the major merozoite surface antigen (PMMSA), have become the predominant merozoite surface antigen being evaluated as a vaccine component. In rodent [433–435] and simian [436,437] experimental model systems, host immune responses elicited by immunization with affinity-purified PMMSA preparations have provided significant levels of protection against challenge infection. In addition, partially protective responses were induced in monkey [438,439] and human [440] trials, by immunization with synthetic peptides derived from the N-terminal portion of *P. falciparum* PMMSA.

The PMMSA is synthesized late in the erythrocytic cycle of infection and is expressed on the surface of mature schizonts and merozoites [428,441]. The 195 kD PMMSA of *P. falciparum* is proteolytically cleaved in a series of steps to yield an N-terminal fragment of 76–83 kD and a glycosylated C-terminal fragment of 40–50 kD, both of which are expressed on the merozoite surface [442,443] noncovalently associated with at least two additional merozoite components of 36 kD and 22 kD [429]. The N-terminal peptide contains an immunogenic, polymorphic, repeated determinant and appears to be shed from the parasite surface at the time of erythrocyte invasion [444–447]. The C-terminal peptide undergoes a further proteolytic cleavage during schizont rupture, producing a 15–20 kD peptide which is retained by the parasite subsequent to the invasion of the red cell [429].

As discussed previously, the terminal sialic acid residues of the erythrocyte glycophorins are the primary candidates for the ligand of a sialic acid-dependent parasite receptor. Evidence suggests that the *P. falciparum* (*Pf*) PMMSA may be involved in this receptor–ligand interaction, as the *Pf* PMMSA appears to bind to sialic acid residues on the surface of human erythrocytes [36]. This *Pf* PMMSA–erythrocyte interaction was shown to be inhibited by a monoclonal antibody directed against a glycosylated domain of glycophorin, as well as a monoclonal antibody recognizing the C-terminal processed fragment of the molecule. This same anti-PMMSA antibody had previously been shown to inhibit the *in vitro* invasion of erythrocytes by *P. falciparum* merozoites [448].

Immunologic and molecular biologic analyses employing several isolates of *P. falciparum* have revealed the polymorphic nature of *Pf* PMMSA. Using panels of monoclonal antibodies, strain-common and strain-restricted B cell epitopes of *Pf* PMMSA have clearly been demonstrated [449,450]. Studies at the nucleic acid sequence level of the genes encoding the *Pf* PMMSA of various strains have provided similar findings. Sequence comparisons of *Pf* PMMSA have revealed the presence of blocks of conserved, semiconserved, and variable

sequences dispersed throughout the molecule [451,452]. These *Pf* PMMSA polymorphisms, however, appear to be quite limited, resulting from recombinational events between two parental alleles. Only one of two possibilities appears to exist for a given segment of the *Pf* PMMSA molecule. For vaccine purposes, recombinant proteins produced in *Escherichia coli* as well as synthetic peptides have been used to map several B cell and T cell epitopes to the nonpolymorphic regions of *Pf* PMMSA [450,453,454]. However, the immunologic relevance of these determinants in the induction of protective antimalarial response remains to be demonstrated.

Studies involving the PMMSA of murine malarial parasites have contributed to the evaluation of this class of molecule as a vaccine antigen. The PMMSA of *P. yoelii* has been shown to be similar to *Pf* PMMSA in terms of the time of expression during blood-stage infection, its surface expression, and post-translational processing [433,455]. In addition, serologic crossreactivity [456] as well as nucleic acid sequence similarity [457–459] have been observed between the *P. falciparum* and *P. yoelii* PMMSA. *In vivo* studies revealed that the immunization of mice with affinity-purified preparations of *P. yoelii* (*Py*) PMMSA converted a virulent *P. yoelii* infection to a nonlethal self-limiting disease [433]. Furthermore, the passive protection of mice against lethal and nonlethal *P. yoelii* infections has been achieved by utilizing a monoclonal antibody recognizing the *Py* PMMSA [95]. The B cell epitope recognized by this antibody has been shown to be disulfide-dependent, mapping to the C-terminal cysteine-rich domain of *Py* PMMSA [460]. However, the expression of this immunologically relevant B cell determinant of *Py* PMMSA appears to be variant-specific, with passive protection correlating with expression [461]. Studies on the PMMSA molecule of *P. chabaudi* have also led to the production of a passively protective monoclonal antibody [96]. This antibody recognized a linear determinant which mapped to a core region of five amino acids derived from the middle portion of the 250 kD *Pc* PMMSA. Interestingly, this region aligns with a variable block of the *Pf* PMMSA.

Rhoptry antigens. A number of molecules appear to be associated with the paired organelles or rhoptries, which are located at the apical end of the merozoite. During the invasion process, the contents of these organelles are released into or onto the erythrocyte membrane and may be responsible for the perturbations of the red cell membrane and cytoskeleton which lead to vacuole formation.

From a vaccine standpoint, results using animal model systems have been encouraging. A monoclonal antibody directed against a 235 kD rhoptry antigen of *P. yoelii* passively protected mice from an otherwise lethal infection [94]. Immunization of BALB/c mice with affinity-purified preparations of the antigen elicited protective immune responses [431]. With *P. falciparum*, the *in vitro* invasion of erythrocytes can be blocked with antibodies which recognize a 41 kD rhoptry antigen [462]. Similar to the murine studies, Saimiri monkeys could be partially protected from challenge infection by immunization with the *P. falciparum* 41 kD rhoptry component [463].

Additional antigens localized to the rhoptries of *P. falciparum* merozoites have been identified and include molecules of 135/155 kD [464,465], 40/80 kD [464,466], 105/130/140 kD [467,468], and 225/240 kD [469]. Although the vaccine potential of this group of antigens is apparent, limited structural data on them are available. As the characterization of these molecules at the molecular level proceeds, their function in the merozoite invasion of erythrocytes and their value as vaccine antigens can be evaluated more critically.

Erythrocyte surface antigens. Through the expression of various antigens on the surface of the infected erythrocyte, the intracellular malarial parasite remains visible to the immune system of the infected host. These parasite-derived red cell surface antigens appear to be important in various aspects of the biology of the parasite. Therefore, appropriate immune responses directed to these erythrocyte surface molecules could result in the death of the intracellular parasite or the destruction of the infected red cell. Three major groups of plasmodial antigens on the *P. falciparum*-infected erythrocyte surface have been considered for vaccine use.

The first is a 155 kD protein, designated Pf155 [470] or RESA [471], which is expressed at the surface of ring-infected erythrocytes. Initially, this antigen is associated with the micronemes at the apical end of the merozoite and is transferred to the erythrocyte membrane during or shortly after invasion. Although Pf155/RESA is not involved in the initial interaction between the merozoite and erythrocyte, antibodies to this antigen can inhibit *P. falciparum* merozoite invasion of erythrocytes *in vitro* [472].

The human anti-Pf155/RESA antibody response is primarily directed against determinants lying within two major blocks of tandemly repeated amino acid residues

[473]. Vaccination of Aotus monkeys with recombinant peptides containing these two immunodominant repeat domains led to partial protection against *P. falciparum* challenge infection [474]. Later studies examining human T lymphocyte responses to Pf155/RESA have mapped several T cell epitopes within repeat as well as nonrepeat domains of the molecule [475,476]. Comparison of the limited number of Pf155/RESA sequences of different *P. falciparum* strains indicate that, unlike the CS protein, these T cell determinants may be nonpolymorphic. Few data are available on the HLA restriction of these epitopes. However, H-2 congenic strains of mice varied significantly in their T cell responses to three different repetitive sequences of Pf155/RESA [477]. As such, it is likely that human leucocyte antigen-dependent responder and nonresponder phenotypes also exist.

An RESA-like antigen of the rodent malarial parasite, *P. chabaudi*, is currently being characterized. This 96–105 kD antigen is similar to the *P. falciparum* molecule in many respects, and serologically shares some cross-reactive epitopes [478–480]. Most significantly, mice immunized with affinity-purified preparations of Pc96/105 can be partially protected against blood-stage malarial infection. Additional studies using this model system should allow a thorough analysis of the immunologic properties and vaccine potential of these erythrocyte surface molecules.

A second group of plasmodial proteins being considered for vaccine use are those present at the erythrocyte membrane believed to be involved in the phenomenon of cytoadherence and sequestration (reviewed by Howard [50]). *Plasmodium falciparum*-infected erythrocytes which contain maturing trophozoites and schizonts are sequestered from the peripheral circulation by specific attachment to capillary EC. This cytoadherence has been associated with protrusions or knobs which are induced by the developing parasite on the erythrocyte surface. Sequestration allows the malaria-infected erythrocyte to avoid passage through, and clearance in, the spleen, while allowing the parasite to develop at a lower, more optimal O_2 level. As mentioned in the section on immunopathology, such binding to venular endothelium may also contribute to the development of cerebral malaria.

The host molecules involved in the attachment of malaria-infected erythrocytes to capillary endothelium include CD36, ICAM-1, and thrombospondin, with different parasite isolates binding to different cell-adhesion receptors. On the infected erythrocyte surface, the knob-like structures induced by the parasite have been considered necessary, but not sufficient, to allow attachment to capillary endothelial cells. Three *P. falciparum* antigens have been associated with these red cell surface protrusions, namely the histidine-rich knob-associated protein PfHRP-1 [481,482] and two erythrocyte membrane proteins PfEMP-1 [483,484] and PfEMP-2 [485,486] or MESA [487].

PfHRP-1 and PfEMP-2/MESA appear to play some structural and/or functional role directly under the knob protrusion [488,489]. The present evidence, however, indicates that these molecules are not exposed on the red cell surface. This may be problematic for vaccine purposes. On the contrary, a portion of PfEMP-1 is expressed on the external surface of the infected erythrocyte [483,490] and current data support its role as the parasite receptor for the endothelial surface or very close association with such a receptor. As yet, monoclonal and/or monospecific polyclonal antibodies are not available which recognize PfEMP-1, limiting further analysis.

One published report documents the cytoadherence of knobless *P. falciparum*-infected red cells to endothelial cells [57]. Of interest, a human monoclonal antibody which recognizes a repeated determinant shared by at least three *P. falciparum* blood-stage antigens (Pf155/RESA, Pf11.1, Ag332) [491] inhibits this binding as well as the binding of knob-positive isolates. Further characterization of these interactions may prove quite useful. Vaccination with the appropriate parasite-derived erythrocyte surface antigen may disrupt the interaction between the infected red cell surface and the capillary endothelium. *Plasmodium falciparum*-infected cells would then be exposed to the specific and non-specific immune mechanisms of the spleen and be potentially eliminated.

The third class of parasite proteins at the erythrocyte surface are those which may be involved in the transport of molecules from the external environment to the intracellular parasite. Speculations are that such molecules exist but only one has been characterized in any detail. Although hemoglobin is present within the infected red blood cell, *P. falciparum* parasites require an exogenous source of iron for intraerythrocytic growth [492]. A 93–102 kD malarial transferrin receptor has been identified and localized to the surface of *P. falciparum*-infected erythrocytes [493,494]. This receptor binds, internalizes, and transports ferrotransferrin to the developing intracellular parasite. A malaria vaccine based on this receptor could disrupt the import of iron to the parasite, suppressing intraerythrocytic growth.

Such a vaccine strategy could be developed only if this plasmodial-derived transferrin receptor is shown to be sufficiently different from the analogous human receptor.

Sexual stages of malaria

A final target for vaccine intervention involves the sexual stages of malaria parasites. This involves the gametocytes which are transmitted from the vertebrate host back to the mosquito vector. An antigametocyte vaccine would be designed to eliminate this stage of the parasite from the bloodstream of the infected individual, preventing uptake by the mosquito vector. In addition, the developmental stages within the mosquito midgut, namely the gamete and zygote stages, could also be targeted for vaccine intervention. Here, specific anti-parasite antibodies ingested during a blood meal might prevent gamete fusion and zygote development within the mosquito midgut. Although of no direct benefit to the infected individual, such a vaccine, even if partially effective, could lead to a decrease in malaria transmission when considering the population at large.

The feasibility of a sexual-stage malaria vaccine has been provided by a number of studies in animal model systems. Antigametocyte transmission-blocking immunity has been induced in mice following immunization with a preparation of *P. yoelii nigeriensis* gametocytes [495]. Of interest, the protective effect was shown to be T cell mediated. In other experiments, serum antibodies from gamete- or gametocyte-immunized animals inhibited the sexual development of the parasite within the mosquito midgut [88,89]. Such antigamete antibodies may block fertilization or directly lyse or inhibit the development of gametes and zygotes.

In the development of transmission-blocking malarial vaccines, several monoclonal antibodies have been produced which block the infectivity of malarial parasites in the mosquito midgut. The antigens expressing the target epitopes of these transmission-blocking antibodies have been identified in *P. gallinaceum* [97], *P. yoelii nigeriensis* [100], and *P. falciparum* [98,100]. In *P. falciparum*, the determinants recognized by these monoclonal antibodies do not appear to be repeated epitopes [496] and are expressed on a gamete antigen of 230 kD and two additional related gamete glycoproteins of 48 kD and 45 kD. Other transmission-blocking monoclonal antibodies recognize a 25 kD antigen located on the surface of developing *P. falciparum* zygotes and ookinetes [101]. A number of these biologically important monoclonal antibodies predominantly recognize disulfide-dependent epitopes. The cloning and sequencing of the 25 kD zygote antigen of *P. falciparum* [497] and *P. gallinaceum* [498] revealed the presence of four epidermal growth factor-like cysteine domains.

The further characterization of these sexual-stage malarial antigens and the mapping of the B cell epitopes recognized by transmission-blocking antibodies will be important for vaccine development. It is encouraging that the present evidence indicates minimal variability among *P. falciparum* isolates in the expression of the determinants recognized by these antibodies [499,500]. Unfortunately, the T cell response to these surface antigens is reminiscent of that seen with the CS protein of malaria sporozoites. Preliminary immunization studies with H-2 congenic strains of mice are indicative of a restricted T cell response to these vaccine candidate antigens [501].

The list of plasmodial antigens under evaluation for vaccine use is impressive but represents only a small subset of the many parasite proteins which interact with the host immune system. Many of these vaccine studies have targeted antigens expressed on the surface membranes of the parasite or the infected erythrocyte. However, it is possible that other parasite molecules, both internal and secreted, may be equally important in the stimulation of the host immune system. For example, heat shock proteins (Hsp) of many infectious agents have been shown to induce strong immune responses (reviewed by Kaufmann [156]). Plasmodial Hsp have been identified and characterized [157–160] but a thorough investigation of the immune responses elicited during malarial infection to Hsp is lacking. This is particularly significant in light of the recent demonstration of elevated levels of γδ T lymphocytes in the peripheral blood of patients with acute falciparum malaria [153,154]. As in other infectious and autoimmune diseases, this population of T cells may be specific for Hsp.

The studies discussed above have focused on the design of an antiparasite vaccine but, as mentioned in the section on immunopathology, the concept of an antidisease vaccine also warrants attention. The induction of certain cytokines in response to specific plasmodial components appears to have a significant impact on the clinical manifestations of malaria. Clark & Chaudhri [208] have provided evidence for a role for TNF in the pathology of malaria, and two soluble antigens of *P. falciparum* have recently been shown to induce TNF release from macrophages [502]. As such, it

may be possible to develop vaccine strategies not for the elimination of the parasite but for a reduction in pathology. A more complete understanding of the elements involved in the immune responses to plasmodial parasites will undoubtedly shape the further development of such malaria vaccines.

ACKNOWLEDGMENTS

The authors wish to thank Ms Mary Ellen Bealor and Mr Sylvester Salas for their expert secretarial assistance. Salary and maintenance support for J.M.-K. and W.P.W. was provided by a grant from the National Institute of Health, No. AI12710.

REFERENCES

1 Warrell DA. Drugs for prevention and treatment of severe malaria. TDR News 1989;29:6.
2 Laitman C. Tropical diseases in media spotlight. TDR News 1990;31:3.
3 Sturcher D. How much malaria is there worldwide. Parasitol Today 1989;5:39–40.
4 Bruce-Chwatt LJ. History of malaria from prehistory to eradication. In Wernsdorfer WH, McGregor I, eds. *Malaria: Principles and Practice of Malariology*, vol. 1. New York: Churchill Livingstone, 1988:1–60.
5 Wyler DJ. Malaria-resurgence, resistance, and research. N Engl J Med 1983;308:875–878.
6 Garnham PCC. Malaria parasites on man: life-cycles and morphology (excluding ultrastructure). In Wernsdorfer WH, McGregor I, eds. *Malaria: Principles and Practice of Malariology*, vol. 1. New York: Churchill LIvingstone, 1988:61–96.
7 Hudson DE, Wellems TE, Miller LH. Molecular basis for mutation in a surface protein expressed by malaria parasites. J Mol Biol 1988;203:707–714.
8 Van der Ploeg LHT, Smits M, Ponnudurai T, Vermeulen A, Meuwissen JHE, Langsley G. Chromosome-sized DNA molecules of *Plasmodium falciparum*. Science 1985;229:658–661.
9 Corcoran LM, Forsyth KP, Bianco AE, Brown GV, Kemp DJ. Chromosome-size polymorphisms in *Plasmodium falciparum* can involve deletions and are frequent in natural parasite populations. Cell 1986;44:87–95.
10 Pologe LG, Ravetch JV. A chromosomal rearrangement in a *P. falciparum* histidine-rich protein gene is associated with the knobless phenotype. Nature 1986;322:474–477.
11 Kemp DJ, Thompson JK, Walliker D, Corcoran LM. Molecular karyotype of *Plasmodium falciparum*: conserved linkage groups and expendable histidine-rich protein genes. Proc Natl Acad Sci USA 1987;84:7672–7676.
12 Foote SJ, Kemp DJ. Chromosomes of malaria parasites. Trends Genet 1989;5:337–342.
13 Shirley MW, Riggs BA, Forsyth KP, *et al*. Chromosome 9 from independent clones and isolates of *Plasmodium fal-*

ciparum undergoes subtelomeric deletions with similar break points *in vitro*. Mol Biochem Parasitol 1990;40:137–145.

14 Corcoran LM, Thompson JK, Walliker D, Kemp DJ. Homologous recombination within subtelomeric repeat sequences generates chromosome-size polymorphisms in *P. falciparum*. Cell 1988;53:807–813.

15 Pologe LG, Ravetch JV. Large deletions result from breakage and healing of *P. falciparum* chromosomes. Cell 1988;55:869–874.

16 Pologe LG, de Bruin D, Ravetch JV. A and T homopolymeric stretches mediate a DNA inversion in *Plasmodium falciparum* which results in loss of gene expression. Mol Cell Biol 1990;10:3243–3246.

17 Anders RF, Coppel RL, Brown GV, Kemp DJ. Antigens with repeated amino acid sequences from the asexual blood stages of *Plasmodium falciparum*. Prog Allergy 1988;41:148–172.

18 McCutchan TF, de la Cruz VF, Good MF, Wellem TE. Antigenic diversity in *Plasmodium falciparum*. Prog Allergy 1988;41:173–192.

19 McGregor IA, Wilson RJM. Specific immunity: acquired in man. In Wernsdorfer WH, McGregor I, eds. *Malaria: Principles and Practice of Malariology*, vol. 1. New York: Churchill Livingstone, 1988:559–619.

20 Cohen S, Lambert PH. Malaria. In Cohen S, Warren KS, eds. *Immunology of Parasitic Infections*. Oxford: Blackwell Scientific Publications, 1982:422–474.

21 Taliferro WH, Taliaferro LG. The effect of immunity on the asexual reproduction of *Plasmodium brasilianum*. J Infect Dis 1944;75:1–32.

22 Hoffman SL, Isenbarger D, Long GW, *et al*. Sporozoite vaccine induces genetically restricted T cell elimination of malaria from hepatocytes. Science 1989;244(4908):1078–1081.

23 Terry RJ. Evasion of host immunity in malaria infections. In Wernsdorfer WH, McGregor I, eds. *Malaria: Principles and Practice of Malariology*, vol. 1. New York: Churchill Livingstone, 1988:639–646.

24 Weidanz WP, Melancon-Kaplan J, Cavacini LA. Cell-mediated immunity to the asexual blood stages of malarial parasites: animal models. Immunol Lett 1990;25:87–98.

25 Cohen S, Butcher GA, Crandall RB. Action of malarial antibody *in vitro*. Nature 1969;223:368–371.

26 Chulay JD, Aikawa M, Diggs C, Haynes JD. Inhibitory effects of immune monkey serum on synchronized *Plasmodium falciparum* cultures. Am J Trop Med Hyg 1981;30:12–19.

27 Phillips RS, Trigg PI, Scott-Finnegan TJ, Bartholomew RK. Culture of *Plasmodium falciparum in vitro*: a subculture technique for demonstrating anti-plasmodial activity in serum from Gambians resident in a malarious area. Parasitology 1972;65:525–535.

28 Mitchell GH, Butcher GA, Voller A, Cohen S. The effect of human immune IgG on the *in vitro* development of *Plasmodium falciparum*. Parasitology 1976;72:149–162.

29 Miller LH, Aikawa M, Dvorak JA. Malaria (*Plasmodium knowlesi*) merozoites: immunity and the surface coat. J Immunol 1975;114:1237–1242.

30 Dvorak JA, Miller LH, Whitehouse WC, Shiroishi T. Invasion of erythrocytes by malaria merozoites. Science 1975;187:748–749.

31 Aikawa M, Miller LH, Johnson J, Rabbege JR. Erythrocyte entry by malarial parasites. A moving junction between erythrocyte and parasite. J Cell Biol 1978;77:72–82.

32 Bannister LH, Butcher GA, Dennis ED, Mitchell GH. Structure and invasive behavior of *Plasmodium knowlesi* merozoites *in vitro*. Parasitology 1975;71:483–491.

33 Ladda R, Aikawa M, Sprinz H. Penetration of erythrocytes by merozoites of mammalian and avian malarial parasites. J Parasitol 1969;55:633–644.

34 Miller LH, Aikawa M, Johnson JG, Shiroishi T. Interaction between cytochalasin B-treated malarial parasites and erythrocytes. Attachment and junction formation. J Exp Med 1979;149:172–184.

35 Miller LH, McAuliffe FM, Mason SJ. Erythrocyte receptors of malaria merozoites. Am J Trop Med Hyg 1977;26:204–208.

36 Perkins ME, Rocco LJ. Sialic acid dependent binding of *Plasmodium falciparum* merozoite surface antigen, Pf 200 to human erythrocytes. J Immunol 1988;141:3190–3196.

37 Perkins ME, Holt E. Erythrocyte receptor recognition of *Plasmodium falciparum* isolates. Mol Biochem Parasitol 1988;27:23–34.

38 Hadley TJ, Miller LH. Invasion of erythrocytes by malaria parasites: erythrocyte ligands and parasite receptors. Prog Allergy 1988;41:49–71.

39 Hadley TJ, David PH, McGinniss MH, Miller LH. Identification of an erythrocyte component carrying the Duffy blood group Fyª antigen. Science 1984;223:597–599.

40 Mason SJ, Miller LH, Shiroishi T, Dvorak JA, McGinniss MH. The Duffy blood group determinants: their role in the susceptibility of humans and animal erythrocytes to *Plasmodium knowlesi* malaria. Br Med J Haematol 1977;36:327–335.

41 Hadley TJ, Klotz FW, Haynes JD, McGinniss MH, Okubo Y, Miller LH. Falciparum malaria parasites invade erythrocytes that lack glycophorin A and B (MkMk): strain differences indicate receptor heterogeneity and two pathways for invasion. J Clin Invest 1988;80:1190–1193.

42 Perkins ME. Erythrocyte invasion by the malarial merozoite: recent advances. Expt Parasitol 1989;69:94–99.

43 Camus D, Hadley TJ. A *Plasmodium falciparum* antigen that binds to host erythrocytes and merozoites. Science 1985;230:553–556.

44 Wertheimer SP, Barnwell JW. *Plasmodium vivax* interaction with the human Duffy Blood Group glycoprotein: identification of a parasite receptor-like protein. Exp Parasitol 1989;69:340–350.

45 Miller LH, Hudson D, Haynes JD. Identification of *Plasmodium knowlesi* erythrocyte binding proteins. Mol Biochem Parasitol 1988;31:217–222.

46 Bignami A, Bastiamelli G. Observations of estivoautumnal malaria. Ref Med 1989;6:1334–1335.

47 Howard RJ, Barnwell JW. Roles of surface antigens on malaria infected red blood cells in evasion of immunity. In Marchalonis JJ, ed. *Contemporary Topics in Immunobiology*, vol. 12. New York: Plenum Publishers, 1984:127–191.

48 Lanners HN, Trager W. Comparative infectivity of a knobless and knobby clones of *P. falciparum* in splenectomized and intact *Aotus trivirgatus* monkeys. Z Parasitenkd 1984;70:739.

49 Howard RJ, Gilladoga AD. Molecular studies related to the pathogenesis of cerebral malaria. Blood 1989;74:2603–2618.

50 Howard RJ. Malarial proteins at the membrane of *Plasmodium falciparum*-infected erythrocytes and their involvement in cytoadherence to endothelial cells. Prog Allergy 1988;41:98–147.

51 Howard RJ, Handunnetti SM, Hasler T, *et al.* Surface molecules on *Plasmodium falciparum*-infected erythrocytes involved in adherence. Am J Trop Med Hyg 1990;43:15–29.

52 Udeinya IJ. *In vitro* and *ex vivo* models of sequestration in *Plasmodium falciparum* infection. Am J Trop Med Hyg 1990;43:2–5.

53 Chulay JD, Ockenhouse CF. Host receptors for malarial-infected erythrocytes. Am J Trop Med Hyg 1990;43:6–14.

54 Rothlein R, Czajkowski M, O'Neill MM, Marlin SD, Mainolfi E, Merluzzi VJ. Induction of intercellular adhesion molecule 1 on primary and continuous cell lines by proinflammatory cytokines. J Immunol 1988;141:1665–1669.

55 David PH, Hommel M, Miller LH, Udeinya IJ, Oligino LD. Parasite sequestration in *Plasmodium falciparum* malaria: spleen and antibody modulation of cytoadherence of infected erythrocytes. Proc Natl Acad Sci USA 1983;80:5075–5079.

56 Magowam C, Wollish W, Anderson L, Leech J. Cytoadherence by *Plasmodium falciparum*-infected erythrocytes is correlated with the expression of a family of variable proteins on infected erythrocytes. J Exp Med 1988;168:1307–1320.

57 Udomsangpetch R, Aikawa A, Berzins K, Wahlgren M, Perlmann P. Cytoadherence of knobless *Plasmodium falciparum* by a human monoclonal antibody. Nature 1989;338:763–765.

58 Biggs BA, Culvenor JG, Ng JS, Kemp DJ, Brown GV. *Plasmodium falciparum*: cytoadherence of a knobless clone. Exp Parasitol 1989;69:189–197.

59 Winograd E, Greenam JR, Sherman IW. Expression of senescent antigen on erythrocytes infected with a knobby variant of the human malaria parasite *Plasmodium falciparum*. Proc Natl Acad Sci USA 1987;84:1931–1935.

60 Winograd E, Sherman IW. Characterization of a modified red cell membrane protein expressed on erythrocytes infected with the human malaria parasite. Possible role as a cytoadherent mediating protein. J Cell Biol 1989;108:23–30.

61 Deans JA, Cohen S. Immunology of malaria. Annu Rev Microbiol 1983;37:25–49.

62 Taylor DW. Humoral immune responses in mice and man to malarial parasites. In Stevenson MM, ed. *Malaria: Host Responses to Infection.* Boca Raton: CRC Press, 1989:1–35.

63 Tobie JE, Abele DC, Wolff SM, Contacos PG, Evans CB. Serum immunoglobulin levels in human malaria and their relationship to antibody production. J Immunol 1966;97:498–505.

64 Curtain CC, Kidson C, Champness DL, Gorman JG. Malaria antibody content of gamma$_2$-7S globulin in tropical populations. Nature 1964;203:1366–1367.

65 Greenwood BM. Possible role of a B cell mitogen in hypergammaglobulinemia in malaria and trypanosomiasis. Lancet 1974;i:435–436.

66 Wyler DJ, Herrod HG, Weinbaum FI. Response of sensitized and unsensitized human lymphocyte subpopulations to *Plasmodium falciparum* antigens. Infect Immun 1979;24:106–110.

67 Ballet JJ, Jaureguiberry G, Deloron P, Agrapart M. Stimulation of T lymphocyte-dependent differentiation of activated human B lymphocytes by *Plasmodium falciparum* supernatants. J Infect Dis 1987;155:1037–1040.

68 Kataaha PK, Facer CA, Mortazavi-Milani SM, Stierle H, Holborow EJ. Stimulation of autoantibody production in normal blood lymphocytes by malaria culture supernatants. Parasite Immunol 1984;6:481–492.

69 Lustig HJ, Nussenzweig V, Nussenzweig RS. Erythrocyte membrane-associated immunoglobulins during malaria infection of mice. J Immunol 1977;119:210–216.

70 Facer CA, Bray RS, Brown J. Direct Coombs antiglobulin reactions in Gambian children with *Plasmodium falciparum* malaria. I. Incidence and class specificity. Clin Exp Immunol 1979;35:119–127.

71 Gilbreath MJ, Pavanand K, MacDermott RP, Wells RA, Ussery MA. Characterization of cold reactive lymphocytotoxic antibodies in malaria. Clin Exp Immunol 1983;51:232–238.

72 Greenwood BM, Muller AS, Valkenburg HA. Rheumatoid factor in Nigerian sera. Clin Exp Immunol 1971;9:161–173.

73 Shaper AG, Kaplan MH, Mody NJ, McIntyre PA. Malarial antibodies and autoantibodies to heart and other tissues in immigrant and indigenous people of Uganda. Lancet 1968;i:1342–1347.

74 Adu D, Williams DG, Quakyi IA, *et al.* Anti-ssDNA and antinuclear antibodies in human malaria. Clin Exp Immunol 1982;49:310–316.

75 Bonfa E, Llovet R, Scheinberg M, De Souza JM, Elkon KB. Comparison between autoantibodies in malaria and leprosy with lupus. Clin Exp Immunol 1987;70:529–537.

76 Jayawardena AN, Kemp JD. Immunity to *Plasmodium yoelii* and *Babesia microti*: modulation by the CBA/N X-chromosome. Bull WHO 1979;57 (suppl. 1):255–259.

77 Taylor DW, Bever CT, Rollwagen FM, Evans CB, Asofsky R. The rodent malaria parasite *Plasmodium yoelii* lacks both types 1 and 2 T-independent antigens. J Immunol 1982;128:1854–1859.

78 Tobie JE, Abele DC, Hill GJ, Contacos PG, Evans CB. Fluorescent antibody studies on the immune response in sporozoite-induced and blood-induced vivax malaria and the relationship of antibody production to parasitemia. Am J Trop Med Hyg 1966;15:676–683.

79 Brown IN. Immunological aspects of malaria infection. Adv Immunol 1969;11:267–349.

80 Cohen S, McGregor IA, Carrington S. Gammaglobulin

and acquired immunity to human malaria. Nature 1961;192:733–737.

81 Coggeshall LT, Kumm HW. Demonstration of passive immunity in experimental monkey malaria. J Exp Med 1937;66:177–190.

82 Phillips RS, Jones VE. Immunity to *Plasmodium berghei* in rats: maximum levels of protective antibody activity are associated with eradication of the infection. Parasitology 1972;64:117–127.

83 Wells RA, Diggs CL. Protective activity in sera from mice immunized against *Plasmodium berghei*. J Parasitol 1976;62:638–639.

84 Freeman RR, Parish CR. *Plasmodium yoelii*: antibody and the maintenance of immunity in BALB/c mice. Exp Parasitol 1981;52:18–24.

85 Murphy JR, Lefford MJ. Host defenses in murine malaria: evaluation of the mechanisms of immunity of *Plasmodium yoelii* infection. Infect Immun 1979;23:384–391.

86 McDonald V, Sherman IW. *Plasmodium chabaudi*: humoral and cell-mediated responses of immunized mice. Exp Parasitol 1980;49:442–454.

87 Mendis KN, Munesinghe YD, de Silva YNY, Keragalla I, Carter R. Malaria transmission-blocking immunity induced by natural infections of *Plasmodium vivax* in humans. Infect Immun 1987;55:369–372.

88 Gwadz RW. Malaria: succesful immunization against the sexual stages of *Plasmodium gallinaceum*. Science 1976;193:1150–1151.

89 Gwadz RW, Green I. Malaria immunization in rhesus monkeys: a vaccine effective against both the sexual and asexual stages of *Plasmodium knowlesi*. J Exp Med 1978;148:1311–1323.

90 Nussenzweig RS, Vanderberg JP, Sanabria Y, Most H. *Plasmodium berghei*: accelerated clearance of sporozoites from blood as part of immune-mechanism in mice. Exp Parasitol 1972;31:88–97.

91 Verhave JP, Meuwissen JHE, Golenser J. Cell-mediated reactions and protection after immunization with sporozoites. Israel J Med Sci 1978;14:611–616.

92 Yoshida N, Nussenzweig RS, Potocnjak P, Nussenzweig V, Aikawa M. Hybridoma produces protective antibodies directed against the sporozoite stage of the malaria parasite. Science 1980;207:71–73.

93 Potocnjak P, Yoshida N, Nussenzweig RS, Nussenzweig V. Monovalent fragments (Fab) of monoclonal antibodies to a sporozoite surface antigen (Pb44) protect mice against malarial infection. J Exp Med 1980;151:1504–1513.

94 Freeman RR, Trejdosiewicz AJ, Cross GAM. Protective monoclonal antibodies recognising stage-specific merozoite antigens of a rodent malaria parasite. Nature 1980;284:366–368.

95 Majarian WM, Daly TM, Weidanz WP, Long CA. Passive immunization against murine malaria with an IgG3 monoclonal antibody. J Immunol 1984;132:3131–3137.

96 Lew AM, Langford CJ, Anders RF, *et al.* A protective monoclonal antibody recognizes a linear epitope in the precursor to the major merozoite antigens of *Plasmodium chabaudi adami*. Proc Natl Acad Sci 1989;86:3768–3772.

97 Harte PG, Rogers N, Targett GAT. Vaccination with purified microgamete antigens prevents transmission of rodent malaria. Nature 1985;316:258–259.

98 Rener J, Graves PM, Carter R, Williams JL, Burkot TR. Target antigens of transmission-blocking immunity on gametes of *Plasmodium falciparum*. J Exp Med 1983;158:976–981.

99 Vermuelen AN, Ponnudurai T, Beckers PJA, Verhave JP, Smits MA, Muewissen JHE. Sequential expression on sexual stages of *Plasmodium falciparum* accessible to transmission-blocking antibodies in the mosquito. J Exp Med 1985;162:1460–1476.

100 Wilson RJM, McGregor IA, Hall P, Williams K, Bartholomew R. Antigens associated with *Plasmodium falciparum* infections in man. Lancet 1969;ii:201–205.

101 Peiris JSM, Premawansa S, Ranawaka MBR, *et al.* Monoclonal and polyclonal antibodies both block and enhance transmission of human *Plasmodium vivax* malaria. Am J Trop Med Hyg 1988;39:26–32.

102 McDonald V, Phillips RS. *Plasmodium chabaudi* in mice. Adoptive transfer of immunity with enriched populations of spleen T and B lymphocytes. Immunology 1978;34:821–830.

103 Gravely SM, Kreier JP. Adoptive transfer of immunity to *Plasmodium berghei* with immune T and B lymphocytes. Infect Immun 1976;14:184–190.

104 Brown KN, Jarra W, Hills LA. T cells and protective immunity to *Plasmodium berghei* in rats. Infect Immun 1976;14:858–871.

105 Cavacini LA, Long CA, Weidanz WP. T-cell immunity in murine malaria: adoptive transfer of resistance to *Plasmodium chabaudi adami* in nude mice with splenic T cells. Infect Immun 1986;52:637–643.

106 Fahey JR, Spitalny GL. Immunity to *Plasmodium yoelii*: kinetics of the generation of T and B lymphocytes that passively transfer protective immunity against virulent challenge. Cell Immunol 1986;98:486–495.

107 Jayawardena AN, Murphy DB, Janeway CA, Gershon RK. T cell-mediated immunity in malaria. I. The Ly phenotype of T cells mediating resistance to *Plasmodium yoelii*. J Immunol 1982;129:377–381.

108 Longenecker BM, Breitenbach RP, Farmer JN. The role of the bursa of fabricius, spleen and thymus in the control of a *Plasmodium lophurae* infection in the chicken. J Immunol 1966;97:594–599.

109 Rank RG, Weidanz WP. Nonsterilizing immunity in avian malaria: an antibody-independent phenomenon. Proc Soc Exp Biol Med 1976;151:257–259.

110 Weinbaum FI, Evans CB, Tigelaar RE. Immunity to *Plasmodium berghei yoelii* in mice. I. The course of infection in T cell and B cell deficient mice. J Immunol 1976;117:1999–2005.

111 Roberts DW, Rank RG, Weidanz WP, Finerty JF. Prevention of recrudescent malaria in nude mice by thymic grafting or by treatment with hyperimmune serum. Infect Immun 1977;16:821–826.

112 Jayawardena AN, Janeway CA Jr, Kemp JD. Experimental malaria in the CBA/N mouse. J Immunol 1979;123:2532–2539.

113 Vanderberg J, Nussenzweig R, Most H. Protective

immunity produced by the injection of X-irradiated sporozoites of *Plasmodium berghei*. V. *In vitro* effects of immune serum on sporozoites. Mil Med 1969;134:1183–1190.

114 Hollingdale MR, Zavala F, Nussenzweig RS, Nussenzweig V. Antibodies to the protective antigen of *Plasmodium berghei* sporozoites prevent entry into cultured cells. J Immunol 1982;128:1929–1930.

115 Hollingdale MR, Nardin EH, Tharavanij S, Schwartz AL, Nussenzweig RS. Inhibition of entry of *Plasmodium falciparum* and *P. vivax* sporozoites into cultured cells: an *in vitro* assay of protective antibodies. J Immunol 1984;132:909–913.

116 Atkinson JP, Glew RH, Neva FA, Frank MM. Serum complement and immunity in experimental simian malaria. II. Preferential activation of early components and failure of depletion of late components to inhibit protective immunity. J Infect Dis 1975;131:26–33.

117 Williams AIO, Rosen FS, Hoff R. Role of complement components in the susceptibility to *Plasmodium berghei* infection among inbred strains of mice. Ann Trop Med Parasitol 1975;69:179–184.

118 Celada A, Cruchaud A, Perrin LH. Opsonic activity of human immune serum on *in vitro* phagocytosis of *Plasmodium falciparum* infected red blood cells by monocytes. Clin Exp Immunol 1982;47:635–644.

119 Tosta CE, Wedderburn N. Immune phagocytosis of *Plasmodium yoelii*-infected erythrocytes by macrophages and eosinophils. Clin Exp Immunol 1980;42:114–120.

120 Shear HL, Nussenzweig RS, Bianco C. Immune phagocytosis in murine malaria. J Exp Med 1979;149:1288–1298.

121 Trubowitz S, Mazek B. *Plasmodium falciparum*: phagocytosis by polymorphonuclear leucocytes. Science 1968;162:273–274.

122 Celada A, Cruchaud A, Perrin LH. Phagocytosis of *Plasmodium falciparum*-parasitized erythrocytes by human polymorphonuclear leucocytes. J Parasitol 1983;69:49–53.

123 Playfair JHL. Immunity to malaria. Br Med Bull 1982;38:153–159.

124 Brown J, Smalley ME. Specific antibody-dependent cellular cytotoxicity in human malaria. Clin Exp Immunol 1980;41:423–429.

125 Shear HL. Variation in expression of antibody-dependent cell-mediated cytotoxicity in rodents with malaria. Infect Immun 1988;56:3007–3010.

126 Carter R, Kumar N, Quakyi I, *et al.* Immunity to sexual stages of malaria parasites. Prog Allergy 1988;41:193–214.

127 Udeinya IJ, Miller LH, McGregor IA, Jensen JB. *Plasmodium falciparum* strain-specific antibody blocks binding of infected erythrocytes to amelanotic melanoma cells. Nature 1983;303:429–431.

128 Weidanz WP, Long CA. The role of T cells in immunity to malaria. Prog Allergy 1988;41:215–253.

129 Webster HK, Brown AE, Wongsrichanalai C, Koncharean S. *38th Annual Meeting of the Am Soc Trop Med Hyg, Honolulu*, 1989, Abstract 25.

130 Melancon-Kaplan J, Weidanz WP. Role of cell-mediated immunity in resistance to malaria. In Stevenson MM, ed. *Malaria: Host Responses to Infection*. Boca Raton: CRC Press, 1989:37–63.

131 Jayawardena AN. Immune responses in malaria. In Mansfield JM, ed. *Parasitic Diseases*, vol. 1. New York: Marcel Dekker, 1981:85–136.

132 Russo DM, Weidanz WP. Activation of antigen-specific suppressor T cells by the intravenous injection of soluble blood-stage malarial antigen. Cell Immunol 1988;115:437–446.

133 Ballet JJ, Druilhe P, Querleux MA, Schmitt C, Agrapart M. Parasite-derived mitogenic activity for human T cells in *Plasmodium falciparum* continuous cultures. Infect Immun 1981;33:758–762.

134 Riley EM, Jepsen S, Andersson G, Otoo LN, Greenwood BM. Cell-mediated immune responses to *Plasmodium falciparum* antigens in adult Gambians. Clin Exp Immunol 1988;71:377–382.

135 Troye-Blomberg M, Pêrlmann P. T cell functions in *Plasmodium falciparum* and other malarias. Prog Allergy 1988;41:253–287.

136 Riley E, Greenwood B. Measuring cellular immune responses to malaria antigens in endemic populations: epidemiological parasitological and physiological factors which influence *in vitro* assays. Immunol Lett 1990;25:139–141.

137 Ho M, Webster HK. T cell responses in acute falciparum malaria. Immunol Lett 1990;25:135–138.

138 Ho M, Webster HK, Green B, Looareesuwan S, Kongcharoen S, White NJ. Defective production of and response to interleukin 2 in human falciparum malaria. J Immunol 1988;141:2755–2759.

139 Weidanz WP. Malaria and alterations in immune reactivity. Br Med Bull 1982;38:167–172.

140 Troye-Blomberg M, Sjöberg K, Olerup O, *et al.* Characterization of regulatory T cell responses to defined immunodominant T cell epitopes of the *Plasmodium falciparum* antigen Pf155/RESA. Immunol Lett 1990;25:129–134.

141 Kern P, Hammer CJ, Van Damme J, Gruss HJ, Dietrich M. Elevated tumor necrosis factor alpha and interleukin 6 serum levels as markers for complicated *Plasmodium falciparum* malaria. Am J Med 1989;87:139–143.

142 Ferrante A, Staugas RM, Bresatz S, *et al.* Production of tumor-necrosis factor alpha and beta by human mononuclear leukocytes stimulated with mitogens and microbial components. Infect Immun 1990;58:3996–4003.

143 Grau GE, Kindler V, Piguet PF, Lambert PH, Vassalli P. Prevention of experimental cerebral malaria by anti-cytokine antibodies. Interleukin 3 and granulocyte macrophage colony-stimulating factor are intermediates in increasing tumor necrosis factor production and macrophage accumulation. J Exp Med 1988;168:1499–1504.

144 Grau GE, Heremans H, Piguet PF, *et al.* Monoclonal antibody against interferon gamma can prevent experimental cerebral malaria and its associated overproduction of tumor necrosis factor. Proc Natl Acad Sci USA 1989;86:5572–5574.

145 Grau GE, Frei K, Piguet PF, *et al.* Interleukin 6 production in experimental cerebral malaria: modulation by anti-

cytokine antibodies and possible role in hypergamma-globulinemia. J Exp Med 1990;172:1505–1508.

146 Scuderi P, Lam KS, Ryan KJ, *et al*. Raised serum levels of tumor necrosis factors in parasitic infections. Lancet 1986; 1:1364–1365.

147 Kwiatkowski D, Cannon JG, Manogue KR, Cerami A, Dinarello CA, Greenwood BM. Tumor necrosis factor production in Falciparum malaria and its association with schizont rupture. Clin Exp Immunol 1989;77:361–366.

148 Grau GE, Bieler G, Pointaire P, *et al*. Significance of cytokine production and adhesion molecules in malarial immunopathology. Immunol Lett 1990;25:189–194.

149 Kwiatkowski D. Tumour necrosis factor, fever and fatality in falciparum malaria. Immunol Lett 1990;25:213–216.

150 Troye-Blomberg M, Romero P, Patarroyo ME, Björkman A, Perlmann P. Regulation of the immune response in *Plasmodium falciparum* malaria. III. Proliferative response to antigen *in vitro* and subset composition of T cells from patients with acute infection or from immune donors. Clin Exp Immunol 1984;58:380–387.

151 Hoffman SL, Piessens WF, Ratinayanto S, *et al*. Reduction of suppressor T lymphocytes in the tropical splenomegaly syndrome. N Engl J Med 1984;310:337–341.

152 Minoprio P, Itohara S, Heusser C, Tonegawa S, Coutinho A. Immunobiology of murine *T. cruzi* infection: the predominance of parasite nonspecific responses and the activation of TCR1 T cells. Immunol Rev 1989;112:182–207.

153 Roussilhon C, Agrapart M, Ballet JJ, Bensussan A. T lymphocytes bearing the $\gamma\delta$ T cell receptor in patients with acute *P. falciparum* malaria. J Infect Dis 1990;162:283–285.

154 Ho M, Webster HK, Tongtawe P, Pattanapanyasat K, Weidanz WP. Increased $\gamma\delta$ T cells in acute *P. falciparum* malaria. Immunol Lett 1990;25:139–142.

155 Janeway CA Jr. Frontiers of the immune system. Nature 1988;333:804–806.

156 Kaufmann SHE. Heat shock proteins and the immune response. Immunol Today 1990;11:129–136.

157 Richman SJ, Vedvick TS, Reese RT. Peptide mapping of conformational epitopes in a human malaria parasite heat shock protein. J Immunol 1989;143:285–202.

158 Kumar N, Zhao Y, Graves P, Folgar JP, Maloy L, Zheng H. Human immune response directed against *P. falciparum* heat shock related proteins. Infect Immun 1990;58:1408–1414.

159 Bianco AE, Favaloro JM, Burkot TR, *et al*. A repetitive antigen of *Plasmodium falciparum* that is homologous to heat shock protein 70 of Drosophila melanogaster. Proc Natl Acad Sci USA 1986;83:8713–8717.

160 Sheppard M, Kemp DJ, Anders RF, Lew AM. High level sequence homology between a *Plasmodium chabaudi* heat shock protein and its *Plasmodium falciparum* equivalent. Mol Biochem Parasitol 1989;33:101–104.

161 Sinigaglia F, Pink JRL. Human T lymphocyte clones specific for malaria (*Plasmodium falciparum*). EMBO J 1985; 4:3819–3822.

162 Pink JRL, Rijnbeek AM, Reber-Liske R, Sinigaglia F. *Plasmodium falciparum*—specific human T cell clones: recognition of different parasite antigens. Eur J Immunol 1987;

17:193–196.

163 Good MF, Quakyi IA, Saul A, Berzofsky JA, Carter R, Miller LH. Human T cell clones reactive with peripheral blood from nonexposed donors. J Immunol 1987;138:306–311.

164 Simitsek P, Chizzolini C, Perrin L. Malaria-specific human T cell clones: cross reactivity with various plasmodia species. Clin Exp Immunol 1987;69:271–279.

165 Sinigaglia F, Matile H, Pink JRL. *Plasmodium falciparum*-specific human T cell clones; evidence for helper and cytotoxic activities. Eur J Immunol 1987;17:187–192.

166 Nardin E, Herrington D, Davis J, *et al*. Conserved repetitive epitope recognized by CD4+ clones from a malaria-immunized volunteer. Science 1989;246:1603–1606.

167 Nardin E. T cell responses in a sporozoite-immunized human volunteer and a chimpanzee. Immunol Lett 1990; 25:43–48.

168 Sinigaglia F, Guttinger M, Romagnoli P, Takacs B. Malaria antigens and MHC restriction. Immunol Lett 1990;25:265–270.

169 Grun JL, Weidanz WP. Immunity to *Plasmodium chabaudi adami* in the B-cell deficient mouse. Nature 1981;290:143–145.

170 Grun JL, Weidanz WP. Antibody-independent immunity to reinfection malaria in B-cell deficient mice. Infect Immun 1983;41:1197–1204.

171 Cavacini LA, Parke LA, Weidanz WP. Resolution of acute malarial infections by T cell-dependent non-antibody-mediated mechanisms of immunity. Infect Immun 1990;58:2946–2950.

172 Chen DH, Tigelaar RE, Weinbaum FI. Immunity to sporozoite-induced malaria infection in mice. 1. The effect of immunization of T and B cell deficient mice. J Immunol 1977;118:1322–1327.

173 Sadoff JC, Ballou WR, Baron LS, *et al*. Oral *Salmonella typhimurium* vaccine expressing circumsporozoite protein protects against malaria. Science 1988;240:336–338.

174 Brake DA, Weidanz WP, Long CA. Antigen-specific, Interleukin 2-propagated T lymphocytes confer resistance to a murine malarial parasite, *Plasmodium chabaudi adami*. J Immunol 1986;137:347–352.

175 Brake DA, Long CA, Weidanz WP. Adoptive protection against *P. chabaudi adami* malaria in athymic nude mice by a cloned T cell line. J Immunol 1988;140:1989–1993.

176 Weidanz WP, Brake LA, Cavacini LA, Long CA. The protective role of T cells in immunity to malaria. In Eisenstein TK, Bullock WE, Hanna N, eds. *Host Defenses and Immunomodulation to Intracellular Pathogens*. New York: Plenum Publishing, 1988:99–111.

177 Romero P, Maryanski JL, Corradin G, Nussenzweig RS, Nussenzweig V, Zavala F. Cloned cytotoxic T cells recognize an epitope in the circumsporozoite protein and protect against malaria. Nature 1989;341:323–326.

178 Romero P, Maryanski J, Cordey A-S, Corradin G, Nussenzweig RS, Zavala F. Isolation and characterization of protective cytolytic T cells in a rodent malaria model system. Immunol Lett 1990;25:27–32.

179 Del Giudice G, Grillot D, Renia L, *et al*. Peptide-primed CD4+ cells and malaria sporozoites. Immunol Lett 1990;

25:59–64.

180 Schofield L, Villaquiran J, Ferreira A, Schellekens H, Nussenzweig RS, Nussenzweig V. Interferon, CD8⁺ T cells and antibodies required for immunity to malaria sporozoites. Nature 1987;330:664–666.

181 Weiss WR, Sedegah M, Beaudoin RL, Miller LH, Good MF. CD8⁺ T cells (cytotoxic/suppressors) are required for protection in mice immunized with malaria sporozoites. Proc Natl Acad Sci USA 1988;85:573–575.

182 Weiss WR, Good MF, Hollingdale MR, Miller LH, Berzofsky JA. Genetic control of immunity to *Plasmodium yeolli* sporozoites. J Immunol 1989;143:4263–4266.

183 Süss G, Eichmann K, Kury E, Linke A, Langhorne J. Roles of CD4- and CD8-bearing T lymphocytes in the immune response to erythrocytic stages of *Plasmodium chabaudi*. Infect Immun 1988;56:3081–3088.

184 Langhorne J, Simon-Haarhaus B, Meding SJ. The role of CD4⁺ T cells in the protective immune response to *Plasmodium chabaudi in vivo*. Immunol Lett 1990;25:101–108.

185 Mazier D, Renia L, Nussler A, *et al*. Hepatic phase of malaria is the target of cellular mechanisms induced by the previous and the subsequent stages. A crucial role for liver nonparenchymal cells. Immunol Lett 1990;25:65–70.

186 Green SJ, Mellouk S, Hoffman SL, Meltzer MS, Nacy CA. Cellular mechanisms of nonspecific immunity to intracellular infection: cytokine-induced synthesis of toxic nitrogen oxides from L-arginine by macrophages and hepatocytes. Immunol Lett 1990;25:15–20.

187 Brown KN, Berzins K, Jarra W, Schetters T. Immune responses to erythrocytic malaria. Clin Immunol Allerg 1986;6:227–249.

188 Bucy RP, Chen CL, Cooper MD. Tissue localization and CD8 accessory molecule expression of T gamma delta cells in humans. J Immunol 1989;142:3045–3049.

189 Brenner MB, Strominger JL, Krangel MS. The γδ T cell receptor. Adv Immunol 1988;43:133–192.

190 Playfair JHL, Jones KR, Taverne J. Cell-mediated immunity and its role in protection. In Stevenson MM, ed. *Malaria: Host Responses to Infection*. Boca Raton: CRC Press, 1989:65–86.

191 Clark IA, Hunt NH, Butcher GA, Crowden WB. Inhibition of murine malaria (*Plasmodium chabaudi*) *in vivo* by recombinant interferon-γ or tumor necrosis factor and its enhancement by butylated hydroxyanisole. J Immunol 1987;139:3493–3496.

192 Shear HL, Srinivasan R, Nolan T, Ng C. Role of IFN-γ in lethal and nonlethal malaria in susceptible and resistant hosts. J Immunol 1989;143:2038–2044.

193 Taverne J, Tavernier J, Fiers W, Playfair JHL. Recombinant tumor necrosis factor inhibits malaria parasites *in vivo* but not *in vitro*. Clin Exp Immunol 1987;67:1–4.

194 Stevensen MM, Tam MF, Nowotarski M. Role of interferon-γ and tumor necrosis factor in host resistance to *Plasmodium chabaudi*: AS. Immunol Lett 1990;25:115–122.

195 Ockenhouse CF, Shulman S, Shear HL. Induction of crisis forms in the human malaria parasite *Plasmodium falciparum* by γ-interferon-activated, monocyte-derived macrophages. J Immunol 1984;133:1601–1608.

196 Kharazami A, Jepsen S, Valerius NH. Polymorphonuclear leukocytes defective in oxidative metabolism inhibit *in vitro* growth of *Plasmodium falciparum*. Evidence against an oxygen dependent mechanism. Scand J Immunol 1984;20:93–96.

197 Cavacini LA, Guidotti M, Parke LA, Melancon-Kaplan J, Weidanz WP. Reassessment of the role of splenic leukocyte oxidative activity and macrophage activation in expression of immunity to malaria. Infect Immun 1989;57:3677–3682.

198 Playfair JHL. Non-specific killing mechanisms effective against blood stage malaria parasites. Immunol Lett 1990;25:173–174.

199 Ohnishi ST, Ohnishi N, Oda Y, Katsuoka M. Effects of membrane acting-drugs on plasmodium species and sickle cell erythrocytes. Cell Biochem Funct 1989;7:105–109.

200 Ohnishi ST, Sadanaga KK, Katsuoka M, Weidanz WP. Effects of membrane-acting drugs on *Plasmodium* species and sickle cell erythrocytes. Mol Cell Biochem 1989;91:159–165.

201 Wyler DJ, Oster CN, Quinn TC. The role of the spleen in malaria infections. In Torrigiani G, ed. *The Role of the Spleen in Immunology of Parasitic Diseases*. Tropical Diseases Research Series No. 1. Basel: Scwabe & Co., 1983:183–204.

202 Pongponratn E, Riganti M, Bunnag D, Hurinasuta T. Spleen in falciparum malaria: ultrastructural study. Southeast Asian J Trop Med Public Health 1987;18:491–501.

203 Lee S-H, Crocker P, Gordon S. Macrophage plasma membrane and secretory properties in human malaria. Effects of *Plasmodium yoelii* blood-stage infection on macrophages in liver, spleen and blood. J Exp Med 1986;163:54–74.

204 Grun JL, Long CA, Weidanz WP. Effects of splenectomy on antibody-independent immunity to *Plasmodium chabaudi adami* malaria. Infect Immun 1985;48:853–858.

205 Playfair JHL, DeSouza JB, Dockrell HM, Agomo PU, Taverne J. Cell-mediated immunity in the liver of mice vaccinated against malaria. Nature 1979;282:731–734.

206 Weiss L, Geduldig U, Weidanz WP. Mechanisms of splenic control of murine malaria: reticular cell activation and the development of a blood spleen barrier. Am J Anat 1986;176:251–285.

207 Clark IA. Cell-mediated immunity in protection and pathology of malaria. Parasitol Today 1987;3:300–305.

208 Clark IA, Chaudhri G. Relationships between inflammation and immunopathology of malaria. In Stevenson MM, ed. *Malaria: Host Responses to Infection*, Boca Raton: CRC Press, 1989:127–146.

209 Grau GE, Taylor TE, Molyneux ME, *et al*. Tumor necrosis factor and disease severity in children with falciparum malaria. N Engl J Med 1989;320:1586–1591.

210 Picot S, Peyron F, Vuillez J-P, Barbe G, Marsh K, Ambroise-Thomas P. Tumor necrosis factor production by human macrophages stimulated *in vitro* by *Plasmodium falciparum*. Infect Immun 1990;58:214–216.

211 Bate CAW, Taverne J, Playfair JHL. Malarial parasites induce TNF production by macrophages. Immunology 1988;64:227–231.

212 Collart MA, Relin D, Vassali J-D, DeKassado S, Vassalli P. γ-interferon enhances macrophage transcription of the tumor necrosis/cachectin, interleukin-1, and urokinase genes, which are controlled by short-lived suppressors. J Exp Med 1986;164:2113–2118.

213 Nedwin GE, Svedersky LP, Bringmann TS, Palladino MA, Goeddel DV. Effect of interleukin-2, interferon-γ and mitogens on the production of tumor necrosis factors α and β. J Immunol 1985;135:2492–2497.

214 Roberts DW, Weidanz WP. Splenomegaly, enhanced phagocytosis, and anemia are thymus-dependent responses to malaria. Infect Immun 1978;20:728–731.

215 Grau GE, Piguet PF, Engers HD, Louis JA, Vassali P, Lambert P-H. L3T4+ lymphocytes play a major role in the pathogenesis of murine cerebral malaria. J Immunol 1986;137:2348–2354.

216 Playfair JHL, Taverne J, Bate CAW, de Souza JB. The malaria vaccine: anti-parasite or anti-disease? Immunol Today 1990;11:25–27.

217 Bate CAW, Taverne J, Playfair JHL. Soluble malarial antigens are toxic and induce the production of tumour necrosis factor in vivo. Immunology 1989;66:600–605.

218 Schofield L, Ferreira A, Nussenzweig V, Nussenzweig RS. Antimalarial activity of alpha tumor necrosis factor and gamma interferon. Fed Proc 1987;46:760.

219 Taverne J, Bate CAW, Playfair JHL. Malaria exoantigens induce TNF, are toxic and are blocked by T-independent antibody. Immunol Lett 1990;25:207–212.

220 Spriggs DR, Sherman ML, Frei E, Kufe DW. Clinical studies with tumor necrosis factor. In Bock G, Marsh J, eds. *Tumor Necrosis Factor and Related Cytokines*, Ciba Foundation Symposium No. 131. Chichester: John Wiley and Sons, 1987:206–227.

221 Galanos C, Freudenberg M. Tumor necrosis factor (TNF), a mediator of endotoxin lethality. Immunobiology 1987; 175:13–14.

222 Raventos-Suarez C, Kaul DK, Macaluso F, Nagel RL. Membrane knobs are required for the micro-circulatory obstruction induced by *Plasmodium falciparum*-infected erythrocytes. Proc Natl Acad Sci USA 1985;82:3829–3833.

223 Macpherson GG, Warrell MJ, White NJ, Looareesunan S, Warrell DA. Human cerebral malaria: a quantitative ultrastructural analysis of parasitized erythrocyte sequestration. Am J Pathol 1985;119:385–401.

224 Poser CM. Disseminated vasculomyelinopathy. A review of the clinical and pathologic reactions of the nervous system in hyperergic diseases. Acta Neurolog Scand 1969; 37:3–44.

225 Adam C, Geniteau M, Gougerot-Pcidalo M, *et al.* Cryoglobulins, circulating immune complexes and complement activation in cerebral malaria. Infect Immun 1981; 31:530–535.

226 Finley RW, McKey LA, Lambert PH. Virulent *P. berghei* malaria: prolonged survival and decreased cerebral pathology in T-cell deficient nude mice. J Immunol 1982; 129:2213–2218.

227 Thomas JD. Clinical and histopathological correlation of cerebral malaria. Trop Geogr Med 1971;23:232–238.

228 Curfs SHAJ, Schetters TPM, Hermsen CC, Jerusalem CR, Van Zon AAJC, Eling WMC. Immunological aspects of cerebral lesions in murine malaria. Clin Exp Immunol 1989;75:136–140.

229 Grau GE, Piguet P-F, Vassalli P, Lambert PH. Tumor-necrosis factor and other cytokines in cerebral malaria: experimental and clinical data. Immunol Rev 1989;112:49–70.

230 Grau GE, Fajardo LF, Piguet P-F, Allet B, Lambert PH, Vassalli P. Tumor necrosis factor (cachectin) as an essential mediator in murine cerebral malaria. Science 1987;237:1210–1212.

231 Abdalla S, Weatherall DJ, Wickramasinghe SN, Hughes M. The anemia of *Plasmodium falciparum* malaria. Br J Haematol 1980;46:171–183.

232 Woodruff AW, Ansdell VE, Petit LE. Cause of anemia in malaria. Lancet 1979;i:1055–1057.

233 Rosenberg EB, Strickland GT, Yabf S-I, Whalen GE. IgM antibodies to red cells and auto-immune anaemia in patients with malaria. Am J Trop Med Hyg 1973;22:146–152.

234 Coleman RM, Remcricca NJ, Ritterhaus CW, Brisetto WH. Malaria: decreased survival of transfused normal erythrocytes in infected rats. J Parasitol 1976;62:138–140.

235 Jeje OM, Kelton JG, Blajchman MA. Quantitation of red cell membrane associated immunoglobulin in children with *Plasmodium falciparum* parasitaemia. Br J Haematol 1983;54:567–572.

236 Facer CA. Direct coombs antiglobulin reactions in Gambia children with *Plasmodium falciparum* malaria. II. Specificity of the erythrocyte bound IgG. Clin Exp Immunol 1980;39:279–288.

237 Srichaikul T, Siriasawakui T. Ferrokinetics in patients with malaria: haemoglobin synthesis and normoblasts *in vitro*. Am J Clin Pathol 1976;59:166–174.

238 Wickramasinghe SN, Abdalla S, Weatherall DJ. Cell cycle distribution of erythroblasts in *P. falciparum* malaria. Scand J Immunol 1982;29:83–88.

239 Clark IA, Chaudhri G. Tumor necrosis factor may contribute to the anaemia of malaria by causing dyserythropoiesis and erythrophagocytosis. Br J Haematol 1988; 70:99–103.

240 Maggio-Price L, Brookhoff D, Weiss L. Changes in hematopoietic stem cells in bone marrow of mice with *Plasmodium berghei* malaria blood. Blood 1985;66:1080–1085.

241 Miller KL, Schooley JC, Smith KL, Kullgren B, Mahimann LJ, Silverman PH. Inhibition of erythropoiesis by a soluble factor in murine malaria. Exp Hematol 1989;17:379–385.

242 Miller KL, Silverman PH, Kullgren B, Mahlmann LJ. Tumor necrosis factor alpha and the anemia associated with murine malaria. Infect Immun 1989;57:1542–1546.

243 Peetre C, Gollberg U, Nilsson E, Olsson I. Effects of recombinant tumor necrosis factor on proliferation and differentiation of leukemic and normal hemopoietic cells *in vitro*. J Clin Invest 1986;78:1694–1700.

244 Klebanoff SJ, Vadas MA, Harlan JM, *et al.* Stimulation of neutrophils by tumor necrosis factor. J Immunol 1986;136: 4220–4225.

245 Aikawa M, Shikano T, Veno N. The effect of monocyte-

released oxygen metabolites on colony formation of erythroid progenitor cells. Acta Haematol Jpn 1986;49: 1140–1146.

246 Trincherieri G, Kobayashi M, Rosen M, Loudon M, Murphy M, Perussia M. Tumor necrosis factor and lymphotoxin induce differentiation of human myeloid cell lines in synergy with immune interferon. J Exp Med 1986;164:1207–1211.

247 Skudowitz RB, Katz J, Lurie A, Levin J, Metz J. Mechanisms of thrombocytopenia in malignant tertian malaria. Br Med J 1973;ii:515–517.

248 Grau GE, Piguet PF, Gretener D, Vesin C, Lambert PH. Immunopathology of thrombocytopenia in experimental malaria. Immunology 1988;65:501–506.

249 Fajardo LF, Tallent C. Malarial parasites within human platelets. J Am Med Assoc 1974;229:1205–1207.

250 Kelton JG, Keystone J, Moore J, et al. Immune-mediated thrombocytopenia of malaria. J Clin Invest 1983;71:832–836.

251 Perrin LH, Mackey L, Ramirez E, Celada A, Lambert PH, Miescher PA. Demonstration d'antigenes parasitaires solubles, d'anticorps correspondant et de complexes immunes lors du traitement de la malaria. Schweiz, Med Wochenschr 1979;109:1832–1834.

252 Kelton JG, Gibbons S. Autoimmune platelet destruction: idiopathic thrombocytopenic purpura. Semin Thromb Hemostat 1982;8:83–104.

253 Houba V. Immunopathology of nephropathies associated with malaria. Bull WHO 1975;52:199–207.

254 Houba V. Immunologic aspects of renal lesions associated with malaria. Kidney Int 1979;16:3–8.

255 Houba V. Specific immunity: immunopathology and immunosuppression. In Wernsdorfer WH, McGregor I, eds. Malaria: Principles and Practice of Malariology, vol. 1. New York: Churchill Livingstone, 1988:621–638.

256 Houba V, Allison AC, Adeniyi A, Houba JE. Immunoglobulin classes and complement in biopsies of Nigerian children with the nephrotic syndrome. Clin Exp Immunol 1971;8:761–774.

257 June CH, Contreras CE, Perrin LH, Lambert PH, Miescher PA. Circulating and tissue-bound immune complex formation in murine malaria. J Immunol 1979;122:2154–2161.

258 Boonpucknavig S, Boonpucknavig V, Bhamarapravati N. Immunopathological studies of Plasmodium berghei infected mice. Arch Pathol 1972;94:322–330.

259 Boonpucknavig V, Boonpucknavig S, Bhamara Pravati N. Plasmodium berghei infection in mice: an ultrastructural study of immune complex nephritis. Am J Pathol 1973;70:89–108.

260 Voller A, O'Neill P, Hymphrey D. Serological indices in Tanzania. II. Antinuclear factor and malaria indices in populations living at different altitudes. J Trop Med 1972;75:136–139.

261 Okerengwo AA, Williams AIO, Osunkoya BO. Serum immunoconglutinin levels in healthy subjects, malarial nephrosis and other conditions. Afr J Med Sci 1979;8:75–78.

262 Bruce-Chawtt LJ. Biometric study of spleen and liver weights in Africans and Europeans with special references to endemic malaria. Bull WHO 1956;15:513–548.

263 Marsh K, Greenwood BM. The immunopathology of malaria. Clin Trop Med Commun Dis 1986;1:91–125.

264 Marsden PD, Crane GG. The tropical splenomegaly syndrome: a current appraisal. Rev Inst Med Trop Sao Paulo 1976;18:54–70.

265 Bryceson A, Fakunle YM, Fleming AF, et al. Malaria and splenomegaly. Trans R Soc Trop Med Hyg 1983;77:879.

266 David-West AS. Relapses after withdrawal of proguanil treatment in tropical splenomegaly syndrome. Br Med J 1974;3:499–501.

267 Fakunle YM, Greenwood BM. Metabolism of IgM in the tropical splenomegaly syndrome. Trans R Soc Trop Med Hyg 1976;70:346–348.

268 Crane GG, Pryor DS, Wells JV. Tropical splenomegaly syndrome in New Guinea. II. Long term results of splenectomy. Trans R Soc Trop Med Hyg 1972;66:733–742.

269 Ziegler JL. Cryoglobulinaemia in tropical splenomegaly syndrome. Clin Exp Immunol 1973;15:65–78.

270 Fakunle YM, Onyewotu IJ, Greenwood BM, Mohammed I, Holborrow EJ. Cryoglobulinaemia and circulating immune complexes in tropical splenomegaly syndrome. Clin Exp Immunol 1978;31:55–58.

271 Ziegler JL, Stuiver PC. Tropical splenomegaly syndrome in a Rivandan Kindred in Uganda. Br Med J 1972;3:79–82.

272 Crane GG. Tropical splenomegaly. Part 2: Oceania. Haematology in tropical areas. Clin Haematol 1981;10:976–982.

273 Fakunle YM, Oduloju AJ, Greenwood BM. T- and B-lymphocyte subpopulations in the tropical splenomegaly syndrome (TSS). Clin Exp Immunol 1978;33:239–245.

274 Piessens WE, Hoffman SL, Wadee A. Antibody-mediated killing of suppressor T lymphocytes as a possible cause of macroglobulinemia in tropical splenomegaly syndrome. J Clin Invest 1985;75:1821–1827.

275 McGregor IA, Barr M. Antibody response to tetanus toxoid inoculation in malarious and non-malarious Gambian children. Trans R Soc Trop Med Hyg 1962;56:364–367.

276 Greenwood BM, Bradley-Moore AM, Palit A, Bryceson ADM. Immunosuppression in children with malaria. Lancet 1972;i:169–172.

277 Williamson WA, Greenwood BM. Impairment of the immune response after acute malaria. Lancet 1978;i:1328–1329.

278 Bradley-Moore AM, Greenwood BM, Bradley AK. Malaria chemoprophylaxis with chloroquine in young Nigerian children. II. Effect on the immune response to vaccination. Ann Trop Med Parasitol 1985;79:563–573.

279 Ho M, Webster HK, Looareesuwan S, et al. Antigen-specific immunosuppression in human malaria due to Plasmodium falciparum. J Infect Dis 1986;153:763–771.

280 Webster HK, Ho M, Looareesuwan S, et al. Lymphocyte responsiveness to a candidate malaria sporozoite vaccine (R32tet$_{32}$) of individuals with naturally acquired Plasmodium falciparum malaria. Am J Trop Med Hyg 1988;38:37–41.

281 Riley EM, Andersson G, Otoo LN, Jepsen S, Greenwood BM. Cellular immune responses to Plasmodium falciparum

antigens in Gambian children during and after an acute attack of falciparum malaria. Clin Exp Immunol 1988;73: 17–22.

282 Brasseur P, Ballet JJ, Druilhe P. Impairment of *Plasmodium falciparum*-specific antibody response in severe malaria. J Clin Microbiol 1990;28:265–268.

283 Loose LD. Characterization of macrophage dysfunction in rodent malaria. J Leuk Biol 1984;36:703–718.

284 Correa M, Narayanan PR, Miller HC. Suppressive activity of splenic adherent cells from *Plasmodium chabaudi*-infected mice. J Immunol 1980;125:749–754.

285 Morges W, Weidanz WP. *Plasmodium yoelii*: the thymus-dependent lymphocyte in mice immunodepressed by malaria. Exp Parasitol 1980;50:188–194.

286 Warren HS, Weidanz WP. Malarial immunodepression *in vitro*: adherent spleen cells are functionally defective as accessory cells in response to horse erythrocytes. Eur J Immunol 1976;6:819–822.

287 Morakote N, Justus DE. Immunosuppression in malaria: effect of hemozoin produced by *Plasmodium berghei* and *Plasmodium falciparum*. Int Arch Allergy Appl Immun 1988;86:28–34.

288 Lelchuk R, Rose G, Playfair JHL. Changes in the capacity of macrophages and T cells to produce interleukins during murine malaria infection. Cell Immunol 1984;84:253–263.

289 Miller LH, Mason SJ, Clyde DF, McGinniss MH. The resistance factor to *Plasmodium vivax* in Blacks. N Engl J Med 1976;295:302–304.

290 Miller LH, McGinniss MH, Holland PV, Sigmon P. The Duffy blood group phenotype in American Blacks infected with *Plasmodium vivax* in Vietnam. Am J Trop Med Hyg 1978;27:1069–1072.

291 Haldane JBS. *Proceedings of the 8th International Congress of Genetics*. 1949:267–272.

292 Allison AC. Protection afforded by sickle cell trait against subtertian malarial infection. Br Med J 1954;1:290–294.

293 Allison AC. The distribution of the sickle cell trait in East Africa and elsewhere and its apparent relationship to the incidence of subtertian malaria. Trans R Soc Trop Med Hyg 1954;48:312–318.

294 Allison AC. Polymorphism and natural selection in human populations. Cold Spring Harbor Symp Quant Biol 1964; 29:137–139.

295 Miller LH, Mason SJ, Dvorak JA, McGinniss MH, Rothman IK. Erythrocyte receptors for (*Plasmodium knowlesi*) malaria: The Duffy blood group determinants. Science 1975;189: 561–563.

296 Barnwell JW, Nichols ME, Rubinstein P. *In vitro* evaluation of the role of the Duffy blood group in erythrocyte invasion by *Plasmodium vivax*. J Exp Med 1989;169:1795–1802.

297 Nichols ME, Rubinstein P, Barnwell J, de Cordoba SR, Rosenfield RE. A new human Duffy blood group specificity defined by a murine monoclonal antibody. Immuno-genetics and association with susceptibility to *Plasmodium vivax*. J Exp Med 1987;166:776–785.

298 Kitchen SF. The infection of reticulocytes by *Plasmodium vivax*. Am J Trop Med 1938;18:347–359.

299 Pasvol G, Weatherall DJ, Wilson RJM. The increased susceptibility of young red cells to invasion by the malarial parasite *Plasmodium falciparum*. Br J Haematol 1980;45:285–295.

300 Miller LH, Haynes JD, McAuliffe FM, Shiroishi T, Durocher JR, McGinniss MH. Evidence for differences in erythrocyte surface receptors for the malarial parasites, *Plasmodium falciparum* and *Plasmodium knowlesi*. J Exp Med 1977;146: 277–281.

301 Pasvol G, Wainscoat JS, Weatherall DJ. Erythrocytes deficient in glycophorin resist invasion by the malarial parasite *Plasmodium falciparum*. Nature 1982;297:64–66.

302 Perkins M. Inhibitory effects of erythrocyte membrane proteins on the *in vitro* invasion of the human malarial parasite (*Plasmodium falciparum*) into its host cell. J Cell Biol 1981;90:563–567.

303 Serjeantson S, Bryson K, Amato D, Babona D. Malaria and hereditary ovalocytosis. Hum Genet 1977;37:161–167.

304 Cattani JA, Gibson FD, Alpers MP, Crane GG. Hereditary ovalocytosis and reduced susceptibility to malaria in Papua New Guinea. Trans R Soc Trop Med Hyg 1987;81: 705–709.

305 Hadley T, Saul A, Lamont G, Hudson DE, Miller LH, Kidson C. Resistance of Melanesian elliptocytes (ovalocytes) to invasion by *Plasmodium knowlesi* and *Plasmodium falciparum* malaria parasites *in vitro*. J Clin Invest 1983;71: 780–782.

306 Kidson C, Lamont G, Saul A, Nurse GT. Ovalocytic erythrocytes from Melanesians are resistant to invasion by malaria parasites in culture. Proc Natl Acad Sci USA 1981; 78:5829–5832.

307 Mohandas N, Lie-Injo LE, Friedman M, Mak JW. Rigid membranes of Malayan ovalocytes: a likely genetic barrier against malaria. Blood 1984;63:1385–1392.

308 Rangachari K, Beaven GH, Nash GB, *et al*. A study of red cell membrane properties in relation to malarial invasion. Mol Bichem Parasitol 1989;34:63–74.

309 Kidson C, Gorman JG. A challenge to the concept of selection by malaria in glucose-6-phosphate dehydrogenase deficiency. Nature 1962;196:49–51.

310 Luzzatto L, Usanga EA, Reddy S. Glucose-6-phosphate dehydrogenase deficient red cells: resistance to infection by malarial parasites. Science 1969;164:839–842.

311 Roth EF Jr, Raventos-Suarez C, Rinaldi A, Nagel RL. Glucose-6-phosphate dehydrogenase deficiency inhibits *in vitro* growth of *Plasmodium falciparum*. Proc Natl Acad Sci USA 1983;80:298–299.

312 Bienzle U, Ayeni O, Lucas AO, Luzzatto L. Glucose-6-phosphate dehydrogenase and malaria. Greater resistance of females heterozygous for enzyme deficiency and of males with non-deficient variant. Lancet 1972;i:107–110.

313 Yoshida A, Roth EF Jr. Glucose-6-phosphate dehydrogenase of malaria parasite *Plasmodium falciparum*. Blood 1987;69:1528–1530.

314 Usanga EA, Luzzatto L. Adaptation of *Plasmodium falciparum* to glucose 6-phosphate dehydrogenase-deficient host red cells by production of parasite-encoded enzyme. Nature 1985;313:793–795.

315 Friedman MJ. Oxidant damage mediates variant red cell resistance to malaria. Nature 1979;280:245–247.

316 Baehner RL, Nathan DG, Castle WB. Oxidant injury of Caucasian glucose-6-phosphate dehydrogenase-deficient red blood cells by phagocytosing leukocytes during infection. J Clin Invest 1971;50:2466–2473.

317 Huheey JE, Martin DL. Malaria, favism and glucose-6-phosphate dehydrogenase deficiency. Experientia 1975;31:1145–1147.

318 Roth EF Jr, Ruprecht RM, Schulman S, Vanderberg J, Olson JA. Ribose metabolism and nucleic acid synthesis in normal and glucose-6-phosphate dehydrogenase-deficient human erythrocytes infected with Plasmodium falciparum. J Clin Invest 1986;77:1129–1135.

319 Chern CJ, Beutler E. Pyridoxal kinase: decreased activity in red blood cells in Afro Americans. Science 1975;187:1084–1086.

320 Solomon LR, Hillman RS. Vitamin B6 metabolism in human red cells. 1. Variations in normal subjects. Enzyme 1978;23:262–273.

321 Martin SK, Miller LH, Kark JA, et al. Low erythrocyte pyridoxal-kinase activity in blacks: its possible relation to falciparum malaria. Lancet 1978;i:524–526.

322 Friedman MJ. Erythrocytic mechanism of sickle cell resistance to malaria. Proc Natl Acad Sci USA 1978;75:1994–1997.

323 Pasvol G, Weatherall DJ. Cellular mechanism for the protective effect of haemoglobin S against P. falciparum malaria. Nature 1978;274:701–703.

324 Friedman MJ, Roth EF, Nagel RL, Trager W. Plasmodium falciparum: physiological interactions with the human sickle cell. Exp Parasitol 1979;47:73–80.

325 Luzzatto L, Nwachuku-Jarrett ES, Reddy S. Increased sickling of parasitized erythrocytes as mechanism of resistance against malaria in the sickle cell trait. Lancet 1970;i:319–321.

326 Luzzatto L. Genetic factors in malaria. Bull WHO 1974;50:195–202.

327 Roth EF Jr, Shear HL, Costantini F, Tanowitz HB, Nagel RL. Malaria in β-thalassemic mice and the effects of the transgenic human β-globin gene and splenectomy. J Lab Clin Med 1988;111:35–41.

328 Brockelman CR, Wongsattayanont B, Tan-Ariya P, Fucharoen S. Thalassemic erythrocytes inhibit in vitro growth of Plasmodium falciparum. J Clin Microbiol 1987;25:56–60.

329 Pasvol G, Weatherall DJ, Wilson RJM. Effects of foetal haemoglobin on susceptibility of red cells to Plasmodium falciparum. Nature 1977;270:171–173.

330 Bruce-Chwatt LJ. DNA probes for malaria diagnosis (Letter). Lancet 1984;i:175.

331 Mackey L, McGregor IA, Lambert PH. Diagnosis of Plasmodium falciparum infection using a solid-phase radioimmunoassay for the detection of malaria antigens. Bull WHO 1980;58:439–444.

332 Avraham H, Golenser J, Gazitt Y, Spira DT, Sulitzeanu D. A highly sensitive solid-phase radioimmunoassay for the assay of Plasmodium falciparum antigens and antibodies. J Immunol Methods 1982;53:61–68.

333 Mackey LJ, McGregor IA, Paounova N, Lambert PH. Diagnosis of Plasmodium falciparum infection in man: detection of parasite antigens by ELISA. Bull WHO 1982;60:69–75.

334 Taylor DW, Parra M, Evans CB, Quakya IA, Howard RJ. The use of a monoclonal antibody "antigen capture" assay for the diagnosis of malaria. In Abstracts of 3rd International Congress on Malaria and Babesiosis, Annecy Edition, Foundation Marcel Merieux, Lyons. 1987:11.

335 Parra ME, Taylor DW. Presence of histidine rich protein 2 (PfHRP-2) in the sera of people infected with Plasmodium falciparum. In Abstracts of the 38th Annual Meeting of the American Society of Tropical Medicine and Hygiene, Honolulu, Hawaii. American Institute of Biological Sciences Washington DC, 1989:212.

336 Fortier B, Delplace P, Dubremetz JF, Ajana F, Vernes A. Enzyme assay for detection of antigen in acute Plasmodium falciparum malaria. Eur J Clin Microbiol 1987;6:596–598.

337 Zavala F, Gwadz RW, Collins FH, Nussenzweig RS, Nussenzweig V. Monoclonal antibodies to circumsporozoite proteins identify the species of malaria parasite in infected mosquitoes. Nature 1982;299:737–738.

338 Burkot TR, Zavala F, Gwadz RW, Collins FH, Nussenzweig RS, Roberts DR. Identification of malaria-infected mosquitoes by a two-site enzyme-linked immunoadsorbent assay. Am J Trop Med Hyg 1984;33:227–231.

339 Burkot TR, Williams JL, Schneider I. Identification of Plasmodium falciparum-infected mosquitoes by a double antibody enzyme-linked immunoadsorbent assay. Am J Trop Med Hyg 1984;33:783–788.

340 Wirtz RA, Burkot TR, Andre RG, Rosenberg R, Collins WE, Roberts DR. Identification of Plasmodium vivax sporozoites in mosquitoes using an enzyme-linked immunoadsorbent assay. Am J Trop Med Hyg 1985;34:1048–1054.

341 Collins FH, Zavala F, Graves PM, et al. First field trial of an immunoradiometric assay for the detection of malaria sporozoites in mosquitoes. Am J Trop Med Hyg 1984;33:538–543.

342 Beier JC, Perkins PV, Wirtz RA, Whitmire RE, Mugambi M, Hockmeyer WT. Field evaluation of an enzyme-linked immunoadsorbent assay (ELISA) for Plasmodium falciparum sporozoite detection in anopheline mosquitoes from Kenya. Am J Trop Med Hyg 1987;36:459–468.

343 Wirtz RA, Burkot TR, Graves PM, Andre RG. Field evaluation of enzyme-linked immunoadsorbent assays for Plasmodium falciparum and sporozoites in mosquitoes (Diptera:Culicidae) from Papau New Guinea. J Med Entomol 1987;24:433–437.

344 Franzen L, Westin G, Shabo R, et al. Analysis of clinical specimens by hybridization with probe containing repetitive DNA from Plasmodium falciparum. Lancet 1984;i:525–528.

345 Aslund L, Franzen L, Westin G, Persson T, Wigzell H, Pettersson U. Highly reiterated non-coding sequence in the genome of Plasmodium falciparum is composed of 21 base-pair tandem repeats. J Mol Biol 1985;185:509–516.

346 Oquando P, Goman M, Mackay M, Langsley G, Walliker D, Scaife J. Characterization of a repetitive DNA sequence

from the malaria parasite, *Plasmodium falciparum*. Mol Biochem Parasitol 1986;18:89–101.

347 Pollack Y, Metzger S, Shemer R, Landau D, Spira DT, Golenser J. Detection of *Plasmodium falciparum* in blood using DNA hybridization. Am J Trop Med Hyg 1985;34:663–667.

348 Enea V. Sensitive and specific DNA probe for detection of *Plasmodium falciparum*. Mol Cell Biol 1986;6:321–324.

349 Barker RH Jr, Subsaeng L, Rooney W, Alecrim GC, Dourado HV, Wirth DF. Specific DNA probe for the diagnosis of *Plasmodium falciparum* malaria. Science 1986;231:1434–1436.

350 Zolg JW, Andrade E, Scott ED. Detection of *Plasmodium falciparum* DNA using repetitive clones as species specific probes. Mol Biochem Parasitol 1987;22:145–151.

351 Holmberg M, Sheton FC, Franzen L, et al. Use of DNA hybridization assay for the detection of *Plasmodium falciparum* in field trials. Am J Trop Med Hyg 1987;37:230–234.

352 McLaughlin GL, Collins WE, Campbell GH. Comparison of genomic, plasmid, synthetic, and combined DNA probes for detecting *Plasmodium falciparum* DNA. J Clin Microbiol 1987;25:791–795.

353 McLaughlin GL, Edlind TD, Campbell GH, Eller RF, Ihler GM. Detection of *Plasmodium falciparum* using a synthetic DNA probe. Am J Trop Med Hyg 1985;34:837–840.

354 Mucenski CM, Guerry P, Buesing M, et al. Evaluation of a synthetic oligonucleotide probe for diagnosis of *Plasmodium falciparum* infections. Am J Trop Med Hyg 1986;35:912–920.

355 Buesing M, Guerry P, Deisanti C, et al. An oligonucleotide probe for detecting *Plasmodium falciparum*: an analysis of clinical specimens from six countries. J Infect Dis 1987;155:1315–1318.

356 McLaughlin GL, Ruth JL, Jablonski E, Steketee R, Campbell GH. Use of enzyme-linked DNA in diagnosis of falciparum malaria. Lancet 1987;1:714–715.

357 Sethabutr O, Brown AE, Gingrich J, et al. A comparative field study of radiolabeled and enzyme-conjugated synthetic DNA probes for the diagnosis of falciparum malaria. Am J Trop Med Hyg 1988;39:227–231.

358 Lanar DE, McLaughlin GL, Wirth DF, Barker RJ, Zolg JW, Chulay JD. Comparison of thick films, *in vitro* culture and DNA hybridization probes for detecting *Plasmodium falciparum* malaria. Am J Trop Med Hyg 1989;40:3–6.

359 Alley AB, Bates MD, Tam JP, Hollingdale MR. Synthetic peptides from the circumsporozoite proteins of *Plasmodium falciparum* and *Plasmodium knowlesi* recognize the human hepatoma cell line HepG2-A16 *in vitro*. J Exp Med 1986;164:1915–1922.

360 Waters AP, McCutchan TF. Rapid, sensitive diagnosis of malaria based on ribosomal RNA. Lancet 1989;1:1343–1346.

361 Lal AA, Changkasiri S, Hollingdale MR, McCutchan TF. Ribosomal RNA-based diagnosis of *Plasmodium falciparum* malaria. Mol Biochem Parasitol 1989;36:67–72.

362 Nussenzweig R, Vanderberg J, Most H. Protective immunity produced by the injection of X-irradiated sporozoites of *Plasmodium berghei*. IV. Dose response,

specificity, and humoral immunity. Mil Med 1969;134:1176–1182.

363 Nussenzweig RS, Vanderberg JP, Most H, Orton C. Protective immunity induced by the injection of X-irradiated sporozoites of *Plasmodium berghei*. Nature 1969;222:488–489.

364 Clyde DF, McCarthy VC, Miller RM, Hornick RB. Specificity of protection of man immunized against sporozoite induced falciparum malaria. Am J Med Sci 1973;266:398–403.

365 Clyde DF, McCarthy VC, Miller RM, Woodward WE. Immunization of man against falciparum and vivax malaria by the use of attenuated sporozoites. Am J Trop Med Hyg 1975;24:397–401.

366 Hollingdale MR, Nardin EH, Tharavanij S, Schwartz AL, Nussenzweig RS. Inhibition of entry of *Plasmodium falciparum* and *Plasmodium vivax* sporozoites into cultured cells; an *in vitro* assay of protective antibodies. J Immunol 1984;132:909–913.

367 Verhave JP, Strickland GR, Jaffe HA, Ahmed A. Studies on the transfer of protective immunity with lymphoid cells from mice immune to malaria sporozoites. J Immunol 1978;121:1031–1033.

368 Yoshida N, Potocnajak P, Nussenzweig V, Nussenzweig RS. Biosynthesis of Pb44, the protective antigen of sporozoites of *Plasmodium berghei*. J Exp Med 1981;154:1225–1236.

369 Nussenzweig V, Nussenzweig RS. Circumsporozoite proteins of malaria parasites. Cell 1985;42:401–403.

370 Cochrane AH, Santoro F, Nussenzweig V, Gwadz RW, Nussenzweig RS. Monoclonal antibodies identify the protective antigens of *Plasmodium knowlesi*. Proc Natl Acad Sci USA 1982;79:5651–5655.

371 Nardin EH, Nussenzweig V, Nussenzweig RS, et al. Circumsporozoite proteins of human malaria parasites *Plasmodium falciparum* and *Plasmodium vivax*. J Exp Med 1982;156:20–30.

372 Santoro F, Cochrane AH, Nussenzweig V, et al. Structural similarities among the protective antigens of sporozoites from different species of malaria parasite. J Biol Chem 1983;258:3341–3345.

373 Zavala F, Cochrane AH, Nardin EH, Nussenzweig RS, Nussenzweig V. Circumsporozoite proteins of malaria parasites contain a single immunodominant region with two or more identical epitopes. J Exp Med 1983;157:1947–1957.

374 Ellis J, Ozaki LS, Gwadz RS, et al. Cloning and expression in *E. coli* of the malarial sporozoite surface antigen gene from *Plasmodium knowlesi*. Nature 1983;302:536–538.

375 Ozaki LS, Svec P, Nussenzweig RS, Nussenzweig V, Godson GN. Structure of the *Plasmodium knowlesi* gene coding for the circumsporozoite protein. Cell 1983;34:815–822.

376 Dame JB, Williams JL, McCutchan TF, et al. Structure of the gene encoding the immunodominant surface antigen on the sporozoite of the human malaria parasite *Plasmodium falciparum*. Science 1984;225:593–599.

377 Arnot DE, Barnwell JW, Tam JP, Nussenzweig RS, Enea V. Circumsporozoite protein of *Plasmodium vivax*: gene

cloning and characterization of the immunodominant epitope. Science 1985;230:815–818.

378 Eichinger DJ, Arnot DE, Tam JP, Nussenzweig V, Enea V. The circumsporozoite protein of *Plasmodium berghei*: gene cloning and identification of the immunodominant epitope. J Mol Cell Biol 1986;6:3965–3972.

379 Galinski MR, Arnot DE, Cochrane AH, Barnwell JW, Nussenzweig RS, Enea V. The circumsporozoite gene of the *Plasmodium cynomolgi* complex. Cell 1987;48:311–319.

380 Lal AA, de la Cruz VF, Welsh JA, Charoenvit Y, Maloy WL, McCutchan TF. Structure of the gene encoding the circumsporozoite protein of *Plasmodium yoelii*. J Biol Chem 1987;262i:2937–2940.

381 Lal AA, de la Cruz VF, Collins WE, Campbell GH, Porcell PM, McCutchan TF. Circumsporozoite protein gene from *Plasmodium brasilianum*: genetic evidence of animal reservoirs for human malaria parasites. J Biol Chem 1988; 263:5495–5498.

382 Lal AA, de la Cruz VF, Campbell GH, Procell PM, Collins WE, McCutchan TF. Structure of the circumsporozoite gene of *Plasmodium malariae*. Mol Biochem Parasitol 1988; 30:291–294.

383 Cochrane AH, Gwadz RW, Ojo-Amaize E, Hii J, Nussenzweig V, Nussenzweig RS. Antigenic diversity of the circumsporozoite proteins in the *Plasmodium cynomolgi* complex. Mol Biochem Parasitol 1985;14:111–124.

384 Sharma S, Svec P, Mitchell GH, Godson GN. Diversity of circumsporozoite antigens and genes from two strains of the malaria parasite *Plasmodium knowlesi*. Science 1985;229: 779–782.

385 Rosenberg R, Wirtz RA, Lanar DE, et al. Circumsporozoite protein heterogeneity in the human malaria parasite *Plasmodium vivax*. Science 1989;245:973–976.

386 Weber JL, Hockmeyer WT. Structure of the circumsporozoite gene in 18 strains of *Plasmodium falciparum*. Mol Biochem Parasitol 1985;15:305–316.

387 Vergara U, Ruiz A, Ferreira A, Nussenzweig RS, Nussenzweig V. Conserved group-specific epitopes of the circumsporozoite proteins revealed by antibodies to synthetic peptides. J Immunol 1985;134:3445–3448.

388 Robson KJH, Hall JRS, Jennings MW, et al. A highly conserved amino-acid sequence in thrombospondin, properdin and in proteins from sporozoites and blood stages of a human malaria parasite. Nature 1988;335:79–82.

389 Goundis D, Reid KBM. Properdin, the terminal complement components, thrombospondin and the circumsporozoite protein of malaria parasites contain similar sequence motifs. Nature 1988;335:82–85.

390 Zavala F, Tam JP, Hollingdale MR, et al. Rationale for development of a synthetic vaccine against *Plasmodium falciparum* malaria. Science 1985;228:1436–1440.

391 Ballou WR, Rothbard J, Wirtz RA, et al. Immunogenicity of synthetic peptide from circumsporozoite protein of *Plasmodium falciparum*. Science 1985;228:996–999.

392 Herrington DA, Clyde DF, Losonsky G, et al. Safety and immunogenicity in man of a synthetic peptide and malaria vaccine against *Plasmodium falciparum* sporozoites. Nature 1987;328:257–259.

393 Young JF, Hockmeyer WT, Gross M, et al. Expression of *Plasmodium falciparum* circumsporozoite proteins in *Escherichia coli* for potential use in a human malaria vaccine. Science 1985;228:958–962.

394 Ballou WR, Hoffman SR, Sherwood JA, et al. Safety and efficacy of a recombinant DNA *Plasmodium falciparum* sporozoite vaccine. Lancet 1987;1:1277–1281.

395 Good MF, Pombo D, Lunde MN, et al. Recombinant human interleukin-2 overcomes genetic nonresponsiveness to malaria sporozoite peptides. Correlation of effect with biological activity of IL-2. J Immunol 1988;141:972–977.

396 Lowell GH, Ballou WR, Smith LF, Wirtz RA, Zollinger WD, Hockmeyer WT. Proteosome-lipopeptide vaccines: enhancement of immunogenicity for malaria CS peptides. Science 1988;240:800–802.

397 Que JU, Cryz SJ Jr, Ballou R, et al. Effect of carrier selection on immunogenicity of protein conjugate vaccines against *Plasmodium falciparum* circumsporozoites. Infect Immun 1988;56:2645–2649.

398 Webster HK, Boudreau EF, Pang LW, Permpanich B, Sookto P, Wirtz RA. Development of immunity in natural *Plasmodium falciparum* malaria: antibodies to the falciparum sporozoite vaccine 1 antigen (R32tet32). J Clin Microbiol 1987;25:1002–1008.

399 Hoffman SL, Oster CN, Plowe CV, et al. Naturally acquired antibodies to sporozoites do not prevent malaria: vaccine development implications. Science 1987;237:639–641.

400 Ferreira A, Schofield L, Enea V, et al. Inhibition of development of exoerythrocytic forms of malaria parasites by γ-interferon. Science 1986;232:881–884.

401 Egan JE, Weber JL, Ballou WR, et al. Efficacy of murine malaria sporozoite vaccines: implications for human vaccine development. Science 1987;236:453–456.

402 Lal AA, de la Cruz VF, Good MF, et al. *In vivo* testing of subunit vaccines against malaria sporozoites using a rodent system. Proc Natl Acad Sci USA 1987;84:8647–8651.

403 Good MF, Berzofsky JA, Maloy WL, et al. Genetic control of the immune response in mice to a *Plasmodium falciparum* sporozoite vaccine. J Exp Med 1986;164:655–660.

404 Russo DM, Sundy JS, Young JF, Maguire HC, Weidanz WP. Cell-mediated immune responses to vaccine peptides derived from the circumsporozoite protein of *Plasmodium falciparum*. J Immunol 1989;143:655–659.

405 Good MF, Maloy WL, Lunde MN, et al. Construction of a synthetic immunogen: use of new T-helper epitope on malaria circumsporozoite protein. Science 1987;235;1059–1062.

406 Good MF, Pombo D, Quakyi IA, et al. Human T-cell recognition of the circumsporozoite protein of *Plasmodium falciparum*: immunodominant T-cell domains map to the polymorphic regions of the molecule. Proc Natl Acad Sci USA 1988;85:1199–1203.

407 Lockyer MJ, Schwarz RT. Strain variation in the circumsporozoite protein gene of *Plasmodium falciparum*. Mol Biochem Parasitol 1987;22:101–108.

408 De la Cruz VF, Lal AA, McCutchan TF. Sequence variation in putative functional domains of the circumsporozoite protein of *Plasmodium falciparum*: implications for vaccine development. J Biol Chem 1987;262:11935–11939.

409 De la Cruz VF, Maloy WL, Miller LH, Lal AA, Good MF, McCutchan TF. Lack of cross-reactivity between variant T cell determinant from malaria circumsporozoite protein. J Immunol 1988;141:2456–2460.

410 Sinigaglia F, Guttinger M, Kilgus J, *et al.* A malaria T-cell epitope recognized in association with most mouse and human MHC class II molecules. Nature 1988;336:778–780.

411 Hoffman SL, Oster CN, Mason C, *et al.* Human lymphocyte proliferative responses to a sporozoite T cell epitope correlates with resistance to falciparum malaria. J Immunol 1989;142:1299–1303.

412 Trager W, Jensen JB. Human malaria parasites in continuous culture. Science 1976;193:673–675.

413 Wilson RJM. Serotyping *Plasmodium falciparum* malaria with S-antigens. Nature 1980;284:451–452.

414 Winchell EJ, Ling I, Wilson RJM. Metabolic labelling and characterization of S-antigens, the heat stable strain-specific antigens of *Plasmodium falciparum*. Mol Biochem Parasitol 1984;10:287–296.

415 Anders RF, Brown GV, Edwards AE. Characterization of an S- antigen synthesized by several isolates of *Plasmodium falciparum*. Proc Natl Acad Sci USA 1983;80:6652–6656.

416 Coppel RL, Cowman AF, Lingelbach KR, *et al.* Isolate-specific S-antigen of *Plasmodium falciparum* contains a repeated sequence of eleven amino acids. Nature 1983; 302:751–756.

417 Cowman AF, Saint RB, Coppel RL, Brown GV, Anders RF, Kemp DJ. Conserved sequences flank variable tandem repeats in two S-antigens of *Plasmodium falciparum*. Cell 1985;40:775–783.

418 Saint RB, Coppel RL, Cowman AS, *et al.* Changes in repeat number, sequence, and reading frame in S antigen genes of *Plasmodium falciparum*. Mol Cell Biol 1987;7:2968–2973.

419 Saul A, Cooper J, Ingram L, Anders RF, Brown GV. Invasion of erythrocytes *in vitro* by *Plasmodium falciparum* can be inhibited by a monoclonal antibody directed against an S-antigen. Parasite Immunol 1985;7:587–593.

420 Jungery M, Boyle D, Patel T, Pasvol G, Weatherall DJ. Lectin-like polypeptides of *P. falciparum* bind to red cell sialoglycoproteins. Nature 1983;301:704–705.

421 Perkins ME. Surface proteins of *Plasmodium falciparum* merozoites binding to the erythrocyte receptor, glycophorin. J Exp Med 1984;160:788–798.

422 Ravetch JV, Kopchan J, Perkins M. Isolation of the gene for a glycophorin-binding protein implicated in erythrocyte invasion by a malaria parasite. Science 1985;227:1593–1597.

423 Kochan J, Perkins M, Ravetch JV. A tandemly repeated sequence determines the binding domain for an erythrocyte receptor binding protein of *P. falciparum*. Cell 1986;44:689–696.

424 Bianco AE, Culvenor JG, Coppel RL, *et al.* Putative glycophorin-binding protein is secreted from schizonts of *Plasmodium falciparum*. Mol Biochem Parasitol 1987;23:91–102.

425 Perkins ME. Stage-dependent processing and localization of a *Plasmodium falciparum* protein of 130 000 molecular weight. Exp Parasitol 1988;65:61–68.

426 Van Schravendijk MR, Wilson RJM, Newbold CI. Possible pitfalls in the identification of glycophorin-binding proteins of *Plasmodium falciparum*. J Exp Med 1987;166:376–390.

427 Haynes JD, Dalton JP, Klotz FW, *et al.* Receptor-like specificity of a *Plasmodium knowlesi* malarial protein that binds to Duffy antigen ligands or erythrocytes. J Exp Med 1988;167:1873–1881.

428 Freeman RR, Holder AA. Surface antigens of malaria parasites. A high molecular weight precursor is processed to an 83 000 M.W. form expressed on the surface of *Plasmodium falciparum* merozoites. J Exp Med 1983;158:1647–1653.

429 McBride JS, Heidrich HG. Fragments of the polymorphic gr 185 000 glycoprotein from the surface of isolated *Plasmodium falciparum* merozoites from an antigenic complex. Mol Biochem Parasitol 1987;23:71–84.

430 Thomas AW, Deans JA, Mitchell GH, Alderson T, Cohen S. The Fab fragments of a monoclonal IgG to a merozoite surface antigen inhibit *Plasmodium knowlesi* invasion of erythrocytes. Mol Biochem Parasitol 1984;13:187–199.

431 Deans JA, Thomas AW, Alderson T, Cohen S. Biosynthesis of a putative protective *Plasmodium knowlesi* merozoite antigen. Mol Biochem Parasitol 1984;11:189–204.

432 Holder AA. The precursor to the major merozoite surface antigens: structure and role in immunity. Prog Allergy 1988;41:72–97.

433 Holder AA, Freeman RR. Immunization against blood-stage rodent malaria using purified parasite antigens. Nature 1981;294:361–364.

434 Freeman RR, Holder AA. Characterization of the protective response of BALB/c mice immunized with a purified *Plasmodium yoelii* schizont antigen. Clin Exp Immunol 1983;54:609–616.

435 Bates MD, Newbold CI, Jarra W, Brown KN. Protective immunity to malaria: studies with cloned lines of *Plasmodium chabaudi chabaudi* in CBA/Ca mice. III. Protective and suppressive responses induced by immunization with purified antigens. Parasite Immunol 1988;10:1–15.

436 Perrin LH, Merkli B, Loche M, Chizzolini C, Smart J, Richle R. Antimalarial immunity in *Saimiri* monkeys. J Exp Med 1984;160:441–451.

437 Siddiqui WA, Tam LQ, Kramer KJ, *et al.* Merozoite surface coat precursor protein completely protects *Aotus* monkeys against *Plasmodium falciparum* malaria. Proc Natl Acad Sci USA 1987;84:3014–3018.

438 Cheung A, Leban J, Shaw AR, *et al.* Immunization with synthetic peptides of a *Plasmodium falciparum* surface antigen induces antimerozoite antibodies. Proc Natl Acad Sci USA 1986;83:8328–8332.

439 Patarroyo ME, Romero P, Torres ML, *et al.* Induction of protective immunity against experimental infection with malaria using synthetic peptides. Nature 1987;328:629–632.

440 Patarroyo ME, Amador R, Clavijo P, *et al.* A synthetic vaccine protects humans against challenge with asexual

blood stages of *Plasmodium falciparum* malaria. Nature 1988; 332:158–161.

441 Holder AA, Freeman RR. The three major antigens on the surface of *Plasmodium falciparum* merozoites are derived from a single high molecular weight precursor. J Exp Med 1984;160:624–629.

442 Lyon JA, Geller RH, Haynes JD, Chulay JD, Weber JL. Epitope map and processing scheme for the 195 000-dalton surface glycoprotein of *Plasmodium falciparum* merozoites deduced from cloned overlapping segments of the gene. Proc Natl Acad Sci USA 1986;83:2989–2993.

443 Lyon LA, Haynes JD, Diggs CL, Chulay JD, Haidaris CG, Pratt-Rossiter J. Monoclonal antibody characterization of the 195-kilodalton major merozoite glycoprotein of *Plasmodium falciparum* malaria schizonts and merozoites: identification of additional processed products and a serotype-restricted repetitive epitope. J Immunol 1987;138: 895–901.

444 Holder AA, Lockyer MJ, Odink KG, *et al.* Primary structure of the precursor to the three major surface antigens of *Plasmodium falciparum* merozoites. Nature 1985;317:270–273.

445 Mackay M, Goman M, Bone N, *et al.* Polymorphism of the precursor for the major surface antigens of *Plasmodium falciparum* merozoites: studies at the genetic level. EMBO J 1985;4:3823–3829.

446 Weber JL, Leininger WM, Lyon J. Variation in the gene encoding a major merozoite antigen of the human malaria parasite *Plasmodium falciparum*. Nucleic Acids Res 1986;14: 3311–3323.

447 Peterson MG, Coppel RL, McIntyre P, *et al.* Variation in the precursor to the major merozoite surface antigens of *Plasmodium falciparum*. Mol Biochem Parasitol 1988;27:291–302.

448 Pirson PJ, Perkins ME. Characterization with monoclonal antibodies of a surface antigen of *Plasmodium falciparum* merozoites. J Immunol 1985;134:1946–1951.

449 McBride JS, Newbold CI, Anand R. Polymorphism of a high molecular weight schizont antigen of the human malaria parasite *Plasmodium falciparum*. J Exp Med 1985; 161:160–180.

450 Gentz R, Certa U, Takacs B, *et al.* Major surface antigen p190 of *Plasmodium falciparum*: detection of common epitopes present in a variety of plasmodia isolates. EMBO J 1988;7:225–230.

451 Tanabe K, Mackay M, Gorman M, Scaife JG. Allelic dimorphism in a surface antigen gene of the malaria parasite *Plasmodium falciparum*. Mol Biol 1987;195:273–287.

452 Peterson MG, Coppel RL, Moloney MB, Kemp DJ. Third form of the precursor to the major merozoite surface antigens of *Plasmodium falciparum*. Mol Cell Biol 1988;8: 2664–2667.

453 Crisanti A, Muller HM, Hilbich C, *et al.* Epitopes recognized by human T cells map within the conserved part of the GP190 of *P. falciparum*. Science 1988;240:1324–1326.

454 Sinigaglia F, Takacs B, Jacot H, *et al.* Nonpolymorphic regions of p190, a protein of the *Plasmodium falciparum* erythrocytic stage, contain both T and B cell epitopes. J

Immunol 1988;140:3568–3572.

455 Holder AA, Freeman RR. Characterization of a high molecular weight protective antigen of *Plasmodium yoelii*. Parasitology 1984;88:211–219.

456 Holder AA, Freeman RR, Newbold CI. Serological cross-reaction between high molecular weight proteins synthesized in blood schizonts of *Plasmodium yoelii*, *Plasmodium chabaudi*, and *Plasmodium falciparum*. Mol Biochem Parasitol 1983;9:191–196.

457 Burns JM Jr, Daly TM, Vaidya AB, Long CA. The 3' portion of the gene for a *Plasmodium yoelii* merozoite surface antigen encodes the epitope recognized by a protective monoclonal antibody. Proc Natl Acad Sci USA 1988;85:602–606.

458 Daly TM, Burns JM Jr, Long CA. Precursor to the major merozoite surface antigen of *Plasmodium yoelii*: cloning and sequencing of the middle 1.9-kb region. Mol Biochem Parasitol 1989;36:283–286.

459 Lewis AP. Cloning and analysis of the gene encoding the 230-kilodalton merozoite surface antigen of *Plasmodium yoelii*. Mol Biochem Parasitol 1989;36:271–282.

460 Burns JM Jr, Marjarian WR, Young JF, Daly TM, Long CA. A protective monoclonal antibody recognizes an epitope in the carboxyl-terminal cysteine-rich domain in the precursor of the major merozoite surface antigen of the rodent malarial parasite, *Plasmodium yoelii*. J Immunol 1989;143:2670–2676.

461 Burns JM Jr, Parke LA, Daly TM, Cavacini LA, Weidanz WP, Long CA. A protective monoclonal antibody recognizes a variant specific epitope in the precursor of the major merozoite surface antigen of the rodent malarial parasite *Plasmodium yoelii*. J Immunol 1989;142:2835–2840.

462 Perrin LH, Ramirez E, Lambert PH, Meischer PA. Inhibition of *P. falciparum* growth in human erythrocytes by monoclonal antibodies. Nature 1981;289:301–303.

463 Perrin LH, Merkli B, Gabra MS, Stocker JW, Chizzolini C, Richle R. Immunization with a *Plasmodium falciparum* merozoite surface antigen induces partial immunity in monkeys. J Clin Invest 1985;75:1718–1721.

464 Campbell GH, Miller LH, Hudson D, Franco EL, Andrysiak PM. Monoclonal antibody characterization of *Plasmodium falciparum* antigens. Am J Trop Med Hyg 1984;33:1051–1054.

465 Holder AA, Freeman RR, Uni S, Aikawa M. Isolation of a *Plasmodium falciparum* rhoptry protein. Mol Biochem Parasitol 1985;14:293–303.

466 Howard RF, Stanley HA, Campbell GH, Reese RT. Proteins responsible for a punctate fluorescence pattern in *Plasmodium falciparum* merozoites. Am J Trop Med Hyg 1984;33:1055–1059.

467 Siddiqui WA, Tam LQ, Kan S, *et al.* Induction of protective immunity to monoclonal antibody-defined *Plasmodium falciparum* antigens requires strong adjuvant in *Aotus* monkeys. Infect Immun 1986;52:314–318.

468 Copper JA, Ingram LT, Bushell GR, *et al.* The 140/130/105 kilodalton protein complex in the rhoptries of *Plasmodium falciparum* consists of discrete polypeptides. Mol Biochem Parasitol 1988;29:251–260.

469 Roger N, Dubremetz JF, Delplace P, Fortier B, Tronchin G, Vernes A. Characterization of a 225 kilodalton rhoptry protein of *Plasmodium falciparum*. Mol Biochem Parasitol 1988;27:135–142.

470 Perlmann H, Berzins K, Wahlgren M, *et al.* Antibodies in malaria sera to parasite antigens in the membrane of erythrocytes infected with early stages of *Plasmodium falciparum*. J Exp Med 1984;159:1686–1704.

471 Brown GV, Culvenor JG, Crewther PE, *et al.* Localization of the ring-infected erythrocyte surface antigen (RESA) of *Plasmodium falciparum* in merozoites and in ring-infected erythrocytes. J Exp Med 1985;162:774–779.

472 Wahlin B, Wahlgren M, Perlmann H, *et al.* Human antibodies to a Mr 155 000 *Plasmodium falciparum* antigen efficiently inhibit merozoite invasion. Proc Natl Acad Sci USA 1984;81:7912–7916.

473 Cowman AF, Coppel RL, Saint RB, *et al.* The ring-infected erythrocyte surface antigen (RESA) polypeptide of *Plasmodium falciparum* contains two separate blocks of tandem repeats encoding antigenic epitopes that are naturally immunogenic in man. Mol Biol Med 1984;2:207–221.

474 Collins WE, Anders RF, Pappaioanou M, *et al.* Immunization of *Aotus* monkeys with recombinant proteins of an erythrocyte surface antigen of *Plasmodium falciparum*. Nature 1986;323:259–262.

475 Kabilan L, Troye-Blomberg M, Perlmann H, *et al.* T-cell epitopes in Pf155/RESA, a major candidate for a *Plasmodium falciparum* malaria vaccine. Proc Natl Acad Sci USA 1988; 85:5659–5663.

476 Rzepczyk CM, Ramasamy R, Ho PCL, *et al.* Identification of T epitopes within a potential *Plasmodium falciparum* vaccine antigen. A study of human lymphocyte responses to repeat and nonrepeat regions of Pf155/RESA. J Immunol 1988;141:3197–3202.

477 Lew AM, Langford CJ, Pye D, Edwards S, Corcoran L, Anders RF. Class II restriction in mice to the malaria candidate vaccine ring infected erythrocyte surface antigen (RESA) as synthetic peptides or as expressed in recombinant vaccinia. J Immunol 1989;142:4012–4016.

478 Gabriel JA, Holmquist G, Perlmann H, Berzins K, Wigzell H, Perlmann P. Identification of a *Plasmodium chabaudi* antigen present in the membrane of ring stage infected erythrocytes. Mol Biochem Parasitol 1986;20:67–75.

479 Wanidworanum C, Barnwell JW, Shear HL. Protective antigen in the membranes of mouse erythrocytes infected with *Plasmodium chabaudi*. Mol Biochem Parasitol 1987;25: 195–201.

480 Snounou G, Viriyakosol S, Holmquist G, Berzins K, Perlmann P, Brown KN. Cloning of genomic fragment from *Plasmodium chabaudi* expressing a 105 kilodalton antigen epitope. Mol Biochem Parasitol 1988;28:153–162.

481 Kilejian A. Characterization of a protein correlated with the production of knob-like protrusions on the membranes of erythrocytes infected with *Plasmodium falciparum*. Proc Natl Acad Sci USA 1979;76:4650–4655.

482 Kilejian A. The biosynthesis of the knob protein and a 65 000 dalton histidine-rich polypeptide of *Plasmodium falciparum*. Mol Biochem Parasitol 1984;12:185–194.

483 Leech JH, Barnwell JW, Miller LH, Howard RJ. Identification of a strain-specific malarial antigen exposed on the surface of *Plasmodium falciparum* infected erythrocytes. J Exp Med 1984;159:1567–1575.

484 Aley SB, Sherwood JA, Howard RJ. Knob-positive and knob-negative *Plasmodium falciparum* differ in expression of a strain-specific malarial antigen on the surface of infected erythrocytes. J Exp Med 1984;160:1585–1590.

485 Greunberg J, Sherman IW. Isolation and characterization of the plasma membrane of human erythrocytes infected with the malarial parasite *Plasmodium falciparum*. Proc Natl Acad Sci USA 1983;80:1087–1091.

486 Stanley HA, Reese RT. *Plasmodium falciparum* polypeptides associated with the infected erythrocyte plasma membrane. Proc Natl Acad Sci USA 1986;83:6093–6097.

487 Coppel RL, Culvenor JG, Bianco AE, *et al.* Variable antigen associated with the surface of erythrocytes infected with mature stages of *Plasmodium falciparum*. Mol Biochem Parasitol 1986;20:265–277.

488 Taylor DW, Parra M, Chapman GB, *et al.* *Plasmodium falciparum* histidine rich protein Pf HRP1: its subcellular localization to the host erythrocyte membrane under knobs using a monoclonal antibody. Mol Biochem Parasitol 1987; 25:165–174.

489 Howard RJ, Lyon JA, Uni S, *et al.* Transport of a Mr 300 000 *Plasmodium falciparum* protein (Pf EMP2) from the intraerythrocytic asexual parasite to the cytoplasmic face of the host cell membrane. J Cell Biol 1987;103:1269–1277.

490 Roberts DD, Sherwood JA, Spitalnik SL, *et al.* Thrombospondin binds falciparum malaria parasitized erythrocytes and may mediate cytoadherence. Nature 1985;318:64–66.

491 Udomsangpetch R, Carlsson J, Wahlin B, *et al.* Reactivity of the human monoclonal antibody 33G2 with repeated sequences of three distinct *Plasmodium falciparum* antigens. J Immunol 1989;142:3620–3626.

492 Pollack S, Fleming J. *Plasmodium falciparum* takes up iron from transferrin. Br J Haematol 1984;58:289–293.

493 Rodriguez MH, Jungery M. A protein on *Plasmodium falciparum* infected erythrocytes functions as a transferrin receptor. Nature 1986;324:388–391.

494 Haldar K, Henderson CL, Cross GAM. Identification of the parasite transferrin receptor of *Plasmodium falciparum* infected erythrocytes and its acylation via 1,2-diacyl-sn-lycerol. Proc Natl Acad Sci USA 1986;83:8565–8569.

495 Harte PG, Rogers NC, Targett GAT. Role of T-cells in preventing transmission of rodent malaria. Immunology 1985;56:1–7.

496 Carter R, Bushell G, Saul A, Graves PM, Kidson C. Two apparently nonrepeated epitopes on gametes of *Plasmodium falciparum* are targets of transmission-blocking antibodies. Infect Immun 1985;50:102–106.

497 Kaslow GC, Quakyi IA, Syin C, *et al.* A vaccine candidate from the sexual stage of human malaria that contains EGF-like domains. Nature 1988;333:74–76.

498 Kaslow DC, Syin C, McCutchan TF, Miller LH. Comparison of the primary structure of the 25 kDa ookinete surface antigens of *Plasmodium falciparum* and *Plasmodium gallinaceum* reveal six conserved regions. Mol Biochem Parasitol 1989;33:283–288.

499 Graves PM, Carter R, Burket TR, Rener J, Kaushal DC,

Williams JL. Effects of transmission-blocking monoclonal antibodies on different isolates of *Plasmodium falciparum*. Infect Immun 1985;48:611–616.

500 Kaslow DC, Quakya IA, Keister DB. Minimal variation in a candidate from the sexual stage of *Plasmodium falciparum*. Mol Biochem Parasitol 1989;32:101–104.

501 Good MF, Miller LH, Kumar S, *et al.* Limited immuno- logical recognition of critical malaria vaccine candidate antigens. Science 1988;242:574–577.

502 Taverne J, Bate CAW, Kwiatkowski D, Jakobsen PH, Play- fair JHL. Two soluble antigens of *Plasmodium falciparum* induce tumor necrosis factor release from macrophages. Infect Immun 1990;58:2923–2928.

15 Babesiosis

Robert J. Dalgliesh

Babesia spp. are tick-transmitted protozoan parasites of wild and domesticated animals in most regions of the world. Among over 100 recognized species [1], at least 18 are pathogenic to domestic and laboratory animals, and to humans [2]. Acute babesiosis is a malaria-like infection, characterized by fever, hemolytic anemia, and hemoglobinuria. Control of parasite multiplication in erythrocytes is essential to the host's well-being. Surviving animals acquire long-lasting specific immunity.

Babesiosis has greatest economic importance in cattle [3]. Bovine babesiosis has historical significance as the first protozoal infection of vertebrates for which an arthropod vector was demonstrated, and for which practical vaccination was developed and widely applied. Humans are susceptible to infection with some species of *Babesia*, but in most cases disease has been related to variable degrees of immunoincompetence in the patient [4].

Advances in immunochemistry and other technology, such as DNA hybridization and monoclonal antibody production, have enabled parasite biology and host–parasite relationships to be studied at increasingly basic levels.

THE PARASITE

Classification

The genus *Babesia* Starcovici is a member of the family Babesiidae Poche, order Piroplasmorida Wenyon, in the phylum Apicomplexa [1]. Major characteristics of *Babesia* include transovarial transmission in the tick, transmission between vertebrate hosts by ticks, and residence within erythrocytes in the vertebrate host, where it divides without forming pigment from host cell hemoglobin. The infecting parasite is called a piroplasm.

The taxonomic positions of *Babesia equi* and *Babesia microti* have been questioned because certain stages of the life-cycle resemble the related piroplasm, *Theileria* [5,6]. Uilenberg [6] suggested that the rodent parasite, *Babesia rodhaini*, may also be a *Theileria*. Nevertheless, all three species are considered as *Babesia* in this review. To date, genomic studies on *Babesia* have been confined to several species (see pp. 354–355, 360). Comparisons between *Babesia* and *Theileria* at the DNA level may clarify taxonomic relationships.

Life-cycles

In the tick

Life-cycles and transmission of *Babesia* by ticks were recently reviewed [5,7]. Each species of *Babesia* is relatively specific for its tick vector. Developmental changes and multiplication occur in mother ticks, in eggs, and in the resultant progeny. Infective sporozoites are released in the saliva of feeding ticks. Environmental temperature affects the development and survival of sporozoites in infected ticks [8,9].

The long-suspected existence of sexual stages in *Babesia* is now confirmed, with the demonstration of syngamy in *Babesia canis* [5] and *B. microti* [10] by electron microscopy, and in *Babesia divergens* by cytophotometric measurement of DNA [11]. Gametocytes developing in erythrocytes within the vertebrate host give rise to gametes that fuse in the lumen of the tick gut.

Transmission

The tick stage which transmits *Babesia* varies between babesial species [5,7]. Infection can be transmitted artificially with blood containing living parasites and with extracts prepared from either feeding or "heat-stimulated" infected ticks (see de Vos *et al*. [2]).

In the vertebrate

Babesia develops and multiplies within erythrocytes. *Babesia microti* may have pre-erythrocytic stages in lymphocytes [12]. Merozoites and trophozoites are the recognized blood forms. Single merozoites invade erythrocytes, transform to the ultrastructurally different trophozoite, and grow and divide into two or, less frequently, four or more oval or pear-shaped bodies. The characteristic location, size, and shape of intra-erythrocytic stages of different *Babesia* species in stained blood films aids diagnosis [5].

In vitro culture

The finding that *Babesia bovis* is amenable to *in vitro* culture in erythrocytes [13] provided a new investigative technique, and has been applied in practical vaccination (see p. 372). The microaerophilous stationary phase method [13] has subsequently been applied, with modifications, to the culture of other economically important species of *Babesia* (reviewed by Canning *et al.* [14]).

Metabolic aspects

Babesia resembles most other parasitic protozoa in being unable to synthesize purines *de novo*. Evidence for dependence on purine salvage pathways, based on the marked uptake of the nucleic acid precursors adenosine and hypoxanthine [15–18], was supported by the demonstration of enzymes which catalyze purine salvage [19]. Detection of radioactivity in the cytoplasm and nucleus of incubated parasites indicated the incorporation of ^{3}H-hypoxanthine in both RNA and DNA [16].

Uptake of the pyrimidines thymidine and uridine, and the pyrimidine precursor orotic acid, varied between species of *Babesia* [15–17]. The variability was tentatively ascribed to differences in permeability between host erythrocytes or in the parasites [16]. Of the six enzymes required for *de novo* pyrimidine biosynthesis, four were detected in *Babesia hylomysci* [20], five in *B. rodhaini* [21], and all six in both *B. bovis* and *Babesia bigemina* [22].

Most of the other enzymes identified in *Babesia* (see Callow & Dalgliesh [23]) are utilized in carbohydrate metabolism, glucose being an important energy source. Barry [24,25] examined aspects of glucose and energy metabolism in *B. rodhaini* and *B. bovis* incubated *in vitro* in relation to parasite survival, and concluded that adenylate energy charge was a useful indicator of infectivity. A comparison of the composition of amino acids produced by *B. rodhaini* and *P. berghei* [26] indicated that hemoglobin was the main source for both parasites. Proteolytic enzymes capable of hemoglobin digestion were detected in bovine *Babesia* [27]. Lactate dehydrogenase, glucose phosphate isomerase and glutamate dehydrogenase were identified in both the rodent species *B. rodhaini* [28] and the bovine species *B. bigemina* [29]. A calcium-dependent protein kinase that apparently phosphorylates a 40 kD parasite protein within the parasite membrane was demonstrated in *B. bovis* [30], although its role remains uncertain.

In *B. bovis* infections at least, parasite-induced changes in the host cell membrane facilitate the entry of glucose [31] and nucleosides [32], apparently for parasite metabolism.

Variation in parasite populations

Antigenic variation between and within strains of the same species of *Babesia* has been recognized for several decades. The findings were obtained mainly from serologic studies and crossimmunity tests in animals (reviewed by de Vos *et al.* [2] and Callow & Dalgliesh [23]). Studies at the molecular level provide supportive evidence for variation. It appears that the parasites collected from naturally infected animals on a unique occasion are a mixture of an unknown number of subpopulations differing genotypically and in biologic characteristics. Selection for different subpopulations may occur during syringe-passage in animals, resulting in host responses different from those caused by the original isolate (see p. 358). Change induced by passaging has particular significance since much published work on the biology of *Babesia* and its host relationships is based on observations made with multipassaged laboratory strains of the parasite, often derived from a single isolate. Re-evaluation of some earlier findings will be necessary in the light of this new knowledge of the parasite.

Much of the information on variation in antigenic proteins and DNA in *Babesia* is derived from studies on a particular strain of *B. bovis*, designated the K strain. The K strain was isolated in Australia from a naturally infected healthy carrier cow. The strain was syringe-passaged in a series of splenectomized calves to reduce virulence, and has been widely used for practical immunization [33]. Stabilates prepared at isolation and during passage were the source of parasites for many studies; derivatives of initial stabilates were coded Kv, and those of stabilates following reduction in virulence,

Ka. Similar coding was used for virulent and avirulent derivatives of other strains (e.g., Cv, Ca) used for comparative purposes in molecular studies.

Proteins and antigens

Protein antigens in geographically separated isolates of *Babesia* have been analyzed by biosynthetic labeling and immunoprecipitation techniques [34–36]. Isolates within each species share many common antigens. Strain-specific immunodominant proteins were also identified in *B. bovis* [35], and peptide mapping indicated absence of amino acid homology between these antigens. Antigenic differences in *B. divergens*, both among isolates collected from cattle within the same herd and among those from widely separated animals, were indicated by variable reactivity to monoclonal antibodies (MAB) [37].

Using the *B. bovis* Ka derivative studied by Kahl *et al.* [35], Gill *et al.* [38] examined molecular characteristics of "cloned lines," derived by a limiting dilution procedure

in splenectomized calves. The "cloned lines" differed from each other and from the original population in the expression of a small number of polypeptides and minor antigens. Differing antibody specificities of calves used to grow the lines were also detected.

Variation in the genome

Comparisons among strains of *B. bovis* have revealed considerable variation at the genomic level. Cowman *et al.* [39] constructed a cDNA library from poly(A)$^+$ RNA of Ka *B. bovis* and used clones designated pK4, pK5, and pK6 as probes to examine variability in the RNA and DNA of *B. bovis*. Differences in the sizes and amount of RNA transcripts were detected among strains derived from different isolates (K and C *B. bovis* strains were similar but differed from L and S). Some cDNA probes also revealed polymorphism in genomic DNA among *B. bovis* strains (Fig. 15.1). Characterization of the rRNA genes of six isolates of *B. bovis* by restriction mapping and DNA hybridization indicated that whereas

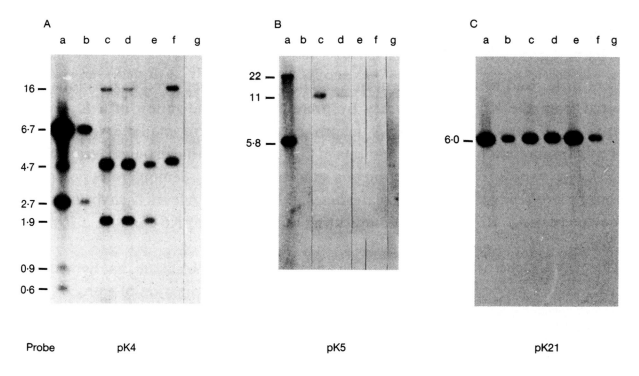

Fig. 15.1 Variability in genomic DNA among *Babesia bovis* strains K, C, and L shown by southern hybridization. DNA (1.0 µg) of each strain was digested with PvuII or EcoR1, fractionated on nitrocellulose, and hybridized with cDNA probes pK4, pK5, and pK21. Tracks a–g in each panel contain Ka, Kv, Ca, Cv, Cnt, Lv, and bovine thymus; a, v, and nt were avirulent, virulent, and nontick-transmissible derivatives of the strains, respectively. Probe pK21 hybridizes with a 6 kb band in all strains and was used to assess track loading. (From Cowman *et al.* [38].)

the gene units were highly conserved, polymorphic flanking regions allowed clear discrimination among the isolates [40] (B.P. Dalrymple, unpublished work). Genomic analysis of 10 *B. bovis* isolates by restriction fragment length polymorphism also showed that most had distinctive patterns (B.P. Dalrymple, W. Jorgensen, A.J. de Vos, & I.G. Wright, unpublished work). The findings reveal an impressive capability for genetic change in *B. bovis*, further evidence for which is presented on p. 359.

Biologic properties

Variation in biologic effects, measured in the host, were demonstrated among putatively cloned populations of *Babesia*. These "clones" were produced by limiting dilution techniques for *B. rodhaini* [41] and *B. bovis* [38] in mice and splenectomized calves, respectively, and also for *B. bovis* in *in vitro* culture [42].

The virulence and immunogenicity of five "clones" from Ka strain *B. bovis* were compared with those of the parent strain [38,43]. Three "clones" and the parent strain were similarly mild in virulence; two "clones" were more virulent, one being markedly so. Heterologous challenge of immunized cattle indicated that differing levels of protective immunity, unrelated to virulence, were induced by the "clones" and none was as protective as the parent strain. Mixing "clones" before immunization did not restore full immunogenicity [43]. None of eight "clones," differing in virulence, were infective for the natural tick vector, *Boophilus microplus*, whereas the parent Ka strain was readily transmissible by the tick [44]. Restoration of high virulence in a relatively avirulent "clone" by syringe-passage in intact cattle did not restore infectivity for ticks [44], thus confirming that the two biologic characteristics are unrelated. Putative clones of a different strain of *B. bovis* were distinguished by different rates of growth *in vitro* [42]. A comparatively fast growing "clone" was apparently less virulent than the parent population [54].

An important step will be to correlate biologic properties of *Babesia*, particularly virulence, with reliable genetic markers. Progress made in this regard includes the detection of a *B. bovis* gene transcribed predominantly in avirulent parasites [39,46] and the identification of a *B. microti* gene apparently associated with virulence [47].

HOST–PARASITE RELATIONSHIPS

Host range and specificity

Familial relationships apparently determine host ranges for *Babesia* in domesticated and larger wild animals. Attempts to produce significant infections with economically important *Babesia* spp. in abnormal hosts have generally been unsuccessful. An exception is *B. divergens*. Hosts susceptible to this bovine parasite include deer, gerbils, splenectomized rats, and splenectomized primates, including humans [2,23].

Babesia spp. of rodents are less host specific. Medical interest in *B. microti* has increased because humans are susceptible to natural infection. *Babesia microti* normally infects at least eight genera of wild rodents. Laboratory infections were established in rats, hamsters, gerbils, and mice, as well as several species of nonhuman primates [2,23].

Epidemiology

Babesia usually coexists with the normal host without causing significant disease. Callow & Dalgliesh [23] suggested that *Babesia*'s prominence as a pathogen stems from human activity in selecting animals for profit or transporting them to unfamiliar environments. A topical example is the world cattle trade, in which European *Bos taurus* cattle, selected for high productivity in temperate countries of origin, are imported into tropical countries, where babesiosis and other tick-borne infections have taken a heavy toll [48,49]. In some countries, disease in imported cattle provided the first evidence for the existence of pathogenic *B. bovis* and *B. bigemina*, which have apparently developed harmonious evolutionary relationships with indigenous *Bos indicus* cattle and the tick vector.

Epidemiologic study has been concentrated on the few species responsible for economic loss in domestic animals, and more recently on *B. microti* because of its prevalence in humans in the northeast of North America.

Qualitative and quantitative laboratory studies on factors affecting the tick transmission of several species of *Babesia* have been reviewed [7]. Interpretation of field transmission based on laboratory studies is difficult for those species with life-cycles in two- and three-host ticks. Recent concepts of field transmission and its relationship to disease patterns in animals derive mainly from studies with *B. bovis* and *B. bigemina* and their one-host tick vector, *Boophilus*. Integration of information obtained from many field and laboratory studies (see Callow [33,50], Joyner & Donnelly [51], and Mahoney [52] for reviews) provides explanations for disease outbreaks in endemic areas. Disease occurs in these areas when hosts, susceptible and apparently not well adapted to a species of *Babesia*, are subjected to fluctuating levels of infection by infected ticks. Animals exposed continuously from birth may not experience disease because natural immunization can occur while maternally derived immunity and subsequent nonspecific protection are present (p. 365). Mathematical models have been used to study the complex relationships among *B. bovis*, *B. bigemina*, their tick vectors, and cattle [53,54].

Interest in *B. microti* infection as a significant zoonosis led to studies of its epidemiology in North America. The incidence of *B. microti* infection in humans steadily increased after the first recorded case in 1969 [55]. A distribution of the disease along the northeastern coast of the USA has been defined. Both *B. microti* and its tick vector (*Ixodes dammini*) have a broad host range, resulting in a complex epidemiology. In the endemic area, immature stages of the three-host tick infest a variety of rodents and other mammals (including humans) but show preference for mice, which act as reservoir hosts for *B. microti* infection. The adult tick, however, prefers the white-tailed deer as a host. Spielman *et al.* [56] correlated the rising incidence of human babesiosis with a growing population of white-tailed deer, which has increased the prevalence of the vector. The relationship is an interesting example of an increased incidence of disease, probably caused by an increasing animal population, and not by a reservoir for infection.

General characteristics of infection

In babesiosis, infection is influenced by parasite factors, such as virulence and dose (number of infecting parasites), and by host factors, such as breed, age, and immunologic competence. Features of babesial infections are described here to facilitate later discussion of the host response.

Natural infections

The severity may vary from hyperacute to subclinical in apparently similar animals exposed to infected ticks. Signs of disease usually begin 7–21 days after the vector commences to feed and generally coincide with detectable parasitemia in peripheral blood. Fever, anorexia,

depression, anemia, and hemoglobinuria are characteristic of acute infection. Nervous signs may develop in terminal infections with some *Babesia*. Control of the acute phase and recovery are indicated by a slowing of the rate of parasite increase in the peripheral blood. Acute signs of disease subside when the parasitemia falls. Patent parasitemias may last as long as 10 days in severe nonfatal infections. Parasites may reach detectable levels again on more than one occasion during the next several weeks and then persist at lower levels for longer periods. Clinical signs rarely occur during parasite recrudescences. Duration of babesial infections in natural hosts varies for different parasites and different hosts (see Callow & Dalgliesh [23]). For example, *Bos indicus* cattle eliminate *B. bovis* more readily than do *Bos taurus* cattle; *Bos taurus* cattle eliminate *B. bigemina* more readily than *B. bovis*.

Artificially induced infections in natural hosts

Infections induced artificially, by injecting parasitized blood or infected tick extracts, are similar to those following natural transmission. In blood-transmitted infections, an inverse relationship between prepatent period and parasite dose allows manipulation of the prepatent period. Splenectomy of the host and intravenous inoculation of the parasite dose allow the most precise definition of the relationship, but even with subcutaneous inoculation of intact cattle the prepatent period shortens as the dose of *B. bovis* is increased [50]. Parasite dose may also influence the severity of infection (see Callow & Dalgliesh [23]).

Accelerated development of infective babesial parasites produced by heating unfed ticks has been used to provide inocula for immunologic studies on *B. bovis* in cattle [57–59]. A disadvantage is that the parasite dose given cannot be accurately quantified.

Infections in unnatural hosts

The severity and course of infection in artificial host–parasite systems vary considerably with the stage of adaptation [60] (see p. 360). With the other known parasite and host variables, this complicates interpretation of findings. Clinical signs in small laboratory animals are generally similar to those in naturally infected hosts, although parasitemia is usually much higher before animals appear sick. Also, the severity of infection induced by identical inocula within a group of animals is more consistent, presumably because of greater genetic uniformity among laboratory animals. Inverse relationships between parasite dose and responses, such as prepatent period, time to reach 1% parasitemia or peak of infection, and survival time of the host, are useful in comparing the infectivity of different inocula (e.g., Dalgliesh *et al.* [61]).

Babesia infections persist in laboratory rats and mice for varying periods (see Callow & Dalgliesh [23]). *Babesia microti* parasitemias in hamsters were high enough to infect ticks for at least 7 months [62]. But transient parasitemias followed by rapid elimination of *B. divergens* occurred in individual gerbils [63,64] and splenectomized rats [60]. This is of interest because these hosts were resistant to challenge, indicating a sterile immunity.

Parasite–host cell relationships

Penetration of erythrocytes

Babesia resembles *Plasmodium* in that parasites enter erythrocytes without disrupting the cell membrane. Penetration is an active process, with five initial phases [5,65]:

1 contact between the merozoite and erythrocyte;
2 orientation of the parasite's apical pole towards the erythrocyte;
3 membrane "fusion" between merozoite and erythrocyte;
4 release of contents of rhoptries;
5 changes in the erythrocyte membrane at the site of attachment, resulting in its invagination and entry of the merozoite.

The invaginated membrane then fuses at its external limits to form a parasitophorous vacuole [65]. The host's cell membrane within the vacuole disintegrates, bringing the parasite in direct contact with the cell's cytoplasm.

Molecular events associated with attachment and penetration are poorly understood. Studies in rats strongly suggested that *B. rodhaini* required activated complement and the presence of C3b receptors on the parasite or rat red cells, or both, for erythrocyte invasion [66]. In another study, however, complement did not facilitate *B. rodhaini* infection in BALB/c mice [67]. Studies of homology between the two parasite strains, and comparisons in similar hosts, are needed to clarify this important issue. There was no apparent requirement for complement by *B. bovis* for the parasite to develop *in vitro* [68].

From ultrastructural studies of *B. microti*, an apparently adhesive surface coat surrounding extracellular merozoites is "shedded" during the entry process [65]. Soluble antigens detected in the supernatant of *in vitro* cultures of *B. bovis* may be derived from this merozoite surface coat [69]. Reactivity of MAB with both parasite and erythrocyte antigens indicated that outer membrane components of *B. rodhaini* [70] and *B. bigemina* [71] could be of host cell origin. Sequence homology in parasite and erythrocyte proteins was also demonstrated [70,72] but no functional relationship between the proteins has been shown [72]. The findings support McHardy's [73] suggestion that *Babesia* might incorporate host antigen. Further study of merozoite surface coats and their relationship with host cells is warranted, surface antigens of parasites being prime targets for immunologic attack.

Change in the host cell

The morphology and behavioral characteristics of host erythrocytes are altered. Viewed microscopically, infected cells may differ from uninfected cells in size, shape, and staining characteristics (see Callow & Dalgliesh [23]). Other changes in erythrocytes attributed to babesial infection include increased permeability to glucose [31,74], increased sedimentation rate [75], variable changes in osmotic fragility [76,77], increased susceptibility to perforation by electric pulses [78], and nucleoside permeation sites induced in the cell membrane [32]. Structural alterations in infected cells have been observed [79–81], and with some species of *Babesia* these may result in selective accumulation of the cells within the host's microcirculation (see p. 363).

Radiolabeling detected new erythrocyte surface proteins, of molecular weight 60 kD and higher, resulting from infection with *B. bovis* [82] and *B. rodhaini* [83]. In both studies, normal erythrocyte surface proteins were lost. The new proteins may have been host and/or parasite-derived molecules absorbed from serum, or cell membrane proteins normally inaccessible to lactoperoxidase-catalyzed radioiodination but exposed during infection [83]. Other changes detected in *B. bovis*-infected erythrocytes include increased malonyldialdehyde and total lipid, decreased vitamin E, sialic acid, and adenosine triphosphate, and the presence of phosphatidyl serine on the outer surface of the cell membrane [84,85]. These lesions may contribute to the increased membrane rigidity and enhanced cytoadherance characteristic of *B. bovis*-infected erythrocytes [84,85].

Different parasites may affect their host cells differently. *Babesia bigemina*-infected cells were osmotically fragile, but *B. bovis*-infected cells decreased in osmotic fragility [77]. *Babesia bovis*-infected cells and some uninfected cells from the same cattle became bound to heparin–sepharose columns, indicating altered membrane protein, but *B. begemina*-infected cells did not [86]. Differing properties of *B. bovis*- and *B. bigemina*-infected erythrocytes, particularly those relating to deformability and cytoadherence, may explain differences in the bovine host's response to the two parasites (see p. 363).

Host-related changes in *Babesia*

Host-induced changes in the parasite influence control of bovine babesiosis. As will be seen, the system of passaging *Babesia* can influence immunogenicity, virulence, and infectivity for tick vectors. As well as being of practical importance, the findings highlight the risk of extrapolating results obtained with parasites maintained in the laboratory to natural host–parasite relationships.

Change while in a single host

The two changes to consider are virulence and anti-

genicity. The relatively low virulence of bovine parasites from clinically recovered carriers compared to that of parasites from acutely infected cattle was first recorded last century and has long been exploited in practical vaccination (see Dalgliesh *et al.* [87]). Serology and cross-immunity testing indicated that *Babesia* is antigenically labile during chronicity in the vertebrate host (see Callow & Dalgliesh [23]) and, for *B. bovis* at least, reverts to a basic antigenic type following development in and transmission by the vector [88].

Effects of unnatural transmission

Babesia's potential for change during syringe-passage is exemplified by studies on *B. bovis* (Fig. 15.2). Continuous syringe-passage in cattle resulted in permanent loss of infectivity for the tick [98]. The effect is associated with abnormal morphology of parasites in the tick gut [97].

Virulence of *B. bovis* is reduced by rapid passage in splenectomized calves [94]. At least 10 *B. bovis* isolates in Australia, South America, and South Africa have been changed in this way [2]. Up to 23 passages may be

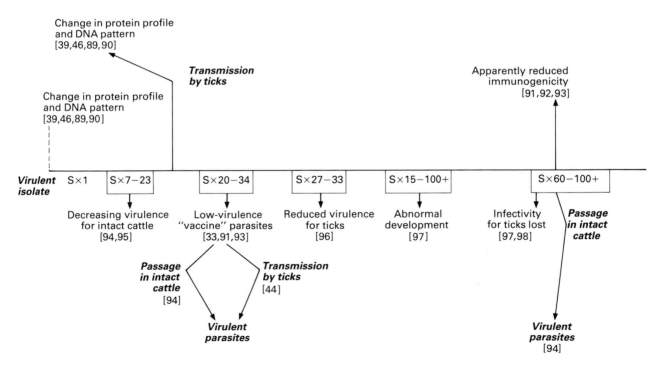

Fig. 15.2 A schematic summary of changes in *Babesia bovis* parasites induced by rapid syringe passage in splenectomized calves and by subsequent passage in intact cattle or ticks. The approximate number of syringe passages (Sx) during which changes were observed is shown along the horizontal line. Key references are shown in square brackets.

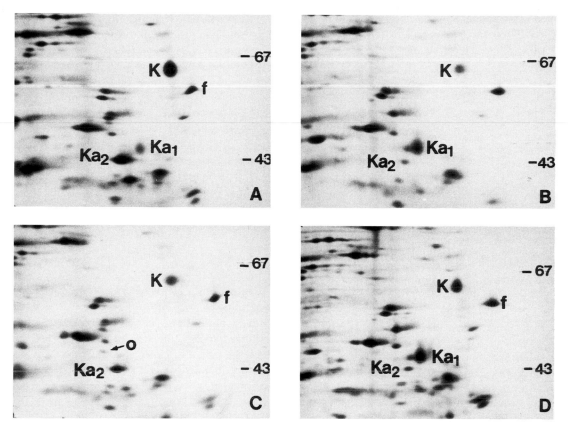

Fig. 15.3 Changes in protein profile of an avirulent (Ka) strain of *Babesia bovis* induced by various passage procedures. Protein patterns were prepared by ³⁵S-labeling and two-dimensional polyacrylamide gel electrophoresis. Panel A: typical pattern for K strain after 20 or more passages in splenectomized calves (Ka). Panel B: Ka after a single tick transmission. Panel C: Ka after serial syringe passage in three nonsplenectomized (intact) calves. Panel D: Ka after three intact passages and a single tick transmission. The letters identify particular proteins. (From Kahl *et al.* [89].)

required to reduce virulence. For unknown reasons, two *B. bovis* isolates were not affected by this form of passaging (R.J. Dalgliesh, W. McGregor, & L.T. Mellors, unpublished work; D.J. Weilgama, unpublished work). Reduced virulence in multipassaged *B. bovis* parasites is not stable. Reversion to virulence followed a single syringe-passage in intact cattle [94] and also transmission by the natural vector [97].

At the molecular level, syringe-passage of the K strain of *B. bovis* (see p. 353) in splenectomized calves resulted in subtle changes in highly acidic polypeptides, seen following ³⁵S-methionine labeling and two-dimensional electrophoresis, and greater expression of at least one dominant antigen (molecular weight 43 kD) [35]. The expression of this and another major acidic protein antigen (molecular weight 47.5 kD) was influenced to different degrees by subsequent passage in intact calves or ticks [89] (Fig. 15.3, proteins labeled Ka1, Ka2). Splenectomized calf passage of a different *B. bovis* strain (C) did not obviously increase expression of the 43 kD antigen, although the pattern of smaller polypeptides (30–35 kD) changed. Thus, protein changes occurred within both strains during passage but were dissimilar.

Both syringe-passage and tick transmission of K strain *B. bovis* induced change in patterns of immunoprecipitated antigens [89] and genomic DNA [39]. Using the cDNA probe pK4 (see p. 354), Cowman *et al.* [39] quantified EcoR1 fragments of DNA, and interpreted an increased abundance of 5.7 and 7.7 kb fragments in syringe-passaged parasites as being the result of selection for an initially minor subpopulation. Multipassaged parasites subsequently transmitted by ticks provided

Table 15.1 Relative amounts of certain DNA fragments, quantified by cDNA probe pK4 (see text), in K strain *Babesia bovis* before and after serial passage in splenectomized calves and tick transmission. (From Cowman *et al.* [39])

DNA fragment size (kb)	K strain derivative*		
	Kv	Ka	Kt
9.7	1.0[†]	1.7	0.6
7.7	_[‡]	0.7	–
5.7	–	0.9	0.1
3.35	0.8	1.7	0.3

* Kv and Ka, *B. bovis* isolate before and after, respectively, serial syringe passage in 22 splenectomized calves; Kt = Ka after a single transmission by *Boophilus microplus* ticks.
[†] The amount of hybridization of pK4 to the 9.7 kb band in Kv was designated as 1 and other values compared with it.
[‡] Level of hybridization not measurable.

only trace amounts of the fragments, suggesting selection of another subpopulation more closely resembling the ancestral isolate (Table 15.1).

Further analysis of the K strain characterized a highly polymorphic gene family within a locus designated BabR (*Babesia* rearranging) [90]. The locus provides a useful marker for detecting population heterogeneity. Sequence analysis identified genetic heterogeneity in multipassaged parasites, and indicated that the Ka strain comprised subpopulations differing in their BabR locus. Cowman *et al.* [90] suggested that putative BabR proteins may be antigenic and subject to selection by the host immune response. Using the same pK4 probe, Carson *et al.* [46] demonstrated passage-induced changes in DNA pattern prepared from parasites derived from the same K stock of *B. bovis*. The findings support the hypothesis that *B. bovis* isolates comprise virulent and avirulent subpopulations, the latter being selected during syringe passage in splenectomized calves [90,94]. Further work with clones is required to determine whether genotypic and phenotypic changes in *B. bovis* result from selection among subpopulations with different amounts of repetitive DNA, as findings with *Plasmodium* species [99,100] might suggest. Differing genetic expression in splenectomized and intact hosts has been suggested as a factor contributing to virulence changes in *B. bovis* [44]. In *B. rodhaini*, sequence diversity in a gene complex possibly analogous to the BabR locus of *B. bovis* [101] indicated that expression of the parasite's surface proteins may change in response to host effects.

Passage-induced change in other species of *Babesia* has not been studied intensively. Attempts to reduce virulence of *B. bigemina* [87] and *B. divergens* [102] by rapid passage in splenectomized calves were unsuccessful. Repeated passage of *B. divergens* in gerbils did not reduce its virulence for splenectomized calves [103]. Syringe passage of *B. bigemina* in either splenectomized or intact cattle altered its normal development in the vector [104]. Maintenance by regular tick transmission in splenectomized rather than intact cattle caused obvious morphologic changes in tick-ingested parasites [104]. Findings from life-cycle studies should therefore be critically evaluated in terms of the laboratory maintenance of the *Babesia* when this is not the natural method.

Adaptation

Adaptation of *Babesia* to unnatural hosts also provides examples of host-induced change in the parasite. The rodent species *B. rodhaini* and *B. hylomysci*, isolated from different species of African tree rats, developed virulence for laboratory mice and fatal infections are now the norm. Characteristics of *B. divergens* (human origin) infection changed during syringe passage in 123 gerbils, there being a 34% increase in maximum parasitemia before death of the host, a 31% increase in the proportion of multiple-infected cells, and differences in parasite morphology [63]. Similar results were obtained during multipassage of *B. divergens* (bovine origin) in splenectomized rats [60]. *Babesia divergens* parasites in laboratory animals are often larger than those in cattle [105].

HUMORAL RESPONSE

Humoral responses may be either beneficial or damaging to the host. Specific antibody against *Babesia* in naïve animals can usually be detected towards the peak of the parasitemia and remains as long as parasites persist in the host. Other chemical responses occur earlier in the infection and may be evident only during the first 1–2 weeks of severe infections. They are therefore classified for discussion as acute-phase responses.

Acute-phase responses

In 1957, Maegraith *et al.* [75] suggested that lesions in dogs with acute *B. canis* infections were indicative of nonspecific inflammation. Since then, support for this conclusion has accumulated from studies of other babesial infections, particularly *B. bovis* in cattle.

Table 15.2 Changes in plasma compounds and their biologic significance during acute *Babesia bovis* infection in cattle

Compound or activity	Direction of change	Known or suggested biologic significance of change	[References]
IgM, IgG (nonspecific)	Decrease	Immunosuppression	[106]
Conglutinin*	Decrease	Erythrocyte aggregation	[107]
Fibronectin	Decrease	Erythrocyte aggregation	[107]
C_3, $C_H50^\dagger$	Decrease	Complement activation	[107, 108]
Active kallikrein Kinin Kininogen	Increase Increase Decrease	Vascular dilatation and increased permeability	[109, 110]
Fibrinogen and related products Plasminogen‡	Increase Decrease	Coagulation disturbances and erythrocyte aggregation	[111–113]
α-Antitrypsin α-Macroglobulin	Increase Decrease	Antiprotease activity	[114, 115]
Malonyldialdehyde Vitamin E Sialic acid	Increase Decrease Decrease	Lipid peroxidation and/or erythrocyte aggregation	[84]

*Bovine nonantibody protein which reacts with C3b.
†Hemolytic complement titer.
‡Decrease in splenectomized cattle; no significant change in intact cattle.

Changes detected in proteins and other compounds in bovine blood during the development of acute infections are shown in Table 15.2. The patterns of change in proteins such as fibronectin, C3, fibrinogen, and α-antitrypsin are suggestive of the acute-phase response induced by various infections and other causes of inflammation or tissue damage [116,117]. Pronounced acute-phase responses are characteristic of severe *B. bovis* infections (see p. 363).

Studies of other species of *Babesia* are relevant. Depletion of whole complement, C2, C3, C4, and C5, but not properdin, occurred in rats infected with *B. rodhaini* [118], indicating activation of the classic pathway only. Activation of both classic and alternative pathways, however, occurred in *B. bovis*-infected cattle [108].

In *B. bigemina* infections in cattle, the hemolytic complement titer (C_H50) and C3 levels were reduced [119]. In other studies with *B. bigemina*, changes indicative of coagulation defects and kallikrein activation also occurred, but were much less pronounced than those observed in *B. bovis* infections and occurred later in the course of infection [120]. The plasma concentration of haptoglobin increased in *B. bigemina*-infected cattle [120]

but decreased in cattle infected with unspecified species of *Babesia* [121].

Increased levels of IgM, IgG, and CIq binding, and decreased C3 and C4 levels were recorded for "acute-phase sera" from humans with *B. microti* infection [4]. The findings may not be comparable with others presented here, as the duration of infection before collection of the serum was apparently unknown.

Responses during recovery

Blood chemistry begins to return to normal with the onset of clinical recovery. Some abnormalities may, however, persist well after the infection is controlled. For example, hemolytic complement activity (C_H50) remained low in *B. bovis*-infected cattle for about 2 weeks after parasitemia had declined [108]. C_H50 levels were below normal in humans tested months after acute *B. microti* infections, although other chronic disorders may have contributed [4].

Host responses not evident during the acute phase of infection may become apparent with recovery. Immunoconglutinin (IK), an autoantibody against fixed complement components C3b and C4b, was detected in the recovery phase of *B. bovis* infections in cattle [107]. It was suggested that IK in serum indicates the presence of soluble immune complexes in babesiosis [122] (see p. 364) and that IK assists in their clearance from the host circulation [107,122].

Animals become serologically positive to *Babesia* during the initial infection. In surviving animals, titers increase for some weeks thereafter. They may gradually diminish over long periods following a single infection [123–125]. A positive serologic test for *B. bovis* is evidence that an animal is infected. Titers rise following challenge [123,126], although this finding may depend on the serologic test used [126].

Increases in both IgM and IgG levels are detectable during initial infections with *B. bovis* and *B. bigemina* in cattle [106,127] and *B. microti* in mice [128,129]. IgM activity may precede IgG activity by several days [58]. In general, IgM levels are lower and decrease with latency of the infection. IgG1 activity, but not IgG2 activity, was demonstrated in *B. bovis* infections [58,59].

CELLULAR RESPONSE

The cellular pathology (including anemia) of babesiosis has been described for various species in previous reviews [23,52,130]. Counts of circulating leukocytes and platelets fall as the acute phase of infection develops and then increase to normal or elevated levels during recovery. Thrombocytopenia is attributable to coagulation defects. Leukocytopenia is due at least in part to diversion of immune cells to lymphoid tissue, particularly in the spleen.

Cell kinetics in lymphoid organs

Splenomegaly is a feature of acute babesiosis. Apart from removal by the spleen of erythrocytes damaged by parasitism, factors contributing to splenomegaly may include the influx and proliferation of lymphocytes, phagocytic monocytes, and natural killer (NK) cells during the developing infection [131–134], and compensatory erythropoeisis during recovery [133].

Inchley *et al.* [133] used ^{125}I-iododeoxyuridine uptake by newly synthesized DNA to compare proliferative responses in the mouse spleen during fatal *B. rodhaini* infections and nonfatal *B. microti* infections. Cell proliferation was comparable during rising infections, but spleen weight increased more rapidly with *B. rodhaini*, suggesting earlier influx of cells occurred in the fatal

infection. Early drug-induced termination of *B. rodhaini* infection was used to show that cell recruitment by the spleen depends on the presence of live parasites, whereas cell proliferation continues after parasites are killed.

Flow cytofluorimetric analysis showed greater proliferation of B cells than T cells in the spleens of *B. microti*-infected mice [135]. Within the T cell population, a greater increase in helper (L3T4) than suppressor/cytotoxic (Ly-2) cells was detected with appropriate monoclonal antibodies [135]. The null-cell component, which may contain NK cell activity implicated in the host response to *B. microti* infection (pp. 363, 366), also contributed to increased splenic cellularity. *Babesia microti* infection induced an early proliferative response in certain lymph nodes of mice, and also induced a pronounced but transient depression of cell division in the thymus [135]. Cell division was suppressed in both tissues in fatal *B. rodhaini* infections.

Phagocytosis and macrophages

Phagocytosis of parasitized and parasite-free erythrocytes has been observed frequently in blood and organ samples from *Babesia*-infected animals. Erythrophagocytosis by both neutrophils and macrophages was obvious in the pathology of rapidly fatal infections [63, 111], suggesting that an initial nonspecific response may occur. Specific, opsonizing antibody enhances the activity [136–138], particularly in specifically primed macrophages [136]. Lymphokines produced by immune spleen cells may also enhance phagocytosis by macrophages [136].

Secretory functions of macrophages appear to be nonspecific. Mouse macrophages were primed to release tumor necrosis factor (TNF) and lymphocyte-activating factor by different stimuli, including infection with either *B. microti* or *Plasmodium vinckei petteri*, and injection of Bacillus Calmette-Guerin (BCG) [139]. An inhibitory effect on *B. bovis* in culture by mononuclear cells, probably macrophages, was increased by prior exposure of the cells to specific antigen [140]. Primed macrophages also inhibited *in vitro* growth of *B. microti*, particularly in the presence of immune serum [141]. Nonspecific priming was not tested in these *in vitro* studies.

T cells

T cell response, measured by antigen-stimulated transformation of peripheral blood lymphocytes, was short-lived in *B. bovis* infections in cattle, being highest at or soon after the peak of parasitemia and becoming undetectable with the onset of latency [142]. Elimination of *B. bovis* from carrier cattle by drug therapy restored specific T cell response [143], suggesting that specifically sensitized T cells may be present during subclinical infection but not available for sampling. Specific T cell response in cattle following inoculation with nonliving culture-derived *B. bovis* antigen persisted at high levels for 6 months [142]. One explanation of these different effects is that specifically sensitized T cells are not available for sampling during active infection, being sequestered near *B. bovis* lying in capillaries; if no such localization or persistence of nonliving antigen occurs, reactive T cells circulate freely and become included in samples of peripheral blood. The high T cell response in cattle receiving nonliving antigen was not indicative of protective immunity [142].

The T cell response to nonspecific mitogens was suppressed during acute *B. microti* infection in humans, although other clinical conditions may have affected the patients' immunologic competence [4]. T cell depletion occurred but a suppressor/cytotoxic (γ T cell) subpopulation was increased. In *B. bovis* infections in cattle, increased activity of cytotoxic T cells occurred during initial and challenge infections [59]. Sections on immunosuppression (p. 364) and immune processes (p. 368) contain further information on T cells.

Because aberrant responses to live vaccines are occasionally seen and are likely to follow immunization with some molecular vaccines, there is a need to define T cell subpopulations that influence immune responses in bovine babesiosis.

Natural killer cells

Natural killer cells are activated in murine babesiosis. Extreme NK cell activity occurred during peak parasitemia and the recovery phase of *B. microti* infections [132]. Ruebush & Burgess [144] also demonstrated NK cell activity, as well as increased interferon production, during acute *B. microti* infection, supporting the suggestion [145] that interferon may mediate NK cell activity. Natural killer cell activity was apparently assumed by other tissues in *B. microti* and *B. rodhaini* infections studied in either surgically splenectomized or congenitally asplenic mice [134]. Natural killer cell activity during the acute infection warrants study because of possible side-effects, such as killing of bystander lymphocytes [145]. Perhaps the transient leukopenia observed during some acute babesial infections is due in part to early and indiscriminate NK cell activity.

IMMUNOPATHOLOGY

General aspects

Damage to erythrocytes is the main pathologic effect of babesiosis. Severity and variability of host responses are thus dependent on the number of erythrocytes destroyed or structurally altered and their distribution within the microcirculation. The pathologic effects induced by those species of *Babesia* without tendency to accumulate in capillaries are those of acute hemolytic anemia, compounded terminally by severe congestion and overload of organs responsible for clearance of hemoglobin and cell debris from the circulation. These events typify acute *B. bigemina* and *B. equi* infections, and also apply to rodent-adapted species such as *B. rodhaini*. However, evidence for coagulation disturbances, a response usually associated with infections by the "accumulating" species *B. bovis* and *B. canis*, was reported for *B. bigemina* [120] and *B. equi* [146]. Thus the more complex pathophysiology of *B. bovis* (see Table 15.2) over *B. bigemina* in the same (bovine) host probably derives from a higher number of altered erythrocytes being present within the microcirculation, in contact with the lymphoid macrophage system, at an earlier stage of infection. All species of *Babesia* might induce similar responses if sufficiently high numbers of parasitized erythrocytes or their remnants are present. With species such as *B. bigemina*, either recovery or death from hemolytic anemia usually intervenes.

Cellular mediators

The trigger for the sequence of pathologic events during infection, particularly those induced by *B. bovis*, has been the subject of considerable study and speculation. Wright *et al.* [147] considered that the stimulus was a parasite esterase which activated the kallikrein–kinin system and the coagulation system. However, neither chemical inhibition of kallikrein [148] nor apparent immunologic inhibition of the parasite's esterase [147] significantly altered the course of acute *B. bovis* infections.

Another suggestion, based mainly on studies with rodent infections, was that endotoxin stimulated release of nonantibody-soluble mediators responsible for much of the pathology in babesiosis and malaria [149]. Low doses of bacterial lipopolysaccharide given during rising parasitemia with *Babesia* in rodents [149] and cattle [23] accelerated the pathologic changes usually seen in fatal infections. There is, however, no evidence for endotoxin in babesiosis. Similarities in the pathogenesis of bacterial endotoxicity and the two hemoparasitic diseases are apparently due to host cellular responses common to all three clinical conditions.

The critical mediators may be TNF [150,151] and oxygen-derived free radicals [152]. Other cytokines and cell products, such as IL-1 [153], thrombospondin [154], and alkenal aldehydes [153], have been implicated. Tumor necrosis factor and IL-1 are likely candidates, as both may induce the acute-phase response [117,155]; TNF enhances the procoagulant activity of endothelial cells [156], thus contributing to cytoadherance and coagulation disturbances. The collective effects of these and other as yet unknown cell products are no doubt significant to the outcome of infection. Further elucidation of their sequential activity and relative importance in control of the infection and host pathology may be possible by correlating their presence with visually determined changes in the host's microcirculation during developing infection. Visual appraisal was used to study progressive vascular pathophysiology of *Plasmodium berghei* infection in golden hamsters [157].

Activation of complement

Components of both classic and alternative pathways of complement are depleted during the acute phase of babesial infections (see p. 361). Suggested mechanisms for activation include protease secretion by the parasite, formation of immune complexes, and the release of hemoglobin [108]. The effects on the host are poorly understood. An association between complement activation and deleterious effects of acute inflammation in babesiosis might be expected, since the complement system coordinates the host's inflammatory response [158,159]. Moreover, some enzymes of the complement system are structurally and functionally similar to certain lymphokines [160], and thus, like lymphokines, may have deleterious effects as well as beneficial immunologic outcomes in babesiosis and other infections.

Immune complexes (IC)

Studies, mainly of *B. rodhaini*, up to 1980 implicated IC in anemia, splenomegaly, and glomerulonephritis of babesiosis (see Callow & Dalgliesh [23]). Subsequently, a cold precipitable IC was demonstrated in the plasma of cattle recovering from acute *B. bovis* infection [161,162]. This IC consisted of immunoglobulins, predominantly IgM, as well as other plasma proteins, C3, and small amounts of parasite antigen. Immunofluorescence of complexes in blood smears and brain tissue from infected cattle suggested that the complex may contribute to aggregation of infected erythrocytes and the inflammatory response in the acute phase of infection [162]. Levels of C1q binding indicative of the presence of IC in human patients with *B. microti* infections were apparently unrelated to either IgM or IgG levels; no significant pathologic effect was associated with persistence of the IC for up to 4 months [4]. Further studies with assays specific for IC are necessary to clarify their importance in babesiosis.

Autoimmune anemia

There is little evidence to suggest that autoantibody-induced anemia is significant in the pathogenesis of natural babesiosis (see Callow & Dalgliesh [23]). Autoantibodies were detected in three splenectomized humans infected with *B. microti* [163], but the pathologic significance of the antibodies was not established. In domestic animals, phagocytosis of apparently parasite-free red cells is sometimes observed, particularly during the acute phase of infection. This may result from nonspecific cellular responses, and some of the red cells removed may be damaged by parasites. Anemia normally parallels parasitemia and reports to the contrary appear to be exceptional (see Mahoney [52] and Zwart & Brocklesby [130]). A false impression of the relationship between peripheral parasitemia and anemia may arise, due to uneven parasite distribution in the circulation, particularly with species such as *B. bovis*, which accumulate in capillaries. However, in a study of *B. bovis* infection in 131 susceptible cattle, the correlation between parasitemia and packed cell volume was highly significant [164]. Thus, from clinical and experimental viewpoints, change in packed cell volume appears to be a reliable indicator of parasite density in the total circulation.

Immunosuppression

Depression of the immune response occurs in *Babesia* infections, although its mechanisms and relevance to the pathogenesis of disease are unclear. The main effects observed in rodent infections have been impaired host responses to other parasites and infectious agents, decreased antibody response to experimental antigens, such as sheep red blood cells, and reduced B and T cell responses to mitogens (see de Vos *et al.* [2]).

In cattle, acute *B. bovis* infection caused decreases in nonspecific IgM and IgG [106], complement components [107,108], and cortisol [165]. Host resistance to other infections and stress may thus be compromised. *Babesia bovis* caused a prolonged, increased susceptibility to the tick *B. microplus* when cattle were exposed to both parasites at the same time [166]. Concurrent infections with *B. bovis* and papular stomatitis virus in cattle apparently exacerbated the severity of both infections (I. Shiels, A.J. de Vos, & H. Prior, unpublished work).

In humans with *B. microti* infection, T and B cell responses to nonspecific mitogens were reduced during the acute phase [4]. T cell responses to nonspecific mitogens were also depressed in dogs infected with *Babesia gibsoni* [167]. However, T cell responses in mice infected with *B. microti*, determined by contact sensitivity to oxazolone and allograft survival, were almost normal [168]. Since helper T cells are essential for the synthesis of antibodies to sheep red blood cells [169], a response markedly reduced by *B. microti* infection, babesial infection apparently affects the response of some T cell subsets more than others.

INNATE RESISTANCE

Absolute resistance

Factors important in absolute innate resistance have not been studied extensively in babesiosis. The spleen has a role, since its removal allows primates, normally refractory to *B. divergens*, to become infected [170]. Other factors are obviously involved because removal or congenital absence of the spleen does not make certain other abnormal hosts susceptible [171]. Congenitally

athymic mice were no more susceptible than intact mice to infection with bovine parasites [171].

Babesia is an obligate parasite of erythrocytes so that biochemical differences in this cell among host species is a consideration. In *in vitro* culture, *B. bovis* was gradually adapted to erythrocytes of sheep, goat, horse, and rabbit [172]. A virulent strain of *B. bovis* was "attenuated" by adapting it to cultures with equine serum as the main component [173]. These adaptations of *B. bovis* to abnormal host constituents may be examples of selection for subpopulations suited to the new environment. The relevance of the changes to host specificity of the parasite is questionable since the manipulations used in adaptation are unlikely to occur in nature.

Age

The age of natural hosts apparently has little or no effect on whether infection becomes established but influences severity in at least some host–parasite relationships. From most studies of bovine babesiosis, cattle less than about 9 months of age are more resistant than older cattle, although this relative resistance may not fully develop until about 2 months of age (see Callow & Dalgliesh [23]). Age-related changes in bovine resistance and their relevance to acquired immunity and epidemiology are shown schematically in Fig. 15.4. The pattern is based mainly on studies of *B. bovis*, for which residual resistance may be present in cattle of 1–2 years of age [174]. With *B. divergens* infection in cattle, no age effect was observed in one experiment [175] whereas an inverse age resistance was demonstrated in another [176]; an analysis of clinical cases of the natural infection in Sweden [177] supported the latter finding. Adult gerbils are more susceptible than immature gerbils to *B. divergens* infection [178]. The effect of age on the resistance of other animal hosts to *Babesia* is variable (see de

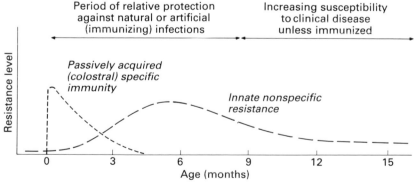

Fig. 15.4 Schematic representation of changes in colostrum-derived immunity and nonspecific resistance against babesiosis (*Babesia bovis* and *Babesia bigemina*) in calves from immune dams and their relevance to immunization.

Vos *et al.* [2]). The relatively high incidence of *B. microti* infections in older people [179] may indicate age-related increased susceptibility. Age exacerbated the immuno-suppressive effect of *B. microti* infections in mice [180].

Genetic factors

Genetic factors may contribute to variable susceptibility among individuals of the same host species but have not been identified. No relationship between BoLA antigens and severity of *B. bovis* infections could be identified in 56 naïve *B. taurus* cattle (A.J. de Vos, C.K. Dimmock, & M. Stear, unpublished work). *Bos indicus* cattle have stronger innate resistance than *B. taurus* cattle to *B. bovis* [181], although fatal infections occasionally occur in susceptible *B. indicus* cattle. Sahiwal cattle, a *B. indicus* type, were relatively resistant to *B. bigemina* [182], but other *B. indicus* types may not be. Crossbred (*B. indicus* × *B. taurus*) cattle were not more resistant than pure *B. taurus* cattle to primary *B. bovis* infection [181].

Variable susceptibility of strains of inbred mice to *B. microti* was not linked to H-2 haplotype [132]. Mice selected for resistance to the nematode parasite *Nematospiroides dubius* did not show enhanced resistance to *B. microti* [129].

Erythrocytes

Variations in surface molecules, hemoglobin type, and age of erythrocytes from different hosts might be expected to affect their suitability for *Babesia* in view of their relevance in malaria. However, apart from studies which provided conflicting evidence for the importance of C3b receptor (p. 357) for *B. rodhaini* infection, no other potential receptors appear to have been identified. Correlations between other erythrocyte factors and resistance to *Babesia* may exist but have not been established. Erythrocytes of *B. taurus* cattle were more suitable than those of *B. indicus* cattle for *in vitro* culture of *B. bovis* (P. Timms, unpublished work). Erythrocytes from newborn calves were unsuitable for *in vitro* culture of the same parasite [183]. A dialyzable, nonantibody component in calf serum also inhibited *in vitro* growth of *B. bovis*, and may contribute to nonspecific resistance of young calves [183].

Lymphoid organs and leucocytes

Splenectomy usually enhances susceptibility in babesial infections. Exceptions are *Babesia felis* infections in cats,

in which the general course of infection was unaltered by splenectomy [184], and those infections that are already rapidly fatal, such as *B. rodhaini* in mice and *B. divergens* in mature gerbils. In immature gerbils, splenectomy removes innate resistance to *B. divergens* and increases susceptibility [178]. Splenectomized humans [179] and nonhuman primates [185] are more susceptible to *B. microti* infection than intact counterparts.

Eugui & Allison [132] correlated variable resistance of inbred strains of mice to *B. microti* with splenic function, measured by increased weight, total cell number, and NK cell activity. Subsequently, Wood & Clark [186] depleted NK cells in mice with strontium-89 without decreasing resistance to *B. microti*. Allison & Eugui [187] explained the difference as being due to the existence of NK cell subsets with different susceptibilities to the isotope. Other workers also considered NK cell activity to be a component of resistance to murine babesiosis [134,144]. The activity was not confined to the spleen [134]. Natural killer cell activity in mice is age related, increasing to a peak at 6–8 weeks of age and then declining [145]. It is possible that similar age-related activity in calves contributes to the pattern shown in Fig. 15.4.

Nonlethal *B. microti* infection increased in severity in nu/nu mice [188] and T cell-depleted mice [189]. Antilymphocyte serum given to hamsters prolonged *B. microti* parasitemia and led to death [190]. The effect was greater than that produced by splenectomy. Ruebush & Hanson [189] suggested that age-related loss of thymic function may contribute to increased susceptibility of elderly people to *B. microti*.

In some experiments, apparent impairment of normal immune responses actually enhanced resistance of naïve mice to lethal *B. rodhaini* infection. Thus, nu/nu mice [191] and intact mice either nonlethally irradiated or treated with the immunosuppressant cyclophospham-ide [192] were more resistant than respective controls to primary infection, perhaps due to the absence or reduced effect of T suppressor cells [192].

ACQUIRED RESISTANCE

Specific resistance

In their natural environment, animals first acquire specific resistance by passive transfer from immune dams to their progeny, probably by way of colostral antibody. Later, active resistance follows recovery from primary tick-transmitted infection. Comparable resistance develops in animals infected artificially, by blood transmission or by inoculation with viable parasites extracted from ticks.

Passive resistance

Hall [193] showed that passive resistance to *B. bovis* persisted in calves for about 6 weeks after birth. Challenged calves showed no clinical signs, anemia or parasitemia, although subclinical infection became established. The resistance was deemed to be antibody mediated. Age-matched calves from nonimmune cows were severely affected by the same challenge. Hall *et al.* [194] obtained similar results with *B. bigemina*, and showed further that passively acquired resistance was restricted to the strain of parasite infecting the dam. The passive resistance against *B. bigemina* was strong enough to prevent infection from becoming established in some of the calves studied. These findings provided the first evidence, albeit indirect, for two important characteristics of immunity against *Babesia*. One is the strain specificity of at least some protective antibodies, and the other is that animals may be strongly resistant without being carriers of infection.

Christensson [195] demonstrated specific antibody (IgG) to *B. divergens* in colostrum-fed calves from immune dams but found no difference in their resistance to subsequent infection compared with calves from nonimmune dams. Low virulence of the infecting parasites may have masked a protective effect of maternal antibody, as infections in both groups of calves were extremely mild.

Specific antibody against *Babesia* in the naïve offspring of immune mothers is detectable by conventional serology (e.g., Weisman *et al.* [196]). Antibody levels are detectable within 3 h after calves receive colostrum and may persist for 4–5 months [196].

In endemic areas, passively acquired resistance complements innate resistance of the young (Fig. 15.4). The observed strain-specific protection against parasites infecting the dam [184] is difficult to explain epidemiologically, because some calves would be exposed to strains different from those infecting the dam. Yet almost all young calves in enzootic areas resist babesiosis.

Resistance acquired by infection

For economic reasons, acquired resistance following infection with *Babesia* has been most intensively studied in cattle. Most recovered cattle possess strong and long-lasting clinical resistance to challenge. The degree of immunity acquired by infection is unrelated to either the virulence of the infecting parasite or the method used to induce infection [124], nor does it depend on infection being maintained. Thus, cattle that naturally eliminate infection exhibit strong resistance to reinfection with the homologous strain for several years and are also resistant to heterologous strain challenge (see de Vos *et al.* [2] and Callow & Dalgliesh [23]). Immunity also persists following elimination of babesial infections in rodents and gerbils [63,192,197,198]. Studies on cattle that are drug-cured of *B. bovis* infection indicated that the degree of antigenic stimulation experienced by the host is more relevant to acquired resistance than the presence of live parasites [199]. Löhr [200] suggested that sterile immunity to *B. bigemina* may decline gradually at a rate related to the host's immune response during infection.

Crossimmunity between certain species of *Babesia* and between *Babesia* and other genera of hemoparasites has been demonstrated. The effect in most cases appears to be due to nonspecific host responses (see p. 368), although there is evidence that at least some species of *Babesia* share common antigens. Increased resistance to *B. bovis* infection in cattle that have recovered from *B. bigemina* infection has been observed, albeit inconsistently (see Callow & Dalgliesh [23]). Serologic crossreactivity between *B. bigemina* and *B. bovis* (e.g., Wright *et al.* [201]), as well as the presence of common polypeptides [201] and proteins with common epitopes [36] in the two species, indicate that the crossimmunity is specific.

Zwart *et al.* [202] observed similar crossprotection and serologic crossreactivity between *B. bigemina* and *B. major*. Sheep exposed to *B. bovis* antigen acquired resistance to *B. ovis*, but in this case the stimulus was crude extract of the bovine parasite administered with an adjuvant [203].

Relapsing infections without adverse effects occur in animals following control of the primary babesial infection. Differing persistence between species of *Babesia* and the number of relapses that occur during chronicity in the same host may be related to differing capacity for antigenic variation [204]. Infected animals experience higher relapse parasitemias when acquired immunity is compromised by splenectomy or treatment with corticosteroids (see Callow & Dalgliesh [23]).

Nonspecific resistance

Agents which stimulate nonspecific resistance to *B. microti* and *B. rodhaini* in rodents include a variety of bacteria and bacterial products, as well as cord factor, zymosan, glucan, muramyl dipeptide, and chloride-oxidized oxyamylase (see de Vos *et al.* [2]). Nonspecific resistance induced in mice was long-lasting and unaffected by age or pregnancy [205]. In gerbils, BCG induced protection against *B. divergens* infection but killed *Propionesbacterium acne* and zymosan did not [206]. Attempts to protect cattle against babesial infections with BCG [207] and *Corynebacterium parvum* vaccine [23, 208] were unsuccessful.

Crossimmunity amongst *B. microti*, *B. rodhaini*, *B. hylomysci*, and other genera of rodent hemoparasites was attributed to nonspecific resistance [205]. In other studies, serologic crossreactivity and crossimmunity among several species of *Babesia* and *Plasmodium* were ascribed to antigens detected in the serum of animals acutely infected with parasites of either genus, but apparently lacking specificity for either the host or the parasite species [209,210]. Similarly, in bovine babesiosis, differentiation of nonspecific and specific reactivity may be complicated by the emergence of host-derived antigens and complexes of host and parasite molecules during infection (see Mahoney & Goodger [211]).

Immune processes

Serum antibodies

Variable protection against babesial infections has followed the passive transfer of immune serum (reviewed by de Vos *et al.* [2], Callow & Dalgliesh [23], Zwart & Brocklesby [130], and Mahoney [212]). Dose of serum, specific Ig content, and host variables (species, strain, age, sex, immunocompetence) probably have affected the outcome of experiments. The most protective antibody appears to be IgG from hyperimmune serum [137]. Mahoney [212] considered that antibody alone was ineffective against *B. bovis* and that opsonization was the basis for protection. *In vitro* studies with several species of *Babesia* [136,138,140,141], in which phagocytosis by macrophages increased in the presence of immune serum, provide a foundation for this view. Goff *et al.* [57] considered that resolution of primary *B. bovis* infection is due mainly to antibody-dependent cytotoxicity but that phagocytosis may become increasingly important in animals exposed to reinfection.

Given the probable strain specificity of protective antibodies [137,194,213], the crossprotection among strains commonly observed in infected animals was tentatively ascribed to common T cell-stimulating carrier moieties on protective antigens of different strains [212]. The carrier moieties may prime T-helper cells, enabling a prompt response against associated molecularly different epitopes.

Immune cell types

Information on the relative importance of different cell types in acquired immunity to babesial infections derives mainly from studies in rodents, particularly adoptive transfer experiments in mice. Prior depletion of T cells from spleen cells transferred from *B. microti*-immune mice did not affect the immunity expressed by recipient mice, and macrophages and NK cells were implicated [131]. In another study, depletion of T cells abrogated the protective activity, leading to the suggestion that T cell lymphokines were responsible for the immunity, acting either directly on the parasite or indirectly by activation of other immunocytes, such as macrophages [214]. Similar conclusions were drawn from studies on sterile immunity in mice that were drug-cured of *B. rodhaini* infection. Sublethal and lethal irradiation of immune mice had little effect on their resistance to challenge infection, and radioresistant T cells and macrophages were considered to be responsible for protection [192]. In nu/nu mice, repeated drug treatment during and following primary infection failed to prevent the occurrence of fatal recrudescence, thus further implicating a T cell-dependent activity in specific acquired immunity [215]. Inoculation with BCG induced protection against *B. rodhaini* in intact mice but not nu/nu mice [216], suggesting nonspecific resistance also is T cell dependent.

In other studies on adoptive immunity, Meeusen *et al.* [128,217,218] examined effects of both B cells and T cells: B cell-enriched subpopulations of immune spleen cells were more protective than T cell-enriched preparations against primary *B. microti* infections, but when the mice were reinfected the recipients of enriched T cell preparations were the more resistant [217]. Pretreatment of transferred B cells with antitheta serum and complement abolished both the primary protective effect and the anamnestic antibody response, suggesting that primed T cells, as well as primed B cells, were necessary for prolonged acquired immunity. B memory cells, rather than already differentiated antibody-forming cells, conferred the better protection [218]. B cell proliferation in the recipient, accompanied by enhanced development of IgG, was necessary to induce immunity [128].

Cellular mediators

Intracellular death of parasites has been described in rodent babesiosis, degenerating "crisis" forms being most obvious at the height of the primary parasitemia [219]. The effect was not antibody related [219], did not depend on contact between effector and target cells [131], and was induced by a variety of nonspecific biologic and synthetic compounds [150]. Clark [150] considered that the parasite killing was mediated by activated macrophages, with TNF and interferon as the likely factors. Infection with *B. microti* and injection with BCG apparently provided the same priming stimulus for subsequent triggering of monokine release [139], suggesting that the process has a role in acquired resistance in babesiosis.

One of TNF's effects may be to sensitize neutrophils and macrophages to secrete reactive forms of oxygen [151], another mechanism implicated both in killing *Babesia* [220] and *Plasmodium* [187] and in host pathology [152,153]. Further evidence is required to support this concept for *Babesia*, as *B. divergens* showed little susceptibility to the action of reactive oxygen forms [206].

Whereas macrophages and their cytokine(s) may be mainly responsible for killing and removing hemoparasites from the host circulation, the trigger appears to be T cell related [151,187,220], and may be interferon-γ (IFN-γ) [151]. Findings that MAB to IFN-γ inhibited the development of resistance against *Toxoplasma gondii* [221] and also suppressed acquired immunity to malaria [222] in rodents suggest that it may have a crucial role in both the acquisition and maintenance of immunity to parasitic protozoa.

IMMUNO- AND MOLECULAR DIAGNOSIS

Detection of parasites and their products

Babesiosis is best diagnosed by using a microscope to search for parasites in Romanowsky-stained smears of blood and organs from diseased animals. Other methods are useful for speciation in mixed infections or with specimens in which parasite morphology is atypical. Both *B. bigemina* and *B. bovis* were identified in organ smears for longer periods after death with a direct

Fig. 15.5 High specificity and sensitivity of DNA probe 19-9B for *Babesia bovis* in blood from live cattle shown by overnight autoradiograph of slot-blotted DNA extracted from 20 μl samples infected with various hemoparasites. The samples were digested with proteinase K, extracted with phenol/chloroform, and probed with plasmid vector PUC 19 DNA with a 1.8 kb insert of DNA containing a repetitive sequence of the *B. bovis* genome (probe 19-9B). The left column shows 2% *B. bovis*-infected blood serially diluted fivefold with uninfected blood. The right column shows in sequence undiluted blood with: (1) *Babesia bigemina*; (2) no parasites; (3) and (4) *B. bovis* (two samples with different parasite concentration); (5) *Theileria buffeli*; (6) *Anaplasma centrale*; (7) *B. bigemina*. The limit of detection of the probe was 0.0006%. The limits of detection of a skilled microscopist examining Giemsa-stained films of the same source material were 0.02% for thin films and below 0.0001% for thick films (examination time 10 min). (From the unpublished work of W.K. Jorgensen & P. Timms, Tick Fever Research Centre, Wacol, Queensland, Australia.)

fluorescent antibody test than with Giemsa staining [223]. The use of specific conjugates can differentiate the two species when poor staining or postmortem degeneration might preclude a precise diagnosis.

McLaughlin *et al.* [224] compared plasmid clones with *B. bovis* DNA inserts with *B. bovis* genomic DNA for specificity and sensitivity in detecting the parasite in blood. The plasmid DNA probes were much better for discriminating against *B. bigemina* and bovine DNA, and their sensitivity compared well with that of the total DNA probe. Dot blot hybridization with overnight exposure detected 100 pg of *B. bovis* DNA, which is that present in 50 μl of blood with 0.01% infected erythrocytes [224]. Another probe for *B. bovis* was compared with the ability of an experienced microscopist to detect *B. bovis* parasites in blood smears (Fig. 15.5).

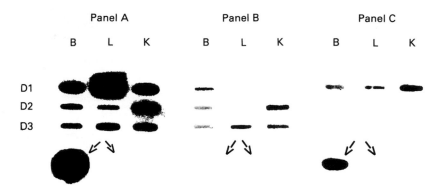

Fig. 15.6 Detection of *Babesia bovis* in decomposing autopsy specimens by overnight autoradiographs of slot-blotted DNA extracted from 20 µl samples of brain (B), liver (L), and kidney (K) from a *B. bovis*-infected calf taken 24 h (D1), 48 h (D2), and 72 h (D3) postmortem. DNA extracts of *B. bovis*-infected blood (left arrow) and parasite-free blood (right arrow) were included as controls. Use of probe 19-9B (see caption for Fig. 15.5) suggests that *B. bovis* DNA is present after 72 h (Panel A), but samples probed with plasmid vector DNA alone (Panel B) indicate that DNA of contaminating bacteria is also being detected by the probe. Samples probed with 1.8 kb insert of *B. bovis* DNA alone (Panel C) are positive for up to 24 h. (From the unpublished work of W.K. Jorgensen, Tick Fever Research Centre, Wacol, Queensland, Australia.)

Examination of thick blood films [225] was more sensitive, but high specificity of the DNA probe (Fig. 15.5) aids diagnosis when speciation on morphologic grounds is difficult. Plasmid vector DNA in probes for detecting *B. bovis* may hybridize with DNA of contaminating bacteria in smears from decomposing organs, thus giving a false positive result (Fig. 15.6). However, *B. bovis* DNA alone specifically detected parasite DNA until at least 24 h postmortem (Fig. 15.6), a period 8 h longer than that for which the direct fluorescent antibody test provided a reliable diagnosis [223]. Other species-specific DNA probes have been prepared for *B. equi* [226] and *B. bigemina* [227].

DNA probes for *Babesia* are unlikely to replace conventional microscopy for practical diagnosis, but their potential for genomic analysis should be fully explored. Methods for distinguishing field and laboratory strains of *Babesia* are needed in diagnostic laboratories servicing endemic areas where living vaccines are used and sometimes incriminated in disease outbreaks. Also, systematic analysis of parasite populations responsible for disease outbreaks may show correlations between genotype and pathogenicity in the field. This information may prove useful in the design of vaccines.

Detection of antibodies

Characteristics and research applications of the various serologic tests developed for babesial antibodies have been reviewed [23,228]. In diagnosing disease outbreaks, detection of specific antibody in recovered animals may provide confirmatory evidence of the causal parasite's identity, should doubt exist. Serologic tests are also applied in establishing the infection status of animals in compliance with trade restrictions within and between countries.

The indirect fluorescent antibody test (IFAT) has been most widely applied, and was used as the standard against which new tests were compared [123,126,229, 230]. Enzyme immunoassays (EIA) or enzyme-linked immunosorbent assays (ELISA) were considered to be less subjective in reading and more sensitive than the IFAT [123]. An EIA detected *B. bovis* antibodies for up to 4 years after a single infection [126]. A slide EIA adapted for field use in detecting *B. bovis* antibody [231] and a dot-EIA for diagnosing canine babesiosis showed promise [230].

Radioimmunoassay (RIA) was highly specific and more sensitive than the IFAT [127] but the requirement for isotopic techniques may limit its use to research. Anamnestic antibody responses were detected with RIA but not with the IFAT in immune cattle following heterologous challenge with *B. bovis* [127].

Rapid slide agglutination tests with antigen adsorbed onto latex particles [229] were developed for use in the field. Inconsistent results and problems in distinguishing between *B. bovis* and *B. bigemina* infections may discourage wide application.

VACCINES

Vaccines against babesiosis will be considered as:
1 those currently available and widely used in certain countries either on a commercial basis or within disease control programs supported by national or international agencies;
2 experimental vaccines which have had limited field application or are presently undergoing laboratory evaluation.

Current vaccines and their deficiencies

Live vaccines

The immunization of cattle against *B. bovis* and *B. bigemina* with blood-based vaccines containing living parasites began almost 100 years ago [50]. Live vaccines against the two parasites are currently produced in several countries by government agencies (see de Vos et al. [2]). Vaccination of cattle with living *B. divergens* parasites has a long history in Sweden [232]. The vaccines are usually effective because a strong heterologous-strain immunity is acquired following a single blood-transmitted infection. The development and general application of live vaccines and other control measures for bovine babesiosis have been reviewed [50,233,234]. The most thoroughly documented vaccine currently available is the Australian *B. bovis* vaccine of which more than 700 000 doses are sold annually [293].

Troublesome characteristics of living babesial vaccines include variable infectivity, severe reactions to vaccination, and an incidence of neonatal hemolytic disease in calves from vaccinated dams. In Australia, low infectivity was overcome by preparing vaccines in splenectomized calves, virulence problems by utilizing syringe-passage procedures, and sensitization of recipients to erythrocyte antigens by reducing the volume of donor blood in the inoculum. The FAO Expert Consultation on Research on Tick-borne Diseases and Their Vectors recommended the methods for wider use in the field [3]. Chemotherapeutic preparations for *Babesia* can be used to suppress severe reactions to vaccination if necessary, although acquired immunity may be compromised by certain treatment regimes applied in conjunction with vaccination [235].

Most currently available vaccines are unfrozen and therefore perishable, and must be used within 6–7 days to be effective. There is thus insufficient time to test them effectively for contaminating viruses and bacteria by culture. A batch of vaccine became contaminated with bovine leucosis virus in Australia [236]. Despite this accident, Australian chilled vaccines have a good safety record because of stringent disease control in cattle used for vaccine preparation.

Techniques for cryopreserving *B. bovis* and *B. bigemina* with reliable recovery of viable parasites have been applied in producing frozen vaccines against the parasites [238–240]. Freezing overcomes the problem of a short shelf-life of chilled vaccines and also allows safety testing to be done before use. The efficacy of Australian vaccine strains of *B. bovis* and *B. bigemina* against field strains in South Africa [95,241] and Sri Lanka (D.J. Weilgama, unpublished observations) indicates that vaccines prepared in one country may be useful in others.

Nonliving vaccines

Successful propagation of *B. canis* in *in vitro* culture was the basis for the first nonliving babesial vaccine to be produced commercially [242]. The vaccine comprises purified, formalin-treated soluble antigen derived from *B. canis* cultures, which is mixed with saponin adjuvant. The vaccination schedule requires two inoculations 3 weeks apart. A reduction in the incidence of canine babesiosis during the first 18 months of field use of the vaccine in France provided evidence of the vaccine's efficacy, but a proportion of vaccine failures caused concern [243]. The presence in France of at least two large types of canine *Babesia*, which do not crossprotect and differ serologically [244], may be relevant to vaccine failures. Lack of crossprotection between *B. canis* from France and South Africa [245] has implications for wider use of the vaccine.

Soluble antigens derived from *in vitro* culture have also been used in producing a combined *B. bovis* and *B. bigemina* vaccine in Venezuela [246]. The parasites are cultured separately and the respective antigens pooled and mixed with saponin before two inoculations 3 weeks apart.

These vaccines lack the troublesome characteristics of living vaccines and have the additional advantage of being amenable to freeze-drying for prolonged storage. However, other factors regarding their further development and wider application must be considered. One is the variable results obtained in laboratory trials with culture-derived *B. bovis* antigens. Whereas protection against homologous (strain) challenge has been a consistent finding, inadequate protection against heterologous challenge was found with antigens derived from some strains of B. bovis [142,247,248]. Moreover, comparisons between culture-derived antigens and a living vaccine strain as protective immunogens heavily favored the latter [248]. Thus, immunity imparted by culture-derived antigens may be marginal. Given the apparent capability of field strains to circumvent immunity provided even by a proven, living vaccine strain (see p. 325), close monitoring of field results over several years will be required to assess fully the usefulness of nonliving culture-derived vaccines.

Another disadvantage is the requirement for their administration on two or more occasions. The necessity for multiple inoculations several weeks apart is a serious constraint to practical vaccination of cattle, and perhaps other animals, in endemic areas where extensive management is practised.

Experimental vaccines—live parasites

Irradiated parasites

The effects of irradiation on blood stages of *Babesia* have been reviewed [2,23,249]. In general, increasing the dose of irradiation either kills or inhibits replication of a progressively larger proportion of the parasites and apparently reduces virulence of the remainder. Parasites totally inhibited by irradiation and used as inoculum confer some protection against homologous challenge, but replication of the irradiated parasites in recipients is necessary for the development of useful acquired immunity. An irradiation dose of between 280 and 350 Gy usually provides an inoculum of suitably modified, infective parasites. Nonlethally irradiated *Babesia* produce mild infections in most recipients and provide varying degrees of protective immunity. Irradiated vaccines against *B. divergens* [250] and *B. bigemina* [251] have been used in the field, but the technique has not been adopted widely. An exception to the generally observed reduced virulence following irradiation was reported for *B. divergens* [102].

Parasites cultured in vitro

Culture techniques are available for short-term and continuous *in vitro* propagation of most economically important species of *Babesia* (see Canning *et al.* [14]). The most reliable and widely applied technique for research is the microaerophilus stationary phase (MASP) method [13]. A *B. bovis* strain adapted to equine serum in MASP

cultures lost virulence [173] without apparent loss of immunogenicity [252]. Maintenance of vaccine strains of *B. bovis* and *B. bigemina* in MASP cultures for 174 days [253] and 100 days [254], respectively, did not alter infectivity, virulence (low), or immunogenicity of the strains tested in recipient cattle. Long-term (18 months) *in vitro* culture of *B. divergens* reduced its virulence for gerbils [255]. *In vitro* culture thus provides a potentially suitable alternative to infected calves as a source of these parasites for live vaccines.

The small volume of infective inoculum provided by MASP cultures is a serious constraint to their practical application [253,254]. Suspension culture techniques developed for *B. bovis* [253,256] have greater potential to provide useful quantities of inoculum, and are now being applied in producing a live *B. bigemina* vaccine in Australia (W.K. Jorgensen, A.J. de Vos, & R.J. Dalgliesh, unpublished work). Current culture methods cannot satisfy the much greater demand for *B. bovis* vaccine. But the goal to replace infected donor animals with *in vitro* culture for producing live *Babesia* vaccines is realistic.

Parasites from abnormal hosts

A novel approach to producing *B. divergens* vaccine for cattle is the use of laboratory gerbils to provide living parasites [257,258]. Recipient cattle were solidly immune to experimental heterologous challenge [257] and apparently to natural infections [258]. Immunizing infections with virulent gerbil-derived *B. divergens* can be moderated by prophylactic drug treatment, although precise dosage is necessary to achieve control without loss of infectivity [257]. The approach has the advantage over cattle-derived parasites that risks from potential contaminants of bovine origin are reduced. However, it is readily applicable only to *Babesia* with low host specificity, and may not be amenable to large-scale production.

Experimental vaccines – nonliving specific antigens

Varying degrees of protection have been obtained with undefined babesial antigens derived from either *in vivo* or *in vitro* sources (see de Vos *et al.* [2]). Currently, the aim is to identify and purify individual antigens that are protective and might be produced in quantity by recombinant DNA techniques, synthesis, or *in vitro* culture. Difficulties have been encountered in separating parasite antigens from host cell products [211,259]

Table 15.3 Murine monoclonal antibodies against bovine *Babesia*

Babesia sp. [Ref.] (Source of parasite)	Code	Ig class	Reactive antigens		Result of biologic test [Ref.]
			Location	Mol. wt (kD)	
B. bovis [261] (Bovine blood)	2C2	IgM	I/C parasite* + rbc cytoplasm[†]	1300	Antigen not protective for cattle
	15B1	IgM	I/C parasite	44	Antigen protective for cattle
	18A5	IgG1	I/C parasite, mainly apical end	180	Antigen not protective for cattle
B. bovis[‡] (Bovine blood)	W11C5	IgG1	I/C parasite and host cell surface	70–200	Recombinant antigen (140 kD) protective for cattle
B. bovis [262] (*In vitro* culture)	BABB35	IgG2	Merozoite surface coat and cytoplasm	37, 42	
	BABB75	IgG2	Merozoite surface coat and possible rhoptry	60	
	BABB90	IgG1	Merozoite surface coat	85	
	BABB93	IgG2	Merozoite cytoplasm	145	
B. bovis [263] (*In vitro* culture)	23.70.174	IgG1	Merozoite surface coat	44	Recombinant antigen induced variable antibody response
B. bigemina [71] (Bovine blood)	14.1	IgG2	Merozoite surface coat	45	Antigen partially protective for cattle [265]
	14.20	IgG1	Merozoite surface coat	55	
	14.29	IgG1	Merozoite surface coat	72	
	14.16	IgG1	Merozoite surface coat	58	
	14.52	IgG1	Merozoite surface coat	16–36	
B. bigemina [34] (*In vitro* culture)	B1B3C4	IgG2	Merozoite external membrane	51–53	
	B1C9A2	IgG1	Merozoite external membrane	36	
B. divergens [37] (Rat blood)	BF1.1	IgM	I/C parasite		Antigen not protective for rats
	BF1.3	IgG1	I/C parasite		Antibody protective for rats
	BF1.18	IgG1	I/C parasite		Antibody protective for rats
B. divergens [264] (*In vitro* culture)	7.8	IgG1	Merozoite, mainly surface coat	50–60	Antibody inhibited merozoite invasion *in vitro*
	7.8	IgG1	Merozoite, mainly surface coat	24–29	Antigen partially protective for gerbils [266]

*I/C, intracellular parasite; stage and more specific location not described.
[†] Reactive antigen observed in red blood cell (rbc) cytoplasm of host cells containing multiple but not single stages.
[‡] Unpublished work of K.R. Gale, B.V. Goodger, I.G. Wright, D.J. Waltisbuhl, & C.M. Dimmock, CSIRO laboratories, Indooroopilly, Queensland, Australia; see p. 374.

and in distinguishing those with protective properties [114,260] from those that induce an antibody response but not resistance [161,260]. Monoclonal antibodies have assisted progress towards the characterization and purification of parasite antigens, some of which have conferred varying degrees of protection (Table 15.3).

Antigens from B. bovis

Wright *et al*. [265] used MAB to isolate three soluble *B. bovis* proteins from a lysate of infected erythrocytes (Table 15.3) and found that the smallest, a 44 kD protein, conferred a degree of protection against homologous strain challenge. The protective antigenic epitope might be contained by a 29 kD polypeptide from the 44 kD native protein [267]. Goff *et al.* [262] considered that three surface proteins (37 kD, 42 kD, 60 kD) of cultured *B. bovis* merozoites, identified by ^{125}I-labeling and precipitation with immune serum and MAB (Table 15.3), were potentially useful antigens. Hines *et al.* [268] also studied merozoite surface proteins of *B. bovis*, using a parasite clone to limit the effects of genetic variation, and confirmed that a 42 kD protein was an immunodominant antigen.

Soluble *B. bovis* antigens released in culture were present in greatest concentration in 24-h cultures [269]. Antigens identified to date have been between 18 and 82 kD [69,269]. The relationship between the 28 kD and 30 kD antigens identified in culture-derived material [269] and the 29 kD fragment characterized by Wright *et al.* [267] warrants investigation.

Initial experiments with recombinant *B. bovis* proteins as immunogens have met with varying degrees of success. Gill *et al.* [270] prepared a cDNA clone coding for a 5–10 kD portion of a 220 kD polypeptide antigen of *B. bovis*. The clone was used subsequently to produce sufficient babesial protein, fused to β-galactosidase, to inoculate each of seven cattle with 0.7 mg of fused protein on four separate occasions [271]. The lack of specific antibody response and poor protection conferred may have resulted from masking of the babesial antigenic epitope by the much larger β-galactosidase molecule [271]. Reduker *et al.* [263] used MAB to screen a λ-GT 11 library for clones expressing surface proteins of *B. bovis*. A fused 44 kD protein, recognized by MAB 23.10.174 (Table 15.3), induced no apparent antibody response in one calf and a highly variable response in four others. The protective quality of the protein was apparently not tested. Gale *et al.* (see Table 15.3) also screened a γ-GT 11 library with MAB and isolated a clone expressing fused 140 kD *B. bovis* protein. The protein, when administered twice with Freund's complete adjuvant, consistently induced specific antibody in cattle, and provided partial protection against heterologous-strain challenge (K.R. Gale *et al.*, unpublished work). Other recombinant proteins of *B. bovis* have been prepared and characterized [272] but not tested in animals.

Kemp *et al.* [273] showed that *B. bovis* genes can be accurately transcribed in the yeast *Saccharomyces cerevisiae*, which thus warrants study as a possible alternative to bacteriophages as expression vectors for candidate antigens.

Antigens from other Babesia spp.

A 45 kD surface protein of *B. bigemina* merozoites, affinity purified with MAB 14.1 (Table 15.3), provided partial homologous-strain protection in calves [265]. Mixtures of culture-derived supernatant antigens of *B. bigemina* (32–86 kD) conferred heterologous-strain immunity in cattle [269]. Similarly derived antigens of another bovine parasite, *B. divergens*, provided poor protection against homologous challenge in gerbils [64]. Affinity-purified merozoite-derived antigens recognized by MAB 7.8 (Table 15.3) showed greater promise by conferring partial homologous-strain protection in gerbils; the MAB inhibited parasite invasion *in vitro* [265,266].

Adjuvants and vaccine administration

An effective adjuvant is needed to ensure the success of nonliving babesial vaccines owing to the relatively weak and short-lived immunogenicity conferred by processed antigens compared with that following the use of living parasites. Freund's complete adjuvant has been used most extensively in studies on experimental vaccines but is unacceptable for commercial use for various reasons, including irritancy. The saponin preparation, Quil A, has been used experimentally, as well as in culture-derived commercial vaccines in France and Venezuela (see p. 371). Glucan, avridine, and muramyldipeptide have also been used experimentally [246]. The paucity of comparative observations on adjuvants in studies on *Babesia* is surprising. However, generally slow progress in finding suitable adjuvants for use with malarial and other parasite antigens is not good news for those wishing to produce subunit vaccines against babesiosis.

CONCLUSIONS

The application of new technology has increased knowledge of *Babesia*, particularly at the molecular level. Perhaps the most disturbing facet of this new information in relation to effective vaccination is the extent to which both phenotypic and genotypic characteristics of the parasite may change with selection pressure (see Fig. 15.2). Although the findings derive mainly from laboratory studies, they must reflect the parasite's potential for change in the field. The genetic diversity observed among field isolates of *B. bovis* (see p. 354) supports this view. The question arises whether field strains of *Babesia* are capable of evading immunity provided either by living vaccine strains or by nonliving antigens. There is evidence to suggest that this has already occurred. The Ka *B. bovis* strain (see p. 353) used successfully for vaccination in Australia has apparently lost efficacy against field strains on several cattle properties where the Ka strain was used regularly for a number of years (A.J. de Vos, unpublished work). Under laboratory conditions, the Ka strain provided poor protection against an isolate from one of these properties but strong protection against isolates from other areas (A.J. de Vos, I. Shiels, & R. Bock, unpublished work). The solution in this case was replacement of the Ka strain with a new, heterologous vaccine strain.

These developments give no cause for optimism that vaccines comprising a small unchanging array of antigens will be useful over an extended period. Whether future vaccine antigens are derived from recombinant DNA techniques, protein synthesis, *in vitro* culture, or other methodologies, field conditions will probably demand periodic change in a vaccine's antigenic makeup to overcome the effects of antigenic shifts in field populations of the parasite.

ACKNOWLEDGMENTS

I am grateful to all colleagues who made available the unpublished results referred to in the preceding text. I thank Dr L.L. Callow, Dr G.F. Mitchell, and Dr A.F. Cowman for their helpful criticisms of the manuscript.

REFERENCES

1 Levine ND. Blood parasites: the piroplasms. In Levine ND, ed. *The Protozoan Phylum Apicomplexa*, vol. II. Boca Raton: CRC Press, 1988:35–45.
2 De Vos AJ, Dalgliesh RJ, Callow LL. *Babesia*. In Soulsby EJL, ed. *Immune Response in Parasitic Infections: Immunology, Immunopathology, and Immunoprophylaxis, vol. III. Protozoa.* Boca Raton: CRC Press, 1987:183–222.
3 McCosker PJ. The global importance of babesiosis. In Ristic M, Kreier JP, eds. *Babesiosis.* New York: Academic Press, 1981:1–24.
4 Benach JL, Habicht GS, Hamburger MI. Immuno-responsiveness in acute babesiosis in humans. J Infect Dis 1982;146:369–380.
5 Mehlhorn H, Schein E. The piroplasms: life cycle and sexual stages. Adv Parasitol 1984;23:37–103.
6 Uilenberg G. Highlights in recent research on tick-borne diseases of domestic animals. J Parasitol 1986;72:485–491.
7 Friedhoff KT. Transmission of *Babesia*. In Ristic M, ed. *Babesiosis of Domestic Animals and Man.* Boca Raton: CRC Press, 1988:23–52.
8 Dalgliesh RJ, Stewart NP. Some effects of time, tempera-

ture and feeding on infection rates with *Babesia bovis* and *Babesia bigemina* in *Boophilus microplus* larvae. Int J Parasitol 1982;12:323–326.
9 Lewengrub S, Rudzinska MA, Piesman J, Spielman A, Zung J. The influence of heat on the development of *Babesia microti* in unfed nymphs of *Ixodes dammini*. Can J Zool 1989;67:1510–1515.
10 Rudzinska MA, Spielman A, Lewengrub S, Trager W, Piesman J. Sexuality in piroplasms as revealed by electron microscopy in *Babesia microti*. Proc Natl Acad Sci USA 1983;80:2966–2970.
11 Mackenstedt U, Gauer M, Mehlhorn H, Schein E, Hauschild S. Sexual cycle of *Babesia divergens* confirmed by DNA measurements. Parasitol Res 1990;76:199–206.
12 Mehlhorn H, Raether W, Schein E, Weber M, Uphoff M. Light and electron microscopic studies on the life cycle of *Babesia microti* and the effects of pentamidine on the erythrocytic stage. Dtsch Tierärztl Wochenschr 1986;93: 400–402, 404–405; Protozool Abstr 1987;11:965.
13 Levy MG, Ristic M. *Babesia bovis*: continuous cultivation in a microaerophilous stationary phase culture. Science 1980;207:1218–1220.
14 Canning EU, Winger CM. Babesiidae. In Taylor AER, Baker JR, eds. *In vitro Methods for Parasite Cultivation.* London: Academic Press, 1987:199–229.
15 Conrad PA. Uptake of tritiated nucleic acid precursors by *Babesia bovis in vitro*. Int J Parasitol 1986;16:263–268.
16 Irvin AD, Young ER. Further studies on the uptake of tritiated nucleic acid precursors by *Babesia* spp. of cattle and mice. Int J Parasitol 1979;9:109–114.
17 Irvin AD, Young ER, Purnell RE. The *in vitro* uptake of tritiated nucleic acid precursors by *Babesia* spp. of cattle and mice. Int J Parasitol 1978;8:19–24.
18 Matias C, Nott SE, Bagnara AS, O'Sullivan WJ, Gero AM. Purine salvage and metabolism in *Babesia bovis*. Parasitol Res 1990;76:207–213.
19 Hassan HF, Phillips RS, Coombs GH. Purine-metabolizing enzymes in *Babesia divergens*. Parasitol Res 1987;73: 121–125.
20 Gero AM, Coombs GH. Pyrimidine biosynthetic enzymes in *Babesia hylomysci*. Int J Parasitol 1982;12:377–382.
21 Holland JW, Gero AM, O'Sullivan WJ. Enzymes of *de novo* pyrimidine biosynthesis in *Babesia rodhaini*. J Protozool 1983;30:36–40.
22 Gero AM, O'Sullivan WJ, Wright IG, Mahoney DF. The enzymes of pyrimidine biosynthesis in *Babesia bovis* and *Babesia bigemina*. Aust J Exp Biol Med Sci 1983;61:239–243.
23 Callow LL, Dalgliesh RJ. Immunity and immunopathology in babesiosis. In Cohen S, Warren KS, eds. *Immunology of Parasitic Infections.* Oxford: Blackwell Scientific Publications, 1982:475–526.
24 Barry DN. Metabolism of *Babesia* parasites *in vitro*. Glucose and energy metabolism of *B. bovis*. Aust J Exp Biol Med Sci 1984;62:53–61.
25 Barry DN. Metabolism of *Babesia* parasites *in vitro*. Change in adenylate energy charge and infectivity of *Babesia rodhaini*-infected erythrocytes. Aust J Exp Biol Med Sci 1984;62:63–71.
26 Barry DN. Metabolism of *Babesia* parasites *in vitro*. Amino acid production by *Babesia rodhaini* compared to

Plasmodium berghei. Aust J Exp Biol Med Sci 1982;60: 175–180.

27 Wright IG, Goodger BV. Proteolytic enzyme activity in the intraerythrocytic parasites *Babesia argentina* and *Babesia bigemina*. Z Parasitenkd 1973;42:213–220.

28 Momen H. Biochemistry of intraerythrocytic parasites. I. Identification of enzymes of parasite origin by starch-gel electrophoresis. Ann Trop Med Parasitol 1979;73:109–115.

29 Vega CA, Buening GM, Rodriguez SD, Carson CA. Concentration and enzyme content of *in vitro*-cultured *Babesia bigemina*-infected erythrocytes. J Protozool 1986;33: 514–518.

30 Ray A, Quade J, Carson CA, Ray BK. Calcium-dependent protein phosphorylation in *Babesia bovis* and its role in growth regulation. J Parasitol 1990;76:153–161.

31 Upston JM, Gero AM. Increased glucose permeability in *Babesia bovis*-infected erythrocytes. Int J Parasitol 1990;20: 69–76.

32 Gero AM. Induction of nucleoside transport sites into the host cell membrane of *Babesia bovis* infected erythrocytes. Mol Biochem Parasitol 1989;35:269–276.

33 Callow LL. *Animal Health in Australia*, vol. 5, *Protozoal and Rickettsial Disease*. Canberra: Australian Government Publishing Service, 1984.

34 Figueroa JV, Buening GM, Kinden DA, Green TJ. Identification of common surface antigens among *Babesia bigemina* isolates using monoclonal antibodies. Parasitology 1990;100:161–175.

35 Kahl LP, Anders RF, Rodwell BJ, Timms P, Mitchell GF. Variable and common antigens of *Babesia bovis* parasites differing in strain and virulence. J Immunol 1982;129: 1700–1705.

36 McElwain TF, Palmer GH, Goff WL, McGuire TC. Identification of *Babesia bigemina* and *Babesia bovis* merozoite proteins with isolate- and species-common epitopes recognized by antibodies in bovine immune sera. Infect Immun 1988;56:1658–1660.

37 Phillips RS, Reid GM, McLean SA, Pearson CD. Antigenic diversity in *Babesia divergens*: preliminary results with three monoclonal antibodies to the rat-adapted strain. Res Vet Sci 1986;42:96–100.

38 Gill AC, Cowman AF, Stewart NP, Kemp DJ, Timms P. *Babesia bovis*: molecular and biological characteristics of cloned parasite lines. Exp Parasitol 1987;63:180–188.

39 Cowman AF, Timms P, Kemp DJ. DNA polymorphisms and subpopulations in *Babesia bovis*. Mol Biochem Parasitol 1984;11:91–103.

40 Dalrymple BP. Cloning and characterisation of the rRNA genes and flanking regions from *Babesia bovis*: use of the genes as strain discriminating probes. Mol Biochem Parasitol 1990;43:117–124.

41 Roberts JA, Tracey-Patte P. *Babesia rodhaini*: immuno-induction of antigenic variation. Int J Parasitol 1975;5: 573–576.

42 Rodriguez SD, Buening GM, Green TJ, Carson CA. Cloning of *Babesia bovis* by *in vitro* cultivation. Infect Immun 1983;42:15–18.

43 Timms P. Development of babesial vaccines. Trans R Soc Trop Med Hyg 1989;83(Suppl.):73–79.

44 Timms P, Stewart NP, de Vos AJ. Study of virulence and vector transmission of *Babesia bovis* by use of cloned parasite lines. Infect Immun 1990;58:2171–2176.

45 Buening GM, Kuttler KL, Rodriguez SD. Evaluation of a cloned *Babesia bovis* organism as a live immunogen. Vet Parasitol 1986;22:235–242.

46 Carson CA, Timms P, Cowman AF, Stewart NP. *Babesia bovis*: evidence for selection of subpopulations during attenuation. Exp Parasitol 1990;70:404–410.

47 Tetzlaff CL, McMurray DN, Rice-Ficht AC. Isolation and characterization of a gene associated with a virulent strain of *Babesia microti*. Mol Biochem Parasitol 1990;40:183–192.

48 Callow LL. Ticks and tick-borne disease as a barrier to the introduction of exotic cattle to the tropics. World Anim Rev 1978;28:20–25.

49 Uilenberg G. Tick-borne livestock diseases and their vectors. 2. Epizootiology of tick-borne diseases. World Anim Rev 1976;17:8–15.

50 Callow LL. Vaccination against bovine babesiosis. In Miller LH, Pino JA, McKelvey JJ, eds. *Immunity to Blood Parasites of Animals and Man*. New York: Plenum Publishers 1977:121–149.

51 Joyner LP, Donnelly J. The epidemiology of babesial infections. Adv Parasitol 1979;17:115–140.

52 Mahoney DF. Babesia of domestic animals. In Kreier JP, ed. *Parasitic Protozoa*. New York: Academic Press, 1977: 1–52.

53 Dallwitz MJ, Young AS, Mahoney DF, Sutherst RW. Comparative epidemiology of tick-borne diseases of cattle with emphasis on modelling. In Howell MJ, ed. *Parasitology quo vadit*? Canberra: Australian Academy of Science, 1986:629–637.

54 Smith RD, Kakoma I. A reappraisal of vector control strategies for babesiosis. Trans R Soc Trop Med Hyg 1989;83:43–52.

55 Western KA, Benson GD, Healy GR, Schultz MG. Babesiosis in a Massachusetts resident. N Engl J Med 1970;283:854–856.

56 Spielman A, Wilson ML, Levine JF, Piesman J. Ecology of *Ixodes dammini*-borne human babesiosis and lyme disease. Ann Rev Entomol 1985;30:439–460.

57 Goff WL, Wagner GG, Craig TM. Increased activity of bovine ADCC effector cells during acute *Babesia bovis* infection. Vet Parasitol 1984;16:5–15.

58 Goff WL, Wagner GG, Craig TM, Long RF. The bovine immune response to tick-derived *Babesia bovis* infection: serological studies of isolated immunoglobulins. Vet Parasitol 1982;11:109–120.

59 Goff WL, Wagner GG, Craig TM, Long RF. The role of specific immunoglobulins in antibody-dependent cell-mediated cytotoxicity assays during *Babesia bovis* infection. Vet Parasitol 1984;14:117–128.

60 Phillips RS. *Babesia divergens* in splenectomised rats. Res Vet Sci 1984;36:251–255.

61 Dalgliesh RJ, Swain AJ, Mellors LT. Bioassay to measure effects of cooling and warming rates and protection by dimethyl sulphoxide on survival of frozen *Babesia rodhaini*. Cryobiology 1976;13:631–637.

62 Piesman J. Intensity and duration of *Borrelia burgdorferi* and *Babesia microti* infectivity in rodent hosts. Int J Parasitol 1988;18:687–689.

63 Liddell KG, Lucas SB, Williams H. *Babesia divergens* infections in the Mongolian gerbil: characteristics of a human strain. Parasitology 1980;82:205–224.

64 Winger CM, Canning EU, Culverhouse JD. Induction of protective immunity to *Babesia divergens* in Mongolian gerbils, *Meriones unguiculatus*, using culture-derived immunogens. Vet Parasitol 1987;26:43–53.

65 Rudzinska MA. Morphologic aspects of host-cell–parasite relationships in babesiosis. In Ristic M, Kreier JP, eds. *Babesiosis*. New York: Academic Press, 1981;87–141.

66 Jack RM, Ward PA. Mechanisms of entry of *Plasmodium* and *Babesia* into red cells. In Ristic M, Kreier JP, eds. *Babesiosis*, New York: Academic Press, 1981;445–457.

67 Seinen W, Stegmann T, Kuil H. Complement does not play a role in promoting *Babesia rodhaini* infections in BALB/C mice. Z Parasitenkd 1982;68:249–257.

68 Levy MG, Kakoma I, Clabaugh G, Ristic M. Studies on the role of complement in the *in vitro* invasion of bovine erythrocytes by *Babesia bovis*. Rev Elev Med Vet Pays Trop 1986;39:317–322.

69 James MA, Levy MG, Ristic M. Isolation and partial characterization of culture-derived soluble *Babesia bovis* antigens. Infect Immun 1981;31:358–361.

70 Snary D, Smith MA. An antigenic determinant common to both mouse red blood cells and several membrane proteins of the parasitic protozoa *Babesia rodhaini*. Mol Biochem Parasitol 1986;20:101–109.

71 McElwain TF, Perryman LE, Davis WC, McGuire TC. Antibodies define multiple proteins with epitopes exposed on the surface of live *Babesia bigemina* merozoites. J Immunol 1987;138:2298–2304.

72 Snary D. Structural homology of membrane proteins of *Babesia rodhaini*. In Chang K-P, Snary D, eds. *Host–Parasite Cellular and Molecular Interactions in Protozoal Infections*. Heidelberg: Springer-Verlag 1987;335–344.

73 McHardy N. Immunization of rats against *Babesia (Nuttallia) rodhaini*. Nature 1967;214:805.

74 Homewood CA, Neame KD, Momen H. Permeability of erythrocytes from mice infected with *Babesia rodhaini*. Ann Trop Med Parasitol 1975;69:429–434.

75 Maegraith B, Gilles HM, Devakul K. Pathological processes in *Babesia canis* infections. Z Tropenmed Parasitol 1957;8:485–514.

76 Dorner JL. A hematologic study of babesiosis of the dog. Am J Vet Clin Pathol 1967;1:67–75.

77 Wright IG. Osmotic fragility of erythrocytes in acute *Babesia argentina* and *Babesia bigemina* infections in splenectomised *Bos taurus* calves. Res Vet Sci 1973;15:299–305.

78 Gneno R, Azzar G, Got R, Roux B. Permeability of membrane of *Babesia canis* infected erythrocytes – influence of an external electric field. Int J Biochem 1986;18:1151–1154.

79 Aikawa M, Rabbege J, Uni S, Ristic M, Miller LH. Structural alterations of the membrane of erythrocytes infected with *Babesia bovis*. Am J Trop Med Hyg 1985;34:45–49.

80 Everitt JI, Shadduck JA, Steinkamp C, Clabaugh G. Experimental *Babesia bovis* infection in Holstein calves. Vet Pathol 1986;23:556–562.

81 Kawai S, Takahashi K, Sonoda M, Kurosawa T. Ultra-

structure of intra-erythrocytic stages of *Babesia ovata*. Jpn J Vet Sci 1986;48:943–949.

82 Howard RJ, Rodwell BJ, Smith PM, Callow LL, Mitchell GF. Comparison of the surface proteins and glycoproteins on erythrocytes of calves before and during infection with *Babesia bovis*. J Protozool 1980;27:241–247.

83 Howard RJ, Smith PM, Mitchell GF. Characterization of surface proteins and glycoproteins on red blood cells from mice infected with haemosporidia: *Babesia rodhaini* infections of BALB/c mice. Parasitology 1980;81:251–271.

84 Commins MA, Goodger BV, Waltisbuhl DJ, Wright IG. *Babesia bovis*: studies of parameters influencing microvascular stasis of infected erythrocytes. Res Vet Sci 1988;44:226–228.

85 Commins MA, Goodger BV, Wright IG, Waltisbuhl DJ. *Babesia bovis* – effects of phospholipid translocation and adenosine tri-phosphate consumption. Int J Parasitol 1990;20:395–396.

86 Goodger BV, Mahoney DF, Wright IG. *Babesia bovis*: attachment of infected erythrocytes to heparin-sepharose columns. J Parasitol 1983;69:248–249.

87 Dalgliesh RJ, Callow LL, Mellors LT, McGregor W. Development of a highly infective *Babesia bigemina* vaccine of reduced virulence. Aust Vet J 1981;57:8–11.

88 Curnow JA. Studies on antigenic changes and strain differences in *Babesia argentina* infections. Aust Vet J 1973;49:279–283.

89 Kahl LP, Mitchell GF, Dalgliesh RJ, *et al*. *Babesia bovis*: proteins of virulent and avirulent parasites passaged through ticks and splenectomized or intact calves. Exp Parasitol 1983;56:222–235.

90 Cowman AF, Bernard O, Stewart N, Kemp DJ. Genes of the protozoan parasite *Babesia bovis* that rearrange to produce RNA species with different sequences. Cell 1984;37:653–660.

91 Callow LL. Some aspects of the epidemiology and control of bovine babesiosis in Australia. J S Afr Vet Assoc 1979;4:353–356.

92 De Vos AJ. Immunogenicity and pathogenicity of three South African strains of *Babesia bovis* in *Bos indicus* cattle. Onderst J Vet Res 1978;45:119–124.

93 Callow LL, Dalgliesh RJ. The development of effective, safe vaccination against babesiosis and anaplasmosis in Australia. In Johnston LAY, Cooper MG, eds. *Ticks and Tick-borne Diseases*. Sydney: Australian Veterinary Association 1980:4–8.

94 Callow LL, Mellors LT, McGregor W. Reduction in virulence of *Babesia bovis* due to rapid passage in splenectomised cattle. Int J Parasitol 1979;9:333–338.

95 De Vos AJ, Bessenger R, Fourie CG. Virulence and heterologous strain immunity of South African and Australian *Babesia bovis* strains with reduced pathogenicity. Onderst J Vet Res 1982;49:133–136.

96 Dalgliesh RJ, Stewart NP, Duncalfe F. Reduction in pathogenicity of *Babesia bovis* for its tick vector, *Boophilus microplus*, after rapid blood passage in splenectomized calves. Z Parasitenkd 1981;64:347–351.

97 Stewart NP. Differences in the life cycles between a vaccine strain and an unmodified strain of *Babesia bovis* (Babes, 1889) in the tick *Boophilus microplus* (Canestrini). J

Protozool 1978;25:497–501.

98 Dalgliesh RJ, Stewart NP. Failure of vaccine strains of *Babesia bovis* to regain infectivity for ticks during long-standing infections in cattle. Aust Vet J 1977;53:429–431.

99 Birago C, Bucci A, Dore E, Frontali C, Zenobi P. Mosquito infectivity is directly related to the proportion of repetitive DNA in *Plasmodium berghei*. Mol Biochem Parasitol 1982;6: 1–12.

100 Corcoran LM, Forsyth KP, Bianco AE, Brown GV, Kemp DJ. Chromosome size polymorphisms in *Plasmodium falciparum* can involve deletions and are frequent in natural parasite populations. Cell 1986;44:87–95.

101 Snary D, Smith MA. Sequence homology of surface membrane proteins of *Babesia rodhaini*. Mol Biochem Parasitol 1988;27:303–312.

102 Taylor SM, Kenny J, Mallon T. The effect of multiple rapid passage on strains of *Babesia divergens*: a comparison of the clinical effects on juvenile and adult cattle of passaged and irradiated parasites. J Comp Pathol 1983;93:391–396.

103 Murphy TM, Gray JS, Langley RJ. Effects of rapid passage in the gerbil (*Meriones unguiculatus*) on the course of infection of the bovine piroplasm *Babesia divergens* in splenectomised calves. Res Vet Sci 1986;40:285–287.

104 Stewart NP, Dalgliesh RJ, de Vos AJ. Effect of different methods of maintenance on the development and morphology of *Babesia bigemina* in the gut of *Boophilus microplus*. Res Vet Sci 1986;40:94–98.

105 Gray JS, Langley RJ, Murphy TM. Morphological comparisons of the bovine piroplasm, *Babesia divergens*, in cattle and jird (*Meriones unguiculatus*) erythrocytes. J Parasitol 1985;71:799–802.

106 James MA, Kuttler KL, Levy MG, Ristic M. Antibody kinetics in response to vaccination against *Babesia bovis*. Am J Vet Res 1981;42:1999–2001.

107 Goodger BV, Wright IG, Mahoney DF. Changes in conglutinin, immunoconglutinin, complement C3 and fibronectin concentrations in cattle acutely infected with *Babesia bovis*. Aust J Exp Biol Med Sci 1981;59:531–538.

108 Mahoney DF, Wright IG, Goodger BV. Changes in the haemolytic activity of serum complement during acute *Babesia bovis* infection in cattle. Z Parasitenkd 1980;62: 39–45.

109 Wright IG. Kinin, kininogen and kininase levels during acute *Babesia bovis* (= *B. argentina*) infection of cattle. Br J Pharmacol 1977;61:567–572.

110 Wright IG, Mahoney DF. The activation of kallikrein in acute *Babesia argentina* infections of splenectomised calves. Z Parasitenkd 1974;43:271–278.

111 Dalgliesh RJ, Dimmock CK, Hill MWM, Mellors LT. *Babesia argentina*: disseminated intravascular coagulation in acute infections in splenectomized calves. Exp Parasitol 1976;40:124–131.

112 Goodger BV, Wright IG. *Babesia bovis* (*argentina*): observations of coagulation parameters, fibrinogen catabolism and fibrinolysis in intact and splenectomized cattle. Z Parasitenkd 1977;54:9–27.

113 Goodger BV, Wright IG, Mahoney DF, McKenna RV. *Babesia bovis* (*argentina*): components of the cryofibrinogen complex and their contribution to pathophysiology of infection in splenectomized calves. Z Parasitenkd 1978;

58:3–13.

114 Goodger BV, Commins MA, Wright IG, Mirre GB. *Babesia bovis*: vaccination of cattle against heterologous challenge with fractions of lysate from infected erythrocytes. Z Parasitenkd 1984;70:321–329.

115 Goodger BV, Wright IG, Mahoney DF. The use of pathophysiological reactions to assess the efficacy of the immune response to *Babesia bovis* in cattle. Z Parasitenkd 1981;66:41–48.

116 Eckersall PD, Conner JG. Bovine and canine acute phase proteins. Vet Res Commun 1988;12:169–178.

117 Pepys MB, Baltz ML. Acute phase proteins with special reference to C-reactive protein and related proteins (pentaxins) and serum amyloid A protein. Adv Immunol 1983;34:141–212.

118 Chapman WE, Ward PA. The complement profile in babesiosis. J Immunol 1976;117:935–938.

119 Adams LG, Wagner GG. Alternations in complement during acute bovine babesiosis. Adv Exp Med Biol 1981; 137:793.

120 Goodger BV, Wright IG. Acute *Babesia bigemina* infection: changes in fibrinogen catabolism. Z Parasitenkd 1977;53: 53–61.

121 Bremner KC. Studies on haptoglobin and haemopexin in the plasma of cattle. Aust J Exp Biol Med Sci 1964;42: 643–656.

122 Thoongsuwan S, Cox HW, Patrick RA. Serologic specificity of immunoconglutinin associated with infectious anemia of rats and its role in nonspecific acquired resistance. J Parasitol 1978;64:1060–1066.

123 Barry DN, Rodwell BJ, Timms P, McGregor W. A microplate enzyme immunoassay for detecting and measuring antibodies to *Babesia bovis* in cattle serum. Aust Vet J 1982;59:136–140.

124 Mahoney DF, Wright IG, Goodger BV. Immunity in cattle to *Babesia bovis* after single infections with parasites of various origin. Aust Vet J 1979;55:10–12.

125 Waltisbuhl DJ, Goodger BV, Wright IG, Commins MA, Mahoney DF. An enzyme linked immunosorbent assay to diagnose *Babesia bovis* infection in cattle. Parasitol Res 1987;73:126–131.

126 Kahl LP, Anders RF, Callow LL, Rodwell BJ, Mitchell GF. Development of a solid-phase radioimmunoassay for antibody to antigens of *Babesia bovis* infected bovine erythrocytes. Int J Parasitol 1982;12:103–109.

127 O'Donoghue PJ, Friedhoff KT, Vizcaino OG, Weyreter H. The detection of IgM and IgG antibodies against *Babesia bigemina* in bovine sera using semi-defined antigens in enzyme immunoassays. Vet Parasitol 1985;18:1–12.

128 Meeusen E, Lloyd S, Soulsby EJL. Antibody levels in adoptively immunized mice after infection with *Babesia microti* or injection with antigen fractions. Aust J Exp Biol Med Sci 1985;63:261–272.

129 Parrodi F, Wright IG, Dobson C. Immunity to *Babesia microti* in male mice selected as resistant or susceptible to *Nematospiroides dubius*. Int J Parasitol 1988;18:539–541.

130 Zwart D, Brocklesby DW. Babesiosis: non-specific resistance, immunological factors and pathogenesis. Adv Parasitol 1979;17:49–113.

131 Allison AC, Christensen J, Clark IA, Elford BC, Eugui EM.

The role of the spleen in protection against murine babesia infections. Tropical Diseases Research Series No. 1. *Proceedings of UNDP/World Bank/WHO Meeting in Geneva, June 12–14 1978*. Basel: Schwabe & Co., 1979:151–182.

132 Eugui EM, Allison AC. Differences in susceptibility of various mouse strains to haemoprotozoan infections: possible correlation with natural killer activity. Parasite Immunol 1980;2:277–292.

133 Inchley CJ, Grieve EM, Preston PM. The proliferative response of mouse lymphoid tissues during infections with *Babesia microti* or *Babesia rodhaini*. Int J Parasitol 1987; 17:945–950.

134 Irvin AD, Young ER, Osborn GD, Francis LMA. A comparison of *Babesia* infections in intact, surgically splenectomised, and congenitally asplenic (Dh/+) mice. Int J Parasitol 1981;11:251–255.

135 Inchley CJ. The contribution of B-cell proliferation to spleen enlargement in *Babesia microti*-infected mice. Immunology 1987;60:57–61.

136 Ishimine T, Nagasawa H, Suzuki N. An in vitro study of monocyte phagocytosis in the peripheral blood of healthy and *Babesia*-infected beagles. Jpn J Vet Sci 1979;41:487–493.

137 Mahoney DF, Kerr JD, Goodger BV, Wright IG. The immune response of cattle to *Babesia bovis* (syn. *B. argentina*). Studies on the nature and specificity of protection. Int J Parasitol 1979;9:297–306.

138 Rogers RJ. Serum opsonins and the passive transfer of protection in *Babesia rodhaini* infections of rats. Int J Parasitol 1974;4:197–201.

139 Wood PR, Clark IA. Macrophages from *Babesia* and malaria infected mice are primed for monokine release. Parasite Immunol 1984;6:309–317.

140 Montealegre F, Levy MG, Ristic M, James MA. Growth inhibition of *Babesia bovis* in culture by secretions from bovine mononuclear phagocytes. Infect Immun 1985;50: 523–526.

141 Bautista CR, Kreier JP. The action of macrophages and immune serum on growth of *Babesia microti* in short-term cultures. Tropenmed Parasitol 1980;31:313–324.

142 Timms P, Stewart NP, Rodwell BJ, Barry DN. Immune responses of cattle following vaccination with living and non-living *Babesia bovis* antigens. Vet Parasitol 1984;16: 243–251.

143 Barry DN, Stewart NP, Timms P. Lymphocyte anergy in bovine babesiosis. In *Programme and Abstracts for 1st Combined Meeting Australian & New Zealand Societies for Immunology, Dec 3–6 1985, Queenstown, New Zealand*, Abstract no. 196.

144 Ruebush MJ, Burgess DE. Induction of natural killer cells and interferon production during infection of mice with *Babesia microti* of human origin. In Heberman RB, ed. *NK Cells and Other Natural Effector Cells*. New York: Academic Press, 1982:1483–1489.

145 Eugui EM, Allison AC. Activation of natural killer cells and its possible role in immunity to intracellular parasites. In Torrigiani G, Bell R, eds. *Immunological Recognition and Effector Mechanisms in Infectious Diseases*. Basel: Schwabe & Co., 1981:161–187.

146 Mahoney DF, Wright IG, Frerichs WM, *et al.* The identification of *Babesia equi* in Australia. Aust Vet J 1977;

53:461–464.

147 Wright IG, Goodger BV, Rode-Bramanis K, Mattick JS, Mahoney DF, Waltisbuhl DJ. The characterisation of an esterase derived from *Babesia bovis* and its use as a vaccine. Z Parasitenkd 1983;69:703–714.

148 Wright IG, Kerr JD. Effect of trasylol on packed cell volume and plasma kallikrein activation in acute *Babesia argentina* infection of splenectomised calves. Z Parasitenkd 1975;46:189–194.

149 Clark IA. Does endotoxin cause both the disease and parasite death in acute malaria and babesiosis? Lancet 1978;ii:75–77.

150 Clark IA. Protection of mice against *Babesia microti* with cord factor, COAM, zymosan, glucan, *Salmonella* and *Listeria*. Parasite Immunol 1979;1:179–196.

151 Clark IA. Cell-mediated immunity in protection and pathology of malaria. Parasitol Today 1987;3:300–305.

152 Clark IA, Hunt NH, Cowden WB. Oxygen-derived free radicals in the pathogenesis of parasitic disease. Adv Parasitol 1986;25:1–44.

153 Buffinton GD, Hunt NH, Cowden WB, Clark IA. Malaria: a role for reactive oxygen species in parasite killing and host pathology. In Rice-Evans C, ed. *Free Radicals, Cell Damage and Disease*. London: Richelieu Press, 1986: 201–220.

154 Parrodi F, Wright IG, Bourne AS, Dobson C. *In vitro* adherence of bovine erythrocytes infected with *Babesia bovis* to thrombospondin and laminin. Int J Parasitol 1989; 19:567–569.

155 Beutler B, Cerami A. The common mediator of shock, cachexia, and tumour necrosis. Adv Immunol 1988;42: 213–231.

156 Nawroth PP, Stern DM. Modulation of endothelial cell hemostatic properties by tumor necrosis factor. J Exp Med 1986;163:740–745.

157 Franz DR, Lee M, Seng LT, Young GD, Baze WB, Lewis GE. Peripheral vascular pathophysiology of *Plasmodium berghei* infection: a comparative study in the cheek pouch and brain of the golden hamster. Am J Trop Med Hyg 1987;36:474–480.

158 Czop JK. Phagocytosis of particulate activators of the alternative complement pathway: effects of fibronectin. Adv Immunol 1986;38:361–398.

159 Ross GD, Medof ME. Membrane complement receptors specific for bound fragments of C3. Adv Immunol 1985; 37:217–267.

160 Young JD-E, Cohn ZA. Cellular and humoral mechanisms of cytotoxicity: structural and functional analogies. Adv Immunol 1987;41:269–332.

161 Goodger BV, Waltisbuhl DJ, Wright IG, Mahoney DF, Commins MA. *Babesia bovis*—analysis and vaccination trial with the cryoprecipitable immune complex. Vet Immunol Immunopathol 1987;14:57–65.

162 Goodger BV, Wright IG, Mahoney DF. Initial characterization of cryoprecipitates in cattle recovering from acute *Babesia bovis (argentina)* infection. Aust J Exp Biol Med Sci 1981;59:521–529.

163 Wolf CFW, Resnick G, Marsh WL, Benach J, Habicht G. Autoimmunity to red blood cells in babesiosis. Transfusion 1982;22:538–539.

164 Callow LL, Pepper PM. Measurement of and correlations between fever, changes in the packed cell volume and parasitaemia in the evaluation of the susceptibility of cattle to infection with *Babesia argentina*. Aust Vet J 1974;50:1–5.

165 Elissalde GS, Wagner GG, Craig TM, Elissalde MH, Rowe L. Hypocholesterolemia and hypocortisolemia in acute and terminal *Babesia bovis* infections. Vet Parasitol 1983; 12:1–11.

166 Callow LL, Stewart NP. Observations on immuno-suppression during primary babesiosis of cattle, expressed against the tick, *Boophilus microplus*. In Johnston LAY, Cooper MG, eds. *Ticks and Tick-borne Diseases*. Sydney: Australian Veterinary Association, 1980:67–72.

167 Kawamura M, Maede Y, Namioka S. Mitogenic responsibilities of lymphocytes in canine babesiosis and the effects of splenectomy on it. Jpn J Vet Res 1987;35: 1–10.

168 Purvis AC. Immunodepression in *Babesia microti* infections. Parasitology 1977;75:197–205.

169 Miller JFA, Mitchell GF. Cell to cell interaction in the immune response. 1. Haemolysin-forming cells in neonatally thymectomised mice reconstituted with thymus or thoracic duct lymphocytes. J Exp Med 1968; 128:801–820.

170 Garnham PCC, Voller A. Experimental studies on *Babesia divergens* in rhesus monkeys with special reference to its diagnosis by serological methods. Acta Protozool 1965;3: 183–187.

171 Irvin AD, Young ER, Osborn GD. Attempts to infect T lymphocyte-deficient mice with *Babesia* species of cattle. Res Vet Sci 1978;25:245–246.

172 Ristic M, Levy MG. A new era of research toward solution of bovine babesiosis. In Ristic M, Kreier JP, eds. *Babesiosis*. New York: Academic Press, 1981:509–544.

173 Yunker CE, Kuttler KL, Johnson LW. Attenuation of *Babesia bovis* by *in vitro* cultivation. Vet Parasitol 1987;24: 7–13.

174 Trueman KF, Blight GW. The effect of age on resistance of cattle to *Babesia bovis*. Aust Vet J 1978;54:301–305.

175 Brocklesby DW, Harness E, Sellwood SA. The effect of age on the natural immunity of cattle to *Babesia divergens*. Res Vet Sci 1971;12:15–17.

176 Christensson DA. Inverse age resistance to experimental *Babesia divergens* infection in cattle. Acta Vet Scand 1989;30: 453–464.

177 Christensson DA, Thorburn MA. Age distribution of naturally occurring acute babesiosis in cattle in Sweden. Acta Vet Scand 1987;28:373–379.

178 Langley RJ, Gray JS. Age-related susceptibility of the gerbil, *Meriones unguiculatus*, to the bovine parasite, *Babesia divergens*. Exp Parasitol 1987;64:466–473.

179 Benach JL, Habicht GS. Clinical characteristics of human babesiosis. J Infect Dis 1981;144:481.

180 Habicht GS, Benach JL, Leichtling KD, Gocinski BL, Coleman JL. The effect of age on the infection and immunoresponsiveness of mice to *Babesia microti*. Mech Ageing Dev 1983;23:357–369.

181 Johnston LAY, Sinclair DF. Differences in response to experimental primary infection with *Babesia bovis* in Hereford, Droughtmaster and Brahman cattle. In Johnston

LAY, Cooper MG, eds. *Ticks and Tick-borne Diseases*. Australian Veterinary Association, 1980:18–21.

182 Löhr KF. Susceptibility of non-splenectomized and splenectomized Sahiwal cattle to experimental *Babesia bigemina* infection. Z Veterinarmed 1973;20:52–56.

183 Levy MG, Clabaugh G, Ristic M. Age resistance in bovine babesiosis: role of blood factors in resistance to *Babesia bovis*. Infect Immun 1982;37:1127–1131.

184 Futter GJ, Belonje PC, Van den Berg A. Studies on feline babesiosis. 3. Haematological findings. JS Afr Vet Assoc 1980;51:271–280.

185 Moore JA, Kuntz RE. *Babesia microti* infections in non-human primates. J Parasitol 1981;67:454–456.

186 Wood PR, Clark IA. Apparent irrelevance of NK cells to resolution of infections with *Babesia microti* and *Plasmodium vinckei petteri* in mice. Parasite Immunol 1982;4:319–327.

187 Allison AC, Eugui EM. A radical interpretation of immunity to malaria parasites. Lancet 1982;ii:1431–1433.

188 Clark IA, Allison AC. *Babesia microti* and *Plasmodium berghei yoelii* infections in nude mice. Nature 1974;252:328–329.

189 Ruebush MJ, Hanson WL. Thymus dependence of resistance to infection with *Babesia microti* of human origin in mice. Am J Trop Med Hyg 1980;29:507–515.

190 Wolf RE. Effects of antilymphocyte serum and splenectomy on resistance to *Babesia microti* infection in hamsters. Clin Immunol Immunopathol 1974;2:381–394.

191 Mitchell GF. Studies on immune responses to parasite antigens in mice. V. Different susceptibilities of hypothalmic and intact mice to *Babesia rodhaini*. Int Arch Allergy Appl Immunol 1977;53:385–388.

192 Zivkovic D, Seinen W, Kuil H, Albero-Van Bemmel CMG, Speksnijder JE. Immunity to *Babesia* in mice. I. Adoptive transfer of immunity to *Babesia rodhaini* with immune spleen cells and the effect of irradiation on the protection of immune mice. Vet Immunol Immunopathol 1983/84; 5:343–357.

193 Hall WTK. The immunity of calves to tick-transmitted *Babesia argentina* infection. Aust Vet J 1963;39:386–389.

194 Hall WTK, Tammemagi L, Johnston LAY. Bovine babesiosis: the immunity of calves to *Babesia bigemina* infection 1968;44:259–264.

195 Christensson DA. Clinical and serological response after experimental inoculation with *Babesia divergens* of newborn calves with and without maternal antibodies. Acta Vet Scand 1987;28:381–392.

196 Weisman J, Goldman M, Mayer E, Pipano E. Passive transfer to newborn calves of maternal antibodies against *Babesia bigemina* and *Babesia berbera*. Refuah Vet 1974;31: 108–113.

197 Cox FEG, Young AS. Acquired immunity to *Babesia microti* and *Babesia rodhaini* in mice. Parasitology 1969;59:257–268.

198 Hussein HS. The nature of immunity against *Babesia hylomysci* and *B. microti* infections in mice. Ann Trop Med Parasitol 1977;71:249–253.

199 Callow LL, McGregor W, Parker RJ, Dalgliesh RJ. The immunity of cattle to *Babesia argentina* after drug sterilisation of infections of varying duration. Aust Vet J 1974;50: 6–11.

200 Löhr KF. Immunity to *Babesia bigemina* in experimentally infected cattle. J Protozool 1972;19:658–660.

201 Wright IG, Goodger BV, Leatch G, Aylward JH, Rode-Bramanis K, Waltisbuhl DJ. Protection of *Babesia bigemina*-immune animals against subsequent challenge with virulent *Babesia bovis*. Infect Immun 1987;55:364–368.

202 Zwart D, Van der Ende MC, Kouwenhoven B, Buys J. The difference between *B. bigemina* and a Dutch strain of *B. major*. Tijdschr Diegeneeskd 1968;93:126–140.

203 Alabay M, Duzgun A, Cerci H, Wright IG, Waltisbuhl DJ, Goodger BV. Ovine babesiosis: induction of a protective immune response with crude extracts of either *Babesia bovis* or *B. ovis*. Res Vet Sci 1987;43:401–402.

204 Mahoney DF, Wright IG, Mirre GB. Bovine babesiosis: the persistence of immunity to *Babesia argentina* and *B. bigemina* in calves (*Bos taurus*) after naturally acquired infection. Ann Trop Med Parasitol 1973;67:197–203.

205 Cox FEG. Non-specific immunization against babesiosis. In *Isotope and Radiation Research on Animal Diseases and their Vectors*. Vienna: International Atomic Energy Agency, 1980:95–105.

206 Langley RJ, Gray JS. Non-specific resistance to *Babesia divergens* in the Mongolian gerbil (*Meriones unguiculatus*). Int J Parasitol 1989;19:265–269.

207 Brocklesby DW, Purnell RE. The failure of BCG to protect calves against *Babesia divergens* infection. Nature 1977;265:343.

208 Corrier DE, Wagner GG. The protective effect of pre-treatment with killed *Corynebacterium parvum* against acute babesiosis in calves. Vet Parasitol 1984;15:165–168.

209 Cox HW, Milar R. Cross-protection immunization by *Plasmodium* and *Babesia* infections of rats and mice. Am J Trop Med Hyg 1968;17:173–179.

210 Sibinovic KH, Sibinovic S, Ristic M, Cox HW. Immunogenic properties of babesial serum antigens. J Parasitol 1967;53:1121.

211 Mahoney DF, Goodger BV. The isolation of *Babesia* parasites and their products from the blood. In Ristic M, Kreier JP, eds. *Babesiosis*. New York: Academic Press, 1981:323–335.

212 Mahoney DF. Studies on the protection of cattle against *Babesia bovis* infection. In Morrison WI, ed. *The Ruminant Immune System in Health and Disease*. Cambridge: Cambridge University Press, 1986:539–554.

213 Callow LL. A note on homologous strain immunity in *Babesia argentina* infections. Aust Vet J 1968;44:268–269.

214 Ruebush MJ, Hanson WL. Transfer of immunity to *Babesia microti* of human origin using T lymphocytes in mice. Cell Immunol 1980;52:255–265.

215 Zivkovic D, Speksnijder JE, Kuil H, Seinen W. Immunity to *Babesia* in mice. II. Cross protection between various *Babesia* and *Plasmodium* species and its relevance to the nature of *Babesia* immunity. Vet Immunol Immunopathol 1983/84;5:359–368.

216 Mitchell GF, Handman E, Howard RJ. Protection of mice against *Plasmodium* and *Babesia* infections: attempts to raise host-protective sera. Aust J Exp Biol Med Sci 1978;56:553–559.

217 Meeusen E, Lloyd S, Soulsby EJL. *Babesia microti* in mice. Adoptive transfer of immunity with serum and cells. Aust J Exp Biol Med Sci 1984;62:551–566.

218 Meeusen E, Lloyd S, Soulsby EJL. *Babesia microti* in mice.

219 Clark IA, Richmond JE, Wills EJ, Allison AC. Intra-erythrocytic death of the parasite in mice recovering from infection with *Babesia microti*. Parasitology 1977;75:189–196.

220 Allison AC. Cellular immunity to malaria and babesia parasites: a personal viewpoint. In Marchalonis JJ, ed. *Contemporary Topics in Immunobiology*, vol. 12. New York: Plenum Publishers 1984:463–490.

221 Suzuki Y, Orellana MA, Schreiber RD, Remington JS. Interferon-γ: the major mediator of resistance against *Toxoplasma gondii*. Science 1988;240:516–518.

222 Scholfield L, Villaquiran J, Ferreria A, Schellekens H, Nussenzweig R, Nussenzweig V. γ Interferon, CD8[+] T cells and antibodies required for immunity to malaria sporozoites. Nature 1987;330:664–666.

223 Johnston LAY, Trueman KF, Pearson RD. Bovine babesiosis: comparison of fluorescent antibody and giemsa staining in post-mortem diagnosis of infection. Aust Vet J 1977;53:222–226.

224 McLaughlin GL, Edlind TD, Ihler GM. Detection of *Babesia bovis* using DNA hybridization. J Protozool 1986;33:125–128.

225 Mahoney DF, Saal JR. Bovine babesiosis: thick blood films for the detection of parasitaemia. Aust Vet J 1961;37:44–47.

226 Posnett ES, Ambrosio RE. Repetitive DNA probes for the detection of *Babesia equi*. Mol Biochem Parasitol 1989;34:75–78.

227 Buening GM, Barbet A, Myler P, Mahan S, Nene V, McGuire TC. Characterization of a repetitive DNA probe for *Babesia bigemina*. Vet Parasitol 1990;36:11–20.

228 Weiland G, Reiter I. Methods for the measurement of the serological response to *Babesia*. In Ristic M, ed. *Babesiosis of Domestic Animals and Man*. Boca Raton: CRC Press, 1988:143–162.

229 James MA, Coronado A, Lopez W, Melendez R, Ristic M. Seroepidemiology of bovine anaplasmosis and babesiosis in Venezuela. Trop Anim Health Prod 1985;17:9–18.

230 Wanduragala L, Kakoma I, Clabaugh GW, Abeygunawardena I, Levy MG, Ristic M. Development of dot-enzyme immunoassay for diagnosis of canine babesiosis. Am J Trop Med Hyg 1987;36:20–21.

231 Kung'u MW, Goodger BV. A slide enzyme-linked immunosorbent assay (SELISA) for the diagnosis of *Babesia bovis* infections and for the screening of *Babesia*-specific monoclonal antibodies. Int J Parasitol 1990;20:341–345.

232 Christensson DA, Moren T. Seroresponse (IgG) after vaccination and natural infection of cattle with *Babesia divergens*. Acta Vet Scand 1987;28:393–402.

233 Irvin AD. Control of tick-borne diseases. Int J Parasitol 1987;17:649–658.

234 Pipano E, Hadani A. Control of bovine babesiosis. In Ristic M, Ambroise-Thomas P, Kreier J, eds. *Malaria and Babesiosis*. Dordrecht: Martinus Nijhoff, 1984:263–303.

235 De Vos AJ, Dalgliesh RJ, McGregor W. Effect of imidocarb dipropionate prophylaxis on the infectivity and immuno-genicity of a *Babesia bovis* vaccine in cattle. Aust Vet J 1986;63:174–177.

236 Rogers RJ, Dimmock CK, de Vos AJ, Rodwell BJ. Bovine leucosis virus contamination of a vaccine produced *in vivo* against bovine babesiosis and anaplasmosis. Aust Vet J 1988;65:285–287.

237 Dalgliesh RJ, Jorgensen WK, de Vos AJ. Australian frozen vaccines for the control of babesiosis and anaplasmosis in cattle—a review. Trop Anim Health Prod 1990;22:44–52.

238 Jorgensen WK, de Vos AJ, Dalgliesh RJ. Infectivity of cryopreserved *Babesia bovis, Babesia bigemina* and *Anaplasma centrale* for cattle after thawing, dilution and incubation at 30°C. Vet Parasitol 1989;31:243–251.

239 Mellors LT, Dalgliesh RJ, Timms P, Rodwell BJ, Callow LL. Preparation and laboratory testing of a frozen vaccine containing *Babesia bovis, Babesia bigemina* and *Anaplasma centrale*. Res Vet Sci 1982;32:194–197.

240 Pipano E. Frozen vaccines against tick fevers of cattle. In Mayer E, ed. *Proceedings Eleventh International Congress Diseases of Cattle, Tel-Aviv.* Haifa: Bregman Press, 1980: 678–681.

241 De Vos AJ, Combrink MP, Bessenger R. *Babesia bigemina* vaccine: comparison of the efficacy and safety of Australian and South African strains under experimental conditions in South Africa. Onderst J Vet Res 1982;49: 155–158.

242 Moreau Y, Laurent N. Antibabesial vaccination using antigens from cell culture fluids: industrial requirements. In Ristic M, Ambroise-Thomas P, Kreier J, eds. *Malaria and Babesiosis.* Dordrecht: Martinus Nijhoff 1984:129–140.

243 Moreau Y, Vidor E, Bissuel G, Dubreuil N. Vaccination against canine babesiosis: an overview of field observations. Trans R Soc Trop Med Hyg 1989;83:95–96.

244 Uilenberg G, Perie NM, Top PDJ, Arends PJ, Kool PJ, Zwart D. Babesiosis in the dog. Trop Geogr Med 1984; 36:102.

245 Uilenberg G, Franssen FFJ, Perie NM, Spanjer AAM. Three groups of *Babesia canis* distinguished and a proposal for nomenclature. Vet Quart 1989;11:33–40.

246 Ristic M, Montenegro-James S. Immunization against *Babesia*. In Ristic M, ed. *Babesiosis of Domestic Animals and Man.* Boca Raton: CRC Press, 1988:163–189.

247 Montenegro-James S, Ristic M, Benitez MT, Leon E, Lopez R. Heterologous strain immunity in bovine babesiosis using a culture-derived soluble *Babesia bovis* immunogen. Vet Parasitol 1985;18:321–337.

248 Timms P, Dalgliesh RJ, Barry DN, Dimmock CK, Rodwell BJ. *Babesia bovis*: comparison of culture-derived parasites, non-living antigen and conventional vaccine in the protection of cattle against heterologous challenge. Aust Vet J 1983;60:75–77.

249 Irvin AD, Brocklesby DW, Purnell RE. Radiation and isotopic techniques in the study and control of piroplasms of cattle: a review. Vet Parasitol 1979;5:17–28.

250 Purnell RE, Lewis D, Brabazon A, Francis LMA, Young ER, Grist C. Field use of an irradiated blood vaccine to protect cattle against redwater (*Babesia divergens* infection) on a farm in Dorset. Vet Rec 1981;108:28–31.

251 Weilgama DJ, Weerasinghe HMC, Perera PSG. Premunisation of calves against babesiosis using irradiated *Babesia bigemina*. In *Livestock Production and Diseases in the Tropics*, Proceedings of 5th International Conference, Kuala Lumpur, Malaysia, Aug 18–22 1986. Abstr Vet Bull 1988;58:946.

252 Kuttler KL, Zaugg JL, Yunker CE. The pathogenicity and immunologic relationship of a virulent and a tissue-culture-adapted *Babesia bovis*. Vet Parasitol 1988;27: 239–244.

253 Timms P, Stewart NP. Growth of *Babesia bovis* parasites in stationary and suspension cultures and their use in experimental vaccination of cattle. Res Vet Sci 1989;47: 309–314.

254 Jorgensen WK, de Vos AJ, Dalgliesh RJ. Comparison of immunogenicity and virulence between parasites from continuous culture and a splenectomised calf. Aust Vet J 1989;66:371–372.

255 Winger CM, Canning EU, Culverhouse JD. A strain of *Babesia divergens*, attenuated after long term culture. Res Vet Sci 1989;46:110–113.

256 Erp EE, Smith RD, Ristic M, Osorno BM. Optimization of the suspension culture method for *in vitro* cultivation of *Babesia bovis*. Am J Vet Res 1980;41:2059–2062.

257 Gray JS, Langley RJ, Brophy PO, Gannon P. Vaccination against bovine babesiosis with drug-controlled live parasites. Vet Rec 1989;125:369–372.

258 Weber C. Epidemiology of *Babesia divergens* infections in cattle in North Germany. 2. Immunization with live vaccine on three farms. Inaugural Dissertation, Tierarzliche Hochschule, Hannover, 1988. Protozool Abstr 1989;13:410.

259 Rodriguez SD, Buening GM, Vega CA, Carson CA. *Babesia bovis*: purification and concentration of merozoites and infected bovine erythrocytes. Exp Parasitol 1986;61: 236–243.

260 Waltisbuhl DJ, Goodger BV, Wright IG, Mirre GB, Commins MA. *Babesia bovis*: vaccination with three groups of high molecular weight antigens from lysate of infected erythrocytes. Parasitol Res 1987;73:319–323.

261 Wright IG, White M, Tracey-Patte PD, *et al. Babesia bovis*: isolation of a protective antigen by using monoclonal antibodies. Infect Immun 1983;41:244–250.

262 Goff WL, Davis WC, Palmer GH, *et al.* Identification of *Babesia bovis* merozoite surface antigens by using immune bovine sera and monoclonal antibodies. Infect Immun 1988;56:2363–2368.

263 Reduker DW, Jasmer DP, Goff WL, *et al.* A recombinant surface protein of *Babesia bovis* elicits bovine antibodies that react with live merozites. Mol Biochem Parasitol 1989;35: 239–248.

264 Winger CM, Canning EU, Culverhouse JD. A monoclonal antibody to *Babesia divergens* which inhibits merozoite invasion. Parasitology 1987;94:17–27.

265 McElwain TF, Perryman LE, Davis WC, McGuire TC. Immunization of cattle with a 45000 M.W. merozoite surface protein provides partial protection against experimental challenge with *Babesia bigemina*. In *Programme for 3rd International Congress on Malaria and Babesiosis, Sept 7–11, 1987, Annecy, France.* International Laveran Foundation, Annecy, 1987:261.

266 Canning EU, Winger CM, Culverhouse JD, Bisson AM, Sjolin SC. Monoclonal antibody derived antigens which inhibit development of *Babesia divergens* in *Meriones unguiculatus.* In *Programme for 3rd International Congress on*

Malaria and Babesiosis, Sept 7–11 1987, Annecy, France. International Laveran Foundation, Annecy, 1987:195.

267 Wright IG, Mirre GB, Rode-Bramanis K, Chamberlain M, Goodger BV, Waltisbuhl DJ. Protective vaccination against virulent *Babesia bovis* with a low-molecular-weight antigen. Infect Immun 1985;48:109–113.

268 Hines SA, McElwain TF, Buening GM, Palmer GH. Molecular characterization of *Babesia bovis* merozoite surface proteins bearing epitopes immunodominant in protected cattle. Mol Biochem Parasitol 1989;37:1–10.

269 Montenegro-James S, Benitez MT, Leon E, Lopez R, Ristic M. Bovine babesiosis: induction of protective immunity with culture-derived *Babesia bovis* and *Babesia bigemina* immunogens. Parasitol Res 1987;74:142–150.

270 Gill A, Timms P, Kemp DJ. cDNA clone encoding a high molecular weight antigen of *Babesia bovis*. Mol Biochem Parasitol 1987;22:195–202.

271 Timms P, Barry DN, Gill AC, Sharp PJ, de Vos AJ. Failure of a recombinant *Babesia bovis* antigen to protect cattle against heterologous strain challenge. Res Vet Sci 1988;45:267–269.

272 Tripp CA, Wagner GG, Rice-Ficht AC. *Babesia bovis*: gene isolation and characterization using a mung bean nuclease-derived expression library. Exp Parasitol 1989;69:211–225.

273 Kemp DJ, Easton KE, Cowman AF. Accurate transcription of a cloned gene from *Babesia bovis* in *Saccharomyces cerevisiae*. Mol Biochem Parasitol 1984;12:61–67.

3 Helminths

16 Schistosomiasis

George R. Newport & Daniel G. Colley

INTRODUCTION

The genus *Schistosoma* (phylum Platyhelminthes, class Trematoda) consists of parasitic flatworms whose definitive habitat is the bloodstream of certain warm-blooded vertebrates. Four species, *Schistosoma haematobium*, *Schistosoma japonicum*, *Schistosoma mansoni*, and *Schistosoma mekongi* are of direct medical importance to humans, affecting the livelihood of around 200–400 million humans and accounting for an estimated 800 000 deaths per year [1–3]. *Schistosoma bovis*, a parasite of cattle, and *Schistosoma mattheei*, a parasite of cattle and sheep, cause significant economic losses that further add to the burden imposed by the parasites. The organisms typically elicit vigorous humoral and cellular immune responses, yet their survival and reproductive capacities in the vertebrate host may occasionally span over 30 years, a remarkable phenotype for an invertebrate. Ample evidence exists, however, to suggest that under certain conditions humans and experimental animals can acquire varying degrees of immunity to the parasites, a finding that has encouraged pursuit of vaccines that sterilize, partially protect, or ameliorate the course of pathogenesis. Pathology is primarily the result of host immune responses towards the parasite's eggs but, as will be discussed, even this can be viewed as a compromise in the host–parasite relationship, for absence of such might be of more serious consequence to the host. It would seem that the parasite's dependence on a relatively long-lasting relationship that ensures transmission has been accomplished by the development, in most patients, of a balanced existence that permits the long-term persistence of both parties.

Schistosome infections are often accompanied by intermittent morbidity, but severe pathology can occur either acutely or over chronic periods. Acute manifestations include a dermatitis, probably due to previous sensitization to larval antigens, and a generalized systemic syndrome referred to as the acute or toxemic phase, or as Katayama fever. The latter resembles serum sickness, may be immune-complex mediated, and is most commonly noted during the initial stages of egg deposition, but may occur earlier in individuals receiving massive infections. Symptoms may include fever, chills, headache, hepatosplenomegaly, and lymphadenopathy, and are often accompanied by high peripheral and tissue eosinophilia. The development of this clinical form of infection most often occurs when persons from non-endemic areas become infected. More commonly, the infection remains relatively asymptomatic, with the life-threatening manifestations of hepatosplenomegaly, portal hypertension, and esophageal varices occurring in 5–10% of those chronically infected.

Key elements relevant to understanding the biology of schistosomes include:

1 they live in the bloodstream, and hence within the full immunologic purview of their definitive host, where they elicit parasite-specific humoral and cellular immune responses;

2 despite the previous point, they may survive for long periods;

3 host immune responses, which consist of protective, irrelevant, and harmful arms, are complex and vary over time and between individuals;

4 evidence exists to indicate that under natural and experimental conditions the host can become resistant to infection in an immunologically mediated manner.

Additionally, the parasites do not replicate in their definitive host and most of the induced pathology is not due to host responses against the worms but rather to eggs entrapped in tissues. In this chapter we review the current understanding of immunologic and molecular aspects of the organisms' relationship with their host, emphasizing the parasite's biology (metabolic and molecular biologic), host humoral and cellular responses, the parasite's antigenic makeup, immunopathogenesis, and

PARASITE BIOLOGY

The life-cycle of schistosomes calls for an alternation of generations, with asexual reproduction taking place in a susceptible snail host and sexual reproduction in a warm-blooded vertebrate. Humans become infected upon contact with water containing cercariae, which are transiently free-living forms of the organism whose locomotion is mediated by a muscular tail lost shortly after infection. This stage possesses a number of external sensory structures that probably regulate a number of tropisms that attract it to areas where the definitive host is likely to be located. In response to fatty acids, the parasite then attaches to the skin, aided by secreted water-insoluble mucopolysaccharides, and searches for surface irregularities [4]. The parasite then penetrates the skin, migrates across dermal and subdermal tissues, and upon entry into the bloodstream via subdermal capillaries 1–3 days later is thought to be passively carried by the circulation [5].

Understanding of the timing of subsequent developmental and migratory events is based primarily on studies in laboratory mice. Once in the bloodstream, migration is entirely intravascular, with the parasites entering all organ systems in proportion to, and in the direction of, cardiac output [5]. Passage from one organ system to another is, for the most part, passive and occurs in a matter of seconds; migration across small blood vessels, on the other hand, takes several hours to days [5]. After approximately 1 week, the parasites are preferentially located in the vasculature of the lungs from where they migrate through the liver into the vasculature surrounding either the urinary bladder or intestine; some of the organisms in the lungs manage to re-enter the systemic circulation from where they eventually are carried back into the liver and back into the lungs, a process that may be repeated on multiple occasions [5]. Most of the parasites that enter the liver do not manage to reach the hepatic vein, probably owing to difficulties in negotiating the sinusoidal system, but those that do so eventually return to the lungs. During the liver phase the parasites undergo a period of rapid growth, they pair, and become sexually mature. The worms then migrate into mesenteric venules, with each mature female eventually laying 200–2000 eggs per day (depending on species). Many of the eggs, which measure approximately 100 μm in length, are swept into the circulation until they reach a physical barrier, such as the liver or lung. Rarely, when adult worm pairs reside in ectopic sites, relatively high numbers of eggs may become lodged in surrounding areas, such as in the spinal cord where they can cause neurologic complications. Eggs that enter the lumen of the intestine or bladder hatch upon contact with water, releasing free-living ciliated miracidia (the second invasive forms). Successful miracidia penetrate the soft tissues of a susceptible snail host, where they transform into asexually reproductive mother sporocysts. The mother sporocysts give rise to several generations of daughter sporocysts, which are sack-like forms containing developing cercariae. Mature cercariae emerge from the snail under appropriate environmental conditions and, again aided by several tropisms, concentrate in areas where the definitive host is likely to be present [6].

Cercaria to schistosomulum transformation

As currently understood, the anatomic and biochemical transition from cercariae to schistosomula is mediated by precursor molecules synthesized during the molluscan phase of development [7], although it has been demonstrated that mature cercariae are not transcriptionally silent [8]. The process is triggered upon penetration of the skin by an increase in ambient osmolarity and involves shedding of a large-surface-area proteoglycan (the glycocalyx), exposing a unit membrane and leading to the formation of a multilayered surface membrane [9]. The latter appears to derive from cytoplasmic membranous whorls synthesized while in the snail, and its formation is inhibited by colchicine, indicating that microtubules play a role in their trafficking [10]. During this period, prominent "glandular" cells whose secretions mediate infection disappear [6]. Metabolically, the parasites change from aerobic to anaerobic energy-yielding pathways, with lactate being the predominant end product [11]. The first proteins synthesized are heat shock proteins, i.e., phylogenetically conserved molecules that are thought to play a role in cellular development and differentiation [12–14]. General protein synthesis first becomes appreciable 8 h after infection and then plateaus to a steady state after about 48 h [15].

During the first 2 weeks after infection of the vertebrate host, the parasites undergo a period of slow growth with anatomic changes, including loss of spines, muscular and connective tissue changes that lead to a more elongated shape, and maturation of a bifurcated blind-ended intestinal tract. Subsequently, during the period coinciding with the first appearance of red blood cells in the gut, they rapidly differentiate, with sexual dimorphisms becoming clear 2–3 weeks later. Surface spines are regained during this period and, for reasons not understood, maturation of the female is strictly dependent on the direct physical presence of a male [16].

Surface

The outer membrane of blood-dwelling trematodes consists of an unusual multilaminate structure consisting of several juxtaposed unit membranes [9]. The structure appears to be preformed by cytoplasmic membrane whorls that join the membrane by fusion, although there is evidence suggesting that the outermost part may contain fused membranes of host origin [17]. The complex does not appear to represent the stacking of the same membrane, as the density of intramembranous particles in, and the protein composition of, each layer is different [18,19]. The topology of the outer layer of adult males resembles that of Swiss cheese, and in some areas there exist protuberances in the form of tubercles and spines; the core of the latter appears to be a crystal of actin filaments [20]. The sponge-like appearance of the surface is thought to be a specialization for increased absorptive area, and the spines are thought to play a role in maintaining position in the bloodstream. The surface of female worms is more uniform in appearance. Below the outer membrane, the epithelium is syncytial, with expected partitions between cells being nonexistent. Nuclei of the syncitial layer are found below layers of circular and longitudinal muscle, connecting with the main cytoplasmic layer via tubular conduits. The syncitial nature of the surface is characteristic of parasitic platyhelminths [18].

Metabolism

As is generally the case for eukaryotic parasites, schistosomes display an abbreviated metabolic plan during their definitive stage, lacking *de novo* pathways for the synthesis of vitamins, purine bases, most amino acids, cholesterol, and fatty acids [11]. Acquisition of these nutrients from the host is carrier-mediated to a large extent, with specific porters being located in the outer surface of the organisms [11]. The rate of glucose metabolism is prodigious, exceeding that of red blood cells and tumor cells by several orders of magnitude [21]. The primary end products of glucose metabolism

appear to be lactate and alanine. The rate-limiting step in glucose metabolism occurs either at the membrane transport level or during the phosphorylation of fructose-6-phosphate. Catabolic reactions of free-living stages of the life-cycle occur by oxidative phosphorylation, ostensibly as an adaptation to an environment, providing lower access to nutrients [22].

As an additional source of nutrients, whose relative contribution is undefined, worms over 1 week of age have a functional blind-ended gut that contains numerous host-derived red blood cells. Approximately every 10 min, the parasites evacuate the contents of the intestine, releasing undigested hemin and various immunogenic molecules that probably play hydrolytic roles or protect the gut epithelium. As the parasites can be maintained *in vitro* for many months in defined media free of red blood cells or hemoglobin, but only a matter of hours in saline, the function of the gut is unclear [23]. It has been suggested that ingestion of erythrocytes may play a role in egg and egg yolk production, a property lost upon maintenance in hemoglobin-free media. In this scheme, hemoglobin is digested by proteinases of worm origin to yield small peptides; these are taken up by the gut epithelium and subsequently shunted to adjacent vitelline cells to serve as yolk and perhaps precursors of the parasite's egg shell [24].

Molecular biology, sexual differentiation, egg production, and stage-specific gene expression

The complexity of the schistosome's life-cycle (coupled with the possibility that it might be regulated by interesting nuances) and the fact that the organisms are one of the very few platyhelminths that display sexual dimorphism have received considerable attention [25]. Sexual determination correlates with the presence of two sex chromosomes, the female being the heterogametic sex. *Schistosoma mansoni* possesses eight chromosomes per haploid genome (the parasite is diploid), composed of 2.3×10^8 bp [26]. Modified bases have not been detected, and 60% of the genome is composed of sequences that reassociate as single copy, a fairly high figure for a eukaryote. Repetitive sequences include sex-specific satellite DNA [27], satellite DNA shared between sexes [26], ribosomal RNA genes [26], and retrotransposon-like sequences [28,29]. The guanosine–cytosine content of the organisms ranges from 43 to 47% (compared to roughly 38–47% in mammals and 30–45% in other invertebrates). Some of the functional schistosome genes sequenced indicate that the parasites contain introns as small as 50 bp that contain typical eukaryotic consensus splice sequences [30]. During development it appears that they delete, rearrange, and amplify genomic segments [28,31], implying plasticity, and such changes may account for the development of resistance to the antischistosomal drug hycanthone [32].

The study of schistosome gene expression and subsequent processing of transcripts, although not of immediate practical importance, has begun to pay dividends of interest to basic science. For example, it has been noted that, like nematodes and trypanosomes [33,34], the organisms trans-splice some transcripts. This is accomplished by ligation of a 36 nucleotide leader, donated by a 90 nucleotide nonpolyadenylated RNA, onto the 5' end of a subset of mRNA [35]. The leader shares no homology with that of trypanosomes and nematodes; however, its predicted secondary structure and the presence of a trimethylguanosine cap and potential Sm-binding site shows similarities. In the case of the hydroxymethylglutaryl coenzyme A reductase transcript, splicing occurs both in *cis* and in *trans*, with either the leader or exon 2 being alternatively spliced onto exon 3 [35]. Although these findings do not clarify the purpose behind trans-splicing of pre-mRNA, they pose interesting considerations in that the only other organisms documented to carry out the reaction are trypanosomes (which do not appear to be capable of *cis*-splicing) and nematodes. Based on phylogenetic considerations, these findings raise the question of whether trans-splicing is more common than appreciated or whether the presence of the phenomenon in three different phyla represents an equal number of separate evolutionary events.

Schistosomes are genetically programmed to survive sequentially within widely different specific habitats. During these transitions, they display both nonstage-specific (glycolysis, DNA replication, transcription and translation, and gluconeogenesis) and stage-specific (respiratory pathways used by larval forms) metabolic strategies. Their developmental plan calls for permutations of:

1 life in aquatic, vertebrate, and invertebrate habitats;
2 shifts in temperature, ambient gas phases, and available nutrients;
3 changes in the identity of hostile environmental pressures.

While the inference of differential gene expression is easy to understand in the case of stage-specific gene expression, many workers have suggested that common metabolic pathways may be mediated by isoenzymes subject to stage-specific appearance. Studies in comparative enzymology have revealed that protein-catalyzed reactions are subject to ionic, pH, and temperature-defined modulation. Temperature is particularly important, as it influences rates of reactions, rates of protein denaturation, affinity towards substrate, and pH dependence. These considerations may critically affect the function of near-equilibrium, regulatory, or membrane-bound enzymes owing to temperature effects on membrane fluidity. The mechanisms whereby these organisms transform from one stage to another are poorly understood, and how they are regulated at the level of gene expression is completely unknown. Several laboratories have recently begun to address the issue, primarily by analyzing stage- and tissue-specific transcripts.

In one study, an expression cDNA library was screened with sera from infected patients [36]. Of 21 cDNA clones expressing parasite antigens, three proved to be developmentally regulated. Northern analyses of adult worm, egg, and cercarial poly A$^+$ RNA revealed that two of the three stage-specific cDNA hybridized to mRNA, which change in length during development. While this observation in itself is not particularly unusual, the observation that the two RNAs appear to be transcribed from the same (single copy) gene is. How, and to what extent, schistosomes manage to produce multiple transcripts from a single gene is unknown; however, preliminary results indicate that noted size variations are due to the use of alternative start sites and alternative RNA processing.

As far as sexual determination is concerned, the most obvious genetic difference between male and female schistosomes is the presence of a unique chromosome (called the W) and the absence of one copy of another (termed the Z) in the female. Sexual determination hence appears to be, as in birds and reptiles (which curiously are representatives of the other two phyla parasitized by schistosomes), due to the presence of a female-specific chromosome, ZZ individuals being male and ZW individuals being female; as simple ZO and ZZW individuals have not been isolated, it is unclear whether the W chromosome acts as an activator of female development or as a repressor of male development, or whether sexual dimorphism is simply determined by gene dosage. The W chromosome is largely heterochromatic in interphase nuclei of all schistosome cells examined to date, being by inference, for the most part, transcriptionally inactive. It is easily identifiable in stained preparations of fixed organisms [37], and cytologic signs of W chromosome activation during female worm development or in the ovary of mature female worms have not been described.

Cytogenetic analyses involving microscopic examination of C-banded preparations indicate that the schistosome W chromosome evolved by a combination of fusions, amplification/divergent drift of repetitive sequences, and in some cases deletion [37,38]. Examination of the karyotype of hermaphroditic and gonochoristic digenetic trematodes suggests that primitive schistosomes were probably hermaphrodites possessing 10–11 pairs of telocentric or subtelocentric chromosomes [37]. Lines of strigeoid trematodes containing 16 pairs of chromosomes are assumed to contain telocentric regions which have translocated to the short arms of subtelomeric chromosomes of other strigeoid species; the alternative explanation that strigeoids containing more than 11 pairs of chromosomes arose by polyploidy is generally considered to be unlikely [37]. As in species of *Spirorchis* (hermaphroditic blood flukes of turtles), ancestral schistosomes probably contained a set of heteromorphic "presex" chromosomes (termed M) which might have resulted from a translocation involving a pair of large telocentric chromosomes and a pair of small or medium-sized chromosomes, one or both of which probably contained a concentration of alleles essential to sexual differentiation. One of these presex chromosomes is thought to have evolved by differentially accumulating one or more sex determining genes [38], possibly as a result of functional loss of these loci in the heteromorphic homolog.

The result of sexual differentiation and focus of pathology—the schistosome egg—is originally laid as a single embryonic cell surrounded by approximately 24 vitelline cells [24]. These are the predominant cell type in female worms, and are thought to contain proteins that are precursors to the eggshell and yolk to nourish the developing embryo. During the first 5 days after deposition, of the embryos differentiate into miracidia. The developing and mature miracidia are metabolically active, receive much of their nutrition from the host, and release excretory–secretory products into their environment. The egg is surrounded by a highly resilient, somewhat pliable, and protease-resistant biopolymer that appears to be composed of proteins crosslinked to each other by quinone bridges. The complex contains pores that allow the influx of nutrients required by developing miracidia and the efflux of macromolecules that lead to granuloma formation in the host. Acid hydrolyzates of purified egg-shells indicate that they are rich in glycine,

aspartate, histidine, and lysine, but contain unexpectedly low levels of tyrosine, the principal amino acid thought to participate after oxidation in crosslinking [39].

Several genes and cDNA encoding putative components of the schistosome eggshell have been sequenced [39]. Their inferred amino acid sequences predict proteins with numerous repetitive domains, in one case Gly–Gly–Gly–Tyr and Gly–Gly–Gly–Cys, and in another Gly–Tyr–Asp–Lys–Tyr [40,41], strengthening the suggestion that the shell is a highly organized conglomerate. The sequences predict the presence of more tyrosine than is actually measured in hydrolysates of the eggshell, suggesting that these residues are post-translationally modified. One suggestion is that they are converted to dihydroxyphenylalanine, and then into highly reactive quinones that form covalent bonds with terminal side-chain amino groups of other amino acids [39]. One eggshell protein rich in glycine and tyrosine contains a carboxy-terminal hydrophobic leader and an amino-terminal sequence of a tandem 16–18 amino acid repeat. Charged residues are asymmetrically distributed, with acidic residues dominating the carboxyl end and basic residues dominating the amino terminal end. The protein's sequence is consistent with a model of short antiparallel strands, in which amino acids with short side-chains lie within the strands and tyrosines and cysteines lie at the bends, where they are available for crosslinking [41]. Examination of the 5' flanking region of one of the genes has revealed the presence of sequences homologous to the cap-site, promoter-element, and tissue- and stage-specific sequence determinants and putative *trans*-acting regulatory proteins of silkmoth chorion genes [25]. Production of the eggshell mRNA appears to be mediated by a small gene family that is under transcriptional control and which, unlike homologous silkworm chorion genes, is not amplified during expression [25].

HOST–PARASITE RELATIONSHIPS

During the course of schistosomiasis, the host must learn to cope with migrating schistosomula; adult, intravascular worms; excretory/secretory products of the worms; and eggs produced by adult worms. It is assumed that excretory/secretory products are principally handled in the spleen and liver. Many of the eggs "intended" by *S. mansoni* and *S. japonicum* worms to exit the body by passage through the vessel and gut walls and then via defecation actually are swept up to the liver where they lodge in presinusoidal capillaries. In *S.*

haematobium infections the worm's "intent" is to get these eggs through the vessel and bladder wall in order to be excreted in the urine. Many end up trapped in the bladder wall and others are deposited in the pulmonary and, to a lesser extent, hepatic vasculatures. Because these internal exposures occur continually, the host must develop systems capable of preventing direct pathogenesis due to these products and yet not allow these continual responses to be expressed as immunopathogenic hypersensitivity states. Thus, while adult worms appear to "evade" host immune responses, or at least their consequences, the host seeks to (and usually does) avoid possible serious consequences due to multiple infection-related insults.

Human schistosomiasis

Many field- and hospital-based cross-sectional and longitudinal studies of the clinical, parasitologic, and epidemiologic aspects of human *S. mansoni*, *S. haematobium*, and *S. japonicum* infections have been published [42–51]. The incidence of infection in most endemic populations is greatest between the ages of 3 and 8 years, with peak prevalence and intensity of infection usually occurring between the ages of 10 and 30 years. Schistosomiasis mansoni and japonica are considered to be long-lived infections, with the mean lifespan of adult worms in well-fed expatriate hosts being in the 5–10-year range [52] and schistosome longevity records reaching more than 30 years have been recorded [53]. The situation in schistosomiasis hematobia endemic areas indicates that most worms may have a lifespan of 1 of 2 years, and continued infection depends more on reinfection [54,55].

The fundamental consideration of the host–parasite relationship in schistosomiasis rests on the chronic endemic nature of the infections. These characteristics form the basis for the interactions of host and parasite, and require of these participants a balanced form of response. It is central to the thoughts of these interactions that in highly endemic areas where over 60–80% of the population may be infected, only 5–10% of these are clinically ill. When dealing with an infection as widespread as schistosomiasis, this still means that of the order of 800 000 individuals die each year as a result [3]. However, the vast majority of schistosome-infected patients are ambulatory, and most are relatively unimpaired by their schistosome infection [49–51]. How the worms and their host achieve and maintain

this usually well-balanced long-term interaction is the essence of the natural host–parasite relationship, affecting all aspects of the interaction and likely to impinge on attempts to alter or prevent infection.

One major contributing factor to the well-balanced long-term interaction is clearly the relatively low level of intensity of infection, as determined by fecal egg output [49–51,56]. Furthermore, it is often observed, and most often only anecdotally reported, that people from nonendemic areas who become infected upon immigration or more temporary sojourns into endemic areas often experience clinically observable acute schistosomiasis and more serious chronic infections [50,56,57]. The bases for this difference between those within endemic areas and those outside of endemic areas are unknown. Genetic differences in antischistosomal immune responsiveness and clinical responses to infection are known [49,50] and, as stated, the intensity of infection often correlates with the severity of a patient's clinical form of schistosomiasis [49–51,56]. The generalization concerning endemic and nonendemic populations seems to go beyond these individual contributing factors.

Maternal/fetal/prenatal interactions

The peak prevalence and intensity of schistosome infections usually occurs between the ages of 10 and 30 years [42–44,47]. This period includes the main childbearing years. It follows that within endemic areas a majority of those who eventually become infected with schistosomes were born to mothers who harbored active schistosome infections during their pregnancy. This is obviously not true of the vast majority of those from nonendemic areas who become infected. It is not known if being born of a schistosome-infected mother alters one's subsequent host–parasite relationship upon becoming infected with schistosomes. However, it has been shown that *in utero* immunologic sensitization through either placental passage of schistosome antigens (the idiotypes on maternal antibodies to schistosome antigens) or antiidiotypic antibodies to those antibodies occurs in both human [58–60] and experimental [61,62] schistosomiasis. It is also known that in other clinical [63–65] and experimental [66,67] settings such influences are excellent, highly efficient ways to alter the subsequent immune responses of offspring [68,69]. It is thus possible that a major difference between the host–parasite relationship that develops between patients from endemic areas and their schistosomes and patients from nonendemic areas and their schistosomes is their different *in utero* experiences related to schistosome antigens and antischistosome-related immune reactants.

Schistosome evasion of the immune response

A fundamental aspect of the host–parasite relationship under discussion is that some time during their maturation the schistosomes express an efficient means for evading hostile immune responses, thus setting the stage for chronic infection. How they accomplish this feat—a question that is one of the most interesting in schistosome research—is unknown. Suggested ways include: molecular mimicry, i.e., presentation of endogenously produced host-like or host-acquired molecules; antigenic variation; stage-specific antigen presentation; suppression of the immune response; innate lack of immunogenicity of critical molecules; anatomic sequestration of relevant epitopes; innate lack of immunogenicity of relevant epitopes; and subversion of immune responses, such as by cleaving bound immunoglobulins, by stimulating blocking antibodies, or by rapid shedding of immunocompromised molecules [70–73]. Whatever the mechanisms, it seems reasonable to assume that the immunologically most accessible parasite antigens are somehow the most protected, and that these molecules have evolved in a manner that maintains function within pressures imposed by the immune response. Genetic studies indicate that the evasive qualities of adult worms are not due to intrinsic differences between individual schistosomes [74].

Antigenic mimicry

A large number of host antigens have been detected on the surface of adult schistosomes. The first demonstration of their existence came from experiments demonstrating rapid destruction of worms raised in mice after transplantation into monkeys immunized with mouse antigens [75]. Since then, various workers have demonstrated the presence of red blood cells A, B, and H, and Lewis antigens (but not rhesus, Duffy, M, N, or S antigens), major histocompatability antigens, immunoglobulins, C1q receptors, antiproteases, and a phospholipase C-sensitive molecule resembling human decay accelerating factor [70,76]. The latter is interesting in that it appears to be bound to the membrane by a

lipid anchor, and its presence may serve to prevent the amplification of C3 deposition [77].

As originally proposed, antigenic mimicry suggests that parasites synthesize and coat themselves with host-like macromolecules, and thus obviate an immune response [70]. Backing this hypothesis, schistosomes appear to synthesize a surface host-like α_2-macroglobulin specific to their host [78]. It should be noted that the organism's production of "host-like" proteins is to be expected, as structural restraints related to function have led to the existence of proteins conserved throughout phylogeny. In fact, as described later, most of the immunogenic schistosome proteins whose function or structure is known in fact share considerable sequence homology with host determinants.

Membrane turnover

One possible way in which schistosomes may evade the immune response is by continual shedding of the surface, particularly of damaged areas [79]. Efforts to demonstrate this phenomenon have yielded conflicting results. Studies using lipophilic fluorescent probes indicate that the schistosome surface, unlike the nematode cuticle, is fluid and that component macromolecules exhibit considerable lateral mobility [80]. Electron microscopic studies have revealed that the tegumental membrane is periodically sloughed off [81], and that areas containing bound immunoglobulins can be shed in a capping-like reaction [82]. Freeze–fracture studies on the effects of antihost antibodies on the surface of the parasite have revealed the formation of complexes composed of four intramembranous particles in the hydrophobic portion of the membrane [82]. In the presence of complement, these complexes concentrate in the pits of the tegument, and in the absence of complement they flow to the spines. As mentioned later, tegumental proteins of adult schistosomes, especially females, are difficult to tag with surface-labeling techniques, suggesting that this compartment may either be masked or constructed in such a way that only nonimmunogenic portions are exposed [83,84].

The controversy behind the authenticity of surface shedding as a means for immune evasion concerns differences in observed half-life of the schistosome membrane. In some workers' hands this measures of the order of 10 min [85], while other workers have measured a considerably longer turnover rate [86,87].

Secretion of immunomodulatory molecules and role of proteases

In vitro studies indicate that schistosomes have Fc receptors, the implication being that they have the potential to bind, and thus render inaccessible, the portion of immunoglobulins necessary for effector cell recognition. At the same time, the organisms have the capacity to hydrolyze immunoglobulins and in the process release a tripeptide (Thr–Lys–Pro) that may interfere with various macrophage functions, such as IgE-mediated cytotoxicity and granuloma formation [88]. Proteases secreted by schistosomula have been shown to stimulate eosinophil activity, to potentiate IgE synthesis, and to increase the density of Fc receptors for IgE on the surface of lymphoid cells [89]. Larval stages of the parasite secrete, and possibly bind on their surface, a serine protease that may interfere with the action of complement [90,91].

Another immunosuppressive molecule from adult schistosomes has been partially characterized and called schistosome-defined inhibitory factor (SDIF). The molecule is heat resistant, dialyzable, and soluble in trichloroacetic acid. It was originally defined by its ability to inhibit mast cell degranulation, and may thus indirectly interfere with mast cell product stimulation of eosinophil functions and increases in permeability [92], both of which are discussed later.

Parasite-derived fractions have also been described that inhibit *in vitro* lymphocyte proliferation [93], *in vivo* immediate hypersensitivity [94], *in vivo* humoral responses [95], and stimulate parasite-specific *in vitro* suppressor cell generation [96]. Parasite-protective effects of these products are presently conjectural.

Disruption of evasive mechanisms during chemotherapy

Whatever the method used by adult schistosomes to evade immunologic destruction, recent studies suggest that several antischistosomal drugs may obviate their effectiveness [97]. Antimony, oxamniquine, hycanthone, and praziquantel all kill fewer worms in immunocompromised animals than in immuno-competent animals [97]. This phenomenon has been best studied using praziquantel, a pyrazino-isoquinoline that provides an excellent means of treating all three schistosome infections. Its mode of action is not precisely understood but involves parasite changes in membrane potential, an influx of calcium, paralysis, a shift to the liver, and damage to the tegument [98].

Antibodies directed towards tegumental antigens are necessary for the *in vivo* vermicidal action of praziquantel, and the drug is less effective in immuno-compromised hosts than in immunologically intact hosts but more effective in animals previously immunized with irradiated cercariae [98]. Depletion of B cells by anti-m treatment results in diminished schistosomidal praziquantel treatment, a defect reversed by adoptive transfer of immune serum (both IgG and non-IgG fractions can substitute) [99]. Prior to death, the tegument of schistosomes loses its ability to avoid binding of parasite-specific immunoglobulins, attachment occurring preferentially to tubercles on the dorsal side of adult males; exposure of normally sequestered surface epitopes occurs, a phenomenon associated with antibody binding to parasite-borne host antigens [99]. Studies employing monoclonal antibodies to tegumental or subtegumental antigens indicate that the altered presentation of epitopes is not a generalized response to the drug. Moreover, a single monoclonal antibody can passively renew the schistosomidal capacities of praziquantel in B cell-depleted mice. Granulocytes enter into the body of worms during treatment, perhaps mediating clearance of debilitated worms. The efficacy of the drug is not reduced in mice deficient in complement or in macrophage function.

A currently entertained idea of the mode of action of praziquantel is that it interacts with components of the schistosome surface and in some manner disrupts its general architecture, leading to conformational changes of component macromolecules and exposure of normally sequestered epitopes to the immune system. Alternatively, the drug's immediate effect of altering membrane potential may somehow directly disrupt the tegument's architecture. Newly exposed epitopes may include regions accessible to phospholipases that can hydrolyze phosphatidylinositol-anchored glycoproteins [100], leading to the release of secondary messenger diacyl glycerol and phosphatidic acid and subsequent changes in membrane potential [98].

Thus, while at exposure concentrations achieved clinically praziquantel and its metabolites are not directly lethal to schistosomes, their induced changes lead to immunologic destruction. Antibodies that participate in worm killing are already present during established infections, and react with a limited number of surface epitopes that are normally sequestered in adult worms. There is evidence that a 27 kD molecule with esterase activity and a 200 kD tubercle protein on male worms are primary targets of these cooperative schistosomidal antibodies [95,101].

INNATE RESISTANCE

Based on susceptibility, experimental hosts for schistosomes can be imprecisely subdivided into permissive and nonpermissive categories. In the former, exemplified by certain strains of mice, hamsters, opossums, and baboons, the parasites survive to maturation and lay viable eggs. In the latter, which includes rabbits, rats, and some primates, the organisms develop at a stunted rate, lay few, no, or deformed eggs, and usually die within a short period of time. The distinction between permissive and nonpermissive hosts is not rigid, but instead represents a spectrum of developments and responses. Permissive and nonpermissive experimental hosts offer different benefits for study, and the host–parasite relationships expressed in this spectrum of schistosome/host combinations differ substantially. Caution must be exercised to confine interpretations of experimental results to the system being analyzed. It should also be noted that even in permissive experimental hosts, a significant proportion of parasites from a primary infection die prior to maturation, and that nothing is known regarding schistosomal survival rates (the ratio of adult worms that develop from a given number of penetrating cercariae) in humans.

The differences in degree of permissiveness of hosts can vary between individuals of the same species. Factors contributing to the variations in host responsiveness to infection are both parasite- and host-defined. Parasite factors determining infectivity and subsequent pathology are poorly understood and may include aspects that affect the parasite's quality, such as vigor, nutritional status within the snail, intensity of infection in the snail, variations in the length of the path taken out of the snail, and elapsed time of emergence from the snail. Host factors determining the outcome of infection may include genetic differences (e.g., major histocompatibility complex (MHC)), physiologic and nutritional status, and nonspecific acquired factors. These are only vaguely appreciated, although it has been demonstrated that the hormonal status of the host has a significant effect on the course of infection [102]. In addition, the fact that schistosome antigens crossreact with those of other parasites [103] raises the possibility that heterologous infection may lead to some degree of immunity.

Nonspecific immunity has been experimentally induced by provoking an inflammatory reaction in the skin immediately prior to infection [104] and following induction of a nonspecific immune response in the lungs by intravenously administered heat-killed *Escherichia coli*

[105]. Similarly, immunization with Bacille Calmette-Guerin (BCG) or cord factor can lead to nonspecific resistance to schistosome infection [106,107]. The overall mechanisms leading to protection in these systems is unclear, but may involve nonspecific macrophage activation and arginase-mediated killing of schistosomula [108].

Certain strains of mice exist whose members respond differentially to either *S. mansoni* or *S. japonicum* infections, with some being permissive and others being resistant to initial exposure. In the latter animals, worm attrition occurs 23–25 days postinfection, before the commencement of egg deposition [109]. Nonpermissive individuals tend to have few to no adult worms in the portal circulation, these instead accumulating in the pulmonary vasculature where they are eventually eliminated in eosinophil-rich inflammatory reactions [110,111]. The latter are preceded by herniation of the worms and eventual rupture of the organ through the tegument [110]. Resistance to infection correlates with a reduction and truncation of peripheral vessels of the liver, with the portal system being more profoundly affected than the arterial system [112]. These changes are thought to prevent physical entrapment of the worms in the liver, permitting their relocation to the lungs where the diameter of associated vessels prevents their re-entry into the liver. The reason for existence of portasystemic anastamoses in roughly half of a strain of mice is unknown, possibilities including prior infection with another parasite or persistent ductus venosus [113]; studies on specific pathogen-free mice suggest that the abonormalities are not due to prior infection with another parasite [112]. It is not clear why ectopically located adult parasites are killed, possibilities including debilitation of the organisms after removal from the nutrient-rich mesenteric vasculature or the existence of more efficient cytotoxic immune mechanisms in the lungs [110–112].

HUMORAL RESPONSES

General comments

Schistosomes present an antigenic mosaic to their host whose immunogenic components can be placed into five overlapping functional categories [114]:

1 inconsequential antigens, or molecules that are immunogenic to the host in an incidental but irrelevant way;

2 immunodiagnostic antigens;

3 antigens that play a role in initiating and/or modulating granuloma formation;
4 parasite protective antigens, or antigens that allow the parasite to evade the immune response;
5 host-protective antigens.

Studies on the immune response to the various stages of the parasites have revealed that it is quite heterogeneous, differing in intensity and specificity from individual to individual [115,116]. This variability is probably due to differences in host lymphocyte and major histocompatability antigen repertoires, as well as lesser understood influences, such as nutritional status and infection with other parasites. It is thus not possible to assign a precise number to how many of the parasite's antigens are immunogenic during the course of natural infection, although this figure probably extends into the low hundreds [116]. The molecules are probably presented to the immune system via multiple routes, such as by degenerating larvae (many invading organisms do not reach maturity, especially after the first host contact with the parasite), eggs encapsulated in tissues, dead eggs (their lifespan is approximately 1 month), and as excretory–secretory products released primarily from the gut, protonephridial canals, and sloughed-off portions of the outer surface of the organisms. The latter process involves the release of blebs [9,79,81] that probably contain cytoplasmic components. Some of the antigens are stage specific, although the majority appear to be shared by at least two stages of the life-cycle. Comparison of total proteins synthesized by the parasites with those which are immunogenic reveal that the latter are a small subset of the former, and that relative abundance is not a factor bridging the two sets of molecules.

Surface antigens

As the exterior of schistosomes directly interfaces with the immune system, it has been generally assumed that its components may represent prime candidates for vaccine development [83,84]; this view has changed over the years, as it has become apparent that not all surface antigens elicit protective responses, and that certain nonsurface antigens do [117,118]. Surface antigens of adult worms and schistosomula have been the most thoroughly characterized, primarily by biochemical, hybridoma, and surface-labeling techniques. Despite intense scrutiny, their functions are generally unknown, possibilities including nutrient transport, host–antigen binding, signal transduction, prevention of blood clotting, and inhibition of the complement cascade. Low-density lipoprotein receptors, glycosyl transferases, an ATPase, alkaline phosphatase, 5-nucleotidase, and a calcium-binding protein have been detected at the biochemical level [119–122], and some of the macromolecules can be phosphorylated, suggesting that they are signal transducers [123,124]. Studies on purified tegument preparations indicate that it contains of the order of 30–50 different proteins [125,126], that these are distributed asymmetrically across the multilaminate coat [19,127], and that those exposed on the exterior are typically glycosylated [128,129].

Examination of surface molecules of *S. mansoni* by hybridoma and recombinant DNA methodologies has revealed the presence of a peculiar array of components whose physiologic function is unclear and in some instances puzzling. For example, the parasites possess a surface-associated glyceraldehyde-3-phosphate dehydrogenase [130], a seemingly unique compartmentalization for a normally cytoplasmic glycolytic enzyme. Other surface antigens characterized at the molecular level include a 38 kD schistosomular glycoprotein whose carbohydrate portion crossreacts with keyhole limpet hemocyanin [131], a cytochrome [132], a molecule demonstrating no homology to other known proteins [133], a protein homologous to a human tumor associated-like protein that is related to lymphocyte receptor CD-37 [134], a calcium-binding protein [119], an acetylcholinesterase [135], several hydrolases [136], alkaline phosphatase [120], a putative lipoprotein receptor [121,122], a protein that crossreacts with an antigen of the human immunodeficiency virus [137], and a protein similar to human stage-specific embryonic antigen 1 [138]. Several of these molecules are currently being tested as vaccine candidates.

As adult worms are relatively impervious to immunologic destruction and most experimentally demonstrated protective responses are directed at developing larvae, surface antigens of the latter have been studied in considerable detail. As schistosomula of *S. mansoni* mature, they alter their surface such that some antigens are retained, others disappear or otherwise become inaccessible to antibodies, and new ones appear, occasionally for short periods of time [84]. Thus 24-hour-old schistosomula expose 15, 20, and 32 kD antigens shared with younger larva, while expressing new ones having molecular weights of 8 and 15 kD. Four days later, exposed antigens have molecular weights of 8, 15, 18, 20, and 32 kD, and a new antigen of 25 kD appears [139].

At the immunologic level, larval *S. mansoni* surface antigens can be subdivided into two classes:

1 those whose immunogenic epitopes are carbohydrate, often crossreacting with antigens of eggs and other species of parasites;

2 those that consist primarily of proteins that may crossreact with antigens of adult worms, but not of eggs or other species of schistosomes [83,140].

In 3-h-old schistosomula, greater than 90% of epitopes consist of carbohydrates associated with antigens having molecular weights of 17, 38, and >200 D [84,140]. The 18, 20, 30, and 32 kD surface antigens contain polypeptide-based epitopes that crossreact with adult worms, but not with eggs and a 15 kD peptide that appears to be schistosomula specific [141,142]. The latter 18, 20, 30, and 32 kD antigens appear to be the major surface epitopes that elicit a humoral response in rodents vaccinated with irradiated cercariae [140].

Several schistosomula surface antigens are attached to the membrane via a glycosyl phosphatidylinositol anchor. These include an acetylcholinesterase, alkaline phosphatase, unidentified serine proteases, and proteins having molecular weights of 18, 32, and 38 kD [95,100,101,135,143]. The latter are also present in cercariae but not 7-day-old schistosomula. Metabolic labeling of adult schistosome proteins with myristate leads to incorporation by more than 15 proteins that partition into detergent phases. One of these is a 200 kD protein exposed after praziquantel treatment [98]. The purpose of these lipid anchors in schistosomes is unclear, one suggestion being that they might be enzymatically hydrolyzed as part of immune evasion [98].

As larval *S. mansoni* mature, their surface is further modulated such that it becomes difficult to label with iodinating agents. Labeling of isolated membranes followed by immunoprecipitation indicates that the dominant antigen has a molecular weight of 25 kD. Other antigens noted have molecular weights of 32, 20, 18, 15, 13, and 8 kD [84,141,142]. Epitopes on these molecules are periodate resistant and crossreact with schistosomular surfaces but not with egg antigens. Several monoclonal antibodies have been isolated that bind to the surface of living schistosomula but not of adult worms; these antibodies, however, do bind to the tegument of sectioned adults, indicating that their respective epitopes become sequestered during development [84,98]. The molecular weights of dominant integral membrane proteins of *S. japonicum* show little similarity to those of *S. mansoni*; however, some of them crossreact at the immunologic level [144].

Muscle proteins

Paramyosin, myosin, and tropomyosin have been demonstrated to be immunogenic proteins of *S. mansoni* [145–147]. The latter two display significant similarities to homologous host proteins, while paramyosin is found in invertebrate catch muscles where, in association with myosin, it forms complexes which permit strong and protracted muscle contraction. Immunocytochemical studies indicate that the schistosome protein is also found in nonfibrillar form within elongated membrane-bound structures within the worm's tegument, although it is apparently not exposed on the surface [145,148]. Preliminary indications are that the amino terminal portion of schistosome myosin may, on the other hand, be exposed on the outer membrane, placed there after an unusual reaction involving proteolytic cleavage (M. Strand, personal communication).

Enzymes involved in energy generation

Empirically determined immunogenic schistosome proteins identified to date are prominently represented by enzymes that play a role in energy-yielding pathways. These include glyceraldehyde-3-phosphate dehydrogenase, triose phosphate isomerase, a guanidino ATP kinase, malate dehydrogenase, and aldolase [118,130, 149–151] (R. Harrison, J. Culpepper, & M. Doenhoff, personal communication). Curiously, the first three enzymes catalyze sequential reactions that form the core of the glycolytic pathway, and which by analogy with homologs in higher eukaryotes may form a macromolecular complex which interacts with actin filaments and membranes [152]. Interestingly, it has been demonstrated that aldolase is a potential vaccine target against *Plasmodium falciparum* and *Ascaris suum* [153,154]. Should these proteins indeed prove protective, their interplay with the immune system would raise several questions. For example, since the enzymes are not rate-limiting to the glycolytic pathway (membrane transport and the three kinases are), why they are sensitive to the immune system is a mystery. Secondly, the enzymes are normally found in the cytoplasm of eukaryotic cells, raising the question of how the immune system senses their presence in intact organisms. Thirdly, the amino acid sequences of the three proteins are very similar to host homologs, making it formally possible that they might elicit an autoimmune response, although preliminary data indicate that, like myosin, they do not do so during the course of normal infection [130,146].

Glutathione- and peroxide-metabolizing enzymes

Another empirically determined class of schistosome immunogens are the isoenzymes of glutathione *S*-transferase. One of these proteins was selected for study on the basis of the observation that mice resistant to *S. japonicum* infection mount a stronger response to the antigen than those that are susceptible [155]. The antigen has been cloned and expressed in bacteria and has been found to be a 26 kD isoenzyme of glutathione *S*-transferase. Unlike the case for the glycolytic enzymes, the protein demonstrates only a modest degree of resemblance to the host homolog. Mice infected with *S. japonicum* produce antibodies which crossreact with *S. mansoni*, and sequencing of the *S. mansoni* homolog has revealed that they are nearly identical in structure [156]. A second isoenzyme of the glutathione *S*-transferase of *S. mansoni* also has prophylactic potential. The protein has a size of 28 kD, shows little sequence similarily to the 26 kD isoenzyme, and induces production of antibodies which do not crossreact with the 26 kD homolog or with homolog of the host [156–158]. The enzymes are normally found in the cytoplasm or microsomal fraction of metazoan cells; however, there are indications that they may also be found on the surface of the parasite [159]. The functional roles of the schistosome isoezymes are unknown, but their presence in the tegument suggests that they may play a role in protecting the organisms from oxygen-mediated damage.

Another stress-related immunogen now characterized is superoxide dismutase. This enzyme probably plays a role in protecting the parasite from oxygen-mediated damage by neutrophils (discussed later), but the importance of the enzyme to diagnosis or host protection remains to be established [160].

Heat shock proteins

Immunodominant antigens presented by schistosomes, and parasites in general [14,161], are strongly represented by a family of macromolecules related to heat shock proteins (Hsp). Heat shock proteins are phylogenetically conserved macromolecules encoded by several families of inducible genes collectively referred to as stress genes. The function of Hsp is incompletely understood, current thought being that they protect cells during stressful conditions by preventing improper aggregation of denatured proteins, removing irreversibly damaged proteins, and promoting refolding of denatured proteins to their native state [162]. Why

parasite Hsp are, despite their phylogenetic conservation, so commonly immunogenic is unknown, but this property is being exploited to develop vaccines against malarial, mycobacterial, and chlamydial infections [160].

The first immunogenic schistosome Hsp described was a 40 kD protein (P40) related to small Hsp and α-crystallins. It is one of the major antigens of eggs and is also found in adult worms. The peptide elicits a strong immunologic response in over 90% of infected humans [163,164]. The antigen is actually a family of four nearly identical proteins that are differentially expressed during the schistosome life-cycle. One form of the peptide is produced by female worms, two in eggs, and four in adult worms [163]. The amino terminal portion of the protein consists of a long stretch homologous to the α-crystallins, the central portion contains a region homologous to the carboxy terminus of small Hsp followed by one related to the amino proximal region homologous to α-crystallins, and the carboxy terminal tail is similar to that of some small Hsp [164,165]. This arrangement suggests that the protein is the result of an early intragenic duplication. The similarity of the schistosome antigen to the α-crystallins and small Hsp suggests that it may form large aggregates which somehow stabilize cellular structures, and hence may be involved in formation of the parasite's eggshell.

The second schistosome Hsp described was a homolog of eukaryotic Hsp70, a relative of the product of the DNAk locus of *E. coli*. Epitope mapping localized the immunogenic portion(s) of the molecule at the carboxy terminus of the protein, the least conserved domain of Hsp70 in general [166]. Similar properties are displayed by the *S. japonicum* Hsp70 cognate [167]. The 70 kD *S. mansoni* and *S. japonicum* immunogens do not crossreact with each other [167], however, the *S. mansoni* antigen may crossreact with other infectious agents [168].

The *S. japonicum* Hsp70, which has been identified as a circulating antigen [169], is found in the tegument and central ganglion of adult worms and in the hatching fluid, surface cells, and nervous system of miracidia [170]. Synthesis of the protein is constitutive in adult worms and sporocysts, although apparently it does not appear to be synthesized by cercariae [12,13]. Transforming schistosomulae produce massive amounts of Hsp70 during a period in which production of nearly every other protein is repressed [13]. The cue for Hsp gene transcription in transforming larva is not heat, as cercariae subjected to heat shock do not synthesize the protein while cercariae placed in mammalian culture media at room temperature do (G.R. Newport, personal observation). Synthesis of other proteins slowly resumes 16 h posttransformation, eventually plateauing to steady-state levels [15].

A third immunogenic Hsp has been identified as being a member of the Hsp90 family. The protein is strongly recognized by 50–60% of infected individuals [164]. The amino acid sequence of the protein demonstrates a high degree of homology with Hsp90 from other organisms, and the protein shares a small stretch of identity with the carboxy terminus of Hsp70 [171].

Studies on the immune response to schistosome Hsp indicate that reactivity may be decisively influenced by the MHC and T cell repertoires of the host. Responses of inbred mice suggest that a failure to respond to Hsp70 is due to a hole in the T cell repertoire or to a deletion of reactive B cell clones [164]; this might represent an example of molecular mimicry, where a parasite molecule strongly resembles a host homolog and is hence tolerated. The lack of responsiveness of some strains of mice to the schistosome Hsp90, by contrast, appears to be mediated by immunosuppression [164].

Proteinases

Adult worm hemoglobinases

Adult schistosomes consume approximately 500 000 red blood cells per day [24]. Contained hemoglobin is digested into small peptides by cysteinyl proteases, referred to as hemoglobinases [24]. Purification of the enzymes has proved difficult, with most preparations yielding activities within the range of 30–35 kD, thus suggesting the presence of multiple forms [172]. The enzymes are highly immunogenic and are currently regarded as promising immunodiagnostic proteins [115,173–176]. The relationship of hemoglobin hydrolysis to the worm's physiology is unknown, although the fact that organisms can survive for long periods in vitro in the absence of the substrate indicates that it is not essential for survival [23]. Several lines of evidence, however, suggest that the host protein may play a role in development, as female worms whose sexual differentiation is stunted in nonpermissive hosts contain little pigment (hematin) in their gut. It has been proposed that products of the hydrolysis of hemoglobin, small peptides and not free amino acids, may be the first step in formation of eggyolk for developing embryos present in eggs deposited by the parasite [24].

Ostensibly, the small peptides are further hydrolyzed during development of the miracidia.

Schistosome hemoglobinases occur as circulating antigens periodically released by regurgitating worms, and the enzymes retain their activity while in the bloodstream despite their acid pH optimum [176]. The role of the enzymes in host pathology is unknown, although they do induce high levels of IgE and elicit an immediate hypersensitivity response [173,175]. Two major adult worm hemoglobinases have been characterized, these having molecular weights of 31 kD (Sm31) and 32 kD (Sm32), respectively. Both proteins are originally synthesized as precursor pre-propeptides that have a hydrophobic leader, with the presumably inactive form of Sm31 having a molecular weight of 42 kD and that of Sm32 of 50 kD [174,177–179]. The precursor of Sm31 appears to be posttranslationally cleaved at both the amino and carboxy termini, yielding the active enzyme. The inferred amino acid sequence of the 42 kD protease shows homology to mammalian cathepsin B [174] while the other protein shows no similarity to proteins in currently available data banks [177], a highly unusual finding for a protease. Interestingly, the latter enzyme is present in the acetabular gland cells of cercariae, where it has been suggested to play a role in skin penetration either by directly hydrolyzing host connective tissue or by activating other protease precursors [179].

The cercarial elastase

Schistosomes gain access into their definitive host by penetration of the skin. The process is to some extent mechanical, enzymatically mediated, and facilitated by secretions of an aggregate variously referred to as the acetabular, cephalic, secretory, poison, or penetration gland complex [180]. Component cells, however, appear to be of mesenchymal origin and are hence not classic glandular cells [180]. Secretion is stimulated by a rise in temperature and the presence of fatty acids normally found in the skin [180]. Secreted products include:

1 water-insoluble mucopolysaccharides [4], which allow the organism to adhere to the skin while it searches for imperfections deemed as favorable ports of entry (joints or loosened edges of squamae, wrinkles and furrows, hair/skin angles, or sites of entry established by other cercariae [6]);

2 calcium, which in addition to serving as a cofactor for secreted cercarial proteases may disturb the architecture of the host extracellular matrix via ionic interactions [181];

3 hydrolytic enzymes such as proteases and glycosidases, which are thought to dissolve the skin barrier [182], mediate removal of the cercarial glycocalyx, and inactivate complement [90,91].

In all, the secretions appear to account for up to one-quarter of the total nitrogen of cercariae. The act of penetration is stressful to the parasite, as inferred by the fact that the rate of attrition during the process ranges from 10 to 75%, depending on host species, and the observation that migrating parasites experience a significant degree of physical damage [183].

Being of nonhost origin, some of the antigens in cercarial secretions (five to 15 proteins and four mucopolysaccharides, which represent the first contact between potentially immunogenic material and the host) elicit a host immune response whose outcome, aside from a hypersensitivity reaction which may result in dermatitis, is currently undefined. The interplay of the macromolecules with the host immune system has been best characterized in studies involving an elastase that facilitates infection by *S. mansoni* cercariae. The enzyme is immunogenic in native form, primarily stimulating a transient IgM response. Studies on experimentally infected mice indicate that, during primary infection, antibodies to the enzyme are detectable within 1 week after exposure to cercariae [184]. Reactivity progressively increases to a maximum at week 9 and then returns to control levels by week 18 [184]. A similar humoral reaction occurs in monkeys. Multiple infections do not enhance the response in nonhuman primates; after each new infection, the response behaves much as in a primary infection [184]. Antibodies mounted against the enzyme by naturally infected humans consist primarily of IgM and to a lesser extent IgG, but no IgE is detectable [184]. The pattern of reactivity induced in the three species suggests that the enzyme may prove to be of use as an indicator of recent exposure to cercariae or as a substrate for early detection of infection. Antibodies to the *S. mansoni* antigen crossreact with elastases of *S. japonicum* and *S. haematobium* but not with that of *Schistosomatium douthitti* [185]. Although well characterized, it should be noted that the cercarial elastase is actually a quantitatively minor component of the protein fraction of cercarial secretions [182]. Dominant peptides in the secretions have unassigned functions, although it is clear that several are immunogenic and that at least one of them elicits an IgE response [186].

The schistosome elastase is a 31 kD calcium-dependent serine protease that is stored as an inactive pre-proenzyme [187]. The protein has a broad substrate specificity that allows it to degrade efficiently the major skin components, such as elastin fibers, proteoglycan, types IV and VIII collagen, desmin, keratin, laminin, and fibronectin [182]. Similar properties are shared by proteases secreted by other tissue-invasive parasites, such as *Pseudomonas aeruginosa*, L3 larvae of *Ankylostoma caninum*, filarial parasites, and *Ascaris* [188]; it is thought that these proteins have placed pressure on the evolution of host antiproteinases, as these have evolved at an accelerated Darwinian rate [189]. Although basically sharing similar catalytic properties, the enzymes of different parasites evolved from separate ancestral genes. For example, the elastase of *S. douthitti*, a parasite of small mammals which is a causative agent of "swimmers itch," derives from a completely different subset of proteolytic enzymes from that of *S. mansoni* (metallo vs. serine protease), although it shares the properties of being an elastase with broad substrate specificity, alkaline pH optimum, and calcium dependency [187].

Other proteases

Schistosomes possess a number of other lesser characterized proteases that are immunogenic and expressed in a stage-specific manner [190,191]. These include neutral proteases [136], aminopeptidases [191], and two proteases secreted by cercariae that play roles in removal of the glycocalyx and inhibition of complement activation [96–98], one of which is probably the elastase [180]. The functions of the other proteases are unknown but may relate to nutrition, interplay with proteolytic components of the blood-clotting cascade, and the cleavage of immunoglobulin in a process that removes the Fc part of the molecule recognized by effector cells and releases immunoregulatory peptides [88].

Two immunogenic peptidases have been identified in schistosome eggs. One is a thiol protease that does not immunologically crossreact with the hemoglobinases but does resemble them in substrate specificity. The enzyme is stored in the escape glands of miracidia and may thus play a role in facilitating the organism's entry into the intestinal lumen or alternatively may be involved in penetration of the intermediate snail host's mantle [192]. The enzyme is released by phospholipases, suggesting the presence of a lipid anchor. The other enzyme is a leucine aminopeptidase released from the parasite's membranes during the hatching process. It is thought that the enzyme may somehow play a role in opening the organism's eggshell [193]. This enzyme is also made in large quantities by recently laid eggs, and is identical to the so-called "7-week antigen" in *S. japonicum* infections [194] (R. Zimmerman & C. Carter, personal communication).

Somatic proteins

A listing of the dominant schistosome immunogens characterized by biochemical or recombinant DNA techniques is presented in Table 16.1. The list most probably represents relatively abundant and/or immunogenic moieties, and will grow quickly in the near future. Most of the antigens listed in Table 16.1 were initially selected for study because they had vaccine or immunodiagnostic potential, because they were developmentally regulated, or because they were dominant immunogens and their identity was retrospectively established as part of their characterization. By inference, it seems likely that many, or perhaps even most, schistosome antigens characterized because of interesting immunologic properties will turn out to have a predictable function based on homologies to proteins of known sequence. How these antigens are presented to the immune system is unknown, but may be related to worm death (most parasites never become established), egg death (eggs live in tissues for only about 1 month), regurgitation, and/or shedding/turnover (e.g., release of membrane blebs).

Polysaccharides and glycoconjugates

Study of immunogenic schistosome polysaccharides has lagged decidedly behind that of proteins, primarily owing to considerations of manipulation and not to a lack of perceived importance. The antigens play prominent roles in the intensity of the immune response, and some members are important to diagnosis, immunity, and pathology. Their functions are poorly understood but may include protection of the gut, hydration of the outer membrane thus facilitating nutrient uptake, osmoregulation, structural purposes, and prevention of blood clot formation [140]. Despite difficulties in isolating the antigens in reasonable quantities, their consideration as vaccine candidates stems from the demonstration that an anti-idiotypic monoclonal antibody, which presumably presents the image of a carbohydrate epitope to the immune system, can stimulate protection against *S. mansoni* infection [196].

Several large proteoglycans have been isolated from the gut of adult worms. The antigens are classified according to their net charge and have received attention

Table 16.1 Listing of schistosome antigens that have been characterized at either the functional level or whose inferred amino acid sequence has been determined by recombinant DNA techniques

---

Protein or glycoprotein schistosome immunogens identified to date [reference]

Stress proteins
26 kD glutathione *S*-transferase [155,156,158]
28 kD glutathione *S*-transferase [156,157]
Superoxide dismutase [160]
A small Hsp [163]
Hsp86 [171]
Hsp70 [166, 167]

Enzymes involved in carbohydrate metabolism
A cytochrome [132]
Aldolase [149] (R. Harrison, J. Culpepper, & M. Doenhoff, personal communication)
Triose phosphate isomerase [118]
Glyceraldehyde-3-phosphate dehydrogenase [130]
Malate dehydrogenase [151]

Muscle proteins
Myosin [146]
Paramyosin [145]
Tropomyosin [147]

Proteolytic enzymes
31 kD hemoglobinase [174,178]
32 kD hemoglobinase [174]
Cercarial elastase [181]
Leucine aminopeptidase [193]
Egg cysteine proteases [192]

Miscellaneous proteins
A human tumor-associated antigen [134]
Alkaline phosphatase [120]
Acetylcholinesterase [135]
ATP guanidine kinase [150]
Embryonic antigen 1-like protein [138]

Antigens of unknown function
A miracidial antigen [195]
25 kD surface protein [133]
Calcium-binding protein I [8]
Calcium-binding protein II [119]
HIV crossreactive 170 kD surface antigen [137]
Repetitive antigen [36]

---

because they represent some of the more promising substrates for diagnosis of ongoing infection. They are released when the parasites regurgitate, and assays have been developed to detect them as circulating antigens. The major known antigens in this category, designated circulating cathodic antigen (CCA) and circulating anodic antigen (CAA), are discussed in the section on immunodiagnosis.

Other prominent proteoglycans of schistosomes are

components of the glycocalyx of the cercarial stage. The glycocalyx is a dense 1–2 mm outer coating consisting of 8–15 nm fibrils that protects the osmotic integrity of the free-living larva [197]. It consists primarily of high-molecular-weight carbohydrates (80%) that activate complement and are responsible for the cercarienhullen reaction, where serum from infected individuals forms a precipitate at the surface of cercariae [197]. The antigen is rapidly shed from the surface of transforming schistosomula and does not reappear for the remainder of the mammalian phase of the life-cycle. Attempts to use it as a vaccine have failed, as immune responses directed at it actually enhance infection, possibly by leading to the production of blocking antibodies [198]. Another polysaccharide, a water-insoluble adhesin, is secreted by cercariae during the initial stages of penetration [4]. The molecule assists attachment to host skin but its immunogenicity has not been tested.

Schistosomula and adult worms contain various O-linked carbohydrate chains linked to peptides either by N-acetylgalactosamine or galactose-N-acetyl galactosamine [199]. The polysaccharide chains do not contain sialic acid, are rich in mannose- and asparagine-linked oligosaccharides, have high amounts of terminal fucose residues, and contain N-acetylgalactosamine in terminal B-linked positions [199]. Approximately 90% of schistosomular surface-reactive antibodies in serum from chronically infected donors are directed at carbohydrate, and it seems likely that the O-linked chains are components of this compartment [199]. The amount of exposed carbohydrate is reduced in lung-stage schistosomula, the least immunogenic stage of the vertebrate phase of the life-cycle. It has been estimated that 70% of the glycosaminoglycans of adult *S. mansoni* are located in the tegumental surface, this being in part accounted for by heparin-like polysaccharides that may prevent the formation of blood clots [200]. The proportion of carbohydrate antigens secreted by developing eggs is undefined; however, a concanavalin A binding fraction of soluble egg antigen (SEA) is a dominant stimulator of granuloma formation and is thus of major importance in pathogenesis [201]. Relatedly, monoclonal antiegg antibodies have been isolated that modify granuloma formation [83,202].

One of the more striking aspects of carbohydrate antigens of schistosomula is that they crossreact extensively with eggs but not with adult worms [140]. Anticarbohydrate antibodies constitute the bulk of the crossreactive humoral response in chronically infected mice, whereas animals vaccinated with irradiated cercariae tend to react more with peptides [140]. The response to carbohydrates appears to be primarily stimulated by the parasite's eggs, is most intense in acute infections, and appears to be down-regulated in chronic infections [83,202]. It has been conjectured that the above observations may account for the slow development of immunity to schistosomes in humans; the line of thought is that acute-stage IgM and IgG2 antibodies stimulated by T cell-independent polysaccharide egg antigens having counterparts in schistosomula would either be ineffective at killing schistosomula or otherwise block the action of protective antibodies [203]. As the infection becomes more chronic, the host would lose or regulate these anticarbohydrate responses [203].

Two glycoproteins have received attention as vaccine candidates whose protective responses are directed at carbohydrate residues. One is a 38 kD surface glycoprotein whose carbohydrate moieties share similarities with keyhole limpet hemocyanin [137]. This antigen, or a crossreactive one, is expressed in schistosomula, on the surface of cercariae, on the ciliary plates of miracidia, in the adult worm, and in the snail intermediate host [137]. Monoclonal antibodies to the antigen partially protect rats after passive transfer, and vaccination with anti-idiotypic antibodies to these also confer protection [196]. During natural infections, however, antibody titers to the antigen do not correlate with protection since part of the response consists of the production of IgM antibodies that do not kill schistosomula and that may actually interfere with the process by competing with the binding of protective IgG antibodies [203–207]. The titers of blocking antibodies in infected humans decline with age, perhaps accounting for a rise in age-related immunity. Antibodies to schistosome carbohydrates in general tend to decline with age, perhaps in the process leading to diminution in the granulomatous response [83,84,202]. The other antigen is a glycoprotein exposed on the surface of schistosomula and in eggs of the parasite [208]. Monoclonal antibodies to one of the carbohydrate chains of the molecule are protective after passive transfer into naïve mice, and bind to the surface of living parasites in a reaction inhibited by the trisaccharide Galβ1–4(Fucx1–3)GlcNAc [138]. This epitope is also found in mammals as part stage-specific antigen 1, an antigen expressed by circulating granulocytes [138]. As antibodies to the antigen are detected in human infections, it has been conjectured that the associated autoimmune response may both account for gradual acquisition of resistance and perhaps for some of the pathology [138].

Excretory–secretory antigens

During all of the mammalian phases of their life-cycle, schistosomes secrete or otherwise excrete a number of potentially immunogenic molecules. As previously mentioned, cercariae empty the contents of the acetabular gland cells and release their glycocalyx. Juvenile and adult worms release surface antigens in the form of blebs that may contain cytoplasmic antigens, gut-associated antigens, and protonephridia-derived antigens. Many of these macromolecules have been, or will be, discussed elsewhere in the text.

Owing to the fact that schistosomula are targets of protective immunity in most experimental models, the excretory–secretory antigens of this stage have received attention. The total number of proteins released by cercariae/schistosomula numbers approximately 40, and the functions of these are, for the most part, unknown [209–211]. Equally little is known about the immune response to antigens released by schistosome larva, except that these include species that elicit hypersensitivity reactions. Possibly of importance, presentation of antigens to the immune system by irradiated or drug-attenuated cercariae (which induce immunity) has been found to occur more efficiently than by unirradiated larvae (which induce a lower degree of immunity) [212–214]. The former organisms persist longer in the skin and draining lymph nodes, thus prolonging release of antigenic material [213,215]. *In vitro* secretions of cercariae/schistosomula have been shown to be protective in rat and monkey models, with parasite-killing ocurring by antibody-dependent mechanisms [216].

Aside from gut-associated hemoglobinases and proteoglycans, little is known about the excretory–secretory antigens of adult worms. Based on analyses *in vitro*, over 10 antigens are released [217,218]. Many of these are immunogenic during the course of human infection, a property that may be useful for immunodiagnosis [217,218]. Owing to the low pH of the schistosome gut [24], it seems unlikely that antigens in this location will prove to be protective targets.

CELLULAR RESPONSES

General comments

Lymphoid cell-mediated immune responses against schistosomal antigenic materials have been studied in relationship to protective immunity [219], and extensively with regard to immunopathology [220–222].

Such studies have been pursued functionally *in vivo*, with results being assayed by reductions in worm burdens [223–225] and egg-focused granuloma formation [220,221,226], and *in vitro* through studies of schistosomular killing [223–225,227–229], lymphoid cell proliferation [230–232], lymphokine production [233–236], and "*in vitro* granuloma" formation [237–240]. There has been a strong emphasis on studies of cell-mediated immunoregulation of the antiegg granulomatous response [220–222]. Most of the reported investigations into cellular responses have been done with crude mixtures of parasite-extracted antigenic preparations and using whole lymphoid cell populations. Considerable information, especially functionally relevant information, has been derived from these relatively undefined studies. In conjunction with this approach, it is heartening that over the last few years dissection of these responses has begun to proceed at the level of specific T cell lines and clones [233,234,241–247] and with more highly purified antigenic materials [248,249]. Nevertheless, the organization of this section on cellular responses cannot be based on recognition of individual antigens, as was possible with humoral responses. Rather, the lymphoid cell responses will be evaluated from the perspective of general functional involvement, and this will be further expanded with regard to immunopathologic consequences in the subsequent section.

In vivo responses

Cellular responses *in vivo* have been measured as delayed-type hypersensitivity (DTH) skin test reactivities in both experimental hosts and patients [220]. Cell-mediated immune reactivity is demonstrable against crude extracts from the various life-cycle stages of schistosomes encountered by the mammalian host (cercariae, schistosomula, adult worms, and eggs), and often against the intact organisms [220]. A unique characteristic of the DTH reactions that develop against schistosomal materials during infection is the strong lymphokine-mediated involvement of eosinophils in the reaction sites [250]. Eosinophil involvement in the reactions to schistosomes and their products has been most studied in regard to SEA [251] but also occurs against other life-cycle stages [252]. It now seems clear that the major involvement of eosinophils in anti-SEA cellular responses closely correlates with the participation of TH-2 lymphoid cells that produce significant amounts of IL-5 [235,253,254]. The differential production of IL-5 in combination with granulocyte macrophage colony stimulating factor (GM-CSF), a mixture known as eosinophil stimulation promoter (ESP) [235], and the lack of, or down-regulation of, cellular responses yielding certain other cytokines produces a set of conditions that leads to eosinophilopoiesis, peripheral blood eosinophilia, and the tissue accumulation of eosinophils in reaction sites and lesions. Related differential stimuli and TH-2 cellular involvements are also probably involved in the unique development of IgE humoral responses, through the mediation of IL-4-assisted TH-2 cell help [255]. The kinetics of the infection-related joint phenomena of eosinophilia and IgE responses are well established [230,256], and the timing of TH-2-dependent lymphokine production is now seen to occur over the same time sequence during mouse infection [253,254]. In this and other helminthic situations, there must exist differential trigger mechanisms that lead to dominance of these particular cellular involvements in these settings. The basis for the apparent selective activation of TH-2 cells, however, remains unclear.

The cumulative total of the host's responses against SEA/eggs during infection or *in vivo* experimental manipulations can be seen in the formation of egg-focused, cell-mediated, eosinophil-rich granulomas [220–222,226]. These lesions are also probably centrally involved in the development of chronic fibrosis in schistosomiasis [222,226]. The induction and regulation of egg-induced granuloma formation will be discussed in detail below.

The *in vivo* involvement of lymphoid responses against larval or worm antigenic moieties has been demonstrated in the skin test studies mentioned previously, and functionally related studies implicate roles for these types of responses in protective antischistosome mechanisms [219,257]. Increased numbers of CD4$^+$ T lymphocytes occur at appropriate times and locations [258] to participate in resistance phenomena and mechanistically their involvement can be traced to both their T helper functions [259–261] and their antigen-induced production of lymphokines, and resultant activation of antischistosomular effector cells [257, 262–264]. Studies of the depletion of CD4$^+$ cells in relationship to the irradiated vaccine model in mice indicate that CD4$^+$ cells are essential in the early development of resistance in once-immunized mice. If mice are multiply immunized (five times), a requirement for T helper/T activation function(s) is no longer required [261]. Thus, sera from multiply immunized (more than

three times) mice can passively transfer protection to naïve mice, while sera from once-immunized mice cannot [261,265].

In vitro responses

The responses of lymphoid cells from experimental hosts and patients with schistosomiasis have been studied extensively [56,220,221,266]. Such studies have been done almost exclusively in relationship to peripheral blood mononuclear cell (PBMC) or spleen cell stimulation by crude schistosomal antigenic extracts, such as SEA or total extracts from adult worms, soluble worm adult antigenic preparation (SWAP), cercariae (CAP), or schistosomula. These preparations, by their general nature, contain considerable overlapping antigenic mosaics of shared moieties. Yet, in lymphocyte blastogenesis studies using patients' PBMC the general patterns of responsiveness observed by independent investigators in many different patient populations to these heterogeneous materials are most often quite distinct [56,220,221]. Patients with acute infections respond strongly and equally to SEA and SWAP, with somewhat lower responses to CAP. Patients with chronic asymptomatic infections (intestinal (INT)) usually present with considerably lower SEA-induced responses, while their SWAP and CAP responses are maintained [56,220,221,266–268]. The overall decreased responsiveness to SEA is also observed in experimental infections [221,230]. These changes in anti-SEA reactivity are based on the development of several immunomodulatory responses, which are termed, in the aggregate, modulation [220–222,226,269]. Immunomodulation was first observed in schistosomiasis as a diminution in the size of newly formed egg-induced granulomas as chronic infection was established [269,270].

Alterations in other cellular functions associated with the immune response, such as the level of natural killer (NK) cells [271–273], IL-1 expression [274], and tumor necrosis factor (TNF) levels [275], have been evaluated in the PBMC and sera of patients. Peripheral blood mononuclear NK cell functional activity is largely unaffected in patients with moderate schistosome infections [271] but can be greatly impaired in heavily infected patients [272,273] and is restored upon curative chemotherapy [276]. IL-1 activity was undetectable in sera from some schistosomiasis mansoni patients [276]. Another study found IL-1 production to be greater by lipopolysaccharide (LPS)-stimulated PBMC from patients than from uninfected controls, and reported a correlation between the level of IL-1 production and the intensity of the patient's infection [274]. Serum levels of TNF activity were higher in hepatosplenic patients than in INT patients, but PBMC from hepatosplenics were less able to produce TNF activity upon stimulation by pokeweed mitogen than PMBC from INT cases [275].

A somewhat unique set of *in vitro* cellular immune responses that evaluates anti-idiotypic interactions [69,277,278] has been studied with PBMC from schistosomiasis patients [277–279] and spleen cells from infected mice [280,281]. In the human studies, T lymphocytes in PBMC from patients or former patients respond by proliferation to certain idiotypes on anti-SEA polyclonal or monoclonal antibodies. These responses to homologous or heterologous anti-SEA antibodies are interpreted as being manifestations of anti-idiotypic T cell reactivity [69,277,278]. These cell stimulatory interactions take place directly, without requirements for processing or MHC presentation of the idiotypes responsible, but with a need for IL-1 and multibinding site stimulatory antibodies [282]. Correlations of these responses with given clinical forms of the infection [278, 281,283] and their participation in immunoregulatory events [279] will be discussed later in relationship to pathogenesis and immunoregulation.

Cellular responses are now being examined in the context of T cell subset categories defined by the abilities of T cell populations and clones to respond by differential production of given lymphokines/cytokines (LK/CK) [255]. Lymphokine/cytokine production by antigen- or mitogen-stimulated cells from experimental hosts or patients is emerging as an important expanding area of research in schistosomiasis [235,253,254,284,285]. At this time, it is difficult to relate *in vitro* LK/CK studies with PMBC from patients to *in vivo* immune events, and the TH-1 vs. TH-2 interpretation now used in mouse studies is less clear in humans. Nevertheless, as appropriate LK/CK assay systems for the human interleukins and/or their mRNA become more available, more studies with this focus will quickly follow.

Studies of the LK/CK responses of *S. mansoni*-infected mice have indicated that SEA-stimulated spleen cells begin to make detectable IL-2 6 weeks after infection [284]. This ability appears to peak at 8 weeks, and then wanes. This pattern is very reminiscent of that seen for the LK activity ESP [230], which is now known to be comprised of a combination of GM-CSF and IL-5 [235]. From *in vivo* depletion studies, IL-5 is known to be largely responsible for peripheral blood eosinophilia in experimental schistosomiasis [236]. Two rather different

sets of observations have been made regarding the likelihood of dominance of the TH-1 vs. TH-2 subsets of T cells in acute and chronic murine schistosomiasis. Boros *et al.* [284] have emphasized the role of IL-2 in the formation of early lesions. They also indicate that the waning of IL-2 production correlates with the onset of modulation, and that granuloma modulation is reversed by rIL-2 administration [284]. In contrast, Sher *et al.* find no evidence of IL-2 production during infection but report high levels of IL-4 production during acute infections [253], or whenever *S. mansoni* eggs are administered [249]. A corroborating report indicates a further lack of evidence for demonstrable IL-2 involvement based on Northern blot analysis of granulomas, liver tissue, lymph nodes, or spleen 8 weeks after infection, although IL-4 mRNA was found to be abundant in these tissues [285]. Infection and assay systems may need to be compared directly and with the same kinetics in order to unravel the differences in these findings. With such analyses should come a more complete understanding of the requisite interdigitation of immune systems needed to produce and regulate granuloma formation.

An *in vitro* cellular response essentially unique to studies of schistosomiasis has been developed and used by Doughty *et al.* [237–240] which allows functional immune analysis. This system is termed "*in vitro* granuloma" formation, and is an adaptation of the *in vivo* assay developed by von Lichtenberg [286] and adapted by Boros & Warren [287] using schistosome eggs or SEA-coated particles injected into the pulmonary vasculature of mice to impact and induce granuloma formation. In the *in vitro* system when either intact eggs or SEA-coupled beads are cocultured with cells from sensitized or infected mice or patients, a reaction is observed which can be quantified with regard to the number and types of cells which are attracted to and cover the egg or suitable bead [237,239,240]. These reactions involve initial macrophage interactions and the recruitment of antigen-specific T cell subsets that further recruit a variety of cell types, such as macrophages, lymphocytes, eosinophils, and fibroblasts, depending on the original cell population used. Both anti-SEA reactivity and its regulation (including anti-idiotypic immunoregulation) can be demonstrated in this system [279]. It allows the dissection of cell subset functions with regard to the sequence of events as they occur with the population of cells available.

IMMUNOPATHOLOGY AND ITS IMMUNOREGULATION

Acute infection

During the course of schistosomiasis the host must deal with migrating schistosomula, adult intravascular worms, worm excretory–secretory products, large numbers of eggs produced by the adult worms, and egg-produced materials. Internal exposures to large quantities of antigens begin upon maturation of the male and female worms, and they continue throughout infection. Primary infections in endemic populations usually lead to little, if any, morbidity during the acute period around patency [57]. In contrast, initial infections of visitors from nonendemic areas [288] and urbanites in endemic countries who visit active transmission sites for the first time [289] often result in a clinical syndrome (1–3 months after infection) of fatigue, fever, sweats, chills, headache, malaise, myalgias, lymphadenopathy, and gastrointestinal discomfort [290,291]. The intensities of infection that lead to clinically acute disease seem extremely variable, in that fecal examinations from these patients can yield very high to very low egg counts. Organomegally at this time is related primarily to lymphoreticular activity, and is transient. Pathophysiologically, human liver needle biopsies demonstrate the presence of large florid egg-focused granulomas that contain high numbers of eosinophils and frequently show Hoeppli precipitates [60]. Patients with this syndrome are usually immunologically hyperreactive [56,267,268,292,293]. This acute toxemic phase, often referred to as the Katayama fever syndrome, symptomatically resembles immune complex disease. At this time, patients' antibody levels are rising in response to large amounts of newly produced egg antigens. This situation would be expected to lead to a transient period of circulating antigen excess and indeed it does yield immune complex formation, with some of the early antibody responses occurring differentially against several carbohydrate antigens [56,266,294]. For example, taking advantage of a shared carbohydrate epitope on schistosomula and keyhole limpet hemocyanin (KLH), an anti-KLH enzyme-linked immunosorbent assay (ELISA) can distinguish between sera from patients with acute (highly reactive) and chronic (low-nonreactive) infections [294]. Complement activation occurs during the acute syndrome, and C1q-binding activity can be seen to parallel both the severity [295] and intensity [296] of disease. It seems clear that immune complexes and complement activation participate in the pathogenesis of this condition.

Another characteristic of clinical [268,295] and experimental [256] acute infection is high peripheral blood eosinophilia. Experimental acute infections have been recently shown to involve transient high-level production of IL-5 by spleen cells from mice infected for 8–16 weeks when stimulated with either SEA or concanavalin A [253]. This major emphasis on IL-5 (and IL-4) production did not occur in mice with single sex infections [253] and another line of similar investigations indicates that the presence of schistosome eggs is sufficient to down-regulate TH-1-dominant LK/CK production (interferon-γ (IFN-γ) and IL-2) and lead to increased TH-2 involvement (IL-4 and IL-5) [254]. Granuloma-bearing liver tissue and antigen-stimulated spleen cells from mice with 8-week infections contain abundant levels of IL-4-specific mRNA, while IL-2 mRNA is not detectable at this time, and by 20 weeks IL-4-specific transcription is greatly diminished [285]. The timings of these occurrences in schistosome-infected mice exactly parallels the rise and fall of anti-SEA IgE-mediated reactivity [242,269] and peripheral blood eosinophilia [230,256]. Removal/inhibition of the eosinophil-active lymphokine [235,255] IL-5 by administration of anti-IL-5 monoclonal antibody to infected mice results in the virtual absence of eosinophils in the peripheral blood and from the granulomas that subsequently develop [236]. Surprisingly, these eosinophil-poor granulomas remain unchanged in size. Normally eosinophils (activated and armed for egg destruction) [297] participate heavily in antiegg reactions, and this is apparently brought about through an IL-5-mediated mechanism. If this action is prevented, it seems that other cellular elements are supplied as needed for granuloma development [236].

The relative absence of Katayama fever manifestations in early infections in endemic populations may be of considerable immunologic interest. It suggests that individuals living in endemic areas handle acute infection in an altered manner from those who develop infection upon visits to endemic areas. It has been proposed that most of those born and raised in an endemic area are not, at the time of their first infection, immunologically naïve with regard to schistosome-related immune responses or antigenic materials [266,298]. It is likely that through their infected mothers they would have been exposed to schistosome-related moieties (antigens, antibodies, idiotypes) (see earlier section on maternal/fetal–perinatal interactions). It seems possible that such exposure may contribute to the expression of a less-fulminant, more-regulated [61,62] response to schistosomal antigens during early initial infection in those native to endemic areas. A note of caution should be raised that essentially all studies comparing responses of acute vs. chronic infections derive their acute patients from nonendemic groups and their chronic patients from endemic populations. While this is a virtual necessity, based on the only circumstances in which one finds clinically acute patients, it may not be totally legitimate to infer progression from one state to the next based on these disparate study populations [266].

Chronic infection

Chronic infections, autoimmune conditions, transplantation situations, and neoplasms are all settings in which the body's immune system is more or less continuously bathed in antigenic stimuli. The antigenic sources can be invading organisms, altered or normal crossreacting self-components, foreign tissue, or tumor-specific antigens. The results of these long-term interactions between a person's immune system and antigens can be immunopathogenic, resulting in severe morbidity and mortality. Fortunately, in the natural setting, the morbidity induced by chronic antigenic exposure is often partially controlled by the development of counterbalancing immunoregulatory responses that hold in check the most severe immunopathologic processes. It has been proposed that chronic parasitic infections may be a primary evolutionary influence that has pushed the immune system to develop such regulatory capacities [299].

In schistosomiasis, the chronic antigenic insults are from excretions and secretions of adult worms and eggs and their products, and the intensity of infection plays a critical role in influencing the severity of the resulting clinical condition. Upon egg deposition, the host response results in the formation of strong, egg-focused, cell-mediated granulomas. Significant colonic polyposis can occur in S. mansoni infections in some endemic areas (notably Egypt), while mild to severe intermittent intestinal disorders are associated with the most commonly occurring syndrome, called chronic INT schistosomiasis. The severe hepatosplenic form of chronic schistosomiasis mansoni and japonica usually occurs in 4–8% of those infected, the remaining >90% of the patients appearing to develop and live with the INT form of infection. Experimental studies imply that unregulated antiegg hepatic granuloma formation leads

to the severe form of the disease, the *sine qua non* of which is Symmers' clay pipestem fibrosis. This periportal fibrosis leads to portal hypertension, collateral circulation, and ascites buildup. Until very late in this progression, patients maintain normal liver function, and the basis of the pathogenesis is primarily a slow, cumulative blockage of portal blood flow, not liver damage [57]. Upper gastrointestinal bleeding from esophageal varices subsequent to portal hypertension is the most common morbid consequence of schistosomal hepatic fibrosis.

Most experimental studies suggest that a direct relationship exists between the formation, fibrosis, and coalescence of granulomas and periportal inflammation, leading to the characteristic fibrosis [299–303]. It is unclear, however, whether this truly represents the complete developmental picture in infected chimpanzees and humans, where the additive periportal fibrotic effect of scattered hepatic egg granulomas is unresolved [304–306]. It has recently been observed that different levels of severity of disease can develop in comparably infected inbred mice (chronic S. mansoni infections of one or two worm pairs in male CBA/J mice) and that these differences correlate with individual (less severe) vs. banded (more severe) patterns of granuloma fibrosis and differences in immunoregulatory (anti-idiotypic) responses (G. Henderson, N. Nix, D. Colley, personal communication). Further analysis of these observations may provide insight into what determines progression down these apparently divergent pathways.

Analysis of the chronic development of Symmers' fibrosis in patients has suffered from the inability to assay the extent of fibrosis by noninvasive means. Happily, the advent and use of portable ultrasonography units has greatly enhanced the ability to diagnose and study the progressive process of the development of Symmers' periportal fibrosis in humans in the field [307–310]. More widespread application of this non-invasive but sensitive tool should assist in field analysis of correlates of fibrotic progression. Its use has already resulted in heartening observations that appropriate curative chemotherapy of hepatosplenic or severe urinary schistosomiasis can be expected to lead to reversal, and sometimes resolution, of severe fibrosis [308,309] or urinary pathology [311].

A predictable pattern of *in vitro* responses to several heterogeneous schistosomal preparations has been reported by different groups working on cellular responses in human schistosomiasis [56,266]. Immune correlations with the presence of different clinical forms of infection have been observed which indicate that establishment of chronic subclinical infections parallels decreased cellular responsiveness to SEA [268], while chronic hepatosplenic disease is associated, at the ambulatory stage, with vigorous SEA-stimulated PBMC proliferation [231]. As will be discussed later in the text, the development of different chronic clinical expressions of schistosomiasis mansoni also correlates with different idiotypic anti-SEA responses in patients [312].

A wide variety of other clinical aspects of chronic schistosomal infections have been observed and reviewed [57]. These conditions include nephropathies, cor pulmonale, cerebral complications, colonic pseudo-polyposis, ectopic sites of adult worm residence, and complications involving schistosomiasis with a variety of other infections and neoplasias. Owing to the chronicity and progressive nature of schistosomal infections, such interdigitation seems inevitable. The import and mechanisms of these interrelationships largely remains to be elucidated.

Granulomas

The immunopathogenesis of experimental schistosomiasis mansoni is largely attributed to cellular immune responses that develop to SEA and result in egg-focused hepatic granuloma formation [226,313]. Relatively few studies have been done on *S. haematobium* egg-induced granulomas, but in mouse lung these responses appear similar in cellularity to those of *S. mansoni* [314], while in mouse liver they involve early marked fibrosis [315]. In *S. japonicum* infections, antiegg granulomatous reactions and their control are considered to result more from an interplay between humoral and cellular reactions [301, 302,316,317]. Regulation of *S. mansoni* granuloma formation is primarily thought to depend on cellular immuno-regulation [220,221,303], or possibly an alteration in the predominant responsive TH population [253,285]. The formation and regulation of schistosome egg-induced granulomas, both *in vivo* [226,294] and *in vitro* [238,279], have been studied extensively, and much of the individual work on the lymphoid and mononuclear phagocytic cells responsible for the production and regulation of these lesions has been reviewed previously [221,226, 266,303,318]. This area remains the central focus of most investigations on schistosomal immunopathology.

The function of the granuloma has for some time [318] been considered primarily to wall off eggs and their potentially noxious secreted products. Recently, an additional role has been proposed independently by two laboratories [319,320]. This hypothesis addresses the question of how intravascularly laid schistosome eggs might cross the vessel wall and gut or bladder wall to make their escape from the body in excreta. It is proposed that these eggs are dependent upon "transport" by motile granulomas across tissues that would otherwise stand as insurmountable barriers to completion of the life-cycle of an intravascular parasite. These thoughts are based on immunodepletion studies in mice, which showed that egg excretion is dependent on an intact immune system [321], and on histo-pathologic observations which suggest a dynamic process of granuloma formation in the gut wall with "excretion" of granuloma/egg units [320]. It has been further proposed that the parasite actually has suborned, or at least exploited, the immune system of the host for this, its own necessary ends [320].

Schistosome egg-induced granuloma formation is a process by which mainly lymphocytes, macrophages, eosinophils, and a few plasma cells, fibroblasts, giant cells, mast cells, and neutrophils actively encase eggs in focused, specific lesions [220,239,301–303,313,316,322, 323]. IL-1 can be directly involved in the production of *in vivo* granulomas [324], and in mice injected with *S. mansoni* eggs the formation of granulomas has been associated with the sequential production by LPS-stimulated granuloma macrophages of IL-1b (2–4 days) and TNF (8–16 days) [324]. Macrophages isolated from hepatic granulomas early in infection are, by multiple criteria, highly activated, while those from later in the infection are considered less so [326–331]. Also, high numbers of early granuloma macrophages suppress lymphocyte functions through an IFN-α/β-like moiety, while both early and late granuloma macrophages have immune accessory capabilities that appeared to be mediated through IL-1 [331]. The source of granuloma macrophages has been studied by enumeration of macrophage progenitors in the bone marrow and in liver granulomas [332]. During acute infection the number of bone marrow macrophage progenitors was very high, but progenitors in the lesions were few in number. This was accompanied by high levels of circulating M-colony stimulating factor (CSF-1), which decreased to normal levels as infections became chronic. At this time the bone marrow progenitors also returned to normal levels, and in concert the number of progenitors in the granulomas increased dramatically, and these lesions secreted high levels of CSF-1.

Miracidia in eggs secrete SEA, the preparation that contains the moieties that can induce and elicit granuloma formation and its regulation [251]. Soluble egg antigen is very heterogeneous, containing multiple proteins, glycoproteins, carbohydrates, and glycolipids. Partial purification of active components in SEA has been undertaken and some dominant antigens have been described that are responsible for humoral and cellular reactivity [248,249]. High-resolution preparative separation and methodologies are now being applied to this problem. The use of agarose-isoelectric focusing separated eluted fractions revealed that only antigens in the pI range of 3.5–5.0 (species of molecular weight >200, 54, and 37 kD) consistently stimulated proliferative responses of lymph node cells from mice infected for 10 weeks. If the donor mice were infected for 25 weeks their cells also responded to antigens in the pI range 6.2–6.4 (species of molecular weight 51 kD) [248]. These cellular recognition patterns were much more restrictive than were parallel studies on humoral reactivity. Electroelution of nine SEA fractions from sodium dodecyl sulfate polyacrylamide gel electrophoresis (SDS-PAGE) separations has shown that T cell-enriched spleen cells from acutely infected mice responded to antigens of all nine molecular weight fractions, but those of chronically infected mice only responded to antigens in the 50–56 and 60–66 kD fractions. Depletion of CD8$^+$ cells from preparations from chronic mice broadened the range of responses, suggesting that modulation is associated with regulation of responses to specific antigens within heterogeneous SEA [249], a result that is consistent with preliminary findings using a T cell Western blot approach to this question (T. Amano & D.G. Colley, personal communication). Continuing endeavors to define the specific antigens primarily responsible for induction, elicitation, and regulation of granuloma formation in experimental systems are needed and their use also needs to be extended to *in vitro* studies with cells of patients.

Numerous studies have examined granuloma formation in *S. mansoni*- or *S. japonicum*-infected, T lymphocyte-depleted, or athymic mice [220,302,303,333]. All reports note that egg-focused granulomas in such animals are dramatically decreased in size and cellularity. Some investigators have observed necrotic lesions instead of granulomas and others have seen only smaller granulomas without concomitant pathologic necrotizing lesions. The overall condition of the mice and the degree of immunosuppression may contribute to such differences. The role(s) of CD4$^+$ T lymphocytes in the promulgation of granuloma formation has been shown

by the dampening effect of anti-L3T4 depletion studies [334] and the restorative effect of adoptive transfers of CD4$^+$ cells to syngeneic athymic mice prior to infection [302]. Studies of the *in vitro* granulomatous response against schistosome eggs or SEA-coated beads, using cells from infected mice or patients, agree with the basic finding that CD4$^+$ cells have a fundamental role in the production of these granulomas [238,239,303].

Cellular responses to SEA that result in LK/CK production were discussed above with regard to both cell-mediated responses and acute infections. Many of these functional responses are considered to be centrally involved in the production of schistosome egg-induced granulomas [234,253,285,303].

The potential involvement of CD4$^+$ cell lines and clones has been demonstrated using anti-SEA-specific clones from the spleens of infected mice [233,242,243] and even T cell lines obtained from granuloma cell populations [234]. Soluble egg antigen-specific helper and suppressor T cell lines have been generated from *S. japonicum* patients' PBMC [244], and have been useful in the definition of the SEA component that might be involved in patients' responses to the crude preparation [245].

Differences in schistosome egg-induced granuloma formation in infected mice have been noted with regard to the duration of infection [270]. It is also known that the anatomic site of egg deposition affects the size and composition of granulomas formed. Acute granulomas formed in the liver and colon are of comparable size but differ in the number of adoptively transferred chronic lymphoid cells needed to regulate them, while granulomas in the ileum are always much smaller than those in the liver or colon [335]. Furthermore, isolated and dispersed ileal granulomas yield primarily macrophages, while those in the liver and colon contain many more T and B lymphocytes [336]. Both chronicity and location make a difference in the types and amounts of extracellular matrix proteins associated with different granulomas [337].

Fibrosis

Hepatic granuloma and periportal-associated fibrosis have been less completely studied in schistosomiasis than has the induction of granuloma formation. Fortunately the tools and approaches for evaluating fibrogenesis are gaining in availability [338] and usage [303,339] in both clinical [339–342] and experimental schistosomiasis [343,344]. Livers of *S. mansoni*-infected mice examined by immunofluorescence with regard to the deposition and patterns of extracellular matrix proteins have demonstrated differences based on the duration of infection that parallel some of the acute to chronic modulatory features of granuloma formation to be discussed under "regulation." At 8 weeks of infection liver granulomas contained low amounts of Type III collagen and fibronectin [337]. The relative quantities of Type I collagen deposits gained equality with Type III as infection progressed, and by 20 weeks of infection, when levels of both Types I and III were high, Type IV was also observed [344]. Ileal granulomas, known to be smaller than those in the colon or liver [253], did not change size during infection, and at 8 weeks contained barely detectable amounts of connective tissue matrix proteins. By 20 weeks, Types I and III collagen attained their highest levels, fibronectin was prominent, but laminin and Type IV collagen were absent. Colonic granulomas were characteristically as large as those in the liver, but contained much less Types I and III collagen at 8 weeks of infection. The formation of these lesions modulated by 20 weeks, when they contained only trace amounts of laminin and Type IV collagen, fibronectin was present in high amounts, and Type III collagen levels were moderate [337]. It is clear that dynamic deposition, accumulation, and turnover of granuloma-associated matrix proteins occurs throughout experimental infection and differs at disparate sites of egg deposition. The early usage of Type III collagen [337,345,346] followed by the accumulation of the more heavily crosslinked, more degradation-resistant, and more rigid Type I collagen in lesions in mice with chronic infections may correlate with enhanced morbidity and less lesion reversibility upon treatment [340,347–349].

Studies by Wyler *et al.* have demonstrated that isolated intact cultured schistosome egg granulomas spontaneously produce several moieties that stimulate fibroblasts to proliferate, migrate, and synthesize collagen and fibronectin [343,350–352]. The implications of such studies are that cells within granulomas produce fibrogenic cytokines which control the type and degree of fibroblast involvement in granuloma (and perhaps periportal) fibrosis. Macrophages isolated from schistosome granulomas produce some of these fibroblast stimulating factors (FSF), as do splenic Lyt-1$^+$ T cells [353,354]. A dual effect was seen with cocultivation of Lyt-1$^+$ cells and primary fibroblasts. Fibroblast proliferation was increased but there was a significant decrease in the incorporation of ^{14}C-proline into collagenase-sensitive protein [354]. Partial characterization of a 17 kD

fibroblast stimulating factor produced by an SEA-specific T cell clone has been accomplished [355]. The differences in fibroblast-active moieties from schistosome granulomas from athymic vs. euthymic infected mice [333] offer potentially interesting insights into the influences of granuloma products on fibrogenesis. Athymic granulomas fail to become fibrotic, but whether the T lymphocyte input into these differences occurs directly due to their products or indirectly by the subsequent influence of their products on other cells (macrophages, eosinophils?) remains uncertain. It has been shown that adoptive transfer of spleen cells from chronically *S. mansoni*-infected mice to those with early infections increases Type I collagen levels in early granulomas, while serum transfers from chronic to early infections augments levels of Type III collagen [356]. The size of the granulomas formed in the recipients was only down-regulated by cell transfers. The independent immunologic effects seen in these studies on inflammation vs. fibrosis demonstrate the complexities of these processes, which are often thought of as uniform but may well be both unique and interdependent.

Studies in inbred mice indicate that considerable strain-dependent differences exist in the degree of fibrosis that develops in murine schistosomiasis japonica [357] and mansoni [358]. In *S. japonicum* infections the degree of fibrosis did not correlate with granuloma size, while in *S. mansoni* infections those strains with the largest granulomas also showed the most hepatic fibrosis. However, even in the latter situation, hepatic fibrosis per egg decreased with increasing infection intensity but granuloma volume was unaffected by intensity, indicating probable differential modulating factors for the two processes [358]. These studies concluded that the regulation of granuloma size and hepatic fibrosis may differ and be affected by genes within and outside of the murine MHC. The suitability of assessing granuloma and periportal fibrosis in the murine system as a model for human Symmers' clay pipestem fibrosis remains unsettled. As a model system, its similarities offer opportunities to ask questions in a sorely needed area of research which should be promoted. Also, as ultrasonography and biochemical, immunologic, and molecular means to study fibrosis in noninvasive ways become more finely tuned and available, they should be rapidly applied to parallel studies of these problems in patients. It is reasonable to question the comparative suitability of assessing granuloma formation and fibrosis in the murine system as a model for severe human schistosome-induced periportal fibrosis of the Symmers' clay pipestem type, and the differences [359] and similarities [300,360] continue to be discussed.

Immunoregulation

Immunoregulation of immunopathogenesis is proposed as a major contributory force in the dynamics of the overall host–parasite relationship in schistosomiasis [69, 231,266,298,361,362]. Other important factors include the intensity of infection [56,363], which is influenced by positive and negative immunoresistance responses [3,204,362], and host genetic differences, which are usually observed as immunogenetic effects [266,298, 364–366]. It is also possible that parasite strain differences could contribute to differences in disease spectra, but only broad comparisons of single field isolates exist to contribute to the knowledge of this potential facet of complexity [221]. The composite result of all such interactions is the relatively well-balanced chronic infection seen in the majority of patients, the less frequently seen intermittent morbidity of gastrointestinal or urogenital distress, and the more uncommon (4–8%) debilitating, sometimes fatal, hepatosplenic disease or severe uropathies [57].

In terms of antiegg antigen cell-mediated responses in patients it has been shown that greater responsiveness to SEA [231,232] and to SWAP [361] correlates with severe hepatosplenic disease. Furthermore, patients with asymptomatic infections can be delineated from those who have hepatosplenic disease based on the expression of certain immunoregulatory, dominant crossreactive idiotypes on their anti-SEA antibodies [278,281,283,298,312]. This implies a relationship between the development or expression of such regulatory modalities and well-tolerated (regulated) infections. The relationships between these various immunoregulatory mechanisms and human asymptomatic chronic infections are only correlative, but cause-and-effect relationships are now under study.

In experimental hosts, acute infections are associated with the formation of large, florid granulomas around eggs deposited in the hepatic or pulmonary vasculatures [269,270]. Upon the establishment of stable chronic infections, the anti-SEA granulomatous responsiveness of the host decreases. This process, termed granuloma modulation, has been extensively studied in the mouse model [220–222,226,298,301,367]. In general, it has been shown that this is an active immune regulatory response that involves both humoral and cellular events, but depends primarily on cellular mechanisms

in *S. mansoni* infections and more on humoral mechanisms in *S. japonicum* infections. Granuloma regulation through suppressor cell mechanisms involving idiotypic/ anti-idiotypic cellular and humoral mechanisms have been described [69,317,368] in both *S. mansoni*- and *S. japonicum*-infected mice, as have systems dependent upon T suppressor factors [368,369] in *S. mansoni* infections.

Current understandings of subpopulations of murine CD4[+] lymphocytes, based on differential LK productivity [255], have required a re-evaluation of some aspects of our understanding of granuloma modulation. As discussed in the section on cellular responses (*in vitro*), this area of investigation is currently very active and unsettled. It appears that exposure to schistosomal eggs leads to an anti-SEA response that may begin very early as a TH-1-type response [334,370,371] but then rapidly becomes largely a TH-2-type response, with high-level production of IL-4, IL-5, and IL-10 mRNA and product [253,285,372] (G. Henderson, personal communication). Although IFN-γ mRNA is produced during acute (8-week) infection [317], controversy exists over product expression [372] and the extent of production of the other major TH-1 product IL-2. Thus the general involvement of TH-1-type responses has been questioned in granuloma formation [253,285,372] (G. Henderson & D. Colley, personal communication). The establishment of chronicity and the parallel supremacy of granuloma modulation is associated by some with a decrease in the level of TH-2 responses [285,372] (G. Henderson & M. Howard, personal communication). However, it is also true that IL-2 production by spleen cells to SEA decreases [284] and exogenous IL-2 both abrogates the *in vitro* production of a T suppressor–effector factor (TseF) [373] and reduces *in vivo* TseF function [369], resulting in an increased size of liver granulomas [284,374]. It should also be remembered that CD4[+] cells do not operate in a vacuum during these changes. CD8[+] cells, which make more IFN-γ than IL-2, increase during the transition phase from acute to chronic infection. This alters the CD4[+]/CD8[+] ratio in both the draining lymph organs (G. Henderson & T. McCorley, personal communication) and within the cell populations that make up the granuloma [374]. Given the complexities of interleukin production and effects, and the multitudes of possible interplays between different immune responses during these infections, it would seem that a complete and coherent understanding of these interweavings will take some time. This area is only studied at this level by a handful of laboratories and yet the activity is brisk, and one can expect that it will continually yield new findings and relationships. It can be predicted that, as with studies of new effector mechanisms in antischistosomal and antileishmanial immunity [372], this area is likely to uncover new basic immunologic findings of general interest.

ACQUIRED RESISTANCE

It has become increasingly clear over the past 20 years that immunologically based immunity to reinfection can develop during the course of natural human infections and that resistance can be induced by vaccination of laboratory animals. Despite this knowledge, acceptable schistosome vaccines have yet to be developed, owing in part to a lack of a definitive interpretation of mechanisms inducing acquired immunity in humans and hence an inability to project information gleaned from experimental animal models to the human situation. The two best studied models of acquired resistance to schistosome infection are laboratory rats and mice either after a primary infection or after vaccination with radiation-attenuated larvae. The course of infection varies significantly between these two hosts and between the two conditions. Immunity in each of these models and systems appears to occur by multifactorial processes whose individual components (effector cells, sites of attrition, and immunoglobulin types) consist of both shared and unique entities. Even within individual models, much disagreement exists about the site of attrition and mechanisms of killing of immunologically destroyed worms.

In general, immunity appears to be dependent on $CD4^+$ but not $CD8^+$ T lymphocytes, involves varying degrees of humoral and cell-mediated immune responses, and does not involve direct cytotoxic participation by lymphocytes [375–377]. A variety of effector cells can participate in the various models, including mast cells, platelets, eosinophils, neutrophils, and macrophages. The cells appear to act by release of reactive oxygen intermediates, lysozomal enzymes, peroxidases, and, when eosinophils are involved, their major basic protein. The cells may also act by entrapping the parasites, thus preventing their migration and development [375,377,378].

Mice

Two forms of acquired immunity have been extensively studied in mice. In the concomitant immunity model, the animals are given a primary infection and subsequently challenged. Concomitant immunity defines a situation where an animal infected with a parasite resists further infection via reactions that do not affect established parasites [75]. In the vaccine model, animals are infected with attenuated larvae that die prior to sexual maturation and are subsequently challenged with normal parasites. Resistance to reinfection observed in the two models appears to operate by different mechanisms, with the concomitant immunity model being dependent on egg deposition, to a large degree portal hypertension, and the presence of adult worms [5]. Moreover, concomitant immunity can protect animals from infection by heterologous species of schistosomes whereas vaccine immunity does not [379]. Immunization of mice by attenuated larvae is not fully understood either, but depends more completely on immunologic mechanisms than on vascular alterations.

Chronically infected mice

Partial resistance of mice to reinfection develops roughly at the time when parasite egg deposition begins, and peaks some 4–6 weeks later [380]. The presence of adult, sexually mature worms and/or viable eggs appears to be necessary for maintaining this balance, and transfer of adult worms into the mesenteric vasculature of a naïve host stimulates this type of immunity, which is directed at immature worms, even though the host has not experienced the cercarial and schistosomular stages of development [381]. The strength of the protective response correlates with egg burden [382]. While monoclonal antibodies to schistosome egg antigens have been isolated that react with the surface of schistosomula and protect *in vivo* upon passive transfer [208], it is not clear that this type of crossreaction is responsible for immunity.

The involvement of various immunologic components in concomitant immunity in infected mice is suggested by several observations. Mice genetically deficient in macrophage function show poor levels of immunity, despite normal granuloma formation in the liver [383], and destruction of neutrophils with a monoclonal antibody suppresses concomitant immunity [384], as does destruction of eosinophils with a polyclonal antibody [385]. Treatment with antileukocyte monoclonal antibodies also inhibits inflammatory reactions in the skin normally associated with reinfection [384]. Many authors have noted that reinfection is associated with anamnestic cellular reactions that appear to encapsulate challenge infections [386]. Histologic examination of these responses generally reveals the presence of very few granulocytes in direct contact with the worms and little evidence of focused leukocyte degranulation, despite the fact that these cells readily kill schistosomula *in vitro* [375,376]. The site of parasite attrition is contested, with the skin [252,387,388], lungs [387,389], and liver [382, 389] having been implicated.

Resistance displayed by infected mice to reinfection is dependent on the presence of an intact thymus during the primary infection but not during the secondary infection [390]. Serum from infected animals cannot correct the defect in thymectomized animals [390]. Resistance can be mediated by antibody, with the active fraction being IgG [391,392]. However, attempts to transfer immunity via parabiosis experiments, where one animal is infected and the other is challenged, have failed [393]. A role, if any, for complement in this model is uncertain; conflicting results having been reported on the effects of cobra venom factor on resistance [394,395].

Significant determinants of concomitant immunity in schistosome-infected mice lack direct immunologic specificity. The phenomenon is dependent on egg deposition and its magnitude is positively correlated to egg burden within tissues, degree of portal hypertension, and number of egg granulomas in the lungs [5,382]. As currently interpreted, development of granulomas in the liver compromises venous blood flow, leading to portal hypertension that is compensated for by an increased supply from the hepatic arteries [5]. Portal hypertension is accompanied by the appearance of portasystemic anastomoses that direct blood flow around the liver, with the size of collateral blood vessels eventually becoming large enough to allow developing worms to exit from the portal system and subsequently reach the lungs. By this time the worms may be too large to negotiate passage through narrow vessels and subsequently die [5].

Whether concomitant immunity seen in animals larger than laboratory rodents (that are, in realistic terms, massively infected) can be attributed to changes in hepatic blood flow is currently unclear. The finding that rhesus monkeys with unisexual infections can become resistant to reinfection argues that nonegg-related nonmechanical mechanisms may be of importance in this species [75,396]. Fibrosis is not a significant aspect of *S. mansoni* infections in baboons—animals that demonstrate resistance to infection—although their livers may show signs of leakiness [397]. The collateral shunting of the circulation observed in mice occurs in infected

humans with severe morbidity, but it should be noted that infection of a mouse with a single egg-laying worm pair leads to what in human terms would be regarded as severe pathology. In all, the concomitant immunity model has received lessening attention as a basis for vaccine development over recent years, in part owing to the complexity of interpreting data [398].

Immunity induced in mice by attenuated larvae

Mice vaccinated with irradiated cercariae reproducibly acquire varying degrees of immunity to challenge infection [382,399]. During the initial stages of invasion, the organisms do not show any obvious differences with unirradiated larvae, which do not directly appear to induce immunity, save for a protracted migration time [111,277]. Attenuated larvae die prior to maturation, indicating that egg-induced changes in the hepatic vasculature are not a factor in defining resistance [277]. The anatomic site of attrition and the developmental stages involved have long been contested, with different laboratories gathering information consistently refuted by others. As in the concomitant immunity model, postulated sites of attrition include the skin, the lungs, and the liver; developmental forms that are targets of immunity include newly transformed schistosomula, lung stage larvae, and hepatic preadult worms [375,376]. The presently emerging picture is that vaccine-induced immunity in the mouse has a strong genetic component and that the strain of mouse and parasite used is of critical importance in determining the site and mechanisms of attrition [375–378].

Resistance of once-vaccinated mice has an absolute requirement for CD4$^+$ T cells but not CD8$^+$ cells [261, 264]. As previously mentioned, removal of the former cells prior to challenge of once-vaccinated animals resulted in significant reductions in immunity while, as previously mentioned, removal of the cells from mice vaccinated five times has no effect, suggesting that upon hyperimmunization the active role of CD4$^+$ cells is no longer required [261]. Paralleling this, immunity observed in once-vaccinated mice can be transferred by adoptive transfer of blood but not by serum or lymphoid cells. Multiple vaccinations, however, result in serum that induces protection in naïve recipients, with IgG being the active fraction [265]. IgG1 and IgM can also be protective [400,401], and except for the use of a given IgE monoclonal antibody in *S. japonicum* infections [402], efforts to show a role for IgE in the mouse have failed [403].

Mouse immunity at skin level is strain-specific, and associated with subcutaneous inflammatory reactions that wall off the parasites [404–406]. The predominant host cells in skin reactions are eosinophils and mononuclear cells, which accumulate in approximately equal numbers [406,407]. Cellular reactions appear to entrap migrating parasites and attendant inflammatory reactions cause the skin to swell by as much as ten times, hence lengthening the infection pathway. Direct damage to the parasites by leukocytes appears to be minimal [406]. For unknown reasons, the most prominent morphologic correlate of schistosomular entrapment is a disruption of the parasite's muscle fibers and lack of damage to the tegumental membrane [406]. Agents that remove leukocytes, such as whole-body irradiation or treatment with appropriate monoclonal antibodies, reduce levels of protection, whereas treatments that interfere with only macrophage function have no effect [384,407]. The route of parasite migration during initial vaccination appears to be important in this model, with highest levels of resistance being expressed when the attenuated larvae spend extended time in the lymph nodes [212–214]. This form of immunity appears to involve specific IgG1 [400], and thus probably depends on TH cell and B cell interactions during the initial stages.

Attrition of schistosomula in the lungs of once-vaccinated mice is associated with the formation of cellular infiltrates consisting mostly of mononuclear cells and the deposition of fibrous protein in the pulmonary interstitium. The reactions are CD4$^+$ T cell-dependent, relatively radiation-insensitive, and adoptive transfer of resistance has to date proved unsuccessful [377,379]. The role of CD4$^+$ cells is currently unclear, possibilities including recruitment of effector cells [405], activation of macrophages [77], and acting as TH cells in antibody production [261].

Immunization with irradiated cercariae leads to an increase in Thy-1 and CD4$^+$ T lymphocytes in the lungs, especially in perivascular areas and in parenchymal aggregates [258,408]. Macrophages showing positive reactions for Mac-1, Class II MHC, macrophage-monocytemarker-1 (MOMA-1), and MOMA-2 also increase, but C3R$^+$ receptors do not, with the overall picture being consistent with a DTH reaction [251,258, 259,378]. Challenge parasites stimulate the same reactions but to a lesser extent [251]. Initially, challenge parasites are located within the vasculature, but later many are found in the alveoli [378]. Little damage has been noted on lung-stage worms, which are generally the least immunogenic stage in the vertebrate host and the least resistant to antibody-dependent cell-mediated damage *in vitro* [378]. It has been hypothesized that much of the attrition noted in vaccinated mice may be due to prolonged migration times in the lung, to unexplained deflections into the alveoli, and to a decreased ability upon parasite aging to re-enter tissues [378]. Development of this form of immunity is dependent on prolonged antigen presentation in the dermal lymph nodes [212,214,409].

Unlike skin-phase immunity, macrophages appear to have a critical role in parasite killing in the lungs. The reaction appears to be a DTH response, where activated macrophages and IFN-γ play dominant roles [72,73,219, 257,410,411]. This form of immunity is not affected by whole-body irradiation [412], and mice with congenital deficiencies in mast cells, IgE, lytic complement, and NK cell function develop normal levels of immunity, suggesting that these mechanisms are not important [403,413]. Strains of mice that demonstrate the highest levels of vaccine-mediated immunity share MHC alleles, although genetic studies indicate that genes outside the MHC locus are also important [73]. P strain mice, which have a defect in T cell lymphokine production and macrophages that are refractory to IFN-γ stimulation, do not develop significant levels of protection [73,384]. This mouse strain has normal T cell proliferative ability, macrophage accessory cell function, and eosinophil response.

Infected and vaccinated rats

Although some strains of rats are permissive hosts for schistosomiasis, strains generally used in the laboratory are not. Despite the fact that the nonpermissive trait does not apply to humans, how rats become immune to schistosome infection has received considerable attention owing to the rationale that immunity in humans may bear some analogy to that in rats, or that an understanding of the phenomenon may suggest ways of stimulating a similar reaction in humans [3,414]. Infection in rats proceeds much the same as in permissive animals during the first month, after which time there occurs a major drop in worm burden. The animal's handling of infection is not strictly due to immunologic mechanisms, since rat parasites develop at a normal rate when transplanted into a permissive host and mouse worms at a stunted rate when placed into a naïve rat [415]. Rejection of worms is, however, to some extent immunologically based, as resistance is not demonstrated by congenitally athymic or thymectomized rats [415–417] and can be adoptively transferred with T lymphocytes [418]. It is relevant to note that in this model eggs are not deposited (and thus do not play a role in stimulating resistance) and that, in addition to surface antigens and excretory–secretory products, the host is exposed to antigens released during the killing of immature worms. This type of immunity can be transfered by implantation of adult worms into the mesenteric vasculature of naïve rats, suggesting that relevant parasite antigens need not be stage-specific [415].

Subsequent infection of rats results in more efficient immune killing of challenge parasites, the efficacy of which is related to the number of parasites used in the initial infection [418]. Resistance is T cell-dependent [419,420] and is not seen in anti-μ-treated animals, indicating that it is to some extent an expression of humoral immunity [421]; confirming this, transfer of serum from infected rats can induce resistance in naïve rats [422,423]. Resistance to secondary infection correlates with definite increases in IgE, IgD, and IgG antibodies, these plateauing after worm rejection, and moderate increases in IgA and IgM [404]. IgG2a and IgE antibodies appear important during the early infection, with IgE predominating subsequently [414]. Antischistosome antibodies can kill schistosomula *in vitro* in the presence of complement, and treatment of rats with cobra venom factor results in diminution of immunity [395]. Specific IgE can also mediate parasite killing by eosinophils, platelets, or macrophages, even when these cells are derived from uninfected rats [424–426]. Treatment of the cells with a monoclonal antischistosomular IgE converts them into potent cytotoxic effectors. Mast cells, but not neutrophils, may also be involved in immunity [413,427]. The targets of these responses are thought to be schistosomular surface antigens and excretory–secretory products, and include KLH cross-reactive surface glycoprotein and glutathione *S*-transferase [3].

Initial efforts to determine the factors responsible for resistance in rats relied heavily on *in vitro* systems involving young schistosomula. These studies suggested that the organisms are sensitive to complement-fixing antibodies, especially in the presence of neutrophils. Current views, however, favor a role for antibody-dependent cell-mediated cytotoxicity (ADCC), with the principal effector cells being eosinophils, mast cells, neutrophils, macrophages, and platelets [3,428]. During primary infections, the first isotype capable of killing the worms are reaginic antibodies of the IgG2a subclass, effector cells being primarily eosinophils. This protective combination wanes in efficiency and, subsequently, IgE antibodies mediate killing by activated macrophages, platelets, and eosinophils. Transfer of the latter cells bearing parasite-specific IgG2a or IgE on their surface induces protection after adoptive transfer into naïve rats [3,422,425].

Studies on infected rats provided the first evidence of direct participation of anaphylactic antibodies in cytotoxic reactions [414]. The interaction appears to involve Fc receptors (FcεRII) on the surface of sub-populations of monocytes, macrophages, eosinophils, and, surprisingly, platelets that specifically bind IgE and which are distinct from IgE receptors on mast cells and basophils (FcεRI) [3].

Rats immunized with irradiated cercariae have also been used as models of acquired immunity to schistosome infection. Here the main site of attrition of challenge infection appears to be the lung [429]. Serum from vaccinated animals can protect naïve recipients from challenge infection, even when administered while the parasites are undergoing the lung phase of migration [429]. Serum from immune rats can also protect naïve recipients infected intravascularly with 5-day-old worms. Immunity in this model is dependent on an intact thymus; however, protection is afforded to nude rats by transfer of serum from vaccinated euthymic rats, indicating that worm elimination does not involve direct participation by T cells or lymphokines [429]. Although the details of the mechanisms responsible for resistance in the vaccinated rat are unclear, there are indications that eosinophils, mast cells, platelets, and neutrophils may not be involved [376,429]. It also appears that IgE may not be important and that the relevant isotype is IgG2a [376,429].

Guinea pigs

Studies on guinea pigs have introduced the idea that immunity to schistosome infection operates in a host-specific, site-specific manner and not a parasite stage-specific manner. When vaccinated with irradiated cercariae, this host kills challenge parasites in the liver. Immunity can be transferred with serum and is most effective when administered around 9 days post-infection. Implantation of 5-day lung worms or 9-day liver worms from mice that kill larvae in the skin into vaccinated guinea pigs leads to parasite death, even though they are at this stage immune to killing mechanisms in mice [389]. The antibody involved in immunity appears to be IgG2, and protection can be induced with mineral oil-activated macrophages [430]. Neither peripheral lymph node cells nor splenocytes from vaccinated donors can adoptively transfer resistance [430].

Humans

For practical reasons, acquired human immunity to schistosomiasis has proved difficult to demonstrate, despite the feeling by some workers that there has never been any doubt about its existence [431]. The age-related prevalence and intensity curves of schistosome-infected humans living in endemic areas have been repeatedly shown to decline in the older age groups, suggesting a refractoriness to superinfection [432–434]. Age-related decreases in worm burden are to some extent due to decreased exposure and worm death [313]; however, studies involving immigrants demonstrate that the duration of exposure is also important [423]. Similarly, studies on residents of endemic areas whose water-contact patterns were carefully monitored strongly suggest that humans gradually become resistant to reinfection by *S. mansoni* and *S. haematobium* [232, 435–438]. Such studies reveal that age-dependent resistance develops over a relatively long period (of the order of 5–10 years) and that it is acquired in a non-uniform way, i.e., despite comparable water contact some children repeatedly become heavily infected whereas others become minimally so [205,238,254]. The slow development of immunity and its interindividual variability may be attributable to prolonged and occasional exposure to small numbers of cercariae, to genetic differences between individuals, or to gradual disappearance of blocking antibodies [205].

Studies on the intensities of reinfection after chemotherapeutic cure of a population whose water-contact patterns were monitored have shown that levels of IgG antibodies promoting *in vitro* eosinophil cytotoxic reactions against schistosomula correlate with levels of antiadult worm and antiegg antibodies, and recent evidence suggests an involvement of IgE [439]. *In vitro* studies suggest that antibodies may act in concert with activated eosinophils [440,441]. Furthermore, studies have shown clear correlations between IgE antibodies and resistance in *S. haematobium* infections [442]. These antibodies accumulate slowly, and their activity may be initially inhibited by IgG4 antibodies. Their production is probably regulated by cytokines; IL-4 stimulates

both IgG4 and IgE production but not other isotypes, and IFN-γ interferes with their production, although the level of IFN-γ necessary to inhibit IL-4-induced synthesis of IgE is lower than that required to stop synthesis of IgG4 (as discussed in Hagan *et al.* [442]).

The intensity of *S. mansoni* reinfection in humans appears not to correlate with overall immunoglobulin levels, in part owing to the presence of blocking antibodies (IgM and possibly IgG4) that bind to schistosomular antigens in ways which hinder the attachment of protective IgG1, IgG2, and IgG3 antibodies [203–207]. Blocking antibodies are thought to be primarily stimulated by carbohydrate antigens of schistosome eggs that crossreact with epitopes present in schistosomula [141,142]. The emergence of resistance with age is conjectured to be due to a gradual reduction in the titers of blocking antibodies and the persistence of protective responses [207].

It is heartening that several field-based studies now indicate that immunologic resistance to reinfection does occur in humans, and the correlations that have been made between the rise or fall of given immune responses and observed protection provide possibilities for given resistance-associated mechanisms that might be goals for vaccine development. However, it should be mentioned that the production of an efficacious immunization scheme need not necessarily mimic those mechanisms observed naturally. In fact, vaccine development may seek to improve on, or even differ from, that which develops due to infection. The data imply that discriminatory production of positive (as opposed to negative immunoregulation) immune responses to any given immunogen may be required. This may necessitate that considerable attention be paid to different immune capabilities. This in turn may require either a more complete understanding of differential immune stimulation or pragmatic protocols that accomplish this same end.

IMMUNO- AND CELLULAR DIAGNOSIS

Antibody detection

Definitive diagnosis of schistosome infection rests on demonstration of parasite eggs in stool or urine samples. The method is time consuming, low-grade infections may be difficult to detect, and owing to daily fluctuations multiple examinations are necessary to gain a true idea of quantitative egg-output. The problems have prompted numerous workers to develop serologic means for detection, with the hope of developing systems that correlate with intensity of infection. Attempts to assay the status of experimental animals or individuals living in endemic areas using antibody detection techniques have proved to be of limited value as they often lack specificity due to crossreactions between schistosomes and other parasites, antischistosome antibody persistence for long periods after chemotherapeutic cure, the interindividual variation in the humoral response to infection, and their inability to relate to the intensity of infection [116,443,444]. The area continues to be active, however, as it is possible that such assays may prove of value in assessing the progress, intensity, and moribidity of infection, and, by correct selection of antigens and isotypes, of immune status and length of infection.

Early antibody systems consisted of crude extracts of cercariae, schistosomula, surface antigens, parenchymal antigens, excretory–secretory products of the various stages, gut antigens, or eggs [443]. These were, in general, found to be inadequate owing to lack of specificity prompting search for more purified substrates. Despite the erratic humoral response to infection, several specificities have been consistently found, primarily directed against eggs and adult worms. These specificities have been studied in various systems, including enzyme-linked immunoassay, radioimmunoassay, immunofluorescence techniques, immunoprecipitation, complement fixation tests, and indirect hemagglutination [443,444].

Soluble egg antigen of *S. mansoni* has been found to contain a number of immunogenic molecules that have been fractionated by various workers [128,445]. Early studies on SEA revealed the presence of three major precipitin lines, designated MSA-1, MSA-2, and MSA-3 [446]. MSA-1 and MSA-2 have been found to be glycosylated (138 kD and 465 kD, respectively), with the former probably being a dimer, and shown to have high specificity when used in radioimmunoassay [446–448]. Other components of SEA showing promise in antibody detection include several glycoproteins [128] and a 70 kD antigen referred to as major egg antigen [449], a 200 kD antigen known as polysaccharide egg antigen [450], and a 20–50 kD antigen called w1 [451]. Attempts to detect specific anti-*S. japonicum* egg antigens using purified fractions have not proved useful, although standardized crude extracts can be used to develop sensitive and specific assay systems. [114].

The immune response to schistosomular and adult worm proteins, as inferred from immunoprecipitation of *in vitro* mRNA translation products, has proved to be quite heterogeneous. This limits the usefulness of this class of macromolecules in antibody detection systems [116]. Partially purified antigens have given better results, examples being a preparation of the microsomal fraction of adult worms [452], excretory–secretory antigens [231,232], and salt-extracted surface antigens [453]. Purified adult worm peptides of potential usefulness include an egg α-crystallin-like protein [163], malate dehydrogenase [151], and the 31 kD and 32 kD hemoglobinases previously discussed [174,175]. Analysis of the worm glycoproteins has introduced a more consistently recognized set of potential immunodiagnostic substrates, some of which are genus-, species-, and even gender-specific [28,129]. Similarly, surface antigens of schistosomula appear to be consistently immunogenic in infected individuals [142].

Antigen detection

Assays to detect circulating schistosome antigens have the advantage of being capable of detecting active infections, and in some cases the presence of infection prior to pathologic manifestations. These assays capitalize on the fact that the parasites secrete/excrete various immunogenic molecules, especially when the worms periodically evacuate the contents of their intestinal tract. Other antigens that have received attention include macromolecules that diffuse from deposited eggs, surface antigens of the worms, and excretory–secretory antigens from unidentified anatomic origin. Specific antigens that are assayable in the blood of infected hosts include a gut-associated anodic proteoglycan [454,455], a gut-associated cathodic proteoglycan [454,455], an *S. japonicum* surface antigen known as Sj23 [456], a 41 kD cercarial antigen [457], an isoenzyme of glutathione *S*-transferase [458], malate dehydrogenase [151], and Hsp70 [459]. The latter three antigens are thought to derive from the protonephridial system of worms, the cecum of worms, and the hatching fluid of eggs, respectively [151,156,170]. Other circulating antigens have been described, but are poorly characterized [460].

The best studied and most widely employed circulating antigens for detection of the presence of live schistosome worms are polysaccharides originally described by Berggren & Weller [461]. These are found lining the gut of the parasite where they are thought to protect it from its own digestive enzymes. The two principal antigens are differentiated on the basis of charge and are called CAA and CCA. Circulating cathodic antigen is composed predominantly of carbohydrate, varying in molecular weight from 66 to 350 kD [454]. Circulating anodic antigen has a molecular weight of 90–130 kD and an additional component of 65–70 kD [454,455,462]. A neutrally charged 10–70 kD antigen consisting of 60% carbohydrate and related to CCA can be detected in urine and milk of infected individuals [463,466]. Similarly, sensitive techniques have been developed to detect CAA in urine samples of patients infected with *S. mansoni* and *S. haematobium* [465]. The drive to develop methods permitting detection of schistosome antigens in serum has led to assays of reasonable sensitivity, and specificity has been improved by use of high-affinity monoclonal antibodies [466].

VACCINE DEVELOPMENT

The ability to confer sterilizing, or even partial, immunity in humans to schistosomes is an obviously desirable objective, especially if this involves development of a long-lasting vaccine that can be safely and inexpensively delivered. This could greatly complement current efforts by the World Health Organization to control the disease via chemotherapy, vector control, and education. Further underscoring the need to develop a vaccine is the realization that in the absence of immunity, individuals treated with antihelminthic drugs tend to become reinfected [217,218]. Scientific and economic considerations, however, make it currently unclear whether this goal can be achieved, and if so whether it can be implemented, although progress to date indicates that development of a vaccine is feasible [3,72,73,412,467,468].

The severity and acceptability of lesions caused by irradiated larval schistosome vaccines is debatable [399]. This, the species specificity of the vaccines, and difficulties of making storable material for use in endemic areas has encouraged research on immunity stimulated by more-defined components. Despite a considerable amount of effort, development of such a vaccine remains an elusive goal. This is in large part due to a current lack of understanding of why the best-developed regimes always yield only partial protection. Theoretical explanations are that a "super antigen" remains to be discovered, that current protocols stimulate blocking antibodies, or that means of vaccination used have been suboptimal. Regardless of the cause, a major part of future work will need to identify ways of enhancing protective immune responses [469]. It should be noted,

however, that the morbidity of infection correlates with worm burden, or more precisely numbers of deposited eggs, and that development of a partially protective vaccine should be considered potentially worthwhile. However, the actual degree of resistance necessary to have a useful impact is debatable.

Most studies relating to schistosome vaccine development have dealt with ways of immunologically killing the parasite's larvae. The fact that the morbidity associated with infection is immunologically based and down-regulated as infection proceeds suggests the possibility of developing complementary antipathology vaccines. The idea would be to either decrease the fecundity of the worms, to facilitate the exit of eggs from the host, or to down-regulate granuloma formation during the early stages of infection. These concepts remain conjectural [470].

Approaches

Development of an antischistosome vaccine of defined components has taken three general directions. One empirical approach has made no assumption about the identity of target antigens and systematically tests the efficacy of fractionated parasite extracts until a protective molecule is purified [117,145]. A variation of this approach has been the testing of randomly generated monoclonal antibodies for *in vitro* larvicidal activity, or ability to passively transfer protection *in vivo*; protective antibodies have then been used to isolate the respective antigens (reviewed by Colley & Colley [468]). Another approach, which we term the rational and others the "Achilles heel" approach, involves the identification of macromolecules vital to the parasite or otherwise exposed but poorly immunogenic, then attempts to direct effective immune responses at them [72,73]. An example would be to interfere immunologically with macromolecules mediating nutrient acquisition, signal transduction, or larval infection and migration. The third, or intermediate, approach has been to make rational assumptions about the anatomic location of suitable targets on the worm, isolate these components, then fractionate and test their protectiveness [209]. Alternatively, the isolated antigen pools have been used to generate monoclonal antibodies. Targets of this approach have included surface antigens, excretory–secretory antigens, and antigens preferentially or exclusively recognized by the immune system of immune hosts [130,155,437,438,471,472].

Since *in vitro* systems for propagating relevant stages

Table 16.2 Nonliving schistosome vaccines experimentally demonstrated to protect animals from challenge infection

Antigen* [reference]	Source†	Route‡	Host	Protection (%)
Schistosomular excretory–secretory products [210,212]	C/S	IP	Rat	40–90
Adult worm excretory–secretory products [474]	A	FP	Rabbit	39–88
Tegument preparation [475]	A	SC	Mouse	20–30
Cercarial extract [476]	C	ID	Mouse	20–35
Schistosomula extract [477]	S	ID	Mouse	36–66
Fasciola hepatica extract [478]	Fh	SC	Mouse	50–80
>200, 160, 38 complex [479]	S	IM	Mouse	31
Protein 155 [480]	A	IM	Mouse	25
Glycoprotein 68 [481]	A	SC	Mouse	30–60
Protein 53 [480]	S	IM	Mouse	25
Protein 45/30 [479]	C	FP	Mouse	30–47
Protein 28 [473]	S	SC	Mouse	38
Protein 22 [473]	S	ID	Mouse	11–27
KLH [131]	KL	IP	Rat	50–70
MAB JM8-36 [185]	Mc		Rat	45
MAB 31-3b6 [482]	Mc	IP/SC	Mouse	16–41
Paramyosin [117]	A	ID	Mouse	30–40
Sm GST [157]	A		Rat	50–70
Sj GST [155]	R	SC	Mouse	30–40
Sm MDH [151]	PP	ID/SC	Mouse	26
Sm TPI [118]	PP	SC	Mouse	38

* KLH, keyhole limpet hemocyanin; MAB, monoclonal antibody; Sm, *Schistosoma mansoni*; GST, glutathione *S*-transferase; Sj, *Schistosoma japonicum*; MDH, malate dehydrogenase; TPI, triose phosphate isomerase.

† C, cercaria; S, schistosomula; A, adult worm; Fh, *Fasciola hepatica*; KL, keyhole limpet; R, recombinant peptide; PP, puified protein.

‡ IP, intraperitoneal; FP, footpad; SC, subcutaneous; ID, intradermal; IM, intramuscular.

of the schistosome life-cycle do not exist, and even the life-cycles of *S. haematobium* and *S. japonicum* are difficult to maintain in the laboratory, vaccine development has relied heavily on recombinant DNA technologies. As such, the best-characterized antigens are proteins or the peptide portion of glycoproteins. Glycoconjugates have received attention disproportionately below their potential importance. Two approaches have been taken to circumvent these problems. In one, protective monoclonal antibodies recognizing carbohydrate epitopes are used to generate anti-idiotypic monoclonal antibodies that will present epitopes resembling the shape of the original antigen [196]. The other involves isolation of relevant monoclonal antibodies and their use to screen panels of small peptides of different amino acid sequence [473]; the idea is that some of these peptides will adopt conformations resembling carbohydrate epitopes of interest.

Target antigens

Undefined antigens

Significant levels of experimentally induced immunologic resistance to primary schistosome infections has been documented to develop in animals immunized with parasite extracts, purified antigens (including recombinant antigens), and specific anti-idiotypic monoclonal antibodies (Table 16.2). Moreover, several different monoclonal antibodies have been isolated that passively confer partial resistance to naïve recipients (Table 16.3).

The least-defined experimental antischistosomal vaccines consist of extracts of several stages of the life-cycle. In the case where cercariae were used as the substrate, a highly successful protocol involved use of alum as the adjuvant, suggesting stimulation of

Table 16.3 Monoclonal antibodies that passively transfer protection to schistosome infection when administered to naïve animals

Molecular weight of target (kD)	Isotype	Schistosome*	Passive transfer	Stages†	Reference
200, 160, 130, 38, 17	IgG1	Sm	Mouse	C,E	[483]
200, 160, 130, 38	IgG1	Sm	Mouse	C,S,M	[209]
200, 45, 30	IgG2b	Sm	Mouse	C,S	[479]
200, 160, 97, 82	IgE	Sj	Mouse	S	[411]
200	IgG3	Sm	Mouse	S,E	[484]
170	IgM	Sm	Rat	S	[137]
155	IgM	Sm	Mouse	S,A,M	[485]
85	IgM	Sm	Mouse	C,S,A,M	[486]
68	IgM	Sm	Mouse	S,A,E	[473]
38	IgG2a	Sm	Rat	C,S,A,M	[487]
32	IgG2a	Sm	Mouse	S	[488]
26	IgE	Sm	Rat	S	[217]
20	IgG1	Sm	Mouse	S	[489]
13–18	IgG3	Sm	Mouse	S	[488]

*Sm, *Schistosoma mansoni*; Sj, *Schistosoma japonicum*.
†C, cercaria; E, egg; S, schistosomula; M, miracidium; A, adult worm.

immediate hypersensitivity responses [476]. In cases where schistosomular extracts were used, best results were obtained by use of BCG [469]. In the latter instance it was found that the route of inoculation was of critical importance, with the intradermal route being the most effective [469]. Interestingly, nonprotective immunization via the intravenous route resulted in a better humoral response to schistosomular surface antigens than did effective intradermal immunization [411,469]. The protective protocol led to schistosome-specific cell-mediated responses involving lymphokine production and macrophage activation [409,411,469]. In murine models where challenge parasites are killed in the lungs, the parasites concentrate in the alveoli and attract predominantly mononuclear leukocytes. Immunity appears to be mediated by activated macrophages and is potentiated by IFN-γ. Other fractions that can confer immunity include adult tegument preparations, extracts of *Fasciola hepatica*, and excretory–secretory products of larval worms (Table 16.2).

Surface antigens

It was reasonable to assume that schistosome vaccines should be directed at molecules exposed by living parasites to the immune system. As a result, much attention has been directed at surface antigens of the parasite, but the results obtained have been confusing. Immunization of animals with purified schistosome surfaces has yielded protection [475], no protection [125], or even an increase in susceptibility [198]. Immunization of mice or rats with adult worm tegument preparations leads to production of high levels of IgG, but titers do not correlate at all with protection [125]; in fact, levels of antitegument antibodies in immunized mice are higher, regardless of protectiveness, than in animals vaccinated with irradiated cercariae [125]. To further complicate matters, antibody titer, and not avidity, appears to correlate with protection in the vaccine mouse model [490].

The hypothesis that schistosome surfaces can confer protection has been best borne out by the study of antitegument monoclonal antibodies [83,468]. Some of these can confer resistance upon passive transfer (Table 16.3) or kill schistosomula *in vitro* in the presence of effector cells [83]. Further evidence that the parasite surface antigens can serve as targets for protective immune responses are reports that purified cercarial surface antigens of 30/45 kD [479] schistosomular surface antigens of 200/160/38, 22, and 28 kD [473], and adult tegumental antigens of 155, 68, and 53 kD [480,481] protect experimental animals from a challenge infection.

Excretory–secretory antigens

Employment of parasite secretions as potentiators of antiparasitic protective immune responses has been suggested as a general strategy for preventing the invasion and migration of tissue-invasive helminths [491]. *In vitro* studies using monoclonal antibodies indicate that the schistosome elastase may represent a reasonable vaccine target, as subsets of antibodies against the antigen have been found that inhibit its enzymatic activity, and related antibodies kill transforming schistosomula *in vitro* [223]. Killing occurs by unknown mechanisms, correlates with damage to the tegument, and is enhanced by the presence of complement [223]. Another monoclonal antibody of unknown specificity has been shown to inhibit schistosome invasion on passive transfer [227]. In general, however, attempts to confer immunity to experimental animals via immunization with cercarial secretions has yielded equivocal results, perhaps due to improper antigen presentation or antigen competition [181].

Excretory–secretory products of schistosomula have also been tested for vaccine potential. Levels of immunity as high as 89% have been demonstrated in rats immunized with secretions obtained from cultured larvae. The active fraction contained proteins in the 22–26 kD range [3,414]. The antigens are targets of an IgE response; resistance can be passively transferred to naïve rats and this is lowered after IgE depletion. The mechanisms of killing are thought to involve ADCC [209–211,414].

Defined antigens

Purified antigens that have been shown to induce partial protection to schistosome infection include isoenzymes of glutathione *S*-transferase, glyceraldehyde-3-phosphate dehydrogenase, paramyosin, malate dehydrogenase, a KLH-like glycoprotein, and triose phosphate isomerase (Table 16.3). Lesser understood protective antigens include surface molecules having molecular weights of 16, 30/45, 53, 65, and 150 kD. Glutathione *S*-transferase and glyceraldehyde-3-phosphate dehydrogenase were isolated on the basis of their preferred recognition by immune hosts. Paramyosin was isolated by extraction and fractionation from crude extracts that were protective [145], and the KLH-like glycoprotein and triose phosphate dehydrogenase were isolated by purification using protective monoclonal antibodies [118,463,487].

Mechanisms explaining the protective responses stimulated by defined schistosome antigens remain unclear. The 28 kD isoenzyme of *S. mansoni* glutathione *S*-transferase appears to act by ADCC mechanisms involving eosinophils [3]. The *S. japonicum* 26 kD iso-

enzyme has been less studied but appears to act in the same way [109]: it has been hypothesized that the enzyme acts in detoxification mechanisms, and that neutralization of the enzyme by antibody may render the parasites more susceptible to oxygen-mediated damage [109]. Paramyosin, on the other hand, appears to act by stimulating DTH-type reactions, resulting in macrophage activation [117]. Thus, killing can occur in several ways, depending on the antigen involved, suggesting that it may prove possible to stimulate several effective pathways by immunizing with a cocktail of antigens. Complicating matters, however, is the demonstration that given adjuvants and routes of presentation are in some instances crucial [469] and a lack of information about the optimal conditions for all candidate antigens (they may also be antigen-dependent) hampers progress. In this regard, adjuvants that have been used in successful trials include none, BCG, alum, and liposomes [468]. Routes of immunization that have lead to protection include intraperitoneal, intradermal, intravenous, and subcutaneous (Table 16.2) [468].

Summary

Taken as a whole, the findings mentioned above add a new dimension to our understanding of schistosome antigens susceptible to effector arms of the immune response. It is puzzling how host antigen-processing cells manage to recognize what may be cytoplasmic components of living schistosomes and subsequently translate this information into an immune response capable of homing in on molecules enclosed by a plasma membrane [117]. Also surprising is the fact that nearly all potentially protective antigens sequenced to date share significant blocks of homology with host molecules, yet they apparently do not elicit an autoimmune response, despite the fact that other autoantibodies are detectable [228]. Whether this is attributable to special modes of presentation of the antigens by the parasites or to an inherent lack of immunogenicity of evolutionarily conserved domains is unknown but of obvious relevance to vaccine development.

The general picture emerging in antischistosomal vaccine development is that it is a feasible end, but that further information is needed. It is clear that the mode of antigen presentation is important, but that this is not a straightforward problem to address. It is probable that further information may have to be empirically derived, primarily by varying adjuvant preparations and routes of administration. Secondly, it is unclear which experimental host most closely resembles humans in immunologic responsiveness. Available data indicate that different mechanisms apparently operate, and sites of killing and modes of resistance differ in the various models. It remains possible that a single target antigen for vaccine development may not exist and that multivalent vaccines or vaccines based on optimal antigen presentations are necessary. While continued progress in experimental systems is needed, it is also time to approach the questions of immunogenicity and standardization in human studies.

ACKNOWLEDGMENTS

The authors would like to acknowledge support from the Department of Veterans Affairs (D.G.C.), NIH Grants AI-11289 and AI 26505 (D.G.C.), The John D. and Catherine MacArthur Foundation (G.R.N.), and the Edna McConnell Clark Foundation (G.R.N.)

REFERENCES

1 Iarotsky LS, Davis A. The schistosomiasis problem in the world: results of a WHO questionnaire survey. Bull WHO 1981;59:115–127.

2 Doumenge JP, Mott KE, Cheung C, et al. Atlas de la repartition mondial des schistosomiases. Talense, CCGET-CNRS. Geneva: OMS/WHO, 1987.

3 Capron A, Dessaint JP, Capron M, Ouma JH, Butterworth AE. Immunity to schistosomes: progress towards vaccine. Science 1987;238:1065–1072.

4 Linder E. Fluorochrome-labelled lectins reveal secreted glycoconjugates of schistosome larvae. Parasitol Today 1986;2:219–221.

5 Wilson RA. Leaky livers, portal shunting and immunity to schistosomes. Parasitol Today 1990; 11:351–355.

6 Stirewalt MA. Schistosoma mansoni: cercariae to schistosomule. Adv Parasitol 1974;12:115–182.

7 Atkinson KH, Atkinson BG. Protein synthesis in vivo by Schistosoma mansoni cercariae. Mol Biochem Parasitol 1981; 4:205–216.

8 Ram D, Grossman Z, Markovics A, et al. Rapid changes in the expression of a gene encoding a calcium-binding protein in Schistosoma mansoni. Mol Biochem Parasitol 1989; 34:167–176.

9 Caufield JP. Cell biology of schistosomes. II. Tegumental membranes and their interaction with human blood cells. In Wyler D, ed. Modern Parasite Biology, Cellular, Immunological, and Molecular Aspects. New York: WH Freeman & Co, 1990:107–125.

10 Wiest PM, Tartakoff AM, Aikawa M, Mahmoud AF. Inhibition of surface membrane maturation in schistosomula of Schistosoma mansoni. Proc Natl Acad Sci USA 1988;85: 3825–3829.

11 Rumjanek FD. Biochemistry and physiology. In Rollinson D, Simpson AJG, eds. The Biology of Schistosomes: from Genes to Latrines. London: Academic Press, 1987:163–183.

12 Blanton R, Loula EC, Parker J. Two heat induced proteins are associated with transformation of Schistosoma mansoni cercariae to schistosomula. Proc Natl Acad Sci USA 1987; 84:9011–9014.

13 Yuckenberg PD, Poupin F, Mansour TE. Schistosoma mansoni: protein composition, and synthesis during early development. Evidence for early synthesis of heat-shock proteins. Mol Biochem Parasitol 1987;63:301–311.

14 Newport G, Culpepper J, Agabian N. Parasite heat shock proteins 87. Parasitol Today 1988;4:306–312.

15 Nagai Y, Gazzinelli G, Moraes GWG, Pellegrino J. Protein synthesis during cercariae–schistosomulum transformation and early development of Schistosoma mansoni larvae. Comp Biochem Parasitol 1977;57B:27–30.

16 Basch PF, Samuelson J. Cell biology of schistosomes. I. Ultrastructure and transformations. In Wyler D, ed. Modern Parasite Biology, Cellular, Immunological, and Molecular Aspects. New York: WH Freeman & Co, 1990: 91–106.

17 Caufield JP, Korman G, Butterworth AE, Hogan M, David JR. Partial and complete detachment of neutrophils and eosinophils from schistosomula: evidence for the establishment of continuity between fused and normal parasite membrane. J Cell Biol 1980;86:64–76.

18 McLaren DJ, Hockley DJ, Goldring OL, Hammond BJ. A freeze fracture study of the developing tegumental outer membrane of Schistosoma mansoni. Parasitology 1978;76: 327–348.

19 McDiarmid SS, Dean LL, Podesta RB. Sequential removal of outer bilayer and apical plasma membrane from the surface epithelial synsytium of Schistosoma mansoni. Mol Biochem Parasitol 1983;7:141–157.

20 Cohen C, Reinhardt B, Castellani L, Norton P, Stirewalt M. Schistosome surface spines are crystals of actin. J Cell Biol 1982;95:987–988.

21 Van Oordt BEP, Tielens AGM, van den Bergh SG. Aerobic to anaerobic transition in the carbohydrate metabolism of Schistosoma mansoni cercariae during transformation in vitro. Parasitology 1989;98:409–415.

22 Gomme J, Albrechtsen S. Problems of interpreting integumental D-glucose fluxes by the integument of Schistosoma mansoni. Comp Biochem Physiol 1988;90A: 651–657.

23 Newport GR, Weller TH. Deposition and maturation of eggs of Schistosoma mansoni in vitro: importance of fatty acids in serum-free media. Am J Trop Med Hyg 1982;31: 349–357.

24 Senft AW, Goldberg MW, Byram JE, Jaivan JS. Schistosomal hemoglobinase: nature of the protease and implications for the host. In van den Bossche H, ed. The Host Invader Interplay. Amsterdam: Elsevier/North Holland Biomedical Press, 1980:427–442.

25 LoVerde PT, Rekosh D, Bobek LA. Developmentally regulated gene expression in Schistosoma. Exp Parasitol 1989;68:116–120.

26 Simpson AJG, Sher A, McCutchan TF. The genome of Schistosoma mansoni: isolation of its DNA, its size, bases,

and repetitive sequences. Mol Biochem Parasitol 1982;6: 125–137.

27 Webster P, Mansour TE, Bieber D. Isolation of a female-specific, highly repeated *Schistosoma mansoni* DNA probe and its use in an assay of cercarial sex. Mol Biochem Parasitol 1989;36:217–222.

28 Spotila LD, Hirai H, Rekosh DM, LoVerde PT. A retrotransposon-like short repetitive DNA element in the genome of the human blood fluke, *Schistosoma mansoni*. Chromosoma 1989;84:421–428.

29 Tanaka M, Iwamura Y, Amanuma H, *et al*. Integration and expression of murine retrovirus-related sequences in schistosomes. Parasitology 1989;99:31–38.

30 Craig SP, Muralidhar MG, McKerrow JH, Wang CC. Evidence for a class of very small introns in the gene for hypoxanthine–guanine phosphoribosyl-transferase in *Schistosoma mansoni*. Nucleic Acids Res 1989;17:1635–1647.

31 Nara T, Iwamura Y, Tanaka M, Irie Y, Yasuraoka K. Dynamic changes of DNA sequences in *Schistosoma mansoni* in the course of development. Parasitology 1990; 100:241–245.

32 Brindley PJ, Lewis F, McCutchan TF, Bueding E, Sher A. A genomic change associated with the development of resistance to hycanthone in *Schistosoma mansoni*. Mol Biochem Parasitol 1989;36:243–252.

33 Nilsen TW. Trans-splicing in nematodes. Exp Parasitol 1989;69:413–416.

34 Agabian N. Trans-splicing of nuclear pre-mRNAs. Cell 1990;61:1157–1160.

35 Rajkovic A, Davis RE, Simonsen JN, Rottman FM. A spliced leader is present on a subset of mRNAs from the human parasite *Schistosoma mansoni*. Proc Natl Acad Sci USA 1990;87:8879–8883.

36 Davis RD, Davis A, Carrol SM, Rajkovic A, Rottman FM. Tandemly repeated exons encode 81-base repeats in multiple, developmentally regulated *Schistosoma mansoni* transcripts. Mol Cell Biol 1988;8:4745–4755.

37 Grossman AI, Short RB, Cain GD. Karyotype evolution and sex chromosome differentiation in schistosomes (*Trematoda, Schistosomatidae*). Chromosoma 1981;84:413–430.

38 Short RB. Sex and the single schistosome. J Parasitol 1983;69:4–22.

39 Cordingley JS. Trematode eggshells: novel protein biopolymers. Parasitol Today 1987;3:341–344.

40 Bobek LA, Rekosh DM, LoVerde PT. Small gene family encoding an eggshell (chorion) protein of the human parasite *Schistosoma mansoni*. Mol Cell Biol 1988;8:3008–3016.

41 Rodrigues V, Chaudhri M, Knight M, *et al*. Predicted structure of a major *Schistosoma mansoni* eggshell protein. Mol Biochem Parasitol 1989;32:7–14.

42 Ongom VL, Bradley DJ. The epidemiology and consequences of *Schistosoma mansoni* infection in West Nile, Uganda. I. Field studies of a community at Panyagoro. Trans R Soc Trop Med Hyg 1972;66:835–851.

43 Cook JA, Baker ST, Warren KS, Jordan P. A controlled study of morbidity of *Schistosoma mansoni* in St Lucian children, based on quantitative egg excretion. Am J Trop Med Hyg 1974;23:625–633.

44 Smith DH, Warren KS, Mahmoud AA. Morbidity in schistosomiasis mansoni in relation to intensity of infection: study of a community in Kisumu, Kenya. Am J Trop Med Hyg 1979;28:220–229.

45 Warren KS, Mahmoud AA, Muruka JF, Whittaker LR, Ouma JH, Siongok YK. Schistosomiasis hematobia in the coast province Kenya. Relationship between egg output and morbidity. Am J Trop Med Hyg 1979;28:864–870.

46 Warren KS, Su DL, Xu ZY, *et al*. Morbidity in schistosomiasis japonica in relation to the intensity of infection. N Engl J Med 1983;309:1533–1539.

47 Costa MMFL, Rocha RS, Magalhaes MHA, Katz N. A clinical epidemiological survey of schistosomiasis mansoni in a hyperendemic area in Minas Gerais State (Comercinho, Brazil). Trans R Soc Trop Med Hyg 1985;79:539–545.

48 King CH, Keating CE, Muruka JF, *et al*. Urinary tract morbidity in schistosomiasis haematobia: associations with age and intensity of infection in an endemic area of Cosat Province, Kenya. Am J Trop Med Hyg 1988;39:361–368.

49 Chen MG, Mott KE. Progress in assessment of morbidity due to *Schistosoma japonicum* infection. A review of recent literature. Trop Dis Bull 1989;85(6):R1–R45.

50 Chen MG, Mott KE. Progress in assessment of morbidity due to *Schistosoma mansoni* infection. A review of recent literature. Trop Dis Bull 1989;85(10):R1–R56.

51 Chen MG, Mott KE. Progress in assessment of morbidity due to *Schistosoma haematobium* infection. A review of recent literature. Trop Dis Bull 1989;86(4):R1–R36.

52 Warren KS, Mahmoud AA, Boros DL, Rall TW, Mandel MA, Carpenter CCJ Jr. Schistosomiasis mansoni in Yemeni in California. Duration of infection, presence of disease, therapeutic management. Am J Trop Med Hyg 1974; 23:902–909.

53 Berberian DA, Paquin HO, Fantauzzi A. Longevity of *Schistosoma haematobium* and *Schistosoma mansoni*. Observations based on a case. J Parasitol 1953;39:517–519.

54 Wilkins HA, Scott A. Variation and stability in *S. haematobium* egg counts: a four year study in Gambian children. Trans R Soc Trop Med Hyg 1978;72:397–404.

55 Wilkins HA, Gall PH, Marshall TF, Moore PJ. Dynamics of *Schistosoma haematobium* infection in a Gambian community. III. The acquisition and loss of infection. Trans R Soc Trop Med Hyg 1984;78:227–232.

56 Nash TE, Cheever AW, Ottesen EA, Cook JA. Schistosome infections in humans: perspectives and recent findings. Ann Intern Med 1982;97:740–754.

57 Von Lichtenberg F. Consequences of infections with schistosomes. In Rollison D, Simpson AJG, eds. *The Biology of Schistosomes: from Genes to Latrines*. London: Academic Press, 1987:185–232.

58 Camus D, Carlier Y, Bina JC, Borojevic R, Prata A, Capron A. Sensitization to *Schistosoma mansoni* antigen in uninfected children born to infected mothers. J Infect Dis 1976;134:405–408.

59 Tachon P, Borojevic R. Mother–child relation in schistosomiasis mansoni: skin test and cord blood reactivity to schistosomal antigens. Trans R Soc Trop Med Hyg 1978;72: 605–609.

60 Eloi-Santos SM, Novato-Silva E, Maselli VM, Gazzinelli G, Colley DG, Correa-Oliveira R. Idiotypic sensitization *in*

utero of children born to mothers with schistosomiasis or Chagas' disease. J Clin Invest 1990;84:1028–1031.

61 Lewert RM, Mandlowitz S. Schistosomiasis: prenatal induction of tolerance to antigens. Nature 1969;224:1029–1030.

62 Hang LM, Boros DL, Warren KS. Induction of immunological hyporesponsiveness to granulomatous hypersensitivity in *Schistosoma mansoni* infection. J Infect Dis 1974; 130:515–522.

63 Gill TH, Repetti CF, Metlay LA, *et al.* Transplacental immunization of the human fetus to tetanus by immunization of the mothers. J Clin Invest 1983;72:987–996.

64 Lee C-J, Takaoka Y, Saito JF. Maternal immunization and the immune response of neonates to pneumococcal polysaccharides. Rev Infect Dis 1987;9:494–510.

65 Weil GJ, Hussain R, Kumaraswami V, Tripathy SP, Philips KS, Ottesen EA. Prenatal allergic sensitization to helminth antigens in the offspring of parasite-infected mothers. J Clin Invest 1983;71:1124–1129.

66 Kresina TF, Nisonoff A. Passive transfer of the idiotypically suppressed state by serum from suppressed mice and transfer of the suppression from mothers to offspring. J Exp Med 1983;157:15–23.

67 Stein KE, Soderstrom T. Neonatal administration of idiotype or antiidiotype primes for protection against *Escherichia coli* K13 infection in mice. J Exp Med 1984; 160:1001–1011.

68 Bona CA. *Regulatory Idiotypes*. New York: John Wiley and Sons, 1987.

69 Colley DG. Occurrence, roles, and uses of anti-idiotypes in parasitic diseases. In Cerny J, Hiernaux J, eds. *Idiotype Network and Diseases*. Washington, DC: ASM, 1990:71–105.

70 Damian RT. Molecular mimicry revisited. Parasitol Today 1987;3:263–266.

71 Pearce EJ, Sher A. Mechanisms of immune evasion in schistosomiasis. Contrib Microbiol Immunol 1987;8:219–232.

72 Sher A, James SL, Correa-Oliveira R, Hieny S, Pearce E. Schistosome vaccines: current progress and future prospects. Parasitology 1989;98:S61–S68.

73 Sher A, James SL. Genetic control of vaccine-induced immunity against a parasitic helminth, *Schistosoma mansoni*. Bioassays 1989;9:163–169.

74 Lewis FA, Hieny S, Sher A. Evidence against the existence of specific *Schistosoma mansoni* subpopulations which are resistant to irradiated vaccine-mediated immunity. Am J Trop Med Hyg 1985;34:86–91.

75 Smithers SR, Terry RJ. Immunity in schistosomiasis. Ann NY Acad Sci 1969;160:826–840.

76 Modha J, Parikh V, Gauldie J, Doenhoff MJ. An association between schistosomes and contrapsin, a mouse serine esterase inhibitor (serpin). Parasitology 1988;96:99–109.

77 Pearce EJ, Hall BF, Sher A. Host-specific evasion of the alternative complement pathway by schistosomes correlates with the presence of a phospholipase C-sensitive surface molecule resembling human decay accelerating factor. J Immunol 1990;144:2751–2756.

78 Damian RT, Greene ND, Hubbard WJ. Occurrence of mouse α_2-macroglobulin antigenic determinants on

Schistosoma mansoni adults, with evidence on their nature. J Parasitol 1973;59:64–73.

79 Pearce EJ, Basch PF, Sher A. Evidence that the reduced surface antigenicity of developing *Schistosoma mansoni* schistosomula is due to antigen shedding rather than host molecule acquisition. Parasite Immunol 1986;8:79–89.

80 Foley M, Kusel JR, Garland PB. Fluorescence recovery after photobleaching: applications to parasite surface membranes. Parasitol Today 1986;2:318–320.

81 Kusel JR, MacKenzie PE, McLaren DJ. The release of membrane antigens into culture by adult *Schistosoma mansoni*. Parasitology 1975;71:247–259.

82 Torpier G, Capron A. Intramembrane particle movements associated with binding of lectins on *Schistosoma mansoni* surface. J Ultrastruct Res 1980;72:325.

83 Simpson AJG, Omer Ali P, Meadows HM, Jeffs SA, Hagan P, Smithers SR. *Schistosoma mansoni* surface proteins Mem Inst Oswaldo Cruz 1989;84:179–187.

84 Simpson AJG. Schistosome surface antigens: developmental expression and immunological function. Parasitol Today 1990;6:40–45.

85 Wilson RA, Barnes PE. The formation and turnover of the membranocalyx on the tegument of *Schistosoma mansoni*. Parasitology 1977;74:61.

86 Saunders N, Wilson RA, Coulson PS. The outer bilayer of the adult schistosome tegument surface has a low turnover rate *in vitro* and *in vivo*. Mol Biochem Parasitol 1987; 25:123–131.

87 Ruppel A, McLaren DJ. *Schistosoma mansoni*: surface membrane stability *in vitro* and *in vivo*. Expt Parasitol 1986; 8:307–318.

88 Auriault C, Joseph M, Tartar A, Bout D, Tonnel AB, Capron A. Regulatory role of a tripeptide (TKP) from the second constant domain of immunoglobulin G. I. Inhibition of rat and human macrophage activities. Int J Immunopharmacol 1985;7:73.

89 Verwaerde C, Auriault C, Neyrinck JL, Capron A. Properties of serine proteases of *Schistosoma mansoni* schistosomula involved in the regulation of IgE synthesis. Scand J Immunol 1988;27:17–24.

90 Marikovsky M, Arnon R, Fishelson Z. Proteases secreted by transforming schistosomula of *Schistosoma mansoni* promote resistance to killing by complement. J Immunol 1988;141:273–278.

91 Marikovsky M, Parizade M, Arnon R, Fishelson Z. Complement regulation on the surface of schistosomula and adult worms of *Schistosoma mansoni*. Eur J Immunol 1990;20:221–227.

92 Mazingue C, Camus D, Dessaint J-P, Capron M, Capron A. *In vitro* and *in vivo* inhibition of mast cell degranulation by a factor from *Schistosoma mansoni*. Int Arch Allergy Appl Immunol 1980;63:178–189.

93 Camus D, Nosseir A, Mazingue C, Capron A. Immunoregulation by *Schistosoma mansoni*. Immunopharmacology 1981;3:193.

94 Mota-Santos TA, Tavares CAP, Gazzinelli G, Pellegrino J. Immunosuppression mediated by adult worms in chronic schistosomiasis mansoni. Am J Trop Med Hyg 1977; 26:727–731.

95 Doenhoff MJ, Modha J, Lambertucci JR. Anti-schistosome

chemotherapy enhanced by antibodies specific for a parasite esterase. Immunology 1988;65:507–510.

96 Colley DG, Lewis FA, Goodgame RW. Immune responses during human schistosomiasis mansoni. IV. Induction of suppressor cell activity by *Schistosoma* antigen preparations and Concanavilin A. J Immunol 1978;120:1225–1232.

97 Doenhoff MJ, Mohda J, Lambeticci JR, McLaren DJ. The immune dependence of chemotherapy. Parasitol Today 1991;7:16–18.

98 Brindley PJ, Sher A. Immunological involvement in the efficacy of praziquantel. Exp Parasitol 1990;71:245–248.

99 Brindley PJ, Sher, A. The chemotherapeutic effect of praziquantel against *Schistosoma mansoni* is dependent on host antibody response. J Immunol 1987;139:215–220.

100 Sauma SY, Strand M. Identification and characterization of glycosylphosphatidylinositol-linked *Schistosoma mansoni* adult worm immunogens. Mol Biochem Parasitol 1990;38:199–210.

101 Brindley PJ, Strand M, Norden A, Sher A. Role of host antibody in the chemotherapeutic action of praziquantel against *Schistosoma mansoni*: identification of target antigens. Mol Biochem Parasitol 1989;34:99–108.

102 Knopf PM, Soliman M. Effects of host endocrine gland removal on the permissive status of laboratory rodents to infection by *Schistosoma mansoni*. Int J Parasitol 1980; 10:197.

103 Correa-Oliveira R, Dusse LM, Viana IR, Colley DG, Carvalho OS, Gazzinelli G. Human antibody responses against schistosomal antigens. I. Antibodies from patients with *Ancylostoma*, *Ascaris lumbricoides*, or *Schistosoma mansoni* infections react with schistosome antigens. Am J Trop Med Hyg 1988;38:348–355.

104 Fauve RM, Dodin A. Influence d' une reaction inflammatoire par le BCG un par un irritant nondigrabable sur la resistance des souris a la bilharziose. C R Acad Sci Paris 1976;282:131.

105 Smith MA, Clegg JA, Jusel JR, Webbe G. Lung inflammation in immunity to *Schistosoma mansoni*. Experientia 1975; 31:595.

106 Bout D, Dupas H, Carlier Y, Afchain D, Capron A. Protection of mice against *Schistosoma mansoni* with BCG. Med Sci: Immunology and Allergy; Microbiology Parasitology and Infectious Diseases; Pharmacology 1977;5:47.

107 Olds GR, Chedid L, Lederer E, Mahmoud AAF. Induction of resistance to *Schistosoma mansoni* by natural cord factor and synthetic lower homologues. J Infect Dis 1980;141:473.

108 Olds GR, Ellner JJ, Kearse LA Jr, Kazura JW, Mahmoud AAF. Role of arginase in killing of schistosomula of *Schistosoma mansoni* J Exp Med 1980;151:1557–1562.

109 Mitchell GF. Glutathione *S*-transferases: potential components of anti-schistosome vaccines? Parasitol Today 1989;5:34–37.

110 Elsaghier AAF, Knopf PM, Mitchell GF, McLaren DJ. *Schistosoma mansoni*: evidence that "non-permissiveness" in 129/Ola mice involves worm relocation and attrition in the lungs. Parasitology 1989;99:365–375.

111 Coulson PS, Wilson RA. Portal shunting and resistance to *Schistosoma mansoni* in 129 strain mice. Parasitology 1989; 99:383–389.

112 Elsaghier AAF, McLaren DJ. *Schistosoma mansoni*: evidence

that vascular abnormalities correlate with the "non-permissive" trait in 129/Ola mice. Parasitology 1989;99: 377–381.

113 Cox FEG. Schistosomiasis: the mouse that wasn't immune. Nature 1990;345:18.

114 Mitchell GF, Cruise KM. Schistosomiasis: antigens and host-parasite interactions. In Pearson TW, ed. *Parasite Antigens: Towards New Strategies for Vaccines*. New York: Marcel Dekker 1985:275–316.

115 Ruppel A, Rother U, Vongerichten H, Lucius R, Diesfeld HJ. *Schistosoma mansoni*: immunoblot analysis of adult worm proteins. Exp Parasitol 1989;60:195–206.

116 Newport G, Hedstrom R, Tarr P, Kallestad J, Klebanoff S, Agabian N. Identification, molecular cloning, and expression of a schistosome antigen displaying diagnostic potential. Am J Trop Med Hyg 1988;38:540–546.

117 Pearce EJ, James SL, Lanar DE, Sher A. Induction of protective immunity against *Schistosoma mansoni* by vaccination with schistosome paramyosin, a nonsurface parasite antigen. Proc Natl Acad Sci USA 1988;85:5678–5682.

118 Shoemaker CS, Gross A, Gebremichael A, Harn DA. cDNA cloning and functional expression of the *Schistosoma mansoni* protective antigen triose phoshate isomerase. Proc Natl Acad Sci USA 1992;89:1842–1846.

119 Havercroft JC, Huggins MC, Dunne DW, Taylor DW. Identification and characterization of Sm20, a 20 kilodalton calcium-binding protein of *Schistosoma mansoni*. Mol Biochem Parasitol 1990;38:211–220.

120 Pujol FH, Cesari IM. Antigenicity of adult *Schistosoma mansoni* alkaline phosphatase. Parasite Immunol 1990; 12:189–198.

121 Rogers MV, Quilici D, Mitchell GF, Fidge NH. Purification of a putative lipoprotein receptor from *Schistosoma japonicum* adult worms. Mol Biochem Parasitol 1990; 41:93–100.

122 Rumjanek FD, Campos EG, Afonso LCC. Evidence for the occurrence of LDL receptors in extracts of schistosomula of *Schistosoma mansoni*. Mol Biochem Parasitol 1988;28:145–152.

123 Podesta R, Karcz S, Ansell M, Silva E. *Schistosoma mansoni*: apical membrane/envelope synthesis, signal transduction and protein phosphorylation. In MacInnis A, ed. *Molecular Strategies for Eradicating Helminth Parasites*. New York: Alan R Liss, 1987:241–255.

124 Kalopothakis E, Rumjanek F, Evans WH. Protein and antigen phosphorylation in the tegument of *Schistosoma mansoni*. Mol Biochem Parasitol 1987;26:39–46.

125 Roberts SM, Boot C, Wilson RA. Antibody responses of rodents to a tegument preparation from adult *Schistosoma mansoni*. Parasitology 1988;97:425–435.

126 Roberts SM, Wilson RA, Ouma JH, *et al*. Immunity after treatment of human schistosomiasis mansoni: quantitative and qualitative antibody responses to tegumental membrane antigens prepared from adult worms. Trans R Soc Trop Med Hyg 1987;81:786–793.

127 Dean LL, Podesta RB. Electrophoretic patterns of protein synthesis and turnover in apical plasma membrane and outer bilayer of *Schistosoma mansoni*. Biochim Biophys Acta 1984;799:106–114.

128 Norden AP, Strand M. *Schistosoma mansoni, S. haematobium,*

and *S. japonicum* identification of genus- and species-specific antigenic egg glycoproteins. Exp Parsitol 1984; 58:333–344.

129 Norden AP, Strand M. *Schistosoma mansoni, S. haematobium*, and *S. japonicum*: identification of genus-, species-, and gender-specific antigenic worm glycoproteins. Exp Parasitol 1984;57:110–123.

130 Goudot-Crozel V, Caillol D, Djabali M, Dessein AJ. The major parasite surface antigen associated with human resistance to schistosomiasis is a 37 kD glyceraldehyde-3-phosphate-dehydrogenase. J Exp Med 1989;170:2065–2080.

131 Grzych JM, Dissous C, Capron M, Torres S, Lambert PH, Capron A. *Schistosoma mansoni* shares with keyhole limpet hemocyanin a protective carbohydrate epitope. J Exp Med 1987;165:865–878.

132 Stein LD, David JR. Cloning of a developmentally regulated tegument antigen of *Schistosoma mansoni*. Mol Biochem Parasitol 1986;20:253–264.

133 Knight M, Kelly B, Rodrigues V, *et al.* A cDNA clone encoding part of the major 25 000 dalton surface membrane antigen of adult *Schistosoma mansoni*. Parasitol Res 1989;75:280–286.

134 Wright MD, Henkler KJ, Mitchell GF. An immunogenic Mr 23 000 integral membrane protein of *Schistosoma mansoni* worms that closely resembles a human tumor associated antigen. J Immunol 1990;144:3195–3200.

135 Espinoza B, Tarrab-Hazdai R, Silman I, Arnon R. Acetylcholinesterase in *Schistosoma mansoni* is anchored to the membrane via covalently attached phosphatidylinositol. Mol Biochem Parasitol 1988;29:171–179.

136 Auriault C, Pierce R, Cesari I, Capron A. Neutral protease activities at different developmental stages of *Schistosoma mansoni* in mammalian hosts. Comp Biochem Physiol 1978; 72B:377–384.

137 Khalife J, Grzych J-M, Pierce R, *et al.* Immunological cross-reactivity between the human immunodeficiency virus type 1 virion infectivity factor and a 170-kD surface antigen of *Schistosoma mansoni*. J Exp Med 1990;172:1001–1004.

138 Ko AL, Drager UC, Harn DA. A *Schistosoma mansoni* epitope recognized by a protective monoclonal antibody is identical to the stage-specific embryonic antigen 1. Proc Natl Acad Sci USA 1990;87:4159–4163.

139 Simpson AJG, Payares G, Walker T, Knight M, Smithers SR. The modulation of expression of polypeptide surface antigens on developing schistosomula of *Schistosoma mansoni*. J Immunol 1984;133:2725–2730.

140 Dunne DW. Schistosome carbohydrates. Parasitol Today 1990;6:45–48.

141 Dunne DW, Grabowska AM, Fulford AJC, *et al.* Human antibody responses to *Schistosoma mansoni*: the influence of epitopes shared between different life-cycle stages on the response to the schistosomulum. Eur J Immunol 1988;18:123–131.

142 Omer Ali P, Mansour M, Woody JN, Smithers SR, Simpson AJG. Antibody to carbohydrate and polypeptide epitopes on the surface of schistosomula of *Schistosoma mansoni* in Egyptian patients with acute and chronic schistosomiasis. Parasitology 1989;98:417–424.

143 Pearce EJ, Sher A. Three major surface antigens of *Schistosoma mansoni* are linked to the membrane by glycosyl-phosphotidylinositol. J Immunol 1989;142:979–984.

144 Rogers MV, Davern KM, Smythe JA, Mitchell GF. Immunoblotting analysis of the major integral membrane protein antigens of *Schistosoma japonicum*. Mol Biochem Parasitol 1988;29:77–88.

145 Lanar DE, Pearce EJ, James SL, Sher A. Identification of paramyosin as the schistosome antigen recognized by intradermally vaccinated mice. Science 1986;234:593–596.

146 Newport GR, Harrison RA, McKerrow J, Tarr P, Kallestad J, Agabian N. Molecular cloning of *Schistosoma mansoni* myosin. Mol Biochem Parasitol 1987;26:29–38.

147 Xu H, Miller S, van Keulen H, Wawrzynski MR, Rekosh DM, LoVerde PT. *Schistosoma mansoni* tropomyosin: cDNA characterization, sequence, expression, and gene product localization. Exp Parasitol 1989;69:373–392.

148 Matsumoto Y, Perry G, Levine RJC, Blanton R, Mahmoud AAF, Aikawa M. Paramyosin and actin in schistosome teguments. Nature 1988;333:76–78.

149 Newport G, Agabian N. Molecular biology of schistosomes and filariae. In Wyler D, ed. *Modern Parasite Biology, Cellular, Immunological, and Molecular Aspects.* New York: WH Freeman & Co, 1990:362–383.

150 Stein LD, Harn DA, David JR. A cloned ATP guanidine kinase in the trematode *Schistosoma mansoni* has a novel duplicated structure. J Biol Chem 1990;265:6582–6588.

151 Bout D, Dupas H, Capron MN, *et al.* Purification, immunological and biochemical characterization of malate dehydrogenase of *Schistosoma mansoni*. Immunochemistry 1978;5:633–638.

152 Srere PA. Complexes of sequential metabolic enzymes. Ann Rev Biochem 1987;56:89–124.

153 Srivastiva IK, Schmidt M, Certa U, Dobel H, Perrin LH. Specificity and inhibitory activity of antibodies to *Plasmodium falciparum* aldolase. J Immunol 1990;144:1497–1503.

154 Mishra NK, Marsh CL. *Asacris suum*: aldolase I: purification and immunologic study. Exp Parasitol 1973;33:89–94.

155 Smith DB, Davern KM, Board PG, Tiu WY, Garcia EG, Mitchell GF. The Mr 26 000 antigen of *Schistosoma japonicum* recognized by resistant WEHI 129/J mice is a parasite glutathione *S*-transferase. Proc Natl Acad Sci USA 1986; 83:8703–8707.

156 Henkle KJ, Davern KM, Wright AMD, Ramos AJ, Mitchell GF. Comparison of the cloned genes of the 26- and 28-kilodalton glutathione S-transferases of *Schistosoma japonicum* and *Schistosoma mansoni*. Mol Biochem Parasitol 1990;40:23–34.

157 Balloul JM, Sondermeyer P, Dreyer D, *et al.* Molecular cloning of a protective antigen against schistosomiasis. Nature 1987;326:149–153.

158 Trottein F, Kieny MP, Verwaerde C, *et al.* Molecular cloning and tissue distribution of a 26-kilodalton *Schistosoma mansoni* glutathione *S*-transferase. Mol Biochem Parasitol 1990;41:35–44.

159 Taylor JB, Vidal A, Torpier G, *et al.* The glutathione transferase and tissue distribution of a cloned Mr 28 000 protective antigen of *Schistosoma mansoni*. EMBO J 1988; 7:465–472.

160 Simurda MC, van Keulen H, Rekosh D, LoVerde PT. *Schistosoma mansoni*: identification and analysis of an

mRNA and a gene encoding superoxide dismutase (Cu/Zn). Exp Parasitol 1988;67:73–84.

161 Newport GR. Parasite heat shock proteins as vaccines. Springer-Verlag Semin Immunol 1991.

162 Pelham H. Heat shock proteins. Coming in from the cold. Nature 1988;332:776.

163 Nene V, Dunne DW, Johnson KS, Taylor DW, Cordingley JS. Sequence and expression of a major egg antigen from *Schistosoma mansoni*: homologies to heat shock proteins and alpha-crystallins. Mol Biochem Parasitol 1986;21:179–188.

164 Cordingley JS, Taylor DW. Schistosome vaccines: strategies for overcoming immune response gene effects. In Laurence L, ed. *Technological Advances in Vaccine Development*. New York: Alan R Liss, 1988:603–614.

165 De Jong WW, Leunissen JAM, Leenen PJM, Zweers A, Versteeg M. Dogfish α-crystallin sequences. Comparison with small heat shock proteins and *Schistosoma* egg antigen J Biol Chem 1988;263:5141–5149.

166 Hedstrom R, Culpepper J, Harrison R, Agabian N, Newport G. A major immunogen in *Schistosoma mansoni* infections is homologous to the heat shock protein Hsp 70. J Exp Med 1987;165:1430–1435.

167 Hedstrom R, Culpepper J, Schinski V, Agabian N, Newport G. Schistosome heat-shock proteins are immunologically distinct host-like antigens. Mol Biochem Parasitol 1988;29:275–282.

168 Moser D, Doumbo O, Klinkert M-Q. The humoral response to heat shock protein 70 in human and murine schistosomiasis mansoni. Parasite Immunol 1990;12:341–352.

169 Fu C, Carter CE. Detection of a circulating antigen in human schistosomias japonica using a monoclonal antibody. Am J Trop Med Hyg 1990;42:347–351.

170 Scallon BJ, Bogitsh BJ, Carter CE. Cloning of a *Schistosoma japonicum* gene encoding a major immunogen recognized by hyperinfected rabbits. Mol Biochem Parasitol 1987;24:237–245.

171 Johnson KS, Wells K, Bock JV, Nene V, Taylor DW, Cordingley JS. The 86 kilodalton antigen from *Schistosoma mansoni* is a heat-shock protein homologous to yeast Hsp-90. Mol Biochem Parasitol 1989;36:19–28.

172 Lindquist RN, Senft AW, Petitt M, McKerrow J. *Schistosoma mansoni*: purification and characterization of the major acidic protease from adult worms. Exp Parasitol 1986;61:398–404.

173 Senft AW, Madidison SE. Hypersensitivity to parasite proteolytic enzyme in schistosomiasis. Am J Trop Med Hyg 1975;24:160–167.

174 Klinkerk M-Q, Felleisen R, Link G, Ruppel A, Beck E. Primary structures of Sm31/32 diagnostic proteins of *Schistosoma mansoni* and their identification as proteases. Mol Biochem Parasitol 1989;33:113–122.

175 Chappell CL, Dresden MH, Gryseels B, Deelder AM. Antibody response to *Schistosoma mansoni* adult worm cysteine proteinases in infected individuals. Am J Trop Med Hyg 1990;42:335–341.

176 Senft AW, Goldberg MW, Byram JE. Hemoglobinolytic activity in serum of mice infested with *Schistosoma mansoni*. Am J Trop Med Hyg 1981;30:96–101.

177 Davis AH, Nanduri J, Watson DC. Cloning and gene expression of *Schistosoma mansoni* protease. J Biol Chem 1987;262:12851–12857.

178 Felleisen R, Klinkerk MQ. *In vitro* translation and processing of cathepsin B of *Schistosoma mansoni*. EMBO J 1990;9:371–377.

179 El Meanawy MA, Aji T, Phillips NFB, *et al.* Definition of the complete *Schistosoma mansoni* hemoglobinase mRNA sequence and gene expression in developing parasites. Am J Trop Med Hyg 1990;43:67–78.

180 McKerrow JH, Newport G, Fishelson Z. Recent insights into the structure and function of a larval proteinase in host infection by a multicellular parasite. Proc Soc Exptl Biol Med 1991;197:119–124.

181 Dresden MH. Proteolytic enzymes of *Schistosoma mansoni*. Acta Leiden 1982;49:81–99.

182 McKerrow JH, Pino-Heiss S, Linquist RL, Werb Z. Purification and characterization of an elastinolytic proteinase secreted by cercariae of *Schistosoma mansoni*. J Biol Chem 1985;260:3703–3707.

183 Clegg JA, Smithers SR. Death of schistosome cercariae during penetration of the skin. II. Penetration of mammalian skin by *Schistosoma mansoni*. Parasitology 1968;58:111–128.

184 Toy L, Pettit M, Wang YF, Hedstrom R, McKerrow JH. The immune response to stage-specific proteolytic enzymes of *Schistosoma mansoni*. In MacInnis AJ, ed. *Molecular Paradigms for Eradication of Helminth Parasites*, vol. 60. UCLA Symposia on Molecular and Cellular Biology. New York: Alan R Liss, 1987:85–103.

185 Amiri P, Sakanari J, Basch P, Newport G, McKerrow JH. The *Schistosomatium douthitti* cercarial elastase is biochemically and structurally distinct from that of *Schistosoma mansoni*. Mol Biochem Parasitol 1988;28:113–120.

186 Minard P, Murrell D, Stirewalt MA. Proteolytic, antigenic and immunogenic properties of *Schistosoma mansoni* secretion material. Am J Trop Med Hyg 1977;26:491–499.

187 Newport GR, McKerrow J, Hedstrom R, *et al.* Cloning of the proteinase that facilitates infection by schistosome parasites. J Biol Chem 1988;263:13179–13184.

188 McKerrow JH. Parasite proteases. Exp Parasitol 1989;68:11–115.

189 Hill RE, Hastie ND. Accelerated evolution in the reactive centre regions of serine proteases. Nature 1987;326:96–99.

190 Doenhoff MJ, Modha J, Curtis RHC, Adeoye GO. Immunological identification of *Schistosoma mansoni* peptidases. Mol Biochem Parasitol 1988;31:233–340.

191 Damonneville M, Auriault C, Pierce J, Capron A. Antigenic properties of *Schistosoma mansoni* aminopeptidases during development in mammalian hosts. Mol Biochem Parasitol 1982;6:265–275.

192 Sung CK, Dresden MH. Cysteinyl proteinases of *Schistosoma mansoni* eggs: purification and partial characterization. J Parasitol 1986;72:891–900.

193 Xu YZ, Dresden M. The hatching of schistosome eggs. Exp Parasitol 1990;70:236–240.

194 Scallon BJ, Carter CE. Cloning of a gene for the *Schistosoma japonicum* 7 week antigen. In MacInnis A, ed. *Molecular Paradigms for Eradicating Helminthic Parasites*. New York: Alan R Liss, 1987:129–136.

195 Scallon BJ, Bogitsh BJ, Carter CE. Characterization of a large gene family in *Schistosoma japonicum* that encodes an immunogenic miracidial antigen. Mol Biochem Parasitol 1989;33:105–112.

196 Grzych JM, Capron M, Lambert PH, Dissous C, Torres S, Capron A. An anti-idiotype vaccine against experimental schistosomiasis. Nature 1985;316:74–76.

197 Caufield JP, Cianci CML, McDiarmid SS, Suyemitsu T, Schid K. Ultrastructure, carbohydrate, and amino acid analysis of two preparations of the cercarial glycocalyx of *Schistosoma mansoni*. J Parasitol 1987;73:514–522.

198 Harn DA, Cianci ML, Caufield JP. *Schistosoma mansoni*: immunization with cercarial glycocalyx preparation increases the adult worm burden. Exp Parasitol 1989;68:108–110.

199 Nyame K, Cummings K, Damian RT. *Schistosoma mansoni* synthesizes glycoproteins containing termial O-linked N-acetylglucosamine residues. J Biol Chem 1987;262:7990–7995.

200 Robertson NP, Cain GD. Isolation and characterization of glycosaminoglycans of *Schistosoma mansoni*. Comp Biochem Physiol 1985;82B:299–306.

201 Weiss JB, Aronstein WS, Strand M. *Schistosoma mansoni*: stimulation of artificial granuloma formation *in vivo*. Exp Parasitol 1987;64:228–236.

202 Simpson AJG, Yi X, Lillywhite J, *et al*. Dissociation of antibody responses during human schistosomiasis and evidence for enhancement of granuloma size by anti-carbohydrate IgM. Trans R Soc Trop Med Hyg 1990;84:808–814.

203 Khalife J, Capron M, Grzych JM, Butterworth AE, Dunne DW, Ouma JH. Immunity in human schistosomiasis mansoni. Regulation of protective immune mechanisms by IgM blocking antibodies. J Exp Med 1986;164:1626–1640.

204 Butterworth AE. Control of Schistosomiasis in man. In Englund PT, Sher A, eds. *The Biology of Parasitism: a Molecular and Immunologic Approach*. New York: Alan R Liss, 1988:43–59.

205 Butterworth AE. Immunology of schistosomiasis. In Wyler D, ed. *Modern Parasite Biology. Cellular, Immunological, and Molecular Aspects*. New York: WH Freeman & Co, 1990:262–288.

206 Butterworth AE, Bensted-Smith R, Capron A, *et al*. Immunity in human schistosomiasis mansoni: prevention by blocking antibodies of the expression of immunity in young children. Parasitology 1986;94:281–300.

207 Butterworth AE, Dunne D, Fulford A, *et al*. Immunity in human schistosomiasis mansoni: cross-reactive IgM and IgG2 anti-carbohydrate antibodies block the expression of immunity. Biochimie 1988;70:1053–1063.

208 Harn D, Mitsuyama M, David JR. *Schistosoma mansoni*. Anti-egg monoclonal antibodies protect against cercarial challenge *in vivo*. J Exp Med 1984;159:1371–1387.

209 Auriault C, Damonneville M, Joseph M, *et al*. Defined antigens secreted by the larvae of schistosomes protect against schistosomiasis: induction of cytotoxic antibodies in the rat and the monkey. Eur J Immunol 1985;15:1168–1172.

210 Auriault C, Dammonneville M, Verwaerde C, *et al*. Rat IgE directed against schistosomula-released products is cytotoxic for *Schistosoma mansoni* schistosomula *in vitro*. Eur J Immunol 1984; 14:132–138.

211 Damonneville M, Auriault C, Verwaerde C, Delanoye A, Pierce R, Capron A. Protection against experimental *Schistosoma mansoni* schistosomiasis achieved by immunization with schistosomula released products antigens (SRP-A); role of IgE antibodies. Clin Exp Immunol 1986;65:244–252.

212 Mountford AP, Coulson PS, Wilson RA. Antigen localisation and the induction of resistance in mice vaccinated with irradiated cercariae of *Schistosoma mansoni*. Parasitology 1988;97:11–25.

213 Mountford AP, Coulson PS, Saunders N, Wilson RA. Characteristics of protective immunity in mice induced by drug-attenuated larvae of *Schistosoma mansoni*. Antigen localisation and antibody responses. J Immunol 1989; 143:989–995.

214 Mountford AP, Wilson RA. *Schistosoma mansoni*: the effect of regional lymphadenectomy on the level of protection induced in mice by radiation attenuated cercariae. Exp Parasitol 1990;71:463–469.

215 Constant SL, Mountford AP, Wilson RA. Phenotypic analysis of the cellular responses in regional lymphoid organs of mice vaccinated with *Schistosoma mansoni*. Parasitology 1991;100:15–22.

216 Verwaerde C, Joseph M, Capron M, *et al*. Functional properties of a rat monoclonal IgE antibody specific for *Schistosoma mansoni*. J Immunol 1987;138:4441–4446.

217 Rotmans JP, van der Voort MJ, Looze M, Mooij GW, Deelder AM. *Schistosoma mansoni*: characterization of antigens in excretions and secretions. Exp Parasitol 1981;52:171–182.

218 Rotmans JP, van der Voort MJ, Looze M, Mooij GW, Deelder AM. *Schistosoma mansoni*: use of antigens from excretions and secretions in immunodiagnosis. Exp Parasitol 1981;52:319–330.

219 James SL, Scott PA. Induction of cell mediated immunity as a strategy for vaccine production against parasites. In Englund PT, Sher A, eds. *The Biology of Parasitism: a Molecular and Immunologic Approach*. New York: Alan R Liss, 1988:249–264.

220 Colley DG. Immune responses and immunoregulation in experimental and clinical schistosomiasis. In Mansfield JM, ed. *Parasitic Diseases*, vol. 1. New York: Marcel Dekker, 1981:1–83.

221 Phillips SM, Colley DG. Immunologic aspects of host responses to schistosomiasis: resistance, immunopathology, and eosinophil involvement. Progr Allergy 1978; 24:49–182.

222 Boros DL. Immunopathology of *Schistosoma mansoni* infection. Clin Microbiol Rev 1989;2:250–269.

223 Pino-Heiss S, Petitt M, Beckstead JH, McKerrow JH. Preparation of mouse monoclonal antibodies and evidence for a host immune response to the preacetabular gland proteinase of *Schistosoma mansoni* cercariae. Am J Trop Med Hyg 1986;35:536–543.

224 Wolowczuk K, Auriault C, Gras-Masse H, *et al*. Protective immunity in mice vaccinated with the *Schistosoma mansoni* P-28-1 antigen. J Immunol 1989;142:1342–1350.

225 Pestel J, Dissous C, Louis J, *et al*. *Schistosoma mansoni*-

specific rat T cell clones. II. Different effects of adult worm-specific T cell clones in immunocompetent and nude infected mice. Eur J Immunol 1989;19:1457–1462.

226 Warren KS. The secret to the pathogenesis of schistosomiasis: *in vivo* models. Immunol Rev 1982;61:189–213.

227 Abdel-Hafez SK, Zodda DM, Pillips SM. The effect of two monoclonal antibodies on *Schistosoma mansoni* cercarial penetration of mouse skin. Folia Parasitol 1983;30:165–218.

228 Thomas MAB, Frampton G, Isenberg DA, *et al.* A common anti-DNA antibody idiotype and anti-phospholipid antibodies in sera from patients with schistosomiasis and filariasis with and without nephritis. J Autoimmun 1989; 2:803–811.

229 James SL, Glaven J, Goldenberg S, Meltzer MS, Pearce E. Tumor necrosis factor (TNF) as a mediator of macrophage helminthotoxic activity. Parasite Immunol 1990;12:1–13.

230 Colley DG. Immune responses to soluble schistosomal egg antigen preparation during chronic primary infections with *Schistosoma mansoni*. J Immunol 1975;115:150–156.

231 Colley DG, Garcia AA, Lambertucci JR, *et al.* Immune responses during human schistosomiasis. XII. Differential responsiveness in patients with hepatosplenic disease. Am J Trop Med Hyg 1986;35:793–802.

232 Colley DG, Barsoum IS, Dahawi HSS, Gamil F, Habib M, El Alamy MA. Immune responses and immunoregulation in relation to schistosomiasis in Egypt. III. Immunity and longitudinal studies of *in vitro* responsiveness after treatment. Trans R Soc Trop Med Hyg 1986;80:952–957.

233 Lammie PJ, Michael AI, Prystowsky MB, Linette GP, Phillips SM. Production of a fibroblast-stimulating factor by *Schistosoma mansoni* antigen-reactive T cell clones. J Immunol 1986;136:1100–1104.

234 Ragheb S, Mathew RC, Boros DL. Establishment and characterization of an antigen-specific T-cell line from liver granulomas of *Schistosoma mansoni* infected mice. Infect Immun 1987;55:2625–2630.

235 Secor WE, Stewart SJ, Colley DG. Eosinophils and immune mechanisms. VI. The synergistic combination of GM-CSF and IL-5 accounts for eosinophil stimulation promoter. J Immunol 1990;144:1484–1489.

236 Sher A, Coffman RL, Hieny S, Scott P, Cheever AW. Interleukin-5 is required for the blood and tissue eosinophilia but not granuloma formation induced by infection with *Schistosoma mansoni*. Proc Natl Acad Sci USA 1990; 87:61–65.

237 Doughty BL, Phillips SM. Delayed hypersensitivity granuloma formation around *Schistosoma mansoni* eggs *in vitro*. I. Definition of the model. J Immunol 1982;128:30–36.

238 Doughty BL, Phillips SM. Delayed hypersensitivity granuloma formation and modulation around *Schistosoma mansoni* eggs *in vitro*. I. Regulatory T cell subsets. J Immunol 1982;128:37–42.

239 Bentley AG, Phillips SM, Kaner RJ, Theodorides VJ, Linette GP, Doughty BL. *In vitro* delayed hypersensitivity granuloma formation: development of an antigen-coated bead model. J Immunol 1985;134:4163–4169.

240 Doughty BL, Ottesen EA, Nash TE, Phillips SM. Delayed hypersensitivity granuloma formation around schistosome eggs *in vitro*. III. Granuloma formation and modulation in

human schistosomiasis. J Immunol 1984;133:993–997.

241 Louis JA, Lima G, Pestel J, Titus R. Murine T-cell responses to protozoan and metazoan parasites: functional analysis of T-cell lines and clones specific for *Leishmania tropica* and *Schistosoma mansoni*. Contemp Top Immunobiol 1984;12:201–224.

242 Lammie PJ, Linette GP, Phillips SM. Characterization of *Schistosoma mansoni* antigen-reactive T cell clones that form granulomas *in vitro*. J Immunol 1985;134:4170–4175.

243 Mak NK, Sanderson CJ. The T-cell mediated immune response to *Schistosoma mansoni*: I. Generation of stage-specific, MHC-restricted proliferative T-cell clones to soluble worm antigens. Immunology 1985;54:625–633.

244 Ohta N, Edahiro T, Tohgi N, Ishii A, Minai M, Hosaka Y. Generation and functional characterization of T cell lines and clones specific for *Schistosoma japonicum* egg antigens in humans. J Immunol 1988;141:2445–2450.

245 Ohta N, Itagaki T, Minai M, Hirayama K, Hosaka Y. *Schistosoma japonicum* egg antigen-specific T cell lines in humans. Induction of suppressor and helper T cell lines and clones *in vitro* in a patient with chronic schistosomiasis japonica. J Clin Invest 1988;81:775–781.

246 Mendlovic F, Arnon R, Tarrab-Hazdai R, Puri J. Genetic control of immune response to a purified *Schistosoma mansoni* antigen. II. Establishment and characterization of specific I-A and I-E restricted T-cell clones. Parasite Immunol 1989;11:683–694.

247 Reynolds SR, Kunkel SL, Thomas DW, Higashi GI. T cell clones for antigen selection and lymphokine production in murine schistosomiasis mansoni. J Immunol 1990; 144:2757–2762.

248 Harn DA, Danko K, Quinn JJ, Stadecker MJ. *Schistosoma mansoni*: the host immune response to egg antigens. I. Partial characterization of cellular and humoral responses to pI fractions of soluble egg antigens. J Immunol 1989; 142:2061–2066.

249 Lukacs NW, Boros DL. Splenic and granuloma T-lymphocyte responses to fractionated soluble egg antigens of *Schistosoma mansoni*-infected mice. Infect Immun 1991; 59:941–948.

250 Rand TH, Clanton JAA, Runge V, English D, Colley DG. Murine eosinophils labeled with indium-111 oxine: localization to delayed hypersensitivity reactions against a schistosomal antigen and to lymphokine *in vivo*. Blood 1983;61:732–739.

251 Boros DL, Warren KS. Delayed hypersensitivity-type granuloma formation and dermal reaction induced and elicited by a soluble factor isolated from *Schistosoma mansoni* eggs. J Exp Med 1970;132:488–507.

252 Savage AM, Colley DG. The eosinophil in the inflammatory response to cercarial challenge of sensitized and chronically infected CBA/J mice. Am J Trop Med Hyg 1980;29:1268–1278.

253 Grzych J-M, Pearce E, Cheever A, *et al.* Egg deposition is the major stimulus for the production of TH2 cytokines in murine schistosomiasis mansoni. J Immunol 1991;146: 1322–1327.

254 Pearce EJ, Caspar P, Grzych J-M, Lewis FA, Sher A. Downregulation of Th1 cytokine production accompanies

induction of Th2 responses by a parasitic helminth, *Schistosoma mansoni*. J Exp Med 1991;173:159–166.

255 Mosmann TR, Coffman RL. TH1 and TH2 cells: different patterns of lymphokine secretion lead to different functional properties. Ann Rev Immunol 1989;7:145–173.

256 Colley DG, Katz SP, Wikel SK. Schistosomiasis: an experimental model for the study of immunopathologic mechanisms which involve eosinophils. Adv Biosci 1973; 12:653–665.

257 James SL, Natoviz PC, Farrarr WL, Leonard JE. Macrophages as effector cells of protective immunity in murine schistosomiasis: macrophage activation in mice vaccinated with radiation-attenuated cercariae. Infect Immun 1984; 44:569–575.

258 Kambara T, Wilson RA. *In situ* pulmonary responses of T cell and macrophage subpopulations to a challenge infection in mice vaccinated with irradiated cercariae of *Schistosoma mansoni*. J Parasitol 1990;76:363–372.

259 Phillips SM, Linette GP, Doughty BL, Byram JE, von Lichtenberg F. *In vivo* T cell depletion regulates resistance and morbidity in murine schistosomiasis. J Immunol 1987; 139:919–926.

260 Phillips SM, Walker D, Abdel-Hafez SK, *et al*. The immune response to *Schistosoma mansoni* infections in inbred rats. VI. Regulation by T cell subpopulations. J Immunol 1987; 139:2781–2787.

261 Kelly EAB, Colley DG. *In vivo* effects of monoclonal anti-L3T4 antibody on immune responsiveness of mice infected with *Schistosoma mansoni*. Reduction of irradiated cercariae-induced resistance. J Immunol 1988;140:2737–2745.

262 Menson EN, Wilson RA. Lung-phase immunity to *Schistosoma mansoni*: flow cytometric analysis of macrophage activation states in vaccinated mice. J Immunol 1989; 143:2342–2348.

263 Menson EN, Wilson RA. Lung-phase immunity to *Schistosoma mansoni*: definition of alveolar macrophage phenotypes after vaccination and challenge of mice. Parasite Immunol 1990;12:353–366.

264 Vignali DAA, Crocker P, Bickle QD, Cobbold S, Waldmann H, Taylor MG. A role for CD4$^+$ but not CD8$^+$ T cells in immunity to *Schistosoma mansoni* induced by 20 krad-irradiated and Ro 11-3128-terminated infections. Immunology 1989;67:466–472.

265 Mangold BL, Dean DA. Passive transfer with serum and IgG antibodies of irradiated cercariae-induced resistance against *Schistosoma mansoni* in mice. J Immunol 1986; 136:2644–2648.

266 Colley DG. Dynamics of the human immune response to schistosomes. In Mahmoud AAF, ed. *Clinical Tropical Medicine and Communicable Diseases, vol. 2, no 2, Schistosomiasis*. London: Baillière Tindel, 1987:315–332.

267 Ottesen EA. Modulation of the host response in human schistosomiasis. I. Adherent suppressor cells that inhibit lymphocyte proliferation responses to parasite antigens. J Immunol 1979;123:1639–1644.

268 Gazzinelli G, Lambertucci JR, Katz N, Rocha RS, Lima MS, Colley DG. Immune responses during human schistosomiasis mansoni. XI. Immunologic status of patients with acute infections and after treatment. J Immunol 1985; 135:2121–2127.

269 Warren KS. Modulation of immunopathology and disease in schistosomiasis. Am J Trop Med Hyg 1977;26:113–119.

270 Andrade ZA, Warren KS. Mild prolonged schistosomiasis in mice. Alterations in host response with time and the development of portal fibrosis. Trans R Soc Trop Med Hyg 1964;58:53–57.

271 Barsoum IS, Freeman GL Jr, Habib M, *et al*. Evaluation of natural killer activity in human schistosomiasis. Am J Trop Med Hyg 1984;33:451–454.

272 Feldmeier H, Gastl GA, Poggensee U, Kortmann C, Daffalla AA, Peter HH. Relationship between intensity of infection and immunomodulation in human schistosomiasis. II. NK cell activity and *in vitro* lymphocyte proliferation. Clin Exp Immunol 1985;60:234–240.

273 Gastl GA, Feldmeier H, Kortmann C, Daffalla AA, Peter HH. Human schistosomiasis: deficiency of large granular lymphocytes and indomethacin-sensitive suppression of natural killing. Scand J Immunol 1986;23:319.

274 Wilson CS, Ellner JJ, Dinarello CA, Keusch GT, El Kholy A. Dissociation of immunoregulatory function of blood monocytes from maturational state and expression of interleukin-1 in humans chronically infected with *Schistosoma mansoni*. Am J Trop Med Hyg 1990;42:234–243.

275 Zwingenberger K, Irschick E, Vergetti Siueira JG, Correia Dacal AR, Feldmeier H. Tumor necrosis factor in hepatosplenic schistosomiasis. Scand J Immunol 1990;31:205–211.

276 Feldmeier H, Gastl GA, Poggensee U, *et al*. Immune response in chronic schistosomiasis haematobium and mansoni. Scand J Immunol 1988;28:147–155.

277 Lima MS, Gazzinelliu G, Nascimento E, Carvalho Parra J, Montesano MA, Colley DG. Immune responses during human schistosomiasis mansoni. XIV. Evidence for anti-idiotypic T lymphocyte responsiveness. J Clin Invest 1986; 78:983–988.

278 Montesano MA, Lima MS, Correa-Oliveira R, Gazzinelli G, Colley DG. Immune responses during human schistosomiasis mansoni. XVI. Idiotypic differences in antibody preparations from patients with different clinical forms of infection. J Immunol 1989;142:2501–2506.

279 Doughty BL, Goes AM, Parra JC, *et al*. Anti-idiotypic T cells in human schistosomiasis. Immunol Invest 1989;18:373–388.

280 Powell MR, Colley DG. Anti-idiotypic T-lymphocyte responsiveness in murine schistosomiasis mansoni. Cell Immunol 1987;104:377–385.

281 Montesano MA, Freeman GL Jr, Gazzinelli G, Colley DG. Expression of cross-reactive, shared idiotypes on anti-SEA antibodies from humans and mice with schistosomiasis. J Immunol 1990;145:1002–1008.

282 Parra JC, Lima MS, Gazzinelli G, Colley DG. Immune responses during human schistosomiasis mansoni. XV. Anti-idiotypic T cells can recognize and respond to anti-SEA idiotypes directly. J Immunol 1988;140:2401–2405.

283 Montesano MA, Freeman GL Jr, Gazzinelli G, Colley DG. Immune responses during human schistosomiasis mansoni. XVII. Recognition by monoclonal anti-idiotypic antibodies of several idiotypes on a monoclonal anti-soluble schistosomal egg antigen and anti-soluble schistosomal egg antibodies from patients with different clinical

forms of infections. J Immunol 1990;145:3095–3099.

284 Mathew RC, Ragheb S, Boros DL. Recombinant IL-2 therapy reverses diminished granulomatous responsiveness in anti-L3T4 treated, *Schistosoma mansoni*-infected mice. J Immunol 1990;144:4356–4361.

285 Henderson GS, Conary JT, Sumnar M, McCurley TL, Colley DG. IL-4, not IL-2 mRNA, is abundant in granulomatous livers and mesenteric lymph nodes of *S mansoni* infected mice. J Immunol 1991;147:992–997.

286 Von Lichtenberg F. Host response to eggs of *S mansoni*. I. Granuloma formation in the unsensitized laboratory mouse. Am J Pathol 1962;41:711–731.

287 Boros DL, Warren KS. Characterization of a model system for infectious and foreign body granulomatous inflammation using soluble mycobacterial, histoplasma and schistosoma antigens. Immunology 1973;24:511–529.

288 Istre GR, Fontaine RE, Tarr J, Hopkins RS. Acute schistosomiasis among Americans rafting the Omo River, Ethiopia. J Am Med Assoc 1984;251:508–510.

289 Farid Z, Mansour N, Kamal K, Girgis N, Woody J, Kamal M. The diagnosis and treatment of acute toxaemic schistosomiasis in children. Trans R Soc Trop Med Hyg 1987; 81:959.

290 Billings FT, Winkenwereder WL, Hunninen AV. Studies on acute schistosomiasis japonica in the Philippine Islands. I. A clinical study of 337 cases with a preliminary report on the results of treatment with Fuadin in 110 cases. Bull Johns Hopkins Hosp 1947;78:21–56.

291 Diaz-Rivera RS, Ramos-Morales F, Koppisch E, *et al*. Acute Manson's schistosomiasis. Am J Med 1956;21:918–943.

292 Ottesen EA, Hiatt RA, Cheever AW, Sotomayor ZR, Neva FA. The acquisition and loss of antigen-specific cellular immune responsiveness in acute and chronic schistosomiasis in man. Clin Exp Immunol 1978;33:38–47.

293 Haitt RA, Sotomayor ZR, Sanchez G, Zombrana M, Knight WB. Factors in the pathogenesis of acute schistosomiasis mansoni. J Infect Dis 1979;139:659–666.

294 Mansour MM, Ali PO, Farid Z, Simpson AJ, Woody JW. Serological differentiation of acute and chronic schistosomiasis mansoni by antibody responses to keyhole limpet hemocyanin. Am J Trop Med Hyg 1989;41:338–344.

295 Lawley TJ, Ottesen EA, Hiatt RA, Gazze LA. Circulating immune complexes in acute schistosomiasis. Clin Exp Immunol 1979;37:221–227.

296 Hiatt RA, Otessen EA, Sotomayor ZR, Lawley TJ. Serial observations of circulating immune complexes in patients with acute schistosomiasis. J Infect Dis 1980;142:665–670.

297 James SL, Colley DG. Eosinophil-mediated destruction of *Schistosoma mansoni* eggs. J Reticuloendo Soc 1976;20:359–374.

298 Sher A, Colley DG. Immunoparasitology. In Paul WE, ed. *Fundamental Immunology*. New York: Raven Press, 1989: 957–983.

299 Mitchinson NA, Oliveira DBG. Chronic infection as a major force in the evolution of the suppressor T-cell system. Parasitol Today 1986;2:312–313.

300 Warren KS. The pathogenesis of "clay-pipe stem fibrosis" in mice with chronic schistosomiasis mansoni, with a note about the longevity of schistosomes. Am J Pathol 1966; 49:477–489.

301 Stavitsky AB. Immune regulation in schistosomiasis japonica. Immunol Today 1987;8:228–233.

302 Cheever AW, Deb S, Duvall RH. Granuloma formation in *Schistosoma japonicum* infected nude mice. The effects of reconstitution with L3T4$^+$ or Lyt2$^+$ splenic cells. Am J Trop Med Hyg 1989;40:66–71.

303 Phillips SM, Lammie PJ. Immunopathology of granuloma formation and fibrosis in schistosomiasis. Parasitol Today 1986;2:296–302.

304 Sadun EH, von Lichtenberg F, Cheever AW, Erickson DG. Schistosomiasis mansoni in the chimpanzee. The natural history of chronic infections after single and multiple exposures. Am J Trop Med Hyg 1970;19:258–277.

305 Von Lichtenberg F, Sadun EH, Cheever AW, Erickson DG, Johnson AJ, Boyce HW. Experimental infection with *Schistosoma mansoni* in chimpanzees. Am J Trop Med Hyg 1971;20:850–893.

306 Kamel IA, Elwi AM, Cheever AW, Mosimann JE, Danner R. *Schistosoma mansoni* and *S. haematobium* infections in Egypt. IV. Hepatic lesions. Am J Trop Med Hyg 1978; 27:931–139.

307 Homeida M, Ahmed S, Dafalla A, Suliman S, Eltom I, Nash T, Bennet JL. Morbidity associated with *Schistosoma mansoni* infection as determined by ultrasound: a study in Gezira, Sudan. Am J Trop Med Hyg 1988;39:196–201.

308 Homeida MA, Dafalla AA, Kardaman MW, *et al*. Effect of antischistosomal chemotherapy on prevalence of Symmers' periportal fibrosis in Sudanese villages. Lancet 1988;2:437–440.

309 Abdel-Wahab MF, Esmat G, Milad M, Abdel-Razek S, Strickland GT. Characteristic sonographic pattern of schistosomal hepatic fibrosis. Am J Trop Med Hyg 1989;40:72–76.

310 Doerhing-Schwerdtfeger E, Abdel-Rahim IM, Mohamed-Ali Q, *et al*. Ultrasonography: evaluation of morbidity. Am J Trop Med Hyg 1990;42:581–586.

311 Hatz C, Mayomban C, de Savigny D, *et al*. Ultrasound scanning for detecting morbidity due to *Schistosoma haematobium* and its resolution following treatment with different doses of praziquantel. Trans R Soc Trop Med Hyg 1990;84:84–85.

312 Colley DG, Montesano MA, Eloi-Santos SM, *et al*. Idiotype net-works in schistosomiasis. In McAdam KPWJ, ed. *New Strategies in Parasitology*. Edinburgh: Churchill Livingstone, 1989:179–190.

313 Warren KS. The immunopathogenesis of schistosomiasis: a multidiciplinary approach. Trans R Soc Trop Med Hyg 1972;66:417–434.

314 Warren KS, Domingo EO. Granuloma formation around *Schistosoma mansoni*, *S. haematobium*, and *S. japonicum* eggs. Size and rate of development, cellular composition, cross-sensitivity, and rate of egg destruction. Am J Trop Med Hyg 1970;19:292–304.

315 Agnew AM, Lucas SB, Doenhoff MJ. The host–parasite relationship of *Schistosoma haematobium* in CBA mice. Parasitology 1988;97:403–424.

316 Warren KS, Grove DI, Pelley RP. The *Schistosoma japonicum* egg granuloma. II. Cellular composition, granuloma size, and immunologic concomitants. Am J Trop Med Hyg 1978;27:271–275.

317 Kresina TF, Olds GR. Concomitant cellular and humoral expression of a regulatory cross-reactive idiotype in acute *Schistosoma japonicum* infection. Infect Immun 1986;53:90–94.

318 Von Lichtenberg F. Studies on granuloma formation. III. Antigen sequestration and destruction in the schistosome pseudotubercle. Am J Pathol 1964;45:75–93.

319 Doenhoff MJ, Hassounah O, Murare H, Bain J, Lucas S. The schistosome egg granuloma: immunopathology in the cause of host protection or parasite survival? Trans R Soc Trop Med Hyg 1986;80:503–514.

320 Damian RT. The exploitation of host immune responses by parasites. J Parasitol 1987;73:1–11.

321 Doenhoff M, Musallam R, Bain J, McGregor A. Studies on the host–parasite relationship in *Schistosoma mansoni*-infected mice: the immunological dependence of parasite egg excretion. Immunology 1978;35:771–778.

322 Boros DL. Granulomatous inflammations. Prog Allergy 1978;24:183–267.

323 Phillips SM, Fox EG. Immunopathology of ʻparasitic diseases: a conceptual approach. Contemp Top Immunobiol 1984;12:241–261.

324 Kasahara K, Kobayashi K, Shikama Y, *et al.* Direct evidence for granuloma-inducing activity of interleukin 1. Induction of experimental pulmonary granuloma formation in mice by interleukin-1-coupled beads. Am J Pathol 1988;130:629–638.

325 Chensue SW, Otterness IG, Higashi GI, Forsch CS, Kunkel SL. Monokine production by hypersensitivity (*Schistosoma mansoni* egg) and foreign body (sephadex bead)-type granuloma macrophages. Evidence for sequential production of IL-1 and tumor necrosis factor. J Immunol 1989;142:1281–1286.

326 Wellhausen SR, Boros DL. Comparison of Fc, C3 receptors, and Ia antigens on the inflammatory macrophages isolated from vigorous or immunomodulated liver granulomas of schistosome-infected mice. J Reticuloendo Soc 1981;30:191–203.

327 Stadecker MJ, Wyler DJ, Wright JA. Ia antigen expression and antigen presentation function by macrophages isolated from hypersensitivity granuloma. J Immunol 1982;128:2739–2744.

328 Chensue SW, Kunkel SL, Higashi GI, Ward PA, Boros DL. Production of superoxide anion, prostaglandins, and hydroxyeicostetraenoic acids by macrophages from hypersensitivity-type (*Schistosoma mansoni* egg) and foreign body-type granulomas. Infect Immun 1983;42:1116–1125.

329 Sunday ME, Stadecker MJ, Wright MJ, Aoki I, Dorf ME. Induction of immune responses by schistosome granuloma macrophages. J Immunol 1983;130:2413–2417.

330 Shook LB, Wellhausen SR, Boros DL, Niederhuber JE. Accessory cell function of liver granuloma macrophages of *Schistosoma mansoni*-infected mice. Infect Immun 1983;42:882–886.

331 Elliott DE, Righhand VF, Boros DL. Characterization of regulatory interferon α and accessory (LAF/IL-1) monokine activities from liver granuloma macrophages of *S. mansoni*-infected mice. J Immunol 1987;138:2653–2662.

332 Clark CR, Chen BD, Boros DL. Macrophage progenitor cell and colony stimulating factor production during granu-

333 Cheever AW, Byram JE, von Lichtenberg F. Immuno-pathology of *Schistosoma japonicum* infection in athymic mice. Parasite Immunol 1985;7:387–398.

334 Mathew RC, Boros DL. Anti L3T4 antibody treatment suppresses hepatic granuloma formation and abrogates antigen-induced interleukin-2 production in *Schistosoma mansoni* infection. Infect Immun 1986;54:820–826.

335 Weinstock JV, Boros DL. Heterogeneity of the granulomatous response in the liver, colon, ileum, and ileal Peyer's patches to schistosome eggs in murine schistosomiasis. J Immunol 1981;127:1906–1909.

336 Weinstock JV, Boros DL. Organ-dependent differences in composition and function observed in hepatic and intestinal granulomas isolated from mice with schistosomiasis mansoni. J Immunol 1983;130:418–422.

337 Grimaud JA, Boros DL, Takiya C, Mathew RC, Emonard H. Collagen isotypes, laminin and fibronectin in granulomas of the liver and intestines of *Schistosoma mansoni* infected mice. Am J Trop Med Hyg 1987;37:335–344.

338 Freundlich B, Bomamaski JS, Neilson E, Jimenez SA. Regulation of fibroblast proliferation and collagen synthesis by cytokines. Immunol Today 1986;7:303–307.

339 Wyler DJ. Fibronectin in parasitic diseases. Rev Infect Dis 1987;9:391–399.

340 Biempica L, Dunn MA, Kamel IA, *et al.* Liver collagen-type characterization in human schistosomiasis. A histological, ultrastructural, and immunocytochemical correlation. Am J Trop Med Hyg 1983;32:316–325.

341 Grimaud JA, Druguet M, Peyrol S, Chevalier O, Hebage D, El Badrawy N. Collagen immunotyping in human liver. Light and electron microscope study. J Histochem Cytochem 1980;28:1145–1156.

342 Zwingenberger K, Feldmeier H, Queiroz JA, *et al.* Liver involvement in human schistosomiasis mansoni. Assessment by immunological and biochemical markers. Parasitol Res 1988;74:448–455.

343 Prakash S, Postlethwaite AE, Stricklin GP, Wyler DJ. Fibroblast stimulation in schistosomiasis. IX. Schistosomal egg granulomas from congenitally athymic mice are deficient in production of fibrogenic factors. J Immunol 1990;144:317–322.

344 Nishimura M, Asahi M, Hayashi M, Takazono I, Tanaka Y, Kohda H, Urabe H. Extracellular matrix in hepatic granulomas of mice infected with *Schistosoma mansoni*. Arch Pathol Lab Med 1985;109:813–818.

345 Wu CH, Giambrone MA, Howard DJ, Rojkind M, Wu GY. The nature of the collagen in hepatic fibrosis in advanced murine schistosomiasis. Hepatology 1982;2:366–371.

346 Parise ER, Summerfield JA, Hahn E, Wiedmann KH, Doenhoff MJ. Basement membrane proteins and type III procollagen in murine schistosomiasis. Trans R Soc Trop Med Hyg 1985;79:663–670.

347 Al Adnani MS. Concomitant immunohistochemical localization of fibronectin and collagen in schistosome granuloma. J Pathol 1985;147:77–85.

348 Olds GR, Griffith A, Kresina TF. Dynamics of collagen accumulation and polymorphism in murine *Schistosoma japonicum*. Gastroenterology 1985;89:617–624.

349 Morcos SH, Khayyal MT, Mansour MM, *et al*. Reversal of hepatic fibrosis after praziquantel chemotherapy of murine schistosomiasis. Am J Trop Med Hyg 1985;34: 314–321.

350 Wyler DJ, Wahl SM, Wahl LM. Hepatic fibrosis in schistosomiasis: egg granulomas sectete fibroblast stimulating factor *in vitro*. Science 1978;202:438–440.

351 Wyler DJ. Regulation of fibroblast functions by products of schistosomal egg granulomas: potential role in the pathogenesis of hepatic fibrosis. In Evered D, Collins GM, eds. *Cytopathology of Parasitic Diseases*. London: Pitman Books, 1983:190–206.

352 Prakash S, Dinarello CA, Danko K, Stadeker MJ, Wyler DJ. Schistosomal egg granuloma-derived fibroblast-stimulating factor is apparently distinct from interleukin 1. Infect Immun 1984;57:679–684.

353 Wyler DJ, Stradecker MJ, Dinarello CA, O'Dea JF. Fibroblast stimulation in schistosomiasis. V. Egg granuloma macrophages spontaneously secrete a fibroblast-stimulating factor. J Immunol 1984;132:3142–3148.

354 Mansour MM, El-ghorab NM, Salah LA, Dunn MA, Woody JN. T lymphocyte subset modulation of hepatic fibroblast function in murine schistosomiasis. Am J Trop Med Hyg 1989;41:454–459.

355 Lammie PJ, Monroe JG, Michael AI, Johnson GD, Phillips SM, Prystowsky MB. Partial characterization of a fibroblast-stimulating factor produced by cloned T lymphocytes. Am J Pathol 1988;130:289–295.

356 Olds GR, El Meneza S, Mahmoud AAF, Kresina TF. Differential immunoregulation of granulomatous inflammation, portal hypertension, and hepatic fibrosis in murine schistosomiasis mansoni. J Immunol 1989;142:3605–3611.

357 Cheever AW, Duvall RH, Hallack TA Jr. Differences in hepatic fibrosis and granuloma size in several strains of mice infected with *Schistosoma japonicum*. Am J Trop Med Hyg 1984;33:602–607.

358 Cheever AW, Duvall RH, Hallak TA Jr, Minker RG, Malley JD, Malley KG. Variation of hepatic fibrosis and granuloma size among mouse strains infected with *Schistosoma mansoni*. Am J Trop Med Hyg 1987;37:85–97.

359 Cheever AW. The intensity of experimental schistosome infections modulates hepatic pathology. Am J Trop Med Hyg 1986;35:124–133.

360 Andrade ZA. Pathogenesis of pipe-stem fibrosis of the liver (experimental observation on murine schistosomiasis). Mem Inst Oswaldo Cruz 1987;82:325–334.

361 Tweardy DJ, Osman GS, El Kholy A, Ellner JJ. Failure of immunosuppressive mechanisms in human *Schistosoma mansoni* infection with hepatosplenomegaly. J Clin Microbiol 1987;25:768–773.

362 Hagan P. The human immune response to schistosome infection. In Rollinson D, Simpson AJG, eds. *The Biology of Schistosomes: from Genes to Latrines*. London: Academic Press, 1987:295–320.

363 Cheever AW. A quantitative post-mortem study of schistosomiasis mansoni in man. Am J Trop Med Hyg 1968;17:38–64.

364 Abdel-Salam E, Abdel-Khalik A, Abdel-Meguid A, Barakat W, Mahmoud AA. Association of HLA Class I antigens (A1, B5, B8, and CW2) with disease manifestations and infection in human schistosomiasis mansoni in Egypt. Tissue Antigens 1986;27:142–146.

365 Ohta N, Hayashi M, Tormis LC, Blas BL, Nosena JS, Sasazuki T. Immunogenetic factors involved in the pathogenesis of distinct clinical manifestations of schistosomiasis japonica in the Philippine population. Trans R Soc Trop Med Hyg 1987;81:142–146.

366 Hirayama K, Matsushita S, Kikuchi I, Iuchi M, Ohta N, Sasazuki T. HLA-DQ is epistatic to HLA-DR in controlling the immune response to schistosomal antigen in humans. Nature 1987;327:426–430.

367 Boros DL. Immunoregulation of granuloma formation in murine schistosomiasis mansoni. Ann NY Acad Sci 1986; 465:313–323.

368 Abe T, Colley DG. Modulation of *Schistosoma mansoni* egg-induced granuloma formation. III. Evidence for an anti-idiotypic, I-J-positive, I-J-restricted, soluble T suppressor factor. J Immunol 1984;132:2084–2088.

369 Perrin PJ, Phillips SM. The molecular basis of granuloma formation in schistosomiasis. III. *In vivo* effects of a T-cell-derived suppressor effector factor and IL-2 on granuloma fromation. J Immunol 1989;143:649–654.

370 Stavitsky AB, Harold WW. Deficiency of interleukin-2 production upon addition of soluble egg antigen to cultures of isolated hepatic granulomas or hepatic granuloma cells from mice infected with *Schistosoma japonicum*. Infect Immun 1989;57:2339–2344.

371 Yamashita T, Boros DL. Changing patterns of lymphocyte proliferation, IL-2 production and utilization, and IL-2 receptor expression in mice infected with *Schistosoma mansoni*. J Immunol 1990;145:724–731.

372 Scott P, Pearce E, Cheever AW, Coffman RL, Sher A. Role of cytokines and CD4$^+$ T-cell subsets in the regulation of parasite immunity and disease. Immunol Rev 1989;112: 161–182.

373 Perrin PJ, Phillips SM. The molecular basis of granuloma formation in schistosomiasis. I. A T cell derived suppressor effector factor. J Immunol 1988;141:1714–1719.

374 Ragheb S, Boros DL. Characterization of granuloma T lymphocyte function from *Schistosoma mansoni* infected mice. J Immunol 1989;142:3239–3246.

375 McLaren DJ. Will the real target of immunity to schistosomiasis please stand up. Parasitol Today 1989;5:279–282.

376 McLaren DJ. Experimental animal models in vaccination against schistosomiasis. Mem Oswaldo Cruz 1989;84(1): 188–194.

377 Vignali DAA, Bickle QD, Taylor MG. Immunity to *Schistosoma mansoni in vivo*: contradiction or clarification? Immunol Today 1989;10:410–416.

378 Wilson RA, Coulson PS. Lung-phase immunity to schistosomes: a new perspective on an old problem. Parasitol Today 1989;5:274–278.

379 Bickle QD, Andrews BJ, Doenhoff MJ, Ford MJ, Taylor MG. Resistance against *Schistosoma mansoni* induced by highly irradiated infections: studies on species-specificity of immunization and attempts to transfer resistance. Parasitology 1985;90:301–312.

380 Smithers SM, Simpson AJG, Yi X, Omer-Ali P, Kelly C, McLaren DJ. The mouse model of schistosome immunity. Acta Trop 1987;44:21–30.

381 Peresan G, Cioli D. Resistance to cercarial challenge upon transfer of *Schistosoma mansoni* into mice. Am J Trop Med Hyg 1980;29:1258–1262.

382 Dean DA. *Schistosoma* and related genera: acquired resistance in mice. Exp Parasitol 1982;55:1–104.

383 James SL, Cheever AW. Comparison of immune responses between high and low responder strains of mice in the concomitant immunity and vaccine models of resistance to *Schistosoma mansoni*. Parasitology 1985;91:301–315.

384 McLaren DJ, Strath M, Smithers SR. *Schistosoma mansoni*: evidence that immunity in vaccinated and chronically infected CBA/Ca mice is sensitive to treatment with a monoclonal antibody that depletes cutaneous effector cells. Parasite Immunol 1987;9:667–682.

385 Mahmoud AAF, Warren KS, Peters PA. A role for the eosinophil in acquired resistance to *Schistosoma mansoni* infection as determined by anti-eosinophil serum. J Exp Med 1975;142:805–812.

386 Delgado VS, McLaren DJ. Variable generation of specific acquired resistance in CBA/Ca mice chronically infected with schistosomiasis mansoni. Parasitology 1989;99:357–362.

387 Smithers SR, Gammage K. Migration of the schistosomula of *Schistosoma mansoni* from the skin, lungs, and hepatic portal system of naive mice and mice previously exposed to *S. mansoni*. Evidence for two phases of parasite attrition in immune mice. Parasitology 1980;80:289–300.

388 Bentley AG, Carlisle SA, Phillips SM. Ultrastructural analysis of the cellular response to *Schistosoma mansoni*. Am J Trop Med Hyg 1981;30:815–824.

389 Delgado VS, McLaren DJ. *Schistosoma mansoni*: evidence that site-dependent responses determine when and where vaccine immunity is expressed in different rodent species. Parasitology 1990;100:57–63.

390 Doenhoff M, Long E. Factors affecting the acquisition of resistance against *Schistosoma mansoni* in the mouse. IV. The inability of T-cell-deprived mice to resist re-infection, and other *in vivo* studies on the mechanisms of resistance. Parasitology 1979;78:171–183.

391 Sher A, Smithers SR, MacKenzie P. Passive transfer of acquired resistance to *Schistosoma mansoni* in laboratory mice. Parasitology 1975;70:347–357.

392 Sher A, Smithers SR, MacKenzie P, Broomfield K. *Schistosoma mansoni*: immunoglobulins involved in passive immunization of laboratory mice. Exp Parasitol 1977; 41:160–166.

393 Dean DA, Bukowski MA, Clark SS. Attempts to transfer the resistance of *Schistosoma mansoni*-infected and irradiated cercariae-immunized mice by means of parabiosis. Am J Trop Med Hyg 1981;30:113–120.

394 Tavares CAP, Gazzinelli G, Mota-Santos TA, Da Silva WD. *Schistosoma mansoni*: complement-mediated cytotoxic activity *in vitro* and effect of decomplementation on acquired immunity in mice. Exp Parasitol 1978;46:145–151.

395 Vignali DAA, Bickle QD, Taylor MG, Tennent G, Pepys MB. Comparison of the role of complement in immunity to *Schistosoma mansoni* in rats and mice. Immunology 1988; 63:55–61.

396 Vogel H, Minning W. Uber die erwor-bene resistenz von *Macacus rhesus* gegenuber *Schistosoma japonicum*. Z Tropenmed Parasitenkd 1953;4:418–505.

397 Sturrock RF, Cottrell BJ, Lucas SW, Reid GD, Seitz HM, Wilson RA. Observations on the implications of pathology induced by experimental schistosomiasis in baboons in evaluating the development of resistance to challenge infection. Parasitology 1988;96:37–48.

398 McHugh SM, Coulson PS, Wilson RA. The relationship between pathology and resistance to reinfection with *Schistosoma mansoni* in mice: a causal mechanism of resistance in chronic infections. Parasitology 1987;94:81–91.

399 Taylor MG, Bickle QD. Towards a schistosomiasis vaccine. Parasitol Today 1986;2:132–134.

400 Delgado V, McLaren DJ. Evidence for enhancement of IgG1 subclass expression in mice polyvaccinated with radiation attenuated cercariae of *Schistosoma mansoni* and the role of this isotype in serum-transferred immunity. Parasite Immunol 1990;12:15–32.

401 Jwo J, LoVerde PT. The ability of fractionated sera from animals vaccinated with irradiated cercariae of *Schistosoma mansoni* to transfer immunity to mice. J Parasitol 1989;75: 252–260.

402 Kojima S, Niimura M, Kanazawa T. Production and properties of a mouse monoclonal IgE antibody to *Schistosoma japonicum*. J Immunol 1987;139:2044–2049.

403 Sher A, Correa-Oliveira R, Hieny S, Hussain R. Mechanisms of protective immunity against *Schistosoma mansoni* in mice vaccinated with irradiated cercariae. IV. Analysis of the role of IgE antibodies and mast cells. J Immunol 1983; 131:1460–1465.

404 Hsu SYL, Hsu HF, Johnson SC, Xu ST, Johnson SM. Histological study of the attrition of challenge cercariae of *Schistosoma mansoni* in the skin of mice immunized by chronic infection and the use of highly X-irradiated cercariae. Z Parasitenkd 1983;69:627–642.

405 McLaren DJ, Smithers SR. Serum from CBA/Ca mice vaccinated with irradiated cercariae of *Schistosoma mansoni* protects naive recipients through the recruitment of cutaneous effector cells. Parasitology 1988;97:287–302.

406 Ward RE, McLaren DJ. *Schistosoma mansoni*: evidence that eosinophils and/or macrophages contribute to skin phase challenge attrition in vaccinated CBA/Ca mice. Parasitology 1988;96:63–84.

407 Delgado VS, McLaren DJ. Evidence that radiosensitive cells are central to skin-phase protective immunity in CBA/Ca mice vaccinated with radiation-attenuated cercariae of *Schistosoma mansoni* as well as in naïve mice protected with vaccine serum. Parasitology 1990;100:45–56.

408 Aitken R, Coulson PS, Wilson RA. Pulmonary leucocytic responses are linked to the acquired immunity of mice vaccinated with irradiated cercariae of *Schistosoma mansoni*. J Immunol 1988;140:3573–3579.

409 Coulson PS, Mountford AP. Fate of attenuated schistosomula administered by different routes, relative to the immunity induced against *Schistosoma mansoni*. Parasitology 1989;99:39–44.

410 James SL. Activated macrophages as effector cells of protective immunity in schistosomiasis. Immunol Res 1986; 5:139–147.

411 James SL, Sher A. Prospects for a non-living vaccine

against schistosomiasis. Parasitol Today 1986;2:134–137.

412 Vignali DA, Bickle QD, Taylor MG. Studies on immunity to *Schistosoma mansoni in vivo*: whole-body irradiation has no effect on vaccine-induced resistance in mice. Parasitology 1988;96:49–61.

413 Dessaint JP, Capron A. Interaction of phagocytic cells with immune complexes of anaphylactic antibodies. In Phillips SM, Escobar A, eds. *The Reticuloendothelial System*. New York: Plenum Publishers, 1986.

414 Capron M, Capron A. Rats mice and men—models for immune effector mechanisms against schistosomiasis. Parasitol Today 1986;2:69–75.

415 Knopf PM, Cioli D. *Schistosoma mansoni*: resistance to an infection with cercariae induced by the transfer of adult worms to the rat. Int J Parasitol 1980;10:13.

416 Cioli D, Malorni W, DeMartino C, Dennert G. A study of *Schistosoma mansoni* in thymectomized rats. Cell Immunol 1980;53:246–256.

417 Phillips SM, Bentley AG, Linette G, Doughty BL, Capron M. The immunologic response of congenitally athymic rats to *Schistosoma mansoni* infection. I. *In vivo* studies of resistance. J Immunol 1983;131:1466–1474.

418 Knopf PM, Nutman TB, Reasoner JA. *Schistosoma mansoni*: resistance to reinfection in the rat. Exp Parasitol 1977; 41:74.

419 Phillips SM, Reid WA, Bruce JI, *et al.* The cellular and humoral immune response to *Schistosoma mansoni* in inbred rats. I. Mechanisms during initial exposure. Cell Immunol 1975;19:99–116.

420 Phillips SM, Walker D, Abdel-Hafez SK, *et al.* The immune response to *Schistosoma mansoni* infections in inbred rats. VI. Regulation by T cell subpopulations. J Immunol 1987; 139:2781–2787.

421 Bazin H, Capron A, Capron M, Joseph M, Dessaint J-P, Pauwels R. Effect of neonatal injection of anti-µ antibodies on immunity to schistosomes (*S. mansoni*) in the rat. J Immunol 1980;124:2373–2377.

422 Capron A, Dessaint J-P, Capron M, Joseph M, Pestel J. Role of anaphylactic antibodies in immunity to schistosomes. Am J Trop Med Hyg 1980;29:849–857.

423 Mangold BL, Knopf PM. Host protective humoral immune responses to *Schistosoma mansoni* infections in the rat. Kinetics of hyperimmune serum-dependent sensitivity and elimination of schistosomes in a passive transfer system. Parasitology 1981;83:559–574.

424 Capron M, Rousseaux J, Mazingue C, Bazin H, Capron A. Rat mast cell–eosinophil interaction in antibody-dependent eosinophil cytotoxicity to *Schistosoma mansoni* schistosomula. J Immunol 1978;121:2518.

425 Capron A, Dessaint J-P, Capron M, Bazin H. Specific IgE antibodies in immune adherence of normal macrophages to *Schistosoma mansoni* schistosomules. Nature 1975;253:474.

426 Capron A, Dessaint J-P, Joseph M, Rousseaux R, Capron M, Bazin M. Interaction between IgE complexes and macrophages in the rat; a new mechanism of macrophage activation. Eur J Immunol 1977;7:315.

427 Ford MJ, Bickle QD, Taylor MG. Immunity to *Schistosoma mansoni* in congenitally athymic, irradiated and mast cell-depleted rats. Parasitology 1987;94:313–326.

428 Kigoni EP, Elsas PP, Lenzi HL, Dessein AJ. IgE antibody

and resistance to infection. II. Effect of IgE suppression and late skin reaction and resistance of rats to *Schistosoma mansoni* infection. Eur J Immunol 1986;16:589–595.

429 McLaren DJ, Smithers SR. *Schistosoma mansoni*: challenge attrition during the lung phase of migration in vaccinated and serum-protected rats. Exp Parasitol 1985;60:1–9.

430 McLaren DJ, Delgado VS, Gordon JR, Rogers MV. *Schistosoma mansoni*: analysis of the humoral and cellular basis of resistance in guinea pigs vaccinated with radiation-attenuated cercariae. Parasitology 1990;100:35–44.

431 Davis A. Historical perspectives. Acta Trop 1987;12:8–12.

432 Clarke V de V. Evidence for the development in man of acquired resistance to infection of *Schistosoma* spp. Cent Afr J Med 1986;12(1):1–30.

433 Kloetzel K, da Silva JR. Schistosomiasis mansoni acquired in adulthood: behaviour of egg counts and the intradermal test. Am J Trop Med Hyg 1967;16:167–169.

434 Bradley DJ, McCullough FS. Egg output and stability and the epidemiology of *Schistosoma mansoni*. Trans R Soc Trop Med Hyg 1973;67:491–500.

435 Hagan P, Blumenthal UJ, Chaudri. Resistance to reinfection with *Schistosoma haematobium* in Gambian children: analysis of their immune responses. Trans R Soc Trop Med Hyg 1987;81:938–946.

436 Wilkins HA, Blumenthal UJ, Hagan P, Hayes RJ, Tulloch S. Resistance to reinfection after treatment of urinary schistosomiasis. Trans R Soc Trop Med Hyg 1987;81:21–35.

437 Butterworth AE, Capron M, Cordingley JS, *et al.* Immunity after treatment of human schistosomiasis mansoni. II. Identification of resistant individuals, and analysis of their immune responses. Trans R Soc Trop Med Hyg 1985;79: 393–408.

438 Dessein AJ, Begley M, Demeure C, *et al.* Human resistance to *Schistosoma mansoni* is associated with IgG reactivity to a 37 kDa larval surface antigen. J Immunol 1988;140:2727–2736.

439 Rihet P, Demeure CE, Bourgois CE, Prata A, Dessein A. Evidence for an association between human resistance to *Schistosoma mansoni* and high antilarval IgE levels. Eur J Immunol 1991;21:2629–2686.

440 Khalife J, Dunne DW, Richardson BA. Functional role of IgG subclasses in eosinophil mediated killing of schistosomula of *S. mansoni*. J Immunol 1989;142:4422–4427.

441 Hagan P, Wilkins HA, Blumenthal UJ, Hayes RJ, Tulloch S. Eosinophilia and resistance to *Schistosoma haematobium* in man. Parasite Immunol 1985;7:625–632.

442 Hagan P, Blumenthal UJ, Dunne D, Simpson AJG, Wilkins HA. Human IgE, IgG4 and resistance to reinfection with *Schistosoma haematobium*. Nature 1991;349:243–245.

443 Mott KE, Dixon H. Collaborative study on antigens for immunodiagnosis of schistosomiasis. Bull WHO 1982;60: 729–753.

444 Maddison SE. The present status of serodiagnosis and seroepidemiology of schistosomiasis. Diagn Miocrobiol Infect Dis 1987;7:93–105.

445 Carter C, Colley DG. An electrophoretic analysis of *Schistosoma mansoni* soluble worm antigen preparation. J Parasitol 1978;64:385–340.

446 Pelley RP, Pelley RJ. *S. mansoni* soluble egg antigen. IV. Biochemistry and immunochemistry of major serological

antigens with particular emphasis on MSA. In van den Bossche H, ed. *Biochemistry of Parasites and Host–Parasite Relationships.* Amsterdam: Elsevier, 1977:283.

447 Pelley RP, Pelley RJ, Hamburger J, Peters PA, Warren KS. *Schistosoma mansoni* soluble egg antigens. I. Identification and purification of three major antigens, and the employment of radioimmunoassay for their further characterization. J Immunol 1976;117:1553–1560.

448 Pelley RP, Warren KS, Jordan P. Purified antigen radioimmunoassay in serological diagnosis of schistosomiasis mansoni. Lancet 1977;ii:78.

449 Hamburger J, Lustigman S, Arap Siongok TK, Ouma JH, Mahmoud AAF. Characterization of a purified glycoprotein from *Schistosoma mansoni* eggs: specificity, stability, and the involvement of carbohydrate and peptide moieties in its serologic activity. J Immunol 1982;128:1864–1869.

450 Boctor FN, Nash TE, Cheever AW. Isolation of polysaccharide antigen from eggs. J Immunol 1979;122:39–43.

451 McClaren ML, Lillywhite JE, Dunne DW, Doenhoff MJ. Serodiagnosis of human *Schistosoma mansoni* infections: enhanced sensitivity and specificity in ELISA using a fraction containing *Schistosoma mansoni* egg antigens. Trans R Soc Trop Med Hyg 1981;75:72.

452 Tsang VC, Hancock K, Maddison SE, Beaty AL, Moss DE. Demonstration of species-specific and cross-reactive components of the adult microsomal antigens from *Schistosoma mansoni* and *S. japonicum* (Mama and Jama). J Immunol 1984;132:2607–2633.

453 Rotmans JP, Mooij GW. KCl-extractable surface antigens of *Schistosoma mansoni*: immunological characterization and applicability in immunodiagnosis. Z Parasitenkd 1982;68: 211–226.

454 Nash TE, Lunde MN, Cheever AW. Analysis and antigenic activity of a carbohydrate fraction derived from adult *Schistosoma mansoni*. J Immunol 1981;126:805–810.

455 Deelder AM, Kornelis D, Van Marck EAE. *Schistosoma mansoni*: characterization of two circulating polysaccharide antigens and the immunological response to these antigens in mouse, hamster, and human infections. Exp Parasitol 1980;50:16–32.

456 Cruise KM, Mitchell GF, Garcia EG, Tiu WU, Hocking RE, Anders RF. Sj23, the target antigen in *Schistosoma japonicum* adult worms of an immunodiagnostic hybridoma antibody. Parasite Immunol 1983;5:37–46.

457 Hayunga EG, Mollegard I, Duncan JF, Sumner MP, Stek M, Hunter KW. Development of circulating antigen assay for rapid detection of acute schistosomiasis. Lancet 1986; 2:716–718.

458 Davern KM, Tiu WU, Samaras N, *et al. Schistosoma japonicum*: monoclonal antibodies to the Mr 26 000 schistosome glutathione *S*-transferase (Sj26) in an assay for circulating antigen in infected individuals. Exp Parasitol 1990; 70:293–304.

459 Fu C, Carter CE. Detection of a circulating antigen in human schistosomiasis japonica using a monoclonal antibody. Am J Trop Med Hyg 1990;42:347–351.

460 Simpson AJG, Smithers SR. Schistosomes surface, egg, and circulating antigens. Curr Top Microbiol Immunol 1985;120:205–239.

461 Berggren WL, Weller TH. Immunoelectrophoretic demon-

stration of specific circulating antigen in animals infected with *Schistosoma mansoni*. Am J Trop Med Hyg 1967;16: 606–612.

462 Kestens L, Mangelschots K, van Marck EAE, Gigase PL, Deelder AM. *Schistosoma mansoni*: impaired clearance of model immune complexes consisting of circulating anodic antigen and monoclonal iGg1 in infected mice. Parasitol Res 1988;74:356–362.

463 Carlier Y, Bout D, Capron A. Further studies on the circulating M antigen in human and experimental *Schistosoma mansoni* infections. Ann Immunol 1978;129:811–818.

464 Santoro F, Carlier Y, Borojevic R, Bout D, Tachon P, Capron A. Parasite "M" antigen in milk from mothers infected with *Schistosoma mansoni* (preliminary report). Ann Trop Med Parasitol 1977;71:121–123.

465 De Jonge N, Fillie YE, Hilberath GW, *et al.* Presence of the schistosome circulating anodic antigen (CAA) in urine of patients with *Schistosoma mansoni* or *S. haematobium* infections. Am J Trop Med Hyg 1989;41:563–569.

466 Deelder AM, De Jonge N, Boerman OC, *et al.* Sensitive determination of circulating anodic antigen in *Schistosoma mansoni* infected individuals by an enzyme-linked immunosorbent assay using monoclonal antibodies. Am J Trop Med Hyg 1989;40:268–272.

467 Capron A, Dessaint JP. Vaccination against parasitic diseases: some alternative concepts for the definition of protective antigens. Ann Inst Pasteur/Immunol 1988;139: 109–117.

468 Colley DG, Colley MD. Protective immunity and vaccines to schistosomiasis. Parasitol Today 1989;5:350–354.

469 James SL. Induction of protective immunity against *Schistosoma mansoni* by a non-living vaccine is dependent on the method of antigen presentation. J Immunol 1985;134:1956–1960.

470 Mitchell GF, Garcia EG, Wood SM, *et al.* Studies on the sex ratio of worms in schistosome infections. Parasitology 1990;101:27–34.

471 Dalton JP, Strand M. *Schistosoma mansoni* polypeptides immunogenic in mice vaccinated with radiation-attenuated cercariae. J Immunol 1987;139:2474–2481.

472 Wright MD, Rogers MV, Davern KM, Mitchell GF. *Schistosoma mansoni* antigens differentially recognized by resistant WEHI 129/J mice. Infect Immun 1988;56:2948–2952.

473 Harn D, Quinn J, Oligino L, *et al.* Candidate epitopes for vaccination against schistosomiasis mansoni. In MacInnis A, ed. *Molecular Paradigms for Eradicating Helminthic Parasites.* New York: Alan R Liss, 1987:55–70.

474 Tendler M, Almeida MSS, Magalhaes Pinto R, Noronha D, Katz N. *Schistosoma mansoni*-New Zealand rabbit model: resistance induced by infection followed by active immunization with protective antigens. J Parasitol 1991; 77:138–141.

475 Smithers SR, Hackett F, Omer Ali P, Simpson AJG. Protective immunization of mice against *Schistosoma mansoni* with purified adult worm surface antigens. Parasite Immunol 1989;11:301–318.

476 Horowitz S, Smolarsky B, Arnon R. Protection against *Schistosoma mansoni* achieved by immunization of mice with sonicated cercariae. Eur J Immunol 1982;12:237–332.

477 Hsu SYL, Hsu HF, Svestka KW, Clarke W. Vaccination

against schistosomiasis in mice with killed schistosomula without an adjuvant. Proc Soc Exp Biol Med 1986;181:454–458.

478 Hillyer GV, Garcia Rosa MI, Alicea H, Hernandez A. Successful vaccination against murine *Schistosoma mansoni* infection with a purified 12 kD *Fasciola hepatica* cross-reactive antigen. Am J Trop Med Hyg 1988;38:103–110.

479 Tarrab-Hazdai R, Levi-Schaffer F, Brenner V, Horowitz S, Eshhar Z, Arnon R. Protective monoclonal antibody against *Schistosoma mansoni*: antigen isolation, characterization, and suitability for active immunization. J Immunol 1985; 135:2772–2779.

480 Smith MA, Clegg JA. Vaccination against *Schistosoma mansoni* with purified surface antigens. Science 1985;227: 535–537.

481 King CH, Lett RR, Nanduri J, *et al.* Isolation and characterization of a protective antigen for adjuvant-free immunization against *Schistosoma mansoni*. J Immunol 1987; 139:4218–4224.

482 Kresina TF, Old GR. Antiidiotype antibody vaccine in murine schistosomiasis mansoni comprising the internal image of antigen. J Clin Invest 1989;83:912–920.

483 Zodda DM, Phillips SM. Monoclonal antibody-mediated protection against *Schistosoma mansoni* infection in mice. J Immunol 1982;129:2326–2332.

484 Kelly C, Simpson AJ, Fox E, Phillips SM, Smithers SR. The identification of *Schistosoma mansoni* surface antigens recognized by protective monoclonal antibodies. Parasite Immunol 1986;8:193–198.

485 Smith MA, Clegg JA, Snary D, Trejdosiewicz AJ. Passive immunization of mice against *Schistosoma mansoni* with an IgM monoclonal antibody. Parasitology 1982;83:83–91.

486 Gregoire RJ, Shi M, Rekosh DM, Lo Verde PT. Protective monoclonal antibodies from mice vaccinated or chronically infected with *Schistosoma mansoni* that recognize the same antigens. J Immunol 1987;139:3792–3881.

487 Dissous C, Grzych JM, Capron A. *Schistosoma mansoni* surface antigen defined by a rat monoclonal IgG2a. J Immunol 1982;129:2232–2234.

488 Bickle QD, Andrews BJ, Taylor MG. *Schistosoma mansoni*: characterization of two protective monoclonal antibodies. Parasite Immunol 1986;8:95–107.

489 Yi X, Omer-Ali P, Kelly C, Simpson AJG, Smithers SR. IgM antibodies recognizing carbohydrate epitopes shared between schistosomula and miracidia of *Schistosoma mansoni* that block *in vitro* killing. J Immunol 1986;137: 3946–3954.

490 Vignali DA, Devey ME, Bickle QD, Taylor MG. The role of antibody affinity and titre in immunity to *Schistosoma mansoni* following vaccination with highly irradiated cercariae. Immunology 1990;69:195–201.

491 Lightowlers MW, Rickard MD. Excretory-secretory products of helminth parasites: effects on host immune responses. Parasitology 1988;96:S123–S166.

17 Cestodes

Marshall W. Lightowlers, Graham F. Mitchell, & Michael D. Rickard

PARASITES

Cestodes, commonly known as tapeworms, are platyhelminth (flat-worms, phylum Platyhelminthes) parasites which are medically and economically import-ant owing to their infection of humans and domesticated animals. In common with other platyhelminths, such as the free-living turbellaria and parasitic trematodes (flukes), the cestodes (class Cestoda) are dorsoventrally flattened acoelomate invertebrates. The mouth and digestive system are completely absent in the cestodes, which, as the sexually mature life-cycle stage, are all parasitic in the intestine or its diverticulae. Sexual re-production takes place in the definitive host which harbors the recognizably tapeworm parasite. The devel-opment of larval stages in an intermediate host is required for completion of the life-cycle in all species with a single exception an intermediate host is optional.

The medically and economically important cestodes belong to the subclass Eucestoda, with the majority of the important species being cyclophyllids (order Cyclophyllidea) of the family Taeniidae (Table 17.1). A comprehensive discussion of immunologic investi-gations which have been carried out with all cestodes is beyond the scope of this chapter. The following serves as an introduction to the more important species and those which have been the subject of more extensive investigation in immunology and molecular biology.

Several cestode species which are not taeniid cestodes are important parasites of humans and/or have been the subject of substantial immunologic study. *Diphyllobothrium latum* (order Pseudophyllidea) is a common tapeworm of carnivores which eat freshwater fish. Humans become infected by ingesting insufficiently cooked freshwater fish infected with larval stages (plerocercoid). Eggs released from the adult worm undergo larval development through two intermediate hosts, the second being a freshwater fish. A proportion

Table 17.1 Principal medically and economically important taeniid species used extensively in laboratory studies

Species	Prinicpal intermediate hosts	Metacestode type	Principal definitive host
Echinococcus granulosus	Sheep, goats, cattle, pigs, and other herbivores (humans)	Unilocular hydatid cyst	Dog
Echinococcus multilocularis	Microtine rodents (humans)	Multilocular hydatid cyst	Fox
Taenia solium	Pigs (humans)	Cysticercus	Human
Taenia saginata	Cattle	Cysticercus	Human
Taenia hydatigena	Sheep	Cysticercus	Dog
Taenia ovis	Sheep	Cysticercus	Dog
Taenia multiceps	Sheep (humans)	Coenurus	Dog
Taenia pisiformis	Rabbit	Cysticercus	Dog
Taenia taeniaeformis	Rodents	Strobilocercus	Cat

of infected patients develop a serious megaloblastic anemia as a result of the intestinal tapeworm com-peting with the host for dietary vitamin B_{12}. Other pseudophyllidean cestodes, thought principally to be *Spirometra* species, cause a disease known as sparganosis. This occurs as a result of people either ingesting a copepod infected with the procercoid larval stage, eating insufficiently cooked flesh (amphibian, reptile, bird, or mammal) infected with plerocercoid larvae, or applying raw infected flesh to an open wound, the eye, or other sites of the body.

Other nontaeniid cyclophyllidean cestodes have been

the subject of a substantial amount of immunologic study. *Hymenolepis nana*, the dwarf tapeworm, is a cosmopolitan parasite of humans and rodents. It is unique among the cestodes in being able to complete its life-cycle without necessarily requiring an intermediate host. Ingested eggs hatch and the oncosphere larva penetrates the intestinal villus where it develops into the cysticercoid stage. Mature cysticercoids emerge into the intestinal lumen and develop into adult worms. Eggs released from the adult may either hatch and, in permissive hosts, repeat the life-cycle, or pass out with the feces and develop into the cysticercoid stage following ingestion by larval fleas or grain beetles. The adult worm develops subsequently following ingestion of the infected invertebrate host by humans or rodents. *Hymenolepis diminuta* is another species normally infecting rats, with arthropods, especially the grain beetle (*Tribolium* spp.) as intermediate hosts. Humans become infected through accidental ingestion of infected intermediate hosts. Unlike *H. nana* the infection must pass through the intermediate host and hence there is not the opportunity for development of massive infections as may occur through autoinfection with *H. nana*. *Mesocestoides corti* is a cestode of birds and mammals. Complete details of the life-cycle are unknown. The cysticercoid-type larva, known as a tetrathyridium, is found usually in the serous cavities, especially the peritoneal cavity, where it multiplies by asexual division. Unidentified *Mesocestoides* species have been described infecting humans. In mice, infection with tetrathyridia has been found to cause some pronounced alterations to the immune system of the host and these have been the subject of experimental analyses.

Members of the cyclophyllidean family Taeniidae constitute the cestode parasites of major importance to humans and the species on which the majority of immunology and molecular biology research has been done. *Taenia* and *Echinococcus* species are the etiological agents of cysticercosis and hydatidosis, respectively, in humans and animals. Neurocysticercosis caused by infection in humans with the larval stage of *Taenia solium* and hydatid infection in humans resulting from *Echinococcus granulosus* and *Echinococcus multilocularis* are serious health problems in many countries of the world. Some of the immunologic characteristics of the host–parasite relationship among this group of parasites favor the development of defined antigen vaccines against infection. The recent development of a highly effective recombinant vaccine against *Taenia ovis* infection in sheep (discussed in detail on pp. 450–451), which is being developed commercially, is the first of these vaccines and provides a useful model for the development of recombinant vaccines against other parasites.

The life-cycle of taeniid cestodes involves carnivorous definitive hosts and herbivorous or omnivorous intermediate hosts. The adult worm attaches via the scolex to the wall of the intestine of the definitive host. Eggs containing fully developed, infective oncospheres are released with the feces. When ingested by a suitable intermediate host the keratinous embryophore breaks down under the influence of intestinal secretions, releasing the activated oncosphere. After penetrating the intestinal epithelium the oncosphere migrates either via the venous or lymphatic system to its site of election. Prior to establishment of the larva there may be some migration through the host tissue, resulting in damage to the organs involved. The encysted larval stage (metacestode) then develops into the adult, recognizably tapeworm, form, following ingestion by a suitable definitive host. Asexual reproduction of the larval stage in hydatid cysts and the coenurus leads to the production of multiple infective protoscoleces derived from an individual infecting oncosphere. Hydatid cysts occur as unilocular cysts when infection is with *E. granulosus*, or multilocular (alveolar) cysts in the case of *E. multilocularis*. Unilocular cysts are characterized by being limited by a cyst wall, with the production of protoscoleces and daughter cysts internally within the fluid-filled lumen of the parent cyst. The protoscoleces and daughter cysts are capable of developing into new cysts following rupture of the parent cyst or injection into another host. Multilocular hydatids are not limited by a cyst membrane, and the spongy mass buds exogenously, leading to the continued infiltration of host tissues and metastasis from the parent cyst. Control of infection with taeniid cestodes in humans may, theoretically, be achieved through alterations in human behavior patterns and public health measures. However, cysticercosis and hydatidosis continue to be serious parasitic diseases of humans and may actually be increasing in prevalence [1–3].

HOST–PARASITE RELATIONSHIP

Relatively little is known of the immunologic relationship between cestode parasites and their definitive hosts. Intimate contact between the scolex and the tissues of the gut wall provides an opportunity for direct immunologic interaction, and the detection of antibodies against scolex antigens in infected hosts indicates that this interaction occurs. Except in the case of *Hymenolepis* species, these immune responses have little host-protective effect against either the continued presence of parasites from an initial infection or the establishment of new parasites from a challenge infection. The appearance of specific antioncospheral antibodies in the circulation of dogs infected with *E. granulosus* or *Taenia* spp. [4,5] suggests that some taeniid eggs may hatch in the gut of the definitive host and penetrate the host tissues sufficiently to induce the antioncosphere immune response.

In the intermediate host there is a greater opportunity for immunologic interaction between the host and the parasite which is encysted in the host tissues or peritoneal cavity. The surface of the metacestode is covered in hair-like projections called microtriches, beneath which there is a syncytial tegument. The microtriches are bound externally by a plasma membrane and an outermost layer, the glycocalyx, composed predominantly of carbohydrate. In *Echinococcus* species, the germinal layer, which forms the boundary of the fluid-filled cavity, has a tegumental structure. Microtriches project outwards into an overlying acellular laminated layer. The presence of a parasite-derived laminated layer which is composed largely of carbohydrate and protein is characteristic of hydatid cysts and forms the cellular barrier between the host and the parasite [6]. In *E. granulosus*, a host-derived fibrous capsule develops around the laminated layer. Although cellular contact between host and parasite may be prevented by the laminated membrane, there is ample descriptive and experimental evidence [6–8] for the passage of intact host macromolecules into hydatid cysts. Host macromolecules have also been found within the metacestode of other taeniid species [8]. The host–parasite interface has been studied closely in the *Taenia taeniaeformis* rat–mouse system [9–11]. Within the first 24 h following infection the oncosphere surface becomes covered in a dense mesh of microvilli, often branched and extending to 40 μm from the parasite surface. After 8 days the multicellular oncosphere is transformed into a fluid-filled vesicle having a syncytial tegument, subtegumental cell bodies, and microtriches. Despite the development of an intense granulocytic infiltrate around some developing metacestodes, Engelkirk & Williams [10] could not find evidence of damage to the parasite and concluded that the vigorous cellular response of the host was ineffective in either containing the expansion of the parasite or compromising the integrity of its surface membrane. Fibroblastic activity subsequently forms a capsule around the metacestode, which survives often for the life of the host. In the taeniid cestodes, an initial infection typically leads to a state of concomitant immunity. In some species there is evidence that immunity to a challenge infection wanes despite the continued presence of metacestodes from the initial infection [12,13]. The expression of an effective immune response against established metacestodes has not been proved, although recent evidence indicates that postoncospheral larvae of *T. taeniaeformis* may be susceptible to immune attack [14]. The length of time for which the metacestodes survive in the intermediate host varies considerably between species, with the majority of larvae in some, e.g., *T. ovis* and *Taenia saginata*, normally not surviving for extended periods. There is no unequivocal evidence that this attrition of viable parasites is immunologically based.

IMMUNOPATHOLOGY

Symptoms of immunopathology are not among the clinical features typically associated with cestode parasitism. However, closer analysis of the immunologic consequences of metacestode infections in humans and experimental animals has revealed substantial alterations in the lymphoid organs and in immunocompetence, manifest in a variety of pathologic consequences.

Membranous glomerulonephritis occurs in patients with hydatid cysts [15,16], with the glomerular lesions associated with deposition of IgG, IgM, and IgA, complement components, and hydatid antigens. Circulating hydatid antigens and immune complexes are frequently associated with unilocular hydatidosis in humans [17–24] and assays for their detection have been found to be valuable for diagnosis of infection [22,24]. Infections with *Echinococcus*, *Taenia*, or *Mesocestoides* species in animals also produce readily detectable levels of circulating parasite antigens and immune complexes [18,25–27]. In addition to the pathology caused by deposition of these complexes in the glomeruli, these released parasite factors are likely to be involved in many of the wide variety of pathologic consequences of metacestode infections, including alterations in immunocompetence (discussed below), induction of malignant sarcoma [28] and gastric and intestinal mucosal hyperplasia [29–33], inhibition of testosterone production [34], growth-promoting effects on the host [35], and alterations in collagen metabolism [36].

Immune mechanisms are implicated in the development of an extremely dense granulomatous infiltration in alveolar echinococcosis which is responsible for destruction of liver parenchyma and clinical complications. The parasite mass is characterized by a peripheral zone of dense fibrous tissue infiltration, leading to irreversible acellular keloid scar-like fibrosis [37]. Vuitton *et al*. [36] have suggested that this fibrosis is mediated by immune mechanisms in the host and note that enhanced collagen deposition could be detected in the liver tissue of patients with very limited *E. multilocularis* lesions in another liver lobe. Experimental alveolar echinococcosis in mice also induces an intense inflammatory infiltration [38]. The characteristics of the development of the infection as either progressive or restrictive is determined by the host strain, with the inflammatory response being implicated in limiting the growth of the parasite in less-permissive hosts [38,39]. One of the correlates of chronic inflammation associated with *E. multilocularis* infection in humans and experimental animals is the deposition of amyloid derived from serum amyloid A protein in the liver and kidney [40,41]. The etiology of this aspect of the disease is unclear and it is not known whether this clinical manifestation of the disease is an immunopathologic consequence of the response to the parasite. In mice the period of amyloidogenesis corresponds to a decrease in the T cell population of peripheral lymphoid organs [42].

An intense inflammatory reaction may also develop around the encysted cysticercus of *Taenia* spp. This reaction typically subsides following the initial establishment of the parasite in the host tissue but may resume at some later time around a proportion of the larvae associated with the death of the parasites [43]. It is not known whether the inflammation and parasite death are related as cause and effect or whether the death of the parasite leads to the release of parasite-derived substances which induce local inflammation. Biologically active peptides produced following activation of complement by parasite factors [44–46] or arachidonic acid metabolites [47] are examples of parasite molecules which could be released after the death of the parasite and contribute to inducing inflammatory responses. Inflammatory reactions surrounding *T. solium* cysticerci in humans have been described and are believed to be associated with the death of the parasite [48]. In the natural intermediate host, the pig, *T. solium* cysticerci are frequently surrounded by a chronic granulomatous reaction containing many plasma cells; however, this reaction does not lead to detectable damage to the encapsulated parasite [49].

Immediate-type hypersensitivity reactions can be demonstrated in hosts parasitized with cestode larvae. A proportion of patients infected with *E. granulosus* hydatid cysts develop signs of nonspecific allergy and the majority are positive in skin tests for specific immediate hypersensitivity [50,51], basophil degranulation assays [52,53], and for the presence of specific circulating IgE [53,54]. Similar responses occur in sheep infected with *E. granulosus* [55]. Allergic sensitization in hydatid patients appears to have a sequel in the development of anaphylaxis-like reactions in some patients following the rupture of a cyst [56] or chemotherapy [57]. Similar reactions have also been shown experimentally following the intravenous injection of hydatid cyst fluid into infected animals [58].

The chronic parasitism which ensues following infection with larval cestodes typically occurs in the face of readily detectable immune responses to the encysted larvae. Several mechanisms have been proposed for the means by which the parasite evades immune attack by the host, and some of these mechanisms have correlates with pathologic changes in the structure of the lymphoid organs and immune competence. *Echinococcus multilocularis* infection in permissive strains of mice results in gross depletion of T-dependent areas of the lymphoid organs and thymic involution correlating with the period of rapid growth of the cyst mass [59]. Alterations to lymphoid tissue architecture also occur in rodents infected with *E. granulosus* [60,61] and *T. taeniaeformis* [31]. These gross alterations in lymphoid organs are frequently associated with altered and often depressed immune responses. In mice infected with *E. multilocularis*, cell-mediated immune responses to parasite antigens are markedly suppressed during proliferative growth of the parasite [62] while immune responses to oxazolone and skin allografts are unaltered, despite the depletion of T cells in the lymphoid tissues. Host neutrophils and macrophages respond vigorously *in vitro* to both parasite and nonspecific chemoattractants during the early phase of *E. multilocularis* infection in mice, but transition to the progressive phase correlates with inhibition of leukocyte chemotaxis [63]. Alveolor echinococcosis in humans is also associated with impaired T cell function and reduction in the numbers of peripheral blood B cells [64]. Mice infected with *E. granulosus* exhibited depressed immune responses to heterologous antigens [65] and mitogens [61]. Immunodepression has been described in mice infected with *Taenia crassiceps* [66,67] and *T. taeniaeformis* [67–69]. Reduced immune responses to sheep red blood cell (SRBC) in *E. granulosus* infection can be transferred to uninfected mice with mesenteric lymph node cells [65]. Similar depression in anti-SRBC responses in *T. crassiceps* infection can be restored *in vitro* by addition of activated peritoneal cells, suggesting that the immunodepression in this case is the result of alterations in the activity of accessory cells [70]. In another study, using *M. corti*, decreased responses to SRBC injected by the intraperitoneal route could be excluded to a retention of antigen in the peritoneal cavity and its destruction in that site [71].

Following infection with *T. taeniaeformis* in mice, a proportion of the early developing larvae are killed in the liver prior to the appearance of detectable levels of antibody, while the remaining larvae continue to develop to mature strobilocerci. During this early period, spleen cells show a pronounced inhibition of mitogen-induced proliferation and IL-2 production *in vitro* [68]. Suppression is less evident in more resistant strains of mice [69]; however, it is not possible to differentiate between this being the cause or effect of the reduced number of viable parasites in these strains.

Polyclonal activation of lymphocytes and auto-antibody production are coming to be recognized as immunopathologic correlates of both hydatidosis and cysticercosis. Investigations by Sealey *et al.* [72] sought to find an explanation for the presence of many plasma cells in chronic granulomatous reactions around *T. solium* cysticerci. An extract of the parasite was found to be mitogenic for mouse lymphocytes *in vitro* which was independent of any contaminating bacterial lipopolysaccharide (LPS). *Echinococcus granulosus* protoscoleces also induce mitogenic activity in naïve mouse lymphocytes [73,74]. This characteristic of cestode parasites has been confirmed and extended by Judson *et al.* [75] who showed that parasite extracts, cyst fluid, or culture supernatants from *E. granulosus*, *Taenia hydatigena*, *Taenia multiceps*, *Taenia pisiformis*, *Monezia expansa*, *Anoplocephala perfoliata*, and *H. diminuta* are mitogenic for ovine lymphocytes. In the case of *E. granulosus* protoscoleces, coculture which allows contact between the murine spleen cells and viable protoscoleces is required to induce proliferation of the lymphocytes [73,76]. Both unprimed T and B cells respond, with T cell activity requiring the presence of macrophages in the culture. In addition, peritoneal macrophages from mice primed by intraperitoneal injection of protoscoleces prior to harvesting presented a mitogenic stimulus to naïve T cells, provided that the cells had identity at the I region of the H-2 complex. Unpublished observations of Cox *et al.* [77] have reported evidence of polyclonal B cell activation in mice within 4 days of infection

with *E. granulosus* protoscoleces. The number of cells secreting antibody to trinitrophenyl- and enzyme-treated autologous red cells was claimed to have increased to a magnitude similar to that induced by other known polyclonal B cell activators, such as LPS. Additional evidence in favor of the induction of polyclonal B cell activation in cestode infections comes from the discovery of autoantibodies in patients infected with *E. granulosus* or *E. multilocularis* (Table 17.2). At present there are insufficient data to allow an assessment of how commonly autoimmune phenomena are associated with cestode infections or the mechanism responsible for the generation of these responses. Sera from patients with collagen diseases and tissue autoantibodies have been reported to crossreact with *Echinococcus* antigens [82]. Pini *et al.* [78] observed the presence of antismooth-muscle antibodies in the sera of hydatid patients but were unable to establish a direct relationship between hydatid antigens and smooth-muscle antigens. Antisera raised in rabbits against the hydatid antigens showed no crossreactivity with smooth-muscle antigens.

The mechanisms which bring about the pathologic alterations in the immune responses of the hosts of cestode larvae have not been defined. Several mechanisms have been proposed and are likely to be involved

Table 17.2 Investigations of the presence of autoantibodies in the sera of patients with hydatidosis

Parasite	Antigens tested	Reactivity with patients' sera	[Ref.]
Echinococcus granulosus	Immunoglobulin	−	[78]
	Nuclear	−	[78]
	Smooth mucle	+	[78]
	Mitochondria	−	[78]
	Erythrocyte	+	[79]
	Immunoglobulin	−	[24]
	MHC class I	+	[80]
	MHC class II	+	[80]
	β_2-Microglobulin	−	[80]
Echinococcus granulosus or *Echinococcus multilocularis*	Double-stranded DNA	+	[81]
	Single-stranded DNA	−	[81]
	Histone	+	[81]
	Actin	+	[81]
	Desmin	+	[81]
	Vinmentin	+	[81]
	Keratin	+	[81]
	Tubulin	−	[81]
	Laminin	−	[81]
	Collagen type 1	−	[81]

at least in part. Principal among those are the direct and indirect effects of excreted or secreted products of parasites. The encysted strobilocerci of *T. taeniaeformis* secrete glycosaminoglycan molecules which are believed to activate the complement cascade within the local microenvironment of the cyst and thereby play an important role in evasion of immune attack [45,83,84]. The wider implications of this secretory product are unknown. Certainly alcian blue staining material has been found streaming through the hepatic tissue at a substantial distance from the parasite in mice early in infection with *T. taeniaeformis* [85]. Cestode protease inhibitors [86–88] may also be involved in suppression of immune responses. A purified proteinase inhibitor from *T. taeniaeformis* larvae inhibits lectin-induced mitogenic responses of rat spleen cells *in vitro* [87] and inhibits IL-2 generation in murine lymphocytes and IL-1-induced proliferation in murine thymocytes [89]. The contribution of this particular molecule (termed taeniaestatin) to the suppression of immune responses in *T. taeniaeformis* is unclear since secretion of the protease inhibitor by the parasite has not been demonstrated. Excretory–secretory products of *T. taeniaeformis* depress lectin-induced proliferative responses of rat splenocytes *in vitro* through the induction of a suppressor cell population [68]. Release of parasite antigens

generating circulating antigen and immune complexes are also likely to lead to a wide spectrum of pathologic and immunologic consequences in addition to the more direct effects in the kidney [26]. The products of arachidonate metabolism by *T. taeniaeformis* larvae may be involved in inducing immunosuppression [47].

The occurrence of autoantibodies to both class I and class II major histocompatibility complex (MHC) gene products in patients with *E. granulosus* infection is particularly interesting in view of the crucial role that these cell surface receptors play in the induction of immune responses. It is tempting to speculate that these antibodies may subvert the host's immune responses in such a way as to promote the long-term survival of the parasite. Should these antibody specificities also be found to be generated in other cestode infections, this possibility would be amenable to experimental analysis. *Echinococcus granulosus* parasites express a gene encoding a protein which exhibits a high degree of amino acid sequence homology with a human protein termed cyclophilin (Fig. 17.1) [90], now known to be peptidyl prolyl *cis–trans* isomerase [91]. The binding of cyclophilin to the fungal metabolite cyclosporin A has profound suppressive effects on T cell function [92,93]. If the parasite were to contain a natural ligand for cyclophilin which had some of the immunosuppressive activities of

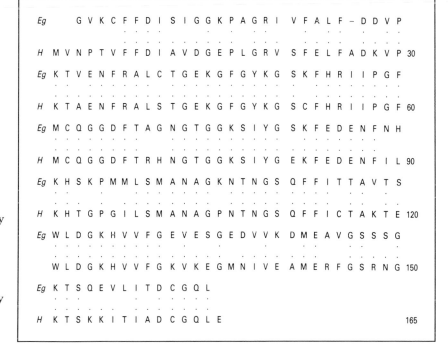

Fig. 17.1 Amino acid sequence homology between an *Echinococcus granulosus* gene (Eg) and human cyclophilin (H) [90]. Amino acid numbering is indicated on the right-hand side for the human protein, beginning from the amino-terminal amino acid. Complete homology is indicated by double points, and single points indicate conservative changes in amino acid.

cyclosporin A, a role for such a ligand could exist in suppressing host immune responses and the survival of the parasite. In view of the detection of autoantibodies in patients with hydatid cysts, it is interesting to note that a mechanism has recently been discovered for the way in which cyclosporin A interferes with the deletion of autoreactive T cell clones during thymic maturation, leading to the appearance of autoimmune disease [94].

INNATE RESISTANCE TO INFECTION

Within an outbred population, variation in innate resistance to parasite infection is important in determining the prevalence and intensity of infection in that population. As such, the factors which influence innate resistance may play crucial roles in the success of parasite control or vaccination programs. For these reasons a detailed knowledge of the factors affecting innate resistance and their mechanisms of action may provide clues as to what determines susceptibility to infection and hence possible mechanisms for inducing resistance.

Innate resistance can be defined as resistance to an initial infection with the parasite. Difficulties arise in determining meaningful measurements of innate resistance. The measure should include not only the number of parasites initially establishing but also their longer term survival as well as their infectivity for the definite host. In some instances there may be no real difference between innate and acquired immunity, since specific immune responses elaborated during the course of an initial infection can influence the number of parasites that establish successfully, as has been shown to occur with *T. taeniaeformis* infection in resistant strains of mice [95].

A wide variety of factors have been shown to influence innate resistance to cestode infections [96]. In addition to numerous host-related factors, parasite strain or isolate may also affect markedly the apparent spectrum of resistance/susceptibility exhibited by host animals. For example, wide variation in the infectivity for mice and rats has been described for different isolates of *T. taeniaeformis* by several workers [97–100].

Physical and biochemical incompatibilities are likely to be major factors in the determination of host range for any parasite species and may also contribute to degrees of resistance evident among individuals within an otherwise susceptible host species. Composition of host bile

has been shown to influence the survival of protoscoleces of *E. granulosus* [101] and has been suggested to be involved in the susceptibility of different host species to taeniid cestodes [102]. Musoke *et al.* [103] suggest that a lack of intestinal proteolytic enzyme activity in young rats may account for the resistance of rats of less than 3 weeks of age to *T. taeniaeformis*. The physiologic state of the host may influence the outcome of cestode infection. Williams *et al.* [100] found that the infectivity of inbred Wistar rats obtained from different suppliers varied markedly in their susceptibility to *T. taeniaeformis* and that this was due to different degrees of stress suffered by the two groups, depending on the length of time the animals had been in transit between the supplier and the laboratory. Host sex and age also affect innate susceptibility to infection in both the intermediate and definitive hosts of cestode parasites [96] and these are important factors to consider in designing experimental investigation of infection/resistance. In an article whose conclusions are succinctly stated in the title ''Evidence that a hydatid cyst is 'seldom as old as the patient,''' Beard [104] examined the epidemiologic data on human hydatid disease during 10 years of the hydatid control campaign in Tasmania, Australia. He concluded that adults were susceptible to hydatid infection and that the period between infection and diagnosis of hydatid cysts was in many cases relatively short. The results of reports on the effect of increasing age on increasing resistance to cestode infection in field animals must be examined with caution in order to differentiate between true age-related resistance and the acquisition with time of prior infection with the same or related parasite species. However, the degree to which host age can influence susceptibility to infection has been demonstrated clearly in the *T. taeniaeformis*/rodent systems. The susceptibility of the rat host to infection is limited to an age window, with both younger and older animals being resistant to infection [103,105]; older mice are also refractory to *T. taeniaeformis* infection [106–108]. In resistant Swiss White mice, Turner & McKeever [107] observed a reduced proportion of oncospheres successfully penetrating the gut of older mice. This was not due to any difference in the proportion of oncospheres hatching in the intestine of the older animals. The histologic characteristics of the host's cellular responses to infection also differed between younger and older animals. The older mice displayed a rapid leukocytic invasion of the vicinity of the parasite within 48 h of infection whereas this infiltrate was absent in younger mice. In this case, however, other factors were also involved in determining resistance/ susceptibility, since even in the younger mice of this strain the oncospheres failed to undergo postoncospheral reorganization and died.

Nonspecific immune mechanisms play an important role in determining the susceptibility/resistance of individuals to most infectious diseases, including parasites, and it is not surprising that there is evidence that these mechanisms are also involved in resistance to cestode larvae. Protection has been demonstrated using Bacillus Calmette-Guerin (BCG) against *E. granulosus* protoscoleces in gerbils [109], *E. multilocularis* protoscoleces in cotton rats [110,111], egg-induced infection with *T. taeniaeformis* in mice [112], and survival/proliferation of *M. corti* tetrathyridia in mice [113]. Other nonspecific immunostimulants have also been found to increase resistance to infection with larval cestodes [114,115]. The mechanism of action of these immunostimulants has been associated with their stimulatory activity on macrophages and other cells of the reticuloendothelial system. At least one of the mechanisms by which BCG treatment leads to resistance to *M. corti* in mice shows specificity for *M. corti*. White *et al.* [113] showed that mice could be rendered resistant to *M. corti* following two immunizations with BCG and that the protection could be adoptively transferred with spleen cells from animals exposed to both parasite and BCG but not from mice exposed to either one alone.

Mitchell *et al.* [95] proposed a mechanism to account for mouse strain variation in susceptibility to *T. taeniaeformis*. Although the susceptibility was measured against a primary infection, the rate of appearance of a specific acquired immune response was found to govern the success or failure of invading larvae to establish in the liver of the host. Susceptible mice produce protective antibody slightly later in the course of infection but after the parasite's protective mechanisms are established. The dose of eggs given to mice may also have a marked effect on the apparent innate susceptibility of a strain of mice to *T. taeniaeformis* [95]. Inoculation of increasing numbers of eggs to relatively resistant C57BL/6 and BALB/c mice results in the development of fewer liver larvae, possibly because the larger number of eggs led to an accelerated immune response and hence the death of a greater proportion of the larvae prior to their developing effective immune evasion mechanisms.

While the rate of production of protective antibodies against a primary infection with *T. taeniaeformis* would appear to account for the death of a proportion of the

invading larvae, other mechanisms must account for the destruction of larvae which occurs in resistant mouse strains within the first few days of infection [116]. Davis & Hammerberg [117] used immunohistochemical techniques to investigate the role played by complement in the destruction of *T. taeniaeformis* larvae prior to 6 days postinfection, when no specific antibody against the parasite could be demonstrated. Deposition of C3 was demonstrated on *in vitro* hatched and activated oncospheres as well as *in vivo* on the surface of 2-day-old liver larvae in both susceptible and resistant strains of mice. Administration of anticomplementary cobra venom factor to the mice before oral infection with eggs resulted in increased survival of larvae, especially in the normally resistant BALB/c mice. The alternative pathway was implicated in the deposition of C3. Polysulfated polysaccharides which activate complement by the alternative pathway have been identified in later larvae [84]. The extent to which complement binding directly damages the developing parasite is questionable, since in Davis & Hammerberg's study larvae in both the susceptible and resistant strains were affected. Damage may be associated with the secondary effects of adherent leukocytes and this mechanism is supported by the results of experiments by Engelkirk *et al.* [118], who demonstrated rapid adherence of leukocytes, principally eosinophils, to the surface of *T. taeniaeformis* larvae. Adherence was dependent on the presence of serum components having the characteristics of complement. Adherence occurred with cells and serum from normal animals and was not enhanced by use of either sera or cells from immune mice. The adherence of these cells was shown to cause substantial damage to the distal tegument of the parasite, suggesting that the early infiltration of cells around larvae in innately resistant mice [107,116] could be directly involved in the killing of early postoncospheral larvae.

Genetic factors related to the MHC immune responses (Ir) genotype as well as other MHC linked or nonlinked genes may influence the innate susceptibility of individuals to larval cestode infection. There is an apparent variability in susceptibility of people to *E. granulosus* infection, where some individuals in close contact with infected dogs apparently fail to contract infection. Investigations into genetic determinants or correlates of susceptibility to hydatidosis or cysticercosis could provide valuable information on whether individuals of particular genotypes display innate susceptibility/ resistance to these important parasites.

IMMUNE RESPONSES AND RESISTANCE TO INFECTION IN THE INTERMEDIATE HOST

In no other helminth infection are the beneficial effects of host antiparasite immune responses so obvious as in larval cestode infections. Most information on immune mechanisms of host resistance comes from the *T. taeniaeformis*/mouse and *T. taeniaeformis*/rat systems, these being natural host–parasite relationships. All evidence points to a dominant role for complement-fixing antibody in mediation of resistance to infection and directed against the oncosphere and early stages of larval development. A role for eosinophils in resistance (e.g., Letonja & Hammerberg [119]), presumably influenced by T cell-derived lymphokines such as IL-5, has been suspected for some time but never demonstrated. The same applies to cell-mediated immunity (CMI), although the very term has become cumbersome and of limited value in recent times. For example, there is obviously a need to differentiate clearly antibody-dependent cell-mediated cytotoxicity (ADCC), cytotoxic $CD8^+$ T cells, and $CD4^+$ T cells of different lymphokine-producing subsets in descriptions of CMI. In this section we will present an update of resistance to infection against larval cestodes, emphasis being on the broad picture; references to the subtleties and nuances as well as the historic perspective can be found in Rickard & Williams [96], and vaccination effects are covered later.

Resistance to reinfection, or concomitant immunity, is very prominent in larval cestode infections and was demonstrated in mice and rabbits in the 1930s using *T. taeniaeformis* [120] and *T. pisiformis* [121], respectively. Similar resistance can be demonstrated in sheep or cattle against *T. hydatigena*, *T. ovis*, or *T. saginata* (reviewed by Rickard & Williams [96] and Lloyd [122]). A variety of crude antigen preparations have been used to demonstrate both humoral immunity and CMI following exposure. Little can be said from all these data except that the serology suggests the existence of dominant stage-specific antigens/epitopes as well as crossreactive if not shared antigens/epitopes between life-cycle stages of the one parasite.

Demonstrations of successful passive transfer with serum again date from the 1930s, even though protective effects of transfer of colostrum or serum from *naturally infected* domesticated animals has often been inefficient [122]. This inefficiency probably reflects the level of infection, and frequency of natural challenge infections, and thus the titer of antibody in the donor. In con-comitant immunity *sensu strictu*, the established parasite maintains immune responses that are effective against the establishing parasite [123,124]. However, it is highly likely that repeated challenges and destruction of oncospheres and invading larvae is a potent means of maintaining high-level resistance to reinfection in cysticercosis.

In the mouse, it is clear that a combination of IgG1 and IgG2 isotypes is an efficient means of transferring resistance to naïve recipients against *T. taeniaeformis*, whereas in rats, IgG2a is probably sufficient [95,103,125]. Serum IgG and colostral IgA can protect mice when given orally [126]. IgA antioncosphere antibodies must block penetration of the intestinal wall by oncospheres, although resistance can be expressed against *Taenia* spp. and *E. granulosus* during penetration or soon after (see Rickard & Williams [96] and Lloyd [122]). Passive transfer of protective serum in rats does not prevent oncospheres from a challenge infection reaching the liver [127]. However, the oncospheres fail to undergo reorganization and die shortly after reaching the liver.

No essential role for IgE antibodies has been demonstrated in resistance to larval cestodes, although such antibodies may be helpful in increasing vascular permeability and thus access of other effector antibodies and/or cells to the invading larva [128]. The idea that IgG1 antibodies to surface antigens of larval cestodes [129,130] may have blocking effects in cysticercosis [131–133] has not been substantiated [134] and Ig complexes are shed readily, at least *in vitro* [135].

Numerous observations with regard to the effects of age, host genotype, sex, and immune status (including injection of nonspecific immune stimulants such as BCG) on resistance to infection can be readily ascribed to differences in the rate at which antibody responses are induced following administration of oncospheres. This is best illustrated in the analysis of mouse strain variation in resistance to first infection. Some mouse strains (e.g., C57BL/6) when exposed to oncospheres of *T. taeniaeformis* develop no or few cysts in the liver, whereas others, such as C3H/He, are highly susceptible [95,108,133,136,137]. Resistance is dominant in F1 hybrids and polygenic genetic control is almost certainly involved rather than oligogenic or unigenic control. Good evidence exists that the essential difference between genetically susceptible and resistant mice is in the rate at which T cell-dependent host-protective IgG antibodies appear in the early stages of infection. At a histologic level this is associated with the accumulation of cells around degenerating larvae in the liver, eosinophils predominating in these lesions. It seems that genetically resistant mouse strains have the capacity to produce sufficiently high titers and/or appropriate types of antibodies before the metacestode develops the capacity to resist aggressive immune attack (see below). The antigens and antibody specificities involved remain unknown, despite some attempts at identification [138–141]. Passive transfer of resistance even from chronically infected genetically susceptible mouse strains can be absolute, even in hypothymic nude recipients of serum. Challenge in such passive transfer studies must be performed soon after serum injection. Passive transfer from mice immunized with oncospheres and totally resistant to challenge infection has been less successful [141]. It was anticipated that challenge in vaccinated mice would lead to a rapid increase in titer of host-protective antibody, although this has not been formally demonstrated. Nevertheless, the possibility that the vaccine need only sensitize for an accelerated host-protective response on challenge with *T. taeniaeformis* oncospheres is appealing. In this regard it is of particular interest that genetically resistant mouse strains challenged in primary infection with a low number of parasites are more susceptible than mice challenged with a high number of parasites. High-dose egg challenge presumably leads to an accelerated antigen dose-dependent immune response compared with the low-dose ("trickle") challenge [95,136].

Several controversial issues in the immunology of larval cestodes can be raised. Firstly, the question of responsiveness and consequences of infection in young animals. Evidence of neonatal unresponsiveness in calves exists, as does evidence for persistent susceptibility to reinfection in infected calves, in some but not all studies (Gallie & Sewell [142,143] and Froyd [144,145]). In lambs, data on neonatal unresponsiveness/tolerance induction is inconclusive (see Lloyd [122] for a summary of the New Zealand work on this point). Comparison of the specific antibody response in sheep and other natural intermediate hosts to infection with *E. granulosus* with the level of response in humans sugests that these hosts may be tolerant or in some other way unable to respond. Immunization with hydatid antigens without adjuvant in infected sheep does, however, stimulate a rapid anamnestic response, even if the antigens are derived from the animal's own cysts [146], suggesting that the hyporesponsiveness is more likely to be due to antigen sequestration.

The dwarf tapeworm of humans, *H. nana*, infects mice and rats and because the life-cycle is very different from most other cestodes the immunology of infection has some unique features. In the mouse, *H. nana* exists as a tissue cysticercoid and as an intestinal adult worm. Evidence for stage-specific immunity is very strong. Thus the invading oncosphere derived from eggs elicits an immune response in mice directed against the tissue phase that derives from eggs (an early response). There is a later response during infection that manifests as resistance to cysticercoid challenge; in all probability this late response is most efficiently induced by luminal worms. Controversial issues of the system, as well as the substantial body of information on the biology of *Hymenolepis* infection, are reviewed by Ito & Smyth [147]. In terms of quantitative immunology and detailed analysis of immune mechanisms of resistance, immuno-chemical analyses have identified dominant stage-specific antigens [148].

Other unusual cestodes are *T. crassiceps* and *Mesocestoides* spp. that are transferrable as metacestodes between mice (in which host they multiply in the peritoneal cavity). Analysis of infection with these parasites in mice has provided interesting data for comparison with the results of studies on larval cestodes of economic or medical importance. With regard to TH-2 cell-dependent (i.e., IL-4- and IL-5-dependent) eosinophilia and antibody isotype responses in mice, *Mesocestoides corti* in particular is proving very useful [149–152].

Evidence that complement-fixing antibody isotypes have protective effects *in vivo*, but only against early larval stages, has led to the notion that established cystic metacestodes protect themselves from such attack by elaboration of anticomplementary activities. Some effort has been expended in analysis of the molecular basis of this form of immune evasion but a very incomplete picture is available. A role for highly sulfated polysaccharides and proteoglycans with local consumption of complement in the aqueous phase has been proposed [84,153]. Larval products can inhibit mitogenic responses in lymphoid cell suspensions (e.g., Burger *et al.* [68] and Leid *et al.* [87]) and a 19.5 kD molecule of *T. taeniaeformis* metacestodes has been identified and termed "taeniaestatin," i.e., a proteinase inhibitor [86,88]. It inhibits IL-2 production [89] and aggregation of inflammatory cells [154]. Whether immunologic (? antibody) neutralization of this molecule will reverse immune evasion in established metacestodes of *Taenia* spp. has yet to be demonstrated. If it does, then the way is open to therapeutic vaccination against larval cestodes.

Because eosinophils and other cell types in the presence of serum (? complement) can attach to, and damage, larval cestodes *in vitro* [118,155,156], it has been proposed that protection against such damage *in vivo* is related to the presence of a "perilarval amorphous layer" [137]. The relationship of this to the polysaccharides and proteoglycans referred to above remains unknown. One thing is clear, larval cestodes are very susceptible to complement attack, at least *in vitro*, the most recent and most comprehensive of a large number of related experiments being conducted by Conder *et al.* [157] using *T. taeniaeformis* oncospheres and more mature metacestodes.

IMMUNE RESPONSES AND RESISTANCE TO INFECTION IN THE DEFINITIVE HOST

In their definitive hosts cestodes are lumen-dwelling parasites. This does not, however, preclude effective stimulation of the host's immune system. Gut presentation can be an effective route of antigen delivery, as evidenced by the effectiveness of living bacterial and viral vaccines and immune responses in some circumstances to nonliving antigens given via the gut [158]. Antigen presentation may also occur directly into the host tissues from the scolex which, particularly in *Echinococcus* species, can penetrate into the crypts of Lieberkühn [159]. Stimulation of immune responses in the definitive host may also occur due to the hatching, activation, and penetration of oncospheres from eggs released from the adult cestodes [4]. The successful rejection of intestinal nematodes under some circumstances attests the potential for these gut-level immune responses to be host-protective [160].

A substantial amount of research has been performed on immunity to *Hymenolepis* species in laboratory rodents which has revealed that host-protective immunity can occur against these lumen-dwelling worms (reviewed by Ito & Smyth [147]). Different host species or strains vary markedly in their suspectibility to infection with the different parasite species. Host-protective immune responses against the adult worm may be expressed as inhibition of growth and fecundity or expulsion of worms. This is ample evidence that host treatments which reduce immunocompetence substantially impair those protective responses (reviewed by Ito & Smyth [147] and Rickard [161]). Impaired immune responses to *H. nana* adult worms in nude mice are restored by thymic grafts [162]. Also, rejection of challenge infections may be enhanced, implicating immunologic mechanisms in the protective response [163–166]. Although there is strong evidence in favor of protective immune responses against the *Hymenolepis* adult worms, the precise identity of these immune mechanisms is unclear. In those host–parasite relationships where there is no or only poor expression of protective immunity it still remains unclear if this can be attributed to immune evasion by the parasite or failure of the host to express the appropriate immune responses [161].

Investigations of specific antibody responses in definitive hosts of other cestodes have confirmed the immunogenicity of these lumen-dwelling parasites. Specific serum antibodies have been found to be produced in response to infection with *D. latum*, *T. saginata*, or *T. solium* in humans [167–172], *Taenia* species in dogs and cats [4,173–175], *E. granulosus* in dogs [5,176–178], *M. expansa* in sheep [179], and *Raillietina cesticillus* in chickens [180,181]. The production of reaginic antibodies in response to infection with adult cestodes has also been indicated by the demonstration of cutaneous hypersensitivity following intradermal injection of parasite antigens in humans [182–184] and dogs [185,186]. Any role for these antibodies in expression of host-protective immunity is obscure. Indeed, the evidence in favor of the development of any acquired resistance to superinfection is poor and, where there is some evidence, immunity is inconsistent and incomplete [187–191].

A similar situation exists with the inconsistent and often poor results obtained in experiments attempting to immunize dogs against infection with *E. granulosus* and *Taenia* species [190–201]. Although "partial protection" has been achieved in some experiments, these results do not hold out much promise for developing effective vaccines against the medically and economically important cestodes in their definitive hosts. There is considerable scope for investigation of the host–parasite immunologic relationship in the taeniid cestodes, particularly at the gut level. A better understanding of the extent and significance of immunologic responses to these parasites and immune evasion mechanisms utilized by adult cestodes may improve the current pessimistic prospects for effective vaccination. Diagnostic techniques (discussed further below) based on the detection of specific antibody in dogs infected with *Taenia* or *Echinococcus* are, however, likely to be of practical value [202–205].

VACCINATION

Why vaccines?

Studies on vaccination against infection with cestode parasites have largely concerned the larval stages of members of the family Taeniidae. The most obvious reasons for this are firstly that this group of parasites contain most tapeworm species which are of economic and public health significance, and secondly that in few host–parasite associations is the phenomenon of adaptive immunity so clearcut. Studies on immunization against the adult stages of these parasites have been much less productive and consequently only a small amount of literature is available [190–201].

Measures for control of these parasites are well known and detailed guidelines have been prepared by international health and agriculture organizations [2,3]. By mathematical modeling, Gemmell et al. [191, 206–210] have analyzed factors important in the transmission and control of various Taenia spp. and Echinococcus spp. They have shown that effective control of the large Taenia spp. in the endemic state will necessitate the development of vaccines for the intermediate host. The adult tapeworm parasites have enormous egg production. Conventional control measures largely rely on reducing the numbers of adult tapeworms so that contamination of the environment with the infective eggs is reduced. Thus, grazing animals no longer become infected with a few parasites when they are young, and so do not acquire immunity to infection. A population of highly susceptible adult animals is thereby generated, and chance contamination from a single infected definitive host can lead to catastrophic "cysticercosis storms" [206]. For effective control, a vaccine is necessary to replace immunity normally acquired by natural infection of the young animal.

Vaccination experiments

Studies on vaccination against larval cestode parasites have been carried out since the 1930s and several comprehensive reviews have been published on this topic [96,211–213]. It is not the intention here to re-review this work, but rather to summarize some of the major advances and to highlight the most recent developments.

Two broad categories of antigen have been used in most vaccination experiments against larval cestodes, i.e., antigens derived from oncospheres or their in vitro products, or from metacestode stages or their in vitro products. The earliest experiments were carried out using metacestode antigens to immunize rats against infection with T. taeniaeformis [214,215]. Strong resistance against challenge infection with eggs developed in immunized rats, and later work, has substantiated these results [216–218]. Later studies with metacestode stages of T. taeniaeformis have been concerned more with the role played by their chemical products in modulating the host immune response rather than with vaccination per se, but some partial characterization of the antigens has been carried out [216]. Vaccination with metacestode-stage antigens has also been achieved with T. pisiformis in rabbits [219,220] and T. solium in pigs [221].

The majority of studies on vaccination have been carried out using cestode oncospheres or their in vitro products, since the pioneering studies of Gemmell et al. in the 1960s used living activated oncospheres as vaccines in sheep against T. hydatigena [222] and T. ovis [223]. These studies laid the emphasis on products secreted by living parasites as being the most likely candidates for host-protective antigens. This philosophy gained credence when it was later shown that antigens released from oncospheres confined in intraperitoneally implanted millipore membrane diffusion chambers immunized rats against T. taeniaeformis and sheep against T. ovis [224]. When techniques were developed for in vitro cultivation of taeniid cestode oncospheres [225], it was shown that products collected in media used to cultivate T. ovis and T. hydatigena oncospheres contained host-protective antigens [226]. These so-called excretion–secretion antigens (ESA) were also shown to be effective in vaccinating cattle against T. saginata [227,228], mice against T. taeniaeformis [229], rabbits against T. pisiformis [230], sheep against T. multiceps [231,232], and sheep against E. granulosus [233]. A series of experiments showed that immunity induced by antigens produced during in vitro cultivation of taeniid cestode oncospheres can be applied as an effective vaccine against T. ovis and T. saginata infection in farm livestock reared under natural farm management conditions [12,234–237]. Subsequent studies showed that oncospheres did not have to be cultured in vitro to collect host-protective antigens, but that freeze–thaw sonicated oncospheres were equally effective [229,238].

Despite the great promise for development of practical cestode vaccines, an apparently insurmountable obstacle remained. It was impossible to envisage that adult cestode parasites obtained from humans or animals could serve as adequate sources of oncosphere antigens to develop commercial vaccines at an economic cost on a sufficient scale.

Development of recombinant cestode vaccines

The advent of recombinant DNA technology provided some hope that host-protective cestode antigens could be produced on a sufficient scale to allow commercial production. It was shown that the host-protective antigens of *T. taeniaeformis* in mice included protein components [139]. Treatment of oncosphere antigens with periodate did not abrogate the host-protective effect, whereas protease treatment caused a drastic reduction. Thus, peptide antigens produced by recombinant DNA methods should be successful in immunization experiments.

Specific antibodies have been the most common reagent used for isolating specific antigen-expressing clones from cDNA libraries expressed in *Escherichia coli*. However, the sheer impracticability of testing vast numbers of cloned antigens in vaccination trials, especially in the large economic animals, is immediately apparent. It is therefore imperative to maximize the possibility that antibodies used for screening cDNA libraries will identify putative host-protective epitopes. A number of approaches have been undertaken with larval cestodes to achieve this objective. Lightowlers *et al.* in a series of papers [139–141] showed that significant host protection against *T. taeniaeformis* in mice could be achieved using a limited subset of *T. taeniaeformis* antigens (probably four) separated by sodium deoxycholate polyacrylamide gel electrophoresis (DOC-PAGE). Antisera to this subset of antigens (DOC-FII) were used to screen cDNA libraries prepared in the bacteriophage λgt11 from mRNA from *T. taeniaeformis* larvae or oncospheres (K.S. Johnson & M.W. Lightowlers unpublished results). Several clones expressing anti-DOC-FII antigen specificities were isolated from this library, and purified β-galactosidase (β-gal) fusion proteins tested in vaccination trials. None of the fusion proteins stimulated host-protective immunity, despite the fact that subsequent studies with fluorescent antibodies against the clones reacted with antigens either on the surface or inside *T. taeniaeformis* oncospheres.

Bowtell *et al.* [239,240] expressed *T. taeniaeformis* recombinant antigens in a λAmp3 cDNA library prepared from mRNA of 28-day-old metacestodes from mice, and isolated several clones using a rabbit anti-*T. taeniaeformis* oncosphere serum. Purified β-gal fusion proteins were

tested in immunization trials but failed to stimulate protective immunity in mice. This was disappointing because two of the isolated antigens were related to antigens thought to be associated with concomitant immunity [138]. A cDNA library has also been prepared from mRNA of *T. ovis* adult worms in the pEx series of plasmids and recombinants expressing antigenic *T. ovis* determinants, as β-gal fusion proteins were isolated using an anti-*T. ovis* oncosphere serum [241]. However, vaccination trials with these antigens have not been reported.

Despite their possible low sensitivity, monoclonal antibodies have been attractive as probes for screening cDNA libraries because of their high specificity, and a mouse monoclonal antibody which passively protects calves against infection with *T. saginata* has been described [242]. This antibody recognizes antigenic epitopes on the surface of *T. saginata* oncospheres and affinity-purifies an oncosphere antigen, which stimulates protective immunity in cattle [243]. This antibody may be useful as a screen for *T. saginata* cDNA libraries, but experiments along these lines have not yet been reported.

There has been a recent report of the successful development of a *T. ovis* recombinant vaccine [244]. In this instance, antibodies used to screen a *T. ovis* oncosphere cDNA library were selected on the basis of their reactivity with two putative host-protective antigens defined in several preliminary experiments, including immunization with sodium dodecyl sulfate polyacrylamide gel electrophoresis (SDS-PAGE) gel cutouts. Oncosphere antigens were separated by SDS-PAGE, electroblotted onto nitrocellulose, and specific antibodies purified by acid elution from the regions of the blot corresponding to the areas containing the host-protective antigens (47 and 52 kD). These affinity-purified antibodies were used to screen the *T. ovis* oncosphere cDNA library, reactive clones were selected and purified, and two β-gal fusion proteins (β-gal 45S and β-gal 45W) purified from transformed *E. coli*. These fusion proteins were not effective in vaccination experiments in lambs, although sera from vaccinated animals contained antibodies against them as well as to native oncosphere antigens in the 47–52 kD region. Because the evidence relating these clones to putative host-protective antigens was so compelling, a new expression system was tested.

The authors [244] chose one of a new series of vectors in which the parasite-encoded portion of the polypeptide is expressed as a fusion protein with glutathione *S*-transferase (GST) of *Schistosoma japonicum* [245]. This allowed the soluble GST fusion proteins, GST-45S and GST-45W, to be purified from *E. coli* sonicates by simple affinity chromatography on glutathione agarose under nondenaturating conditions. When these GST fusion proteins were tested in vaccination trails in sheep, it was shown that a total dose of 50 μg (three injections) of GST-45W in the adjuvant saponin stimulated up to 94% protection against challenge infection with *T. ovis* eggs. This is the highest level of protection thus far reported for a recombinant antigen in a natural host–parasite system. The full nucleotide sequence of the antigen was reported [244] and it encodes a sequence of 238 amino acids, giving a calculated molecular weight of 25 830 D.

The future

The *T. ovis* recombinant antigen GST-45W [244] forms a sound basis for developing a commercial recombinant vaccine to be used in control of this parasite in the field. There are obviously a number of further avenues to be pursued. The fusion protein is unstable and manipulation of the vector and/or the parasite cDNA may assist in overcoming this problem. Also, maximal expression of fusion protein must be achieved to allow commercial development at an economical price per dose.

Crossimmunity is known to occur between many of the taeniid cestode species (Table 17.3) and, specifically, *T. ovis* has been shown to have crossimmunity with *T. hydatigena* [222,226] and *E. granulosus* [246,247]. It is highly likely, therefore, that the host-protective antigens of these different species may share determinants in common so that the cDNA of *T. ovis* GST-45W could be used in hybridization experiments to identify homologous cDNA coding for protective antigens in cDNA libraries of the other species. Also, GST-45W itself may provide useful crossprotection against some of the other species.

Very little information is available concerning the source and nature of host-protective oncosphere antigens and the precise manner in which they react with the immune system of the host. It has been shown that the antigens are particulate or bound to membranes [229] but that they could be solubilized using sodium deoxycholate [258,259]. The particulate or membrane-bound nature of the antigens led to speculation that they may be associated with the secretion granules contained in the so-called penetration glands of the oncospheres [229]. It is interesting that preliminary results using immunogold labeling (D. Heath, unpublished

work) show that an anti-*T. ovis* GST-45W monoclonal antibody binds to the secretion granules of *T. ovis* oncospheres as well as to sites on the oncosphere surface. The availability of totally characterized host-protective cloned antigens should allow detailed studies to clarify aspects such as the biologic role of these antigens in the parasite, the precise locations of B cell and T cell epitopes on the antigens, and their interaction with the host immune system. It would be simpler for these experiments to have a cloned *T. taeniaeformis* antigen available because of the greater ease of carrying out experiments in mice rather than in sheep. A search for such an antigen is now under way, but even if this were not successful the *T. ovis* GST-45W should yield substantial important information.

IMMUNO- AND MOLECULAR DIAGNOSIS

Serologic diagnosis plays a vital role in the differential diagnosis of both hydatidosis and cysticercosis. The identification of space-occupying lesions or the development of nonspecific pathologic symptoms may indicate hydatidosis or cysticercosis as potential causes; however, it is usually necessary to discriminate these from several other potential etiologies. The appropriate therapy, specific surgical techniques to be used, and/or choice of drug therapy depend critically on an accurate primary diagnosis. In hydatid disease this is especially important, given the potential for protoscoleces released through rupture of a parent cyst to establish secondary infection.

Hydatid diagnosis in humans

In most parts of the world, the majority of patients presenting with *E. granulosus* or *E. multilocular* infection are found to have readily detectable levels of specific serum antibody against the parasite. Although cross-reactivity occurs with other parasitic infections, particularly in some tests, accurate diagnosis is possible in most cases [260,261]. Serologic techniques for the identification of specific antibodies or antigens in patient serum have largely replaced the use of intradermal injection and assessment of immediate hypersensitivity (Casoni test) as the preferred method of diagnosis. This has occurred principally because of the occurrence of an unacceptable proportion of false-positive reactions in the Casoni test [262,263].

There are several potential applications for serologic diagnosis of human hydatid disease. While test sen-

Table 17.3 Summary of published data on crossprotection between taeniid parasites

Challenge species	Immunizing species	Immunizing regimen/antigen	Protection afforded (%)	[Ref.]
Echinococcus granulosus	*Taenia hydatigena*	Onc. (V), IM	31.3*	[246]
		Eggs (V), IM	0	[246]
		Prior infection[†]	38.5	[247]
	Taenia ovis	Onc. (V), IM	38.9*	[246]
		Eggs (V), IM	19.8	[246]
		Prior infection & eggs (V), IM	66.7	[247]
	Taenia pisiformis	Eggs (V), IM	0	[246]
		Onc. (V), IM	0	[246]
Taenia hydatigena	*Echinococcus granulosus*	Prior infection	49.9	[247]
	Taenia ovis	Prior infection	31.3	[247]
			38.1	[131]
			73.6[‡]	[248]
		Coinfection[§]	66.3	[131]
		Eggs (NV), IM	23.9	[222]
		Eggs (V), IM	39.7	[222]
			0[‡]	[249]
		Onc. (V), IM	40.4*	[222]
			0*,[‡]	[249]
	Taenia pisiformis	Eggs (NV), IM	28.9	[222]
		Onc. (V), IM	22.6	[222]
		Eggs (V), IM	23.0	[222]
			35.4*,[‡]	[250]
Taenia ovis	*Echinococcus granulosus*	Prior infection	32.4	[247]
	Taenia hydatigena	Prior infection	63.8[‡]	[248]
			47.4	[131]
			65.7	[247]
		Coinfection	46.1*	[131]
		Eggs (V), IM	54.3*	[223]
			66.7[‡]	[249]
		Onc. (V), IM	48.0*	[223]
			38.2[‡]	[249]
		Onc. ES (NV)**, IM	96.8	[226]
	Taenia pisiformis	Eggs (V), IM	21.5	[223]
		Onc. (V), IM	11.7	[223]
			72.5*,[‡]	[250]
Taenia pisiformis	*Taenia hydatigena*	Eggs (V), oral	88.8	[251]
		Eggs (V), IM	19.3	[252]
		Onc. (V), IM	28.1*	[252]
			47.5*,[‡]	[250]
		Onc. (V), DC[††]	79.8	[251]
	Taenia ovis	Eggs (V), oral	22.9	[251]
		Eggs (V), IM	0	[252]
		Onc. (V), IM	40.6*	[252]
			46.3*,[‡]	[250]
		Onc. (V), DC	0	[251]
		Onc. ES (NV), IM	0	[251]
Taenia saginata	*Taenia crassiceps*	Metacestodes (V), IP	28.3	[253]
		Metacestodes (NV), IP	22.6	[253]
	Taenia hydatigena	Eggs (V), oral	15.6	[253]
		Onc. (V), IM	98.8	[254]
			25.7	[253]

continued

Table 17.3 *Continued*

Challenge species	Immunizing species	Immunizing regimen/antigen	Protection afforded (%)	[Ref.]
Taenia saginata	*Taenia hydatigena*	Onc. (V), IV	26.8	[253]
		Onc. ES (NV), IM	56.0*	[227]
			60.3	[238]
			77.8††	[237]
	Taenia ovis	Onc. ES (NV), IM	64.9*	[227]
			21.1	[253]
	Taenia taeniaeformis	Onc. ES (NV), IM	28.4*	[228]
Taenia taeniaeformis	*Taenia crassiceps*	Metacestode (V), IP	73.5	[228]
		Metacestode (NV), IM	54.3	[228]
	Taenia hydatigena	Eggs (V), oral	88.4	[255]
		Onc. (NV), oral	12.6	[255]
		Onc. (NV), SC	30.4	[255]
	Taenia pisiformis	Eggs (V), oral	41.3	[256]
			94.5	[255]
		Onc. (NV), oral	45.9	[255]
		Onc. (NV), SC	70.6	[255]
		Adult worm (V), IP	83.6*	[256]
		Metacestode (NV), SC	56.3*	[257]
	Taenia saginata	Gravid proglottid (NV), IM	35.3	[228]
		Adult worm (NV), SC	0	[257]
		Adult worm (V), SC	0	[257]
		Adult worm (NV), IP	33.3	[256]
		Onc. ES (NV), IM	95.9	[228]
		Onc. ES (NV), oral	96.3	[228]
	Taenia serialis	Metacetode (NV), SC	88.5*	[257]
	Taenia solium	Adult worm (V), SC	31.0*	[257]

DC, diffusion chamber; ES, excretory−secretory; IM, intramuscular; IP, intraperitoneal; IV, intravenous; NV, nonviable; Onc., oncosphere; SC, subcutaneous; V, viable. Unless otherwise stated, protection is assessed as the percentage reduction in the number of metacestodes detected following a challenge infection irrespective of the state of development or viability of the metacestodes. Where the results stated in a particular publication include repeat experiments or several immunizing regimens (e.g., antigen dose, route of administration, host strain, etc.) only the highest level of protection achieved is included.

* Original data indicate that the immunization resulted in a decreased viability of metacestodes established following the challenge infection compared with controls.

† Prior infection via the normal route of infection as distinct from atypical infection which may occur as a result of other immunization protocols, e.g., development of cestodes intramuscularly, in draining lymph nodes, etc., following intramuscular injection of oncospheres.

‡ Percentage protection calculated on transformed data, see original reference.

§ Egg infection with "challenge species" and "immunizing species" given simultaneously.

** Supernatant from *in vitro* culture of oncospheres.

†† Activated oncospheres implanted intraperitoneally in diffusion chambers.

‡‡ Immunized with culture supernatant plus freeze−thaw sonicate of cultured oncospheres.

sitivity and specificity remain imperfect, these applications have different requirements with respect to the practical application of the techniques. The foremost requirement for serology is for primary diagnosis of infection on an individual patient basis, where other symptoms suggest the possibility of hydatid disease. In this situation it is particularly important that false-negative results are minimized such that inappropriate biopsies or surgical interventions are avoided. False positives are more important when the test is used for seroepidemiologic purposes. Where the disease may be rare in a population, say 5 per 100 000, a false-positive rate of 1 in 1000 sera results in a 20 times over-estimate of the prevalence of infection.

Serology is also of value for postoperative or post-chemotherapy surveillance. The continuing production of specific antibodies is useful as an indication of continuing infection after surgery as a result of the presence of additional cysts or the development of secondary cysts from protoscoleces released prior to or during surgery. It can also be used to implicate the continued presence of viable parasites after chemotherapy. Care must be taken in interpreting results of serologic testing after treatment of the patient because release of antigen during an operation or as a result of the death of a cyst from an anthelmintic may temporarily boost antibody production to high levels. Nonviable parasite material can also provide a source of antigenic stimulation to the host. For these reasons, post-treatment serology is of most value when the general trend in antibody titers is viewed. Rising or continuing steady-state positive reactions assessed every few months for 18 months or 2 years following treatment are more likely indicators of continuing infection than a single positive serology reaction after 2 years [260,264].

There have been several comprehensive reviews of the literature on diagnosis of hydatid disease [260,261, 265,266]. What follows is an overview of the topic, including the most recent developments. Essentially, every serologic test capable of identifying the presence of specific antibody has been used with some success for diagnosis of hydatid disease in humans. Specific IgG, IgM, IgA, and IgE may be present in the sera of hydatid patients, with the predominant antibody class being IgG [17,264,267]. The convenience of enzyme-linked immunosorbent assay (ELISA) in producing accurate and objective estimates of antibody levels has led to this being the favored serologic technique in recent times [268–279]. Very high sensitivities have been claimed for diagnosis, frequently greater than 90%; however, these claims need to be interpreted with care. Sera which are particularly important are those from patients who have yet to receive surgical or anthelmintic treatment, i.e., a primary diagnosis. Where information is derived from laboratories performing routine diagnosis, it is essential that those found to be serologically negative are confirmed, as much as is practicably possible, not to be infected. In this situation, patients with positive serology are more likely to be followed up to confirm infection. This leads to bias in the estimates in favor of recording a high apparent sensitivity for the test, since positive reactions will always be scored while some or many false negatives are overlooked. A recent study of *E. granulosus* diagnosis in which substantial effort was invested into identifying most or all false-negative results, 76% of primary diagnosis sera were found to have titers of specific antibody against cyst fluid antigens which were indicative of infection (Table 17.4). This can be regarded as the maximum test sensitivity since it was not possible to be certain that all false-negative cases were identified.

Crossreactivity does occur in serologic tests for

Table 17.4 Sensitivity of immunoelectrophoresis (IEP) and enzyme-linked immunosorbent assay (ELISA) for primary diagnosis of *Echinococcus granulosus* infection in humans*

Sera category	Number of sera	Percentage sera positive in	
		ELISA	IEP
All Sera	78	72	74
Positive ELISA	56	100	91
Positive IEP	57	74	100
Negative ELISA	22	0	27
Negative IEP	21	24	0
Negative ELISA and IEP	19	0	0

* M.W. Lightowlers, unpublished observations using serological techniques described by Rickard *et al.* [275] and Varela-Díaz & Colforti [280]. Criteria for positivity were: IEP, presence of Arc5 or three or more other bands; ELISA, titer 1:200 or greater.

hydatid infection, particularly with sera from patients with other helminth parasites [275]. False-positive reactions can be eliminated, with the exception of patients infected with other taeniid species, by the identification of specific antibodies [280] to the lipoprotein cyst fluid antigen termed antigen 5. This immunodominant molecule was first identified in immunoelectrophoresis (IEP) by Chordi & Kagan [281] and subsequently by Capron *et al.* [282–284] who coined the terms antigen 5 for this antigen and Arc5 for the precipitation band formed between this antigen and specific antibody in IEP. Antibodies reacting with this antigen and producing Arc5 in IEP are not detected in the sera of patients with diseases or parasitic infections other than hydatidosis or cysticercosis. The antigen is, however, common to *E. granulosus*, *E. multilocularis*, *E. vogeli*, and *T. solium* and hence IEP Arc5 cannot be used to differentiate between these infections [285–288]. The relative specificity afforded by detection of IEP Arc5 makes this test valuable in the diagnosis of hydatidosis, particularly where clinical and radiologic data can be used to differentiate hydatidosis from cysticercosis. Immunoelectrophoresis is a technique which uses large amounts of antigen (equivalent to 10 ml of fluid from fertile ovine cysts for each serum tested) and large amounts of serum (0.3–0.6 ml per sample) [280]. Modifications of the technique have been developed which utilize fewer resources, such as counterimmunoelectrophoresis on cellulose acetate membranes [289]. Double diffusion against a positive control serum which is specific for antigen 5 can also be used [288,290], particularly now that Arc5-forming monoclonal antibodies are becoming available [291,292].

Substantial progress has been made by Gottstein and colleagues in the development of a routine serologic test for species-specific diagnosis of alveolar hydatidosis in humans [269–273]. An *E. multilocularis* antigen preparation was depleted of determinants crossreacting with *E. granulosus* by affinity depletion with antibodies raised against *E. granulosus* coupled to a solid support. The run-through fraction, termed Em2, was found to have increased specificity for reactivity in ELISA with sera from *E. multilocularis* patients [269]. The antigen was found to be composed predominantly of a single parasite-derived antigen together with contaminant host-derived proteins [270]. Affinity purification of the antigen with anti-Em2 antibodies raised in the same host species from which the parasite was derived was successful in purification of a single predominant antigen (termed Em2a) of 54 kD, p*I* 4.8, which showed species specificity for *E. multilocularis* in immunoblots and ELISA [271]. An international study of sera from 50 *E. granulosus* patients and 32 *E. multilocularis* patients with Em2 ELISA assigned only three *E. granulosus* sera incorrectly as being infected with *E. multilocularis* [272]. Sera from *E. granulosus* patients which showed relatively strong reactivity with Em2 were not assessed to determine whether they had specific antibodies against the 54 kD antigen. An analysis of these critical sera would indicate more accurately the true specificity of this antigen for *E. multilocularis*. Sera from 12 000 blood donors from northern Switzerland, where *E. multilocularis* infection is endemic, were screened using Em2 ELISA [273]. Five individuals were positive in the assay, one of which was subsequently shown to be infected with *E. multilocularis*. The additional four individuals who were Em2 ELISA-positive showed no signs of infection on ultrasound or computer-assisted tomography scanning. Unless these patients are subsequently shown to be infected, this represents a false/positive ratio of four to each correct diagnosis. When followed up with clinical examination, such surveys are valuable for the early detection of infection when resources are available for such studies. Adaptation of the techniques to simple kit form or some other method which could be used on a decentralized basis would increase the value of the test in the endemic region. Difficulties with supply of adequate quantities of antigen for differential diagnosis of *E. multilocularis* infection may be overcome by the use of antigen expressed in *E. coli* using recombinant DNA techniques; one potentially suitable antigen has been cloned, hyper-

expressed, and tested in ELISA [293–295], although this is apparently not related to Em2a antigen [293]. Genes encoding *E. multilocularis* antigens with possible potential for use in serodiagnosis have also been cloned and expressed by another group [296]. All results reported for these antigens, one of which has added to nomenclature confusion by being designated EM2, have utilized only pooled serum samples, hence the potential utility of these antigens is yet to be assessed.

The source of antigens used for diagnosis of *E. granulosus* is usually cyst fluid from fertile cysts in naturally infected animal intermediate hosts. The fluid contains significant amounts of host-derived serum proteins as well as parasite-derived macromolecules. There are two parasite-derived antigens which are dominant in terms of their immunogenicity in hydatid patients. Since their first identification by Chordi & Kagan [281], who adopted the terms antigen 4 and antigen 5, there have been several different terms applied to the antigens, including that of Capron et al. [282,284] who applied the term antigen 5 to the same antigen which had previously been designated antigen 4 by Chordi & Kagan [281]. Oriol et al. [297], who used physicochemical techniques to characterize these antigens, adopted the term antigen A for the antigen which forms "Arc5" in IEP, and antigen B for the other major cyst fluid antigen. The nomenclature and antigen characteristics have been reviewed recently and a standardized nomenclature proposed [260] and subsequently adopted by others [298]. The antigen forming Arc5 in IEP is termed antigen 5 and the second major antigen, which also forms a characteristic band in IEP, is termed antigen B. The previously used nomenclatures and characteristics of these two antigens are summarized in Table 17.5. Arc5 does not occur in IEP with human sera other than from patients infected with taeniid cestodes. Interestingly, immunoprecipitation analysis of radiolabeled hydatid antigens shows nonspecific precipitation of the components of antigen 5 [292,298]. Immunoprecipitation or immunoblotting with monoclonal antibodies directed against phosphorylcholine (PC) indicate that the same, nonspecifically precipitated components bear PC epitopes [292,298]. Nonspecific reactivity with these components is quantitatively less than the specific recognition of the antigen 5 components by hydatid sera and these naturally occurring anti-PC antibodies do not form Arc5 in IEP. Nevertheless, Shepherd & McManus [298] were able to reduce the background binding to hydatid antigens in ELISA by addition of 10 mM phosphorylcholine to the serum

Table 17.5 Synonyms and characteristics of the two major cyst fluid antigens of *Echinococcus granulosus*

Characteristic	Antigen 5	[Reference]	Antigen B	[Reference]
Synonym	Antigen A	[297,299]	Antigen B	[297,299]
	Antigen 4	[281,300–304]	Antigen 5	[281,300–304]
	Arc5, F_5, fraction 5	[282,305,306]		
	Antigen 5			
Principal composition	Lipoprotein	[297]	Lipoprotein	[297]
Antigen stability at 100°C	Labile	[297,304–306]	Stable	[297,304]
Relative concentration in sheep				
hydrated cyst fluid	1	[303]	10	[303]
Estimated molecular weight (D)				
Native antigen	>400 000	[300]	160 000	[297]
	100 000–300 000	[307]	120 000	[299]
	>200 000 + 60 000	[305]	150 000	[300]
In SDS-PAGE	67 000	[292,302]	Three components	[302]
(nonreduced)	66 000 + 56 000	[291]	10 500–20 000	
	39 000	[291]	16 000, 24 000, 32 000	[292]
	38 000 + 24 000	[302]	plus higher molecular weight molecules differing in size by 8000	

diluent, particularly with sera from patients with other helminth infections.

Serologic diagnosis of hydatidosis suffers from problems associated with obtaining antigen from infected intermediate hosts, namely poor quality control of antigen batches, leading to variation in antigen concentration and specificity and limitation in the supply of antigen. These problems may be overcome by the use of antigens expressed *in vitro* through the use of recombinant DNA techniques. Several cDNA cloned antigens of *E. granulosus* have been expressed in *E. coli* and tested in ELISA with sera from hydatid patients (M.W. Lightowlers, unpublished observations). Two particular clones have potential for practical application in diagnostic tests; however, as is also the case with the native cyst fluid antigens, they do not discriminate between infections with the various taeniid species infecting humans.

Circulating hydatid antigens, free or as immune complexes, occur frequently in hydatid patients and their detection provides a useful addition to the other serologic techiques for diagnosis [17–24]. Detection of circulating hydatid antigen has been found to be particularly valuable for diagnosis of infection in the Turkana people of Kenya who, as a group, have comparatively low levels of specific antibody detectable with routine techniques [308]. Detection of circulating antigens is also valuable for postoperative surveillance of hydatid patients, where the continued presence of parasite antigens could be expected to be indicative of continuing infection.

Hydatid diagnosis in other intermediate hosts

Much effort has been expended in the quest to develop a sensitive and specific diagnostic test for identification of *E. granulosus* infection in domestic livestock. An effective test would be valuable in hydatid eradication campaigns. Successful campaigns lead to a situation where relatively few herds or flocks remain infected and infections in dogs are virtually eliminated. This situation cannot lead rapidly to eradication, however, because there is no practical means by which infection in livestock can be ascertained. Serologic techniques for accurate diagnosis of infection on an individual animal basis have never been successful.

In contrast with the presence of readily detectable specific antibody against *E. granulosus* infection in humans, domesticated animals are typically serologically negative or have low titers of serum antibody (reviewed by Rickard & Lightowlers [260]). A compounding problem is the ubiquitous presence of taeniid cestode parasitism with species known to produce antibodies which crossreact with *E. granulosus* antigens. Claims in the published literature of high sensitivity and specificity in hydatid serology need to be examined with care. Particularly important is the choice of noninfected controls for the tests. These must be animals with similar histories with respect to parasitism with species other than *E. granulosus*, especially *T. ovis*, *T. hydatigena*, or *T. saginata*. Comparison of serologic titers for hydatid infection in sheep with results from noninfected controls of substantially different age, breed, and infection with other parasites can lead to misinterpretation of the significance of the antihydatid antibodies. Attempts to affinity-deplete crossreactive antigens with antisera or monoclonal antibodies to the heterologous species antigen have been largely unsuccessful [309–311].

The reasons for the apparent difference between the immune response to the parasite in humans and other intermediate hosts are obscure. Frequently, patients whose sera are submitted for hydatid serology have come to the attention of medical practitioners because of some problem caused by the cyst. This may involve rupture or other pathologic symptoms indicative of overt reactivity to the parasite. Animals, on the other hand, are effectively tested at random. For this reason, the sampling of the infected human population may bias towards higher reactivity in this group. An alternative explanation is that the "natural" intermediate host is by some means immunologically tolerant to hydatid antigens. Lightowlers *et al.* [146] investigated this hypothesis by immunizing infected sheep or control noninfected sheep with small doses of cyst fluid antigen intramuscularly without adjuvant. The infected animals responded anamnestically to the antigen, in contrast to the controls, indicating that the animals were capable of an immune response to the cyst fluid antigens. Cyst fluid released from the animal's own cysts at laparotomy also induced rapid antibody production. It was suggested that antigen may be more effectively sequestered in the cysts in domesticated animals. The recent discovery that infected sheep have detectable levels of circulating hydatid antigens [312] contradicts the sequestration hypothesis.

Diagnosis of *Taenia solium* cysticercosis

Cysticercosis of the central nervous system and skeletal muscles is a major health problem in many parts of Latin America, Asia, and Africa. As many as 1.9% of all general autopsies performed in Mexico have been found to have *T. solium* cysticerci in the central nervous system [313]. Many patients infected with the parasite have readily detectable levels of specific antibody in both the serum and cerebrospinal fluid (CSF) (reviewed by Flisser *et al.* [313] and Schantz [314]). The predominant immunoglobulin class is IgG, although some patients also have detectable levels of IgM, IgA, and IgE [315–317]. Detection of these antibodies using a variety of *in vitro* assays provides a valuable aid to confirmation of clinical diagnosis. The published success rates for

serologic diagnosis cover a broad spectrum [313,315, 318–325]. However, the tests used routinely to date typically fail to detect antibody in a significant proportion of patients with surgically confirmed infection. Serious problems also exist with crossreactivity of antibodies in the sera from patients with other helminthic diseases, especially hydatidosis.

Variations in technique undoubtedly contribute to the range of success rates reported for serodiagnosis of cysticercosis in humans. Schantz [314], however, believes that a more significant factor is that different groups of patients have been the subject of the studies. Patients with calcified [326] or few cysts [327] and the relatively asymptomatic patients [328,329] are more likely to be negative in serologic tests. Patients suffering substantial symptoms of the infection [329] and with cysts in the early stages of degeneration [330] are more likely to be serologically positive. Clinical signs are frequently associated with the development of inflammatory reactions around degenerating parasites [331] which may cause the release of parasite antigens and boost antibody production. There is, however, no simple correlation between damage to cysts, antibody titer, and clinical findings [326]. The proportion of serologically negative patients at primary diagnosis can be high, as many as 25% of sera and 45% CSF samples in a recent study [327]. If indeed many of these patients have no antibody to the parasite, assays for antibody, no matter how effective, will not be useful for diagnosis. What is required for these patients is some specific assay not based on detection of antibody. Detection of parasite antigens in CSF [332] may provide a solution for some of these patients. The cysticercotic patients investigated by Estrada & Kuhn [332] were positive in assays for both CSF antigen and specific antibody and the extent to which this approach to diagnosis might assist in the detection of antibody-negative patients remains to be determined. Recently, Harrison *et al.* [333] (and reported in Flisser *et al.* [334]) have successfully applied the use of a monoclonal antibody raised against a *T. saginata* antigen for the detection of *T. solium* antigens in the CSF of patients with neurocysticercosis.

The second limitation to the success of serologic tests for diagnosis of *T. solium* cysticercosis is the lack of species specificity of crude antigen preparations derived from either the cysticercus or adult worm. Crossreactivities occur with sera from patients with a variety of parasitic infections, particularly hydatidosis [292,320,327]. This is not so much a problem in the clinical situation where other evidence is likely to implicate cysticercosis over other parasitic causes of positive serology. These crossreactivities limit the value of these tests for seroepidemiologic studies. Attempts have been made to improve the specificity of the tests. Diwan *et al.* [322] improved specificity by using the ratio of reactivity in ELISA with a crude cysticercus extract and an extract of pig muscle as a control for background reactivity with contaminant host-derived antigens in the parasite preparation. Larralde *et al.* [325] found that the vesicular fluid gave superior results compared with extracts from whole cysticeri. Nascimento *et al.* [335] had most success with an extract from the cysticercus scolex and were subsequently able to isolate, by monoclonal antibody affinity chromatography, a derivative of this extract with improved specificity [336]. Similar techniques have also been used by Kim *et al.* [337] in the purification of an antigen which improves specificity in serologic specificity for cysticercosis. Partial purification of cysticercus antigens by chromatofocusing techniques has been used in the isolation of a preparation which shows reduced crossreactivity in ELISA, although false positives continue to occur with sera from patients with hydatidosis [320]. A similar situation exists in the case of a cysticercus antigen termed antigen B, which has been shown to be a dominant antigen eliciting antibody responses in cysticercosis patients [315]. This antigen has been partially characterized and found to have fibronectin-like properties [338]. Crossreactivity with sera from hydatid patients also occurs with this antigen [339].

Analyses of the antigenic constituents of *T. solium* using immunoblotting techniques have been useful in identifying species-specific components [317,325,340]. Gottstein *et al.* [340] have identified a pair of cysticercus antigens with estimated molecular weights of 8 and 26 kD which are frequently detected in blots with sera from cysticercosis patients and show no crossreactivity with sera from patients with either *E. granulosus* or *E. multilocularis* infection. In a followup study, none of 147 sera from patients with a variety of potentially crossreactive infections was found to have antibodies to the 8 and 26 kD antigens [327]. Reactivity with one or both of these antigens in sera or CSF from cysticercosis patients were 92% and 100%, respectively, when only sera which were positive in ELISA with crude cysticercus antigen were tested. The potential for this technique to detect positive reactions in some of the samples which are negative in ELISA was not evaluated. The discovery of these antigens represents an important advance in developing a specific diagnostic test for cysticercosis

because it is the first time that a serologic technique has been found which unequivocally differentiates between *T. solium* cysticercosis and hydatidosis. Application of these antigens in other less-sophisticated serologic techniques and *in vitro* production of the antigens by recombinant DNA techniques would make diagnosis based on these antigens more readily suited for seroepidemiology and widescale screening for infection.

Results of studies on the antibody response of patients with *T. solium* cysticercosis and "crossreactivity" with sera from patients with taeniasis raise an important question in the epidemiology of human cysticercosis which remains to be answered—the extent to which a patient who has the adult worm in the small intestine (as a result of eating undercooked pork) is susceptible to autoinfection. This could occur directly by the hatching of eggs released in the intestine, possibly in conjunction with reverse peristalsis or vomiting to bring the eggs into specific areas of the gut for initiation of hatching and activation, if this is strictly necessary, or indirectly through fecal/oral contamination with eggs. An interesting observation which may be relevant to this question is the surprising finding that of 33 sera from individuals with *T. solium* taeniasis examined for antibodies to cysticercus antigens, none was positive [336]. The vital role that immune responses play in determining susceptibility to infection with this group of parasites was discussed earlier in this chapter. *Taenia solium* is an unusual taeniid cestode in that it is the only species in which the same host can, under normal circumstances, become infected with both the adult and larval forms. For this reason there has been no investigation in any model system of the effects of antiadult worm immune responses on the *subsequent* establishment of infection in the same individual. Taeniid cestodes are certainly immunogenic in the definitive host [4,173,174,186,204] and oral dosing with antigens derived from the oncosphere or metacestode stages can induce a high level of protection against subsequent challenge infection with eggs [217,228,255]. Antigens derived from the adult worm strobila have not been used in immunization trials by the oral route although they are very effective when given parenterally [215,257, 341]. Initial infection with the adult worm of *T. solium* may protect against subsequent infection with *T. solium* eggs. Definitive hosts of taeniid cestodes can be repeatedly infected with the parasites and there is scant evidence for the development of protective immunity to the adult worms [187–191,342]. Thus, patients found infected with both the adult and larval forms of *T. solium* may have become infected with the adult worm after they had ingested eggs.

Diagnosis of *Taenia saginata* cysticercosis

Taenia saginata is a widespread zoonosis in which humans are the obligate definitive host. Infection with the parasite is common where raw or poorly cooked beef is eaten and where cattle feed becomes contaminated with human feces. Contamination may be direct or through the outflow from sewage treatment works. High-level infections occur at times from contamination of cattle in feed lots [343,344]. Development of a reliable technique for antemortem diagnosis of infection would be useful in controlling the transmission of this parasite and in monitoring parasite control programs.

The limitations in test sensitivity and specificity encountered with serologic tests for diagnosis of other larval cestode infections also occur with tests for *T. saginata* cysticercosis to such an extent that reliable diagnosis of infection on an individual animal basis is not possible. The principal problems stem from:

1 the absence of detectable antibody in the sera of cattle with only light infections;

2 antigenic crossreactivity between antibodies to *T. saginata* and antibodies in the sera of cattle exposed to other helminth parasites, particularly *Fasciola hepatica* and *T. hydatigena*;

3 difficulties with obtaining an adequate supply of *T. saginata* parasite material for serologic tests.

This latter difficulty has been overcome to some extent by the utilization of crossreacting antigens derived from other taeniid cestode species [345–348]. Immuno-chemical analysis has confirmed the extent of antigen crossreactivity [349,350]. Naturally, the limitations imposed on diagnosis by serologic crossreactivity due to exposure to other taeniid species are likely to be exacerbated by using antigens from an heterologous species. The problems with obtaining significant quantities of *T. solium* antigens from human infections or from the small cysticercus stage in cattle muscle place severe limitations on practical application of diagnosis of *T. saginata* cysticercosis. Techniques currently being applied successfully for the *in vitro* production of other taeniid cestode antigens may provide a solution to this problem, although to date there have been no reports of production of *T. saginata* antigens by recombinant DNA approaches.

The early literature on diagnosis of *T. saginata*

cysticercosis has been reviewed by Geerts *et al.* [351]. Careful analysis of antibody titers in experimentally and naturally infected cattle by Craig & Rickard [352], using control animals given monospecific infections with heterologous parasites, confirmed the poor sensitivity of "crude" antigens from *T. saginata* adult worm strobila for detection of naturally acquired infection and the high level of crossreactivity, especially with sera from cattle infected with *F. hepatica*. Other approaches, such as examination of hypersensitivity reactions following inoculation of parasite antigen, are also unsuccessful [353]. Antibodies can be detected in the sera of infected cattle using IEP and counter-IEP with antigens derived from *T. saginata* adult worm strobila, however, the limit of test sensitivity is approximately 50 cysticerci, with animals having fewer cysts being negative in the tests [354,355]. Antigenic crossreactivity has also been shown between the major diagnostic antigen in IEP and antigens from other taeniid species [355]. Attempts have been made to affinity-purify diagnostic antigens and affinity-deplete crossreacting specificities. These have resulted in a reduced test sensitivity and absolute specificity was not obtained [309]. A lipoprotein antigen of *T. hydatigena* cyst fluid has been identified which can be used in purified form for detection of anti-*T. saginata* antibodies [347]. The antigen apparently does not crossreact with sera from cattle infected with *F. hepatica*; however, false positives were found and the authors indicate that this may be due to exposure of the cattle to *T. hydatigena*.

A significant advance in this field has recently been made by Harrison *et al.* [333] who have identified specific circulating parasite antigens in the sera of infected cattle. In the assay, a monoclonal antibody reactive with a repetitive carbohydrate epitope present on the surface and in secretions of *T. saginata* cysticerci is used to capture the circulating antigen. The sensitivity of the assay is 200 viable cysticerci, which is less than that claimed for other serologic tests based on the detection of antibody. However, an advantage is that the assay results are negative when the cysticerci are dead while anti-*T. saginata* antibody levels may still be high. In this way the assay provides a method for detection of infection with viable parasites.

Identification of tapeworms or tapeworm eggs and diagnosis of infection in the definitive host

Diagnosis of taeniid cestode infection in dogs and humans can often be readily achieved by identification of the characteristic proglottids and/or eggs in the feces.

Differentiation of which species of parasite is involved is more difficult, requiring careful investigation of the morphology of the adult worm and/or proglottid architecture and this may not be completely reliable because of the considerable variation in segment morphology within each species [356]. Identification of the taeniid cestode species with eggs alone is not possible for any species. Accurate diagnosis is important where *E. granulosus* may be present in dogs and to differentiate infection with *T. solium* and *T. saginata* in humans. Isoenzyme analyses or proglottid proteins can be used to distinguish between *T. solium* and *T. saginata* [357], but this would not be a suitable technique for identifying *E. granulosus* infection *per se* in dogs because, if the small worms or proglottids can be found, they are diagnostic in themselves. Isoenzymes are useful for differentiating different *E. granulosus* strains [358–360]. Cloned genomic DNA sequences of *E. granulosus* have also been used for identification of *E. granulosus* strains by differential hybridization to genomic DNA [361,362]. Similar techniques have been applied to the differentiation of *T. saginata* and *T. solium* based on hybridization with cloned DNA sequences derived from *Schistosoma mansoni* ribosomal RNA genes or cloned fragments of genomic DNA from *T. solium* [363]. Cloned mitochondrial DNA fragments may also have potential for use in strain/speciation studies [364]. Should these techniques be sufficiently sensitive they could be applied to the identification of taeniid eggs, either in feces or in environmental contamination studies. Some progress has been made in this direction through the use of a monoclonal antibody with specificity for the oncosphere antigens of *E. granulosus* [365]. Practical application of this technique may be hampered by the necessity to hatch the eggs prior to incubation with the monoclonal antibody, particularly with old fragile eggs isolated from environmental contamination. The monoclonal antibody can be used successfully to identify eggs obtained from the perianal area of naturally infected dogs, providing one method for specific diagnosis of infection with this parasite. A separate approach which may be valuable in species-specific diagnosis of taeniid cestodes is through the detection of coproantigens [366].

Detailed and carefully controlled experiments by Jenkins & Rickard [4,173,178] on the antibody response of dogs to infection with *Taenia* and *Echinococcus* have resulted in the development of serologic tests for diagnosis of infection. Practical application of the tests has shown that naturally acquired infection can be reliably diagnosed in the majority of infected dogs [5,174,202,203] and widescale application of serologic diagnosis of *T. ovis* infection is to be implemented in New Zealand in the near future (D. Heath, personal communication). The major difficulty in widescale application of these tests is in the supply of adequate quantities of antigen. Immunochemical analysis of *E. granulosus* protoscolex antigens has identified species-specific antigenic molecules [204], one of which has been successfully cloned and expressed by Gasser *et al.* [205] using recombinant DNA techniques. The recombinant protein is detected by 20% of sera from naturally infected dogs and shows no crossreactivity with sera from dogs known to be infected with other parasites but known not to have been infected with *E. granulosus*. Continuing research by Gasser *et al.* is seeking to identify additional clones for use as a cocktail of recombinant antigens to improve the sensitivity of diagnosis.

CONCLUDING REMARKS

Studies on the immunobiology of cestode infections have provided valuable insights into the immunity to, and immune evasion by, both tissue-dwelling and gut lumen-dwelling parasites. Taeniid metacestodes have provided models for investigation of host-protective immune responses where immunity to reinfection is a prominent feature of the natural host–parasite relationship. Knowledge of the immunobiology has been exploited in research on vaccination, leading to the development of the first highly effective recombinant vaccine against a parasite with the production of the *T. ovis* vaccine. Application of molecular biology has also been successful in the *in vitro* production of antigens for improved serodiagnosis of alveolar hydatidosis in humans. Recent progress in serologic diagnosis of infection in the definitive host indicates that cloned antigens are soon likely to permit practical serodiagnosis of *E. granulosus* infection in dogs.

Crossprotection between taeniid cestode species may allow the development of the *T. ovis* vaccine to be rapidly translated into vaccines for other taeniid species. Cloned genes which show species specificity in DNA hybridization have been used successfully for parasite identification and strain typing with some cestode species. Application of signal amplification using polymerase chain reaction technology with species- or strain-specific primer sequences could improve these techniques and possibly allow identification of taeniid eggs to species or strain level.

Although great progress has been made in understanding the immunobiology of cestode infections, several areas provide challenging topics for research. In particular, more information is needed on the factors which influence susceptibility/resistance to infection and immune evasion by the parasites, especially in humans. Accurate diagnosis of E. granulosus infection in domestic livestock continues to remain elusive and probably requires a radical change from standard serologic assays. Assays for the presence of antigen-specific circulating T cells, through detection of interferon-γ production, is one possible approach. Finally, immunization against infection with taeniid cestodes in their definitive hosts has not been particularly successful to date. A significant degree of resistance to infection has been shown to develop against E. granulosus in dogs following several challenge infections with the worms, indicating the potential for development of at least limited host-protective immunity. More research is required into these host-resistance mechanisms towards the development of practical immunization protocols suitable for use in hydatid control programs.

REFERENCES

1 Matossian RM, Rickard MD, Smyth JD. Hydatidosis: a global problem of increasing importance. WHO 1977;55: 499–501.

2 Eckert J, Gemmell MA, Soulsby EJL, eds. FAO/UNEP/ WHO Guidelines for Surveillance, Prevention and Control of Echinococcosis/Hydatidosis. WHO, Geneva, 1981.

3 Gemmell M, Matyas Z, Pawlowski Z, Soulsby EJL, eds. Guidelines for Surveillance Prevention and Control of Taeniasis/ Cysticercosis. WHO, Geneva, 1983.

4 Jenkins DJ, Rickard MD. Specific antibody responses to Taenia hydatigena, T. pisiformis and Echinococcus granulosus infection in dogs. Aust Vet J 1985;62:72–78.

5 Gasser RB, Lightowlers MW, Obendorf DL, Jenkins DJ, Rickard MD. Evaluation of a serological test system for the diagnosis of natural Echinococcus granulosus infection in dogs using E. granulosus protoscolex and oncosphere antigens. Aust Vet J 1988;65:369–373.

6 Coltorti EA, Varela-Díaz VM. Echinococcus granulosus: penetration of macromolecules and their localization on the parasite membranes of cysts. Exp Parasitol 1974;35: 225–231.

7 Coltorti EA, Varela-Díaz VM. Penetration of host IgG molecules into hydatid cysts. Z Parasitenkd 1975;48:47–51.

8 Hustead ST, Williams JF. Permeability studies on taeniid metacestodes. I. Uptake of proteins by larval stages of Taenia taeniaeformis, T. crassiceps and Echinococcus granulosus. J Parasitol 1977;63:314–321.

9 Engelkirk PG, Williams JF. Taenia taeniaeformis (cestoda) in the rat: ultrastructure of the host–parasite interface on days 1 to 7 postinfection. J Parasitol 1982;68:620–633.

10 Engelkirk PG, Williams JF. Taenia taeniaeformis (cestoda) in the rat: ultrastructure of the host-parasite interface on days 8 to 22 postinfection. J Parasitol 1983;69:828–837.

11 Bortoletti G, Ferretti G. Morphological studies on the development of Taenia taeniaeformis larvae in susceptible mice. Int J Parasitol 1985;15:365–375.

12 Rickard MD, White JB, Boddington EB. Vaccination of lambs against infection with Taenia ovis. Aust Vet J 1976;52: 209–214.

13 Gemmell MA, Johnstone PD. Factors regulating tapeworm populations: estimation of the duration of acquired immunity by sheep to Taenia hydatigena. Res Vet Sci 1981; 30:53–56.

14 Bogh HO, Rickard MD, Lightowlers MW. Studies on stage-specific immunity against Taenia taeniaeformis metacestodes in mice. Parasite Immunol 1988;10:255–264.

15 Ibarrola AS, Sobrini B, Guisantes J, et al. Membranous glomerulonephritis secondary to hydatid disease. Am J Med 1981;70:311–315.

16 Vialtel P, Chenais F, Desgeorges P, Couderc P, Micouin C, Cordonnier D. Membranous nephropathy associated with hydatid disease. N Engl J Med 1981;304:610–611.

17 Richard-Lenoble D, Smith MD, Loisy M, Verroust PJ. Human hydatidosis: evaluation of three serodiagnostic methods, the principal subclass of specific immunoglobulin and the detection of circulating immune complexes. Ann Trop Med Parasitol 1978;72:533–560.

18 Zvolinskene V. Determination of circulating hydatid antigens in the sera of patients with hydatidosis and their diagnostic significance. Acta Parasitol Litu 1981;19:56–62 (in Russian).

19 Leikina ES, Kovrova EA, Krasovskaya NN. Detection of circulating antigens in the bloodstream of patients with hydatid disease, alveococcosis and trichinosis. Med Parazitol Parazit Bolezni 1982;60:7–15 (in Russian).

20 D'Amelio R, Pontesilli O, Palmisano L, et al. Detection and partial characterization of circulating immune complexes in hydatid disease. J Clin Microbiol 1983;18:1021–1026.

21 Pini C, Pastore R, Valesini G. Circulating immune complexes in sera of patients infected with Echinococcus granulosus. Clin Exp Immunol 1983;51:572–578.

22 Gottstein B. An immunoassay for the detection of circulating antigens in human echinococcosis. Am J Trop Med Hyg 1984;33:1185–1191.

23 Craig PS, Nelson GS. The detection of circulating antigen in human hydatid disease. Ann Trop Med Parasitol 1984; 78:219–227.

24 Craig PS. Detection of specific circulating antigen, immune complexes and antibodies in human hydatidosis from Turkana (Kenya) and Great Britain, by enzyme-immunoassay. Parasite Immunol 1986;8:171–188.

25 Sogandares-Bernal F, Race MC, Dennis MV, Voge M. Circulating antigens in infections of mice by tetrathyridia of Mesocestoides corti Hoeppli, 1925. Z Parasitenkd 1981;64: 157–167.

26 Ali-Khan Z, Siboo R. Immune complexes in experimental

alveolar hydatidosis. Tropenmed Parasitol 1983;34:187–192.

27 Craig PS. Circulating antigens, antibodies and immune complexes in experimental *Taenia pisiformis* infection of rabbits. Parasitology 1984;1:121–131.

28 Bullock FD, Curtis MR. A study of the reactions of the tissues of the rat's liver to the larvae of *Taenia crassicollis* and the histiogenesis of cysticercus sarcoma. J Cancer Res 1924;8:446–481.

29 Bullock FD, Curtis MR. Spontaneous tumors of the rat. J Cancer Res 1930;14:1–115.

30 Cook RW, Williams JF. Pathology of *Taenia taeniaeformis* infection in the rat: gastrointestinal changes. J Comp Pathol 1981;91:205–217.

31 Cook RW, Trap AL, Williams JF. Pathology of *Taenia taeniaeformis* infection in the rat: hepatic, lymph node and thymic changes. J Comp Pathol 1981;91:219–226.

32 Rikihisa Y, Lin YC. *Taenia taeniaeformis*: increased cell growth and neutral mucus production in the gastric mucosa of the rat infected with the larvae. Exp Parasitol 1984;58:147–155.

33 Rikihisa Y, Letonja T, Pratt N, Lin YC. *Taenia taeniaeformis*: characterization of larval metabolic products and growth of host gastric cells *in vitro*. Exp Parasitol 1984;58:230–238.

34 Rikihisa Y, Lin YC, Fukaya T. *Taenia taeniaeformis*: inhibition of rat testosterone production by excretory-secretory product of the cultured metacestode. Exp Parasitol 1985;59:390–397.

35 Shiwaku K, Hirai K, Torii M, Tsuboi T. Evidence of the growth factor in mouse serum infected with *Spirometra erinacei* plerocercoids. Z Parasitenkd 1986;72:83–87.

36 Vuitton DA, Guerret-Stocker S, Carbillet JP, Mantion G, Miguet JP, Grimaud JA. Collagen immunotyping of the hepatic fibrosis in human alveolar echinococcosis. Z Parasitenkd 1986;72:97–104.

37 Miguet JP, Monange C, Carbillet JP, *et al*. L'echinococcose alvéolaire du foie. A propos de 20 cas observe's en Franche-Comté. II. Etude anatomo-pathologique. Arch Fr Mal Appar Dig 1976;65:23–32.

38 Ali-Khan Z, Siboo R. Pathogenesis and host response in subcutaneous alveolar hydatidosis. I. Histogenesis of alveolar cyst and a quantitative analysis of the inflammatory infiltrates. Z Parasitenkd 1980;62:241–254.

39 Ali-Khan Z. Host-parasite relationship in echinococcosis. I. Parasite biomass and response in three strains of inbred mice against graded doses of *Echinococcus multilocularis* cysts. J Parasitol 1974;60:231–235.

40 Ali-Khan Z, Jothy S, Alkarmi T. Murine alveolar hydatidosis: a potential experimental model for the study of AA-amyloidosis. Br J Exp Parasitol 1983;64:599–611.

41 Ali-Khan Z, Rausch RL. Demonstration of amyloid and immune complex deposits in renal and hepatic parenchyma of Alaskan alveolar hydatid disease patients. Ann Trop Med Parasitol 1987;81:381–392.

42 Ali-Khan Z, Sipe JD, Du T, Riml H. *Echinococcus multilocularis*: relationship between persistent inflammation, serum amyloid A protein response and amyloidosis in four mouse strains. Exp Parasitol 1988;67:334–345.

43 Leid RW, Williams JF. Helminth parasites and the inflammatory system. In Scheer BT, Florkin MA, eds. *Chemical Zoology*. New York: Academic Press, 1979:229–271.

44 Hammerberg B, Musoke AJ, Williams JF. Activation of complement by hydatid cyst fluid of *Echinococcus granulosus*. J Parasitol 1977;63:327–331.

45 Hammerberg B, Williams JF. Interaction between *Taenia taeniaeformis* and the complement system. J Immunol 1978;120:1033–1037.

46 Pericone R, Fontana L, DeCarolis C, Ottaviani P. Activation of alternative complement pathway by fluid from hydatid cysts. N Engl J Med 1980;302:808–809.

47 Leid RW, McConnell LA. Thromboxane A2 generation by the larval cestode, *Taenia taeniaeformis*. Clin Immunol Immunopathol 1983;28:67–76.

48 Cardenas Y, Cardenas J. Cysticercosis of the nervous system. II. Pathologic and radiologic findings. J Neurosurg 1962;19:635–640.

49 Willms K, Merchant MT. The inflammatory reaction surrounding *Taenia solium* larvae in pig muscle: ultrastructural and light microscopic observations. Parasite Immunol 1980;2:261–275.

50 Williams JF. An evaluation of the Casoni test in human hydatidosis using an antigen solution of low nitrogen concentration. Trans R Soc Trop Med Hyg 1972;66:160–164.

51 Yarzábal LA, Schantz PM, Lopez-Lemes MH. Comparative sensitivity and specificity of the Casoni intradermal and the immunoelectrophoresis tests for the diagnosis of hydatid disease. Am J Trop Med Hyg 1975;24:843–848.

52 Laynadier F, Luce H, Abrego A, Huguier M, Dry J. Human basophil degranulation test in diagnosis of hydatidosis. Br Med J 1980;280:1251–1252.

53 Vuitton DA, Bresson-Hadni S, Lenys D, *et al*. IgE-dependent humoral immune response in *Echinococcus multilocularis* infection: circulating and basophil-bound specific IgE against *Echinococcus* antigens in patients with alveolar echinococcosis. Clin Exp Immunol 1988;71:247–252.

54 Afferni C, Pini C, Misiti-Dorello P, Bernardini L, Conchedda M, Vicari G. Detection of specific IgE antibodies in sera from patients with hydatidosis. Clin Exp Immunol 1984;55:587–592.

55 Schantz PM. Homocytotropic antibody to *Echinococcus* antigen in sheep with homologous and heterologous larval cestode infection. Am J Vet Res 1973;34:1179–1181.

56 Romero-Torres R, Campbell JR. An interpretive review of the surgical treatment of hydatid disease. Surg Gynecol Obstet 1965;121:851–864.

57 Schantz PM, Van den Bossche H, Eckert J. Chemotherapy for larval echinococcosis in animals and humans—report of a workshop. Z Parasitenkd 1982;67:5–26.

58 Schantz PM. *Echinococcus granulosus*: acute systemic allergic reactions to hydatid cyst fluid in infected sheep. Exp Parasitol 1977;43:268–285.

59 Ali-Khan Z. Cellular changes in the lymphoreticular tissues of C57L/J mice infected with *Echinococcus multilocularis* cysts. Immunology 1978;34:831–839.

60 Riley EM, Dixon JB, Kelly DF, Cox DA. The immune response to *Echinococcus granulosus*: sequential histological observations of lymphoreticular and connective tissues

during early murine infection. J Comp Pathol 1985;95:
93–104.

61 Riley EM, Dixon JB, Jenkins P, Ross G. *Echinococcus granulosus* infection in mice: host responses during primary and secondary infection. Parasitology 1986;92:391–403.

62 Ali-Khan Z. Cell mediated immune response in early and chronic alveolar murine hydatidosis. Exp Parasitol 1978; 46:157–165.

63 Alkarmi T, Behbehani K. *Echinococcus multilocularis*: inhibition of murine neutrophil and macrophage chemotaxis. Exp Parasitol 1989;69:16–22.

64 Vuitton DA, Lasségue A, Miguet JP, *et al.* Humoral and cellular immunity in patients with hepatic alveolar echinococcosis. A two year follow-up with and without flubendazole treatment. Parasite Immunol 1984;6:329–340.

65 Allan D, Jenkins P, Connor RJ, Dixon JB. A study of immunoregulation of BALB/c mice by *Echinococcus granulosus equinus* during prolonged infection. Parasite Immunol 1981;3:137–142.

66 Good AH, Miller KL. Depression of the immune response to sheep erythrocytes in mice infected with *Taenia crassiceps* larvae. Infect Immun 1976;14:449–456.

67 Nichol CP, Sewell MM. Immunosuppression by larval cestodes of *Babesia microti* infections. Ann Trop Med Parasitol 1984;78:228–233.

68 Burger CJ, Rikihisa Y, Lin YC. *Taenia taeniaeformis*: inhibition of mitogen induced proliferation and interleukin-2 production in rat splenocytes by larval *in vitro* product. Exp Parasitol 1986;62:216–222.

69 Letonja T, Hammerberg C, Schurig G. Evaluation of spleen lymphocyte responsiveness to a T-cell mitogen during early infection with larval *Taenia taeniaeformis*. Parasitol Res 1987;73:265–270.

70 Miller KL, Good AH, Mishell RI. Immunodepression in *Taenia crassiceps* infection: restoration of the *in vitro* response to sheep erythrocytes by activated peritoneal cells. Infect Immun 1978;22:365–370.

71 Mitchell GF, Handman E. Studies on the immune responses to larval cestodes in mice: a simple mechanism of non-specific immunosuppression in *Mesocestoides corti*-infected mice. Aust J Exp Biol Med Sci 1977;55:615–622.

72 Sealey M, Ramos C, Willms K, Ortiz-Ortiz L. *Taenia solium*: mitogenic effect of larval extracts on murine B lymphocytes. Parasite Immunol 1981;3:299–307.

73 Dixon JB, Jenkins P, Allan D. Immune recognition of *Echinococcus granulosus*. I. Parasite-activated primary transformation by normal murine lymph node cells. Parasite Immunol 1982;4:33–41.

74 Judson DG, Dixon JB, Skerritt GC, Stallbaumer M. Mitogenic effect of *Coenurus cerebralis* cyst fluid. Res Vet Sci 1984;37:128.

75 Judson DC, Dixon JB, Skerritt GC. Occurrence and biochemical characteristics of cestode lymphocyte mitogens. Parasitology 1987;94:151–160.

76 Dixon JB, Jenkins P, Allan D, Connor RJ. Blastic stimulation of unprimed mouse lymphocytes by living protoscolices of *Echinococcus granulosus*: a possible connection with transplant immunity. J Parasitol 1978;64:949–950.

77 Cox DA, Dixon JB, Marshall-Clark S. Transformation induced by *Echinococcus granulosus* protoscoleces in

unprimed murine spleen cells: identity and MHC restruction of participating cell types. Immunology 1986;57: 461–466.

78 Pini C, Pastore R, Valesini G. Circulating immune complexes in sera of patients infected with *Echinococcus granulosus*. Clin Exp Immunol 1983;51:572–578.

79 Ben Izhak O, Tatarsky I. Positive direct antiglobulin test associated with echinococcosis: a case report. J Trop Med Hyg 1985;88:389–390.

80 Ameglio F, Saba F, Bitti A, *et al.* Antibody reactivity to HLA classes I and II in sera from patients with hydatidosis. J Infect Dis 1987;156:673–676.

81 Mori H, Wernli B, Weiss N, Franklin RM. Autoantibodies in humans with cystic or alveolar echinococcosis. Trans R Soc Trop Med Hyg 1986;80:978–980.

82 Kagan IG, Norman L, Allain DS, Goodchild CG. Studies on echinococcosis: non-specific serologic reactions of hydatid-fluid antigen with serum of patients ill with diseases other than echinococcosis. J Helminthol 1960;64: 635–640.

83 Hammerberg B, Williams JF. Physico-chemical characterization of complement-interacting factors from *Taenia taeniaeformis*. J Immunol 1978;120:1039–1045.

84 Hammerberg B, Dangler C, Williams JF. *Taenia taeniaeformis*: chemical composition of parasite factors affecting coagulation and complement cascades. J Parasitol 1980;66:569–576.

85 Bogh HO, Lightowlers MW, Sullivan D, Mitchell GF, Rickard MD. Stage-specific immunity to *Taenia taeniaeformis* infection in mice: an histological study of the course of infection in mice vaccinated with either oncosphere or metacestode antigens. Parasite Immunol 1990;12:153–162.

86 Nemeth I, Juhasz S. A trypsin and chymotrypsin inhibitor from the metacestodes of *Taenia pisiformis*. Parasitology 1980;80:433–446.

87 Leid RW, Suquet CM, Perryman LE. Inhibition of antigen- and lectin-induced proliferation of rat spleen cells by a *Taenia taeniaeformis* proteinase inhibitor. Clin Exp Immunol 1984;57:187–194.

88 Suquet C, Green-Edwards C, Leid RW. Isolation and partial characterization of a *Taenia taeniaeformis* metacestode proteinase inhibitor. Int J Parasitol 1984;14:165–172.

89 Leid RW, Suquet CM, Bouwer HGA, Hinrichs DJ. Interleukin inhibition by a parasite proteinase inhibitor, taeniaestatin. J Immunol 1986;137:2700–2702.

90 Lightowlers MW, Haralambous A, Rickard MD. Amino acid sequence homology between cyclophilin and a cDNA-cloned antigen of *Echinococcus granulosus*. Mol Biochem Parasitol 1989;36:287–289.

91 Takahashi N, Hayano T, Suzuki M. Peptidyl-prolyl *cis-trans* isomerase is the cyclosporin A-binding protein cyclophilin. Nature 1989;337:473–475.

92 Lafferty KJ, Borel JF, Hodgkin P. Cyclosporin-A (CsA): models for the mechanism of action. In Kahan BD, ed. *Cyclosporine Biological Activity and Clinical Applications*. Orlando: Grune & Stratton Inc, 1984.

93 Hess AD, Esa AH, Colombani PM. Mechanisms of action of cyclosporine: effects on cells of the immune system and on subcellular events in T cell activation. Trans Proc 1988; 20(Suppl. 2):29–40.

94 Shi Y, Sahai BM, Green DR. Cyclosporin A inhibits activation-induced cell death in T-cell hybridomas and thymocytes. Nature 1989;339:625–626.

95 Mitchell GF, Rajasekariah GR, Rickard MD. A mechanism to account for mouse strain variation in resistance to the larval cestode, *Taenia taeniaeformis*. Immunology 1980; 39:481–489.

96 Rickard MD, Williams JF. Hydatidosis/cysticercosis: immune mechanisms and immunization against infection. Adv Parasitol 1982;21:229–296.

97 Oliver L. Natural resistance to *Taenia taeniaeformis*. I. Strain differences in susceptibility of rodents. J Parasitol 1962;48: 373–378.

98 Heath DD, Elsdon-Dew R. The *in vitro* culture of *Taenia saginata* and *Taenia taeniaeformis* larvae from the oncosphere, with observations on the role of serum for *in vitro* culture of larval cestodes. J Parasitol 1972;2:119–130.

99 Ambu S, Kwa BH. Susceptibility of rats to *Taenia taeniaeformis* infection. J Helminthol 1980;54:43–44.

100 Williams JF, Shearer AM, Ravitch MM. Differences in susceptibility of rat strains to experimental infection with *Taenia taeniaeformis*. J Parasitol 1981;67:540–547.

101 Smyth JD, Haselwood GAD. The biochemistry of bile as a factor in determining host specificity in intestinal parasites, with particular reference to *Echinococcus granulosus*. Ann NY Acad Sci 1963;113:234–260.

102 Weinman CJ. Cestodes and acanthocephala. In Jackson CJ, Herman R, Singer I, eds. *Immunity to Parasitic Animals*, vol. 2. New York: Appleton-Century-Crofts, 1970:1021–1059.

103 Musoke AJ, Williams JF, Leid RW, Williams CSF. The immunological response of the rat to infection with *Taenia taeniaeformis*. IV. Immunoglobulins involved in passive transfer of resistance from mother to offspring. Immunology 1975;29:845–853.

104 Beard TC. Evidence that a hydatid cyst is seldom "as old as the patient". Lancet 1978;ii:30–32.

105 Greenfield SH. Age resistance of the albino rat to *Cysticercus fasciolaris*. J Parasitol 1942;28:207–211.

106 Dow C, Jarrett WFH. Age, strain and sex differences in susceptibility to *Cysticercus fasciolaris* in the mouse. Exp Parasitol 1960;10:72–74.

107 Turner HM, McKeever S. The refractory responses of the White Swiss strain of *Mus musculus* to infection with *Taenia taeniaeformis*. Int J Parasitol 1976;6:483–487.

108 Mitchell GF, Goding JW, Rickard MD. Studies on the immune response to larval cestodes in mice. I. Increased susceptibility of certain mouse strains and hypothymic nude mice to *Taenia taeniaeformis* and analysis of passive transfer of resistance with serum. Aust J Exp Biol Med Sci 1978;55:165–186.

109 Thompson RCA. Inhibitory effect of BCG on development of secondary hydatid cysts of *Echinococcus granulosus*. Vet Rec 1979;99:273.

110 Rau ME, Tanner CE. BCG suppresses growth and metastasis of hydatid infections. Nature 1975;256:318–319.

111 Reuban JM, Tanner EE, Rau ME. Immunoprophylaxis with BCG of experimental *Echinococcus multilocularis* infection. Infect Immun 1978;21:135–139.

112 Thompson RCA, Penhale WJ, White TR, Pass DA. BCG induced inhibition and destruction of *Taenia taeniaeformis* in mice. Parasite Immunol 1982;4:93–99.

113 White TR, Thompson RCA, Penhale WJ. Studies on BCG immunotherapy in mice infected with *Mesocestoides corti*. Int J Parasitol 1988;18:389–393.

114 Toye PG, Jenkins CR. Protection against *Mesocestoides corti* infection in mice treated with zymosan or *Salmonella enteritidis* IIRX. Int J Parasitol 1982;12:399–402.

115 Reuben JM, Tanner CE. Protection against experimental echinococcosis by non-specifically stimulated peritoneal cells. Parasite Immunol 1983;5:61–66.

116 Letonja T, Rikihisa Y, Hammerberg C. Differential cellular response of resistant and susceptible rodents to the early stages of infection of *Taenia taeniaeformis*. Int J Parasitol 1984;14:551–558.

117 Davis SW, Hammerberg B. Activation of the alternative pathway of complement by larval *Taenia taeniaeformis* in resistant and susceptible strains of mice. Int J Parasitol 1988;18:591–597.

118 Engelkirk PG, Williams JF, Signs MM. Interactions between *Taenia taeniaeformis* and host cells *in vitro*: rapid adherence of peritoneal cells to strobilocerci. Int J Parasitol 1981;11:463–474.

119 Letonja T, Hammerberg C. *Taenia taeniaeformis*: early inflammatory response around developing metacestodes in the liver of resistant and susceptible mice. II. Histochemistry and cytochemistry. J Parasitol 1987;73:971–979.

120 Miller HM. Immunity of the white rat to super infestation with *Cysticercus fasciolaris*. Proc Soc Exp Biol 1931;28:467–468.

121 Kerr KB. Immunity against a cestode parasite—*Cysticercus pisiformis*. Am J Hyg 1935;22:169–182.

122 Lloyd S. Cysticercosis. In Soulsby EJL, ed. *Immune Responses in Parasitic Infections: Immunology, Immunopathology and Immunoprophylaxis: vol. 2, Trematodes and Cestodes*. Boca Raton: CRC Press, 1987:183–212.

123 Smithers SR, Terry RJ. Immunity in schistosomiasis. Ann NY Acad Sci 1969;160;826–840.

124 Mitchell GF. A note on concomitant immunity in host–parasite relationships: a successfully transplanted concept from tumour immunology. Adv Cancer Res 1989;54:319–332.

125 Musoke AJ, Williams JF. The immunological response of the rate to infection with *Taenia taeniaeformis*. V. Sequence of appearance of protective immunoglobulins and mechanism of action of 7S2a antibodies. Immunology 1975;29: 855–866.

126 Lloyd S, Soulsby EJL. The role of IgA immunoglobulins in the passive transfer of protection to *Taenia taeniaeformis* in the mouse. Immunology 1978;34:939–945.

127 Heath DD, Pavloff P. The fate of *Taenia taeniaeformis* oncospheres in normal and passively protected rats. Int J Parasitol 1975;5:83–88.

128 Musoke AJ, Williams JF, Leid RW. Immunological response of the rat to infection with *Taenia taeniaeformis*. VI. The role of immediate type hypersensitivity in resistance to reinfection. Immunology 1978;34:565–570.

129 Mitchell GF, Marchalonis JJ, Smith PM, Nicholas WL, Warner NL. Studies on immune responses to larval cestodes in mice. Immunoglobulins associated with the

larvae of *Mesocestoides corti*. Aust J Exp Biol Med Sci 1977; 55:187–211.

130 Chapman CB, Knopf PM, Hicks JD, Mitchell GF. IgG1 hypergammaglobulinaemia in chronic parasitic infections in mice. Magnitude of the response in mice infected with various parasites. Aust J Exp Biol Med Sci 1979;57:369–387.

131 Varela-Díaz VM, Gemmell MA, Williams JF. Immunological responses of the mammalian host against tapeworm infections. XII. Observations on antigen sharing between *Taenia hydatigena* and *Taenia ovis*. Exp Parasitol 1972;32:96–101.

132 Rickard M. Hypothesis for the long-term survival of *Taenia pisiformis* cysticerci in rabbits. Z Parasitenkd 1974;44:203–209.

133 Gibbens JC, Harrison LJ, Parkhouse RM. Immunoglobulin class responses to *Taenia taeniaeformis* in susceptible and resistant mice. Parasite Immunol 1986;8:491–502.

134 Toye PG, Ey PL, Jenkin CR. Activation of complement by tetrathyridia of *Mesocestoides corti*: enhancement of antibodies from infected mice and lack of effect on parasite viability. J Parasitol 1984;70:871–878.

135 Siebert AE, Blitz RR, Morita CT, Good AH. *Taenia crassiceps*: serum and surface immunoglobulins in metacestode infections in mice. Exp Parasitol 1981;51:418–430.

136 Conchedda M, Ferretti G. Susceptibility of different strains of mice to various levels of infection with the eggs of *Taenia taeniaeformis*. Int J Parasitol 1984;14:541–546.

137 Bortoletti G, Conchedda M, Ferretti G. Damage and early destruction of *Taenia taeniaeformis* larvae in resistant hosts, and anomalous development in susceptible hosts: a light microscopic and ultrastructural study. Int J Parasitol 1985;15:377–384.

138 Bowtell DDL, Mitchell GF, Anders RF, Lightowlers MW, Rickard MD. *Taenia taeniaeformis*: immunoprecipitation analysis of protein antigens of oncospheres and larvae. Exp Parasitol 1983;56:416–427.

139 Lightowlers MW, Mitchell GF, Bowtell DDL, Anders RF, Rickard MD. Immunization against *Taenia taeniaeformis* in mice: studies on the characterization of antigens from oncospheres. Int J Parasitol 1984;14:321–332.

140 Lightowlers MW, Rickard MD, Mitchell GF. Immunization against *Taenia taeniaeformis* in mice: identification of oncospheral antigens in polyacrylamide gels by Western blotting and enzyme immunoassay. Int J Parasitol 1986;16:297–306.

141 Lightowlers MW, Rickard MD, Mitchell GF. *Taenia taeniaeformis* in mice: passive transfer of protection with sera from infected or vaccinated mice and analysis of serum antibodies to oncospheral antigens. Int J Parasitol 1986;16:307–315.

142 Gallie GJ, Sewell MMH. The survival of *Cysticercus bovis* in resistant calves. Vet Rec 1972;91:481–482.

143 Gallie GJ, Sewell MMH. The serological response of calves infected neonatally with *Taenia saginata* (*Cysticercus bovis*). Trop Anim Health Prod 1974;6:163–171.

144 Froyd G. The artificial infection of calves with oncospheres of *Taenia saginata*. Res Vet Sci 1961;2:243–247.

145 Froyd G. The artificial infection of cattle with *Taenia saginata* eggs. Res Vet Sci 1964;5:434–440.

146 Lightowlers MW, Rickard MD, Honey RD. Serum antibody following parenteral immunization with hydatid cyst fluid in sheep infected with *Echinococcus granulosus*. Am J Trop Med Hyg 1986;35:818–823.

147 Ito A, Smyth JD. Adult cestodes—immunology of the lumen-dwelling cestode infections. In Soulsby EJL, ed. *Immune Responses in Parasitic Infections: Immunology, Immunopathology and Immunoprophylaxis: vol. 2, Trematodes and Cestodes*. Boca Raton, CRC Press, 1987:115–163.

148 Ito A, Itoh M, Andreassen J, Onitake K. Stage-specific antigens of *Hymenolepis microstoma* recognized in BALB/c mice. 1989;11:453–462.

149 Abraham KM, Teale JM. The contribution of parasite specific T cells to isotype restriction in *Mesocestoides corti*-infected mice. J Immunol 1987;15:2530–2537.

150 Lammas DA, Mitchell LA, Wakelin D. Adoptive transfer of enhanced eosinophilia and resistance to infection in mice by an *in vitro* generated T-cell line specific for *Mesocestoides corti* larval antigens. Parasite Immunol 1987;9:591–601.

151 Sanderson CJ, Strath M. Isolation of specific antigen-reactive T-cell clones from nude (nu/nu) mice infected with *Mesocestoides corti*. Immunology 1985;54:275–279.

152 Mitchell GF. Effector cells, molecules and mechanisms in host-protective immunity to parasites. Adv Immunol 1979;38:209–223.

153 Williams JF, Picone J, Engelkirk P. Evasion of immunity by cestodes. In van den Bossche H, ed. *The Host Invader Interplay*. Amsterdam: Elsevier, 1980:205–216.

154 Leid RW, Grant RF, Suquet CM. Inhibition of neutrophil aggregation by taeniaestatin, a cestode proteinase inhibitor. Int J Parasitol 1987;17:1349–1353.

155 Kwa BH, Liew FY. Studies on the mechanisms of long-term survival of *Taenia taeniaeformis* in rats. J Helminthol 1978;52:1–6.

156 Letonja T, Hammerberg B. Third component of complement, immunoglobulin deposition and leukocyte attachment related to surface sulfate on larval *Taenia taeniaeformis*. J Parasitol 1983;69:637–644.

157 Conder GA, Picone J, Geary AM, deHoog J, Williams JF. Lytic effects of normal serum on isolated postoncospheral and metacestode stages of *Taenia taeniaeformis*. J Parasitol 1983;69:465–472.

158 Stokes CR. Induction and control of intestinal immune responses. In Newby TJ, Stokes CR, eds. *Local Immune Responses of the Gut*. Boca Raton: CRC Press, 1984:98–141.

159 Thompson RCA, Dunsmore JD, Hayton AR. *Echinococcus granulosus*: secretory activity of the rostellum of the adult cestode in situ in the dog. Exp Parasitol 1979;48:144–163.

160 Rothwell TLW. Immune expulsion of parasitic nematodes from the alimentary tract. Int J Parasitol 1989;19:139–168.

161 Rickard MD. Immunity. In Arme C, Pappas P, eds. *The Biology of the Eucestoda*. London: Academic Press, 1983:539–579.

162 Isaak DD, Jacobson RH, Reed ND. Thymus dependence of tapeworm (*Hymenolepis diminuta*) elimination from mice. Infect Immun 1975;12:1478–1479.

163 Hopkins CA, Subramanian G, Stallard H. The development of *Hymenolepis diminuta* in primary and secondary infections in mice. Parasitology 1972;64:401–412.

164 Howard RJ. The growth of secondary infections of *Hymenolepis microstoma* in mice: the effect of various primary infection regimes. Parasitology 1976;72: 317–323.

165 Andreassen J, Hopkins CA. Immunologically mediated rejection of *Hymenolepis diminuta* by its normal host, the rat. J Parasitol 1980;66:898–903.

166 Hopkins CA, Andreassen J, Barr IF. Duration of immunological memory evoked by adult tapeworms. Parasitology of 1980;81:xl–xli.

167 Kondo K, Yoshimura H, Ohnishi Y, Nishida K, Kamimura K. Immunological studies on diphyllobothriasis. I. Immunoglobulin and precipitation tests using ochterlony and immunoelectrophoresis in the patients. Jpn J Parasitol 1977;26:265–270.

168 Deschiens R, Renaudet R. La reaction de fixation du complement dans les teniasis a *Taenia saginata*. Bull Soc Pathol Exot (Paris) 1941;34:17–25.

169 Machnicka-Roguska B, Zwierz C. Haemagglutination reaction in people infected with *Taenia saginata*. Wiad Parazytal 1964;10:467–468.

170 Machnicka B, Zwierz C. The immunologic reactivity of the sera of people infected with *Taenia saginata* to *Cysticercus bovis* antigens. Bull Acad Polon Sci 1974;22:259–261.

171 Capron A, Wattre P, Capron M. Value of the immunological diagnosis of taeniasis. Lille Med 1973;18: 513–516.

172 Kosmiderski S, Polak S, Burczek R. Styrene latex used for the detection of antibodies to *Taenia*. Pol Tyg Leka 1971;29: 1271–1272.

173 Jenkins DJ, Rickard MD. Specificity of scolex and oncosphere antigens for the serological diagnosis of taeniid cestode infections in dogs. Aust Vet J 1986;63: 40–42.

174 Heath DD, Lawrence SB, Glennie A, Twaalfhoven H. The use of excretory and secretory antigens of the scolex of *Taenia ovis* for the serodiagnosis of infection in dogs. J Parasitol 1985;71:192–199.

175 Nascimento E. *Taenia taeniaeformis*: aspects of the host parasite relationship. Mem Inst Oswaldo Cruz 1982;77: 319–323.

176 Chordi A, González-Castro J, Tormo J. Aportacion al estudio de les helminthiasis intestinales en los perros. II. Resultados de la prueba de floculatión con "Benthid" y de la de fijación de complemento en perros con *Echinococcus granulosus*. Rev Ibér Parasitol 1962;22:285–290.

177 Movsesijan M, Mladenović Z. The possibility of using different development stages of *E. granulosus* for detection of specific antibodies against their parasite. Vet Glas 1971; 25:159–163.

178 Jenkins DJ, Rickard MD. Specific antibody responses in dogs experimentally infected with *Echinococcus granulosus*. Am J Trop Med Hyg 1986; 35:345–349.

179 Machnicka-Roguska B. Studies on *Monezia expansa* antigens, IV. Serological examinations of sheep from an endemic area. Acta Parasitol (Pol) 1972;xx:409–419.

180 Gray JS. Studies on the course of infection of the poultry cestode *Raillietina cesticillus* (Molin 1858) in the definitive host. Parasitology 1972;65:243–250.

181 Gray JS. Studies on host resistance to secondary infec-

tions of *Raillietina cesticillus*, Molin 1858, in the fowl. Parasitology 1973;67:375–382.

182 Ramsdell SG. A note on skin-reaction in *Taenia* infestation. J Parasitol 1927;14:102–105.

183 Machnicka-Roguska B, Zwierz C. Intradermal test with antigenic fractions in *Taenia saginata* infection. Acta Parasitol 1970;17:293–299.

184 Slusarski W, Zapart W. Diagnostic value of intradermal test with acid soluble protein fractions of *Taenia* infections in man. Acta Parasitol 1971;19:445–455.

185 Turner EL, Dennis FW, Berberian DA. The value of the Casoni test in dogs. J Parasitol 1935;21:180–182.

186 Williams JF, Pérez-Esandi MV. Reaginic antibodies in dogs infected with *Echinococcus granulosus*. Immunology 1971; 20:451–455.

187 Miller HM. Superinfection of cats with *Taenia taeniaeformis*. J Prev Med 1932;6:17–29.

188 Clapham PA. Studies on *Coenurus glomeratus*. J Helminthol 1940;18:45–52.

189 Vukovic V. Infection and superinfection of the dog with *Taenia hydatigena*. Arch Sci Biol 1949;1:258–261.

190 Rickard MD, Coman BJ, Cannon RM. Age resistance and acquired immunity to *Taenia pisiformis* infection in dogs. Vet Parasitol 1977;3:1–9.

191 Gemmell MA, Lawson JR, Roberts MG. Population dynamics in echinococcosis and cysticercosis: biological parameters of *Echinococcus granulosus* in dogs and sheep. Parasitology 1986;92:599–620.

192 Turner EL, Berberian DA, Dennis EW. Successful artificial immunization of dogs against *Taenia echinococcus*. Proc Soc Exp Biol Med 1933;30:618–619.

193 Turner EL, Berberian DA, Dennis EW. The production of artificial immunity in dogs against *Echinococcus granulosus*. J Parasitol 1936;22:14–28.

194 Forsek Z, Rukavina J. Experimental immunization of dogs against *Echinococcus granulosus*. I. Preliminary findings. Veterinaria Sarajevo 1959;8:479–482.

195 Gemmell MA. Natural and acquired immunity factors interfering with development during the rapid growth phase of *Echinococcus granulosus* in dogs. Immunology 1962;5:496–503.

196 Movsesijan M, Mladenovic Z. Active immunization of dogs against *Echinococcus granulosus*. Vet Glas 1970;24: 189–193.

197 Smyth JD, Gemmell M, Smyth MM. Establishment of *Echinococcus granulosus* in the intestine of normal and vaccinated dogs. In Singh KS, Tandau BK, eds. *HD Srivastava Commemoration Volume*. Division of Parasitology, Indian Veterinary Research Institute. Izatnagar: Utlar Pradesh, 1970:167–178.

198 Herd RP, Chappel RJ, Biddell D. Immunization of dogs against *Echinococcus granulosus* using worm secretory antigens. Int J Parasitol 1975;5:395–399.

199 Rickard MD, Parmeter SN, Gemmell MA. The effect of development of *Taenia hydatigena* larvae in the peritoneal cavity of dogs on resistance to a challenge infection with *Echinococcus granulosus*. Int J Parasitol 1975;5:281–283.

200 Herd RP. Resistance of dogs to *Echinococcus granulosus*. Int J Parasitol 1977;7:135–138.

201 Heath DD, Parmeter SN, Osborne PJ. An attempt to

immunise dogs against *Taenia hydatigena*. Res Vet Sci 1980;29:388–389.

202 Heath DD, Lawrence SB, Oudemans G. A blind test of the serological response of dogs to infection with *Taenia ovis*. NZ Vet J 1988;36:143–145.

203 Gasser RB, Lightowlers MW, Rickard MD, Lyford RA, Dawkins HJS. Serological screening of farm dogs for *Echinococcus granulosus* infection in an endemic region. Aust Vet J 1990;67:145–147.

204 Gasser RB, Lightowlers MW, Rickard MD. Identification of protein components of *Echinococcus granulosus* protoscolex antigens for specific serodiagnosis of *E. granulosus* infection in dogs. Parasite Immunol 1989;11:279–291.

205 Gasser RB, Lightowlers MW, Rickard MD. A recombinant antigen with potential for Serodiagnosis of *Echinococcus granulosus* infection in dogs. Int J Parasitol 1990;20:943–950.

206 Lawson JR, Gemmell MA. Hydatidosis and cysticercosis: the dynamics of transmission. Adv Parasitol 1983;22:261–308.

207 Roberts MG, Lawson JR, Gemmell MA. Population dynamics in echinococcosis and cysticercosis: mathematical model of the life cycle of *Echinococcus granulosus*. Parasitology 1986;92:621–641.

208 Roberts MG, Lawson JR, Gemmell MA. Population dynamics in echinococcosis and cysticercosis: mathematical model of the life cycle of *Taenia hydatigena* and *T. ovis*. Parasitology 1987;94:181–197.

209 Gemmell MA, Lawson JR, Roberts MG. Population dynamics in echinococcosis and cysticercosis: evaluation of the biological parameters of *Taenia hydatigena* and *T. ovis* and commparisons with those of *Echinococcus granulosus*. Parasitology 1987;94:161–180.

210 Gemmell MA. A critical approach to the concepts of control and eradication of echinococcosis/hydatidosis and taeniasis/cysticercosis. Int J Parasitol 1987;17:465–472.

211 Williams JF. Recent advances in the immunology of cestode infections. J Parasitol 1979;65:337–349.

212 Williams JF. Cestode infections. In Cohen S, Warren KS, eds. *Immunology of Parasitic Infections*, 2nd edn. Oxford: Blackwell Scientific Publications, 1982:676–714.

213 Rickard MD. Immunization against infection with larval taeniid cestodes using oncospheral antigens. In Flisser A, Willms K, Laclette JP, Larralde C, eds. *Cysticercosis. Present State of Knowledge and Perspectives*. New York: Academic Press, 1982:633–646.

214 Miller HM Jr. The production of artificial immunity in the albino rat to a metazoan parasite. J Prev Med 1931;5:429–452.

215 Campbell DH. Active immunization of albino rats with protein fractions from *Taenia taeniaeformis* and its larval form *Cysticercus fasciolaris*. Am J Hyg 1936;23:104–113.

216 Kwa BH, Liew FY. Immunity in taeniasis-cysticercosis. I. Vaccination against *Taenia taeniaeformis* in rats using purified antigen. J Exp Med 1977;146:118–131.

217 Ayuya JM, Williams JF. Immunological response of the rat to infection with *Taenia taeniaeformis*. VII. Oral and parenteral immunization with parasite antigens. Immunology 1979;36:825–834.

218 Rajasekariah GR, Mitchell GF, Rickard MD. Immunization of mice against infection with *Taenia taeniaeformis* using

various antigens prepared from eggs, oncospheres, developing larvae and strobilocerci. Int J Parasitol 1980;10:315–324.

219 Miller HM, Kerr KB. Attempts to immunize rabbits against a larval cestode, *Cysticercus pisiformis*. Proc Soc Exp Biol 1932;29:670–671.

220 Heath DD. Resistance to *Taenia pisiformis* larvae in rabbits. I. Examination of the antigenically protective phase of larval development. Int J Parasitol 1973;3:485–489.

221 Molinari JL, Meza R, Suarez B, Palacios S, Tato P. *Taenia solium*: immunity in hogs to the cysticercus. Exp Parasitol 1983;55:340–357.

222 Gemmell MA. Immunological responses of the mammalian host against tapeworm infections. I. Species specificity of hexacanth embryos in protecting sheep against *Taenia hydatigena*. Immunology 1964;7:489–499.

223 Gemmell MA. Immunological responses of the mammalian host against tapeworm infections. III. Species specificity of hexacanth embryos in protecting sheep against *Taenia ovis*. Immunology 1965;8:281–290.

224 Rickard MD, Bell KJ. Immunity produced against *Taenia ovis* and *T. taeniaeformis* infection in lambs and rats following *in vivo* growth of these larvae in filtration membrane diffusion chambers. J Parasitol 1971;57:571–575.

225 Heath DD, Smyth JD. *In vitro* cultivation of *Echinococcus granulosus*, *Taenia hydatigena*, *T. ovis*, *T. pisiformis* and *T. serialis* from oncosphere to cystic larva. Parasitology 1970;61:329–343.

226 Rickard MD, Bell KJ. Successful vaccination of lambs against infection with *Taenia ovis* using antigens produced during *in vitro* cultivation of the larva stages. Res Vet Sci 1971;12:401–402.

227 Rickard MD, Adolph AJ. Vaccination of calves against *Taenia saginata* infection using a "parasite-free" vaccine. Vet Parasitol 1976;1:389–392.

228 Lloyd S. Homologous and heterologous immunization against metacestodes of *Taenia saginata* and *Taenia taeniaeformis* in cattle and mice. Z Parasitenkd 1979;60:87–96.

229 Rajasekariah GR, Mitchell GF, Rickard MD. *Taenia taeniaeformis* in mice: protective immunization with oncospheres and their products. Int J Parasitol 1980;10:155–160.

230 Rickard MD, Outteridge PM. Antibody and cell-mediated immunity in rabbits infected with the larval stages of *Taenia pisiformis*. Z Parasitenkd 1974;44:187–201.

231 Edwards GT, Herbert IV. Preliminary investigations into the immunization of lambs against infection with *Taenia multiceps* metacestodes. Vet Parasitol 1982;9:193–199.

232 Verster A, Tustin RC. Immunization of sheep against the larval stage of *Taenia multiceps*. Onderstepoort J Vet Res 1987;54:103–105.

233 Heath DD, Parmeter SN, Osborn PJ, Lawrence SB. Resistance to *Echinococcus granulosus* infection in lambs. J Parasitol 1981;67:797–799.

234 Rickard MD, Adolph AJ, Arundel JH. Vaccination of calves against *Taenia saginata* infection using antigens collected during *in vitro* cultivation of larvae: passive protection via colostrum from vaccinated cows and

vaccination of calves protected by maternal antibody. Res Vet Sci 1977;23:365–367.

235 Rickard MD, Boddington EB, McQuade N. Vaccination of lambs against *Taenia ovis* infection using antigens collected during *in vitro* cultivation of larvae: passive protection via colostrum from vaccinated ewes and the duration of immunity from a single vaccination. Res Vet Sci 1977; 23:368–371.

236 Rickard MD, Arundel JH, Adolph AJ. A preliminary field trial to evaluate the use of immunization for the control of naturally acquired *Taenia saginata* infection in cattle. Res Vet Sci 1981;30:104–108.

237 Rickard MD, Brumley JL, Anderson GA. A field trial to evaluate the use of antigens from *Taenia hydatigena* oncospheres to prevent infection with *T. saginata* in cattle grazed on sewage irrigated pasture. Res Vet Sci 1982; 32:189–193.

238 Rickard MD, Brumley JL. Immunization of calves against *Taenia saginata* infection using antigens collected by *in vitro* incubation of *T. saginata* oncospheres or ultrasonic disintegration of *T. saginata* and *T. hydatigena* oncospheres. Res Vet Sci 1981;30:99–103.

239 Bowtell DDL, Saint RB, Rickard MD, Mitchell GF. Expression of *Taenia taeniaeformis* antigens in *Escherichia coli*. Mol Biochem Parasitol 1984;13:173–185.

240 Bowtell DDL, Saint RB, Rickard MD, Mitchell GF. Immunochemical analysis of *Taenia taeniaeformis* antigens expressed in *Escherichia coli*. Parasitology 1986;93:599–610.

241 Howell MJ, Hargreaves JJ. Cloning and expression of *Taenia ovis* antigens in *Escherichia coli*. Mol Biochem Parasitol 1988;28:21–30.

242 Harrison LJS, Parkhouse RME. Passive protection against *Taenia saginata* infection in cattle by a mouse monoclonal antibody reactive with the surface of the invasive oncosphere. Parasite Immunol 1986;8:319–332.

243 Harrison LJS, Joshua GWP, Parkhouse RME. Identification of protective antigens in *Taenia saginata* cysticercosis. *Proc VI Int Congresson Parasitology*. Brisbane: University of Queensland Press, 1986:277.

244 Johnson KS, Harrison GBL, Lightowlers MW, et al. Vaccination against ovine cysticercosis using a defined recombinant antigen. Nature 1989;338:585–587.

245 Smith DB, Johnson KS. Single step purification of polypeptides expressed in *Escherichia coli* as fusions with glutathione *S*-transferase. Gene 1988;67:31–40.

246 Gemmell MA. Immunological responses of the mammalian host against tapeworm infections. IV. Species specificity of hexacanth embryo in protecting sheep against *Echinococcus granulosus*. Immunology 1966;11: 325–335.

247 Heath DD, Lawrence SB, Yong WK. Cross-protection between the cysts of *Echinococcus granulosus*, *Taenia hydatigena* and *T. ovis* in lambs. Res Vet Sci 1979;27: 210–212.

248 Gemmell MA. Hydatidosis and cysticercosis. I. Acquired resistance to the larval phase. Aust Vet J 1969;45:521–524.

249 Gemmell MA. Hydatidosis and cysticercosis. 3. Induced resistance to the larval phase. Aust Vet J 1970;46:366–369.

250 Gemmell MA. Immunological responses of the mammalian host against tapeworm infections. XI. Antigen

251 Rickard MD, Coman BJ. Studies on the fate of *Taenia hydatigena* and *Taenia ovis* larvae in rabbit, and cross-immunity with *Taenia pisiformis* larvae. Int J Parasitol 1977;7:257–267.

252 Gemmell MA. Immunological responses of the mammalian host against tapeworm infection. II. Species specificity of hexacanth embryos in protecting rabbits against *Taenia pisiformis*. Immunology 1965;8:270–280.

253 Gallie GJ, Sewell MMH. Attempted immunisation of calves against infection with the cysticercus stage of *Taenia saginata*. Trop Anim Health Prod 1981;13:213–216.

254 Wikerhauser T, Zukovic M, Dzakula N. *Taenia saginata* and *Taenia hydatigena*: intramuscular vaccination of calves with oncospheres. Exp Parasitol 1971;30:36–40.

255 Rickard MD, Rajasekariah GR, Mitchell GF. Immunisation of mice against *Taenia taeniaeformis* using antigens prepared from *T. pisiformis* and *T. hydatigena* eggs or oncospheres. Z Parasitenkd 1981;66:49–56.

256 Miller HM. Acquired immunity against a metazoan parasite by use of non-specific worm material. Proc Soc Exp Biol 1932;29:1125–1126.

257 Kan K. Immunological studies of *Cysticercus fasciolaris*. Keio Igaka 1934;14:663–687.

258 Rajasekariah GR, Rickard MD, Mitchell GF, Anders RF. Immunization of mice against *Taenia taeniaeformis* using solubilized oncospheral antigens. Int J Parasitol 1982;12: 111–116.

259 Rajasekariah GR, Rickard MD, O'Donnell IJ. *Taenia pisiformis*: protective immunization of rabbits with solubilized oncospheral antigens. Exp Parasitol 1985;59: 321–327.

260 Rickard MD, Lightowlers MW. Immunodiagnosis of hydatid disease. In Thompson RCA, ed. *The Biology of Echinococcus and Hydatid Disease*. London: George Allen & Unwin, 1986:217–249.

261 Schantz PM, Gottstein B. Echinococcosis (hydatidosis). In Walls KF, Schantz PM, eds. *Immunodiagnosis of Parasitic Diseases*, vol. 1. Helminthic diseases. Orlando: Academic Press, 1986:69–107.

262 Schantz PM, Ortiz-Valqui RE, Lumbreras H. Nonspecific reactions with the intradermal test for hydatidosis in persons with other helminth infections. Am J Trop Med Hyg 1975;24:849–852.

263 Yarzábal LA, Schantz PM, López-Lemes MH. Comparative sensitivity and specificity of the Casoni intradermal and immunoelectrophoresis tests for the diagnosis of hydatid disease. Am J Trop Med Hyg 1975;24:843–848.

264 Gottstein B, Eckert J, Woodtli W. Determination of parasite-specific immunoglobulins using the ELISA in patients with echinococcosis treated with mebendazole. Z Parasitenkd 1984;70:385–389.

265 Kagan IG, Agosin M. *Echinococcus* antigens. Bull WHO 1968;39:13–24.

266 Schantz PM. Echinococcosis. In Steele J, Arambulo P, eds. *Handbook of Zoonoses*, vol. 1. Section C. Boca Raton: CRC Press, 1982:231–277.

267 Dessaint JP, Bout D, Wattre P, Capron A. Quantitative

determination of specific IgE antibodies to *Echinococcus granulosus* and IgE levels in sera from patients with hydatid disease. Immunology 1975;29:813–823.

268 Guisantes JA, Rubio MF, Diaz R. Application of an enzyme-linked immunosorbent assay (ELISA) method to the diagnosis of human hydatidosis. Bull Pan Am Health Org 1981;15:260–266.

269 Gottstein B, Eckert J, Fey H. Serological differentiation between *Echinococcus granulosus* and *E. multilocularis* infections in man. Z Parasitenkd 1983;69:347–356.

270 Gottstein B. Isolation of an antigen fraction from *Echinococcus multilocularis* with high species-specificity in ELISA and its identification with the Western blotting technique. In Avrameas S, Dreut P, Masseyeff R, Feldmann G, eds. *Immunoenzymatic Techniques.* Amsterdam: Elsevier, 1983:299–302.

271 Gottstein B. Purification and characterization of a specific antigen from *Echinococcus multilocularis*. Parasite Immunol 1985;7:201–212.

272 Gottstein B, Schantz PM, Todorov T, Saimot AG, Jacquier P. An international study on the serological differential diagnosis of human cystic and alveolar echinococcosis. Bull WHO 1986;64:101–105.

273 Gottstein B, Lengeler C, Bachmann P, *et al.* Seroepidemiological survey for alveolar echinococcus (by Em2-ELISA) of blood donors in an endemic area of Switzerland. Trans R Soc Trop Med Hyg 1987;81:960–964.

274 Merioua A, Bout D, Capron A. Evaluation of ELISA and RAST using purified antigens for diagnosis of hydatidosis. Pathol Biol 1984;32:15–22.

275 Rickard MD, Honey RD, Brumley JL, Mitchell GF. Serological diagnosis and post-operative surveillance of human hydatid disease. II. The enzyme-linked immunosorbent assay (ELISA) using various antigens. Pathology 1984;16:211–215.

276 Knobloch J, Lederer I, Mannweiler E. Species-specific immunodiagnosis of human echinococcosis with crude antigens. Eur J Clin Microbiol 1984;3:554–555.

277 Knobloch J, Biedermann H, Mannweiler E. Serum antibodies in patients with alveolar echinococcosis before and after therapy. Trop Med Parasitol 1985;36:155–156.

278 Coltorti EA. Standardization and evaluation of an enzyme immunoassay as a screening test for the seroepidemiology of human hydatidosis. Am J Trop Med Hyg 1986;35:1000–1005.

279 Wattal C, Malla N, Khan IA, Agarwal SC. Comparative evaluation of enzyme-linked immunosorbent assay for the diagnosis of pulmonary echinococcosis. J Clin Microbiol 1986;24:41–46.

280 Varela-Díaz VM, Coltorti EA. *Techniques for the Immunodiagnosis of Human Hydatid Disease.* Buenos Aires: Pan American Zoonosis Centre, 1976.

281 Chordi A, Kagan IG. Identification and characterization of antigenic components of sheep hydatid fluid by immunoelectrophoresis. J Parasitol 1965;51:63–71.

282 Capron A, Vernes A, Biguet J. Le diagnostic immuno-électrophorétique de l'hydatidose. In *Le Kystehydatique du foie.* Lyon: SIMEP, 1967:27–40.

283 Capron A, Biguet J, Vernes A, Afchain D. Structure antigénique des helminthes. Aspects immunologiques des

relations hote-parasite. Pathol Biol (Paris) 1968;16:121–138.

284 Capron A, Yarzabal L, Vernes A, Fruit J. Le diagnostique immunologique de l'echinococcose humaine. Pathol Biol (Paris) 1970;18:357–365.

285 Varela-Díaz VM, Eckert J, Rausch RL, Coltorti EA, Hess U. Detection of *Echinococcus granulosus* diagnostic arc5 in sera from patients with surgically-confirmed *E. multilocularis* infection. Z Parasitenkd 1977;53:183–188.

286 Yarazábal LA, Bout DT, Naquira FR, Capron A. Further observations on the specificity of antigen 5 of *Echinococcus granulosus*. J Parasitol 1977;63:495–499.

287 Varela-Díaz VM, Coltorti EA, D'Alessandro A. Immunoelectrophoresis tests showing *Echinococcus granulosus* arc5 in human cases of *Echinococcus vogeli* and cysticercus-multiple myeloma. Am J Trop Med Hyg 1978;27:554–557.

288 Schantz PM, Shanks D, Wilson M. Serologic cross-reactions with sera from patients with echinococcosis and cysticercosis. Am J Trop Med Hyg 1980;29:609–612.

289 Hira PR, Shweiki HM, Siboo R, Behbehani K. Counterimmunoelectrophoresis using an arc5 antigen for the rapid diagnosis of hydatidosis and comparison with the indirect hemagglutination test. Am J Trop Med Hyg 1987;36:592–597.

290 Coltorti EA, Varela-Díaz VM. Detection of antibodies against *Echinococcus granulosus* Arc5 antigens by double diffusion test. Trans R Soc Trop Med Hyg 1978;72:226–229.

291 Di Felice G, Pini C, Afferni C, Vicari G. Purification and partial characterization of the major antigen of *Echinococcus granulosus* (antigen 5) with monoclonal antibodies. Mol Biochem Parasitol 1986;20:133–142.

292 Lightowlers MW, Liu D, Haralambous A, Rickard MD. Subunit composition and specificity of the major cyst fluid antigens of *Echinococcus granulosus*. Mol Biochem Parasitol 1989;37:171–182.

293 Vogel M, Gottein B, Muller N, Seebeck T. Production of a recombinant antigen of *Echinococcus multilocularis* with high immunodiagnostic sensitivity and specificity. Mol Biochem Parasitol 1988;31:117–126.

294 Müller N, Vogel M, Gottstein B, Scholle A, Seebeck T. Plasmid vector for overproduction and export of recombinant protein in *Escherichia coli*: efficient one-step purification of a recombinant antigen from *Echinococcus multilocularis* (Cestoda). Gene 1989;75:329–334.

295 Müller N, Gottstein B, Vogel M, Flury K, Seebeck T. Application of a recombinant *Echinococcus multilocularis* antigen in an enzyme-linked immunosorbent assay for immunodiagnosis of human alveolar echinococcosis. Mol Biochem Parasitol 1989;36:151–160.

296 Hemmings L, McManus DP. The isolation, by differential antibody screening, of *Echinococcus multilocularis* antigen gene clones with potential for immunodiagnosis. Mol Biochem Parasitol 1989;33:171–182.

297 Oriol R, Williams JF, Pérez-Esandi MV, Oriol C. Purification of lipoprotein antigens of *Echinococcus granulosus* from sheep hydatid fluid. Am J Trop Med Hyg 1971;20:569–574.

298 Shepherd JC, McManus DP. Specific and cross-reactive antigens of *Echinococcus granulosus* hydatid cyst fluid. Mol Biochem Parasitol 1987;25:143–154.

299 Oriol C, Oriol R. Physicochemical properties of a lipoprotein antigen of *Echinococcus granulosus*. Am J Trop Med Hyg 1975;24:96–100.

300 Pozzuoli R, Musiani P, Arru E, Piantelli M, Mazzarella R. *Echinococcus granulosus*: isolation and characterization of sheep hydatid fluid antigens. Exp Parasitol 1972;32:45–55.

301 Pozzuoli R, Piantelli M, Perucci C, Arru E, Musiani P. Isolation of the most immunoreactive antigens of *Echinococcus granulosus* from sheep hydatid fluid. J Immunol 1975;115:1459–1463.

302 Piantelli M, Pozzuoli R, Arru E, Musiani P. *Echinococcus granulosus*: identification of subunits of the major antigens. J Immunol 1977;119:1382–1386.

303 Musiani P, Piantelli M, Lauriola L, Arru E, Pozzuoli R. *Echinococcus granulosus*: specific quantification of the two most immunoreactive antigens in hydatid fluid. J Clin Pathol 1978;31:475–478.

304 Lauriola L, Piantelli M, Pozzuoli R, Arru E, Musiani P. *Echinococcus granulosus*: preparation of monospecific antisera against antigens in sheep hydatid fluid. Zentralbl Bakteriol Parasitkde Abt I, Abt Orig A 1978;240:251–257.

305 Bout D, Fruit J, Capron A. Purification d'un antigène spécific de liquide hydatique. Ann Inst Pasteur/Immunol 1974;125C:775–788.

306 Varela-Diáz VM, Coltorti EA, Ricardes MI, Guisantes JA, Yarzabal LA. The immunoelectrophoretic characterization of sheep hydatid cyst fluid antigens. Am J Trop Med Hyg 1974;23:1092–1096.

307 Dottorini S, Tassi C. *Echinococcus granulosus*: characterization of the major antigenic component (Arc5) of hydatid fluid. Exp Parasitol 1977;43:307–314.

308 Craig PS, Zeyhle E, Romig T. Hydatid disease: research and control in Turkana. II. The role of immunological techniques in the diagnosis of hydatid disease. Trans R Soc Trop Med Hyg 1986;80:183–192.

309 Craig PS, Rickard MD. Studies on the specific immunodiagnosis of larval cestode infections of cattle and sheep using antigens purified by affinity chromatography in an enzyme-linked immunosorbent assay (ELISA). Int J Parasitol 1981;11:441–449.

310 Craig PS, Hocking RE, Mitchell GF, Rickard MD. Murine hybridoma-derived antibodies in the processing of antigens for the immunodiagnosis of hydatid infection in sheep. Parasitology 1981;83:303–317.

311 Lightowlers MW, Rickard MD, Honey RD, Obendorf DL, Mitchell GF. Serological diagnosis of *Echinococcus granulosus* infection in sheep using cyst fluid antigen processed by antibody affinity chromatography. Aust Vet J 1984;61:101–108.

312 Judson DG, Dixon JB, Clarkson MJ, Pritchard J. Ovine hydatidosis: some immunological characteristics of the seronegative host. Parasitology 1985;91:349–357.

313 Flisser A, Pérez-Montfort R, Larralde C. The immunology of human and animal cysticercosis: a review. Bull WHO 1979;57:839–856.

314 Schantz PM. Improvements in the serodiagnosis of helminthic zoonoses. Vet Parasitol 1987;25:95–120.

315 Flisser A, Woodhouse E, Larralde C. Human cysticercosis:

316 Correa D, Dalma D, Espinoza B, *et al*. Heterogeneity of humoral immune components in human cysticercosis. J Parasitol 1985;71:535–541.

317 Grogl M, Estrada JJ, MacDonald G, Kuhn RE. Antigen-antibody analyses in neurocysticercosis. J Parasitol 1985; 71:433–442.

318 Arambulo PV, Walls S, Kagan IG. Serodiagnosis of human cysticercosis by microplate ELISA. Acta Trop 1978;35:63–67.

319 Coker-Vann MR, Subianto DB, Brown P, *et al*. ELISA antibodies to cysticerci of *Taenia solium* in human populations in New Guinea, Oceania and Southeast Asia. Southeast Asian J Trop Med Hyg 1981;12:499–505.

320 Coker-Vann M, Brown P, Gajdusek C. Serodiagnosis of human cysticercosis using a chromatofocused antigenic preparation of *Taenia solium* cysticerci in an enzyme-linked immunosorbent assay (ELISA). Trans R Soc Trop Med Hyg 1984;78:492–496.

321 Chopra JS, Kaur U, Mahajan RC. Cysticerciasis and epilepsy: a clinical and serological study. Trans R Soc Trop Med Hyg 1981;75:518–520.

322 Diwan AR, Coker-Vann M, Brown P, *et al*. Enzyme-linked immunosorbent assay (ELISA) for the detection of antibody to cysticerci of *Taenia solium*. Am J Trop Med Hyg 1982;31:364–369.

323 Miller B, Goldberg MA, Heiner D, Myers A, Goldberg A. A new immunologic test for CNS cysticercosis. Neurology 1984;34:695–697.

324 Knobloch J, Delgado E. Immunodiagnosis of cysticercosis: standardization of ELISA and its application to field conditions. Trop Med Parasitol 1985;36:157–159.

325 Larralde C, Laclette JP, Owen CS, *et al*. Reliable serology of *Taenia solium* cysticercosis with antigens from cyst vesicular fluid: ELISA and hemagglutination tests. Am J Trop Med Hyg 1986;35:965–973.

326 Espinoza B, Ruiz-Palacios G, Rovar A, Sandoval MA, Plancarte A, Flisser A. Characterization by enzyme-linked immunosorbent assay of the humoral immune response in patients with neurocysticercosis and its application in immunodiagnosis. J Clin Microbiol 1986;24:536–541.

327 Gottstein B, Zini D, Schantz PM. Species-specific immunodiagnosis of *Taenia solium* cysticercosis by ELISA and immunoblotting. Trop Med Parasitol 1987;38:299–303.

328 McCormick GF, Zee CS, Heiden J. Cysticercosis cerebri: review of 127 cases. Arch Neurol 1982;39:534–539.

329 Corona T, Pascoe D, Gonzalez-Barranco D, Abad P, Landa C, Estanol B. Anticysticercosis antibodies in serum and cerebrospinal fluid in patients with cerebral cysticercosis. J Neurol 1986;49:1044–1049.

330 Chang KH, Kim WS, Cho SY, Han MC, Kim CW. Comparative evaluation of brain CT and ELISA in the diagnosis of neurocysticercosis. Am J Neuroradiol 1988; 9:125–130.

331 Marquez-Monter H. Cysticercosis. In Marcial-Rojas, ed. *Pathology of Protozoa and Helminthic Diseases*. Baltimore: Williams & Wilkins, 1971.

332 Estrada JJ, Kuhn RE. Immunochemical detection of

antigens of larval *Taenia solium* and anti-larval antibodies in the cerebrospinal fluid of patients with neurocysticercosis. J Neurol Sci 1985;71:39–48.

333 Harrison LJS, Joshua GWP, Wright SH, Parkhouse RME. Specific detection of circulating surface/secreted glycoproteins of viable cysticerci in *Taenia saginata* cysticercosis. Parasite Immunol 1989;11:351–370.

334 Flisser A, Overbosch D, Knapen F van. C-now: report of a workshop on neurocysticercosis. Parasitol Today 1989;5: 64–66.

335 Nascimento E, Nogueira PM, Tavares CA. Improved immunodiagnosis of human cysticercosis with scolex protein antigens. Parasitol Res 1987;73:446–450.

336 Nascimento E, Tavares CA, Lopes JD. Immunodiagnosis of human cysticercosis (*Taenia solium*) with antigens purified by monoclonal antibodies. J Clin Microbiol 1987; 25:1181–1185.

337 Kim SI, Kang SY, Cho WY, Hwang ES, Cha CY. Purification of cystic fluid antigen of *Taenia solium* metacestodes by affinity chromatography using monoclonal antibody and its antigenic characterization. Korean J Parasitol 1987;24: 145–158.

338 Plancarte A, Flisser A, Larralde C. Fibronectin-like properties in antigen B from the cysticercus of *Taenia solium*. Cytobios 1983;36:83–93.

339 Espinoza B, Flisser A, Palacios GR, Larralde C. The enzyme-linked immunosorbent assay (ELISA) for the diagnosis of human cysticercosis using antigen B purified from the cysticerci of *T. solium*. Mol Biochem Parasitol 1982; (Suppl.):235–236.

340 Gottstein B, Tsang VC, Schantz PM. Demonstration of species-specific and cross-reactive components *Taenia solium* metacestode antigens. Am J Trop Med Hyg 1986;35: 308–313.

341 Miller HM Jr. Further studies on immunity to a metazoan parasite. *Cysticercus fasciolaris*. J Prev Med 1932;6:37–46.

342 Gemmell MA, Souslby EJL. The development of acquired immunity to tapeworms and progress towards active immunization with special reference to *Echinococcus* spp. Bull WHO 1968;39:44–45.

343 Slonka GF, Moulthrop JI, Dewhirst LW, Hotchkiss PM, Vallaza B, Schultz MG. An epizootic of bovine cysticercosis. J Am Vet Med Assoc 1975;166:678–681.

344 Slonka GF, Matulich W, Morphet E, Miller CW, Bayer E. An outbreak of bovine cysticercosis in California. Am J Trop Med Hyg 1978;27:101–105.

345 Geerts S, Kumar V, Ceulemans F, Mortelmans J. Serodiagnosis of *Taenia saginata* cysticercosis in experimentally and naturally infected cattle by enzyme-linked immunosorbent assay. Res Vet Sci 1981;30:288–293.

346 Geerts S, Kumar V, Aerts N, Ceulemans F. Comparative evolution of immunoelectrophoresis, counterimmunoelectrophoresis and enzyme linked immunosorbent assay for the diagnosis of *Taenia saginata* cysticercosis. Vet Parasitol 1981;8:299–307.

347 Rhoads ML, Murrell KD, Dilling GW, Wong MM, Baker NF. A potential diagnostic reagent for bovine cysticercosis. J Parasitol 1985;71:779–787.

348 Kamanga-Sollo EI, Rhoads ML, Murrell KD. Evaluation of

an antigenic fraction of *Taenia hydatigena* metacestode cyst fluid for immunodiagnosis of bovine cysticercosis. Am J Vet Res 1987;48:1206–1210.

349 Parkhouse RME, Harrison LJS. Cyst fluid and surface associated glycoprotein antigens of *Taenia* spp. metacestodes. Parasite Immunol 1987;9:263–268.

350 Joshua GWP, Harrison LJS, Sewell MMH. Excreted/secreted products of developing *Taenia saginata* metacestodes. Parasitology 1988;97:477–487.

351 Geerts S, Kumar V, Vercruysse J. *In vivo* diagnosis of bovine cysticercosis. Vet Bull 1977;47:653–664.

352 Craig PS, Rickard MD. Evaluation of "crude" antigen prepared from *Taenia saginata* for the serological diagnosis of *T. saginata* cysticercosis in cattle using the enzyme-linked immunosorbent assay, ELISA. Z Parasitenkd 1980; 61:287–297.

353 Hilwig RW, Cramer JD. *In vivo* cross-reactivity of *Taenia saginata* and *Taenia crassiceps* antigens in bovine cysticercosis. Vet Parasitol 1983;12:155–164.

354 Geerts S, Kumar V, Aerts N. Antigenic components of *Taenia saginata* and their relevance to the diagnosis of bovine cysticercosis by immunoelectrophoresis. J Helminthol 1979;53:293–299.

355 Geerts S, Vervoort T, Kumar V, Ceulemans F. Isolation of fraction 10 from *Taenia saginata* and evaluation of its specificity for the diagnosis of bovine cysticercosis. Z Parasitenkd 1981;66:201–206.

356 Vester A. Redescription of *Taenia solium* Linnaeus, 1758 and *Taenia saginata* Goeze, 1782. Z Parasitenkd 1967;29: 313–328.

357 Le Riche PD, Sewell MMH. Differentiation of *Taenia saginata* and *Taenia solium* by enzyme electrophoresis. Trans R Soc Trop Med Hyg 1977;71:327–328.

358 Le Riche PD, Sewell MMH. Identification of *Echinococcus granulosus* strains by enzyme electrophoresis. Res Vet Sci 1978;25:247–248.

359 McManus DP, Smyth JD. Isoelectric focusing of some enzymes from *Echinococcus granulosus* (horse and sheep strains) and *E. multilocularis*. Trans R Soc Trop Med Hyg 1979;73:259–265.

360 Kumaratilake LM, Thompson RCA. Biochemical characterisation of Australian strains of *Echinococcus granulosus* by isoelectric focussing of soluble proteins. Int J Parasitol 1984;14:581–586.

361 Rishi AK, McManus DP. Genomic cloning of human *Echinococcus granulosus* DNA: isolation of recombinant plasmids and their use as genetic markers in strain characterization. Parasitology 1987;94:369–383.

362 McManus DP, Simpson AJ. Identification of the *Echinococcus* (hydatid disease) organisms using cloned DNA markers. Mol Biochem Parasitol 1985;17:171–178.

363 Rishi AK, McManus DP. Molecular cloning of *Taenia solium* genomic DNA and characterization of taeniid cestodes by DNA analysis. Parasitology 1988;97:161–176.

364 Yap KW, Thompson RCA, Rood JI, Pawlowski ID. *Taenia hydatigena*: isolation of mitochondrial DNA, molecular cloning and physical mitochondrial genome mapping. Exp Parasitol 1987;63:288–294.

365 Craig PS, Macpherson CNL, Nelson GS. The identification

of eggs of *Echinococcus* by immunofluorescence using a specific anti-oncospheral monoclonal antibody. Am J Trop Med Hyg 1986;35:152–158.

366 Deplazes P, Gottstein B, Stingelin Y, Eckert J. An ELISA for detection of *Taenia* antigen in fecal samples of dogs. Cited by Eckert J. New aspects of parasitic zoonoses. Vet Parasitol 1989;32:37–55.

18 Filariasis

James W. Kazura, Thomas B. Nutman, & Bruce M. Greene

INTRODUCTION

On the order of half a billion people are infected with filarial parasites of various species. Although clinical manifestations vary considerably from parasite species to parasite species, and even from person to person infected with the same parasite species, certain common biologic properties can be identified that are relevant to the discussion to follow. First, filarial infections are long term, with insect-borne infective larvae penetrating into subcutaneous tissues, migrating, and developing into adult male or female worms over a period of months. During this developmental process, molting occurs from the third larval stage (infective larvae from the insect) to the fourth larval stage, and another molt may precede development into juvenile parasites. Adult male and female worms live in lymphatic vessels or subcutaneous and deeper tissues, reproduce sexually, and survive for about 10 years or even longer. The adult females produce millions of microfilariae which dwell in the lymphatics, the bloodstream, or in the dermal, subcutaneous, and ocular tissues.

Because of the chronic antigenic challenge associated with long-term infection, the host mounts a complex humoral and cell-mediated immune response against the parasite. An additional biologic feature of primary importance is the multicellular nature of filarial parasites. This results in an extraordinary diversity of antigens to which the vertebrate host is exposed. Although the pathogenesis of disease is poorly understood, it is believed that the majority of clinical manifestations result directly from the host immune response.

In general, the corollary that disease is associated directly with parasite burden is true, although many infected persons defy this rule.

A final common theme of great importance is the lack of replication of filarial parasites from the microfilaria (L1) stage to the stage of adult male and female para-sites, i.e., a single microfilaria can yield at most one infective larva, and likewise a single infective larva can yield at most one adult male or female parasite. In reality there is tremendous attrition from the L1 to the adult stage, occurring both in the insect vector and in the vertebrate host. This feature has very significant implications for the potential for success of vaccine development efforts.

Thus, it is evident that the host is exposed to a wide array of parasite antigens, and therefore it is not surprising that there is, in fact, such a great diversity of host responses and of clinical manifestations. Further understanding of the precise determinants of these variations should follow from a combined approach based upon, on the one hand, animal experimentation and, on the other, careful clinical characterization of infected individuals and examination of the immune response to defined, as opposed to crude mixes of, antigens.

LYMPHATIC FILARIASIS (*WUCHERERIA BANCROFTI, BRUGIA MALAYI,* AND *BRUGIA TIMORI* INFECTIONS)

Infection with the lymphatic-dwelling nematodes *Wuchereria bancrofti, Brugia malayi,* and *Brugia timori* affects more than 100 million individuals in Africa, Asia, islands of the southern and western Pacific Ocean, and South America [1]. The frequency of these infectious diseases is expected to increase as urbanization and poor sanitation in many developing countries result in spread of breeding sites of the mosquito vectors.

The major clinical outcomes of lymphatic filariasis include elephantiasis of the extremities, genital lymphatic inflammation, and tropical pulmonary eosinophilia. As is the case for other chronic nematode (e.g., onchocerciasis) and trematode infections (e.g., schistosomiasis), disease manifestations occur in some but not all parasitized individuals, i.e., a proportion of residents of endemic areas remain asymptomatic despite repeated exposure to infective stages of the helminths. These differences in symptomatology are believed to reflect variability in parasite-specific immune reactivity and possibly cumulative worm burdens. Innate and acquired resistance to infective stages of the helminth and bloodborne microfilariae are dependent on, as yet, poorly understood cellular and humoral immunity to specific parasite antigens.

Since the second edition of this book (1982), there have been significant advances in our understanding of the molecular and cellular events which underlie and regulate immune reactivity and pathologic responses to filariae. This increase in knowledge is, in large part, based on the application of tools and concepts derived from other disciplines to the study of lymphatic filariasis. For example, examination of the roles of IL-5 and IL-4 in regulation of filarial-induced eosinophilia and IgE responses was facilitated by cloning of cytokines and demonstration of their bioactivities in other systems.

This review will summarize recent observations of human and experimental animals models of lymphatic filariasis that are pertinent to the peculiar immunologic features of this infection. These features most notably include:

1 The spectrum of T helper cell and filarial-specific antibody isotype responses;
2 Possible genetic basis and role of tolerance in regulating the spectrum of immune responses;
3 Evidence for and possible mechanisms of protective immunity.

The salient biologic features of lymphatic filariae which distinguish these infectious agents from other classes of microbial pathogens and the clinical and epidemiologic features of human infection will be described first, to impart a more general understanding of the host–parasite relationship.

Parasite

Life-cycle and morphology

Brugia and *Wuchereria* species are characterized by five morphologically distinct developmental stages which exist in tissues of the mammalian host and obligatory arthropod vector. Adult male and female worms primarily inhabit the lumen of afferent lymphatics, especially those of the lower extremities and male genitalia. This parasite stage also resides in pulmonary vessels of experimental animals, such as the gerbil

(*Meriones unguinculatus*). Adult-stage worms are thread-like and measure 40–100 mm in length and 0.1–0.3 mm in width (females are larger than males and *Wuchereria* larger than *Brugia*). Fertilized female worms release embryonic microfilariae or first-stage larvae (L1), which circulate in the bloodstream. Microfilariae are approximately 250 μm in length and enclosed by an acellular sheath which is a remnant of the vitelline membrane formed during *in utero* development. Following ingestion in the bloodmeal of the mosquito vector, which include *Anopheles*, *Aedes*, *Culex*, and *Mansonia* species, microfilariae exsheath, penetrate the midgut of the insect, and migrate to the thoracic musculature. L1 molt to form second-stage larvae (L2) and eventually infective third-stage larvae (L3), which live in the mouthparts of the mosquito. Development of L1 to L3 occurs only in female mosquitoes having undergone oviposition and takes 10–14 days. L3 are released from the mouthparts when the mosquito takes its bloodmeal. The infective larvae migrate to the skin puncture site and into the lymphatics of the host. The mechanisms which attract L3 to the puncture site and enable it to penetrate host dermis are not well understood. Proteolytic enzymes such as collagenase have been implicated in the latter process [2]. Following skin penetration, two additional molts occur to form fourth-stage larvae (L4) and adult worms. Sexually mature adult filariae appear after 9–12 months. The host and parasite molecules which regulate molting and differentiation of various specialized organ systems of *Brugia* and *Wuchereria* (e.g., peptide growth factors and ecdysteroids) and determine the host specificity for animal and human filariae are not known.

Structure and biochemistry

Adult worms. Considerable effort has been devoted to characterization of the biochemical nature of the filarial surface because of its presumed importance in stimulating host immunity. Adult worms are covered by a cuticle made by the underlying hypodermis. The most abundant structural component of the nematode cuticle is collagen. Because of the insolubility of these proteins, it has generally been presumed that they are poorly immunogenic. However, recent studies indicate that sera of subjects with brugian filariasis contain antibodies to nematode collagen which are crossreactive with human collagen [3]. It is thus plausible that immune reactivity engendered by parasite collagens may be injurious to host tissues by virtue of crossreactivity with structural components of host lymphatics, such as basement membrane. Numerous other immunogenic "surface" molecules have been identified using adult worms extrinsically radio-iodinated at tyrosine or biosynthetically labeled with ^{35}S-methionine and ^{3}H-proline [4,5]. The functions of these molecules have not been elucidated.

Microfilariae. Microfilariae are enclosed by an acellular sheath, a complex structure which is shed following ingestion by the mosquito vector. The sheath is anionic and composed of acid mucopolysaccharides and phospholipids. Lectin-binding studies have been performed to characterize the chemical nature of the glycoproteins. These demonstrate that *N*-acetylglucosamine residues are detectable on the surface of the sheath of mature microfilariae but not during intrauterine development of the organisms. Chitin is also synthesized during formation of the sheath. It has not been established whether this molecule is also synthesized by blood-stage microfilariae [6].

Loss of the sheath represents an important developmental event in the filarial life-cycle since it precedes the first molt in the mosquito vector. The molecular basis of this process is poorly understood. Studies in which exsheathment has been monitored in *Anopheles* indicate that larvae lose this structure at variable periods between the time the parasite is ingested and penetrates the midgut wall of the insect [7,8]. Observations of exsheathment *in vitro* are consistent with the possibility that external calcium ion concentration and proteases are involved in initiation of this process. For example, microfilariae incubated in papain or the calcium chelator ethylenediaminetetraacetic acid (EDTA) have been found to exsheath [9]. It is not known if a recently described microfilarial calcium-binding protein [10] serves a similar chelating function. The relevance of these observations to exsheathment and subsequent molting in the mosquito remain to be established.

Excretory–secretory products. Soluble or excretory–secretory products released by living organisms have been shown to be highly immunogenic molecules in many nematode infections. In the cases of adult filariae and microfilariae, excretory–secretory products include a complex mixture of "shed" surface molecules and "internal" molecules which are not readily detectable on the surface of the organisms [11–13]. Recent studies demonstrate that acetylcholinesterase is abundant in products released by adult *Brugia* maintained *in vitro* [14]. This enzyme is anchored to the surface membranes

of other eukaryotes by a glycophosphatidyloinositol (GPI) linkage to fatty acids rather than a conventional stretch of hydrophobic amino acids. It will be of interest in the future to determine if other antigens contained in released products of filariae have this structure and whether they are cleaved by parasite- or host-derived GPI-specific enzymes such as phosphatidylinositol phospholipase C [15]. Few studies have concentrated on characterization of the surface or excretory products of L3 and L4, largely because sufficient numbers of organisms are not available.

Lipids. With respect to lipid metabolism, there is increasing evidence that adult worms and microfilariae synthesize and possibly interconvert biologically active arachidonic acid metabolites. *Brugia malayi* take up long-chain fatty acids such as arachidonic and linoleic acids, rapidly esterify them into phosphatidylinositol, and incorporate them into lipid bodies [16,17]. Recent investigations have established that microfilariae are capable of converting these arachidonate stores to the prostanoids PGE_2 and 6-keto-$PGF_{1\alpha}$, the stable hydrolysis product of prostacyclin [18]. These observations are noteworthy in that eicosanoids modulate a variety of other cellular and immune interactions in the intravascular space (i.e., platelet–endothelium, neutrophil–endothelium) [19]. It is thus possible that intravascular microfilariae impair adherence to and avoid damage by cellular elements in the bloodstream by virtue of their capacity to generate these lipid mediators. The ability of the antifilarial drug diethylcarbamazine to lower rapidly the number of microfilariae circulating in peripheral blood may also be related to inhibition of filarial and/or pulmonary endothelial cyclooxygenase activity (N. Kanesa-Thasan, J. Douglas, & J. Kazura, unpublished observations) and possibly modification of platelet leukotriene metabolism [20].

Nucleic acids. Molecular biologic studies of *Wuchereria* and *Brugia* have largely been applied research related to cloning of antigens related to vaccine development or of importance for immunodiagnosis. These are described below in the context of their relevance to various aspects of the immune response. However, Nilsen *et al.* [21,22] have recently made observations which provide insight into basic aspects of gene regulation in filariae. These investigators observed that RNA processing by *B. malayi* and other parasitic nematodes involves a *trans*-splicing reaction as well as conventional *cis*-splicing. In the former process, a 22-nucleotide leader sequence identical to that described initially for the free-living nematode *Caenorhabditis elegans* and intestinal roundworm *Ascaris lumbricoides* is transcribed from the 5S rRNA gene cluster. This transcript is subsequently spliced to the 5′ end of a subset of mRNA containing a highly specific splice acceptor site. The development of a cell-free extract from *Ascaris suum* embryos that is capable of accurately and efficiently synthesizing spliced leader RNA of *Brugia* should facilitate studies aimed at defining the nucleotide sequences which regulate spliced leader RNA transcription and the intermediate steps of this intermolecular reaction.

Host–parasite relationship

Microfilarial periodicity

Brugia and *Wuchereria* microfilariae have circadian rhythms whereby the density of organisms in the peripheral bloodstream varies with the time of day. The intensity of parasitemia is generally highest concomitant with peak biting times of local vectors, a characteristic which facilitates transmission. Nocturnally periodic forms in which the peak intensity of microfilaremia occurs between 2400 and 0400 h are most common. Nonperiodic and subperiodic varieties exist in several areas of the South Pacific.

The mechanisms underlying microfilarial periodicity are poorly understood. Daily activity of the infected individual is the major factor that correlates with appearance of the parasites in the peripheral blood. Thus, if the sleep–wake cycle is reversed, peak intensity of parasitemia tends to shift to the time of sleeping. The reduction in mixed pulmonary venous oxygen tension which occurs during sleep in the recumbent position has been suggested by an unknown sensing mechanism to "drive" microfilariae from the pulmonary vasculature to the peripheral circulation [23].

Disease manifestations

Endemic populations. Disease caused by lymphatic filarial infection is characterized by a broad spectrum of clinical manifestations [24]. Within a population of residents in an endemic area, there is at one extreme individuals who are asymptomatic with no obvious signs or symptoms attributable to filarial infection. Most commonly, asymptomatic people are microfilaremic. A smaller proportion of asymptomatic people lack detectable microfilariae in the bloodstream and are often referred to as "endemic normals." These individuals may have developed complete or partial resistance against L3, adult worms, and/or microfilariae [25].

The initial clinical manifestations of lymphatic filarial infection are most frequently fever and/or retrograde lymphangitis of the extremities and male genitalia (for *W. bancrofti*). These signs usually last for a period of one to several weeks and are most common in teenagers and young adults. They may recur with increasing frequency in some people and eventually result in persistent lymphedema of the extremities and genital lymphatic incompetence. Such people with "chronic disease" are usually amicrofilaremic [26], although microfilaremia has been observed in residents of endemic areas where transmission is high [27]. Immunologic studies comparing asymptomatic microfilaremic (mf+) to amicrofilaremic (mf−) individuals with chronic disease demonstrate a relatively lower degree of filarial antigen-specific T cell responses and lower titers of antibodies to the microfilarial surface in the former group. The possible mechanisms underlying these phenomena will be discussed below.

At the extreme opposite, asymptomatic mf+ people are subjects with tropical pulmonary eosinophilia (TPE). Tropical pulmonary eosinophilia occurs more frequently in males than females and is characterized by recurrent nocturnal wheezing and amicrofilaremia. If the disease is not treated with the antimicrofilarial agent diethylcarbamazine, interstitial pulmonary fibrosis and restrictive lung disease may develop [28]. People with TPE have extremely elevated filarial-specific and total IgE (>5000 U/ml), blood eosinophilia, and rapid resolution of pulmonary symptoms following administration of diethylcarbamazine [29]. The pathogenesis of TPE is likely related to marked immediate hypersensitivity reactions to microfilariae trapped in the pulmonary vasculature [30,31].

Nonendemic populations. Brief travel by residents of nonendemic to endemic areas is rarely associated with development of clinical disease or patent infection. However, exposure of such individuals to large numbers of infective larvae over a short period of time may result in a hyperresponsive state with acute lymphangitis. This situation was observed in American servicemen stationed in endemic areas of the South Pacific during World War II [32]. These individuals did not develop microfilaremia and the clinical manifestations resolved several months after they left the endemic area. There is no clearcut evidence that chronic lymphatic obstruction, such as elephantiasis, develops in this setting.

Epidemiology

The frequency of exposure to mosquitoes transmitting infective larvae is the paramount factor which determines prevalence of infection in an endemic area [33]. Variability in age- and sex-specific prevalences of infection in different endemic areas is thus related primarily to unique environmental and/or social conditions which determine abundance of the mosquito vector and contact with humans.

A general pattern of infection occurs in stable human populations in most endemic areas. Active infection diagnosed by the presence of bloodborne microfilariae appears initially in 5–10-year-old age groups. The peak frequency of parasitemia occurs in adults between the ages of 20 and 40 years. The proportion of mf+ people (microfilarial carrier rate) remains the same or decreases in the fourth to sixth decades of life. As is the case with many chronic helminthic infections, it is not possible in humans to determine unequivocally whether these decreases in microfilarial carrier rates and intensity of parasitemia represent changes in exposure to infective larvae or acquired resistance to microfilariae or other stages of the organism.

Animal models

A significant impediment in advancing our understanding of immune responses to filarial infection has been the lack of convenient experimental animal models that mimic accurately the human infection. Nevertheless, several models have been useful for dissecting selected aspects of immunity that are relevant to the acquisition of resistance and development of immunopathologic reactions.

Jirds have been the most widely utilized as a source of adult worms and microfilariae. These rodents are permissive for the development of *B. malayi* and *Brugia pahangi* L3 to fecund adult female worms [34]. Acquired resistance and development of lymphatic thrombi and granulomas following sensitization to parasite antigens have also been described in this host [35]. Although some workers have examined immune regulation in jirds, investigations exploiting this animal model are severely limited by the lack of well-defined immunologic reagents, such as phenotypic markers for T cells and antibodies to specific Ig isotypes.

Mice are not susceptible to *Brugia* infection but have been used to examine acquired resistance to microfilariae and L3 [36,37]. In addition, athymic nude mice are permissive for development of *B. pahangi* L3 to sexually mature adult worms [38]. Reconstitution of these animals with syngeneic T cells results in the development of lymphatic granulomas. This host–parasite interaction may be useful for studies directed at identifying antigens that induce immunopathology.

Cats and ferrets infected with *B. malayi* are permissive for L3 maturation to fecund adult worms. In addition, these mammals develop protective immunity and lymphatic lesions reminiscent of those observed in human infection [39,40]. The widespread use of cats and ferrets as well as primates, such as the Patas monkey (susceptible to *B. malayi* infection) and green leaf monkey (susceptible to *W. bancrofti* infection), has been hampered by their limited supply and high cost.

Immunology of lymphatic filariasis

Humoral immunity

IgE and IgG subclasses. Humans with lymphatic filariasis develop an extraordinarily complex serum Ig response that typically includes elevated levels of filarial-specific IgG, IgM, and IgE. Total or polyclonal IgE is also frequently increased [40]. IgE plays a central role in mediating immediate hypersensitivity reactions in nonparasitic allergic diseases, such as asthma and hayfever. Emphasis has therefore been placed on ascertaining whether disease and/or parasitologic status of people with lymphatic filariasis correlates with the level of this specific Ig isotype and whether these antibodies are directed against specific filarial antigens. Filarial-specific IgE and IgG antibodies are generally higher in mf− adults with lymphatic disease than asymptomatic mf+ adults [41]. Similar sized antigens are recognized by IgE and IgG antibodies when assessed by one-dimensional immunoblots [42]. There are, however, differences between mf− and mf+ people with respect to filarial-specific IgG subclass responses. Mf+ and mf− people from India have similar levels of IgG1, IgG2, and IgG3 antibodies. In contrast, filarial-specific IgG4 is 17 times higher in the mf+ group than the mf− group [43]. This observation may be relevant to the lack of pathologic manifestations among mf+ people in that IgG4, unlike other IgG subclasses, is monovalent. It thus has the capacity to form small immune complexes with circulating parasite antigens and to block IgE-dependent allergic responses [43,44]. The former process provides a mechanism to abrogate the formation of larger immune complexes that might become fixed in tissues and the latter means a decrease release of mast cell-derived mediators of immediate hypersensitivity reactions. To determine whether quantitative differences in antibody subclasses between mf− and mf+ people were related to differences in reactivity to specific parasite antigens, the immunoblot patterns for each subclass were compared. IgG1 and IgG2 antibodies of mf+ and mf− people bound to bands corresponding to molecules of ~15 to greater than 200 kD. In contrast, distinct patterns of IgG3 and IgG4 antibody binding were noted. IgG3 in mf− people recognized primarily antigens of >68 kD whereas sera of mf+ individuals bound to fewer bands. The opposite pattern was noted for IgG4 antibodies. It is not known if these qualitative differences are predictive of the development of lymphatic disease or of parasitologic status. Similar analyses of sera obtained from children residing in endemic areas and adults with acute lymphangitis will be necessary to examine this possibility. Further, utilization of techniques for measurements of *in vitro* antibody synthesis by B cells will allow more precise assessment of actual production of a specific Ig isotype than is possible from a simple serum determination [45], which is reflective not only of production but also clearance, catabolism, and distribution between the intravascular and extravascular spaces.

Cytokine regulation of Ig isotype response. Progress in understanding the role of specific T helper cell (TH) cytokines in regulation of Ig isotype switching [46] has allowed dissection of the cellular mechanisms that result in increased parasite-specific IgG4 and IgE. Based on investigations of subjects with *Loa loa* and *Onchocerca volvulus* infection, it is likely that B cell production of polyclonal and filarial-specific IgE is positively regulated by IL-4 and suppressed by interferon-γ (IFN-γ) [47]. Studies of BALB/c mice sensitized to *B. malayi* antigens and challenged with live microfilariae indicate that increased production of IL-4 correlates directly with increased frequencies of polyclonal and filarial-specific IgE and IgG1 but no change in IgG2a-producing B cells (E. Pearlman, J.W. Kazura, S.-S. Chen, unpublished observations). These experimental approaches will markedly facilitate examination of the possible roles of specific parasite antigens in eliciting immediate hypersensitivity reactions and tolerance in determining the spectrum of lymphatic filariasis.

Cell-mediated immunity

As is the case for filarial-specific antibody levels, there are marked differences in antigen-driven T cell responses between mf− and mf+ people. These observations initially relied on ³H-thymidine uptake by peripheral blood mononuclear cells (PBMC) as a measure of T cell reactivity and showed that proliferation was significantly greater for PBMC of mf− adults than mf+ adults [48,49]. Indeed, the latter group demonstrated little or no reactivity to lysates of microfilariae or adult-stage *B. malayi*. Similar findings were reported subsequently for the lymphokines IL-2 and IFN-γ [50]. The possible mechanisms underlying the low degree of antigen-specific T cell proliferation for mf+ people have not been completely resolved. Piessens *et al.* [51] demonstrated in Indonesian subjects the existence of adherent mononuclear cells and T suppressor cells. Depletion of these populations from PBMC resulted in reversal of the apparent defect in T cell proliferation. Factors present in the sera of mf+ subjects may also have the capacity to suppress antigen-specific as well as mitogen-driven T cell responses *in vitro* [52]. Phosphocholine, which is abundant in the secreted products of filariae and other nematodes, has a direct inhibitory effect on T cell proliferation [53]. The observation that administration of the microfilaricidal drug diethylcarbamazine results in reversal of the defective T cell proliferation is also consistent with the possibility that products released by microfilariae contribute to this phenomenon [54]. On the other hand, there is increasing evidence that the low degree of T cell reactivity in mf+ individuals is related to antigen-specific tolerance. Nutman *et al.* [45,50] have demonstrated by limiting dilution analysis for frequency of antigen-specific T cells and by measurements of *in vitro* antibody production that PBMC of mf+ asymptomatic Bancroftian filariasis subjects contain fewer antigen-responsive cells than mf− subjects. Further, estimates of the proportions of CD4+ helper/inducer and CD8+ suppressor/cytotoxic T cells made by fluorescent-activated cell sorter analysis demonstrate no significant differences between mf+ and mf− people [55]. An increase in the proportion of CD8+ T cells might be expected if these cells were important in a suppressive network. Possible reasons for the difference in results of these studies and those of Piessens *et al.* might be related to the heterogeneity of the endemic populations studied, differences in the T cell antigens contained in parasite lysates, and the endpoint used to quantify T cell responses (proliferation vs. cytokine and Ig production *in vitro*).

Experimental hosts. With respect to experimental animals, Lammie *et al.* [56–58] have examined the kinetics of antigen-specific lymphoid cell responses in jirds infected with *B. pahangi*. These investigators observed that lymphoid cells isolated from the periaortic lymph nodes and spleens of animals during the prepatent period of infection proliferate vigorously in response to parasite lysates. The onset of patency was associated with decreased T cell reactivity to parasite antigens and mitogens for cells obtained from the spleen but not the lymph nodes. This anatomically restricted antigen- and mitogen-driven hyporesponsiveness was reversed by depletion of histamine-bearing cells and associated with reduced production of an IL-2-like cytokine and possibly other T cell growth factors [58]. Although the jird models of *B. pahangi* and *B. malayi* infection provide a convenient animal system to examine issues related to modulation of cell-mediated immunity, their utility for immunologic studies will be limited until defined reagents for T cell phenotypes and Ig isotopes are available.

T and B cell epitopes

It should be emphasized that detergent and/or aqueous extracts of filariae were utilized as "antigen" in the studies described above, as well as others of humans and experimental animals. These preparations are exceedingly complex in terms of their protein and carbohydrate compositions. Furthermore, molecules with suppressive activity, such as phosphocholine, may also obfuscate interpretation of experimental results, dependent on measurements of proliferation or cytokine production [52,53]. For example, Lal *et al.* [59] demonstrated that there are specific fractions of *Brugia* lysates separated by ion exchange chromatography that preferentially drive T cell responses *in vitro*. Highly purified native molecules or cloned antigens will ultimately be required to identify T or B cell epitopes. There are several recently described molecules which promise to fulfill this need. A cDNA corresponding to a 548 amino acid 62 kD *B. malayi* antigen expressed by microfilariae and adult-stage worms has been sequenced and ex-

pressed as a fusion protein [60]. This molecule is antigenic in humans, as assessed by its capacity to bind IgG antibodies in *W. bancrofti*-infection in humans (J.W. Kazura, F.H. Hazlett, E. Pearlman, K.P. Forsyth, & M. Alpers, unpublished observations). Examination of T cell proliferative responses of mice immunized with the recombinant antigens are consistent with the presence of a T cell epitope located between amino acids 401 and 489 [61]. The immunogenicity of *Brugia* paramyosin has also been recently demonstrated in mice [62] and humans (J.W. Kazura & J. Nanduri, unpublished observations). Selkirk *et al.* [63] have noted that sera of amicrofilaremic Malayan filariasis subjects contain antibodies to a 70 kD cognate heat shock protein produced in an expression cDNA library. Finally, Arasu *et al.* [64] described a fusion protein corresponding to a clone designated λBm19 that was selected on the basis of its reactivity with sera of *B. malayi*-infected subjects. The fusion protein induced proliferation of PBMC of subjects with loaisis and onchocerciasis but not healthy uninfected individuals.

Tolerance and neonatal sensitization

Humans. The low degree of parasite-specific immune reactivity observed in some endemic residents despite the presence of patent infection suggests that tolerance may develop as a consequence of pre- or postnatal exposure to filarial antigens. There are, however, few studies of humans which directly address this issue. Microfilariae have occasionally been observed in the bloodstream of newborns of *W. bancrofti*-infected women [65]. The frequency of this event, and whether it truly represents transplacental migration or occurs because of placental trauma at the time of birth, is not known. Evidence that newborns have been exposed to microfilariae or circulating parasite antigens accrues primarily from studies of filarial-specific antibodies in cord blood. Dissanayake & de Silva [66] detected antifilarial IgM antibodies in 12 of 340 cord blood samples obtained in an area of Sri Lanka where *W. bancrofti* is endemic. This study did not describe the microfilarial status of the mother (an antibody assay was used as an indicator of past or current infection) and the animal parasite *Setaria digitata* was used as a source of antigen. It is thus possible that crossreactive non-*Wuchereria* antibodies were measured. In the most convincing report to date, Weil *et al.* [67] quantified the level of

IgE and filaria-specific IgE antibodies in cord blood of
W. bancrofti-infected mf+ mothers. More than 80% of
samples had total IgE levels greater than that reported
for non-Indian controls. Moreover, 33 of 57 samples
contained filarial-specific IgE. There was not a direct
quantitative relationship between maternal and cord
blood filarial-specific IgE, suggesting that IgE antibodies
were derived from the fetus and not obtained as a
consequence of transfer of maternal blood at the time
of birth. Filarial antigen-induced histamine release
from cord blood cells was also demonstrated, further
suggesting that IgE was of fetal origin. There are no
published reports of filarial antigens being detected in
cord blood.

With respect to breast milk, recent studies conducted
by Petralanda et al. [68] demonstrate the existence of
O. volvulus antigens and the capacity of such antigens
(or anti-idiotypic antibodies) to induce suppression of
mitogen and purified protein derivative (PPD)-induced
lymphocyte proliferation of an uninfected donor. It has
not been established if milk of W. bancrofti or B. malayi-
infected women contains antigens and if these influence
antigen-specific immune reactivity of infants. Because
the neonatal intestinal tract is relatively "permeable" to
macromolecules as well as maternal lymphoid cells, it
will be also important to address the possibility that
transfer of parasite antigens occurs by the oral route.

Several additional observations suggest that exposure
to filarial antigens in utero or infancy diminishes the
degree of immune reactivity to parasite antigens following infection. American solders stationed in hyperendemic areas of Bancroftian filariasis during World
War II developed acute and recurrent lymphangitis
and other "allergic" phenomenon (e.g., eosinophilia)
not generally seen in men of the same age who were
life-long residents of the endemic area [69]. Biopsy of
affected sites showed inflammatory reactions consisting
of macrophages and eosinophils surrounding degenerating filarial worms. Departure from the endemic area
was associated with resolution of symptoms. More
recently, transmigrants from nonendemic to endemic
areas of Indonesia were observed to develop accelerated
lymphatic inflammation relative to endemic residents
[70]. Finally, Nutman et al. [71] recently described the
immune correlates of the "hyperresponsive" syndrome
in expatriates who enter areas where L. loa infection
is endemic. Elevation of filarial-specific IgE and IgG,
vigorous lymphocyte proliferative responses to Brugia
lysates relative to other antigens (e.g., tetanus toxoid),

and increased levels of CD4+ T cells compared to control
North Americans were observed.

Experimental animals. There is abundant evidence
in experimental animals that intact helminths (e.g.,
Dirofilaria immitis microfilariae) or parasite antigens
pass from the infected mother to her offspring in utero [65].
Haque et al. [72] reported that *Dipetolenema viteae*
microfilariae migrate transplacentally and induce filarial
antigen-specific lymphoid cell unresponsiveness as
measured by proliferation. Further studies by the same
group suggested that serum factors such as microfilarial
molecules in association with IgG inhibited antigen-
specific lymphoid cell proliferation in vitro. Klei et al. [35]
compared the parasitologic and immunopathologic
responses (lymphatic thrombi) to B. pahangi in male
offspring of infected and control jirds. No differences
in adult worm burdens were observed; however, the
frequency of microfilaremia was greater in offspring of
infected than uninfected control mothers. Lymphatic
granulomatous reactions and thrombi were also less
severe in offspring of infected mothers. A report by
Schrater et al. [73] also indicates that offspring of Brugia-
infected jirds have a propensity to develop higher levels
of microfilaremia following infection compared to
control animals. These studies are consistent with the
possibility that in utero exposure to antigens results in
tolerance manifested by reduced immunopathology and
lack of resistance to microfilariae.

Genetic factors

Humans. Knowledge of host genetic influences on the
immune response and pathologic manifestations of
lymphatic filarial infection is the key to improving our
understanding of the spectral nature of this infectious
disease as well as the nature of protective immunity.
Only a few studies have addressed these issues in
humans. Ottesen et al. [74] in Polynesia noted familial
clustering of infection as determined by microfilaremia.
There was no linkage between parasitologic status
and any of the human leucocyte antigen (HLA)-A or
HLA-B loci examined. An association between HLA-B15
and elephantiasis has been reported from an endemic
area of South India and Sri Lanka [75].

Experimental animals. Because of practical difficulties
involved in studying human populations, more exten-

sive investigations of host genetic factors have been performed using rodent models of filarial immunity. Although euthymic mice are not permissive for development of L3 to adult worms, the availability of multiple inbred strains has made possible the examination of susceptibility to other parasite stages and its relation to immune reactivity. Fanning & Kazura [76] injected 14 inbred strains of mice with *B. malayi* microfilariae and monitored the level and duration of parasitemia. A wide range was observed. BALB/c and C57BL/6 harbored the organisms for the longest period of time. No linkage to a specific H-2 haplotype was discerned. CBA/n *xid* mice, which are relatively deficient in development of IgM antibodies, especially to phosphocholine [77], have been compared to the CBA/Ca strain which lacks the *xid* defect, in terms of duration of microfilaremia. In the cases of *B. malayi* and *Acanathacheilonema viteae*, prolonged durations of microfilaremia were observed in animals with the *xid* defect, [77–79]. More recently, Kwan-Lim & Maizels [80] have examined the possible influence of MHC on serologic responses to *B. malayi* adult worms placed in the peritoneal cavity. The development of IgG antibodies to a 29 kD surface glycoprotein of adult worms and a 40 kD microfilarial surface antigen was restricted to strains of the H-2^k haplotype, whereas antibodies to 24 and 66 kD antigens of adult *B. malayi* were noted only in mouse strains of the H-2^d haplotype. Extension of this experimental approach to analyses of the immune response to molecularly defined T and B cell epitopes will provide important information relevant to the role of specific filarial antigens in resistance and immunopathology.

Protective immunity

Humans. Acquired resistance to filarial infection and/or specific stages of the parasite has been difficult to document unequivocally in humans for both practical and theoretical reasons. Inherent problems in measurement of acquired resistance to L3 and/or microfilariae in humans include an inability to quantify directly adult worm burdens, difficulty in accurately measuring exposure to mosquitoes carrying infective larvae, and the as yet ill-defined influences of genetic and maternal factors on resistance. Despite these problems, epidemiologic surveys in several areas of the world where transmission is stable suggest that a proportion of adults may be partially resistant to lymphatic filarial

infection and/or acquire resistance to microfilariae. In most endemic areas, the density of parasitemia generally increases up to the age of 20 years, presumably secondary to accumulation of fecund adult female worms. A proportion of individuals older than 30 years (10–30%) either remain or become amicrofilaremic despite continued exposure to mosquitoes bearing L3. These people may be partially resistant to superinfection and/or express a relatively high degree of resistance to blood-borne microfilariae [25]. Future studies in this area will be aided by the development of serologic assays to quantify adult worm-specific molecules so as to provide an indirect measure of worm burden.

Experimental animals. Immunization with attenuated parasites induces partial resistance to subsequent challenge with L3 in several experimental hosts. Injection of jirds with irradiated L3 prior to *B. malayi* or *B. pahangi* infection results in reduction of the number of parasites that develop into mature adult worms [25]. Cats, dogs, jirds, and rhesus monkeys injected with attenuated L3 before challenge with various animal filariae also develop resistance, manifest by 40–80% decreases in the number of adult parasites recovered at necropsy [81,82]. Published data related to acquired immunity in other primate hosts susceptible to *B. malayi* infection (i.e., Patas monkeys) are not available. Extracts of mf or adult worms may also elicit protective responses. Our studies of jirds and mice immunized with aqueous extracts of *B. malayi* mf indicate that subcutaneous injection of these preparations without adjuvant results in 50% reductions in adult worm burdens and parasitemia, respectively [83].

Protective antigens. Filarial molecules of ~25, 29, 35–38, 43, 62, 70–72, 97, and ~110 kD have been implicated as being potentially protective [35,84–86]. Because of the difficulties in purifying these molecules in quantities sufficient for protein sequencing and animal immunization experiments, identification of the specific epitopes involved in induction of resistance and examination of the immune mechanisms by which they elicit protection has been reported to date for only three of these molecules. These include a 62 kD [22], paramyosin (97 kD) [62], and as yet uncharacterized 26–28 kD molecules that elicit enhanced clearance of L3 from the peritoneal cavities of BALB/c mice [86]. The optimal mode of sensitization, delivery system, and efficacy of combinations of vaccine candidates have not been defined.

ONCHOCERCIASIS

Onchocerciasis is viewed as an immune-mediated disease, in which the host response to the parasite, particularly microfilariae in the skin and ocular tissues, leads to tissue damage. However, present knowledge and understanding of the immune response to *O. volvulus* infection in humans, and of the immuno-pathology of the disease, is quite limited. Efforts are underway to develop a vaccine to prevent infection by eliciting an immune response against developing larvae and against critical parasite proteins.

Parasite

Infection in humans begins with inoculation of infective larvae into the skin by the bite of the female blackfly (*Simulium* species). Infective larvae develop into adult worms over a period estimated at several months; the adult worms then coil up into roughly spherical bundles, typically containing two to three females and one to two males, but with wide variations in these numbers. The gravid female releases microfilariae which then migrate out of the nodule and throughout the tissues of the host, concentrating in the dermis. Microfilariae are found in the skin after a prepatent period of 7–34 months following introduction of infec-tive larvae. Transmission of infection to other individ-uals is initiated by the bite of a female fly, which, along with a blood meal, ingests microfilariae from the host skin. Some of the ingested microfilariae migrate from the gut of the blackfly into the thoracic muscles and develop into infective larvae over a period of 6–8 days. These infective larvae then migrate to the head of the fly where they may be transmitted to a second host in the process of taking a blood meal. Infective larvae are 440–700 μm in length (mean 600 μm) and 19–28 μm in width. Adult *O. volvulus* females are 23–70 mm in length and 275–325 μm in width.

Humoral immune response

Infection in humans leads to the formation of antibodies against multiple antigens of the parasite. Because the onset of the infection is usually unknown, and there are few studies in children, data are not available which define the evolution of the antibody against response. However, it is clear that with chronic infection, anti-bodies against a multiplicity of parasite antigens develop [87,88]. In addition, antibodies of various classes, espe-cially IgG, IgM, and IgE, with no apparent specificity for *O. volvulus* are elevated and it is likely that *O. volvulus* infection leads to polyclonal B cell activation [89].

While it has not been possible to evaluate the immune response soon after infection in human onchocerciasis, recent data based upon a chimpanzee model of infection has yielded new insights into the dynamics of the immune response following infection [90,91]. Within 1 month following infection, antigens of developing larvae elicit both a humoral and a cell-mediated immune response against crude parasite antigen. This is mani-fested by IgG recognition of multiple bands on Western blot, and by a brisk lymphocyte blastogenic response to soluble adult worm extract [90]. There is a parallel rise in IgE class antibodies against parasite antigen [91].

With regard to functional significance, there are no data which clarify a possible protective role for anti-body in the human immune response against infection with *O. volvulus*. However, *in vitro* experiments show that sera from chronically infected persons promote granulocyte adherence to microfilariae and infective larvae [92–94] and destruction of microfilariae in such a system has been demonstrated [92]. Of interest also is the observation that sera which strongly promote attachment of eosinophils to microfilariae *in vitro* have been correlated with the presence of corneal punctate opacities [94], suggesting that antibody in such sera may be responsible for some of the acute inflammatory manifestations of disease in the eye and the skin (e.g., papular dermatitis). Degranulation of eosinophils *in vivo* in areas of the skin surrounding microfilariae, with release of eosinophil granule proteins [95], not only suggests that eosinophils may be a major effector cell in antibody-dependent killing of microfilariae but also raises the possibility that local and perhaps even dis-tant tissue damage may result from repeated, massive eosinophil degranulation (see below). Similar mech-anisms may be operative in the eye.

The IgE response in onchocerciasis deserves special mention. Many, although not all, people with *O. volvulus* infection have elevated IgE levels [96]. A greater proportion of IgE antibody, as compared to IgG anti-body (which is also usually elevated with *O. volvulus* infection), is demonstrably directed against parasite antigens [97]. Further, the mean levels of IgE are extremely high in *O. volvulus* infection, probably higher than in other filarial infections [96]. The significance of high levels of IgE is unclear. Perhaps relevant in this regard is the report of IgG blocking antibodies in generalized onchocerciasis [96], which may serve to modulate IgE-mediated responses under most circumstances. It seems likely that IgE antibody may contribute to acute inflammatory complications of *O. volvulus* infection, including some ocular lesions [98].

Antigen–antibody complexes are found in the sera of a high percentage of *O. volvulus*-infected people [99–101]. The functional significance of circulating antigen–antibody complexes in onchocerciasis is uncertain, but it is possible that immune complex deposition leads to acute inflammation and tissue damage in some infected people. One study reported a correlation between parasite antigen-specific circulating immune complexes and disease complications [101]. Further, immune complexes may contribute to down-regulation of the host immune response.

Cell-mediated immunity

It is clear that *O. volvulus*-infected people have impaired cell-mediated immunity. There is decreased reactivity to tuberculin skin testing and an increased prevalence of lepromatous leprosy in populations infected with *O. volvulus* [87,102–104]. Further, in Africans, delayed hypersensitivity reactions and lymphocyte reactivity to streptococcal antigens appear to be suppressed [87,105]. More importantly, there is diminished reactivity, both by skin testing and *in vitro* lymphocyte activation assays, to *O. volvulus* antigens [105–107]. This phenomenon appears to be more prominent in people with microfilariae distributed throughout the skin and little cutaneous reaction as compared to those with highly reactive skin disease and few microfilariae demonstrable [106,108,109]. Furthermore, lymphocyte reactivity in infected young people (12–16 years) was greater than that in older infected individuals, suggesting that there is acquired down-regulation of the cellular immune response [106].

The precise mechanisms by which dampening of the cell-mediated immune response occurs is as yet unknown. One study showed that addition of exogenous IL-2 restored reactivity to parasite antigen in about half of infected people [106].

Because the cell-mediated immune response may exert a regulatory effect on the function of diverse cell types (e.g., granulocytes, fibroblasts), variations in cell-mediated immunity may be of central importance in determining person-to-person and temporal variations in the tissue response to infection and, therefore, in disease manifestations (see below). The direct relation between lymphocyte response to *O. volvulus* antigen

and skin disease mentioned above would be consistent with this hypothesis.

Immunopathology

The pathology of onchocerciasis [110–112] is manifested primarily by onchocercomata (containing adult worms), and in the skin, lymph nodes, and eye. Definition of pathologic changes associated with *O. volvulus* infection remains incomplete, particularly for ocular lesions, as pathologic material is very scarce and often shows end-stage disease.

Over a period of months to years, adult worms become encased in a rim of host tissue, thus forming the characteristic subcutaneous nodules (onchocercomata). This rim is composed of hyalinized and vascularized scar tissues surrounding the adult worms and variable amounts of chronic inflammatory elements, including fibrin, plasma cells, neutrophilic and eosinophilic granulocytes, lymphocytes, giant cells, and Russell bodies. Some nodules contain only necrotic material with liquefied and sometimes partially calcified remnants of adult worms.

In the skin, microfilariae are most frequently seen in the upper dermis, characteristically with no evidence of a surrounding tissue reaction. Aside from microfilariae in the dermis, early changes include localization of inflammatory cells around vessels and dermal appendages and an increase in dermal fibroblasts and mast cells. Subsequent changes include hyperkeratosis, focal parakeratosis, and acanthosis with melanophores and increased mucin in the upper dermis, accompanied by dilated lymphatics and tortuous upper dermis vessels. This process ultimately leads to loss of elastic fibers and progressive fibrosis. End-stage disease is characterized by advanced atrophy of the epidermis with loss of rete ridges; only very thin layers of epidermis and keratin remain overlying the dermis.

The pathology of lymph nodes from Africans infected with *O. volvulus* consists of scarring of lymphoid areas, with sinus histiocytosis and infiltration with inflammatory cells. In contrast, lymph nodes from infected Yemenites show follicular hyperplasia.

In the ocular tissues, the bulbar conjunctiva shows an infiltrate of plasma cells, eosinophils, and mast cells with hyperemia and dilatation of vessels and some vessel thickening and perivascular fibrosis. Punctate keratitis has been shown on corneal biopsy to consist of a collection of inflammatory cells surrounding degenerating microfilariae [113]. Sclerosing keratitis shows scarring, chronic inflammation, and vascularization. The pathology of anterior uveitis and chorioretinitis reflects a low-grade chronic nongranulomatous inflammatory process.

Thus, the pathologic changes in onchocerciasis suggest a chronic inflammatory response to the presence of the parasite, especially microfilariae, in the tissues. However, the precise pathogenesis of disease manifestations in onchocerciasis is largely unknown. Punctate keratitis has been simulated in several experiments utilizing animal models in which microfilariae are injected into the ocular tissues [98,114–118]. However, in humans, these lesions are transitory and resolve completely, and any direct relation of microfilariae to other tissue lesions remains speculative. As discussed further below, an animal model of sclerosing keratitis in guinea pigs was recently described, in which animals presensitized with *O. volvulus* antigen and subsequently injected intrastromally in the cornea with soluble antigen showed clinical and histopathologic features similar to human onchocercal sclerosing keratitis [119]. Fractionation of the challenge soluble antigen suggested that the inducing substance was of intermediate molecular weight.

Because of their vast numbers and wide distribution throughout the body, as well as the clinical correlation between microfilaria counts and complications, microfilariae or their products are thought to cause, directly or indirectly, most of the disease manifestations. It is believed that the host immune response directed against microfilariae contributes directly (e.g., through tissue-destructive products of granulocytes, including eosinophils [95]) or indirectly (e.g., through products of activated mononuclear cells) to the pathologic consequences of infection. The guinea pig model mentioned above is consistent with this hypothesis. It has also been suggested that the reaction to microfilariae may induce pathologic autoantibody [118]. While the immune response appears to cause most of the disease manifestations, it is possible that products of microfilariae, e.g., elastase or other proteases, may also be of importance. Much additional research is needed in this area.

A striking feature of human onchocerciasis is that only a relatively small percentage of people with microfilariae in the skin and ocular tissues show major disease. Onchocercal skin disease occurs in 5–60% of infected adults in different areas, and onchocercal sclerosing keratitis occurs in <1%–10% of adults. Thus, in striking contrast to bacterial disease, for example, infection does not lead to obvious disease in all infected people who have microfilariae distributed throughout various tissues. Examination of effector cells from infected individuals [92,94] and of antibodies [92] has failed to reveal a defect that could explain the lack of response. However, a consistent observation has been the lack of a robust cell-mediated immune response to parasite-derived antigens in most infected adults [106]. Since this hyporesponsiveness appears to be acquired in adolescence or adulthood, any relation to prenatal or early childhood sensitization [120,121] seems doubtful. It can be hypothesized that acquired down-regulation of cell-mediated responses to parasite-derived antigens [106] may explain the lack of *in vivo* reactivity to microfilariae in the tissues, and further that this may represent an adaptive, protective mechanism that prevents tissue damage related to a vigorous host response to microfilariae.

In contrast to most infected people who have little or no active onchodermatitis at any one time, individuals with severe hyperreactive onchodermatitis and enchanced peripheral cell-mediated immune responses to onchocercal antigen have an extensive inflammatory cell infiltrate of the upper dermis composed of plasma cells, eosinophils, and lymphocytes [122,123]. This latter subgroup of patients also have increased numbers of circulating activated peripheral lymphocytes, as assessed by expression of HLA-DR, transferrin receptors, and CD25 (IL-2 receptor) on circulating lymphocytes, when compared to individuals with microfilariae positivity and no disease [124]. Furthermore, increased numbers of CD4+ cells, increased numbers of cells with the natural killer (NK) phenotype, and an increased CD4/CD8 ratio were found in the hyperreactive onchodermatitis patients [124]. Less is known of the late skin lesions associated with chronic microfilardermia without clinical disease, which is manifested histologically by severe dermal fibrosis; an inflammatory cell infiltrate clearly precedes end-stage disease, although quantification of the intensity of the systemic immune response has never been done in this latter clinical subgroup.

The major cause of blindness in areas with severe onchocerciasis is sclerosing keratitis. This complication is associated with invasion of the cornea with microfilariae, and is believed to follow diffuse stromal keratitis associated with the presence of microfilariae [110]. It seems likely that microfilariae enter the cornea from the skin by way of the conjunctiva. Those in the uveal tract and aqueous may well be blood-borne, although direct invasion along the vessels and nerves also occurs. Microfilariae may be found in the conjunctiva, cornea, posterior sclera, anterior and posterior chambers, vitreous, uveal tract, inner retina, and optic nerve and its sheath.

Sclerosing keratitis begins peripherally, typically adjacent to the limbus either nasally or temporally, appearing as a white haziness in the anterior third of the stroma [125]. This is followed by inward migration of limbal pigment and superficial thin vessels. The process extends around the cornea and spreads centrally. Histologically, there is an infiltrate of chronic inflammatory cells with occasional eosinophils and neutrophils in a fibrovascular pannus of the superficial stroma at the level of Bowman's membrane [125]. Limbitis with perivascular plasma cells is usually present. The process may extend into the deep stroma. The hallmarks of onchocercal sclerosing keratitis are vascularization, fibroblast proliferation, and chronic inflammatory infiltrates. Although in early sclerosing keratitis microfilariae are frequently abundant [110], they are seen only infrequently or are absent altogether in advanced disease.

Experimentally, it has been difficult to study definitively onchocercal sclerosing keratitis because there is no established animal model. Although chimpanzees can be infected [90], there are prohibitions against performing invasive procedures and sacrificing these subhuman primates. Horse and cattle infections with *Onchocerca* species are not associated with sclerosing keratitis. Injection of microfilariae of *O. volvulus* and other onchocercids into rabbits and rodents has yielded useful information, however.

Chorioretinitis is the most common form of blinding onchocercal ocular disease in the forest areas of endemic onchocerciasis [115]. The pathogenesis of chorioretinitis is poorly defined.

The earliest change is granular atrophy of the retinal pigment epithelium. Intraretinal pigment is often seen. Chorioretinal atrophy and subretinal fibrosis with neovascularization are the other structural changes recognized clinically. Histopathologically, there is a chronic nongranulomatous chorioretinitis with infiltration by lymphocytes, plasma cells, and eosinophils. These changes are believed to be immune-mediated and to result directly or indirectly from invasion of microfilariae and the associated immune response. However, little information is available regarding the human immune response and its role in onchocercal chorioretinitis.

Many theories for the pathogenesis of the posterior pole lesions of onchocerciasis have been proposed, including a circulating toxin secreted by adult worms [116], an inflammatory reaction around dead microfilariae with perivascular infiltration and endothelial proliferation leading to occlusion of the smaller blood vessels [117], and lesions caused by the disintegration of entrapped microfilariae in the choroid, retina, and sclera [118].

Immune mechanisms, such as immune complex deposition [119] and antiretinal antibodies [118], have been proposed. Recent clinical observations of patients with onchocerciasis revealed live intraretinal microfilariae and intraretinal hemorrhages, cotton wool spots, and vasculitis [126]. The significance of the retinal findings in relation to subsequent chorioretinal changes is unknown, but both clinical and histopathologic observation suggest that active infiltration of the retina induced by microfilariae may chronically and progressively lead to widespread chorioretinal damage.

A recent study of experimental ocular onchocerciasis in cynomolgus monkeys suggests that the nonhuman primate eye may be an accurate model, producing ocular lesions resembling human onchocerciasis [127]. In this model, intravitreal injections of approximately 10 000 *Onchocerca lienalis* microfilariae led to an intense posterior and anterior chamber reaction which precluded a view of the fundus. In a subsequent study [128], intravitreal injections of 0, 10, 50, and 500 live microfilariae to two normal monkeys and two monkeys that had received prior immunization with subcutaneous injections of *O. lienalis* microfilariae led to disc edema, venous engorgement, retinal vasculitis, intraretinal hemorrhage, and progressive retinal pigment epithelial disturbances in the posterior segment. Histopathologically, perivascular infiltrates with eosinophils and loss of pigment in the retinal pigment epithelium were seen. This study showed that the posterior segment lesions of onchocerciasis can be reproduced in cynomolgus monkeys by intravitreal injection of small numbers of microfilariae. This closely resembles the clinical situation, since both histopathologic studies [129] and clinical observations [126] have shown that microfilariae are only occasionally encountered in the vitreous, retina, and choroid, in spite of the extensive chorioretinal damage which has been observed.

It is unclear whether the changes in the posterior segment in the monkey model are initiated by live microfilariae, either mechanically or by the release of toxic excretory products or antigens, or by dead or dying microfilariae, either directly by the release of toxic or antigenic molecules or indirectly through local or circulating inflammatory or immune products. It seems likely that the local release of toxic products or antigens by dying microfilariae is the key initiating event. However, precise immune mechanisms are ill-defined and in the monkey model the fact that preimmunization with microfilariae did not result in an increase in subsequent response to intravitreal microfilarial challenge [128] raises the real question of whether the observed pathologic changes are immune-mediated.

Innate and acquired resistance

There is no clear evidence to support the existence in humans of any form of resistance to infection in nature. Patients cured of onchocerciasis by treatment with suramin become reinfected following re-exposure [130]. Furthermore, repeated challenge of chimpanzees with living infective larvae over an 8-year period revealed no evidence of acquired resistance [131].

In a recent study, immune parameters were examined in 28 Guatemalans who, despite exposure to infection, were found not to harbor microfilariae and had negative Mazotti tests [132]. In this putatively immune population, the production of IL-2 by PBMC in response to *O. volvulus* antigen stimulation was significantly greater than in the comparison group with microfilariae in the skin and in a nonendemic control group. In contrast, parasite-specific antibody (IgG, IgE) was not increased in this group. These results raise the possibility that cell-mediated immunity in particular may be associated with protective immunity.

Immuno- and molecular diagnosis

Attempts to develop sensitive and specific immunoassays to detect *O. volvulus* infection have been made through the years, with little success [88]. The major problem has been crossreactivity with other filarial species of humans which are coendemic with *O. volvulus* in many areas. A recent report [133] suggests that by utilizing a low-molecular-weight fraction of crude antigen, the problem of specificity can be largely solved. Seventy-three sera from individuals with proven *O. volvulus* infection showed negligible or no reactivity in a microenzyme-linked immunosorbent assay (micro-ELISA) to the low-molecular-weight fraction, while crossreactivity with crude antigen was common. The assay also showed impressive sensitivity, diagnosing 32 of 33 children less than 16 years of age, and predicting

conversion from skin biopsy negative to positive 1–4 years before microfilariae were found in the skin in four of nine children. Finally, a cloned antigen has been described that performs similarly to the low-molecular-weight fraction [134]; this should facilitate development of practical immunodiagnostic assays.

DNA probes have been constructed that are capable of distinguishing different species of *Onchocerca* and different forms of *O. volvulus* [135]. Such probes may prove useful in determining the origin of larvae in vectors, or even in human skin, but the sensitivity and specificity of these DNA hybridization-based assays as field diagnostic techniques in onchocerciasis remains to be determined.

Vaccines

Development of an effective, long-lasting, vaccine that would prevent infection would clearly represent a major advance in control of onchocerciasis. This would be particularly useful in areas where it is not possible to treat with ivermectin on a yearly basis, which unfortunately is the case for a large proportion of the world in which onchocerciasis is endemic.

Although, as noted above, there is no evidence of a protective immune response developing in nature, there is some optimism concerning the possible development of a vaccine based upon results in animal filarial infections and the power of the newer techniques of molecular immunology and molecular biology. Attempts at vaccine development have focused upon two major approaches:

1 development of an immune response directed against developing larvae, especially L3 and L4;
2 development of an immune response against molecules that are critical for survival of the parasite.

A large number of recombinant molecules have been produced.

With regard to the former approach, larva-associated antigens have been identified and isolated from an expression cDNA library [136]. These antigens were identified utilizing rabbit antiserum raised against third-stage larvae of *O. volvulus*. Another approach is based upon utilizing the immune response in animals challenged with irradiated larvae to identify antigens that may be associated with protective immunity. This approach derives from results in other species, including *D. immitis*, for example, in which irradiated larvae elicit an impressive protective immune response [137]. In addition, an antigen of *O. volvulus*-infective larvae (molecular weight 133 000) has been identified, and a cDNA clone encoding part of this antigen has been isolated [138,139]. Cloned antigens that are recognized by sera from putatively immune people have been identified, and additional studies are under way to examine the structure of these antigens and the possible relationship between them and immunity. One cloned antigen, OW-10, has been described in which a T cell epitope has been localized [140].

Clearly, vaccine development is at an early stage. Unfortunately, ultimate testing will have to rely on the chimpanzee model [90], which is extraordinarily expensive and time-consuming. Thus, much preliminary and indirect data concerning potential immunogens must be collected in anticipation of trials in chimpanzees and ultimately in humans. It is not known whether a humoral or cell-mediated immune response against potential immunogens is more important. Therefore, vaccine development attempts will seek to identify immunogens that will induce both an antibody and cell-mediated immune response in subhuman animal models. The potential for success of any of these approaches is unpredictable.

LOIASIS

Loa loa is a filarial parasite causing chronic infection that is characterized by occasional angioedematous swelling and often by the migration of adult worms in subcutaneous tissues and across the eye. It is limited to, and highly endemic in, equatorial West and Central Africa [141]. Although nonhuman primates can be infected with *L. loa*, the available evidence suggests that humans are the only significant reservoir of loiasis [142,143].

Loa loa, the causative agent of loiasis (the African eyeworm), was probably first observed in 1589 by Pigafetta, but it was not until 1777 that Francois Guyot designated the term ''loa'' for the species of filaria, pathogenic for humans, that was commonly found in West Africa [144]. However, the *L. loa* organism was first described in detail in 1904 [145].

The parasite

Life-cycle and morphology

Tabanid flies (deerflies or mango flies) of the genus *Chrysops* are the vector and the necessary intermediate host in which microfilariae taken up from the blood develop over a period of 10 days into the infective larvae;

these infective larvae emerge from the mouth parts and, in a process that remains poorly understood, enter the human host. After undergoing two molts (a process that occurs over a period of 6 months to 4 years), they appear as adult parasites residing in the subcutaneous tissues.

The adult *L. loa* is a semitranslucent thread-like parasite; the females are 50–70 mm by 450–600 μm and the males are 30–34 mm by 350–400 μm [146]. A cuticle surrounds the adults from which chitinous bosses protrude. The *L. loa* microfilariae are sheathed and 185–300 μm by 5–8 μm. Their most distinguishing feature is the terminal elongated nucleus at the tip of the tapered posterior tail [147].

Surface characterization

Each stage of the *L. loa* life-cycle has distinct biochemical, ultrastructural [148,149], and immunologic attributes. At an ultrastructural level, the microfilariae have a thick, coarse electron-dense mat [148,150] and characteristic lateral cuticular ridges, both of which are unique to *L. loa*. Biochemical and immunochemical characterization of immunoreactive *L. loa* surface molecules have recently been reported [151–153]. Interestingly, the blood microfilariae contain two major surface molecules, a 23 kD protein and one of 67 kD which appears to be human albumin. Uterine microfilariae appear to have this 23 kD surface molecule along with major 40 and 40–67 kD molecules. Minor microfilarial surface molecules also have been noted and are awaiting further characterization. Adult parasites, in contrast, have an immunodominant 29–31 kD surface glycoprotein along with less abundant proteins of multiple relative molecular masses [151].

Host–parasite relationship

The *L. loa* parasite (like all the filarial parasites), in its evolutionary adaptation to the human host, has likely evolved mechanisms for avoiding rejection by the immune system. Because these parasites must survive within the definitive host for a period of time that is sufficiently long to ensure continued transmission of the infective stages to their intermediate host, chronic infections are the rule. This chronicity of infection causes the release of large quantities of parasite antigens, which have profound immunopathologic consequences; they may be deposited in host tissue as immune complexes or they may induce by themselves both immediate and delayed-type hypersensitivity (DTH) reactions.

In addition, chronic stimulation by parasite antigens may result in immunosuppression or immunologic tolerance—a mechanism that likely promotes the parasite's survival. Indeed, this type of mechanism has been implicated both in asymptomatic patients with lymphatic filariasis [24] as well as in the immunologically less responsive *L. loa*-infected patients seen among natives of highly endemic areas [71,154].

Humoral responses

Humoral immune responses induced by *L. loa* infection include those that are parasite-specific and those that are polyclonal. Hypergammaglobulinemia is a common finding in loiasis both in temporary residents of and those native to *Loa*-endemic areas [155]. Further, polyclonal IgE elevation is found in the majority of patients [71,156].

More important are the parasite-specific antibody responses that *L. loa* infection engenders. Using a variety of immunologic techniques [157] (including immunofluorescent antibody hemagglutination, complement fixation, and ELISA [158]), with either *L. loa* or related filarial parasite extracts as the antigen target, the presence of antifilarial antibody in individuals with *Loa* is a constant finding. In a cross-sectional, age-related study in a hyperendemic region for *Loa*, IgM and IgE antibody against *L. loa* appeared within the first year of life and peaked between the ages of 2 and 3 years. Furthermore, 95% of this population had *Loa*-specific antibodies by age 5 years, suggesting that parasite-specific responses are found universally in this infection [158]. In addition, using nonhuman primates experimentally infected with *L. loa*, antibody was clearly shown to be induced [159]; further, the antibody levels were inversely correlated with levels of circulating microfilariae [159], a finding felt to occur in humans as well [156]. Indeed, antibody to a 23 kD surface microfilarial protein has been identified as possibly mediating microfilarial clearance [151].

At a quantitative level, antifilarial antibodies are clearly higher in infected expatriates than in those infected individuals native to endemic areas, who in turn had higher levels than those exposed but not infected [154]. At a qualitative level, using adult antigen from *L. loa* [160] in immunoblotting, there was a difference in the antigen recognition patterns between amicrofilaremic (occult or "resistant") patients and those with asymptomatic microfilaremia. Those individuals who were able to clear their microfilariae recognized a high-molecular-weight antigen (~160 kD) which was not

recognized by those with microfilaremia; conversely, microfilaremic individuals exclusively recognized an 18 kD antigen [160].

Cellular responses

Relatively little is known about the cellular response to *L. loa* infection. While eosinophilia is an almost constant finding in loiasis and several different cell types have been shown to adhere to and kill *Loa* microfilariae *in vitro* (in the presence of specific antibody [20,161]), the nature of the cell-mediated responses remains largely unstudied. Nevertheless, immediate-type IgE-mediated mast cell responses were clearly demonstrated by deliberately feeding infected *Chrysops* on a previously sensitized individual [162]. Furthermore, a DTH reaction probably also occurred in this setting 48 h later. Lymphocyte responses to parasite antigen have also been examined, and *Loa*-infected individuals had marked responsiveness to filarial antigens (compared to non-parasite antigens) and an increase in the antigen-specific T cell precursor frequency [71]. Recently, however, parasite antigen-specific cellular responses (lymphocyte proliferation) have been studied in a group of 50 infected individuals from West Africa; in all infected individuals there was clearly a measurable response, although there was little relationship between clinical status and the ability to mount a lymphocyte response [154].

Immunopathology

Various pathologic manifestations of loiasis have been described, the most characteristic of which is a localized angioedema found predominantly on the extremities but sometimes extending to nearby joints or peripheral nerves and causing symptoms related to these structures. Nephropathy [163–168], cardiomyopathy [169–172], retinopathy [173,174], arthritis [175,176], peripheral neuropathy [177,178], lymphedema and lymphadenitis [179,180], and encephalopathy [181–185] have also been described, albeit uncommonly, in this disease.

It appears that important differences exist between the clinical presentation of loiasis in visitors to filaria-endemic areas and in people who are native to these same regions; indeed, the infection in temporary residents of endemic regions is generally characterized by a greater predominance of allergic symptoms, frequently recurring episodes of angioedema, high peripheral blood eosinophilia, and greater debilitation and pathology [71, 153]. Studies examining the immunologic correlates of this phenomenon suggest that the intensity and the nature of the immune response are critical to the development of pathology [154].

The nature of the pathology seen in loiasis remains poorly defined. Immune complex deposition has been demonstrated in the nephropathy associated with loiasis; this same mechanism has also been postulated as being responsible for the post-treatment hematuria commonly seen in this disease [71].

Blood eosinophilia is a hallmark of *L. loa* infection in humans [71,154,155]. While it is tempting to hypothesize that the toxic cationic proteins (such as major basic protein, eosinophil-derived neurotoxin, or eosinophil cationic protein) released by eosinophils are mediating the peripheral neuropathy and cardiomyopathy seen in loiasis (as in the idiopathic hypereosinophilia syndrome), direct proof is lacking.

Additional studies suggest that granulomatous reactions must also play a role in the pathology seen in this infection. These types of reactions have been well-documented in the spleens of naturally and experimentally infected primates [186,187] and possibly in humans as well [188]. In addition, both granulomatous and fibrotic reactions around microfilariae have been described at other sites, including the lymph nodes and liver in both biopsy specimens and postmortem examinations in humans [189,190].

Resistance

Innate and nonspecific resistance

Lower vertebrates and nonprimate mammals cannot support the life-cycle of *L. loa*. While a number of nonhuman primates can be experimentally infected with *L. loa* [142,191], only the baboon, the drill (*Mandrillus* spp.), and the green monkey (*Cerocebus* spp.) appear to be definitive hosts. Interestingly, however, the simian *L. loa* has a microfilarial periodicity which is nocturnal [142] (as opposed to the diurnal periodicity of the human parasite). While the genetic background of the host is also likely to be important in susceptibility to *L. loa* infection, no clearcut answer has emerged. In one small study there was no evidence of HLA-A, -B, -C, -DR, or -DQ types or susceptibility to acquisition of infection [71].

Acquired resistance

Direct evidence that resistance to infection or reinfection with *L. loa* occurs does not exist. However, as with other filarial infections, the existence of naturally developed protective immunity is suggested by the presence, in highly endemic areas, of some individuals who appear to remain free of infection despite high levels of local transmission.

Immuno- and molecular diagnosis

As with all the filariae that infect humans, diagnosis only presents a problem in the absence of demonstrable microfilariae. Serology that is *Loa*-specific has not been developed and assays to detect circulating *Loa* antigen also have not been developed, although filarial-specific serology is available. Unfortunately, individuals in *Loa*-endemic areas often have coinfections with other filarial parasites, making *Loa*-specific diagnostics an important priority.

A molecular biologic approach will likely provide both the tools needed to speciate among the filariae as well as the quantities of relevant antigens needed for diagnosis. Of note is the recent identification of repetitive 356 bp sequences of *Loa*-genomic DNA containing 37 copies of a hexamer, many of which are arranged in longer repeated motifs. This repetitive sequence appears to be interspersed in the *Loa* genome, with a copy number of ~5500 per haploid genome. The repetitive unit is similar in organization and sequence to eukaryotic satellite DNA, based on unusual guanosine cytosine content and small (hexamer) repeated elements. Oligonucleotides derived from this repeated sequence have been used successfully as *Loa*-specific in dot blot analyses [192]. Using non-radioactive techniques, these *Loa*-specific DNA probes have been able to detect one larva in an infected *Chrysops* or one microfilaria in a milliliter of blood and can clearly distinguish *L. loa* from the other filariae.

REFERENCES

1 Sasa M. *Human Filariasis—a Global Survey of Epidemiology and Control*. Baltimore: University Park Press, 1976.
2 Petralanda I, Yarzabal L, Piessens WF. Studies on a filarial antigen with collagenase activity. Mol Biochem Parasitol 1986;19:51–59.
3 Selkirk ME, Nielsen L, Kelly C, Partono F, Sayers GA, Maizels RM. Identification, synthesis and cross-reactivity of cuticular collagens from the filarial nematodes *Brugia malayi* and *Brugia pahangi*. Mol Biochem Parasitol 1988; 32:229–240.
4 Maizels RM, Gregory WF, Kwan-Lim GE, Selkirk ME. Filarial surface antigens: the major 29 kilodalton glycoprotein and a novel 17–200 kilodalton complex from adult

Brugia malayi parasites. Mol Biochem Parasitol 1989;32: 213–228.

5 Maizels RM, Partono F, Oemijati S, Denham DA, Ogilvie BM. Cross-reactive surface antigens on three stages of *Brugia malayi, B. pahangi* and *B. timori*. Parasitology 1983; 87:249–259.

6 Fuhrman JA, Urioste SS, Hamill B, Spielman A, Piessens WF. Functional and antigenic maturation of *Brugia malayi* microfilariae. Am J Trop Med Hyg 1987;36:70–74.

7 Schrater AF, Rossignol PA, Hamill B, Piessens WF, Spielman A. *Brugia malayi* microfilariae from the peritoneal cavity of jirds vary in their ability to penetrate the mosquito midgut. Am J Trop Med Hyg 1982;31:292–296.

8 Fuhrman JA. Biochemistry of microfilarial exsheathment. Exp Parasitol 1990;70:363–366.

9 Devaney E, Howells RE. The exsheathment of *Brugia pahangi* microfilariae under controlled conditions *in vitro*. Ann Trop Med Parasitol 1979;73:227–233.

10 Fuhrman JA, Piessens WF. A stage-specific calcium-binding protein from microfilariae of *Brugia malayi*. Mol Biochem Parasitol 1989;35:249–258.

11 Kaushal NA, Hussain R, Nash TE, Ottesen EA. Identification and characterization of excretory–secretory products of *Brugia malayi* adult filarial parasites. J Immunol 1982; 129:338–446.

12 Maizels RM, Sutanto I, Denham DA. Secreted and circulating antigens on the filarial parasite *Brugia pahangi*: analysis of *in vitro* released components and detection of parasite products *in vivo*. Mol Biochem Parasitol 1985; 17:277–291.

13 Egwang TG, Kazura JW. Immunochemical characterization and biosynthesis of major antigens of Iodo-bead surface-labeled *Brugia malayi* microfilariae. Mol Biochem Parasitol 1987;22:159–174.

14 Rathaur S, Robertson BD, Selkirk ME, Maizels RM. Secretory acetylcholinesterases from *Brugia malayi* adult and microfilarial parasites. Mol Biochem Parasitol 1987;26: 257–265.

15 Low MG. Biochemistry of the glycosyl-phosphaditylinositol membrane protein anchors. Biochem J 1987;244:1–13.

16 Longworth DL, King DC, Weller PF. Rapid uptake and esterification of arachidonic acid and other fatty acids by microfilariae of *Brugia malayi*. Mol Biochem Parasitol 1987; 23:275–284.

17 Longworth DL, Monaham-Earley RA, Dvorak AM, Weller PF. *Brugia malayi*: arachidonic acid uptake into lipid bodies of adult parasites. Exp Parasitol 1988;65:251–257.

18 Liu LX, Serhan CN, Weller PF. Intravascular filarial parasites elaborate cyclooxygenase-derived eicosanoids. J Exp Med 1990;172:993–996.

19 Plescia OJ, Racis S. Prostaglandins as physiological immunomodulators. Prog Allergy 1988;44:153–190.

20 Cesbron J-Y, Capron A, Vargaftig BB, *et al.* Platelets mediate the action of diethylcarbamazine or microfilariae. Nature 1987;325:533–536.

21 Takacs AM. Denker JA, Perrine KG, Maroney PA, Nilsen TW. A 22-nucleotide spliced leader sequence in the human parasitic nematode *Brugia malayi* is identical to the trans-spliced leader exon in *Caenorhabditis elegans*. Proc Nat Acad Sci USA 1988;85:7932–7936.

22 Maroney PA, Hannon GJ, Nilsen TW. Transcription and cap methylation of nematode spliced leader RNAs in a cell-free system. Proc Nat Acad Sci USA 1990;87:709–713.

23 Hawking F, Gammage K. The periodic migration of microfilariae of *Brugia malayi* and its response to various stimuli. Am J Trop Med Hyg 1968;17:724–729.

24 Ottesen EA. Immunologic aspects of lymphatic filariasis and onchocerciasis. Trans R Soc Trop Med Hyg 1984; 78:9–18.

25 Phillip M. Davis TB, Storey N. Carlow CK. Immunity in filariasis: perspectives for vaccine development Ann Rev Microbiol 1988;42:685–716.

26 Ottesen EA. Immunopathology of lymphatic filariasis in man. Springer Semin Immunopathol 1984;2:373–385.

27 Kazura JW, Spark R, Forsyth K, *et al.* Parasitologic and clinical features of bancroftian filariasis in a community in East Sepik Province, Papua New Guinea. Am J Trop Med Hyg 1984;33:1119–1123.

28 Webb JKG, Job CK, Gault EW. Tropical eosinophilia: demonstration of microfilariae in lung, liver, and lymph nodes. Lancet 1960;i:835–842.

29 Pinkston P, Vijayan VK, Nutman TB, *et al.* Acute tropical pulmonary eosinophilia. Characterization of the lower respiratory tract inflammation and its response to therapy. J Clin Invest 1987;80:216–225.

30 Neva FA, Ottesen EA. Tropical (filarial) eosinophilia. N Engl J Med 1978;298:1129–1131.

31 Egwang TE, Kazura JW. The BALB/c mouse as a model for immunological studies of microfilariae-induced pulmonary eosinophilia. Am J Trop Med Hyg 1990;43:61–66.

32 Coggeshall LT. Filariasis in the serviceman. Retrospect and prospect. J Am Med Assoc 1946;131:8–12.

33 Piessens WF, Partono F. Host–vector–parasite relationships in human filariasis. Semin Infect Dis 1980;3:131–152.

34 McCall JW, Malone JB, Ah H-S, Thompson PE. Mongolian jirds (*Meriones unguiculatus*) infected with *Brugia pahangi* by the intraperitoneal route: a rich source of developing larvae, adult filariae and microfilariae. J Parasitol 1973; 59:436–449.

35 Klei TR, Blanchard DP, Coleman SU. Development of *Brugia pahangi* infections and lymphatic lesions in male offspring of female jirds with homologous infections. Trans R Soc Trop Med Hyg 1986;80:214–216.

36 Carlow CKS, Phillips M. Protective immunity to *Brugia malayi* in BALB/c mice: potential of this model for the investigation of protective antigens. Am J Trop Med Hyg 1987;37:597–604.

37 Kazura JW, Davis RS. Soluble *Brugia malayi* microfilarial antigens protect mice against challenge by an antibody-dependent mechanism. J Immunol 1982;128:1792–1796.

38 Vincent AL, Vickery AC, Lotz MJ, Desai U. The lymphatic pathology of *Brugia pahangi* in nude (athymic) and thymic C3H/HeN mice. J Parasitol 1984;70:48–56.

39 Crandall RB, Crandall CA, Hines SA, Doyle TJ, Nayar JK. Peripheral lymphedema in ferrets infected with *Brugia malayi*. Am J Trop Med Hyg 1987;37:138–149.

40 Nanduri J, Kazura JW. Clinical and laboratory aspects of filariasis. Clin Microbiol Rev 1989;2:39–50.

41 Hussain R, Ottesen EA. IgE responses in human filariasis. III. Specificites of IgE and IgG antibodies compared by

immunoblot analysis. J Immunol 1985;135:1415–1421.

42 Ottesen EA, Skavaril F, Tripathy SP, Poindexter RW, Hussain R. Prominence of IgG4 in the IgG antibody response to human filariasis. J Immunol 1985;134: 2707–2711.

43 Hussain R, Grugl M, Ottesen EA. IgG antibody subclasses in human filariasis: differential subclass recognition of parasite antigens correlates with different clinical manifestations of infection. J Immunol 1987;139:2794–2798.

44 Hussain R, Ottesen EA. IgE responses in human filariasis. IV. Parallel antigen recognition by IgE and IgG4 subclass antibodies. J Immunol 1986;136:1859–1866.

45 Nutman TB, Kumaraswami V, Pao L, Narayanan PR, Ottesen A. An analysis of in vitro B cell immune responsiveness in human lymphatic filariasis. J Immunol 1987; 138:3954–3961.

46 Coffman RL, Seymour BW, Lebman DA, et al. The role of helper T cell products in mouse B cell differentiation and isotype regulation. Immunol Rev 1988;102:5–28.

47 King CL, Ottesen EA, Nutman TB. Cytokine regulation of antigen-driven immunoglobulin production in filarial parasite infection in humans. J Clin Invest 1990;85:1810–1815.

48 Ottesen EA, Weller PF, Heck L. Specific cellular immune unresponsiveness in human filariasis. Immunology 1977; 33:413–424.

49 Piessens WF, McGreevy PB, Piessens PW, et al. Immune response in human infections with Brugia malayi. Specific cellular unresponsiveness to filarial antigens. J Clin Invest 1980;65:172–179.

50 Nutman TB, Kumaraswami V, Ottesen EA. Parasite-specific anergy in human filariasis: insights after analysis of parasite antigen-driven lymphokine production. J Clin Invest 1987;79:1516–1522.

51 Piessens WF, Partono F, Hoffman SL, et al. Antigen-specific suppressor T lymphocytes in human lymphatic filariasis. N Engl J Med 1982;307:144–148.

52 Wadee AA, Vickery AC, Piessens WF. Characterization of immunosuppressive proteins of Brugia malayi microfilariae. Acta Trop 1987;44:343–352.

53 Lal RB, Kumaraswami V, Steel C, Nutman TB. Phosphocholine-containing antigens of Brugia malayi nonspecifically suppress lymphocyte function. Am J Trop Med Hyg 1990;42:56–64.

54 Piessens WF, Ratiwayanto S, Piessens PW, et al. Effect of treatment with diethylcarbamazine on immune responses to filarial antigens in patients infected with Brugia malayi. Acta Trop 1981;38:227–234.

55 Lal RB, Kumaraswami V, Krishnan N, Nutman TB, Ottesen EA. Lymphocyte subpopulations in Bancroftian filariasis: activated (DR+) CD8+ T cells in patients with chronic lymphatic obstruction. Clin Exp Immunol 1989; 77:77–82.

56 Lammie PJ, Katz SP. Immunoregulation in experimental filariasis. I. In vitro suppression of mitogen-induced blastogenesis by adherent cells from jirds chronically-infected with Brugia pahangi. J Immunol 1983;130:1381–1385.

57 Lammie PJ, Katz SP. Immunoregulation in experimental filariasis. II. Responses to parasite and nonparasite anti-

gens in jirds with Brugia pahangi. J Immunol 1983;130: 1386–1389.

58 Leiva LE, Lammie PJ. Regulation of parasite antigen-induced T cell growth factor activity and proliferative responsiveness in Brugia pahangi-infected jirds. J. Immunol 1989;142:1304–1309.

59 Lal RB, Lynch TJ, Nutman TB. Brugia malayi antigens associated with lymphocyte activation in filariasis. J Immunol 1987;139:1652–1657.

60 Nilsen TW, Maroney PA, Goodwin RG. Cloning and characterization of a potentially protective antigen in lymphatic filariasis. Proc Natl Acad Sci USA 1988;85:3606–3607.

61 Kazura JW, Maroney PA, Pearlman E, Nilsen TW. Protective efficacy of a cloned Brugia malayi antigen in a mouse model of microfilaremia. J Immunol 1990;145:2260–2264.

62 Nanduri J, Kazura JW. Paramyosin-enhanced clearance of Brugia malayi microfilaremia in mice. J. Immunol 1989; 143:3359–3363.

63 Selkirk ME, Denham DA, Partono F, Maizels RM. Heat shock cognate 70 is a prominent immunogen in Brugian filariasis. J Immunol 1989;143:299–308.

64 Arasu P, Nutman TB, Steel C, et al. Human T cell stimulation, molecular characterization, and in situ mRNA localization of a Brugia malayi recombinant antigen. Mol Biochem Parasitol 1989;36:223–232.

65 Loke YW. Transmission of parasites across the placenta. Adv Parasitol 1982;21:155–228.

66 Dissanayake S, de Silva LVK. IgM antibody to filarial antigens in human cord blood: possibility of transplacental infection. Trans R Soc Trop Med Hyg 1980;74:542–544.

67 Weil GJ, Hussain R, Kumaraswami V, Tripathy SP, Phillips KS, Ottesen EA. Prenatal allergic sensitization to helminth antigens in offspring of parasite-infected mothers. J Clin Invest 1983;75:1124–1129.

68 Petralanda I, Yarzabal L, Piessens WF. Parasite antigens are present in breast milk of women infected with Onchocerca volvulus. Am J Trop Med Hyg 1988;38:372–379.

69 Wartman WB. Filariasis in American armed forces in World War II. Medicine 1947;26:181–189.

70 Partono F, Purnomo, Pribaldi W. Epidemiological and clinical features of Brugia timori in a newly established village, Karakuak, west Flores, Indonesia. Am J Trop Med Hyg 1978;27:910–915.

71 Nutman TB, Reese W, Poindexter W, Ottesen EA. Immunologic correlates of the hyperresponsive syndrome in loaisis. J Infect Dis 1988;157:544–550.

72 Haque A, Cuna W, Pestel J, Capron A, Bonnel B. Tolerance in rats by transplacental transfer of Dipetalenema viteae microfilariae: recognition of putative tolerogen(s) by antibodies that inhibit antigen-specific lymphocyte proliferation. Eur J Immunol 1988;18:1167–1172.

73 Schrater AF, Spielman A, Piessens WF. Predisposition to Brugia malayi microfilaremia in progency of infected gerbils. Am J Trop Med Hyg 1983;32:1306–1308.

74 Ottesen EA, Mendell NR, MacQueen JM, Weller PF, Amos DB, Ward FE. Familial predisposition to filarial infection — not linked to HLA-A or B-locus specificities. Acta Trop 1981;38:205–212.

75 Chan SH, Dissayanake S, Mak JW, et al. HLA and filariasis

in Sri Lankans and Indians. Southeast Asian J Trop Med Public Health 1984;15:281–292.

76 Fanning MM, Kazura JW. Genetic association of murine susceptibility to *Brugia malayi* microfilaremia. Parasite Immunol 1983;5:305–314.

77 Thompson JP, Crandall RB, Crandall CA, Neilson JT. Microfilaremia and antibody responses in CBA/H and CBA/N mice following injection of microfilariae of *Brugia malayi*. J Parasitol 1981;67:728–737.

78 Haque A, Worms MJ, Ogilvie BM, Capron A. *Dipetalonema viteae*: microfilariae production in various mouse strains and nude mice. Exp Parasitol 1980;49:398–406.

79 Storey N, Wakelin D, Behnke JM. The genetic control of host responses to *Dipetalonema viteae* (Filarioidea) infections in mice. Parasite Immunol 1985;7:349–355.

80 Kwan-Lim G, Maizels RM. MHC and non-MHC-restricted recognition of filarial surface antigens in mice transplanted with adult *Brugia malayi* parasites. J Immunol 1990;145:1912–1920.

81 Denham DD. Vaccination against filarial worms using radiation-attenuated vaccines. J Nucl Med Biol 1980;7:105–111.

82 Grieve RB, Abraham D, Mika-Grieve M, Seibert BP. Induction of protective immunity in dogs to infection with *Dirofilaria immitis* using chemically-abbreviated infections. Am J Trop Med Hyg 1988;39:373–379.

83 Kazura JW, Cicirello H, McCall JW. Induction of protection against *Brugia malayi* infection in jirds by microfilarial antigens. J Immunol 1986;136:1422–1426.

84 Kazura JW, Cicirello H, Forsyth KP. Differential recognition of a protective filarial antigen by antibodies from humans with bancroftian filariasis. J Clin Invest 1986;77:1985–1992.

85 Freedman DO, Nutman TB, Ottesen EA. Protective immunity in Bancroftian filariasis. Selective recognition of a 43-kD larval stage antigen by infection-free individuals endemic area. J Clin Invest 1989;3:14–22.

86 Hammerberg B, Nogami S, Nakagaki K, Hayashi Y, Tanaka H. Protective immunity against *Brugia malayi* infective larvae in mice. II. Induction by a T cell-dependent antigen isolated by monoclonal antibody affinity chromatography and SDS-PAGE. J Immunol 1989;143:4201–4207.

87 Greene BM, Gbakima AA, Albiez EJ, Taylor HR. The humoral and cellular immune response to *Onchocerca volvulus* in man. Rev Infect Dis 1985;7:789–795.

88 Williams JF, El Khalifa M, Mackenzie DC, Sisley B. Antigens of *Onchocerca volvulus*. Rev Infect Dis 1985;7:831–836.

89 Nutman TB, Withers AS, Ottesen EA. *In vitro* parasite antigen-induced antibody responses in human helminth infections. J Immunol 1985;135:2794–2799.

90 Greene BM. Primate model for onchocerciasis research. CIBA Found Symp 1987;127:236–243.

91 Soboslay PT. *Doctoral Dissertation*. Germany: University of Tübingen, 1989.

92 Greene BM, Taylor HR, Aikawa M. Cellular killing of microfilariae of *Onchocerca volvulus*: eosinophil and neutrophil-mediated immune serum-dependent destruction. J Immunol 1981;127:1611–1618.

93 Mackenzie DC. eosinophil leucocytes in filarial infections. Trans R Soc Trop Med Hyg 1980;74:51–58.

94 Williams JF, Ghalib HW, MacKenzie CD, Elkhalifa MY, Ayuyn JM, Kron MA. Cell adherence to microfilariae of *O. volvulus*: a comparative study. Ciba Found Symp 1987;127:146–163.

95 Kephart GM, Gleich GJ, Connor DH, Gibson DW, Ackerman SJ. Deposition of eosinophil granule major basic protein onto microfilariae of *Onchocerca volvulus* in the skin of patients treated with diethylcarbamazine. Lab Invest 1984;50:51–61.

96 Ottesen EA. Immediate hypersensitivity responses in the immunopathogenesis of human onchocerciasis. Rev Infect Dis 1985;7:796–801.

97 Weiss N, Speiser F, Hussain R. IgE antibodies in human onchocerciasis. Application of a newly developed radioallergosorbent test (RAST). Acta Trop 1981;38:353–362.

98 Donnelly JJ, Rockey JH, Taylor HR, Soulsby EJL. Onchocerciasis: experimental models of ocular disease. Rev Infect Dis 1985;7:820–825.

99 Paganelli R, Ngu JL, Levinsky RJ. Circulating immune complexes in onchocerciasis. Clin Exp Immunol 1980;38:570–575.

100 Greene BM, Taylor HR, Brown EJ, Humphrey RL, Lawley TJ. Ocular and systemic complications of diethylcarbamazine therapy for onchocerciasis: association with circulating immune complexes. J Infect Dis 1983;147:890–897.

101 Sisley BM, Mackenzie CD, Steward MW, *et al*. Associations between clinical disease, circulating antibodies and Clq-binding immune complexes in human onchocerciasis. Parasite Immunol 1987;9:447–463.

102 Buck AA, Anderson RT, Kawata K, Hitchcock JC. Onchocerciasis: some new epidemiologic and clinical findings. Am J Trop Med Hyg 1969;18:217–230.

103 Rougemont A, Boisson-Pontal ME, Pontal PG, Gridel F, Sangare S. Tuberculin skin tests and BCG vaccination in hyperendemic area of onchocerciasis. Lancet 1977;i:309.

104 Prost A, Nebout M, Rougemont A. Lepromatous leprosy and onchocerciasis. Br Med J 1979;1:589.

105 Greene BM, Fanning MM, Ellner JJ. Non-specific suppression of antigen-induced lymphocyte blastogenesis in *Onchocerca volvulus* infection in man. Clin Exp Immunol 1983;52:259–265.

106 Gallin M, Edmonds K, Ellner JJ, *et al*. Cell-mediated immune responses in human infection with *Onchocerca volvulus*. J Immunol 1988;140:1999–2007.

107 Bartlett A, Turk J, Ngu JL, Mackenzie DC, Fuglsang H, Anderson J. Variation in delayed hypersensitivity in onchocerciasis. Trans R Soc Trop Med Hyg 1978;72:372–377.

108 Ngu JL. Immunological studies on onchocerciasis. Acta Trop 1978;35:259–265.

109 Bryceson ADM. What happens when microfilariae die? Trans R Soc Trop Med Hyg 1976;70:397–399.

110 Anderson J, Font RL. Ocular onchocerciasis. In Binford CH, Connor DH, eds. *Pathology of Tropical and Extraordinary Diseases*, vol. 2. Section 8. Washington, DC: Armed Forces Institute of Pathology, 1976:373–381.

111 Gibson DW, Heggie C, Conner DH. Clinical and pathologic aspects of Onchocerciasis. In Somers SC, Rosen PP, eds.

Pathology Annual, vol. 15. Part 2. New York: Appleton-Century-Crofts, 1980:195–240.

112 Connor DH, George GH, Gibson DW. Pathologic changes of human onchocerciasis: implications for future research. Rev Infect Dis 1985;7:809–819.

113 Rodger FC. The dissolution of microfilariae of *Onchocerca volvulus* in the human eye and its effect on the tissues. Trans R Soc Trop Med Hyg 1959;53:400–403.

114 Duke BOL, Anderson J. A comparison of the lesions produced in the cornea of the rabbit eye by microfilariae of the forest and Sudan-savanna strains of *Onchocerca volvulus* from Cameroon. I. The clinical picture. Trop Med Parasitol 1972;23:354–368.

115 Anderson J, Duke BOL. A comparison of the lesions produced in the cornea of the rabbit eye by microfilariae of the forest and Sudan-savanna strains of *Onchocerca volvulus* from Cameroon. II. The pathology. Trop Med Parasitol 1973;24:385–396.

116 Duke BOL, Garner A. Reactions to subconjunctival inoculation of *Onchocerca volvulus* microfilariae in pre-immunized rabbits. Trop Med Parasitol 1975;26:435–448.

117 Sakla AA, Donnelly JJ, Lok JB, *et al*. Punctate keratitis induced by subconjunctivally injected microfilariae of *Onchocerca lienalis*. Arch Ophthalmol 1986;104:894–898.

118 Donnelly JJ, Xi MS, Haldar JP, *et al*. Autoantibody induced by experimental onchocerca infection. Invest Ophthalmol Visual Sci 1988;29:827–831.

119 Gallin MY, Murray D, Lass JH, Grossniklaus HE, Greene BM. Experimental interstitial keratitis induced by *Onchocerca volvulus* antigens. Arch Ophthalmol 1988;106:1447–1452.

120 Brinkmann UK, Krämer P, Presthus GT, Sawadogo B. Transmission *in utero* of microfilariae of *Onchocerca volvulus*. Bull WHO 1976;54:3570.

121 Petralanda I, Yarzabal L, Piessens WF. Parasite antigens are present in breast milk of women infected with *Onchocerca volvulus*. Am J Trop Med Hyg 1988;38:372–379.

122 Büttner DW, Von Lear G, Mannweiler E, Büttner M. Clinical parasitological and serological studies on onchocerciasis in the Yemen Arab Republic. Trop Med Parasitol 1982;33:201–212.

123 Gibson DW, Connor DH, Brown HL, *et al*. Onchocercal dermatitis: ultrastructural studies of microfilarial and host tissue, before and after treatment with diethylcarbamazine (Hetrazan). Am J Trop Med Hyg 1976;25:74–87.

124 Brattig NW, Tischendorf FW, Albiez EJ, Büttner DW, Berger J. Distribution pattern of peripheral lymphocyte subsets in localized and generalized form of onchocerciasis. Clin Immunol Immunopathol 1987;44:159.

125 World Health Organization. *The Pathogenesis and Treatment of Ocular* Onchocerciasis, TDR/FIL/SWG(8)/82.3. WHO, Geneva, 1982.

126 Murphy RP, Taylor HR, Greene BM. Chorioretinal damage in onchocerciasis. Am J Ophthalmol 1984;98:519–521.

127 Donnelly JJ, Taylor HR, Young EM, Khatami M, Lok JB, Rockey JH. Experimental ocular onchocerciasis in cynomulgus monkeys. Invest Ophthalmol Visual Sci 1986;27:492–499.

128 Semba RD, Donnelly JJ, Rockey JH, Lok JB, Sakla AA,

Taylor HR. Experimental ocular onchocerciasis in cynomolgus monkeys. II. Chorioretinitis elicited by intravitreal *Onchocerca lienalis* microfilariae. Invest Ophthalmol Visual Sci 1988;29:1642–1651.

129 Paul EV, Zimmerman LE. Some observations on the ocular pathology of onchocerciasis. Hum Pathol 1970;1:581.

130 Duke BOL. Reinfections with *Onchocerca volvulus* in cured patients exposed to continuing transmission. Bull WHO 1968;38:307–309.

131 Duke BOL. Experimental transmission of *Onchocerca volvulus* from man to a chimpanzee. Trans R Soc Trop Med Hyg 1962;56:271.

132 Ward DJ, Nutman TB, Zea-Flores G, Portocarrero C, Lujan A, Ottesen EA. Onchocerciasis and Immunity in Humans: enhanced T cell responsiveness to parasite antigen in putatively immune individuals. J Infect Dis 1988;157:536–543.

133 Weiss N, Karam M. Evaluation of a specific enzyme immunoassay for onchocerciasis using a low-molecular-weight antigen fraction of *Onchocerca volvulus*. Am J Trop Med 1989;40:261–267.

134 Lobos E, Weiss N, Karam M, Taylor HR, Ottesen EA, Nutman TB. An immunogenic *Onchocerca volvulus* antigen: a specific and early marker of infection. Science 1991;251:1603–1605.

135 Erttmann KD, Unnasch TR, Greene BM, *et al*. A DNA sequence specific for forest form *Onchocerca volvulus*. Nature 1987;327:415–417.

136 Unnasch TR, Gallin M, Soboslay PT, Erttmann KD, Greene BM. Isolation and characterization of expression cDNA clones encoding antigens of *Onchocerca volvulus*. J Clin Invest 1988;82:262–269.

137 Wong MM, Guest MF, Lavoipierre MJ. Dirofilaria immitis: fate and immunogenicity of irradiated infective stage larvae in beagles. Exp Parasitol 1974;35:465–474.

138 Lucius R, Schulz-Key H, Büttner DW, *et al*. Characterization of an immunodominant *Onchocerca volvulus* antigen with patient sera and a monoclonal antibody. J Exp Med 1988;167:1505–1510.

139 Lucius R, Erondu N, Kern A, Donelson JE. Molecular cloning of an immunodominant antigen of *Onchocerca volvulus*. J Exp Med 1988;168:1199–1204.

140 Colina KF, Perler FB, Matsumura I, Meda M, Nutman TB. The identification of an *Onchocerca*-specific recombinant antigen containing a T cell epitope. J Immunol 1990;145:1551–1556.

141 Fain A. Current problems of loiasis. Bull WHO 1978;56(2):155–167.

142 Duke BOL, Wijers DJB. Studies on loiasis in monkeys. I. The relationship between human and simian *Loa* in the rain forest zone of the British Cameroons. Ann Trop Med Parasitol 1958;52:158–175.

143 Eberhard ML, Orihel TC. Development and larval morphology of *Loa loa* in experimental primate hosts. J Parasitol 1981;67(4):556–564.

144 Gruntzig J. The first description of *Loa-loa* infestation of the eye by Pigafetta – a historical error (author's translation). Klin Monatsbl Augenheilkd 1976;169(3):383–386.

145 Guyot F. In *Fonds Academie Royale de Chirugie, Carotn No. 55. Dossier: Maladies de la tete No. 4 bis No. 129*. Paris:

Academie Nationale ded Medicine Bibliotheque, 1981.

146 Looss A. Zur Kentniss des Baues der Filaria Loa Guyot. Zoologische Jahrbucher 1904;3(20):549–557.

147 Connor DH, Neafie RC, Meyers WM. Loiasis. In Binford CH, Connor DH, eds. *Pathology of Tropical and Extraordinary Diseases*. Washington: Armed Forces Institute of Pathology, 1976:356–359.

148 Kozek WJ, Orihel TC. Ultrastructure of *Loa loa* microfilaria. Int J Parasitol 1983;13(1):19–43.

149 Franz M, Melles J, Buttner DW. Electron microscope study of the body wall and the gut of adult *Loa loa*. Z Parasitenkd 1984;70(4):525–536.

150 McClaren DJ. Ultrastructural studies on microfilariae (Nematoda: Filarioidea). Parasitology 1972;65:317–332.

151 Pinder M, Dupont A, Egwang TG. Identification of a surface antigen on *Loa loa* microfilariae the recognition of which correlates with the amicrofilaremic state in man. J Immunol 1988;141(7):2480–2486.

152 Egwang TG, Akue JP, Dupont A, Pinder M. The identification and partial characterization of an immunodominant 29–31 kilodalton surface antigen expressed by adult worms of the human filaria *Loa loa*. Mol Biol Parasitol 1988;31:263–272.

153 Egwang TG, Dupont A, Akue JP, Pinder M. Biochemical and immunochemical characterization of surface and excretory–secretory antigens of *Loa loa* microfilariae. Mol Biol Parasitol 1988;31:251–262.

154 Klion AD, Massougboudji A, Sadeler BC, Ottesen EA, Nutman TB. Loiasis in endemic and non-endemic populations: immunologically-mediated differences in clinical presentation. J Infect Dis. 1991;163:1318–1325.

155 Nutman TB, Miller KD, Mulligan M, Ottesen EA. *Loa loa* infection in temporary residents of endemic regions: recognition of a hyperresponsive syndrome with characteristic clinical manifestations. J Infect Dis 1986;154(1):10–18.

156 Carme B, Mamboueni JP, Copin M, Noireau F. Clinical and biological study of *Loa loa* filariasis in Congolese. Am J Trop Med Hyg 1989;41:331–337.

157 Ambroise-Thomas P, Peyron F. Filariasis. In Walls KW, Schantz PM, eds. *Immunodiagnosis of Parasitic Diseases*, vol. I. New York: Academic Press, 1986:233–236.

158 Goussard B, Ivanoff B, Frost E, Garin Y, Bourderiou C. Age of appearance of IgG, IgM, and IgE antibodies specific for *Loa loa* in Gabonese children. Microbiol Immunol 1984; 28(7):787–792.

159 Grieve RB, Eberhard ML, Jacobson RH, Orihel TC. *Loa loa*: antibody responses in experimentally infected baboons and rhesus monkeys. Tropenmed Parasitol 1985;36:225–229.

160 Egwang TG, Dupont A, Leclerc A, Akue JP, Pinder M. Differential recognition of *Loa loa* antigen by sera of human subjects from a loiasis endemic zone. Am J Trop Med Hyg 1989;41:664–673.

161 Haque A, Cuna W, Bonnel B, Capron A, Joseph M. Platelet mediated killing of larvae from different filarial species in the presence of *Dipetalonema viteae* stimulated IgE antibodies. Parasite Immunol 1985;7(5):517–526.

162 Grewe W, Gordon RM. The immediate reaction of the mammalian host to the bite of uninfected Chrysops and of Chrysops infected with human and monkey *Loa*. Ann Trop Med Parasit 1959;53:334.

163 Ngu JL, Chatelnat F, Leke R, Noumbe P, Youmbiss J. Nephropathy in Cameroon evidence for filarial derived immune complex pathogenesis in some cases. Clin Nephrol 1985;24(3):128–134.

164 Barsotti P, Brandimarte C, Feriozzi S, *et al*. Glomerulonefrite in corso di filariosi e malaria. Studio di un caso. Minerva Nefrol 1983;30:25–31.

165 Katner H, Beyt BE Jr, Krotoski WA. Loiasis and renal failure. South Med J 1984;77(7):907–908.

166 Malik STA, McHug M, Morley AR, Ngu J, Zureshi M, Wilkinson R. Filariosis (*Loa-loa*) associated with membranous glomerulonephritis demonstration of filarial antigen. Kidney Int 1981;20:157.

167 Pillay VK, Kirch E, Kurtzman NA. Glomerulopathy associated with filarial loiasis. J Am Med Assoc 1973;225(2): 179.

168 Zuidema PJ. Renal changes in loiasis. Folia Med Neerl 1971;14:168–172.

169 Andy JJ, Bishara FF, Soyinka OO, Odesanmi WO. Loasis as a possible trigger of African endomyocardial fibrosis: a case report from Nigeria. Acta Trop 1981;38(2):179–186.

170 Ive FA, Willis AJP, Ikeme AC, Brockington IF. Endomyocardial fibrosis and filariasis. Q J Med 1967;144:495–515.

171 Brockington IF, Olsen EGJ, Goodwin JF. Endomyocardial fibrosis in Europeans resident in tropical Africa. Lancet 1967;i:583–588.

172 De la Heeran Herrera A, Gonzalez Garrido EA, Pila Perez R, Leon Diaz R. Miocardioptia en un paciente con loaiasisi Presentacion de un caso. Rev Cubana Med Trop 1981; 33:201–206.

173 Toussaint D, Danis P. Retinopathy in generalized *Loa-loa* filariasis. A clinicopathological study. Arch Ophthalmol 1965;74(4):470–476.

174 Renard G, Morand L, Lacombe B, Offret G. A case of retinol filariasis. J Fr Ophthalmol 1978;1:41–46.

175 Doury P, Saliou P, Charmot G. Articular effusions with eosinophils. Apropos of a case report. Sem Hop 1983; 59(22):1683–1685.

176 Bouvet JP, Therizol M, Auquier L. Microfilarial polyarthritis in a massive *Loa loa* infestation. A case report. Acta Trop 1977;34(3):281–284.

177 Sarkany I. Loiasis with involvement of peripheral nerves. Trans St John's Hosp Dermatol Soc 1959;82:49–51.

178 Schofield FD. Two cases of loiasis with peripheral nerve involvement. Trans R Soc Trop Med Hyg 1955;49:588–589.

179 Grove DI, Schneider J. Arm lymphedema associated with filariasis. Arch Intern Med 1981;141:137.

180 Paleologo FP, Neafie RC, Connor DH. Lymphadenitis caused by *Loa loa*. Am J Trop Med Hyg 1984;33(3):395–402.

181 Downie CGB. Encephalitis during treatment of loiasis with diethylcarbamazine. JR Army Med Corp 1966;112:46–49.

182 Kivits M. Quatre cas d'encephalite mortelle avec invasion du liquide cephaloarachidien par des microfilaires *Loa loa*. Ann Soc Belg Med Trop 1952;32:235–242.

183 Pays JF, Ecalle JC, Cornet A, Brumpt L. Neuropsychic manifestations of loaiasis. A clinical case of potomania. Bull Soc Pathol Exot 1976;69(3):265–272.

184 Van Bogaert L, Dubois A, Janssen PG, Radermercker J,

Tverdy G, Wanson M. Encephalitis in *Loa loa* filariasis. J Neurol Neurosurg Psychiatr 1955;18:103–119.

185 Stanley SL Jr, Kell O. Ascending paralysis associated with diethylcarbamazine treatment of M. *Loa loa* infection. Trop Doct 1982;12(1):16–19.

186 Orihel TC, Eberhard ML. *Loa loa*: development and course of patency in experimentally-infected primates. Trop Med Parasitol 1985;36(4):215–224.

187 Duke BOL. Studies on loiasis in monkeys. III. The pathology of the spleen in drills (*Mandrillus leucophaeus*) infected with *Loa*. Ann Trop Med Parasitol 1960;54:141–146.

188 Negesse Y, Lanoie LO, Neafie RC, Connor DH. Loiasis: Calabar: swellings and involvement of deep organs. Am J Trop Med Hyg 1985;34(3):537–546.

189 Woodruff AW. Destruction of microfilariae of *Loa loa* in the liver in loiasis treated with banocide. Trans R Soc Trop Med Hyg 1951;44:4.

190 Klotz O. Nodular fibrosis of the spleen associated with Filaria *Loa*. Am J Trop Med 1930;10:57–64.

191 Orihel TC, Moore PJ. *Loa loa*: experimental infection in two species of African primates. Am J Trop Med Hyg 1975;24(4):606–609.

192 Klion AD, Raghavan N, Brindley PJ, Nutman TB. Cloning and characterization of a species-specific repetitive DNA sequence from *Loa loa*. Mol Biochem Parasitol 1991;45:295–306.

19 Nematodes

Derek Wakelin, William Harnett, & R. Michael E. Parkhouse

INTRODUCTION

Nematodes are among the commonest of all parasites and are responsible for diseases of major importance in humans and in domestic animals. From the standpoint of immunoparasitology it is convenient, although not entirely logical, to consider separately those species which live exclusively within the body tissues and those which live exclusively, or during the adult phase, within the intestine. The former, the filarial worms, are the subject of Chapter 18 and will not be considered in any detail here. The latter, the gastrointestinal nematodes, have a very high prevalence in humans, infecting perhaps one-quarter of the world's population; they also cause serious economic losses in cattle, sheep, and other stock (Table 19.1).

Four worms account for the majority of human infections: *Ascaris lumbricoides*, *Trichuris trichiura*, *Necator americanus*, and *Ankylostoma duodenale*. Some excellent reviews summarizing their incidence have recently appeared [1–6]. In spite of the widespread occurrence of these parasites, they have been greatly neglected, by both researcher and research funding authorities. Indeed, a WHO initiative was formulated only in the last decade [4]. Like most helminthiases, the gastrointestinal nematodes do not attract the medical headlines, largely because they are Third World diseases and provoke low mortality. At the same time, however, they are typically associated with considerable misery, through general debilitation, pathology, and direct and indirect associations with malnutrition. Sadly, however, we are largely ignorant of the nature and interplay of the factors caused by these parasites which operate to the detriment of global human health. Investigation is urgently required at all conceivable levels, so that these parasites can be understood and controlled, but perhaps most urgently at the level of the immunologic interactions between host and parasite.

NEMATODES AS TARGETS OF IMMUNITY

It is characteristic of many of these infections that they are persistent, with reinfection occurring readily, often throughout the lifetime of the host. This has led to the assumption that protective immunity against these species is absent, or at best ineffective. Indeed, despite the abundant evidence that gastrointestinal nematodes are highly immunogenic in humans, there is still no

Table 19.1 Major nematodes of humans and domestic animals (excluding filarial spp.)

Nematodes	Stage associated with disease	Location of disease
In humans		
Angiostrongylus	Larva	CNS
	Adult	Intestine
Anisakis	Larva	Intestine
Ascaris	Larva	Liver, lung
	Adult	Intestine
Capillaria	Adult	Intestine
Hookworms	Larva	Skin
	Adult	Intestine
Strongyloides	Larva	Tissues (general)
	Adult	Intestine
Toxocara	Larva	Tissues (especially CNS, eye)
Trichinella	Larva	Muscles
	Adult	Intestine
Trichuris	Adult	Intestine
In domestic animals		
Dictyocaulus	Adult	Lungs
Gastrointestinal trichostrongyles	Adult	Intestine

unequivocal evidence for protective immunity, although recent epidemiologic data now point to this as a determinant of prevalence and intensity. The situation is quite different in domestic animals and in laboratory rodents, where immunity can be readily demonstrated and has been extensively investigated. These studies have made considerable progress in recent years, contributing both to our understanding of the mechanisms of antinematode immunity as such, as well as adding more generally to our knowledge of the initiation and expression of immune and inflammatory responses. The challenge now is to explain this immunity in molecular terms and to apply this knowledge to understanding and eventually controlling these infections.

A number of recent reviews have summarized the broad field of immunity to nematodes, some dealing specially with intestinal species [7–12]. The reader is referred to these for an overview of current work and for useful summaries of older papers. Rather than attempt to duplicate the cover provided by these reviews, this section will focus on well-studied laboratory systems where immunologic and molecular studies have made substantial progress.

Biology

The nematodes included within this chapter show considerable diversity in size, biology, and life history. Entry into the host may occur by ingestion of infective eggs or larvae (*Ascaris, Toxocara*, gastrointestinal (GI) trichostrongyles, *Trichuris*), by skin penetration of larvae (hookworms, *Strongyloides*), or by ingestion of other infected hosts (*Angiostrongylus, Capillaria, Trichinella*). Development may be wholly restricted to the intestine itself or the worm may undertake extensive tissue migrations. The infecting stages of all species are relatively small (<1 mm), but the size of the adult worms varies considerably, from a few millimeters (*Strongyloides, Trichinella*) to several centimeters (*Ascaris*). Common to all is the possession of a collagenous cuticle, which provides the worm with a substantial degree of protection both against the host's physicochemical environment and against some of the host's protective mechanisms. The precise structural organization of this layer varies between species, but in all that have been studied it expresses antigenic molecules to which the host responds strongly; it may also be the target for attack by macrophages, granulocytes, and platelets via antibody-dependent cell-mediated cytotoxicity (ADCC) reactions. Relatively little is known of the metabolism of these nematodes but it is well documented that they release a variety of molecules, some of which elicit strong host immune responses whereas others apparently down-regulate responsiveness. These molecules are associated with feeding, tissue penetration, excretion, reproduction, and molting. At all stages, nematodes are generally highly mobile and display precise behavioral responses to changes in their environment. Others (e.g., *Trichinella* and *Toxocara*) have larval stages which localize within tissues and become the focus of inflammatory reactions.

Habitats

As Table 19.1 shows, nematodes of clinical importance are approximately equally divided between those where pathogenicity is associated with stages located in parenteral sites and those where disease is caused by the adult worm in the intestine. In numerical terms the latter is by far the most significant group, but many of the most serious pathologic consequences of infection are associated with the former group. The range of immune responses elicited by nematodes and the consequences for the host, whether associated with resistance or with pathology, are to some extent determined by the site of infection. Larval stages within tissues are in intimate contact with the efferent and afferent arms of the immune response and are vulnerable, at least in theory, to antibody, complement, ADCC, and direct damage from cellular mediators. In immunologic terms, therefore, the host–parasite relationships of such species are similar to those involving other tissue-located helminths; differences in the outcome of these relationships can be expected to follow from the different structural characteristics of nematodes. With the intestinal stages the situation is more complex. Some species (*Ascaris*) are purely lumen-dwelling, others live in the lumen but feed on the mucosa (hookworms, GI trichostrongyles), and others penetrate into the mucosa. The relationship of these species to the host's immune system is therefore likely to be different from that of the tissue phases and will reflect the peculiarities of the intestinal immune system.

It is characteristic of these nematodes that they elicit pronounced antibody responses, involving all major isotypes. An increase in parasite-specific IgE and potentiation of nonspecific IgE is a particular feature of almost all infections; levels of selected IgG subclasses (e.g., IgG1 in the mouse) may also show significant increases. Intestinal parasite-specific IgA responses have been recorded in many infections. Intestinal worms provoke profound inflammatory changes, in part the result of direct damage to the tissues but primarily the consequence of T cell-mediated responses. These changes alter both the structure and the physiology of the intestine, they dramatically change the environment of the worms, and they bring a number of immunologic and inflammatory effectors into immediate contact with the parasites. Amine-containing cells (mast cells and basophils) and eosinophils are prominent in the cellular infiltrate which infection provokes.

The immune response, both cellular and humoral, is the critical factor in determining the clinical and pathologic course of the disease. A thorough understanding of parasite antigens and of the immunologic responses that they elicit in the host is thus a prerequisite for understanding the factors which determine whether the outcome of the host–parasite interaction will be resistance or susceptibility, enhanced responsiveness or immunosuppression, asymptomatic, or pathologic. In essence, this involves determining which parasite components are immunogenic, defining the range of immune responses and effector mechanisms called into play, and identifying the interactions within this complex network of immunologic phenomena.

Antigens

The variety of antigens presented by nematodes to their hosts can be analyzed at a number of different levels, from the biologic to the molecular [13]. In terms of their origin they can be considered at the levels of:

1 the parasite stage concerned;
2 the antigenic compartments within a stage (e.g., surface, excretion–secretion (ES), or somatic);
3 the antigen components within a compartment;
4 the epitopes of a single defined antigenic component.

They can also be considered in immunologic or immunochemical terms, i.e., in relation to allergenicity, isotype-response specificity, T cell stimulation/suppression, or to their molecular composition (proteins, glycoproteins, glycolipid, carbohydrate). Most importantly, perhaps, they can be considered in terms of the function that they have in the life of the parasite, whether they are enzymes, antienzymes, receptors, metabolites, structural elements, molting hormones, pheromones, etc.

Many earlier studies of nematode antigens were necessarily restricted to the use of crude and complex homogenates, prepared from the most accessible parasite stage. More recently, the application of immuno-

chemical and molecular techniques has enabled much more precise antigen analysis to be undertaken and has provided the means of studying immune responses to precisely defined molecules. Among the most useful of these techniques have been surface and metabolic labeling, sodium dodecyl sulfate polyacrylamide gel electrophoresis (SDS-PAGE), immunoblotting (including T cell blotting), and monoclonal antibody production for identification, localization, and affinity chromatography. These techniques have been applied now to most of the major nematodes of humans and domestic animals, as well as to the familiar laboratory model systems. There have been several useful reviews of progress in this field [14–19] so we shall not go over it here. To illustrate the approaches undertaken and some of the significant results, attention will be given primarily to work on *Trichinella spiralis*.

Antigens of *Trichinella spiralis*

Trichinella spiralis was the first nematode in which protective immune responses were associated with defined antigenic components [20]. Silberstein & Despommier [20] isolated molecules of molecular weight 48 and 50–55 kD from the stichosomal granules of the muscle larval stage and showed that these elicited high levels of protective immunity when injected into mice. Many subsequent studies have confirmed these observations and it is now generally recognized that stichosomal glycoproteins of this size represent major immunogens. There has been some disagreement over the precise molecular weights of these antigens, no doubt reflecting the glycosylation of the molecules and different techniques of preparation, but a consensus has now been established. Despommier's original 48 kD antigen has been redefined as being equivalent to the 43/45 kD molecules described by himself in later papers [21] and by Gamble & Graham [22]. The 50–55 kD entity is considered equivalent to the more recently described 45–50 kD and 49–53 kD molecules of Despommier *et al.* [23] and Gamble & Graham [22], respectively. The range of hosts which can be protected by these antigens has now been extended to include pigs [24].

Characterization of these molecules has been carried out and partial N-terminal amino acid sequences determined [23]. These sequences are quite distinct, but the molecules themselves show extensive serologic cross-reactivity. This is lost when the molecules are deglycosylated using glycopeptidase, showing that the immunodominant epitopes are associated with *N*-linked carbohydrates. The deglycosylated molecules have stable molecular weights of 32 (43 kD) and 33 (45–50 kD), respectively. *In vivo* studies in mice have confirmed the immunodominance of carbohydrate epitopes [25]. Infected mice show a biphasic antibody response, one set of antigens being recognized 13 days after infection and another set after 35 days. The latter includes ES antigens and elicits predominantly IgG1 responses, 80% of which react with a larval stage-specific determinant shared by virtually all of the antigens concerned. A monoclonal antibody against this shared determinant allowed affinity purification of antigens capable of protectively immunizing mice. Immunochemical studies have shown that six major molecular species from the muscle larvae share the common determinant [26]. Their molecular weights range between 43 and 68 kD and collectively they comprise about 3% by weight of the total protein in larval homogenate preparations. The immunodominant epitope is associated with *N*- and *O*-linked oligosaccharides and was shown not to be phosphorylcholine.

At least two of the major immunogens in larval ES material have been cloned, the 43/45 kD and the 53 kD, but published details are available only for the latter [27]. Screening of a cDNA library with antisera from infected pigs allowed selection of positive clones which were then rescreened using a rabbit anti-ES antibody. One clone selected coded for a 123 kD fusion protein, antibody against which crossreacted with the 53 kD antigen in immunoblots and identified stichosomal granules in worm sections. mRNA for this molecule is expressed in both larval and adult worms, but the molecule itself is present only in the former. The gene sequence is present as multiple copies, and the possibility has been raised that if these genes are not 100% homologous they may control the synthesis of all the antigenically related 45, 49, and 53 kD antigens.

Many of the stichosomal granule components are released by the worm, and the major immunogens are present in ES material. They also appear in material stripped by detergent from the cuticular surface [28]. Which of these components is of primary importance in stimulating immune responses *in vivo* is not certain and nor is the precise relationship between origin and expression of the molecules. The kinetics of stichosomal granule release imply that secreted antigens will provide a rapid and powerful stimulus immediately after entry of larvae into the intestinal mucosa [29]. Nevertheless, purified surface antigens from the muscle larvae also stimulate a substantial immunity [30].

The surface antigens of *T. spiralis* have been the subject of many papers. It was with this species that the phenomenon of stage specificity was first demonstrated [31], a characteristic now described in very many nematodes. The infective larva, adult worm, and newborn larva all show quite distinct and restricted patterns of surface antigens, and these elicit sequential and distinct antibody responses [32]. Monoclonal antibodies raised against surface antigens have been used both to define their precise localization [33] and, in the case of newborn larvae, to infer a role in protective immunity [34]. The genetic control of the expression of these stage-specific molecules is an important field of study, in which only limited progress has so far been made [17]. Cloning of *T. spiralis* antigens has likewise been relatively slow. Zarlenga & Gamble [27] have successfully cloned a 145 kD antigen from the muscle larval stage. Antibodies raised against polyacrylamide gel electrophoresis (PAGE)-purified recombinant antigen recognized a doublet of 53 kD in antigen extracts of larvae, i.e., in the molecular weight range known to contain protective molecules and antigens suitable for specific diagnosis.

Applications of Trichinella spiralis antigen preparations

In addition to the obvious uses of defined antigens in the analysis of protective responses or of vaccination, the availability of such antigens makes it possible to study other aspects of the host–parasite relationship. Two examples can be used to illustrate these. The first concerns the level of complexity in the serologic response to infection, in terms of the variable immunoglobulin isotype profile elicited by each antigen or epitope. The result of infection with any parasite is a complex array of antibody specificities, presenting a problem in elucidating which, if any, are relevant to resistance, to pathology, or to immunodiagnosis. Stage-specific labeled surface and secreted molecules of *T. spiralis* have been used in isotype-specific analysis of the antibody response to individual antigens, providing a model of the kind of analysis necessary with other species [35]. The important results from this study were that there is independent variation in each immunoglobulin isotype to each epitope of every antigen compartment of each stage, and that this variation additionally reflects genetic variation in the host animal. The data suggest that predictive diagnosis of the course, and outcome of infection (in terms of resistance or pathology) may be possible if precisely defined assay systems can be set up, e.g., by using individual stage-specific antigens and a specific immunoglobulin isotype (see also the section on immunodiagnosis pp. 511–519).

The second area of application of *T. spiralis* antigens concerns studies using related species, subspecies, and isolates of this parasite. These can be differentiated by looking for DNA restriction fragment length polymorphisms [36]; they also show distinct patterns of infectivity and pathology in laboratory animals; analysis of immune responses may therefore be able to throw light on the immunologic determinants of resistance and susceptibility or of pathologic responses. Equally, such analyses can provide useful models of the likely efficacy of candidate vaccine antigens against the variant population of parasite that may be encountered in geographically widely dispersed species. In this context, comparisons of *T. spiralis* with *Trichinella pseudospiralis* and of *T. spiralis* with *T. spiralis nativa* and *T. spiralis nelsoni* have been undertaken by several workers [37–39]. Differences in antigen profiles (in all compartments) and in antibody responses have been shown among these parasites, and it should theoretically be possible to associate these with the well-defined differences in pathology, infectivity, and immunity that are known to occur [40], although this has not yet been achieved. Tests using isolate-specific antigens to immunize against homologous and heterologous isolates of *T. spiralis* have shown that good crossimmunity exists, all preparations eliciting >50% protection [41]. However, with some combinations, protection is only just greater than 50%, whereas in others it is >90%. One broader implication of this result is that standard antigen preparations may not immunize equally well when used against a parasite that occurs in widely dispersed populations.

Trichinella spiralis-isolate antigens have also been used in comparative studies of T cell proliferative responses [42]. Since it is becoming increasingly clear that the repertoire and balance of the T cell subsets elicited by parasitic infections play key roles in determining the outcome of infection [43], analyses of this kind using defined but variable antigen moieties are likely to be essential in establishing predictive models.

Antigens of other nematodes

General reviews of recent progress in antigen analyses of other species are available in the references cited earlier (p. 498). Certain of these species have been studied in considerable detail and selected examples are used here to illustrate particular areas of current research.

Ascaris spp.

Ascariasis is characteristically associated with the development of allergic responses, the worms themselves being potent sources of allergens even to research workers handling them. Many allergenic molecules have been described in *Ascaris* [44], but interest has recently been focused on a 10/14 kD molecule, derived from the body fluid, identified as ABA-1 [45]. This molecule is found in the larval and adult stages of both *A. lumbricoides* (from humans) and *Ascaris suum* (from pigs). The N-terminal amino acid sequence of ABA-1 has been determined and is identical in both species [46]. The isolation and characterization of this molecule opens the way for determining the relationship between molecular structure and allergenicity, a property associated with antigens from many nematodes. This work will be facilitated by the observation that there is major histocompatibility complex (MHC)-associated genetic control of IgE antibody recognition of this molecule [47], a finding which may have important implications for the immunopathology of ascariasis in humans. Certainly it has been shown that humans also show marked variation in their ability to recognize this molecule [48].

Hookworm

The surface and secretions of a nematode provide the major interface between the host and the living parasite. From this it follows that host-protective mechanisms, parasite evasion strategies, and parasite-induced immunopathology must all act principally through the molecules present in these compartments. Somatic antigens can be recognized only when the worm is dead or when its intact surface has been breached by some other protective immune process. Protective responses involving molecules of somatic origin can therefore operate only secondarily. When the parasite dies, the host is presented with a multiplicity of antigens, some of which may trigger pathologic reactions. A number of these points have been addressed in relation to hookworm infections.

Hookworms are one of the few nematode groups where a direct relationship has been established between a functionally significant enzyme and protective immunity [49]. *Ancylostoma caninum* releases a histolytic anticoagulant protease that is inactivated by immune serum. In immune dogs, adult worms have a reduced capacity to feed on blood, presumably because of defective enzyme activity. This protease has been isolated and purified as a 37 kD molecule [50]. Similar molecules also occur in the related species *N. americanus*, a cause of hookworm disease in humans [51], and it is tempting to think of these as candidate antigens for potential vaccines. Cloning of these molecules is being actively pursued. *Necator* also releases proteases during its penetration process and these also can be considered as targets for protective responses.

Infections with hookworms in humans are characteristically long-lasting, implying a failure of, or an active interference with, protective responses. It is possible that the worms have surface molecules which protect more vulnerable underlying structures from immune attack. If these protective molecules could themselves be attacked, the worms might then be open to direct damage [52]. Isolated collagen-like proteins involved in maintaining the structural integrity of the cuticle are indeed susceptible to digestion by purified proteases from mast cells, but in the intact undamaged worm this does not occur [53].

Toxocara canis

In addition to being responsible for an important zoonosis, *T. canis* provides an excellent laboratory system with which to study the antigenic basis of host immune and pathologic responsiveness [54]. Infections are readily established in mice, and infective larvae can be maintained for long periods *in vitro*, during which time large amounts of ES antigens are released. *Toxocara canis* was one of the first nematodes in which active (i.e., metabolically dependent) transcuticular release of antigen was demonstrated [55] and it has served as a model for studies of this process.

The ES antigens of *T. canis* are known to act as targets for protective host responses (particularly involving ADCC), they contain allergenic components, induce eosinophilia, and activate complement. All of the major molecules are glycoproteins. Collectively, ES antigens contain more than 40% carbohydrate—*N*-acetylgalactosamine and galactose being the principal sugars [56]. Monoclonal antibodies against ES antigens have been used to follow the developmental expression of particular molecules and to define particular epitopes [57]. The antibodies were predominantly against carbohydrate epitopes, and each reacted with several distinct ES components. Similar multiple reactivities with monoclonal reagents are known in other ascarid species, but in this group they do not appear to be based upon shared phosphorylcholine determinants, as is the case with

other nematodes, particularly the filariae. In addition to the crossreactivity seen between ES molecules from *T. canis* itself, there is substantial crossreactivity between anti-*T. canis* monoclonals and ES antigens from the related *Toxocara cati* [58]. One anticarbohydrate monoclonal was species-specific, however, and this forms the basis for species-specific diagnostic assays (see later section on immunodiagnosis pp. 511–519).

Trichostrongyles of sheep

Attempts to vaccinate against trichostrongyle parasites of sheep have largely used the approach shown to be successful in the vaccines used against *Dictyocaulus* and *Ancylostoma*, i.e., direct infection with irradiation-attenuated larvae. Despite the efficacy of these vaccines, only limited success has been achieved against the major trichostrongyles *Haemonchus* and *Trichostrongylus* (see section on vaccination pp. 508–511). As a consequence there has been renewed interest in the potential of more-defined immunogens.

Electron microscope studies of *Haemonchus contortus* in the 1970s identified an unusual extracellular polymeric protein—contortin—which is closely associated with the microvillar membrane of the intestinal cells of the worm [59]. In its monomeric form, contortin has a molecular weight of about 60 kD. It may function as an immobilized, renewable anticoagulant. The blood-feeding habit of this species means that its intestinal surface is repeatedly exposed to plasma components of the host, including antibodies. If these were able to interact with antigens present on that surface, the ability of the worm to feed and thus to grow and reproduce might be severely impaired (a situation analogous to that known to occur in ticks feeding on immune hosts, and which forms the basis of current antitick vaccination strategies).

Contortin and a number of other intestinal membrane antigens (e.g., H11), have now been isolated and used successfully to immunize lambs against infection with *H. contortus* [60]. Vaccinated animals developed circulating antibodies against these antigens and were considerably more resistant to infection than controls. Genes for contortin and H11 have been cloned, and there is now a clear vaccine potential for the recombinant antigens.

Although it is possible to protect lambs against infection with *Trichostrongylus colubriformis* by using an irradiated larval vaccine, strong genetic influences define the degree of immunity expressed (see later section on vaccination). Earlier observations that SDS-PAGE-separated fractions of larval homogenate were protectively immunogenic [61,62] have been followed by more precise identification and gene cloning [63]. Sodium deoxycholate extracts of larval homogenate showed a relatively restricted antigen profile, with four major bands. One of these bands (41 kD) has been partially sequenced and identified as worm tropomyosin. When injected into guinea pigs, the 41 kD component gave substantial protection against infection. cDNA clones for this molecule have been isolated and characterized.

Protective antigens have also been described in ES products of adult and larval *T. colubriformis*, and two have been isolated and cloned. One is a 30 kD glycoprotein [64] and the other is an 11 kD nonglycosylated component [65]. The glycoprotein (designated ESgp30) can be isolated after binding to lentil lectin and gives 59–69% protection against infection in the guinea pig model, i.e., approximately the same level as treated whole ES material. Purifed ESgp30 was used for N-terminal sequencing, and an oligonucleotide probe based on this sequence was used to screen a cDNA library in λgt11. A number of positive clones was identified and one expressed in *Escherichia coli* as a 15 kD polypeptide. ESgp30 shows some sequence homology with valosin, a porcine intestinal peptide, and thus the question is raised whether *T. colubriformis*-secreted material exerts a (possibly protective) physiologic influence over host intestinal function. The 11 kD peptide also gives a high degree of immunity when used to vaccinate guinea pigs at 20 µg per animal [65]. A reverse-complement 23-base oligonucleotide based on sequenced peptides was used to screen a cDNA library, detecting 36 clones, one of which, when expressed in *E. coli*, gave a product recognized by specific rabbit antibody. Part of the native protein shares a significant homology with a 34-amino acid sequence in a human interferon-γ (IFN-γ)-induced protein, raising the possibility that the worm-derived molecule may play an immunomodulatory role. Immune inactivation of this molecule may therefore interfere with the ability of the worm to survive. No antibody response to this molecule is seen in the infected host, suggesting either that it is unrecognized or that it is recognized only at the T cell level.

The work with *H. contortus* and *T. colubriformis* described in the preceding paragraphs suggests a novel approach to the search for candidate vaccine antigens. This is the identification and use of molecules which, in the living worm, are inaccessible to, or unrecognized by, the host's immune system but when given in a vaccine elicit strong protective immune responses. There would appear to be no reasons why this approach

should not work equally well with many other species.

Work with the species described above provides a good example of the approaches (and pitfalls) in defining and producing reagents capable of providing accurate and unequivocal diagnosis and in the production of potential candidate vaccine antigens, whatever the target species. Knowledge of stage, age, and species specificity of particular surface and secreted components is essential, as is an understanding of their potential to produce undesirable as well as desirable immune responses. Equally important is the ability to obtain large quantities of the molecules chosen. *Toxocara canis* is particularly good at releasing large quantities of antigen over a long period of time in simple media; other species are much less cooperative in this respect, and obtaining adequate amounts of their antigens presents problems. When the antigens required are peptides, large-scale production via recombinant DNA technology is feasible. For non-protein (i.e., noncloneable) antigens the preparation of anti-idiotype reagents to complementary monoclonal antibodies provides an alternative approach. Such anti-idiotype reagents can find immediate application as diagnostic tools, as substitutes for antigen, and they may also provide potential vaccines.

HOST RESPONSE: IMMUNITY AND IMMUNOPATHOLOGY

Innate and acquired resistance to infection

The species listed in Table 19.1 must represent a relatively small fraction of the total number of nematode parasites to which humans and domestic animals are exposed. It is possible to conclude from this that there is a substantial innate resistance to many potential infections. The nature of this resistance is poorly understood, but in a majority of cases probably reflects physiologic incompatibilities between host and parasite rather than an innate resistance expressed through other means. However, mechanisms of innate resistance involving inflammatory cells and their mediators may well be important in defense against penetrating larval stages, e.g., in protecting humans against the infective larvae of animal hookworms. Many nematode larvae are known to have the property of activating the alternative complement pathway [66] and this would provide the means by which parasites could be attacked by inflammatory cells independently of immune recognition. Nevertheless, it seems improbable that such attack is ever immune independent and it is likely that resistance to infection with any nematode capable of survival in the body is primarily expressed through components of the adaptive immune system. Identification and analysis of these components has been most fully undertaken in a number of laboratory systems (Table 19.2) where the degree of resistance expressed can be accurately quantified by counting the numbers of worms present after challenge. Data from these systems will be discussed in detail to provide a framework for consideration of immunity in other hosts.

T cell dependence of acquired immunity

In all cases protective immune responses against nematodes are initiated and mediated through the activity of T cells, being reduced or absent in thymus-deprived and T cell-deficient animals. In addition, immune T cells, purified by adherence to nylon wool, by positive or negative selection using specific antibodies, or selected by repeated antigenic stimulation *in vitro*, have been used successfully to transfer immunity adoptively into naïve animals.

The identity of the T cell subset involved in mediating resistance has been defined in relatively few systems, notably using *T. spiralis* in mice and rats and *Nippostrongylus brasiliensis* in mice [71]. In the former in mice, it has been shown that only CD4$^+$ (L3T4$^+$, Ly-2$^-$) T helper cells can transfer immunity [72]. *In vivo* depletion of this subset, using monoclonal antibody against the L3T4 molecule, removes the ability of mice to control a primary infection, the remaining CD8$^+$ (L3T4$^-$, Ly-2$^+$) T cells apparently being ineffective. Antigen-specific T cells, maintained *in vitro* by repeated stimulation with muscle larval homogenate antigen, were also shown to be exclusively CD4$^+$ and these transferred immunity adoptively in very small numbers ($< 2 \times 10^5$/mouse)

Table 19.2 Major laboratory systems used in analysis of acquired immunity to nematodes

Species	Host	[References]
Nippostrongylus brasiliensis	Mouse, rat	[9]
*Nematospiroides dubius**	Mouse	[8]
Strongyloides ratti	Mouse, rat	[67]
Trichuris muris	Mouse	[68]
Trichinella spiralis	Mouse, rat	[69,70]
Trichostrongylus colubriformis	Guinea pig	[5]

* The taxonomically correct name for this species is *Heligmosomoides polygyrus*. Although recent papers use this name, the majority of those cited in this chapter do not; the more common name is therefore retained to avoid confusion.

[73]. CD4$^+$ T helper cells (W3/25$^+$, OX8$^-$) are also known to be responsible for transfer of immunity against *T. spiralis* in rats and there is again evidence that these cells are effective in relatively small numbers [74]. In both mice and rats, parasite-specific T helper cells are generated very rapidly after infection, within 2–4 days, and much evidence points to their origin in the intestinal mucosa itself. As far as *in vivo*-generated T cells are concerned, effective transfer is achieved only with dividing cells, although this is not the case with *in vitro*-generated cells. This difference may simply indicate that, in the former, division identifies those cells responding specifically to infection, while after prolonged *in vivo* culture only parasite-specific cells are present.

Although similar analyses of the T cell subsets responsible for immunity have not been undertaken in other model systems, it seems reasonable to assume that CD4$^+$ T cells, or their equivalent, will be involved. This conclusion is based not only on the pivotal role that this subset is known to play in immune responses generally, but also on the similarity of components of resistance brought into play by infection in all host systems.

The demonstration that the T helper subset is essential for the expression of acquired immunity provides a basis for identifying the mechanisms involved but clearly does not narrow the options very greatly, as T helper cells regulate antibody-mediated, cell-mediated, and inflammatory responses, all of which might contribute to protection. Central to this regulation is the release of lymphokines, and recent studies have examined both the kinetics and the range of cytokines released by T cells involved in immunity to *T. spiralis*. Splenic and mesenteric lymph node T helper cells from mice known to express the high-responder phenotype release more IL-2 after stimulation with antigen than cells from low responders and are present in greater numbers [75]. Antigen-specific T helper cells from *in vitro*-maintained lines released IL-2, IL-3, and IFN-γ in response to antigenic stimulation and these cytokines were also released by mesenteric node cells taken early (day 4) after infection in high-responder mice [76].

Antibody-mediated immunity

Evidence that antibody is functionally involved in resistance comes from three sources: experiments involving passive or maternal transfer of immunity, those which correlate the kinetics of infection with the kinetics of serologic responses, and those which analyze antibody–parasite interactions *in vitro*.

Transfer of immunity with immune sera has been achieved in all of the host–parasite systems listed in Table 19.2, but with varying degrees of effectiveness. For example, high levels of resistance can be transferred against the intestinal stages of *N. brasiliensis* and *Trichuris muris*, but transfer is less effective against *Nematospiroides dubius* and *T. spiralis*, where a cooperative interaction with some cell-mediated function seems necessary to maximize worm expulsion. In the case of *T. spiralis* both maternal and passive transfer of antibody are effective in eliciting a rapid expulsion of worms from neonatal rats but this does not occur in adults [77]. Transfer of immune serum itself, or of immune B cells, also suggests a role for antibody in reducing the growth and fecundity [78] of *T. spiralis*, possibly through an effect upon feeding and metabolism, although infection is not markedly enhanced in B cell-suppressed mice [79]. The isotype responsible for successful passive transfer has most often been identified as IgG, although there is evidence that other isotypes can function protectively. Given the characteristic association of IgE with nematode infections [44], and the importance of IgA as a mucosal antibody, it is surprising that there is so little direct evidence for their involvement in immunity against nematodes, particularly the intestinal species. In addition to data obtained with *T. muris* (see below), a number of correlative studies have implicated IgA in immunity to *T. spiralis* [35] and to *N. brasiliensis* [80], the former identifying surface antigens of the adult worm as an important target. No clearcut effector role for this isotype has been defined. Perhaps the most likely functions of IgA may be to limit the motility and invasiveness of worms, which may become trapped in the mucus layer covering the mucosal surface [81], or to interact with released molecules that function in tissue penetration and external digestion. The kinetics of serum IgE responses have rarely suggested a causal link with immunity to intestinal stages, although this idea is still firmly entrenched in the literature. Obviously, circulating levels of IgE do not accurately reflect the amounts of antibody bound to amine-containing cells in the mucosa, which is presumably the crucial site for the expression of antiworm immunity. Nevertheless there are a few reports of successful passive transfer of immunity where the evidence suggests a functional role for IgE [82].

Transfer of immunity using monoclonal antibodies provides a more precise way of identifying effective isotypes and their target antigens than the use of polyvalent infection-induced antisera. This approach has only recently begun to be applied to the study of nematode infections. IgG1 and IgG2c rat monoclonal antibodies generated against muscle larval antigens transferred the ability rapidly to expel *T. spiralis* to 10–14-day-old suckling rats [83], the level of immunity transferred being greater than 70%. IgG1 monoclonals recognizing antigens in adult and larval extracts were found to transfer immunity to recipient mice against *N. dubius* [84], but this was reflected primarily in effects against growth and fecundity rather than worm survival. These results parallel those previously obtained in this system with transfer of antisera [8,85], and may reflect interference by antibody with molecules released in ES material. Transfer of an IgA monoclonal antibody, which recognizes antigens present in the stichocytes of *T. muris*, resulted in a substantial loss of worms from infected mice, the first direct demonstration of protective activity against nematodes associated with this isotype [86].

Antibody-dependent cellular cytotoxicity

Many *in vitro* studies have shown that nematode larvae can be immobilized or killed when incubated with antibody, complement, and a variety of cells [66,87]. However, the relevance of *in vitro* correlates of immunity to the degree of immunity expressed *in vivo* remains controversial, especially for those species located in the intestine. Data from such studies are perhaps more relevant for species with parenteral stages, where worms are more accessible to ADCC as well as to other forms of antibody-mediated attack and where the effectors of immunity can be more readily released onto their targets. It is now well established that the nematode cuticle is immunogenic, that surface antigens elicit strong antibody responses in the host and that antigen-antibody interaction provides the means by which cells can adhere to the cuticle via their Fc and C' receptors [14].

The isotypes so far known to mediate ADCC are those of the IgG subclasses. Of the several studies identifying this isotype, only one has presented unequivocal evidence for functional activity *in vivo*, Ortega-Pierres *et al.* [34] achieved killing of *T. spiralis in vitro* with a monoclonal IgG1-antibody that recognized a 64 kD surface antigen. Passive transfer of this antibody into mice significantly reduced the number of muscle larvae developing from intravenously injected newborn larvae, showing that, in this case, *in vitro* data did correlate with *in vivo* function. Whether such ADCC is relevant to the immunity generated and operative during infection is less certain. The site at which immunity operates against newborn larvae released by adult females *in vivo* is not known, although data are available which suggest that larvae are inhibited or destroyed within the intestine itself, an effect that is transferable with immune sera [88].

The involvement of IgE in ADCC dependent on eosinophils and other cells, which is well documented for schistosomes [89], has not yet been demonstrated in nematodes (other than filaria) although it may underlie the observations that muscle larval burdens increased in *T. spiralis* in rats treated with anti-E antibody [90]. Although a number of cells have been shown to function in ADCC, most attention has been focused on eosinophils. These cells appear to be powerfully cytotoxic *in vitro*, and *in vivo* depletion of these cells increased the number of *T. spiralis* muscle larvae that established after infection [91]. Eosinophil-mediated killing of *T. spiralis* larvae occurs through release of the granule-associated mediators directly onto the cuticular surface [92]. Major basic protein [93] and peroxidase [94] have both been shown to damage larvae, although the latter authors found that myeloperoxidase was more toxic. It is interesting that an antibody-dependent neutrophil-mediated immunity has also been demonstrated in mice infected with *N. dubius* [95], the cells presumably being involved in ADCC reactions against the mucosal larval stages.

Both neutrophils and eosinophils (together with mast cells) contribute to the cellular infiltrates induced by skin penetrations by *Strongyloides ratti* [96]. Degranulating eosinophils have been seen around invading larvae in immune hosts [97] and it is assumed that these cells contribute to the expression of acquired immunity. Eosinophils also form a prominent component of the granulomata around *T. canis* larvae but, although there is circumstantial evidence for an involvement in resistance [98] this is not borne out by *in vitro* studies. These have shown *T. canis* to be relatively resistant to ADCC as a consequence of turnover of cuticular material [99]. Similarly, eosinophils form a major component of the cellular reaction around larvae and young adults of *Angiostrongylus cantonensis* in nonpermissive hosts. Large amounts of peroxidase and other material are released onto the surface of the worms and may contribute to worm death [100]. Antibody-dependent cell-mediated cytoxicity reactions may also contribute to the trapping and killing of migrating larvae, e.g., of hookworms and *N. brasiliensis*, as they pass through the lungs of immune animals. Certainly bronchial-alveolar lavage studies have documented marked leukocyte responses in both systems during challenge infections [101,102].

Inflammation

The inflammatory reactions elicited by nematode infections have a complex etiology [103]. Direct physical and chemical damage to host tissues from the movement, growth, and metabolism of the worms may itself initiate an inflammatory response that is independent of immunity. It is likely, however, that inflammation is primarily mediated through the host's immune response to antigens released from the worms. The immunologic components which mediate inflammation include antibodies, complement components, monocyte-myeloid cells, sensitized T cells, and a variety of soluble mediators. Each of these components may exert a direct antiparasite effect and in this way contribute to the expression of immunity. In addition, the inflammatory changes they induce affect both the structure and the function of host tissues, drastically altering the environment in which worms live and thus acting as powerful effectors of host resistance.

Inflammatory reactions around larval nematodes within tissues not only focus immune effectors, with the possibility of direct damage, but they may also encapsulate and localize the worms. However, in many cases worms escape from these reactions or survive despite them, particularly during initial infections, good examples of this being seen in *T. canis* [104] and in *N. dubius* [105]. Subsequent infections may be more severely affected because both quantitative and qualitative changes in the response lead to an earlier and more intense reaction.

Relatively little is known about control of the development and expression of immunologically mediated inflammation during infections with tissue nematodes, although it is clear that T cell-dependent phenomena are important components. Much more is understood about inflammation during infections with intestinal species. Most data relate to experimental systems using *N. brasiliensis*, *S. ratti*, and *T. spiralis* in mice and rats, where the importance of inflammatory changes as a final effector mechanism in worm expulsion has been demonstrated using manipulative adoptive transfer protocols [106]. The effectiveness of immune lymphocytes in transfer is abolished, or severely reduced, by prior irradiation of the recipients. Restoration of the expression of lymphocyte function in these animals requires provision of bone marrow stem cells, presumably to reconstitute inflammatory capacity.

The major changes occurring in the intestines of infected animals are summarized in Table 19.3. It is

Table 19.3 Inflammatory changes in the intestines of animals infected with parasitic nematodes

---

Cellular infiltration—mast cells, eosinophils, neutrophils, plasma cells, macrophages

Increase in intraepithelial lymphocytes, expression of Ia antigens

Raised levels of leukocyte enzymes, amines, prostaglandins, leukotrienes; goblet cell hyperplasia

Increased mucus secretion

Villous atrophy, decreased epithelial cell transit time

Altered permeability and vascular flow

Net secretion of fluid into lumen

Altered gut motility

---

possible that individually, as well as collectively, all of these changes contribute to a reduction in the worm's ability to survive in the intestine, but it has proved difficult to manipulate the inflammatory process selectively in order to determine their relative importance. Recent research interest has focused upon three particular aspects—mast cells, intestinal mucus, and soluble mediators—and these will be discussed briefly.

Mast cells

Circumstantial evidence linking the activity of mucosal mast cells (MMC) with protective responses against intestinal nematodes has been available for many years (extensively reviewed by Miller [9] and Rothwell [11]). The kinetics of mastocytosis and worm expulsion often show a striking correlation, suggesting a causal relationship, but this is not always the case. Although some studies using mast cell deficient mice have confirmed this correlation, others have failed to do so. Worm expulsion has been restored only when MMC responses have been reconstituted by transfer of bone marrow or of cultured MMC [107]. One interpretation of these apparently contradictory data is that under certain circumstances and in certain host–parasite combinations, MMC function can be a major effector mechanism of resistance but can be replaced by other mechanisms. This interpretation is supported by evidence which shows that when immune expulsion of worms is accelerated by adoptive transfer of T cells, mastocytosis is also enhanced [108], i.e., transfer has emphasized this component of the intestinal response and increased its contribution to resistance.

All early work on MMC responses employed histologic

methods to obtain quantitative data. The identification of MMC-specific serine proteinases and the development of immunoassays for them [109] have now made it possible to measure the degree of mast cell activity, as these enzymes are released when MMC are triggered. Studies with both *N. brasiliensis* and *T. spiralis* have shown that the kinetics of MMC-specific serine proteinase release correlate with worm expulsion [110], strengthening the case for a functional involvement of MMC. However, detailed studies in mice with *T. spiralis* show that this involvement is most direct and most important in mice that are high responders and is much less clearcut in low responders [111]. This again supports the view that MMC are one of several components in worm expulsion, and sometimes, but not always, the dominant component.

The precise way in which MMC effect worm expulsion is still not understood and nor are the ways in which their activity is triggered. IgE-mediated activation remains the likeliest explanation, although correlative studies of the kinetics of IgE levels have not proved particularly helpful in this respect. Indeed, studies in mice made incapable of IgE production by injection of anti-μ heavy chain antibody [112], and in strains genetically deficient in IgE [113], have failed to show significant differences in control of either *N. brasiliensis* or *T. spiralis*. The latter finding appears to contradict the results from *T. spiralis*-infected rats made IgE deficient [90], but may reflect a differential effect upon immunity acting intestinally against adults and systemically against larvae. Direct activation by "degranulators" released from worms remains an additional possibility. Arming of resident MMC by antigen-specific T cell-derived factors has also been proposed as a mechanism for initial inflammatory changes that follow infection [114]. Whatever the trigger, it can be assumed that the major event in MMC-mediated inflammation is the release of mediators which initiate local changes in the worms' environment. Detailed physiologic studies using preparations of guinea pig intestine have documented how MMC activation induces profound functional changes in the mucosa, affecting the membrane properties of enterocytes [115].

Mucus

Nematode infections have long been known to cause a change in the quantity and nature of mucus present in the intestine. Renewed interest in mucus as a component of antiworm immunity comes from studies in immune animals, in which the rapid elimination of challenge infections appeared to involve a trapping of incoming larvae, resulting in their exclusion from the mucosa and their removal from the intestine [81]. Although increased release of mucus is known to follow challenge with antigen in primed animals, and goblet cell hyperplasia is under T cell control, it is not clear how these phenomena operate during primary infections.

The basis of trapping, whether entirely a consequence of physicochemical changes in mucus or due to the action of mucus-associated antibody, is uncertain. A further possibility is the incorporation into mucus of mediators derived from inflammatory cells, and their direct activity against invading worms. One mediator for which there has been experimental evidence for such mucus-related activity is SRS-A (now known to be the leukotrienes LTC4, LTD4, and LTE4) [116].

Soluble mediators

The complexity of the intestinal inflammatory response is such that many soluble mediators are released into the mucosa and into the lumen. Implication of these in the process of worm expulsion is largely circumstantial and is based on correlations between worm loss and mediator release or on interference with expulsion by manipulations designed to inhibit the release or function of specific mediator molecules [11]. With the exception of mucus-related SDS-A and of earlier studies using 5-HT and prostaglandins, there is little evidence that mediators play a direct antiparasite role, and it has to be concluded that their function lies primarily in alteration of the intestinal environment. However, this may not be the case for free oxygen radicals released by leukocytes during infection. There is a close correlation between loss of *N. brasiliensis* and production of free radicals, inhibition of free-radical release by incorporation of butylated hydroxyanisole in the diet being associated with reduced worm expulsion [117,118]. Significantly, *N. brasiliensis* has been shown to be vulnerable to damage by free radicals generated *in vitro*, and this may reflect low levels of scavenging enzymes such as superoxide dismutase, catalase, and glutathione reductase [119]. *Nematospiroides dubius*, a species which establishes chronic infections in mice and which seems relatively resistant to intestinal inflammation, has much higher levels of these enzymes.

Initiation and regulation of immunity

Until recently, research activity has been primarily concerned with analysis of the components of the immune responses against nematodes through which resistance is expressed. Developments in immunology itself, the current emphasis upon vaccination against parasitic infections, and the realization that parasites can promote their own survival by depressing host responses have led to a growth of research into the initiation and regulation of immunity. These are key fields, where understanding is essential if interventionist strategies are ever likely to make it possible for immune responses to be manipulated in favor of the host.

The central role of CD4$^+$ T cells in initiating protective immune responses during nematode infections (see p. 503) has focused attention on the role of antigen-presenting cells (APC) and on the interaction of antigens with class II molecules. Relatively little is known of the former in relation to nematode infections, although it can be assumed that the quite distinct populations of APC which function in systemic, enteral, and dermal locations may result in the initiation of qualitatively different responses against parasites which live in these sites. Important work by Wassom suggested that significant differences in immunity to *T. spiralis* in mice resulted from antigen presentation via one or other of the sets of class II molecules on the APC [120]. Presentation via IA molecules was associated with enhanced resistance to infection, whereas IE-mediated presentation was associated with decreased resistance.

One interpretation of these data was the induction of suppressor T cells after IE presentation, but an alternative is differential stimulation of the TH-1 and TH-2 subsets of the CD4$^+$ population [43], as has been proposed in the case of *Leishmania* infections. Certainly the two subsets differ in their release of cytokines, with properties relevant to the development of effective antinematode responses [121].

Data consistent with a down-regulation of immunity after IE-mediated antigen presentation have also been obtained by Wassom from studies with *N. dubius* [122], but conflicting data with this worm were described by Behnke & Wahid [123]. Thus it is not clear how general are the effects associated with this phenomenon—indeed data with *T. muris* in mice also failed to support a depressive role for IE-related presentation [124]. However, the important point made by this and related work is that differences in antigen presentation associated with differences in class II molecules can lead to quantitative and qualitative differences in response. A further issue is the relationship of the T cell receptor to antigen recognition and induction of immunity. Recent evidence that epithelial T cells express a γδ TCR rather than an αβ TCR may imply some important differences in recognition and response [125], although such cells are predominantly CD8[+].

If resistance to many nematodes, and particularly the intestinal species, is indeed dependent upon the development of immune-mediated inflammatory changes, then the regulation of these changes at the T cell level will be through release of the appropriate cytokines. However, the final degree of inflammation will depend both upon the level of these cytokines and on the capacity of bone marrow stem cells to respond. To date, relatively few workers have been concerned with these aspects of nematode infection. Grencis *et al.* [76] showed that antigen-specific cells capable of releasing IL-2 on stimulation were present within the mesenteric lymph node of mice by day 4 after infection with *T. spiralis*. Maximal release of both IL-2 and IL-3 occurred at this time. Comparative studies of IL-2 production in high- and low-responder strains infected with this parasite confirmed this pattern of release, and additionally showed that high-responder NFS mice produced two to three times more IL-2 than low-responder B10.BR mice at peak [75]. NFS mice also had proportionately more IL-2 receptor positive cells at day 5 postinfection, which agrees with the higher frequency of *T. spiralis*-specific cells present in the node at this time. Nematode infections appear to be remarkably efficient in stimulating TH-2 populations, thus releasing the cytokines which selectively regulate eosinophila (IL-5), IgG1, and IgE production (IL-4), as well as mastocytosis (IL-3) [126]. Recent studies have shown that these responses are manipulable *in vivo* [127] by using injected monoclonal anti-IL antibodies, and this opens the way for a more detailed analysis of host responses to infection.

Studies on the eosinophilia accompanying infection with *T. spiralis* have shown that the differences in degree of response seen between different mouse strains are not related to lymphocyte-dependent release of IL-5 but reflect a genetically determined stem cell responsiveness [128]. The relevance of eosinophilia to the intestinal phase of immunity is not clear, although the release of eosinophil-related mediators may well play a contributory role in the generation of intestinal inflammation, but its relevance to expression of immunity against tissue stages is likely to be much greater. It is also known that mouse strain-dependent differences in mucosal mastocytosis following infection with *T. spiralis* are likewise dependent on stem cell response rather than on IL-3 production [129].

Modulation of host resistance

Detailed analysis of the cellular and genetic mechanisms that influence the initiation and regulation of host resistance will not only contribute to a greater understanding of the overall response to nematode parasites but it will help to explain some of the ways in which certain species modulate host responsiveness. Thus, it is already known that worms such as *N. dubius* up-regulate IgG1 production but down-regulate mastocytosis, as well as inducing a more general immune suppression [8]. *In vitro* studies indicate that at least one of these effects (depressed mastocytosis) may reflect a defective production of the cytokines necessary for stem cell proliferation and differentiation [130]. Whether this modulation reflects the release of particular factors by the worm is not known, although there is evidence for the production of low-molecular-weight material with immunosuppressive properties [131]. Molecules with similar properties have also been identified in other species, including the trichostrongyle, *Ostertagia ostertagi* [132].

VACCINATION

Introduction

There is some irony in the fact that the only large-scale commercially successful vaccine against nematodes, that against *Dictyocaulus viviparus* in cattle (DICTOL), was produced largely by empirical means [133]. Neither the mechanisms underlying the immunity induced by vaccination nor the target antigens were known when the vaccine was developed, and that situation is essentially still unchanged. The vaccine uses larvae attenuated by irradiation as the antigenic stimulus, and similar approaches have been used in the other vaccines which have been used on any scale (against lungworm in sheep and hookworm in dogs). The use of attenuated larvae has been dictated by the difficulties encountered in supply of antigen and by failures to immunize successfully with nonliving material. Nevertheless it is clear that despite the success of DICTOL, and the efficacy of other live vaccines in protecting against the pathologic consequences of infection, future vaccines will require the use of defined nonliving antigens to meet both the demands of supply and the necessary safety standards [134,135]. At the present time, work towards the development of such vaccines has been largely restricted to experimental systems in which the relevant parameters can be systematically defined [136]. Among these are the following important objectives:

1 identification of those components of the antigenic stimulus presented by the parasite which induce protection but not pathology;

2 identification of the parasite stages most susceptible to immune control;

3 measurement of the likely variation in important antigens within widely dispersed infections;

4 analysis of genetic and environmental factors which prevent full responsiveness to vaccination;

5 development of optimal protocols for administration of vaccines.

Experimental systems

Experimental studies relevant to the development of vaccines have been carried out with a number of the parasites already discussed in the context of immunity. Successful immunization with attenuated larvae or with nonliving antigens has been achieved in several model systems, particularly useful data coming from work with *H. contortus, N. brasiliensis, N. dubius, T. spiralis, T. colubriformis*, and *T. muris*.

Attenuated larvae induce substantial immunity against both *N. dubius* [137] and *T. colubriformis* [138]. The degree of immunity (100%) achieved in the former is all the more striking when compared with the absence of immunity resulting from normal infections in mice. The effectiveness of X-irradiated infective larvae is associated with their death in the intestinal wall and with the absence of subsequent immune suppression from adult stages. Experiments have shown very clearly that viable adult worms very effectively prevent the expression of immunity elicited by attenuated larvae. X-irradiated larvae also have been used as an experimental vaccine against *T. colubriformis* in sheep. The degree of response to the vaccine is determined by the host's genetically determined ability to develop and express protective resistance but is substantial in responder individuals.

Nonliving antigens of *N. dubius* appear rather ineffective in stimulating immunity, and an interesting comparison has been made of the immunogenicity of ES antigens of *N. dubius* and *N. brasiliensis* in this respect [139]. Greater resistance to challenge can be induced with *N. dubius* antigens when used with pertussigen as an adjuvant [140]. Homogenate antigens of *T. colubriformis* stimulate resistance to challenge in the guinea pig model and some progress has been made in identifying components with particular activity. Recently a deter-

gent-soluble fraction of infective larvae, containing a relatively simple set of antigens, was also found to elicit immunity, a 41 kD molecule, identified as worm tropomyosin, being primarily responsible for the protective activity [62] (cf. paramyosin as a protective antigen in schistosome infections [141]).

Most progress in identification of protective antigens has been made using *T. spiralis*. Mice have been successfully vaccinated against infection with a variety of antigen preparations derived from all stages of the life-cycle, including newborn larvae [142]. Homogenates, ES, and surface antigen preparations [28,30] from infective larvae have all been found to be particularly effective. Purified antigens, with molecular weights between 43 and 55 kD, which originate in the larval stichocytes elicit high levels of protection in relatively small amounts [20]. Protection was assessed in terms of reduced burdens of both adult intestinal worms and muscle larval worms after challenge. Immunization studies with this parasite have been extended into the pig. Here, antigens derived from infective larvae were substantially less protective than those from newborn larvae [24], a result which may correlate with the rather different patterns of host response seen during infections in this host.

Mice are readily vaccinated against *T. muris* using a variety of homogenate and ES antigen preparations, and some of these reciprocally immunize against *T. spiralis* [143], suggesting at least a common origin for the antigens concerned. One antigen of molecular weight 43 kD appears particularly effective in immunization. In all cases, antigens taken from adult worms appear to immunize effectively against earlier stages in the developmental cycle. In *T. muris* as with *T. spiralis*, response to vaccination is strain-variable, high-responder mice expressing more effective vaccine-induced resistance than low responders [144,145]. Significantly, prior infection of low responders is associated with a much poorer subsequent response to immunization, implying a degree of parasite-induced modulation.

Almost all experiments involving use of nonliving antigens for vaccination have used parenteral routes of injection, with the antigen incorporated into adjuvant material. In the case of *T. spiralis* and *T. muris*, effective immunization can be achieved with simple depot adjuvants such as Freund's incomplete adjuvant. Very few studies have achieved success with oral administration of nonliving antigens, although this obviously would be the preferred route of administration in many circumstances. Recently, however, success in protecting mice with orally administered liposome-entrapped adult-derived antigens of *N. brasiliensis* has been reported [146]. Three doses of 150 µg antigen given 15–19 days before challenge resulted in >95% reduction of adult worm burdens 7 days later.

Vaccination: limitations and constraints

Successful vaccination against *H. contortus* and other trichostrongyles in sheep has now been achieved using defined components of parasite intestinal cells (see p. 501).

The major constraints on successful vaccination are those relating to identification, production, and presentation of antigens, as discussed above. However, even when these problems have been overcome there remain a number of limitations upon the use of vaccines against parasitic nematodes. These can be considered under three headings:
1 target effector mechanism;
2 genetic variation in host and parasite;
3 host immune competence.

Target effector mechanisms

It will be clear that, with very few exceptions, the mechanisms responsible for effective immunity against nematodes are known only in general terms. In consequence it is at present not possible to think in terms of designing vaccines to elicit precisely the responses required for resistance. A number of particular problems present themselves. If IgA does play a role in resistance, for which there is some evidence, how can this isotype be best simulated? (Similar constraints may also apply to other isotypes.) If orally administered vaccines are preferred, how can the IgA response and other mucosally restricted responses be elicited without risking activating the powerful networks that down-regulate sensitivity to intestinal antigens? [147]. If inflammatory responses are important effectors against both systemic and intestinal nematodes, how can these be provoked without risking an unacceptable degree of host pathology? An important aspect here may be the screening of potential vaccine antigens for allergenicity. At least one antigen known to protect against *T. spiralis* also has allergenic properties [148]; this may also be true of molecules which may protect against *Ascaris* [47]. If vaccine antigens elicit responses directed against or triggered by enzymes released by nematodes, care will obviously have to be taken to minimize crossreactivity with similar host molecules. Certainly enzyme molecules are known

to be important components of ES antigens [18], and there is some evidence for their being targets of host responses.

Genetic variation in host and parasite

As Mitchell has pointed out, the goal of vaccine strategies is to protect the majority of a genetically diverse host population against the majority of a genetically diverse parasite population [149]. Epidemiologic data show that in most host populations infected with nematodes the parasites are distributed in an aggregated manner [150]. This implies that the majority of hosts are able to resist infection and could therefore respond to vaccination. Those individuals that fail to resist infection naturally are likely to be low responders or nonresponders to vaccination as well. Crucial questions are therefore the importance of being able to protect this minority and the reasons why this minority has defective resistance.

In human populations low responders or nonresponders are the individuals who may suffer most from the pathologic consequences of infection and who may make a disproportionate contribution to transmission of infection through release of eggs or larvae into the environment. In domestic animals, where populations are more uniform genetically, low-responder status may be more frequent and thus a more serious cause of pathology and a more important determinant of transmission. In both cases, therefore, ability to protect the minority becomes important both to reduce pathology and to reduce transmission.

Analysis of the reasons why individuals may fail to respond immunologically to the parasite itself or to antigens of the parasite administered as a vaccine are discussed in Chapter 2 and will be dealt with only briefly here. Failure to recognize crucial antigens at the lymphocyte level is less likely to explain low response to infection but may become important when vaccine antigens of restricted heterogeneity, and particularly subunit vaccines, are employed [134]. No examples in terms of vaccine-induced resistance are yet documented for this group of worms, although well known in other host–parasite systems. However, there are already clear examples of MHC-restricted recognition of important antigens in a number of model systems (e.g., *A. suum* [47], *T. muris* [151]). Major histocompatibility complex-related down-regulation of responsiveness, as seen in *T. spiralis* infections, might also reduce the level of resistance following vaccination [152].

Genetically determined poor responsiveness to vaccines against nematodes is perhaps more likely to reflect variation at the level of nonimmunologic effects, e.g., those involved in inflammatory responses. Studies with *T. spiralis* have shown that levels of "vaccine" (larval antigen)-induced responsiveness are much higher in strains that respond vigorously to infection than in those which do not [145]. Analysis shows that this difference does not affect T cell function, measured by *cytokine* production, but is reflected in parameters of inflammatory responsiveness (eosinophilia, mucosal mastocytosis). It is known that variation in these parameters reflect genetically determined differences in stem cell production and response [128,129]. Similar relationships between inflammatory responsiveness and vaccine-induced resistance have also been demonstrated using *T. colubriformis* in sheep [153], which in turn correlate with data obtained using infections with *T. colubriformis* in high- and low-responder guinea pigs [154]. The efficiency of vaccination with ES antigens of *T. muris* is also host-strain dependent, low-responder B10 background mice taking considerably longer than high-responder strains to eliminate a challenge infection [144].

Genetical variation in parasite populations, although recognized as a difficulty in the development of vaccines against malarial parasites [155] and to a lesser degree against schistosomes, has received relatively little attention in the context of nematodes. One model system which is amenable to detailed analysis is that using *T. spiralis* in mice. The parasite is widely distributed in nature and many different isolates have been described. Molecular heterogeneity has been demonstrated in these isolates by the techniques of isoenzyme electrophoresis and immunochemistry, and it is clear that there is differential recognition of certain antigens by hosts. Experiments using reciprocal immunization and challenge with isolates have shown that while all isolates elicit immunity, some do so more effectively than others and some are more affected by vaccine-induced immunity than others [41]. Thus, heterogeneity associated with geographic variation in the target population may be an important consideration in the use of vaccines against widely distributed species.

Host immune competence

The two important determinants of host immune competence are genetic and environmental. A given level of competence can only be expressed optimally under appropriate conditions; many environmental factors can operate as constraints to reduce the level of immunity expressed and these are likely to influence responsiveness to vaccination as much as to infection. It has been demonstrated experimentally, with *T. colubriformis* in sheep [156] and with *N. dubius* in mice [157], that inadequate levels of host nutrition severely reduce the efficacy of vaccination. In the latter, reduction of the protein content of the diet from 16% to 2% resulted in a reduction in vaccine-induced protection of between 50% and 75%. Concurrent infections, known to reduce vaccine efficiency in many other situations, are also likely to affect vaccination against nematodes. Perhaps most important is the evidence that homologous infection can also bring this about. Thus, the presence of adult *N. dubius* in mice suppresses the expression of resistance conferred by administration of irradiated larvae. Similarly, mice exposed to adult infections of *T. muris* remain unresponsive to vaccination with adult ES antigen, even when the adult worms are removed prior to injection of the antigen [144]. Both nutritional and infection-related constraints are therefore likely to create severe problems should it ever be possible to vaccinate against the nematodes endemic in Third World populations.

IMMUNODIAGNOSIS

The need for immunodiagnosis

Of the nonfilarial nematodes that are important parasites of humans, convenient and satisfactory diagnosis by parasitologic means is restricted to *A. lumbricoides*, *T. trichiura*, and the hookworms *Ankylostoma* and *Necator*. Diagnosis is achieved in these cases by the detection of eggs in stool specimens [158]. Strongyloidiasis may also be successfully diagnosed by this approach but the number of larvae passed in the feces is of a much lower order [159]. This, allied to variation in the presence of numbers of larvae released each day, means that diagnosis requires thorough investigation of several stool samples acquired on different days [160,161]. Even then, an active infection may be missed. It is also worth pointing out that the problem of sensitivity can extend to hookworm infection, where over 50% of known positive cases may score negative following examination of a single stool specimen. Parasitologic diagnosis of other species, which do not have stages in the feces, depends on locating the worms within the host body. Unfortunately, unlike the filarial nematodes, these parasites do not circulate in the bloodstream. Their anatomic locations are usually such that finding them is invariably more difficult and more inconveniencing. In the cases of *T. canis*, *T. spiralis*, and *Capillaria hepatica*, which are found in the tissues, muscles, and liver respectively, biopsies must be taken. Detection of *A. cantonensis* is dependent on lumbar puncture and examination of cerebrospinal fluid. Fiber-optic gastroscopy or investigative surgery is required when infection with *Anisakis simplex* is suspected. In addition to the inconvenience associated with these procedures, they are not always practical or successful. In the case of *T. canis* infection, tissue biopsies are very rarely available and multiple sections may be required [162]. *Trichinella spiralis* may not be detected when infection levels are low or when muscle tissue is taken too early in infection. In consequence, diagnosis of infection with these nematodes must rely on a combination of factors, including likelihood of exposure, clinical signs and symptoms, and serologic findings. Many of the clinical signs and symptoms, e.g., diarrhea, abdominal pain, muscle pain, and weakness, are associated with infection with several helminths (and indeed numerous other agents). The same can be said for serologic findings, such as elevated IgE and blood eosinophil count. For this reason there is a need to develop and employ assays which offer specific diagnosis of particular nematode infections. The problems associated with the development of such assays, the nature of tests which have been and are being employed at present, and recent improvements and future strategies are described below.

Problems associated with the development of diagnostic tests

Choice of assay

The majority of immunodiagnostic tests for nematode infection can be placed into one of three categories: serologic assays which measure antibody, assays which measure circulating parasite products, and intradermal tests. The first encompasses a wide range of approaches and, primarily as a consequence of their greater ease of use, they have received by far the most attention. Modern versions, such as the enzyme-linked immunosorbent assay (ELISA) and radioimmunoassay (RIA), have the potential to achieve high levels of sensitivity and specificity. This, allied to their simplicity and economy with respect to reagents, dictates that they represent the method of choice for the foreseeable future. Their one drawback is that they cannot unequivocally differentiate between existing and past infection. An

increase in antibody titer can, however, be used as evidence of recent infection and also of a strong likelihood of existing infection. In addition, detection of antibodies as evidence of previous exposure to the parasite allows calculations to be made of exposure rates within a given population. The fact that the quantity, quality (Ig class), and kinetics of the antibody response to infection may vary, sometimes because of genetic differences between hosts, has two important consequences for the construction of diagnostic tests based on the detection of antiparasite antibodies. Firstly, the probability of a given antigenic determinant being recognized by all sera from infected individuals at all times during and after infection must be low, and thus antigen mixtures will probably be required for diagnostic tests with 100% sensitivity. Secondly, a thorough knowledge of the kinetics and classes of antibodies synthesized within a population will allow the construction of prognostic tests, or "antibody windows" indicative, for example, of early vs. late infections, susceptible vs. resistant hosts, or the probability of pathologic consequences in a particular host. Thus, an immunochemical dissection of humoral antiparasite responses may help in the understanding of protective vs. pathologic responses.

Existing and previous infection can be differentiated by assays which measure circulating parasite products. Such assays have been much explored in the diagnosis of filarial nematodes [163,164] and have recently begun to be used for immunodiagnosis of nonfilarial nematodes [165,166]. One drawback is that host antibody may interfere with antigen detection. This has been demonstrated with respect to filarial nematodes [167–170] and with *T. canis* [165,170].

Intradermal tests have been applied to the diagnosis of several nematode parasites with varying degrees of success. In general, this form of assay is less sensitive and convenient than ELISA or RIA. These factors, allied to the current awareness of the dangers associated with the introduction of parasite material of human origin into the body, suggest that future developments in this area may be minimal.

Specificity

Many immunodiagnostic tests described to date suffer from a low level of specificity. Thus, although antibody responses to infection may be detected, the organism in question often cannot be unequivocally identified. In situations where one particular parasite is strongly suspected and no other is likely to be present, e.g., *T.*

spiralis in outbreaks in the Western World, it may be reasonable to confirm a diagnosis in this way, although clearly it is inferior to absolute identification. In other situations, however, such a "nonspecific" diagnosis is of little value. For example, diagnosis of infection in Third World patients, who are likely to be harboring several nematode parasites, or cases of visceral larva migrans, which has more than one causative agent, requires much greater specificity [158].

The suboptimal specificity observed in certain immunodiagnostic assays is primarily a consequence of sharing of antigenic epitopes among nematode parasites [10]. The majority of epitopes present in any one species are probably not specific to that species. Many immunodiagnostic tests are based on the use of whole worm extracts, which are likely to contain a number of shared determinants. In recent years, attention has turned to the use of antigen preparations which contain a more restricted range of epitopes, particularly those obtained from the parasite surface and/or ES. These two parasite compartments appear to contain more specific epitopes than the somatic structures [171,172]. Shared epitopes clearly do exist in the surface and ES, however [173–176], and assays based on detection of somatic antigens may have comparable specificity [177].

Sensitivity

Immunodiagnosis of most diseases is moving away from techniques based on immunoprecipitation and hemagglutination to assays with, in general, higher sensitivities, such as ELISA and RIA. Thus, although assays of the former type are in some cases still routinely employed in nematode immunodiagnosis (e.g., in the detection of *T. spiralis* [178]), most recent developments relate to the latter. Enzyme-linked immunosorbent assay and/or RIA have now been employed for the detection of antibodies to the majority of major human nonfilarial nematodes, including *T. canis* [179,180], hookworm [181, 182], *Strongyloides stercoralis* [161,183], *T. spiralis* [178, 184], *Gnathostoma* species [185], *A. simplex* [186], and *A. cantonensis* [187,188]. The levels of sensitivity attained have invariably been high, reaching 100% in some cases [182], although there is room for improvement in some areas. This improvement may not come from technical advances, as it appears that both ELISA and RIA can offer a satisfactory level of sensitivity. It is more likely to come from careful assay design, in particular from selection of a suitable target antigen. The effect of different antigens or antigen preparations on assay sensitivity has

been investigated and has been shown, in some cases, to exert considerable influence [189,190]. Clearly, what is required in all cases is an epitope recognized by all infected individuals and which is available for a sufficiently long period (i.e., not merely expressed for a short time during development) to allow for variation in the rapidity of host response. Since humans also vary with respect to the nature of their antibody response to nematode infection [191], more than one epitope may be required.

Immunodiagnosis of specific infections

Trichinosis

There are probably more assays for the detection of *T. spiralis* than for any other infectious organism. Certainly the list is extensive and detailed reviews on their development, use, and performance are available [178,192, 193]. Several of the assays are routinely employed in diagnostic centers in the USA, including the indirect hemagglutination test, bentonite flocculation test, fluorescent antibody test, counterimmunoelectrophoresis, and ELISA.

The application of these antibody detection assays to the diagnosis of acute infection with *T. spiralis* involves investigation of serum samples at different time points, to determine whether an increase in antibody titer is present. In spite of the general use of target antigens obtained from the infective parasite stage (muscle larva), the earliest appearance of detectable antibody is in the region of 3 weeks postinfection [178]. This is unfortunate, as the critical stage of the disease, which can be treated with drugs, is at 2 weeks. The role of immunodiagnosis in cases of acute trichinosis is therefore essentially confirmatory. Diagnosis of the disease is based at this stage on a combination of factors, including history of exposure to raw or poorly cooked meat, clinical signs and symptoms, nonspecific serologic tests such as enumeration of eosinophils, and muscle biopsy.

In addition to their role as confirmatory tests in acute trichinosis, serologic assays based on antibody diagnosis are of value in analysis of patients who are suspected of having been exposed to *T. spiralis* but show no clinical signs. Thus, Ivanoska *et al.* [166] were able, by the use of an indirect immunofluorescent test, to detect antibody in 46% of 39 such cases. Antibody assays clearly also have a role in diagnosis of mild cases of trichinosis, where symptoms may be particularly vague and/or similar to a range of other infections, and in cases where the clinical signs and symptoms may be atypical. A recent example of the latter concerned human infection in the Arctic region, where the classic symptoms of brief diarrhea followed by fever, myalgia, muscle weakness, and edema were replaced by prolonged diarrhea without fever and only brief muscle symptoms [194].

Immunodiagnosis of trichinosis has also employed measurement of circulating antigen, in particular by the immunoradiometric assay [166]. This was found to be considerably less sensitive than antibody detection by immunofluorescence with respect to confirmed cases (47% vs. 100%) and also produced a lower detection rate for suspected cases (13% vs. 46%). It was noted, however, that a few antibody-negative cases were found to be antigen-positive. Antigen was observed in this study to be present occasionally in serum before the appearance of antibody, and was detectable in samples from the first time point available, 21 days postinfection. This raises the possibility that antigen could, in fact, be detectable at an earlier period and indeed antigen has been found as early as day 13 in *T. spiralis*-infected mice [195]. Early detection of circulating antigens could provide a rapid screening assay for suspected trichinosis, and hence it may be worth testing serum samples obtained on first examination from suspected cases in an immunoradiometric or similar assay.

Immunodiagnostic techniques are also applied to the detection of *T. spiralis* in pigs. Here, the priority is different, namely to estimate the level of infection in populations earmarked for human consumption. As with human trichinellosis, a number of antibody detection tests have been evaluated [178]. It appears likely that ELISA will become the method of choice, as, in addition to its merits of high sensitivity, good specificity, and economy of reagents, it can be easily adapted for automated rapid diagnosis. A number of ELISA tests have been described, differing mainly in the preparation and nature of the antigens used. A level of sensitivity approaching 100% was observed in one study which employed ES antigens [196]. In a similar but independent study, however, the use of ES resulted in a number of false positives: this was not the case when an extract of the large-particle fraction of muscle larvae was used as antigen [189]. The chances of obtaining false positives are apparently influenced by the age of the ES [197]. Excretions–secretions collected during the first 24 h appears to have the highest level of specificity. Antigen preparations obtained by culture after 48 and 72 h yield an increasing rate of false-positive reactions and this appears to be due to the presence of additional mole-

cules, perhaps somatic components derived from dead or dying worms. These findings should be considered when ES is being employed in the development of novel assays for other parasites.

Toxocariasis

Infection with *T. canis* is typically associated with young children. Common laboratory findings in these cases include persistent eosinophilia, leukocytosis, elevated anti-A or anti-B isohemagglutinin titer, and elevated serum IgG (see Beaver *et al.* [158] and Glickman *et al.* [180]). Presumptive diagnosis of *T. canis* infection can be based on these findings, together with a history of pica and exposure to dogs. Definitive diagnosis is dependent upon locating the parasite in tissue biopsy. The lack of specificity of laboratory findings, allied to the inconvenience and low success rate in examining tissue biopsies, points to the need for a convenient immunodiagnostic test.

As with *T. spiralis* infection, a wide range of immunologic tests has been applied to the diagnosis of *T. canis* (see Glickman *et al.* [180]). Again, the major problems relate to sensitivity and specificity. The problems concerning sensitivity reflect the lack of a standard definition of clinically diagnosed toxocariasis and the common failure to obtain a parasitologic diagnosis. They are thus rather different from those associated with other infections in that an increase in assay sensitivity cannot be generated by an improvement in assay methodology or reagents. Specificity is compromised by variation in the extent of elevated antibody titers to *T. canis* within and among different populations (reviewed by Glickman *et al.* [180]). It is also, of course, influenced by the degree of crossreactivity of the antigens employed. This is, perhaps, of particular importance in the case of *T. canis* infection, as visceral larva migrans may also be caused by other nematodes, such as *T. cati* and *Baylisascaris* spp., which show particularly extensive antigenic crossreactivity with *T. canis* [48,170,174,198].

As with any other nematode infection, the key to specific diagnosis is the use of an appropriate antigen. Early work on the nature of somatic antigens revealed them to be complex and neither stage- nor species-specific (reviewed by Glickman *et al.* [180]). After adsorption and purification, antigens were isolated which were thought to be genus-specific [199,200]. Similar work by Welch *et al.* [201] also provided a much more specific antigen preparation. Excretion–secretion products of the L3 stage were found to be a more convenient source of "specific" antigens requiring no purification/ adsorption. These were reported to be genus-specific, using immunofluorescence to identify immunoprecipitates at the oral, anal, and excretory pores of living larvae exposed to sera from infected animals or humans [171]. Development of a long-term *in vitro* cultivation method for *T. canis* larvae [202] paved the way for the incorporation of parasite ES into a routine assay. The first tests incorporating ES were the hemagglutination test and the soluble antigen fluorescent test: both were satisfactory, possessing high sensitivity and genus level specificity [203]. Later, ES was successfully incorporated into an ELISA [179]. This test was, in fact, introduced for reference immunodiagnosis and seroepidemiology in the UK. It has since been evaluated in a number of laboratories and is now generally accepted as the method of choice for detection of antibody to *T. canis* [180]. It should be noted, however, that most of the analysis designed to indicate the level of crossreactivity has been undertaken using serum samples from Caucasian subjects. When serum samples are taken from individuals from tropical regions, in whom the level and variety of exposure to helminths is likely to be considerably greater, the existence of significant crossreactivity is revealed [176]. This problem can be overcome, however, by preadsorption of serum samples with extracts prepared from heterologous parasites, particularly *A. lumbricoides*, and this is now recommended as a routine step in the assay procedure [204].

A second approach to the problem of crossreactivity is to focus on truly species-specific ES epitopes. A monoclonal antibody, Tcn-2, has been described which recognizes certain ES components of *T. canis*, but not of the very closely related *T. cati* [48]. Similarly, a monoclonal antibody raised against *T. canis* ES recognizes the *T. canis* larval surface but not that of *T. cati* [205]. This probably represents a second species-specific anti-ES monoclonal antibody. Tcn-2 has been successfully used in an inhibition assay, in which binding to its target antigen can be prevented by antibodies of the same specificity in human serum samples [165].

Although an ELISA based on detection of antibody to ES is now being used in laboratories in Canada, Europe, Japan, and the USA, differences in assay procedure and also in interpretation mean that results are difficult to compare with confidence. It is now apparent that the composition of *T. canis* larval ES may vary slightly with respect to batch, time spent in culture, and laboratory [206,207], although Speiser & Gottstein [208] showed that the sensitivity and specificity obtained with differ-

ent batches of ES prepared and tested in different laboratories could be both comparable and reproducible. This has now been confirmed by Glickman *et al.* [180], who suggest that *T. canis* ES should be standardized, thereby allowing the establishment of uniform criteria for interpretation of results. One way of approaching this would be to employ recombinant antigen, and progress in this direction has been reported [209].

The sensitivity of the ELISA based on detection of antibody to ES is lower at any antibody titer cutoff point for diagnosis of ocular than for visceral larva migrans [180,203]. This is because patients with ocular disease appear to have significantly lower levels of serum IgG antibody against ES. The reason for this is unknown, but it may reflect a lower larval burden or a longer time period between infection and serodiagnosis [210]. The latter suggestion is strengthened by the observation that a reduction in the time period leads to an increase in antibody titer [211].

Reliable serologic diagnosis of ocular larva migrans is needed to differentiate it from a variety of other important ocular diseases, in particular retinoblastoma [212]. One approach is to measure the antibody content of the humor, which is generally higher than in serum [213–215]. A second incorporates the measurement of circulating immune complexes and this may also be of value [216]. A third approach which is perhaps of most interest involved measurement of IgE antibodies directed against the parasite by the radioallergosorbent test (RAST) [215]. This was demonstrated to have a greater sensitivity than measurement of IgG by the standard ELISA, thereby making it possible to recognize clinical cases which are missed by the latter. Both tests should be employed for diagnosis when ocular disease caused by *T. canis* is suspected.

Existing tests for measurement of antibody suffer from the common drawback that they offer little information on the nature of the infection. Thus it cannot be determined whether a positive serologic finding reflects acute, chronic, clinically normal, or past infection. It has been suggested [180] that characterization of the sequential immunoglobulin class and subclass specific response may be of value. IgM responses have been observed in *T. canis* infection [203] and these may reflect recent infection [180]. It has been found, however, that IgM levels are persistently elevated during experimental infection of mice [170]. This is an unusual observation, but if mimicked in infected humans would preclude estimation of the age of infection based on relative IgM and IgG levels.

In experimentally infected rabbits there is variation in the kinetics of the antibody responses when these are measured against individual epitopes [165]. This raises the possibility that the longevity or status of infection in humans may be determined by measuring antibody to particular epitopes.

Diagnosis of active infection can, of course, be approached by measurement of circulating parasite products. Circulating antigen has, in fact, been detected in the sera of infected individuals, including some in whom antibody could not be measured [165]. However, studies undertaken in experimental animals indicate that host antibody may interfere with antigen detection [165,170]. High antibody titer may result in failure to detect antigen, or antigen may only be detectable at low levels. Robertson et al. [165] suggest that diagnosis of T. canis infection should employ measurement of both antigen and antibody.

Measurement of circulating antigen may also give an indication of the intensity of infection. Experiments in mice have indicated that there is a linear association between the number of eggs that an animal receives and the amount of antigen detectable in its blood [170]. In a separate study it was observed that serum antibody levels were proportional to the number of eggs given [217].

Strongyloidiasis

Strongyloidiasis has received less diagnostic attention than trichinellosis or toxocariasis, possibly because of its lower pathogenicity. The medical importance of this organism has increased considerably in recent years, however, as it has become apparent that it can be life-threatening in immunocompromised hosts. Problems associated with parasitologic diagnosis of strongyloidiasis have already been mentioned: the number of larvae released is small, infections are often light or moderate, and clinical symptoms are absent or vague and nonspecific. A degree of specificity is associated with larva currens, a creeping skin eruption associated with migrating larvae, but the frequency of its appearance varies with the individual and in some infected populations it appears to be absent. Laboratory tests are of little value in the diagnosis of strongyloidiasis. High eosinophilia and elevated total IgE are common, but, as mentioned earlier, both are associated with a number of helminth infections.

Several serologic tests have been used in the immunodiagnosis of strongyloidiasis, although fewer than with trichinellosis or toxocariasis. Most studies have been concerned with immunofluorescent antibody tests or ELISA [160,161]. Good sensitivity and specificity have been obtained with both, although there is some cross-reactivity with sera from filariasis patients [160,190,218]. Enzyme-linked immunosorbent assay is now generally accepted as the method of choice and is in routine use in at least one laboratory [161]. IgE antibodies against *S. stercoralis* larval antigen have also been measured by ELISA [219]. The sensitivity observed was very low, IgE being detected in only 21.2% of patients. More success was obtained when this isotype was measured by the RAST. Here, 89.5% of parasitologically proven infected patients had specific IgE antibody [220]. Measurement of IgG antibodies by ELISA in the same study produced an even higher sensitivity value (93%). The number of known positive cases missed by either assay was about 3% and hence it was suggested by the authors that the tests should be used in combination. The RAST for IgE also appears to demonstrate a very low level of cross-reactivity, although this was not investigated with respect to filarial worms.

IgE antibodies directed against *S. stercoralis* have also been measured by intradermal tests [160,161,219]. These have generally been satisfactory, although specificity and sensitivity have not been rigorously evaluated. In spite of this, in one case the intradermal test was shown to be much more sensitive for measurement of IgE than ELISA [219]. This, taken in combination with the rapidity of the skin test, suggests that it could have a role to play in routine diagnosis.

Serologic measurement of specific IgE has been shown to vary considerably in both infected individuals and populations [220]. With respect to the latter, Asians were found in general to have the highest values, followed by Latin Americans and then North Americans. These results might be explained by a combination of the source of parasite antigen being a patient in Thailand and the American group possessing the largest number of patients treated with steroids. The first explanation implied the existence of strain-specific antigens, which should be taken into account in the future development and evaluation of diagnostic tests. Also, although antigens of the more readily available rat parasite, *S. ratti*, have been substituted for those of *S. stercoralis*, it has recently been demonstrated that their use results in a reduction in assay sensitivity and in geometric mean antibody titer for both the ELISA and indirect hemagglutination assay (IHA) [190]. Information on the antigens of *Strongyloides* species as a whole is lacking and it was only in 1988 that the first paper appeared in the literature concerning characterization of the surface and ES antigens of the filariform larva of *S. stercoralis* [221].

Serologic assays described to date and intradermal tests both measure antibody. As stated previously, such tests cannot differentiate between existing and previous infection, and this is clearly a problem with respect to *S. stercoralis*. This is particularly the case in endemic regions [161]. For this reason, existing immunoassays are of more value when applied to individuals who have acquired infection when visiting such areas. More reliable serologic diagnosis in endemic areas may await the development of an assay for circulating antigens.

Anisakiasis

The major problem in developing a specific serologic assay for diagnosis of anisakiasis concerns the extensive crossreactivity of *Anisakis* species with other ascaridoid nematodes which are extremely common parasites of humans, such as *A. lumbricoides* and *T. canis*. The crossreactivity has recently been examined at the molecular level in both somatic and ES products [175]. Immunodiagnosis of this infection has followed the standard path of moving from techniques such as electrophoresis and immunodiffusion to the present-day methods, such as ELISA and RAST [207,222], and from whole worm antigen extracts to less complex reagents [186]. Although these moves have in general provided better serologic diagnoses, problems with crossreactivity and sensitivity still remain. Thus, although demonstrating 100% sensitivity, Desowitz *et al.* [222] have shown that patients other than those infected with *A. simplex* also recognize larval antigen in the RAST. These patients included individuals with schistosomiasis or *Ascaris* infections, visceral larva migrans, and asthmatic children. The antigen used in this study was a whole worm extract. It may be assumed that the use of an antigen source such as ES might improve the situation, although no difference was found with regard to specificity when crude or ES antigen were compared in the RAST [37]; assay sensitivity was also extremely low (40%, $n = 5$). Sensitivity was increased to 100% if IgG antibodies were measured by ELISA, but this was associated with an increase in crossreactivity.

Clearly, successful serodiagnosis will require the use of a more specific antigen preparation: ES has been shown to be of insufficient specificity in serologic testing and indeed this can be explained by the molecular analysis of Kennedy *et al.* [175]. In this particular case, there-

fore, a specific epitope may have to be sought in some other parasite fraction.

Gnathostomiasis

Humans have been shown to act as a paratenic host for both *Gnathostoma spinigerum* and, more rarely, *Gnathostoma hispidum* [158]. Infection with *G. spinigerum* usually results from eating inadequately processed fish which contain encapsulated third-stage larvae. *Gnathostoma hispidum* has been shown to be acquired by eating live juvenile loaches [223]. Diagnosis is currently based on a number of clinical criteria.

Serologic immunoassay clearly has a role to play in the diagnosis of gnathostomiasis, as witnessed by a number of recent investigations employing ELISA [177,185]. Prior to this, a number of conventional immunoassays had been tested but, with the exception of precipitation tests [224,225], had lacked sensitivity and/or specificity. Two studies have compared the ELISA test with other techniques. In the first, concerning infection with *G. hispidum* [185], the ELISA was found to be marginally superior to a double diffusion test with respect to sensitivity (73% vs. 64%) but considerably inferior with regard to specificity (67% vs. 100%). This study used antigen from the related *Gnathostoma doloresi* (from wild boars), and, although the double diffusion test was superior with regard to specificity, it was pointed out that it used much larger amounts of the antigen. The second study compared the ELISA, IHA, and purified protein test (PPT) for the diagnosis of infection with *G. spinigerum* [177]. The ELISA was the most sensitive assay (seven out of eight positive cases) and the PPT the least (three out of eight). Results were independent of the source of antigen, both somatic and ES being tested. The specificity of the ELISA was found to be extremely high ($>96\%$) following investigation of serum samples from individuals harboring a wide range of parasites. Interestingly, in the report of Maleewong *et al.* [177], antibody titers were found to be lower in ocular than in visceral disease. As noted earlier, this is generally the case with respect to *T. canis* infection [180].

Angiostrongyliasis

Few reports relating to the immunodiagnosis of *A. cantonensis* have appeared in the recent literature, despite the fact that this infection is an important cause of eosinophilic meningoencephalitis in Southeast Asia. Attempts to provide assays of improved specificity by undertaking antigen purification have been described in the literature. The problem of crossreactivity, in particular with *Toxocara* [188,266] still has to be resolved by further antigenic analysis, including analysis of individual surface and ES components. In relation to the latter, information is now becoming available on the nature of circulating antigens and the timing and persistence of their complementary antibody responses [227]. This may help not only in the design and evaluation of techniques which measure antibody to ES but also in the planning of assays which detect circulating antigen.

Investigations undertaken to date have employed a number of standard immunoassays, virtually all of which have included antigen subjected to some degree of purification. These include skin tests, immunodiffusion and indirect hemagglutination [228], immunofluorescence [229], and ELISA [188,230,231].

Ascariasis

The availability of generally satisfactory parasitologic diagnosis in combination, perhaps, with the parasite's relatively low pathogenicity has rendered serodiagnosis of *Ascaris* infection a comparatively unexplored area. A few papers have appeared in the literature concerning purified or ES antigens and their use in assays [232–234]. Recent evidence indicates that the parasite ES is highly crossreactive [175,235]. If serodiagnostic tests are ever to be used as a convenient alternative to parasitologic diagnosis, studies on antigen purification and assay development are necessary.

Hookworm

As with *Ascaris*, immunodiagnosis of hookworm infection has received little attention because, in general, parasitologic diagnosis is relatively satisfactory. Earlier studies, which included skin tests [182,236] and serologic assays, such as complement fixation and fluorescent antibody tests [237], appear to have suffered from a high degree of false positivity. More recently, the ELISA [181] and RIA (measuring IgE) [182] have been evaluated and appear to demonstrate promising specificity. Excretion–secretion antigens were used in both studies.

Future developments

The problems associated with parasitologic diagnosis of a number of the major nematodes of humans invite the development and routine employment of immunodiagnostic assays. A considerable amount of research has been undertaken in this area, particularly during the last three or four decades. This has followed the pattern demonstrated in developments in diagnosis of other infectious agents, i.e., as they have become available, new techniques and methods have been adapted to the needs of nematode diagnosis. The methods of choice at present for antibody detection are predominantly ELISA and RIA: both have the attributes of potentially high sensitivity and specificity, allied to economy of reagents and simplicity and convenience of use. They have been applied, in many cases successfully, to detection of a number of nematode parasites and in some cases are available as routine laboratory tests. Of equal importance to advances in methodology has been the increasing awareness of the need to employ specific antigen. Researchers are thus moving away from the use of whole parasite extracts to concentrate on less heterogeneous compartments, such as the surface and ES, or they are purifying individual antigens. The latter strategy would appear to offer the best prospects for success, as it is becoming clear that the surface and ES antigens of some parasites are more crossreactive than was originally considered. Specific diagnosis of *T. canis* is likely to benefit from the recent introduction of an assay based on the use of a monoclonal antibody directed against a single unique epitope, and this may point the way for future developments in diagnosis of other nematode parasites. Antibody diagnosis has also benefited from the development of assays focusing on the IgE antibody response, particularly the RAST. In some cases, e.g., in the diagnosis of VLM, these have produced a level of sensitivity even higher than that achievable with IgG detection by ELISA. It has been observed that the use of both assays on the same sample can help to reduce the incidence of false negatives and hence they should be used in combination. Another suggestion is the combined use of assays detecting antibody and antigen, and for the same reason. The idea of diagnosing nematode infection by measurement of circulating antigen has been proposed frequently in recent years. The method has now been applied to trichinellosis and toxocariasis, but results obtained demonstrate inferior sensitivity to antibody diagnosis and problems with interference by host antibody. Clearly the difficulties of differentiating existing and past infection, which were thought to be resolvable by this technique, must await further developments. One obvious development would be the introduction of assays having as their target molecules of low or zero antigenicity. Finally, some nematodes, in spite of the absence of suitable parasitologic diagnosis, have received little immunodiagnostic attention to date. These tend to be species the extent of whose medical importance has only recently become apparent, e.g., *Capillaria hepatica* and *philippinensis*. It can be expected that in the near future modern assays such as ELISA will be adapted for their diagnosis.

CONCLUSIONS

After a long period of relative neglect the gastrointestinal and other nonfilarial nematodes of humans are becoming a major focus for immunoparasitologic work. In part, this renewal of interest has been epidemiology-led. Field-based studies and the application of mathematical modeling techniques have uncovered a fascinating interplay of environmental, parasitologic, immunologic, and genetic factors as determinants of the prevalence and intensity of infections in endemic areas. These studies have focused attention on many aspects of host–parasite relationships in humans that have previously been studied only in laboratory models or in the context of veterinary parasitology. What has been gratifying, despite the continuing lack of hard, direct evidence for substantial immunity in human infections with these nematodes, is the relevance and predictive value of much of the experimental data generated by laboratory studies. These have provided a sound basis and a rationale for approaching the key questions of the antigens and the effector mechanisms necessary for host-protective, nonpathogenic immune responses in humans. Further progress will clearly come rapidly with the application of immunologic, immunochemical, and molecular techniques to these questions. We can expect to see definition of major immunogens, cloning of the genes for these molecules, analysis of immune responses at the levels of antigen–receptor interactions and cytokine release, and identification of the molecules and responses which parasites use to evade immunity. As has already been demonstrated in experimental systems, we now have the potential to manipulate host responsiveness selectively, given the present knowledge of genetic constraints on antigen recognition and response. Knowledge of defined antigens and their ready availability is already improving the sensitivity and specificity of diagnosis; it is also contributing to the development of assays for detection of circulating parasite products, thus providing a means of identifying and quantifying active infection. Antigen availability will

also make possible a rational approach to immunoprophylaxis. The apparent absence of effective immunity to the major gastrointestinal nematodes need no longer be seen as an insuperable obstacle to successful vaccination. The realization that vaccines can make use of target antigens that normally go unrecognized by the host, which has come from recent work with veterinary vaccines, the present awareness that nematodes actively interfere with host immune response capacity, and the understanding of the extent of immunogenetic influences upon the expression of resistance all combine to create a climate in which positive manipulation of mechanisms for effective resistance can be seen as an achievable goal.

REFERENCES

1 APCO. *Collected Papers on the Control of Soil-transmitted Helminthiases, vols I and II.* Tokyo: The Asian Parasite Control Organization, 1980.
2 Bundy DAP, Cooper ES. *Trichuris* and trichuriasis in humans. Adv Parasitol 1989;28:107–173.
3 Crompton DWT, Nesheim MC, Pawlowski ZS, eds. *Ascariasis and its Public Health Significance.* London: Taylor & Francis, 1985.
4 Davis A. This wormy world. World Health 1984; March: 2–3.
5 Peters W. The relevance of parasitology to human welfare today. Symp Br Soc Parasitol 1978;16:25–40.
6 Walsh JA, Warren KS. Selective primary health care. N Engl J Med 1979;301:967–974.
7 Befus AD, Bienenstock J. Induction and expression of mucosal immune responses and inflammation to parasitic infections. Contemp Top Immunobiol 1984;12:71–108.
8 Behnke JM. Evasion of immunity by nematode parasites causing chronic infections. Adv Parasitol 1987;26:1–71.
9 Miller HRP. The protective mucosal response against gastrointestinal nematodes in ruminants and laboratory animals. Vet Immunol Immunopathol 1984;6:167–259.
10 Almond NM, Parkhouse RME. Nematode antigens. Curr Top Microbiol Immunol 1985;120:173–203.
11 Rothwell TLW. Immune expulsion of parasitic nematodes from the alimentary tract. Int J Parasitol 1989;19:139–168.
12 Soulsby EJL, ed. *Immune Responses in Parasitic Infections: Immunology, Immunopathology and Immunoprophylaxis, vol. 1. Nematodes.* Boca Raton: CRC Press, 1987.
13 Pritchard DI. Antigens of gastrointestinal nematodes. Trans R Soc Trop Med Hyg 1986;80:728–734.
14 Phillip M, Rumjanek FD. Antigenic and dynamic properties of helminth surface structures. Mol Biochem Parasitol 1984;10:245–268.
15 Kennedy MW, ed. *Parasite Genes, Membranes and Antigens,* London: Taylor & Francis, 1990.
16 Macinnis AJ, ed. *Molecular Paradigms for Eradicating Helminthic Parasites.* UCLA Symposia on Molecular and Cellular Biology, vol. 59. New York: Alan R Liss, 1987.

17 Parkhouse RME, Harrison LJS. Antigens of parasitic helminths in diagnosis, protection and pathology. Parasitology 1989;99:S5–S19.

18 Lightowlers MW, Rickard MD. Excretory–secretory products of helminth parasites: effects on host immune responses. Parasitology; 1988;96:S123–S166.

19 Maizels RM, Selkirk ME. Biology and immunochemistry of nematode antigens. In Englund PT, Sher AF, eds. *The Biology of Parasitism: a Molecular and Immunologic Approach.* New York: Alan R Liss, 1988.

20 Silberstein DS, Despommier DD. Antigens from *Trichinella spiralis* that induce a protective response in the mouse. J Immunol 1984;132:898–904.

21 Despommier DD, Gold AM, Buck SW, Capo V, Silberstein D. *Trichinella spiralis:* secreted antigen of the infective L1 larva localizes to the cytoplasm and nucleoplasm of infected host cells. Exp Parasitol 1990;71:27–38.

22 Gamble HR, Graham CE. Monoclonal antibody-purified antigen for immunodiagnosis of trichinosis. Am J Vet Res 1984;46:67–74.

23 Gold AM, Despommier DD, Buck SW. Partial characterization of two antigens secreted by L1 larvae of *Trichinella spiralis.* Mol Biochem Parasitol 1990;41:187–196.

24 Marti HP, Murrell KD, Gamble HR. *Trichinella spiralis:* immunization of pigs with newborn larval antigens. Exp Parasitol 1987;63:68–73.

25 Denkers EY, Wassom DL, Krco CJ, Hayes CE. The mouse antibody response to *Trichinella spiralis* defines a single, immunodominant epitope shared by multiple antigens. J Immunol 1990;144:3152–3159.

26 Denkers EY, Wassom DL, Hayes CE. Characterization of *Trichinella spiralis* antigens sharing an immunodominant, carbohydrate-associated determinant distinct from phosphorylcholine. Mol Biochem Parasitol 1990;41:241–250.

27 Zarlenga DS, Gamble HR. Molecular cloning and expression of an immunodominant 53-kDa excretory–secretory antigen from *Trichinella spiralis* muscle larvae. Mol Biochem Parasitol 1990;42:165–174.

28 Grencis RK, Crawford C, Pritchard DI, Behnke JM, Wakelin D. Immunization of mice with surface antigens from the muscle larvae of *Trichinella spiralis.* Parasite Immunol 1986;8:587–596.

29 Despommier DD. The stichocyte of *Trichinella spiralis* during morphogenesis in the small intestine of the rat. In Kim CW, ed. *Trichinellosis.* Proceedings of the Third International Conference on Trichinellosis. New York: Intext Educational Publishers, 1974.

30 Ortega-Pierres G. Protection against *Trichinella spiralis* induced by purified stage-specific surface antigens of infective larvae. Parasitol Res 1989;75:563–567.

31 Phillipp M, Parkhouse RME, Ogilvie BM. Changing proteins on the surface of a parasitic nematode. Nature 1980;287:538–540.

32 Jungery M, Ogilvie BM. Antibody response to stage-specific *Trichinella spiralis* surface antigens in strong- and weak-responder mouse strains. J Immunol 1982;129:839–843.

33 McLaren DJ, Ortega-Perres MG, Parkhouse RME. *Trichinella spiralis:* immunocytochemical localization of surface and intracellular antigens using monoclonal antibody probes. Parasitology 1987;94:101–114.

34 Ortega-Pierres G, Mackenzie CD, Parkhouse RME. Protection against *Trichinella spiralis* induced by a monoclonal antibody that promotes killing of newborn larvae by granulocytes. Parasite Immunol 1984;6:275–284.

35 Almond NM, Parkhouse RME. Immunoglobulin class specific responses to biochemically active defined antigens of *Trichinella spiralis.* Parasite Immunol 1986;8:391–406.

36 Chambers AE, Almond NM, Knight M, *et al.* Repetitive DNA as a tool for the identification and comparison of nematode variants: application to *Trichinella* isolates. Mol Biochem Parasitol 1986;21:113–120.

37 Almond NM, McLaren DJ, Parkhouse RME. Comparison of the surface and secretions of *T. pseudospiralis* and *T. spiralis.* Parasitology 1986;93:163–176.

38 Gamble HR, Murrell KD. Conservation of diagnostic antigen epitopes among biologically diverse isolates of *Trichinella spiralis.* J Parasitol 1986;72:921–925.

39 Bolas-Fernandez F, Wakelin D. Infectivity, antigenicity and host responses to isolates of the genus *Trichinella.* Parasitology 1990;100:491–494.

40 Bolas-Fernandez F, Wakelin D. Infectivity of *Trichinella* isolates in mice is determined by host immune responsiveness. Parasitology 1989;99:83–88.

41 Bolas-Fernandez F, Wakelin D. Immunization against geographical isolates of *Trichinella spiralis* in mice. Int J Parasitol. (In press.)

42 Wassom DL, Dougherty DA, Dick TA. *Trichinella spiralis* infections of inbred mice: immunologically specific responses induced by different *Trichinella* isolates. J Parasitol 1988;74:283–287.

43 Wassom DL, Kelly EA. The role of the major histocompatibility complex in resistance to parasite infections. Crit Rev Immunol 1990;10:31–52.

44 Jarrett EEE, Miller HRP. Production and activities of IgE in helminth infection. Prog Allergy 1982;31:178–233.

45 McGibbon AM, Christie JF, Kennedy MW, Lee TDG. Identification of the major *Ascaris* allergen and its purification to homogeneity by high-performance liquid chromatography. Mol Biochem Parasitol 1990;39:163–172.

46 Christie JF, Dunbar B, Davidson I, Kennedy MW. N-terminal amino acid sequence identity between a major allergen of *Ascaris lumbricoides* and *Ascaris suum* and MHC-restricted IgE responses to it. Immunology 1990;69:596–602.

47 Tomlinson LA, Kennedy MW, Christie JF, Fraser EM, McLaughlin D, McIntosh AE. MHC restriction of the antibody repertoire to secretory antigens, and a major allergen, of the nematode parasite *Ascaris.* J Immunol 1989;143:2349–2356.

48 Kennedy MW, Tomlinson LA, Fraser EM, Christie JF. The specificity of the antibody response to somatic antigens of *Ascaris:* heterogeneity in infected humans, and MHC (H-2) control of the repertoire in mice. Clin Exp Immunol 1990;80:219–224.

49 Thorson RE. Proteolytic activity in extracts of the esophagus of adults of *Ancylostoma caninum,* and the effect of immune serum on this activity. J Parasitol 1956;42:21–25.

50 Hotez PJ, Le Trang N, McKerrow JH, Cerami A. Isolation and characterization of a proteolytic enzyme from the adult hookworm *Ancylostoma caninum*. J Biol Chem 1985; 26:7343–7348.

51 Pritchard DI. *Necator americanus*—antigens and immunological targets. In Warren KS, Schad GA, eds. *Hookworm Disease: Current Status and New Directions*. London: Taylor & Francis, 1990.

52 Pritchard DI, McKean PG, Rogan MT. Cuticle preparations from *Necator americanus* and their immunogenicity in the infected host. Mol Biochem Parasitol 1988;28:275–284.

53 McKean PG, Pritchard DI. The action of a mast cell protease on the cuticular collagens of *Necator americanus*. Parasite Immunol 1989;11:293–297.

54 Robertson BD, Rathaur S, Maizels RM. Antigenic and biochemical analyses of the excretory–secretory molecules of *Toxocara canis* infective larvae. In Geerts S, Kumar V, Brandt J, eds. *Helminth Zoonoses*. Dordrecht: Martinus Nijhoff Publishers, 1987:167–173.

55 Smith HV, Quinn R, Kusel JR, Girdwood RWA. The effect of temperature and antimetabolites on antibody binding to the outer surface of second stage *Toxocara canis* larvae. Mol Biochem Parasitol 1981;4:183–193.

56 Meghji M, Maizels RM. Biochemical properties of larval excretory–secretory glycoproteins of the parasite nematode *Toxocara canis*. Mol Biochem Parasitol 1986;18:155–170.

57 Maizels RM, Kennedy MK, Meghji M, Robertson BD, Smith HV. Shared carbohydrate epitopes on distinct surface and secreted antigens of the parasite nematode *Toxocara canis*. J Immunol 1987;139:207–214.

58 Kennedy MW, Maizels RM, Meghji M, Young L, Qureshi F, Smith HV. Species-specific and common epitopes on the secreted and surface antigens of *Toxocara cati* and *Toxocara canis* infective larvae. Parasite Immunol 1987; 9:407–420.

59 Munn EA. A helical, polymeric extracellular protein associated with the luminal surface of *Haemonchus contortus* intestinal cells. Tissue Cell 1977;9:23–34.

60 Smith TS, Munn EA. Strategies for vaccination against gastrointestinal nematodes. In Miller HRP, ed. *Immunity to and Diagnosis of Internal Parasitism*. Rev Sci Tech Off Int Epiz 1990;9:577–595.

61 O'Donnell IJ, Dineen JK, Rothwell TLW, Marshall RC. Attempts to probe the antigens and protective immunogens of *Trichostrongylus colubriformis* in immunoblots with sera from infected and hyperimmune sheep and high- and low-responder guinea pigs. Int J Parasitol 1985;15:129–136.

62 O'Donnell IJ, Dineen JK, Wagland BM, Letho S, Werkmeister JA, Ward CW. A novel host-protective antigen from *Trichostrongylus colubriformis*. Int J Parasitol 1989; 19:327–335.

63 Frenkel MJ, Savin KW, Bakker RE, Ward CW. Characterization of cDNA clones coding for muscle tropomyosin of the nematode *Trichostrongylus colubriformis*. Mol Biochem Parasitol 1989;37:191–200.

64 Savin KW, Dopheide TAA, Frenkel MJ, Wagland BM, Grant WN, Ward CW. Characterisation, cloning and host-protective activity of a 30-kDa glycoprotein secreted by the parasitic stages of *Trichostrongylus colubriformis*. Mol Biochem Parasitol 1990;41:167–176.

65 Dopheide TAA, Tachedjian M, Phillips C, Frenkel MJ, Wagland BM, Ward CW. Molecular characterisation of a protective, 11-kDa excretory–secretory protein from the parasitic stages of *Trichostrongylus colubriformis*. Mol Biochem Parasitol 1991;45:101–108.

66 Mackenzie CD, Jungery M, Taylor PM, Ogilvie BM. Activation of complement, the induction of antibodies to the surface of nematodes and the effect of these factors and cells on worm survival *in vitro*. Eur J Immunol 1980;10:594–601.

67 Genta RM, Walzer PD. Strongyloidiasis. In Walzer PD, Genta RM, eds. *Parasitic Infections in the Compromised Host*. New York: Marcel Dekker, 1989.

68 Wakelin D, Lee TDG, Immunobiology of *Trichuris* and *Capillaria* infections. In Soulsby EJL, ed. *Immune Responses in Parasitic Infections: Immunology, Immunopathology and Immunoprophylaxis, vol. 1. Nematodes*. Boca Raton: CRC Press, 1987:61–88.

69 Despommier DDD. The immunobiology of *Trichinella spiralis*. In Soulsby EJL, ed. *Immune Responses in Parasitic Infections: Immunology, Immunopathology and Immunoprophylaxis, vol. 1. Nematodes*. Boca Raton: CRC Press, 1987; 1:43–60.

70 Wakelin D, Denham DA. The immune response. In Campbell Wc, ed. *Trichinella and Trichinois*. New York: Plenum Publishers, 1983:265–308.

71 Katona IM, Urban JF, Finkelman F. The role of L3T4$^+$ and Ly2$^+$ T cells in the IgE response and immunity to *Nippostrongylus brasiliensis*. J Immunol 1988;140:3206–3211.

72 Grencis RK, Riedlinger J, Wakelin D. L3T4-positive T lymphoblasts are responsible for transfer of immunity to *Trichinella spiralis* in mice. Immunology 1985;56:213–218.

73 Riedlinger J, Grencis RK, Wakelin D. Antigen-specific T-cell lines transfer protective immunity against *Trichinella spiralis in vivo*. Immunology 1986;58:57–61.

74 Korenaga M, Wang CH, Bell RG, Zhu D, Ahmad A. Intestinal immunity to *Trichinella spiralis* is a property of OX8$^-$ OX22$^-$ T-helper cells that are generated in the intestine. Immunology 1989;66:588–594.

75 Zhu D, Bell RG. IL-2 production, IL-2 receptor expression and IL-2 responsiveness of spleen and mesenteric lymph node cells from inbred mice infected with *Trichinella spiralis*. J Immunol 1989;142:3262–3267.

76 Grencis RK, Riedlinger J, Wakelin D. Lymphokine production by T cells generated during infection with *Trichinella spiralis*. Int Arch Allergy Appl Immunol 1987; 88:92–95.

77 Appelton JA, McGregor DD. Characterization of the immune mediator of rapid expulsion of *Trichinella spiralis* in suckling rats. Immunology 1987;62:477–484.

78 Grencis RK, Wakelin D. Immunity to *Trichinella spiralis* in mice. Factors involved in direct anti-worm effects. Wiad Parazytol 1983;29:387–399.

79 Almond NM, Parkhouse RME. *Trichinella spiralis*: B-cell suppression does not 'exacerbate disease in mice. J Parasitol 1987;73:848–850.

80 Wedrychowicz H, Maclean JM, Holmes PH. Secretory IgA responses in rats to antigens of various developmental

stages of *Nippostrongylus brasiliensis*. Parasitology 1984; 89:145–157.

81 Miller HRP. Gastrointestinal mucus, a medium for survival and for elimination of parasitic nematodes and protozoa. Parasitology 1987;94:S77–S100.

82 Gabriel BW, Justus DE. Quantitation of immediate and delayed hypersensitivity responses in *Trichinella*-infected mice. Int Arch Allergy Appl Immunol 1979;60:275–285.

83 Appleton JA, Schain LR, McGregor DD. Rapid expulsion of *Trichinella spiralis* in suckling rats: mediation by monoclonal antibodies. Immunology 1988;65:487–492.

84 East IJ, Washington EA, Brindley PF, Monroy GF, Scott-Young N. *Nematospiroides dubius*: passive transfer of protective immunity to mice with monoclonal antibodies. Exp Parasitol 1988;66:7–12.

85 Ey PL. *Heligmosomoides polygyrus*: retarded development and stunting of larvae by antibodies specific for excretory/secretory antigens. Exp Parasitol 1988;65:232–243.

86 Roach TIA, Else KJ, Wakelin D, McLaren DJ, Grencis RK. Antigen recognition and transfer of immunity against *Trichuris* muris in mice by IgA monoclonal antibodies. Parasite Immunol 1991;13:1–12.

87 Butterworth AE. Cell-mediated damage to helminths. Adv Parasitol 1984;23:144–235.

88 Wang CH, Bell RG. *Trichinella spiralis*: intestinal expression of systemic stage-specific immunity to newborn larvae. Parasite Immunol 1987;9:465–475.

89 Capron A, Dessaint JP, Capron M, Joseph M, Ameison JC, Tonnel AB. From parasites to allergy: a second receptor for IgE. Immunol Today 1986;7:15–18.

90 Dessein AJ, Parker WL, James SL, David JL. IgE antibody and resistance to infection. I. Selective suppression of the IgE antibody response in rats diminishes the resistance and the eosinophil response to *Trichinella spiralis* infection. J Exp Med 1981;153:423–436.

91 Grove DI, Mahmoud AAF, Warren K. Eosinophils and resistance to *Trichinella spiralis*. J Exp Med 1977;145:755–759.

92 McLaren DJ, Mackenzie CD, Ramalho-Pinto FJ. Ultrastructural observations on the *in vitro* interaction between rat eosinophils and some parasitic helminths (*Schistosoma mansoni*, *Trichinella spiralis* and *Nippostrongylus brasiliensis*). Clin Exp Immunol 1977;30:105–118.

93 Wassom DL, Gleich G. Damage to *Trichinella spiralis* newborn larvae by eosinophil major basic protein. Am J Trop Med Hyg 1979;28:860–863.

94 Buys J, Wever R, Ruitenberg EJ. Myeloperoxidase is more efficient than eosinophil peroxidase in the *in vitro* killing of newborn larvae of *Trichinella spiralis*. Immunology 1984; 51:601–607.

95 Pentilla JA, Ey PL, Jenkins CR. Infection of mice with *Nematospiroides dubius*: demonstration of neutrophil-mediated immunity *in vivo* in the presence of antibodies. Immunology 1984;53:147–153.

96 McHugh TD, Jenkins T, McLaren DJ. *Strongyloides ratti*: studies of cutaneous reactions elicited in naive and sensitized rats and of changes in surface antigenicity of skin-penetrating larvae. Parasitology 1989;98:95–103.

97 Moqbel R. Histopathological changes following primary, secondary and repeated infections of rats with *Strongyloides ratti*, with special reference to tissue

eosinophils. Parasite Immunol 1980;2:11–27.

98 Sugane K, Oshima T. Eosinophilia, granuloma formation and migratory behaviour of larvae in the congenitally athymic mouse infected with *Toxocara canis*. Parasite Immunol 1982;4:307–318.

99 Badley JE, Grieve RB, Rockey JH, Glickman LT. Immune-mediated adherence of eosinophils to *Toxocara canis* infective larvae: the role of excretory–secretory antigens. Parasite Immunol 1987;9:133–143.

100 Yoshimura K, Uchida K, Sato K, Oya H. Ultrastructural evidence for eosinophil-mediated destruction of *Angiostrongylus cantonenesis* transferred into the pulmonary artery of non-permissive hosts. Parasite Immunol 1984; 6:105–118.

101 Egwang TG, Gauldie J, Befus AD. Bronchoalveolar leucocyte responses during primary and secondary *Nippostrongylus brasiliensis* infection in the rat. Parasite Immunol 1984;6:191–201.

102 Wells C, Behnke JM. Acquired resistance to the human hookworm *Necator americanus* in mice. Parasite Immunol 1988;10:493–505.

103 Leid RW, Williams JF. Helminth parasites and the host inflammatory system. Chem Zool 1979;11:229–271.

104 Lloyd S. Immunobiology of *Toxocara canis* and visceral larva migrans. In Soulsby EJL, ed. *Immune Responses in Parasitic Infections: Immunology, Immunopathology and Immunoprophylaxis. vol. 1. Nematodes*. Boca Raton: CRC Press, 1987:299–324.

105 Behnke JM, Parish HA. *Nematospiroides dubius*: arrested development of larvae in immune mice. Exp parasitol 1979;47:116–127.

106 Wakelin D, Grencis RK. Immunological responses to intestinal parasite infection. In Miller K, Nicklin S, eds. *Immunology of the Gastrointestinal Tract*, vol. II. Boca Raton: CRC Press, 1987.

107 Reed ND. Function and regulation of mast cells in parasite infections. In Galli SJ, Austen KF, eds. *Mast Cell and Basophil Differentiation and Function in Health and Disease*. New York: Raven Press, 1989.

108 Alizadeh H, Wakelin D. Mechanism of rapid expulsion of *Trichinella spiralis* from mice. In Kim CW, Ruitenberg EJ, Teppema JS, eds. *Trichinellosis*. Chertsey: Reedbooks, 1981.

109 Miller HRP, Huntley JF, Newlands GFJ, *et al*. Mast cell granule proteases in mouse and rat: a guide to mast cell heterogeneity and activation in the gastrointestinal tract. In Galli SJ, Austen KF, eds. *Mast Cell and Basophil Differentiation and Function in Health and Disease*. New York: Raven Press, 1989.

110 Woodbury RG, Miller HRP, Huntley JF, Newlands GFJ, Palliser AC, Wakelin D. Mucosal mast cells are functionally active during the spontaneous expulsion of primary intestinal nematode infections in the rat. Nature 1984;312:450–452.

111 Tuohy M, Lammas DA, Wakelin D, *et al*. Functional correlations between mucosal mast cell activity and immunity to *Trichinella spiralis* in high and low responder mice. Parasite Immunol 1990;12:675–686.

112 Jacobson RH, Reed ND, Manning DD. Expulsion of *Nippostrongylus brasiliensis* from mice lacking antibody production potential. Immunology 1977;32:867–874.

113 Watanabe N, Katakura K, Kobayashi A, Okamura K, Ovary Z. Protective immunity and eosinophilia in IgE-deficient SJA/9 mice infected with *Nippostrongylus brasiliensis* and *Trichinella spiralis*. Proc Natl Acad Sci USA 1988;85:4460–4462.

114 Parmentier HK, de Vries C, Ruitenberg EJ, van Loveren H. Involvement of serotonin in intestinal mastocytosis and inflammation during a *Trichinella spiralis* infection in mice. Int Arch Allergy Appl Immunol 1987;83:31–38.

115 Harari Y, Russell D, Castro GA. Anaphylaxis-mediated epithelial C^+ secretion and parasite rejection in rat intestine. J Immunol 1987;138:1250–1255.

116 Douch PGC, Harrison GBL, Buchanan LL, Greer KS. *In vitro* bioassay of sheep gastrointestinal mucus for nematode paralysing activity mediated by a substance with some properties characteristic of SRS-A. Int J Parasitol 1983;13:207–212.

117 Smith NC. The role of free oxygen radicals in the expulsion of primary infections of *Nippostrongylus brasiliensis*. Parasitol Res 1989;75:423.

118 Smith NC, Bryant C. The effect of antioxidants on the rejection of *Nippostrongylus brasiliensis*. Parasite Immunol 1989;11:161–167.

119 Smith NC, Bryant C. The role of host generated free radicals in helminth infections: *Nippostrongylus brasiliensis* and *Nematospiroides dubius* compared. Int J Parasitol 1986; 16:617–622.

120 Wassom DL, Krco CJ, David CS. I–E expression and susceptibility to parasite infection. Immunol Today 1987; 8:39–43.

121 Mossman TR, Cherwinski H, Bond MW, Giedlin MA, Coffman RL. Two types of murine helper T cell clone. I. Definition according to profiles of lymphokine activities and secreted proteins. J Immunol 1986;136:2348–2357.

122 Enriquez FJ, Brooks BO, Cypess RH, David CS, Wassom DL. *Nematospiroides dubius*: two H-2-linked genes influence levels of resistance to infection in mice. Exp Parasitol 1988;67:221–226.

123 Behnke JM, Wahid F. Immunological relationships during primary infections with *Heligmosomoides polygyrus* (*Nematospiroides dubius*): H-2 linked genes determine worm survival. Parasitology 1991;103:157–164.

124 Else KJ, Wakelin D, Wassom DL, Hauda KM. The influence of genes mapping within the major histocompatibility complex on resistance to *Trichuris muris* infections in mice. Parasitology. (In press.)

125 Bonneville M, Janeway CA, Ito K, *et al*. Intestinal intraepithelial lymphocytes are a distinct set of $\gamma\delta$ T cells. Nature (London) 1988;336:479–481.

126 Zakroff SGH, Beck L, Platzer EG, Spiegelberg HL. The IgE and IgG subclass responses of mice to four helminth parasites. Cell Immunol 1989;119:193–201.

127 Coffman RL, Seymour BWP, Hudak S, Jackson J, Rennick D. Antibody to interleukin-5 inhibits helminth-induced eosinophilia in mice. Science 1989;245:308–310.

128 Lammas DL, Mitchell LA, Wakelin D. Genetic control of eosinophilia. Analysis of production and response to eosinophil-differentiating factor in strains of mice infected with *Trichinella spiralis*. Clin Exp Immunol 1989;77:137–143.

129 Reed ND, Wakelin D, Lammas DA, Grencis RK. Genetic

130 Reed ND, Dehlawi MS, Wakelin D. Medium conditioned by spleen cells of *Nematospiroides dubius*-infected mice does not support development of cultured mast cells. Int Arch Allergy Appl Immunol 1988;85:113–115.

131 Monroy FG, Dobson C, Adams JH. Low molecular weight immunosuppressors secreted by adult *Nematospiroides dubius*. Int J Parasitol 1989;19:125–127.

132 Cross DA, Klesius PH. Soluble extracts from larval *Ostertagia ostertagi* modulating immune function. Int J Parasitol 1989;19:57–61.

133 Urquhart GM. Application of immunity in the control of parasitic disease. Vet Parasitol 1980;6:217–239.

134 Mitchell GF. Problems specific to parasite vaccines. Parasitology 1989;98:S19–S28.

135 Barbet AF. Vaccines for parasitic infections. Adv Vet Sci Comp Med 1989;33:345–375.

136 Miller HRP. Vaccination against intestinal parasites. Int J Parasitol 1987;17:43–51.

137 Hagan P, Behnke JM, Parish HA. Stimulation of immunity to *Nematospiroides dubius* in mice using larvae attenuated by cobalt 60 irradiation. Parasite Immunol 1981;3:149–156.

138 Dineen JK, Gregg P, Lascelles AK. The response of lambs to vaccination at weaning with irradiated *Trichostrongylus colubriformis* larvae: segregation into "responders" and "non-responders". Int J Parasitol 1978;8:59–63.

139 Day KP, Howard RJ, Prowse SJ, Chapman CB, Mitchell GF. Studies on chronic versus transient intestinal nematode infections in mice. I. A comparison of responses to excretory/secretory (ES) products of *Nippostrongylus brasiliensis* and *Nematospiroides dubius* worms. Parasite Immunol 1979;1:217–239.

140 Mitchell GF, Munoz JJ. Vaccination of genetically susceptible mice against chronic infection with *Nematospiroides dubius* using pertussigen as adjuvant. Aust J Exp Biol Med Sci 1983;61:425–434.

141 Pearce EJ, James SL, Henry S, Lanar D, Sher A. Induction of protective immunity against *Schistosoma mansoni* by vaccination with schistosome paramyosin (Sm97), a nonsurface parasite antigen. Proc Natl Acad Sci USA 1988;85: 5678–5682.

142 Silberstein DS. Antigens. In Campbell WC, ed. *Trichinella and Trichinosis*. New York: Plenum Press, 1983.

143 Lee TDG, Grencis RK, Wakelin D. Specific cross immunity between *Trichinella spiralis* and *Trichuris muris*: immunization with heterologous infections and antigens, and transfer of immunity with heterologous immune mesenteric lymph node cells. Parasitology 1982;84:381–389.

144 Else KJ, Wakelin D. Genetically determined influences on the ability of poor responder mice to respond to vaccination against *Trichuris muris*. Parasitology 1990;100:479–489.

145 Wakelin D, Mitchell LA, Donachie AM, Grencis RK. Genetic control of immunity to *Trichinella spiralis* in mice. Response of rapid- and slow-responder strains to immunization with parasite antigens. Parasite Immunol 1986; 8:159–170.

146 Rhalem A, Bourdieu C, Luffau G, Pery P. Vaccination of mice with liposome-entrapped adult antigens of

Nippostrongylus brasiliensis. Ann Inst Pasteur/Immunol 1988;139:157–166.

147 Mowat AMCI. The regulation of immune responses to dietary protein antigens. Immunol Today 1987;8:93–98.

148 Durham CP, Murrell KD, Lee CM. *Trichinella spiralis*: immunization of rats with an antigen fraction enriched for allergenicity. Exp Parasitol 1984;57:297–306.

149 Mitchell GF. Immunoregulation and the induction and expression of host-protective immune responses to parasites. In Ogra PL, Jacobs DM, eds. *Regulation of the Immune Response*. Basel: Karger, 1983:278–287.

150 Anderson RM, May RM. Helminth infections of humans: mathematical models, population dynamics and control. Adv Parasitol 1985;24:1–101.

151 Else KJ, Wakelin D. Genetic variation in the humoral immune response of mice to the nematode *Trichuris muris*. Parasite Immunol 1989;11:77–90.

152 Wassom DL, Dougherty DA, Krco CJ, David CS. H-2-controlled, dose-dependent suppression of the response that expels adult *Trichinella spiralis* from the small intestine of mice. Immunology 1984;53:811–818.

153 Dawkins HJS, Windon RG, Eagleson GK. Eosinophil responses in sheep selected for high- and low-responsiveness to *Trichostrongylus colubriformis*. Int J Parasitol 1989;19:199–207.

154 Rothwell TLW, Abeydeera LR, Geczy AF. Relationship between basophils and eosinophils in cutaneous basophil hypersensitivity reactions in guinea pigs and susceptibility to *Trichostrongylus colubriformis* infection. Int J Parasitol 1988;18:347–351.

155 Good MF, Kumar S, Miller LH. The real difficulties for malaria sporozoite vaccine development: nonresponsiveness and antigenic variation. Immunol Today 1988;9:351–355.

156 Wagland BM, Steel JW, Windon RG, Dineen JK. The response of lambs to vaccination and challenge with *Trichostrongylus colubriformis*: effect of plane of nutrition on, and the inter-relationship between, immunological responsiveness and resistance. Int J Parasitol 1984;14:39–44.

157 Slater AFG, Keymer AE. The influence of protein deficiency on immunity to *Heligmosomoides polygyrus* (nematoda) in mice. Parasite Immunol 1988;10:507–522.

158 Beaver PC, Jung RC, Cupp EW. *Clinical Parasitology*, 9th edn. Philadelphia: Lea & Febiger, 1984.

159 Nishigori M. On various factors influencing the development of *Strongyloides stercoralis* and autoinfection. Taiwan Igakkai Zasshi 1928;277:31–33.

160 Genta RM. Strongyloidiasis. In Walls KM, Schantz PM, eds. *Immunodiagnosis of Parasitic Diseases, vol. 1. Helminthic Diseases*. Orlando: Academic Press, 1986:183–199.

161 Grove DI. Diagnosis. In Grove DI, ed. *Strongyloidiasis: a Major Roundworm Infection of Man*. London: Taylor & Francis, 1989.

162 Woodruff AW. Toxocariasis. Br Med J 1970;3:663–669.

163 Dissanayake S, Ismail NN. Immunodiagnosis of bancroftian filariasis. In *Ciba Foundation Symposium 127—Filariasis*. Chichester: John Wiley and Sons, 1987.

164 Harnett W. Molecular approaches to the diagnosis of *Onchocerca volvulus* in man and the insect vector. In

165 Robertson BE, Burkot TR, Gillespie SH, Kennedy MW, Wambai Z, Maizels RM. Detection of circulating parasite antigen and specific antibody in *Toxocara canis* infections. Clin Exp Immunol 1988;74:236–241.

166 Ivanoska D, Cuperlovic K, Gamble HR, Murrell KD. Comparative efficacy of antigen and antibody detection tests for human trichinellosis. J Parasitol 1989;75:38–41.

167 Dissanyake S, Forsyth KP, Ismail MM, Mitchell GF. Detection of circulating antigen in bancroftian filariasis by using a monoclonal antibody. Am J Trop Med Hyg 1984;33:1130–1140.

168 Hamilton RG, Hussain R, Ottesen EA. Immunoradiometric assay for detection of filarial antigens in human serum. J Immunol 1984;133:2237–2242.

169 Weiss N, Van den Ende M-C, Albiez EJ, Barbiero VK, Forsyth K, Prince AM. Detection of serum antibodies and circulating antigens in a chimpanzee experimentally infected with *Onchocerca volvulus*. Trans R Soc Trop Med Hyg 1986;80:587–591.

170 Bowman DD, Mika-Grieve M, Grieve RB. Circulating excretory–secretory antigen levels and specific antibody responses in mice infected with *Toxocara canis*. Am J Trop Med Hyg 1987;36:75–82.

171 Hogarth-Scott RS. Visceral larva migrans—an immunofluorescent examination of rabbit and human sera for antibodies to the ES antigen of the second stage larvae of *Toxocara canis*, *Toxocara cati* and *Toxoscaris leonina*. Immunology 1966;10:217–223.

172 Suzuki T. Studies on the immunological diagnosis of anisakiasis. Jpn J Parasitol 1968;17:212–220.

173 Parkhouse RME, Philipp M, Ogilvie BM. Characterization of surface antigens of *Trichinella spiralis* infective larvae. Parasite Immunol 1981;3:339–352.

174 Boyce WM, Branstetter BA, Kazacos KR. Comparative analysis of larval excretory–secretory antigens of *Baylisascaris procyonis*, *Toxocara canis* and *Ascaris suum* by Western blotting and enzyme immunoassay. Int J Parasitol 1988;18:109–113.

175 Kennedy MW, Tierney J, Ye P, *et al*. The secreted and somatic antigens of the third stage larva of *Anisakis simplex*, and antigenic relationship with *Ascaris suum*, *Ascaris lumbricoides* and *Toxocara canis*. Mol Biochem Parasitol 1988;31:35–46.

176 Lynch NR, Wilkes LK, Hodgen AN, Turner KJ. Specificity of *Toxocara* ELISA in tropical populations. Parasite Immunol 1988;10:323–337.

177 Maleewong W, Morakote N, Thamasonthi W, Charuchinda K, Tesana A, Khamboonruang C. Serodiagnosis of human gnathostomiasis. Southeast Asian J Trop Med Public Health 1988;19:201–205.

178 Despommier DD. Trichinellosis. In Walls KW, Schantz PM, eds. *Immunodiagnosis of Parasitic Diseases, vol. 1. Helminthic Diseases*. Orlando: Academic Press, 1986:163–181.

179 De Savigny O, Voller A, Woodruff AW. Toxocariasis: serological diagnosis by enzyme immunoassay. J Clin Pathol 1979;32:284–288.

180 Glickman LT, Schantz PM, Grieve RE. Toxocariasis. In

Walls KW, Schantz PM, eds. *Immunodiagnosis of Parasitic Diseases, vol. 1. Helminthic Diseases.* Orlando: Academic Press, 1986:201–231.

181 Ogilvie BM, Bartlett A, Godfrey FC, Turton JA, Worms MJ, Yeates RA. Antibody responses in self infections with *Necator americanus.* Trans R Soc Trop Med Hyg 1978;72: 66–71.

182 Ganguly NK, Mahajan RC, Sehgai R, Shetty P, Dilawart JB. Role of specific immunoglobulin E to excretory–secretory antigen in diagnosis and prognosis of hookworm infections. J Clin Microbiol 1988;26:739–742.

183 Tribouley-Duret J, Tribouley J, Appriou M, Megroud RN. Application du test ELISA au diagnostia de la strongyloidosa. Ann Parasitol Hum Comp 1978;53: 641–648.

184 Engrall E, Ljungstrom I. Detection of human antibodies to *Trichinella spiralis* by enzyme linked immunosorbent assay, ELISA. Acta Pathol Microbiol Scand Sect C 1975;83: 231–237.

185 Tada I, Araki T, Matsuda H, Araki K, Akahane H, Mimori T. A study on immunodiagnosis of gnathostomiasis by ELISA and double diffusion with special reference to the antigenicity of *Gnathostoma doloresi.* Southeast Asian J Trop Med Public Health 1987;18:444–448.

186 Poggensee U, Schommer G, Jansen-Rosseck R, Feldmeier H. Immunodiagnosis of human anisakiasis by use of larval excretory–secretory antigen. Zentralbl Bakteriol Hyg A 1989;270:503–510.

187 Cross JH. Clinical manifestations and laboratory diagnosis of eosinophilia meningitis syndrome associated with angiostrongyliasis. Southeast Asian J Trop Med Public Health 1978;9:161–170.

188 Chen SN Enzyme-linked immunosorbent assay (ELISA) for the detection of antibodies to *Angiostrongylus cantonensis.* Trans R Soc Trop Med Hyg 1986;80:398–405.

189 Rapic D, Dzakula N, Matic-Piantanida D. Evaluation of different antigens in a seroepizootiological survey of trichinellosis by enzyme-linked immunosorbent assay. Vet Parasitol 1986;21:285–289.

190 Gam AA, Neva FA, Krotoski WA. Comparative sensitivity and specificity of ELISA and IHA for serodiagnosis of strongyloidiasis with larval antigens. Am J Trop Med Hyg 1987;37:157–161.

191 Almond NM, Parkhouse RME, Chapa-Ruiz MR, Garcia-Ortigoza E. The response of humans to surface and secreted antigens of *Trichinella spiralis.* Trop Med Parasitol 1986;37:381–384.

192 Kagan IG. Serology of trichinosis. In Gould SE, ed. *Trichinosis in Man and Animals.* Springfield: Thomas, 1970: 257.

193 Ljungstrom I. Immunodiagnosis in man. In Campbell WC, ed. *Trichinella and Trichinosis.* New York: Plenum Press, 1983:715–757.

194 MacLean JD, Viallet J, Law C, Staudt M. Trichinosis in the Canadian Arctic: report of five outbreaks and a new clinical syndrome. J Infect Dis 1989;160:513–520.

195 Smith HV, Kennedy MW. Soluble antigens and antibodies in the sera of mice infected with *Trichinella spiralis,* detected by a modified double counter-immunoelectrophoresis technique. J Helminthol 1984; 58:71–78.

196 Murrell KD, Anderson WR, Schad GA, *et al.* Field evaluation of the enzyme-linked immunosorbent assay for swine trichinosis: efficacy of the excretory–secretory antigen. Am J Vet Res 1986;47:1046–1049.

197 Gamble HR, Rapic D, Marinculic A, Murrell KD. Evaluation of excretory–secretory antigens for the serodiagnosis of swine trichinellosis. Vet Parasitol 1988;30:131–137.

198 Boyce WM, Asai DJ, Wilder JK, Kazacos KR. Physico-chemical characterization and monoclonal and polyclonal antibody recognition of *Baylisascaris procyonis* larval excretory–secretory antigens. J Parasitol 1989;75:540–548.

199 Jeska EL. Antigenic analysis of a metazoan parasite, *Toxocara canis.* 1. Extraction and assay of antigens. Exp Parasitol 1967;10:38–50.

200 Jeska EL. Purification and immunochemical analysis of a genus-specific cuticular antigen of *Toxocara canis.* J Parasitol 1969;55:465–471.

201 Welch JS, Symons MH, Dobson C. Immunodiagnosis of parasitic zoonoses: purification of *Toxocara canis* antigens by affinity chromatography. Int J Parasit 1983;13:171–178.

202 De Savigny D. *In vitro* maintenance of *Toxocara canis* larvae and a simple method for the production of *Toxocara* ES antigen for use in serodiagnostic tests for visceral larva migrans. J Parasitol 1975;61:781–782.

203 De Savigny D, Tizard IR. Toxocaral *larva migrans:* the use of larval secretory antigens in haemagglutination and soluble antigen fluorescent antibody tests. Trans R Soc Trop Med Hyg 1977;71:501–507.

204 Glickman LT, Grieve RB, Lauria SS, Jones DL. Serodiagnosis of ocular toxocariasis: a comparison of two antigens. J Clin Pathol 1985;38:103–107.

205 Bowman DD, Mika-Grieve M, Grieve RB. *Toxocara canis:* monoclonal antibodies to larval excretory–secretory antigens that bind with genus- and species-specificity to the cuticular surface of infective larvae. Exp Parasitol 1987;64:458–465.

206 Badley JE, Grieve RB, Bowman DD, Glickman LT, Rockey JH. Analysis of *Toxocara canis* larval excretory–secretory antigens: physicochemical characterisation and antibody recognition. J Parasitol 1987;73:593–600.

207 Clemett RS, Williamson HJE, Hidajat RR, Allardyce RA, Stewart AC. Ocular *Toxocara canis* infections: diagnosis by enzyme immunoassay. Aust N Z J Ophthalmol 1987;15: 145–150.

208 Speiser F, Gottstein B. A collaborative study on larval excretory/secretory antigens of *Toxocara canis* for the immunodiagnosis of human toxocariasis with ELISA. Acta Trop 1984;41:361–372.

209 Sugane K, Irving DO, Howell MJ, Nicholas WL. *In vitro* translation of mRNA from *Toxocara canis* larvae. Mol Biochem Parasitol 1985;14:275–281.

210 Glickman LT, Schantz PM. Epidemiology and pathogenesis of zoonotic toxocariasis. Epidemiol Rev 1981;3:230–250.

211 Schantz PM, Myer D, Glickman LT. Clinical, serologic and epidemiological characteristics of ocular toxocariasis. Am J Trop Med Hyg 1979;28:24–28.

212 Shields JA. Ocular toxocariasis: a review. Surv Ophthalmol 1984;28:361–381.

213 Glickman LT, Cypress R, Hiles D, Gessner T. *Toxocara* specific antibody in the serum and aqueous humor of a

patient with presumed ocular and visceral toxocariasis. Am J Trop Med Hyg 1979;28:29–35.

214 Biglin AW, Glickman LT, Lobes LA. Serum and vitreous *Toxocara* antibody in nematode endophthalmitis. Am J Ophthalmol 1979;88:898–901.

215 Genchi C, Tinelli M, Brunello F, Falagiani P. Serodiagnosis of ocular toxocariasis: a comparison of specific IgE and IgG. Trans R Soc Trop Med Hyg 1986;80:993–994.

216 Aguila C, Cuellar C, Fenoy S, Guillen JL. Comparative study of assays detecting circulating immune complexes and specific antibodies in patients infected with *Toxocara canis*. J Helminthol 1987;61:196–202.

217 Kayes SG, Omholt PE, Grieve RB. Immune responses of CBA/J mice to graded infections with *Toxocara canis*. Infect Immun 1985;48:697–703.

218 Daffalla AA. The indirect fluorescent antibody test for the serodiagnosis of strongyloidiasis. J Trop Med Hyg 1972; 75:109–111.

219 Sato Y, Otsuru M, Takara M, Shiroma Y. Intradermal reactions in strongyloidiasis. Int J Parasitol 1986;16:87–91.

220 McRury J, De Messias IT, Walzer PD, Huitger T, Genta RM. Specific IgE responses in human strongyloidiasis. Clin Exp Immunol 1986;65:631–638.

221 Brindley PJ, Gam AA, Pearce EJ, Poindester RW, Neva FA. Antigens from the surface and excretions/secretions of the filariform larva of *Strongyloides stercoralis*. Mol Biol Parasitol 1988;28:171–180.

222 Desowitz RS, Raybourne RB, Ishikura H, Kliks MM. The radioallergosorbent test (RAST) for the serological diagnosis of human anisakiasis. Trans R Soc Trop Med Hyg 1985;79:256–259.

223 Akahane H, Iwata K, Miyazaki I. Studies on *Gnathastoma hispidum fedchenko*, 1872 parasitic in loaches imported from China. Jpn J Parasitol 1982;31:507–511.

224 Yamaguchi T. Immunological studies on human gnathostomiasis II. Precipitin test. J Kurume Med Assoc 1952;15:26–34.

225 Furunc O. An immunological study on gnathostomiasis. (Precipitin ring tests and Sarles phenomenon.) Acta Med Jpn 1959;29:2802–2822.

226 Suzuki T, Sato Y, Yamashita T, Sekikawa H, Otsuru M. *Angiostrongylus cantonensis*: preparation of a specific antigen using immunosorbent columns. Exp Parasitol 1975;38: 191–201.

227 Fujii T. Immunoblot analysis of the circulating antigens occurring in serum of rats infected with *Angiostrongylus cantonensis*. Parasitol Res 1988;74:476–483.

228 Sato Y, Otsuru M, Asato R, Kinjo K. Immunological observations on seven cases of eosinophilia meningo-encephalitis probably caused by *Angiostrongylus cantonensis* in Okinawa, Japan. Jpn J Parasitol 1977;26:209–220.

229 Welch JS, Dobson C, Campbell GR. Immunodiagnosis and seroepidemiology of *Angiostrongylus cantonensis* zoonoses in man. Trans R Soc Trop Med Hyg 1980;74:614–623.

230 Cross JH. Clinical manifestations and laboratory diagnosis of eosinophilia meningitis syndrome associated with angiostrongyliasis. Southeast Asian J Trop Med Public Health 1978;9:161–170.

231 Tharavanij S. Immunology of angiostrongyliasis. In Cross JH, ed. *Studies on Angiostrongyliasis in Eastern Asia and Australia*, special publication. Taipei, Taiwan: US Naval Medical Research Unit No. 2, 1979.

232 Tanaka K, Miyachi Y, Tsuji M, Miyoshi A. Radio-immunoassay for *Ascaris* specific protein and its clinical approach to gastrointestinal diseases. Nippon Shokakibyo Gakkai Zasshi 1978;75:1832–1838.

233 Mukerji KI, Saxena RP, Chatak S, Saxena KC, Chandra R, Srivastava VK. Partially purified human *Ascaris* antigen in immunodiagnosis of ascariasis. Indian J Exp Biol 1980; 18:905–909.

234 De Savigny D, Stevenson W. Unpublished observations cited by Ogilvie BM, de Savigny D. In Cohen S, Warren KS, eds. *Immunology of Parasitic Infections*, 2nd edn. Oxford: Blackwell Scientific Publications, 1982:715–757.

235 Kennedy MW, Qureshi F, Haswell-Elkins M, Elkins DB. Homology and heterology between the secreted antigens of the parasitic larval stages of *Ascaris lumbricoides* and *Ascaris suum*. Clin Exp Immunol 1987;67:20–30.

236 Vinayak VK, Singh T, Naik SR. Evaluation of intradermal test in ancylostomiasis. Indian J Med Res 1977;66:737–744.

237 Ball PAJ, Bartlett A. Serological reactions to infection with *Necator americanus*. Trans R Soc Trop Med Hyg 1969;63: 362–369.

4 Synopsis of Parasitology

20 Protozoa

Wallace Peters

In the 10 years since the second edition of this book appeared, the clinical importance of parasitic protozoa has taken on a new dimension. The advent of the human immunodeficiency virus (HIV) and the later stages of infection with this organism—acquired immunodeficiency syndrome (AIDS)—have highlighted the existence of several protozoa formerly accepted as being of relatively minor significance, as well as exposing several previously unrecognized species of human pathogen. Today cryptosporidiosis, for example, is almost a household word, whereas even 10 years ago it was a term familiar mainly to veterinarians and the more academic protozoologists. Toxoplasmosis now kills a significant proportion of sufferers from AIDS, frequently after producing severe cerebral pathology associated with gross personality disorders. Visceral leishmaniasis increasingly is being identified in individuals in whom the organisms had probably lived a more or less commensal existence prior to the superposition in such persons of HIV. Paradoxically, in homosexual males in whom HIV is most common, at least in the "developed" countries, it has been shown that the high prevalence of infection with *Entamoeba histolytica* is due mainly to infection with nonpathogenic variants of this organism and that infection even with virulent ameba does not seem to be aggravated by the virus. Evidence gathered to date in malaria endemic countries seems to indicate that the incidence and severity of this infection, too, is not influenced by concomitant HIV infection.

Ten years ago it appeared that the origin of relapses in malaria had been resolved with the discovery of the hypnozoite stage in parenchymal cells of the liver, yet the controversy lingers on even now. The final details of the life-cycles of other protozoa also await elucidation, e.g., those of the coccidian parasite *Cryptosporidium*, while an academic battle still rages for the proprietary rights to an important killer of AIDS victims, *Pneumocystis carinii* and the intestinal organism *Blastocystis hominis*,

between protozoologists and mycologists, hence their only brief mention in this chapter. Meanwhile the biochemical tools used to resolve such problems have become more sophisticated. To the invaluable range of techniques for determining species affiliations by characterizing their isoenzymes have been added monoclonal antibodies and a range of methods for identifying and comparing sequences of DNA and RNA of various types. Electron microscopy has extended our knowledge of the fine structure of protozoa, a topic of particular value in throwing light on the intimate host–parasite interface of intracellular parasites and their techniques for evading the damaging effects of host immune responses to their presence. As Sadun [1] pointed out, the successful access of a parasite to a new host does not necessarily imply the successful establishment of that parasite at the site of access. The invader may have to undergo an extensive change in its form and migrate within the host in order to reach its ultimate and optimal site of residence. This may be required not only for its survival in that host but also to enable it to be carried over into a fresh host, either via the free environment or through the intermediary of a vector.

As in the second edition, the various protozoa are here discussed in a zoologically systematic order which roughly parallels that in which the immunologic aspects are dealt with in earlier chapters of this work.

PHYLUM SARCOMASTIGOPHORA

Subphylum Sarcodina (the amebae)

This phylum holds equal rank with the Ciliophora, of which there is only one representative in humans (the parasitic ciliate *Balantidium coli*), and the Apicomplexa, which includes the malaria parasites, piroplasmids, and coccidians. The Sarcomastigophora contain two important subphyla, the Sarcodina (or amebae) and the

Table 20.1 The Ciliophora and Sarcodina of medical importance

Phylum, subphylum	Superclass	Order	Genera and species
Apicomplexa (see Table 20.5)			
Ciliophora			
Rhabdophora		Vestibulifera	*Balantidium coli*
Sarcomastigophora			
Mastigophora	Rhizopodea	Amoebida	*Entamoeba coli*
			Entamoeba histolytica*
			Entamoeba gingivalis
			Endolimax nana
			Iodamoeba bütschlii
			Acanthamoeba culbertsoni [†]
			Acanthamoeba spp. [†]
			Blastocystis hominis [‡]
		Schizopyrenida	*Naegleria fowleri* [†]

[*] Contains numerous "zymodemes," most of which are nonpathogenic.
[†] Pathogenic "free-living" species. The other species are obligatory parasites or commensals.
[‡] Classified in order Amoebida, suborder Blastocystina [2].

Mastigophora (or flagellates). The relationship of these groups to each other is summarized in Table 20.1.

Superclass Rhizopodea, order Amoebida

Entamoeba histolytica. The order Amoebida contains all the amebae found in the intestinal tract of humans, extending from the mouth to the anus, of which the only undoubtedly pathogenic species is *E. histolytica*. Other amebae that inhabit the lower intestine and bear a close resemblance to *E. histolytica* and to each other have been called *Entamoeba hartmanni*, *Entamoeba polecki*, *Entamoeba dysenteriae*, and *Entamoeba minuta*. They are readily distinguished from the other species listed in Table 20.1 by morphologic characters that are detailed in every textbook of parasitology. However, the isoenzyme studies of Sargeaunt *et al.* [3] indicate that *E. polecki* and *E. minuta* are almost certainly no more than growth stages of *E. histolytica*, while the taxonomic status of *E. dysenteriae* for the pathogenic zymodemes of *E. histolytica* is still disputed. It is now widely accepted that few of the zymodemes of *E. histolytica* are pathogenic (see p. 531).

Amebiasis is passed from human to human by feco-oral transmission, generally through the intermediary of contaminated food (usually unwashed fruit, green vegetables, or salad) or occasionally via water, milk, or other liquids. The infection is also prevalent in homosexual males. Notorious examples where poor hygiene standards facilitate transmission are centers for the mentally handicapped in which outbreaks of epidemic proportions may occur. Earlier investigations on the influence of genetic factors, diet, and environmental factors as well as the geographic distribution of amebiasis have been handicapped until recently by the inability to separate pathogenic from nonpathogenic organisms, and much of the earlier literature on these topics must be interpreted with caution. A pathogenic infection is defined as one in which trophozoites invade the tissues where they can readily be seen to be phagocytosing host erythrocytes.

The life-cycle of *E. histolytica* and closely related amebae is summarized in Fig. 20.1 which, however, omits recent data which indicate that genetic exchange probably occurs at some stage through an unidentified form of sexual reproduction [3]. Ingested cysts pass through the stomach and upper intestine where the cyst walls are digested by proteolytic enzymes to liberate from each cyst four actively growing trophozoites in the region of the lower ileum. The trophozoites feed mainly on bacteria in the gut lumen where they reproduce repeatedly by binary fission. As they pass down the gut they again form a cyst wall which they will require to protect them from dessication once they pass to the exterior in the feces. They also accumulate a food reserve in the form of glycogen and, later, crystalline aggregates of ribosomes which form the so-called "chromatoid bodies." The nucleus has a characteristic appear-

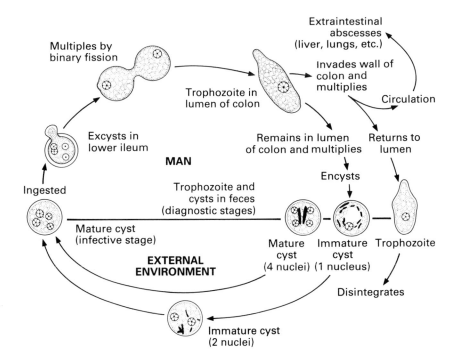

Fig. 20.1 Diagram of the life-cycle of *Entamoeba histolytica*.

ance with a small, central, DNA-rich karyosome and chromatin granules containing RNA uniformly lining the inner surface of the nuclear membrane. The nucleus divides into two and then four daughter nuclei. The trophozoites of pathogenic strains (zymodemes) ingest erythrocytes and invade the glands, mucosa, and sub-mucosa of the large intestine (Fig. 20.2) causing pro-gressive destruction characterized by undermined "flask-shaped" ulcers. Attachment to and invasion of host tissues seems to be facilitated by parasite proteases that degrade fibronectin [4].

If a small blood vessel of the mesenteric plexus is invaded the trophozoites can metastasize to the liver where they may establish an infective focus with the formation of abcesses containing sterile pus. Amebic abcesses are unusual in that they provoke little cellular reaction on the side of the host so that, for example, an amebic abcess does not become walled off by fibrous tissue but remains somewhat diffuse at the edges where active trophozoites are readily detected. Secondary spread may occur from the liver either directly, e.g., across the diaphragm to the lungs, or metastatically to the brain or other tissues. Cyst formation does not occur in extraintestinal sites.

Trophozoites of *E. histolytica* have a complex structure (Fig. 20.3) (see also Martínez-Palomo [5]) including the usual cellular organelles, with the exception of rough endoplasmic reticulum and mitochondria, the latter being redundant in these facultative anaerobes [6]. Electron microscopy shows that they are mainly associ-ated with the uroid, numerous, thread-like filopodia, the function of which is unknown, although they are apparently not related to cytotoxicity [7].

The diagnosis of *E. histolytica* is normally made by detecting trophozoites or cysts in fecal preparations, but the finding of cysts alone is not necessarily indicative of the existence of invasive amebiasis. The organisms are readily grown from cysts or trophozoites on Robinson's medium [8]. Amebae harvested from such cultures, as well as those in fresh feces, can be identified by standard microscopy in unstained or stained preparations, and the various zymodemes can be typed biochemically [9]. Most pathogenic *E. histolytica* are associated with Group II (Fig. 20.4). Immunodiagnostic procedures to confirm the existence of intestinal or extraintestinal amebiasis are discussed in Chapter 10. They include several types of enzyme-linked immunosorbent assay (ELISA) to detect amebic antigens in feces but these still require further evaluation. No DNA probe is yet available.

Pathogenic "free-living" amebae. In addition to the amebae of the gastrointestinal tract, the order Amoebida also contains free-living but potentially pathogenic

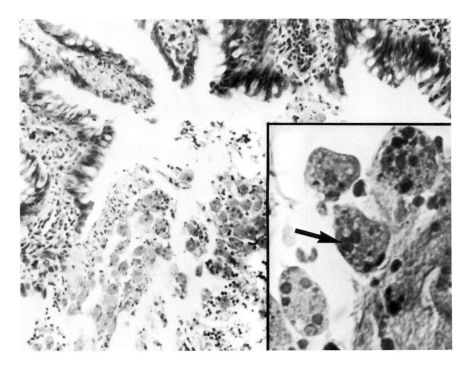

Fig. 20.2 Trophozoites of *Entamoeba histolytica* invading the submucosa of the colon. The inset shows individual trophozoites containing ingested erythrocytes (arrow). Original magnification, ×90; inset ×360.

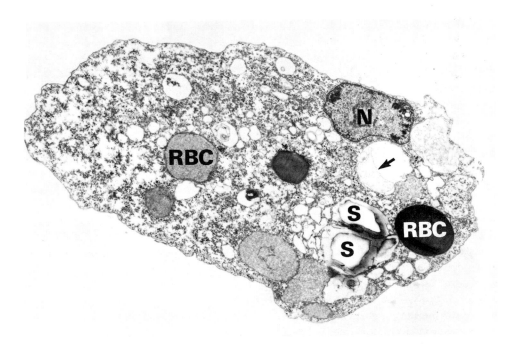

Fig. 20.3 Electron micrograph of *Entamoeba histolytica*. N, nucleus; RBC (white lettering), freshly ingested red blood cell; RBC (black lettering), partially digested red cell; S, starch granule; arrow indicates residual membrane left after digestion of an erythrocyte. Original magnification, ×4000. (Courtesy of Dr D. S. Ellis, London School of Hygiene and Tropical Medicine.)

		Origin PGM +	Origin +	Origin ME +
E. histolytica	Gp I			
E. histolytica	Gp II			
E. histolytica	Gp III			
E. histolytica	Gp IV			
E. histolytica	-like			
E. moshkovskii	GP II			
E. moshkovskii	GP II			
E. moshkovskii	EC clone			
E. invadens	Gp I			
E. invadens	Gp II			
E. chattoni				

Fig. 20.4 Diagram to show the comparative band positions of three isoenzymes in various *Entamoeba* of the *histolytica* type. ME, L-malate−NADP$^+$ oxidoreductase; GPI, glucose phosphate isomerase; PGM, phosphoglucomutase. Pathogenicity is associated with group II *E. histolytica*. (From Sargeaunt *et al.* [3].)

organisms in the genus *Acanthamoeba*, *Acanthamoeba culbertsoni* being one of the most notorious (see review by Rondanelli [10]). They are ameboflagellates which possess a free-living flagellate stage and which form tough-walled cysts that are almost ubiquitous. There are also potentially pathogenic, free-living species in the genus *Naegleria* which belongs to the order Schizopyrenida. Organisms of both genera have been identified in material from patients with meningoencephalitis (now called primary amebic meningoencephalitis) [11] and, more recently, in people with severe keratoconjunctivitis and corneal ulceration, particularly associated with the use of "soft" contact lenses. *Acanthamoeba* infections may be contracted by bathing in warm waters, e.g., in thermal baths or swimming pools in warm climates, the organisms apparently entering the brain via the nose and the cribriform plate. Hence infection tends to be seasonal or localized to certain health resorts in temperate climates, but perennial in the subtropics and tropics where the infection is very likely to be mistaken for bacterial or viral meningitis or encephalitis. Apart from the typical structure of the trophozoite nucleus in stained preparations of biopsy or autopsy material, and of the cysts in pus (Fig. 20.5), the species can be identified by the use of isoenzyme electrophoresis [12].

Infection with this group of amebae responds poorly to chemotherapy, amphotericin B being one of the few active drugs. Infection of the brain is usually fatal.

Blastocystis hominis. This common resident of the human intestine has usually been considered to be a harmless commensal fungus. Recent investigations indicate that it is, in fact, an unusual anaerobic protozoon [2] in the order Amoebida which may, in a few individuals, cause serious bowel pathology

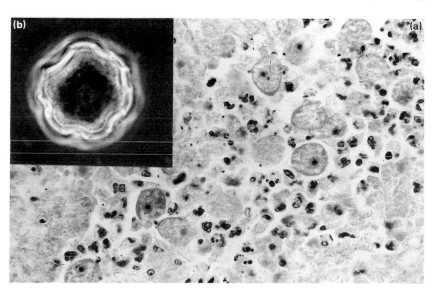

Fig. 20.5 Pathogenic "free-living" amebae. (a) Trophozoites of an *Acanthamoeba*, identified serologically as *Acanthamoeba castellanii* in a section of brain taken at postmortem from a Zambian patient who died from AIDS. The nuclear structure is well seen in this section. Original magnification, ×640. (b) Cyst of *Acanthamoeba polyphaga* as seen in material from an infected human cornea. Original magnification, ×2700. (Specimens courtesy of Dr D. C. Warhurst, London School of Hygiene and Tropical Medicine.)

and diarrhea. More research is required to determine the clinical importance of *B. hominis*, especially in immunocompromised individuals.

Subphylum Mastigophora, class Zoomastigophora

The human parasites of this subphylum fall into two general groups, as shown in Table 20.2. The parasitic flagellates that are found in the intestinal tract and genitalia belong to three different orders, while those in the blood and tissues are classified in the order Kinetoplastida. In the first group are five species, *Giardia lamblia* (which should probably be referred to as *Giardia intestinalis* or *duodenalis*) occurring in the duodenum and jejunum, *Chilomastin mesnili* in the cecum and colon, *Trichomonas hominis* in the distal ileum and cecum, *Trichomonas tenax* in the mouth, and *Trichomonas vaginalis* in the female and male urogenital tracts. Although only *G. lamblia* and *T. vaginalis* are pathogenic in humans, both are extremely common and cosmopolitan [13].

Order Diplomonadida

Giardia lamblia. About forty species of *Giardia* have been described from a wide variety of mammalian hosts,

Table 20.2 The Mastigophora of medical importance

Class	Order	Genus and species
Zoomastigophorea		
No vectors	Retortomonadida	*Chilomastix mesnili*
	Diplomonadida	*Giardia lamblia*
	Trichomonadida	*Trichomonas vaginalis*
		Trichomonas hominis
		Dientamoeba fragilis
Insect vector	Kinetoplastida	*Trypanosoma* spp.
		(see Table 20.3)
		Leishmania spp.
		(see Table 20.4)

including humans, but their precise identification is difficult and their taxonomic classification still obscure. Biochemical taxonomic techniques, including DNA mapping and probes, are beginning to make order now that methods have become available for the axenic culture of *Giardia* trophozoites excysted from feces (see review by Erlandsen & Meyer [14]).

Infection occurs when cysts are swallowed, commonly in contaminated water. Waterborne epidemics accounted for over 20 000 cases in the USA between 1965 and 1981. Feco-oral infection, direct or via contaminated food, is often sporadic, especially among travellers to

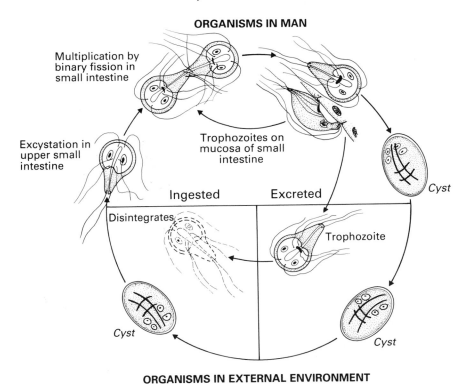

ORGANISMS IN MAN

Multiplication by binary fission in small intestine

Excystation in upper small intestine

Trophozoites on mucosa of small intestine

Ingested

Excreted

Disintegrates

Trophozoite

Cyst

Cyst

Cyst

ORGANISMS IN EXTERNAL ENVIRONMENT

Fig. 20.6 Diagram of the life-cycle of *Giardia lamblia*. (From Meyer & Jarroll [16].)

warmer climates. The significance of animal reservoirs is still being investigated, the beaver in North America and cats in Australia [15] being two of the species that are currently under suspicion. After the cyst has been swallowed its wall is digested away and the liberated trophozoite enters the duodenum and jejunum where it attaches to the surface of the mucosa by means of a sucking disc (Fig. 20.6).

The trophozoites undergo binary fission by a rather complicated process which involves disappearance of the median bodies and sucking disc, nuclear division, and then finally reformation of the disc and other organelles in the daughter cells. The trophozoites appear to feed on intestinal contents rather than on the epithelial cells to which they are attached, often in huge numbers (Fig. 20.7).

Their presence leads to damage to the epithelium, readily seen as a flattening of the villi which is associated with functional changes such as diarrhea, pain, and an intestinal malabsorption syndrome in predisposed individuals. Poorly nourished children in parts of the tropics and subtropics appear to be especially vulnerable. Most infections in well-nourished people, however, are probably asymptomatic (about 80% according to Manson-Bahr & Bell [17]) and are only recognized by the incidental finding of cysts in the feces. The immunologic responses of the host are reviewed in

Chapter 11.

Individual flagellates that pass to the distal part of the jejunum and ileum encyst and the nuclei undergo binary fission twice, producing a four-nucleated cyst which is passed in the feces. The cysts can survive in cold, even chlorinated water for up to 2 months but are killed by dessication, heat, and extreme cold. Sewage-contaminated water supplies are probably a common source of infection.

The diagnosis of giardiasis is generally made by identifying the characteristic trophozoites and/or cysts in the feces. In more difficult cases a duodenal capsule may be swallowed and the trophozoites sought on the enclosed cord. Immunodiagnostic methods include the detection of antigen in feces by a simple ELISA procedure [18]. Diagnostic DNA probes are being developed but are not yet generally available. Once diagnosed, giardiasis usually responds readily to treatment with metronidazole or related compounds, although some infections may prove refractory.

Order Trichomonadida

Trichomonas vaginalis. As Honigberg [13] pointed out, trichomoniasis is one of the most widespread, predominantly venereal diseases of humans, the clinical

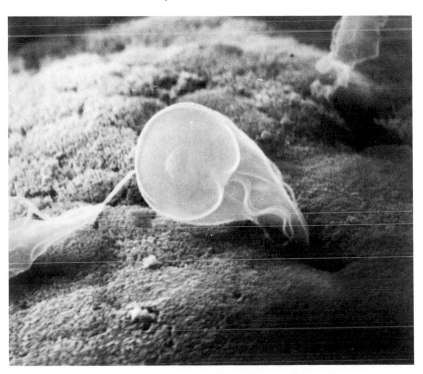

Fig. 20.7 Scanning electron micrograph of *Giardia lamblia* on the surface of a section of human ileum removed at biopsy. Original magnification, ×5300. (Courtesy of Dr G. Rapeport, Western Infirmary, Glasgow.)

importance of which, especially in the female genital tract, is often seriously underrated. It has been estimated that over 180 million people, predominantly women, are infected with this protozoan. The morphology of the flagellate as seen, for example, in smears of vaginal secretion or in culture (Fig. 20.8) varies considerably, as does the size, depending upon the manner of fixing and staining the preparation, as well as the angle from which it is viewed. This is not surprising since *T. vaginalis*, depicted in most publications as roughly ovoid parasites, are actually rather ameboid organisms that spread over surfaces such as the squamous epithelial cells of the vagina and urethra with the aid of pseudopodia which probably help to anchor them against the current. It is likely that the flagellates secrete exotoxins or digestive enzymes, as well as causing mechanical damage with the axostyles that help to attach them to the epithelial surfaces.

Infection is almost invariably by venereal contact, the male partner frequently harboring an asymptomatic infection of the genitourinary tract, especially in the urethra and prostate, or in the subprepucial sac. The flagellates cause edema and inflammation of the squamous epithelium, particularly of the vagina. They divide by simple binary fission and their actively moving flagellae are readily spotted in fresh wet smears of the abundant, foamy secretion that is associated with the vaginitis. They may also be found in centrifuged urine specimens from either sex and in urethral secretions or semen of infected males. The trophozoites are susceptible to most external environmental conditions as they do not encyst, so that transmission other than by direct contact must be very exceptional.

Since the immunologic aspects of trichomoniasis are not dealt with in this volume, the reader is referred to the review by Ackers [19]. Immunologic tests at present have little place in diagnosis. A number of investigators have been able to culture *T. vaginalis* from urine and semen of known male contacts but the small numbers of organisms in the male suggests that their infectivity to a susceptible female must be very high. Trichomonal vaginitis is very frequently associated with an abnormal vaginal flora in which *Candida albicans* may predominate, adding to the severity of the vaginitis and complicating the treatment.

At least a dozen serotypes of *T. vaginalis* have been

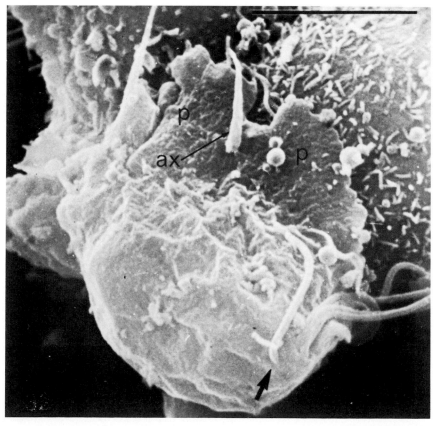

Fig. 20.8 Scanning electron micrograph of an ameboid stage of *Trichomonas vaginalis* moving over the surface of an epithelial cell in tissue culture. The axostyle (ax) is seen projecting from the posterior part of the parasite which has two broad, flat pseudopodia (p) extending over the epithelial cell. One of the anterior flagella (arrow) does not arise from the usual site, i.e., the flagellar pocket, in this specimen. (Bar = 5 μm). (Courtesy of Dr J.P. Heath, Strangeways Research Laboratory, Cambridge.)

identified but, so far, no correlation has been found between them and the level of pathogenicity of the organisms [20]. Antitrichomonal antibody was demonstrated in the blood and cervical or vaginal secretions of a proportion of infected women and men by complement fixation, direct agglutination, and the indirect immunofluorescent antibody test (IFAT), but the titers were variable and the results inconsistent. IgA antibodies to *T. vaginalis* were identified in the vaginal secretions but not in male contacts [21]. A newly developed enzyme immunoassay and a latex agglutination test have proved to be far more reliable [22]. It is still uncertain whether humoral or cell-mediated immunity is the most important in protection against *T. vaginalis* infection. There is still much to be learned also about the interrelationships of host susceptibility, intercurrent infection, parasite strain or, possibly, even parasite species, and pathogenicity. Although little use has been made, so far, of biochemical techniques for parasite characterization as compared, for example, with the protozoa causing amebiasis or leishmaniasis, Soliman *et al.* [23] identified five groups among as few as 32 strains of *T. vaginalis* on the basis of the electrophoretic characterization of only four isoenzymes.

Order Kinetoplastida: genus Trypanosoma

The protozoa in this order that parasitize humans, in the two genera *Trypanosoma* and *Leishmania*, unlike those discussed above, have evolved complex life-cycles to permit their transmission cyclically from vertebrate to vertebrate through the intermediary of an invertebrate vector. The parasites have evolved a specialized DNA-containing organelle, the kinetoplast from which the name of the order derives. This is intimately involved in the radical structural and functional modifications that the organisms undergo, not only when passing between vertebrate and invertebrate but also within each type of host. Within the mammalian host the parasites are either intracellular or live in the bloodstream or tissue fluids, moving from one site to another during the course of their complicated cycles. Reproduction is primarily by binary fission, which occurs in both types of host. Only recently has evidence been obtained to indicate that genetic exchange can take place in trypanosomes of the subgenus *Trypanozoon* [24,25] and in *Leishmania* [26], probably during the invertebrate cycle. The kinetoplast contains DNA which appears to control, among other things, cyclic metamorphoses of the mitochondrion and flagellum.

The pathogenic trypanosomes of medical and veterinary importance are summarized in Table 20.3. Two complexes in the subgenus *Trypanozoon* are responsible for sleeping sickness in West Africa and the more acute East African form, namely *Trypanosoma brucei gambiense* and *T. brucei rhodesiense*, *T. brucei brucei* itself being restricted to various domestic and wild animals [27]. The two species found in humans in the New World are *Trypanosoma (Schizotrypanum) cruzi*, which causes Chagas' disease, and *Trypanosoma (Herpetosoma) rangeli*, which is probably nonpathogenic.

Subgenus Trypanosoma (Trypanozoon): Trypanosoma brucei complex. On the basis, originally, of their infectivity to humans and/or animals, *T. brucei* was divided into three subspecies: *T. brucei brucei*, which is essentially an animal parasite, *T. brucei gambiense*, the causative agent of sleeping sickness in humans in West Africa, and *T. brucei rhodesiense*, which is a zoonosis acquired from game animals that produces a fulminating parasitemia in humans in East Africa. These subdivisions have been largely substantiated by extensive surveys of isolates from humans and animals, employing the biochemical characterization of trypanosome zymodemes by electrophoresis [28], while more recent work with DNA probes shows promise for the identification of the organisms both in their mammalian hosts and vectors [29]. A simplified life-cycle of these organisms is shown in Fig. 20.9. Trypanosomes possess a number of unique biochemical pathways, several of which are associated with specialized organelles. The glycosome, for example, contains a number of the enzymes required for the functioning of the glycolytic cycle [30]. *Trypanosoma brucei sensu lato* undergoes cyclic changes from a largely anaerobic type of respiration in the vertebrate to an aerobic mechanism in the invertebrate host, a switch that is associated with marked changes in the structure of the mitochondrion which, in turn, is the site of enzymes associated with anaerobic respiration. It has only recently been shown that the trypanosomes also possess a unique pathway associated with polyamine metabolism, based not on glutathione as is that in other eukaryotic cells, but on trypanothione. Such unique biochemical characteristics offer exceptional opportunities for the development of selectively targeted antitrypanosomal drugs [31] and one such compound, α-difluoromethylornithine, has already proved of value in the therapy of advanced *T. b. rhodesiense* [32] and *T. gambiense* infection in humans [33]. The African trypanosomes are transmitted from mammal to mammal

Table 20.3 Trypanosomes of medical and veterinary importance

	Genus (subgenus)	Species	Host species	Disease
In Africa				
Salivaria	*Trypanosoma (Duttonella)*	*vivax*	Antelopes, ruminants, equines, dogs	Souma
		uniforme	Antelopes, ruminants	(Pathogenic)
	Trypanosoma (Nannomonas)	*congolense*	Antelopes, ruminants, equines, pigs, dogs	(Pathogenic)
		simiae	Pigs, warthogs, camels	(Pathogenic)
	Trypanosoma (Trypanozoon)	*brucei brucei*	Antelopes, domestic mammals	Nagana
		brucei rhodesiense	Antelopes, humans	Sleeping sickness (acute form)
		brucei gambiense	Humans, pigs	Sleeping sickness (chronic form)
		evansi[*]	Bovines, equines, camels, dogs, etc.	Surra
		equiperdum[†]	Equines	Dourine
	Trypanosoma (Pycnomonas)	*suis*	Domestic and wild pigs	(Pathogenic)
In South America				
Salivaria	*Trypanosoma (Duttonella)*	*vivax*[‡]	Bovines	(Pathogenic)
	Trypanosoma (Herpetosoma)	*rangeli*[§]	Many wild animals, humans	(Nonpathogenic)
Stercoraria	*Trypanosoma (Schizotrypanum)*	*cruzi*[§]	Humans, armadillos, opposums, dogs, etc.	Chagas' disease

All species transmitted by tsetse flies except:
[*] by tabanid flies;
[†] by coitus;
[‡] by various biting flies;
[§] by reduviid bugs.

by tsetse flies, the riverine *Glossina palpalis* being the principal vector in West Africa and the savanna species *Glossina morsitans* being a major vector in East Africa. Slender (long and flat) trypomastigotes circulating in the blood and present in tissue fluid are picked up by the female flies when they take a blood meal. Within the fly the parasites move passively via the crop to the midgut inside the blood meal which, in turn, is kept in place within a short-lived sac secreted by the fly, the peritrophic membrane. Here, the parasites divide by binary fission to produce slender trypomastigotes which were believed to move posteriorly, skirting round the peritrophic membrane, passing between it and the midgut epithelium, and regaining the foregut (cardia). From there it was thought that they enter the hyopharynx and then move in a retrograde fashion along the salivary ducts into the salivary glands where they undergo a complex metamorphosis to the epimastigote form. Some midgut forms pass directly through the midgut epithelium to enter the hemocoele from which, presumably, they gain access directly to the salivary glands [34]. There they multiply as epimastigotes, eventually transforming yet again to the (metacyclic) trypomastigote stage. In this form they enter a new host when the fly takes a further meal (see caption to Fig. 20.9).

Once in the new host the trypomastigotes commence a further cycle of reproduction by binary fission in the tissue spaces, the accumulation of parasites and, perhaps, toxic byproducts resulting in the swelling known as a trypanosomal chancre. From this site trypomastigotes of various shapes (short and stumpy, long and slender, and intermediate) spill over into the circulation as well as being distributed throughout the tissues in the lymphatic circulation. While the East African parasites tend to produce massive parasitemia and toxemia, the West African form of infection tends to be more chronic, but finally trypomastigotes penetrate the brain where focal perivascular infiltration occurs

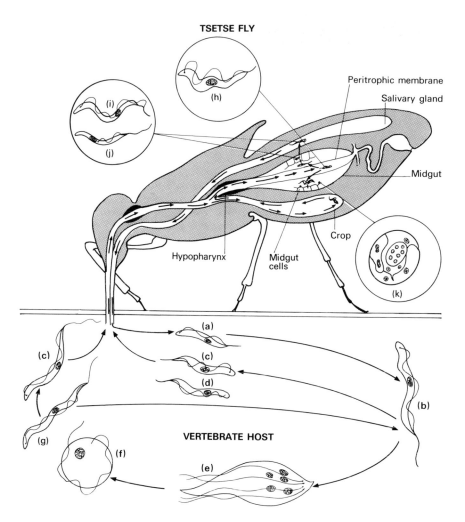

TSETSE FLY

Peritrophic membrane

Salivary gland

Midgut

Crop

Hypopharynx

Midgut cells

VERTEBRATE HOST

Fig. 20.9 Diagram of the life-cycle of *Trypanosoma brucei*. Metacyclic trypomastigotes (a) divide in tissues and blood to form long, flat trypomastigotes (b). These also divide in blood and extracellularly in tissues. (They are infective both to vertebrate and invertebrate hosts.) They mature to form stumpy (c) and posterior-nucleated forms (d). (These are degenerate and may not be infective to the tsetse fly.) Long, flat trypomastigotes (b) in the tissues may also form multiple division forms (e) and spheromastigotes (f). From the tissues very long, slender trypomastigotes (g) and spheromastigotes (f) emerge to form more long, flat forms (b). In the tsetse fly, blood containing forms (b), (c), (d), and (g) enters the crop where it remains for about 30 min before passing into the midgut within the peritrophic membrane. After 9–12 days the trypomastigotes, which have become broader (h), enter the ectoperitrophic space and then penetrate epithelial cells of the midgut wall. Development and division occur in these cells, in which giant cells (k) are also found. (It is possible that genetic exchange may occur just prior to or during this phase of parasite development.) Trypomastigotes (i) and epimastigotes (j) pass from the gut to enter the salivary glands, possibly by penetrating the hemocoel. Epimastigotes (j) transform to metacyclic trypomastigotes (a) in the salivary glands and are injected into a fresh vertebrate host when the fly next bites. (Courtesy of Dr W.E. Ormerod, London School of Hygiene and Tropical Medicine.)

around cerebral capillaries. In time, severe brain damage is caused by cerebral edema, punctate hemorrhages, and a generalized meningoencephalitis, leading to the familiar syndrome of sleeping sickness. There is evidence that the parasites undergo a further developmental cycle in the ependymal cells of the choroid plexus. Raseroka & Ormerod [35] have proposed that this is the site in which the organisms can evade the action of several kinds of antitrypanosomal drugs and from which they can later emerge to give rise to relapses. Eventually, slender trypomastigotes of the "long and flat" type are picked up by *Glossina* which thus perpetuate the cycle. The nature of the immune response of the mammalian host to invasion by the African trypanosomes and the intricate genetic devices that enable them to evade the host defense mechanisms are reviewed by Steinert & Pays [36] and in Chapter 12.

Subgenus Trypanosoma (Schizotrypanum): Trypanosoma cruzi. The life-cycle of *T. cruzi*, the causative agent of Chagas' disease and the only species of this subgenus that is pathogenic for humans, afflicting between 15 and 20 million people out of about 65 million in endemic areas of the New World, is entirely different from that of the African trypanosomes. This zoonosis is transmitted to humans through a number of species of large biting bugs, the Reduviidae, which because they frequently bite the face of sleeping people are known as "kissing bugs." The metacyclic trypomastigotes pass out in the feces of the bug onto the skin since the insect excretes as it feeds. The parasites gain entry into the site of the bite through the puncture wound, probably when the victim scratches. Within the skin the trypomastigotes enter tissue macrophages in which they are able to survive and multiply unscathed. Their entry may be facilitated by their initial binding to host cell fibronectin [4]. Trypomastigote multiplication at the initial site leads to the formation of a local swelling or trypanosomal chancre which, if near the eye, produces marked local edema, the so-called Romaña's sign. As the infection progresses (1–2 weeks), numerous trypomastigotes enter macrophages in different parts of the body, as well as heart muscle and unstriated muscle of the gastrointestinal tract. In the macrophages they are able to evade the normally protective action of lysosomal enzymes with which they become surrounded. They transform into amastigotes, which proceed to multiply by binary fission until the host macrophages are filled with small colonies of parasites. As the colony grows and the macrophage becomes disrupted the parasites are freed, some transforming back into trypomastigotes which invade the circulation and others entering fresh macrophages. Severe myocarditis in the acute phase of the disease may be fatal or it may settle down spontaneously, but the damage caused often leads to the chronic myocarditis of Chagas' disease in later life. Similarly, damage to intestinal smooth muscle and cells of the autonomic nervous system may lead to loss of muscular control of the organs and progressive, passive dilatation, the megasyndrome of Chagas' disease. In 83% of 230 sera from patients with Chagas' disease, Ribeiro dos Santos *et al*. [37] demonstrated antineuronal

antibodies which they considered may be responsible for some of these pathologic changes.

When trypomastigotes are picked up together with a blood meal by a reduviid bug, the organisms pass to the midgut where they transform into epimastigotes. After a number of replications in this form, new metacyclic trypanosomes develop in the rectal ampulla. These are discharged when the bug next feeds (Fig. 20.10). On penetrating the new mammalian host the trypomastigotes again reproduce at the site of the bite, later to invade the bloodstream and metastasize throughout the body. The reservoir hosts of *T. cruzi* are various wild animals, notably opossums and armadilloes, in which the parasites apparently do not induce the same pathologic changes as in humans. Moreover, it has been shown recently in at least one species of opossum, *Didelphis marsupialis*, that *T. cruzi* invades and multiplies within the lumen of the anal glands, not in the amastigote or trypomastigote form classically found within other vertebrate tissues but as the epimastigote, the stage normally seen within the invertebrate vector [39]. It has been suggested that the parasites reach the glands directly from the intestinal tract of animals that eat infected bugs. The discovery of this phase of the life cycle opens up a fascinating insight into the evolution of the kinetoplastids [40].

Remarkable regional differences have long been recognized in the severity and localization of the pathologic changes produced in humans by *T. cruzi*. Cardiopathies are, for example, particularly severe in Panama where megasyndromes are uncommon. In the south of Brazil, on the other hand, the latter is seen very frequently. This feature of Chagas' disease, as well as morphologic variations in the trypanosomes isolated from humans, wild animals, and Reduviid bugs, led to the suggestion that *T. cruzi* might indeed be a complex of sibling species, each with its own epidemiology, pathogenic potential, host range, etc. The application of biochemical taxonomic techniques has given some justification for these suspicions. Miles [41] in Brazil and Miles *et al*. [42] in Chile described several clearly defined zymodemes, at least three of which are pathogenic for humans. Restriction enzyme analysis of a wide range of isolates has broadly confirmed these groupings. Tibayrenc *et al*. [43] suggested that *T. cruzi* is a highly polymorphic diploid organism. To date, no clear correlation has been revealed between such taxonomic criteria and pathogenicity for humans.

Laboratory procedures for the diagnosis of Chagas' disease include the unusual technique of "xeno-

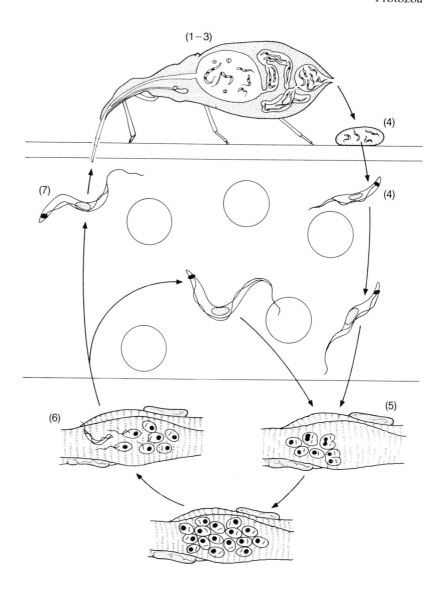

Fig. 20.10 Diagram of the life-cycle of *Trypanosoma cruzi* in the reduviid bug and in the vertebrate host. (1–4) represent the metamorphosis from ingested amastigotes to epimastigotes, then to metacyclic trypomastigotes in the bug. In (5) and (6) metacyclic trypomastigotes are shown converting to amastigotes in muscle, then converting back through epimastigotes to circulating trypomastigotes (7) in the blood. (From Geigy & Herbig [38].)

diagnosis.'' Uninfected Reduviid bugs are allowed to feed on patients with a suspected infection. Any organisms they take up multiply in the bugs' tissues in which they may subsequently be detected after a suitable interval. Serologic tests currently depend on the detection of antibodies, and specific *T. cruzi* antigens are being sought (see Chapter 13). A purified 60 kD surface glycoprotein of *T. cruzi* epimastigotes has proved to be a more sensitive antigen than whole promastigotes [44].

Order Kinetoplastida: genus Leishmania

The complex of visceral, cutaneous, and mucocutaneous disorders known as the leishmaniases, caused by para-sites belonging to the even more complex genus *Leishmania* of the Kinetoplastida, was reviewed in 1987 [45]. The classification (Table 20.4) and accurate identi-fication of leishmanial isolates are essential to the under-standing of the epidemiology, pathology, immunology, chemotherapy, or indeed any aspect of infections caused by these organisms. They have now been divided into two subgenera, *Leishmania sensu stricto* and *Viannia*, by Lainson & Shaw [46], primarily on the basis of their sites of reproduction in the insect vectors, which are small biting ''sandflies'' of the genera *Phlebotomus* in the Old World and *Lutzomyia* in the New World. Zymodeme characterization (see, for example, Rioux [47]), mono-clonal antibodies, and both kinetoplast and genomic

Table 20.4 Parasites of the genus *Leishmania* infecting humans

Subgenus	Species	Distribution	Disease*
Leishmania	*Leishmania major*	USSR to western India, Middle East, Africa north and south of Sahara, Sudan	CL, MCL
	Leishmania tropica	USSR to western India, Middle East, Kenya	CL, R
	Leishmania aethiopica	Ethiopia, Kenya	CL, DCL
	Leishmania donovani	India, China	KA, PKDL
	Leishmania infantum	Mediterranean basin, Middle East, USSR, China	KA, CL
	Leishmania donovani sensu lato	Sudan, Ethiopia, Saudi Arabia	KA
	Leishmania chagasi	South America	KA
	Leishmania mexicana	Central America	CL
	Leishmania pifanoi	Venezuela	CL, DCL
	Leishmania garnhami	Venezuela	CL
	Leishmania venezuelensis	Venezuela	CL
	Leishmania amazonensis	Brazil (Amazon basin), Trinidad	CL, DCL
	Leishmania (Leishmania) sp.	Dominican Republic	DCL
Viannia	*Leishmania braziliensis*	Brazil, Colombia, Peru, Paraguay, Ecuador, Bolivia	CL, Espundia
	Leishmania guyanensis	Guyanas, northern Brazil	Pian bois
	Leishmania panamensis	Panama	Pian bois
	Leishmania peruviana	Peruvian highlands	Uta
	Leishmania lainsoni	Brazil—Parà	CL

* CL, cutaneous leishmaniasis (Oriental sore, etc.); DCL, disseminated cutaneous leishmaniasis (leishmaniasis diffusa); KA, visceral leishmaniasis (kala-azar); MCL, mucocutaneous leishmaniasis (other than S. American Espundia); PKDL, post kala-azar dermal leishmaniasis; R, leishmaniasis recidivans.

DNA analysis [48] are revealing both homologies and disparities between the many species and subspecies of *Leishmania* and have recently indicated the presence of genetic exchange in this genus which has, until now, been considered to lack a sexual cycle [28]. Visceral leishmaniasis has taken on a new clinical dimension with the advent of AIDS, since numerous infections with the parasites are surfacing in immunodeprived subjects in whom the organisms were apparently present in a subclinical or cryptic condition. The cutaneous and mucocutaneous forms of leishmaniasis, most of which are zoonoses, also are increasing in parallel with ecologic changes associated with agricultural and other types of land development.

Leishmania possess the same general structure as other members of the Kinetoplastida, with two major stages in the life-cycle (Fig. 20.11): the amastigote (Fig. 20.12), which occurs in vertebrate macrophages; and the promastigote, which develops in the vector. The life-cycle is simpler than that of the African and South American trypanosomes.

Metacyclic promastigotes in the foregut of the sandfly enter the wound when the fly takes a blood meal. It is not yet certain whether they immediately transform

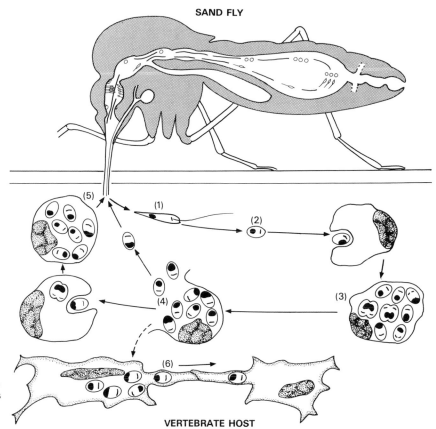

Fig. 20.11 Diagram of the life-cycle of *Leishmania* in the sandfly and in the vertebrate host. The lower part of the figure represents the promastigotes (1) injected by the fly, invading tissue macrophages in which the parasites metamorphose to amastigotes which subsequently divide by binary fission to form ''cell nests'' (2–5). In some cases (6), amastigotes may pass into fresh host cells during division of the latter. (From Geigy & Herbig [38].)

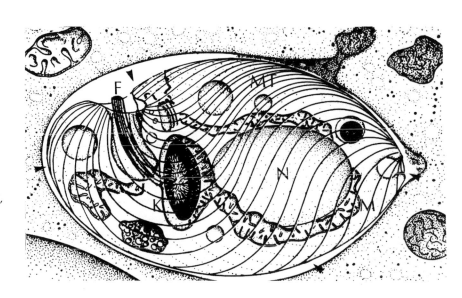

Fig. 20.12 Diagram to illustrate the morphology of an amastigote of *Leishmania* inside a parasitophorous vacuole of a macrophage. N, nucleus; K, kinetoplast; F, flagellum in the flagellar pocket; M, mitochondrion; MT, microtubules forming a cytoskeletal structure; arrows indicate the cavity of the parasitophorous vacuole. (From Gardener [49].)

into amastigotes under the influence of the warmer vertebrate milieu or whether they first enter macrophages with which they come into contact after their deposition in the tissues. Like *T. cruzi*, this seems to be facilitated by parasite binding to fibronectin on the surface of the host cells [4]. In either case, as amastigotes they grow and multiply by binary fission within the macrophages, forming small "cell nests." In time, the host macrophages become disrupted and the amastigotes are released and rapidly picked up by fresh macrophages in which the cycle is repeated. Certain species of *Leishmania* develop optimally at the temperature of the body surface and others at the slightly higher temperature of the internal organs, spleen, liver, lymph glands, or bone marrow. In either site of predilection the parasites occupy macrophages of the reticuloendothelial system.

It has classically been considered that cutaneous and mucocutaneous disease result from infection with *Leishmania tropica* and *Leishmania major* in the Old World and the *Leishmania mexicana* and *Leishmania braziliensis* complexes in the New World, visceral disease (kala-azar) being associated with *Leishmania donovani* in India and probably *Leishmania infantum sensu lato* elsewhere. However, the rigid distinction between dermatotropic and viscerotropic species of *Leishmania* no longer holds, several zymodemes of *L. infantum*, for example, having been shown to be responsible for cutaneous and, indeed, mucosal infections [50].

The remarkable manner in which *Leishmania* amastigotes survive within the macrophages, which is considered at length in Chapter 14, makes an interesting contrast to another intramacrophage protozoon, *Toxoplasma gondii*, which is discussed below. It is likely that they are protected from host enzymes in the parasitophorous vacuole by a special lipophosphoglycan [51]. When amastigotes in macrophages of the blood or skin are picked up by further sandflies, the parasites enter the abdominal midgut where they rapidly transform and reproduce in the promastigote stage. In the midgut they remain loosely attached by intertwining their flagellae through the microvilli lining the midgut epithelium (Fig. 20.13a). Parasites of the subgenus *Leishmania* migrate forwards to become attached to the chitinous lining of the esophageal valve by hemidesmosomes (Fig. 20.13b) and they become more rounded [52]. In the subgenus *Viannia* the parasites move posteriorly where they become attached in a similar manner to the hindgut. In either anterior or posterior site they undergo further asexual reproduction

and then finally move forward into the esophagus and the pharynx from where the small promastigotes can be pumped into a new host along with "salivary fluid" via the proboscis. In its simplest form the life-cycle is summarized in Fig. 20.11.

Although rather less is known of the biochemistry of *Leishmania* than of *Trypanozoon*, the data available indicate that in this genus, too, a cyclic switch in the form of the mitochondrion is paralleled by a cyclic switch from a largely anaerobic type of respiration in the vertebrate to aerobic respiration in the invertebrate host (see reviews by Chang & Bray [53] and Peters & Killick-Kendrick [45].

With the probable exception of *L. donovani* in India and *L. tropica* in a number of Old World countries, the leishmaniases of humans are zoonoses, the parasites normally developing in various wild reservoir hosts, dogs and wild canids being the major hosts for the visceralizing forms of *L. infantum sensu lato*. In the Mediterranean basin a high infection rate has been found in domestic dogs by serologic testing, e.g., 40% in the area of Marseilles [54] (see also review by Lainson & Shaw [46]).

PHYLUM CILIOPHORA (THE CILIATES)

The only species of ciliate parasitic in humans is *B. coli*, which normally inhabits the large intestine of the domestic pig. If humans are infected by consuming food contaminated with pig feces containing cysts, the trophozoites (Fig. 20.14), which may range from 30 to 300 μm in length and breadth, may invade the crypts and epithelium of the large intestine, causing ulceration not unlike that of amebiasis caused by *E. histolytica*, but stopping short of invasion of the muscular layers. A marked cellular reaction occurs round the ciliates, consisting mainly of lymphocytes and eosinophils. Balantidiasis is a rare infection in humans and tends to be focal in distribution. A good review was published by Zaman [55]. As distinct from most of the organisms discussed up to this point, the ciliates possess a sexual phase in their life-cycles.

PHYLUM APICOMPLEXA

Parasites of at least five families (Table 20.5) infect humans, species of two, genera *Plasmodium* and *Babesia*, being transmitted by invertebrate vectors. All the apicomplexan parasites possess complicated life-cycles, including a sexual phase. Malaria caused by four species of *Plasmodium* which exists in parts of the tropics and

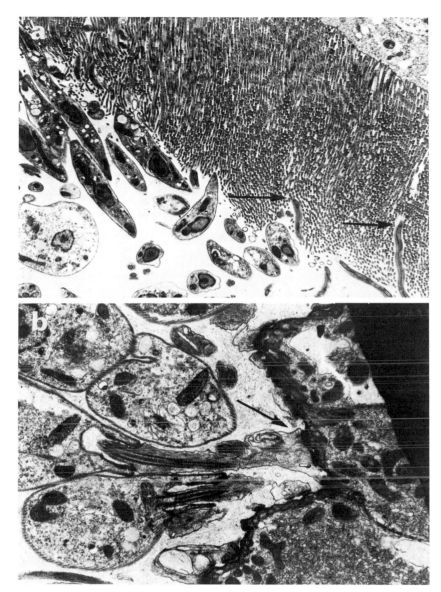

Fig. 20.13 Electron micrographs of promastigotes of *Leishmania amazonensis* in the abdominal midgut (a) and attached by hemidesmosomes to the esophageal valve (b) of the sandfly *Lutzomyia longipalpis*. In (a) the arrows indicate cross-sections of flagellae which are intertwined with the microvilli of the midgut epithelium. The arrow in (b) indicates the expanded flagellar tip of a promastigote forming a hemidesmosomal attachment to the chitin of the esophageal valve. Original magnification of a, ×9600; original magnification of b, ×15000. (Courtesy of Professor D.H. Molyneux.)

subtropics occupied by half the world's population must rank as the most serious parasitic disease of humankind in terms of morbidity and mortality. On the contrary, babesiosis is relatively rare in humans, although it is a major veterinary problem in many countries. Coccidian parasites, while probably very common in humans, are frequently unrecognized except when the immune status is compromised. Toxoplasmosis, for example, is a serious hazard to women of child-bearing age while cryptosporidiosis has only been recognized since its presence was found to cause intractable diarrhea in

patients with advanced AIDS in whom toxoplasmosis, too, is often a terminal complicating factor.

Subclass Coccidiasina, order Eucoccidiorida

Suborder Eimeriorina

Sarcocystosis. Included in this suborder are parasites of three families. In the Sarcocystidae, *T. gondii* has a remarkably wide host range which includes humans in

Table 20.5 The Apicomplexa of medical importance

Subclass, order, suborder	Family	Genus and species
Subclass Piroplasmasina	Babesiidae	*Babesia microti** *Babesia bovis* *Babesia divergens* *Babesia* spp.
	Theileriidae	*Theileria* spp.[†]

Subclass Coccidiasina
  Order Eucoccidiorida
    Suborder
    Haemospororina
	Plasmodiidae	*Plasmodium vivax*
		Plasmodium ovale
		Plasmodium malariae
		Plasmodium (Laverania) falciparum

Suborder Eimeriorina	Sarcocystidae	*Toxoplasma gondii* *Sarcocystis hominis*[‡] *Sarcocystis suihominis*[§] *Sarcocystis* spp.[¶]
	Eimeriidae	*Isospora belli*
	Cryptosporidiidae	*Cryptosporidium* spp.

* Rare in human host.
[†] Only of veterinary importance, not found in human host.
[‡] Man definitive, cattle intermediate host.
[§] Man definitive, pig intermediate host.
[¶] *Sarcocystis lindemanni* consists of a number of unidentified species for which humans are an intermediate host.

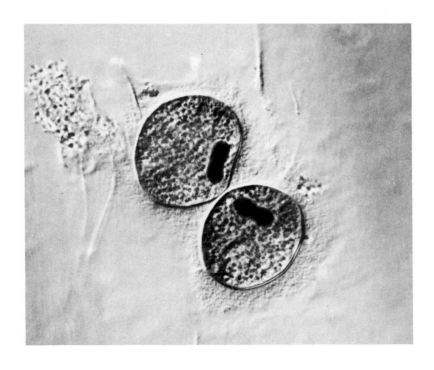

Fig. 20.14 Cysts of *Balantidium coli* in human feces (×380).

whom several of the asexual stages develop. A number of species of *Sarcocystis* in this family and *Isospora belli* within the family Eimeriidae are true parasites of humans, while in the latter family species of *Eimeria* parasitic in fish can cause spurious infections when their oocysts, released from the intestines of such small fish as sardines, pass unchanged through the human gut. The exact identity (or identities) of the species of *Cryptosporidium* that are now known to be common causes of transient diarrhea in otherwise normal children or severe diarrhea in immunocompromised individuals have not been finally established. Sarcocystosis in humans is usually cryptic, either occurring in the intestine where *Sarcocystis hominis* (cycling between humans and cattle) or *Sarcocystis suihominis* (cycling between humans and the pig) may be present, or as cysts in striated muscle. The latter, formerly called "*S. lindemanni*," are now considered to be dead-end stages of unidentified *Sarcocystis* species of other animals that have been consumed.

The life-cycle of *Sarcocystis* is illustrated by the example of *S. hominis* which circulates between humans, the definitive host, and cattle, the intermediate host (Fig. 20.15). The sarcocysts of this species occur in the skeletal and cardiac muscle of cattle where, unlike *Sarcocystis cruzi* (which circulates between the dog and cattle), they appear to elicit little systemic or local tissue response [57]. When infected meat is eaten either raw or half-cooked, the sarcocysts are digested and the merozoites (also called bradyzoites by some workers, e.g., Dubey [58] and Mehlhorn & Frenkel [59]) are released into the small intestine where they invade the epithelium to take up a subepithelial position in cells of the lamina propria. There they develop into male and female (micro- and macro-) gametes. Sexual reproduction then takes place with the formation of oocysts which, while migrating towards the gut lumen, sporulate *in situ* to form two sporocysts, each containing four sporozoites. These are shed slowly through the intestinal epithelium and passed into the feces where they are usually detected as free sporocysts (Fig. 20.16).

If sporocysts are ingested by cattle, sporozoites are

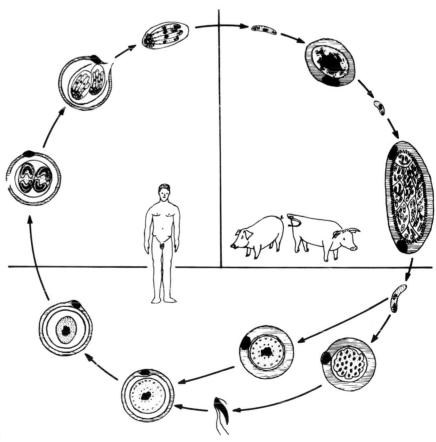

Fig. 20.15 Diagram of the life-cycle of *Sarcocystis hominis* and *Sarcocystis suihominis* in the definitive host (humans) and intermediate host (bovine and pig, respectively). (From Mehlhorn *et al.* [56].)

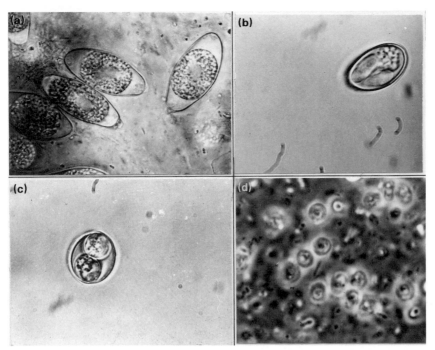

Fig. 20.16 Oocysts and sporocysts of coccidia of medical importance in unstained fecal preparations: (a) unsporulated oocysts of *Isospora belli*; (b) sporocyst of *Isospora belli*; (c) sporulated oocyst of *Toxoplasma gondii* in cat feces; (d) oocysts of *Cryptosporidium* species in human feces. Original magnification, all ×800. ((a–c) Courtesy of Professor J.J. Laarman and Dr W. Tadros; (d) courtesy of Mr J. Williams.)

released in the intestine from where they invade the circulation to enter reticuloendothelial cells of various tissues, especially blood vessels. There they undergo a stage of asexual schizogony, the daughter cells (merozoites) eventually becoming located in muscle where they form large well-defined sarcocysts with their large numbers of daughter merozoites by endodyogeny. Heydorn [60] found that the ingestion of raw *Sarcocystis*-infected beef gave rise to nausea, abdominal pain, and diarrhea within a few hours, with exacerbation of these symptoms about a fortnight later when there was peak production and shedding of sporocysts. Much more severe symptoms were produced by the consumption of raw pork containing the sarcocysts of *S. suihominis* [61]. However, the literature contains conflicting reports about the causal relationship between the ingestion at least of *S. hominis* and such symptoms. The infected individual may continue to shed sporocysts of *S. hominis* for up to 6 months, the parasites apparently eliciting little or no cellular reaction in the intestine. Nevertheless, Tadros & Laarman [62], by means of an IFAT, demonstrated antibodies to a *Sarcocystis* antigen in the sera of people who were shedding *Sarcocystis* sporozoites, the youngest being a 9-month-old baby. Infection in humans appears to be very common. Circulating, passively acquired antibodies were detected in infants of up to 3 months of age and no crossreactions were

observed with *T. gondii*. It is uncertain whether repeated infection induces protective immunity in normal individuals but, as implied above, severe symptoms may develop in sufferers from AIDS. The IFAT also detects sarcocyst infection in cattle and other domestic animals, so that the finding of a positive IFAT does not prove that a human infection with these organisms is necessarily limited to the intestine. Muscle infection with the 10 or more species of *Sarcocystis* already suspected to infect humans as an intermediate host [63] cannot be excluded but such infections are usually only found accidentally at autopsy. Cases diagnosed by muscle biopsy have been associated with systemic disturbances, such as myositis and fever, although histologic examination reveals little or no cellular reaction around the sarcocysts.

Toxoplasmosis. *Toxoplasma gondii* is the only protozoon parasite that appears to lack both tissue and host specificity. Human infection is usually acquired by the ingestion of bradyzoites from cysts present in raw or undercooked meat of various domestic animals. The definitive host is the cat, which sheds oocysts that become infective after they sporulate, with the production of two sporocysts each containing four sporozoites. These normally infect rodents or birds, which act as an intermediate host in which the parasites form, firstly, tachyzoites in the acute stage and then bradyzoites and cysts as the host develops cellular immunity [64]. Humans can also become infected by the accidental ingestion of infective sporocysts from cat feces. A particular hazard is the congenital transmission of toxoplasmosis to the fetus from a mother with acute infection. The life-cycle is summarized in Fig. 20.17.

Toxoplasmosis is very common in humans, serologic surveys having revealed that, for example, up to 50% of a randomly selected population had evidence of past infection [65]. Probably most acute infections are unrecognized and masquerade as transient febrile conditions resembling influenza, although at least one epidemic has been ascribed to infection from cats [66]. The awareness of the potential danger in pregnancy makes the diagnosis of acute toxoplasmosis important and the diagnostic procedures and immunologic aspects are dealt with elsewhere in this book (Chapter 15). With such a prevalent but generally cryptic infection, it is hardly surprising that severe immunosuppression such as that ocurring in late HIV infection often unmasks and unleashes acute toxoplasmosis which then becomes a

terminal parasitosis causing distressing symptoms in many patients with AIDS, commonly cerebral.

From a biologic viewpoint, *T. gondii* presents many intriguing aspects and problems. In its life-cycle the parasite undergoes a typical coccidian development in the epithelial cells of the small intestine of the cat in which the sexual phase and oocyst development occur. The oocysts may infect almost any species of animal and some birds in which they undergo extraintestinal development similar in some respects to that of *Sarcocystis*. The sporozoites liberated in the intestine invade first intestinal cells, in which they undergo endodyogeny to yield tachyzoites, and then a variety of host cells, in which they form small intracellular pseudocysts. In these develop further tachyzoites which, in turn, are disseminated to other tissues in which, as immunity develops, larger cystic forms containing large numbers of daughter cells (bradyzoites) are produced. These cysts are commonly found in the brain and muscles, especially of small mammals such as mice which are the normal prey of cats. Strangely enough, the cat itself can also become host to the extraintestinal stages. The bradyzoites are resistant to proteolytic enzymes of the cat's intestine so that, when cysts are ingested by the cat together with the brain or other tissues of mice and other animals, the parasites are released in the cat's small intestine. There, several asexual reproductive cycles take place, after which gametocytes and oocysts (Fig. 20.16) form. These are liberated into the intestinal lumen by rupture of the host cells. Some tachyzoites enter the lamina propria, where they form pseudocysts, and subsequently other tissues, where cysts are formed in the cat as cellular immunity develops so that the cat, therefore, can serve both as definitive and intermediate host.

As noted earlier, *T. gondii* tachyzoites develop intracellularly. In contrast to *Leishmania*, another intracellular protozoon (see p. 544), they have evolved a mechanism whereby they can survive the destructive action of host cell enzymes by preventing the fusion of host cell lysosomes with the parasitophorous vacuole in which they develop [67]. This interesting phenomenon is discussed in Chapter 15. Nevertheless, the presence of *T. gondii* in the intermediate host is associated with a marked humoral and especially cellular response which is manifested by fever, lymphadenopathy, and the production of immune IgG. These responses, however, fail to destroy the tissue cysts which can survive in chronically infected and (depending on the number and location of the cysts) often asymptomatic hosts for many

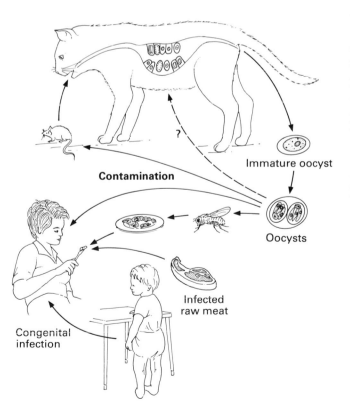

Fig. 20.17 Simplified diagram to show the life-cycle and routes of transmission of *Toxoplasma gondii*.

years, perhaps even the lifespan of small mammals such as rodents which thus remain infective to predatory cells.

Isosporosis. Unlike parasites of the genus *Sarcocystis*, which has an obligatory two-host cycle, *Isospora* is, at least in some species, apparently direct in a single host, passing from individual to individual through contamination with infective sporocysts. One species occurs in humans, *Isospora belli*. The large oocysts (Fig. 20.16) contain two sporocysts, each with four sporozoites. Oocysts are usually shed unsporulated in the feces. When sporulated oocysts are ingested, the sporozoites emerge to penetrate the epithelial cells of the crypts and villi of the upper small intestine and occasionally the lamina propria. There they undergo asexual schizogony, the merozoites reinvading new epithelial cells. Gametogony also occurs with the formation of oocysts which are then released into the gut lumen and shed in the feces.

Infection with *I. belli* can cause acute symptoms including fever, diarrhea, and, in more chronic cases, steatorrhea related to the flattening of the intestinal mucosa resembling that seen in sprue. A second species, *Isospora natalensis*, has been reported from two patients in South Africa. Because of its rarity, little immunologic information has been obtained on infection with *I. belli* but it, like other coccidian infections, can become a serious infection in immunocompromised subjects.

Cryptosporidiosis. While it is possible that several species in the genus *Cryptosporidium* (family Cryptosporidiidae) infect humans, the members of this genus, like *T. gondii*, appear to be able to infect a broad range of host species and, at present, only four species are accepted as valid [68]. *Cryptosporidium muris*, which was originally found in rodents, also infects other mammals, including cattle, and is considered to be the species occurring in humans. The presence of the oocysts (Fig. 20.16) of this coccidian in human feces was generally missed because of their minute size and only the introduction of simple staining techniques (e.g., the Ziehl–Neelsen acid-fast technique or rhodamine fluorescent staining) in the last few years has enabled a realistic assessment of their prevalence rates to be made [69]. Standard formol-ether or zinc sulfate concentration techniques are unsatisfactory and should be replaced by sugar flotation. Serologic techniques are being developed (see review by Cook [70]). The clinical significance of cryptosporidiosis was first highlighted by the detection of massive infections,

not only in the gut but also lining the pharynx of patients with AIDS and other immunocompromising conditions. It is now evident that *Cryptosporidium* may be present in the intestine of as many as 5% of normal children but in a far higher proportion of those with transient self-limiting diarrhea. Stehr-Green *et al.* [71], detected oocysts in the feces of one-third (28/84) of the children and one-quarter (4/18) of staff members during an outbreak of diarrhea in a daycare center in Florida. Oocysts may continue to be shed by such individuals for up to 2 months after the symptoms cease.

The life-cycle of *Cryptosporidium* differs in several aspects from the coccidian cycles discussed above. In the definitive host the parasites are located beneath the limiting membrane of epithelial cells lining the surface usually of some part of the gastrointestinal tract but occasionally, in severely immunocompromised individuals, the pharynx. The parasites which are extra-cytoplasmic (Fig. 20.18) undergo two cycles of asexual schizogony, the first producing six or eight merozoites and the second four. The latter form macro- and micro-gametes, then gametocytes (probably 16 for each microgamete), which fuse to give rise to zygotes. From these may develop thick-walled oocysts that are shed into the lumen of the gut and are passed in the feces. They are only 2–5 µm in diameter and contain four sporozoites. Infection of a new host (usually by feco-oral contamination) occurs when these oocysts are swallowed, the sporozoites being released in the intestine to penetrate beneath the surface membrane of new host cells. Other zygotes appear to produce thin-walled oocysts which rupture before being shed, thus releasing their sporozoites to cause autoinfection in the host. The incubation period is probably about 5 days.

Cryptosporidiosis is particularly serious in immuno-compromised individuals since there is, at present, no effective form of chemotherapy to control the infection (see review by Tzipori [72]). Patients with AIDS may suffer intractable, watery diarrhea with abdominal pain.

Suborder Haemospororina (malaria parasites)

Plasmodium species of humans. The family Plasmodiidae contains three species that are widely infective to humans in the subgenus *Plasmodium*, namely *Plasmodium vivax* and *Plasmodium ovale* which produce two hepatic stages, and *Plasmodium malariae* which produces only one, as well as one species in the subgenus *Laverania*, namely *Plasmodium falciparum*. Humans can also be infected by

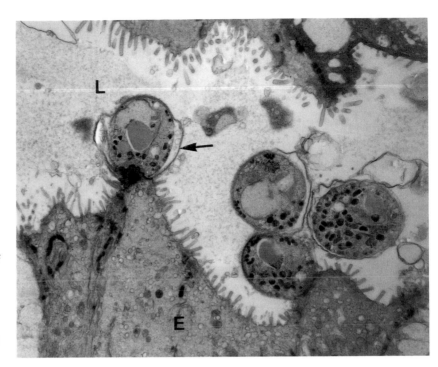

Fig. 20.18 Electron micrograph of *Cryptosporidium* in a rectal biopsy from a patient with AIDS. Note the membrane of host cell origin (arrow) that surrounds a developing oocyst which is attached by a convoluted structure to a rectal epithelial cell (E). L, lumen of rectum. Original magnification, ×4500. (Courtesy of Dr D.S. Ellis, London School of Hygiene and Tropical Medicine.)

the simian parasites *Plasmodium cynomolgi bastianellii*, *Plasmodium knowlesi*, and *Plasmodium simium*. Although human malaria parasites were first described in 1880 by Laveran, their transmission through *Anopheles* mosquitoes was only elucidated in 1897 by Ross (see Garnham [73]). The complete life-cycles of the four parasites have still not been completely unraveled, although techniques for the continuous cultivation of the intraerythrocytic stages first described by Trager & Jensen [74] and Haynes *et al.* [75] have greatly facilitated such studies. More recently the intrahepatic stages of rodent and human malaria parasites have been cultured *in vitro* (see review by Peters [76]). The generalized life-cycle of the malaria parasites is shown in Fig. 20.19.

Plasmodium falciparum is the organism responsible for malignant tertian fever. Humans are infected with this, as with the other species of *Plasmodium*, by the inoculation of sporozoites from the salivary glands of female anopheline mosquitoes when they take a blood meal. The sporozoites circulate passively in the peripheral circulation for about 30 min until they enter, by a process still incompletely defined (see review by Meis & Verhave [77]), into hepatocytes in which they undergo asexual reproduction to form large (50–60 μm diameter) intracellular schizonts (the pre-erythrocytic schizonts). These contain of the order of 30 000 merozoites, each about 0.7 μm in diameter,

which mature in 5–6 days. As they mature the schizonts appear to stimulate no cellular reaction on the part of the host but, as they rupture releasing the merozoites into the liver sinusoids, phagocytic cells invade the site and remove residual material. The merozoites invade erythrocytes with which they come into contact by a complex process, the nature of which has only been demonstrated in recent years [78]. The process involves, firstly, random contact, followed by orientation so that the apical complex of the merozoite is directed towards the host cell membrane. The parasite then inserts into this membrane a histidine-rich substance produced in the rhoptries that changes the physical characteristics of the red cell surface, thus facilitating the invagination by the parasite of its host cell [79] with the aid of a "moving junction," pushing a layer of host membrane ahead of it and, finally, the sealing off of the point of entry and fusion of host and parasite membranes (Fig. 20.20). Calmodulin also appears to play a role in erythrocyte invasion [81].

Within the parasitophorous vacuole thus formed, some of the organelles of the young trophozoite degenerate leaving the familiar intracellular trophozoite which proceeds to grow and later undergo asexual schizogony. At least three different receptors may be involved in the attachment of merozoites to the erythrocyte surface (see Hadley & Miller [82] and

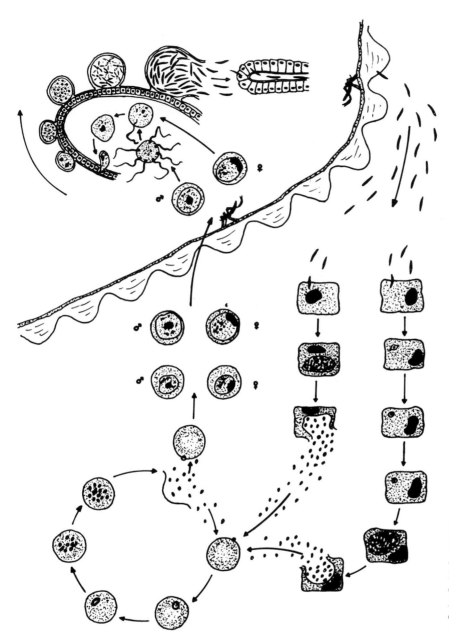

Fig. 20.19 Diagram of the life-cycle of *Plasmodium vivax*. Note that relapses due to the production of secondary exoerythrocytic schizonts from hypnozoites in liver parenchymal cells occur only in *Plasmodium vivax* and *Plasmodium ovale*. This cycle does not occur in *Plasmodium falciparum* or *Plasmodium malariae*.

Chapter 16 for a more detailed account of the molecular mechanisms involved).

The trophozoite feeds on the stroma of the host cell which is ingested by means of a newly formed organelle, the cytostome, through which small portions of stroma are segregated in phagocytic vesicles. Inside these vesicles the hemoglobin is digested, providing some of the amino acid requirements of the parasite and leaving an insoluble pigment, hemozoin. Early in its develop-

ment the trophozoite induces changes in the surface of the erythrocyte, including the formation of fine knob-like protrusions (Fig. 20.21). These contain a histidinerich protein [83] which has affinities with the histidinerich material (HRP I) found in the rhoptries of the developing schizont and its contained merozoites [84,85]. At the same time, the red cells containing trophozoites become sequestered in capillaries of the deep circulation, including the brain, where they tend to

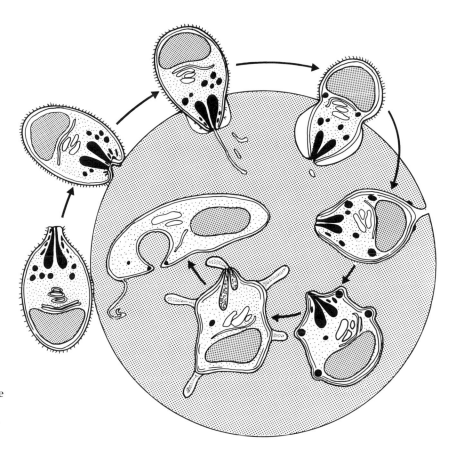

Fig. 20.20 Diagram to show the sequence of events from invasion of *Plasmodium* to the formation of the asexual trophozoite. (From Bannister *et al.* [80].)

remain during the remainder of the schizogonic cycle. It has been suggested that this is a survival mechanism that *P. falciparum* has evolved to avoid being picked out by macrophages during the passage of its host cells through the spleen [86]. When the schizonts mature, the host cells rupture releasing some 16 merozoites, the hemozoin granules which by then have coalesced into a residual body, together with other residual parasite and host cell debris. Schizont rupture is associated with fever, probably due to the release of pyrogenic material, some of which is certainly antigenic. Asexual schizogony is associated both with the development of antiparasitic immunity (humoral and cellular) and antitoxic antibody production (see Chapter 16). The sequestration of *P. falciparum* in capillaries of the brain results in damage to the endothelium and a sequence of pathologic changes, possibly involving the production of cytotoxines by host macrophages [87], which can culminate in the host's death. Parasite sequestration in other organs, e.g., the placenta, may also have serious consequences. The heavy asexual parasitemia that develops in falciparum malaria brings in its train

massive hemolysis and anemia which can result in peripheral vascular collapse and death, hence the term "malignant malaria" (see reviews by Boonpucknavig *et al.* [88] and White [89].

Liberated merozoites rapidly invade fresh host erythrocytes in the general circulation, thus continuing the cycle. Triggered by an as yet undefined stimulus, some merozoites develop into gametocytes which mature in approximately 10 days. The crescent-shaped gametocytes survive for 11–12 days but have an infective "half-life" of only 2–3 days [90]. Although much information has accumulated on the biochemistry of the asexual erythrocytic stages (see Sherman [91]), little is known about the gametocytes and other stages in the life-cycle.

When gametocytes are taken into the mosquito stomach together with a blood meal, the male and female gametocytes undergo a rapid series of transformations, first breaking loose from the enveloping host cell. The male (micro-) gametocyte becomes rounded, then forms eight uniflagellated microgametes. Fertilization occurs with the formation of a motile

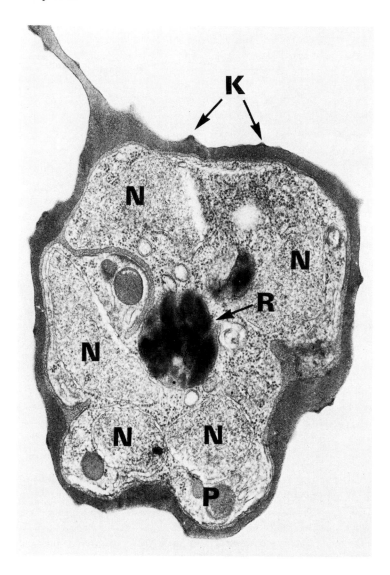

Fig. 20.21 Electron micrograph of an intraerythrocytic schizont of *Plasmodium falciparum*. N, nuclei of the developing merozoites; P, paired organelles (rhoptries); R, hemozoin granules in the residual body; K, knob-like protrusions on the surface of the host erythrocyte. Original magnification, ×26 000. (Courtesy of Professor R.E. Sinden, Imperial College, London)

zygote known as the ookinete. This passes through the mosquito midgut to become established under the peripheral lining membrane of the gut where asexual reproduction results in the formation of oocysts containing several hundreds of sporozoites. This process takes 10 days or longer, depending upon the ambient temperature. On maturing, the oocysts rupture releasing sporozoites into the body cavity of the mosquito from where they migrate towards the salivary glands which they penetrate. They undergo a short period of maturation within the acinar cells of the salivary glands and then enter the salivary duct where they remain until injected into a new host.

Plasmodium vivax and *P. ovale* differ from *P. falciparum* in the life-cycle as well as in their morphology and physiology. Both have a 48-h cycle in the intra-erythrocytic phase, and infection is thus associated with a tertian fever. (The cycle is also 48 h in *P. falciparum* but the presence of asynchronous crops of parasites leads often to an irregular pattern of fever.) In both species two types of sporozoites are produced, one that establishes a normal pre-erythrocytic schizogony in hepatocytes and the other that gives rise to dormant unicellular parasites, now called "hypnozoites." These do not develop in the early stages of infection but, after several months or years, commence asexual schizogony which terminates in the development of periodic relapses of parasitemia. Vivax and ovale malaria, therefore, are characterized by their chronicity unless the infections are treated with a hypnozoitocidal drug (primaquine being the only one currently available). Hypnozoites, which were first described in the simian relapsing parasite *P. cynomolgi* by Krotoski *et al.* [92], have since been observed in *P. vivax* both in chimpanzees and in tissue culture [93]. What processes determine when a batch of hypnozoites will mature are completely unknown at present, although one must assume that they involve host immune factors and, perhaps, changes in the parasites' antigenicity. After a period of about 8 days of pre-erythrocytic schizogony, merozoites of *P. vivax* and *P. ovale* are released to enter erythrocytes and commence the asexual erythrocytic cycle and gametocytogenesis. Vivax malaria is peculiar in that the merozoites can only penetrate immature erythrocytes and then only ones on which the Duffy factor is present, the geographic distribution of this infection thus being related to that of the Duffy antigen. *Plasmodium vivax* is, for example, very limited in West Africa where a high proportion of the indigenous population are Duffy negative. Sequestration of infected erythrocytes is much less marked in these infections than in *P. falciparum*. Moreover, gametocytogenesis commences early in the infection so that all stages of the intraerythrocytic parasites may be seen in the peripheral circulation at any time. The gametocytes undergo a cycle that leads to maximum infectivity to mosquitoes about 36 h after their formation, usually around midnight, after which their

infectivity rapidly declines [94]. Relapses from *P. vivax* usually cease after about 3 years in untreated individuals, whereas those from *P. ovale* may continue for decades.

Plasmodium malariae develops more slowly than the other species in all stages. The 72-h cycle of asexual erythrocytic schizogony results in a quartan fever which is relatively benign but which is sometimes associated with severe immunopathologic changes in the kidney or liver (see Chapter 4). Pre-erythrocytic schizogony occupies about 2 weeks and the schizonts contain less than 2000 merozoites. The intraerythrocytic schizonts produce only about eight merozoites, in contrast to *P. falciparum* and *P. vivax* (about 16), but the same as *P. ovale*. Although infection with *P. malariae*, if untreated, may persist for decades, it is no longer believed that this is due to the existence of a secondary exoerythrocytic stage. Garnham [95] suggested that it may be due to the persistence of the occasional intraerythrocytic asexual forms that manage to evade host immune attack. If this is the case, the mechanism of this evasion, by antigenic variation or by sequestration in an immunologically privileged site, for example, remains to be elucidated. It is noteworthy that *P. malariae*, like *P. falciparum*, also induces a type of "knob" formation on its host erythrocytes. The characters of the four malaria parasites of humans are reviewed and compared in Table 20.6.

Table 20.6 Comparative characters of malaria parasites of humans (stained thin smears)

Stage or period	*Plasmodium vivax*	*Plasmodium ovale*	*Plasmodium malariae*	*Plasmodium falciparum*
Ring	Relatively large; usually one prominent chromatin dot, sometimes two; often two rings, sometimes more, in one cell	Compact; one chromatin dot; double infection uncommon	Compact; one chromatin dot; double infection uncommon	Small, delicate; sometimes two chromatin dots; multiple red cell infection common; appliqué forms
Large trophozoite	Large; markedly ameboid; abundant chromatin; prominent vacuole; pigment in fine rodlets	Small; compact; vacuole inconspicuous; pigment coarse	Small; compact; often band-shaped; vacuole inconspicuous; pigment coarse	Medium size; usually compact; vacuole inconspicuous; rare in peripheral blood after half-grown; pigment granular
Young schizont (presegmenter)	Large; somewhat ameboid; dividing chromatin masses numerous; pigment in fine rodlets	Medium size; compact; chromatin masses few; pigment coarse	Small; compact; chromatin masses few; pigment coarse	Small; compact; chromatin masses numerous; single pigment mass; rare in peripheral blood
Mature schizont (segmenter)	Schizonts and merozoites large; pigment coalescent	Merozoites larger than in *P. malariae*; irregular rosette	Schizonts smaller but merozoites larger; forms rosette	Smaller merozoites; single pigment mass

Continued on p. 556

Table 20.6 *Continued*

Stage or period	*Plasmodium vivax*	*Plasmodium ovale*	*Plasmodium malariae*	*Plasmodium falciparum*
Number of merozoites	12–24, usually 12–18	6–12, usually 8	6–12, usually 8	8–26, usually 8–18
Microgametocytes (usually smaller and less numerous than macrogametocytes)	Spherical; compact; no vacuole; undivided chromatin; diffuse coarse pigment; cytoplasm stains light	Similar to *P. vivax* but somewhat smaller; never abundant	Similar to *P. vivax* but smaller and less numerous	Crescents usually sausage-shaped; chromatin diffuse; pigment scattered large grains; nucleus rather large; cytoplasm stains paler blue
Macrogametocytes	Spherical; compact; larger than microgametocyte; smaller nucleus; pigment stains darker blue	Similar to *P. vivax* but somewhat smaller; never abundant	Similar to *P. vivax* but smaller and less numerous	Crescents often longer and more slender; chromatin central; pigment more compact; nucleus compact; cytoplasm stains darker blue
Alterations in the infected red cell	Enlarged and decolorized; Schüffner's dots usually seen	Enlarged; decolorized; prominent Schüffner's dots appear early; infected cells may be oval-shaped with fimbriated ends	Cell may seem smaller; fine stippling (Ziemann's dots) occasionally seen	Normal size but may have "brassy" appearance; Maurer's dots (or "clefts") common; Garnham's bodies occasionally seen
Length of asexual stage (h)	48	49–50	72	48
Prepatent period, minimal (days)	8	9	14, average 28	5, average 8–12
Usual incubation period (days)	8–31, average 14	11–16	28–37, average 30	7–27, average 12
Interval between parasite patency and gametocyte appearance (days)	3–5	5–6; appearance irregular and numbers few	10–14; appearance irregular and numbers few	8–11
Development period in mosquito	10 days at 25°C	16 days at 25°C	25–28 days at 22–24°C	10–12 days at 27°C

The classic method of diagnosing malaria by searching for parasites on Romanowsky-stained thick and thin blood films is being supplanted in epidemiologic surveys by the application of highly sensitive DNA probes which can be interpreted by enzyme or radiometric methods. These are still less sensitive than direct examination by a skilled microscopist, but less time consuming [96].

Nonhuman Plasmodium species. In his classic monograph on the malaria parasites, Garnham [73] described about 130 species in the genus *Plasmodium*, and since then several others have been described. Three groups have been widely used in immunologic studies, namely certain species of *Plasmodium* occurring in birds, rodents, and monkeys [97]. They are listed in Table 20.7, which summarizes and compares their main biologic characters. The life-cycle of avian parasites differs greatly from that of mammalian species since pre- and secondary exoerythrocytic schizogony takes place in cells of the reticuloendothelial system, and these can be reinvaded by parasites from the intraerythrocytic phase of the cycle (Fig. 20.22).

The most widely employed avian malaria parasite is *Plasmodium gallinaceum*, a parasite first described by Brumpt in 1935 [98] and isolated by him from a chicken in Ceylon in 1958. This strain, which proved a valuable tool for immunologic research from that time, has fallen

Table 20.7 Comparison of biologic characters of some commonly used nonhuman *Plasmodium* species

Species	Usual laboratory host(s)	Minimum duration of PE cycle (hours)	Site of PE cycle	Duration of E cycle (h)	Site of 2ᵉ EE cycle(s)	Usual laboratory vector(s)
Plasmodium gallinaceum	Chick	72–75	Skin macrophages and fibroblasts (two generations)	36	RES; capillary endothelium, especially in brain	Aedes aegypti
Plasmodium lophurae	Duckling	<5 days (? 60 h)		? 24		
Plasmodium fallax	Turkey	40–42		?	RES; fibroblasts of marrow, liver, spleen, etc.	Culex pipiens
Plasmodium relictum	Canary	65		36		
Plasmodium cathemerium		72		24		
Plasmodium berghei	Mouse, rat, hamster	50	Hepatocyte	24	Nil	Anopheles stephensi
Plasmodium yoelii		43				
Plasmodium vinckei	Mouse	61				
Plasmodium chabaudi		52				
Plasmodium knowlesi	Rhesus	5 days	Hepatocyte	24	Nil	Anopheles spp.
Plasmodium cynomolgi		8 days		48	Hepatocyte	
Plasmodium inui		? 10 days		72	Nil	

E, erythrocytic; EE, exo-erythrocytic; PE, pre-erythrocytic; RES reticulo-endothelial system.

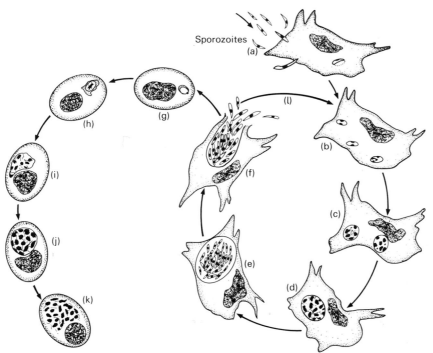

Fig. 20.22 Diagram of the life-cycle of the avian malaria parasite *Plasmodium gallinaceum*. a, sporozoites; b–f, development of exoerythrocytic schizonts in fixed tissue macrophages; g, erythrocyte invaded by cryptozoite; h–k, development of intraerythrocytic schizonts. Note that cryptozoites may reinvade tissue macrophages (l) or endothelial cells in which secondary exoerythrocytic schizogony takes place, with subsequent reinvasion of erythrocytes by metacryptozoites. Reinvasion of reticuloendothelial cells occurs from merozoites that originate in intraerythrocytic schizonts. (From Wright *et al.* [104].)

out of favor with the availability of more readily manipulated rodent species. Its life-cycle is illustrated in Fig. 20.22. The parasite is readily transmitted by *Aedes aegypti*, which itself can easily be reared in large numbers in the laboratory. The life-cycle of *P. gallinaceum* in the vector is essentially the same as that of human malaria parasites.

The discovery of *Plasmodium berghei* in African thicket rats by Vincke & Lips in 1948 [99] opened the way to an immense volume of immunologic and other research by providing a readily manageable and economic mammalian model. Subsequent discoveries of other rodent malaria parasites have extended the range so that they now provide the most widely used of all mammalian malaria models [96,100]. Rodent parasites differ from the human malaria parasites in having a very rapid pre-erythrocytic schizogony which varies from about 48–72 h, depending upon the parasite species and strain and the host species. Under natural conditions, two types of pre-erythrocytic schizogony are found in *Plasmodium yoelii* in its wild host: a normal, rapidly developing schizont (Fig. 20.23); and a more slowly developing form that Landau & Boulard [101] suggested was responsible for relapses of parasitemia in this infection. I. Landau (personal communication) now considers that the slow form can be compared to the

hypnozoite of the primate relapsing malarias. The immunologic aspects of rodent malaria were reviewed by Nussenzweig *et al.* [102].

One of the interesting aspects of the immune response to malaria in the vertebrate host is that it appears to be stage-specific (see Chapter 16). Although the pre-erythrocytic stage in the hepatocytes does not appear to produce a cellular response, at least until the schizonts are mature, Guérin-Marchand *et al.* [103] have produced evidence showing that a specific liver-stage antigen is produced by *P. falciparum*. This is not surprising when one considers their intimate intracellular location (Fig. 20.22) and the likelihood that they secrete enzymes or other products into their host cells [105].

Two species of simian malaria have been widely utilized in immunology research: *P. knowlesi* with a 24-h asexual erythrocytic cycle and *P. cynomolgi* with a 48-h cycle. Both parasites infect rhesus monkeys, the first producing a fulminating infection that leads rapidly to the death of untreated animals [106] and the latter a benign, relapsing type of malaria with secondary exoerythrocytic schizogony due to the presence of hypnozoites [107]. For the past two decades the South American Owl monkey "*Aotus trivirgatus*" (now known to consist of a complex of species in this genus) and a squirrel monkey, *Saimiri sciureus*, have been used

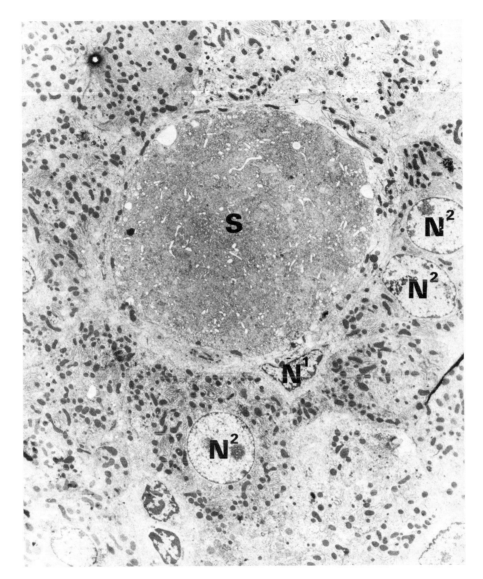

Fig. 20.23 Electron micrograph of a 43-hour-old, pre-erythrocytic schizont of *Plasmodium yoelii* in a hepatocyte of an albino rat. S, schizont; N[1], nucleus of host hepatocyte; N[2], nuclei of uninfected, neighboring hepatocytes. Original magnification, ×2400. (Courtesy of Professor C. Seureau & Dr A. Szöllösi, 1981.)

increasingly for studies on *P. falciparum* and *P. vivax*. Unlike the situation in its human host, *P. falciparum* does not sequester in these simian hosts and produces a massive parasitemia, with all developmental erythrocytic stages appearing in the peripheral circulation. While *P. vivax* produces a benign, self-limiting infection, hypnozoite formation has not yet been reported to follow sporozoite-induced infections in these monkeys.

Subclass Piroplasmasina

Genus Babesia

The genus *Babesia*, which is widely distributed throughout the world, contains a number of species of major veterinary importance and a small number that can produce zoonotic infection in humans. The infection that they produce in some hosts may be chronic

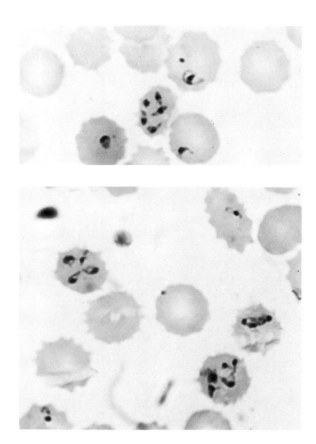

Fig. 20.24 *Babesia bigemina* in the blood of a splenectomized man who died with overwhelming parasitemia. Original magnification, ×1200. (Courtesy of Dr I.A. Cook, Raigmore Hospital, Inverness.)

and apparently cause little damage to the host (e.g., *Babesia microti* in voles), whereas others give rise to life-threatening, hemolytic disease resembling malaria, e.g., *Babesia canis* in domestic dogs. Unlike *Plasmodium*, *Babesia* are transmitted by ticks wherein the parasites undergo a complicated life-cycle which includes a sexual phase. On rare occasions such fulminating infections, which are usually caused by species of *Babesia* that are enzootic in cattle or wild ruminants (Fig. 20.24), are reported from people who have, for other reasons, previously been splenectomized [108]. Several series of nonfatal cases of *Babesia* infection in immunologically competent patients were reported from Nantucket Island on the eastern seaboard of the USA from 1970 onwards [109]. These infections were attributed to *B. microti* which, while normally a parasite of voles and small rodents, is transmitted to humans by the nymph of the tick *Ixodes dammini* which feeds indiscriminately on rodents, deer, or humans. Two other infections with

piroplasms in immunologically compromised individuals have been attributed to the genus *Entopolypoides*, a *Babesia*-like intraerythrocytic parasite formerly recognized only in monkeys [110].

A further focus of human babesiosis was reported by Osorno *et al.* [111] from a locality on the Gulf coast of Mexico where 30% of human serum samples proved to be seropositive. *Babesia* were isolated in hamsters injected with blood from three people, all of whom were healthy and asymptomatic. It seems very likely that a number of species of *Babesia* may be widespread in the form of a cryptic zoonosis. However, some crossreaction serologically with *P. falciparum* may cause confusion [108] (see Chapter 17).

Infection with *Babesia bigemina* and *Babesia bovis* can cause serious disease in domestic cattle, particularly when *Babesia*-free stock are introduced into enzootic areas [112]. One to two weeks after cattle are bitten by infected ticks parasites may be detected in blood smears. *Babesia bovis*-infected erythrocytes, like those containing. *P. falciparum*, become sequestered in capillaries of the brain and other organs, especially the lung, which often leads, in addition to fever, hemoglobinemia, and hemoglobinuria followed by jaundice and general loss of condition, to severe central nervous system disorders and respiratory distress. As in the case of falciparum malaria, much of the pathophysiology associated with such infections appears to be mediated by the production of cytokines such as tumour necrosis factor, a fall in plasma fibronectin, and possibly the production of fibrin–fibronectin complexes, associated with the cascade of changes in the blood and reticuloendothelial system that follow their production [104]. Animals that recover are resistant to subsequent homologous infection. Dogs may develop a similar disease when infected with *B. canis*, and immunocompromised people when infected with bovine parasites. It is interesting that, so far, babesiosis does not seem to have been reported in people whose immune system is depressed by HIV infection.

The life-cycle of *Babesia* (Fig. 20.25) parallels in some ways that of *Plasmodium*. Infective stages found in the salivary gland cells of various genera of ixodid ticks (e.g., *Ixodes, Dermacentor, Boophilus, Rhipicephalus*) are injected into the vertebrate when the tick feeds [113]. The parasites of most species apparently invade erythrocytes directly without the intervention of a pre-erythrocytic phase. However, at least in their natural hosts, *Babesia equi* and *B. microti* are transmitted trans-stadially, i.e., after infecting larvae or nymphs they develop to a

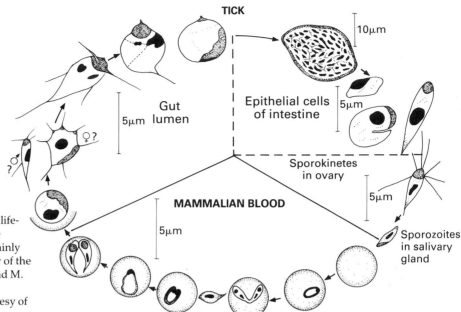

TICK

10μm

Gut lumen

5μm

Epithelial cells of intestine

5μm

♀?

♂?

Sporokinetes in ovary

5μm

MAMMALIAN BLOOD

5μm

Sporozoites in salivary gland

Fig. 20.25 Diagram of the probable life-cycle of *Babesia* in the tick and in the vertebrate host. (Drawing based mainly on observations on the morphology of the tick stages by Drs S. Wattendorff and M. Weyreter, Institute of Parasitology, Veterinary College, Hanover, courtesy of Professor K.T. Friedhoff.)

sporozoite stage which is passed to a new mammalian host by the succeeding stage of the tick. They then enter lymphocytes in which they undergo schizogony, the progeny of which then infect erythrocytes [114]. These species are not transmitted transovarially. Most other species of *Babesia* appear to follow a transovarial route. Within the erythrocytes the babesial trophozoites feed in a different manner from *Plasmodium*, probably by taking in red cell stroma by pinocytosis, and they do not produce hemozoin granules as do the malaria trophozoites. They form pyriform bodies which reproduce asexually by "budding," giving rise to two or four daughter cells, depending upon the species. This process takes about 8 h. The daughter merozoites usually cause the host cells to lyse. They then enter fresh erythrocytes by a process reminiscent of that described for *Plasmodium* but differing in that complement appears to play a more important role in the process. Some of the trophozoites are believed to form gametocytes. Within the gut of the tick, "strahlenkörper" (which are almost certainly gametes) appear within the first 20 h [115]. After about 2 days, when the intestinal lumen becomes filled with enlarging intestinal epithelial cells, these forms are replaced by oval and spherical forms without "strahlen" which are probably zygotes. Ookinetes apparently develop from the zygotes and elongate to form "vermicules" which invade the gut epithelial cells in which they undergo sporogony. Unlike the intra-

erythrocytic stages, the late intraluminal stages contain multiple micropores (cytostomes) which appear to function as feeding organelles. From these multiple fission bodies, sporokinetes are released to migrate through the tissues of the tick, invading the salivary glands of the nymphs and larvae, as well as other tissues, including the ovaries of the adults. In the ovum they remain dormant until the egg hatches and the young nymph takes its first feed. They then rapidly multiply by sporogony, producing large numbers of sporokinetes, some of which invade the salivary glands. When the young tick starts to feed on a new animal, the sporokinetes within the salivary gland cells undergo a different type of asexual sporogony that culminates, within 5 days, in the production of large numbers of long pyriform sporozoites which can infect the host during the prolonged blood feed, invading host erythrocytes and recommencing the cycle. Thus the cycle continues through the new generation of ticks, which become the main reservoir of infection through transovarial transmission.

Genus Theileria

This genus contains at least nine species that are parasitic in cattle, buffalo, sheep, goats, and camels, almost exclusively in various tropical and subtropical countries, as well as reindeer (*Theileria tarandi*) in the northern

USSR [116]. Theileriosis is an extremely serious economic problem in such areas as East Africa where *Theileria parva* causes heavy losses of domestic cattle. The disease in such animals is characterized by fever (East Coast fever), lymph node hypertrophy, hemorrhages, and pulmonary edema. The presence of bile pigments and hemoglobin in the urine has given rise to the term "redwater fever." It has been estimated that over 1 billion animals worldwide are infected with the different species of *Theileria*, three being especially pathogenic in Bovidae, namely *T. parva*, *T. lawrencei*, and *T. bubalis* [117].

Like the babesioses, the theilerioses are transmitted from animal to animal by the bite of male and female ixodid ticks. *Rhipicephalus appendiculatus* is the principal vector of *T. parva*. Piroplasms in the erythrocytes which are taken up when the tick feeds as larva, nymph, or adult are liberated in the gut where, much as in *Babesia*, gametocytes or gametes are freed to form zygotes within the gut epithelial cells. The zygotes eventually migrate as kinetes to the salivary glands where they undergo schizogony over a period of several days (4 days in adult ticks). The daughter cells (oval or round bodies about 1.5 μm in diameter) are then available to infect a new host when the tick feeds again during its next stage of development. The parasites do not survive after engorgement has been completed and there is no transovarial transmission. However, adult ticks that do not feed may remain infective for several months.

When the infective particles are injected with the saliva into a new host they appear to migrate to the local lymph nodes and from there to lymphatic tissue in other organs. They enter the cytoplasm of lymphoblasts and lymphocytes in which they undergo asexual schizogony, forming macroschizonts ("Koch's blue bodies"). These are bodies about 2–16 μm in diameter with a clear blue-staining cytoplasm (with Giemsa stain) and about eight large nuclei. They appear to feed on host cell cytoplasm through a cytostome. The parasites act as powerful mitogenic agents, stimulating lymphopoiesis so that lymph nodes become hypertrophied and, in bone marrow, other blood elements are displaced. Enormous numbers of macroschizonts are produced until as many as 80% of all lymphoblasts and lymphocytes may be parasitized. Fever develops 1–3 weeks after the ticks feed. At a later stage of the infection, a smaller body, the microschizont, develops not in the lymphocytes as much as in macrophages. In the microschizonts are produced about 50–120 micromerozoites, each about 1 μm long. When these are released on rupture of the

host cells, they invade erythrocytes in which they form elongated "bacillary" or "bayonet" forms, not dissimilar to the intraerythrocytic forms of *Babesia* but smaller than most of those species. Like the latter, they possess a cytostome through which they ingest red cell stroma, and they do not produce hemozoin. They divide asexually to produce up to four merozoites, some of which appear to become gametocytes that are infective to the next tick that feeds. It is very likely that the main reservoir of infection is the small number of these forms that survive in the circulation of animals that recover from the acute stages of infection. The manner in which *Theileria* survives in cells of the reticuloendothelial system and the immune mechanisms of the vertebrate host are discussed in Chapter 18. A number of features of the life-cycle of *Theileria* still remain obscure. It is believed, for example, that the macroschizonts increase in number by dividing together with their host cells, but it is uncertain whether "free" macroschizonts can divide or whether it is time or host temperature that determines at which point in the infection microschizonts are formed [113]. Nor is it certain what determines the level of pathogenicity of a particular species of *Theileria*. The removal of infected red cells by the spleen and liver causes a variable degree of anemia but other pathologic changes may be due partly to toxic effects on the bone marrow. A reduction in thrombocytes causes a general hemorrhagic diathesis and, for example, changes associated with cerebral hemorrhages may give rise to a syndrome known as "turning sickness" in cattle.

PHYLUM MICROSPORA

The Microsporidia are parasites that infect all classes of vertebrates [118] but only in recent years have they been suspected of causing pathologic changes in humans. Two young children have been found to be infected with parasites of the genus *Encephalitozoon*, of which one of the best-known species is *Encephalitozoon cuniculi*, a pathogen of rabbits. Serologic sampling has demonstrated that antibodies to this parasite are present in many individuals who have been resident in tropical areas in the absence of associated pathologic conditions [119]. Recently, however, a new microsporidian, now named *Enterocytozoon bienusi*, has been discovered in the intestinal enterocytes of people with AIDS, associated with intractable diarrhea. At the time of writing, this infection has been reported in at least eight patients. The possibility must be considered that human infection with microsporidians, like a number of other protozoan parasites (e.g., *T. gondii*), is common but that the infections remain cryptic unless the host's immunity is seriously compromised. The question is difficult to answer, since microsporidia are rarely spotted in normal histologic preparations of biopsy or autopsy material and their presence and structure are usually only determined by electron microscopy of tissue samples. This is clearly a field that demands extensive investigation.

PATHOGENS OF DISPUTED TAXONOMIC STATUS

Pneumocystis carinii

Prior to the advent of the current pandemic of HIV infection, human infection with *P. carinii* was considered to be a rare event limited to patients whose immunity had been seriously compromised by the use of such substances as corticosteroids or of cytotoxic agents used in the therapy of various neoplasms. The picture has now changed dramatically, with *Pneumocystis* pneumonia being one of the most prevalent opportunistic infections in patients with AIDS in the developed countries and one of those commonly causing death. It is a eukaryote which forms eight-nucleated spores in the pulmonary alveoli of a wide range of vertebrates and, like several protozoa referred to above, it is likely that it is a widespread commensal in many immunologically competent animals, including humans.

The life-cycle of *P. carinii* is only known in the intrapulmonary stage. In the alveoli there are two cycles: a sexual cycle in which haploid trophozoites fuse to form diploid zygotes; and an asexual cycle in which two meioses and one mitosis culminate in the production of eight haploid bodies within a cyst. Heavy infections in immunocompromised people are associated with patchy pneumonic consolidation. The mode of infection is uncertain but is probably by the inhalation of infective trophozoites within spores [120].

The question of the taxonomic status of *P. carinii* appears to have been resolved by the finding that the 16S-like rRNA of the organisms is phylogenetically related to the Fungi and not to the Protozoa [121].

The diagnosis of pneumocystosis formerly depended upon observing the cysts in sputum smears stained by Giemsa or in silver-stained lung biopsy or autopsy specimens. Diagnosis has now been greatly facilitated by the development of monoclonal antibodies that can be used to identify the organisms in material obtained either by bronchoalveolar lavage or, more simply, in

sputum induced by the inhalation of a saline aerosol (e.g., Elvin *et al.* [122]).

ACKNOWLEDGMENT

The author wishes to thank Mrs L.J. Williams for drawing Figs 20.5, 20.9, 20.10, 20.11, 20.15, 20.21, and 20.23.

REFERENCES

1 Sadun EH. Life cycles of common parasites of medical importance. In Cohen S, Sadun EH, eds. *Immunology of Parasitic Infections*, 1st edn. Oxford: Blackwell Scientific Publications, 1976:469–488.

2 Zierdt CH, Donnolley CT, Muller J, Constantopoulos G. Biochemical and ultrastructural study of *Blastocystis hominis*. J Clin Microbiol 1988;26:965–970.

3 Sargeaunt PG, Williams JE, Neal RA. A comparative study of *Entamoeba histolytica* (NIH: 200, HK9, etc.), *E. histolytica*-like and other morphologically identical amoebae using isoenzyme electrophoresis. Trans R Soc Trop Med Hyg 1980;74:469–474.

4 Wyler DJ. Fibronectin in parasitic diseases. Rev Infect Dis 1987;9(4):S391–S399.

5 Martínez-Palomo A, ed. *Amebiasis*. Amsterdam: Elsevier, 1986.

6 Albach RA, Booden T. Amoebae. In Kreier JP, ed. *Parasitic Protozoa*, vol. II. New York: Academic Press, 1977:455–506.

7 Lushbaugh WB, Miller JH. The morphology of *Entamoeba histolytica*. In Ravdin JI, ed. *Amebiasis. Human Infection by Entamoeba histolytica*. New York: John Wiley & Sons, 1988: 41–68.

8 Robinson GL. The laboratory diagnosis of human parasitic amoebae. Trans R Soc Trop Med Hyg 1968;62:285–294.

9 Sargeaunt PG. Zymodemes of *Entamoeba histolytica*. In Ravdin JI, ed. *Amebiasis. Human Infection by Entamoeba histolytica*. New York: John Wiley & Sons, 1988:370–387.

10 Rondanelli EG, ed. *Infectious Diseases. Color Atlas Monographs. 1. Amphizoic Amoebae. Human Pathology*. Padua: Piccin, 1987.

11 Griffin JL. Pathogenic free-living amoebae. In Kreier JP, ed. *Parasitic Protozoa*, vol. II. New York: Academic Press, 1977:507–549.

12 Visvesvara GS, Healy GR. Disc electrophoretic patterns of esterase isoenzymes of *Naegleria fowleri* and *N. gruberi*. Trans R Soc Trop Med Hyg 1980;74:411–412.

13 Honigberg BM. Trichomonads of importance to human medicine. In Kreier JP, ed. *Parasitic Protozoa*, vol. II. New York: Academic Press, 1977:275–454.

14 Erlandsen SL, Meyer EA, eds. *Giardia and Giardiasis. Biology, Pathogenesis and Epidemiology*. New York: Plenum Publishers, 1984.

15 Meloni BP, Lymbery AJ, Thompson RCA. Isoenzyme electrophoresis of 30 isolates of *Giardia* from humans and felines. Am J Trop Med Hyg 1988;38:65–73.

16 Meyer EA, Jarroll EL. Giardiasis. Am J Epidemiol 1980;111: 1–12.

17 Manson-Bahr PEC, Bell DR. *Manson's Tropical Diseases*,

19th edn. London: Baillière Tindall, 1987.

18 Green EL, Miles MA, Warhurst DC. Immunodiagnostic detection of *Giardia* antigen in faeces by a rapid visual enzyme-linked immunosorbent assay. Lancet 1985;ii:691–693.

19 Ackers JP. Immunology of amebas, *Giardia* and trichomonads. In Nahmias AJ, O'Reilly RJ, eds. *Comprehensive Immunology. Vol. 9. Immunology of Human Infection Part II: Viruses and Parasites; Immunodiagnosis and Prevention of Infectious Diseases*. New York: Plenum Medical Books, 1982: 403–443.

20 Ackers JP. Immunologic aspects of human trichomoniasis. In Honigberg BM, ed. *Trichomonads Parasitic in Humans*. New York: Springer-Verlag 1990:36–52.

21 Ackers JP, Catterall RD, Lumsden WHR, McMillan A. Absence of detectable local antibody in genitourinary tract secretions of male contacts of women infected with *Trichomonas vaginalis*. Br J Vener Dis 1978;54:168–171.

22 Carney JA, Unadkat P, Yule A, Rajakumar R, Lacey CJN, Ackers JP. New rapid latex agglutination test for diagnosing *Trichomonas vaginalis* infection. J Clin Pathol 1988; 41:806–808.

23 Soliman MA, Ackers JP, Catterall RD. Isoenzyme characterisation of *Trichomonas vaginalis*. Br J Vener Dis 1982; 58:250–256.

24 Tait A. Evidence for diploidy and mating in trypanosomes. Nature 1980;287:536–538.

25 Jenni L, Marti S, Schweizer J, *et al*. Hybrid formation between African trypanosomes during cyclical transmission. Nature 1986;322:173–175.

26 Evans DA, Kennedy WPK, Elbihari S, Chapman CT, Smith V, Peters W. Hybrid formation within the genus *Leishmania*? Parassitologia 1988;29:165–173.

27 Gibson WC. Will the real *Trypanosoma b. gambiense* please stand up. Parasitol Today 1986;2:255–257.

28 Gibson WC, Marshall TF de C, Godfrey DG. Numerical analysis of enzymes polymorphism: a new approach to the epidemiology and taxonomy of trypanosomes of the subgenus *Trypanozoon*. Adv Parasitol 1980;18:175–246.

29 Ole-Moi Yoi OK. Trypanosome species-specific DNA probes to detect infection in tsetse flies. Parasitol Today 1987;3:371–374.

30 Opperdoes FR. Biochemical peculiarities of trypanosomes, African and South American. Br Med Bull 1985;41: 130–136.

31 Henderson GB, Fairlamb AH. Trypanothione metabolism: a chemotherapeutic target in trypanosomatids. Parasitol Today 1987;58:250–256.

32 Schechter PJ, Sjoerdsma A. Difluoromethylornithine in the treatment of African trypanosomiasis. Parasitol Today 1986;2:223–224.

33 Doua F, Boa FY, Schechter PJ, *et al*. Treatment of human late stage gambiense trypanosomiasis with α-difluoromethylornithine (Eflornithine): efficacy and tolerance in 14 cases in Côte d'Ivoire. Am J Trop Med Hyg 1987;37:525–533.

34 Evans DA, Ellis DS. Recent observations on the behaviour of certain trypanosomes within their insect hosts. Adv Parasitol 1983;22:1–42.

35 Raseroka BH, Ormerod WE. The trypanocidal effect of

drugs in different parts of the brain. Trans R Soc Trop Med Hyg 1986;80:634–641.

36 Steinert M, Pays E. Selective expression of surface antigen genes in African trypanosomes. Parasitol Today 1986;2: 15–19.

37 Ribeiro dos Santos R, Furtado CCVG, Ramos de Oliveira JC, Martins AR, Köberle F. Antibodies against neurons in chronic Chagas' disease. Tropenmed Parasitol 1979;30: 19–23.

38 Geigy R, Herbig A. *Erreger und Uberträger tropische Krankheiten*. Basle: Verlag für Recht und Gesselschaft AG, 1955.

39 Deane MP, Jansen AM. From a mono to a digenetic life-cycle: how was the jump for flagellates of the family Trypanosomatidae? Mem Inst Oswaldo Cruz 1988;83: 273–275.

40 Deane MP, Lenzi HL, Jansen AM. *Trypanosoma cruzi*; vertebrate and invertebrate cycles in the same mammal host, the opossum *Didelphis marsupialis*. Mem Inst Oswaldo Cruz 1984;79:513–515.

41 Miles MA. *Trypanosoma* and *Leishmania*: the contribution of enzyme studies to epidemiology and taxonomy. In Oxford GS, Rollinson DJ eds. *Protein Polymorphism: Adaptive and Taxonomic Significance*, London: Academic Press, 1983: 37–57.

42 Miles MA, Apt BW, Widmer G, Povoa MM, Schofield CJ. Isozyme heterogeneity and numerical taxonomy of *Trypanosoma cruzi* stocks from Chile. Trans R Soc Trop Med Hyg 1984;78:526–535.

43 Tibayrenc M, Ward P, Moy A, Ayala FJ. Natural populations of *Trypanosoma cruzi*, the agent of Chagas' disease, have a complex multiclonal structure. Proc Natl Acad Sci USA 1986;83:115–119.

44 Schechter M, Flint JE, Voller A, Gohl F, Marinkelle CJ, Miles MA. Purified *Trypanosoma cruzi* specific glycoprotein for discriminative serological diagnosis of South American trypanosomiasis (Chagas' disease). Lancet 1983;2:939–941.

45 Peters W, Killick-Kendrick R, eds. *The Leishmaniases in Biology and Medicine, vol. I. Biology and Epidemiology*. London: Academic Press, 1987.

46 Lainson R, Shaw JJ. Evolution, classification and geographical distribution. In Peters W, Killick-Kendrick R, eds. *The Leishmaniases in Biology and Medicine, vol. I. Biology and Epidemiology*. London: Academic Press, 1987:1–120.

47 Rioux JA, ed. *Leishmania. Taxonomie—phylogenèse. Applications éco-épidémiologiques*. Montpellier: Institut Méditerranéen d'Etudes Épidémiologiques et Écologiques (IMEEE), 1986.

48 Barker DC, Gibson JJ, Kennedy WPK, Nasser AAAA, Williams RH. The potential of using recombinant DNA species-specific probes for the identification of tropical *Leishmania*. Parasitology 1986;91:S139–S174.

49 Gardener PJ. Pellicle-associated structures in the amastigote stages of *Trypanosoma cruzi* and *Leishmania* species. Ann Trop Med Parasitol 1974;68:167–176.

50 Moreno G, Rioux J-A, Lanotte G, Pratlong F, Serres E. Le complexe *Leishmania donovani* s.l. Analyse enzymatique et traitement numérique. In Rioux J-A, ed. *Leishmania. Taxonomie et phylogenèse*. Montpellier: IMEEE, 1985:105–117.

51 King DL, Chang Y, Turco SJ. Cell surface lipo- phospho-glycan of *Leishmania donovani*. Mol Biochem Parasitol 1987; 24:47–53.

52 Molyneux DH, Killick-Kendrick R. Morphology, ultra-structure and life cycles. In Peters W, Killick-Kendrick R, eds. *The Leishmaniases in Biology and Medicine, vol. I. Biology and Epidemiology*. London: Academic Press, 1987;121–176.

53 Chang K-P, Bray RS. *Leishmaniasis*. Amsterdam: Elsevier, 1985.

54 Ranque J, Quilici M, Belleudy P, Dunan S. Les réservoirs de virus de la leishmaniose viscérale en Provence. Med Trop 1978;38:405–409.

55 Zaman V. *Balantidium coli*. In Kreier JP, ed. *Parasitic Protozoa*, vol. II. New York: Academic Press, 1977:633–653.

56 Mehlhorn H, Heydorn AO, Senaud J, Schein E. Les modalités de la transmission des protozoaires parasites des genres *Sarcocystis* et *Theileria*, agents de graves maladies. Ann Biol 1979;18:98–120.

57 Markus MB. *Sarcocystis* and sarcocystosis in domestic animals and man. Adv Vet Sci Comp Med 1978;22:159–193.

58 Dubey JP. *Toxoplasma, Hammondia, Besnoitia, Sarcocystis* and other tissue cyst-forming coccidia of man and animals. In Kreier JP, ed. *Parasitic Protozoa*, vol. III. New York: Academic Press, 1977:101–237.

59 Mehlhorn H, Frenkel JK. Ultrastructural comparison of cysts and zoites of *Toxoplasma gondii, Sarcocystis muris,* and *Hammondia hammondi* in skeletal muscles of mice. J Parasitol 1980;66:59–67.

60 Heydorn AO. Sarcosporidieninfiziertes Fleisch als mögliche Krankheitursache für den Menschen. Arch Lebensmittelhyg 1977;28:27–31.

61 Piekarski G, Heydorn AO, Aryeetey ME, Hartlapp J-H, Kimmig P. Klinische, parasitologische und serologische Untersuchungen zur Sarkosporidiose (*Sarcocystis suihominis*) des Menschen. Immun Infekt 1978;6:153–159.

62 Tadros W, Laarman JJ. *Sarcocystis* and related coccidian parasites: a brief general review, together with a dis-cussion on some biological aspects of their life cycles and a new proposal for their classification. Acta Leiden 1976;44: 1–107, figs 1–27.

63 Beaver PC, Gadgil RK, Morera P. *Sarcocystis* in man: a review and report of five cases. Am J Trop Med Hyg 1979;28:819–844.

64 Frenkel JK. Pathophysiology of toxoplasmosis. Parasitol Today 1988;4:273–278.

65 Krick JA, Remington JS. Toxoplasmosis in the adult. N Engl J Med 1978;298:550–553.

66 Teutsch SM, Juranek DD, Sulzer A, Dubey JP, Sikes RK. Epidemic toxoplasmosis associated with infected cats. N Engl J Med 1979;300:695–699.

67 Jones TC, Masur H. The survival of *Toxoplasma gondii* and other microbes within host-cell cytoplasmic vacuoles. In Van den Bossche H, ed. *The Host–Parasite Interplay*. Amsterdam: Elsevier, 1980:157–164.

68 Levine ND. Taxonomy and review of the Coccidian genus *Cryptosporidium* (Protozoa, Apicomplexa). J Protozool 1984;31:94–98.

69 Navin TR, Juranek DD. Cryptosporidiosis: clinical, epidemiologic, and parasitologic review. Rev Infect Dis 1984;6:313–327.

70 Cook GC. *Cryptosporidium sp. and Other Intestinal Coccidia. A Bibliography*. London: Bureau of Hygiene and Tropical Diseases, 1987.

71 Stehr-Green JK, McCaig L, Remsen HM, Rains CS, Fox M, Juranek DD. Shedding of oocysts in immunocompetent individuals infected with *Cryptosporidium*. Am J Trop Med 1987;36:338–342.

72 Tzipori S. Cryptosporidiosis in perspective. Adv Parasitol 1988;27:63–129.

73 Garnham PCC. *Malaria Parasites and Other Haemosporidea*. Oxford: Blackwell Scientific Publications, 1966.

74 Trager W, Jensen JB. Human malaria parasites in con-tinuous culture. Science 1976;193:673–675.

75 Haynes JD, Diggs CL, Hines FA, Desjardins RE. Culture of human malaria parasites, *Plasmodium falciparum*. Nature 1976;263:767–769.

76 Peters W. *Chemotherapy and Drug Resistance in Malaria*, 2nd edn. London: Academic Press, 1987.

77 Meis JFGM, Verhave JP. Exoerythrocytic development of malarial parasites. Adv Parasitol 1988;27:1–61.

78 Aikawa M, Miller LH. Structural alteration of the erythrocyte membrane during malarial parasite invasion and intraerythrocytic development. In *Malaria and the Red Cell*. CIBA Foundation Symposium 94. London: Pitman, 1983:45–63.

79 Sam-Yellowe TY, Shio H, Perkins ME. Secretion of *Plasmodium falciparum* rhoptry protein into the plasma membrane of host erythrocytes. J Cell Biol 1988;106: 1507–1513.

80 Bannister LH, Butcher GA, Dennis ED, Mitchell GH. Structure and invasive behaviour of *Plasmodium knowlesi* merozoites *in vitro*. Parasitology 1975;71:483–491.

81 Matsumoto Y, Perry G, Scheibel LW, Aikawa M. Role of calmodulin in *Plasmodium falciparum*: implications for erythrocyte invasion by the merozoite. Eur J Cell Biol 1987;45:36–43.

82 Hadley TJ, Miller LH. Invasion of erythrocytes by malaria parasites: erythrocyte ligands and parasite receptors. In Perlmann P, Wigzell H, eds. *Progress in Allergy, vol. 41. Malaria Immunology*. Basel: Karger, 1988:49–71.

83 Kilejian A. Homology between a histidine-rich protein from *Plasmodium lophurae* and a protein associated with the knob-like protrusions on membranes of erythrocytes infected with *Plasmodium falciparum*. J Exp Med 1980;151: 1534–1538.

84 Kilejian A, Jensen JB. A histidine-rich protein from malaria and its interaction with membranes. Bull WHO 1977;55: 191–197.

85 Howard JH. Malarial proteins at the membrane of *Plasmodium falciparum*-infected erythrocytes and their involvement in cytoadherence to endothelial cells. In Perlmann P, Wigzell H, eds. *Progress in Allergy, vol. 41 Malaria Immunology*. Basel: Karger, 1988:98–147.

86 Barnwell JW, Howard RJ, Miller LH. Influence of the spleen on the expression of surface antigens on parasitized erythrocytes. In *Malaria and the Red Cell*. CIBA Foundation Symposium 94. London: Pitman, 1983:117–136.

87 Clark IA, Hunt NH, Cowden WB. Oxygen-derived free

radicals in the pathogenesis of parasitic disease. Adv Parasitol 1986;25:1–44.

88 Boonpucknavig V, Srichaikul T, Punyagupta S. Clinical pathology. In Peters W, Richards WHG, eds. *Antimalarial Drugs I. Biological Background, Experimental Methods, and Drug Resistance*. Berlin: Springer-Verlag, 1984:127–176.

89 White NJ. Pathophysiology. In Strickland GT, ed. *Clinics in Tropical Medicine and Communicable Diseases*, vol. 1, *No. 1. Malaria*. London: Saunders, 1986:55–90.

90 Smalley ME, Sinden RE. *Plasmodium falciparum* gametocytes: their longevity and infectivity. Parasitology 1977;74:1–8.

91 Sherman IW. Biochemistry of *Plasmodium* (malarial parasites). Microbiol Rev 1979;43:453–495.

92 Krotoski WA, Krotoski DM, Garnham PCC, *et al*. Relapses in primate malaria: discovery of two populations of exoerythrocytic stages. Preliminary note. Br Med J 1980;1: 153–154.

93 Hollingdale MR, Collins WE, Campbell CC, Schwartz AL. *In vitro* culture of two populations (dividing and nondividing) of exoerythrocytic parasites of *Plasmodium vivax*. Am J Trop Med Hyg 1985;34:216–222.

94 Hawking F, Worms MJ, Gammage K. 24- and 48-hour cycles of malaria parasites in the blood: their purpose, production and control. Trans R Soc Trop Med Hyg 1968; 62:731–760.

95 Garnham PCC. The continuing mystery of relapses in malaria. Protozool Abstr 1977;1:1–12.

96 Sethabutr O, Brown AE, Gingrich J, *et al*. A comparative field study of radiolabeled and enzyme-conjugated synthetic DNA probes for the diagnosis of falciparum malaria. Am J Trop Med Hyg 1988;39:227–231.

97 WHO. The biology of malaria parasites. WHO Tech Rep Ser 1987;763:743.

98 Brumpt M. Paludisme aviaire: *Plasmodium gallinaceum* n.sp. de la poule doméstique. C R Acad Sci 1935;200: 783–786.

99 Vincke LH, Lips M. Un nouveau *Plasmodium* d'un rongeur sauvage du Congo *Plasmodium berghei* n.sp. Ann Soc Belg Med Trop 1948;28:97–104.

100 Killick-Kendrick R, Peters W, eds. *Rodent Malaria*. New York: Academic Press, 1978.

101 Landau I, Boulard Y. Life cycles and morphology. In Killick-Kendrick R, Peters W, eds. *Rodent Malaria*. New York: Academic Press, 1978:53–84.

102 Nussenzweig RS, Cochrane AH, Lustig HJ. Immunological responses. In Killick-Kendrick R, Peters W, eds. *Rodent Malaria*. New York: Academic Press, 1978:247–307.

103 Guérin-Marchand C, Druilhe P, Galey B, *et al*. A liver stage-specific antigen of *Plasmodium falciparum* characterized by gene cloning. Nature 1987;329:164–167.

104 Wright IG, Goodger BV, Clark IA. Immunopathophysiology of *Babesia bovis* and *Plasmodium falciparum* infections. Parasitol Today 1988;4:214–218.

105 Seureau C, Szöllösi A, Boulard Y, Landau I, Peters W. Aspects ultrastructureaux de la rélation hôte-parasite entre le schizonte de *Plasmodium yoelii* et la cellule hépatique de rat. Protistologica 1980;16:419–426.

106 Schmidt LH, Fradkin R, Harrison J, Rossan RN. Differences in the virulence of *Plasmodium knowlesi* for *Macacus irus* (*fascicularis*) of Philippine and Malayan origins. Am J Trop Med Hyg 1977;26:612–622.

107 Schmidt LH, Cramer DV, Rossan RN, Harrison J. The characteristics of *Plasmodium cynomolgi* infections in various Old World primates. Am J Trop Med Hyg 1977; 26:356–372.

108 Ristic M, Lewis GE. *Babesia* in man and wild and laboratory-adapted mammals. In Kreier JP, ed. *Parasitic Protozoa*, vol. IV. New York: Academic Press, 1977:53–76.

109 Healy GR, Spielman A, Gleason N. Human babesiosis: reservoir of infection on Nantucket Island. Science 1976; 192:479–480.

110 Wolf RE, Gleason NN, Schoenbaum SC, Western KA, Klein CA, Healy GR. Intraerythrocytic parasitosis in humans with *Entopolypoides* species (Family Babesiidae). Association with hepatic dysfunction and sérum factors inhibiting lymphocyte response to phytohemagglutinin. Ann Intern Med 1978;88:769–773.

111 Osorno BM, Vega C, Ristic M, Robles C, Ibarra S. Isolation of *Babesia* spp. from asymptomatic human beings. Vet Parasitol 1979;2:111–120.

112 Mahoney DF. *Babesia* of domestic cattle. In Kreier JP, ed. *Parasitic Protozoa* IV. New York: Academic Press, 1977:1–52.

113 Huff MF. In Boyd CA. Life cycles of malaria parasites with special reference to the newer knowledge of preerythrocytic stages, ed. *Malariology*, vol. I. Philadelphia: WB Saunders, 1949:60.

114 Young AS, Morzaria SP. Biology of *Babesia*. Parasitol Today 1986;2:211–219.

115 Weber G, Friedhoff KT. Preliminary observations on the ultrastructure of supposed sexual stages of *Babesia bigemina* (Piroplasmea). Z Parasitenkd 1977;53:83–92.

116 Barnett SF. *Theileria*. In Kreir JP, ed. *Parasitic Protozoa*, vol. IV. New York: Academic Press, 1977:77–113.

117 Ristic M, McIntyre I, eds. *Diseases of Cattle in the Tropics*. The Hague: Martinus Nijhoff, 1981.

118 Canning EU, Lom J, Dykova I. *The Microsporidia of Vertebrates*. London: Academic Press, 1986.

119 Canning EU, Hollister WS. Microsporidia of mammals— widespread pathogens or opportunistic curiosities? Parasitol Today 1987;3:267–273.

120 Matsumoto Y, Yoshida Y. Advances in *Pneumocystis* biology. Parasitol Today 1986;2:137–142.

121 Edman JC, Kovacs JA, Masur H, Santi DV, Elwood HJ, Sogin ML. Ribosomal RNA sequence shows *Pneumocystis carinii* to be a member of the Fungi. Nature (London) 1988;334:519–522.

122 Elvin KM, Björkman A, Linder E, Heurlin N, Hjerpe A. *Pneumocystis carinii* pneumonia: detection of parasites in sputum and bronchoalveolar lavage fluid by monoclonal antibodies. Br Med. J 1988;297:381–384.

21 Helminths

Ralph Muller

The spread of acquired immunodeficiency syndrone (AIDS) has not had the dramatic impact on helminthologic studies as it has on protozoologic studies; there is no previously obscure parasite which has now emerged as an important human pathogen. This reflects the fact that, with very few exceptions, helminths do not multiply inside the body. Unlike other infectious organisms in which pathogenicity depends on a single infection, in helminth infections it is usually related to the magnitude of intensity of the worm load, which often depends on the accumulation of parasites following repeated reinfection.

One of the few helminths which can replicate in man is *Strongyloides* but, although hyperinfection with this parasite sometimes occurs in immunocompromised patients, including a few with AIDS [1], surprisingly few such opportunistic infections have been reported from AIDS patients [2,3]. It appears that the impairment to cell-mediated immunity in AIDS patients does not favor helminth infections as it does intracellular protozoan infections. On the other hand, disseminated strongyloidiasis does seem to be associated with the presence of HTLV-I infection [4].

The forecast in the previous edition that the role of immunologists in devising immunologic tests would increase has certainly been realized. With such large organisms, the purification and utilization of specific antigens from most helminths has until recently been an impossible task. However, the field is being revolutionized by the use of monoclonal antibody and DNA hybridization techniques. The detection of circulating parasite antigens which will be able to distinguish between current and past infection is also rapidly becoming a feasible possibility for many helminths. In addition, these new techniques have stimulated research on the discovery and amplification of antigens which evoke mechanisms of resistance in the host and have increased optimism that it will be possible to produce vaccines against some helminth infections.

The basis of host variations in susceptibility and resistance to helminth infections has been stimulated by a great increase in the availability of well-defined strains of laboratory hosts in recent years. Many human helminths or analogous models can be studied in laboratory mice and, for some of these, genetically determined variations in susceptibility and resistance of different mouse strains have been demonstrated [5] (Table 21.1).

The statement in the previous edition that very little is known about the metabolism of many medically important helminths is still partly true [34–39] and no additional species have been successfully maintained *in vitro* throughout their life-cycles [40,41]. There have, however, been notable advances in chemotherapy against almost all trematode and cestode infections and against geohelminth infections and onchocerciasis [42,43]. The extent of synergy between drugs and immunologic reactions is a field of increasing interest [44].

One recent sphere of interest has been the role of oxygen-derived free radicals produced by phagocytes in the killing mechanisms against helminths, as against other infectious organisms, and in the pathogenesis of helminth diseases [45]. There is also evidence that many helminths produce antioxidant enzymes [46].

PHYLUM PLATYHELMINTHES

Class Digenea (the trematodes)

Digenetic trematodes, often known as flukes, are all parasitic and members inhabit the intestinal tract, bile ducts, lungs, or blood of humans. Most are flat, hermaphroditic, organisms which produce large numbers of eggs. All have complicated life-cycles with an alternating sexual cycle in humans without multiplication and an asexual multiplicative cycle in a gastropod snail intermediate host. Many trematodes also have a second intermediate host. The common human

Table 21.1 List of helminths used as model systems in the mouse in which genetically determined variation in response has been demonstrated. (From Wakelin [5])

Phylum Platyhelminthes

Class Digenea
Fasciola hepatica [6]
*Schistosoma japonicum** [7]
*Schistosoma mansoni** [8,9]

Class Cestoda
Echinococcus multilocularis [10]
Hymenolepis microstoma [11]
Hymenolepis nana [12]
Mesocestoides corti [13]
Taenia crassiceps [14]
*Taenia taeniaeformis** [15–17]

Phylum Nematoda

*Acanthocheilonema viteae** [18,19]
Ascaris suum [20]
Aspiculurus tetraptera [21]
Brugia malayi [22]
Brugia pahangi [23]
*Heligmosoides polygyrus** [24]
Litomosoides carinii [25]
Necator americanus [26]
Nippostrongylus brasiliensis [27]
Onchocerca lienalis [28]
Strongyloides ratti [29]
*Trichinella spiralis** [30–32]
Trichuris muris [33]

* Parasites for which there has been most analysis of genetic control of variation or of mechanisms of resistance.

trematodes and some of their biologic characteristics are listed in Table 21.2. It can be seen from the table that nearly all trematode infections are zoonoses with other mammals serving as reservoir hosts [47]. The most important and widely distributed human trematode infection is schistosomiasis but clonorchiasis and paragonimiasis are of local importance in areas of Southeast Asia.

The classification used here is a recent one and the divisions above the family level are not yet universally accepted [48].

Order Strigeiformes, family Schistosomatidae

Schistosoma species (blood flukes). The schistosomes are the cause of the most important helminth diseases of humans and were first recovered from the vesical veins at autopsy by Theodor Bilharz in Cairo in 1851. It is

likely that there are well over 200 million cases in the world [49], mostly in agricultural communities, and in Africa also in fishermen.

There are five species of *Schistosoma* which parasitize humans, four of these (*Schistosoma mansoni*, *Schistosoma japonicum*, *Schistosoma mekongi*, and *Schistosoma intercalatum*) cause intestinal schistosomiasis while the fifth (*Schistosoma haematobium*) causes urinary schistosomiasis.

Schistosoma mansoni is widely distributed in Africa and also occurs in foci in the Middle East and South America and the Caribbean; *S. japonicum* is confined to Southeast Asia, principally China, Sulawesi, and the Philippines; *S. haematobium* is very widely distributed in Africa and also has foci in the Middle East; *S. intercalatum* has limited foci in Central and West Africa; *S. mekongi* is a recently described species confined to the Mekong delta [50,51].

The adult schistosomes differ from all other human trematodes because the sexes are separate (diecious); the slimmer female resides permanently in the gynecophoric canal formed by the fleshy ventral folds of the more robust body of the male, the latter having well-developed oral and ventral suckers and measuring about 12 mm in length.

Adults occupy the venules of the posterior mesenteric veins, except those of *S. haematobium* which occupy the venules of the vesical plexus. The site of the adults determines the type of pathology caused. Female schistosomes produce fewer eggs than most other human trematodes: less than 200 per day for *S. mansoni* and *S. haematobium* and about 1500 per day for *S. japonicum*. The numbers are important because in schistosomiasis it is the eggs which produce the pathology, and the severity of disease caused is proportional to the egg load in the tissues.

The life-cycles of the human schistosomes are illustrated in Fig. 21.1.

For oviposition the female worms move into very small venules and deposit eggs which penetrate through the walls of the venule and the large intestine into the lumen and are then passed out with the feces, or in the case of *S. haematobium*, through the wall of the bladder and into the lumen and then out with the urine. The passive movement of eggs through the mucosa into the feces or urine depends on miracidial secretions which are released through the eggshell. The secretions stimulate a cellular response and softening of the tissues, a process which may also be immunologically mediated. When an egg, already containing a micracidium larva,

Table 21.2 Trematodes of medical importance

	Habitat of first intermediate snail host	Mode of infection to humans	Definitive hosts*
Order Strigeiformes			
Family Schistosomatidae			
Schistosoma mansoni	Slow-flowing rivers, canals, and lakes (*Biomphalaria*)	Active penetration of skin by cercaria	Humans, rodents, baboons
Schistosoma japonicum	Banks of rivers and canals (*Oncomelania*)	Active penetration of skin by cercaria	Humans, dogs, rats, cattle
Schistosoma haematobium	Ponds or margins of lakes (*Bulinus*)	Active penetration of skin by cercaria	Humans
Schistosoma intercalatum	Ponds or margins of lakes (*Bulinus*)	Active penetration of skin by cercaria	Humans
Order Opisthorchiiformes			
Family Opisthorchiidae			
Clonorchis sinensis	Slow-flowing rivers, canals, and ponds (*Bulimas, Parafossarulus*)	Cercariae encyst as metacercariae in freshwater fish	Humans, carnivores
Opisthorchis viverrini	Slow-flowing rivers, canals, and ponds (*Bithynia*)	Cercariae encyst as metacercariae in freshwater fish	Humans, carnivores
Family Heterophyidae			
Heterophyes heterophyes	Lakes, ponds, streams, and canals (*Pirenella, Cerithidea*)	Cercariae encyst as metacercariae in freshwater fish	Humans, carnivores
Metagonimus yokagawai	Lakes, ponds, streams, and canals (*Semisulcospira*)	Cercariae encyst as metacercariae in freshwater fish	Humans, carnivores
Order Plagiorchiiformes			
Family Troglotrematidae			
Paragonimus westermani	Fast-flowing mountain streams (*Semisulcospira, Oncomelania, Thiera*)	Cercariae encyst as metacercariae in freshwater crustaceans	Humans, carnivores
Paragonimus spp.	Fast-flowing mountain streams (various)	Cercariae encyst as metacercariae in freshwater crustaceans	Carnivores (humans)
Order Echinostomatiformes			
Family Fasciolidae			
Fasciola hepatica	Damp pastures (*Lymnaea*)	Cercariae encyst as metacercariae on vegetation	Sheep, cattle (humans)
Fasciolopsis buski	Ponds and slow-flowing streams (*Polypylis*)	Cercariae encyst as metacercariae on vegetation	Humans, hogs

* Parentheses indicate reservoir host.

reaches fresh water it hatches in a few minutes releasing the free-swimming ciliated miracidium. The miracidium swims actively, searching for a suitable gastropod snail host; when it finds one, it penetrates through the epithelium by the use of glands which secrete proteolytic enzymes.

Inside the digestive gland of the snail a process of asexual multiplication takes place, each miracidium

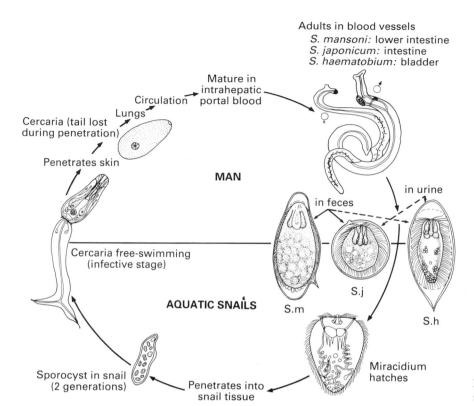

Adults in blood vessels
S. mansoni: lower intestine
S. japonicum: intestine
S. haematobium: bladder

Mature in intrahepatic portal blood

Circulation

Lungs

Cercaria (tail lost during penetration)

Penetrates skin

MAN

in urine

in feces

Cercaria free-swimming (infective stage)

S.j

S.m

S.h

AQUATIC SNAILS

Miracidium hatches

Sporocyst in snail (2 generations)

Penetrates into snail tissue

Fig. 21.1 Diagram of the life-cycles of schistosome species.

becoming transformed into a primary sporocyst containing numerous germinal cells, each of which is released inside the body of the snail and develops into a secondary sporocyst. Within each secondary sporocyst many thousands of the next free-living stage, the cercaria, develop and are released into the water. The whole process takes 4–8 weeks, depending on the species, and all cercariae produced from a single miracidium are of the same sex. Many of these fork-tailed cercariae are released each day and swim around in the water for a few hours but can quickly penetrate through human skin if they come into contact with a new host. The snail hosts of the different species of schistosomes are quite distinct: *S. mansoni* develops in the relatively large discoid aquatic snail *Biomphalaria* which is found particularly in flowing water such as streams and canals or in lakes and can release hundreds of cercariae every day for months; the globose snail *Bulinus*, which acts as host for *S. haematobium* and *S. intercalatum*, lives principally in ponds; the very small amphibious snail *Oncomelania*, which transmits *S. japonicum*, is associated with the banks of rivers and rice paddies and produces only a few cercariae per day.

Cercariae attach to the skin and, by means of secretions from lytic enzymes and by a vigorous wriggling movement, lose their tails and penetrate the dermis. The immature schistosomes or schistosomules develop from the cercariae, which lose their surface coat to adapt to the change in osmotic pressure, and the ''unit membrane'' of the tegument is replaced by two closely apposed trilaminate membranes (Fig. 21.2). Schistosomules also lose their sensitivity to complement and the ability to produce the envelope of the ''cercarienhullenreaktion'' which surrounds cercariae incubated in immune serum [52]. The larvae are carried in the lymph to the lungs. They remain in the lungs for a few days before moving down the pulmonary veins to the left side of the heart and being pumped around the body to their next destination—the hepatic portal system. They still measure only a few millimeters and can elongate to pass through anastomoses between the arterial system and the portal system along the gut wall or reach the liver via the hepatic artery. In the liver the parasites take 3–4 weeks to reach maturity and during this time the male and female worms come together. They migrate together down the portal veins against the bloodstream until they reach their egg laying site in the small venules. In humans, *S. haematobium* migrates through the anastomoses between the portal and systemic veins in the pelvis.

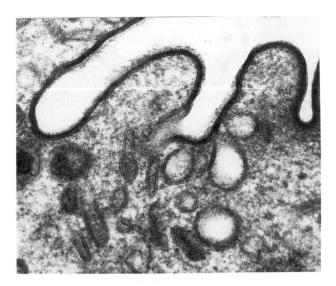

Fig. 21.2 Electron micrograph of the tegument and double outer membrane of *Schistosoma mansoni*. Original magnification, ×83 500. (Courtesy of Dr D.L. McLaren.)

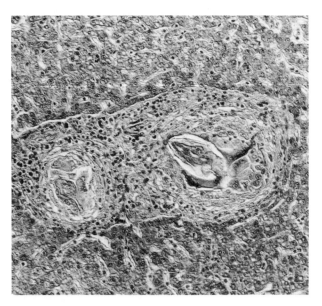

Fig. 21.3 Section of liver with schistosome circumoval granulomas (*Schistosoma mansoni*). There is a mixed leukocytic infiltrate around the eggs and presence of multinuclear cells. Original magnification, ×200.

The adult worms have a mean life expectancy of around 5 years, depending on the strain and species of parasite, but some may continue to produce viable eggs in people who are no longer exposed to transmission for more than 20 years.

Penetration of cercariae after primary exposure produces a transient dermatitis (swimmer's itch). Infections with *S. japonicum* can cause hypersensitivity reactions at the time when eggs are first produced, with lymphadenopathy, splenomegaly, diarrhea, lung eosinophilia, and liver tenderness (Katayama fever).

Beginning about 10–12 weeks after infection there is rapid production of eggs which penetrate through the mucosa of the colon, or the bladder in the case of *S. haematobium*. In chronic infections, many eggs get trapped in the tissues with miracidial secretions stimulating immune responses until death in about 4 weeks. Eggs become surrounded by inflammatory cells, leading to the formation of granulomas or pseudotubercles (Fig. 21.3), with eventual fibrosis [53]. These are the cause of all the important pathologic effects of schistosomiasis and in general the severity of disease is proportional to the number of eggs in the tissues. In intestinal schistosomiasis (which is caused by all species except *S. haematobium*) the colon becomes thickened and the mucosa may have many small ulcers surrounding granulomas. In Egypt in particular, papillomas and inflammatory polyps develop in *S. mansoni* infections and can even lead to obstruction of the lumen of the colon. However, in the majority of patients the phase of intestinal damage is accompanied by few symptoms, except perhaps for headache and bloody intermittent diarrhea.

Schistosomiasis hematobium, in which the eggs cause granulomatous lesions of the bladder and ureters, is usually a mild infection, the only sign in children being a recurrent painless hematuria, but occasionally granulomas in the ureter produced by only a few worms can have a devastating effect on the urinary system. In heavy infections, however, the bladder wall becomes fibrosed and has a reduced capacity. Bladder cancer is also more common where prevalence is high, although other factors also appear to be involved.

In heavy infections with *S. japonicum* and *S. mansoni*, many eggs are carried to the liver and the granulomas formed around them can lead to an extensive coarse periportal fibrosis (Symmers' clay pipestem fibrosis) 5–15 years after infection. The resulting portal hypertension causes liver and spleen enlargement and possibly ascites; it also leads to gross enlargement of the esophageal and gastric veins (esophageal varices), which sometimes burst. The periportal fibrosis (without significant destruction of liver cells), which is the result of an initial chronic perisinusoidal inflammation, is most severe in *S. japonicum* infections because of the high numbers of eggs.

Eggs can reach other sites in the body and in Egypt

and Brazil those of *S. mansoni* deposited in the small pulmonary arterioles cause pulmonary hypertension and corpulmonale. Cerebral and spinal cord granulomatous lesions are caused principally by *S. japonicum*, probably because the eggs are smaller and almost spineless (Fig. 21.1).

Clinical manifestations are likely to be more severe with schistosomiasis japonicum. The Katayama syndrome may cause fever, abdominal pain, and diarrhea, followed in a few years in a minority of patients by severe hepatosplenic schistosomiasis with dysentery, emaciation, edema of the limbs, anemia, and hepatic and splenic enlargement. Cardiopulmonary disease may develop and in both *S. japonicum* and *S. mansoni* infections esophageal varices may lead to the development of a collateral circulation. Carcinoma of the colon is also associated with *S. japonicum* infections.

The most important methods of diagnosis are still parasitologic by finding the characteristic eggs in a sample of feces (except for *S. haematobium*) examined under the microscope. Various concentration techniques can also be used, such as sedimentation, Kato thick smear, formol-ether sedimentation, MIFC method, and filtration staining methods. For *S. haematobium* infections most eggs are passed out in the urine and the deposit of centrifuged or sedimented urine, preferably obtained near midday, can be examined microscopically. A filtration staining method can also be used. A rough guide to the presence of infection can also be obtained by the dipstick method of measuring hematuria.

Rectal biopsies examined microscopically as a crush preparation between two microscope slides can also show the presence of eggs (even of *S. haematobium*).

Many immunodiagnostic methods have been investigated but most are more useful for seroepidemiologic surveys rather than for individual diagnosis [54,55]. Recently, large numbers of antigens have been characterized. The enzyme-linked immunosorbent assay (ELISA) test using antigens purified from soluble egg homogenates (e.g., CEF6) appears to be highly sensitive and very specific, particularly for the diagnosis of *S. mansoni* [56].

The schistosomes have continued to attract numerous research studies in the last few years partly because of the importance of the human diseases they cause but also because *S. mansoni* and more recently *S. japonicum* can be maintained with relative ease in inbred mice and laboratory-reared snails, and many aspects of the host–parasite relationship appear to follow a similar chronic course to human infections [57].

There have also been recent advances in understanding the immunology of human infections and in applying recombinant DNA techniques to the cloning of schistosome antigens [58]. As stated by Butterworth [59], "immunologists have been promising a schistosome vaccine for sufficiently long that an amused or angry scepticism among administrators of practical control programs is entirely understandable." However, there have been great advances in the last few years. Immunization with irradiation-attenuated larvae has been shown to confer a high degree of immunity in rodents, primates, and cattle. Such a vaccine against a cattle species, *Schistosoma bovis*, has been successfully field tested and is now a practical proposition [60]. While the use of living larvae cannot be envisaged in humans, many protective antigens are being identified in mice and the production of various monoclonal antibodies that confer passive protection *in vivo* has indicated that antibodies against a single antigen can confer protection [61–63]. Recombinant DNA techniques have discovered various molecules which confer protection in mice [64, 65]. While it is unlikely that any vaccine would confer complete immunity, it should be able to greatly lessen disease and reduce transmission rates. One of the current concerns is whether any vaccine should contain single or multiple epitopes [66].

Optimism that a vaccine against human schistosomiasis is feasible has also been bolstered by recent long-term human epidemiologic studies carried out with *S. haematobium* in the Gambia [67] and with *S. mansoni* in Kenya, Egypt, and Brazil [68,69]. These studies have examined intensities of reinfection after treatment in individuals whose level of contact with contaminated water can be observed. These studies show that although water contact undoubtedly decreases with age, there is also the slow development of an additional age-dependent resistance to infection, probably partly caused by a decline in inappropriate "blocking" antibodies [59]. Until these studies, it was by no means certain that any protective immunity does develop in humans. The protective effect of IgE in humans against reinfection with *S. haematobium* has also been demonstrated recently [70].

Order Opisthorchiiformes, family Opisthorchiidae

Clonorchis sinensis (human liver fluke). The human liver flukes *Clonorchis sinensis* and the closely related *Opisthorchis viverrini* are endemic in areas of Southeast Asia where freshwater fish are eaten uncooked, with about 28 million cases worldwide.

The hermaphroditic adults measure about 15 mm in length and 4 mm in width and are very flattened. They inhabit the bile ducts, and eggs produced in large numbers are passed out in the feces. For further development the contained miracidia have to reach freshwater colonized by suitable species of snails. Once inside the snail the usual digenean process of asexual multiplication takes place and the emergent cercariae penetrate many different kinds of freshwater fish; when infected fish are eaten uncooked, the immature worms excyst in the intestine and pass up the main bile duct.

In the early stages of infection the biliary epithelium becomes inflamed and in heavy infections there may be proliferation accompanied by production of mucus; this is followed by an encapsulating fibrosis of the ducts. Smaller portal vessels also become fibrosed so that portal hypertension and splenomegaly may ensue. Carcinoma of the bile ducts is also associated with *Clonorchis* and *Opisthorchis* infection.

About three-quarters of cases have less than 100 worms and suffer no symptoms at all. A few have heavy infections of over 1000 worms and suffer from acute liver pain, fever, recurrent cholangitis with stones, pancreatitis, and a slight jaundice.

Diagnosis is almost entirely by examination of feces for eggs.

Both *Clonorchis* and *Opisthorchis* have animal reservoir hosts, dogs and cats for the former and civet cats for the latter. There has been some recent experimental work on pathology and immunology in hamsters [71].

Family Heterophyidae

Heterophyes heterophyes. These are tiny trematodes, adults measuring 2 mm in length, which reside in the small intestine attached to the mucosa or between the villi. Human infection occurs in some Mediterranean countries, particularly the delta of the Nile, and in five countries of Southeast Asia [72].

Eggs which are passed out in the feces release a miracidium in freshwater and there is the usual multiplicative asexual cycle in snails. Emerging cercariae encyst in the muscles or under the scales of freshwater fish. Infection is contracted by eating raw or undercooked fish.

Most cases of infection are asymptomatic. In heavy infections there may be mild inflammation and necrosis of the mucosa resulting in abdominal tenderness, colicky pain, and mucus diarrhea.

Diagnosis is by finding the small eggs in the feces.

A similar parasite *Metagonimus yokogawai* is found in humans only in Southeast Asia. Heterophyidiasis is a zoonotic infection, the natural hosts for this group of trematodes being fish-eating mammals or birds.

Order Plagiorchiiformes, family Troglotrematidae

Paragonimus westermani (lung fluke). Many species of lung flukes occur in carnivores throughout the world but most of the estimated 5 million human infections are caused by *P. westermani* in Southeast Asia [73].

Adult *Paragonimus* are hermaphroditic trematodes which inhabit pulmonary cysts, usually in pairs. They measure about 12 mm in length by 6 mm in width.

Eggs which are produced by the adult worms are conveyed up the trachea and passed out in sputum or are swallowed and escape in the feces. Provided that an egg reaches freshwater, a miracidium larva develops inside it in weeks. The miracidium hatches out, enters a suitable snail, and undergoes the usual multiplicative processes inside the snail before the next free-living stages, the cercariae, emerge. Cercariae penetrate into a freshwater crab or crayfish and, when the infected crustacean is eaten uncooked, the metacercarial larvae excyst and reach the lungs after penetrating through the intestinal wall, the diaphragm, and the lung capsule.

Pairs of developing worms in the lungs provoke inflammatory and granulomatous reactions around them to form a cyst with an opening into a bronchiole. Cysts contain a brown purulent fluid and there may be up to 25 of them, each measuring about 20 mm in diameter. In chronic infections the cyst walls become fibrotic and eventually calcify. Eggs can also become surrounded by pseudotubercles. Worms which migrate out of the lungs along the soft tissues to the brain become foci for abscesses and can cause symptoms resembling epilepsy, a cerebral tumor, or an embolism. Paralysis can also result.

Most clinical cases of pulmonary paragonimiasis, who exhibit a dry cough with chest pain and bloody sputum (hemoptysis), are found in young adults but cerebral cases are often found in children.

Diagnosis is usually by finding the characteristic eggs in the feces or sputum. Serologic tests such as ELISA are fairly specific and become negative a few months after treatment.

There have been few recent experimental studies on *Paragonimus* or *Clonorchis*, principally because the life-cycles cannot be satisfactorily maintained in the laboratory away from an endemic area. The success of chemotherapeutic treatment with praziquantel against all trematode infections of humans, except rather unexpectedly *Fasciola* infections, has also tended to discourage both research and epidemiologic studies on these organisms [74,75].

Order Echinostomatiformes, family Fasciolidae

Fasciola hepatica (sheep and cattle liver fluke). *Fasciola* is a common parasite of sheep and cattle kept on damp pastures in many parts of the world but only occasionally does it infect humans. Most human cases have been from South America, Cuba, North Africa, and from western France and the UK in years with a wet summer.

The hermaphroditic adults measure about 30 mm in length and 12 mm in width and are much flattened dorsoventrally. They inhabit the bile ducts and eggs are passed out in the feces. Miracidium larvae develop inside the eggs on wet pastures, hatch out, penetrate into suitable amphibious snail hosts, and undergo a phase of asexual multiplication inside the snail. The cercariae which emerge from the snail form metacercarial cysts on certain vegetation—usually wet grass, but watercress or radishes in the case of human infection.

The larvae ingested on vegetation excyst in the duodenum and reach the bile ducts by penetrating the duodenal wall and eating their way through the liver tissues (Fig. 21.4). Human cases usually present a few weeks after infection with symptoms of nausea, fever, and acute abdominal pain, accompanied by a very high eosinophilia [76]. The chronic phase, which commences when the parasites reach the bile ducts in about 5 weeks, is characterized by a marked biliary fibrosis and is often accompanied by hepatomegaly. In chronic infections diagnosis can be made by searching for eggs in the feces, but in the early stages immunodiagnostic methods are necessary and highly purified specific antigens which will distinguish *Fasciola* from *S. mansoni* or *Paragonimus* infections have been described and large quantities of specific antigens have been produced by cotransformation in mouse tissue culture cells [77].

Fasciola can be maintained in both mice and rats and are widely used for pathologic, chemotherapeutic, and immunologic studies [78].

Fasciolopsis buski. This is the largest trematode infecting humans, adults measuring up to 75 mm in length and

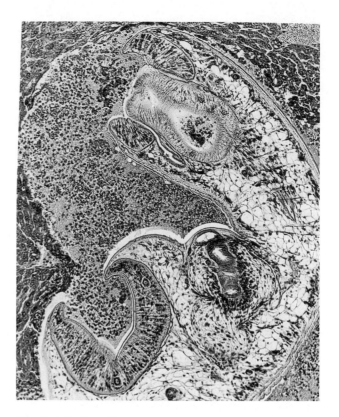

Fig. 21.4 Section of immature *Fasciola hepatica* migrating through the liver in an experimental infection; oval (left) and ventral suckers are visible. There is dense leukocytic infiltration. Original magnification, ×450. (Courtesy of Mr K.W. Iles.)

inhabiting the small intestine. The largest endemic foci of infection are in Central and South China, with smaller foci in other parts of Asia [79].

The life-cycle is similar to that of *Fasciola* except that the snails are aquatic; human infection is contracted from ingesting metacercarial cysts on the fruits and roots of water plants (e.g., water caltrop, water chestnut, and lotus).

In most cases infections are light and are symptomless. in heavy infections the adult worms cause traumatic damage to the mucosa at the site of attachment with an excessive production of mucus. Clinically there is nausea, diarrhea, and intense griping pains. A characteristic facial edema and other signs of generalized allergy can also occur.

Diagnosis is by finding eggs in the feces.

Fasciolopsiasis is a zoonosis, hogs acting as reservoir hosts.

Class Cestoda (the tapeworms)

All adult tapeworms or cestodes of humans inhabit the intestinal tract and are flat ribbon-shaped worms, lacking a gut and consisting of a chain of separate segments or proglottids. They all attach to the intestinal wall by an anterior holdfast region, usually equipped with suckers and hooks (Fig. 21.5).

Although infections with adult tapeworms are relatively benign, with clinical symptoms often only appearing once the infection has been diagnosed, infection with larval tapeworms can be very serious and in some parts of the world cysticercosis (infection with the cysts of *Taenia solium*) and echinococcosis (infection with larvae of *Echinococcus granulosus*), are major public health problems.

The principal human tapeworms are listed in Table 21.3.

Family Taeniidae

Adult taenias of humans are located in the ileum with their scolex embedded in the mucosa and the rest of the organism, measuring about 5 m, hanging free in the lumen.

Taenia solium. This is the pork tapeworm, with adults in humans and larvae in the hog. Distribution is worldwide where pork is eaten undercooked, with an estimated 3 million cases in the world.

The life-cycle is shown in Fig. 21.6.

Humans are an obligatory part of the life-cycle and acquire infection by ingesting the encysted, hol-

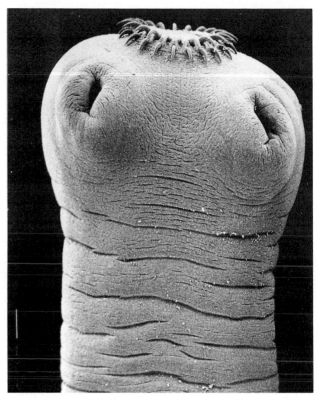

Fig. 21.5 Scanning electron micrograph of the scolex of *Taenia solium* (scale = 100 μm). (Courtesy of Dr L.M. Gibbons.)

low, second-stage larva containing an invaginated protoscolex, known as a cysticercus or bladder worm, in improperly cooked pork or pork products. The larval parasite grows to maturity in 2–3 months and the posterior gravid segments filled with eggs are shed into

Table 21.3 Cestode infections

		Distribution	Definitive host	Intermediate host*
Order Cyclophyllidea				
Family Taeniidae	*Taenia solium*	Cosmopolitan	Humans	Hogs (humans)
	Taenia saginata	Cosmopolitan	Humans	Cattle
	Echinococcus granulosus	Cosmopolitan	Dogs	Sheep, cattle, camels, goats (humans)
	Echinococcus multilocularis	Nearctic, central Europe	Foxes	Rodents (humans)
Family Hymenolepidiidae	*Hymenolepis nana*	Cosmopolitan	Humans	None
	Hymenolepis diminuta	Cosmopolitan	Rats (humans)	Fleas, flour beetles
Order Pseudophyllidea	*Diphyllobothrium latum*	Finland, North America, East Europe, and SE Asia	Humans, dogs	Copepods then fish

* Parentheses indicate reservoir host.

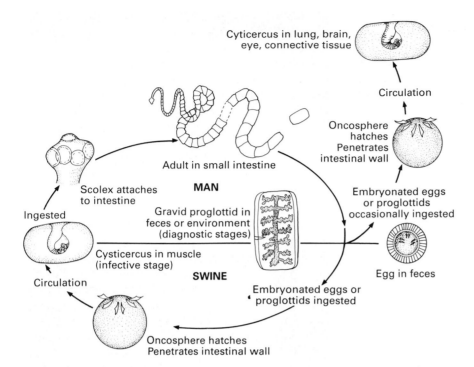

Fig. 21.6 Diagram of the life-cycle of *Taenia solium*.

Within the figure:
- Cyticercus in lung, brain, eye, connective tissue
- Circulation
- Adult in small intestine
- Oncosphere hatches Penetrates intestinal wall
- Scolex attaches to intestine
- **MAN**
- Ingested
- Gravid proglottid in feces or environment (diagnostic stages)
- Embryonated eggs or proglottids occasionally ingested
- Cysticercus in muscle (infective stage)
- Circulation
- **SWINE**
- Egg in feces
- Embryonated eggs or proglottids ingested
- Oncosphere hatches Penetrates intestinal wall

the lumen of the bowels and escape out of the anus, often under their own volition.

Thousands of eggs are liberated from the segments onto soil and are ingested by hogs. Inside the intermediate host the oncosphere larvae released from eggs penetrate the intestinal mucosa and are carried to the muscles, lungs, brain, heart, and connective tissues, where they develop into cysticerci measuring about 8 × 5 mm.

The pathology caused by the presence of adult tapeworms is virtually nil but infection with larvae can be serious (see below).

Diagnosis is by finding the characteristic gravid proglottids expelled from the anus. Eggs can be found in the feces in about 50% of infected individuals.

Taenia saginata. This is the beef tapeworm in which humans are an obligatory definitive host and cattle the intermediate host. Otherwise, the life-cycle is similar to *T. solium* except that it is a more cosmopolitan infection since beef is more widely eaten and also more often eaten undercooked. There are an estimated 45 million cases worldwide.

The gravid proglottids passed out of the anus can usually be distinguished from those of *T. solium* although the eggs are identical.

Cysticercosis. Cysticercosis is human infection with the cysticerci of *T. solium* and is endemic everywhere; infection with the adult tapeworm occurs particularly in Mexico, South America, Indonesia, and South Africa [80]. It is contracted by ingesting eggs on salad vegetables; those of *T. saginata* cannot develop in humans.

Cysts in most parts of the body are benign and often calcified cysts are only recognized on a radiograph. Cysts induce a granulomatous reaction, becoming surrounded first by an intense inflammatory reaction and then by a fibrous capsule and may be palpable if located in the subcutaneous tissues.

Cysts in the brain can produce a variety of symptoms—motor, sensory, or mental—depending on their location. Epilepsiform attacks of the Jacksonian type often occur a few years after infection and other signs and symptoms include increased intracranial pressure, hypertension, transient hemiplegia, and visual and aural symptoms [81,82]. This is the most common parasite of the central nervous system of humans. Cysts in the eye do not become encapsulated but can result in pain and blurring of vision.

Diagnosis of cysticercosis is often by radiography, with computerized tomography (CT) proving to be particularly useful for cerebral cysticercosis. An ELISA test using purified antigens from cysts has shown high sensitivity and good specificity [83] and encouraging results have been obtained by employing a specific monoclonal antibody.

The tapeworm *Taenia taeniaeformis*, with adults in cats and cysts in rodents, is proving useful as a model system, particularly for studies on genetics of resistance [15–17].

Echinococcus granulosus. Echinococcosis is caused by infection with the enormous cystic larval stage of the minute adult tapeworm of the dog and other carnivores. This tapeworm is unusual in that its cystic stage keeps on growing in the intermediate host, which can be humans, with the production of thousands of larvae (protoscolices) from a single infective egg.

The life-cycle is summarized in Fig. 21.7.

Eggs which are morphologically identical to those of *Taenia* and equally resistant to environmental conditions are passed out in the feces of dogs and normally infect sheep, cattle, goats, or camels by ingestion on pasture [84]. Infection now occurs in most parts of the world apart from the Pacific islands and appears to be spreading [85]. The highest human rates of infection (0.22%) occur in the nomadic Turkana people of northern Kenya, who have a very close relationship with dogs which are used as nursemaids to infants [86].

Inside the intermediate host an oncosphere larva emerges from the egg and develops in the liver, lungs, or other organs of the herbivore into a large slow growing and thick-walled hydatid cyst. The inner layer of the cyst produces many thousands of protoscolices which bud off into the contained fluid, each capable of developing into an adult tapeworm if the cyst is ingested by another dog. Inside the main cyst there may be numerous thin-walled daughter cysts produced, each containing many protoscolices (Fig. 21.8). When dogs ingest the hydatid cysts, usually from eating discarded offal, adult worms measuring 5 mm mature in the small intestine in about 7 weeks.

Humans become accidental intermediate hosts by ingesting eggs which may be present on contaminated foodstuffs or by contact with dogs. Cysts develop slowly in humans. First the young larva develops into a hollow bladder which becomes surrounded by a host-produced fibrous capsule and after 5 months the cyst measures about 10 mm in diameter. Cysts continue growing and can contain many million of protoscolices in 15–20 liters of fluid. Those in humans are most commonly found in the liver (about 50% of cases) and cause compression of the liver cells which can lead to biliary stasis and cholangitis if they rupture into the biliary tract. Lung cysts (40% of total) are more spherical than liver cysts

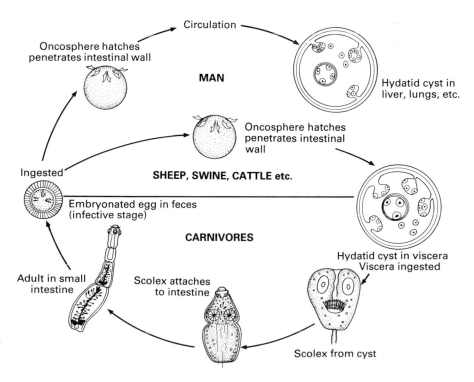

Fig. 21.7 Diagram of the life-cycle of *Echinococcus granulosus.*

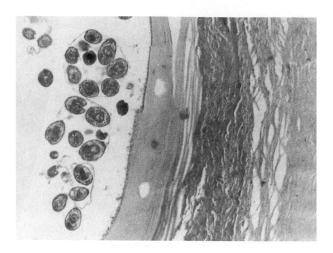

Fig. 21.8 Section of wall of hydatid cyst of *Echinococcus granulosus*. Thin-walled daughter cysts containing protoscolices are budding off from the thin germinal layer. The outer wall of the cyst is laminated and there is a fibrotic host reaction. Original magnification, ×110.

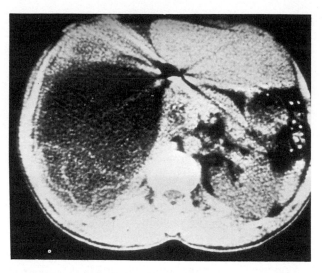

Fig. 21.9 Computerized tomography scan of abdomen with large cyst of *Echinococcus granulosus* in liver (dark area on left). (Courtesy of Dr A. Bryceson.)

and their rupture can result in hemoptysis from bursting of the pulmonary capillaries. When developing in bones the cysts fill the marrow cavities and may cause bone erosion but do not elicit a host reaction.

Infection often occurs in childhood but pathologic changes, caused almost entirely by pressure effects, may only become apparent in later life, except for cysts in the brain or eyes. Many cysts die and calcify in the tissues without being diagnosed.

The cyst is filled with hydatid fluid, daughter cysts, and hydatid sand (consisting of thousands of minute free protoscolices); if it ruptures into the abdominal or pleural cavity there can be acute anaphylactic shock, and each released protoscolex is capable of developing into a new cyst.

Diagnosis is usually by radiology or CT scanning when the appearance of a cyst is characteristic (Fig. 21.9). Ultrasound is also very satisfactory and portable machines have proved useful in northern Kenya [86]. Latex agglutination and indirect hemagglutination tests are simple to perform but not very specific [86]. A purified antigen (fraction 5) isolated from hydatid fluid can specifically measure IgE antibodies by immuno-electrophoresis or ELISA [87]. However, in Kenya false negatives are common and detection of circulating antigen appears more promising [88].

Strain variation is a feature of *Echinococcus*, with differences in ability to develop *in vitro*, in isoenzyme patterns, in energy metabolism, and with DNA hybridization in restriction site analysis [89–93]. These differences are reflected in infectivity to various intermediate hosts; the dog/horse strain in the UK for instance probably not being infective to humans.

Echinococcus multilocularis. This is a similar parasite in which the minute adult tapeworms occur in the intestines of foxes and dogs but in which the larvae normally develop in small holarctic rodents in a matter of weeks rather than years. The cysts do not have a thick outer layer and form a spongy mass of proliferating pseudomalignant vesicles in the liver.

Humans usually become infected by ingesting eggs on wild berries or vegetables or from fur trapping. Cysts in humans do not contain protoscolices but there is a tendency for them to spread and human cases are often fatal. Infection is most common in arctic regions but there are also occasional cases in central Europe and new foci have recently been identified in China, India, Iran, and the USA (in addition to Alaska) [92].

Experimentally, almost the entire life-cycles of both species of *Echinococcus* have been duplicated *in vitro* and this has enabled many physiologic and biochemical, but not so far immunologic, problems to be elucidated [93]. However, immune mechanisms and possibilities for the production of vaccines have recently been reviewed [94].

Family Hymenolepidiidae

Hymenolepis nana. This is the dwarf tapeworm of humans, adults measuring only 15–40 mm with about 200 proglottids and a minute hooked scolex. It is cosmopolitan in distribution, usually in children, and particularly in warm countries and in southern and eastern Europe: there are estimated to be 43 million cases worldwide.

Hymenolepis nana is the only tapeworm known which does not require an intermediate host; eggs passed in the feces of an infected individual can develop in another individual when ingested. The onchosphere larva released from an egg enters a villus of the ileum, transforms in the mucosa into a solid cysticercoid larva, and is released back into the lumen to mature into an adult tapeworm. Thus, although adult worms live for only a few months, repeated reinfection can occur.

Light infections are generally asymptomatic but in a small proportion of cases there may be over 2000 worms present and these cause abdominal pain, headache, dizziness, anorexia, diarrhea, nausea, and vomiting. The larvae cause superficial damage to the mucosa and can cause disturbances to protein digestion.

Diagnosis is parasitologic by finding the characteristic eggs in the feces.

Hymenolepis nana var. *fraterna* is a common parasite of the mouse and has been widely used for biochemical and physiologic studies. In immunosuppressed mice, cysticercoids have developed aberrantly and invaded many organs but this has not been reported from humans [95].

Hymenolepis diminuta. This is a rather larger tapeworm of rats which utilizes fleas or beetles as intermediate hosts and is a not uncommon parasite of humans in some poor areas plagued by rodents. It is also a widely used laboratory species [96].

Order Pseudophyllidea

Diphyllobothrium latum. This is the broad fish tapeworm of humans, the adult worms measuring 3–10 m in length and with up to 4000 proglottids. It occurs in temperate countries with many lakes, such as Finland, the USSR, Central Europe, Japan, the Great Lakes regions of the USA and Canada, and Chile, and it has been estimated that there are about 9 million cases worldwide.

Eggs are passed out in the feces and have to reach fresh water for further development. A larva develops inside the egg, hatches out, and is ingested by microcrustacea in the water (*Diaptomus* or *Cyclops*). These are in turn ingested by fishes, and larvae (plerocercoids) develop in the muscles of the fish. When infected fish such as pike or trout are eaten raw by humans, the adult tapeworms mature in the ileum, attaching to the mucosa by two sucking grooves on the scolex.

Clinical effects are usually minor but the most notorious is a pernicious-type megaloblastic (macrocytic) anemia that was once particularly common in Finland but is now rare, probably because of a better winter diet and health care and more effective treatment. The anemia is caused by a competitive uptake of vitamin B12, and a biochemical pathway involving this vitamin has been described from a related tapeworm [97].

Diagnosis is by finding eggs in the feces.

PHYLUM NEMATODA (THE NEMATODES OR ROUNDWORMS)

The nematodes are one of the most successful groups of animals, occupying all available habitats, whether aquatic, terrestrial, or as parasites of plants and animals. They have accomplished this with a remarkable uniformity of structure. About one dozen species are natural parasites of humans, although many more animal species sometimes cause zoonotic infections; they range in size from 1 to 700 mm and have separate sexes (Table 21.4). All are elongated, cylindrical, organisms with a mouth and an anus and have a body covered with a tough, multilayered, proteinaceous cuticle. Until the last few years this cuticle was thought to be inert but it is becoming apparent that it is a dynamic layer, continuously shedding surface components and in some species permeable to nutrients [98–100]. Collagen represents about 80% of nematode cuticle and at least 20 collagen genes have been identified in *Ascaris*, scattered throughout the genome [101]. Work on gene regulation in nematodes is gleaning a lot from detailed studies carried out on the free-living nematode *Caenorhabditis elegans* [102,103].

The soil-transmitted intestinal helminths, known as geohelminths (*Ascaris*, hookworms, *Strongyloides*, and *Trichuris*), are the most prevalent human parasites and among helminths are second only to the schistosomes in world importance as causes of human suffering. Typically these nematodes all cause chronic infections and their contribution to malnutrition has in the past either been dismissed as unimportant or alternatively

Table 21.4 Nematodes of medical importance

	Mode of transmission	Normal definitive host
Subclass Secernentea		
Order Ascaridida		
Ascaris lumbricoides	Soil transmitted	Humans
Toxocara canis	Soil transmitted	Dogs
Anisakis simplex	Ingested in fish	Marine mammals
Pseudoterranova decipiens	Ingested in fish	Marine mammals
Order Strongylida		
Ancylostoma duodenale	Soil transmitted	Humans
Necator americanus	Soil transmitted	Humans
Angiostrongylus cantonensis	Ingested in snails, slugs, salad	Rodents
Angiostrongylus costaricensis	Ingested in slugs, salad	Rodents
Order Rhabditida		
Strongyloides stercoralis	Soil transmitted	Humans
Order Oxyurida		
Enterobius vermicularis	Direct contamination	Humans
Order Spirurida		
Superfamily Filarioidea		
Wuchereria bancrofti	Insect transmitted	Humans
Brugia malayi	Insect transmitted	Humans, monkeys
Onchocerca volvulus	Insect transmitted	Humans
Loa Loa	Insect transmitted	Humans
Mansonella spp.	Insect transmitted	Humans
Superfamily Dracunculoidea		
Dracunculus medinensis	Ingested in microcrustacean	Humans
Superfamily Gnathostomatoidea		
Gnathostoma spinigerum	Ingested in microcrustacean	Dogs, cats
Subclass Adenophorea		
Superfamily Trichuroidea		
Trichuris trichiura	Soil transmitted	Humans
Trichinella spiralis	Ingested in meat	Many scavenging mammals
Capillaria philippinensis	Ingested in fish	Birds (?)

regarded as self-evident; it is only in the past few years that there has been a resurgence of interest in these nematodes that is unmatched since the efforts of the Rockefeller Commission in the 1920s and their population dynamics and effect on nutrition are being scientifically evaluated [104–108]. One feature that is typical of almost all human helminth infections is the stability of helminth populations within human communities, which indicates that density-dependent checks on parasite establishment, survival, and fecundity play an important role in transmission dynamics [109–111].

The great majority of individuals infected with geo-helminths harbor only a few worms but with a small proportion of "wormy people" having very high worm burdens. This predisposition to heavy infection has often been attributed to behavioral or social factors but from laboratory studies appears also to be significantly affected by genetic and nutritional factors [112]. The intriguing question of why in endemic areas such a high proportion of the population can apparently control the numbers of worms present but cannot eliminate them completely is also being addressed [113,114].

Tissue nematodes all have an indirect life-cycle involving another host. In most cases this is an arthropod vector, but *Trichinella* larvae are ingested in the flesh of other mammals.

Subclass Secernentea, order Ascaridida

Ascaris lumbricoides. The large roundworm of humans is one of the most common and most widespread of human infections, with an estimated 1.5 billion cases world-wide. It is a long established human parasite having been reported from humans in antiquity and was first named by Linneaus. The large stout adult nematodes measuring 200×4 mm have separate sexes and inhabit the lumen of the small intestine.

The adult female produces about 240 000 fertilized but undeveloped eggs every day, which are passed out in the feces. Worms live for up to 1 year so each female can pass out 65 million eggs in its lifetime. The eggs are very resistant to disinfectants, composting, or cold.

The life-cycle is illustrated in Fig. 21.10.

If feces containing eggs is deposited on soil, a larva develops inside the egg in about 2 weeks under warm, moist conditions. When these infective eggs are swallowed, either in soil around the house in the case of young children or on salad vegetables, the larvae hatch in the small intestine, penetrate the mucous membranes, and are carried by the bloodstream to the liver, heart, and lungs. In the lungs they penetrate into the alveoli,

ascend the bronchi and trachea, and are swallowed again to reach maturity in the intestine in approximately 2 months. The migrating lung larvae produce numerous lesions, sometimes giving rise to Loeffler's syndrome, characterized by fever, cough, dyspnea, and a high eosinophilia.

The majority of cases have only a few worms (the average number is six) and cause a minimum of distress but the presence of numbers over 100 can result in digestive disorders and, particularly in children, a protein energy malnutrition [115]. Intestinal obstruction results in severe abdominal pain with nausea and vomiting, worms in the bile duct to cholangitis and gallstones with severe colicky pains, nausea, fever, and a slight jaundice, and worms in the pancreatic duct to pancreatitis. While it is obvious that intestinal obstruction or bile duct invasion are likely to have important clinical effects, and are a common cause of acute abdominal emergencies in children, recent studies have shown unequivocally that even moderate infections in children can impair lactose absorption [116] and cause intestinal hurry. Growth retardation in children reported from many countries is probably caused by a reduction in food intake [117]. Worldwide, it has been estimated that ascariasis is the cause of over 1 million cases of disease annually, with 20 000 deaths [118].

Diagnosis is almost invariably by the presence of

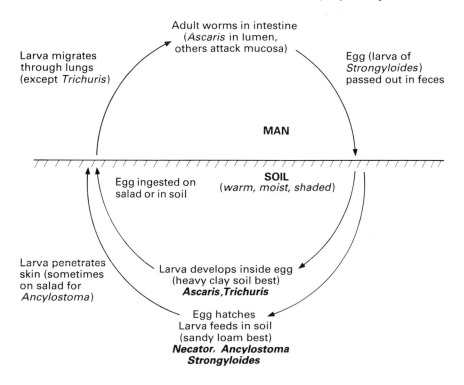

Fig. 21.10 Diagram of the life-cycles of soil-transmitted helminths.

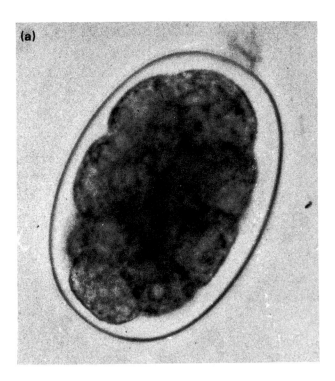

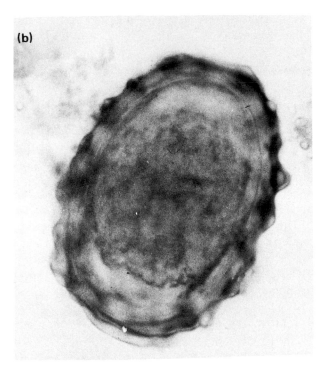

Fig. 21.11 The egg of (a) hookworm and (b) *Ascaris lumbricoides* as passed in feces. Original magnification, ×13 000).

numerous eggs in the feces, although sometimes adult worms can be visualized radiologically as a clear space after a barium meal (Fig. 21.11).

While *Ascaris lumbricoides* will only infect humans, the closely related *Ascaris suum* of hogs is being used as a model [119], particularly for nutritional studies in the natural host and immunologic studies in rodents, although larvae do not mature in these latter hosts.

Toxocara canis. Eggs of the common and cosmopolitan dog roundworm, *Toxocara canis*, passed in the feces of young dogs are widely distributed in the soil of most parts of the world and, when ingested by children, larvae can hatch in the intestine and reach the blood-¨tream. *Toxocara* larvae do not develop further in humans but being thinner than the larvae of *Ascaris* can sometimes pass through the sinusoids of the liver and disperse to the kidneys, brain, or eyes. Larvae usually end up in the liver where they elicit a strong inflammatory response, each becoming surrounded by an eosinophilic microabscess which eventually fibroses to give a granuloma. Occasionally larvae in the eye can cause serious damage. Overt disease is rare, usually consisting of long-lasting liver enlargement with hypereosinophilia and fever (visceral larva migrans).

However, toxocaral antibodies shown by ELISA are present in an appreciable proportion of children in many developed and developing countries. Seropositivity levels are over 50% in some developing countries but the existence of any associated morbidity has not been assessed [120]. The cat roundworm, *T. cati*, is another important cause of visceral larva migrans.

Anisakis spp. (causes herring worm disease) and Pseudoterranova decipiens (causes codworm disease). Both are natural parasites of fish-eating mammals. Human infections are contracted by ingesting larvae in the fish intermediate hosts, and occur in countries where fish are eaten raw, such as the Netherlands and Japan. The larvae cause abscesses in the stomach or intestinal wall [121].

Diagnosis is usually by examination of biopsy specimens or by endoscopy [122].

Order Strongylida

Hookworms. Two species infect humans, both attaching to the wall of the small intestine by their anterior ends. Infection occurs in most countries with a warm, moist climate and it has been estimated that there are about 1 billion cases worldwide [123,124].

1 *Ancylostoma duodenale* (the "Old World" hookworm). This is the larger of the two species, the female worms measuring about 12 mm in length and 0.4 mm in maximum diameter.

Thin-shelled eggs are produced by female worms at the rate of about 20 000 per day and are passed out in the feces (Fig. 21.10). If they reach well-aerated shady, sandy, soil at a temperature above 25°C, larvae hatch out and feed and develop to the third infective stage in under 1 week. Infective larvae can survive in the upper few millimeters of soil for a few weeks but in order to develop further must penetrate through the skin of another person. Alternatively, larvae may be swallowed on salad vegetables or possibly passed from mother to infant in breast milk. After skin penetration, larvae reach the circulation, penetrate into the alveoli of the lungs, and migrate up the trachea and down the esophagus, as for *Ascaris*. With oral infection there is apparently no larval migration through the lungs.

In areas of India at least, *Ancylostoma* larvae also undergo a period of hypobiosis or arrested development in the body [123,125]. This is an adaptation to climatic changes and ensures that the maturation of adults and production of eggs coincides with the most favorable ecologic conditions following the monsoon.

Adults mature in the small intestine in 5 weeks to many months. The adult worms attach to the internal mucosa by their anterior ends and cause small hemorrhages. The worms suck blood; they also change their position frequently and secrete an anticoagulant [126] so that in heavy infections there is a daily blood loss of about 100 ml and also a protein-losing enteropathy [127]. Hookworms, like many other intestinal nematodes, secrete acetylcholinesterase and it is possible that through its action on parasympathetic control of mucus secretion it may serve to interfere with or delay worm expulsion [128].

In the early stages of infection there may be pruritis at the site of entry of the larvae (ground itch) and the migration through the lungs can cause cough with sore throat and nausea (Wakana disease). Early symptoms of adult infection are epigastric pain resembling peptic ulcer, diarrhea with blood and mucus, anorexia, and eosinophilia [129]. A hypochromic microcytic anemia develops in chronic, heavy infections (e.g., over 1000 adult worms) or even in relatively light ones where dietary iron intake is low (e.g., 100 worms). In children, stunting of growth is characteristic of severe, chronic infections [107].

Diagnosis of infection is based primarily on fecal examination for eggs; concentration methods, such as the Kato thick smear technique, can be used but usually many eggs are present. Eggs of *Ancylostoma* and *Necator* are identical (Fig. 21.11). Quantitative egg counts are sometimes used to obtain an estimate of intensity of infection; these assume that there is a direct relationship between egg output and worm numbers, but this does not appear to be true for very heavy or very light infections [130,131].

2 *Necator americanus* (the "New World" hookworm). This is a similar parasite to *Ancylostoma*, except that the adults are slightly smaller and have a pair of cuticularized cutting plates in the mouth to bite off pieces of mucosa rather than teeth (Fig. 21.12) and infection is always percutaneous. *Necator* is also more strictly tropical than *Ancylostoma* and is the dominant species in the Americas, Central Africa, Southeast Asia, and the Pacific.

Recently human hookworms have been established in laboratory rodents and this should facilitate research on antigen characterization, immunity, and possibly also vaccine development [123,132,133]. Other strongylid nematode species of rodents which have been widely used as laboratory models for physiologic, biochemical, and immunologic studies are *Nippostrongylus brasiliensis* and *Heligmosoides polygyrus* (=*Nematospiroides dubius*) [24,27].

Angiostrongylus cantonensis.   This is a normal parasite of the rat, with adults occupying the pulmonary arteries. Larvae are passed out in feces and develop further when ingested by slugs or snails which act as intermediate hosts. When a human eats an infected mollusk, either as an edible snail or as a slug on salad vegetables, the third-stage larvae penetrate the intestine and are carried via the circulation to the brain. Here they cause an eosinophilic meningitis with intense inflammation of the brain and meninges under the arachnoid. There can be severe headache, back and neck stiffness, vomiting, and vertigo, which usually diminish in a few weeks but can be prolonged [134]. Occasionally there is eye involvement without meningitis. Human infection is sporadic on Pacific islands and in Southeast Asia, West Africa, and the Caribbean [135].

Immunodiagnostic tests include indirect fluorescent antibody and ELISA [136,137].

Angiostrongylus costaricensis.   The normal definitive host of this species is the cotton rat, *Sigmodon hispidus*, and

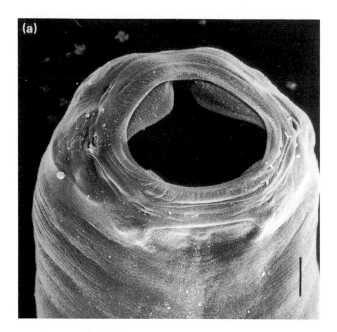
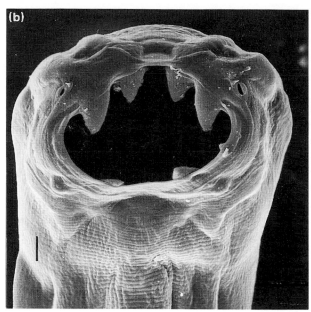

Fig. 21.12 Scanning electron micrographs of the anterior ends and buccal capsules of: (a) *Necator* with cutting plates; (b) *Ancylostoma* with two pairs of teeth! (Courtesy of Dr L.M. Gibbons.)

other rodents. Human infections were first described 20 years ago and it has now been found in several Latin American countries, but is usually diagnosed only from the routine examination of pathology specimens. However, recent findings suggest that it causes much more clinical disease than is realized [138].

In man, adult parasites are located in the ileocecal branches of the anterior mesenteric artery or sometimes vessels in the liver. Both adults and eggs produce inflammation and the intestine shows areas of necrosis and thickening of the wall. There is often a tumor-like mass in the large bowel. Liver lesions resembles those caused by *T. canis*.

Infections are most common in children who play with small slugs, which are the intermediate hosts.

Diagnosis is by an agglutination reaction using polystyrene beads or by ELISA.

Order Rhabditida

Strongyloides stercoralis. *Strongyloides* is the smallest nematode infecting humans. Parthenogenetically reproducing females measuring 2×0.05 mm live embedded in the mucosa of the small intestine. Infection is cosmopolitan with a similar distribution to hookworm but everywhere appears to be less common with an estimated 90 million cases worldwide. Prevalence rates

are usually below 1% in temperate climates but may exceed 25% in some areas of the tropics [139,140].

The life-cycle is shown in Fig. 21.13.

Strongyloides shows great developmental plasticity. In susceptible hosts, females produce eggs which develop into larvae in the lower gut and these larvae are passed out in the feces or develop and molt twice on the perianal skin into infective third-stage larvae which reinfect the same individual (autoinfection). Larvae from feces feed on microflora and develop to the infective stage in a few days in soil; these can penetrate through the skin to undergo a similar migration to hookworm larvae. However, under certain conditions, probably affected by the immunologic status of the host [141], the first-stage larvae in the soil develop into free-living males and females. After mating the females produce eggs and the third-stage larvae which develop can penetrate into a new host.

The majority of individuals infected with *Strongyloides* are asymptomatic and also appear to have few intestinal lesions. However, in a proportion of cases the submucosa becomes edematous and the mucosa becomes flattened and atrophic. Adults and larvae are present in the tissues and in very severe cases the epithelium becomes ulcerated with rigidity of the intestinal wall because of the edema and fibrosis. The patient suffers from intermittent abdominal pain, diarrhea alternating

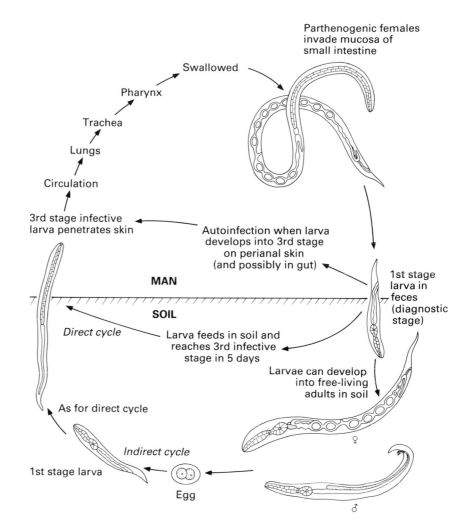

Parthenogenic females
invade mucosa of
small intestine

Swallowed

Pharynx

Trachea

Lungs

Circulation

3rd stage infective
larva penetrates skin

Autoinfection when larva
develops into 3rd stage
on perianal skin
(and possibly in gut)

MAN

1st stage
larva in
feces
(diagnostic
stage)

SOIL

Direct cycle

Larva feeds in soil and
reaches 3rd infective
stage in 5 days

Larvae can develop
into free-living
adults in soil

♀

As for direct cycle

Indirect cycle

1st stage larva

Egg

♂

Fig. 21.13 Diagram of the life-cycle of
Strongyloides stercoralis.

with constipation, and weight loss [142]. Mesenteric lymph nodes are likely to be enlarged. The enteritis may result in severe malabsorption and electrolyte imbalance, sometimes leading to death.

Because of the occurrence of autoinfection, an individual may be infected for a lifetime [143,144]; typical of these chronic infections are recurrent itchy rashes, usually in the perianal region where the larvae penetrate ("larva currens").

In patients on immunosuppresive therapy or who are severely malnourished, but not usually in AIDS patients [2,3], hyperinfection can ensue. In this often fatal condition, larvae penetrate through the gut into the lymphatics and may be found in many organs, particularly the lungs.

It is clear that much remains to be elucidated concerning the complex relationship between the host defence mechanisms and this unusual parasite and it is

not yet possible to formulate a general hypothesis to explain the circumstances under which severe disease ensues.

Diagnosis is usually by observing first-stage larvae in a fecal sample. Larvae are often difficult to find and concentration methods may be necessary, or larvae can be trapped on a swallowed brushed nylon string which is then withdrawn and examined (enterotest). Not surprisingly, most immunodiagnostic tests work well and the ELISA has high sensitivity and specificity, even in immunosuppressed patients. However, it is not clear how much past infection contributes to immunoreactivity and serologic tests for antibody need to be viewed with caution in people who reside in an endemic area [145].

Another species, not yet definitely identified, so termed *Strongyloides* cf. *fuelleborni* since it is morphologically identical to *S. fuelleborni* of African primates and sometimes humans, causes overwhelming infections characterized by an enormously swollen abdomen in young infants in areas of Papua New Guinea [146]. Infection can be differentiated from that due to *S. stercoralis* since eggs rather than larvae are passed in the feces.

Strongyloides stercoralis can be kept experimentally in dogs, cats, or primates but most laboratory studies have used *Strongyloides ratti* in rats or more recently in mice [147,148].

Order Oxyurida

Enterobius vermicularis (pinworm or seatworm). Enterobiasis has a worldwide distribution and is the only parasitic worm common in temperate regions, particularly in children. Adults are small nematodes (10 mm × 0.4 mm) with a pointed tail and inhabit the lumen of the cecum and appendix. The life-cycle is direct. Mature female worms crawl out of the anus usually at night and deposit sticky eggs on the skin in the perianal regions.

A first-stage larva develops in the egg in about 4 h after deposition and eggs containing these infective larvae are transmitted to a new host via the hands or in house dust: it is thus not a soil-transmitted nematode. The most common symptom of infection is anal pruritis which can lead to secondary bacterial infections following scratching. Migrating worms also invade the female reproductive tract and very occasionally worms block the lumen of the appendix leading to appendicitis. As with many other helminth infections, there appears to be a predisposition to infection so that a small proportion of infected individuals harbor a large number of worms, while most have just a few [149].

Infection is normally diagnosed by swabbing the anal region with sticky transparent tape and then examining this for eggs under the microscope.

Humans are the only host for this and a similar recently described species (*Enterobius gregorii*) [150], but another oxyurid of rodents, *Syphacia obvelata*, has been used as a laboratory model. Biologically this group is interesting since it appears to be the only endoparasitic haplodiploid taxon, i.e., females are diploid while males are haploid and are derived from unfertilized eggs [151].

Order Spirurida

Suborder Spirurina, Superfamily Filarioidea. Lymphatic filariasis. This is an important and spreading helminth disease of humans with about 90 million cases world-wide [152], transmitted by mosquitoes, and involving two species of filarial parasite.

The filarial nematode, *Wuchereria bancrofti*, infects only humans and is widely distributed in the tropics in Africa, Asia, the Pacific, South America, and the Caribbean. About two-thirds of the estimated total of infections come from China, India, and Indonesia. The long, thin adult worms (females 90 × 0.25 mm, males about half this size) inhabit the lymphatics of the axillary, inguinal, and genital regions. Female worms produce sheathed prelarval stages known as microfilariae which reach the bloodstream. In most parts of the world, microfilariae circulate in the peripheral circulation only at night—an adaptation to night-biting *Anopheles* and *Culex* mosquitoes. However, in the eastern Pacific region microfilariae are found in the blood during the day and are transmitted by day-biting *Aedes* mosquitoes.

The life-cycle is shown in Fig. 21.14 and the possible course of the disease in Fig. 21.15. In a proportion of individuals infected in childhood no symptoms ever appear. However, there may be recurrent episodes of fever, nausea, headaches, and possibly a rash lasting for about 2 weeks. This filarial fever may last for many years, accompanied by adenolymphangitis with painful inflammation of the lymph trunks and microfilariae disappearing from the blood. It is probable that the lymph vessels become blocked by an allergic reaction to adult worms but it is not known what triggers this. This results in leakage of lymph into the tissues. After each attack the edema may become harder and more permanent, with growth of new tissue and progressive fibrosis leading to a disfiguring and grotesque elephantiasis. In Sri Lanka and western Malaysia a linkage between human leucoyte antigen and elephantiasis has been observed with familial clustering [153].

People who move into an area show hyperreactivity and develop acute disease faster than indigenous populations.

In males, worms often block the spermatic lymph vessels, which leads to hydrocele. In some infected individuals microfilariae do not appear in the blood but infection is followed by pulmonary disorders such as fever, breathlessness, cough, and chest pain, accompanied by hypereosinophilia. This condition, known as tropical pulmonary eosinophilia (TPE), is caused by a hyperresponsive reaction to the microfilariae accumulated in the lung capillaries.

Microfilaremia is a state of long-term antigen-specific immunosuppression and host immunoregulation must play a decisive role in this chronic infection where adults

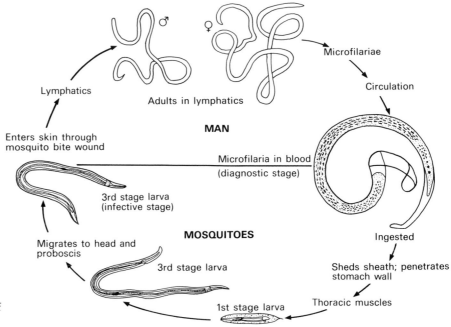

Fig. 21.14 Diagram of the life-cycle of *Wuchereria bancrofti*.

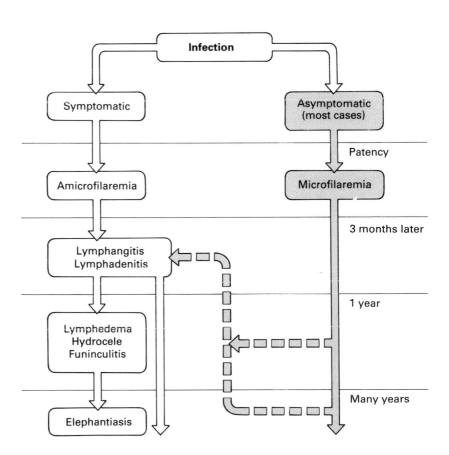

Fig. 21.15 Diagram of the possible course of disease in lymphatic filariasis. (From R. Muller, J.R. Baker *Medical Parasitology*. Lippincott, 1989.)

can live for many years [154]. The ways in which this state may change to one of hyperreactivity has been challengingly reviewed by von Lichtenberg [155].

The usual method of diagnosis is to examine microscopically a stained, thick, dehemoglobinized blood film made from a drop (about 20 mm^3) of finger prick blood for microfilariae (Fig. 21.16). The blood must be obtained at night for the nocturnally periodic strain of parasites. For amicrofilaremic cases, 10 ml of blood can be obtained in a hypodermic syringe and passed through a nuclepore filter; this is then stained and examined.

Many serologic tests have been utilized but most lack specificity [156]. One problem is that antigen preparations have usually been obtained from animal species of filariae and most tests have been useful only for epidemiologic surveys. IgE antibodies tend to be more specific and monoclonals are being investigated [157].

Tests for the detection of circulating antigens are being investigated but are hindered by the masking of antigen by antibody in immune complexes; current tests pick up less than half of amicrofilaremic cases known to be positive on clinical grounds.

Brugia malayi is a closely related parasite which occurs only in Southeast Asia. Adults are found principally in

the lymphatics of the lower limbs so elephantiasis is usually restricted to the legs. *Brugia* also has a nocturnally periodic form with microfilaremia at night and a sub-periodic form; the latter is found in forest primates as reservoir hosts and is a zoonotic infection in humans in cleared forest areas. There is a closely related animal species, *Brugia pahangi*, which might infect humans. Another species, *Brugia timori*, occurs in humans on some Indonesian islands.

The diagnosis of *Brugia* is similar to that of *Wuchereria*, the stained microfilariae being capable of differentiation under the microscope. Recently, a specific repeated-sequence DNA probe has been cloned and characterized which can distinguish *B. malayi* from all other filariae apart from *B. timori* [158].

Animal models of lymphatic filariasis and of other human filariae include *B. pahangi* in cats and jirds and two rodent species, *Acanthocheilonema viteae* in jirds and ticks and *Litomosoides carinii* in cotton rats and mites.

Onchocerca volvulus. This is the cause of an important filarial disease of humans found principally in West, Central, and East Africa, with smaller foci in Central and South America, and is estimated to have about 17 million cases worldwide [159].

An adult female *Onchocerca* measures 400 × 0.3 mm and a male 30 × 0.2 mm. Adults inhabit the subcutaneous tissues; in long-standing infections they often congregate so that intermingled worms become surrounded by permanent connective tissue nodules containing lymphocytes, plasma cells, and eosinophils, usually on bony prominences.

The life-cycle is similar to that of *Wuchereria* except that the intermediate host is a small biting fly, the buffalo gnat or blackfly *Simulium*. This breeds in well-oxygenated flowing water, often large rivers in Africa, which determines the epidemiology of the disease. The sheathless microfilariae produced by adult female worms wander through the dermal layers of the skin rather than being carried in the bloodstream. The flies have scarifying mouthparts and are pool feeders; ingested microfilariae reach the thoracic muscles and develop into infective third-stage larvae in about 8 days. Larvae enter a new host through the puncture wound made by the fly and develop into adults in 1 year. The longevity of the adult worms appears to be about 11 years. This is a very important biologic character because various United Nations' agencies have been undertaking a control campaign in the savanna area of the Volta river basin involving 11 West African countries since 1975. The campaign has been based entirely on control of flies by

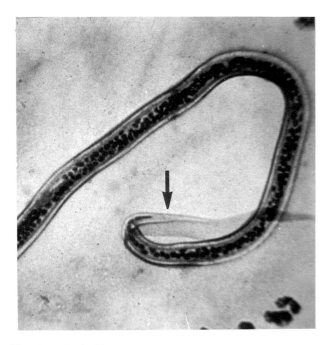

Fig. 21.16 Stained blood smear with sheathed microfilaria of *Wuchereria bancrofti*. Arrow points to the tail devoid of nuclei, which is a diagnostic feature. Original magnification, ×1200.

application of larvicides and it was only in 1986 that the microfilarial counts of infected individuals began to fall rapidly to zero [159].

In contrast to lymphatic filariasis, the clinical effects of onchocerciasis are caused by microfilariae rather than by adults. Dead microfilariae provoke an inflammatory response and the first clinical effect in children is usually a pruritic skin rash in the region near to the usually impalpable adults (Fig. 21.17). In heavy infections this leads in months or years to intradermal edema, thickening of the skin, and loss of elastic fibers (Fig. 21.18) with sometimes hanging groin, in which the inguinal lymph glands hang down in pockets of skin. Eventually there is skin atrophy, giving a prematurely aged appearance.

The most serious effects are produced when microfilariae reach the eye. This can either be from the skin surrounding the eye or via minute capillaries into the retina. Microfilariae which cross the conjunctiva and die produce a punctate keratitis, which can be accompanied by photophobia and conjunctivitis. This is a potentially reversible stage but is later followed by increasing conversion of the cornea into connective tissue, leading to reduced vision and blindness. In Africa, these anterior eye lesions are much more common in savanna than in forest regions and this is the rationale for the Volta river control campaign being confined to savanna areas, where blindness levels were previously up to 15%. Posterior lesions consist of areas where all retinal elements have disappeared except for the retinal vessels and result in complete blindness or tubular vision.

It has long been recognized that the clinical and

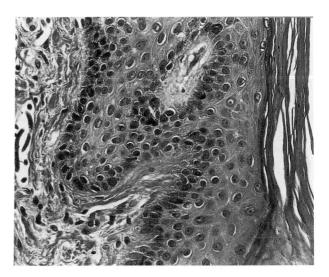

Fig. 21.18 Section of skin with microfilariae of *Onchocerca volvulus*. There is marked hyperkeratosis, thickening of the epidermis (acanthosis), edema of the dermis with dilated lymphatics, and presence of inflammatory cells. Original magnification, ×500.

pathologic manifestations of onchocerciasis vary in different areas and regions of the world and in Africa at least six strains of parasite can be identified, based on characteristics of the disease and on transmission studies [160]. The success of the control program, which is being carried out only in savanna areas, depends on the finding that there appears to be complete incompatibility between the savanna strain of parasite and forest flies, and vice versa.

Since 1989 mass treatment with a chemotherapeutic agent Mectizan (ivermectin, MSD) is supplementing *Simulium* control. This drug kills microfilariae and prevents their release from the uterus of the female for up to 12 months after treatment. It does not appear to have any effect against other human filarial parasites and has been licensed only for use in onchocerciasis. What is unprecedented is that the compound has been donated free by the manufacturers [161].

Diagnosis is by staining and identifying microfilariae microscopically obtained by teasing a bloodless skin snip, obtained with a razor blade or corneoscleral punch, in a drop of saline.

A wide range of immunodiagnostic tests have been used but lack specificity [162].

A DNA sequence from the forest form of parasite has been isolated which hybridizes with DNA from other forest isolates but not with that from savanna parasites, and other probes are being developed [163].

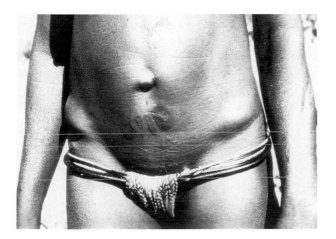

Fig. 21.17 Child with nodules containing adult *Onchocerca volvulus* in hip region. The skin also shows characteristic thickening. (Courtesy of Dr S.J. Anderson and H. Fuglsang.)

Onchocerca volvulus cannot be maintained in any laboratory host. However, there are numerous species which infect large herbivores, and cattle species are being used in chemotherapeutic studies [164]. Adult *O. volvulus* can also be obtained by digesting in collagenase [165], but, although worms appear to be morphologically [166] and antigenically [167] unaffected they do not survive well *in vitro* and must have serious physiologic damage. Worms carefully dissected out from nodules can only be maintained for up to 9 months, even after cryopreservation [168]. Microfilariae injected into laboratory mice survive for long periods and can be used for immunologic and chemotherapeutic studies [169].

Loa loa. Infection with this filarial parasite is restricted to forest areas of Central and West Africa. The adult males and females (measuring 50 × 0.5 mm) are found together in the subcutaneous tissues. Females produce sheathed microfilariae which circulate in the peripheral bloodstream during the day. If these are picked up in a blood meal by various species of the large biting tabanid fly, *Chrysops*, larvae develop inside the fly and in under 2 weeks are infective to a new host when the fly bites again. Immature worms migrate through the connective tissues and mature in 4–6 months in experimental primate hosts. Worms migrate through the connective tissues, most commonly of the wrists and ankles, where they provoke transient nonpitting edematous ("Calabar") swellings, which disappear after a few days to reappear elsewhere. An intense generalized pruritis with arthralgia and fatigue is common. There are also high eosinophilia and raised IgE levels [170,171]. Worms sometimes cross the eye under the conjunctiva and can be extracted. High levels of microfilaremia can result in encephalitis and have been associated with endomyocardial fibrosis and low fertility [172].

Diagnosis is by identifying the characteristic sheathed microfilariae in a stained, thick, dehemoglobinized blood film collected during daytime. However, a majority of infected individuals do not have circulating microfilariae and serologic tests are necessary. Unlike infections with *Wuchereria* or *Brugia*, it does not appear that this occult state is associated with increased pathology, except in temporary residents of an endemic area. For immunodiagnosis, ELISA using antigens from adult worms is reasonably specific.

A 23 kD protein antigen, which appears to be located on the microfilarial sheath and is perhaps identical to surface antigens identified in lymphatic filariasis, might be the target antigen of host-protective antibody-dependent responses.

Mansonella spp. Three species are found in humans, all transmitted by small biting midges, *Culicoides*. *Mansonella ozzardi* occurs in Central and South America and the Caribbean, *Mansonella perstans* in Central and West Africa and in areas of South America, and *Mansonella streptocerca* in Central and West Africa. The first two species can be the cause of allergic manifestations such as pruritis, articular pains, and headache, while the microfilariae of *M. streptocerca* reside in the skin and cause similar effects to mild onchocerciasis.

Diagnosis is by finding characteristic microfilariae in blood films for the first two species and in skin snips for the last. ELISA has been used for detection of *M. perstans* [173].

None of the species can be maintained in convenient animal models.

Suborder Camallanina, Superfamily Dracunculoidea. Dracunculus medinensis. This is the longest nematode parasite of humans, related to the filariae. The female worm measures up to 700 mm in length and at patency emerges from the subcutaneous tissues, usually of the foot, releasing many thousands of larvae when the limb is immersed in water. The males measure only 25 mm, remain in the tissues, and are never seen. Infection is confined to the poorest rural communities in West, Central, and East Africa, India, and Pakistan. Of the estimated 3 million cases in the world, 2.5 million are in Nigeria [174] (the first active case search in Nigeria in 1987/88 discovered 654 000 cases).

This worm infection appears to be a long-established human parasite, mentioned in ancient Egyptian and Indian writings and recovered from Egyptian mummies; however, it is almost never fatal and has been regarded more as a nuisance than as an infection worthy of serious study. In the present decade attitudes have changed and the eradication of dracunculiasis from the world was formally adopted as a resolution of the World Health Assembly in 1986 and is now targeted for 1995. This is the first human parasitic disease for which such an ambitious program has been envisaged, apart from the abortive attempts against malaria in the 1950s and hookworm in the 1920s.

Larvae released into water are ingested by microcrustaceans (cyclops) and develop to the infective stage inside them. When cyclops are ingested in drinking water obtained from ponds or open step wells in parts of India, the released larvae penetrate the intestinal wall and reach the subcutaneous tissues where the female worms become patient in about 12 months. Dracunculiasis is the only water-associated disease dependent entirely on contamination of drinking water for transmission.

When a female worm emerges it provokes the formation of a blister which bursts and a portion of the worm is extruded. In about half of all cases the whole worm is extruded in a few weeks and the lesion rapidly heals. However, in the other half of cases the track of the worm becomes secondarily infected and morbidity may last for months; about 12 000 people in Nigeria alone suffer permanent disability [175]. An unusual feature of this infection is that there appears to be no protective immunity and one individual can be infected many times in a lifetime.

It is difficult to assess the economic benefits of eradicating a disease which chiefly affects communities living at subsistence level, but in an area of eastern Nigeria it has been estimated that on average 11 mandays are lost for each person involved in rice production annually at a cost of $20 million [176].

In Africa, control campaigns are still at the surveillance stage but in India control efforts have been under way since 1983 and incidence figures have dropped from 32 792 cases in 1984 to 12 023 cases in 1988. Control is based primarily on the provision of safe drinking water sources, such as boreholes and safe wells, but also includes measures aimed at preventing contamination of water sources, personal filtering of water, and in some situations chemical treatment of ponds and wells; there are no effective anthelmintics. Recent success in maintaining a closely related species, *Dracunculus insignis*, in laboratory-reared ferrets at the Centers for Disease Control in Atlanta should enable research on chemotherapy and early immunodiagnosis to be pursued [177]. Studies in this model have shown the ineffectiveness of ivermectin against developing worms [178].

Superfamily Gnathostomatoidea. Gnathostoma spinigerum. This is a natural parasite of fish-eating carnivores such as the dog and cat, residing in nodules in the stomach.

The life-cycle involves a freshwater microcrustean (cyclops) as first intermediate and then a freshwater fish, amphibian, or reptile as second intermediate host. Human infection is confined to Asia and occurs when fish containing third-stage larvae are eaten raw or undercooked [179]. The parasite does not mature in humans but the larvae penetrate through the intestinal wall to the subcutaneous tissues where they can produce a migrating edema that can periodically reoccur for many years.

Within two days of ingestion of parasitized fish there is likely to be acute epigastric discomfort with nausea and pruritis accompanied by a very high eosinophilia. This subsides and is followed a few weeks later by the creeping eruption. Occasionally, larvae enter the central nervous system giving rise to an eosinophilic myelenoencephalitis [134].

Diagnosis is by clinical means, including the use of a CT scan for cerebral cases or by ELISA for determination of specific IgG or IgE antibodies.

Subclass Adenophorea

Superfamily Trichuroidea

Trichuris trichiura (whipworm). Trichuriasis is a cosmopolitan soil-transmitted (geohelminth) infection but is of clinical significance principally in warm/moist, tropical, and subtropical regions, with a similar distribution to ascariasis. From examination of coprolites, infection also appears to have been universal in temperate regions in prehistoric and historic times and it is likely that the low current prevalences reflect an increase in standards of sanitation rather than climatic factors. Estimates of the number of infected people in the world range from 500 [180] to 800 [108] million, with about 1 in 5000 infected people suffering morbidity [118] (although recent studies show that the proportion should be about 10 times as great [108]).

An adult *Trichuris trichiura* measures about 30 mm in length with a narrow anterior portion which is buried superficially in the mucosa of the cecum or appendix and a stouter posterior portion which hangs free in the lumen.

Very resistant unembryonated eggs are passed out in the feces and for further development require warm, moist, and well-oxygenated conditions in soil. A first-stage larva develops inside each egg in about 3 weeks under optimum conditions and when ingested with salad vegetables, or in infants with soil, the hatched larva develops in the gut without undergoing any somatic migration (Fig. 21.10).

In the great majority of infections no clinical symptoms can be attributed to the presence of *Trichuris* but in heavy infections there may be dysentery, abdominal pain, anorexia, and weight loss, with rectal prolapse in children occasionally resulting from the straining consequent on irritation and hyperemia. Prevalence rates are often high with many individuals already infected by the age of 2 years. Recently, more effective chemotherapeutic agents have enabled studies to be carried out on worm burdens. Such studies indicate that in most individuals intensity of infection rises to a peak between the ages of 4 and 10 years and then falls to reach a much lower plateau at about 25 years old. However, as with other geohelminths, there appears to be a clear predisposition to infection, so that people harboring many worms who are cured become reinfected at a similarly high level as before.

The anterior portions of the worms cause petechial hemorrhages of the mucosa which becomes edematous and hyperemic and in heavy infections this can lead to a chronic, insidious, colitis similar in some ways to Crohn's disease [108].

Hypochromic anemia and stunting of growth are associated with chronic whipworm dysentery in children [108]. The whipworm of the pig, *Trichuris suis*, is very similar to the human species and could probably be profitably used for studies on the effects of this worm on nutrition. The mouse species, *Trichuris muris*, has been extensively used recently in immunologic studies.

Trichinella spiralis. This nematode has received a great deal of attention from experimental parasitologists and from immunologists, mainly because it has a short development cycle in the usual laboratory animals and requires no intermediate host. It continues to be the subject of comprehensive monographs following those of the late E.S. Gould [181].

The life-cycle is illustrated in Fig. 21.19.

The natural hosts of *Trichinella* are carrion-feeding carnivores and scavengers, such as polar bears and walruses in the Arctic, foxes and rats in temperate countries, and jackals, hyenas, and warthogs in the tropics [182]. The small adult worms (3 × 0.06 mm) live partially embedded in the mucosa of the ileum (Fig. 21.20). They are short-lived, surviving for only about 1 month before disappearing, but each female worm produces up to 2000 larvae which penetrate the mucosa and reach the skeletal muscles via the lymphatics and blood vessels. In the muscles the larvae grow and become surrounded by a cyst. After about 5 weeks a hyaline capsule forms around the larva with an inflammatory infiltrate of lymphocytes and eosinophils (Fig. 21.21), but after about 6 months the larvae die and the cyst calcifies. Trichiniasis is thus a self-limiting infection lasting only a matter of weeks or months and light infections are usually asymptomatic. However,

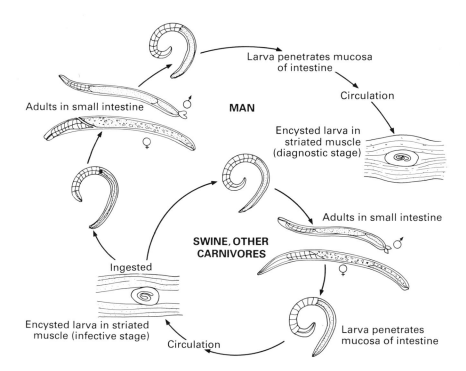

Fig. 21.19 Diagram of the life-cycle of *Trichinella spiralis.*

Fig. 21.20 Electron micrograph of a section through the cuticle and epicuticle of *Trichinella spiralis*. Original magnification, ×82 000. (Courtesy of Dr D.L. McLaren.)

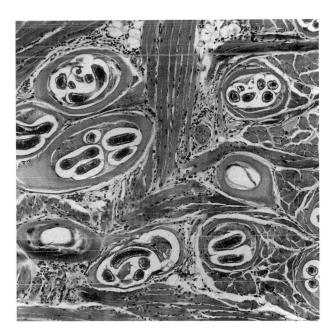

Fig. 21.21 Section of muscle from a fatal case of *Trichinella spiralis* infection. Many muscle fibers have been replaced by cysts. There is marked edema with infiltration of inflammatory cells. Original magnification, ×110.

in heavy infections with over 1000 larvae per gram of muscle, there is severe myositis and edema and symptoms resemble those of typhoid but with a high eosinophilia; death can occur from exhaustion or heart failure.

It has long been recognized that different geographic isolates have marked variation in biologic characteristics, particularly host susceptibility, and the single species was split into four: *T. spiralis* in domestic hogs; *Trichinella nativa* in the Arctic and *Trichinella nelsoni* in the tropics, both of which are not infective to hogs or to rats; and the rarely found *Trichinella pseudospiralis* which has almost no cyst and primarily infects birds. However, this division is not generally accepted although it is probable that *T. pseudospiralis* has progressed to full speciation. Isoenzymes studies on 12 enzyme systems of isolates from many different hosts and geographic localities have provided three which can act as markers for *T. pseudospiralis* but not for the other "species" [183]. It is possible that these are a series of subspecies in various stages of speciation which will probably disappear or become part of the normal variation of *T. spiralis* [184].

Human infections in temperate regions are usually contracted from eating undercooked pork or pork products, when hogs are fed unsterilized scaps of other animals in swill or garbage or sometimes hogs become infected from eating dead rats which have been infected from eating pork scraps in garbage. However, many of the recent outbreaks in Europe have involved horsemeat.

Serologic tests for diagnosis become positive 2–4 weeks after infection, by which time treatment against the adults in the gut is likely to be too late. ELISA, indirect fluorescence antibody tests, bentonite flocculation, and latex agglutination can all be used and refined antigens are now becoming available [185].

Trichinella spiralis has very low host specificity and can be maintained easily in rodent hosts in the laboratory. It has been widely used as a model system to study mechanisms of resistance in both the intestinal and tissue phases [186].

Capillaria philippinensis. This previously unknown nematode was reported as causing a new disease syndrome in the Philippines in 1963. Subsequently cases have been reported also from Thailand and Iran [187].

The small adults (females measuring 4 mm in length) burrow into the mucosa of the small intestine, causing atrophy of the villi. Eggs which are produced can hatch into larvae in the intestine so this is an unusual nematode which can multiply inside the body.

Freshwater or brackish water fish act as intermediate hosts; the normal definitive hosts are probably birds and human infection is a zoonosis.

Clinical manifestations can be severe, with many voluminous stools and malabsorption, and resemble tropical sprue. Muscle wasting, emaciation, and weakness occur in many patients and, until effective chemotherapy was discovered, mortality was over 20%, principally from heart failure.

Diagnosis is by finding eggs in the feces.

REFERENCES

1 Maayan S, Wormser GP, Widerhorn J, Sy ES, Kim YH, Ernst JA. *Strongyloides stercoralis* hyperinfection in a patient with the acquired immune deficiency syndrome. Am J Med 1987;83:945–948.

2 Pawlowski ZS. Intestinal helminthiases and human health: recent advances and future needs. Int J Parasitol 1987;17:159–167.

3 Genta RM, Walzer PD. Strongyloidiasis. In Walzer PD, Genta RM, eds. *Parasitic Infections in the Immunocompromised Host*. New York: Marcel Dekker, 1989:463–525.

4 Coulaud JP, Bartczak S. Auguillulose et deficience immunitaire. Med Afr Noire 1988;35:637–640.

5 Wakelin DM. Helminth infections. In Wakelin DM, Blackwell JM, eds. *Genetics of Resistance to Bacterial and Parasitic Infection*. London: Taylor & Francis, 1988:153–224.

6 Chapman CB, Mitchell GF. *Fasciola hepatica*: comparative studies on fascioliasis in rats and mice. Int J Parasitol 1982;12:81–91.

7 Kaji R, Kamijo T, Yano A, Kojima S. Genetic control of immune responses to *Schistosoma japonicum* antigen. Parasite Immunol 1983;5:25–35,217–221.

8 Colley DG, Freeman GL. Differences in adult *Schistosoma mansoni* worm burden requirements for the establishment of resistance to reinfection in inbred mice. II. Am J Trop Med Hyg 1983;32:543–549.

9 Kee KC, Taylor DW, Cordingley JS, Butterworth AE, Munro AJ. Genetic influence on the antibody response to antigens of *Schistosoma mansoni* in chronically infected mice. Parasite Immunol 1986;8:565–574.

10 White TR, Thompson RCA, Penhale WJ. A comparative study of the susceptibility of inbred strains of mice to infection with *Mesocestoides corti*. Int J Parasitol 1982;12:29–33.

11 Oldfield NF. Studies on the interaction of *Hymenolepis microstoma*, the mouse bile duct tapeworms, and inbred strains of mice of diverse genetic backgrounds. Diss Abs Int B 1986;46:2231.

12 Ito A, Kano S, Hioki A, Kasnya S, Ohtomo H. Reduced fecundity of *Hymenolepis nana* due to thymus-dependent immunological responses in mice. Int J Parasitol 1986;16:81–85.

13 White TR, Thompson RCA, Penhale WJ. The effects of selective immunosuppression on resistance to *Mesocestoides corti* in strains of mice showing high and low initial susceptibility. Z Parasitenkd 1983;69:91–104.

14 Joysey HS. Experimental infection in high and low responder Biozzi mice with *Taenia crassiceps* (Cestoda). Int J Parasitol 1986;16:217–221.

15 Mitchell GF. Genetic variation in resistance of mice to *Taenia taeniaeformis*: analysis of host-protective immunity and immune evasion. In Flisser A, Willms K, Laclette JP, Larralde C, Ridaura C, Beltran F, eds. *Cysticercosis: Present State of Knowledge and Perspectives*. New York: Academic Press, 1982:575–584.

16 Conchedda M, Ferreti G. Susceptibility of different strains of mice to various levels of infection with the eggs of *Taenia taeniaeformis*. Int J Parasitol 1984;14:54–546.

17 Gibbens JC, Harrison LJS, Parkhouse RME. Immunoglobulin class responses to *Taenia taeniaeformis* in susceptible and resistant mice. Parasite Immunol 1986;8:491–802.

18 Haque A, Worms MJ, Ogilvie BM, Capron A. *Dipetalonema viteae*: microfilariae production in various mouse strains and in nude mice. Exp Parasitol 1980;49:398–407.

19 Storey N, Behnke JM, Wakelin D. Immunity to *Dipetalonema viteae* (Filarioidea) infections in resistant and susceptible mice. Acta Trop 1987;44:43–54.

20 Kennedy MW, Gordon AMS, Tomlinson LA, Qureshi F. Genetic (major histocompatability complex?) control of the antibody repertoire to the secreted antigens of *Ascaris*. Parasite Immunol 1987;9:269–273.

21 Dunn MC, Brown HW. Comparison of the susceptibility of CFI, C57 Brown, C3H, C57 Black and DBA 2 strains of mice to *Aspiculurus tetraptera*. J Parasitol 1963;49(2):32–33.

22 Fanning MM, Kazura JW. Genetic association of murine susceptibility to *Brugia malayi* microfilaraemia. Parasite Immunol 1983;5:305–316.

23 Vickery AC, Nayar JK. *Brugia pahangi* in nude mice: protective immunity to infective larvae is Thy 1.2^+ cell dependent and cyclosporin A resistant. J Helminthol 1987;61:19–27.

24 Behnke JM, Robinson M. Genetic control of immunity to *Nematospiroides dubius*: a 9-day anthelmintic abbreviated immunizing regime which separates weak and strong responder strains of mice. Parasite Immunol 1985;7:235–253.

25 Wenk P. Der invasionweg der metazyklischen larven von *Litomosoides carinii* 1931 (Filariidae). Z Parasitenkd 1967;28:240–263.

26 Wells C, Behnke JM. The course of primary infections with *Necator americanus* in syngeneic mice. Int J Parasitol 1989;18:47–51.

27 Mitchell LA, Wescott RB, Perryman LE. Kinetics of expulsion of the nematode, *Nippostrongylus brasiliensis*, in mast-cell deficient W/Wᵛ mice. Parasite Immunol 1983;4:1–12.

28 Townson S, Bianco AE. Experimental infection of mice with the microfilariae of *Onchocerca lienalis*. Parasitology 1982;85:283–293.

29 Dawkins HJS, Grove DI, Dunsmore JD, Mitchell GF. *Strongyloides ratti*: susceptibility to infection and resistance to reinfection in inbred strains of mice as assessed by excretion of larvae. Int J Parasitol 1980;10:125–129.

30 Alizadeh H, Murrell KD. The intestinal mast cell response to *Trichinella spiralis* infection in mast-cell deficient W/Wᵛ

mice. J Parasitol 1984;70:707–773.

31 Bell RG, Adams LS, Ogden RW. *Trichinella spiralis*: genetics of worm expulsion in inbred and F1 mice infected with different worm doses. Exp Parasitol 1984;58:345–355.

32 Wakelin D, Mitchell LA, Donachie AM, Grencis RK. Genetic control of immunity to *Trichinella spiralis* in mice. Response of rapid- and slow-responder strains to immunization with parasite antigens. Parasite Immunol 1986;8:159–170.

33 Else K, Wakelin D. Genetic variation in the humoral immune response of mice to the nematode *Trichuris muris*. Parasite Immunol 1989;11:77–90.

34 Barrett J. *Biochemistry of Parasitic Helminths*. London: Macmillan, 1981.

35 Barrett J. Amino acid metabolism in helminths. Adv Parasitol 30 1991;30:39–107.

36 Bryant C, Behm C. *Biochemical Adaptation in Parasites*. London: Chapman & Hall, 1989.

37 Coles GC. Recent advances in schistosome biochemistry. Parasitology 1984;89:603–637.

38 Smyth JD, McManus DP. *The Physiology and Biochemistry of Cestodes*. Cambridge: Cambridge University Press, 1989.

39 Barret J. Parasitic helminths. In Bryant C, ed. *Metazoan Life without Oxygen*. London: Chapman & Hall, 1991:146–164.

40 MacInnes AJ, ed. *Molecular Paradigms for Eradicating Helminthic Parasites*. UCLA Symposium on Molecular and Cellular Biology, vol. 60. New York: Alan R Liss, 1987.

41 Taylor AER, Baker JR, eds. *In vitro Methods for Parasite Cultivation*. London: Academic Press, 1987.

42 Campbell WC. The chemotherapy of parasitic infections. J Parasitol 1986;72:45–61.

43 Vanden Bossche H. How anthelmintics help us to understand helminths. Parasitology 1985;90:675–685.

44 Brindley PJ, Sher A. Immunological involvement in the efficacy of praziquantel. Exp Parasitol 1990;71:245–248.

45 Clark IA, Hunt NH, Cowden WB. Oxygen-derived free radicals in the pathogenesis of parasitic disease. Adv Parasitol 1986;25:1–44.

46 Callahan HL, Crouch RK, James ER. Helminth anti-oxidant enzymes: a protective mechanism against host oxidants? Parasitol Today 1988;4:218–225.

47 Nelson GS. Parasitic zoonoses. In Englund PT, Sher A, eds. *The Biology of Parasitism: a Molecular and Immunologic Approach*. New York: Alan R Liss, 1988:13–41.

48 Brooks DR, O'Grady RT, Glen DR. Phylogenetic analysis of the Digenea (Platyhelminthes: Cercomeria) with comments on their adaptive radiation. Can J Zool 1985;63:411–443.

49 Warren KS. The global impact of parasitic diseases. In Englund PT, Sher A. *The Biology of Parasitism: a Molecular and Immunologic Approach*. New York: Alan R Liss, 1988:3–12.

50 Jordan P, Webbe G. *Schistosomiasis: Epidemiology, Treatment and Control*. London: Heinemann Medical, 1982.

51 Bruce JI, Sornmani S, Asch HL, Crawford KA. *The Mekong Schistosome*, Suppl 2. Ann Arbor, MI: Ann Arbor, 1980.

52 Wilson RA. Development and migration in the mammalian host. In Rollinson D, Simpson AJG, eds. *The Biology of Schistosomes: from Genes to Latrines*. London: Academic Press, 1987:117–146.

53 Von Lichtenberg F. Consequences of infections with schistosomes. In Rollinson D, Simpson AJG, eds. *The Biology of Schistosomes: from Genes to Latrines*. London: Academic Press, 1987:185–232.

54 Walls K, Schantz P, eds. *Immunodiagnosis of Parasitic Diseases. I. Helminthic Diseases*. New York: Academic Press, 1986.

55 Mott KE, Dixon KE. Collaborative study on antigens for immunodiagnosis of schistosomiasis. Bull WHO 1982;60:729–753.

56 Doenhoff MJ, Dunne DW, Lillywhite JE. Serology of *Schistosoma mansoni* infections after chemotherapy. Trans R Soc Trop Med Hyg 1989;83:237–238.

57 Cheever AW. *Schistosoma japonicum*: the pathology of experimental infection. Exp Parasitol 1985;59:1–11.

58 Rollinson D, Walker JK, Simpson AJG. The application of recombinant DNA technology to problems of helminth identification. Parasitology 1986;91:553–571.

59 Butterworth AE. Control of schistosomiasis in man. In Englund PT, Sher A, eds. *The Biology of Parasitism: a Molecular and Immunologic Approach*. New York: Alan R Liss, 1988:45–59.

60 Taylor MG. Live vaccines for bovine schistosomiasis. In Staines NA, Doenhoff MJ, eds. *Immunoparasitology*. London: British Society of Immunology, 1986:161–165.

61 Simpson AJG, Cioli D. Progress towards a defined vaccine for schistosomiasis. Parasitol Today 1987;3:26–28.

62 Grzych JM, Capron M, Lambert PH, Dissous C, Torres S, Capron A. An anti-idiotype vaccine against experimental schistosomiasis. Nature 1985;316:74–76.

63 Capron M, Capron A. Rats, mice and man—models for immune effector mechanisms against schistosomiasis. Parasitol Today 1986;2:69–74.

64 Balloul JM, Sondermeyer P, Dreyer D, *et al.* Molecular cloning of a protective antigen against schistosomiasis. Nature 1987;326:149–153.

65 Sher A. Strategies for vaccination against parasites. In Englund PT, Sher A, eds. *The Biology of Parasitism: a Molecular and Immunologic Approach*. New York: Alan R Liss, 1988:169–182.

66 James SL. *Schistosoma* spp.: progress towards a defined vaccine. Exp Parasitol 1987;63:247–252.

67 Wilkins HA, Blumenthal UJ, Hayes RJ, Tulloch S. Resistance to reinfection after treatment of urinary schistosomiasis. Trans R Soc Trop Med Hyg 1987;81:29–35.

68 Butterworth AE, Dalton PR, Dunne DW, *et al.* Immunity after treatment of human schistosomiasis mansoni. I. Study design, pretreatment observation and the results of treatment. Trans R Soc Trop Med Hyg 1984;78:108–123.

69 Butterworth AE, Fulford AJC, Dunne DW, Ouma JH, Sturrock RF. Longitudinal studies on human schistosomiasis. Philos Trans R Soc London B 1988;321:495–511.

70 Hagan P, Blumental UJ, Dunne D, Simpson AJG, Wilkins HA. Human IgE, IgG4 and resistance to reinfection with *Schistosoma haematobium*. Nature 1990;349:243–245.

71 Flavell DJ, Flavell SU. *Opisthorchis viverrini*: pathogenesis of infection in immunodeprived hamsters. Parasite Immunol 1986;8:455–466.

72 Velasquez CC. Heterophyes. In Hillyer GV, Hopla CE,

eds. *Parasitic Zoonoses*, vol. 3. Florida: CRC Press, 1982: 99–107.

73 Kawashima K, ed. *Paragonimus in Asia: Biology, Genetic Variation and Speciation*. Research Report 1. Fukuoka, Japan: Kyushu University, 1987.

74 Cross JH. Chemotherapy of intestinal trematodiasis in man. In Vanden Bossche H, Thienpont D, Janssens PG, eds. *Chemotherapy of Intestinal Helminths*. Berlin: Springer-Verlag, 1985:541–556.

75 Farag HF, Ragab M, Salem A, Sadek N. A short note on praziquantel in human fascioliasis. J Trop Med Hyg 1986; 86:79–80.

76 Gonzalez JF, Perez O, Rodriguez G, Arus E, Lastre M. Fasciolasis humana epidemica, Cuba 1983. GEN 1985;39: 276–281.

77 Beardsell PL, Howell MJ. Production of parasite antigen by co-transformation. Parasitol Today 1987;3:28–29.

78 Hanna REB, Trudgett AG, Anderson A. *Fasciola hepatica*: development of monoclonal antibodies against somatic antigens and their characterization by ultrastructural localization of antibody binding. J Helminthol 1988;62: 15–28.

79 Rim HJ. Fasciolopsiasis. In Hillyer GV, Hopla CE, eds. *Parasitic Zoonoses*, vol. 3. Florida: CRC Press, 1982:89–97.

80 Mahajan RC. Geographical distribution of human cysticercosis. In Flisser A, Willms K, Ladette JP, Larralde C, Ridaura C, Beltran F, eds. *Cysticeriosis: Present State of Knowledge and Perspectives*. New York: Academic Press, 1982:39–46.

81 Flisser A. Neurocysticercosis in Mexico. Parasitol Today 1988;4:131–137.

82 Palacios E, Rodriguez-Carbajal J, Taveras JM. *Cysticercosis of the Central Nervous System*. Springfield Thomas, 1983.

83 Coker-Vann M, Brown P, Gadjusek DC. Sero-diagnosis of human cysticercosis using a chromato-focused antigenic preparation of *Taenia solium* cysticerci on an enzyme-linked immunosorbent assay (ELISA). Trans R Soc Trop Med Hyg 1984;78:492–496.

84 Thompson RCA, ed. *The Biology of Echinococcus and Hydatid Disease*. London: Allen & Unwin, 1986.

85 Schwabe CW. Current status of hydatid disease: a zoonosis of increasing importance. In Thompson RCA, ed. *The Biology of Echinococcus and Hydatid Disease*. London: Allen & Unwin, 1986:81–113.

86 Craig PS, Zeyhle E, Romig T. Hydatid disease: research and control in Turkana. II. The role of immunological techniques for the diagnosis of hydatid disease. Trans R Soc Trop Med Hyg 1986;80:183–192.

87 Di Felice G, Pini C, Afferni C, Vicari G. Purification and partial characterization of the major antigen of *Echinococcus granulosus* (antigen 5) with monoclonal antibodies. Mol Biochem Parasitol 1986;20:133–142.

88 Craig PS, Nelson GS. Detection of circulating antigen in human hydatid disease. Ann Trop Med Parasitol 1984;78: 219–227.

89 Rausch RL. Life-cycle patterns and geographic distribution of *Echinococcus* species. In Thompson RCA, ed. *The Biology of Echinococcus and Hydatid Disease*. London: Allen & Unwin, 1986:44–80.

90 Muller R, Baker JR. The demands of medicine and veterinary sciences (for identification at the sub-specific level). In Hawksworth DL, ed. *Prospects in Systematics*. Oxford: Clarenden Press, 1988:377–395.

91 Thompson RCA, Lymberry AJ. The nature, extent and significance of variation within the genus *Echinococcus*. Adv Parasitol 1988;27:210–258.

92 McManus DP, Smyth JD. Hydatidosis: changing concepts in epidemiology and speciation. Parasitol Today 1986;2: 163–168.

93 Howell MJ. Cultivation of *Echinococcus* species *in vitro*. In Thompson RCA, ed. *The Biology of Echinococcus and Hydatid Disease*. London: Allen & Unwin, 1986:143–163.

94 Rickard MD, Williams JF. Hydatidosis/cysticercosis: immune mechanisms and immunization against infection. Adv Parasitol 1982;21:229–296.

95 Lucas SB, Hassounah O, Muller R, Doenhoff MJ. Abnormal development of *Hymenolepis nana* larvae in immunosuppressed mice. J Helminthol 1980;54:75–82.

96 Arai HP, ed. *Biology of the Tapeworm Hymeolepis diminuta*. New York: Academic Press, 1980.

97 Arme C, Bridges JF, Hoole D. Pathology of cestode infections in the human host. In Arme C, Pappas PW, eds. *Biology of the Eucestoda*, vol. 2. London: Academic Press, 1983:499–538.

98 Lightowlers MW, Rickard MD. Excretory–secretory products of helminth parasites: effects on host immune responses. Parasitology 1988;96:S123–S166.

99 Wright KA. The nematode cuticle—its surface and the epidermis: function, homology, analogy—a current consensus. J Parasitol 1987;73:1077–1083.

100 Bettschart B. Structure, molecular biology and immunology of the cuticle of parasitic nematodes. Acta Trop 1990;47: 251–407.

101 Kingston IB. Nematode collagen genes. Parasitol Today 1990;7:11–15.

102 Ogilvie BM, Selkirk ME, Maizels RM. The molecular revolution and nematode parasitology: yesterday, today and tomorrow. J Parasitol 1990;76:607–618.

103 Rajan TV. Molecular biology of human lymphatic filariasis. Exp Parasitol 1990;70:500–503.

104 Crompton DWT. The prevalence of ascariasis. Parasitol Today 1988;4:162–169.

105 Crompton DWT, Nesheim MC, Pawlowski ZS. *Ascariasis and its Public Health Significance*. London: Taylor & Francis, 1985.

106 Pawlowski ZS. Strategies for the control of ascariasis. Ann Soc Belg Med Trop 1984;64:125–134.

107 Stephenson LS. *Impact of Helminth Infections on Human Nutrition*. London: Taylor & Francis, 1987.

108 Bundy DAP, Cooper ES. *Trichuris* and trichuriasis in humans. Adv Parasitol 1989;28:107–173.

109 Bundy DAP. Population ecology of intestinal helminth infections in human communities. Philos Trans R Soc London B 1988;321:405–420.

110 Anderson RM, May RM. Helminth infections of humans: mathematical models, population dynamics, and control. Adv Parasitol 1985;24:1–102.

111 Anderson RM. The population dynamics and epidemiology of intestinal nematode infections. Trans R Soc Trop Med Hyg 1986;80:686–696.

112 Anderson RM. The population biology and genetics of resistance to infection. In Wakelin DM, Blackwell JM, eds. *Genetics of Resistance to Bacterial and Parasitic Infections*. London: Taylor & Francis, 1988:223–263.

113 Wakelin D. Evasion of the immune response: survival within low responder individuals of the host population. Parasitology 1984;88:637–657.

114 Behnke JM. Evasion of immunity by nematodes causing chronic infections. Adv Parasitol 1987;26:2–72.

115 Nesheim MC. Nutritional aspects of *Ascaris suum* and *A. lumbricoides* infections. In Crompton DWT, Nesheim MC, Pawlowski ZS, eds. *Ascariasis and its Public Health Significance*. London: Taylor & Francis, 1985:147–160.

116 Carrera E, Nesheim MC, Crompton DWT. Lactose maldigestion in *Ascaris* infected preschool children. Am J Clin Nutr 1984;39:255–264.

117 Crompton DWT. Nutritional aspects of infection with gastro-intestinal helminths. Trans R Soc Trop Med Hyg 1986;80:697–705.

118 Walsh JA. Estimating the burden of illness in the tropics. In Warren KS, Mahmond AAF, eds. *Tropical and Geographical Medicine*, vol. 1. New York: McGraw-Hill Book Co., 1984:1073–1085.

119 Komuniecki R, Komuniecki, PR. *Ascaris suum*: a useful model for anaerobic mitochondrial metabolism and the aerobic-anaerobic transition in developing parasitic helminths. In Bennet E-M, Behm C, Brant C, eds. *Comparative Biochemistry of Parasitic Helminths*. London: Chapman & Hall, 1989:1–12.

120 Schantz PM. *Changing Patterns of Parasitic Diseases*. Proc 6th Int Congress on Parasitology. Canberra: Australian Academy of Science, 1986:697–710.

121 Ishikura H, Kikuchi K. *Intestinal Anisakiasis in Japan*. Tokyo: Springer-Verlag, 1990.

122 Oshima T. Anisakiasis—is the sushi bar guilty? Parasitol Today 1987;3:44–48.

123 Schad GA, Warren KS, eds. *Hookworm Disease*: *Current Status and New Directions*. London: Taylor & Francis, 1990.

124 Crompton DWT. Hookworm disease: current status and new directions. Parasitol Today 1989;5:1–2.

125 Gibbs HC. Hypobiosis in parasitic nematodes. Adv Parasitol 1986;25:129–174.

126 Hotez PT, Cerami A. Secretion of a proteolytic anticoagulant by *Ancylostoma duodenale* hookworms. J Exp Med 1983;157:1594–1603.

127 Weatherall DJ, Wasi P. Anaemia. In Warren KS, Mahmond AAF, eds. *Tropical and Geographical Medicine*, vol. 1. New York: McGraw-Hill Book Co., 1984:61–69.

128 Philipp M. Acetylcholinesterase secreted by intestinal nematodes: a re-interpretation of its putative role of 'biochemical holdfast'. Trans R Soc Trop Med Hyg 1984;78: 138–139.

129 Clive BL, Little MD, Bartholomew RK, Halsey NA. Larvicidal activity of albendazole against *Necator americanus* in human volunteers. Am J Trop Med Hyg 1984;33:387–394.

130 Hall A. Intestinal helminths of man: the interpretation of egg counts. Parasitology 1982;85:605–613.

131 Anderson RM, Schad GA. Hookworm burdens and faecal egg counts: an analysis of the biological basis of variation.

Trans R Soc Trop Med Hyg 1985;79:812–825.

132 Behnke JM, Paul V, Rajasekarinh GR. The growth and migration of *Necator americanus* following infection of neonatal hamsters. Trans R Soc Trop Med Hyg 1986;80: 146–149.

133 Carr A, Pritchard DI. Identification of hookworms (*Necator americanus*) antigens and their translation *in vitro*. Mol Biochem Parasitol 1986;19:251–258.

134 Jaroonvesama N. Differential diagnosis of eosinophilic meningitis. Parasitol Today 1988;4:262–266.

135 Cross JH. Public health importance of *Angiostrongylus cantonensis* and its relatives. Parasitol Today 1987;3:367–369.

136 Welch JS, Dobson C, Campbell GR. Immunodiagnosis and seroepidemiology of *Angiostrongylus cantonensis* zoonoses in man. Trans R Soc Trop Med Hyg 1980;74:614–623.

137 Cross JH, Chi JCH. Enzyme linked immunosorbent assay for the detection of *Angiostrongylus cantonensis* antibodies in patients with eosinophilic meningitis. Southeast Asian J Trop Med Public Health 1982;13:73.

138 Morera P. Abdominal angiostrongyliasis: a problem of public health. Parasitol Today 1985;1:173–175.

139 Pawlowski ZS. Epidemiology, prevention and control. In Grove DI, ed. *Strongyloidiasis: a Major Roundworm Infection of Man*. London: Taylor & Francis, 1989:233–249.

140 Grove DI, ed. *Strongyloidiasis: a Major Roundworm Infection of Man*. London: Taylor & Francis, 1989.

141 Schad GA. Morphology and life history of *Strongyloides stercoralis*. In Grove DI, ed. *Strongyloidiasis: a Major Roundworm Infection of Man*. London: Taylor & Francis, 1989:85–104.

142 Onile B, Komsafe F, Oladiran B. Severe strongyloidiasis presenting as occult gastro-intestinal tract malignancy. Ann Trop Med Parasitol 1985;79:301–304.

143 Pelletier LL Jr, Baker CB, Gam AA, Nutman TB, Neva FA. Diagnosis and evaluation of treatment of chronic strongyloidiasis in ex-prisoners of war. J Infect Dis 1988; 157:573–576.

144 Hill JA. Strongyloidiasis in ex-Far East prisoners of war. Br Med J 1988;296:753.

145 Grove DI. Diagnosis. In Grove DI, ed. *Strongyloidiasis: a Major Roundworm Infection of Man*. London: Taylor & Francis, 1989:175–197.

146 Ashford RW, Barnish G. Strongyloidiasis in Papua New Guinea. In Pawlowski ZS, ed. *Clinics in Tropical Medicine and Communicable Diseases: Intestinal Helminths*. London: Ballière-Tindall and Saunders, 1987:765–773.

147 Genta RM, Ottesen EA, Neva FA, Walzer PD, Tanowitz HB, Wittner M. Cellular responses in human strongyloidiasis. Am J Trop Med Hyg 1983;32:990–994.

148 Dawkins HJS. *Strongyloides ratti* infections in rodents: value and limitations as a model of human strongyloidiasis. In Grove DI, ed. *Strongyloidiasis: a Major Roundworm Infection of Man*. London: Taylor & Francis, 1989;287–332.

149 Haswell-Elkins MR, Elkins DB, Anderson RM. Evidence for predisposition in humans to infection with *Ascaris*, hookworm, *Enterobius* and *Trichuris* in a south Indian fishing community. Parasitology 1987;95:323–337.

150 Chittenden AM, Ashford RW. *Enterobius gregorii* Hugot

1983: first report in the UK. Ann Trop Med Parasitol 1987;81:195–198.

151 Adamson ML. Evolutionary biology of the Oxyurida (Nematoda): biofacies of a haplodiploid taxon. Adv Parasitol 1989;28:175–228.

152 WHO. *Lymphatic Filariasis*. Technical Report Series No. 702. WHO, Geneva, 1984.

153 Chan SH, Dissanayake S, Mak JW, *et al.* HLA and filariasis in Sri Lankans and Indians. Southeast Asian J Trop Med Public Health 1984;15:281–286.

154 Ottesen EA. Immunological aspects of lymphatic filariasis and onchocerciasis in man. Trans R Soc Trop Med Hyg 1984;78:9–17.

155 Von Lichtenberg F. Inflammatory responses to filarial connective tissue parasites. Parasitology 1987;94:S101–S122.

156 Ambroise-Thomas P, Peyron F. Filariasis. In Walls KW, Schantz PM, ed. *Immunodiagnosis of Parasitic Diseases*, vol. 1. Orlando: Academic Press, 1986:233–254.

157 Haque A, Capron A. Filariasis: antigens and host–parasite interactions. In Pearson JW, ed. *Parasite Antigens*. New York: Marcel Dekker, 1986:317–402.

158 McReynolds LA, DeSimone SM, Williams SA. Cloning and comparison of repeated DNA sequences from the human filarial parasite *Brugia malayi* and the animal parasite *Brugia pahangi*. Proc Natl Acad Sci USA 1985;83:797–801.

159 WHO. *WHO Expert Committee on Onchocerciasis*. WHO, Geneva, 1987.

160 Muller R. Identification of *Onchocerca*. In Taylor AER, Muller R, eds. *17th Symposium of the British Society for Parasitology*. Oxford: Blackwell, 1979:175–206.

161 Bradshaw H. Onchocerciasis and the Mectizan donation programme. Parasitol Today 1989;5:63–64.

162 Mackenzie CD, Burgess PJ, Sisley BM. Onchocerciasis. In Walls KW, Schantz PM, eds. *Immunodiagnosis of Parasitic Disease*. Orlando: Academic Press, 1986:255–290.

163 Erttman KD, Unnansch TR, Greene BM, *et al.* A DNA sequence specific for forest form *Onchocerca volvulus*. Nature 1987;327:415–417.

164 Townson S. The development of a laboratory model for onchocerciasis using *Onchocerca gutturosa: in vitro* culture, collagenase effects, drug studies and cryopreservation. Trop Med Parasitol 1988;39:475–479.

165 Schulz-Key H. The collagenase technique: how to isolate and examine adult *Onchocerca volvulus* for the evaluation of drug effects. Trop Med Parasitol 1988;39:423–440.

166 Buttner DWH, MaDonald A. The fine structure of adult *Onchocerca volvulus* recovered by collagenase digestion. Trop Med Parasitol 1985;36:171–174.

167 Prod'Hon J, Lucius R, Kern A, Hebrard G, Diesfeld HJ. Studies on the protein and antigen composition of individual female *Onchocerca volvulus* after collagenase digestion. Trop Med Parasitol 1985;36:238–240.

168 Townson S, Shay K, Dobinson A, Connelly C, Comley J, Zea-Flores G. *Onchocerca gutturosa* and *O. volvulus*: studies on the viability and drug responses of cryopreserved adult

worms *in vitro*. Trans R Soc Trop Med Hyg 1989;83:664–669.

169 Townson S, Dobinson A, Connelly C, Muller R. Chemotherapy of *Onchocerca lienalis* microfilariae in mice: a model for the evaluation of novel compounds for the treatment of onchocerciasis. J Helminthol 1988;62:181–194.

170 Pinder M. *Loa loa*—a neglected filaria. Parasitol Today 1988;4:279–284.

171 Duke BOL. Loiasis. Recent Adv Trop Med 1984;1:171–178.

172 Anon. *Loa loa*—a pathogenic parasite. Lancet 1986; ii(8507):554.

173 Van Hoegaerden M, Chabaud B, Akue JP, Ivanoff B. Filariasis due to *Loa loa* and *Mansonella perstans*: distribution in the regions of Okondja, Haut-Ogoone Province, Gabon, with parasitological and serological follow-up over one year. Trans R Soc Trop Med Hyg 1987;81:441–446.

174 Watts S. Dracunculiasis in Africa. Am J Trop Med Hyg 1987;37:119–125.

175 Smith GS, Blum D, Huttly SRA, Okeke N, Kirkwood BR, Feachem RG. Disability from dracunculiasis: effect on mobility. Ann Trop Med Parasitol 1989;83:151–158.

176 De Rooy C. *Guinea Worm Control as a Major Contribution to Self-sufficiency in Rice Production in Nigeria*. Lagos, Nigeria: Unicef, 1987.

177 Eberhard ML, Ruiz-Tiben E, Wallace SV. *Dracunculus insignis*: experimental infection in the ferret, *Mustela putorius furo*. J Helminthol 1988;62:265–270.

178 Eberhard ML, Brandt FH, Ruiz-Tiben E, Hightower A. Chemoprophylactic drug trials for treatment of dracunenliasis using the *Dracunculus insignis*-ferret model. J Helminthol 1990;64:79–86.

179 Daengsvang S. *A Monograph on the Genus Gnathostoma and Gnathostomiasis in Thailand*. Tokyo: SEAMIC, 1980.

180 WHO. *Prevention and Control of Intestinal Parasitic Diseases*. WHO, Geneva, 1987.

181 Campbell WC, ed. *Trichinella and Trichinosis*. New York: Plenum Publishers, 1983.

182 Campbell WC. Trichinosis revisited—another look at modes of transmission. Parasitol Today 1988;4:83–86.

183 Flockhart HA. *Trichinella* speciation. Parasitol Today; 1986; 2:1–3.

184 Dick TA. The species problem in *Trichinella*. In Stone AR, Platt HM, Khalil LF, eds. *Concepts in Nematode Systematics*. London: Academic Press, 1983:351–360.

185 Ljungstrom I. Immunodiagnosis in man. In Campbell WC, ed. *Trichinella and Trichinosis*. New York: Plenum Publishers, 1983:403–442.

186 Wakelin DM, Denham DA. The immune response. In Campbell WC, ed. *Trichinella and Trichinosis*. New York: Plenum Publishers, 1983:265–308.

187 Cross JH, Bhaibulaya M. Intestinal capillariasis in the Philippines and Thailand. In Croll NA, Cross JH, eds. *Human Ecology and Infectious Diseases*. New York: Academic Press, 1983:104–136.

Index